SUBSURFACE GEOLOGIC METHODS

METHODS

(A Symposium)

THIRD PRINTING—SECOND EDITION

Compiled and Edited by

L. W. LeROY

Associate Professor of Geology, Colorado School of Mines

Single Copy $7.00

COLORADO SCHOOL OF MINES
Department of Publications
Golden, Colorado
1951

First Edition, June 1, 1949.
First Printing, Second Edition, June 1, 1950.
Second Printing, Second Edition, June 1, 1951.
Third Printing, Second Edition, July 1, 1955.

Printed by The A. B. Hirschfeld Press,
Denver, Colorado.

Engravings by Cocks-Clark Engraving Co.,
Denver, Colorado.

CONTRIBUTORS TO *SUBSURFACE GEOLOGIC METHODS*

L. W. LeRoy, Associate Professor of Geology, Colorado School of Mines, Golden, Colorado, *Compiler* and *Editor*

George W. Johnson, Department of English, Colorado School of Mines, Golden, Colorado, *Editor*

O. E. Barstow, Dowell Incorporated, Tulsa, Oklahoma

C. M. Bryant, Dowell Incorporated, Tulsa, Oklahoma

John G. Caran, Core Engineer, San Antonio, Texas

Kirk Carlsten, Mechanical Engineer, Eastman Oil Well Survey Company, Denver, Colorado

S. R. B. Cooke, Department of Metallurgy and Mineral Dressing, University of Minnesota, Minneapolis, Minnesota

James G. Crawford, Chemical and Geological Laboratories, Casper, Wyoming

Jack De Ment, De Ment Laboratories, Portland, Oregon

H. G. Doll, Schlumberger Well Surveying Corporation, Houston, Texas

George H. Fancher, Department of Petroleum Engineering, University of Texas, Austin, Texas

R. D. Ford, Schlumberger Well Surveying Corporation, Houston, Texas

M. G. Frey, Department of Geology, University of Cincinnati, Cincinnati, Ohio

John W. Gabelman, Colorado Fuel and Iron Corporation, Pueblo, Colorado

R. J. Gill, Geologic Survey Company, Wichita, Kansas

R. G. Hamilton, Schlumberger Well Surveying Corporation, Houston, Texas

P. N. Hardin, Dowell Incorporated, Tulsa, Oklahoma

W. E. Hassebroek, Halliburton Oil Well Cementing Company, Duncan, Oklahoma

Sigurd Kermit Herness, Department of Geology, Colorado School of Mines, Golden, Colorado

John M. Hills, Consulting Geologist, Midland, Texas

H. A. Ireland, Department of Geology, University of Kansas, Lawrence, Kansas

J. Harlan Johnson, Department of Geology, Colorado School of Mines, Golden, Colorado

F. WALKER JOHNSON, Creole Petroleum Corporation, Maracaibo, Venezuela

Paul F. Kerr, Department of Geology, Columbia University, New York, New York

Truman H. Kuhn, Department of Geology, Colorado School of Mines, Golden, Colorado

J. L. Kulp, Department of Geology, Columbia University, New York, New York

H. L. Landua, Humble Oil & Refining Company, Houston, Texas

Arthur Langton, Baroid Sales Division, Los Angeles, California

Julian W. Low, The California Company, Denver, Colorado

V. J. Mercier, Mountain Iron and Supply Company, Wichita, Kansas

Carl A. Moore, Department of Geology, University of Oklahoma, Norman, Oklahoma

J. B. Murdoch, Jr., Eastman Oil Well Survey Company, Denver, Colorado

P. B. Nichols, Geolograph Company, Incorporated, Oklahoma City, Oklahoma

W. D. Owsley, Halliburton Oil Well Cementing Company, Duncan, Oklahoma

L. L. Payne, Hughes Tool Company, Houston, Texas

Gordon Rittenhouse, Department of Geology, University of Cincinnati, Cincinnati, Ohio

George L. Robb, United States Bureau of Reclamation, Denver, Colorado

N. Cyril Schieltz, United States Bureau of Reclamation, Denver, Colorado

G. Frederick Shepard, General American Oil Company of Texas, Dallas, Texas

L. L. Sloss, Department of Geology, Northwestern University, Evanston, Illinois

W. Alan Stewart, Department of Geology, Colorado School of Mines, Golden, Colorado

Harrison E. Stommel, Department of Geophysics, Colorado School of Mines, Golden, Colorado

E. F. Stratton, Schlumberger Oil Well Surveying Corporation, Houston, Texas

H. F. Sutter, Baroid Sales Division, National Lead Company, Houston, Texas

Wilfred Tapper, Halliburton Oil Well Cementing Company, Duncan, Oklahoma

John D. Todd, Consulting Petroleum Geologist, Houston, Texas

Paul D. Torrey, Lynes Incorporated, Houston, Texas

R, Maurice Tripp, Consulting Engineer, South Lincoln, Massachusetts

Warren R. Wagner, Department of Geology, Colorado School of Mines, Golden, Colorado

W. A. Wallace, Halliburton Oil Well Cementing Company, Houston, Texas

PREFACE TO THE FIRST EDITION

During the past fifteen years numerous papers pertaining to methods applied in deciphering subsurface geologic conditions have appeared in periodicals such as *World Oil*, the *Oil and Gas Journal*, the *Petroleum World*, the *Bulletin* of the American Association of Petroleum Geologists, the *Bulletin* of the Geological Society of America, the *Journal of Paleontology*, the *Journal of Sedimentary Petrology*, the *Transactions* of the American Geophysical Union and publications of the American Institute of Mining and Metallurgical Engineers. Commercial organizations have issued in pamphlet form discussions of techniques in the field of subsurface geology that have been highly enlightening.

Such text and reference books as *Petroleum Production Engineering* (McGraw-Hill) by Uren, *Manual of Sedimentary Petrography* (Appleton-Century) by Krumbein and Pettijohn, *Methods of Study of Sediments* (McGraw-Hill) by Twenhofel and Tyler, *Sedimentary Petrography* (Nordeman) by Milner, *Sedimentary Rocks* (Harper) by Pettijohn and *Examination of Fragmental Rocks* (Stanford University Press) by Tickell treat to a degree certain methods applicable to the establishment of subsurface values.

The need for a publication that would bring together information relating to subsurface geologic techniques that has hitherto been unavailable or scattered has long been realized by geologists and petroleum engineers. An attempt to satisfy this need is undertaken in the present text, *Subsurface Geologic Methods*. Had it not been for the liberal cooperation and interest of the contributors to this volume, such a compilation would not have been possible. Special recognition and appreciation are extended to those contributing sections of this text and to the organizations of which they are a part.

Many illustrations used in this symposium have been obtained from published sources, credit being given for each.

The section entitled "Multiple-Differential Thermal Analysis" has been reprinted by permission of the *American Mineralogist;* those entitled "Induction Logging" and "Dipmeter Surveys" are by permission of the American Institute of Mining and Metallurgical Engineers. The sections "Deep-Well Camera" and "Character of Pores in Oil Sand" are reprinted through the courtesy of *World Oil*.

Plans are to revise the symposium periodically as new methods are introduced or as the methods discussed are modified. Certain topics in this compilation may not be treated as completely as some individuals may wish; those desiring further detail should refer to the articles and publications cited. Criticisms and suggestions will be gratefully received by the editor, and these will be considered for subsequent editions.

It is hoped that *Subsurface Geologic Methods* will assist geologists and petroleum engineers in the field as well as educational institutions that offer formal courses in subsurface geology.

June 1, 1949

PREFACE TO THE SECOND EDITION

The first edition of *Subsurface Geologic Methods*, published in June of 1949, had a gratifying reception as a textbook for universities offering courses in subsurface geology. Owing to the unexpected rapid depletion of the supply of this volume, it was deemed necessary to revise and enlarge the book as a second edition.

The first printing of second edition, published in June, 1950, includes several new sections which are of concern either directly or indirectly to the subsurface geologist. These additions cover the following subjects: secondary-recovery methods, evaluation of petroleum properties, geochemical methods, micrologging, drill-stem testing, mud chemistry, cementing problems, acidization, shale-density analysis and graphic methods in mining. A series of questions has been added at the termination of each chapter to permit improved instruction in university work. An index has also been included in this volume.

Grateful acknowledgment is hereby given to the Department of Publications of the Colorado School of Mines for making possible the publication of this volume and to George W. Johnson, Acting Director of Publications, for his efforts and interest in the editing of the present volume and for supervising the numerous engraving and binding problems.

Special acknowledgment is given to the American Association of Petroleum Geologists for permission to reprint John M. Hills' paper, "Sampling and Examination of Well Cuttings," and to the American Institute of Mining and Metallurgical Engineers for the courtesy of permitting the publication of H. G. Doll's paper, "The Microlog."

Acknowledgment is also given William F. Dukes, student at the Colorado School of Mines, for the excellent and accurate drafting of 162 illustrations in both the first and second editions.

To former President Ben H. Parker of the Colorado School of Mines sincere appreciation is extended for his wholehearted support of this symposium.

I am greatly indebted to the contributors of this volume whose cooperation has made *Subsurface Geologic Methods* possible.

The demand for Subsurface Geologic Methods has exhausted the present supply and it has become necessary to issue a second printing.

June 1, 1951   L. W. LeRoy

CONTENTS*

Chapter   *Page*

1. Introduction *by L. W. LeRoy* 1

2. Stratigraphic, Structural, and Correlation Considerations
   by L. W. LeRoy 12
   Unconformities *by W. Alan Stewart* 32

3. Comments on Sedimentary Rocks *by L. W. LeRoy* 71

4. Subsurface Laboratory Methods 84
   Micropaleontologic Analysis *by L. W. LeRoy* 84
   Calcareous Algae *by J. Harlan Johnson* 95
   Detrital Mineralogy *by Gordon Rittenhouse* 116
   Insoluble Residues *by H. A. Ireland* 140
   Petrofabric Analysis *by Warren R. Wagner* 157
   Micro (Petrographic) Analysis *by Warren R. Wagner*
       and *John W. Gabelman* 172
   Size Analysis *by L. W. LeRoy* 184
   Settling Analysis *by L. W. LeRoy* 193
   Stain Analysis *by L. W. LeRoy* 193
   Shape Analysis *by L. W. LeRoy* 199
   Electron-Microscope Analysis *by Carl A. Moore* 202
   X-Ray Analysis *by N. Cyril Schieltz* 211
   Multiple-Differential Thermal Analysis *by Paul F. Kerr*
       and *J. L. Kulp* 240
   Water Analysis *by James G. Crawford* 272
   Core Analysis *by John G. Caran* 295
   Fluoroanalysis in Petroleum Exploration *by Jack De Ment* 320
   Shale Density Analysis *by F. Walker Johnson* 329

5. Subsurface Logging Methods 344
   Sampling and Examination of Well Cuttings *by John M. Hills* 344
   Electric Logging *by E. F. Stratton* and *R. D. Ford* 364
   Induction Logging and Its Application to Logging of Wells
       Drilled with Oil-Base Mud *by H. G. Doll* 393
   The Microlog *by H. G. Doll* 399
   Radioactivity Well Logging *by V. J. Mercier* 419
   Caliper and Temperature Logging *by Wilfred Tapper* 439
   Well Logging by Drilling-Mud and Cuttings Analysis
       by Arthur Langton 449
   Drilling-Time Logging *by G. Frederick Shepard* 455
   Driller's Logging *by L. W. LeRoy* 475
   Drilling-Time Logging *by P. B. Nichols* 478
   Spectrochemical Sample Logging *by L. L. Sloss* and
       S. R. B. Cooke 487
   Composite-Cuttings-Analysis Logging *by R. J. Gill* 495

6. Miscellaneous Subsurface Methods 504
   Controlled Directional Drilling *by J. B. Murdoch, Jr.* ... 504
   Oil-Well Surveying *by J. B. Murdoch, Jr.* 548
   Oriented Cores *by Kirk Carlsten* 591

---

* An index of authors and a subject index are included at the back of this volume.

CONTENTS—Continued

Chapter   *Page*

6. Miscellaneous Subsurface Methods—Continued
   Magnetic Core Orientation *by M. G. Frey*................................... 596
   Coring Techniques and Applications *by H. L. Landua*............... 609
   Application of Dipmeter Surveys *by E. F. Stratton* and
      R. G. Hamilton 625
   Design and Application of Rock Bits *by L. L. Payne*.................. 643
   Deep-Well Camera *by O. E. Barstow* and *C. M. Bryant*.............. 664
   The Electric Pilot in Selective Acidizing, Permeability
      Determinations, and Water Locating *by P. N. Hardin*........ 676
   The Porosity and Permeability of Clastic Sediments and
      Rocks *by George H. Fancher*... 685
   Drilling Fluid Chemistry *by H. F. Sutter*.................................. 713
   Hydrafrac Treatment *by W. E. Hassebroek*............................... 723
   Formation Testing *by W. A. Wallace*.. 731
   Oil-Well Cementing *by W. D. Owsley*....................................... 746
   Well Acidization *by L. W. LeRoy*.. 750
   Geochemical Methods *by R. Maurice Tripp*.............................. 760

7. Secondary Recovery of Petroleum *by Paul D. Torrey*.................. 775

8. Valuation and Subsurface Geology *by John D. Todd*.................. 792

9. Duties and Reports of the Subsurface Geologist
      by George W. Johnson.. 810

10. Graphic Representations *by L. W. LeRoy*................................. 856

11. Subsurface Maps and Illustrations *by Julian W. Low*................ 894

12. Subsurface Methods as Applied in Mining Geology
       by Truman H. Kuhn... 969

13. Subsurface and Office Representation in Mining Geology
       by Sigurd Kermit Herness.. 989

14. Subsurface Methods as Applied in Geophysics
       by Harrison E. Stommel...1038

15. Geologic Techniques in Civil Engineering *by George L. Robb*.........1120

16. Sources of Well Information..1150

CHAPTER 1

INTRODUCTION
L. W. LeROY

Subsurface geology involves interpretation of the stratigraphic, structural, and economic values below the earth's surface. These interpretations are based on information obtained from bore holes, geophysical data, and projected surface information. As stated by Jakosky,[1] "The subsurface is an extremely complex three-dimensional region, which the interpreter must diagnose with only the help of the very limited type of data obtained from our present techniques."

The subsurface geologist is required to have: (1) a basic background in geologic structure, (2) a broad and fundamental knowledge of rock types, (3) a three-dimensional concept of geologic phenomena, and (4) an understanding of the economics of the problem. He must be able to coordinate and accurately integrate these related phases. He must realize the importance of structure to stratigraphy and stratigraphy to structure. He must be aware that stratigraphic and structural problems cannot be solved without first evaluating the rocks and their relationships. To place rock types in their proper categories, emphasis must be made on the various techniques and methods which permit more exacting classification. Electrical, radioactive, caliper and other logging data cannot be accurately interpreted until the lithologic aspects of the strata have been established. The rocks and structures developed within them must be treated from the three-dimensional viewpoint because both change in space. Folding intensities and characteristics and lithologic patterns exhibit variations which must be recognized before appraising the geologic impress. Subsurface geology demands a creative imagination, an analytical and systematic approach, and a multiple-hypotheses manner of thinking. The geology of tomorrow depends largely on the assertiveness of the subsurface geologist of today.

Subsurface geology as applied in the petroleum industry has made rapid advances since 1925. In many areas it has attained greater importance than surface geology; the discovery of most future oil fields of the world will undoubtedly be attributed to subsurface geologic studies. Oil operators today fully realize the exigency of appraising subsurface conditions in exploration and development programs.

Methods applied in evaluating subsurface conditions vary in nature and complexity, depending on the character of the rocks penetrated, the type of equipment available, the quality and quantity of data desired, and the time allocated to the solution of the problem.

---

[1] Jakosky, J. J., *Whither Exploration:* Am. Assoc. Petroleum Geologists Bull., vol. 31, no. 7, p. 1121, July 1947.

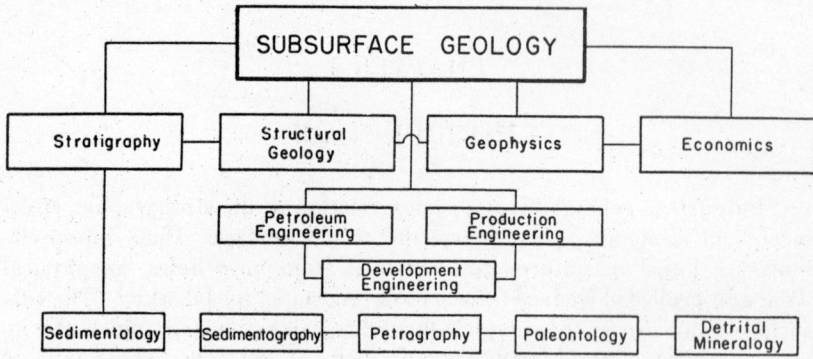

FIGURE 1. Subsurface geology and its relationship to other sciences.

The subsurface geologist should be familiar with all techniques that assist in determining subsurface phenomena and should recognize limitations of these methods.

Some 25,000 wells drilled each year add new sources of geological data requiring close study by specialists in petrology, paleontology, stratigraphy, geochemistry, geophysics, and structure who have the responsibility of determining correct regional correlations, suitable classification of the rock sequences, isopach and paleogeologic maps, faunal and floral zones, facies changes, vertical and lateral extent of oil- and gas-bearing zones, character

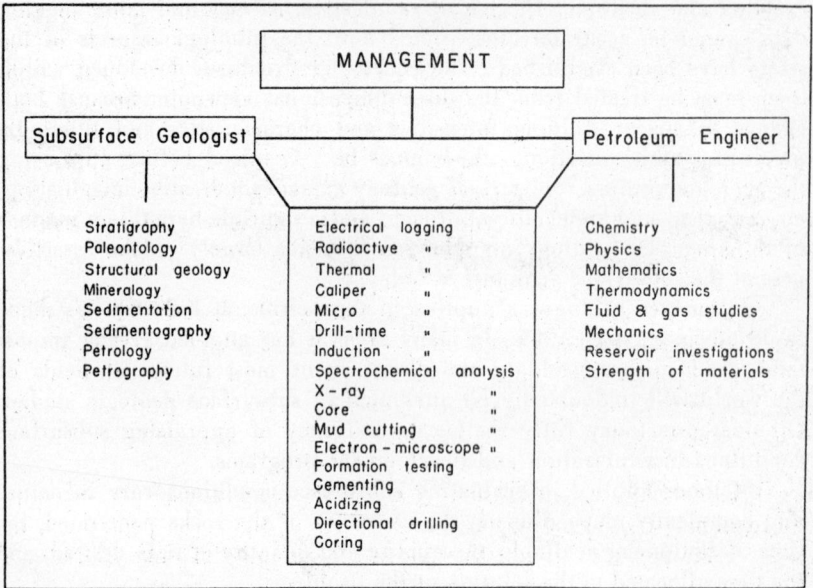

FIGURE 2. Correlation between some of the duties of the subsurface geologist and petroleum engineer.

FIGURE 3. Illustration of the complexity of subsurface conditions across the Permian Basin of west Texas. (*The Lamp*, 1948.)

of reservoir rocks and their fluids, time and place of progressive structural developments as well as the present normal strike and rate of dip for each "layer of geology." . . . Not only is there need of a greater store of geological information but also of well established criteria by which such data may be used more effectively. Much progress has been made toward the development and use of methods for recognition of favorable regions, localities, and well locations, but more definite analysis is no doubt possible as well as highly desirable.[2]

The subsurface geologist is scheduled to play a major role in fulfilling these requirements.

In the early history of drilling, little attention was given to the detailed characteristics of subsurface conditions; as a result, many geologic data were lost or so imperfectly recorded that reliable interpretations are now difficult to make. Present-day programs require systematic and accurate subsurface recordings. These requirements have vastly improved such problems as (1) exact structure contouring; (2) accurate definition of fault patterns and their relationship to oil-producing intervals; (3) the location and evaluation of unconformities; (4) facies changes and thickness trends; (5) the interpretation of geophysical data; (6) the origin, migration, and accumulation of hydrocarbons; (7) building, bridge, and dam foundations and tunneling; (8) locating and outlining ore bodies; (9) improvement of surface drainage systems; (10) evaluation of groundwater patterns; and (11) interpretation of soil data.

The decipherment of certain of these problems offers little difficulty and thus requires only a few techniques. Other subsurface conditions may be of such complexity that voluminous data are needed before logical and satisfactory answers become evident.

Noble[3] recently made the following statement:

Much of our future supply of oil will be found by close teamwork between various oil-finding groups, utilizing all of our present known prospecting tools and exploration methods. To a great extent this effort will consist of detailed subsurface studies in the search for stratigraphic traps and accumulations on the flanks and extensions of known structures; and deeper drilling wherever possible. There are, however, some rather extensive areas having good oil possibilities but about which we know very little because of a cover which masks the underlying geological conditions and which we cannot effectively penetrate with any of our present methods because of physical or economic considerations.

These masks include overthrust segments, blankets of young volcanics, thick alluvial and glacial deposits, multiple unconformities, and even water, which covers favorable stratigraphic and structural conditions on the continental shelves. Noble points out the oil and gas possibilities underlying the thrust sheets, as at Turner Valley in western Alberta, and similar conditions in Montana and the Pacific Coast and Midcontinent

[2] Cheney, M. G., The Geological Attack: Am. Assoc. Petroleum Geologists Bull., vol. 30, no. 7, pp. 1079-1080, July 1946.
[3] Noble, E. B., Geological Masks and Prejudices: Am. Assoc. Petroleum Geologists Bull., vol. 31, no. 7, pp. 1109-1117, July 1947.

regions. Attention must be given the possibilities buried beneath late Tertiary volcanics of the Rocky Mountain and Columbia Plateau areas and in Central and South America. The evaluation of these problems rests with the subsurface geologist.

DUTIES OF THE SUBSURFACE GEOLOGIST

The duties and responsibilities of the subsurface petroleum geologist are numerous and varied. He must be diversely trained and have an accurate sense of geologic and economic values. He must be able to present his data concisely, to adhere to his convictions, and to cooperate fully with the petroleum, production, and development engineer, the field geologist, the geophysicist, the management, and all other personnel that contribute to the solution of the subsurface problem. Frequently in exploration work he may be called upon to devote many continuous hours to special assignments; or he may be designated to obtain information from a "no dope" well, wherein he must investigate the casing program, the acidizing and shooting procedure, the record of mud, chemicals, and bit sales, and logging activities.

SOME MAJOR SUBSURFACE PROBLEMS

Subsurface problems in the fields of petroleum exploration and development are varied and numerous. The solution of some of these problems is relatively simple, whereas, the solution of others demands voluminous data that must be carefully screened and integrated. Some of the more important problems commonly encountered in subsurface work are as follows:

1. *Correlation of Surface to Subsurface Stratigraphic Units:* It has been demonstrated that both recent and ancient deposits of the stable shelf areas frequently are lithologically and faunally in discord with deposits of the unstable shelf. The intracratonic basin sediments and their organic elements vary widely with those that accumulated under geosynclinal conditions. Thus, wells drilled in a geosynclinal facies penetrate sections which require correlation with equivalent though lithologically and faunally dissimilar marginal strata. The establishment of correlations of this type are essential before the tectonic and sedimentational history of the region can be properly evaluated.

2. *Reef Investigations:* During the past few years considerable attention has been given reef production in Texas and Canada. Fanatical attempts have been made to improve and devise new methods applicable to the discovery of gas and petroleum reservoirs of this type. Seismograph exploration has been extremely instrumental in many of the reef production discoveries. Detailed lithologic, paleontologic, and well-logging data must be coordinated in reef investigations, as the reef elements (porosity,

permeability, composition, texture) change rapidly both vertically and horizontally. Reef trends and development are not constant; thus, it is necessary cautiously to outline drilling and leasing programs. Acidizing, shooting, and formation testing are
critical subsurface problems in reef production problems.

3. *Secondary Recovery:* Oil companies today are concerned more
than ever before with the problem of obtaining greater yield of
oil and gas per acre. Repressuring, water flooding, acidizing
and shooting of wells, directional drilling, proper water shutoffs, systematization of drilling, proper well spacing and controlled production are of major concern in secondary recovery
programs. The subsurface geologist and engineer must combine their efforts to secure best results in secondary recovery
problems.

4. *Interpretation of Well-logging Data:* Since 1930 the electrical
log has been very successful in evaluating features of the penetrated strata. Since this date radioactive, caliper, thermal, drill-
time, induction and micrologging have been introduced into
subsurface investigations. Information obtained from these
logs is based on the characteristics of the rocks—their composition, texture, and fluid and gas content. Many profile anomalies
cannot be adequately explained; thus, more attempt should be
made to evaluate these idiosyncrasies by detailed analysis.

5. *Acidizing and Shooting:* What interval to acidize and to shoot
is a problem of major concern to operators drilling in carbonate
sections. Before either or both methods are initiated, the lithologic and structural aspects of the strata should be adequately
known and this information integrated into other logging data.

6. *Prediction of Drilling Difficulties:* The oil operator and contractor are vitally interested in knowing the difficulties and
magnitude of difficulties prior to commencement of drilling. A
sandstone-shale section would present different problems than
a carbonate section or a section containing numerous beds of
salinifereous material. A rock sequence containing considerable
bentonite might alter an entire drilling and casing program.
Other problems may be encountered during penetration of fault
surfaces, unconformities and vuggulated strata in which circulation could not be maintained.

7. *Improved Subsurface Mapping:* Subsurface data may be conveniently represented by contour-type maps. These maps are
based on structural, isopachous, isochor, isothermal, isosperm,
isochron, isopotential, lithofacies, and depth pressure information. Viscosity, fluid and gas density maps may also be prepared. Paleogeologic data in many areas are commonly plotted

and shown in map form. All such maps permit an improved understanding of subsurface conditions.

8. *Unconformities:* The character, extent of erosional surfaces, and the relationship of such surfaces to adjacent strata are often much improved by subsurface information. These surfaces must be accurately defined before stratigraphic and economic values can be evaluated.

9. *Onlap and Offlap:* Onlap and offlap problems require the three-dimensional approach. Subsurface studies permit determination of rate, dimension, and trend of these depositional conditions.

10. *Miscellaneous Problems:* Other subsurface problems confronting the geologist and engineer in addition to those mentioned above include: cementing, setting of casing, hole caving, fishing, stabilization of drilling fluids, perforating, formational water variations, porosity and permeability changes, coring, and testing.

TRAINING OF THE SUBSURFACE GEOLOGIST

Courses in subsurface geology, as given in some universities in the United States, vary according to geographic location, facilities, and instructional personnel. Prior to these courses the student should have a thorough background in petrology, petrography, structural geology, field geology, petroleum geology, mineralogy, stratigraphy, sedimentation, paleontology, and geophysics.

A sequence of subject material in a formal university course in subsurface geology is suggested.

Lithologic Studies: Lithologic types including shales, limestones, dolomites, sandstones, and other lithologic varieties should be examined and studied under the binocular microscope. Each lithology should be represented by chip fragments in a reservoir-type slide and viewed by the student during instruction. This method is extremely applicable and time-saving in preparing the student for subsequent well-logging assignments.

Well-Logging Methods: The various types of well-logging methods should be briefly summarized. These methods should include lithologic, electric, radioactive, drill-time, caliper, and thermal logging. The instrumentation, use, and limitations of these methods should be treated.

Theoretical Electrical Profile Interpretation: After the student has become familiar with lithologic and electrical relationships, he should be required to plot from tabulated data (mimeographed) a percentage log from which an interpretive log is prepared. On the basis of the latter log theoretical electric profiles may be drawn. This problem demands that the student think in terms of both lithology and its probable electrical reflection.

Preparation of Well Log: The student, once having become familiar with the basic lithologic varieties and logging methods, should be assigned a well section to log. This work involves the examination and plotting of a lithologic log from ditch cuttings and cores. The lithology should first be plotted on a percentage basis, and subsequently the lithic boundaries should be adjusted and an interpretive log prepared. Color symbols should be used to represent lithologies (pl. 11). In addition to interpreting and recording the lithologic sequence of a well, each student should have available an electric-log profile with which to make comparisons. Upon completion of the log a final report of the well should be required.

Contouring: Several electric-log series from oil fields should be studied and correlated, and subsurface structural and isopachous maps prepared. Cross sections should also be incorporated. In addition to this problem, a number of theoretical contour problems should periodically be submitted to the class to improve structural interpretation.

Correlation: Paired electric, radioactive, and lithologic logs from various fields should be correlated in detail, and the results carefully drafted and discussed.

Principles of Stratigraphy and Correlation: The fundamental concepts of stratigraphy and correlation should be treated and should include such topics as facies changes, unconformities, onlaps and offlaps, and any other criteria that control the correlation of strata.

Laboratory Methods: The various techniques followed in subsurface or stratigraphic laboratories should be broadly outlined and reviewed and should include such topics as detrital mineralogy, insoluble residues, stain tests, micropaleontology, and thin-section, screen, and sedimentation analyses.

Miscellaneous Topics: Such topics as directional drilling, acidizing, cementing, secondary recovery, formation testing, and coring should be briefly reviewed with emphasis placed on the geologic aspects.

In the graduate school more specialized subsurface problems should be emphasized. The subject matter should include structure contouring, fault problems, correlation interpretation, paleogeologic and lithofacies mapping, isopachous studies, and the preparation of subsurface geologic reports. These courses should be presented with the intention of introducing to the student the "work pressure" factor that prevails in economic programs.

Those students having a geologic-engineering background are favorably adapted for subsurface geologic investigation. This does not imply that such training is essential to develop good subsurface personnel; it cannot be denied, however, that the basic sciences of mathematics, physics, chemistry, hydraulics, thermodynamics, and descriptive geometry provide favorable attributes.

To train an individual for subsurface geology as applied today in the petroleum industry should require at least eight years after termination of his advanced academic work. A minimum of two years in surface geologic mapping is a necessity. Subsurface phenomena would be difficult to interpret adequately without exposed stratigraphic and structural relationships having first been observed and studied. The student of subsurface geology should spend at least three years in stratigraphic and paleontologic laboratories in order to become familiar with fundamental techniques applied in subsurface problems. This period of training should be divided among the Pacific Coast, the Gulf Coast, and the Mid-continent areas. One year should be devoted to geophysical work, another to logging methods, and the last to petroleum and production engineering.

FUTURE OF SUBSURFACE GEOLOGY

Most of the oil fields and ore deposits in the future will be discovered from the application and coordination of subsurface investigations.

. The number and complexity of subsurface methods now employed require specialists. There are those who devote their efforts to lithologic studies, to paleontologic investigations, and to logging methods, and others to interpretation of structural conditions. The field of geophysics offers an unlimited opportunity for the subsurface geologist whose responsibilities are to synchronize electrical and gravity data with stratigraphy and structure. It is essential that each specialist know the position and relationship of his field to the general subsurface problem involved.

Subsurface geologic methods and approaches as employed by the petroleum industry have not yet been fully accepted or utilized by the mining industry in exploration and exploitation. In the early history of mining only the surface or near-surface deposits were exploited and developed. Today it is required that possibilities of deeper ore concentrations be considered. Cram [4] comments: "Mineral geologists must be permitted by the mineral industry to spread their wings on a full-time basis, not on a consulting basis, if the nation's undiscovered reserve of minerals is to be developed."

Subsurface investigations are scheduled to occupy an important position in engineering geology. Before a civil engineer can properly design any structure, he should be versed in the materials upon which his structure is to rest or in which his work is to be carried out. In any engineering project, preliminary investigations are required and should be undertaken with two objectives in view: (1) the evaluation of subsurface conditions at and in the immediate vicinity of the proposed site that may affect the work program, these conditions generally involving local geologic structure and distribution and character of undergound water; and (2) the determination of the character of rock types expected during the progress of construction.

Many projects involving tunnels, excavation, earth movements, bridge, dam, and building foundations, and water supply require the

---

[4] Cram, I. R., *Geology Is Useful:* Am. Assoc. Petroleum Geologists Bull., vol. 32 no. 1, pp. 7-8, Jan. 1948.

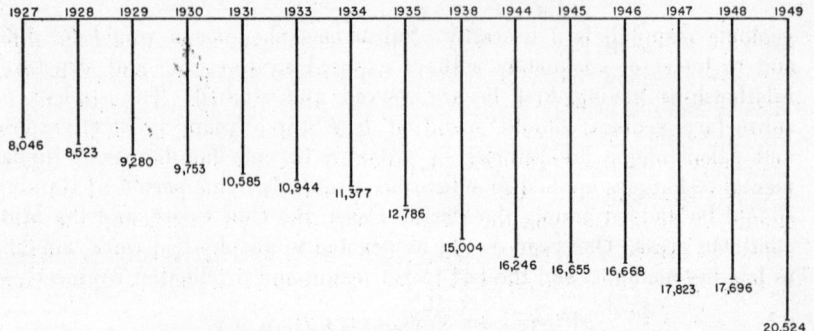

FIGURE 4. Drilling-depth records from 1927 to 1949. (Modified from *World Oil*.)

services of a geologist who has the ability to coordinate and evaluate sub-surface data.

1927 Chansler-Canfield-Midway Oil Company's Olinda 96, Olinda, California.
1928 Texon Oil and Land Company's University 1-B, Big Lake, west Texas.
1929 Shell Oil Company's Nesa 1, Long Beach, California.
1930 Standard Oil Company of California's Mascot 1, Midway, California.
1931 Penn-Mex Fuel Company's Jardin 35, State of Vera Cruz, Mexico.
1933 North Kettleman Oil & Gas Company's (later taken over by Union Oil Company) Lillis-Welch 1, Kettleman Hills, Kern County, California.
1934 General Petroleum Corporation's Berry 1, Belridge, Kern County, California.
1935 Gulf Oil Corporation's McElroy 103, Gulf-McElroy, Upton County, west Texas.
1938 Continental Oil Company's KCLA-2, Wasco, Kern County, California.
1944 Standard Oil Company of California's KCL 20-13, South Coles Levee, Kern County, California.
1945 Phillips Petroleum Company's Schoeps 3, wildcat, Brazos County, Gulf Coast, Texas.
1946 Pacific Western Oil Corporation's National Royalties 1, Miramonte area, Kern County, California.
1947 Superior Oil Company of California's Weller 51-11, wildcat, Caddo County, Oklahoma.
1948 Standard Oil Company of California's Maxwell 1, Ventura County, California.
1949 Superior Oil Company's Pacific Creek Unit 1, Sublette County, Wyoming.

COMPARATIVE USE OF SOME SUBSURFACE TECHNIQUES

At present, lithologic- and electric-logging methods are most extensively used in correlating the substrata. Radioactive and drill-time logging are becoming increasingly important and will play a greater role in future subsurface evaluations. Radioactive logging (gamma and neutron) has proved its dependability in limestone and dolomite sections by its ability to indicate porous strata which indirectly reflect possible petroliferous zones. Controlled mechanical means of recording penetration rates have greatly enhanced the value of the drill-time log and have placed data of this category on a firm basis.

Of the micropaleontologic criteria employed in correlating the substrata, the use of Foraminifera is the most widely applied, with that of ostracods second in application. Insoluble-residue work is adaptable locally and has been of major assistance in correlating carbonate sections. Detrital mineralogy has been minimized in stratal correlations; locally, however, it has its value. Table 1 indicates the relative usage of the more common methods of correlating subsurface strata in various areas.

TABLE 1

RELATIVE IMPORTANCE OF THE MORE COMMON METHODS USED
IN CORRELATING SUBSURFACE STRATA IN VARIOUS AREAS

	Lithology	Electric logging	Radioactive logging	Drill-time logging	Paleontology		Insoluble residue	Detrital mineralogy
					Foraminifera	Ostracoda		
Pacific Coast	2-3	1	3	5	1			5
Gulf Coast	3	1	3	5	1	2-3		5
Permian Basin	1	1	3	2	1		2	
Midcontinent (Kansas, Oklahoma)	1	1	3	2-3	2		4	
Illinois Basin	1	1	2-3	2-3		4	3	
Rocky Mountain region	2	1	3	6			6	5
Western Venezuela	1-2	1	4	4	2			2
Eastern Venezuela	1-2	1	4	4	2			
Burma	2	1			1			
Trinidad, B. W. I.	2	1		5	1			
Central Sumatra	2	2-3			1	3		

QUESTIONS

1. Define "subsurface geology."

2. Why has it been necessary to improve subsurface geologic practices?

3. Subsurface methods vary in nature and complexity. Why?

4. Present-day petroleum-exploration programs require systematic and accurate subsurface interpretation. These requirements have improved what problems?

5. What are some of the essential duties of the subsurface geologist?

6. Review table 1 which concerns the relative importance of methods used for correlating strata in the subsurface.

7. Compare the drilling records from 1927 to 1949.

8. What is meant by "geologic masks"? Give several examples.

CHAPTER 2

STRATIGRAPHIC, STRUCTURAL, AND CORRELATION CONSIDERATIONS

L. W. LeROY

Stratigraphy incorporates the study of the character, sequence, relationship, distribution, and origin of sedimentary rocks. As expressed by Kay,[1] "Stratigraphy is veritably the interpretation of the record of progressive movements (crustal) evidenced in sedimentation." Stratigraphy constitutes the basis of correlation, and correlations are primary requisites for surface and subsurface mapping and for the evaluation of structural and sedimentational patterns.

The basic principles of stratigraphic geology have remained more or less unmodified, although methods and techniques applied in the solution and presentation of stratigraphic problems have greatly improved during the last decade. In recent years emphasis has been placed on graphic representation of stratigraphic data. The construction of lithofacies, isopachous, log, paleogeographic, paleogeomorphic, paleoclimatologic, paleogeologic, and palinspastic maps has contributed enormously in outlining stratigraphic trends and has introduced new avenues of stratigraphic approach. The recent contributions by Krumbein,[2] and Dapples, Krumbein, and Sloss [3] [4] are examples of new lines of thought in stratigraphic compilation. Block diagrams are being used frequently in order to illustrate third-dimensional effects and to assist nongeologic personnel better to understand stratigraphic concepts. Such diagrams are particularly useful in preparing regional reconnaissance reports.

Large sums of money are being spent annually by oil companies on stratigraphic research and in the training of specialists in such fields as micropaleontology, lithology, detrital mineralogy, insoluble residue, and well logging. Some companies prefer training individuals who will subsequently devote their major efforts to restricted phases of stratigraphic geology; other companies favor personnel familiar with diversified stratigraphic procedures.

Many mineral concentrations occur in sedimentary rocks. They may be of either primary or secondary origin. Silver chloride has been noted in the cross-laminated Painted Desert sandstone in southwestern Utah. In western Colorado and southeastern Utah uranium and vanadium minerals

---

[1] Kay, Marshall, *Analysis of Stratigraphy:* Am. Assoc. Petroleum Geologists Bull., vol. 31, no. 1, pp. 161-181, Jan. 1947.
[2] Krumbein, W. C., *Lithofacies Maps and Regional Sedimentary-Stratigraphic Analysis:* Am. Assoc. Petroleum Geologists Bull., vol. 32, pp. 1909-23, 1948.
[3] Dapples, E. C., Krumbein, W. C., and Sloss, L. L., *Tectonic Control of Lithologic Associations:* Am. Assoc. Petroleum Geologists Bull., vol 32, pp. 1924-47, 1948.
[4] Sloss, L. L., Krumbein, W. C., and Dapples, E. C., *Integrated Facies Analysis:* Geol. Soc. America Mem. 39, pp. 91-123, 1949.

occur disseminated throughout Jurassic and Triassic sandstones. Possible minor sedimentary copper deposits are widely distributed in Texas, New Mexico, Colorado, Utah, and Arizona. Manganese minerals (sulphides, oxides, carbonates, and silicates) are widespread in sedimentary sections west of the Mississippi Valley. Rich deposits of phosphate occur in certain Permian strata in Idaho and Wyoming, in the Silurian and Devonian of Tennessee, and in the Tertiary of the Carolinas and Florida. Vast reserves of potash (polyhalite and sylvite) are associated with Permian strata in New Mexico and west Texas. Other stratigraphically controlled deposits of nonmetallics include clay, borates, sulphur, gypsum, magnesite, barite, celestite, strontianite, diatomite, limestone, dolomite, slate, and marble. Lead and zinc deposits occur in limestone, dolomite, and calcareous shale in various parts of the world. The distribution of the Clinton iron ores (Silurian) of the Appalachian States and the hematite deposits of the Lake Superior region are governed mainly by stratigraphic fabric.

The distribution, degree of localization, and value of many pyrometasomatic ore deposits are largely dependent upon the type of sedimentary section intruded. Rarely are these deposits found in argillaceous strata, sandstones, and shales, whereas limestones, dolomites, and calcareous shales are more reactive and thus are most adaptable for mineral concentrations.

Many deposits formed under mesothermal conditions (mineralization at 200°–300° C.) occur in sedimentary rocks. Examples of mesothermal replacement deposits involving sedimentary strata are known in the Cordilleran region of the United States and elsewhere in the world.

Hypothermal or deep-seated (mineralization at 300°–500° C.) ore accumulations are commonly associated with highly metamorphosed sediments.

Unconformities and variations in porosity, permeability, competency, composition planes, texture, and chemical composition of the host rock are some of the controlling factors governing the development and location of ore bodies.

From the foregoing it is obvious that the stratigraphy of a mineralized area should be carefully evaluated during prospecting and development stages. The structural aspects of a region are equally important. Structural irregularities cannot be satisfactorily evaluated without knowledge of stratigraphic relationships. Similarly, stratigraphic values may be erroneously recorded if structural conditions are inadequately known.

SUBDIVISIONS OF STRATIGRAPHIC GEOLOGY

Stratigraphic geology may be subdivided into two major categories, macrostratigraphy and microstratigraphy. The former involves field observation and interpretation of exposed stratigraphic sequences, whereas the latter implies laboratory approach and routine and detailed evaluation of stratal successions. To obtain the maximum value from stratigraphic investigations both informational sources must be harmoniously synchronized. Field and laboratory personnel attacking a stratigraphic problem should be thoroughly aware of their limitations.

The macrostratigrapher operates under diverse conditions, depending on the nature of the assignment (detailed, semidetailed, or reconnaissance), the character and desired quality of results, the quality of assisting personnel, the time allotted to the problem, and the field environment. He should be aware that inadequate field control promotes erroneous microstratigraphic conclusions, which may introduce unnecessary and excessive expenditure for the operating company. The macrostratigrapher is responsible for field mapping, the interpretation of structural anomalies, the selection and definition of type outcrop sections, the collection of stratigraphically controlled representative samples, the orientation of facies variations, and the establishment of formational units. It is essential that the macrostratgrapher periodically became acquainted with problems confronting the microstratigrapher.

The microstratigrapher controls and coordinates laboratory procedures essential for stratigraphic refinements. He must be versed in basic geologic and stratigraphic principles, their applications, limitations, and interrelationships, as well as be thoroughly familiar with methods required to decipher stratigraphic problems. The laboratory should be systematically organized and the personnel sufficiently trained to minimize the time factor, as the macrostratigrapher is invariably concerned with knowing the results of the analyses of his samples as soon as possible. The microstratigrapher should have knowledge of Foraminifera, ostracodes, diatoms, Radiolaria, and Mollusca, from which depositional-environmental deductions may be made. The basic fundamentals of lithology, detrital mineralogy, thin and polished sections, stain tests, insoluble-residue techniques, and porosity and permeability tests should be ably and efficiently applied whenever the occasion demands. The microstratigrapher should be aware of the principles and significance of electric, radioactive, thermal, caliper, and drill-time logging as these methods have either a direct or indirect bearing on the interpretation, evaluation, and correlation of stratigraphic sequences.

NATURE AND CLASSES OF STRATIGRAPHIC UNITS

In 1933 a stratigraphic code, commonly cited as the "Ashley et al. report,"[5] was published for the purpose of minimizing inconsistencies in stratigraphic terminology. For thirteen years this code served as a basis for stratigraphic standardization. In 1946 representatives of the Association of American State Geologists, the American Association of

---

[5] Ashley, G. H., et al., Classification and Nomenclature of Rock Units: Geol. Soc. America Bull., vol. 44, pp. 423-459, 1933; Am. Assoc. Petroleum Geologists Bull., vol. 17, pp. 843-868, 1933; vol. 23, pp. 1068-1069, 1939.

Petroleum Geologists, the Geological Survey of Canada, the Geological Society of America, and the United States Geological Survey met at Chicago under the chairmanship of R. C. Moore to discuss reorganization and improvement of the Ashley report. As a result of this meeting the American Commission on Stratigraphic Nomenclature was founded. The purposes of the commission

. . . are to develop a statement of stratigraphic principles, to recommend procedures applicable to classification and nomenclature of stratigraphic units, to review problems in classifying and naming stratigraphic units, and to formulate expression of judgment.[6]

Distinction between time, time-rock, and rock units must be recognized by geologists before satisfactory stratigraphic concepts can be harmoniously discussed. Renz [7] ably clarifies this necessity by saying:

During the last decade, a number of geologists and stratigraphers in the United States have strongly advocated adopting uniformity in stratigraphic nomenclature and following more closely the original definitions of the terms to be used. Incorrect application of terms in stratigraphic geology causes confusion and misunderstanding, thereby impeding or even nullifying a clear conception of the stratigraphic conditions of areas to be studied from available publications. . . . In setting up the stratigraphy of a given area, a clear distinction has to be made between the classification of rocks into lithogenetic units of various magnitudes, such as groups, formations, and members, and the classification of the same rock sequences into time-stratigraphic units delimited by the vertical ranges of fossil life; such time-stratigraphic units are termed "series," "stages," "zones," etc. The corresponding time units, such as epoch, age, and secule (moment), express the interval of time during which these stratigraphic units were deposited.

The establishment of lithogenetic units is the domain of the field geologist who maps them in the field according to the physical expression of the rocks only, without special reference to the stratigraphic range of the fossils they may contain. The paleontologist and biostratigrapher, on the other hand, build up their classification into time-stratigraphic units by studying the vertical range of fossil life. The classifications arrived at independently by the field geologist and by the biostratigrapher may, but often do not, coincide. In general, lithogenetic units have a rather limited geographic extent and are useful for correlation over comparatively small areas only. On the other hand, time-stratigraphic units are prone to exceed the geographic extent of lithologic units and, therefore, are more useful for regional or even interregional correlations.

To promote better understanding of these stratigraphic terms, the American Commission on Stratigraphic Nomenclature has recommended the following three classes of stratigraphic units: (1) "time units" for divisions of geologic time, (2) "time-rock units" for divisions of rocks segregated on the basis of their relation to determined segments of geologic time, and (3) "rock units" for divisions of rocks segregated on the basis

[6] *Organization and Objectives of the Stratigraphic Commission:* Am. Assoc. Petroleum Geologists Bull., vol. 31, no. 3, pp. 513-518, Mar. 1947. Prepared by R. C. Moore.
[7] Renz, H. H., *Stratigraphy and Fauna of the Agua Salada Group, State of Falcón, Venezuela:* Geol. Soc. America Mem. 32, pp. 1-219, 1948.

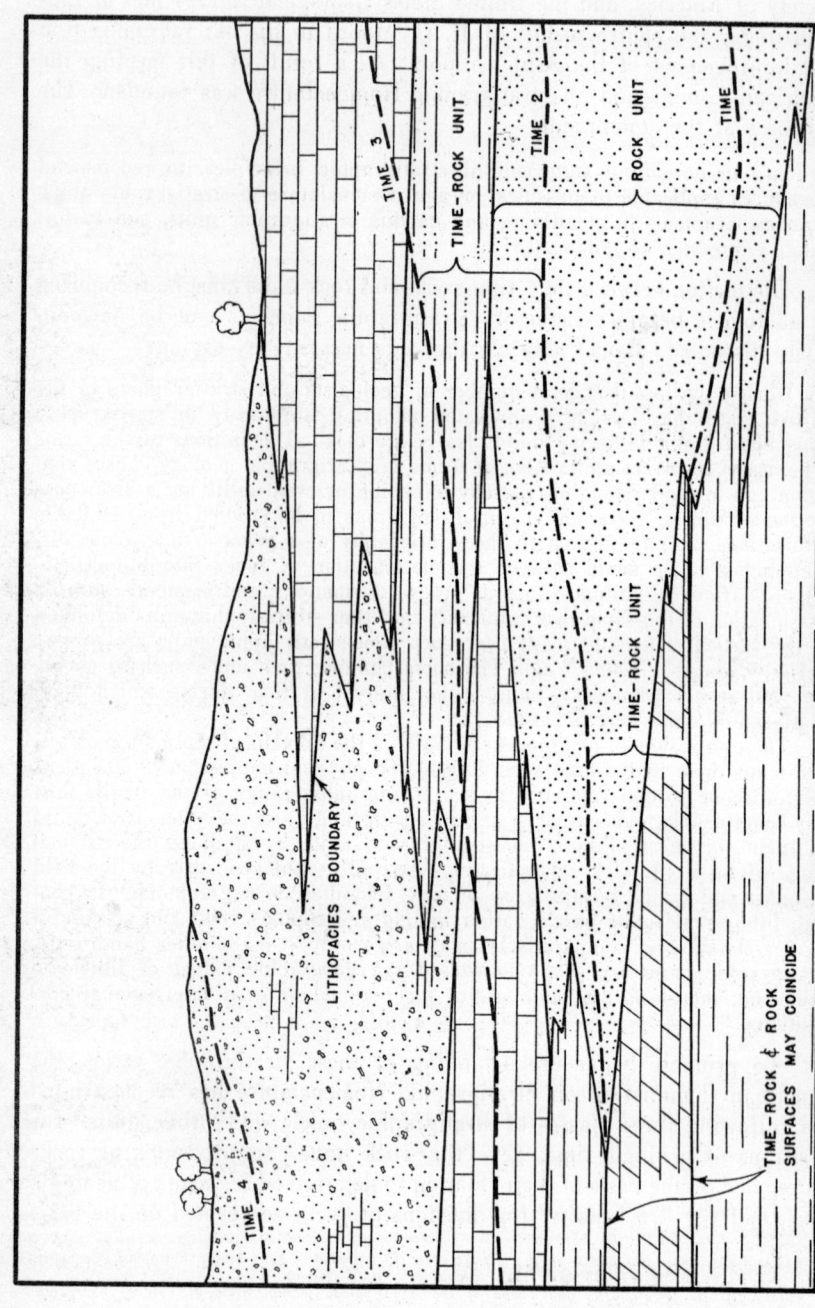

FIGURE 5. Diagrammatic sketch showing relationship of rock and time-rock units. Heavy dashed lines (Time 1, 2, etc.) represent surfaces of contemporaneous deposition. Observe how these surfaces transect lithofacies boundaries.

of objective characteristics deemed to have significance in classification, a differentiation not based on time relations. (See figs. 5 and 6.)

Time units, involving eras, periods, epochs, and ages, are defined indirectly and somewhat indefinitely on the basis of time-rock units. They represent time spans.

Time-rock (time-stratigraphic) units, including systems, series, and stages, have time boundaries only and are represented by sediments deposited during time intervals. Stratal thickness is not involved. The boundaries of time-rock units are essentially established on the basis of paleontology. Hedberg [8] comments:

> Fossils, of course, constitute one of the best means of both correlating and dating rocks; because of the more or less orderly evolutionary sequence of life forms on the earth (worked out, however, only through relation of fossil occurrences to the succession and superposition of strata), they constitute by far the most effective means of setting up a *chronological* system of time-stratigraphic divisions. However, there are limits to the resolving power of fossils as chronological indicators. While sediments differing in age by twenty million years, for example, may be readily placed in their correct sequence by an experienced paleontologist, smaller differences in time become progressively more difficult to place correctly, and a limit is finally reached where facies variations and other factors completely mask the changes in fossil record due to difference in age.
>
> Numerous other features besides order of superposition and paleontology can contribute evidence of relative age. Among these are radioactive measurements, relations to diastrophic events, evidences of volcanic activity, climatic changes, unconformities, sedimentary cycles, transgressions, and regressions. Many of these may, in special cases, become of outstanding importance and exceed in value all other means. However, only fossils (and perhaps radioactive measurements) are of much service in determining complete and worldwide *geochronological* sequences. . . . *In short, it is desirable to be able to express as a time-stratigraphic unit the sediments equivalent in age to the time scope of any recognizable features of sedimentary rocks which may be useful as a stratigraphic measuring stick.*

Relationships between time-rock and rock boundaries are difficult to establish, and many are impossible to evaluate accurately.

Time surfaces may be defined by (1) careful study of stratigraphic sections containing lithologies and faunas common to two or more controlled stratal sequences; (2) "walking out" of key beds such as ash, bentonite, and limestone; (3) correlation of benthonic faunas possessing wide ecologic valence; (4) application of pelagic faunas and floras; (5) widely dispersed detrital minerals; (6) vertical limits of faunal sequences; and (7) biologic evolutionary changes.

Rock units (lithogenetic units), including the group, formation, member, lentil, tongue, stratum, and layer, are defined on the basis of lithology. These units are essential in geologic mapping, description, expressing structural conditions, and deciphering the geologic history of

[8] Hedberg, H. D., *Time-Stratigraphic Classification of Sedimentary Rocks:* Geol. Soc. America Bull., vol. 59, pp. 447-462, May 1948.

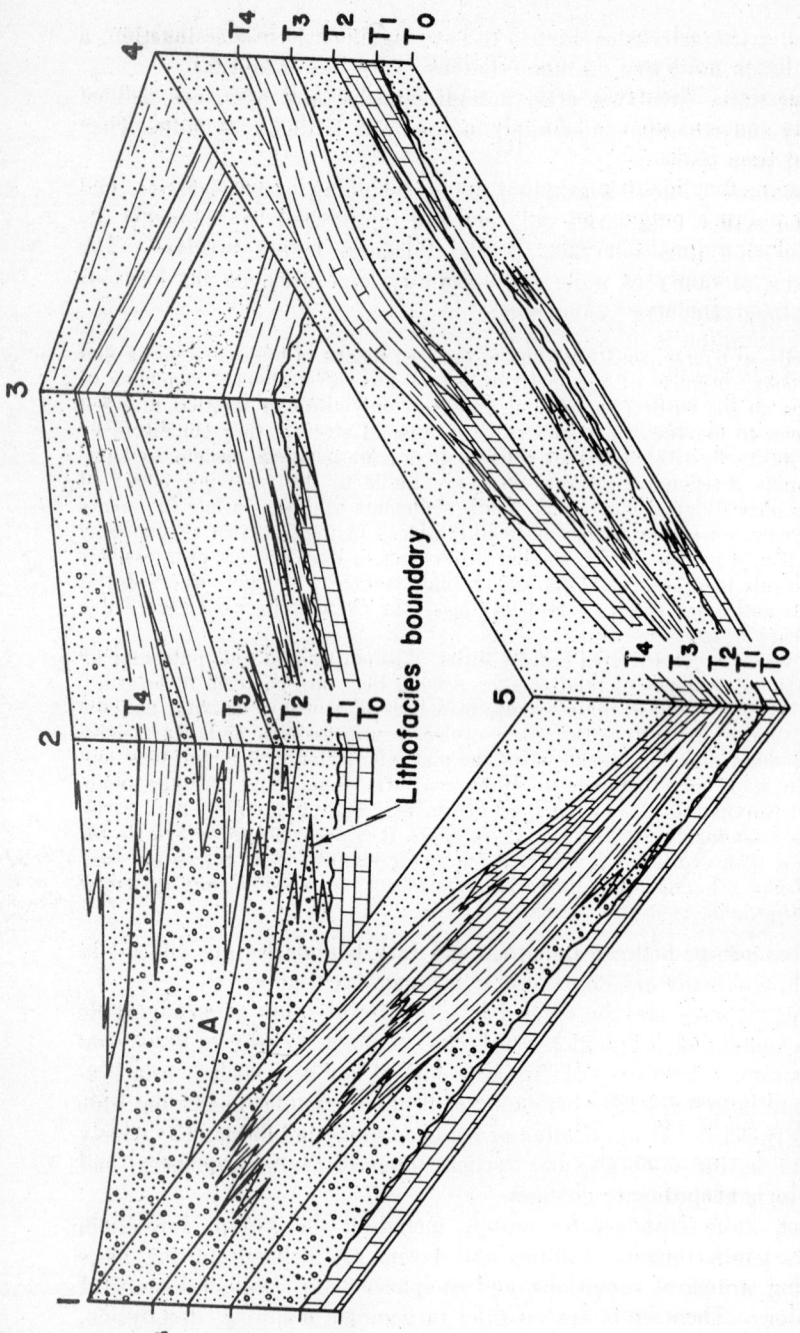

FIGURE 6. Sketch demonstrating lateral gradation and intertonguing of strata (facies changes). Sediments included between the time surfaces (T_0, T_1, T_2, T_3 and T_4) represent time-rock units. Formation A typically exemplifies a rock unit transecting time surfaces. From this diagram it is obvious that relationships between time-rock and rock units must be established before the geologic history of a region can be interpreted accurately.

expressing structural conditions, and deciphering the geologic history of a region. Rock units may be objectively shown on a map and in stratigraphic sections. Their boundaries do not necessarily coincide with time boundaries. Several rock units may be incorporated with a time-rock unit.

It has been indicated [9] by the American Commission on Stratigraphic Nomenclature that in treating a sedimentary formation (rock unit), the following points should be considered: (1) It must contain no apparent evidence of an appreciable break in deposition. (2) In its simplest form it consists dominantly of one general type of rock. (3) Recognition of the same formation in different areas is only justified when its essential lithologic definition is applicable. (4) The upper or lower contacts of a sedimentary formation may laterally transgress horizons of a neighboring formation. (5) The top and bottom of a sedimentary formation are defined either by a change in lithology or by evidence of an appreciable

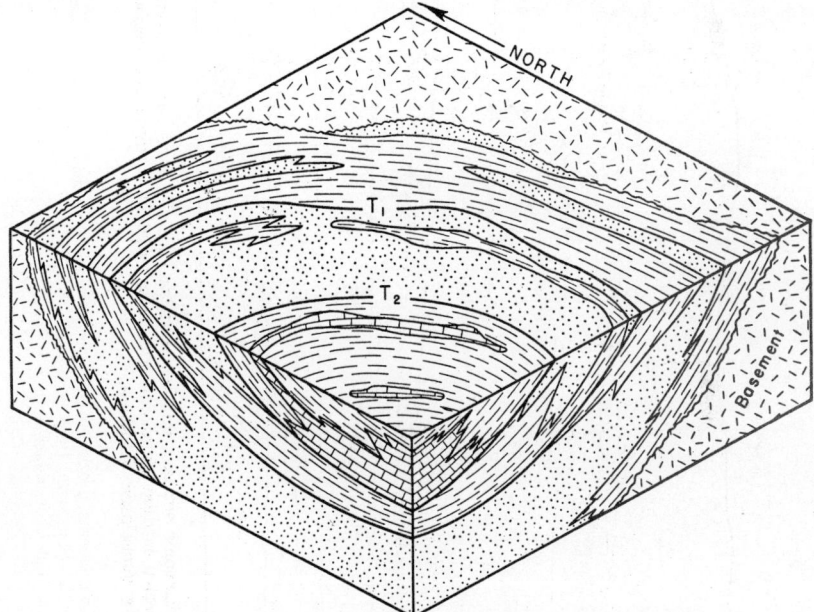

FIGURE 7. Time-space relationships of sedimentary deposits. T_1 and T_2 represent time surfaces. Changes in sedimentary rock types must always be considered in three dimensions.

interval of nondeposition. (6) A formation may hold one or more faunas or floras. (7) A sedimentary formation may include minor developments of volcanic rocks. (8) Pyroclastic materials, whether deposited in water or on land, are to be regarded as volcanic sediments and, hence, as constituents of sedimentary formations.

In naming surface rock units the following points should be con-

[9] *Rules of Geologic Nomenclature of the Geological Survey of Canada:* Am. Assoc. Petroleum Geologists Bull., vol. 32, no. 3, pp. 366-381, Mar. 1948. Prepared by R. C. Moore.

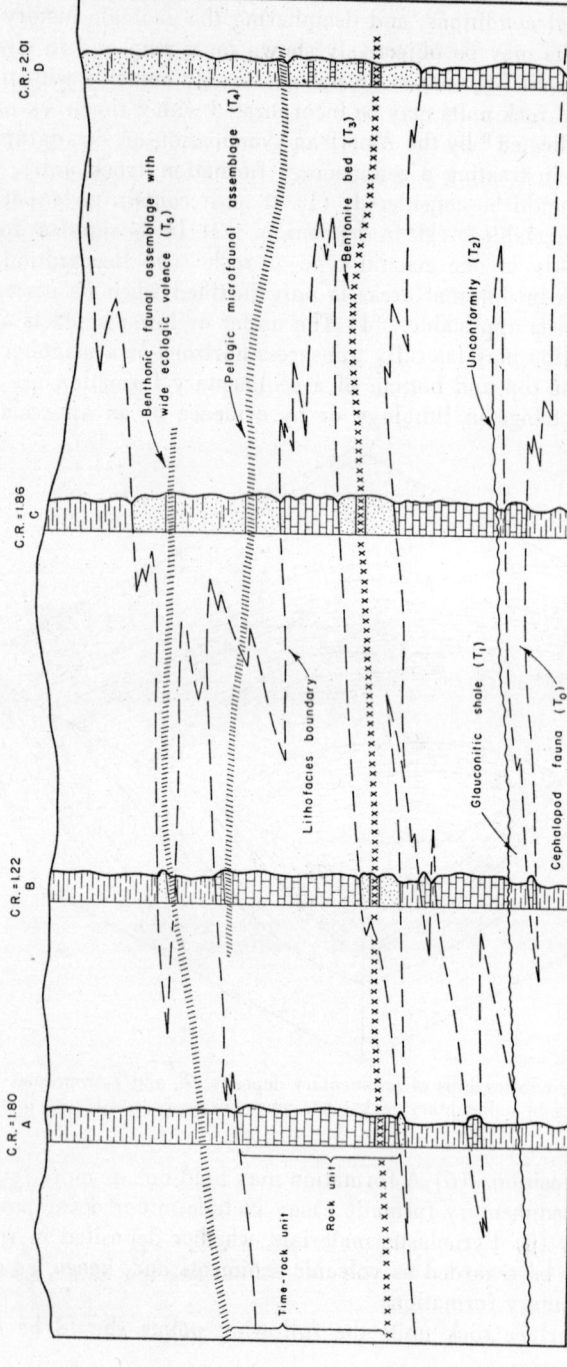

FIGURE 8. Use of control stratigraphic sections and key horizons (T_1, T_2, etc.) for integrating sedimentary facies relationships. Clastic ratio (C. R.) for each section is determined and the value used in deciphering regional paleogeologic problems. Clastic ratio = clastic thickness divided by non-clastic thickness.

sidered: (1) the description of the type locality or area; (2) the detailed and summary lithologic description; (3) a statement pertaining to thickness variations; (4) a statement concerning the surface and subsurface distribution of the unit; (5) a discussion of stratigraphic relationships; (6) comments on detrital mineralogy and paleontology; (7) a statement pertaining to physiographic expression; (8) a discussion of correlation; (9) comments on environment of deposition; and (10) economic aspects. Similar procedure should be followed in defining groups, members, lentils, or other rock units.

In naming subsurface rock units the American Commission on Stratigraphic Nomenclature [10] recommends that:

Subsurface units may be given formal names when such names are necessary for adequate presentation of the geologic history of the region or when the subsurface section is materially different from equivalent exposed beds.

(a) Subsurface units, recognized and named from well logs, including sample logs, core records, and electrical and other mechanically recorded logs, with the assistance of paleontological determinations differ from surface units in that the "type locality" cannot be visited and restudied from time to time by subsequent workers. With the advent and wide dissemination of the electrical log and the systematic saving of representative well cuttings, particularly in areas of deep drilling, it frequently has become possible, in recent years, to bring the "type locality" into the office or laboratory of any truly interested worker. In fact, the log portion of the "type locality" can and should be published. This condition lessens the need for restricting the number of names of subsurface units to an irreducible minimum. If a subsurface type section is definitely better and more typical than that available at any surface type locality, the subsurface section may be designated as type. Otherwise, a surface type locality shall be designated.

(b) Names applied to subsurface units shall be governed by the same restrictions and regulations as prevail for exposed units. [See Articles 7, 9, 10, and 11 of the original paper.]

(c) When it becomes possible to correlate a named subsurface unit with a named surface unit, and when the surface and subsurface facies are sufficiently similar that two names are unnecessary, the name of the surface unit is to be applied, even though the subsurface name has priority, unless much more extensive usage of the subsurface name renders its retention preferable or necessary.

(d) When beds are discovered which are equivalent and in similar facies to a named subsurface unit, the name of the subsurface unit shall have priority.

(e) When it is found that a subsurface unit has been named for but miscorrelated with a named surface unit, a new name shall be given the subsurface unit or it shall be renamed for its true correlative on the surface. In rare instances, exceptionally widespread use of the name by the subsurface unit may make it advisable to permit the "pirating" of the name by that unit and thus force the renaming of the surface unit from which it derived its name. Such "pirating" should be held to a minimum and should be accepted for publication only after a favorable ruling by the American Commission on Stratigraphic Nomenclature.

---

[10] *Naming of Subsurface Stratigraphic Units:* Am. Assoc. Petroleum Geologists Bull., vol. 32, no. 3, pp. 369-370, Mar. 1948. Prepared by W. V. Jones and R. C. Moore.

(*f*) In proposing a new name for a subsurface unit, it is desirable to describe for the type section the following features:

(1) Location of the type locality well; name of operating company or individual; date of drilling; results and present status of the well; elevation of surface at the well and depths to top and bottom of the new unit.

(2) If all data needed to establish the type section properly cannot be furnished from one well, two or more wells shall be used and the data called for under (1) shall be furnished for each well so used.

(3) As complete a section as possible shall be described in detail from cores of the new unit. Where sample logs are available, critical portions of them shall be included in written or graphic form or both. The boundaries and subdivisions, if any, of the new unit shall be indicated clearly in these logs and core records.

(4) Where electrical or other mechanically recorded logs are available, the critical parts of such logs, preferably of several wells located in a single area, shall be published in the article proposing the new named unit. The boundaries and subdivisions, if any, of the new unit shall be marked plainly on these published logs, which shall be on a scale large enough to permit full appreciation of all details of the new unit.

(5) Fauna and flora. Diagnostic fossils of the new unit shall be described in detail and, if possible, figured. Description and figuring of diagnostic fossils are essential if the new unit is a time-rock unit.

(6) Nature of underlying and overlying units.

(7) Correlation and position in the general stratigraphic scale.

(8) Present location of the cuttings or samples.

(9) Present location of the fossils.

(10) Critical parts of written driller's logs of all wells used, unless these are considered to be so inaccurate that their inclusion would be confusing.

(*g*) The cuttings and the fossils, accompanied by copies of all available types of logs should be placed in some official, permanent depository. As a rule, the appropriate state geological survey will serve as such a depository.

(*h*) The editorial staffs of all publishing agencies are urged to insist that the provisions of this article be followed in detail whenever a subsurface unit is being given a name.

FACIES CONCEPT

Although the principle of facies and facies changes has long been recognized by stratigraphers, it has only been during the past ten years that more serious consideration has been given the concept. Those interested in sedimentary facies are referred to the following: "Intertonguing Upper Cretaceous Deposits" by W. S. Pike, Jr., (Geol. Soc. America Mem. 24, 1947), "Sedimentary Facies in Geologic History" (Symposium) (Geol. Soc. America Mem. 39, 1949) and "Sedimentary Facies in Gulf Coast" by S. W. Lowman (Am. Assoc. Petroleum Geologists Bull., vol. 33, no. 12, pp. 1939-1997, Dec. 1949).

Neglect of the facies concept in stratigraphic geology has led to many questionable and ill-founded correlations. The recognition and evaluation of facies changes are cardinal to the proper establishment of the stratigraphic and structural fabric of any area.

The term "facies" has been variously interpreted by stratigraphers,

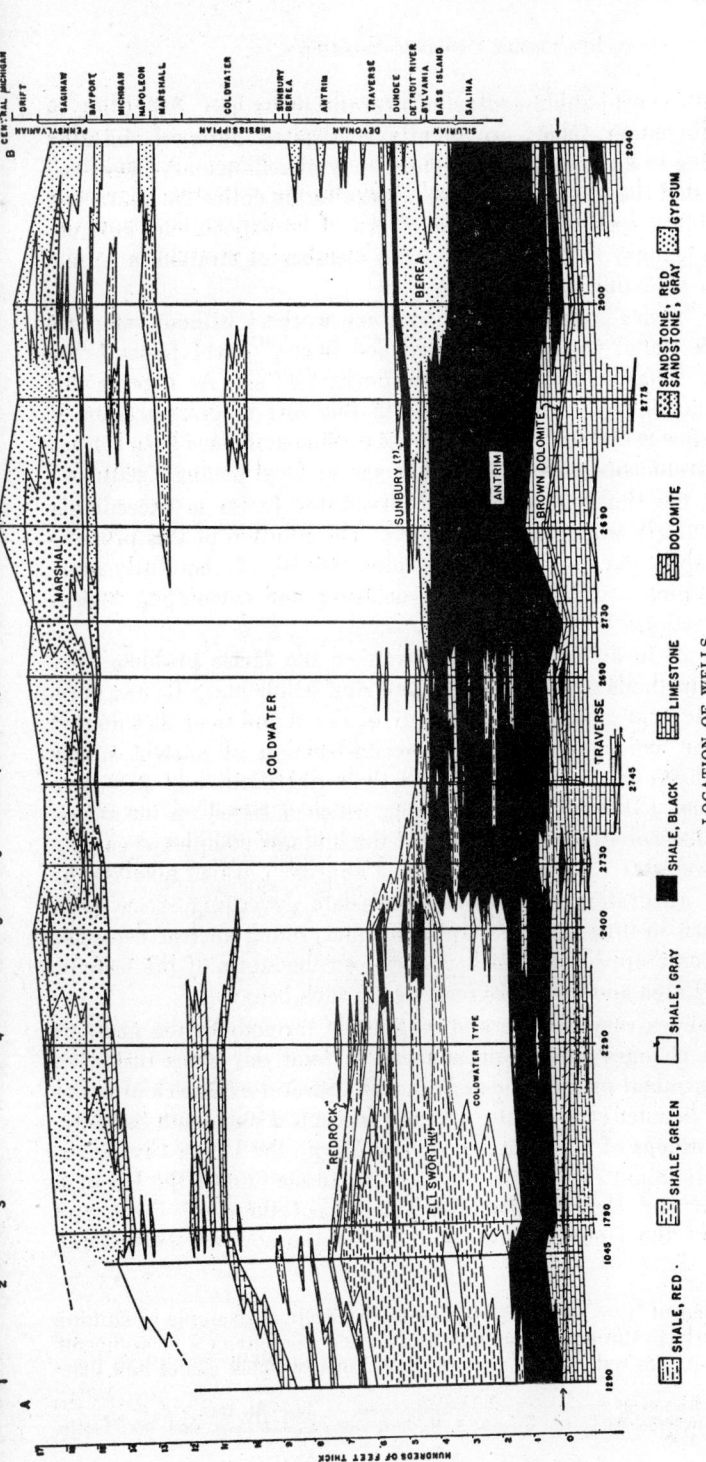

LOCATION OF WELLS

No.	Permit	County	Sec.	Twp.	R.	Name of Well	Company	Elevation, Feet
1	5538	Allegan	10	3 N.	15 W.	James Overbeck 1	Haze & McGill	637.1
2	6817	Ottawa	5	5 N.	14 W.	Huyser 2	Freeman Oil	700.1
3	6293	Ottawa	27	7 N.	13 W.	M. Wisniswski 1	Smith Petroleum	661.3
4	6115	Kent	10	9 N.	11 W.	Davis 1	Wicklund	778.
5	3144	Kent	29	9 N.	10 W.	Plank 1	Associated Petr.	859.
6	4477	Montcalm	31	10 N.	7 W.	Hansen 1	Hirzel Leland	887.
7	3081	Montcalm	29	10 N.	6 W.	Russell 1	Old Dutch	850.
8	2395	Montcalm	2	10 N.	5 W.	Barger 1	Eastern Gulf	787.
9	2889	Montcalm	2	10 N.	5 W.	Fisher 1	Eastern Gulf	787.
10	2471	Gratiot	17	11 N.	2 W.	Walter 1	Mountain Oil	753.
11	3210	Gratiot		12 N.	1 W.	Redman 1	Gulf	696.
12	5422	Midland	9	13 N.	2 E.	Hepinstall 1	Weller	633.
13	4080	Bay	1	14 N.	4 E.	Marston 1	Gulf	595.

FIGURE 9. Facies interpretation along an east-west cross section in Allegan and Bay Counties, Michigan. (From Tarbell. Reproduced permission Am. Assoc. Petroleum Geologists.)

and, as a result, considerable confusion prevails in its use. According to Moore,[11] "sedimentary facies are areally segregated parts of differing nature belonging to any genetically related body of sedimentary deposits." He also states that the term "lithofacies" "denotes the collective characters of any sedimentary rock which furnish record of its depositional environment." A sedimentary facies may involve a member or stratum, a formation or group, or a time-rock unit.

The term "facies" has been used by some workers without reference to stratigraphic units; for example, "red bed facies," "reef facies," "lagoonal facies," "marine facies," and "evaporite facies." As regards this usage, Moore comments, "There is value in this sort of classification of sedimentary deposits and in comparing rocks presumed to have been formed under like environments, without regard to age or local geologic settings."

To prove the time equivalency of dissimilar facies is exceedingly difficult, particularly when data are limited. The solution of this problem lies in (1) establishing regional stratigraphic trends; (2) carefully planning areal mapping programs; and (3) detailing and correlating control stratigraphic sections.

Sloss [12] et al in an excellent discussion on the facies problem have suggested four methods of approach in analyzing sedimentary facies: (1) a paleogeographic approach, wherein a study of facies and their distribution patterns in time and space attempts a reconstruction of ancient source areas and depositional environments and their distribution in past geographic patterns; (2) a biologic approach, which is based on the reconstruction of paleoecology from the study of the biologic complex occurring in fossiliferous strata; (3) an oceanographic approach, which involves the collection and integration of environmental data governing recent sedimentation, which in turn may be helpful in interpreting ancient deposits; and (4) a tectonic approach, which is based on the study of the tectonic behavior of any area and the facies response to such behavior.

Facies changes vary in type and magnitude throughout the geologic column. These changes may be pronounced in relatively short distances, others may be gradual over extended distances. Several examples of facies variations may be cited: the Middle Tertiary of central and south Sumatra, the Upper Cretaceous of northern and central Egypt, the Upper Cretaceous of the Rocky Mountain region, the Permian of Russia and of the Permian Basin of Texas and New Mexico, the Tertiary of the Gulf Coast and Pacific Coast of the United States, and the Tertiary of northern South America.

The Devonian of New York State presents an excellent example of shifting facies, particularly in the post-Onondaga parts where the changes of sediments have been traced from red beds in eastern New York to black shales and lime-

[11] Moore, R. C., *Meaning of Facies*, Geol. Soc. America Mem. 39, pp. 1-34, 1949.
[12] Sloss, L. L., Krumbein, W. C., and Dapples, E. C., *Integrated Facies Analysis*, Geol. Soc. America Mem. 39, pp. 91-123, 1949.

stones in Ohio. In general, the change is from red sands and conglomerates to gray, fine-grained sandstones to dark siltstones, thin to dark-gray and black shales, and finally to calcareous shales and limestones containing coral plantations and bioherms. The pattern of these shifts is now so well known that relationships not yet recognized in parts of the Appalachian geosyncline can be anticipated.[13]

According to Weller:[14]

Facies variation is not peculiar to the Mississippian system [North America], but the problems which result from certain types of rapid lateral variation in sediments, in the fields of both practical stratigraphy and stratigraphic nomenclature, have received more attention in the Lower Mississippian and Upper Devonian rocks of Ohio, Pennsylvania, Indiana, and Kentucky than in other parts of the stratigraphic column and other regions of the continent. The work of Hyde,[15] Stockdale,[16][17] Chadwick,[18] and Caster[19] is particularly important. Both Hyde and Stockdale recognized certain major formations, each of which, they presumed, was deposited contemporaneously throughout its extent. Each formation was then divided vertically into different facies developments, which were given geographic names. Finally the facies were divided horizontally into members which were also partially or completely named. This system has proved to be flexible and convenient for description. The facies names are, in effect, synonyms of the formation names but have only local significance. Although this system introduces a large number of names, many of them can be ignored by persons having no interest in the detailed stratigraphy of these formations.

Caster was more concerned with the interrelationships of rock-stratigraphic and time-stratigraphic units. To units of more or less uniform lithologic characters which transgress time lines he applied the term "magnafacies"; these are rock-stratigraphic units and correspond to the original lithologic formations of the northern Appalachian region. He used the terms "stage," "formation," or "stratigraphic unit" for time-stratigraphic units. *The magnafacies and stages intersect each other, and for the intersected strata Caster introduced the term "parvafacies."* According to this system, which is an elaboration of conclusions reached earlier by Chadwick and others, each magnafacies consists of a succession of parvafacies of similar lithologic character but unequal age, and each stage consists of a succession of parvafacies of similar age but different lithologic character. Geographic names were introduced for all so that strata at any place have three major names: a more or less local parvafacies name, extensive magnafacies, and stage names. In addition, named members are also recognized. This system is of considerable theoretical interest and may be very useful in the detailed study and description of the strata deposited near the margin of an expanding delta. It is not likely, however, to be widely applicable to other types of facies problems.

[13] Cooper, A. G., et al., *Correlation of the Devonian Sedimentary Formations of North America:* Geol. Soc. America Bull., vol. 53, no. 12, pp. 1729-1794, 1942.

[14] Weller, J. M., et al., *Correlation of the Mississippian Formations of North America:* Geol. Soc. America Bull., vol. 59, no. 2, pp. 91-196, 1948.

[15] Hyde, J. E., *Stratigraphy of the Waverly Formations of Central and Southern Ohio:* Jour. Geology, vol. 23, pp. 655-682, 757-779, 1915.

[16] Stockdale, P. B., *The Borden (Knobstone) Rocks of Southern Indiana:* Indiana Dept. Cons. Pub. 98, 1931.

[17] Stockdale, P. B., *Lower Mississippian Rocks of the East-Central Interior:* Geol. Soc. America Special Paper 22, 1939.

[18] Chadwick, G. H., *Faunal Differentiation in the Upper Devonian:* Geol. Soc. America Bull., vol. 46, pp. 305-342, 1935.

[19] Caster, K. E., *The Stratigraphy and Paleontology of Northwestern Pennsylvania,* pt. 1: Am. Paleontology Bull., vol. 21, no. 71, 1934.

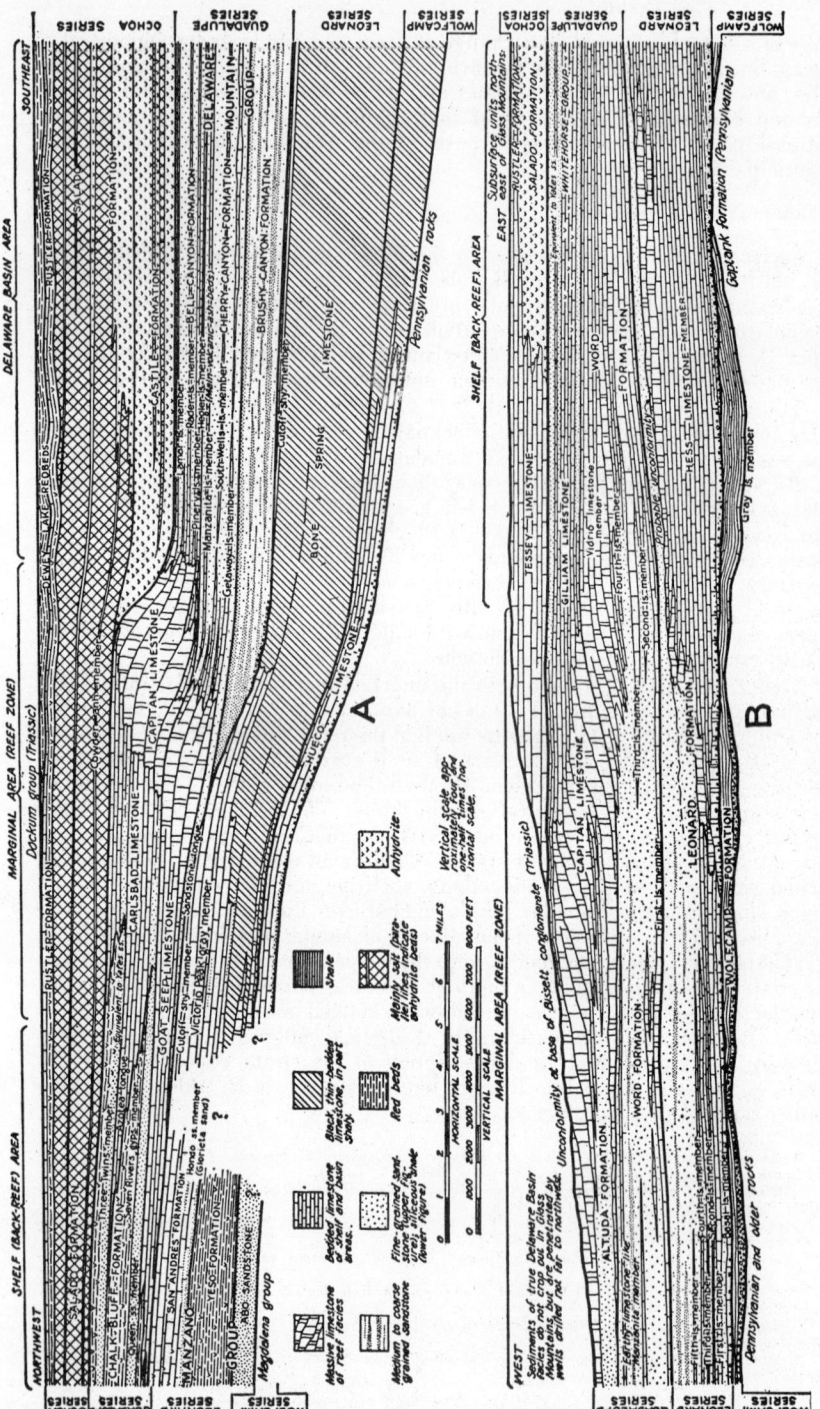

FIGURE 10. Lateral variations within the Permian in the Guadalupe Mountains (A) and in the Glass Mountains (B). Observe the change from organic limestones to clastic sediments. (From King. Reproduced permission Am. Assoc. of Petroleum Geologists.)

In a classic paper on the sedimentary facies of the Gulf Coast, Lowman [20] says:

As the search for oil turns more and more toward a search for new provinces and new trends, we realize the inadequacy of present methods of evaluating the significance of sedimentary properties and their patterns of distribution. . . . The problem of discovering sedimentary criteria for the recognition of petroleum provinces and trends surely involves the study of sedimentation, stratigraphy, and structural history.

Lowman emphasizes that the logical place to begin a fundamental investigation of facies would appear to be in the environments of deposition and states:

In a search for sedimentary criteria by which we may identify environment of deposition of sedimentary rocks, it may be useful to classify the sedimentary variables as "static" and "dynamic." Among the static variables may be listed those properties which retain their original depositional characteristics (although they may be somewhat modified by postdepositional processes) as follows: (1) gross mineral composition of the solid rock, for example, limestone, shale, or sandstone; (2) size, shape, and distribution of detrital grains; (3) fossils; (4) sedimentary structures; and (5) possibly some mass effects related to chemical composition, such as radioactivity and magnetic properties. Among the dynamic variables might be included: (1) minor changes in gross composition, for example, limestone to dolomite; (2) cement; (3) chemical composition of rock fluid; (4) character of those parts of the mineral assemblage which are susceptible to postdepositional change through differential loss of some minerals and authigenic gain of others; and (5) some physical mass properties such as density and porosity. Stated in another way, the static variables are those which should be useful in making interpretations of depositional environments. The dynamic class, on the other hand, should serve in making interpretations of post-depositional change.

Dapples et al [21] express a fundamental concept that cannot be minimized in facies interpretation. It is believed by these writers that:

The tectonic behavior of the depositional area is the most important factor in the control of lithofacies, and the environment of deposition (littoral, nertic, et cetera) plays a part which depends on the length of time environmental conditions can affect the material before it is buried. The source area, except in special instances, appears to be a less important factor. The tectonic behavior of the depositional area itself includes several factors, among which are the geographic distribution of tectonic elements, and the intensity of the tectonism in each.

Pettijohn [22] comments that:

The fundamental cause of the observed differences in lithology and associated phenomena has been the rate of sedimentation which, in turn, is controlled by the related rates of elevation and depression of the source area and the

---

[20] Lowman, S. W., *Sedimentary Facies in Gulf Coast:* Am. Assoc. Petroleum Geologists Bull., vol. 33, no. 12, pp. 1939-1997, Dec. 1949.
[21] Dapples, E. C., Krumbein, W. C., and Sloss, L. L., *Tectonic Control of Lithologic Associations:* Am. Assoc. Petroleum Geologists Bull., vol. 32, no. 10, pp. 1924-47, 1948.
[22] Pettijohn, F. J., *Sedimentary Rocks*, p. 436, Harper's Geoscience Series, 1949.

basin of sedimentation, respectively. Tectonics is indeed the soul of the matter. Therefore, the first breakdown of environments must be tectonics.

The importance of facies development and interrelationships have too often been neglected and minimized in favor of the structural impress in selecting potential petroliferous areas. The definition and evaluation of facies changes are as important in outlining favorable petroliferous areas as are the structural anomalies developed within them.

The question arises: What is the importance of evaluating facies changes? There are several answers:

1. Lateral changes commonly involve variations in porosity and permeability, which control selective accumulation of liquid and gaseous hydrocarbons.

2. Selection of areas containing proper facies and facies relationships may reduce exploratory costs for an oil company. The facies factor has many times been more important in controlling oil and gas accumulations than structural conditions.

3. Paleogeologic data may be more accurately interpreted when lateral variations and relationships of the sediments are properly integrated.

4. Before faunal and stratal sequences can be properly arranged chronologically, a knowledge of facies relationships must be known.

5. Variations in rock types may possibly control localization of ore deposits.

6. Migrating lithofacies boundaries may be incorrectly interpreted as structural reflections.

7. Changes in facies may complicate engineering problems, such as tunneling, excavating, and estimating costs.

Facies changes result from fluctuation of sea level, climatic variations, modification of topographies, diastrophic readjustments of the hinterland, changes in oceanic currents, and drainage patterns, erosional cycles, readjustment in deposition basins, and migration of shore lines.

When one is introduced into an unfamiliar province the approach to the solution of stratigraphic problems is first to generalize the stratigraphy and structure of the area and second to give consideration to the details of the section. This stage is followed by repeated regeneralization and redetailing. Overemphasis of either generalizing or detailing may retard progress or promote erroneous interpretations.

Reliable stratigraphic information demands accurate field control. Without adequate structural information, stratigraphic sequences and facies variations of deposits cannot be properly allocated. An unrecognized fault or nonrecognition of dissimilar facies equivalents may increase or decrease the normal thickness of stratal sequence by thousands of feet. The failure to recognize unconformities may introduce conflicting stratigraphic interpretations.

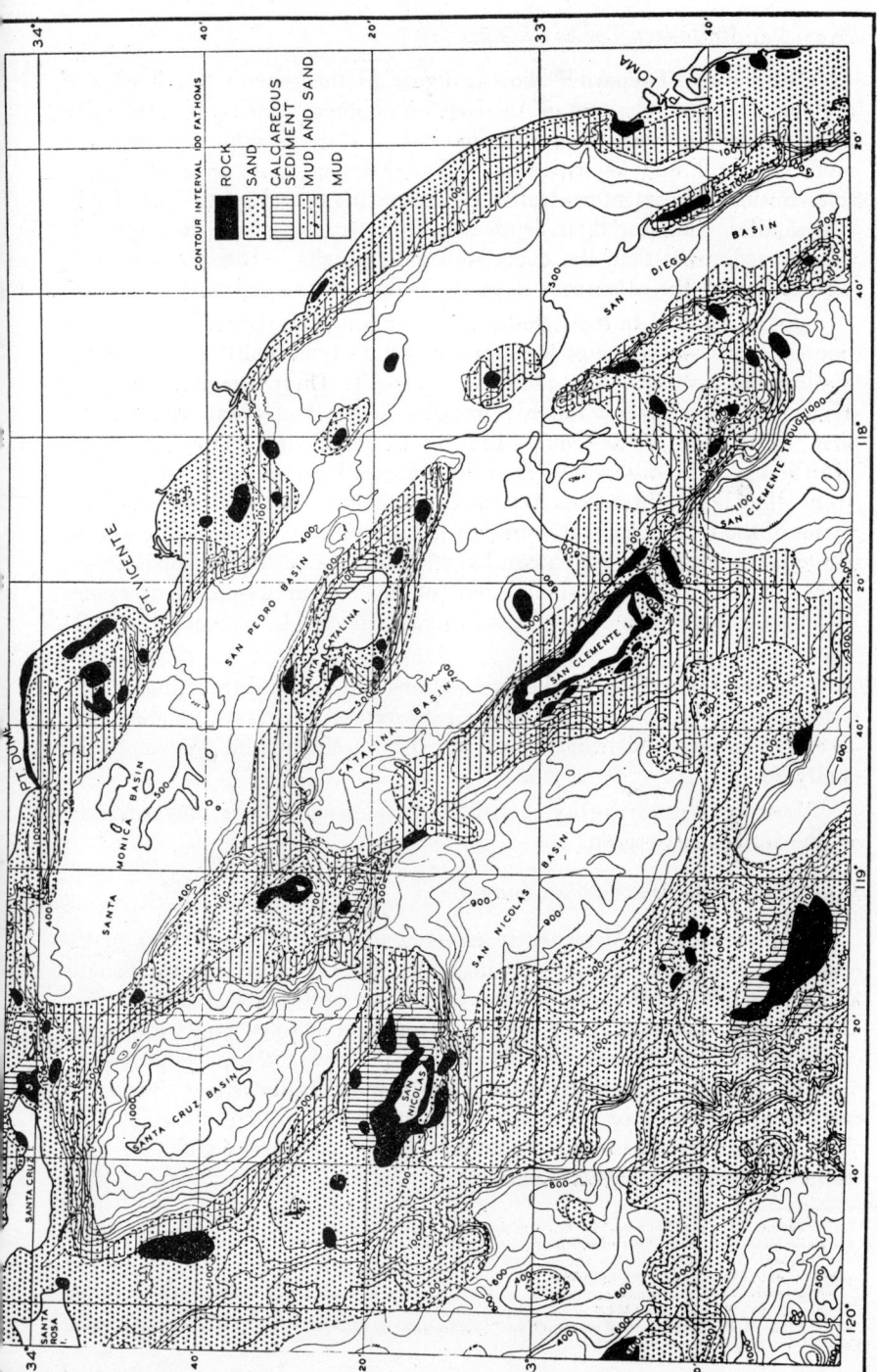

FIGURE 11. Subsea topography and distribution of sediments off the southern California coast. Exemplifies lateral variations of accumulating sediments. (From Revelle and Shepard, *Recent Marine Sediments*, Am. Assoc. Petroleum Geologists.)

Examples of Modern Facies Changes

Revelle and Shepard [23] show in figure 11 the general distribution of sediments and rock bottom off the coast of southern California. The types of these recent sediments fall into four categories: (1) sand, (2) sand and mud, (3) mud, and (4) calcareous. The distribution of these deposits is to a considerable extent controlled by submarine topography. "The ridges and saddles, whatever their depth and distance from shore, have notably coarser sediments than the depression and troughs." (See fig. 11) The distribution of the calcareous deposits in the area are extremely variable.

Tracey [24] et al in their studies of the Bikini reef discuss the distribution of corals and algae of the more important types and the relationship of channels, caverns, pools, and detrital deposits. They recognize a number of distinct facies zones roughly parallel to the reef front. It is stated that "Differences in the composition of the reef surface and in organic growth are also observable laterally—along lines parallel to the reef front—but these differences are less striking than the banding." Specific facies of the reef include (1) a marginal zone on the windward side *(Lithothamnium* abundant), (2) a coral-algal zone (inside the marginal zone), (3) a reef flat (forms the major part of the reef and consists of eroded coral and algae with some foraminiferal sand), (4) a beach, and (5) lagoonal.

As ancient reefs in carbonate sections are now being seriously considered by oil companies in their exploration programs, modern reef development and distribution should be carefully and systematically analyzed.

Lowman [25] in a worthy contribution has discussed the modern facies of the Gulf Coast region.

Examples of Ancient Facies Changes

Many examples of ancient facies changes may be cited. One of the classic examples of lateral variations in strata is that reported on by King [26] within the Permian of the Guadalupe and Glass Mountains (fig. 10), of west Texas and New Mexico. Three well-defined facies are recognized in the Guadalupe series: (1) shelf (back-reef), (2) marginal (reef), and (3) basinal. The shelf facies is represented by limestone, evaporites, and minor amounts of sandstone. The marginal facies is dominated by the Capitan reef limestone. Sandstones, shales, and occasional thin limestones comprise the basinal facies.

[23] Revelle, R., and Shepard, F. P., *Sediments off the California Coast: Recent Marine Sediments, a Symposium:* Am. Assoc. Petroleum Geologists, p. 245-281, 1939.
[24] Tracey, J. I., Ladd, H. S., and Hoffmeister, J. E., *Reefs of Bikini, Marshall Islands:* Geol. Soc. America Bull., vol. 59, pp. 861-878, 1948.
[25] Lowman, S. W., *Sedimentary Facies in Gulf Coast:* Am. Assoc. Petroleum Geologists Bull., vol. 33; no. 12, pp. 1939-96, Dec. 1949.
[26] King, P. B., *Permian of West Texas and Southeastern New Mexico:* Am. Assoc. Petroleum Geologists Bull., vol. 26, no. 4, pp. 535-763, 1942.

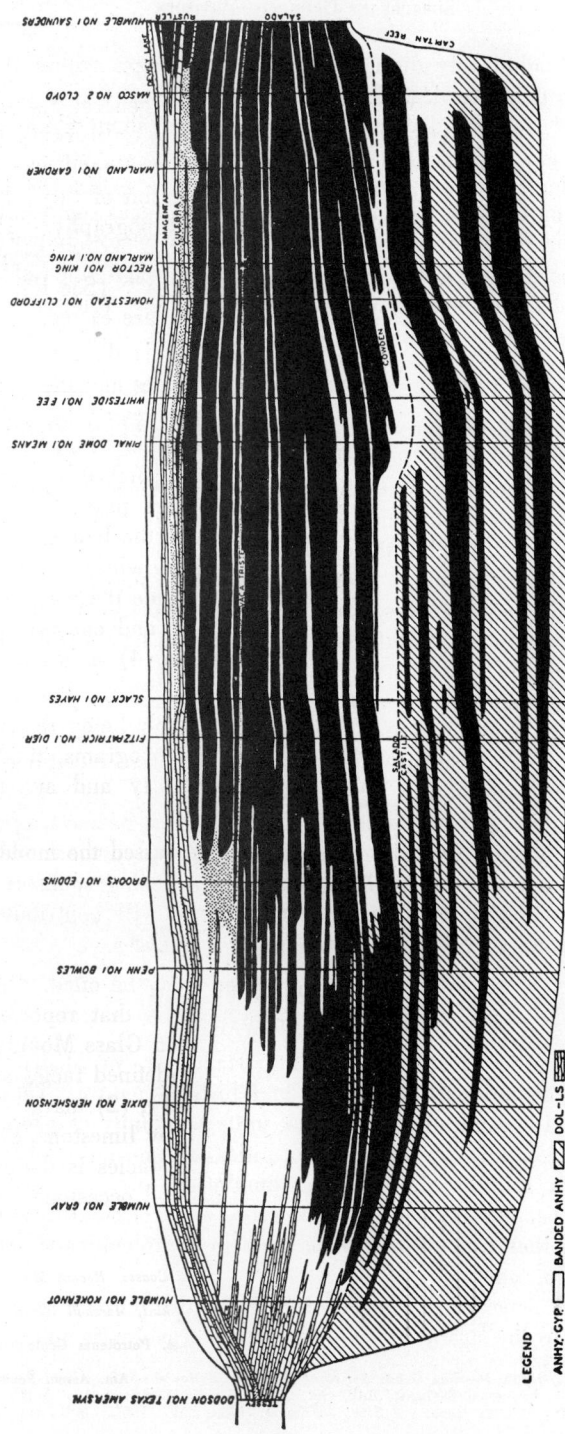

FIGURE 12. Diagrammatic south-north section of Ochoa (Permian) rocks in the Delaware basin of New Mexico and Texas. (From Adams. Reproduced permission Am. Assoc. Petroleum Geologists.)

Adams [27] admirably discusses the facies changes within the uppermost Permian deposits of southwestern United States.

Dunbar [28] reports that the Kazanian series about Perm (Russia) is represented by a bright-red, unfossiliferous sandstone and siltstone some 500 feet thick. This is the nonmarine Ufimian facies, so called for the city of Ufa, which is on the strike south of Perm. In conspicuous contrast, the equivalent beds in the environs of Kazan are richly fossiliferous white limestone.

The Upper Cretaceous of the Rocky Mountain region of Colorado and Wyoming offers excellent examples of facies changes wherein the section becomes more arenaceous and less marine from east to west.

IMPROVING INTERPRETATION OF STRATIGRAPHIC SEQUENCES

Confusion and disagreement in stratigraphic studies often arise as a result of one or more of the following: (1) projection of stratigraphic nomenclature of one area into another without sufficient intervening control; (2) failure to distinguish between rock and rock-time units; (3) publication of stratigraphic material by those who are inexperienced or who have drawn conclusions based on inadequate data; and (4) failure to recognize the importance of structural and depositional idiosyncrasies.

The evaluation of stratigraphic sequences has been greatly improved by application and coordination of the following studies: (1) detailing of surface and subsurface sections; (2) subsurface logging techniques; (3) detailed paleontologic and mineralogic analysis; (4) improved laboratory and field procedures; and (5) better understanding of sedimentational principles.

Projects devoted to the study of clay minerals, bacteria, organic and carbonate sediments, formation waters, recent and ancient sedimentary processes, diagenesis, and well logging, as now being sponsored by the American Association of Petroleum Geologists, will contribute greatly in more accurately evaluating stratigraphic sequences.

UNCONFORMITIES

W. ALAN STEWART

Unconformities are of prime importance in problems of stratigraphic, sedimentational, structural, and historic nature. They are the natural basis for separating both rock units and units of geologic time. The economic importance of unconformities is constantly increasing because of their relationship to oil accumulation and to ore deposition. The occurrence, development, extent, type, and relationships of unconformities are therefore of major concern in petroleum and mining development programs.

Since the detection of unconformities in the subsurface is of paramount value to the stratigrapher and economic geologist, much emphasis

[27] Adams, J. E., *Upper Permian Ochoa Series of Delaware Basin, West Texas and Southeastern New Mexico:* Am. Assoc. Petroleum Geologists Bull., vol. 28, no. 11, pp. 1596-1625, Nov. 1944.
[28] Dunbar, C. O., *Permian faunas: A Study In Facies:* Geol. Soc. America Bull., vol. 52, p. 313-32, 1941.

is given to the criteria for unconformity recognition. Considerable space has been devoted to the use of unconformities in interpreting the geologic record. A number of examples of unconformity-type oil fields are described and illustrated. A section on the influence of unconformities on ore deposits is also included.

An unconformity has been defined by Twenhofel [29] as a surface of erosion or nondeposition separating two groups of strata. Unconformities that are surfaces of erosion are more common than surfaces of nondeposition, possibly because they are more easily recognized. Surfaces of nondeposition indicate a long time break in sedimentation and may leave little evidence of their occurrence in the geologic column. Such a break would involve the establishment of a profile of depositional equilibrium for a long period of time.

The time value of an unconformity is the interval of geologic time represented by the break. This ranges from the time required to deposit a single formation to the enormous time interval represented by the unconformity between the pre-Cambrian and Pleistocene glacial deposits. Twenhofel [30] aptly likens all unconformities to the branches of a tree, the trunk of which is the present land surface and whose branches spread out to intercalate in all directions through the strata. Thus every unconformity directly or through another unconformity intersects the present land surface.

The relief on surfaces of unconformity may vary from peneplanation to several thousand feet. The relief on the Ep-Algonkian surface of the Grand Canyon is at least 800 feet, while Carboniferous sedimentary rocks around Boston, Massachusetts, rests unconformably on a surface of at least 2,100 feet. The amount of relief on an unconformity is no indication of the time value of the unconformity. A surface of great relief might represent only a small fraction of the time involved in the formation of a peneplained surface of slight relief. Conversely, a surface of small relief may not have passed through the initial stages of subaerial or submarine erosion, and may involve only a short period of geologic time. The physical appearance of an unconformity should not be used to gage its time value. Fossil evidence, if present on both sides of an unconformity, is the most reliable criterion of time value.

Nomenclature

The various kinds of unconformities depend for their classification upon the attitude of the strata on both sides of the unconformity, the genesis of the rocks involved, and their areal extent.

Disconformity. In a disconformity, the strata on both sides of the unconformity are parallel. The older strata have lain undisturbed in position during the break in deposition. The younger beds have then

---

[29] Twenhofel, W. H., *Treatise on Sedimentation*, Baltimore, Md., Williams and Wilkins Co., 1926.
[30] *Ibid.*, p. 447.

been deposited with their stratification parallel to that of the older beds. This type of unconformity has been called a parallel unconformity by Lahee,[31] and they have been aptly named stratigraphic unconformities by Grabau.[32]

The term "diastem" may be used to designate a disconformity whose time value is less than that of a formation.

Disconformities may be difficult to detect, especially if there is little relief on the surface. Evidences of erosion, if present, may simplify the task. If no erosion is detected, then fossils are the best means of determining disconformity.

Nonconformity. In a nonconformity, the strata below the unconformable contact are at an angle with those above. This type may be de-

FIGURE 13. Cores taken two feet apart from a well section in central Sumatra. Observe difference in dip; this difference may be attributed to variance in competency of lithologic types, to irregularity in deposition, and perhaps to preconsolidation slumping movement. If these two sections were taken as spot cores several hundred feet apart stratigraphically a nonconformity could possibly be inferred.

scriptively called an "angular unconformity." According to Grabau,[33] this is the true unconformity where there is a discordance of strata, and he has called them "structural unconformities" to contrast with his stratigraphic unconformities.

[31] Lahee, F. H., *Field Geology*, 4th ed., New York, McGraw-Hill Book Co., 1941.
[32] Grabau, A. W., *Guide to the Geology of the Schoharie Valley in Eastern New York:* New York State Mus. Bull. 92, 1906.
[33] *Idem.*

The term "nonconformity" has been used extensively in the literature to include unconformable contacts of sedimentary rocks with metamorphic or plutonic rocks that have been subjected to erosion before deposition of sediments. Billings [34] says that the term is utilized most satisfactorily for unconformities where the older rock is plutonic in origin, and uses "angular unconformity" to designate the unconformable contacts of discordant strata. Willis,[35] however, suggests that the term be used in a broader sense to include any unconformity that is not a dis-

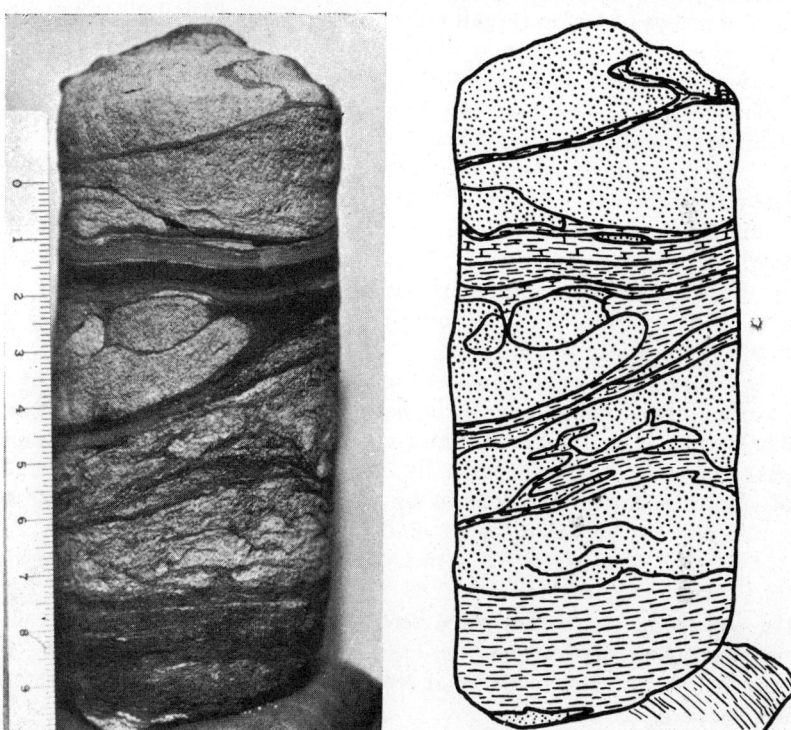

FIGURE 14. Core section showing the effect of sliding movement in sediments during the hydroplastic stage. Structural interpretations cannot be depicted from data of this type.

conformity. This is probably the most satisfactory application of the term.

A nonconformity or angular unconformity involves a surface separating two or more stratal units at an angle of discordance, which may vary from 90 degrees to less than 1 degree. If the latter variation exists, it is sometimes difficult to determine whether the discordance angle is

[34] Billings, M. P., *Structural Geology*, p. 243, New York, Prentice-Hall, 1942.
[35] Willis, Bailey and Robin, *Geologic Structures*, 3d ed., rev., New York, McGraw-Hill Book Co., 1934.

attributable to deformation or to variance in initial depositional dip without extensive and accurate field mapping. It is even more difficult to make this determination if interpretations must be made on limited subsurface evidence. (See figs. 13, 14.)

Local Unconformity. A local unconformity is similar to a disconformity, but differs in that it is of small areal extent and represents a comparatively short time interval. It is commonly the result of stream action in continental deposits. During times of flood, streams may scour out channels in their flood plains, which may be many feet wide and deep and thousands of feet to several miles in length. As the flood subsides, the channel may fill up, or it may fill up days or even years later. The stratification above such an unconformity is parallel to that below, making it a disconformity. The term "local unconformity" is used to indicate its small areal and short time value.

An example of a local unconformity in a marine section would be the wave-truncated surfaces of some buried bioherms. These surfaces would develop on the seaward side of the reef, as the bioherms built upward and shoreward during a marine transgressive overlap. The bench cut by wave action would represent an erosion surface of limited extent. If this surface was then buried by further advances of the sea, a local unconformity would result.

The erosion involved in local unconformities is accomplished by a cessation of the upbuilding of the deposits and is called "contemporaneous erosion." Local unconformities may be produced by contemporaneous erosion in marine deposits by changes in strength or direction of marine currents; in lake deposits by similar changes in currents; and in eolian deposits by variations in wind.

Blended Unconformity. A surface of erosion may be covered with a thick mantle of residual sands and gravels that grade downward into the rock from which they were derived. If they are later overlain by sediments, they may become the basal member of a new formation. In such a case, no distinct surface of separation can be seen, and the feature is a blended unconformity.

Genesis of Unconformities

Continental Unconformities. Continental unconformities are those whose surface is cut by a continental agent, succeeding deposits being of continental origin. The agent of deposition may be fluvial, eolian, or glacial.

Aqueous: Fluvial unconformities are found in valley-flat, or in alluvial-fan or deltaic deposits.

Unconformities in valley flats may arise from scour-and-fill and from the meandering of aggrading streams. In scour-and-fill, unconformities may be abundant and the relief is apt to be locally great, and correlation difficult or impossible. The relations are those of a disconformity, but as initial dips may be high, a pseudononconformable relationship may exist.

An aggrading stream building a flood plain of construction will occupy at one time or another every position in the deposit. This is accomplished through migration of meanders or by the stream breaking natural levees and seeking the lower land of the back swamps. As a flood plain often contains swamps and lakes, the unconformity developed may be between fluvial, paludal, or lacustrine deposits. Unconformities developed by aggrading streams have disconformable contacts and have a short to moderately long time value.

Alluvial fans are deposited by stream distributaries on surfaces of varied origin, over which they extend outward from the uplands. The buried surface becomes an unconformity that may represent a long or short time value and may be either a disconformity or a nonconformity.

Unconformities in deltas may be either continental or marine. A delta may advance or retreat according to the balance between supply and deposition. This advance and retreat leaves unconformable surfaces which may separate fluvial, paludal, lacustrine, or marine sediments.

Eolian: The work of the wind may result in unconformable contacts of either type. Sand may be deposited by the wind on surfaces not previously affected to produce unconformities on burial. Winds may erode earlier eolian deposits and later bury the eroded surface. Discordance between cross laminations may give such a contact the appearance of a nonconformity. The relief on eolian unconformities may be none to very great and the time value small to large.

Glacial: The advance of a glacier will be accompanied by erosion, which tends to decrease the relief of the land surface over which it is passing. Deposition of morainal material on this surface may result in an unconformity of either type.

Marine Unconformities. Marine unconformities are those found in marine deposits or between marine deposits above and deposits of another environment below.

Overlap: Deposition of the basal sediments of a formation on an erosion surface are not likely to be laid down uniformly and contemporaneously over a wide area. Actually, deposition will begin in a few favorable places and spread to other areas as formations become thicker. Thus succeeding strata will overlap the older rocks below the erosion surface.

As the land is submerged by the rapid rise of an encroaching sea, a condition of overlap is created which has been called the "unconformity of progressive overlap" by Grabau.[36] The coarse debris of the land surface forms a basal conglomerate, and there may be buried soils. As deposits are made in the advancing sea, each depth is characterized by

---

[36] Grabau, A. W., *Principles of Stratigraphy*, p. 723, 1913.

sediments related to depth. Thus a coast line will be characterized by roughly parallel belts of beach sands, muds, and limy oozes. As the sea moves in on the land, these three types of deposits will retain their relative positions. Muds will be deposited on sands and oozes on muds. A vertical cross section through a marine transgressive overlap will show coarse clastics overlain by finer clastics. The relief on a surface of transgressive overlap is essentially that of the submerged surface. Figure 15 shows a marine transgressive overlap.

The slow rise of the sea may result in a different kind of progressive overlap. Erosion may take place before deposition and a wave-cut surface extending a long distance from shore may be developed. As the sea level continues to rise, the wave-cut surface is brought below the base level of erosion, and sediments are deposited. There will be fine-grained

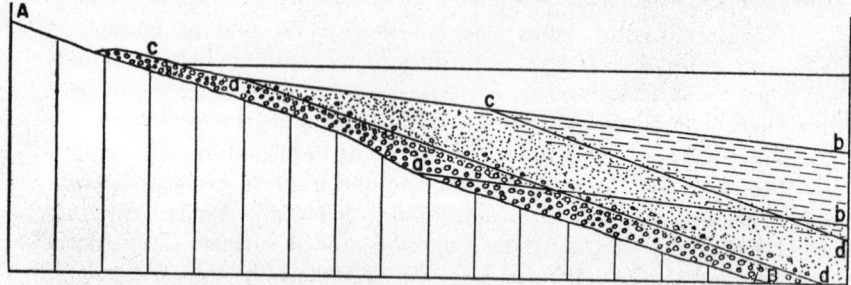

FIGURE 15. A marine transgressive overlap; lines parallel to ocean floor are called ."time lines" (ab), for they run through material deposited contemporaneously; lines essentially parallel to gravel, sand, and mud deposits are called "formation lines" (cd). A-B—Erosion surface overlain unconformably by overlapping sediments.

sediments, and the basal conglomerate produced by rapid submergence will not be present.

Unconformities resulting from overlap may be found in lacustrine and alluvial fan deposits as well as marine.

Offlap: Offlap or regressive overlap is produced when the sea recedes from the land. The beach zone will migrate over earlier offshore muds and the muds over earlier deposited oozes. If sufficient time has elapsed between deposition of offlap beds and the earlier deposited sediments, a marine regressive unconformity may be formed. This is likely to be a disconformity which might be difficult to distinguish from conformable deposition.

Other Marine Unconformities: The rise of sea level may be interrupted by periods of stability during which the sea bottom may be built to the base level of deposition for a considerable distance from shore. Equilibrium established over a long period of time will result in a disconformity of nondeposition.

A falling sea may lower the base level of erosion so that previously deposited sediments will be removed, and the erosion surface will become an unconformity should deposition follow this erosion.

Recognition of Unconformities

There are a number of signposts that are useful in detecting the presence of unconformities. Some of these, such as a basal conglomerate, have become traditions, although experience shows that they may be the exception rather than the rule. There are probably more unconformities without a basal conglomerate than with one.

Unconformities are best recognized from direct observation in a single outcrop. Such an outcrop might be observed in a road cut, a

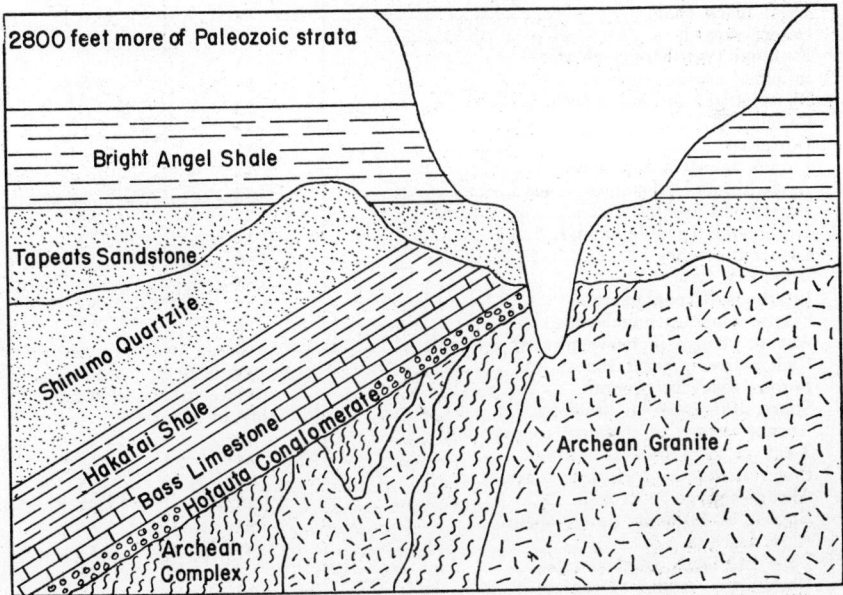

2800 feet more of Paleozoic strata

Bright Angel Shale

Tapeats Sandstone

Shinumo Quartzite

Hakatai Shale

Bass Limestone

Hotauta Conglomerate

Archean Complex

Archean Granite

FIGURE 16. Diagrammatic cross section of inner part of Grand Canyon, showing Ep-Algonkian and Ep-Archean unconformities.

quarry, a surface exposure on canyon walls, a ravine, or a cliff. For example, probably the most striking exposure of an unconformity can be seen in the Grand Canyon of the Colorado River, in the walls of which there are two unconformities exposed. Sharp [37] has described them thoroughly; the diagrammatic section shown in figure 16 is from his article.

In the subsurface the recognition of unconformities depends on the ability of the geologist to recognize any of the many criteria for unconformity from the sparse evidence afforded by well cuttings, cores, and

[37] Sharp, R. P., *Ep-Archean and Ep-Algonkian Erosion Surfaces, Grand Canyon, Arizona:* Geol. Soc. America Bull., vol. 51, pp. 1235-1270, 1940.

electric logs. A single criterion suggesting unconformity may only complicate the problem, as almost all such criteria can be explained by faulting or some sedimentary process or variation. The more associated criteria than can be established for a given horizon, the greater the probability

TABLE 2

CRITERIA FOR RECOGNITION OF UNCONFORMITIES

Criterion	*Associated with unconformities of indicated origin*	
	Subaerial	*Submarine*
Basal conglomerate	xx	
Basal black shale	xx	
Desert varnish	xx	
Residual (weathered) chert	xx	
Silicified erosion surface	xx	
Phosphatized erosion surface	xx	
Caliche	xx	
Duricrust	xx	
Porous zones in limestone	xx	
Asphaltic and oil-stained zones	xx	
Buried soil profiles	xx	
Lag gravels (pebble bands)	xx	xx
Glauconite zones	x	xx
Iron-oxide zones	xx	x
Interbedded conglomerate	xx	xx
Clastic zones in nonclastics	xx	xx
Abrupt change in heavy-mineral assemblages	xx	x
Radioactive mineral zones	xx	?
Porous zones in general	xx	xx
Sharp differences in lithology	xx	x
Abrupt change in chemical composition	xx	?
X-ray patterns in well cuttings	xx	xx
Concretionary and pisolitic zones	xx	x
Abrupt change in fauna	xx	xx
Gaps in evolutionary development	xx	xx
Algal biscuits	xx	xx
Bone and tooth conglomerates	?	xx
Undulatory surface of contact	xx	x
Edgewise conglomerate		xx
Phosphatic pellets or nodules		xx
Manganiferous zones		xx
Pyritiferous zones		xx
Corrosion surfaces		xx
Borings of littoral marine organisms		xx
Lateral spreading of coral reefs		xx

of its being a surface of unconformity. Pyrite crystals by themselves may suggest unconformity, but they might also be a normal dissemination through a conformable series. If they are associated with glauconite and phosphatic pellets, then a marine unconformity is probably present.

Krumbein [38] has tabulated criteria cited by various workers for recognition of unconformities. Pettijohn [39] relisted Krumbein's criteria with

[38] Krumbein, W. C., *Criteria for the Subsurface Recognition of Unconformities:* Am. Assoc. Petroleum Geologists Bull., vol. 26, no. 1, pp. 36-62, Jan. 1942.
[39] Pettijohn, F. J., *Sedimentary Rocks,* p. 147, New York, Harper & Brothers, 1949.

modifications as shown in table 2. A common or an exclusive association is indicated by xx, an occasional or rare association by x, and an association that probably does not occur by a blank. A number of these criteria are included in the subsequent discussion.

Structural Features. Discordance in Dip: Lack of parallelism in beds on both sides of an unconformable contact may be readily observed in vertical sections such as the face of a cliff. Discordance alone should not be accepted as positive proof without further observation, especially if the exposure is a small one. Faulting, contemporaneous erosion, or large-scale cross bedding may produce discordance.

The presence of dip discordance in a sporadically cored subsurface section may suggest, but does not prove, angular unconformity. A number

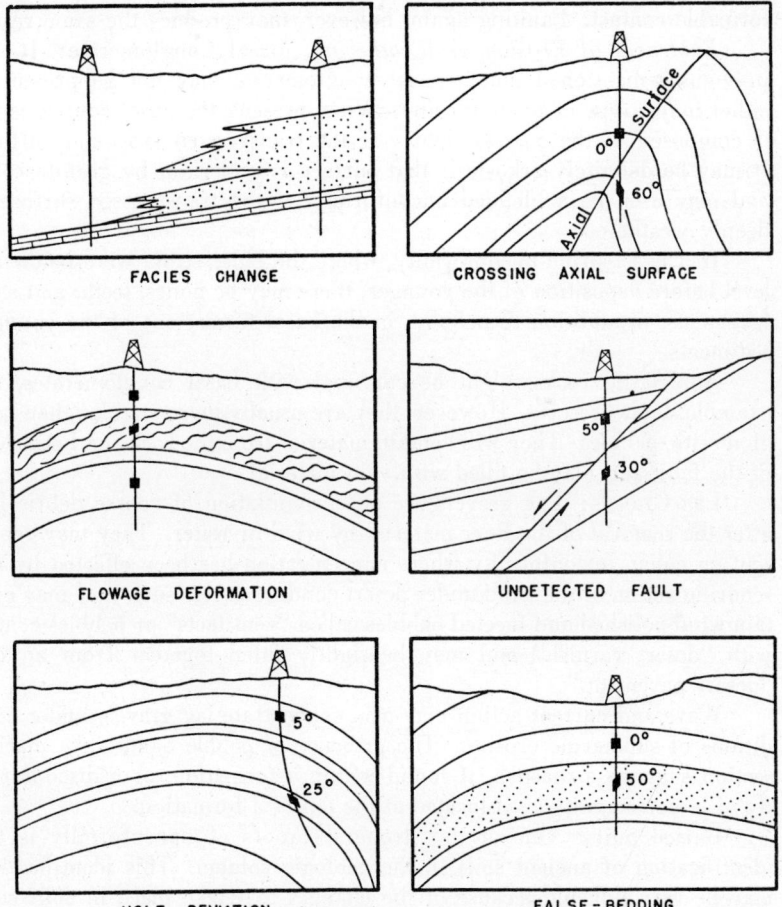

FIGURE 17. Diagrammatic sketches showing conditions involving spot cores the dip variances of which may suggest unconformity.

of features that might give dip variation are undetected faults, hole deviation, cross bedding, flowage, and crossing of the axial surface of a fold. Figure 17 illustrates several situations in which a nonconformity would be apparent from cores taken at intervals, if angular discordance were the only criterion.

Truncation of Faults or Intrusives: Faults and intrusive bodies in older strata may be truncated by erosion and end abruptly at unconformable contacts. There may be a definite contrast in the number of intrusives and the faulting present above and below an unconformity. Again caution should be used in using this criterion, as faulting may produce the same contrasts.

Degree of Folding or Metamorphism: There may be definite contrasts in the degree of deformation or metamorphism on both sides of an unconformable contact. Faulting again, however, may produce the same result.

Evidences of Erosion or Weathering. Basal Conglomerate: It was previously mentioned that a basal conglomerate may not be present in either major type of unconformity. When present, the basal conglomerate is composed of the coarser debris of the transgressed erosional surface. It may be definitely arkosic if that surface is underlain by granite rocks and may even be a blended unconformity if the transgressed surface is deeply weathered.

If it is a marine unconformity, where the older strata were below sea level before deposition of the younger, there may be bones, teeth, and shell fragments, in addition to pebbles, in the basal formations of the younger sediments.

Autoclastic rocks might be confused with basal conglomerates, for example, fault breccias. However, they are usually more angular than conglomeritic pebbles. They will contain material from the rocks on both sides of the fault and may be filled with vein material.

Lag Gravels: Lag gravels are an accumulation of coarse debris left after the removal of the finer material by wind or water. They may represent an eolian unconformity where concentration has been effected by the scouring action of the wind under desert conditions. As such they may contain wind-polished and faceted pebbles called "ventifacts" or pebbles coated with "desert varnish" and may be tightly fitted together from an old "desert pavement."

Wave and current action may also concentrate lag gravels under conditions of submarine erosion. The presence of pebble bands in a marine section suggests a diastem, if found within a formation, or a disconformity of larger magnitude, if located at the top of a formation.

Buried Soils: One of the strongest proofs of unconformity is the identification of ancient soils in the geologic column. This identification may be very difficult because of the changes that take place in soils subsequent to burial. They may be partly or completely incorporated into the basal part of the new formation and may lose the zoned characteristics typical of recent soils while being reworked by an advancing sea. Krumbein [40] in his excellent article enumerates a number of criteria useful in recognizing ancient soil horizons.

A concentration of iron oxides may be indicative of old soil profiles. In the Midcontinent, thin beds of red shale in normal sections of the Pennsylvanian are considered to be ancient soil profiles. During the formation of lateritic soils, there is a strong concentration of iron, which may show up as a tough, porous "duricrust." Zones of concretions may be found in fossil soil horizons. Caliche, for example, is often concretionary. It has been suggested that variegated shales might be the end products of modifications in long-buried soils. Basal black shales may be formed when soils with a high humus content are incorporated into the first formation of a marine transgressive overlap.

Certain residual deposits should also be considered under this heading. The weathering of limestones with the selective solution of the calcium carbonate may leave a residual soil rich in silica, phosphate, or clay. Soils formed from the weathering of cherty limestones might be recognized by masses of broken and cemented chert fragments. Concentrations of clay or calcium phosphates at the top or within a limestone section are good signposts of unconformity.

Porous Limestone: In areas of limestone bedrock, a considerable porosity may be developed by the solvent effect of ground water. Solution action may develop porosity ranging from microvugular to cavernous, depending on such factors as precipitation, rate of erosion, and length of exposure to erosion. Although other agents and processes may be responsible for limestone porosity, solution channels, caverns, and a decrease in porosity in depth without change in lithology are good evidences of groundwater action and hence of unconformity. Further, the porosity not only is a criterion for unconformity but may also be the locus for commercial oil accumulation.

Angular Coal Fragments: An interesting proof of a long erosional break in Pennsylvania sedimentation in eastern Kansas has been advanced by John L. Rich.[41] He found angular coal fragments in the lower few feet of a channel sandstone in the base of the Lawrence shale near Ottawa, Kansas. Many fragments have square ends indicating jointing before burial. The coal must have advanced at least to the lignite stage. The source of the coal fragments must be from underlying Weston shale, also Pennsylvanian, which shows marked crumpling below the unconformity with the Lawrence shale. If coal of Weston age had become lignite before its burial in Lawrence shale, the unconformity between these two must represent a long time interval.

Silicified Shell Fragments: Weathering of some Paleozoic limestones with subsequent development of residual clay has been observed

[40] Op. cit.
[41] Rich, J. L., Angular Coal Fragments as Evidence of a Long Time Break in Pennsylvanian Sedimentation in Eastern Kansas: Geol. Soc. America Bull., vol. 44, no. 4, pp. 865-870, Aug. 31, 1933.

to result in a peculiar type of silicification of the included fossils. The shells in the limestone matrix are wholly calcareous, while those in the overlying residuum are wholly silicified. On the surface of the shells occur many of the so-called "beekite rings." These are small doughnut-like circlets of bluish-gray to white, opaque to translucent quartz. Finding of silicifed shell fragments with beekite rings indicates that an erosional zone has been entered and marks an unconformity in a series of limestones even if there is no lithologic evidence.

Color Contrasts: A sharp contrast above and below an unconformity in the colors of the sediments often marks a disconformable contact where there is little other evidence. A bright red, yellow, or purple below the unconformity indicates possible weathering and oxidation of iron—or manganese-bearing minerals.

Chemical Sediments. Concentrations of chemical sediments on unconformity surfaces may be indicative of erosional breaks or cessation of sedimentation.

Manganese: Many manganese deposits are found associated with the basal strata overlying unconformities. Some of these concentrations are due to accumulation of residual materials on a weathered surface. The oxides of manganese are very stable, and accumulations of them are present on deeply eroded surfaces. Other accumulations of manganese may be found at the base level of deposition during times of little or no deposition of clastic sediments.

Phosphates: Phosphate nodules are indicative of cessation or extreme slowness of deposition. The source of phosphorus may be shells and other organic material. Marine currents working over a base level of deposition may remove parts of the sediments, concentrating the phosphates. Limestones or limy muds are particularly susceptible; the carbonates are removed, and the phosphates are concentrated.

Glauconite: Glauconite deposits are formed by diagenetic processes and require periods of nondeposition or very slow deposition for their formation. A concentration of glauconite indicates but does not prove unconformity.

Caliche: Caliche deposits, if identifiable as such in the geologic column, are indicative of unconformities. Caliche is formed by the evaporation of carbonate-bearing capillary waters that come to the surface in semi arid regions. These deposits may be a yard or more in thickness and conform to the surface. They form on uplands as well as lowlands. On submergence, these deposits may be buried beneath a blanket of sediments and, if identified, mark an unconformity.

Pyrite: Pyrite may be concentrated at or near surfaces of disconformity in marine sediments. Precipitation of iron sulphide apparently takes place along ocean bottoms where profiles of deposition have been established. Pyrite crystals may be disseminated through carbonaceous shales and organic limestones where no unconformity exists and are, therefore, not conclusive proof of breaks in sedimentation. If they are found in association with other chemical sediments such as manganese or phosphate nodules or glauconite, an unconformity is strongly suggested. The presence of pyrite at contacts of contrasting lithologies is good evidence of unconformity.

Fossil Evidence: Probably the most reliable means of detecting unconformities in the geologic column are the index fossils found in formations on both sides of the unconformity. Many extensive time breaks in sedimentation, especially disconformities, if unaccompanied by erosion or deformation, may leave no other proof of their presence in a series of stratified rocks. There are probably many unrecognized unconformities because of a lack of fossils for comparison.

Importance of Unconformities

Unconformities are of paramount value to the stratigrapher in separating units of rocks and of geologic time. They are of great importance to the oil geologist because of the vast amounts of petroleum that may be trapped at or near their surfaces.

Stratigraphic. Geologists have long used diastrophism as the ultimate basis for subdividing geologic time and the geologic column. The geologic record shows evidence of rhythmic changes in sea level. These changes are controlled by recurrent or cyclic periods of disturbance. A geologic cycle usually starts with an unconformity and ends with an unconformity. The diastrophism that closes one cycle is usually accompanied by mountainbuilding and deepening of the ocean basins. Not only are there wide spread breaks in sedimentation, but vast areas are exposed to erosion. As land areas are worn down and sediments are carried into the ocean basins, the seas will slowly rise and cover the lower parts of the continents with shallow seas. Sediments will be deposited unconformably on the submerged erosion surface. An examination of the geologic record shows a great succession of marine inundations of the continents separated by unconformities marking the retreat of marine waters. Since the oceans are freely connected, a change in sea level will affect all, and major breaks will be worldwide.

The cycles of earth movements do not follow a smooth pattern. Between major periods of diastrophism will be irregular periods of disturbance involving local emergence and submergence of land areas. In regions affected by these minor breaks there will be a marked hiatus and unconformity.

Breaks in the Fossil Record: The fossil record affords us the best direct means of evaluating the time breaks indicated by unconformities. The evolution of plant and animal forms is always greatly accelerated during periods of crustal disturbance. Some forms adapt themselves to the changes in environment and survive with variations in structure, while others fail to adjust themselves and perish. The next invasion of the sea traps within its sediments a new assemblage of fossil flora and fauna. Great breaks in the physical records are accompanied by breaks in the biologic record. Fossils are, therefore, the most dependable means of detecting the time break necessary to establish an unconformable break in the geologic column.

Subdivisions of the Geologic Record: The greatest breaks in the geologic record result from widespread continental emergence. These may be accompanied by large-scale mountain building with much deformation of previously formed sediments. These revolutions separate geologic time into eras and are marked by worldwide unconformities. Within each era are periods of crustal disturbance strong enough to cause widespread retreats of the seas from the continent masses. These produce breaks in the geologic record, which are widespread but not universal. The unconformities produced by those breaks separate rocks into systems, and the corresponding time units called periods. Smaller and more local breaks divide periods into epochs and systems into series of rocks.

Oil Reservoirs. A number of conditions may be responsible for oil accumulation near unconformable contacts. Soluble rocks such as limestone when exposed to erosion may become porous by solution. Subsequent deposition of an impervious shale above the unconformable contact may trap oil in large quantities in the limestone. The West Edmond field of Oklahoma produces from the Bois d'Arc limestone member of the Hunton formation. An unconformity overlain by Pennsylvanian shales seals the oil, which is trapped in truncated Bois d'Arc limestone. This limestone was made porous by solution during the weathering of the upturned formations.

Insoluble residual materials may collect on an old erosional plain or against the slope of an old high. These accumulations of porous materials form potentially good reservoirs.

The rocks immediately above an unconformity are often shallow-water sands and gravels. These may be highly porous and suitable for reservoir rocks.

"Bald-Headed" Structures: "Bald-headed" structures are anticlines or domes that have been eroded so that the producing formations are removed from the tops of the structures. Subsequent submergence and deposition have sealed the truncated formations below the unconformity. Oil is found in the flank sands of these structures. The Oklahoma City pool, a cross section of which is shown in figure 18, is an excellent example of this sort of pool. Simpson formations were stripped from the top of the structure by pre-Pennsylvanian erosion. Production from this formation was found on the flanks of the structure. The Billings dome in Noble County, Oklahoma, was barren at the top owing to erosion of Simpson formations. It was not until 17 years later that geologists recommended drilling on the flanks of the structure, where large production was found in the "Wilcox" sand. The Nemaha buried ridge of Kansas and Oklahoma is perhaps the largest "bald head" of all.

Shoestring Sands: Very long sand lenses a few hundred feet wide, a few score feet thick, and seveıal thousand feet to several miles long, are called "shoestring sands," when they are found buried in mud deposits or in shale formations. They may be fillings of old stream channels or buried offshore bars. If the former, the cross section of the deposit will have its greatest width at the top and will have a base convex

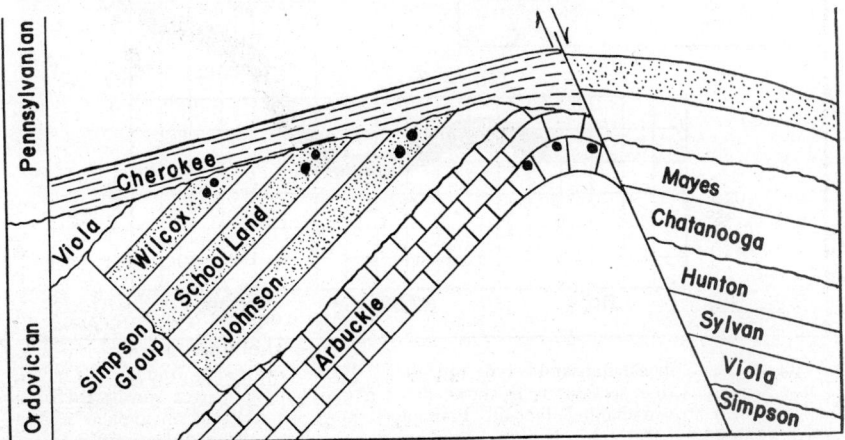

FIGURE 18. Idealized section west-east across Oklahoma City field showing relation of unconformities to oil and gas production. A typical "bald-headed" structure.

downward. In the latter case, the cross section will have a flat base and a top convex upward. These sands will have disconformable relationships with the surrounding rocks.

An example of a shoestring-sand oil pool is the Bush City field in Anderson County, Kansas. Oil is trapped in a sand body 13 miles long, about one-fourth mile wide, and buried 30 to 40 feet below the top of the Cherokee shale of Pennsylvanian age. The sand has been folded into minor anticlines and synclines. Production is from both because of the water-free nature of the sand. A typical cross section and plan of the sand is shown in figure 19.

Disconformities: Oil may be found at the surfaces of disconformities where impervious beds overlie an erosion surface of some relief and will be found in structural adjustment in the highs of the older strata. Figure 20 shows diagrammatically how petroleum has been trapped in hills of the Wilcox formation buried under the impervious clays of the Cane River formation; both are Tertiary.

Regional Unconformities: Unconformities whose surfaces can be traced over wide areas are termed "regional unconformities." The intersection of two unconformable surfaces against the flanks of the Sabine

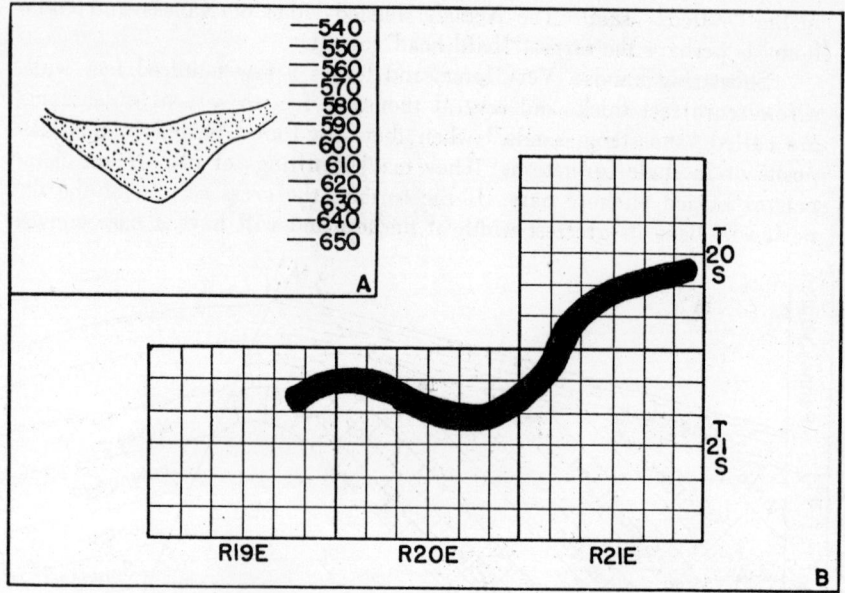

FIGURE 19. "Shoestring-sand" type oil field; Bush City field, Anderson County, Kansas. Cross section in *A* shows flat top and bottom convex downward of a typical stream-channel deposit. Plan in *B* shows windings of an ancient stream channel.

uplift are responsible for the vast quantities of oil in the East Texas field. The Woodbine formation of Upper Cretaceous age has an unconformity at the top of the formation and is overlain nonconformably by the Eagle Ford in the Mexia and Balcones fault zones and by the Austin chalk in east Texas (fig. 21). The base of the Woodbine is marked by another unconformity, below which are beds of the Comanche group. These two unconformities are of regional extent and intersect on the flank of the Sabine uplift in east Texas. Oil from the Woodbine is found in the fault zone, where it is trapped at a fault surface and below the upper unconformity. Figure 21 is an idealized west-east section through the Tyler

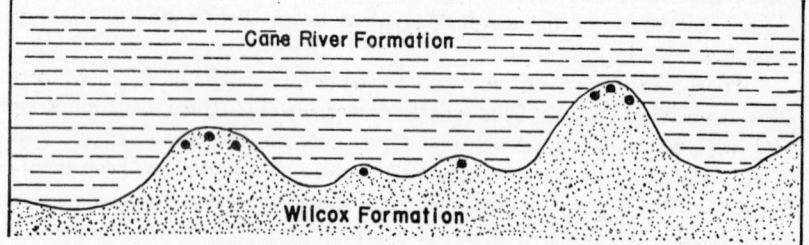

FIGURE 20. Idealized diagram of Urania field, Louisiana, where production comes from oil trapped in the highs of old erosion surface disconformably overlain by impervious shales. Both the Cane River and Wilcox formations are Eocene.

basin showing the function of a regional unconformity in localizing production in the Woodbine sand.

Major oil structures are often completely obscured by nonconformable deposition of later sediments. The geologic history of the older rocks may have been such that great oil and gas accumulations were formed below regional unconformities. For example, the Central Kansas uplift is obscured by a uniform cover of north- and west-dipping Permian and

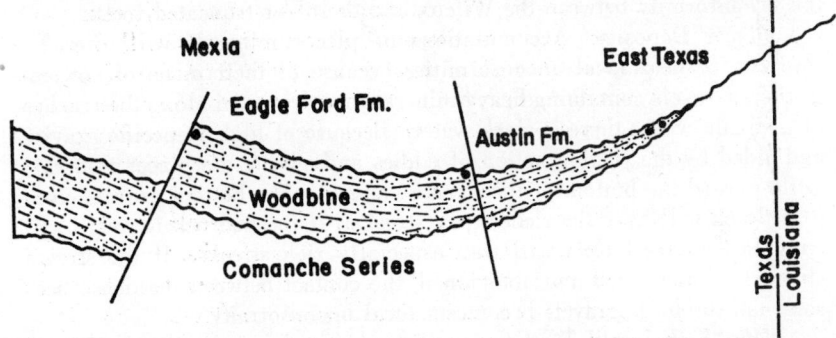

FIGURE 21. Idealized section west-east of Woodbine sand in Tyler basin of northeast Texas showing relation of unconformities to production. In Mexia field to west, production is from oil trapped between a fault and an unconformity. In East Texas field to east, two unconformities intersect on flank of Sabine uplift to form oil trap. This section illustrates importance of regional unconformity to production of oil.

Cretaceous rocks. Again, the Bend arch of Texas is obscured by overlapping and west-dipping Pennsylvanian strata.

Two areas offer great possibilities for new geologic conditions concealed below unconformities. The Comanche rocks of northern Louisiana, northeast Texas, and southern Arkansas, where 5,000 to 10,000 feet of sediments are folded, tilted, and completely overlapped unconformably by Upper Cretaceous rocks. Pre-Carboniferous rocks of west Texas have been folded and eroded with great lateral changes in porosity and are widespread beneath a cover of Carboniferous rocks.

Ore Deposits

Types of deposits that may be found associated with unconformities are residual and placer ores. Surfaces of unconformity may also influence the localization of hydrothermal deposits.

Residual Deposits: Erosion surfaces of regional extent may be the loci for residual deposits of bauxite, manganese ores, lateritic iron ores, and phosphates. If this surface is then buried completely or partly, an unconformity results, the tracing and mapping of which may lead to the discovery of commercial concentrations of ore.

The bauxite deposits of the southern Appalachian states[42] are an example of residual accumulations on a partly buried erosion surface. These deposits were formed during the Eocene when the climate was favorable for lateritic weathering. The margins of the Appalachians had been leveled by the Highland Rim peneplain, with a karst topography developing in areas of limestone outcrop. Bauxite accumulated in the sinkholes and was incorporated into the basal beds of the Wilcox formation. The deposits are now located on remnants of the dissected peneplain, at the unconformity between the Wilcox and the older truncated rocks.

Placer Deposits: Accumulations of placer minerals will often be found at or near local unconformities because of their mode of concentration. Gravels containing heavy minerals are deposited along the reaches of a stream where there is slack water. Because of higher specific gravity and aided by the jigging action of eddies and swirls, the placer minerals settle toward the bottom of the stream channel. The bottom gravels close to bedrock will have the richest pay streaks. If the bedrock is rough and irregular, natural riffles will trap especially rich streaks. If the stream channel is later filled and abandoned, the contact between the dense bedrock and the loose gravels becomes a local unconformity.

The "high level" Tertiary, auriferous gravels of the Sierra Nevada described by Lindgren[43] are an excellent illustration. Buried Tertiary gravels containing placer gold are exposed high on the sides of Quaternary valleys. These gravels were deposited in stream channels cut in bedrock during the Eocene, with the richest pay streaks in the deep gravels two to three feet off the bottom. The stream channels were then buried by lean gravels and volcanics which completely filled the valleys. After elevation of the Sierra Nevada in the Pliocene and Pleistocene, new streams eroded canyons an average of 2,600 feet below the early Tertiary stream bottoms. The gold-bearing gravels have been worked by drift mining and hydraulic procedure.

Hydrothermal Deposits: The influence of unconformities in controlling ore formation by hydrothermal solutions is that of modification and not primary control. A review of the literature revealed no districts or mines in which an unconformity was considered the primary control in ore deposition. Unconformities, if present in the section, were rarely considered as having any modifying influence at all.

Bell[44] mentions the apparent effect of the slope of an unconformable surface on ore deposition in the Hallnor mine, second-largest mine in the east Porcupine area of Ontario. An angular unconformity of considerable erosional irregularity separates a series of younger sediments from older lava flows. A factor that has proved useful in prospecting for new ore zones is the slope of the lava-sediment contact. At the west end of the

[42] McKinstry, H. E., *Mining Geology*, New York City, Prentice-Hall, Inc., 1948.
[43] Lindgren, W., *Tertiary Gravels of the Sierra Nevada*, U. S. Geol. Surv. Prof. Paper 72, 1911.
[44] Bell, A. M., *Hallnor Mine, Structural Geology of Canadian Ore Deposits:* Canadian Inst. of Min. Metallurgy Trans., 1948.

property at the upper levels, a hill of lava projects stratigraphically into the sediments and is marked by a vertical contact. In the lower levels the contact flattens where an embayment of sediments cuts into the old lava surface. The No. 1 vein system occurs opposite the steep contact. Below this the ore is absent along the flatter-dipping contact but recurs where the unconformity starts to steepen.

Conclusion

Because of the stratigraphic and economic value of unconformities, the geologist should become adept in using the criteria by which unconformities may be detected. Undiscovered reservoirs of petroleum of the magnitude of the East Texas field may lie beneath the anonymity of regional unconformities on the Gulf Coast, in west Texas, and in some of the Rocky Mountain States. Unconformities may prove to be important factors in controlling and localizing ore deposition in many mining areas where their potentialities have been ignored.

FAULTS

Geologists are aware of the need of fault control before stratigraphic problems can be solved. Fault patterns and their relationships may be extremely complicated (fig. 22) and may play such a role that both surface and subsurface structural and stratigraphic trends can be evaluated only after voluminous data become available for interpretation.

A fault represents a surface along which one rock segment has moved with respect to the other. The magnitude of faults ranges from millimeters of displacement to several thousands of feet or even miles. Faults are developed in all rock types; their displacement may increase or decrease with depth or vice versa, and their relative movement may be vertical, horizontal, or rotational.

The apparent movement along a fault is a function of many variables, and depends not only on the net slip, but also on the strike and dip of the fault, the dip and strike of the disrupted stratum, and the attitude of the surface on which the observations are made.[45]

Common fault types include strike faults, in which the strikes are more or less parallel to the strike of the rocks involved; oblique faults, in which the strikes are diagonal to the strike of the rocks involved; longitudinal faults, in which the strikes are roughly parallel to the strike of the regional structural fabric; and transverse faults, in which the strikes are perpendicular or oblique to the strike of the regional structural fabric. Other fault varieties are en échelon, peripheral, and radial. High-angled faults are those with surfaces that dip greater than 45 degrees. Low-angled faults dip less than 45 degrees.

A thrust fault (reverse) or thrust is a fault along which the hanging wall has moved up relative to the footwall. A gravity fault (normal) is a

---

45 Billings, M. P., Structural Geology, p. 137, New York, Prentice-Hall, Inc., 1942.

fault along which the hanging wall has moved down relative to the foot-wall.[46]

Billings [47] lists the following criteria to aid in the recognition of faults: (1) the discontinuity of structures, (2) the repetition or omission of strata, (3) features characteristic of fault planes, (4) silicification and mineralization, (5) sudden changes in sedimentary types and (6) physiographic data. Only certain of these criteria may be applied in evaluating

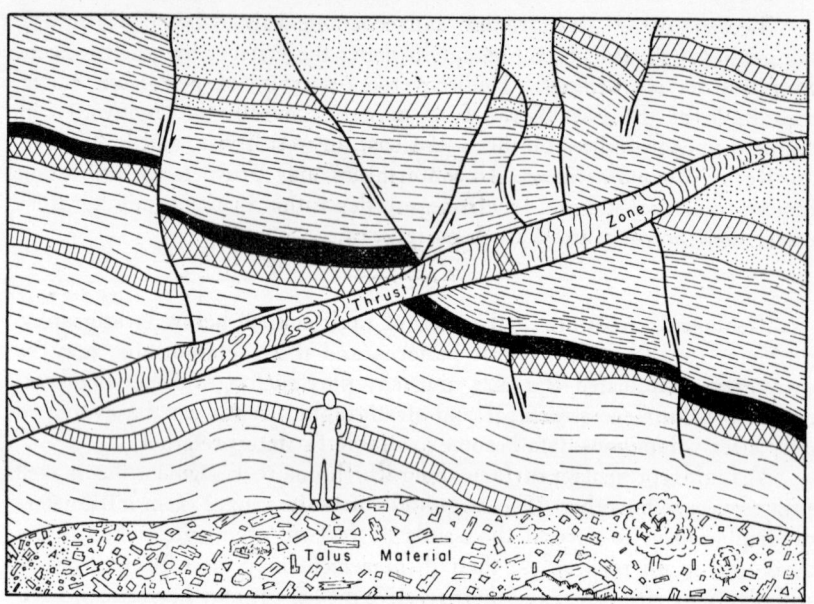

FIGURE 22. Complexly faulted Miocene shales exposed in a road cut in Grimes Canyon, Ventura County, California. This sketch clearly demonstrates complex structural conditions which may develop in the subsurface as a result of thrust faulting. Note structural relations above and below thrust zone. Decipherment of this type of geologic problem requires close coordination between the structural and stratigraphic geologists. Drawing was sketched from a projected 35-mm. Kodachrome slide. Graphic symbols used do not represent true lithology but illustrate only minor compositional variations in the shale section.

the presence of a fault in the subsurface. To establish the type and character of faulting in well sections the following indications have proved applicable: (1) anomalous profiles of lithologic, electric, and radioactive logs; (2) a change in dip and disturbed phases as exhibited by cores; (3) mineralogic and paleontologic irregularities; (4) lost circulation; (5) a caving hole; (6) an increase in penetration rate; (7) an increase in porosity and permeability; (8) slickensided fragments in ditch samples; (9) a sudden increase of drilling-mud temperature; (10) poor core recovery; and (11) an abrupt deviation of the hole from the vertical.

[46] Billings, M. P., *op cit.*, p. 152.
[47] Billings, M. P., *op. cit.*, p. 155.

Many subsurface faults are not reflected at the surface owing to "dying out" or to truncation by buried erosional surfaces (fig. 23). Conversely, faults intersecting the surface may be absorbed by incompetent beds in the subsurface and thus may have restricted vertical downward extension.

It should be recognized that faults involve surfaces and not planes. Many irregularities and curvatures which these surfaces may assume make their position and trend in the subsurface difficult to define. These ir-

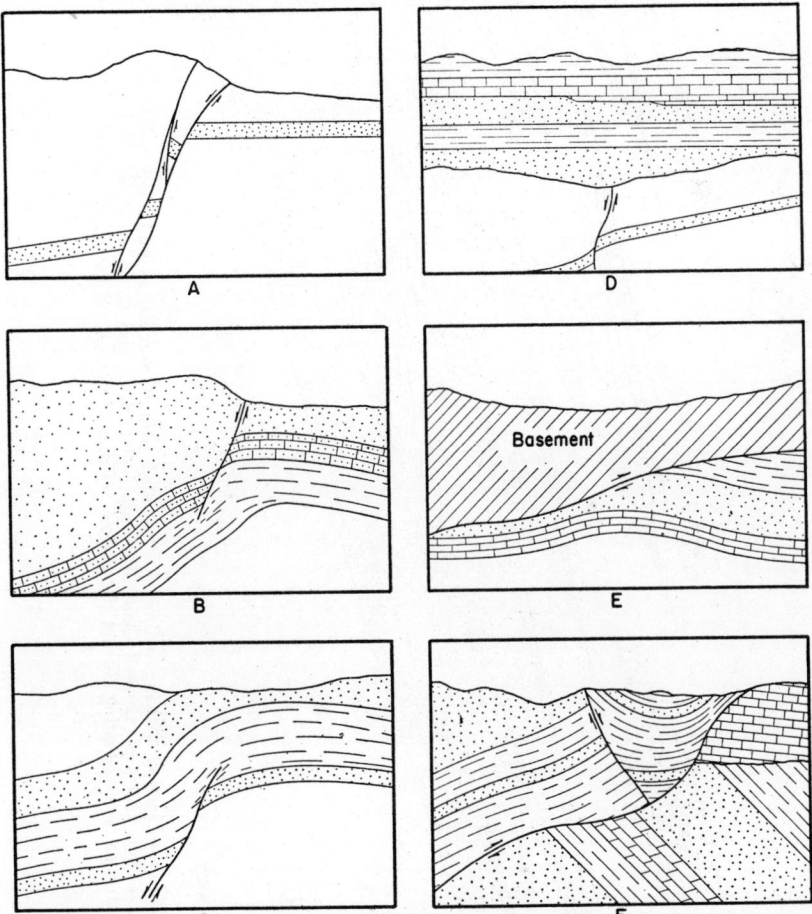

FIGURE 23. Various types of faults. A—fault surfaces may be extremely irregular; the proper interpretation of details of fault zones in the subsurface requires considerable data. B—Many faults disappear with depth owing to displacement absorption by incompetent strata. C—Surface folds may be replaced by vertical displacement with depth. D—Fault systems may be truncated by erosional surfaces. E—Favorable petroliferous structure may underlie overthrust sheets. F—Major thrust faulting may create accompanying tensional fault patterns.

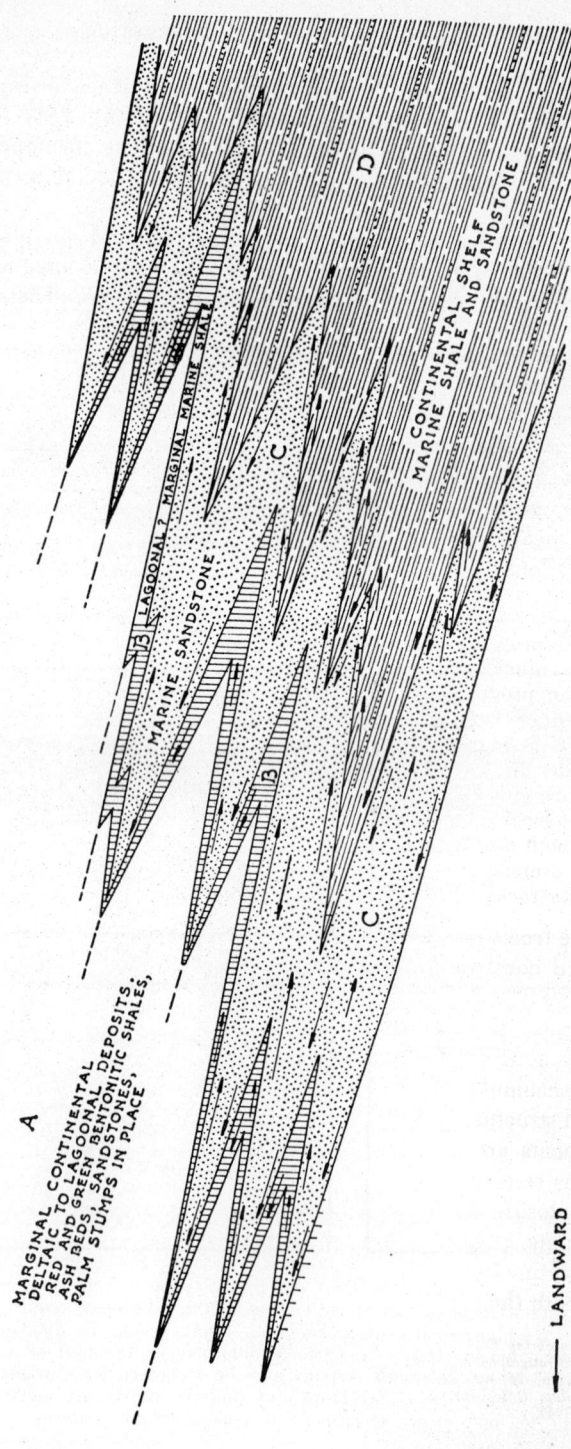

FIGURE 24. Idealistic cross section showing sediments associated with transgressions and regressions of the sea in Starr and Zapata Counties, Texas. *C* and *D* show facies of exposed Cook Mountain beds; *A*, *B*, and *C*, facies of Yegua and Fayette beds. Arrows indicate the direction in which the sea was migrating during deposition. (From Patterson. Reproduced permission Am. Assoc. Petroleum Geologists.)

regularities complicate and, in many instances, control the accumulation pattern of oil and gas and other concentrations of minerals. Fault relationships should always be viewed in three dimensions. Fault surfaces may be contoured and fault gaps outlined.[48]

Marine Onlap and Offlap

Transgressions and regressions of the sea across continental segments produce sedimentary onlaps and offlaps. These depositional features are of local or regional magnitude and have been responsible in many areas for conditions permitting the accumulation of large quantities of oil and gas. Problems of onlap and offlap of varying complexity are commonly encountered in stratigraphic work.

Patterson [49] has clearly demonstrated the transgressive and regressive relationships of the Cook Mountain, Yegua, and Fayette formations (upper Eocene) developed across Webb, Zapata, and Starr Counties, Texas. These relationships are idealistically shown in figure 24.

Swain's definitions [50] of various terms relating to transgressions and regressions follow:

1. Onlap: The progressive pinching-out, toward the margins of a depositional basin, of the sedimentary units of a conformable sequence of rocks.
2. Offlap: The progressively offshore degression of the updip terminations of the sedimentary units of a conformable sequence of rocks.
3. Overstep: The regular truncation of older units of a complete sedimentary sequence by one or more later units of the sequence. The resulting unconformity may be either marginal to the basin of deposition, or within the basin as a result of local uplifts. If more than one unit rests on those beneath the unconformity, both overstep and onlap are involved.
4. Complete overstep: The entire blanketing with unconformable relationship of the older rocks of a basin by younger rocks.

To obtain the true trend and magnitude of marine onlaps and offlaps, one must give full consideration to the third-dimensional concept of the region.

Oil and Gas Traps

Oil and gas accumulations (fig. 25) occur under favorable structural, stratigraphic, and structural-stratigraphic conditions. Examples of various types of entrapments are graphically illustrated in figures 26-32. Prospective oil and gas reservoirs must be considered from a three-dimensional viewpoint. The closure factor should under no circumstances be minimized. To evaluate closure, whether it be controlled by structural or stratigraphic conditions or a combination of both, requires the interpretative ability of both the surface and subsurface geologist.

---

[48] Reiter, W. A., *Contouring Fault Planes:* Oil World, July 14, 1947.
[49] Patterson, J. M., *Stratigraphy of Eocene between Laredo and Rio Grande City, Texas:* Assoc. Petroleum Geologists Bull., vol. 26, no. 2, pp. 256-274, Feb. 1942.
[50] Swain, F. M., *Onlap, Offlap, Overstep* and *Overlap,* Am. Assoc. Petroleum Geologists Bull., vol. 33, no. 4, pp. 634-635, 1949.

Clapp [51] in 1910 proposed the first detailed classification of natural oil and gas accumulations and modified [52] this classification in 1917. Finally in 1929 [53] he presented the following classification:

I. Anticlinal structures
   1. Normal anticlines
   2. Broad geanticlinal folds
   3. Overturned folds
II. Synclinal structures
III. Homoclinal structures
   1. Structural "terraces"
   2. Homoclinal "noses'"
   3. Homoclinal "ravines"
IV. Quaquaversal structures (domes)
   1. Domes on anticlines
   2. Domes on homoclines and monoclines
   3. Closed salt domes
   4. Perforated salt domes

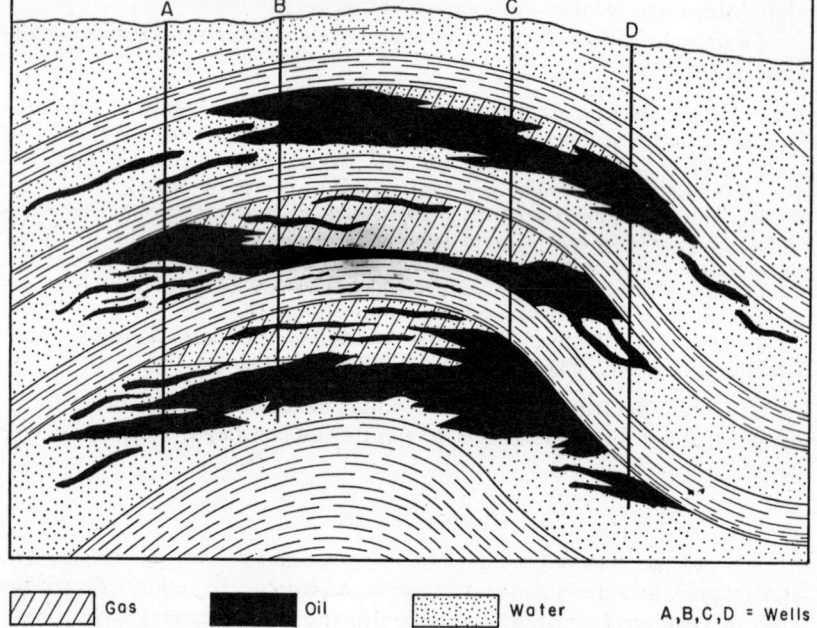

///// Gas   ▓ Oil   ▒ Water   A,B,C,D = Wells

FIGURE 25. Diagram showing irregular distribution of gas, oil and water in an asymmetric anticline. Such relationships require orderly development of field to insure maximum and efficient recovery. Rate and pattern of encroachment of edge and bottom waters are of primary concern to operators.

[51] Clapp, F. G., *A Proposed Classification of Petroleum and Natural Gas Fields:* Econ. Geol., vol. 5, pp. 503-521, 1910.
[52] Clapp, F. G., *Revision of the Structural Classification of Petroleum and Natural Petroleum and Natural Gas Fields:* Geol. Soc. American Bull., vol. 28, pp. 553-602, 1917.
[53] Clapp, F. G., *Role of Geologic Structure in the Accumulation of Petroleum,* in *Structure of Typical American Oil Fields, II:* Am. Assoc. Petroleum Geologists Bull., pp. 671-672, 1929.

      5. Domal structures caused by igneous intrusions
V. Unconformities
VI. Lenticular sands (on structure)
VII. Crevices and cavities irrespective of other structure
      1. In limestones and dolomites
      2. In shales
      3. In igneous rocks
VIII. Structures due to faulting
      1. On upthrown and downthrown sides
      2. Overthrusts
      3. Fault blocks

Wilson [54] in 1934 proposed a very logical and well-organized classification based on local deformation of strata, variation of rock porosity, and combinations of the two. His classification is as follows:

I. Closed reservoirs
  A. Reservoirs closed by local deformation of strata
    1. Reservoirs closed by folding
      a. Reservoirs in closed anticlines and domes
      b. Reservoirs in closed synclines and basins
    2. Reservoirs developed by offsetting of strata by faulting of homoclinal structure
    3. Reservoirs defined by combinations of folding and faulting
    4. Reservoirs formed through disturbance of strata by intrusions
      a. Intrusions of salt
      b. Intrusions of igneous rock
    5. Reservoirs developed in fault and joint fissures and in crushed zones
  B. Reservoirs closed because of varying porosity of rock. No deformation of strata required other than regional tilting
    1. Reservoirs in sandstone caused by lensing of sandstone or by varying porosity in sandstone
    2. Lensing porous zones in limestones and dolomites
    3. Lensing porous zones in igneous and metamorphic rocks
    4. Reservoirs in truncated and sealed strata
      a. Closed by overlap of relatively impervious rock
      b. Closed by seal of viscous hydrocarbons
  C. Reservoirs closed by combination of folding and varying porosity
  D. Reservoirs closed by combination of faulting and varying porosity
II. Open reservoirs
  None of economic importance.

Sanders [55] proposed a broad subdivision of trap types: (1) structural traps, (2) stratigraphic traps, and (3) combinations of structural-stratigraphic traps. Heroy [56] recognized four groups of oil and gas traps: (1) depositional traps, (2) diagenetic traps, (3) deformational traps and (4) combination traps.

In 1945 Wilhelm [67] in a very worthy discussion, subdivided petroleum

[54] Wilson, W. B., *Proposed Classification of Oil and Gas Reservoirs*, Problems of Pet. Geol. (Tulsa, Am. Assoc. Petroleum Geologists, pp. 433-445, 1934).
[55] Sanders, C. W., *Stratigraphic Type Oil Fields and Proposed New Classification of Reservoir Traps*, Am. Assoc. Petroleum Geologists Bull., vol. 27, pp. 539-550, 1943.
[56] Heroy, W. B., *Petroleum Geology*, Geology, 1888-1938, Fiftieth Anniversary Volume, Geol. Soc. America Bull., pp. 534-539, 1941.
[67] Wilhelm, O., *Classification of Petroleum Reservoirs*, Am. Assoc. Petroleum Geologists Bull., vol. 29, pp. 1537-1579, 1945.

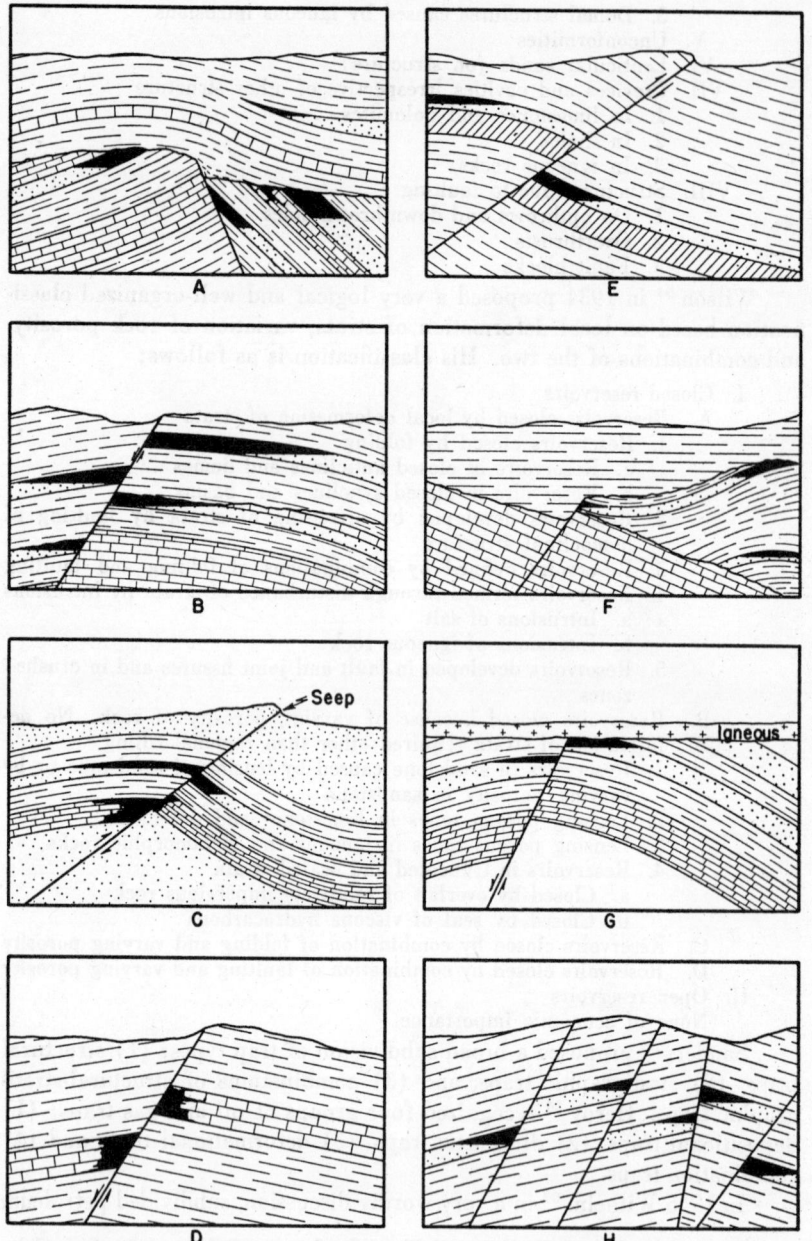

FIGURE 26. *A*—Accumulation in porous beds below an erosional surface overlain by impervious strata. A common variety of fault and lens trap is also shown. The reservoirs represent both stratigraphic and stratigraphic-structural types. *B, C, D, E*—Accumulations attributed mainly to structural control. *F, G*—Structural and stratigraphic-structural relationships that account for accumulations. *H*—Accumulation that has been controlled by structural conditions.

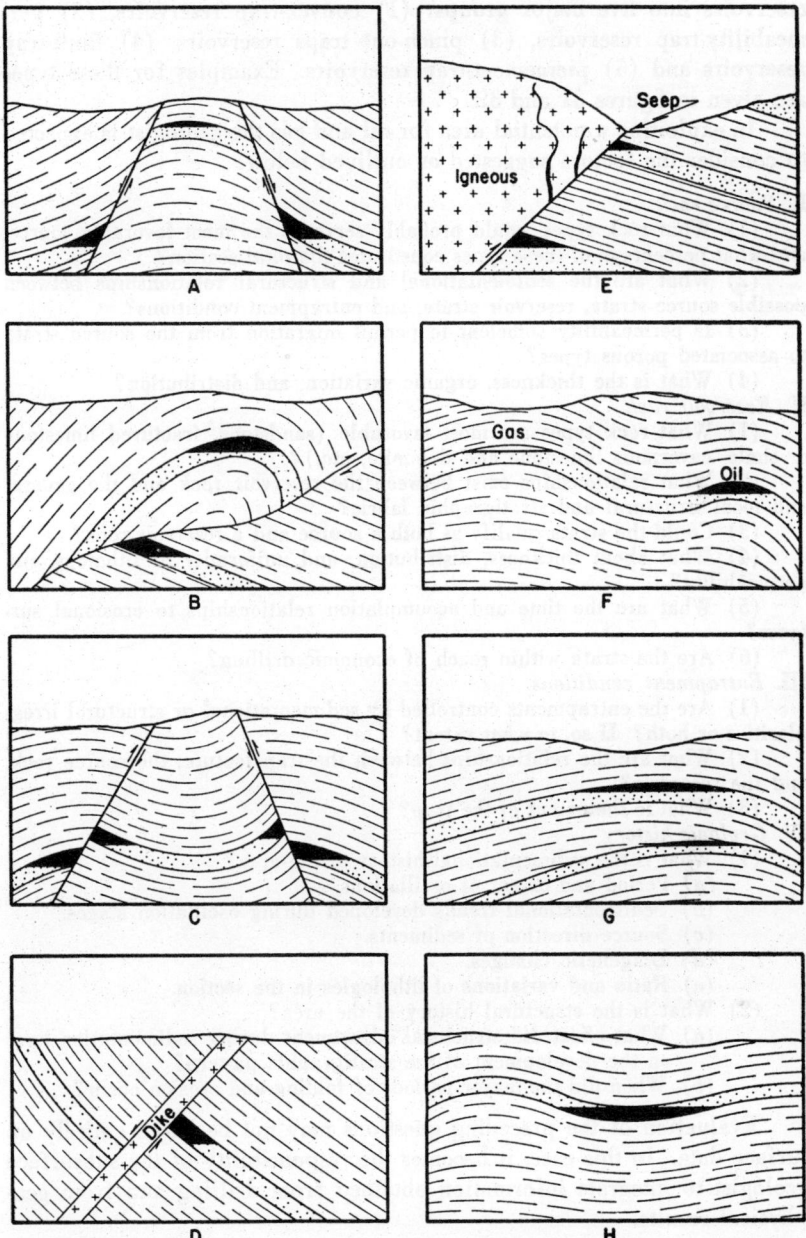

FIGURE 27. Various types of structural traps; combination of folding and faulting involved. *G* illustrates migration of the crestal high as a result of convergence of section.

reservoirs into five major groups: (1) convex-trap reservoirs, (2) permeability-trap reservoirs, (3) pinch-out traps reservoirs, (4) fault-trap reservoirs and (5) piercement-trap reservoirs. Examples for these types are given in figures 31 and 32.

In exploiting a potential area for oil and gas the geologist is expected to consider the factors suggested as outlined below:

I. *Source rock*

(1) What rock types would probably serve as the most favorable source, and what percentage of these types constitutes the total section?

(2) What are the sedimentational and structural relationships between possible source strata, reservoir strata, and entrapment conditions?

(3) Is permeability sufficient to permit migration from the source strata to associated porous types?

(4) What is the thickness, organic variation, and distribution?

II. *Reservoir rock*

(1) What rock types are most favorable (sandstone, fractured limestone or shale, cavernous limestone and dolomite, etc.)?

(2) What relationships exist between the reservoir rock and the general and local structural and stratigraphic fabric?

(3) Could the strata qualify as both a source and a reservoir rock?

(4) What about thickness, distribution, and uniformity of lithology and permeability?

(5) What are the time and accumulation relationships to erosional surfaces?

(6) Are the strata within reach of economic drilling?

III. *Entrapment conditions*

(1) Are the entrapments controlled by sedimentational or structural irregularities or both? If so, to what extent?

(2) What are the relationships between the trap feature, the source rock, and the reservoir?

(3) What is the extent of the trap?

IV. *Geologic history*

(1) What is the sedimentational history of the area?

    (*a*) Period and extent of oscillations.

    (*b*) Sedimentational trends developed during oscillation stages.

    (*c*) Source direction of sediments.

    (*d*) Diagenetic changes.

    (*e*) Ratio and variations of lithologies in the section.

(2) What is the structural history of the area?

    (*a*) What effect did structural adjustment during sedimentation have on the development of the stratigraphic pattern?

    (*b*) When did the major periods of folding and erosion occur?

Evaluation of the preceding questions may not be based entirely on surface data. In this case, it becomes the responsibility of the subsurface geologist to integrate information obtained from drilling and from geophysical results.

CORRELATION CONSIDERATIONS

There is constant demand from the stratigraphic geologist for accurate evaluation of sedimentary units, their lithologic, paleontologic,

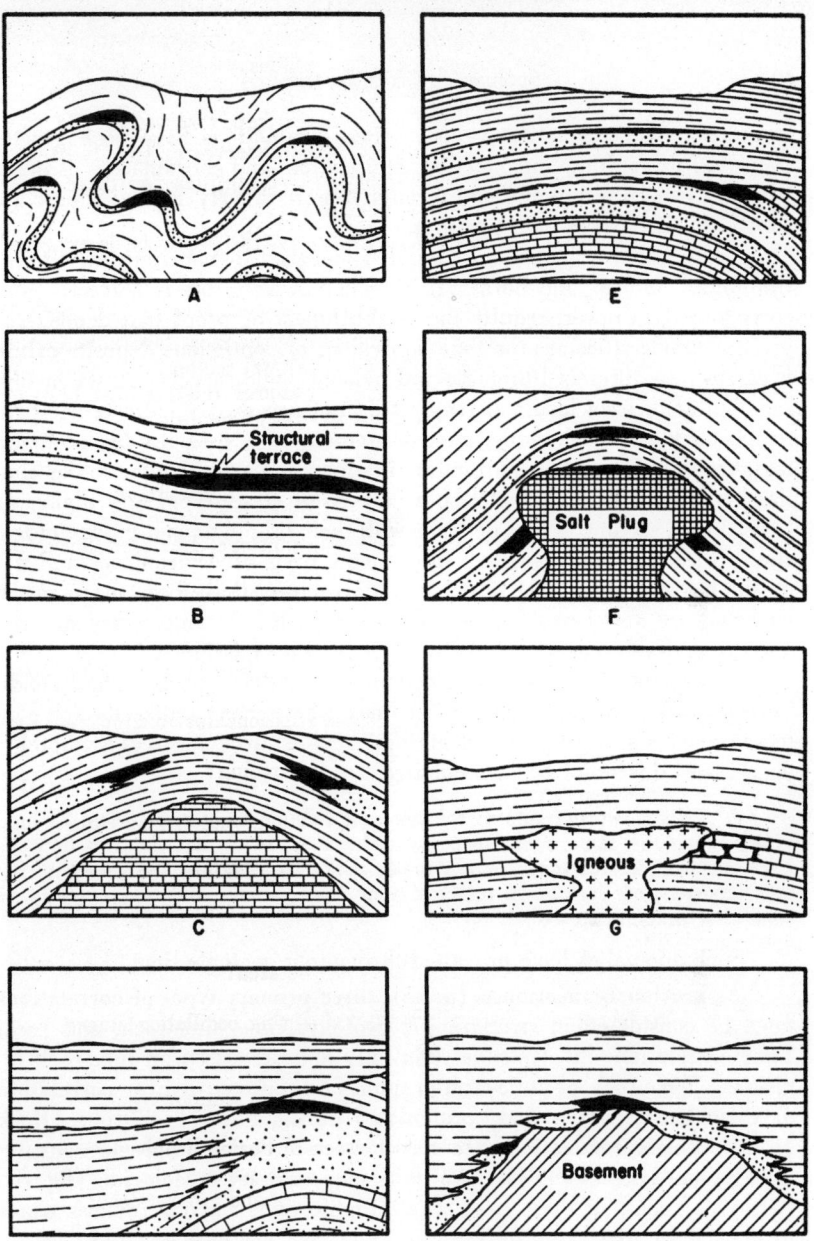

FIGURE 28. *A, B*—Structural traps. *C*—Differential compaction over buried competent rocks, which produces flexures in overlying strata. The reservoirs shown are controlled by both the structural and stratigraphical factor. *D, E*—Combination of structural and stratigraphic conditions accounting for the accumulations. *F*—Salt intrusions commonly arch the overlying strata thus permitting accumulation of oil and gas at various points in the structure. *G*—Igneous intrusions into sedimentary sections, which sometimes induce secondary permeability in the host strata and develop favorable reservoir conditions. *H*—Progressive overlap which gives rise to favorable stratigraphic reservoir traps; accumulation may be erratically distributed.

structural, and correlation values. These evaluations have been greatly improved through the introduction, application, and coordination of many new and revised techniques and through a more orderly and systematic approach to subsurface investigations.

Correlation, as commonly implied, consists in matching up similar lithologies, faunas, and floras. In certain instances this is correct; other correlations, however, require the establishment of proof that deposits of given characteristics are the time equivalent of contiguous deposits exhibiting entirely different lithologic and paleontologic aspects. Strata involving similar or even identical features do not necessarily indicate age contemporaneity. For example, an environment in a stratigraphic sequence may have given rise to a particular lithology and microfauna; under similar conditions these characteristics may occur stratigraphically higher or lower in the section. These deposits with their lithologic and paleontologic similarities may be correlated only on the basis of environment and not on the basis of equivalent time. Such correlations have been made in the past and are now being made, and they have led to unnecessary or even disastrous development and exploration recommendations.

To exemplify the preceding comments, one has only to review conditions prevailing along modern coast lines, where many unlike though contemporaneous sedimentary and biologic realms are evident. Twenhofel [58] emphasizes this point by commenting:

. . . When two deposits of the geologic column have been found to hold pretty much the same organisms, it has been assumed that the two deposits have synchronous relations. It is equally if not more valid to assume that the two deposits were laid down under similar environments and may actually be somewhat different in age.

Such anomalies have prevailed throughout geologic time.

As previously mentioned (p. 15), three primary types of correlations must be considered in stratigraphic geology: time, time-rock, and rock. Time correlations involve time span only and are based on time-rock divisions. Time-rock or time-stratigraphic correlations involve correlation of variable rock types that accumulated during intervals of time. Rock or lithogenetic correlations are based essentially on lithologic constitution. Boundaries of the last two units may transect or coincide (fig. 6). Neglect of defining these basic stratigraphic units only introduces chaos to the science of stratigraphy.

Some correlations, particularly those involving the time-rock type, generally require voluminous data before being satisfactorily and adequately established. This problem is most critical in areas in which the strata exhibit extreme and rapid vertical and lateral facies changes. In areas where stratal sequences are reasonably uniform and constant, as in certain parts of the Paleozoic section of the Midcontinent region, correla-

---

[58] Twenhofel, W. H., *Report of Committee on Paleoecology*, Nat. Research Council, Oct. 1935.

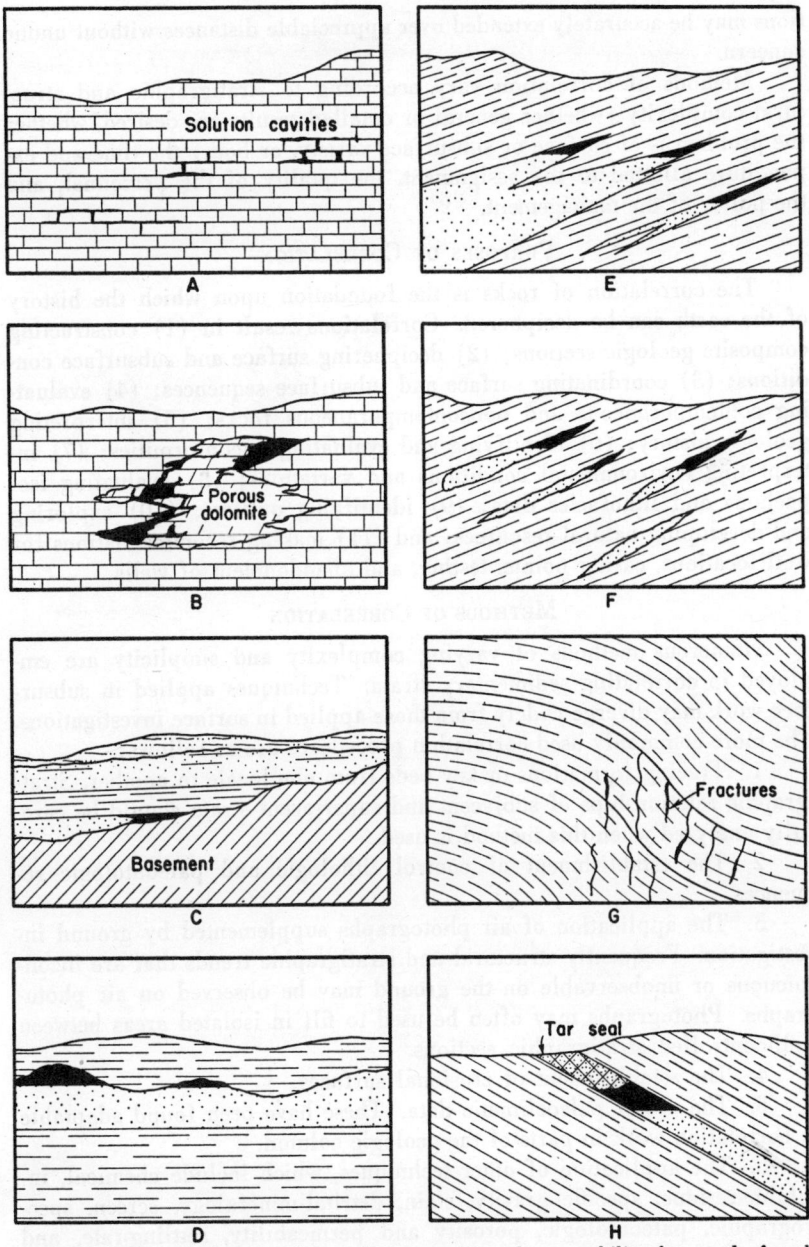

FIGURE 29. *A, B*—Primary and secondary porosity and permeability frequently found in limestones and dolostones. Such conditions favor oil and gas concentrations. These accumulations are difficult to discover and develop and their reserves difficult to predict. *C-F*—Typical stratigraphic traps. *C*—progressive overlap. *D*—unconformity. *E*—updip lensing. *F*—sand lentils. *G*—Accumulation in fold fractures of relatively impermeable strata. *H*—Updip tar seal.

tions may be accurately extended over appreciable distances without undue concern.

Methods of correlation vary according to stratigraphic and structural complexity (whether general or detailed results are desired, whether the problem is of surface or subsurface variety, or both) the time and expenditure allotted to the assignment, the quality of the personnel, and the policy of the management.

PURPOSES OF CORRELATION

The correlation of rocks is the foundation upon which the history of the earth can be deciphered. Correlations result in (1) constructing composite geologic sections; (2) deciphering surface and subsurface conditions; (3) coordinating surface and subsurface sequences; (4) evaluating contemporaneous and noncontemporaneous rocks; (5) interpreting geologic history; (6) identifying and evaluating unconformities; (7) interpreting environmental conditions and variations; (8) evaluating isopachous and lithofacies data; (9) identifying outliers; (10) exploring and developing natural resources; and (11) making recommendations for well locations, casing points, testing, and abandonment of wells.

METHODS OF CORRELATION

Numerous methods of varying complexity and simplicity are employed in correlating sedimentary strata. Techniques applied in subsurface work may diverge widely from those applied in surface investigations. The more commonly used correlation procedures are as follows:

1. Tracing formations or key beds from one locale to another. Stratigraphic relationships of subjacent and superjacent strata should be carefully analyzed when this method is used.

2. The establishment of control lithologic and paleontologic sequences.

3. The application of air photographs supplemented by ground investigation. Frequently structural and stratigraphic trends that are inconspicuous or unobservable on the ground may be observed on air photographs. Photographs may often be used to fill in isolated areas between well-controlled stratigraphic sections.

4. The establishment of erosional surfaces.

5. The use of paleoclimatic data. These have been found adaptable to correlation work in parts of the geologic column.

6. The application of other techniques, which include chemical, insoluble-residue, specific gravity, stain, detrital-mineralogy, screen, spectrographic, paleontologic, porosity and permeability, settling-rate, and X-ray analysis.

7. The use of well-logging methods, such as electric, radioactive, thermal, caliper, mud, and drill-time. These methods have greatly assisted in establishing more reliable correlations in the subsurface.

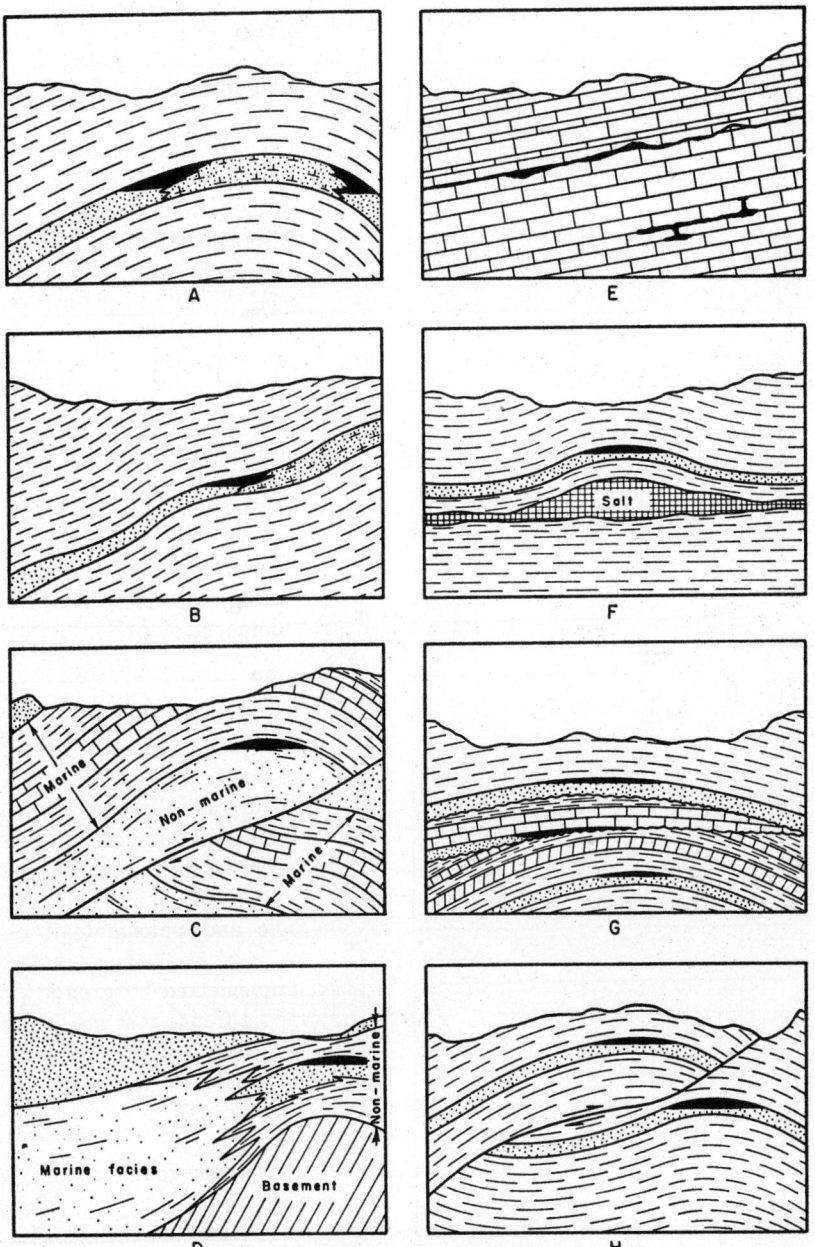

FIGURE 30. *A*—Accumulation on flank of an anticline as a result of crestal cementation of producing strata. *B*—Trap resulting from terracing and updip cementation. *C*—Anticlinal accumulation of an overthrust sheet. *D*—Structural accumulation in the updip nonmarine section. *E*—Accumulation along an unconformity and in solution cavities within a carbonate section. *F*—Anticlinal accumulation above a non-piercement salt mass. *G*—Accumulation in a complexly folded region. *H*—Accumulation in thrust-fault areas.

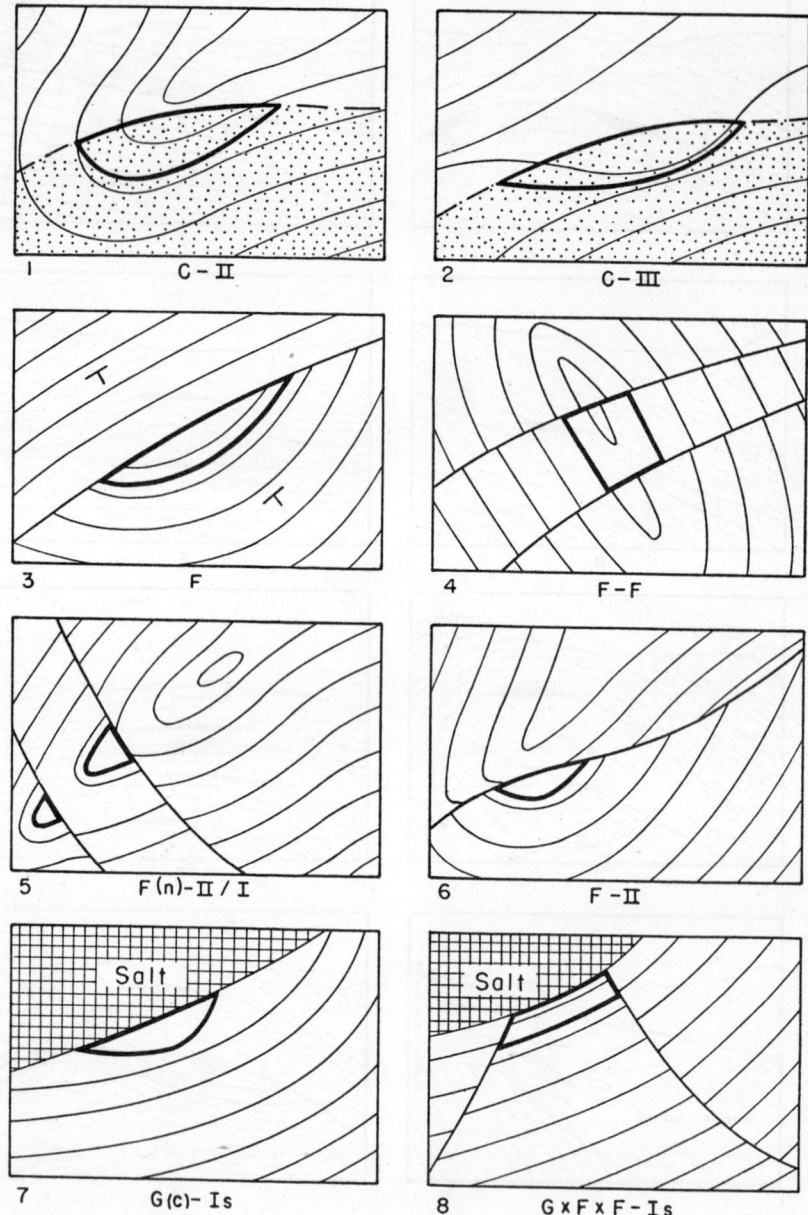

FIGURE 31. Plan views of oil concentration. Numbers 1 and 2 represent pinch-out-trap reservoirs. Numbers 3, 4, 5, and 6 are typical fault-segment reservoirs. Numbers 7 and 8 are salt-piercement type reservoirs. Wilhelm's reservoir classification indices are given below each drawing. (Adapted from Wilhelm. Am. Assoc. Petroleum Geologists.)

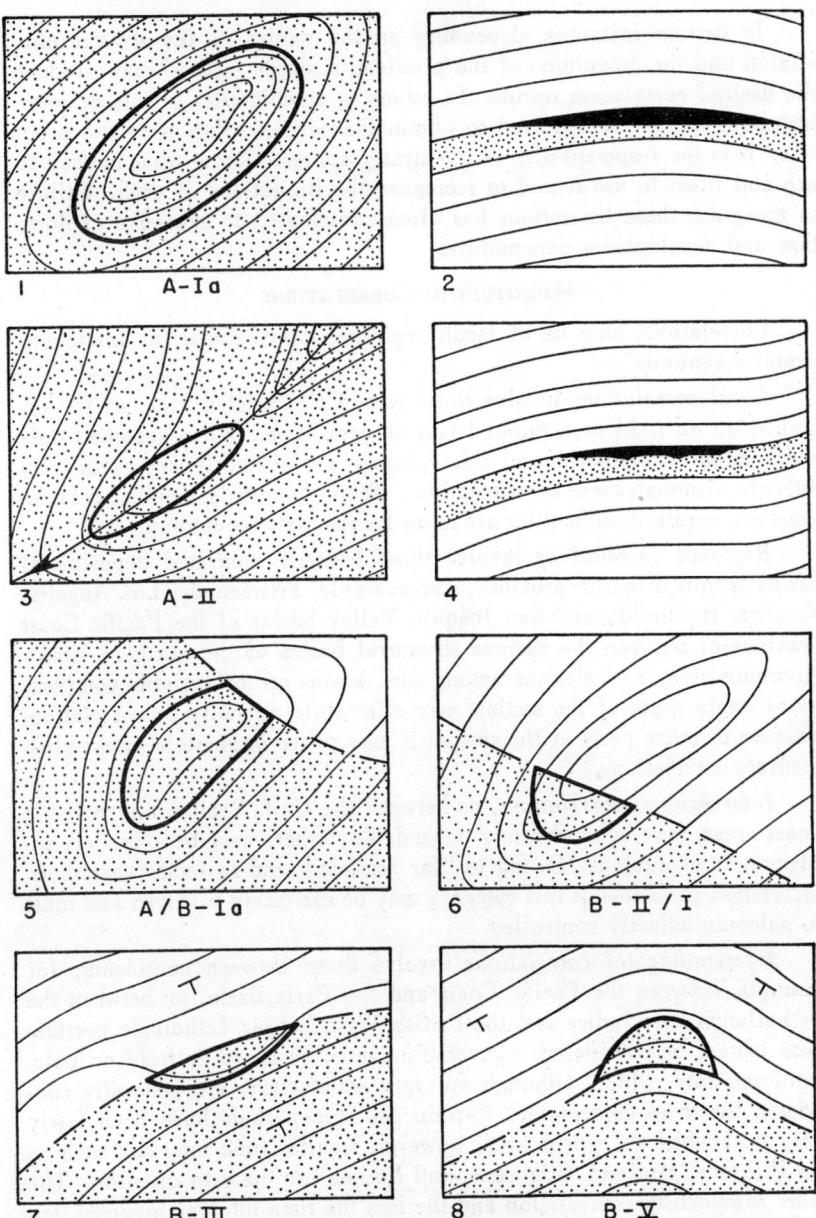

FIGURE 32. Plan views (except 2 and 4) of oil concentration. Numbers 1 and 2 are simple convex trap reservoirs. Numbers 5, 6, 7, and 8 are permeability trap reservoirs. Wilhelm's reservoir classification indices are given below each drawing. (Adapted from Wilhelm. Am. Assoc. Petroleum Geologists.)

In certain instances, depending on the nature of the stratigraphic section and the magnitude of the problem, a single method may produce the desired correlation results. In others, a combination of one or more techniques might be required to obtain the correct solution of the problem. It is the responsibility of the stratigrapher to know what method to use and when to use it and to recognize the limitations of each. Failure to recognize these limitations has often contributed to excessive exploration and development expenditure.

MAGNITUDE OF CORRELATIONS

Correlations may be of local, regional, interregional, or intercontinental magnitude.

Local correlations involve those within a restricted area, as for example, an oil field or a minor basin of deposition. When sufficient data are available, correlations of this category are generally not excessively difficult, although there are exceptions. When extremely detailed correlations are required, difficulties are more frequently encountered.

Regional correlations involve those between separated depositional basins within a major province; for example, between the Los Angeles, Ventura, Humboldt, and San Joaquin Valley basins of the Pacific Coast province or between the various structural basins of the northern Rocky Mountain area. Correlations among such basins are of variable complexity. Certain parts of the section may offer little difficulty in correlation, whereas in other parts of the section it may prove impossible to establish accurate correlations.

Interregional correlations, as between the Gulf, Pacific, and Atlantic Coast areas, may be moderately dependable; others may be extremely inadequate and dubious. Owing to "far removal" and to facies variations, correlation problems of this category may be extremely involved and must be paleontologically controlled.

Intercontinental correlations involve those between continents; for example, between the Pacific Coast and the Paris Basin, or between the Netherlands East Indies and the British West Indies. Lithologic correlations cannot be considered. Correlations must be based entirely on paleontologic data, which, although not very satisfactory, are generally considered the most dependable. Certain geologic periods have been fairly well established the world over; however, smaller time intervals such as the Pliocene, Miocene, Oligocene, and Eocene are open to question. The more extended the correlation and the less the time interval involved, the more inexact is the chronologic value of the correlation.

CORRELATION INDICATORS

The selection of correlation indicators depends on the magnitude and character of the lithologic and paleontologic aspects of rock and time-rock units. Some indicators applicable for local or even regional correlation work include coprolites; fish teeth and scales; pollen; spores; oöliths; chert phases; glauconite, bentonite, and ash layers; detrital minerals; limestone, coal, and anhydrite beds; cyclothems; and lithic and paleontologic sequences. For interregional and intercontinental correlations pelagic Foraminifera, fusulinids, orbitoids, and ammonites have been used with varying success. Evolutionary trends in certain species and genera are notably applicable in certain long-range correlation problems.

CORRELATION DIFFICULTIES

Some of the more commonly encountered difficulties in correlation work are (1) the discontinuity of outcrops; (2) lateral variations in thickness and lithology; (3) the interval variation between key strata; (4) the presence of unrecognized unconformities and faults; (5) the lack of lithologically and paleontologically controlled sections; (6) the multiplicity of time-rock and rock nomenclature; and (7) erroneously compiled and interpreted data obtained from the literature.

Upon entering a new area, these factors should be carefully analyzed. All possible lithologic, paleontologic, and structural data should be screened and coordinated with the intention of defining intervals and surfaces which will foster improved correlations.

QUESTIONS

1. Define stratigraphy.
2. Distinguish between microstratigraphy and macrostratigraphy.
3. What are the purposes of the American Commission on Stratigraphic Nomenclature?
4. Define "time unit," "time-rock unit," and "rock unit."
5. Carefully read Hedberg's statement pertaining to stratigraphic units.
6. On what basis may "time surfaces" be defined?
7. What considerations are necessary for naming a sedimentary formation?
8. What recommendations are made by the American Commission on Stratigraphic Nomenclature for naming and defining subsurface units?
9. Define "sedimentary facies" and "lithofacies."
10. What is the importance of evaluating sedimentary-facies changes?
11. What are the two major types of unconformities and what importance is attached to these features in stratigraphic geology?
12. Give ten criteria for recognition of unconformities.
13. Faults may be recognized on what criteria?
14. What is meant by marine onlap and offlap?
15. Study carefully the schematic diagrams illustrating the various. types of oil and gas traps.

16. In exploiting a new area for oil and gas, what major factors should the geologist consider?
17. For what purposes are correlations of sedimentary rocks made?
18. Give five basic procedures followed in correlating sedimentary rocks.
19. Discuss briefly "magnitude of correlations."
20. Give five difficulties commonly encountered in correlating data.

CHAPTER 3

COMMENTS ON SEDIMENTARY ROCKS

L. W. LeROY

No attempt is herein made to present a complete synopsis of the major types of sedimentary rocks. However, brief mention of the various methods of study applicable in evaluating each type is made.

The classification of sedimentary rocks is difficult because of intergradations of textures and compositions. Various classification outlines have been proposed and suggested, though none are fully complete. In general these classifications fall into two categories: descriptive and genetic. The former involves classification without knowledge of origin, whereas the latter requires data concerning origin. Neither approach can be completely divorced from the other. Two workable classification charts are given, one by Van Tuyl [1] (fig. 33) and one by Shrock [2] (fig. 34).

For a comprehensive and monumental dissertation on sedimentary rocks, Pettijohn's *Sedimentary Rocks*, published by Harper and Brothers in 1949, should be consulted.

TYPES OF SEDIMENTARY ROCKS

Conglomerates and Breccias

A conglomerate or its unconsolidated equivalent (gravel) is composed mainly of rounded granules, pebbles, and boulders exceeding two millimeters in diameter. Roundness, sphericity, and flatness of these components show considerable variation. An accumulation of fragments exhibiting high angularity is commonly termed a "breccia."

Stratification and degree of sorting shown by the coarse clastics range from poor to excellent. Cross-lamination and imbrication patterns are frequently developed.

Conglomerates and breccias have a wide range in color which is controlled by the type of finer matrix, the composition of the fragments, and the degree of weathering.

In some clastics the variety of the pebbles represented is relatively simple (oligomictic), whereas others contain a complicated and diversified pebble suite (polymictic). Pebble composition is a useful attribute for decipherment of the origin of the deposit and for interpreting conditions under which the deposit accumulated.

Limonite, calcite, silica, clay, and a combination of two or more of these minerals are common bindents.

---

[1] Van Tuyl, F. M., Professor and Head of Geology Department, Colorado School of Mines, Golden, Colorado.
[2] Shrock, R. R. Associate Professor of Geology, Massachusetts Institute of Technology, Cambridge, Massachusetts.

MECHANICAL OR CLASTIC		ORGANIC		CHEMICAL	REPLACEMENT PRODUCTS	ALTERATION PRODUCTS
UNCONSOLIDATED	CONSOLIDATED	UNCONSOLIDATED	CONSOLIDATED			
Pebbles and Gravel Quartz Pebbles Limestone Pebbles Chert Pebbles Rhyolite Pebbles Etc. Angular Fragments Quartz Fragments Limestone Fragments Chert Fragments Etc. Sand with Pebbles Sand Sand with Feldspar Sand with Ferromagnesian Minerals Sand with Mica Sand with Glauconite Clay Muds Calcareous Muds Dolomitic Muds	Conglomerate (Psephite) Quartz Conglomerate Limestone Conglomerate Chert Conglomerate Rhyolite Conglomerate Etc. Breccia Quartz Breccia Limestone Breccia Chert Breccia Etc. Conglomeratic Sandstone Sandstone (Psammite) Calcareous Sandstone Siliceous Sandstone Ferruginous Sandstone Argillaceous Sandstone Arkose Graywacke Micaceous Sandstone Glauconitic Sandstone Shale or Pelite Calcareous Shale Clay Shale (Argillite) Siliceous Shale Ferruginous Shale Carbonaceous Shale Oil Shale Clastic Limestone Hydraulic Limestone Waterlimes Clastic Dolomite	Calcareous Shells and Fragments Siliceous Shells and Fragments Diatomaceous Ooze Sponge Spicules Radiolarian Ooze Phosphatic Shells, Bones and Guano Peat	Organic Limestone Crinoidal Limestone Shell Limestone Coralline Limestone Fusulina Limestone Chalk Etc. Diatomaceous Earth (Tripolite) Sponge Cherts Radiolarian Cherts Rock Phosphate Coal	Calcareous Oolite Oolitic Phosphate Oolitic Iron Ore (in part) Siliceous Oolite (in part) Chemical Limestone Chemical Dolomite Rock Salt Rock Gypsum Rock Anhydrite Borax Deposits Bog Iron Deposits Bog Manganese Deposits Potash Salt Deposits Soda Deposits Travertine Siliceous Sinter (Geyserite) Novaculite?	Siliceous Oolite (in part) Chert (in part) Flint (in part) Phosphatized Limestone Sedimentary Iron Ore (in part)	Dolomitized Limestone Limestone after Siderite Anhydrite from Gypsum Gypsum from Anhydrite Etc.

Figure 33. Classification of sedimentary rocks. (From F. M. Van Tuyl)

Nature of sediments			Sedimentary rocks	
Dominantly fragmental	Angular particles more than 2 mm. in greatest dimension	Rubble composed of sharpstones	Sharpstone	CONGLOMERATE
	Rounded particles more than 2 mm. in greatest dimension	Gravel composed of roundstones	Roundstone	
	Angular and rounded particles of rocks and minerals ranging in greatest dimension from 2 mm. to 0.06 mm.	Volcanic fragments = Tuff	Tuffstone	SANDSTONE
		Mixture of rock and mineral fragments	Graywacke	
		Quartz + Feldspar	Arkose	
		Quartz + other minerals in large amount	Normal	
		Quartz + other minerals in small amount	Quartzose	
	Rock and mineral particles ranging in greatest dimension from 0.06 mm. to 0.001 mm. and colloidal particles less than 0.001 mm. in greatest dimension	Volcanic ash	Ashstone	SHALE
		Silt particles — 0.06 to 0.001 mm.	Siltstone	
		Clay materials less than 0.01 mm.	Claystone	
		Silt + Clay + Water = Mud	Mudstone	
Partly fragmental	Fe^{II} and Fe^{III} compounds precipitated inorganically and organically as concretions, nodules and layers Impurities commonly present in the layers	Iron concretions	Concretionary	IRONSTONE
		Iron compounds + mud, silica, etc.	Precipitated	
	Siliceous inorganic fragments less than 0.06 mm. in greatest dimension	Inorganic fragments	Fragmental	SILICASTONE
	Siliceous organic hard parts and their fragments	Diatom frustules, radiolarian skeletons and sponge spicules		
	Silica precipitated as oölites, pisolites, etc.	Siliceous concretions	Concretionary	
	Silica precipitated from suspensions and solutions	Chert, flint, sinter, etc.	Precipitated	
	Plant structures—spores, fronds, leaves, wood, etc. Inorganic sediment Waxes, resins, etc., from decomposition of plants	Plant debris; inorganic impurities Plant fluids	COAL	
	Calcite and Aragonite fragments Calcareous organic hard parts—shells, exoskeletons, plates, spines, and fragments		Fragmental	LIMESTONE
	Organically and inorganically precipitated concretions		Concretionary	
	Inorganically precipitated $CaCO_3$—Evaporation, etc. Organically precipitated $CaCO_3$—(1) by NH_3 from decomposition; (2) loss of CO_2 to plants; etc.		Precipitated	
	Dolomite fragments Dolomitized organic hard parts		Fragmental	DOLOSTONE
	Dolomitic concretions		Concretionary	
	Inorganically precipitated dolomite Organically precipitated dolomite		Precipitated	
Possibly fragmental	Fragments of anhydrite, gypsum, halite, alkali, nitrate caliche, etc.		Fragmental	SALINA-STONE
	Evaporites—minerals precipitated during evaporation of saline waters	Anhydrite Gypsum Chlorides Nitrates Other rare salts	Precipitated	Anhydrock Gyprock

FIGURE 34. Classification of sedimentary rocks. (From Shrock, *Sequence of Layered Rocks.*)

The thickness of conglomerates and breccias ranges from inches to thousands of feet. Extreme variations may occur in relatively short distances horizontally. These rocks may grade in all directions into time equivalent though lithologically dissimilar strata. They frequently represent the basal phase of many formations and thus serve as criteria for recognizing unconformable relationships. Intraformational, coarse clastics have also been observed in many major lithologic units.

A systematic analysis of the coarse clastics should include such studies as (1) type of pebbles, (2) ratio of pebble types, (3) degree of sorting, (4) fabric patterns, (5) alteration of pebbles, (6) type and distribution of bindents, (7) thickness trends and variations, (8) shape of pebbles (roundness, sphericity, flatness), and (9) relationship to adjacent deposits. The results of these investigations should be recorded whenever possible in graphic form (histograms, percentage curves, composition triangles, and fabric diagrams).

Sandstones

Sands and sandstones represent the medium-grained clastic sediments and are composed of grains of various rocks and minerals ranging from 1/16 to 2 millimeters in diameter.

Lithologists frequently assume sandstones to be composed primarily of quartz. This concept should be discouraged, as detailed examinations reveal the composition of sandstones to be extremely diverse and complex as exemplified by Pettijohn:[3]

Rough inspection of 50 thin sections of sandstone chosen at random from the University of Chicago collection shows than 45 percent are graywackes and subgraywackes, 35 percent orthoquartzites, and 20 percent are arkoses.

A *graywacke*, according to Pettijohn, is composed of large very angular grains, mainly quartz, feldspar, and rock fragments (chiefly chert, phyllite, and slate). The grains are set in a prominent-to-predominant "clay" matrix which was, on low-grade metamorphism, converted to a mixture of chlorite and sericite and partially replaced by carbonate.

An *orthoquartzite* is a sedimentary quartzite developed as a result of excessive silicification without the impress of metamorphism.

An *arkose* or *arkosic sandstone*, according to the Committee on Sedimentation, contains 25 percent or more of feldspar derived from the disintegration of acid igneous rock of granitoid texture.

The terms "graywacke" and "arkose" have recently received considerable attention in studies of the relationship between tectonism and sedimentation. Arkoses are considered to be typically developed in intracratonic basins, whereas graywackes accumulate dominantly within geosynclinal downwarps. The contrasting features of these two rock types are given by Pettijohn:[4]

---

[3] Pettijohn, F. J., *Sedimentary Rocks*, p. 229, New York, Harper and Bros. 1949.
[4] Pettijohn, F. J., *op. cit.*, p. 261.

Arkose usually is coarser-grained, lighter in color (pink or light gray). It may be strongly cross-bedded, is cleaner sorted, i.e. is without interstitial clay or silt and consequently, is bound by introduced mineral cement (usually carbonate). Arkose is closely associated with coarse conglomerates that contain much granite debris, and it is usually terrestrial in origin. Graywacke is the opposite in nearly all particulars. It is finer-grained, dark in color, rarely cross-bedded, but with marked graded bedding if any bedding is visible, with an interstitial paste having the composition of a slate, and hence without mineral cement. Graywacke is interbedded with black, pyrite slates and associated with pillow lavas and chert. It is usually marine.

Sandstones grade in texture from very coarse (2 to 1 mm.) to very fine ($\frac{1}{8}$ to $\frac{1}{16}$ mm.). Sorting may be in certain instances exceptionally high, whereas in others it is extremely low. The coefficient of sorting may be determined by construction of a cumulative-frequency curve.

The colors of sandstones have a wide range (white, gray, red, green, to black) depending on composition of the grains, type and amount of cement, and degree of weathering. Cementing materials commonly consist of limonite, hematite, carbonate silica, organic material, anhydrite and gypsum, and clay. Pure quartz sandstones are indicative of stable shelf-depositional conditions.

Bedding characteristics of sandstones may be massive to thinly laminated. Bedding surfaces may be sharp to gradational, parallel or converging.

Accessory components comprising the heavy-mineral fraction of arenaceous rocks consist of such minerals as pyrite, hornblende, pyroxene, olivine, magnetite, leucoxene, garnet, tourmaline, zircon, mica, staurolite, anatase, and glauconite.

Sandstone varieties should be designated by a qualifying adjective whenever possible; i.e., glauconitic, micaceous, garnetiferous, pyritic, and hornblendic.

Studies of arenaceous rock types should treat: textural aspect, grain fabric, type and distribution of cement, character of grains (surface features, alteration, shape, inclusions), grain composition, and variations in porosity and permeability. These values may be determined by thin sections, screen analysis, heavy-liquid separation, polished surfaces, and chemical analysis. In subsurface investigations fluid and gas contents may be evaluated by core analysis and by data obtained from various well-logging methods.

Siltstone

Siltstones, the indurated equivalent of silts, are fine-grained clastics represented by particles ranging in size from 1/16 to 1/256 millimeter. These rocks vary considerably in color and structure. They frequently contain various compounds of organic matter and as a rule yield normal suites of heavy accessory minerals. In many instances rocks of this category are classified as sandy (arenaceous) or silty shales.

Shale and Mudstone

Shales and mudstones constitute the finest of the clastic materials. Particle sizes range below 1/256 millimeter. The primary constituents of these rocks are represented by the complex clay minerals which are extremely difficult to determine, especially without reverting to advanced petrographic procedure. Silica is the dominant element in shales and mudstones. It is present either as free silica (quartz) or in the form of silicates. Alumina is next in importance to silica. Other elements include titanium, iron, manganese, calcium, sodium, potassium, and phosphorous. The common clay minerals include kaolinite, montmorillonite, and illite.

Many lithologists arbitrarily designate all argillaceous clastics as shales. This terminology should be applied to only those rocks exhibiting fissility or lamination. Rocks failing to show these structural features should be termed mudstone or claystone depending on the plasticity value.

The porosity of argillaceous rock types may range up to 50 percent, although the average value, according to Pettijohn, is 13 percent.

Hybrid types of argillaceous rocks include marlstones (50 to 80 percent carbonate), clay ironstone (rich in siderite), porcellanite (high in opaline silica), and black shale (exceptionally rich in disseminated organic matter).

Important accessory minerals in shales and mudstones include mica, glauconite, pyrite, silt, and sand. Cementing materials may be of a siliceous, ferruginous, calcareous, carbonaceous, or bituminous nature.

In descriptive work these sediments should be qualified by proper adjectives; i.e., glauconitic, carbonaceous, dark gray shale.

During the past few years considerable attention has been given to clay mineralogy. This study involves such methods as elutriation, thin-section, rate of settling, X-ray, specific gravity, spectroscopy, differential thermal analysis, electron microscope, staining and chemical analysis.

Limestone and Dolostone

Limestones and dolostones may be of clastic or chemical origin or may be developed as a result of both processes of deposition. The chemical composition of these rock types varies considerably. Rarely does there occur a pure limestone, $CaCO_3$, or dolomite, $CaMg(CO_3)_2$, because of the inclusion of detrital materials. These two carbonate types invariably grade into each other. In certain instances limestones and dolomites contain an abundance of organic remains, whereas in other cases the remains of organisms are absent or nearly so. The dominant organic remains within carbonate rocks include those of algae, mollusks, corals, echinoids, Bryozoa, ostracodes, and Foraminifera.

The partial or total alteration and replacement (metasomatism) of limestone strata have been recognized and have resulted in dolomitization, chertification, and phosphatization. Diagenetic changes frequently obliterate original textures and structures and develop new ones. Carbonates accumulate under various environmental conditions: stable shelf, mildly unstable shelf, intracratonic basin, and geosyncline.

The color of carbonate rocks varies from white to black; crystallinity ranges from fine to coarse. Limestones and dolostones may be thick-bedded to finely laminated, vuggulated, oolitic, or pisolitic. Foreign constituents as shown by insoluble-residue work include chert, clay, quartz, pyrite, glauconite, arenaceous Foraminifera, and silicified fragments.

During the past several years, oil companies have shown intense interest in reef limestones and dolostones as possessing great oil and gas potentials. Systematic surveys of all possible reef areas are being made and criteria are being collected to evaluate these deposits more accurately.

The following techniques have been followed in the study of carbonate rocks: thin-section, polished surface, insoluble residue, chemical-stain test, spectrochemical, porosity and permeability, relative solubility, petrofabrics, and chemical analysis. In addition to these methods, electrical, radioactive, and micrologging have added considerably to our knowledge of subsurface carbonate characteristics. Detailed carbonate investigations contribute to improvement of more efficient production and well-completion methods.

Evaporites

The sediments composing the evaporites are primary precipitates which have resulted for the most part from the evaporation of saline solutions. The best-developed evaporite sections in North America are found in the Permian Basin of west Texas and eastern New Mexico, and in the Silurian salt basins of Michigan and New York. The evaporites are represented by the sulphates (anhydrite, gypsum), chlorides(mainly halite), and minor carbonates. The sodium and potassium sulphates and the nitrates are also included in the evaporites. Rock salt, anhydrite, and gypsum are the most commonly encountered.

Rock-salt beds, ranging from inches to 80 feet in thickness, are generally associated with anhydrite and gypsum. They are frequent in red-bed sections. Their purity, color, and texture are variable. Crystallinity of the salt varies from fine to coarse. Great masses of this material have flowed and produced salt domes in the Gulf Coast area of Texas and Louisiana and in Germany, Iran, and Russia.

Anhydrite and gypsum occur as bedded deposits mainly in nonmarine stratal sequences and are commonly associated with dolostone and shale phases. Frequently these rocks exhibit normal lamination or corrugated lamination. Anhydrite generally ranges in color from white to dark gray and assumes a fine to coarse crystallinity. The structural and textural features are sometimes appreciably modified by conversion to gypsum upon hydration. Gypsum frequently occurs in argillaceous strata as transparent crystals of selenite.

Anhydrite and gypsum beds frequently serve as dependable correlation markers in red-bed sections. In certain instances they are quite continuous and uniform. The Blaine anhydrite in the Permian section of southeastern Colorado is an example.

The study of evaporite rocks should include chemical analysis; textural, structural, and spectrochemical investigations; thin-section analysis; and rhythmic characteristics.

Carbonaceous Rocks

Carbonaceous rocks are represented by three types of residue: humus, peat, and sapropel. *Humus* is produced within the upper part of the soil phase. *Peat* originates from partial decay of plant material under fresh-water swamp conditions. *Sapropel* (high in fatty and protein substances) results from concentration of complex organic compounds which accumulate on the bottoms of lakes, lagoons, and quiet-water embayments.

Coal is by far the most common variety of carbonaceous deposit and is classified according to degree of coalification and physical characteristics. Lignite is the lowest grade and anthracite the highest. Some of the more common ingredients of coal include vitrain, fusain, clarian, and durain. The chemical composition of coal is extremely variable. The main constituents in coal are carbon, hydrogen and oxygen, nitrogen, sulphur, and water.

Sections containing coal beds generally present many complex problems to the stratigrapher and structural geologist owing to their extreme diversity of lithologies and depositional irregularities.

Considerable attention in recent years has been given to cyclothemic sequences in coal sections. In the Midcontinent region, in the Illinois basin, and in Kansas, coal cyclothems and megacyclothems have been used advantageously in establishing correlations and decipering structure.

The investigation of coals includes such analyses as thin section, moisture, volatile matter, fixed carbon, and ash.

Sapropelic deposits have been considered by some geologists to be responsible for the formation of petroleum; however, there is no agreement as to the physico-chemical processes of conversion.

To evaluate sapropelic rocks properly, studies involving thin-section, heat analysis, and ether solubility should be made.

Miscellaneous Rock Types

1. Siliceous rocks: The most common siliceous rocks are represented by the microcrystalline cherts and flints. Colors are extremely variable: white, green, red, brown, and black. Chert and flint are common constituents in the carbonate rocks and occurs as lenses, nodules, thin beds, and fracture fillings. The saturation of argillaceous strata with opaline silica produces siliceous shales and *porcellanites*. The presence of abundant diatoms and volcanic ash produces this type of deposits. Cher-

tified layers in carbonate sections have served many times as dependable correlation criteria.

2. Ferruginous rocks: The ferruginous sediments may be classified as carbonates (siderite), iron silicates (glauconite), ferric oxides and hydroxides (hematite, limonite), and sulphides (pyrite, marcasite).

Iron carbonate (siderite) is commonly associated with argillaceous cherty strata in the form of beds, lentils, and concretions. Siderite also occurs in the carbonate rocks (limestones, dolostones).

Chamosite is the principle iron-bearing silicate in sedimentary rocks, principally the finer-grained clastics. Glauconite constitutes an important silicate in many types of marine strata (shales, sandstones, and limestones). This mineral should be recorded in all stratigraphic investigations because of its correlation value and depositional environment index. Glauconite varies in color from pale green to greenish-black. It occurs mainly as rounded to elliptically shaped granules commonly exhibiting shrinkage fractures.

Limonite and hematite are common oxide bindents of sedimentary clastics. These minerals occur as disseminations or in some cases as oolites. In the Silurian strata of the Appalachian region sedimentary hematites are commercially important.

Pyrite and marcasite occur in all types of sedimentary rocks in varying percentages. They may be present as nodules, crystal aggregates, and disseminations. Pyrite is particularly common in highly organic, black shales. Occasional beds of pyrite have been observed.

3. Manganiferous rocks: Manganese occurs in minor amounts in all sedimentary rocks. The oxides, hydroxides, and carbonates are the chief mineral types.

4. Phosphatic rocks: Phosphate-bearing rocks are commonly referred to as phosphorites. Phosphatic materials are found primarily in shales and limestones, and may be either of primary or secondary origin, or both. Color of phosphates varies from brown to black, although by leaching it may assume lighter hues. The material may be bedded or may occur as concentrically banded oölites and nodules. Nodules range up to several centimeters in diameter. Shell fragments are frequently phosphatized. The origin of phosphate deposits appears to be related to animal remains (bones, guana).

TEXTURE OF SEDIMENTARY ROCKS

The *texture* of a rock refers to the size, shape, and arrangement of the individual particles. Textural values are extremely important in the description of a rock. The size of clastic particles may be expressed by the terms coarse (gravel), medium (sand), and fine (clay) depending on their dimensions as determined by selective screening or by actual measurement. The limits of various grain sizes are more or less arbitrary.

Wentworth's subdivision of grade size [5] has been widely accepted by most sedimentologists. Particle-size percentages may be graphically represented by histograms and frequency curves, and such values as median, coefficient of sorting, skewness, and kurtosis determined. These values permit a quantitative representation of particle size.

The shape (sphericity, roundness, and flatness) of particles may be determined by various methods and numerical values given. Such terms as "angular" (showing very little or no abrasion), "subangular" (showing some effect of wear), "subrounded" (showing considerable abrasion), and "rounded" (exhibiting conspicuous wear) are commonly used to designate degree of angulation. Some medium-grained clastics exhibit pronounced uniformity in grain angularity, whereas others show considerable variation of angularity.

Close examination of the clastics reveals in certain instances distinct fabric pattern of the grains. Petrofabric diagrams (figs. 69 and 70) are helpful for illustrating these orientation trends.

In addition to the size, shape, and arrangement of grains, special attention should be given the characteristics of grain surfaces, such as degree of polish, smoothness, striation, and pitting. These features have definite genetic significance.

Permeability and porosity are controlled in large part by the texture of the rock. Since these two factors are of primary importance to the oil geologist in production problems, textural attributes of producing strata should be carefully evaluated.

Textures of carbonate and evaporite rocks range from fine to coarse crystalline. Textures of chemical sediments are clearly outlined by Pettijohn.[6] Such terms as "macrocrystalline" (granoblastic, over 0.75 mm.), "mesocrystalline" (porphyroblastic, 0.20 to 0.75 mm.), "microcrystalline" (0.01 to 0.20 mm.), and "cryptocrystalline" (less than 0.01 mm.) are discussed.

According to DeFord,[7] "the grade scales for clastic rock are not suitable for carbonate rock, because the names of the scale units imply the clastic origin of the rock." DeFord recommends the terms and size limits given in figure 35.

STRUCTURES IN SEDIMENTARY ROCKS

Structures developed in and exhibited by sedimentary strata are numerous and varied. Special attention should be given these features, because of their usefullness in deciphering depositional environments, stratigraphic succession, and structural anatomy. Two types of structures in sedimentary rocks are recognized: inorganic and organic. The former includes such features as ripple marks, swash marks, current marks, pit-

[5] Wentworth, C. K., *Fundamental Limits to the Sizes of Clastic Grains:* Science, vol. 77, pp. 633-634, 1933.

[6] Pettijohn, F. J., *op. cit.*, p. 73.

[7] DeFord, R. K., *Grain Size in Carbonate Rock:* Am. Assoc. Petroleum Geologists Bull., vol. 8, pp. 1921-1926, 1946.

and-mound, rain-drop impressions, gas pits, crystal imprints, mud cracks, bedding, cross-lamination, scour and fill, imbrication, laminar corrugation, intra- and interstitial flow, unconformities and diastems, nodules, geodes, septaria, stylolites, cone-in-cone, and veinlets. Organic structures are represented by tracks and trails, borings, and petrifactions.

Sedimentary structures and their significance are treated in full by Pettijohn [8] and by Shrock.[9]

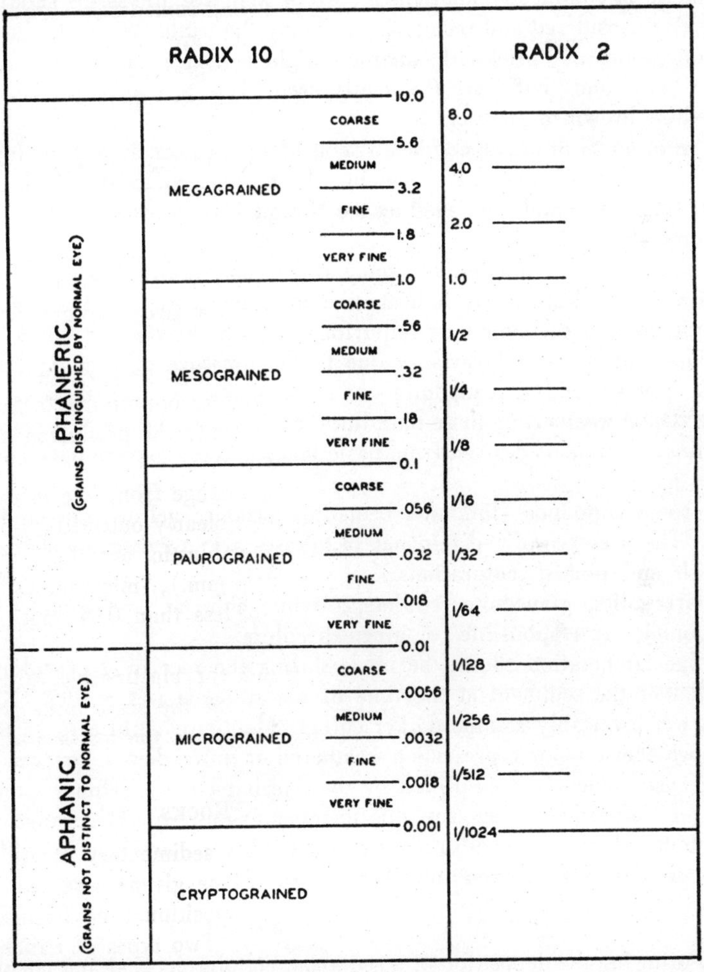

FIGURE 35. Tentative scale for carbonate rocks, grain diameters in millimeters, plotted logarithmically. Scale based on radix 2 (used by Wentworth) added for comparison. (From DeFord. Reproduced permission Am. Assoc. Petroleum Geologists.)

[8] Pettijohn, F. J., *op. cit.*, pp. 120-168.
[9] Shrock, R. R., *Sequence in Layered Rocks*, New York, McGraw-Hill Book Co., Inc., 1948.

COLOR OF SEDIMENTARY ROCKS

The color of sedimentary rocks is controlled by the grain size, by the composition of the grain, and by the chemical pigmentation. Colors may be primary or secondary, or both.

Several proposals have been made in order to standardize colors of sedimentary rocks more adequately. DeFord's statement,[10] "The description by geologists of the colors of rock outcrop and rock cuttings from test wells is completely anarchistic: every man for himself," should be carefully considered and recognized as being the truth. What one individual designates as "brick-red" another might consider "orange-red." The terms "tan" and "buff" are frequently used. Tan, to one person, may be a medium-brown to another.

In order to improve standardization of rock colors among geologists, it is recommended that the rock-color chart prepared by the Rock-Color Chart Committee and published by the National Research Council in 1948 be carefully followed.

Evaporites, carbonates, and some argillaceous sediments, which range from white to light gray, indicate the total or nearly total absence of bituminous and carbonaceous impurities. The dark coloring (dark gray to black) in rocks is invariably due to the presence of organic matter, black iron sulphides, manganiferous constituents, or dark detrital minerals. Upon weathering, these dark hues may become lighter as a result of leaching. Sediments derived from basic igneous rocks assume dark coloration.

Iron compounds (limonite, hematite) produce yellow, tan, and red hues. The presence of red feldspar in arkoses are largely responsible for reddish and pinkish colorations.

Greenalite, glauconite, epidote, olivine, chlorite, and ferrous iron compounds are responsible for greenish colors.

Special notation should be made during the recording of colors as to whether the sediment at the time of recording is wet or dry. Rocks when wet invariably assume darker colors. Mention should also be made as to whether a color represents a weathered or unweathered surface. The latter case is clearly exemplified by the Apishapa shale (Upper Cretaceous) of Colorado. In outcrop this member assumes a pale orange to a light-buff color. In the subsurface or on fresh exposure it is gray-black. Many similar examples may be cited.

DIAGENESIS OF SEDIMENTARY ROCKS

Following the deposition of a sediment, certain physical and chemical processes are initiated which tend to adjust the sediment to its environment. These processes are varied and complex, and their relationships are unknown in many instances. Certain modifications of the sediments

---

[10] DeFord, R. K., *Rock Colors:* Am. Assoc. Petroleum Geologists Bull., vol. 28, no. 1, pp. 128-137, Jan. 1944.

occur prior to lithification, whereas others are not evident until long after burial.

Compaction, cementation, and metasomatism are largely responsible for the rearrangement and replacement of sedimentary constituents.

Compaction is most obvious in the fine-grained clastics (shale, mudstone, claystone). Porosity is substantially minimized and original fabric drastically modified. New minerals may be formed by closer packing and by increased pressures and temperatures produced as a result of weight of overburden. Compaction generally has little or no diagenetic effect on the coarse clastics.

Cementation is responsible for many modifications of a sediment. Porosity and permeability are reduced, new minerals formed or the original minerals replaced in whole or in part. Secondary silica deposited around quartz grains and giving rise to secondary facets is a common phenomenon.

Metasomatism, involving replacement and alteration, is a common process in sedimentary rocks, particularly in the carbonates. Calcite is frequently replaced by dolomite and silica. Several stages of replacement may be involved.

QUESTIONS

1. In evaluating a conglomerate, what features should be considered?
2. What is the difference between a graywacke, an orthoquartzite, and an arkose?
3. Define *shale* and *mudstone*.
4. What methods of study may be followed in analyzing a shale or mudstone? Limestone or dolostone?
5. What are the most common evaporites?
6. State the difference between humus, peat, and sapropel.
7. What is porcellanite, glauconite, and chert?
8. Define *texture*.
9. What is a petrofabric diagram?
10. Give an example of a metasomatic change in sedimentary rocks.

CHAPTER 4

SUBSURFACE LABORATORY METHODS
MICROPALEONTOLOGIC ANALYSIS
L. W. LeROY

Prior to 1925, micropaleontology played an insignificant role in stratigraphic and paleontologic investigations. It was not until that year that the science was recognized and appreciated as a valuable tool in surface and subsurface problems of the petroleum industry. All major and many minor oil companies now sponsor micropaleontologic laboratories.

The economic micropaleontologist is essentially a microstratigrapher. His time is not only devoted to the paleontologic aspects of strata, but also to lithology, to detrital mineralogy, and to the many other techniques which aid in the solution of stratigraphic problems.

Methods followed by micropaleontologists are extremely variable and are controlled by the type of problem (surface or subsurface), the time allocated to the problem, the quality of personnel involved, and company policy. In some areas only major faunal divisions of sections are desired, whereas in other areas it is necessary to introduce detailed investigations before the problem under consideration can be properly solved. The micropaleontologist should be familiar with the field geologist's assignment and should visit field operations whenever it is deemed necessary in order to coordinate the laboratory work properly. He should systematize laboratory routine so that data may be obtained as soon as possible for final analysis. It should be remembered that most micropaleontological problems are of the "pressure" variety, and that not uncommonly management desires results before the project is started. The micropaleontologist must be versatile, having a knowledge of all varieties of microfaunas as well as being able to evaluate their significance and recognize their use limitations.

Micropaleontology has aided materially in evaluating unconformities, structural conditions, and facies changes, in dating and correlating strata, and in interpreting depositional environments. The science has its limitations, and this fact should be recognized; otherwise, incompetent conclusions and interpretations may be introduced. Micropaleontologic data should be coordinated with all other available stratigraphic information.

Only those microfossils that have been used by paleontologists in the oil industry are discussed here briefly. Some types are more applicable in the solution of stratigraphic problems than are others. Those herein considered are Foraminifera, ostracodes, Radiolaria, conodonts, otoliths, fish scales, calcareous algae, diatoms, spores and pollen, and grass seeds.

FIGURE 36. Assemblages of Foraminifera from the Late Tertiary of the South Pacific region. These single-celled micro-organisms have had world-wide use in correlating strata (×15).

FORAMINIFERA

The Foraminifera (fig. 36) are single-celled, microscopic animals belonging to the phylum Protozoa. These forms, ranging in diameter from 0.01 mm. to 50.0 mm., attain their best development in marine environments, although some genera and species occur profusely in brackish- and even fresh-water habitats. Their tests (shells) are extremely variable in structure and composition. Certain genera secrete chitinous, arenaceous, and siliceous tests, although most of them produce calcareous structures. Tests of the Early Paleozoic Foraminifera are structurally simple; those of the Mesozoic and Cenozoic exhibit more complexity and variety. Foraminifera occur abundantly locally in the Late Paleozoic deposits and even

EXPLANATION OF PLATE 1
Figure
1- 3. *Cibicides telisaensis* LeRoy × 82. Fig. 1, ventral view. Fig. 2, dorsal view. Fig. 3, peripheral view.
4- 6. *Anomalina* sp. A LeRoy × 75. Figs. 4, 5 opposite sides. Fig. 6, peripheral view.
7- 9. *Cibicides dorsopustulosus* LeRoy × 43. Fig. 7, dorsal view. Fig. 8, ventral view. Fig. 9, peripheral view.
10-12. *Cibicides foxi* LeRoy × 75. Fig. 10, dorsal view. Fig. 11, ventral view. Fig. 12, peripheral view.
13-15. *Anomalina* sp. A LeRoy × 65. Figs. 13, 15, opposite sides. Fig. 14, peripheral view.
16-18. *Cancris auriculus* (Fichtel and Moll) × 47. Fig. 16, dorsal view. Fig. 17, ventral view. Fig. 18, peripheral view.
19-21. *Valvulineria* aff. *inaequalis* (d'Orbigny) × 75. Fig. 19, dorsal view. Fig. 20, ventral view. Fig. 21, peripheral view.
22-24. *Eponides praecintus* (Karrer) × 35. Fig. 22, dorsal view. Fig. 23, ventral view. Fig. 24, peripheral view.
25-27. *Quinqueloculina* sp. H LeRoy × 29. Figs. 25, 26, opposite sides. Fig. 27, apertural view.
28-30. *Valvulineria araucana* (d'Orbigny) car. *malagaensis* Kleinpell × 47. Fig. 28, dorsal view. Fig. 29, ventral view. Fig. 30, peripheral view.
31-33. *Baggina inflata* LeRoy × 47. Fig. 31, dorsal view. Fig. 32, ventral view. Fig. 33, peripheral view.
34-36. *Globorotalia barissanensis* LeRoy × 73. Fig. 34, peripheral view. Fig. 35, dorsal view. Fig. 36, ventral view.
37, 38. *Globigerinella aequilateralis* (Brady) × 48. Fig. 37, side view. Fig. 38 peripheral view.
39, 40. *Globigerina siakensis* LeRoy × 54. Fig. 39, dorsal view. Fig. 40, ventral view.
41, 42. *Globigerinoides trilocularis* (d'Orbigny) × 37. Fig. 41, ventral view. Fig. 42, dorsal view.
43, 44. *Globigerina baroemoenensis* LeRoy × 41. Fig. 43, dorsal view. Fig. 44, ventral view.
45, 46. *Globigerinoides sacculiferus* (Brady) var. *irregularus* LeRoy, n. var. × 39. Opposite views.

more so in strata of Mesozoic and Cenozoic age. It is not uncommon to find limestones and marlstones of the Pennsylvanian and Permian composed almost entirely of the remains of these organisms (fusulinids). Many Cretaceous chalks and Eocene limestones contain multitudes of foraminiferal tests as well as numerous argillaceous deposits of the Late Tertiary (fig. 36), which accumulated under tropical or subtropical conditions.

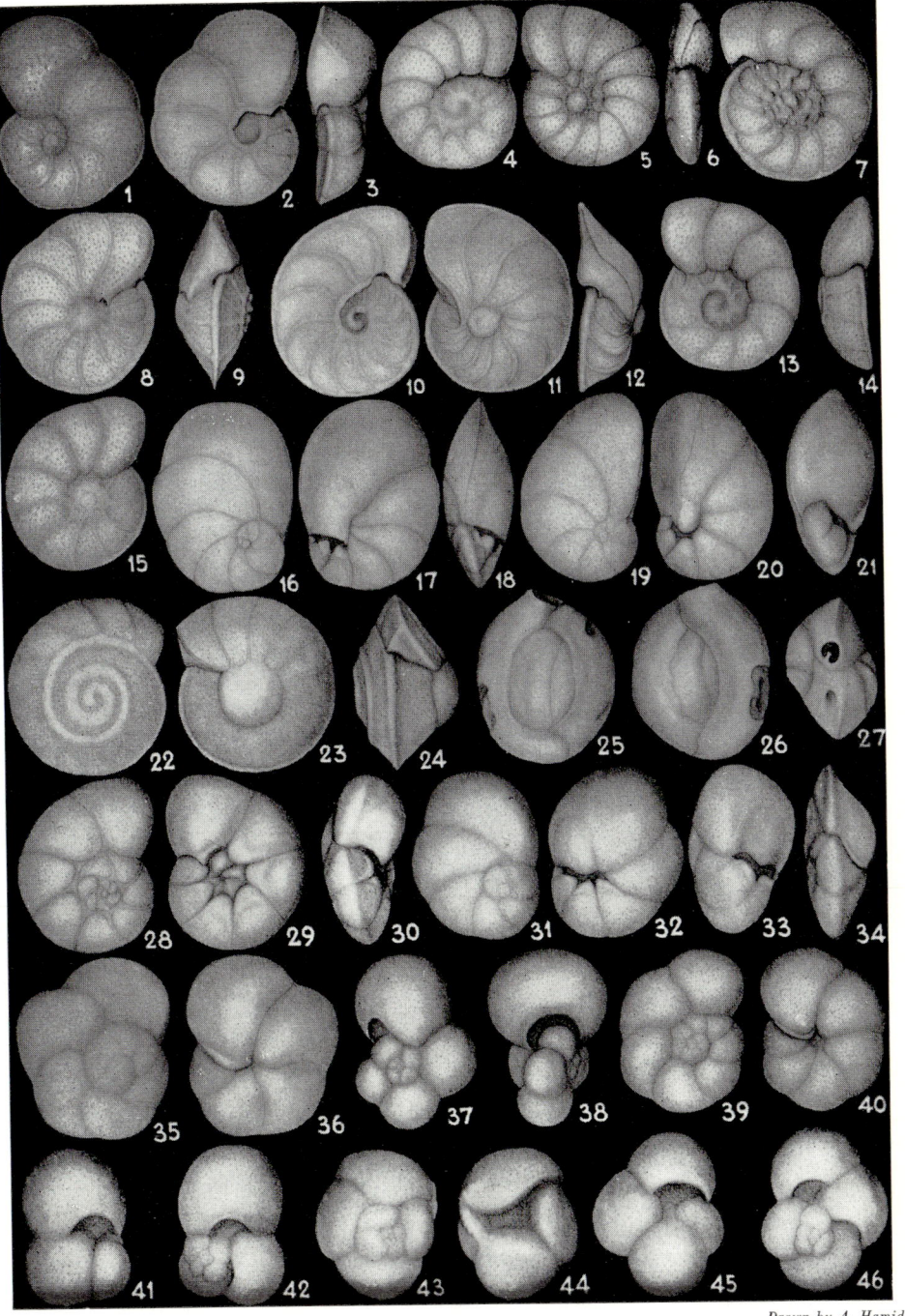

Drawn by A. Hamid

PLATE 1. Tests (shells) of the small Foraminifera. (See p. 86 for explanation.)

PLATE 2. The large foraminifer *Cyclocypeus* from the East Indies. Thin section shown in lower microphotograph. Top six views 8x; bottom view 18x. (From Tan. Wet. Meded.)

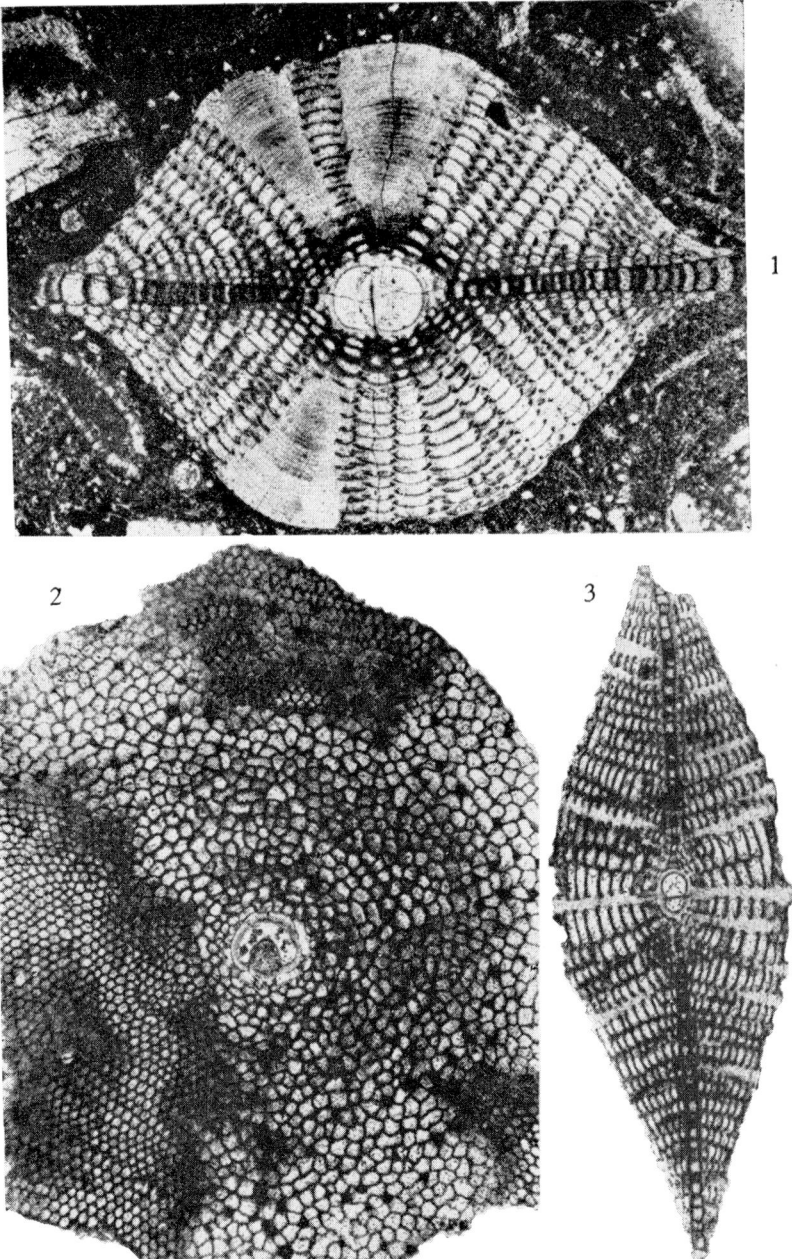

PLATE 3. Thin sections of a large foraminifer *(Lepidocyclina)* 30x. The internal structure of these forms must be studied before identification may be established. (From Scheffen. Wet. Meded.)

According to Cushman [1] "The habits and physiologic characters of the animal and its relationships to the environment are very incompletely known." The Foraminifera reproduce both sexually (resulting in microspheric varieties) and asexually (resulting in megalospheric varieties). The tests of these two generations vary considerably in size and structure and must be considered in speciation studies. Some forms are bottom-dwelling (benthonic), some are floating (pelagic) types, whereas others prefer attachment (sissile).

Benthonic assemblages adjust to temperature. At any given locale, shallow-water suites may vary radically from their deeper-water neighbors. Shallow-water forms (genera and species) of tropical environs are conspicuously divergent from shallow-water assemblages of higher latitudes. These variables must be considered in correlation interpretations, as two assemblages possessing identical time values may be quite dissimilar in composition. Pelagic suites, more restricted in genera and species than benthonic assemblages, offer the best possibilities for long-range correlations because of their nondependency on bottom ecology and the temperature-depth factor.

The tests of Foraminifera and other micro-organisms possessing hard parts are released from sediments by various means. A procedure commonly applied is first to examine the rock sample carefully and record any pertinent information such as lithology, color, minerals, and cementations, which may assist in evaluating the ecology of the contained fauna. If the sediment is not indurated, it is crushed to $\pm\frac{1}{4}$-inch fragments, soaked in water (boiled if necessary) for several hours, and then washed through a series of screens (80-, 100-, 150-mesh) in order to remove the fine clastics. The washed residue may then be examined either under water or dried by means of a binocular microscope ($\times$ 30 to $\times$ 60). If the ratio of the microfauna to the detrital material is not large, the assemblage may be concentrated by heavy liquids or by gravity methods. Swirling the material under water in an examination dish frequently aids in segregation. In the event the host sediment is highly cemented (calcareous, siliceous, or ferruginous) or compacted, the tests may be released by excessive cracking of the material. If this method is not feasible, thin-section or polished-surface studies are required.

When an assemblage is once released and concentrated, the laborious procedure of examination and recording follows. In the initial stages of micropaleontologic work, it is essential that all genera, species, and species varieties be accurately determined and their relative abundances tabulated. It is on the basis of these data that distribution charts (figs. 37, 38, and 39) are prepared, faunal zones defined, sections subdivided, index species determined, and correlations established.

Foraminiferal correlations are based on (1) single species or genus,

[1] Cushman, J. A., Foraminifera, Their Classification and Economic Use: p. 3, Cambridge, Mass., Harvard University Press, 1948.

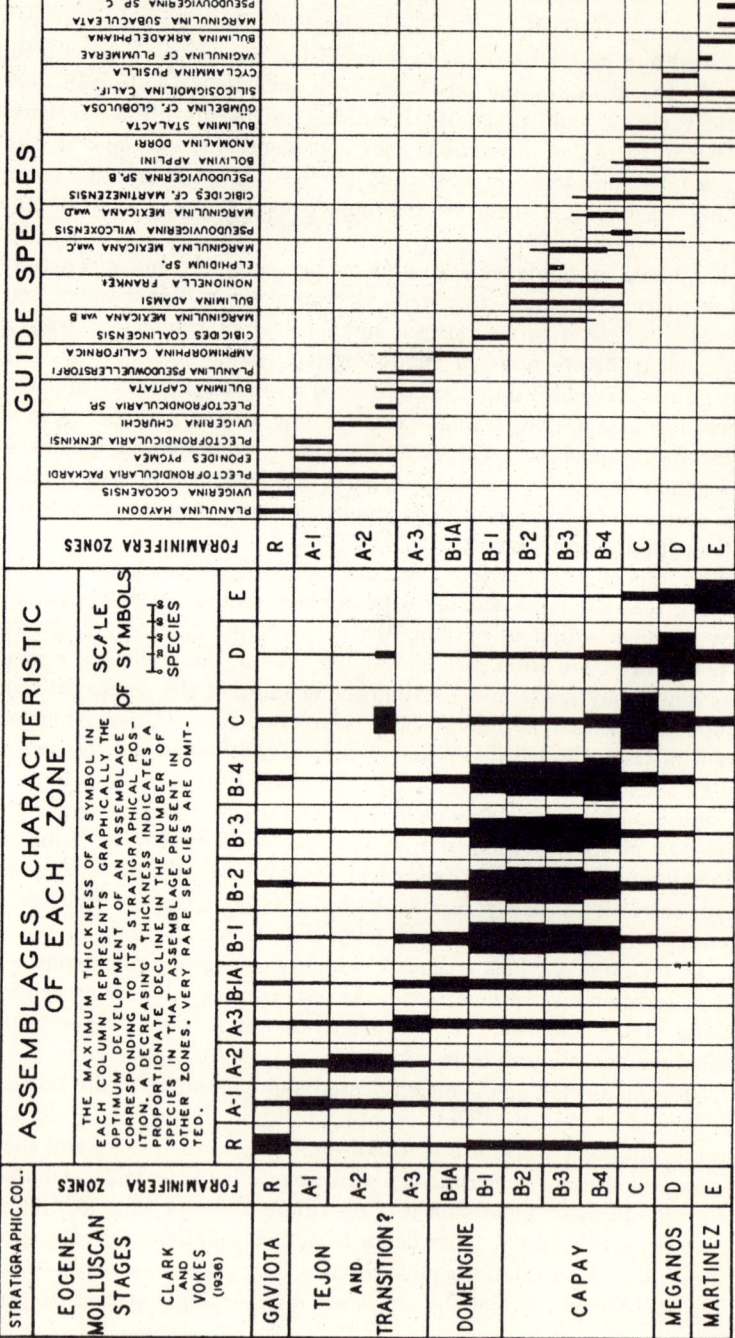

FIGURE 37. Chart showing distribution of Eocene foraminiferal assemblages. (After Laiming.)

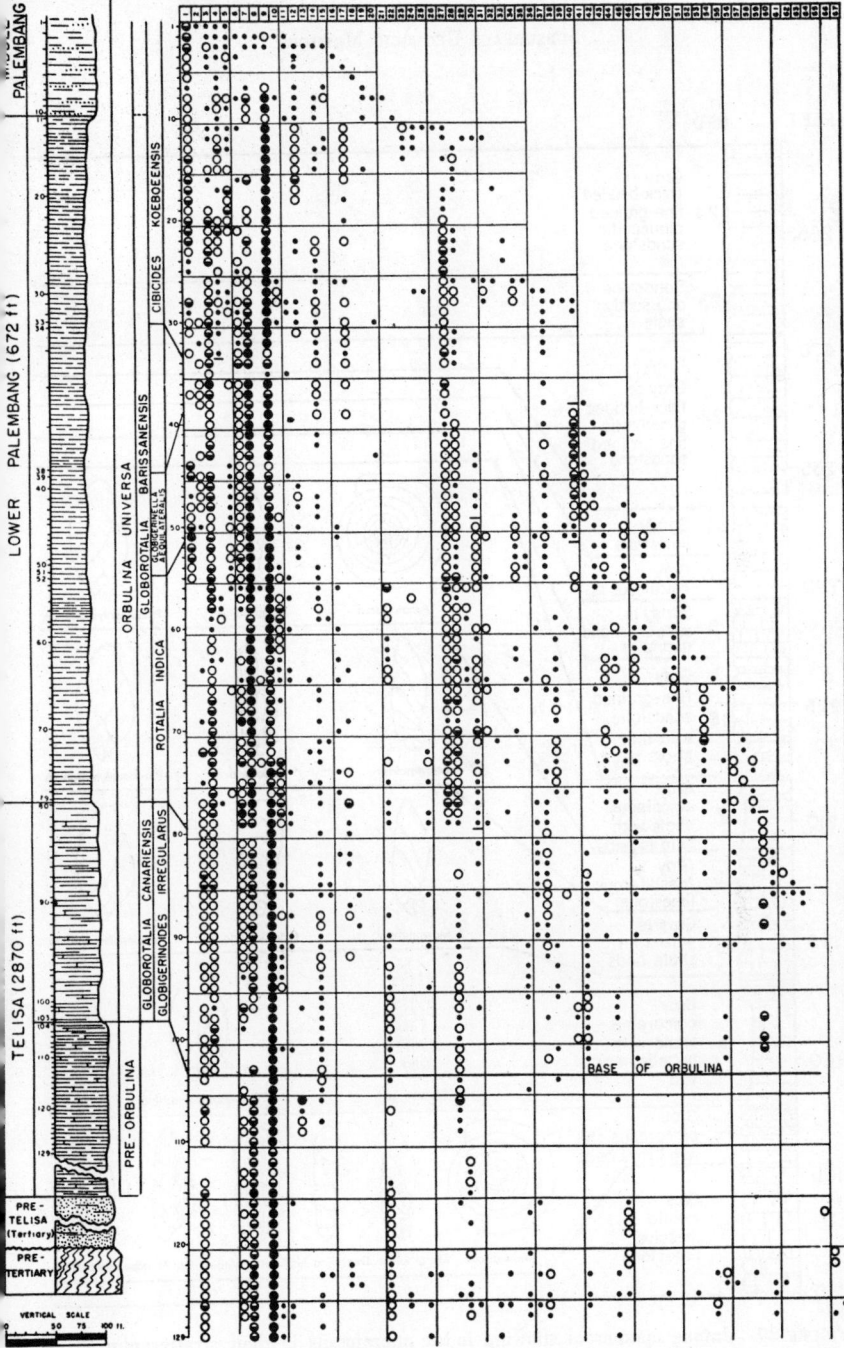

FIGURE 38. A foraminiferal distribution chart. Species are recorded across the top from left to right. Symbols represent relative abundance of each species at various stratigraphic positions. Such charts are essential for subdividing homogeneous marine shale sections into paleontological zones.

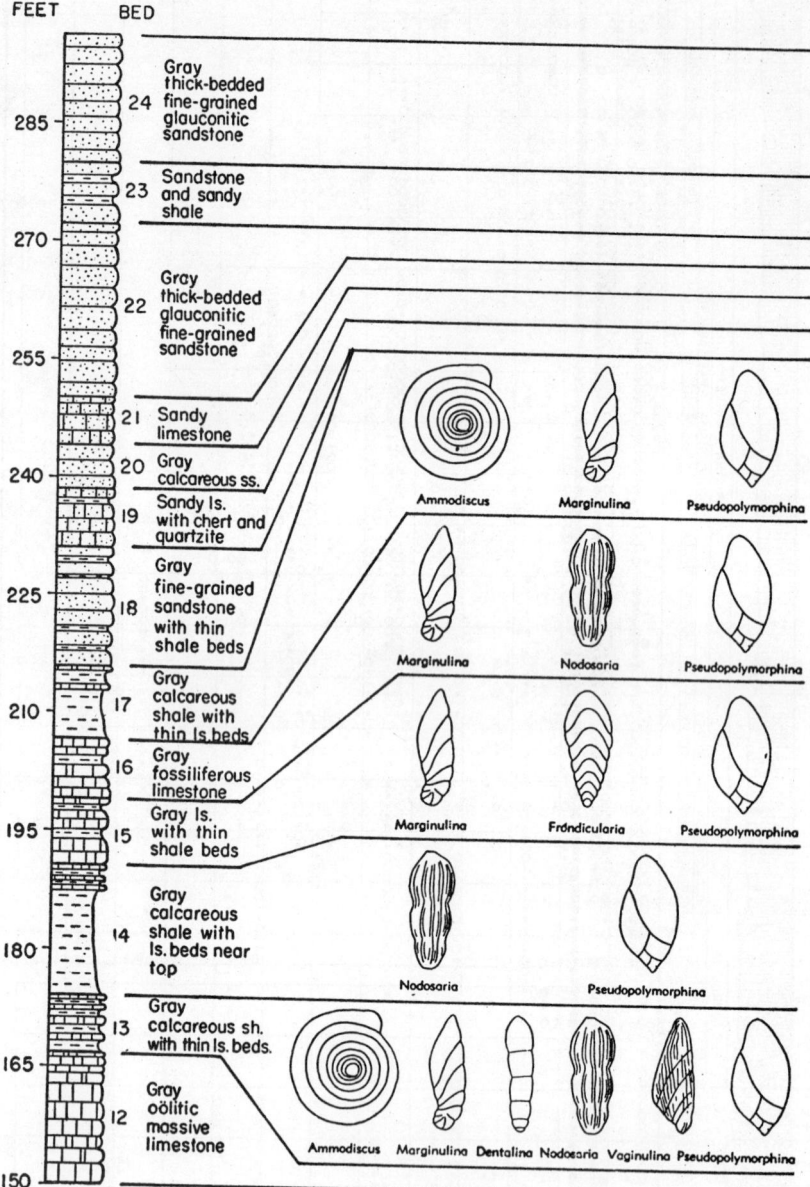

FIGURE 39. Unique manner of showing index microfossils in their stratigraphic position. (From Lalicker. Univ. Kansas Paleontological Contrib.)

(2) one or more species or genera, (3) general assemblages, (4) relative abundance of one or more than one species or genus, (5) ratio of various species and genera, (6) evolutionary development, and (7) faunal sequence.

The recognition of key species requires considerable detailed work. The index value of these forms must be repeatedly tested and compared in several stratigraphic sequences. For example, it required the writer two years to determine ten key species from a 375-species fauna of the Middle Tertiary deposits of central Sumatra. A given species may have an index value in one area, whereas its value as such may be slight or even worthless in adjacent areas. The responsibility to recognize and determine these peculiarities rests with the micropaleontologist.

It is general laboratory practice to develop a species type or control set for reference and comparison. Type species are either identified specifically or are given work numbers. The latter procedure is most applicable when a reference is not available for establishing species determinations.

In working foraminiferal faunas and interpreting their significance, it is necessary constantly to keep in mind the facies concept. Natland[2] demonstrated that recent foraminiferal assemblages off the coast of southern California differ in composition from shallow to deep water. He recognized a similar faunal sequence vertically in the Pliocene section of the Ventura Basin, California. The significance of this relationship exemplifies that these various assemblages have transected time horizons, a characteristic that many faunas undoubtedly possess. This problem constantly confronts micropaleontologists when long-range correlations are based on benthonic assemblages. Unlike faunas may be of similar age, although similar faunas may not possess the same time value.

Lowman[3] in his studies of the distribution of recent Foraminifera of the Gulf of Mexico has shown the extreme lateral variation of this group from fresh-water to brackish-water to open-sea environments. The object of this investigation was to improve "our ability to use fossil foraminifera as criteria of depositional environments, and, second, to help disentangle environmental and evolutionary factors in the three-dimensional distribution of fossil faunas."

A unique method employed by the Bataafsche Petroleum Maatschappij (Royal Dutch-Shell) in preparing micropaleontologic facies logs is described by Ten Dam.[4] This method is based primarily on the quantitative occurrence of certain genera or groups of genera of Foraminifera. Ten Dam summarizes the general procedure as follows:

The microfauna of each washed sample is divided into benthonic and

[2] Natland, M. L., *The Temperature and Depth Distribution of Some Recent and Fossil Foraminifera in the Southern California Region:* Scripps Inst. Oceanography Bull., vol. 3, no. 10, pp. 225-230, 1933.
[3] Lowman, S. W., *Sedimentary Facies of Gulf Coast:* Am. Assoc. Petroleum Geologists, vol. 23, no. 12, pp. 1939-1997, 1949.
[4] Ten Dam, A., *Micropaleontological Facies-Logs:* The Micropaleontologist, vol. 1, no. 4, pp. 13-15, Oct. 1947.

planktonic elements. The planktonic elements are the Globigerinidae and certain other genera. The rest of the microfauna will be almost entirely benthonic. The necessary micropaleontological data for compiling a facies-log are: (1) approximate number of the planktonic specimens; (2) approximate number of benthonic specimens; (3) approximate number of the specimens of genera or species indicating more or less brackish waters; (4) approximate number of the specimens of the genera indicating deep water or cold water; and (5) approximate number of the specimens of large Foraminifera.

De Sitter [5] and Ten Dam gave a key list of some ecologic conclusions based on facies logs. These are as follows: (1) A great number of planktonic specimens indicates a good connection with the open ocean; a growing number of specimens indicates opening of the basin to the open sea; a diminishing number indicates closing of the basin. Absence of plankton may indicate very shallow water. (2) A large number of benthonic species is an indication of favorable living conditions. (3) A small number of benthonic species may indicate limited or unhealthy living conditions. (4) The development of very large foraminiferal faunas indicates shelf conditions and clear and shallow water. Combined with much plankton it may indicate submarine ridges. Without plankton it probably signifies nearness to the coast or location in a more or less closed basin. (5) A microfauna with many specimens of *Cassidulina* or other species that seem to imply deep water may be an indication of deep water, generally with favorable bottom conditions.

None of these factors, according to Ten Dam, are in themselves de-

EXPLANATION OF FIGURE 40

FIGURE 40A. An assemblage from the Santa Barbara formation ("upper Pliocene") at Bath House Beach, city of Santa Barbara, Santa Barbara County, California. Note *Cythereis pennata* LeRoy (19) and *C. kewi* LeRoy (18).

FIGURE 40B. A recent brackish-water assemblage from Sunset Lagoons, Orange County, California. *Ammocytheridea* sp. (21) is very closely related to *Cytheridea beaconensis* LeRoy.

FIGURES 40C and 40D. A recent assemblage from the intertidal zone, Mussel Rock, Monterey Bay, California. *Hemicythere palosensis* LeRoy (23).

FIGURES 40E and 40F. An assemblage from the Lomita facies ("Pleistocene"), Hilltop quarry, Palos Verdes Hills, Los Angeles County, California.

  4. *Hemicythere? californiensis* LeRoy.
  6. *Bairdia verdesensis* LeRoy.
  7. *Cythereis glauca* Skogsberg.
  16. *Loxoconcha lenticulata* LeRoy.
  17. *Cytherelloidea californica* LeRoy.
  18. *Cythereis kewi* LeRoy.
  19. *Cythereis pennata* LeRoy.
  20. *Caudites fragilis* LeRoy.
  21. *Ammocytheridea* sp.
  22. *Ammocytheridea* sp.
  23. *Hemicythere palosensis* LeRoy.

cisive, but a combination of several factors may lead to a solution. Carefully interpreted facies logs of a series of field and well sections may give some concept of lateral and vertical facies changes.

[5] De Sitter, Geologie en Mijnbouw, new ser., vol. 3, no. 8, pp. 225-237, 1941.

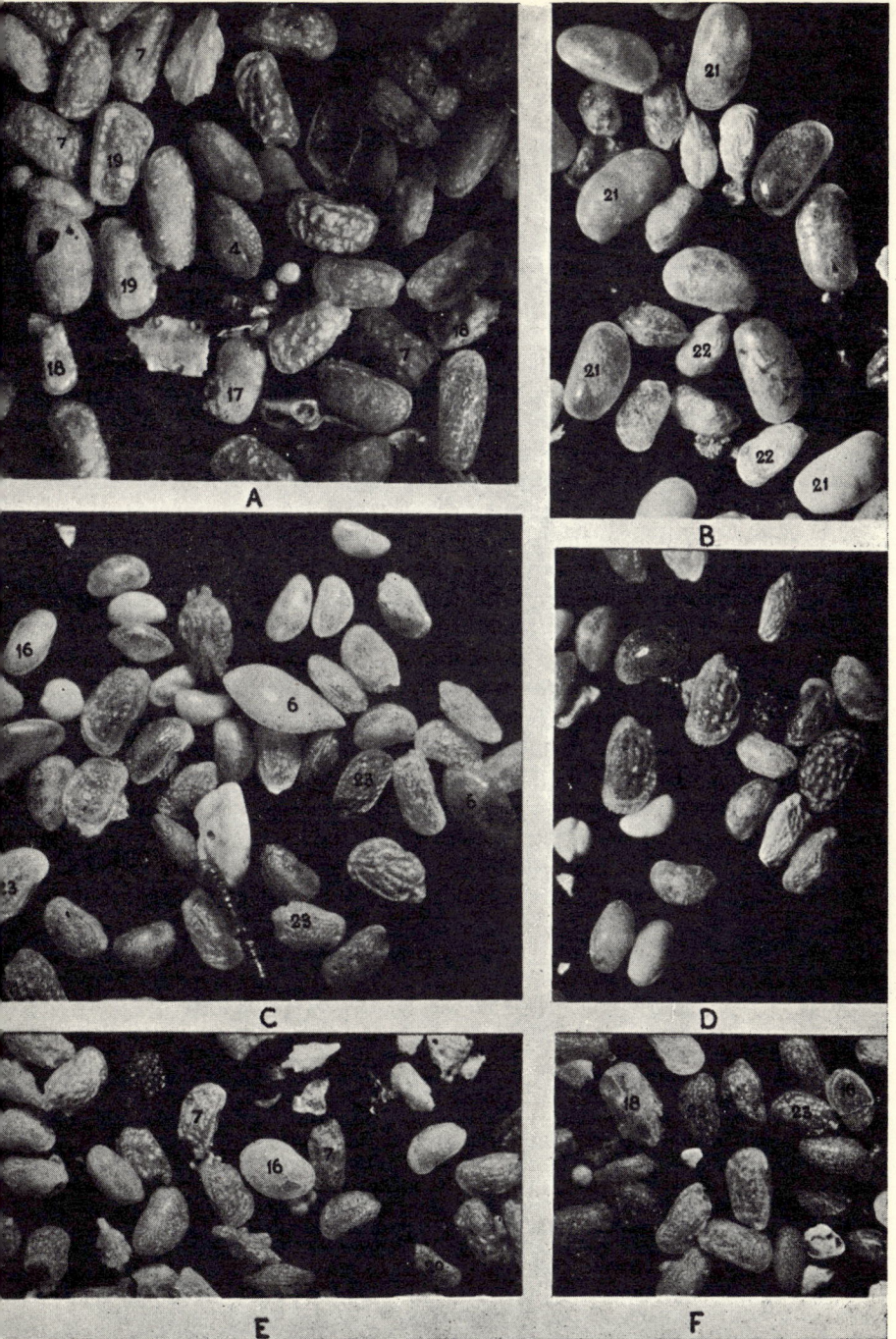

FIGURE 40. Assemblages of marine ostrocodes (magnification ± 15). These organisms occur most profusely in shallow-water environments. They also occur abundantly in fresh-and brackish-water habitats. In many areas they serve as index fossils. (For explanation see page 92:)

Supplementing the preceding data, information on other microfaunas and microfloras, lithologies, and mineralogy is also recorded and has been found to improve ecologic interpretations.

OSTRACODES

Next in importance to Foraminifera in economic micropaleontologic analysis are the ostracodes (fig. 40). These organisms, which are small, bottom-dwelling, bivalved crustaceans range in length from 0.5 mm. to 2.0 mm. They thrive under variable conditions, marine, brackish, and fresh water. The most diversified assemblages are those which inhabit shallow-marine environments. W. T. Rothwell (Richfield Oil Corp., California), in his studies on the distribution of living ostracodes of Newport Bay, California, show them to exist in the following salt-water environments: (1) tidal flat, (2) marsh channel, (3) lagoon channel, (4) bay-mouth and subtidal channel, and (5) open-sea rocky-tide pool. He further comments that the plants appear to be a major biologic factor influencing the distribution of the ostracodes at this locale. Further investigations by Rothwell of recent forms from samples across the San Pedro Channel and from collections along the California coast, indicate that this group of animals adjust to various water depths as do the Foraminifera.

Ostracode carapaces (shells) in some sediments may constitute the bulk of the material. Geologically ostracodes extend upward from the Ordovician. Those of the Paleozoic assume carapace characteristics that are pronouncedly different from those of the Mesozoic and Cenozoic.

The external or surface pattern of the valves is extremely variable and complex. Since most species molt (periodically shed their valves and then form new ones), surface features are not constant throughout their

EXPLANATION OF PLATE 4

Figure
1– 4. *Hemicythere? californiensis* LeRoy var. *hispida* LeRoy × 29. Fig. 1, right valve. Fig. 2, dorsal view. Fig. 3, ventral view. Fig. 4, interior view of valve showing hinge structure.
5– 9. *Bairdia verdesensis* LeRoy × 24. Fig. 5, right valve. Fig. 6, dorsal view. Fig. 7, ventral view. Figs. 8, 9, inside view of left and right valves.
10–13. *Caudites fragilis* LeRoy × 60. Fig. 10, right valve. Fig. 11, dorsal view. Fig. 12, ventral view. Fig. 13, inside view of right valve.
14–18. *Hemicythere palosensis* LeRoy × 42. Fig. 14, right valve. Fig. 15, dorsal view. Fig. 16, ventral view. Figs. 17, 18, interior views of right and left valves.
19–23. *Loxoconcha lenticulata* LeRoy × 42. Fig. 20, dorsal view. Fig. 21, ventral view. Figs. 22, 23, inside views of right and left valves.
24–27. *Cythereis kewi* LeRoy × 37. Fig. 24, right valve. Fig. 25, dorsal view. Fig. 26, ventral view. Fig. 27, inside view of left valve.
28–30. *Cytheropteron minutum* LeRoy × 37. Fig. 28, oblique view of right valve. Fig. 29, right valve. Fig. 30, posterior view.
31–39. Characteristic muscle-scar patterns of left valves. Fig. 31, *Hemicythere? californiensis* var. *hispida.* Fig. 32, *Hemicythere? californiensis.* Fig. 33, *Brachycythere lincolnensis.* Fig. 34, *Hemicythere palosensis.* Fig. 35, *Basslerites delreyensis.* Fig. 36, *Brachycythere driveri.* Fig. 37, *Bairdia verdesensis.* Fig. 38, *Paracypris pacificus.* Fig. 39, *Hemicythere jollaensis.*

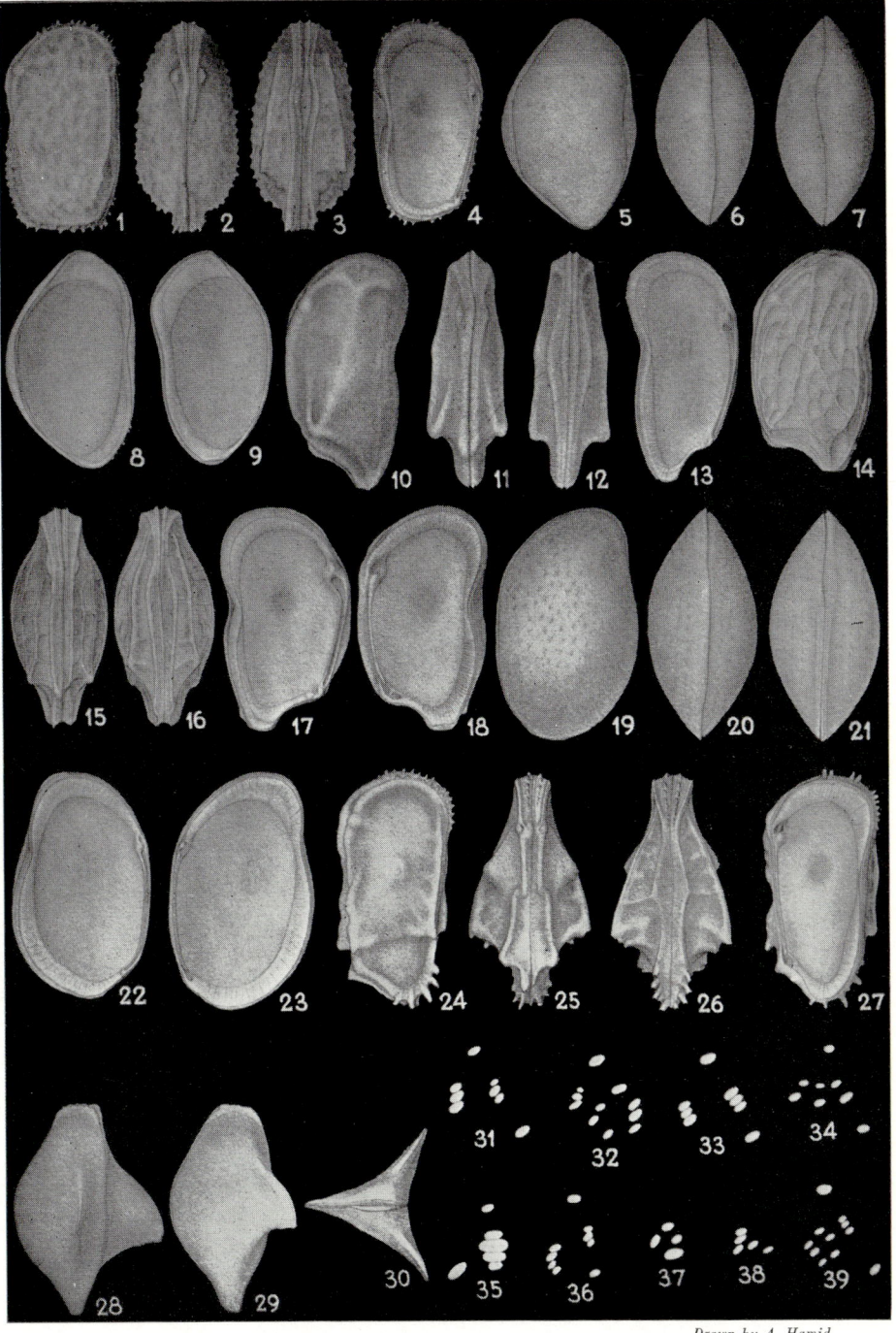

Drawn by A. Hamid

PLATE 4. Typical marine ostracodes; note the complex hinge structure, surface ornamentations, and muscle-scar patterns.

life cycle. Each series of molt valves generally differs somewhat from the preceding one. This factor must be carefully considered by micropaleontologists in speciation work, as classification of fossil ostracodes depends primarily on shell characteristics. Living forms are commonly identified on the basis of the animal appendages.

The most important elements of fossil ostracodes in descriptive work are (1) general shape and outline (side, dorsal, ventral views), (2) relative size and overlap of the valves, (3) surface ornamentation, (4) hinge characteristics, (5) muscle-scar pattern, and (6) characteristics of the interior marginal zones.

Before a species can be properly described, it is also necessary that the male and female individuals be distinguished. This distinction is not always possible because of the rarity of valves and slight differences in shell structure of the opposite sexes.

Fossil carapaces may be obtained from the enclosing sediment by extraction methods mentioned under "Foraminifera," although the screen series generally involves only the 60- and 100-mesh units.

In determining the vertical and horizontal distribution of ostracodes in stratigraphic sequences, the same procedure is followed as in foraminiferal studies. Charts are prepared showing species occurrences and abundences plotted against stratigraphic position. These data are then evaluated in terms of zonal intervals.

CALCAREOUS ALGAE [6]

Algae are seaweeds. Some have the ability to secrete lime around or within their tissue, and, hence, may be preserved as fossils. These are known as the "calcareous algae." During many times in the geologic past in many places they have grown so luxuriantly as to build or largely build extensive deposits of limestone.[7]

Frequently fragments or even entire specimens occur in well cuttings. They can be separated and concentrated in the same manner as most other microfossils. Most of them are about the size and shape of fusulinids (fig. 41).

There are a number of groups of calcareous algae, but only three, the coralline algae (Corallinaceae), the siphonous algae (Dasycladaceae), and the Charophyta (Chara) are of economic value at present as microfossils.

Corallinaceae

The Corallinaceae (fig. 41, no. 5; fig. 42) are a family of the red algae. They develop a very large number of growth forms, the most common being (1) forms with thin crusts, (2) forms with crusts from which rise mammillary protuberances or small stubby branches and (3) branching forms. The branches may be quite extended and may develop as

---
[6] The section "Calcareous Algae" was prepared by J. H. Johnson.
[7] Johnson, J. H., *Limestones Formed by Plants:* Mines Mag., vol. 33, pp. 526-533, 1943.

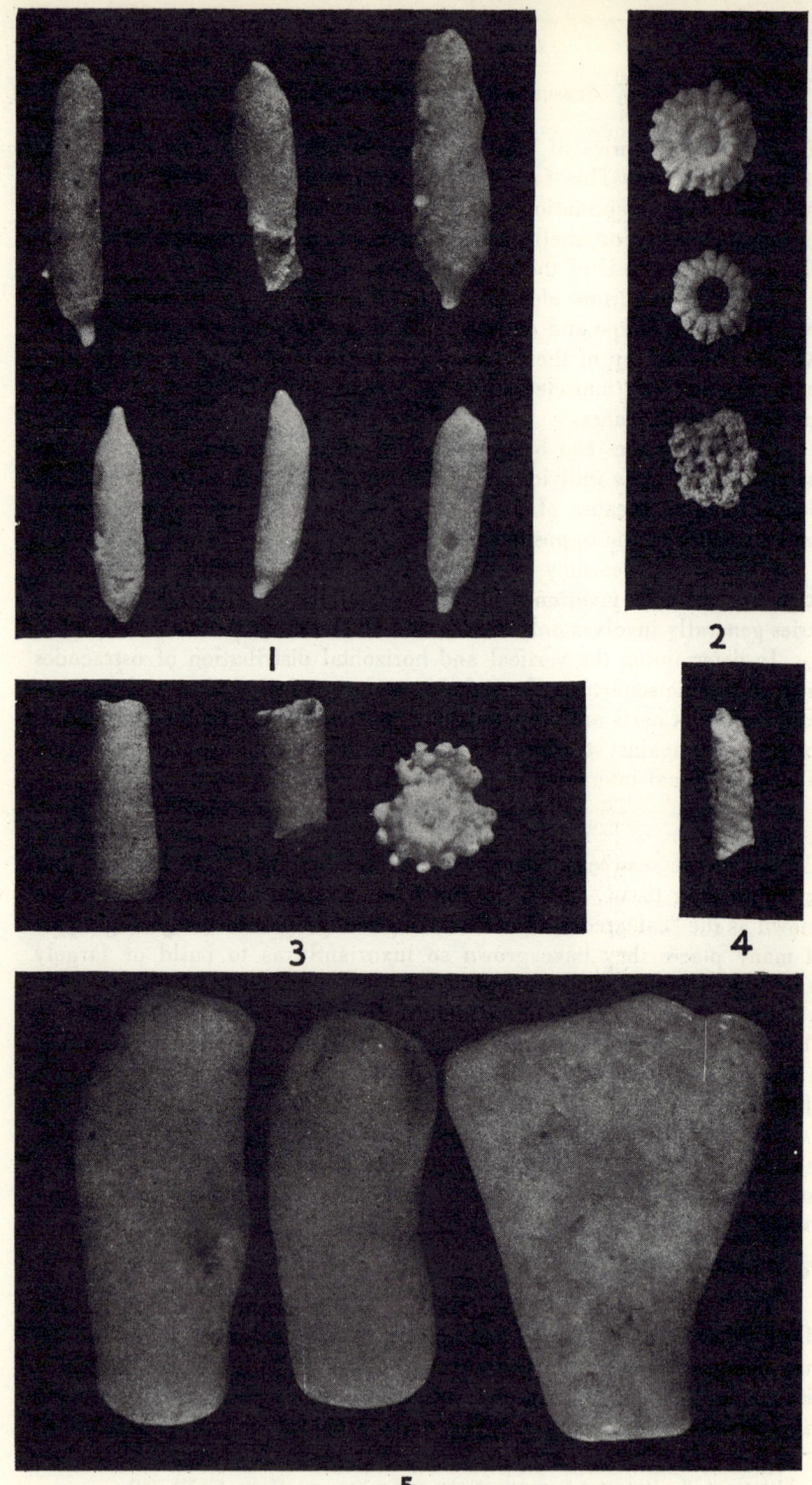

FIGURE 41. Typical fragments from calcareous algae (washed samples).

long, slender needles, may be branching stag-horn types, or may be broad and thick, resembling the horn of a moose.

The coralline algae secrete lime within and between the cell walls, thus showing definite microstructure (fig. 42). The various genera are separated on the basis of arrangement of cells in the tissue and the character and arrangement of the spore cases (conceptacles). Individual species within a genus are separated on the basis of cell and spore-case dimensions.[8]

Coralline algae develop in all seas from the poles to the equator and from tide level almost to the maximum depth of light penetration. Certain genera and individual species have specialized to accommodate certain ecologic environs. Coralline algae are recorded from rocks of all geologic ages from the Ordovician to the present. They are important after Late Cretaceous. Although they grow abundantly in most of the seas of the world, they attain their greatest development in the tropics, particularly in and around the reefs.

In the atolls of the Marshall Islands algae are very important as builders of reef limestones; locally they form up to 80 percent of the reef rock. In the fringing reefs of the Marianas, corals predominate, but algae still play a very important part both as contributors to the limestone and by acting as binding agents in the reefs.

Coralline algae produce distinctive fossils, and, because of their microstructure, they can be exactly identified in thin sections. Except in limited areas, they have not yet been sufficiently studied to determine the geologic range of the species. In those areas, however, where they have been studied, most of the species appear to have restricted time ranges and, hence, have possibilities for serving as guide fossils. They can give considerable ecologic information.

Dasycladaceae

The Dasycladaceae are small, bushy plants belonging to the green algae (fig. 41, nos. 1-4). They consist of a central stem with regularly spaced whorls of primary branches radially arranged. These may bear secondary and tertiary branches. Calcium carbonate is deposited around the central stem and primary branches, forming a shell of variable thickness. The type of fossil obtained and the amount of structure it exhibits depends upon the degree of calcification. The fossils usually appear as small rods or club-shaped fragments. A few are spherical; some are disc- or umbrella-shaped They commonly range in length from about one-eighth of an inch to more than three-fourths of an inch, although a few may grow much larger. Some have characteristic shapes and consequently are easily recognized, whereas others need to be sectioned for identification. Typically the structure consists of a mold of the central stem from which whorls of tiny branches develop at regular intervals.

---

[8] Lemoine, Mme. P., *Algues calcaires fossiles de l'Algérie:*. Materiaux pour la Carte Géol. de l'Algérie, 1e ser., no. 9, 128 pp., 3 pl., 1939.

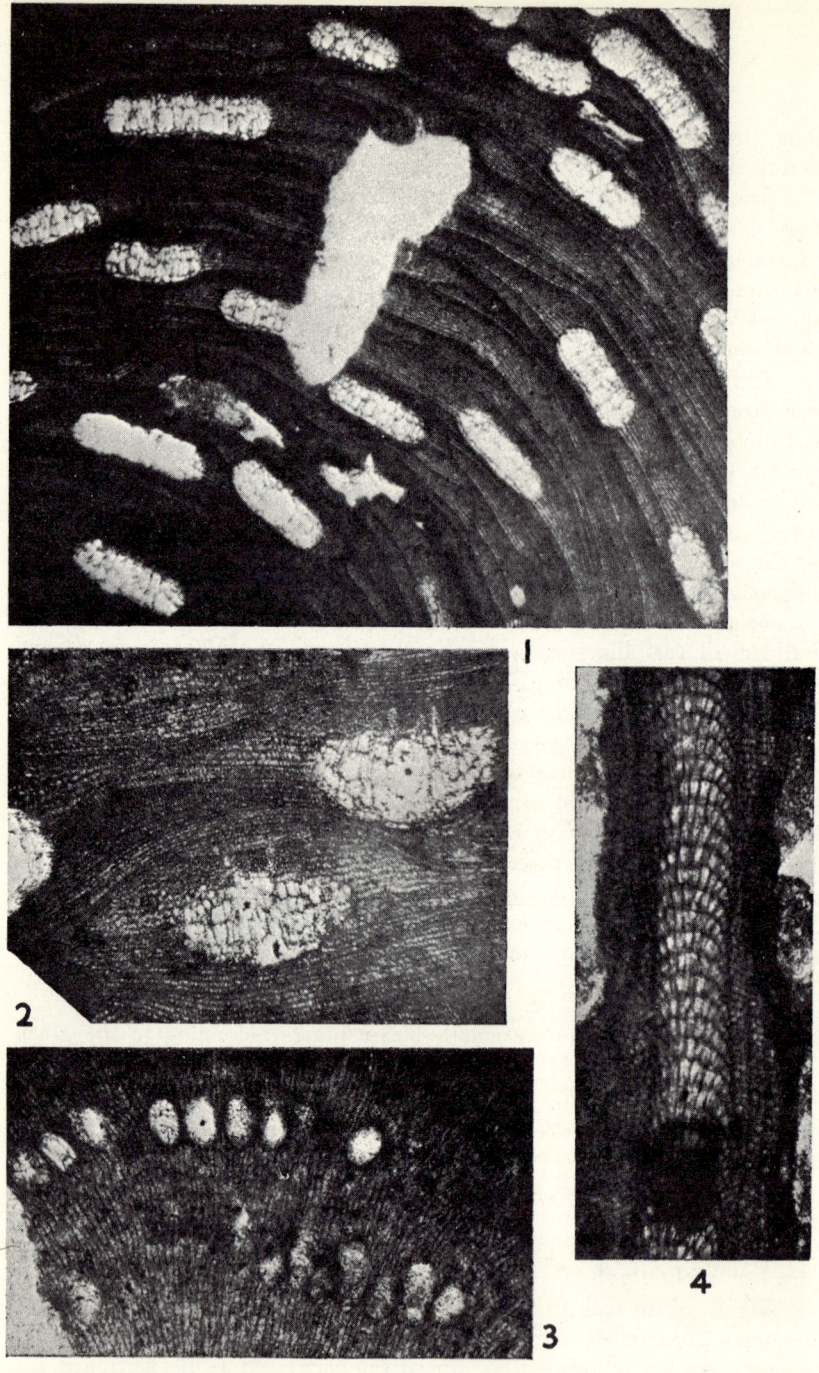

FIGURE 42. Microstructure of coralline algae in thin sections. No. 1, *Lithothamnium concretum* Howe (×50), Oligocene West Indies (after Howe). No. 2, *Mesophyllum sancti dionysii* Lemoine (×50), Miocene, Algeria (after Lemoine). No. 3, *Archaeolithothamnium floridum* (×75) Johnson and Ferris, Eocene, Florida. No. 4, *Lithophyllum* aff. *prelichenoides* Lemoine (×75), Miocene, Borneo.

These branches show on the surface of the fossil as rows of small openings or clusters of openings in regular rows circling around the fossil.

Ecologically these algae grow in shallow water, on mud flats, or in sheltered portions of reefs. Fossil forms occur in many marls, calcareous shales, and limestones.

Geologically they are known in rocks ranging in age from the Early Ordovician to the present, although they appear to have made their greatest development during the Triassic and Jurassic.

They have been extensively studied in certain areas of Europe,[9] where it has been demonstrated that the individual species have short geologic ranges; consequently, they can be used as guide fossils.[10] Very few have been described from North America, although a careful search will undoubtedly show them to be present in many localities.

Charophyta

The Charophyta are a distinctive, isolated group of algae (fig. 43) usually classed with the green algae (Chlorophyta). Calcium carbonate is usually precipitated around the tips of the branches and spore cases. These structures are the only portions of many of the plants that are calcified. The spore cases (oögonia) form a distinctive fossil. With their spiral ornamentation they are readily recognized, as no other microfossil even closely resembles them in structural constitution.

At the present time the Charophyta inhabit only fresh- and brackish-water environs, but during the Paleozoic and possibly during the early Mesozoic it appears that some forms may have lived in shallow-water, near-shore, marine environments. Their remains occur in great numbers in many of our continental deposits,[11] and for some such formations they are distinctive microfossils.

According to Peck and Recker,[12]

Considering the difficulty of establishing fine divisions in the classification of the Charophyta, it is believed that they will never be of value for the discrimination of small stratigraphic units. Yet they are of real value in period differentiation: continental deposits of Jurassic, Lower Cretaceous, Cretaceous, and Lower Cenozoic ages can be readily differentiated on the basis of Charophyta oögonia.

Geologically, their remains have been noted in rocks ranging in age from the Devonian to the present.

General

Fossil algae as a group require a great deal of further study. Recent work has shown them to be widespread and abundant. They have useful

[9] Morellet, L., and Morellet, J., *Les Dasycladaceés du Tertiaire Parisien:* Soc. géol. France Mém. vol. 21, no. 1, p. 43, 3 pls., 1913; *Tertiary Siphoneous Algae in the W. K. Parker Collection:* British Mus. Nat. History, 55 pp., 7 figs., 6 pls., 1939.
[10] Pia, J., *Die Siphoneae verticillatae vom Karbon bis zuir Kreide:* Zool.-bot. Gessel. Wien Abh., vol. 11, pt. 2, 1920.
[11] Peck, R. E., *Fossil Charophyta:* Am. Midland Naturalist, vol. 36, no. 2, pp. 275-278, 1946.
[12] Peck, R. E., and Recker, C. C., *Eocene Charophyta from North America:* Jour. Paleontology, vol. 22, no. 1, pp. 85-90, 1948.

FIGURE 43. Washed samples of typical oögonia of various genera of the Charophyta. (After Peck.)

possibilities, both as time-index fossils and as indicators of environment of deposition.

Calcareous algae often coat and fill voids of other fossils. Certain algae occur symbiotic with other organisms and produce growths that make characteristic and often rather common fossils (e.g., *Archimedes*).

Fossil algae are among the oldest fossils known; in fact, they are the only group that has been found in considerable numbers in pre-Cambrian rocks.

DIATOMS

Diatoms are microscopic unicellular plants belonging to the phylum Thallophyta. The skeleton or frustule, composed of opaline silica, commonly consists of two shallow, disk-shaped halves (example: *Craspedodiscus*), one of which fits into the other similar to a flat pillbox. Other forms are elongate and bilaterally symmetrical (example: *Glyphodiscus*) with respect to an axial strip. Individual specimens may range in diameter from 0.1 to 0.15 mm. Magnifications up to × 200 are generally required to study this group of microflora. It has been estimated that a cubic inch of some diatomites (deposits composed essentially of diatoms) contains as many as two billion frustules.

Classification of fossil diatoms is based on the size, shape, and surface ornamentation of the frustule. The intricacy and complexity of surface ornamentation of certain species are fantastic and remain surprisingly uniform. (See fig. 44).

Although diatoms live under a wide range of environmental conditions, they are highly selective as regards sunlight, temperature, salinity, turbidity, and percentage of silica in the water. Some species live only in fresh water, whereas others thrive only in saline water. Some forms attach themselves to objects, although most are of the planktonic variety.

The fats secreted by diatoms are considered by some petroleum geologists to be the source of much of the oil and gas being produced from the Miocene sediments of California.

Diatomite has a world-wide distribution, although the size of the individual deposits is not large. Marine Miocene diatomaceous deposits occur in California, Maryland, Virginia, Algeria, and Denmark. Large fresh-water deposits are found in North America, Europe, and Japan. The largest deposit in the world is that at Lompoc, California. This deposit consists of 1,400 or more feet of stratified diatomite in the Monterey series of the upper Miocene and covers an area of 4,000 acres. Estimated reserves have been placed at 100,000,000 tons. The more important commercial uses of diatomite are as mineral filters, paint filler, insulation, building blocks, and abrasives.

Diatoms have had a relatively short geologic history ranging from the Late Cretaceous to the Recent. They were common in the Cretaceous, became prolific in the Tertiary, and reached their highest development

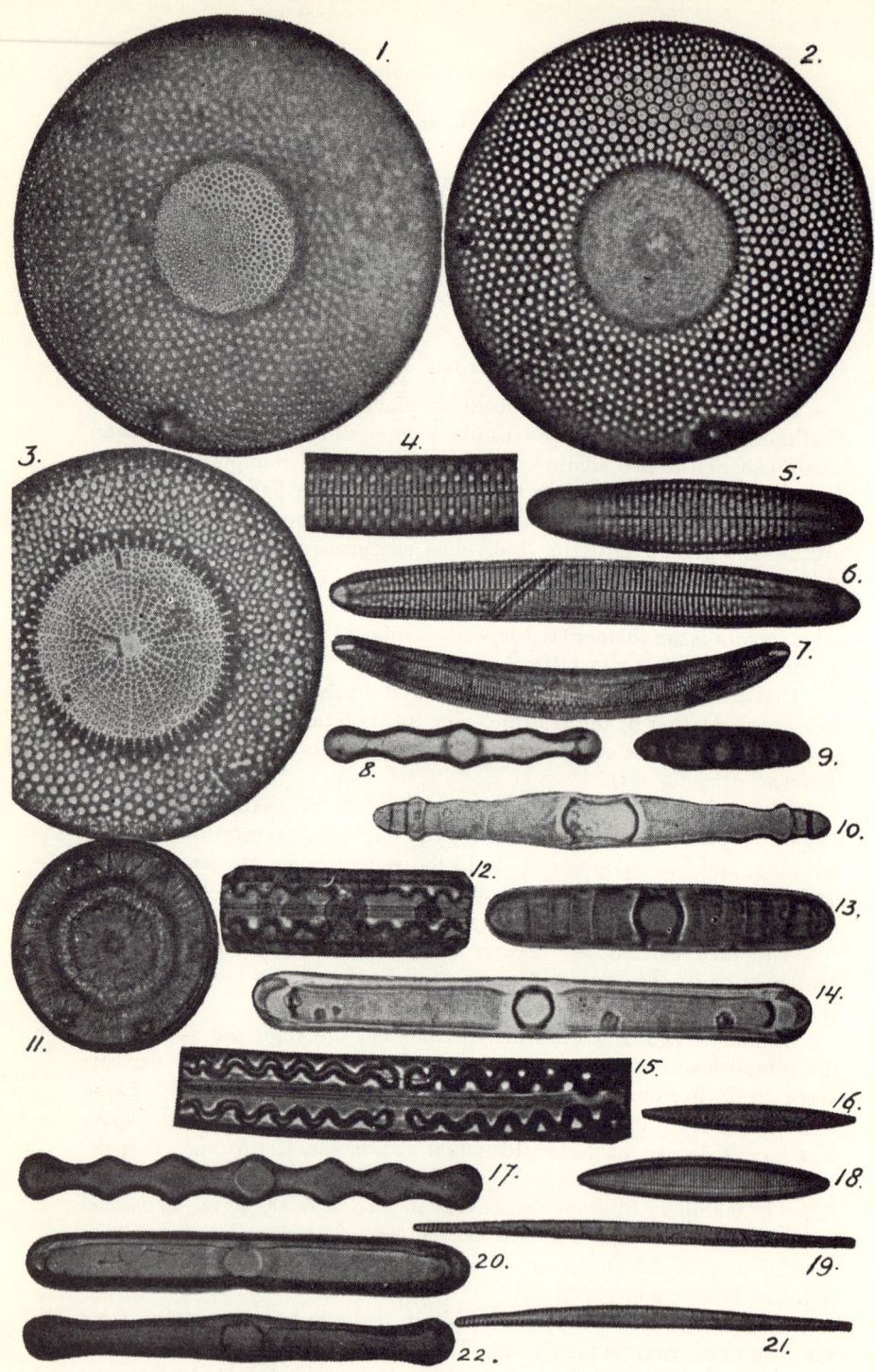

FIGURE 44. Typical marine diatoms (single-celled plants possessing a siliceous skeleton). Diatoms occur in all types of aqueous environments. They may be satisfactorily used in correlation work, owing to their pelagic adaptability. (After Reinhold.)

during the Miocene. Following Miocene time they decreased rapidly but were still present in large numbers. It is impossible to predict how much diatomaceous ooze is now being deposited on the present ocean floor.

Marine diatoms flourish in all latitudes and at all seasons of the year, in the warmer and coldest seas. It is well known that they are so abundant in frigid zones as sometimes to colour the seas and to tinge with a particular hue the blocks of floating ice.

They are capable of surviving in conditions so diverse, it is difficult to believe that any fixed laws of geographical distribution can be discovered with respect to them; on the contrary, it might be supposed that the continuity of adjacent seas, the surface and submarine currents, the movement of tidal waves, the existence of periodical and other winds, the traffic of ships and the movement of fishes would all tend to facilitate or bring about the mingling of local floras.[13]

Diatoms offer great possibilities for establishing local and world-wide correlations because of their pelagic character and the siliceous nature of their skeletons. Attempts have been made in California and in the Netherlands East Indies (Java) to use diatoms for subdividing certain parts of the Tertiary sequence.

Cleaning the diatoms out of different samples is a tedious business. To clean a sample of a fossil marl thoroughly, when the particles are solidly cemented together perhaps with some volcanic material enclosing the fragile microfossils, is not a small task and requires even from the most experienced cleaner much skill and patience. Different chemical agents may be used, depending on the chemical character of the matrix; too strong solvents may destroy the diatoms entirely, which happens sometimes in the most unexpected manner, as any diatomist knows. After the texture of the rock has been loosened and something like a dispersion has been obtained, the material may be sifted through very fine sieves, 150- and 300-mesh, and subjected to several washing operations, decanting after a fixed time. The well-known sulphuric-acid treatment with either potassium chlorate or nitrate for bleaching is in almost any case inevitable. The whole process cannot be hurried through and must take its time. It may last several weeks. However cumbersome the process of cleaning may be, well-cleaned samples are extremely important and save much time in the inspection and sorting. Only in the case of well-cleaned samples, a complete review of the fossil content is possible and the diatoms accessible to further study and to be photographed and arranged in neatly mounted slides, which allow a close and thorough inspection. Fragments have as a rule to be disregarded except in instances where a view of the internal structure is wanted. When incomplete tests are inspected, errors will heap up, for studying rare and unknown diatoms from badly conserved and imperfectly cleaned fragments leads to false determinations, which may remain undetected in case of a study purely in search of new and strange forms, but when the establishment of fossil lists for stratigraphic use is planned, wrong determinations may seriously influence the result as to the zonal distribution of the fossils and the geologic age of the samples.[14]

[13] Castracane, Fr. D. A., Report on the Scientific Results of the Voyage of H.M.S. Challenger During the Years 1773-76: Botany, vol. 2, p. 9, 1886.
[14] Reinhold, Th., Fossil Diatoms of the Neogene of Java and Their Zonal Distribution: Geol.-mijnb. genootsch, Nederland en Kolonien Verh., Geol. ser. Deel 12, pp. 43-133, 21 pls., 1937.

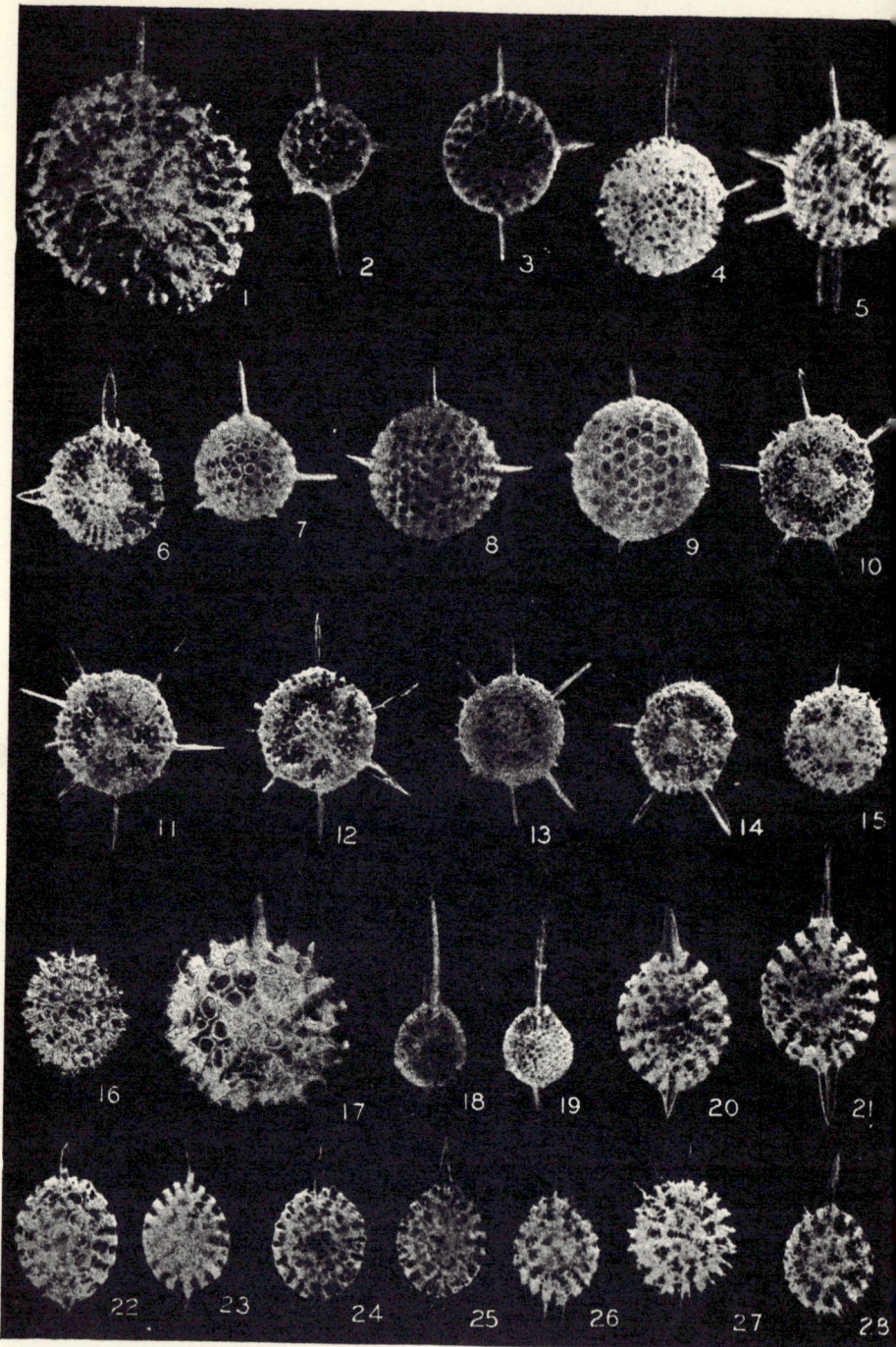

FIGURE 45. Skeletal remains of Radiolaria. These micro-organisms (single-celled animals) occur only in marine habitats and at all latitudes. (After Clark and Campbell, Geol. Soc. America Spec. Paper 39, 1942.)

RADIOLARIA

Radiolaria are single-celled, pelagic, open-sea, marine Protozoa having diameters of about 0.1 to 0.5 millimeters. The skeleton or shell (fig. 45), consisting of silica which is secreted by the protoplasm, makes up an important part of the animal. Little attention has been given to fossil radiolarian faunas because of the complex classification, general scarcity, and difficulties encountered in procuring well-preserved specimens. The classification of this group of organisms is based upon the structure and composition of the hard parts.

The majority of the many forms developed within the Radiolaria are adaptive to flotation and to lowering the rate of sinking in the water. The presence of long spines, many of which are multiplied, of surface spines and thorns, and the high development of the hat or cuplike shape of the lattice-shell in some are all responses to a reduction in the rate of sinking. These structures also allow less effort on the part of the animal to keep it well within the photosynthetic zone. . . . Radiolaria occur in all the seas of the world, in all climatic zones, and at all depths. However, the Pacific Ocean appears to be the richest both quantitatively and qualitatively in these creatures, excelling both the Indian and Atlantic oceans.[15]

Radiolaria should be more seriously considered in the future by micropaleontologists as a basis for establishing long-range correlations because of their pelagic character and the siliceous nature of their skeleton.

CONODONTS

Conodonts are small, tooth-shaped, single- or multiple-pointed, or platelike fossils occurring locally and in considerable number in Paleozoic shales (fig. 46). These structures have been interpreted by various workers as fish remains, annelid jaws, and gastropod teeth. In greatest dimension they range from one to two millimeters. For practical purposes, according to Ellison,[16] conodonts can be divided into four groups as follows: (1) fibrous, (2) simple cones, (3) blades and bars, and (4) platforms. Ellison states further:

The fibrous and simple cone groups have many genera that serve as excellent guides to the Ordovician. The blade and bar group is mainly long ranging. However, a few genera serve as guides in the Ordovician, Devonian, and Mississippian. The genera in the platform group are best guides to beds younger than Silurian. Many of these are remarkably restricted.

Ellison [17] concludes that (1) the composition of conodonts is the same as the mineral matter in fossil and modern vertebrate hard parts; (2) conodonts may be classified as fish or lower vertebrates on the basis of their composition, size, shape, assemblage associations, internal struc-

[15] Clark, B. L., and Campbell, A. S., *Eocene Radiolarian Faunas from the Mt. Diablo Area, California*: Geol. Soc. America Special Paper 39, pp. 1-112, 9 pls., 1942.

[16] Ellison, S. P., Jr., *Conodonts as Paleozoic Guide Fossils*: Jour. Paleontology, vol. 30, no. 1, pp. 93-110, 1946.

[17] Ellison, S. P., Jr., *The Composition of Conodonts*: Jour. Paleontology, vol. 18, no. 2, pp. 133-140, 1944.

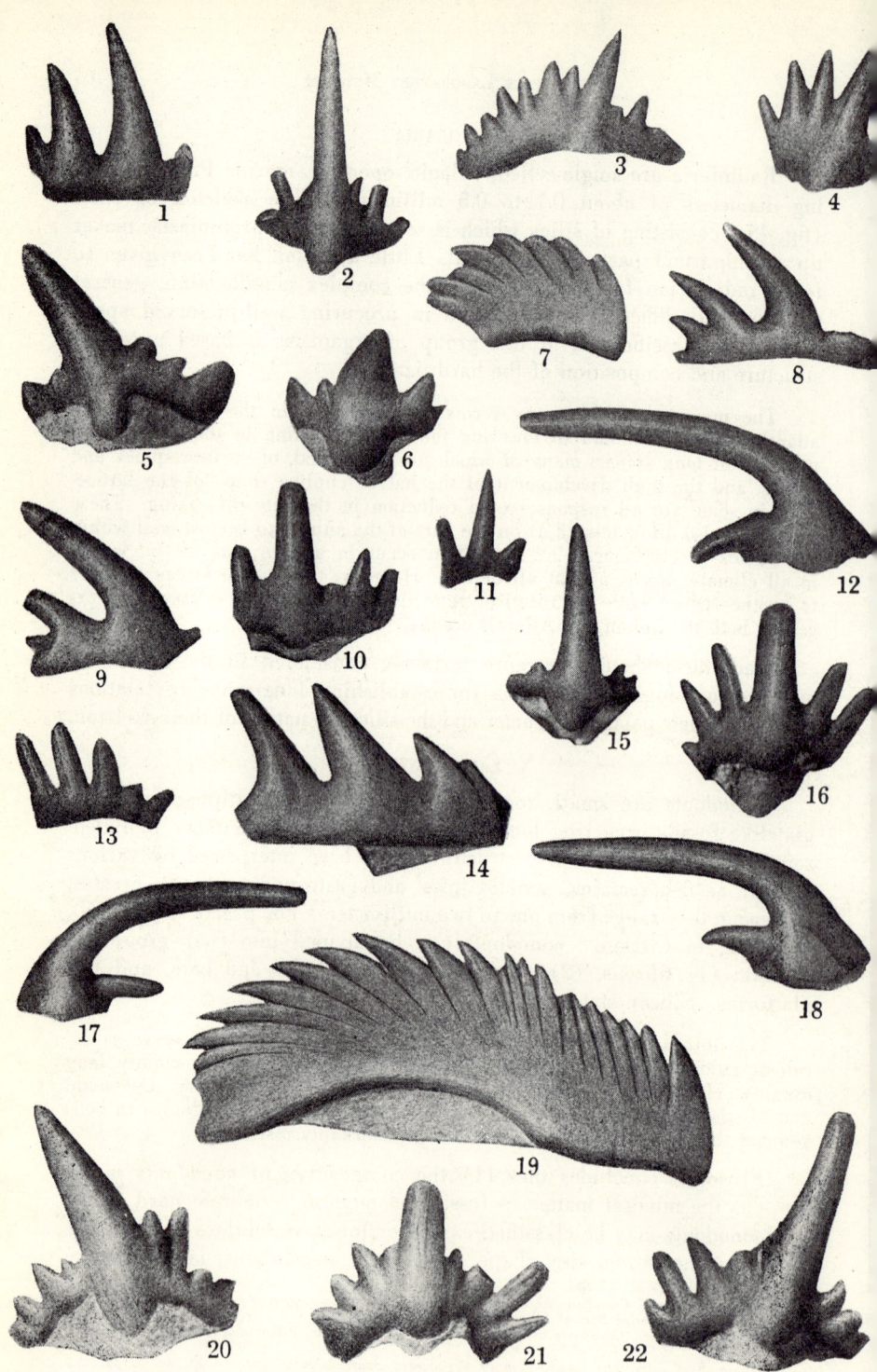

FIGURE 46. Structure of conodonts. Conodonts have been widely used for correlation in the Paleozoic section. (After Youngquist and Cullison, *Jour. Paleontology*, 1946.)

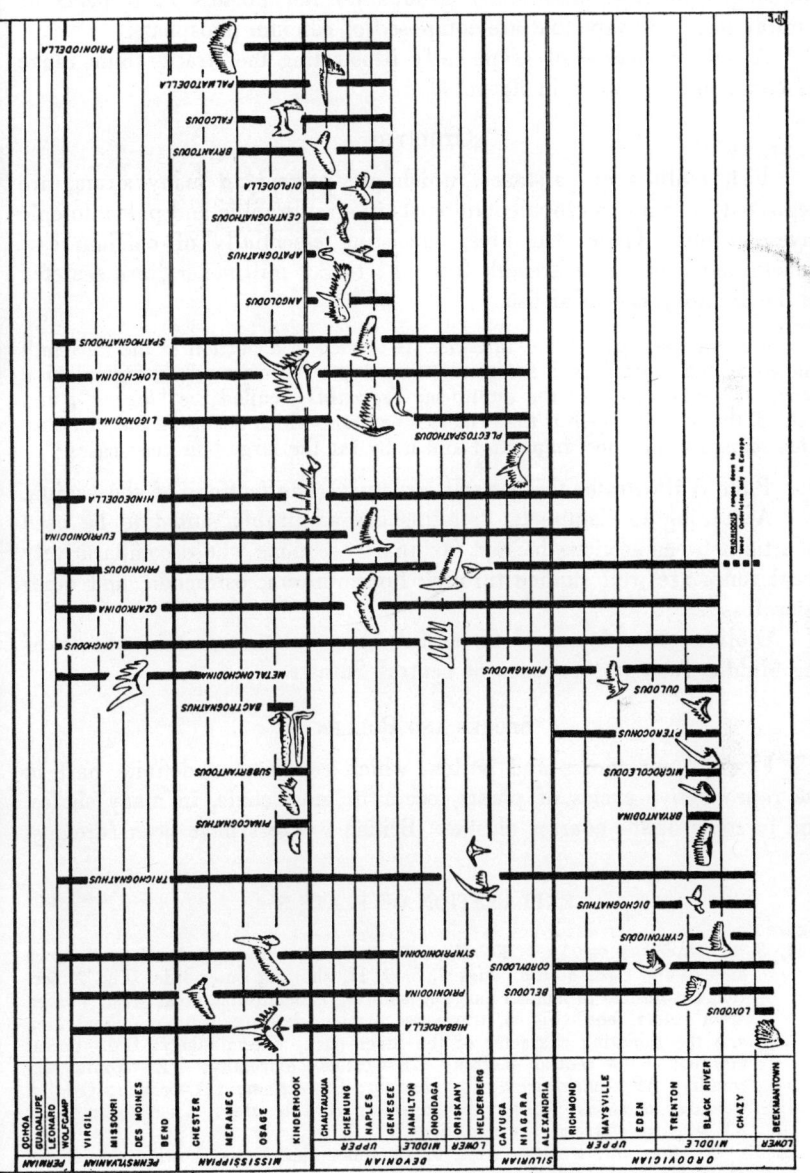

FIGURE 47. Method of recording stratigraphic ranges of conodonts. (After Ellison, Am. Assoc. Petroleum Geologists.)

ture, associated bone material and jaw parts, and stratigraphic occurrence; and (3) the assignment of conodonts to other zoological groups is challenged because these other groups do not possess hard parts of similar size and shape that are composed of calcium phosphate.

A unique method of graphically illustrating the stratigraphic range of conodonts is shown in figure 47.

OTOLITHS

Fish otoliths or earstones, which are present in many strata, are neglected by many micropaleontologists in stratigraphic and paleontologic investigations. These structures, consisting essentially of calcium carbonate and ranging in breadth from 0.1 to 3.0 millimeters, are secreted in the auditory system of fish.

A small one, termed the "*lapillus*," is formed in a portion of the labyrinth known as the "*utriculus*"; a second, termed the "*asteriscus*," is formed in a posterior prolongation of the otolith-sac *(sacculus)*, called the "*lagena*"; and a third, the *sagitta*, which is the principal earstone which occurs in the *sacculus*. This *saggita* is the most important and is by far the largest in most cases.[18]

Plate 5 illustrates the general structural implications of the sagitta.

According to Campbell, "otoliths are admirably suited to be used as a tie between sections located far apart—sections whose comparatively local zones are well studied through Foraminifera, ostracoda, and other microfossils."

Otoliths were found to be useful in correlating certain phases of the Middle Teritiary sequence of central Sumatra.

SPORES AND POLLEN

Fossil plant spores and pollen, which represent a definite part in the reproductive cycles of plants, occur in most coals, in many shales, and in some of the coarser clastics. British workers have been foremost

EXPLANATION OF PLATE 5

Figure

1–11, 3, 6. *Otolithus* var. A × 20. Figs. 1, 4, 7, 10, showing variation of inner side of the right otolith. Figs 2, 5, 8, 11, showing outer side. Fig. 2, section cut normal to median axis and showing elongate umbilical area. Viewed from caudal end. Fig. 6, transverse section showing ovate umbilical area and the radiating character of the inner part. (Terminology: DM = dorsal margin; VM = ventral margin; CE = caudal extremity; FE = frontal extremity; AR = *antirostrum*; R = *rostrum*; O = *ostium*; CA = *cauda*; O + CA = *sulcus acusticus*; CS = *crista superior*; A = *the area*; E = *excisura ostii*; VF = *ventral furrow*; U = *umbilicus*.)

9, 12, 13, *Otolithus* var. B × 20. Figs. 9, 12, inner side views showing the closed character of the *sulcus acusticus*. Observe the prominent projections along the ventral margin, which constitutes a distinctive feature of the variety. Fig. 13 illustrates the radial character as well as the concentric growth lines.

---

[18] Campbell, R. B., *Fish Otoliths, Their Occurrence and Value as Stratigraphic Markers:* Jour. Paleontology, vol. 3, no. 3, pp. 254-279, Sept. 1929.

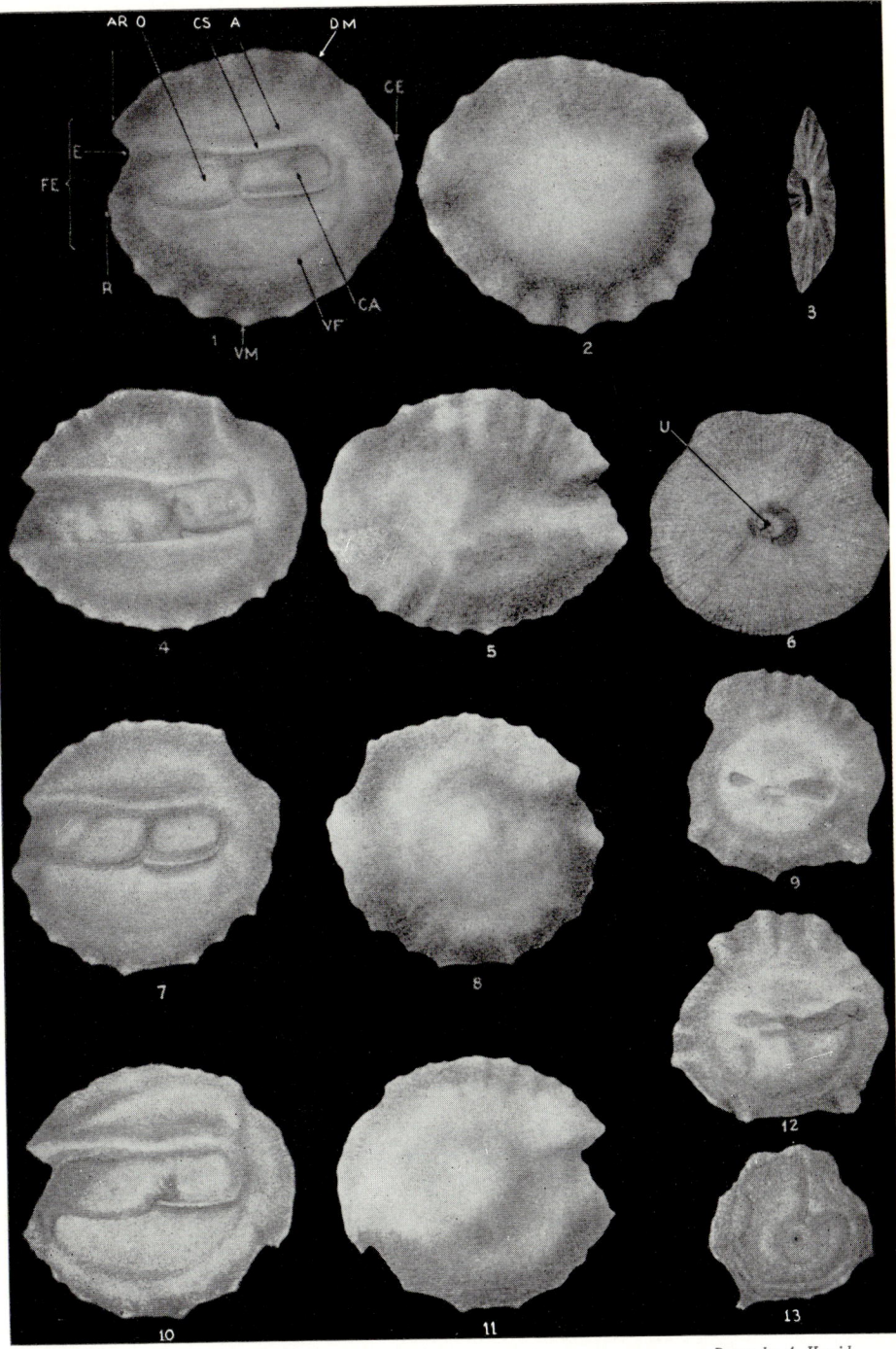

Drawn by A. Hamid

PLATE 5. Showing the structural constitution of fish otoliths (ear bones).
Locally these structures have been used in correlation work.

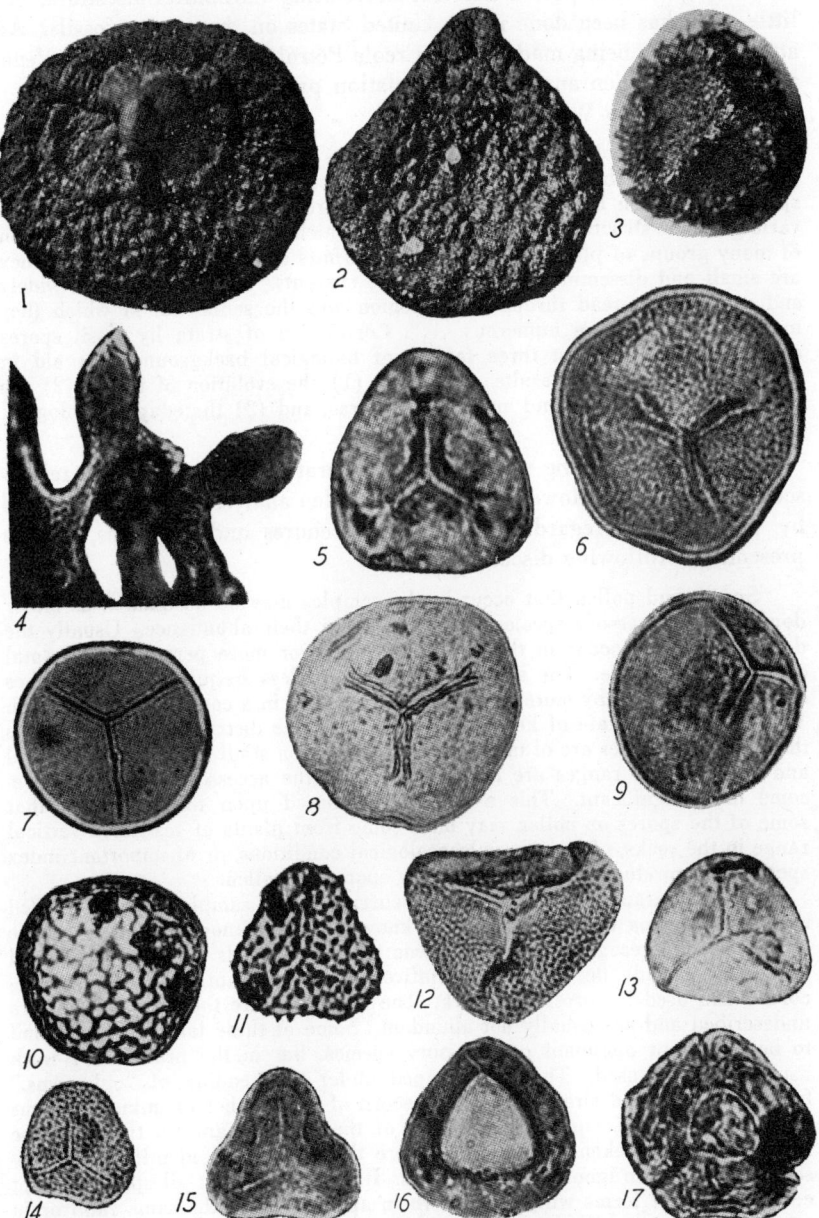

FIGURE 48. Carboniferous plant spores from Daggett County, Utah. Numbers 1-4, *Triletes*; 5-10, *Punctati*; 11-15 *Granulati*; 16-17, *Denso-sporites*. All about 400X, except 1-3 (30X). (From Schemel, *Jour. Paleontology.*)

in using spore and pollen data for correlating Carboniferous strata, but little work has been done in the United States on these microfossils. An attempt is now being made by the Creole Petroleum Corporation in Venezuela to use pollen analysis in correlation problems.

According to Wilson,[19]

Spores and pollen meet the requirements for correlation work and have proved their worth, at least in a limited capacity. The chemical nature of spores and pollen is such that most species are resistant to decay. They are varied in their structure and ornamentation, which allows the easy identification of many groups of plants and frequently permits specific determination. They are small and disseminated mainly by air currents. Many species are widely and uniformly spread throughout a region and the sediments in which they may be preserved are numerous. . . . Correlation of strata by fossil spores and pollen has at least three factors of biological background that aid in the accomplishment of results. These are (1) the evolution of floras, (2) the geographic distribution and migration of floras, and (3) the edaphic ecological relations of plants.

Methods pertaining to collection, maceration of samples, and microscopic techniques followed in spore and pollen analysis are clearly treated by Wilson.[20] As regards correlation procedures and problems, Wilson presents the following discussion:

Spores and pollen that occur in the samples may be designated as either dominant or accessory species depending upon their abundance. Usually the dominant species occur in the frequency of five or more percent of the total count in the sample. The accessory species are less frequent and sometimes are not represented by more than a single specimen in a count of one thousand. With the present state of knowledge concerning the distribution of the fossils the dominant species are of most value in correlation studies, but as the vertical and geographical ranges are better understood the accessory species will become more significant. This assumption is based upon the possibility that some of the spores or pollen may have come from plants of restricted vertical range in the rocks, of specific paleoecological conditions, or of important index species that produced comparatively few spores or pollen.

For stratigraphic work the fossils ocurring in the samples may be divided into the following three groups: (1) knowns, (2) unknowns, and (3) those broken beyond recognition. The knowns are those fossils that are recognized as species already described or tentatively given descriptions. They are recorded and used in correlation work. The unknowns are those fossils that are undescribed and are usually not abundant. Some of these later may be found to be important dominant or accessory species, but in the preliminary work usually are not used. They are lumped under the heading of "unknowns." In the counts it is desirable to have a record of the number of unknown forms and to prepare descriptive illustrations of the more diagnostic types. Those fossils that are broken to the point where they cannot be identified with described species are ignored in the counts. It is assumed that all species except certain resistant forms will break down in approximately the same ratio or in specific ratios that will not materially affect the final count.

The number of spores and pollen that should be counted for satisfactory correlation work appears to be based on the use to which the study is directed,

[19] Wilson, L. R., *The Correlation of Sedimentary Rocks by Fossil Spores and Pollen:* Jour. Sedimentary Petrology, vol. 16, no. 3, pp. 110-130, 1946.
[20] Wilson, L. R., *ibid.*

and the number of species present in samples. In peat investigations, pollen of tree species are usually the only ones used and the conclusions are usually based upon a count of 200 fossils. In peat seldom more than a dozen plant species are present and frequently the total number of tree pollen species is not more than six in any one level. In coal and shales studies the paleontologist uses all types of spores and pollen and seldom restricts his studies to tree pollen. For this reason the number of species is often several times the number encountered in peat. Consequently, it is desirable to examine and count a greater number of individuals. What this count should be is not established and differs widely among workers. . . .

Graphic treatment of the counts is probably the best method of demonstrating plant microfossil correlations. Line graphs were early used in peat pollen studies to show paleoecological succession, but these have been largely replaced by bar graphs of several types. . . .

Two types of graphs can be constructed for correlation purposes. These are channel-sample graphs and horizon-segment graphs. The former is the better type for direct correlation purposes, if the strata involved are not more than several feet thick. In thick coal seams, channel sample correlation becomes difficult. Where possible, a coal seam should be divided into segments separated by shale or pyrite partings. These partings are often of extensive areal extent and make good natural boundaries. In shale deposits lithological types should be used to limit the extent of the sample. In order to determine paleoecological succession within a rock member, contiguous thin vertical samples must be studied. The graphed results will show the percentage of the fossils at successive levels. Close correlation of comparable measured horizons has not yet shown exact percentages of identical species, but successional trends are usually indicated, if the section is viewed as a whole, and as such have paleoecological and stratigraphic value.

How similar the percentages of each fossil species must be in strata to indicate correlation is a question of considerable pertinence. Experiments with seams of coal have shown that the dominant fossils will frequently vary between five and ten percent in a 200-fossil count, if the samples are collected several miles apart, or the maceration process has not been uniform. In the first instance, areal distribution of the ancient vegetation apparently is a factor, or the coal seam was thicker or more completely represented at one locality than at another. In the second instance, where uniformity of maceration is not attained, fossils with various thicknesses of spore or pollen coat, or great differences in size, will appear in variable percentages. An effective method of combating such problems is to divide the sample in the course of its preparation and allow additional time for each portion of the sample. Uniformity can be checked with test slides by using the corrosion of certain species of fossils as an index for uniformity of maceration.

In conclusion, it might be stated that results have been attained with fossil spores and pollen which show conclusively that all coals and shales thus far studied can be assigned within the limits of geological periods, and that various strata can be separated from each other by their spore and pollen facies, or specific abundance of various species. It would seem that plant microfossils have great future scientific and economic value as the science develops and broadens.

Grass Seeds

Several years ago Elias [21] published a noteworthy paper pertaining to prairie-grass seeds of the Late Tertiary deposits of the Prairie States.

[21] Elias, M. K., *Tertiary Prairie Grasses and Other Herbs from the High Plains:* Geol. Soc. America Special Paper 41, pp. 1-176, 1942.

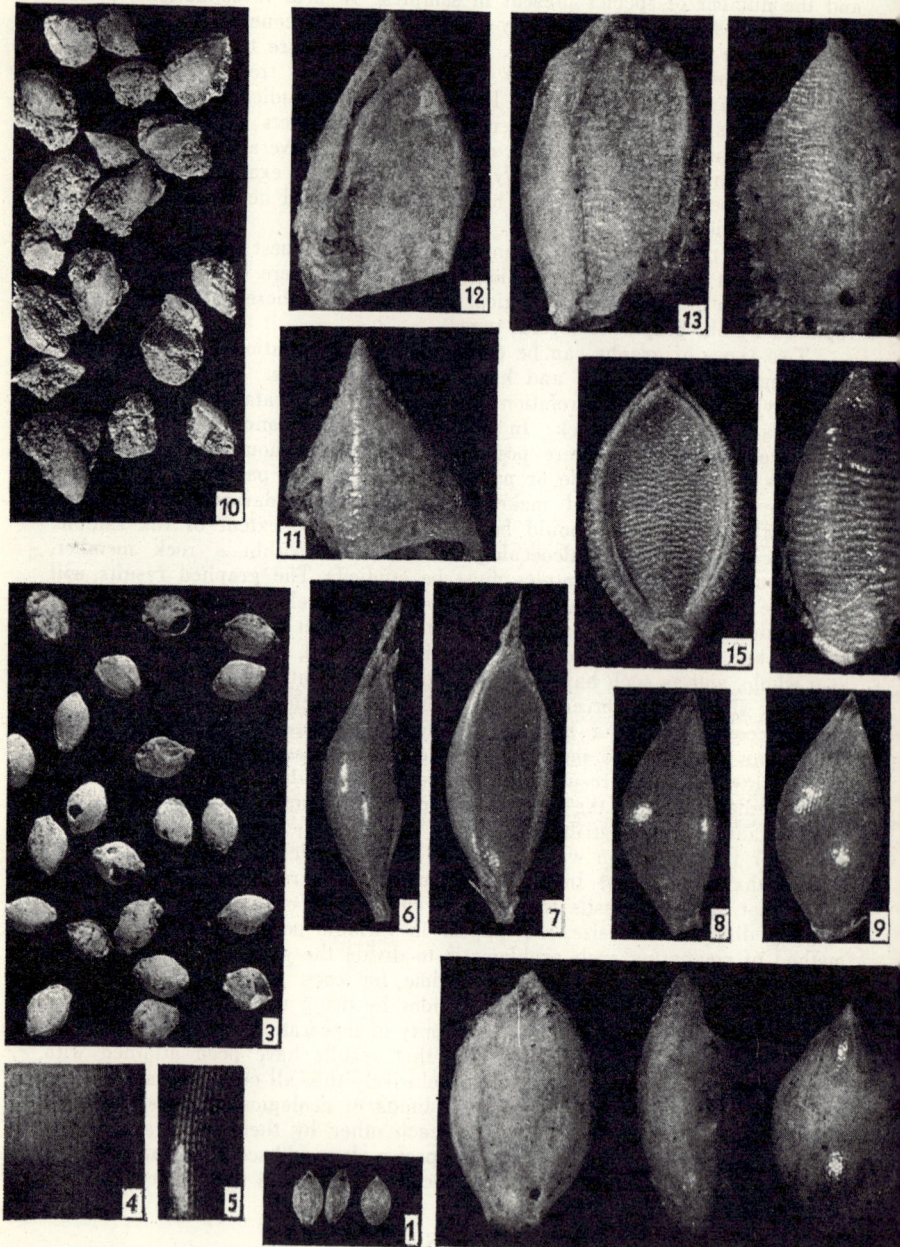

FIGURE 49. Fossil and recent grass seeds. (From Elias, Geol. Soc. America *Special Paper* 1942.) (For explanation see page 113.)

(See fig. 49.) In abstract he says:

The most common among these fossil seeds show close relation to the most typical modern prairie grasses. . . . Comparative study of the fossil and living forms reveals evolutionary trends of the seeds of prairie grasses. The rather small and generalized Miocene ancestor gave rise to greatly diversified Pliocene and Recent species. The seeds of these include small and very large, very slender, and very stout forms, all of them variously adapted for protection against drought and for more efficient dispersal. Abundance, good preservation, and rapid ecologic and evolutionary changes make grass seeds the best index fossils for subdivision of the continental Late Tertiary rocks.

FISH SCALES

Fish remains, especially fish scales, commonly occur in sediments. In the Pacific Coast Tertiary section, according to David,[22] a

. . . great number of fish scales can be used characteristically as index or marker fossils. Others occur in well-marked abundance zones, indicating in that way horizons of definite age and making it possible to correlate quite accurately. . . . Scales of different kinds of fishes show innumerable variations of their ornamentation, of the designs and angles formed by the fingerprintlike impressions that mark the scales. . . . Herrings and round herrings are the most abundant scales through the Tertiary. . . . A number of very distinct scales are of importance for paleoecological determinations. . . . These fossil

EXPLANATION OF FIGURE 49

Magnification

Nos. 1, 3, and 10, × 3; No. 4, × 33; No. 5, × 29; Nos. 6–9, 11–16, × 15.
Fossil *Panicum elegans* nebraskense; *Panicum elegans*; *Setaria chasea*. Living Panicum *angustifolium*; *Setaria chasea*. Living *Panicum angustifolium*; *Setaria crusgalli*; *S. glauca*.

Nos.

1, 2. *Panicum elegans* Elias, early mutation *nebraskense* Elias n. mut.; basal part of *Biorbia fossilia* zone, Ash Hollow formation, Ogallala group; east of railroad bridge, 1½ miles west of Wauneta, Chase County, Nebraska; middle Pliocene.

3, 4. *Panicum elegans* Elias. 3, Autotypes, about seventy feet below the top of the local section of Ogallala, two miles east-northeast of Ogallala, SE¼, sec. 33, T. 14 N., R. 38 W., Keith County, Nebraska. Both from upper part of *Biorbia fossilia* zone. Ash Hollow formation, Ogallala group; middle Pliocene. 4, Holotype, showing epidermis of *palea*, from about forty feet below algal *(Chlorellopsis)* limestone (at top of Ogallala group), sec. 21, T. 14 S., R. 39 W., Wallace County, Kansas.

5. Living *Panicum angustifolia* Elliott; eastern United States.

6–8. Living *Echinochloa crusgalli* (Linné). Beauvois; from Mexico.

10–14. *Chaetochloa chasea* Elias, n. sp.; syntypes; west of United States Department of Agriculture Experiment Station at North Platte, Lincoln County, Nebraska. Upper portion of Ogallala group above *Biorbia fossilia* zone. Collected by H. E. Weakly. 11, showing tripartite apex of the *lemma*; back view. 12, showing frontal view of upper part of hull, *palea* parting from *lemma*; uppermost middle Pliocene or upper Pliocene. 15-16, living *Chaetochloa glauca*; from near Almera, Himalaya Mountains. Note slight parting of *lemma* and *palea* in No. 15; also similar type of sculpture of both *lemma* and *palea* with *Setaria cahsea* Elias.

[22] David, L. R., *How Fossil Fish Remains Have Been Used in Pacific Coast Stratigraphy:* Petroleum Eng., vol. 18, no. 8, pp. 104-113, May 1947.

fishes are highly significant in the study of stratigraphy, sedimentation, and paleontology.

From these comments it is obvious that paleoichthyology has a definite place in stratigraphic and correlation work, and in the future should be considered more seriously by micropaleontologists.

Suggested Micropaleontologic Studies

For more adequate interpretation of micropaleontologic data, a thorough understanding of present-day biotopes and ecologic relationships is required. Integrated detailed investigations of modern faunas and floras would furnish much information that could be of great value to paleontologists. Contributions by Natland,[23] Norton,[24] Phleger,[25] and Cushman [26] in this field are basic. Similar studies should be instigated and liberally supported by the oil companies and educational institutions.

A systematic review of the biologic literature would undoubtedly reveal information that would be of considerable value to the paleontologist.

Detailed examination and recording of microfaunas (Foraminifera, ostracodes, diatoms, Radiolaria, spores, pollen, and the like) in controlled stratigraphic sections should be continued with maximum effort. More study should be devoted to phylogeny, taxonomy, and habitats of recent microfaunas and microfloras.

Pelagic microfaunas and microfloras are scheduled to play an important role in the future in establishing long-range geologic correlations. To evaluate fossil suites more adequately, studies of related recent types should be started. Geographic distribution patterns of these "floaters" in modern seas should be prepared and analyzed in relation to water and wind currents, salinity, turbidity, food supply, temperature, and land barriers.

More data are required on the rate of accumulation of remains in micro-organisms in recent sediments. Detailed investigations of microfossils of the various periods and epochs are critically needed and should be periodically evaluated and published. Total faunas should be illustrated and each species carefully described. Too frequently have incomplete faunas been described in the literature during the past. Published faunas should be considered stratigraphically as well as paleontologically.

Commercial Micropaleontologic Laboratories

The routine of micropaleontologic laboratories depends on the problem involved (surface or subsurface, detailed or reconnaissance, research

---

[23] Natland, M. L., *The Temperature and Depth Distribution of Some Recent and Fossil Foraminifera in the Southern California Region:* Scripps Inst. Oceanography Tech. Ser. Bull., vol. 3, no. 10. pp. 225-230, 1 table, 1933.

[24] Norton, R. D., *Ecologic Relations of Some Foraminifera:* Scripps Inst. Oceanography Tech. Ser. Bull., vol. 2, pp. 331-388, 1930.

[25] Phleger, F. B., Jr., *Foraminifera of Submarine Cores from the Continental Slope:* Geol. Soc. America Bull., vol. 50, pp. 1395-1422, 1939.

[26] Cushman, J. A., *A Study of the Foraminifera Contained in Cores from the Bartlett Deep:* Am. Jour. Sci., vol. 239, pp. 128-142, 1941.

or economic), the personnel, the expenditure permitted, and the work output. Methods applied in a laboratory of one area may vary considerably from those followed in a laboratory of another. Laboratory procedures of various companies may differ widely within the same area. Some companies support only a one-man laboratory; others support laboratories having personnel up to 120 men working in three eight-hour shifts. Operations for a medium-sized laboratory are shown in figure 50.

Routine operations of oil-company laboratories are much the same in many parts of the world in spite of considerable diversity in the nature of the

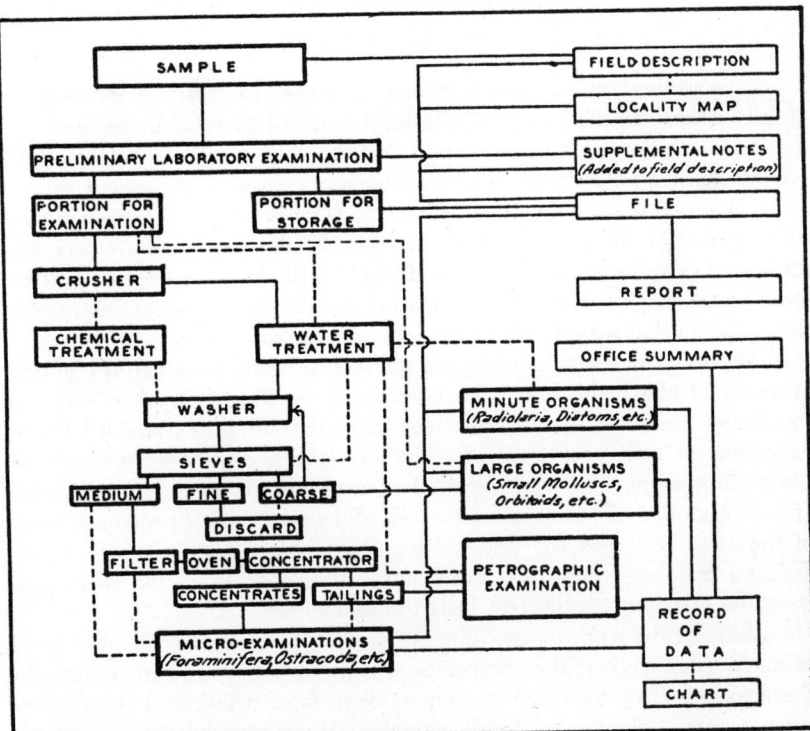

FIGURE 50. Flow chart of operations in a medium-sized micropaleontologic laboratory. (After Driver.)

problems confronting the paleontologists and their differences in viewpoint. Whatever the area and whatever the details of procedure, the operations of these laboratories may be divided into two parts: the preparation of material for examination, and the work of the microscopist in studying and reporting on the material. Systematic orderliness, fine equipment, and special techniques help speedily to achieve the primary objective—the accurate correlation of strata.

The procedure in a commercial laboratory is not automatic, stereotyped, unthinking mass production of data. It is, on the contrary, research by trained persons who are thoroughly familiar not only with their objectives but also with the difficulties that confront them. . . . It is true that oil company micro-paleontologists have invented some useful equipment but most of it is of minor importance. Principally they have adopted inventions made by others; they have modified equipment already manufactured; they have organized material and personnel for systematic, rapid study. In other ways, however, their contributions have been of great importance; by their refined correlations of strata, they have aided in the discovery of oil and in the unraveling of the complicated history of the earth.[27]

DETRITAL MINERALOGY

GORDON RITTENHOUSE

This section outlines briefly the methods used to study the mineralogy of subsurface samples, some of the limitations of these methods, and the uses to which the results may be put. Much has been written on the methods of studying and interpreting sedimentary rocks. Agreement is far from unanimous on either the best methods of study or the meaning of the information after it has been obtained. In this section the writer has attempted to emphasize those methods that he thinks have been or will be most useful in practical subsurface work and to present those interpretations of data that seem to him most logical and reasonable.

The study of the mineralogy of subsurface samples is helpful in one or more of the following: (1) correlating or identifying key beds or producing horizons, (2) determining the extent, thickness, and lithologic variations of beds or groups of beds, (3) selecting the best methods of completing wells, (4) selecting the best methods for secondary recovery, (5) determining the source or sources of various beds or various parts of the same bed, and (6) determining the geologic history as it can be deduced from the composition and physical properties of the rocks and from the relations of different beds to one another.

Much of the work on the mineralogy of subsurface samples has been primarily descriptive: the composition, texture, color, and other obvious characteristics are described in varying degrees of detail, and the rock is given a name. Much exceedingly valuable information has been obtained and will continue to be obtained in this way. One has only to consider the great progress that has been made in subsurface geology during the past three decades to appreciate the value of such descriptive work. Although geophysical logging methods have recently supplanted sample examination in part, subsurface samples will continue to be used extensively in the future.

As stratigraphic traps or combined stratigraphic and structural traps become increasingly important sources of oil and gas, interpretation of subsurface samples in terms of sources of sedimentary materials, the

---

[27] Schenck, H. G., and Adams, B. C., *Operations of Commercial Micropaleontological Laboratories:* Jour. Paleontology, vol. 17, no. 6, pp. 554-583, 1943.

conditions of transportation and deposition, and post-depositional changes may be expected to increase in importance. Sedimentary rocks have many properties from which their past history may be deduced with varying degrees of success. Most rocks are made up of a large number of grains and these grains have the individual properties of size, shape, roundness, surface texture, orientation in space, and a variety of other properties that are dependent on and can be grouped under composition. In the aggregate the particles have mass properties. These include average size and a spread about that average, average shape, roundness, and orientation, and a spread about these averages. Porosity, permeability, color, mud cracks, bedding, ripple marks, and the kind and degree of induration are some of the many other mass properties.

Of these properties, some, like composition, may be inherited from the original source material; some, like shape and roundness, were developed as the particles were transported from the source to their place of deposition; some, such as orientation, size, and bedding, reflect the environment of deposition; and some, like porosity and permeability, at least in part reflect post-depositional changes. It is with the study and interpretation of these individual and mass properties that this section is primarily concerned.

CHOICE OF METHODS

Many methods are used to study the mineralogy of subsurface samples. Some properties of the samples can be determined by megascopic examination, supplemented by simple chemical and physical tests. Other properties can be studied best with a binocular or petrographic microscope. Certain minor constituents may be concentrated as insoluble or heavy-mineral separates before binocular or petrographic examination.

In selecting methods of study careful consideration must be given to the objectives of the investigation and to the type of data required to achieve these objectives. Methods that provide satisfactory information for one problem or one area may be entirely unsuited to another problem or another area. The hazards of unthinking application of techniques and methods cannot be overemphasized.

Naturally the type, number, size, and location of samples that are available are controlling factors in outlining a program of investigation and in choosing methods of analysis. Rotary cuttings, cable-tool cuttings, and cores may yield different types of data. Given certain samples, however, choice of methods depends on the answers to two questions, namely, what methods will provide the data needed to solve the problem under consideration, and which method will yield the information with the minimum expenditure of time and money.

As an example, many correlation problems may be solved by rapid megascopic or binocular examination of a few samples or a series of samples. Where this is possible, further extensive laboratory examination

118 SUBSURFACE GEOLOGIC METHODS

to obtain information about grain size or heavy-mineral content is time wasted insofar as correlation is concerned, although such examinations may yield valuable information on other problems. In general, one should start with the simplest and fastest procedure and try successively more complex and slower procedures until one is found that will provide the needed information. A little forethought and a preliminary examination of a few representative samples may not only save time and money but determine the ultimate success or failure of an investigation.

CHARACTERISTICS OF DIFFERENT TYPES OF SUBSURFACE SAMPLES

Subsurface samples are usually of three types, namely, rotary cuttings, cable-tool cuttings, and cores. The way in which these three types of samples are procured limits the amount of data than can be obtained from them and the interpretation of that data. The brief and perhaps over-simplified explanation that follows outlines some of the major points to be considered in studying these three types of samples.

In rotary drilling the cutting action is provided by the abrasion and downward pressure of a steel bit attached to a hollow drill pipe. Drilling mud is pumped down through the drill pipe and returns to the surface in the space between the drill pipe and the sides of the hole. The drill cuttings carried to the surface in the drilling mud are usually separated from the mud as it passes over a vibrating wire screen or through a shaker. Samples of the accumulated cuttings are taken at intervals. Rotary cuttings have the following characteristics that limit their usefulness:

1. Some rocks, particularly poorly cemented sands or bentonitic shales that break into small fragments or disintegrate into mud, pass through the screen or shaker and are not present or are present in only small amounts in the samples. Sometimes more representative cuttings can be obtained by diverting a part of the mud and cuttings into a container and washing out the drilling mud after the cuttings have settled.

2. Because rotary wells usually are not cased until drilling is completed, cavings from above may form a very large porportion of the sample, especially when mud consistency has not been carefully controlled.

3. Large and small cuttings, or cuttings of different specific gravity, tend to be carried upward at different rates, and as a result the cuttings from different horizons are mixed.

4. Owing to the time required for the cuttings to be carried from the bottom of the hole to the surface, the samples come from a somewhat higher horizon than is being drilled at the time they are collected. A depth correction may or may not be made at the well. A rule-of-thumb correction of ten feet per 1,000 feet of depth is commonly used.

5. The drill cuttings are usually less than three-eighths inch in diameter. Hills,[28] whose recent paper on examination of subsurface samples

---

[28] Hills, J. M., *Sampling and Examination of Well Cuttings:* Am. Assoc. Petroleum Geologists Bull., vol. 33, no. 1, pp. 73-92, Jan. 1949. (See pp. 344-364 of this Symposium).

gives more information than is possible here, notes that much larger cuttings can be obtained by reverse circulation.

In cable-tool drilling, a chisel-shaped bit attached to a cable is alternately raised and dropped. After drilling five feet or so, the bit is removed from the hole and the accumulated cuttings are bailed out of the hole with an elongated bucket having a valve at the lower end. The bailer is emptied into a bucket or barrel and the cuttings are washed free of mud. The hole is cased past places where excessive caving occurs or where excessive water enters the hole. Cable-tool cuttings have the following characteristics that limit their usefulness:

1. Some contamination of the cuttings from the uncased parts of the hole occurs. Usually some material from five to ten feet above is included because the hole is widened as drilling proceeds. In general, cable-tool cuttings are less contaminated and more representative of the interval drilled than are rotary cuttings.

2. Stretching of the drilling cable results in inaccuracies in depth measurements up to twenty feet or more. Steel-line measurements at intervals provide a method of correction.

3. The cuttings are usually less than three-eighths inch in diameter.

4. Poorly cemented pay sands may be blown out of the hole by escaping gas, and consequently samples, if any, from such horizons may not be representative.

5. Electric logs cannot be made on cased holes and therefore are not available to supplement the sample studies.

6. In some areas cable-tool drillers are not so careful in collecting and washing samples as are rotary drillers.

Cores may be collected with rotary, cable-tool, or core-drilling apparatus. The chief disadvantages of coring are (1) high cost, (2) selective loss of weaker or more soluble rocks, and (3) storage and transportation of the cores. Improvements in coring methods are reducing costs and increasing core recovery.

MEGASCOPIC AND BINOCULAR EXAMINATION

To a certain extent, the old saying "the closer you look, the less you see" is applicable to some properties of sedimentary rocks. The color, texture, and composition of cuttings and cores and the bedding, cross-bedding, and porosity in cores can often be determined better and faster by megascopic examination and simple chemical and physical tests than in other ways.

By placing ten to twenty cable-tool or rotary samples on sheets of paper or in cardboard or metal trays and observing their appearance megascopically, major changes in color, texture, and composition are readily apparent if the samples have been washed clean of drilling mud. The boundaries of beds or formations that differ markedly from those above or below often can be determined quite accurately. After the major breaks have been picked, attention should be directed to the character of the sediments in each unit and to the nature of the transition between units. Some of these characteristics can be determined by megascopic examination; others can be determined better with the binocular microscope. The following discussion applies to both megascopic and binocular examination of samples.

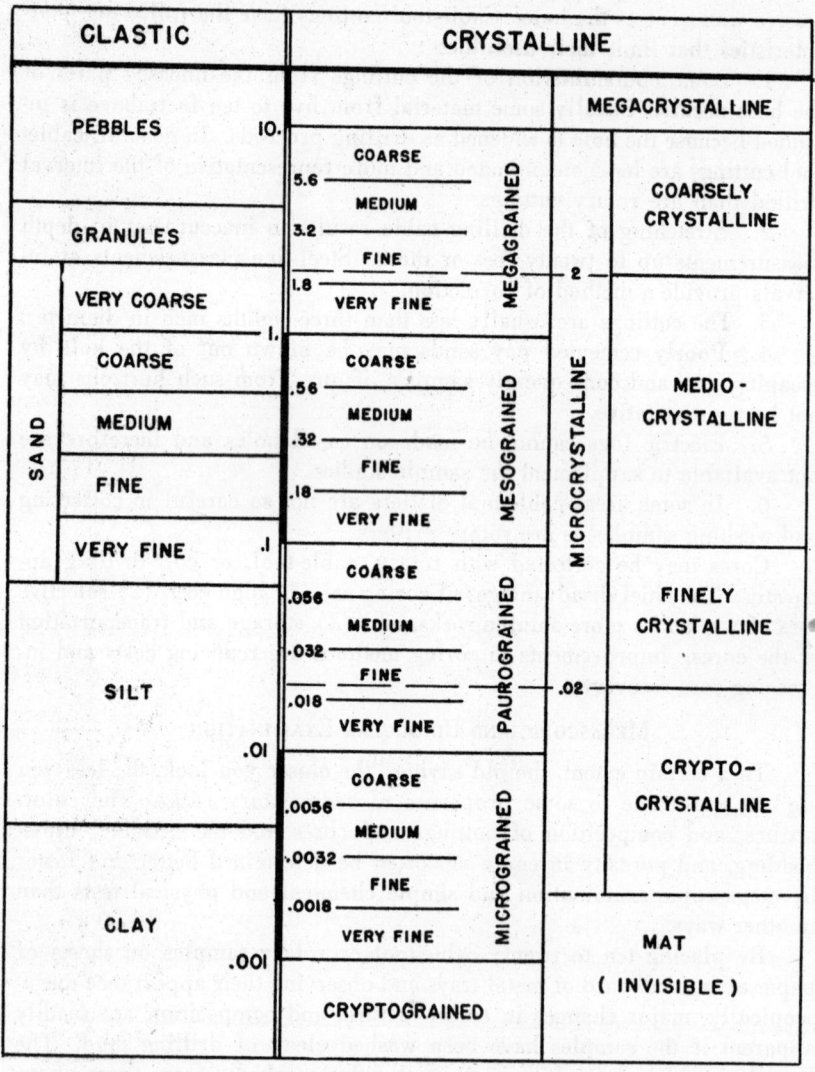

FIGURE 51. Suggested grade scales for clastic and crystalline sedimentary rocks.

Color

In the past color description of sedimentary rocks has been a more or less "every man for himself" proposition. Not only have different men described the same rock differently, but the same man often cannot duplicate his own descriptions. Recently, DeFord [29] discussed previous work and made suggestions that have been partly responsible for the preparation and publication of a "Rock Color Chart." [30] Although the writer has not yet had the opportunity of using this chart, it or one of the other color standards recommended by DeFord should be used in describing color.

Grain Size

Most sediments are composed of clastic particles that have been transported to their place of deposition, of crystals that have grown *in situ*, or a combination of clastic particles and crystals. Rocks containing fossils that have not been transported might be considered as another group. No size terms for either clastic or crystalline sedimentary rocks have been generally accepted for subsurface work. Thus, it is important, whatever method of size classification is adopted, to indicate what that scale of measurement is.

For clastic rocks, exclusive of clastic carbonates, the terms "conglomerate," "sand" (sandstone), "silt" (siltstone), "clay" (claystone), and "shale" are widely used, but different workers define and use them differently. The Wentworth grade scale, shown in figure 51, has probably been used most widely, and has the advantage of separating sand and silt at the one-sixteenth-millimeter size. Above this size the individual grains are clearly visible to the naked eye; below this size they are indistinct or invisible. This size also is approximately the point of separation between wash-load and bed material in some present-day streams and may have rather widespread genetic significance in ancient deposits. One-sixteenth millimeter also is approximately the size that separates the productive from nonproductive sands in some oil and gas fields.

For most subsurface work, comparison with a standard set of samples or sieve separates mounted in glass vials or on microscope slides (fig. 52) will permit adequate description of size. The description should indicate the average size of grain and the spread or "sorting" about that average. In combinations of sand with pebbles, silt, or clay, the main constituent should determine the rock name, and the minor constituent should be used as a modifier, i.e., silty fine sand, fine sandy silt, etc.

The size of crystalline sediments is commonly recorded as "coarse," "medium," "fine," "very fine," "cryptocrystalline," or "lithographic," but

[29] DeFord, R. K., *Rock Colors* (review): Am. Assoc. Petroleum Geologists Bull., vol. 28, no. 1 pp. 128-137, Jan. 1944.

[30] Goddard, E. N., Chairman Rock-Color Chart Comm., *Rock Color Chart:* U. S. Geol. Survey Spec. Pub., 1948. (For sale by Nat. Research Council, Div. of Geology and Geography, 2101 Constitution Ave., Washington 25, D. C., $5.50.)

no general agreement exists as to what sizes these terms include. Recently the tentative size scales shown in figure 51 have been proposed by De Ford [31] and Hills [32] (West Texas Geological Society) for carbonate rocks. These are recommended for consideration and trial.

For rocks that are combinations of clastic grains and crystals, the dominant constituents should determine the main name, and the minor constituent should determine the modifier. To each, the size terms in figure 51 may be applied. A fine sandy, medium-paurograined dolomite would be a dolomite with crystals between .056 and .032 mm. in diameter that contained less than 50 percent of fine sand. In those combinations in which this terminology may be cumbersome, especially when the composition of the sand is indicated, the modifier may be added as a qualifying phrase. In those cases where more accurate designation of the proportion of the constituents is desirable, the estimated percentages of them may be indicated.

As Hills has noted, experience has shown that full descriptions of subsurface samples have saved much time and money, whereas meager descriptions have necessitated one or more re-examinations of the cuttings. Hills also recommends that wildcat wells and pay sections be described in more detail than field wells or long sections of comparatively insignificant beds between key beds.

Shape and Roundness

Shape (sphericity) and roundness of sedimentary grains are independent properties and should not be confused. Sphericity is a measure of the approach to spherical form and may be expressed roughly as a ratio between the diameters of the grains. In contrast, roundness is a measure of the angularity of the corners and edges of the grains.

In most subsurface samples the particles or crystals are monomineral, and their shape is largely dependent on their composition. Therefore, separate-shape terms usually are not necessary when the mineral composition and texture are indicated. For rock fragments, the shape terms suggested by Krynine,[33] as slightly modified by the writer, are indicated. The methods used for more detailed measurements of shape and roundness than ordinarily needed are outlined elsewhere in this section.

Equant—Length of grain is less than $1\frac{1}{2}$ times its width and thickness.

Prismatic—Length of grain is $1\frac{1}{2}$ to 3 times its width and thickness.

Tabular—Length and width of grain are $1\frac{1}{2}$ to 3 times its thickness.

Acicular—Length of grain is more than 3 times its width and thickness.

Platy—Length and width of grain are more than 3 times its thickness.

[31] DeFord, R. K., op. cit.
[32] Hills, J. M., op. cit.
[33] Krynine, P. D., The Megascopic Study and Field Classification of Sedimentary Rocks: Jour. Geology, vol. 56, no. 1, pp. 130-165, Jan. 1948.

Three or more of the terms, "well-rounded," "rounded," "sub-rounded," "subangular," and "angular," are generally used to describe roundness or angularity in megascopic and binocular examinations. There has been no general agreement as to the meanings of these terms. For

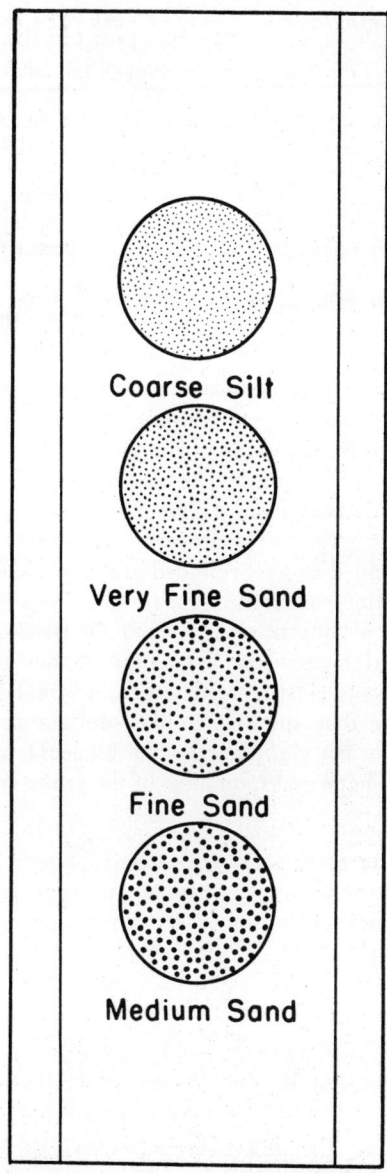

FIGURE 52. Textural standard for sample work. Dots do not represent true grade size.

most megascopic and binocular examinations of subsurface samples, the following definitions as proposed by Russel and Taylor [34] are recommended:

Angular—Showing very little or no evidence of wear. Edges and corners are sharp.

Subangular—Showing definite effects of wear. The grains still have their original form, and the faces are practically untouched, but the edges and corners have been rounded off to some extent though the angles between the faces may still be sharp.

Subrounded—Showing considerable wear. The edges and corners are rounded off to smooth curves, and the area of the original faces is considerably reduced, but the original shape of the grain is still distinct.

Rounded—Original faces almost completely destroyed, but some comparatively flat faces may be present. There may be broad re-entrant angles between remnant faces. All original edges and corners have been smoothed off to rather broad curves.

Well-rounded—No original faces, edges, or corners remain. The entire surface consists of broad curves; flat areas are absent. However, the original shape of the grain may be suggested by its present form.

It is particularly important to note whether the sedimentary particles have about the same rounding, or whether there is a mixture of angular and well-rounded grains. "Mixed" rounding of the grains of a sand generally indicates that the grains have not had the same past history and may have been derived from different sources. Because rounding of quartz sand grains is a very slow process, particularly for grains of fine or very fine size, rounded or well-rounded grains probably indicate derivation from a pre-existing sediment.

The deposition of quartz cement in a quartzose sandstone usually modifies the original shape and roundness of the grains. Shape and roundness measurements must be made on the original grains to be significant. Any sandstone that sparkles in the sunlight or is seen to reflect light from flat faces when viewed under the binocular microscope usually has had the original shape and roundness of the grains modified by cement.

Surface Texture

The surface textures of sand grains may be seen megascopically or inferred from the general appearance of the rock but usually can be seen more clearly through a binocular microscope. A classification of surface textures prepared by Williams [35] and recommended here is as follows:

Luster (grains also may be smooth or rough).
  A. Dull
  B. Polished
Relief (grains may be dull or polished).
  A. Smooth

---

[34] Russell, R. D., and Taylor, R. E., *Roundness and Shape of Mississippi River Sands:* Jour. Geology vol. 45, no. 2, pp. 225-267, Apr.-May 1937.
[35] Williams, Lou, *Classification and Selected Bibliography of the Surface Textures of Sedimentary Fragments:* Nat. Research Council Comm. on Sedimentation Rept. 1936-1937, pp. 114-128, 1937.

B. Rough
    1. Striated
    2. Faceted
    3. Frosted
    4. Etched
    5. Pitted

One must be careful in interpreting the meaning of surface textures. If transportation has not been sufficiently long or rigorous, a pre-existing surface texture may be modified only slightly and consequently may be an inheritance from a pre-existing rock. Also the original surface texture of detrital grains may be modified either by solution or by the addition of cement of the same composition after deposition. The presence in a sand of a wide range of surface textures probably means a mixed source for the sand.

Orientation.

Although it is an important property, the arrangement of sedimentary particles in space does not lend itself to megascopic or binocular examination or description in most subsurface work. The small size of clay and silt particles, the small size and lack of known orientation of cuttings, and the tendency of many sands to break down to individual grains make orientation study of rotary and cable-tool cuttings difficult or impossible. In oriented cores that contain pebbles or coarse sand the orientation of the particles may give clues concerning the direction of flow of the depositing currents. The imbricate or shingled arrangement of flattened pebbles shows the direction of current movement.

Dapples and Rominger [36] have shown that the long axes of quartz sand grains tend to parallel the direction of flow in streams, and that the largest ends of the grains are toward the current. Similar measurements should be possible along the bedding planes of some cores. Particular attention should be given to elongated grains. The orientation of at least 25 randomly selected grains should be determined.

Recent work on the directional permeability in sands [37] suggests that primary orientation may be important in primary production of oil, in secondary-recovery operations, and in working out the paleogeography and geologic history of some sands. Much basic work on deposits of known environment is needed. For the finer sands and silts and for the limestones, dolomites, and other crystalline sediments petrofabric analysis will be needed.

Composition

Although more than 100 minerals have been identified in sedimentary rocks, only about 20 minerals or families of minerals commonly are

[36] Dapples, E. C., and Rominger, J. F., *Orientation Analysis of Fine-Grained Clastic Sediments: A Report of Progress:* Jour. Geology, vol. 53, no. 4, pp. 246-261, July 1945.
[37] Johnson, W. E., and Hughes, R. V., *Directional Permeability Measurements and Their Significance:* Producers Monthly, vol. 13, no. 1, pp. 17-25, Nov. 1948.

present in quantities that exceed one percent of the rock. If the distinguishing properties of these minerals are learned, the mineral composition of most subsurface samples can be determined by megascopic or binocular examination.

Ordinarily the grains in coarse clastic rocks (sands and siltstones) will be composed of one or more of the following: quartz, detrital chert, feldspar, mica (usually muscovite), calcite, dolomite, glauconite, and collophane. Fragments of sedimentary, igneous, and metamorphic rocks also are important constituents of some clastic sediments. The crystalline sedimentary rocks may contain calcite, dolomite, anhydrite, gypsum, halite, and chert. The chief cements in clastic and crystalline sedimentary rocks are quartz, chert, calcite, and dolomite. Less frequently the cements may be pyrite, hematite, limonite, and opal. The dominant constituents of some mudstones and shales are the kaolin, illite, and montmorillonite (bentonite) groups of clay minerals and other fine-grained, platy minerals like sericite and chlorite.

Probably the best method for a beginner to learn these minerals is to obtain and study a series of samples in which they are present, preferably a series that has been worked previously by a good mineral man. The writer knows of no table that is designed primarily for identification of the common minerals as they are observed in cuttings. Observations ordinarily may be made on some or all of the following:

Color—Not very diagnostic because some minerals may occur in several colors, and color may be given by a small amount of cement or matrix. The green color of glauconite, however, is distinctive.

Hardness—Not exceptionally useful because of the small size of the cuttings. Hardness may be determined with a probe under the binocular microscope.

Grain size—Very useful.

Grain shape—Useful.

Reaction in acid—Very useful, particularly under the binocular microscope. Effervescence in cold dilute hydrochloric acid separates calcite from other carbonates. Effervescence of powder in cold dilute hydrochloric acid or of fragments in hot dilute hydrochloric acid separates other carbonates. Gypsum and anhydrite are slowly soluble in hot dilute hydrochloric acid but do not effervesce. Etching for about 30 seconds in cold dilute hydrochloric acid often brings out structures and textures. Etching gives dolomite a flat-gray appearance that is distinctive and often makes the rhombic structure visible.

Sparkle—Usually indicates (a) quartz cement, (b) dolomite, or (c) mica. Mica reflections will be only from the bedding plane.

Drilling appearance—Usually more important in indicating permeability or in differentiating beds or formations than in identifying individual minerals.

Shaly parting—Useful.

Slaking and swelling in water—Useful in differentiating clay-mineral groups. Kaolins ordinarily do not slake. Very marked slaking and swelling differentiate some bentonites from other bentonites and illites.

Smell—Earthy smell suggests kaolin; no smell suggests illites and montmorillonites.

In cores the identification of minerals is the same as in other hand specimens. It is possible, however, to bring out structures and textures

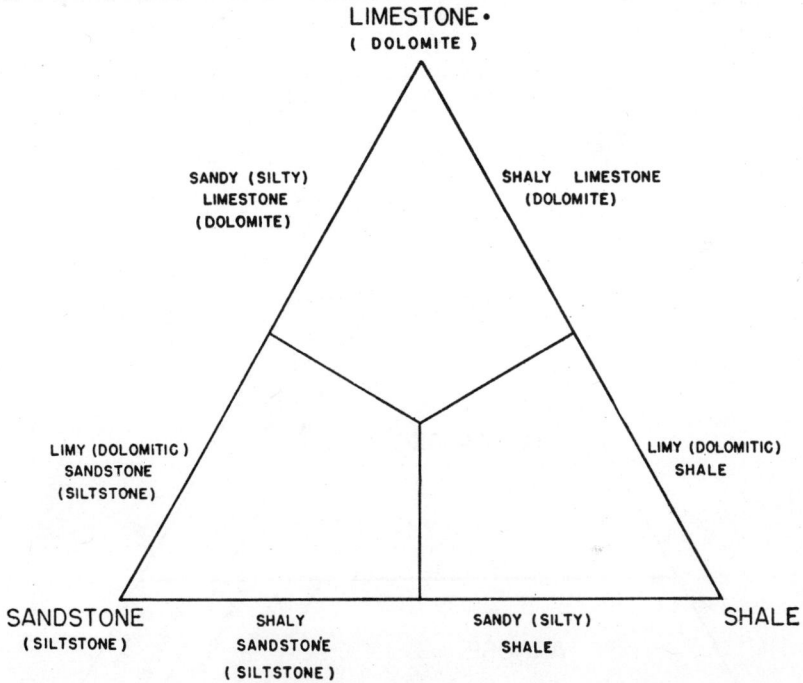

FIGURE 53. Classification of the common-sedimentary rocks. (Modified from Pirsson and Schuchert.)

of cores by preparing a polished section that can be examined mega-scopically or when wet with a binocular microscope. By using a carborundum stone or carborundum powder on a glass plate, one can quickly prepare such polished sections.

Texture

Texture which may be defined as the intimate grain-to-grain relationships of a rock, represents the sum total of such properties as grain-size distribution, grain shape and roundness, fabric, pore shape, and cementation, as distinguished from mineralogic composition. Composition, in so far as it controls or partly controls the other properties, may be considered a factor in texture.

Texture may be indicated in part by the rock name and in part by modifiers of the rock name. Thus the term "breccia" implies angularity of the component grains, whereas the term "sand" has no roundness implication but may be modified by such terms as "angular" or "subangular."

Structure

Structure is somewhat similar to texture in meaning but is applied to such large-scale characteristics of the rock as bedding, cross-bedding,

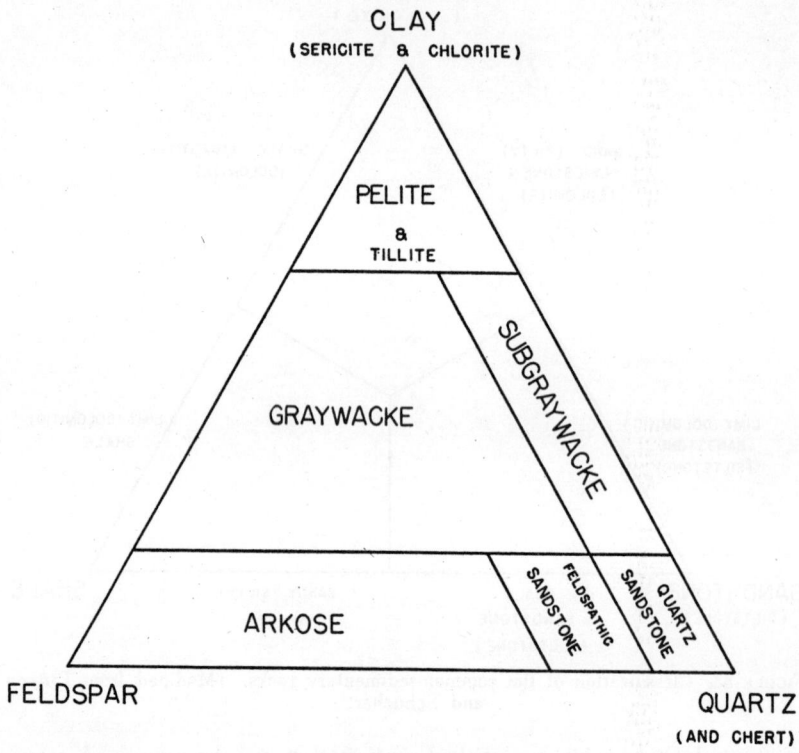

FIGURE 54. Classification of sandstones. (Modified from Pettijohn.)

jointing, and folding, which ordinarily are seen best in outcrops but may be observed in some cores or hand specimens. Except for lamination, structures seldom can be identified in drill cuttings.

Rock Types

No standard classification of rocks has been generally accepted for megascopic and binocular subsurface work. A relatively simple system that meets the needs of most subsurface work and follows the general usage of many subsurface men is that shown in figure 53. Subdivision of the sandy and silty groups on the basis of mineralogic composition is desirable in light of the recent trend that attempts to tie up tectonics and sedimentation. Figure 54, modified from Pettijohn,[38] is a mineralogic classification of sands and silts that merits consideration by subsurface workers. Clastic limestones, that is, those in which the grains have been transported to their place of deposition, should be distinguished from limestones in which no clastic texture can be recognized.

To the main rock name, modifiers are added to permit full description of the sample. As suggested by Krynine,[39] it is possible to apply to sediments the basic standardized descriptive sequence that has been used for many years for igneous rocks, namely, color, subtexture, varietal minerals and cement, and finally the main rock name. If necessary, terms describing structure may follow color. Thus one could have a "gray, cross-bedded, fine-grained, glauconitic, dolomitic, quartz sandstone." For ease in plotting sample data and in picking out changes in lithology, it is convenient to capitalize the main rock name and place it first, and to abbreviate as much as possible, as "QTZ SS, gy, x-b, f-gr., glauc., dol.," for the example above. In commercial work it is not possible to give as complete descriptions as might be desirable in purely scientific research, but the description should be made as complete as possible in the time available.

Porosity and Permeability

Porosity is the percentage of total volume of a rock not occupied by mineral components. Pores may differ in size and may be connected or isolated. In magascopic or binocular examination, porosity usually can be seen best in clean, dry samples.

Permeability is the fluid-transmitting capacity of a porous material. Permeability is not necessarily a function of porosity. Clays, for example, may be very porous but relatively impermeable. In megascopic or binocular examination a rough measure of permeability may be made by observing how rapidly a drop of water will soak into a dry fragment. The presence of individual sand grains or oölites, rather than clusters of such grains, suggests slight cementation and consequently high porosity and permeability.

Subsurface samples may be examined wet or dry. Both methods have advantages and disadvantages and when time permits a combination of both wet and dry study is advisable. Study of dry samples is faster and permits better observation of gross texture, porosity, permeability, sparkling, and some color differentiations. Study of wet samples is advantageous when the samples are not clean. Some textures and colors also can be seen better in wet samples than in dry samples. Calcareous samples that have been etched with dilute hydrochloric acid are usually studied when wet. It is important to indicate whether color has been

---

[38] Pettijohn, F. J., *Sedimentary Rocks*, p. 526, New York, Harper & Brothers, 1949.
[39] Krynine, P. D., *op. cit.*

determined from a wet or a dry sample. For binocular examination of both wet and dry samples magnifications of 12 to 24 diameters are commonly used. Higher magnification may be used for special purposes. For most low-power study, flourescent is superior to incandescent lighting. The percentage of various constituents is usually estimated.

Hills [40] notes that there are two principal ways of describing samples:

. . . The first of these is the interpretative system, in which the geologist picks out the cuttings which he believes to be representative of the formations penetrated and describes the entire sample as composed of this rock The rest of the sample is assumed to be cavings. This kind of description brings out formational changes and is of greatest value in areas where the various formations are of wide extent and relatively constant character, as in the Paleozoic of the Midcontinent region. In areas of rapid lateral gradation in the lithologic character of formations, as in the Permian Basin of west Texas, this method results in masking of lateral variations and misinterpretations of the nature of the stratigraphic column. This, of course, results in miscorrelation of the well logs.

In regions of pronounced lateral gradation, it has been found that a second method of sample description is most satisfactory. This is the percentage description, where the geologist describes all material in the sample, disregarding obvious foreign substances and cavings. This system, though making it difficult to determine formational boundaries from the sample log, shows the gradations of the beds and often enables one to trace a horizon through different sedimentary facies.

Hills also discusses many other points on sample examination that could not be covered in this section.

Converted Binocular Microscope

Recently the writer has used two polaroid plates to convert a binocular microscope into a low-power polarizing microscope. The maximum magnification that can be obtained is low, being essentially the same as the low power of a petrographic microscope. This permits study of sand but not of silt and clay. The writer first tried the conversion by mounting sand grains in clove oil on one lens of a pair of clip-on polaroid sunglasses, placing the other lens above and at right angles to the first, and fastening the two lenses in place with drafting tape. Light was transmitted from below the glass stage of the microscope, and the grains were moved by pressing on the upper lens. In fine and medium sands, quartz grains, which are single crystals, were differentiated from chert grains and rock fragments, which are composed of many smaller crystals. Microcline and plagioclase feldspars could be recognized by their twinning. The high birefringence and relief of carbonates separated them from quartz, feldspar, and chert. It should be possible to differentiate anhydrite, which has strong birefringence (0.044) and an index of refraction higher than clove oil, from gypsum, which has weak birefringence (0.010 —quartz is 0.009) and an index of refraction lower than clove oil.

---

[40] Hills, J. M., *op. cit.*

The color of the sunglasses was a disadvantage. It should be possible to obtain two nearly colorless polaroid discs. Place one on the microscopic stage, mount the grains in oil on a glass microscope slide above it, and mount the second polaroid disc on a cardboard support that could be slipped in above the slide as needed. With this arrangement, however, rotation of the grains might be difficult.

HEAVY MINERALS

General

Heavy minerals are minerals of high specific gravity (2.86 to 2.96), which occur in minor amounts in all sands and sandy limestones. Even though present in small amounts, such heavy minerals as tourmaline, zircon, hornblende, and staurolite may be exceedingly useful in correlating sands, outlining petrographic provinces, indicating sources and past history of the source material and helping to decipher geologic history. To facilitate their examination, heavy minerals are separated from the quartz and other light minerals with which they are associated.

The study of heavy minerals has the following disadvantages: (1) The preparation and study of the samples are time-consuming, and consequently the costs are relatively high. (2) The ability to use a petrographic microscope is necessary. (3) An understanding of principles that govern the size distribution and post-depositional modification of heavies is necessary for the correct interpretation of the results. A lack of such understanding has been partly responsible for the present low opinion of heavy-mineral studies in the oil industry.

Preparation of Sample

The heavy minerals of the entire sample or of one or more sieve separates may be studied. If the sieve separates are used, particular care should be given in cleaning the sieves, so that contamination will be kept to a minimum. When marked differences in amounts or kinds of heavy minerals occur in different samples, it may be well to sieve and discard a preliminary sample. In carbonate rocks or in sandstones with carbonate particles or carbonate cement, the carbonate should be removed prior to sieving by boiling in dilute hydrochloric acid.

Indurated rocks must be broken down to free the heavy-mineral grains. A test sample of 5 to 100 grams may be soaked in water overnight and then, after the excess water has been removed by siphoning or decantation, the sample is boiled for ten minutes in 3 to 4 normal hydrochloric acid to remove carbonates and iron oxides. The sample is transferred to a 1,000-cc. beaker, and water is added to fill the beaker. After being stirred, the mixture is allowed to settle one minute, and the upper 800 cc. of water and sediment suspension is siphoned off. This washing process is repeated until the water is clear after the one-minute settling period. This procedure removes the fine silt and clay (less than about 0.01 mm.).

If aggregates of particles still persist, the sample may be placed in the sieves and shaken by hand for about one minute. The sediment in each sieve is removed to a square of heavy brown paper, and the aggregates are broken down by rolling the end of an iron pestle gently over the material. At intervals the sand is resieved by hand to remove disaggregated particles. The completeness of disaggregation is checked by examination with a hand lens or binocular miscroscope. This method of disaggregation causes little breakage of heavy minerals, even if the sandstone is very quartzitic. The heavy-mineral grains, being of different composition, come loose like peas from a pod. The sample is recomposed and sieved, the procedures outlined in the discussion of that subject in this section being followed.

Acetylene tetrabromide of specific gravity, about 2.93, or bromoform of specify gravity, about 2.86, is placed in a glass funnel, to the stem of which is attached a piece of rubber tube closed by a pinch clamp. The sand sample or sieve separate is introduced, and the mixture is stirred at intervals until the heavy minerals have settled into the stem of the funnel. The pinch clamp is opened, and the heavy fraction is washed onto a filter paper in a second glass funnel. After the excess heavy liquid has been filtered into a receptacle and returned to the stock bottle, the filter paper is washed several times with alcohol. The light minerals and the remaining heavy liquid are then drained onto another filter paper, the heavy liquid is filtered off, and the light minerals are washed with alcohol. Both heavy and light minerals are dried in an oven at 205° F. The alcohol-heavy liquid washings are saved for recovery of the heavy liquid.

When a large number of heavy-mineral separations are to be made, mass-production methods can be used. The writer has employed two batteries of six separation units each, and one assistant could make as many as 24 separations a day. Most of these were of sand of one-eighth- to one-sixteenth-millimeter size, requiring about two hours of alternate stirring and settling to effect a satisfactory separation. For coarser sands that settle more rapidly, more separations a day would be possible.

Acetylene tetrabromide and bromoform are the most commonly used heavy liquids. The writer prefers tetrabromide because its greater specific gravity reduces the number of altered grains, rock fragments, carbonates, micas, and chlorites in the heavy-mineral fraction. The acetylene tetrabromide or bromoform is reclaimed from the alcohol washings by shaking it with an excess of water and decanting the alcohol-water mixture. This process is repeated several times, and the acetylene tetrabromide clarified by filtering.

Attention to two apparently minor details will save much time in making heavy-mineral separations. When the heavy minerals settle, most of them come to rest on the sloping sides of the funnels. If the suspension is stirred vigorously, most of these heavy minerals go into suspension and settle again on the funnel slopes. Consequently the suspension should be stirred gently in such a way that the heavy minerals are worked down the slope of the funnel and into its stem. Then a vigorous stirring will free additional heavy minerals that are trapped with the light minerals at the top of the funnel.

Selection of a proper filter paper is also important. The most porous paper that permits the heavy minerals to be caught and recovered should be used. The writer uses Whatman No. 4. A less porous paper, which filters more slowly, may double the time required for a heavy-mineral separation.

The writer weighs the heavy-mineral fractions of most samples to the nearest half-milligram on an analytic balance. This weighing permits computation of the hydraulic ratio [41] if that is needed. For most routine work, weighing the heavy-mineral separates will not be necessary.

The heavy-mineral fractions are split to 1,000 to 1,500 grains with an Otto miscrosplit [42] and mounted in Canada balsam on 1- by 2-inch glass microscope slides. Often more than one mount is needed, especially when much pyrite or barite is present in the heavy-mineral fraction. The sample number, size grade, and slide number are scratched on each slide with a diamond pencil. This provides a permanent mount, to which reference can be made at a later date. For easier identification, splits of the heavy-mineral fraction may be mounted in oils of various refractive indices.

Mineral Identification

Heavy minerals are usually studied with a petrographic microscope, using medium power (5 × ocular and an 8-mm. objective) for most work. The properties most useful in rapid identification of the mineral grains are opaqueness or translucence, appearance in reflected light (particularly for opaque minerals, rutile, and tourmaline), color, relief as compared to the mounting medium, pleochroism, inclusions and alteration products, crystal form, grain shape, cleavage, birefringence, extinction angle, and isotropic or anisotropic character. Milner's [43] book probably is the best for mineral identification. Krumbein and Pettijohn [44] and Russell [45] also have tables that may be used. These identification tables list many properties of the minerals, but those given above are most useful.

If a large number of samples is to be studied, time and effort can be saved by mounting the heavy-mineral suites in oils of various indices of refraction and identifying all species that are present. Ordinarily minerals that are very rare in the heavy-mineral separate may be neglected, as their presence or absence in any amount is a matter of chance. Then simple criteria can be set up by which each mineral can be recognized at sight or

---

[41] Rittenhouse, Gordon, *Transportation and Deposition of Heavy Minerals:* Geol. Soc. America Bull., vol. 54, no. 12, pp. 1725-1780, Dec. 1, 1943.

[42] Otto, G. H., *Comparative Tests of Several Methods of Sampling Heavy Mineral Concentrates:* Jour. Sedimentary Petrology, vol. 3, no. 1, pp. 30-39, April 1933.

[43] Milner, H. B., *Sedimentary Petrography*, 3d ed., 666 pp., London, Thomas Murby & Co., 1940.

[44] Krumbein, W. C., and Pettijohn, F. J., *Manual of Sedimentary Petrography*, 549 pp., New York, D. Appleton-Century Co., 1938.

[45] Russell, R. D., *Tables for the Determination of Detrital Minerals:* Nat. Research Council Comm. on Sedimentation Rept., 1940-1941, pp. 6-8, 1942. (Separate copies of tables, 50 cents.)

by one or two rapid microscopic tests. With these criteria 90 percent or more of most mineral suites can be identified by observation in plane-polarized light. The speed of counting the grains on a slide may be doubled or trebled and the eyestrain greatly reduced, if an assistant records the grain counts as they are made by the observer.

Use of Heavy-Mineral Data

The type of heavy-mineral data that should be obtained depends primarily on the objectives of the investigation, the samples that are available, and the heavy-mineral content of the samples. The heavy-mineral distribution in sandstones is closely related to the size distribution of the light minerals in the sample and consequently the proportion of heavy minerals, and sometimes their presence or absence in a particular sample depends on the grain size of the sample. Some heavy minerals are removed by solution and others are deposited after deposition. These factors must be considered in determining the type of data to be obtained and how the data should be interpreted.

Ordinarily, if different kinds of heavy minerals occur in two samples, for example a hornblende-epidote-ilmenite suite in one and a staurolite-kyanite-magnetite suite in another, the samples reflect different sources for the two samples. If the same kind of heavy minerals are present in two samples but they are present in different proportions, different sources are not indicated unless all of the minerals are of about the same specific gravity. Major differences in ratios between varieties of the same mineral or between minerals of about the same specific gravity usually indicate different sources. Even in the last case, authigenic minerals must not be used, and the possibility that some minerals may be removed by solution must be given consideration. Pettijohn[46] and Dryden and Dryden [47] give the order of stability of minerals in sediments (both chemically and mechanically weathered) and weathered rock respectively as follows:

	Pettijohn	Dryden and Dryden	
Most persistent			
Anatase (authigenic)	Epidote	Zircon	100
Muscovite	Hornblende	Tourmaline	80?
Rutile	Andalusite	Sillimanite	40
Zircon	Topaz	Monazite	40
Tourmaline	Sphene	Chloritoid	20?
Monazite	Zoisite	Kyanite	7
Garnet	Augite	Hornblende	5
Biotite	Sillimanite	Staurolite	3
Apatite	Hypersthene	Garnet	1
Ilmenite	Diopside	Hypersthene	1–
Magnetite	Actinolite		
Staurolite	Olivine		
Kyanite	Least persistent		

---

[46] Pettijohn, F. J., *Persistence of Heavy Minerals and Geologic Age:* Jour. Geology, vol. 49, no. 6, pp. 610-625, Aug.-Sept. 1941.
[47] Dryden, Lincoln, and Dryden, Clarissa, *Comparative Rates of Weathering of Some Common Heavy Minerals:* Jour. Sedimentary Petrography, vol. 16, pp. 91-96, 1946.

Large apparent differences in heavy-mineral composition may be due to the selective removal of certain minerals. In the example below removal of the less stable minerals would leave a very different mineral suite.

	Before removal (percent)	After removal (percent)
Hornblende	20	0
Garnet	30	0
Staurolite	25	0
Tourmaline	15	60
Zircon	5	20
Rutile	5	20

The types of minerals present in a sediment are indicative of the sources from which they were derived. Andalusite, kyanite, staurolite, and sillimanite probably indicate a metamorphic source. Ilmenite, zircon, rutile, apatite, olivine, titanite, and some varieties of tourmaline probably indicate an igneous source.. Any heavy minerals that are very well rounded suggest a sedimentary source.

Errors in Heavy-Mineral Analysis

Heavy-mineral analysis, like other types of analyses, is subject to various errors that must be considered. in interpreting the results. These errors are due to (1) the composite nature of the samples, (2) contamination, (3) grain breakage, (4) the misidentification of the grains, and (5) the size of the final sample. These errors have been discussed by Rittenhouse.[48]

GRAIN ROUNDNESS

The roundness of detrital grains may be an important criterion for identifying producing horizons, outlining petrographic provinces, and deciphering geologic history. A short discussion of roundness, giving particular emphasis to its interpretation, is given here.

Roundness and sphericity (shape) are independent properties of sediment particles and must not be confused. Roundness is a measure of the angularity of the corners and edges of a grain. In contrast, sphericity is a measure of the approach to spherical form and may be expressed roughly as a ratio between the length and breadth of a grain. Thus in figure 55, grain A has high roundness and high sphericity. Grain B has high roundness but the sphericity is much lower; the length of the grain is much greater than its breadth. Grain C has low roundness—the corners are very sharp; but it is nearly equidimensional and consequently has a fairly high sphericity.

[48] Rittenhouse, Gordon, *Analytical Methods as Applied in Petrographic Investigations of Appalachian Basin;* U. S. Geol. Survey Circ. 22, 20 pp., 1948.

The three lines of four grains each in figure 55 are representative of three roundness classes that the writer has used for rapid roundness studies. Roundness varies with grain size, and consequently roundness measurements must be made on grains of the same size. Because differences and similarities in roundness have been found to be most significant for the very fine sand size, that size is recommended for initial study.

Roundness of detrital mineral grains must be measured on the original grains. In many indurated or partly indurated sandstones the original shape of the quartz, feldspar, and carbonate grains has been modified by the deposition of quartz, feldspar, or carbonate on them. Consequently, the present shape and roundness of such grains have no significance. When the original grain outlines can be recognized in these sections, the

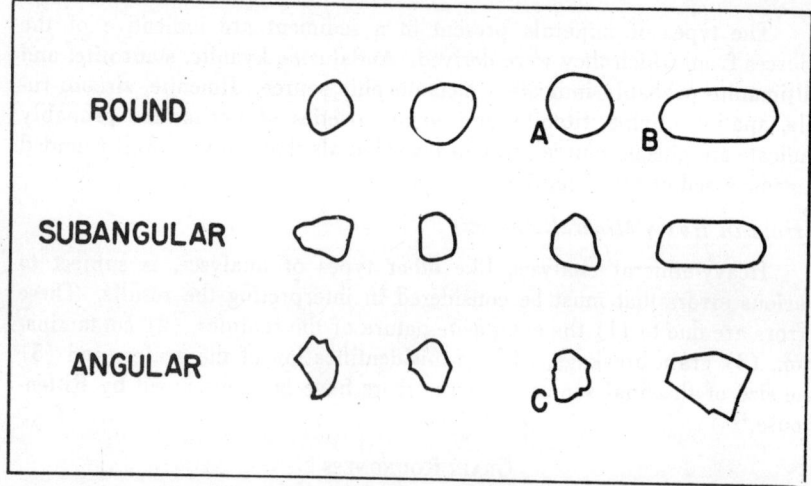

FIGURE 55. Representative round, subangular, and angular grains.

roundness can be determined. When an indurated or partly indurated rock is crushed, the heavy minerals, being of different composition, tend to break out of the rock along their original boundaries. Consequently, heavy minerals can be used for roundness measurements in many cases in which the major constituents of the rock are badly broken.

The number of grains on which roundness measurements should be made depends on the difference in roundness in the samples being studied. If the differences are large, fewer grains need be measured than if the differences are small. Probably a minimum of 25 to 50 grains should be measured in any case. The errors due to the size of the final sample can be determined from figure 56. Roundness studies are subject to the same types of errors as are heavy-mineral analyses.

When quantitative measurements are made on a number of samples of a formation and the results are plotted on a triangular diagram, the data commonly will occupy a restricted part of the diagram. Other formations that differ in grain roundness will be restricted to other parts of the diagram. Then the identity of any unknown sample can be determined. The roundness of tourmalines of very fine-sand size in three formations in Ohio is shown in figure 57.

The rounding of tourmaline, quartz, and other hard minerals of very

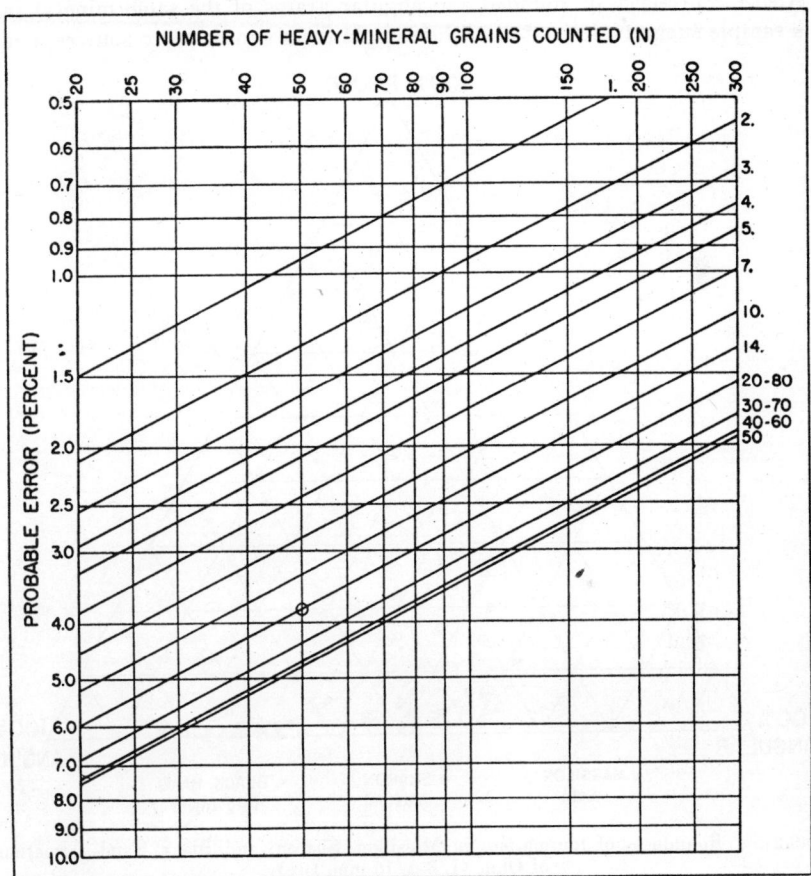

FIGURE 56. Curves for determining probable errors in heavy-mineral, shape, and roundness studies. The probable errors are expressed as percent of the total number of all grains; for example, with 20 percent frequency and 50 grains counted, the probable error is 3.8 percent.)

fine-sand size appears to be very slow. In the Appalachian Basin the sands have the same roundness over thousands of square miles. There is no observable gradation of the type that would indicate progressive change in roundness due to wear as the sediment was transported along a river or a beach. Along certain lines the roundness changes abruptly. The sands on the two sides of such lines appear to have been derived from different sources. Thus roundness has been used to outline petrographic provinces.

Because rounding of grains of very fine-sand size occurs very slowly, the presence of rounded grains of that size in a sediment strongly suggests the derivation of that sediment from a pre-existing sedimentary source. Also the presence of rounded and angular grains of the same mineral in a sample suggests derivation of the sample from two or more sources, one

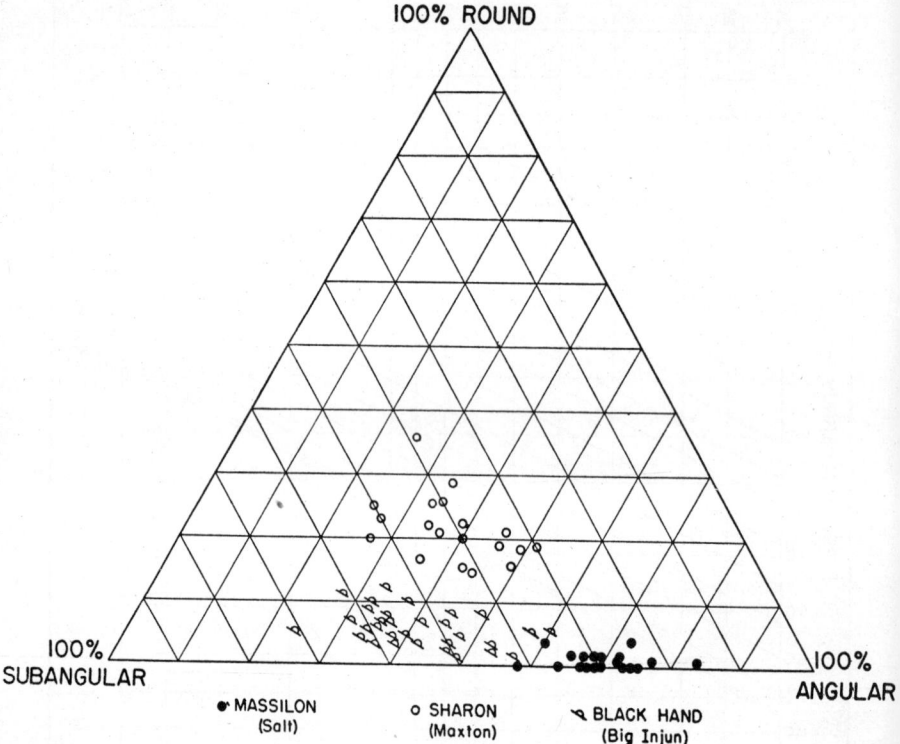

FIGURE 57. Roundness of tourmalines in Massillon, Sharon, and Black Hand formations of Ohio (1/8–1/16 mm. size).

of which is sedimentary. In figure 57 the Sharon is such a mixed sand. The use of roundness has been discussed in more detail by Rittenhouse.[49]

MINOR MINERALS

Some sands contain small percentages of grains that are distinctive in color or appearance. A group of sands in the same area or different parts of a single sand may contain such distinctive grains in different pro-

⁴⁹ Rittenhouse, Gordon, *Grain Roundness—A Valuable Geologic Tool:* Am. Assoc. Petroleum Geologists Bull., vol. 30, no. 7, pp. 1192-1197, July 1946.

portions. Quantitative measurement of these differences may provide useful criteria for correlation or differentation. Because distinctive grains may be present in a ratio to the total of 1 to 1,000, 1 to 100,000, or even less, the determination of the proportion by count would be very tedious and time-consuming. By using a good sample splitter and a binocular microscope, however, the relative proportions can be determined much more rapidly. The procedure is as follows:

The sample is disaggregated and sieved to size the sand. Sizing is necessary because the proportion of distinctive grains may differ with size, and different ratios would occur in coarse and fine sands. The sieve separate is weighed to the nearest 0.01 gram on a triple-beam balance. Using an Otto microsplit a random sample is split out. The number of splits necessary to obtain the test sample is recorded; that is, the test sample is 1/64, 1/128, or other fraction of the original weighed-sieve separate.

A 5- by 8-inch file card is folded into the shape of an M and the test sample is scattered as evenly as possible along the trough of the M to form a line of grains. The card is placed under the microscope, and the number of distinctive grains in the test sample is recorded as the card is moved across the microscope stage. Only distinctive grains are counted; the others are ignored. The number of distinctive grains is calculated. This calculation can be shown better by example than by a word description; for example, the sieve separate weighs 6.02 grams, the test sample represents 1/256 of the sieve separate, and 17 of the distinctive grains were in the test sample. The number of distinctive grains per gram is $\frac{17 \times 256}{6.02} = 723$. It should be noted that in this example each distinctive grain recognized in the test sample represents 43 grains per gram. Consequently, a difference of at least 600 grains per gram more or 400 grains per gram less would be necessary to be considered significant. If twice as large a sample had been counted and twice as many distinctive grains had been recognized, differences of about 400 grains more or 300 grains less would be considered significant.

One requisite of the distinctive grains is that they must be approximately the same specific gravity and shape as the quartz grains. They cannot be flaky grains like the micas or heavy minerals like the tourmalines or garnets. In the Appalachian Basin a variety of quartz that contained small wormlike inclusions of green chlorite gave useful information.

THIN SECTIONS

Thin sections can be used to very good advantage in studying detrital mineralogy. They have the following advantages: (1) Much higher magnifications are possible than with a binocular microscope, thus permitting clear inspection of the smaller features of the rock. (2) The minerals may be identified by means of their optical properties. (3) The sections, being cut through the mineral grains, cement, and pores, give a view of these internal characteristics of the rock and their relationships to one another that cannot be obtained with a binocular microscope. Thin sections have the disadvantage of requiring the use of petrographic microscope and a geologist skilled in its operation. Also the preparation of the sections, especially of cuttings that must be cemented together before being sectioned, requires some time and skill.

INSOLUBLE RESIDUES

H. A. IRELAND

An insoluble residue may be defined as the material remaining after rock fragments have been digested in acid. Hydrochloric acid is generally used, but acetic acid is occasionally used if the preservation of delicate fossils or other structures is desired. Residues, such as shale, pyrite, gypsum, anhydrite, and glauconite, are not siliceous; therefore, the term "siliceous residues" cannot be applied correctly. The chief residues are quartz and various types of chert, with chert the most diagnostic for identification and correlation.

McQueen and Martin in 1931 published methods of preparation, terminology, and practical application of insoluble residues to surface and subsurface correlation and identification of calcareous rocks. The work of Martin is not known so well as that of McQueen although it is a significant contribution. The use of insoluble residues was not widespread prior to 1938. After 1940 rapid advances were made with the application of residue work to petroleum geology. The United States Geological Survey and many state surveys now have many publications based wholly or in part on insoluble residue work. Most of the residue work in Texas was developed independently of that of McQueen, and a diversity of nomenclature resulted. In 1946 Ireland called a conference of active workers from the central United States, which resulted in the publication of standardized terminology and a chart, which is published herein in modified form (See table 3.)

Preparation of Residues

Types of Samples

The materials treated for insoluble residues are well cuttings, cores, and outcrop samples. The most desirable outcrop samples are channel samples or a composite mixture of each exposed stratum within a five-foot or other closed interval. Point-to-point correlation is rarely possible, since there is very little probability of sampling exactly the equivalent point some distance away. A six-inch layer outcropping within a five-foot interval will not represent the whole interval, and it cannot be correlated with the equivalent interval a mile away, which may have a six-inch layer exposed a foot above or below the one in the first outcrop. Only zones or intervals may be correlated successfully. Outcrop samples of unweathered chips, without lichen, soil, or other extraneous matter are desirable.

Oil- or water-well cuttings and cores are the most widely used materials for residues. Cable-tool cuttings are the best samples, as they contain a minimum amount of caved material, and each sample represents a composite of the rock within the sampled interval.

Rotary-tool cuttings are the most common well samples and are generally the only type of samples available from deep wells. They are also the worst samples. Caving is very common because long sections of the drill hole are frequently open. If shale beds or loosely aggregated materials lie above a given sample, caving may reduce the amount of indigenous material of the sample to such a small percentage that an insufficient amount of residue or none will be left after solution. Such samples may require the use of forceps for picking out chips of the indigenous material for solution. Drilling time, electric logs, and a thorough knowledge of the section facilitates the identification of the indigenous material.

Well cores must be split and a fragment taken from each inch or short interval, and the whole mixed for the equivalent of a five-foot sample, or a shorter interval if the lithology changes. Otherwise, inconclusive point-to-point correlation would be necessary.

The observable amount of indigenous material in a sample having eighty to ninety percent shale caving may be increased by placing two or more unit volumes of the sample in acid, and, after solution, sieving out as many unit values less one. Thus, if three units were used, two units would be sieved out after solution. This will leave less than one unit volume, which will contain a minimum amount of caved material but several times more residue from the indigenous material. Large fragments of chert or other insoluble material considered indigenous may be picked out with forceps from the sieve and added to the residue.

Amount of Sample

The volume or weight of sample used to make a residue depends on the purpose of the study, the type of samples used, and individual judgment. Seven grams is an ample amount of sample for ordinary uses. This weight is an average for the volume contained in a one-dram vial, 45 by 15 mm. The same volumes of ten different homogenous samples ranging from very fine to very course fragments of limestone, shale, sand, and chert were weighed, and the average of seven grams was determined. The volume-weight of seven grams reduces considerably the time for the preparation of residues. Small samples of less-than-unit volume must be weighed if percentage determinations are desired. The use of a small scoop sized for a unit volume or a tip balance saves time.

Many workers do not use percentages, but the percentages of residue are valuable in many cases for correlation and identification of beds.

Samples from rotary tools can rarely be used satisfactorily for percentage determinations unless caving intervals have been cased.

Siliceous limestone, tripolitic or cotton chert, and calcareous shale may lose up to half their weight but retain unit volume after solution. Such samples should be weighed before and after acidification, if the percentage of residue compared to the original volume is needed. If the volume of the vial is used as the unit of the original sample, the percentage of nonporous residues may be scaled or observed through the glass vial. Pure limestone or dolomite samples from cores or outcrops may leave only a few grains of residue, and it may be necessary to use two, three, or even five units of the original sample to obtain sufficient residue for examination and determination.

Solution of Samples

Samples are generally dissolved in commercial hydrochloric (muriatic) acid. It is inexpensive, easily obtained, and effective. The acid should be diluted with water to at least fifty percent but to no less than ten percent. Warming will hasten the reaction, but undesirable precipitates may form. Many complex reactions occur between caved material, constituents of the indigenous material, and the impurities in the muriatic acid. Iron, gypsum, and other precipitates frequently coat, stain, and contaminate many types of residues. Many samples will not dry clean if left in the spent acid and precipitates longer than six to eight hours.

Chemically pure (CP) hydrochloric acid has advantages for special work where outcrop samples are used, where precipitates or impurities are undesirable, or when solution is extended over several days. Acetic acid is best for liberation of delicate, fragile, lacy material or for microscopic organisms. Delicate residues may be preserved by using very dilute hydrochloric acid, but the time required for solution is lengthened.

Beakers are the best receptacles for solution of the samples. They are preferred because the lip facilitates washing and decanting, residues may be easily removed, and a glazed spot is provided on the side for identification of each sample. Molded tumblers or other cheap glassware may be used, but the breakage due to heat while drying samples equals or exceeds the greater original cost of heat-resistant glassware.

The procedure for residue preparation is simple. Samples of unit quantity are placed in a glass receptacle properly identified on a slip of paper under a pyrex dish or by any consistent regular arrangement. Samples are then digested in acid, washed, dried, labelled, and stored for examination. The use of several stainless-steel trays, or other type of tray, holding forty to fifty beakers, facilitates the bulk movement of samples to the hood for acid application, washing, drying, or other operations involving the handling of large numbers of samples.

The first application of acid should be small to prevent foaming caused by the rapid effervescence of powder and fine material. The foaming may easily cause the overflow and loss of considerable material. A few minutes after the initial application, additional acid may be added, but only experience will tell how much, generally not more than one-third to one-half the capacity of the receptacle. After several hours of digestion the samples should be washed once or twice to remove spent acid, precipitates, and undesirable material. The second application of acid will generally complete the digestion, although one application may be sufficient. Small applications of acid will digest samples which obviously are chert, sand, or shale. Incomplete digestion will leave dolomite pellets with rough, jagged surfaces and rounded pellets of limestone. When samples are incompletely dissolved, individual euhedral dolomite rhombs may be a large part of the residue. Final washing should be thorough to remove all traces of acid and prevent scum, caking, or coating on the residues.

Clay and fine silt are generally decanted in routine work. Little or no work has been done with the fine residues, and their value for correlation and identification is yet to be determined. Only outcrop or core samples can be used for study of clay and silt residues, because caving and other contamination of well samples obscure diagnostic features and makes uncertain the identification of indigenous fine clastic material.

Residues may be dried in an oven, on a hot plate, or on a sand bath. Dry residues are brushed into a pan or funnel for transfer into glass vials, which may be labelled on the cap, cork, or a paper sticker. Permanent storage requires a painted label or glazed surface on the side of the vial, because silverfish enjoy eating the glue from stickers. One-dram vials hold ample residue for study and require very small storage space. Trays, drawers, original vial boxes, or special boxes are suggested methods of storage.

DESCRIPTION OF RESIDUES

The most common insoluble residues are chert, chalcedony, disseminated silica, clastic and crystalline quartz, aluminous matter, and replaced fossils. Anhydrite, gypsum, feldspar, glauconite, hematite, pyrite, fluorite, and sphalerite are the most common minerals, but other insoluble minerals are found.

Table 3 is a modified arrangement of the original chart published with the paper on standardized terminology.[50] The terminology is based on description rather than genesis of the residues because genesis of many constituents is unknown, vague, or controversial. Many possible types of residues are given a place in the table, although their existence has not been confirmed. Each term is clear-cut and restrictive, and within certain limits a residue fragment may be pigeon-holed. It should be emphasized that types of residues grade into other types, and, as some specific fragments may not be easily placed, workers may place a fragment under a different but related type in the classification.

---

[50] Ireland, H. A., *et al.*, *Terminology for Insoluble Residues:* Am. Assoc. Petroleum Geologists Bull., vol. 31, no. 8, pp. 1479-1490, Aug. 1947.

TABLE 3

CHART FOR INSOLUBLE RESIDUES

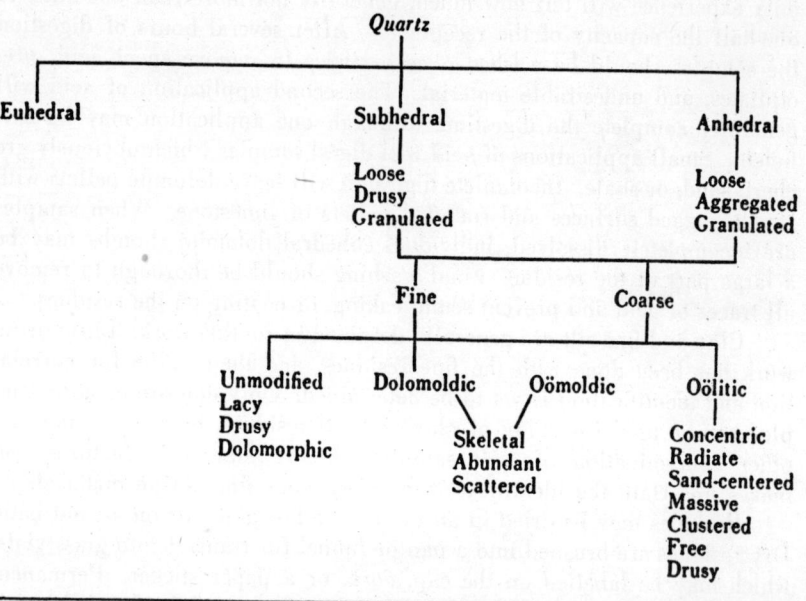

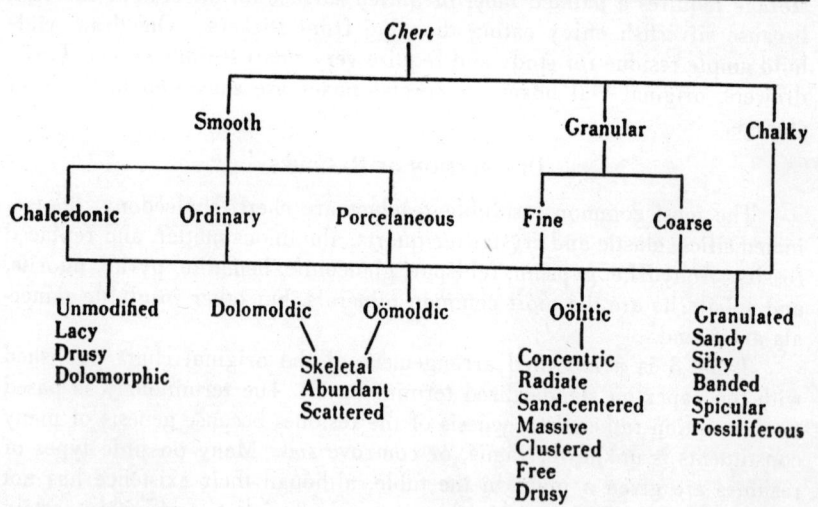

TABLE 3—Continued

Argillaceous material

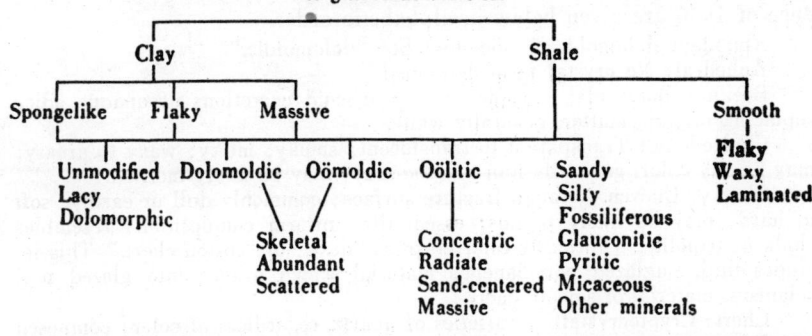

```
                        Argillaceous material
        ┌──────────────────────┴──────────────────────┐
      Clay                                           Shale
  ┌──────┼──────┐                            ┌────────┼────────┐
Spongelike Flaky Massive                                     Smooth

  Unmodified Dolomoldic Oömoldic  Oölitic    Sandy           Flaky
  Lacy                                        Silty          Waxy
  Dolomorphic                                 Fossiliferous  Laminated
                   Skeletal      Concentric   Glauconitic
                   Abundant      Radiate      Pyritic
                   Scattered     Sand-centered Micaceous
                                 Massive      Other minerals
```

Arenaceous material

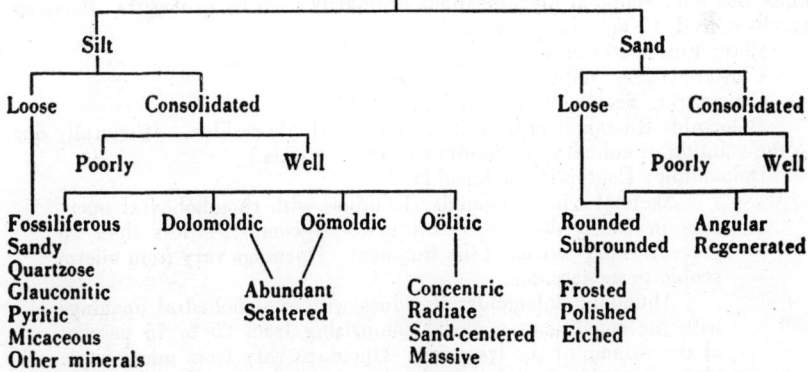

```
                         Arenaceous material
        ┌──────────────────────┴──────────────────────┐
      Silt                                           Sand
  ┌──────┴──────┐                              ┌────────┴────────┐
Loose      Consolidated                     Loose        Consolidated
         ┌──────┴──────┐                              ┌──────┴──────┐
       Poorly        Well                           Poorly        Well

Fossiliferous Dolomoldic Oömoldic  Oölitic   Rounded          Angular
Sandy                                        Subrounded       Regenerated
Quartzose
Glauconitic        Abundant       Concentric  Frosted
Pyritic            Scattered      Radiate      Polished
Micaceous                         Sand-centered Etched
Other minerals                    Massive
```

Anhydrite

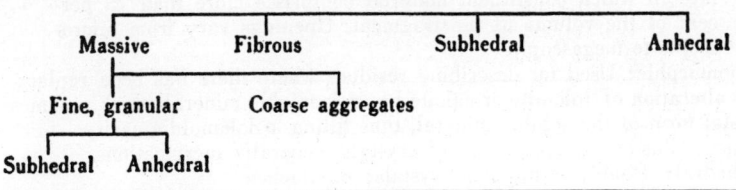

```
                            Anhydrite
        ┌──────────┬──────────────┬──────────────┐
     Massive    Fibrous        Subhedral      Anhedral

Fine, granular    Coarse aggregates

Subhedral  Anhedral
```

Gypsum

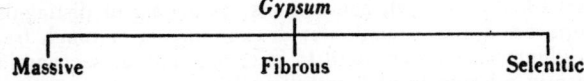

```
                    Gypsum
        ┌───────────┼───────────┐
     Massive     Fibrous     Selenitic
```

UNCLASSIFIED ACCESSORY RESIDUES

Sulphur, pyrite, marcasite, sphalerite, millerite, magnetite, hematite, limonite, feldspar, muscovite, biotite, chlorite, glauconite, barite, celestite, other insoluble minerals, fossils, pellets, beekite.

Terminology for Insoluble Residues

Definitions of special terms as agreed upon by the Residue Conference of 1946 are given below in alphabetic order.

Abundant dolomolds or oömolds: See "dolomoldic."

Anhedral: No crystal form developed.

Beekite: Botryoidal, subspherical, or discoid accretions of opaque silica replacing organic matter, generally white.

Chalcedonic: Transparent to translucent; smoky; milky; waxy to greasy; may be any color, generally buff or blue-gray; may be finely mottled.

Chalky: Uneven or rough fracture surface; commonly dull or earthy; soft to hard; may be finely porous; essentially uniform composition; resembles chalk or tripolite. (Formerly referred to as "dead" or "cotton chert." This includes dull, unglazed porcelaneous material which grades into glazed porcelaneous material of smooth chert.)

Chert: Cryptocrystalline varieties of quartz, regardless of color; composed mainly of petrographically microscopic fibers of chalcedony and/or quartz particles whose outlines range from easily resolvable to nonresolvable with binocular microscope at magnifications ordinarily used by geologists. Particles rarely exceed 0.5 mm. in diameter.

Clay: Fine material of clay size.

Clustered: See "oölith."

Concentric: See "oölith."

Dolomold: Rhombohedral cavities in an insoluble residue. (Generally due to the solution of euhedral dolomite or calcite crystals.)

Dolomoldic: Containing dolomolds.

> Skeletal with dolomolds: Residues with rhombohedral openings in which the constituent material comprises less than 25 percent of the volume of the fragment. Openings vary from microscopic to megascopic.

> Abundant dolomolds: Residues with rhombohedral openings with the constituent material comprising from 25 to 75 percent of the volume of the fragment. Openings vary from microscopic to megascopic.

> Scattered dolomolds: Residues having rhombohedral openings in which constituent material comprises more than 75 percent of the volume of the fragment. Openings vary from microscopic to megascopic.

Dolomorphic: Used for describing residues where there has been replacement or alteration of dolomite or calcite by an insoluble mineral which assumes the crystal form of the soluble mineral, thus filling a dolomoldic cavity.

Drusy: Clusters or aggregates of crystals, generally incrustations.

Euhedral: Doubly terminated crystals; unattached.

Free: See "oölith."

Granular: Chert; compact, homogenous; composed of distinguishable relatively uniform-size grains, granules, or druses; uneven or rough fracture surface; dull to glimmering luster; hard to soft; may appear saccharoidal. (This type is frequently referred to as "crystalline.")

Granulated: Grains or granules partly cemented or loosely aggregated; saccharoidal; grades from angular to drusy; fine to coarse; particles rarely larger than 0.5 mm. in diameter.

Lacy: Residues with irregular openings in which the constituent material comprises less than 25 percent of the volume of the fragment.

Massive: See "oölith." Used also to include fine or coarse granular anhydrite or gypsum.

Mottled: Residue fragments with two or more colors or different material interspersed and irregularly shaped with the boundaries between either sharp or gradational; often appears flocculated; grades into speckled residue.

Oölite: Composed of an aggregation of oöliths.

Oölith: Spheroidal bodies with nucleus or central mass enclosed by one or more surrounding layers of the same or different material; may be any color and of many kinds of material, generally less than 1.0 mm. in diameter. Those over 2.0 mm. are pisoliths.

Concentric: Peripheral layers around a small, undetermined nucleus.

Clustered: Attached oöliths without solid matrix.

Drusy: Oölith covered with subhedral quartz; may be free or clustered.

Free: Unattached oölith.

Massive: Interior of granular, smooth, or chalk-textured material comprising nearly the entire mass of the spheroid.

Radiate: Fibers radiating from small or large nucleus; may have several peripheral layers.

Sand-centered: Nucleus, a quartz sand grain.

Oömold: Spheroidal opening representing the former presence of oöliths.

Oömoldic: Containing oömolds.

Skeletal with oömolds: Same definition as for "dolomoldic."

Abundant oömolds: Same definition as for "dolomoldic."

Scattered oömolds: Same definition as for "dolomoldic."

Ordinary: smooth chert with even fracture surface; all colors, chiefly white, gray, or brown; may be mottled; approaches opaque; generally homogeneous, but may have slight evidence of granularity or crystallinity; grades into chalcedonic or granular chert.

Porcelaneous: Chert with smooth fracture surface; hard; opaque to subtranslucent; typically china-white resembling chinaware or glazed porcelain; grades to chalky.

Pseudoölithic: Rounded pellets with no peripheral layers or sharp distinction between pellets and matrix.

Quartz: Clear, colorless quartz; not detrital.

Radiate: See oölith."

Regenerated: Used in reference to quartz sand grains with secondary regrowth of crystal faces oriented with the original axis of the grain.

Rounded: Spheroidal or ellipsoidal sand grains, coarse to fine, may be polished, frosted, or etched.

Sand: Grains of sand size, chiefly quartz, but may be composed entirely or partly of other minerals.

Sand-centered: See "oölith."

Scattered: See "dolomoldic" and "oömoldic."

Silt: Grains of silt size, chiefly quartz, but may be composed entirely or partly of other minerals.

Skeletal: See "dolomoldic" and "oömoldic."

Smooth: Major type of chert with conchoidal to even fracture; surface devoid of roughness; may be botryoidal; homogeneous; no distinctive structure, crystallinity, or granularity.

Spicular: Containing inclusions of sponge spicules. Free spicules have been noted.

Speckled: Disseminated fine spots of color or material different from that of the matrix and having relatively sharp boundaries.

Subhedral: Crystal forms partly developed; may be loose, drusy, or granulated.

Subrounded: Polygonal grains or fragments but with well-rounded edges and corners.

Unmodified: Residue uniform with no modifying characteristics.

The most common residues are chert and sand, with chert rated as the most diagnostic. Texture, color, transparency, luster, and crystallinity are the chief factors for the differentiation of chert. Inclusions and modifying characteristics are secondary factors. Chalcedonic and ordinary chert are the most abundant of the smooth cherts. The term "granular chert" is applied to obviously crystalline chert or that with observable grains. Smooth and granular cherts grade into each other and into chalky chert. The chalky types are those of which the original internal structure and filled interstices have been affected by weathering and probably by circulating water. Tripolitic chert when placed in acid leaves a very fine, porous, chalky chert because of the solution of disseminated calcium carbonate. All the cherts may be dolomoldic, the dolomolds ranging in type from scattered to skeletal and in size from fine to very large.

The color of chert is an important diagnostic feature. It is prevalently colorless, white, gray, tan, and brown, but all colors are found. Many residues from beds in Missouri, Kansas, Oklahoma, and Texas have sudden color changes which mark boundaries of zones or formations. The smooth, brown chert of the Lower Devonian in west Texas is difficult to differentiate from that in the Upper Ordovician, and drilling to an underlying boundary is necessary in many places for positive identification.

Organisms may be replaced by silica or other insoluble matter and may be identified in the residues, especially small forms and Foraminifera, which generally are not broken by the drill. Molds of organisms are common where soluble shells or fragments have been imbedded in an insoluble matrix. Beekite occurs most commonly in replaced megascopic fossils found in outcrop samples.

Quartz may be euhedral and authigenic or subhedral and anhedral from veins, cavity filling, or interstitial openings. Quartz sand of various types from rounded to angular may be found as scattered inclusions or as a dominant feature in a sandy calcareous rock. Secondary enlargement or regrowth of quartz crystals around sand grains is a common occurrence in the Lower Paleozoic. The crystal growth in some sandstones is distorted and interlocked with adjacent grains in such a manner that a tight, nonporous formation results. Feldspar, mica, glauconite, and other minerals are common as residue constituents of sandstone, although quartz is the chief constituent. Calcareous material interstitially mixed with very fine quartz in silt and clay sizes results in a very fine porous residue.

Glauconite is abundant in sands and is scattered throughout many calcareous beds. It is a good marker for many beds in the Paleozoic, chiefly in the Mississippian, Middle Devonian, Middle and Lower Silurian, and

Upper Cambrian. Few of the lower Ordovician Beekmantown beds have glauconite, and the appearance of glauconite generally marks the top of the Cambrian.

Pyrite is a common insoluble residue seen as small to large, euhedral crystals in limestone, dolomite, and shale. It also occurs spongelike, disseminated, and in veins and cavities. Pyrite has little value as a diagnostic residue, but it has a secondary value as an inclusion in chert or shale. When pyrite occurs in abundance it may serve as a marker bed and often identifies a zone of circulating water or an unconformity.

Interstitial spaces due to primary or secondary permeability, alteration, or replacement in calcareous rocks may become filled with silica, pyrite, or other insoluble material. Solution of the matrix leaves fragile, lacy networks that are generally destroyed by acid effervescence and washing. These residues are the extreme upper limit of skeletal dolomolds, pyrimolds, and oömolds. Residues from veins or fractures are curved or tabular flakes. Vein fillers or cement for brecciated residues include gilsonite, silica, pyrite, and sphalerite.

Siliceous limestones have residues that are generally earthy, finely porous, and dark-colored. These residues are especially noteworthy because examination of such samples before solution gives no clue to the type of residue. The residues from siliceous limestone also appear to be 100-percent insoluble by volume, but they may be 50-percent insoluble by weight, owing to the removal of the interstitial lime.

Siliceous oölites are common and may be found free, clustered, or in a matrix. An oölite, to be identified as such, must have a nucleus and at least one concentric layer or shell. Nuclei may range in size from very minute to one occupying nearly all of the interior mass. Most oöliths have several shells. Oöliths are classified according to the interior structure as concentric, massive, radiate, or sand-centered. Clustered or free oöliths may be frosted with a crust or minute drusy quartz or may have a smooth, siliceous shell. Silica may replace calcareous oöliths and cause them to be preserved as residues. Oöliths have many colors and frequently occur embedded in different-colored matrices. All types of chert have oöliths, although in chalky chert they are rare.

Chert in many cases has included sand grains, which may be confused with oöliths. Shells are absent, however, and the clear quartz of the said grain may be observed.

Pseudoölites or "shadow oölites" resemble oölites and may resemble included quartz sand grains. The boundary between the matrix and the oölith is indistinct, however, and the central portion, which cannot be identified as quartz is only a shade lighter or darker than the other portions. Pseudoölites may be oöliths or sand grains that have been resorbed, thus destroying any formerly existing boundaries.

Dolomolds occur chiefly in chert residues from dolomites, rarely in chert from limestone. Dolomolds are common in shale residues and are present in some pyrite and glauconite residues. Natural dolomolds resulting from weathering are common on certain types of outcrop samples. In dolomite the dolomolds are assumed to be the impressions from dissolved dolomite rhombs, but in shale the cavities are likely to be a result of dissolved interstitial calcite. Disseminated abundant fine dolomite or calcite crystals in chert, silt, or shale will leave a very finely porous residue, too fine to be observed except under high magnifications. The residue of a sample with large quantities of dolomite rhombs will have an intersecting lacework of fragile skeletal dolomolds, while a sample with a few rhombs will leave scattered dolomolds in the insoluble matrix. Dolomolds may be large or small, but generally all in any one fragment will be essentially the same size.

Use of Insoluble Residues

The study of insoluble residues is a supplement to and not a substitute for lithologic sample examination. The cost of preparing and filing residues and the longer time necessary for the more detailed and careful examination of them are factors that must be considered. The mass characteristics of the major constituents of insoluble residues generally have enough similarity horizontally and vary enough vertically to serve for identification and correlation of lithologic units within a thick section of calcareous rock.

Lithologic similarities of thick sections of nonfossiliferous calcareous rocks prevent their subdivision into thinner zones for more-detailed correlation and identification and structural mapping. Insoluble constituents having diagnostic characteristics may be obscured by the volume of the fragments in a lithologic sample and by being embedded in a solid matrix. These constituents are liberated, concentrated, and exposed by solution of the matrix. Diagnostic material such as Foraminifera, some types of chert, dolomolds, disseminated pyrite, fossil replacements, euhedral crystals, mineral or clastic inclusions, and silt aggregates are not observed or recognized until they become residues.

Residues reflect clastic conditions, sea-bottom environment, current action, and adjacent land-mass conditions, which may supply various types of source materials. These factors may change independently over short or long periods of time. If the source of material and the conditions of deposition or precipitation of calcareous matter remain fairly constant for a long time, no significant lithologic variations would result that might serve to identify a stratum. A slight change involving the source, type, or amount of clastic furnished to a lime-depositing environment might not affect greatly the lithologic appearance of a sediment, but such material when left as a residue would be diagnostic and serve for correlation and identification. The amount of silica, iron, or salts in the sedimentary basin might change and give pyrite, siliceous limestone, various types of chert, and other minerals or constituents of diagnostic value;

all of which might be independent of clastic material or changes in land-mass conditions or source material.

Circulating water and replacement and alteration of constituents before and after lithification would change the original residues. These changes, if of sufficient magnitude, might be observed in a lithologic examination of samples, but only the study of residues would show the small changes that might be useful in a detailed subdivision of beds. Correlation using residues of secondary origin could only be used locally or as far as the effect of the modifying conditions could be traced.

Correlations for distances greater than fifty miles are risky, unless some significant wide-range constituent can be determined, because the residues will change as the sedimentary environment changes. Obviously correlation using any specific zone of residue types would be less reliable in a basinward or landward direction than laterally in a direction at right angles.

Correlation of individual thin beds may be difficult because of lateral and vertical changes of the sedimentary-environment time. The subdivision of a thick calcareous section and the inclusion of nondiagnostic thin beds into zones make correlation possible. Identification of the zones is based on such factors as sequence of beds, position in the section, percentage of residue, association of types of residues, and dominant characteristics with chief reliance on dominant characteristics. A distinctly significant residue may identify certain zones, although other residue constituents may be present, and even though the diagnostic residue is not the dominant one. An assemblage of residue constituents often determines the correlation or identification just as an assemblage of fossils serves for determination. Both microscopic and macroscopic fossils replaced with insoluble material are valuable in some zones.

Positive identification of some subdivisions is difficult with only a few samples, unless a significant break or change in residue occurs within the interval examined. For example, assume that a limestone 1,400 feet thick is divided into six zones having intervals of 350, 100, 200, 400, 300, and 50 feet. If only ten 5-foot samples were available from zone 1 at the top, it would be difficult or impossible to identify their positions in the zone, although the zone itself could be identified. If the samples overlapped into zone 2, then the boundary could be recognized, and it could be stated than 300 feet or more of zone 1 was absent.

Many cherts are alike in color and texture, and similar cherts in two different zones would prevent identification, unless an associated residue was diagnostic or a zone boundary was passed. The similarity of the brown, smooth chert in the Lower Devonian and the Upper Ordovician in west Texas has been mentioned previously. If a chert in zone 1 was similar to a chert in zone 4 in the section postulated in the last paragraph, the two zones could easily be confused. If the set of ten samples was identified as belonging to zone 1, but actually zones 1, 2, 3, and part of 4 had been eroded, an error of at least 650 feet in correlation would result. The correct identification of zone 4 would show a structural upfold. If the samples were identified as zone 4 and the producing bed was zone 2, the absence of zone 2 would be concluded and deeper drilling prevented.

The foregoing discussion shows the necessity of having some associated diagnostic residue or a zone boundary included in the sample interval for positive identification. Knowledge of the similarity of two zones would call for careful drilling and a postponement of identification until the underlying zone was encountered. With lithologic examination no zones could be identified.

Pyrite, regenerated sand grains, a sandy chert or sandy zone, a shale break, or a detritus often present clews to a formational change, which in some cases can be confirmed by other evidence.

The use of residues is not restricted to the laboratory. A microscope, a jug of acid, and a half-dozen beakers may be carried to the field. Water from a drilling well may be used for washing, and heat from an automobile-engine head or a drilling well will dry the samples for examination on location. Obviously, a geologist attempting such work must be familiar with residue zones and sequences, as the necessary samples for comparison would not likely be available.

New workers with residues should be well aware that successful correlation by residues comes only after a thorough knowledge of residue types, principles of secondary replacement, and facies changes and the examination of many samples. Experience with residue material is prerequisite to the successful correlation and identification of beds. Of course, the foregoing statement is true for lithologic examination, but an inexperienced geologist can soon learn the surficial characteristics of rock fragments and correctly correlate, but he would find it difficult to correlate with residues without experience or the supervision of one experienced in residue work.

The use of residues for correlation has been successful in the thick calcareous sections of most Paleozoic rocks but has had little success in the thick Permian section of west Texas and New Mexico. Residue work has been especially useful in subsurface work and petroleum geology in Texas, Oklahoma, Kansas, Missouri, and Illinois and has contributed much to geologic science in the states between the Appalachian and the Rocky Mountains. The beds receiving the most attention have been Upper Cambrian and Lower Ordovician, but Silurian, Devonian, and Mississippian beds have been extensively studied.

The space allotted here would be inadequate to give worth-while descriptions of the subdivisions of the thick sections of calcareous rocks in the various parts of the United States. Anyone concerned would profit more to confer with workers familiar with local areas and sections.

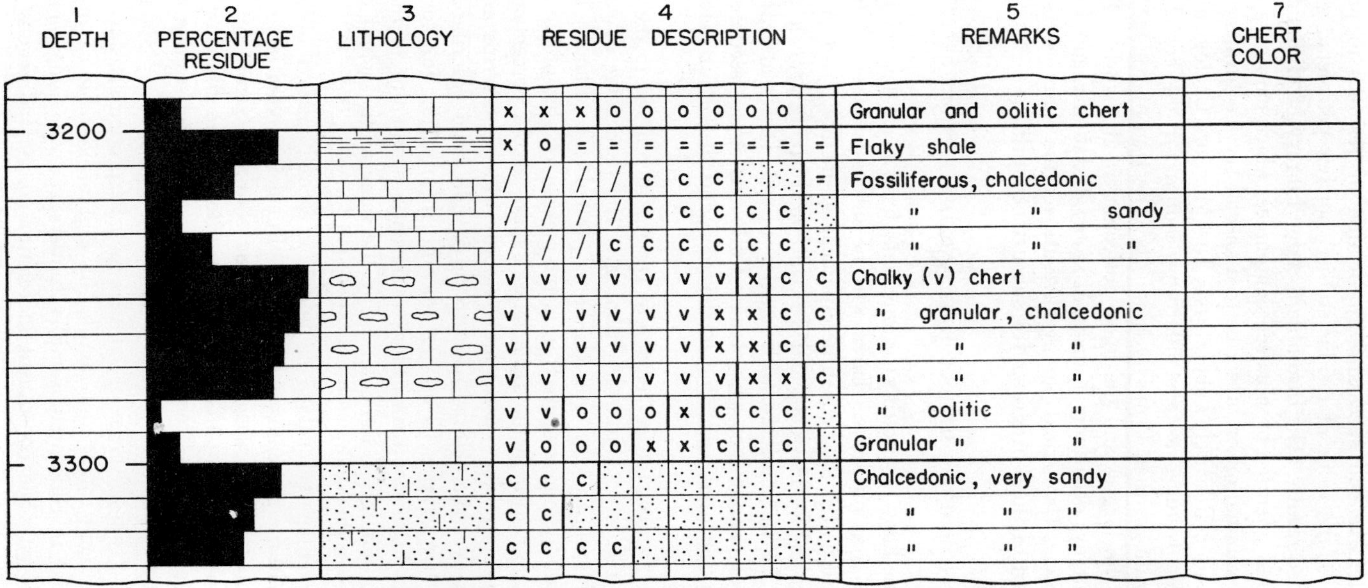

1 DEPTH	2 PERCENTAGE RESIDUE	3 LITHOLOGY	4 RESIDUE DESCRIPTION	5 REMARKS	7 CHERT COLOR

Depth markers: 3200, 3300

Remarks:
Granular and oolitic chert
Flaky shale
Fossiliferous, chalcedonic
"   "   sandy
"   "   "
Chalky (v) chert
"   granular, chalcedonic
"   "   "
"   "   "
"   oolitic   "
Granular   "
Chalcedonic, very sandy
"   "   "
"   "   "

FIGURE 58. Constituent-percentage method of plotting residue data.

The extensive and successful application of residues to petroleum geology proves the value of residue studies, but few petroleum geologists have published results. The Insoluble Residue Library of Midland, Texas, is financed and operated by nineteen companies, which employ specialists for residue examination or subscribe to a special service furnished by the Midland Residue Research Laboratory. The Missouri Geological Survey uses insoluble residues as a standard procedure for the correlation of formations younger than Pennsylvanian. Its collection of residue samples is probably the largest and the finest collection in the world. The state geological surveys of Illinois, Kansas, Missouri, Oklahoma, and Texas and the United States Geological Survey have utilized the study of residues as a regular part of their programs for subsurface work.

PLOTTING RESIDUE DATA AND DESCRIPTIONS

Many methods of plotting residue data have been devised for individual needs and purposes. Three will be discussed here. The writer uses a method called the "constituent-percentage method," which is illustrated in figure 58. The data and description are plotted on a strip log 100 feet to the inch printed with a grid ruling. This scale allows the comparison of residue logs with standard oil-well logs or sections. Other intervals may be used according to the need and desire for detailed description.

Column 2 shows the percentage of residue in relation to the original sample, and column 3 shows the lithology. The percentage of each constituent in reference to the total residue is plotted in column 4. Thus the percentages of the constituents from a ten-percent residue of the original sample will be shown in the same lateral space as the percentages from a ninety-percent residue. Color and symbols with superscrips and overprints in ink over the colored background in column 4 describe and distinguish the constituents. The most specific information for correlation work appears in this column. Lines representing the color of the cherts are placed in column 7, the color being the same as the actual color of the chert, except that white chert is designated by green.

The percentage-percentage method is a second method of plotting. By this method the percentage of each constituent is plotted in proportion to its percentage of the original sample as shown in figure 59. Superscripts and overprints in ink and color similar to the first method are used for differentiation of the constituents. An expanded scale is required and percentages over 75 are eliminated.

Residues from all types of samples may be plotted satisfactorily by this method except rotary-tool samples having considerable cavings. The caved material in rotary-tool samples hinders accurate judgment of the percentage of any one constituent in relation to the indigenous portion

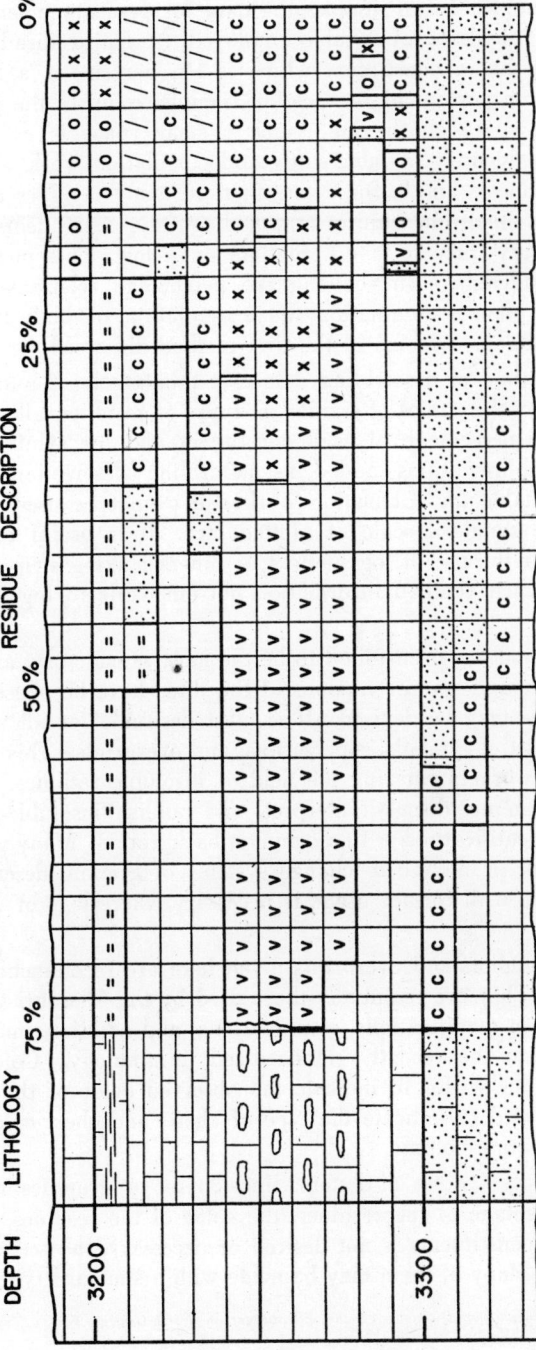

FIGURE 59. Percentage-percentage method of plotting insoluble residues. Plotting is based upon the same data and the same section as in figure 58.

of the original sample. In all samples the data for percentage-percentage plotting must be calculated or a table employed. A major disadvantage of the method is that a constituent which is ten percent of a residue, which in turn is ten percent of the original sample, requires the plotting of a 0.01-percent space on the log. Such a small space is difficult to plot as well as to identify in later study and correlation work. Though a ten-percent residue is ample for determination, many residues are less than five percent. Small but significant and diagnostic constituents would be obscure and very difficult to differentiate on a log. This method of plotting has an advantage in showing the percentage relations of the constituents at a glance, eliminating the examination of both the percentage and constituent columns of the first method discussed.

The third method of assembling data is a tabulation. A number of columns, headed by the names of significant types of residues, allow space for tabulating the percentage of each constituent and for symbols with superscripts or abbreviations added as modifying descriptions. This method is not suitable for correlation work and cannot be used in conjunction with the standard scale of plotted logs. It is useful only for tabulating data for the use of log plotters or others making strip logs or for consultation when detailed information not amenable to log plotting is needed.

The first method has been found to be the most satisfactory, and it is recommended, although several versions of the three have been used, and other combinations may be devised. It is most desirable for all workers to use a standard set of symbols, superscripts, and overprints. This applies especially to new workers entering the field of insoluble residues. Workers could then examine, discuss, interpret, and publish insoluble-residue correlations and identifications with a common background. Many workers will find it difficult or unwise to change systems of graphic description, because consistence with former usage is necessary where logs or records are involved.

The sets of symbols and overprints given here are recommended for standardized use. They are essentially those used by the Missouri Geological Survey. Certain modifications, combinations, and additions make the symbols conform to the recently standardized terminology. Color was used formerly by the writer to indicate the observed color of the chert; but color is now used to indicate the type of chert, and chert colors are indicated in a separate column.

Hendricks [50a] uses a set of letters, lines, bars, or graphics without color. If the percentage of the residues, the color of the residues, or the percentage of the constituents is not desired or necessary, the set of symbols is adaptable. Many of them may be made with a standard typewriter.

[50a] Hendricks, Leo, *Subsurface Divisions of the Ellenburger in North-Central Texas:* Texas Univ. Bur. Econ. Geology Bull. 3945, pp. 923-968, 1940.

PETROFABRIC ANALYSIS
WARREN R. WAGNER

Although petrofabrics has not been widely used, it is a valuable geologic tool that may be applied with marked success in deciphering complicated structural conditions (surface and subsurface). The detail to which it is carried depends upon the nature of the problem; its application, however, may produce results on broad reconnaissance studies or on large-scale detailed problems. The greatest value is probably obtained in the latter.

The idea that the petrofabric data for an entire area may be obtained from one thin section and presented by making one or two orientation (petrofabric) diagrams seems to be prevalent with some geologists. One thin section an inch square may be the key to a particular part of the problem, but the making of orientation diagrams alone does not get the answer; it is their interpretation and their correlation with all other petrofabric data of the area under study that complete the picture.

The limited definition accorded the term "petrofabrics" by some workers may have been responsible for this. The term has been defined in various ways, and numerous students of the subject give it somewhat different connotations. One of the most complete definitions is to be found in Ingerson's short paper "Why Petrofabrics?"[51]

But the word as it should be understood in petrofabrics is much more comprehensive. In this sense, it is analogous to the fabric of a building, that is, its entire make-up from the structure of the individual bricks and the mortar that holds them together, to the steel framework that binds the whole into a unit. In other words, "fabric" includes all of the spatial relation of a rock from the space-lattice of the individual mineral grains through cleavage, fractures, joints, schistosity, lineation, and fold-axes.

Therefore, petrofabrics to be complete should be approached as a field problem supplemented by laboratory work.

As defined above, all the details of a structural unit are brought together. True enough, to acquire all data may require a tremendous amount of tedius labor. But then, is not the price paid for any item controlled by the returns? At the present moment, this tool is being employed chiefly by those more academically minded, although its application to economic problems is advancing steadily.

The metamorphic and igneous rocks have received the greatest attention, because the features and principles of petrofabrics are best developed in them. Many of these principles probably hold true for the less-deformed sedimentary rocks but are not so pronounced and thus have received less attention.

The purpose of this discussion is not to present the techniques and methods of petrofabrics in detail but to summarize the rudiments with references to literature on various phases.

[51] Ingerson, Earl, *Why Petrofabrics?*: Carnegie Inst. Washington, Geophysics Lab., Paper 1081, 1944.

FIELD WORK AND ITS PRESENTATION

Some of the more important papers on petrofabric methods and presenting information on petrofabrics gathered in the field are the following: "The Application of Recent Structural Methods in the Interpretation of the Crystalline Rocks of Maryland," by Cloos,[52] a volume that contains not only Cloos' paper on methods but also a number of supporting papers by his students; "Lineation," by Cloos,[53] dealing exclusively with lineation, its formation, interpretation, and mapping; "Oölite Deformation in the South Mountain Fold, Maryland," also by Cloos,[54] which is a detailed account of the use of deformed oöids in interpreting the structure of an area; and "Structural Petrology of Deformed Rocks," by Fairbairn,[55] which is a detailed account of petrofabrics, its interpretation, and presentation, chapters 1 through 7 dealing with the theoretical aspects of the subject.

The presentation of petrofabric features is accomplished through the use of symbols. The United States Geological Survey has recently published a "New List of Map Symbols" that may be obtained free of charge from the Geological Map Editor, United States Geological Survey, Washington 25, D. C. These symbols were submitted by a committee composed of Ernst Cloos, L. B. Pusey, W. W. Rubey, and E. N. Goddard, Chairman. Reproduced on pages 160 to 163 are the symbols from this list that are most frequently used in mapping petrofabric data.

Plastic flow in crystalline material takes place by intergranular lattice displacement and rotation and migration of materials without destroying the cohesion between the grains of the rock subjected to deformation. Rocks thus deformed are called "tectonites," whereas all others are classed as "nontectonites."

In order to carry out the correlation between the field data and microscopic data, the structures of the rock are referred to three coordinates or axes a, b, and c, each normal to the other. Figure 60 illustrates a plunging anticline with its elements tied to these axes. From this diagram it is seen that a is the direction of movement or transport, b is the fold axis, and c is the vertical component. The a–c plane is normal to the b direction.

While field mapping is in progress, oriented specimens are collected from the area for later microscopic study. These are so marked (fig. 61) that, once in the laboratory, they are reoriented to their proper position in space; thus, the microdata may be correlated with the a, b, and c structure axes.

Figures 62 to 67 show tight folds in bedded, schistose quartzite of the

[52] Cloos, Ernst, *The Application of Recent Structural Methods in the Interpretation of the Crystalline Rocks of Maryland:* Maryland Geol. Survey, vol. 13, 295 pp., 1937.
[53] Cloos, Ernst, *Lineation:* Geol. Soc. America Mem. 18, 1946.
[54] Cloos, Ernst, *Oölite Deformation in the South Mountain Fold, Maryland:* Geol. Soc. America Bull., vol. 68, pp. 843-918, 1947.
[55] Fairbairn, H. W., *Structural Petrology of Deformed Rocks,* Cambridge, Mass., Addison-Wesley Press, Inc., 1942.

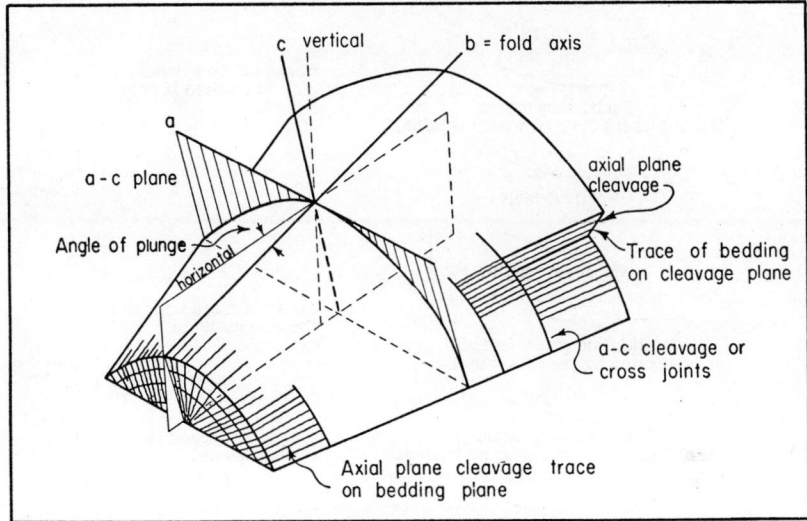

FIGURE 60. Illustrating a plunging anticline with its elements tied to axes *a*, *b*, and *c*.

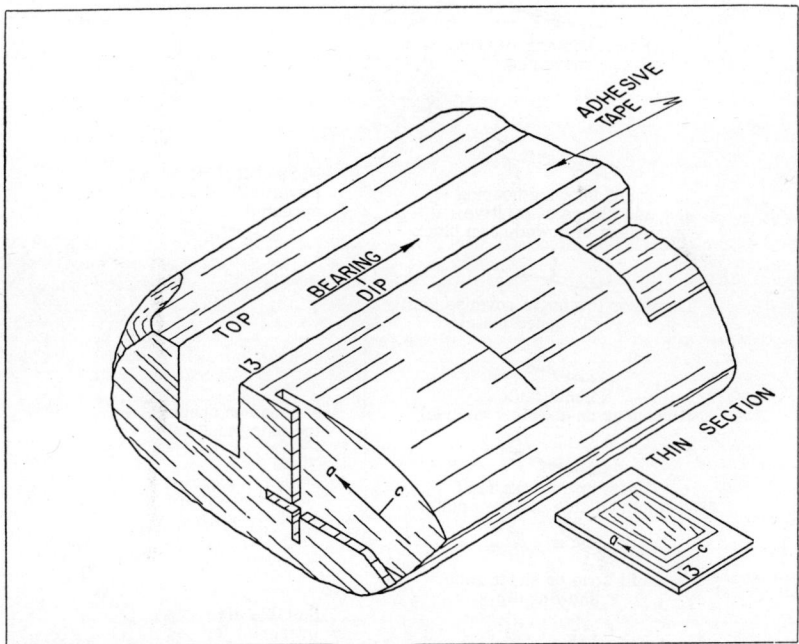

FIGURE 61. Oriented· hand specimen labeled to show its geographic location and space orientation. The top of the specimen, coordinates, and number are marked on adhesive tape. Some workers prefer to scratch this information directly on the rock. Oriented thin sections are cut from the specimen in the laboratory and marked as shown. The section illustrated is cut parallel to the *a-c* plane and normal to the *b* axis.

FAULTS

Fault, showing dip
(Dashed where approximately located)

Barbs on dip symbol
may be omitted if pre-
ferred.

Vertical fault

Concealed fault

Doubtful or probable fault
(Dotted where concealed)

Question mark indicates
uncertainty as to
existence of fault

Fault, showing bearing
and plunge of grooves, striations,
or slickensides

Plunge measured in
vertical plane.

High angle fault
(U, upthrown side; D, downthrown side)

Normal or reverse
fault.

Fault, showing relative
movement

Fault, showing bearing
and plunge of relative
movement of downthrown block

Normal fault is shown.
Reverse fault would
appear thus:

Thrust or low angle reverse fault
(T, upper plate)

Normal fault
(Hachures on downthrown side)

For use on special
tectonic maps only.

Thrust or reverse fault
(Saw-teeth on side of upper plate)

Fault zone or shear zone,
showing dip

Fault breccia

Suitable also as an
overprint for
mylonitized zones and
broad areas of fault
breccia.

FOLDS
(May be shown in color where
structure is unusually
complex)

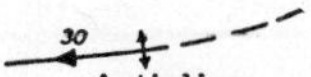

If crest line of the
fold is mapped rather
than the trace of the
axial plane, the wording
should be "showing crest
line."

Anticline
(Showing trace of axial plane and
bearing and plunge of axis.
Dashed where approximately located)

Concealed anticline

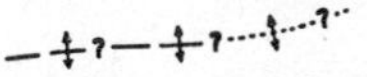

Doubtful or probable anticline
(Dotted where concealed)

Solid, dashed and
dotted as on
anticline

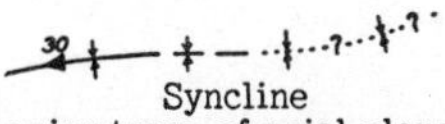

If position of trough is
mapped rather than trace
of axial plane, the word-
ing should be "showing
position of trough."

Syncline
(Showing trace of axial plane and
bearing and plunge of axis)

Overturned anticline
(Showing trace of axial plane, direction of dip
of limbs, and bearing and plunge of axis)

Overturned syncline
(Showing trace of axial plane
and direction of dip of limbs)

Plunge measured in
vertical plane

Plunge of minor anticline

Plunge of minor syncline

To be used where beds
are too tightly folded
to show individual folds
separately.

Plunge of fold axes

Horizontal fold axes

BEDDING

⌐₃₀

Strike and dip of beds

⌐₄₀

Strike and dip of overturned beds

⊢₉₀

Strike of vertical beds

⊕

Horizontal beds

₆₀ʃ   ₆₀ʃ

Generalized strike and dip
of crumpled, plicated,
crenulated, or undulating beds

₆₀⟋←₃₀

Strike and dip of beds
and plunge of slickensides

It is suggested that this
symbol be used only where
the beds are known to be
right-side up. If it is
not known which side is
up, the following symbol
is suggested: ⟋₃₀

The position of the 90
can be used to indicate
the up side of the beds.
If so, this should be
stated in the Explanation.

FOLIATION AND CLEAVAGE

⌐₆₅

Strike and dip of foliation

⊢

Strike of vertical foliation

┼

Horizontal foliation

₇₅⟋

Strike and dip of cleavage

₉₀⟋

Strike of vertical cleavage

╫

Horizontal cleavage

To be used for either
primary or secondary
foliation.' For
distinguishing between
various types of planar
structures, the following
additional symbols are
suggested:

⫻ ⫻ ⫻ ⫻ ⫻

The type of cleavage
mapped should be
specified in the
Explanation

LINEATIONS
(Includes flow lines, alinement
of minerals, inclusions, streakings, etc.)

These can also
be used for
special type's of   Bearing and plunge of lineation
lineation such as
intersection of
planes, wrinklings,
etc., but such uses   Strike and dip of foliation
should be so stated   and plunge of lineation
in the Explanation.

Vertical lineation

Horizontal lineation

JOINTS

Strike and dip of joints

Strike of vertical joints

Horizontal joints

Point of observation
is at base of arrow.

Plunge measured in the
vertical plane. If the
lineation is measured in
the plane of the foliation
it is suggested that the
term rake be used and
that the symbol be shown
thus:

It is recommended that
the term pitch be aban-
doned as it has been so
widely used in both senses
and appears on many pub-
lished maps indicating
the vertical angle.

* On the previous list of survey map symbols, the term "pitch" was applied to lineations measured in the vertical plane. However, many comments have been received urging that the term "plunge" be used instead, as "plunge" was originally defined by Lindgren in this sense and is so defined in the text books of Lindgren and Billings. It was also pointed out that "plunge" has rarely been used in any other sense, and that an increasing number of geologists are adopting Lindgren's definition. For these reasons, the Map Symbol Committee decided to adopt the term "plunge" for the angle measured in the vertical plane. This is the measurement that is usually recorded on geologic maps, but for special structural problems, some geologists prefer to record the angle measured in the plane of the foliation, fault or vein, or in the axial plane of the fold. Lindgren and Billings used "pitch" for this angle. However, "pitch" has been so widely used in both senses and has appeared on so many published maps indicating the vertical angle, that its continued use is likely to lead to further confusion. Therefore, after wide discussion with structural geologists, the committee decided to abandon the term "pitch" and to suggest the use of the term "rake" for the angle measured in the plane of the structure. This term has been occasionally used to describe the inclination of ore bodies, but it has never been clearly defined.

Belt series in the Avery district of Shoshone County, Idaho. The elements of these folds can be readily related to the three above-mentioned axes.

The folds shown are comparatively small and simple; the features such as lineation, jointing, and schistosity are pronounced. Careful detailed mapping may demonstrate these on large, complicated structures. The use of such components correlated with microfabrics permits the reconstruction of complexly deformed units.

Laboratory Techniques and Presentation of Data

Once the macrocomponents are determined, the microscope is resorted to for further aid. The following publications present the labora-

Figure 62. Tight fold in white schistose quartzite. Schistosity is produced by parallel flakes of muscovite. The lineation is parallel to *b* and is shown in figure 63.

tory techniques of determining microfabric and the methods of its presentation: "Laboratory Technique of Petrofabric Analysis," by Ingerson,[56] which begins with the study of the hand specimen after it has been oriented in the field and carries the reader through the steps of microscopic analysis of the specimen and recording and presentation of the data; "Structural Petrology of Deformed Rocks," by Fairbairn,[57] chapters 8, 9, and 10, giving much the same information as Ingerson's paper but with a somewhat different presentation; and "Federow Method (Universal-Stage) of Indicatrix Orientation," by Haff,[58] a paper presenting the universal-stage and its operation.

In the laboratory the oriented hand specimens collected as described above are studied megascopically or with the binocular or by both methods to determine the selection of coordinates. The most prominent structure plane such as schistosity is taken as the a–b plane. The direction for the thin sections are determined by the choice of the three reference axes. Generally three sections are cut normal to a, b, and c (fig. 61). If there is doubt as to the selection of the axes, random thin sections may be cut and studied in order to discover any preferred orientation.

Quartz, calcite, and the micas are the minerals commonly used for the construction of orientation diagrams. In quartz the attitude of the optic axis is determined; in calcite, the optic axis or the poles (the normal to) of twin or glide planes are mapped; in the micas the poles of the cleavage (010-plane) planes are plotted. The measurements of these features are carried out on the universal stage mounted on a petrographic microscope; the results are plotted on a Schmidt net (fig. 68). This net is the equal-area, azimuth projection of the lower half of a sphere.

If quartz orientations are being studied, the direction of the c-axis is determined on the universal-stage. This axis is brought into coincidence with the microscope axis; its original attitude may then be read from the graduated circles of the U-stage. These values are a bearing and an angle of inclination, which are then plotted on the net. Both are represented by a point on the diagram that in a three-dimensional solid would be the point at which the c axis pierces the lower hemisphere. If mica books are to be oriented, however, the cleavage planes are brought into line with the plane containing the microscope axis, and the attitude of the poles (the normals to) of the cleavage planes is plotted in much the same manner as the quartz c axes.

[56] Ingerson, Earl, *Laboratory Technique of Petrofabric Analysis* (pt. 2 of *Structural Petrology* by Knopf, E. B., and Ingerson, Earl): Geol. Soc. America Mem. 6, pp. 209-262, 1938. (A separate of this is Carnegie Inst. Washington, Geophysics Lab., Paper 959.)

[57] Fairbairn, H. W., *Structural Petrology of Deformed Rocks*, pp. 106-131, Cambridge, Mass., Addison-Wesley Press, Inc., 1942.

[58] Haff, J. C., *Federow Method (Universal-Stage) of Indicatrix Orientation:* Colorado School of Mines Quart., vol. 37, no. 3, pp. 3-28, July 1942.

Normally two or three hundred grains are oriented and plotted for each thin section. Upon completion of plotting, the concentration of points is contoured. That is, the concentrations according to percentages are separated by lines having values such as 1, 2, 3, and 4 percent. The

FIGURE 63. Detail of crest of fold in figure 62. The scale is parallel to the *b* axis. The lineation is caused by the intersection of the schistosity and a bedding plane. The joints normal to the *b* axis are *a-c* joints.

area for each percentage is given a pattern, and the greatest density is generally shown in solid black. This procedure completes the orientation diagram except for marking the reference plane, which must be clearly shown.

Petrofabric diagrams are classed as "elemental" when a single mineral is studied in one thin section, and as "collective" when the plotted points from a number of elemental diagrams are combined into a single diagram.

In general, statistical diagrams of quartz show (fig. 69) either isolated maxima or a distribution of axes in bands that may or may not
contain maxima. Fairbairn,[59] has an illustration in his book that shows
the types of quartz diagrams possible from thin sections cut normal to the
reference axes *a*, *b*, and *c*.

FIGURE 64. Looking south along *b* axis of a tight, slightly overturned fold in bedded,
schistose quartzite.

The micas (biotite, muscovite) ordinarily show a complete or partial
girdle (fig. 70) around *b* parallel to the *a–c* plane with a maximum at *c*.
Figure 71 is a schematic drawing of the fold shown in figure 64.
This fold is in a schistose, white quartzite, in which parallel muscovite
flakes produce the schistosity. The simplified petrofabric diagrams super-

[59] Fairbairn, H. W., *Structural Petrology of Deformed Rocks*, p. 8, Cambridge, Mass., Addison-Wesley
Press, 1942.

imposed normal to the *a*, *b*, and *c* fold axes are the types one may expect from the statistical study of the muscovite books.

FIGURE 65. Lineation on west or upper limb of fold shown in figure 64. The horizontal lineation is due to intersection of bedding (dipping 45° out of the picture) and schistosity (dipping 50° out of picture). The lineation almost normal to it is caused by quartz filling tension joints normal to the *b* axis and opened by stretching along that axis. The fold shown in figure 64 was discovered by careful mapping of the lineation and schistosity.

POSSIBLE APPLICATIONS

Detailed structural studies to be complete should make thorough use of petrofabrics. The individual techniques of petrofabrics, however, may be used separately on special problems.

Ingerson [60] gives a list of possible applications of petrofabrics. There

---

[60] Ingerson, Earl, *Why Petrofabrics?*: Am. Geophys. Union Trans., vol. 25, pp. 636-652, 1944.

are additional uses, and with the advance of the science still others will be found.

From magnetically oriented drill cores (see "Magnetic-Core Orienta-

FIGURE 66. Looking along *b* axis of an isoclinal fold in the same white quartzite as shown in figures 62-65. The schistosity and the bedding planes intersect at a low angle and produce prominent lineation as shown in detail in figure 67.

tion") oriented thin sections can be obtained. The statistical analysis of such thin sections tied in with other known petrofabric data will help work out subsurface structures.

Hohlt,[61] in a recent paper on limestone porosity, made statistical studies of carbonate rocks with the idea of correlating mineral orientation and porosity. He concluded that orientation is related to dolomitization.

---

[61] Hohlt, R. B., *The Nature and Origin of Limestone Porosity:* Colorado School of Mines Quart., vol. 43, no. 4, Oct. 1948.

The study of sedimentary rocks by petrofabric methods has been comparatively neglected. Ingerson [62] indicates that valuable information on current direction may be gained from the fabric study of sediments.

FIGURE 67. Pronounced lineation parallel to *b* axis of fold. The angle this lineation makes with the horizontal gives the plunge of the fold. The joints normal to the lineation are *a-c* joints.

Wayland [63] has shown that the tendency of the long axis of quartz grains is to be parallel to the *c* axis and that the *c* axis lies nearly parallel to the bedding plane in sandstones.

Size and shape of fragmental materials in clastic sedimentary rocks

[62] Ingerson, Earl, *Fabric Criteria for Distinguishing Pseudo-Ripple Marks from Ripple Marks:* Geol. Soc. America Bull., vol. 51, pp. 557-570, 1940.

[63] Wayland, R. G., *Optical Orientation in Elongate Clastic Quartz:* Amer. Jour. Sci., vol. 237, pp. 99-109, 1939.

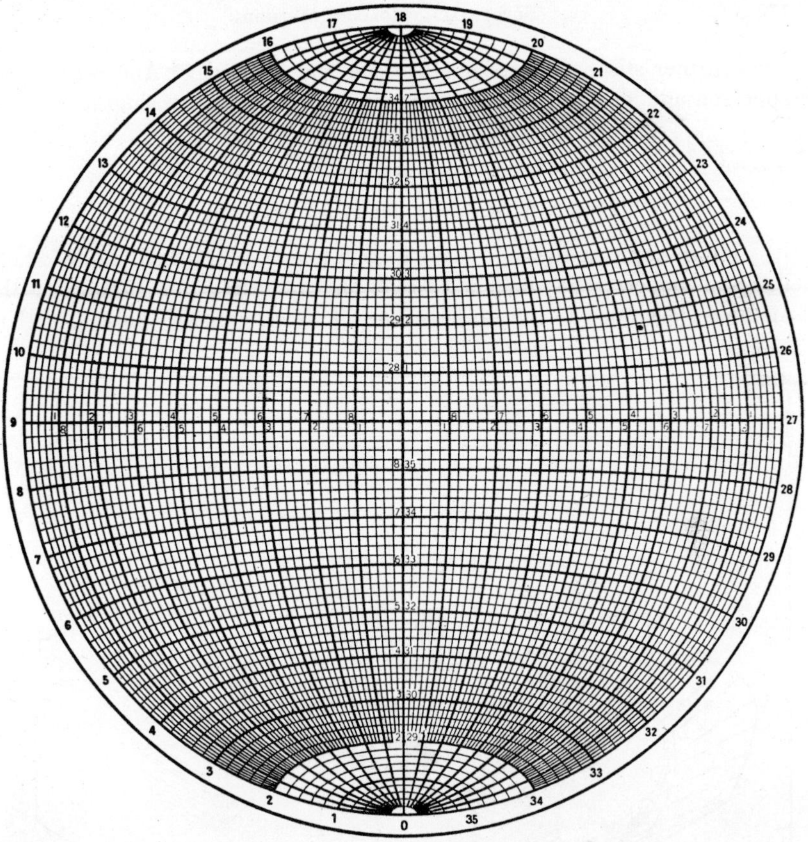

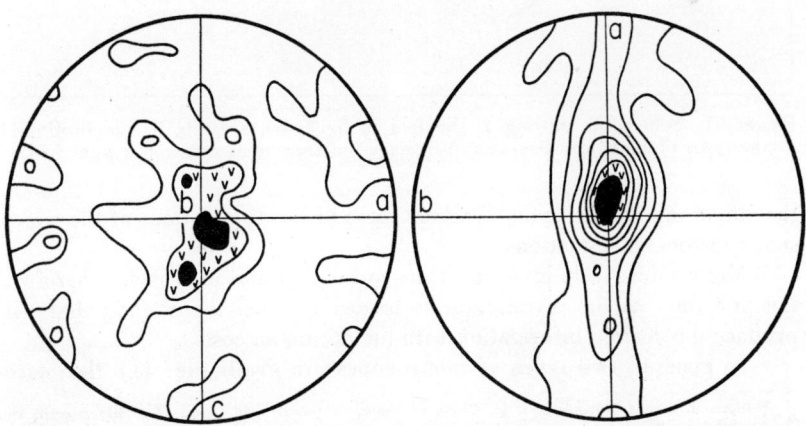

FIGURE 68. Reproduction of a Schmidt net of 20-centimeter diameter for use in petrofabric work. (After Cloos.)

FIGURE 69. 470 quartz axes showing maxima about *b*. Contours 4-2-1 percent. (Modified from Fairbairn.)

FIGURE 70. 134 cleavage poles of biotite forming a girdle parallel to *a-c* with maximum about *c*. Contours 8-6-4-2-1 percent. (After Fairbairn.)

need further study. Papers by Ingerson and Ramisch,[64] Anderson,[65] and Ingerson and Tuttle [66] point to the origin of quartz grain shapes.

MICRO (PETROGRAPHIC) ANALYSIS
WARREN R. WAGNER and JOHN W. GABLEMAN

Microscopic analysis of rocks is such a well-established field of study that this paper is not intended to be an exhaustive presentation of the methods developed to the present time. The writers wish only to call to

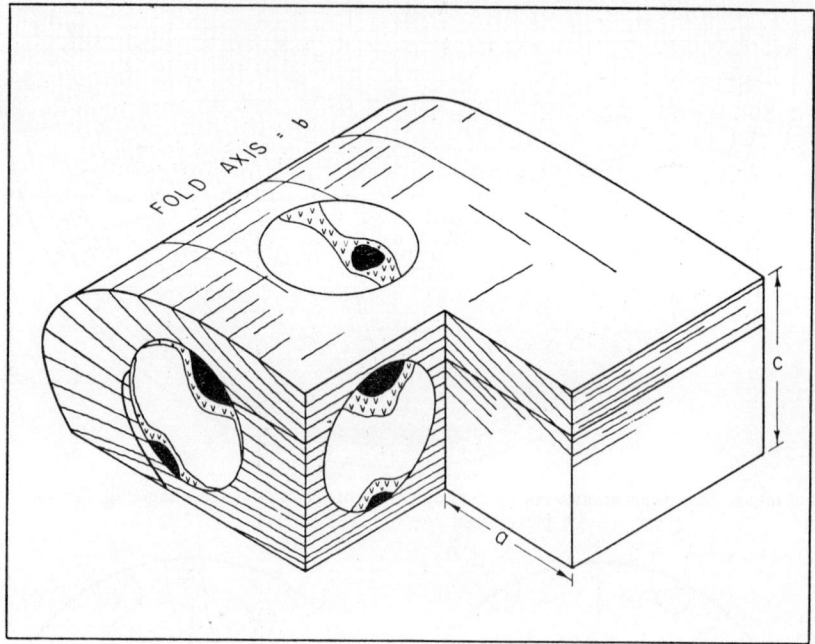

FIGURE 71. Schematic drawing of the fold in figure 64, showing in simplified form the types of statistical diagrams to be expected from mica flakes in a tight fold.

the attention of students and workers some of the applications of the micro-scope to rock examinations.

Many micro-techniques are time-consuming and expensive; therefore, the first duty of the investigator is to decide upon the method that will produce the desired information with the minimum cost.

In general, two types of microscopes are availiable: (1) the petro-

---

[64] Ingerson, Earl, and Ramisch, J. L., *Origin of Shapes of Quartz Sand Grains:* Am. Mineralogist, vol. 27, pp. 595-606, 1942.

[65] Anderson, J. L., *Deformation Planes and Crystallographic Directions in Quartz:* Geol. Soc. America Bull., vol. 56, pp. 409-430, 1945.

[66] Ingerson, Earl, and Tuttle, O. F., *Relations of Lamellae and Crystallography of Quartz and Fabric Direction in Some Deformed Rocks:* Am. Geophys. Union Trans., vol. 26, pt. 1, pp. 95-105, 1945.

graphic miscroscope and (2) the binocular microscope. The former makes use of polarized light for the identification of minerals and the detailed study of thin sections, whereas the latter uses ordinary transmitted or incident light and is normally used for observing the larger features such as lithology, texture, and structure.

Some rivalry appears to exist among workers as to the superior quali-

FIGURE 72. Basic equipment required for preparing a rapid heavy-mineral concentrate of a rock.

ties of the two types of microscopes. Each type has its field of usefulness. Numerous problems are solved most economically with the aid of both kinds of microscopes. Any well-equipped microscopic laboratory should contain both petrographic and binocular microscopes, and should have trained technicians trained to operate them.

On pages 119-120, Rittenhouse discusses some of the uses of the binocular microscope on sedimentary rocks. His article deals primarily with detrital mineralogy, whereas this paper is chiefly concerned with thin-section investigation, although a rapid method of heavy-mineral separation is presented.

Rapid Method of Heavy-Mineral Separation [67]

One of the major drawbacks to heavy-mineral studies is the time required in making clean separations. A competent worker with the proper setup can make a complete heavy-mineral separation, including the permanent micro-mount, in a lapsed time of 10 minutes by following the steps outlined and illustrated below.

Necessary Equipment

Figure 72 shows the equipment required in the procedure for making a single separation. The articles shown are bromoform, a standard 3-inch evaporating dish, stainless-steel teaspoon, filter paper, product to be concentrated, wash bottle with alcohol, adjustable wooden rack, and glass funnels and beakers for filtering. The setup may be varied to fit the size of the problem. If a number of separations are to be carried on, the process may be speeded up by having the workbench arranged for multiple equipment.

Procedure

Figure 73, steps *a*, *b*, *c*, *d*, *e*, and *f*, illustrate the stages of making a separation. *Step a:* The prepared and weighed sample is placed in the evaporating dish with sufficient bromoform (diluted to desired density) to float the light fraction freely. The material is then stirred thoroughly with the teaspoon in order that the "heavies" may sink. *Step b:* After the sample is stirred, it is allowed to settle for one-half to one minute, and the larger part of the floating light fraction is spooned off into the filter. *Step c:* The remaining "lights" are then carefully poured off in successive order with a few seconds of gentle circular motion (panning) between each pouring. *Step d:* The filter containing the bromoform and "lights" is then held above the evaporating dish to allow the bromoform that is filtering out to wash down the sides of the dish. This step is repeated until the separation is complete and the spout or pouring side of the dish is free from the light fraction. *Step e:* The final concentrate is washed from the dish into the second filter with alcohol from the wash bottle. The "heavies" are then washed clean of bromoform, filtered, and dried on a hot plate. *Step f:* The final, dried, heavy fraction is then divided and mounted as desired.

If proper caution is exercised throughout the procedure, a clean separation is obtained. Bromoform is expensive and care should be taken to avoid wastage.

Preparation of Thin Sections

A thin section is a slice of rock or mineral 0.03 millimeter thick mounted on a glass slide for examination under the petrographic microscope. The nonopaque, rock-forming minerals have a high degree of trans-

---

[67] Adapted from procedure taught the senior author by Dr. J. L. Anderson, Department of Geology, Johns Hopkins University.

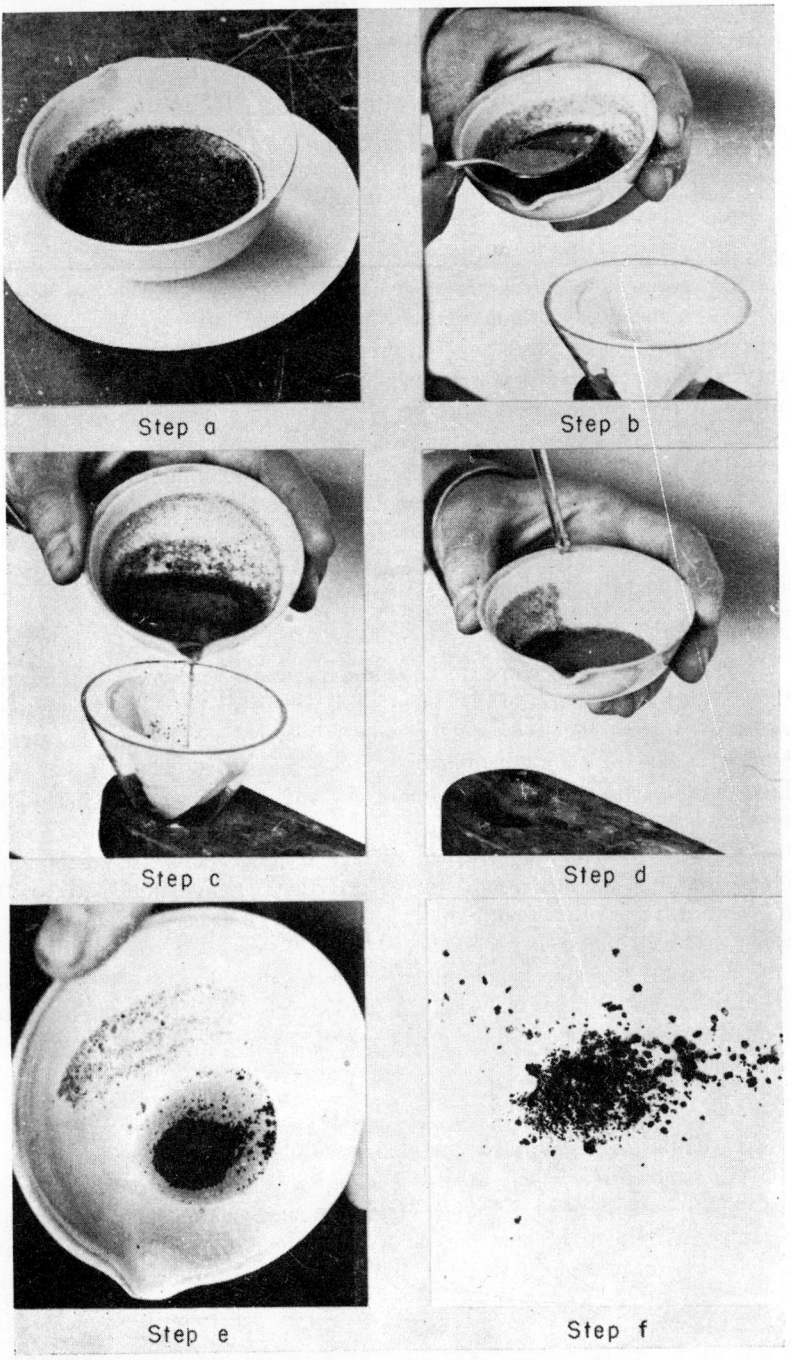

Step a

Step b

Step c

Step d

Step e

Step f

FIGURE 73. Procedure followed in making a rapid heavy-mineral separation.

parency in thin slices, and their different reactions to transmitted, polarized light constitutes the basis of optical mineralogy. Because the optical properties of the minerals differ with the thickness of the section, 0.03 millimeter has been selected as an arbitrary standard thickness.

Thin sections may be made of well-consolidated rocks, of friable or less well-consolidated rocks, of individual minerals, of fragments of minerals or rock, or of a heavy-mineral concentration. The friable or poorly consolidated material is impregnated with a bonding cement before sectioning, whereas the fragmental material or heavy-mineral concentrate requires a special technique that will be described later.

The equipment necessary for preparing thin sections in the laboratory include a diamond power saw, power-lap wheels of cast iron, glass plates, a hot plate, and a microscope. Materials needed are mounting cement, abrasives, glass mounting slides, and cover glasses.

The number of laps and grades of grinding compound used depends upon the technique to be followed. At least two grinding laps are required, one each for course and fine abrasives. The technique briefly outlined below is used at the Colorado School of Mines; it may be varied to suit the needs of individual specimens.

Sawing and Grinding

From the rock specimen to be studied, saw a piece 3 to 5 millimeters thick with the two parallel, flat faces. Trim the edges until the final slice measures approximately 30 x 22 x 3 mm. Now choose the smoothest side and, with 80- to 100-mesh abrasive on the course lap wheel, grind this face until all traces of the saw marks disappear. Wash the slice thoroughly, transfer to the 320-mesh abrasive on the fine lap wheel, and polish the ground surface. The polished surface is to be mounted next to the glass slide and, therefore, must be perfectly flat. A final polish with 600-mesh abrasive is often required. The worker must learn from experience when this final polish is necessary.

When the grinding is completed, wash the slice thoroughly to remove all abrasive and foreign material. A tooth brush is helpful for this. The shaped and polished piece now has one flat, smooth surface ready for mounting.[68]

Mounting the Slice

Place the rock slice on a hot plate with the polished side up and exclude all moisture. When drying is complete, place a standard petrographic glass slide (45 x 25 mm.) beside the specimen and allow the temperature of both to become the same. (If a controlled-temperature hot plate is available, keep it at about 300° F.) Cover the upper surface of

---

[68] The abrasives are kept in kitchen-size salt shakers. Abrasive is applied to the wet lap wheel as needed. Some workers mix it with water in a bottle provided with a glass tube through the cork in order that the contents of the bottle may be shaken on to the lap as required. Experience is necessary before the operator can obtain the right mixture of water and abrasive that will produce the most efficient cutting.

both the rock slice and the glass slide with an even, thin layer of raw Canada balsam and cook. If the rock slice is porous, apply an excess of balsam. The hot plate should be kept level so that the balsam may flow evenly in all directions.

The success of mounting the specimen lies in cooking the balsam. If it is insufficiently done, the rock slice will not adhere to the glass slide; if it is overdone, the balsam will be too brittle and will break away when ground. The object is to cook the balsam to the point that, when it is cooled, one may barely dent it with the thumbnail. While cooking is in progress, the cement is continually tested by taking a small amount on the end of a toothpick and biting it between the front teeth. When balsam is properly cooked, it will stick to the teeth, barely begin to pull, and then abruptly break. In other words, it becomes "tacky." [69]

When the cement is properly cooked, place the rock slice on the glass slide with the two balsam-covered sides in contact. Remove the mount from the hot plate to an asbestos pad. With the eraser end of a pencil, work the rock slice around, meanwhile applying considerable pressure in order that the air bubbles and excess balsam will be squeezed from between the slice and the slide. Center the slice and allow the cement to set while pressure is applied. As the mount cools, test the balsam for hardness with the thumbnail. Cooling should be allowed to proceed normally, as sudden changes in temperature tend to pull the rock slice from the glass. The mount is now ready to be ground to the desired thinness.

Grinding to 0.03 Millimeter

The final grinding to 0.03 millimeter is carried out in three stages. The mount is ground on the coarse lap with 80- to 100-mesh abrasive until it is about 0.10 millimeter thick. It is then cleansed carefully and transferred to the fine lap wheel with 320-mesh abrasive. During this stage, with proper caution, a well-cemented slide may be ground almost to the desired thinness. The writers ordinarily carry the grinding with 320-mesh abrasive to a point where such colored minerals as hornblende or biotite are fairly transparent when held before a light source. Further grinding by mechanical means becomes hazardous; therefore, the last stage of grinding is done on a glass plate with 600-mesh abrasive in water. The section is moved with a gentle, circular motion on the plate. If the slide has become wedge-shaped, more pressure is applied on the thicker portion to bring the entire rock slice to the same thickness. As the slide is now very thin, constant checking for thickness under the microscope is necessary. Quartz is the mineral commonly used as an index, and the section has reached 0.03 millimeter when quartz shows a faint, straw-yellow, interference color.

---

[69] Other cements are available, but the writers have had the most consistent results with raw Canada balsam.

Mounting the Cover Glass

The concluding operation in preparing a thin section is to mount the cover glass. The simplest method of mounting the cover glass is to use cooked balsam dissolved in xylene. To cement the cover glass, place a thin, even layer of prepared balsam on the surface of the section. Put the cover glass in place and press down sufficiently with the eraser end of a pencil to remove all air bubbles and excess balsam. After removing the squeezed-out balsam, the slide is complete. Several days may be required for the xylene to evaporate and for the balsam to set; but with proper precautions, the section can be used immediately. Until the balsam is dry, the slide should be stored in a flat position.

Cover glasses are obtainable in various sizes and thicknesses. For most purposes the No. 1 thickness is preferable.

Thin Sections of Fragments or Heavy Concentrates

The investigator may find that fragmental materials and heavy-mineral concentrates contain minerals which defy identification by physical or optical means in their present state. If the material is such that a thin section may aid identification, it is not difficult to make one from even rather finely divided substances. The process and materials used are much the same as those used in making an ordinary thin section, except that no sawing is required and the mounting procedure is repeated for the second time.

To make a thin section of a heavy-mineral concentrate, place a petrographic, glass, mounting slide on the hot plate. Put raw Canada balsam on it and cook. Immediately before the balsam is sufficiently cooked, sprinkle an excess amount of the heavy concentrate into the hot balsam and permit the grains to settle. An excess is required as some material will be lost in a later transfer. For this step, the cooking of the balsam is not critical as this material is merely a holding mount. Remove the slide from the hot plate and allow to cool.

The next step is grinding a flat, permanent-mounting surface on the mineral grains with a 150- to 200-mesh abrasive. The amount of grinding necessary depends upon the grain size; if the grains are of uniform size, one should grind about halfway through them. The technician must learn the required amount to be ground by experimenting. An inspection under a low-power microscope may prove useful in determining whether or not the surface on the grains is suitable. If not, a transfer to 320- or even 600-mesh abrasive may be essential. (The entire grinding procedure may be carried out on glass plates with the desired size abrasive.)

Assuming that the ground surface is now flat and polished for mounting, the slide is cleaned of all foreign material. Now moisten a cloth with xylene and slowly dissolve the excessive cement from between the grains; leave only sufficient balsam to hold the grains in place.

The next step is remounting the grains on a new petrographic slide with the flat surface on the glass. This operation requires some practice, as it must be accomplished quickly and at a given time. To accomplish the remounting, a glass slide covered with an even, thin layer of balsam is placed on the hot plate and the balsam is cooked within an "instant" of the required consistency. The first mount is then laid upon the new slide with the flat surface of the grains in contact with the newly cooked balsam. It is allowed to remain there until the undissolved balsam melts. (The time required here is short; the new balsam should be correctly cooked upon completion of the operation. One must remove rapidly.) The slides are now removed to an asbestos pad; the upper one (the temporary mount) is pressed down, moved with a circular motion, and gently slid in the direction of the narrowest dimension off the lower slide. A small quantity of mineral grains may adhere to the first mount, but their loss is not serious as an excess of grains were originally used. For the final step, the slide is allowed to cool. Then it is ground to the required 0.03 millimeter in thickness, and the cover glass is mounted as recommended for an ordinary thin section.

Manual Preparation

In the foregoing procedure for thin-section preparation, a laboratory with power equipment was assumed. Thin sections are often needed when this equipment is not available. Excellent sections may be prepared manually from rock chips or even small well cuttings by carrying out the grinding operations on glass plates. The manual procedure is often to be preferred with friable material, very small chips, or soft rocks such as limestones and shales. For base-camp use, a small, compact grinding outfit can be assembled with which a careful worker can make thin sections of any desired rock.

THIN-SECTION STUDY

A complete petrographic analysis requires the use of every available tool. The use of thin sections is merely one phase in the breakdown of a rock into its component parts. It is, however, an important phase, and usually the choice of the final method of study to be employed in the study of a given rock is determined from thin sections or a combination of thin section, polished surface, and hand specimen. The finer-grained sediments such as shales, siltstones, and mudstones yield comparatively little information to the investigator in thin section, whereas, the medium-grained clastic rocks offer a fertile field for micro-study. The flow sheet (fig. 74) shows the organization and methods that may be used in a modern petrographic laboratory.

Some of the more important data and information that may be acquired from thin-section studies of sedimentary rocks are summarized in the paragraphs below.

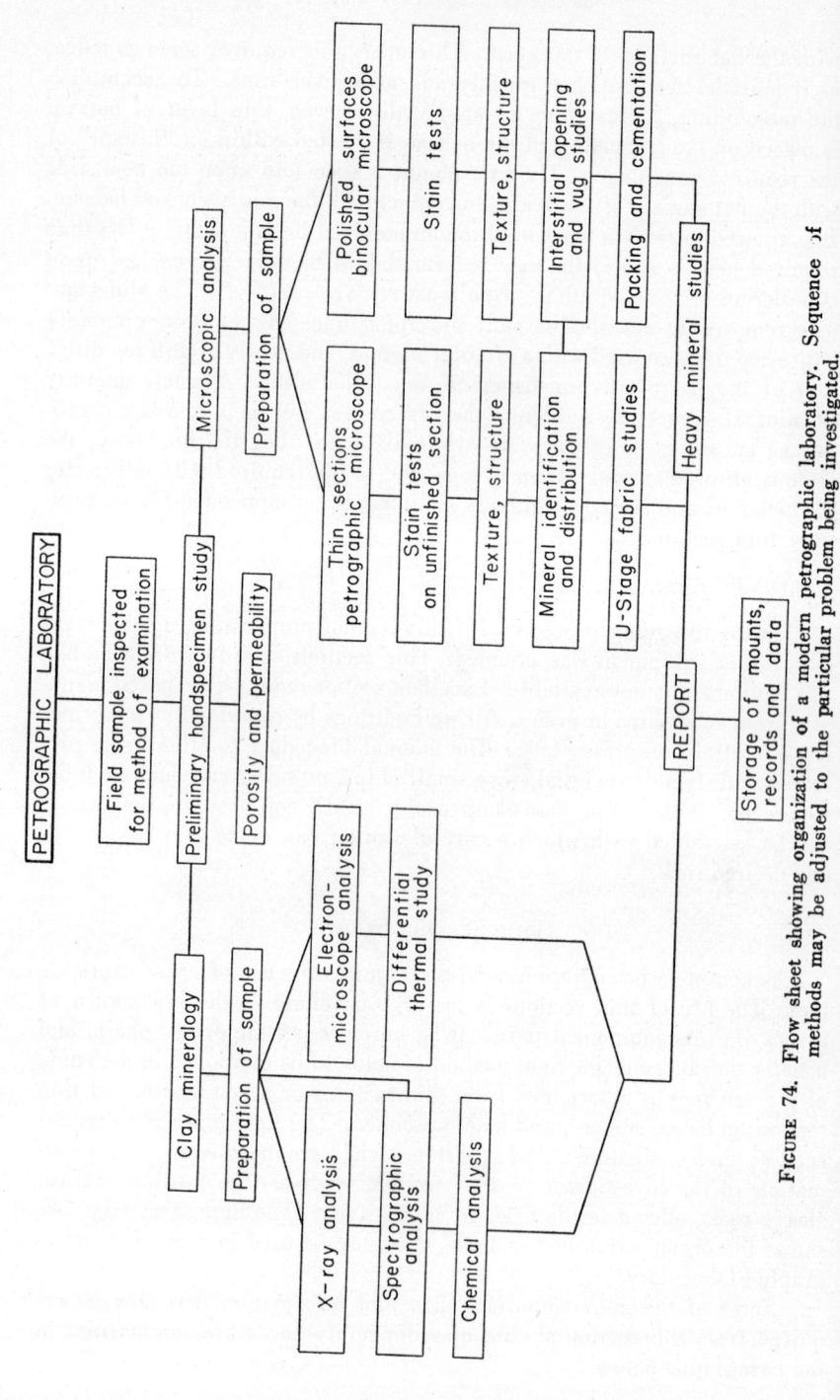

FIGURE 74. Flow sheet showing organization of a modern petrographic laboratory. Sequence of methods may be adjusted to the particular problem being investigated.

Rock Classification [70]

One of the first aims of the petrographer is to classify the rock as to type, i.e., sandstone, calcareous sandstone, argillaceous sandstone, etc. The typing or naming of the rock in itself yields much information about its internal makeup; and, incidentally, the thin section will usually reveal sufficient information on the clastic materials of any size for classification purposes.

Cementation

Waldschmidt,[71] in his paper on cementing materials of sandstones in the Rocky Mountain region, has worked out a sequence of deposition for the various binding materials. He also shows their relation to porosity and permeability in sandstones having various combinations of these cementing minerals. His order of deposition of cementing minerals is as follows:

One cementing mineral	*Two cementing minerals*	*Three cementing minerals*	*Four cementing minerals*
Quartz	1st Quartz 2nd Calcite	1st Quartz 2nd Dolomite 3rd Anhydrite or 1st Quartz 2nd Dolomite 3rd Anhydrite	1st Quartz 2nd Dolomite 3rd Calcite 4th Anhydrite

Waldschmidt's conclusions are based on the study of 111 sections. His paper is an excellent illustration of the use of thin sections in this type of study.

Porosity and Permeability [72] [73]

Many of the principles governing porosity and permeability go beyond the realm of thin-section study; yet, much of the information as to why and how certain rocks have as much or little porosity and permeability may be learned from the study of thin sections. Grain size and arrangement or packing, authigenic mineralization, alteration, interstitial materials, and cementation all exert considerable influence on the amount of pore space available in a rock.

A single thin section presents a two-dimensional view; therefore, for

[70] Pettijohn, F. J., *Sedimentary Rocks,* New York, Harper and Brothers, 1949.

[71] Waldschmidt, W. A., *Cementing Materials in Sandstones and Their Probable Influence on Migration and Accumulation of Oil and Gas:* Am. Assoc. Petroleum Geologists Bull., vol. 25, no. 10, pp. 1839-1879, 1941.

[72] Graton, L. C., and Fraser, H. J., *Systematic Packing of Spheres with Particular Relation to Porosity and Permeability:* Jour. Geology, vol. 43, pt. 1, pp. 785-909, 1935.

[73] Fraser, H. J., *Experimental Study of the Porosity and Permeability of Clastic Sediments:* Jour. Geology, vol. 43, pt. 1, pp. 910-1010, 1935.

complete studies three sections each normal to the other should be cut from a specimen. It is also an excellent idea to polish the surface of the specimen from which each section was originally sawed and to study these under the microscope. If the data obtained from these studies are

FIGURE 75. Photomicrographs of thin sections from four sandstones of Morrison formation near Golden, Colorado, illustrating morphological differences. Note chalcedonic cementation in upper right.

recorded on a block diagram, any directional variation in space relationships shows up more readily.

Thin-Section Correlation

Consolidated sedimentary rocks are often sufficiently individualistic enough that a careful megascopic inspection is adequate for correlation purposes. On the other hand, in a thick, monotonous series of shales, of limestones, of sandstones, or of combinations of these, megascopic identification and correlation are frequently uncertain or impossible. Identification and differentiation of a specific horizon may be accomplished by a microscopic study of thin sections. Possible microscopic evidences that may be used include textures, structures, cementing minerals, detrital material, degree of crystallization, amount of recrystallization, finer organic forms, and other morphological criteria. The photomicrographs of figure 75 illustrate a few of the microscopic features of four different sandstones.

Vug and Opening Studies

Normal-sized thin sections are limited in area. Ordinarily, in petrographic work the slides used are 45 x 25 millimeters, and the finished rock section rarely covers more than two-thirds of this area. For vug and opening studies, larger areas are required. The writers have made and used to advantage thin sections mounted not only on a 2 x 2-in. kodachrome, slide-cover glasses, but also on standard $3\frac{1}{4}$ x 4-in. lantern, slide-cover glasses. These sections are usually somewhat thicker than 0.03 millimeter, but they may be projected in an ordinary slide projector equipped with polaroids. If the sections are cut in parallel series, a fair picture of the size and continuity of "vugation" is presented.

Storage of Sections

Problems of storage of specimens and sections for future reference continuously plague the research worker. Each man has his own solution. One space-saving device is to mount the thin section, heavy-mineral concentrate, and a chip sample from the same specimen on a three-inch biological glass slide rather than on the conventional petrographic slide.

Summary

Thin sections are not the answer to all sedimentary problems, but they disclose the internal view of a rock from which the research man may add to his present knowledge of the rock; and at the same time, they aid in determining his future mode of attack.

SIZE ANALYSIS

L. W. LeROY

Size analysis permits comparison of grain-size similarities and dissimilarities of sands and dissaggregatable sandstones. With regard to results obtained by size analysis, Twenhofel and Tyler [74] comment:

Statistical analyses certainly permit rapid and easy comparison of large numbers of sediments and render it simple to point out similarities and differences. The best that may be stated is that the significances of the studies are not apparent. Statistical studies certainly permit extensive use of mathematical formulae which are of interest to those mathematically inclined. The writers have found these formulae of great interest but not particularly useful so far as interpretation of the sediments is concerned.

According to Pettijohn [75] the purposes and significance of size analyses are as follows:

(1) The improvement of classification and the precision of nomenclature of clastic sediments, (2) the study of the influence of grain-size distribution on porosity and permeability, (3) the study of relations between the dynamics of stream flow and the transportation of particulate materials, (4) quantitative studies of facies changes and correlation problems, and (5) identification of the agent or environment responsible for the origin of the sediment.

In the Lake Maraciabo Basin of western Venezuela, size-analysis data locally reflect the contact between the El Milagro (Pleistocene) and Onia (Pliocene) formations and between the Onia and La Villa (Miocene) (fig. 76). The Lyons (Permian) and Fountain (Pennsylvanian) contact east of the Front Range of Colorado may be locally differentiated (fig. 77). The classification of soils has been based on the grade-size principle. The procedure is utilized as a basis for the computation of the most efficient size of casing perforations in petroleum-production problems and for the selection of gravel-packing installations in water wells.

Fine-grained clastics (particles less than 0.088 mm. in diameter) may be graded by decantation or by elutriation methods. Pipette and hydrometer techniques have also been employed for fractionating fine materials.

The Wentworth size classification given in table 4 (p. 187) has been widely adopted for defining grain-size fractions of clastic sediments.

PREPARATION OF SAMPLE

A 300-gram sample is disaggregated by carefully crushing the sample in a mortar with a rubberized pestle. Grinding should be minimized to prevent excessive grain breakage. Checking the aggregate periodically by microscope is essential to insure normal and complete disaggregation.

The material is placed in a nest of U. S. or Tyler sieves. The sieve series may be hand-shaken or placed on the "Ro-Tap," a mechanical vibra-

---

[74] Twenhofel, W. H., and Tyler, S. A., *Methods of Study of Sediments*, p. 120, New York, McGraw-Hill Book Co., Inc., 1941.

[75] Pettijohn, F. S., *Sedimentary Rocks*, p. 30, New York, Harper & Brothers, 1949.

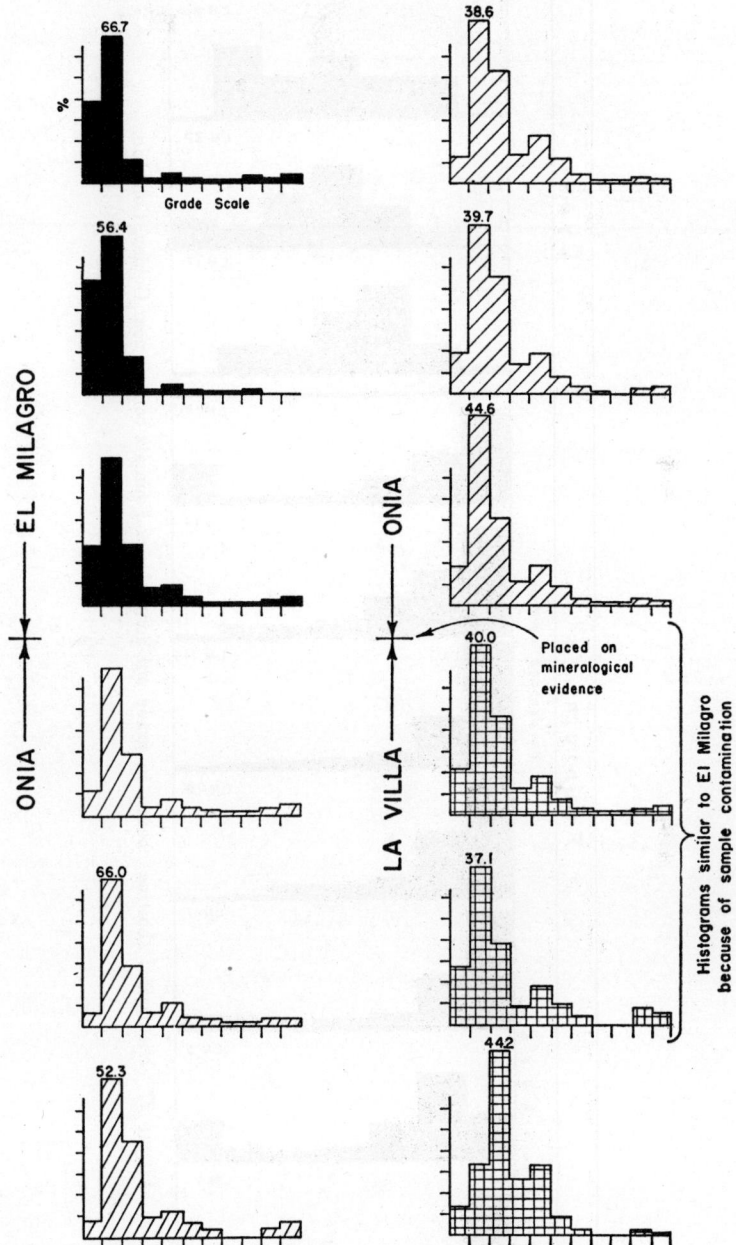

FIGURE 76. Histograms demonstrate grade changes across formational boundaries. The data shown are based on continuous ten-foot ditch samples from depths of about 1,200 feet. Maracaibo Basin, Venezuela.

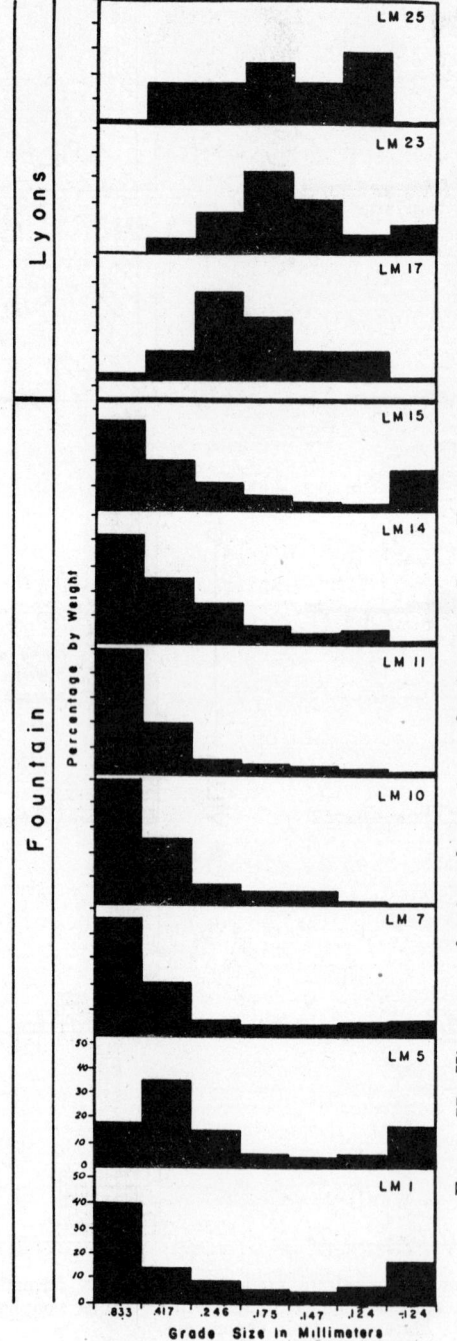

FIGURE 77. Histograms showing change in grade size across Lyons-Fountain contact near Golden, Colorado.

TABLE 4
WENTWORTH SIZE SCALE

Type of sediment	Size limit (mm.)	Sediment
Rudaceous	+256 256–64 64–4 4–2	Boulder Cobble Pebble Granule
Arenaceous	2–1 1–1/2 1/2–1/4 1/4–1/8 1/8–1/16	Very coarse sand Coarse sand Medium sand Fine sand Very fine sand
Siltaceous	1/16–1/256	Silt
Argillaceous	–1/256	Clay

TABLE 5
COMPARISON OF TYLER AND U. S. SIEVES

Tyler sieves Mm.	Mesh	U.S. sieves Mesh	Wentworth classification
3.96	5	5	
3.33	6	6	Granule
2.79	7	7	
2.36	8	8	
1.98	9	10	
1.65	10	12	Very coarse sand
1.40	12	14	
1.17	14	16	
0.991	16	18	
0.833	20	20	Coarse sand
0.701	24	25	
0.589	28	30	
0.495	32	35	
0.417	35	40	Medium sand
0.351	42	45	
0.295	48	50	
0.246	60	60	
0.208	65	70	Fine sand
0.175	80	80	
0.147	100	100	
0.124	115	120	
0.104	150	140	Very fine sand
0.088	170	170	
0.074	200	200	
0.061	250	230	Silt

tor equipped with an automatic clock for time control. Results are more complete if the material first is placed on the "Ro-Tap" and the final separation then completed by hand. After the separation stage, the weight of material retained on each screen is determined, and the results are tabulated as shown in table 6.

TABLE 6

Results of Size Analysis Graphically Shown in Figure 80

Sample No.: P-64   Lithology: Clear quartz sand   Formation: Dakota
Date: Nov. 26, 1948   Sample weight: 500 grams   Analyzed by: W. Stuart

Mesh	Mm.	Grams retained	Percentage retained	Accumulative percentage
28	.589	0.3	0.06	0.06
32	.495	0.2	0.04	0.10
48	.295	2.4	0.48	0.58
60	.246	9.9	1.98	2.56
100	.147	374.0	74.80	77.36
115	.124	43.3	8.67	86.03
150	.104	38.3	7.66	93.69
170	.088	17.5	3.50	97.19
200	.074	6.7	1.34	98.53
-200		7.4	1.47	100.00
		500.0	100.00	

PLOTTING OF DATA

Three methods of graphic representation are followed in illustrating size-analysis data, the histogram, the simple-frequency curve, and the cumulative-frequency curve. In each, weight percentages of the various grades are plotted against dimension with the former represented on the vertical axis and the latter on the horizontal axis.

Histogram Plot

The histogram method requires cross-section paper, with the largest grade size being placed on the left (figs. 78 and 80). Each grade percent is designated by rectangular blocks, the height of which represents the percentage by weight of the respective grade. Histograms are useful for rapid visual comparison. The shape of the diagram is controlled by the number and percentage of grade fractions involved. For comparative work a uniform scale should be employed. Another type of histogram involves various percentage blocks laid end on end along the horizontal axis with each grade block graphically symbolized (fig. 81).

Simple-Frequency Curve

The simple-frequency curve is commonly constructed on cross-sectional paper (fig. 80). The weight percentage is plotted on the vertical axis and the grade value on the horizontal. Points are connected by a

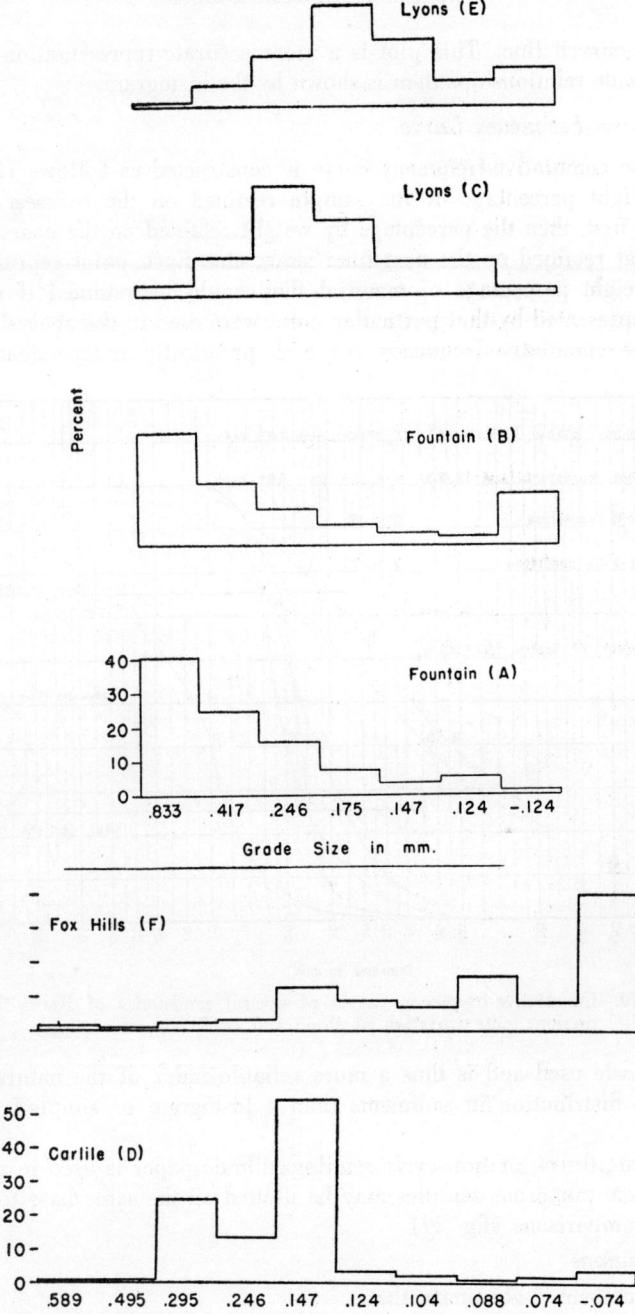

FIGURE 78. Typical histograms of various sandstones of Rocky Mountain region. Refer to figure 79 for corresponding cumulative-frequency-curve data.

smooth curved line. This plot is a more accurate representation of sediment-grade relationships than is shown by the histogram.

Cumulative-Frequency Curve

The cumulative-frequency curve is constructed as follows (fig. 79): The weight percentage of the sample retained on the coarsest sieve is plotted first, then the percentage by weight retained on the coarsest sieve plus that retained on the next finer sieve, etc. Each point represents the total weight percentage of material that would be retained if only the sieve represented by that particular point were used in the analysis.

The cumulative-frequency curve is practically independent of the

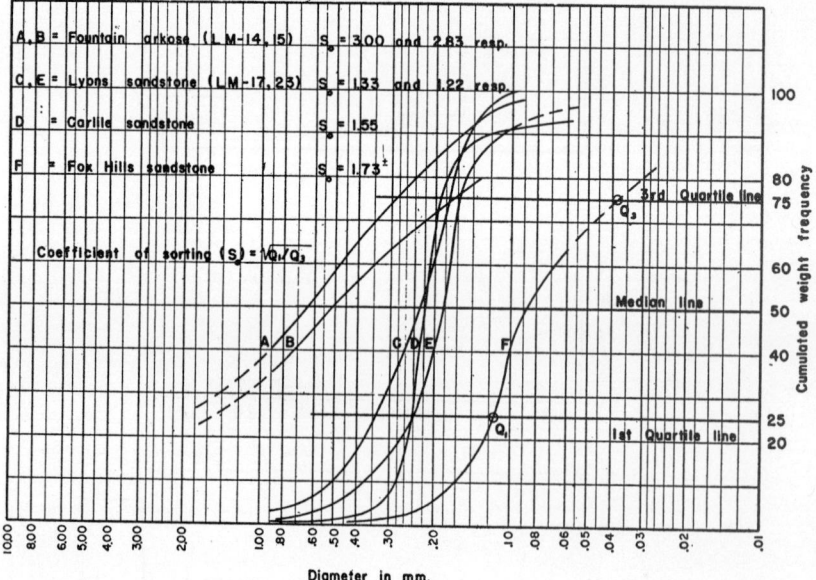

Diameter in mm.

FIGURE 79. Cumulative-frequency curves of several sandstones of Rocky Mountain region; note variation of slope and coefficient of sorting.

grade scale used and is thus a more reliable index of the nature of the particle distribution in sediments than a histogram or simple-frequency curve.

Two-, three-, or four-cycle semilogarithmic paper is used in plotting. Data from numerous samples may be plotted on the same base to permit direct comparisons (fig. 79).

Computations

Krumbein [76] comments that

Some workers have used cumulative curves in a purely descriptive manner, similar to the use of histograms. That is, the slopes of the curves, their spread,

[76] Krumbein, W. C., *Graphic Presentation and Statistical Analysis of Sedimentary Data* in *Recent Marine Sediments*, p. 564, Am. Assoc. Petroleum Geologists, 1939.

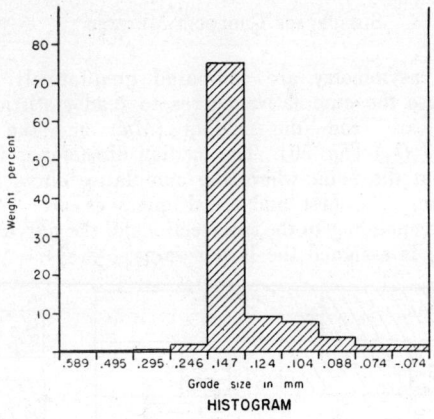

HISTOGRAM

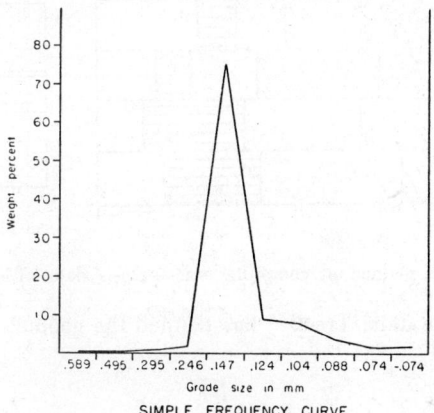

SIMPLE FREQUENCY CURVE

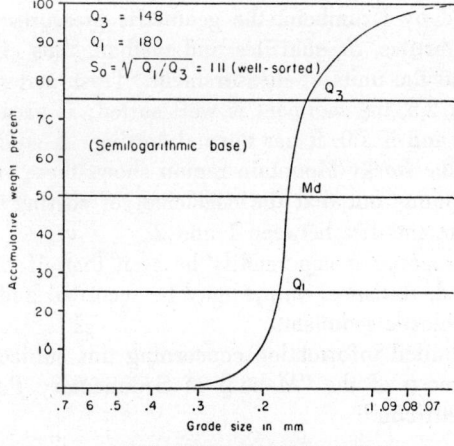

CUMULATIVE FREQUENCY CURVE

FIGURE 80. Histogram, simple-frequency curve, and cumulative-frequency curve, of the Dakota sandstone tabulated in table 6.

and the degree of asymmetry are compared qualitatively. A majority of workers, however, use the cumulative curves to read statistical values. Generally three values are read—the median (Md) and the first and third quartiles (Q_1) and (Q_3) [fig. 80]. The median diameter is found by reading the diameter value at the point where the cumulative curve is intersected by the fifty-percent line. The first and third quartiles are determined as the diameter values corresponding to the intersections of the curve with the 25- and 75-percent lines. Q_1 is assigned the larger value.

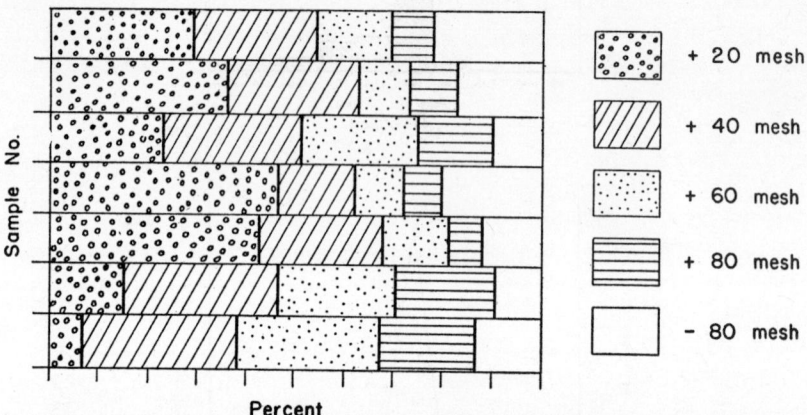

Percent

FIGURE 81. Graphic method of compiling size-analysis data of several samples.

Applying these data, Trask [77] has defined the geometric coefficient of sorting as:

$$S_0 = \sqrt{\frac{Q_1}{Q_3}}$$

As mentioned by Krumbein, the geometric measures are essentially ratios between quartiles, or quartiles and median, thus eliminating both the size factor and the units of measurement. Trask states that if the S_0 value is less than 2.5, the sediment is well sorted; if greater than 4.5, it is poorly sorted; and if 3.0, it has normal sorting. A study of a number of sandstones of the Rocky Mountain region shows these values to be too high. Hough [78] points out that the coefficient of sorting for most near-shore marine sediments lies between 1 and 2.

From the foregoing it can readily be seen that, if this general approach is followed, statistical values may be recorded and compared for various granular clastic sediments.

For more detailed information concerning this subject the reader is referred to chapter 6 of the "Manual of Sedimentary Petrography" by Krumbein and Pettijohn.[79]

[77] Trask, P. D., *Origin and Environment of Source Sediments of Petroleum*, pp. 71-72, Houston, Tex., Gulf Publishing Co., 1932.
[78] Hough, J. L., *Sediments of Buzzaris Bay, Massachusetts*: Jour. Sedimentary Petrology, vol. 10 p. 26, 1940.
[79] Krumbein, W. C., and Pettijohn, F. J., *Manual of Sedimentary Petropraphy*, New York, Appleton Century, 1938.

SETTLING ANALYSIS

L. W. LeROY

Stratigraphic sections involving fine-grained clastics (siltstones, claystones, and shales) in most areas are difficult to subdivide and correlate, particularly if paleontologic and lithologic data are inadequate. Skeeters [27] has suggested a method that is based on the rate of settling of minute particles, through a liquid medium, which may be of some correlative value. This technique does not involve numerical determination of particle size but instead considers the settling rate of the particles and resistivity characteristics of the supernatant liquid.

The procedure of this investigation is as follows: The argillaceous sample, after thorough drying, is pulverized to −120-mesh and placed in a four-foot vertical glass tube (2.5 inches in diameter) into which compressed air is introduced through a stopper in the base. One hundred grams of material is added to 2,000 cc. of water. This mixture is then air-agitated for 15 minutes, after which the height of the settling material is measured at five-minute intervals and the results plotted.

During the settling stage and at five-minute intervals, electrical-resistance values of the supernatant turbid liquid are measured between two electrodes spaced half an inch apart and suspended two feet below the fluid surface.

The settling and resistivity results of several Pierre and Fox Hills shales are shown in figure 82. Skeeters [80] concluded that

> The height-of-settling-surface curves show the greatest promise of applicability to correlation. The resistance curves show considerably more similarity between runs on the same shale and are of sufficient variation between shales to offer some promise of possible value in correlation.

STAIN ANALYSIS

L. W. LeROY

The application of stain solutions to polished surfaces and to thin sections of rocks permits the rapid identification of certain minerals and assists in establishing the distribution and mutual relationship of the minerals.

Stain results depend on such factors as the texture and structure of the rock, the purity and relationships of the minerals, and the uniformity of the applied procedure. If more exact mineralogic determinations are desired, optical investigations should supplement the stain tests.

Frequently the subsurface geologist during an examination of well cuttings and cores is concerned with distinguishing between aragonite, calcite (limestone), dolomite (dolostone), quartz, feldspar, and certain basic types of clay minerals. Some of the methods applicable for rapidly de-

---

[80] Skeeters, W. W., unpublished research report, Colorado School of Mines, 1942.

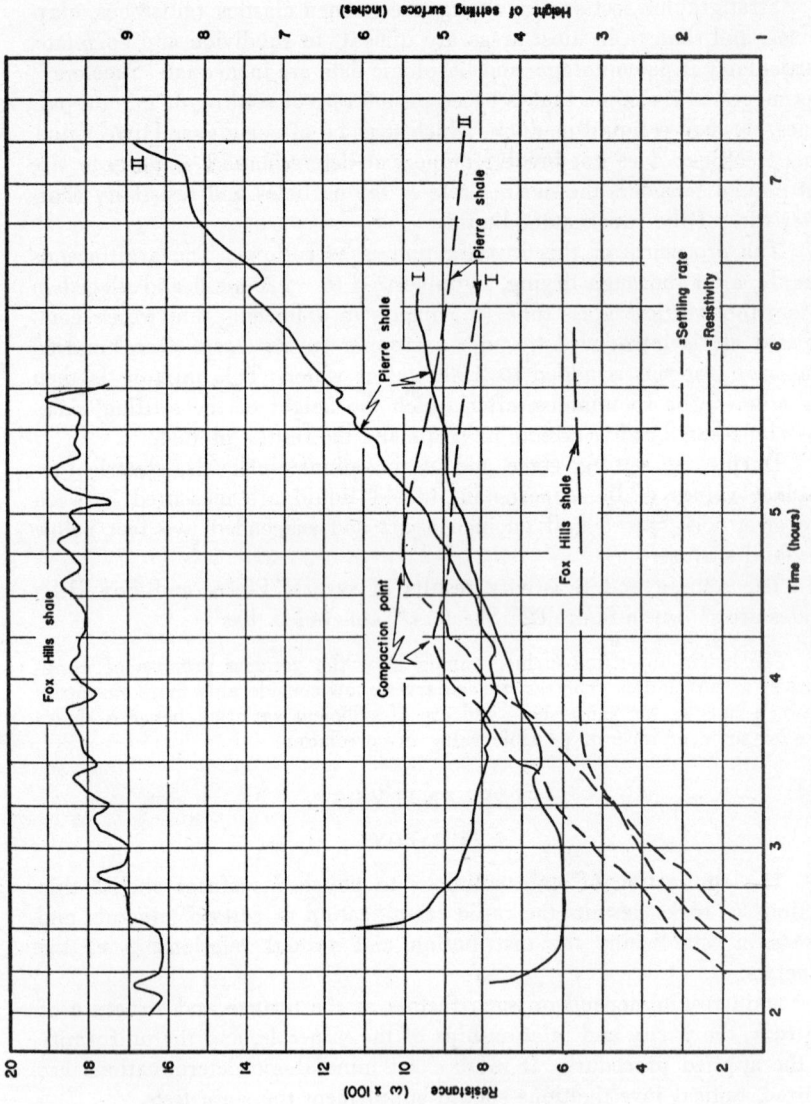

FIGURE 82. Compaction and resistivity curves of Pierre and Fox Hills shales near Golden, Colorado. (Adapted from Skeeters).

termining these minerals and for evaluating their interrelationships in a rock are here briefly outlined.

CALCITE AND ARAGONITE

Meigen [81] developed a method of distinguishing aragonite from calcite by immersing a polished rock surface or thin section of each for 20 minutes in a solution of boiling cobalt nitrate and observing the resulting color. Aragonite stains a light purple in the initial stages but upon continued boiling assumes a violet hue. Calcite attains a similar color only after several hours of immersion. For fine-grained rocks the two minerals are difficult to differentiate owing to the spreading of the stain.[82] Aragonite grains treated with cobalt nitrate, when immersed in a solution of ammonium sulphide, become coated with a film of black cobalt sulphide.

CALCITE AND DOLOMITE

Several staining methods are used for distinguishing calcite from dolomite. The results of these stains differ, some being more dependable and exacting than others. These tests are given in the order of preference.

Fairbanks Method [83]—In the Fairbanks method the solution to be used is prepared by mixing 0.24 grams of haematoxylin, 1.6 grams of aluminum chloride, and 22 cc. of water and bringing the mixture to a boil; the solution is cooled and, after the additions of a small quantity of hydrogen peroxide, filtered. Calcite, upon being immersed in the solution, rapidly stains dark purple, whereas dolomite remains unaffected. The advantage of this test is its rapidity and dependability. The polished surface or thin section is allowed to remain in the stain only thirty seconds; and is then removed and carefully washed in water. Boiling the rock in the stain solution is not required. This method is exceptionally favorable for evaluating limestone and dolostone fragments in well samples.

Copper Nitrate Method—Calcite boiled in a concentrated solution of copper nitrate assumes a medium-green color; dolomite is not affected. The color may be fixed by immersing the sample in ammonia. This test is effective and yields consistent results.

Silver Chromate Method—The polished surface or thin section is immersed three or four minutes in a boiling ten-percent solution of silver nitrate. The nitrate is then washed free from the sample, which is subsequently treated with a saturated solution of potassium chromate. Calcite and aragonite grains are stained reddish-brown; dolomite retains its original color. This method is accredited to Lemberg [84] and may be considered as giving dependable results.

    [81] Meigen, W., *Eine einfach Reaktion zur Unterscheidung von Aragonit Kalkspath:* Centralb. f. Min., etc., pp. 577-578, 1901.
    [82] Twenhofel, W. H., and Tyler, S. A., *Methods of Study of Sediments,* p. 129, New York, McGraw-Hill Book Company, Inc., 1941.
    [83] Fairbanks, E. E., *A Modification of Lemberg's Staining Methods:* Am. Mineralogist, vol. 10, pp. 126-127, 1925.
    [84] Lemberg, J., *Zur microchemischen Untersuchung einiger Minerale:* Zeitschr. geol. Gesell., Band 44, pp. 224-242, 1892.

Lemberg Method [85]—The solution to be used is prepared by boiling for 20 to 30 minutes a mixture of 4 grams of $AlCl_2$, 6 grams of logwood, and 60 grams of water; the mixture is filtered and the filtrate diluted with 1,000—1,200 cc. of water. Calcite when immersed in the solution is stained light purple after five to ten minutes of boiling; dolomite remains unchanged. This reaction causes a film of aluminum hydroxide on the calcite grains and this film absorbs the logwood dye.

Potassium Ferricyanide Method—The potassium ferricyanide method was developed by Heeger.[86] It consists of immersing the rock sample first in a dilute solution of hydrochloric acid (1:100) containing a few drops of postassium ferricyanide. If the dolomite contains ferrous iron, the mineral assumes a deep-blue color; calcite is not affected. This test is considered satisfactory only if the dolomite contains ferrous iron; otherwise, it fails to produce results.

IDENTIFICATION OF FELDSPARS

Twenhofel and Tyler [87] summarized the method of distinguishing quartz from feldspar as follows:

A few drops of hydrofluoric acid are placed on a thin section, or on grains mounted in Canada balsam with their upper surfaces exposed, and allowed to remain one or two minutes before being gently washed off. The acid produces a thin, gelatinous film of aluminum fluorosilicate on the feldspar and other aluminous minerals but leaves the quartz clear. After washing, the specimen is immersed in a water-soluble organic dye for about five minutes and then again washed. Fuchsine, methylene blue, safranine, or malachite green may be used as a stain. . . . The depth of color retained on staining is greatest with anorthite; becomes successively lighter with less calcic feldspars; and is lightest with orthoclase or microcline.

The degree of staining is improved if the grains are exposed to the fumes of hydrofluoric acid for two or three minutes. Care should be exercized in washing after staining, as the stain is easily removed from the corroded feldspar grains.

According to Gabriel and Cox,[88] the potash feldspars may be identified by exposing the rock to hydrofluoric-acid fumes and then staining it with a diluted solution of sodium cobalt nitrite, which is prepared by adding 15 cc. of glacial acetic acid and 25 cc. of water to 12.5 grams of $Co(NO_3)_2.6H_2O$ and 20 grams of $NaNO_3$. The potash feldspars assume a strong yellow color from the formation of potassium cobalt nitrite. Quartz and plagioclase grains are not affected.

Potash feldspars (orthoclase, microcline) may be differentiated from the calcic plagioclase feldspars (laboradorite, bytownite, and anorthite)

[85] Lemberg, J., *Zur microchemischen Untersuchung von Calcit, Dolomit, un Predazzpit*: Zeitschr. Géol. Gesell., Band 39, pp. 489-492, 1887.

[86] Heeger, J. E., *Ueber die Mikrochemische Untersuchung fein verteiler Carbonate im Gesteinsschliff*: Centralbl. Mineralogie 1913, pp. 44-51, 1913.

[87] Twenhofel, W. H., and Tyler, S. A., *op. cit.*, p. 131.

[88] Gabriel, A., and Cox, E. P., *A Staining Method for the Quantitative Determination of Certain Rock Minerals*: Am. Mineralogist, vol. 14, pp. 290-292, 1929.

by the following procedure: (1) Pulverize the sample or disaggregate the sandstone to minus-80-mesh; (2) boil the material for one minute in hydrofluoric acid; (3) wash the acid-treated sample gently in distilled water; (4) boil the washed sample for ten minutes in a water-saturated solution of eosine y; and (5) carefully wash the sample and remove the excess dye solution.

The plagioclase grains are coated with a medium- to dark- to orange-red film. Orthoclase and quartz are not stained. The combination of this test and the cobalt-nitrite test for orthoclase serves as a basis for rapidly estimating the feldspathic content of sands and sandstones. The sodic feldspars (albite, oligoclase, and andesine) are not noticeably affected by the eosine test.

Eosine dye may be used to identify nephelite and cancrinite (orthosilicates). Nephelite assumes a light-pink discoloration, whereas cancrinite attains a much darker pink. The discoloration is produced within the grain and not as an exterior film as with the calcic plagioclases. Sodalite is not affected by the dye.

CLAY-MINERAL STAIN TESTS

In recent years considerable work has been devoted to clay mineralogy. Mineralogic analyses have indicated three important groups of the clay minerals (kaolinite, montmorillonite, and illite). Identification of these clay-mineral species within each group is extremely difficult owing to their minute size. Chemical, optical, X-ray, electron-diffraction, and differential thermal-dehydration methods are required for precise determination.

Several dye tests are employed for assisting in differentiating various clay groups. Extreme care should be exercised in applying these tests, for results may be extremely variable because of impurities, complex mineralogic associations, and inconsistent preparation procedure. The stain results may be observed in reflected light under either a petrographic or binocular microscope at magnifications from 30 to 120 diameters.

Benzidine Test—A saturated water solution of the organic compound benzidine (or benzidine hydrochloride) produces a blue coloration in contact with clay minerals of the montmorillonite and illite groups, although the benzidine solution itself is slightly pink. The sample is not treated with hydrochloric acid prior to application of the stain solution. It has been reported that manganese dioxide and organic matter may cause formation of a blue coloration in the absence of bentonite, and that ferrous iron or other reducing agents may prevent the development of coloration.[89] Gypsum has a pronounced effect on the benzidine test. This effect may be minimized by first boiling the material in water, pouring off the fine fraction, thoroughly drying it at 105° C., and then applying the stain solution.

---

[89] McConnell, Duncan, *Notes on Properties and Testing of Bentonites:* U. S. Bur. Reclamation, Denver, Laboratory Rept. Pet-44B, 1946.

Crystal-Violet Test—The crystal-violet dye solution (25 cc. of nitro-benzene, 0.1 gram of crystal violet) causes acid-treated montmorillonite first to stain green and then greenish yellow or orange yellow. Illite assumes a rather dark-green color. Kaolinite merely absorbs the violet stain.

Safranine y Test—Another stain applicable for identifying clays of the montmorillonite and illite groups is the safranine y (nitrobenzene saturated with safranine y). McConnell [90] summarizes this test as follows:

(1) A small representative sample (about 20 grams) is selected, crushed, and placed in a beaker; (2) strong hydrochloric acid is added in amounts four or five times the volume of earth material. If significant amounts of carbonates are present the quantity of acid is proportionally increased. The sample in acid is retained at elevated temperatures for an hour or two. Suitable temperatures can be obtained by placing the sample on top of a small laboratory oven; (3) the acid-treated sample is washed five times, using 200 milliliters of distilled water for each washing. The earth material is then transferred to a filter paper, which is placed in a dish and oven-dried at about 105° C.; (4) the dried material is examined and one or more samples are removed from the filter paper for staining. Considerable care must be exercised in the selection of this sample (or samples) because stratification invariably takes place in the funnel during washing; (5) three or four drops of nitrobenzene saturated with safranine y are added to the mineral powder and the quantities of colorless, red, purple, and blue grains are estimated. The quantity of blue and purple grains compared with the total number is an indication of the amount of bentonitic material present.

This test is apparently capable of giving anomalous results in rare instances but is probably subject to interferences no more frequently than the benzidine test.

Kaolinite is unaffected by safranine y. Minerals of the montmorillonite group become blue when the dye is applied, whereas illite grains tend to exhibit a more bluish-purple to purplish hue.

Malachite-Green Test—After being acidized with hydrochloric acid the clay minerals of the kaolinite group, when in contact with malachite-green solution (25 cc. of nitrobenzene, 0.1 gram of malachite green), become a bright apple-green. The montmorillonite and illite minerals commonly become pale yellow or greenish yellow.

SUMMARY OF CLAY-STAIN RESULTS

In table 7 results of clay-stain tests are given in summary.

For favorable results in clay-stain tests the following precautions should be observed.

1. The acidization (HCl) procedure should be complete and uniform. Best results are obtained if the material is pulverized and passed through a 200-mesh screen.

2. After step (1), the sample should be thoroughly washed free of the acid with distilled water; otherwise, consistent stain results cannot be obtained.

---

[90] McConnell, Duncan, *op. cit.*

3. After step (2), the acidized material must be completely dehydrated by drying for several hours at temperatures of about 105° C.

4. About one milligram of the sample material should be used when the stain solutions are applied.

5. The treated sample should remain in the stain solution for five minutes in order to obtain the best coloration results.

6. The reflected light source should be controlled.

Clay mineralogy offers possibilities for serving as a means of correlating and subdividing homogeneous argillaceous and carbonate sections. The latter rock types involve insoluble residues. Clay mineralogy can also be useful in evaluating changes in the porosity and permeability of sands and sandstones.

The presence of montmorillonite clay types is extremely detrimental

TABLE 7

SUMMARY OF CLAY-STAIN RESULTS

Mineral group	Safranine "y" (acidized sample)	Malachite green (acidized sample)	Crystal violet (acidized sample)	Benzidine (unacidized sample)
Kaolinite	Red	Green	Violet	No reaction
Montmorillonite	Blue	Yellow to greenish yellow	Yellow, greenish yellow, or orange yellow	Blue (variable hues)
Illite	Bluish purple to purple	Yellow	Dark green	

in many engineering projects because of their swelling properties. Sediments containing these minerals should be completely analyzed before construction of buildings, highways, dams, and other projects in order to predict the reaction of the earth materials.

SHAPE ANALYSIS

L. W. LeROY

Various procedures are followed in determining the sphericity, roundness, and flatness values of sedimentary particles.

Krumbein and Pettijohn [91] give the following factors controlling the shape of sedimentary grains and fragments: (1) the original shape of the fragment; (2) the structure of the fragment, as cleavage or bedding; (3) the durability of the material; (4) the nature of the geologic agent; (5) the nature of action to which the fragment is subjected and the violence of that action (rigor); and (6) the time or distance through which the action is extended.

Varying degrees of sphericity ("measured by the ratio of s/S, where

[91] Krumbein, W. C., and Pettijohn, F. J., *Manual of Sedimentary Petrography*, p. 278, New York, D. Appleton-Century Co., 1938.

s is the surface area of a sphere of the same volume as the fragment, and *S* is the actual area of the object. For a sphere the ratio is 1. For all other solids the ratio has a value less than one") and roundness (a measure of the angularity of the edges and corners) of detrital grains have served in correlating certain strata. In the Rangely oil field of northwestern Colorado and adjacent areas, the Entrada and Navajo (Jurassic) sandstones are differentiated from adjacent lithic units by the rounded and frosted character of the quartz grains.

Rittenhouse [92] has used the degree of roundness of tourmaline and zircon in correlating various strata in the Appalachian Basin. He states:

In the Appalachian Basin roundness of heavy minerals is extremely valuable as a criterion for differentiating various Mississippian and Pennsylvanian oil and gas sands, for outlining petrographic provinces, and for inter-

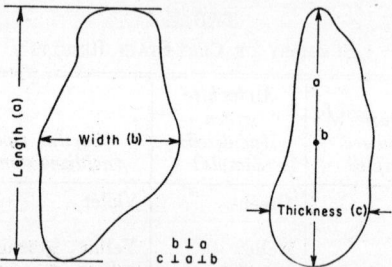

FIGURE 83. Measurements of pebbles required in determining sphericity, roundness, and flatness values; *a* and *b* are determined from maximum image orientation; *c* value is normal to *a* and *b*.

preting geologic history. Roundness is particularly significant in the basin because fossils are rare and the heavy-mineral suite is restricted. . . .

Pettijohn comments: [93]

The roundness of a clastic particle sums up its abrasion history. Sphericity, on the other hand, more largely reflects the conditions of deposition at the moment of accumulation, though to a more limited extent sphericity is modified by the abrasion processes.

According to Fraser,[94] the absolute size of the grain, nonuniformity in the size of the grain, the proportions of various sizes of grains, and the shape of the grain control porosity of unconsolidated deposits. He further states:

Regularities in shape should result in a larger possible range in porosity, as irregular forms may theoretically be packed either more tightly or more loosely than spheres. The degree of rounding generally varies for different

[92] Rittenhouse, Gordon, *Grain Roundness—A Valuable Geologic Tool*: Am. Assoc. Petroleum Geologists Bull., vol. 30, no. 7, pp. 1192-97, July 1946.
[93] Pettijohn, F. J., *Sedimentary Rocks*, p. 53, New York, Harper and Brothers, 1949.
[94] Fraser, H. J., *Experimental Study of the Porosity and Permeability of Clastic Sediments*: Jour. Geology, vol. 43, no. 8, pp. 910-1010, 1935.

grain sizes in any natural deposit, because of differences in the mineralogical composition of different grades. . . . It is difficult to determine the effect of shape of grain on porosity, because of the difficulty of obtaining angular particles of the same size. . . . Angularity may either increase or decrease porosity; most often it increases porosity. The only type of "angularity" found to cause a decrease in porosity is that in which the grains are mildly and uniformly disk-shaped.

Factors affecting permeability (in addition to temperature, hydraulic gradient, and coefficient of permeability) include uniformity and range of grain-size, shape of grain, nature and uniformity of packing, surface conditions of the grains, stratification, consolidation, and cementation of the material.[95]

Krumbein [96] has given an interesting discussion on determining spher-

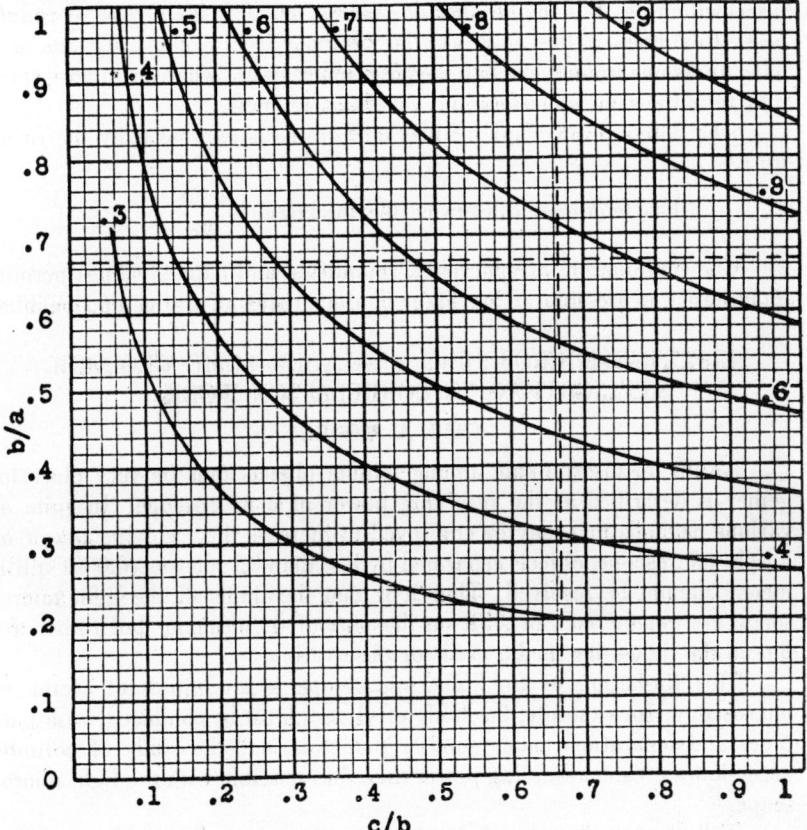

Figure 84. Chart for determining sphericity of pebbles; a, b, and c, values represent the long, intermediate, and short dimensions of the pebble. (From Krumbein, *Jour. Sedimentary Petrology*.)

[95] Fraser, H. J., *op. cit.*, p. 959.
[96] Krumbein, W. C., *Measurement and Geological Significance of Shape and Roundness of Sedimentary Particles:* Jour. Sedimentary Petrology, vol. 11, no. 2, pp. 64-72, Aug. 1941.

icity values of pebbles. Three-dimensional values are recorded, maximum (*a*), intermediate (*b*), and minimum (*c*) (fig. 83). The ratios *b:a* and *c:b* are calculated. From these ratios the sphericity index is obtained from a control chart (fig. 84).

Roundness values of grains and pebbles as determined by Wadell [97] are computed from a maximum-plane image (projected cross section) in which the summation of the radius of the individual corners is divided by the number of corners and this value divided by the radius of the maximum-inscribed circle. Roundness is expressed by the formula:

$$P=\frac{\Sigma r}{N}\div R$$

where *r* is the radius of each corner, *R* is the radius of the maximum inscribed circle, *N* is the number of corners, and *P* is the degree of roundness. Roundness values may also be obtained from a chart. Grains may have the same roundness but varying sphericities, whereas other grains may have the same sphericity but varying roundness.

The flatness ratio [98] of pebbles and grains is expressed by the formula:

$$F=(\text{length}+\text{width}\div\text{twice the thickness})=\frac{a+b}{2c}$$

The combination of sphericity, roundness, and flatness values permits quantitative expression of the shape characteristics of grains and pebbles.

ELECTRON-MICROSCOPIC ANALYSIS—SOME GEOLOGIC APPLICATIONS IN CORRELATION WORK

CARL A. MOORE

Ability to discriminate among minute objects that lie very close together is described as the resolving power of a microscope. In spite of various methods that may be employed to increase the resolving power of a light microscope, objects separated by less than 0.1 micron (0.0001 millimeter) cannot be resolved. Thus it is that the limits of the light microscope are not the lack of skill on the part of the designer but rather are due to the light—the media used for observation.

This limitation of the light microscope is an important factor in microscopy; for example, the study of viruses must be conducted with particles and separations much smaller than this, and the study of colloids necessitates greater resolving power than that possible with the light microscope.

With the introduction of the electron microscope, this limit on resolving power has been greatly decreased, since the magnification is no longer

---

[97] Wadell, Hakon, *Sphericity and Roundness of Rock Particles:* Jour. Geology, vol. 41, pp. 310-331, 1933.

[98] Wentworth, C. K., *The Shapes of Beach Pebbles:* U. S. Geol. Survey Prof. Paper 121-C, pp. 75-83, 1922.

limited by the wave length of visible light. Theoretically, the electron microscope should be capable of resolving powers as small as atomic dimensions. In actual practice, however, the microscope has not been perfected to that extent. Nevertheless magnifications of over 100,000 diameters are practical with the electron microscope, as compared with a useful limit of 2,000 diameters for the light microscope.

DESCRIPTION OF THE MICROSCOPE

Figure 85 is a comparison of the optical microscope with the electron microscope showing equivalent parts: magnetic fields are equivalent to lenses; both have specimen levels; and both have photographic plates for

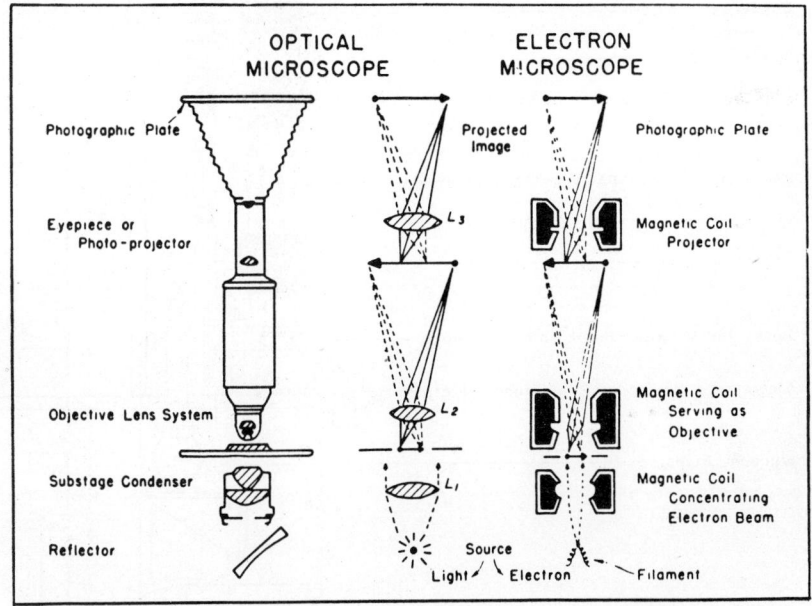

FIGURE 85. Comparison of light microscope and magnetic electron microscope. (From Burton and Kohl.)

pictures. For comparison, the electron microscope is diagrammed upside down.

A simplified drawing of the R.C.A. compound magnetic electron microscope, type EMB, is shown in figure 86. Focusing is accomplished by varying the lens power. The specimen mount is the movable stage. As the stage is inside the vacuum portion of the microscope, it is moved by means of fine screws and a metal flexible bellows.

The electron beam is concentrated on the specimen by the magnetic field produced in the condenser-lens coil. After passing through the specimen, the electrons are focused by the objective-lens coil into an intermediate image, and the projection-lens coil produces a further magnified image on the fluorescent screen in the final viewing chamber.

To facilitate the initial adjustment of the specimen, a port is provided for viewing the intermediate image on a fluorescent screen close to the

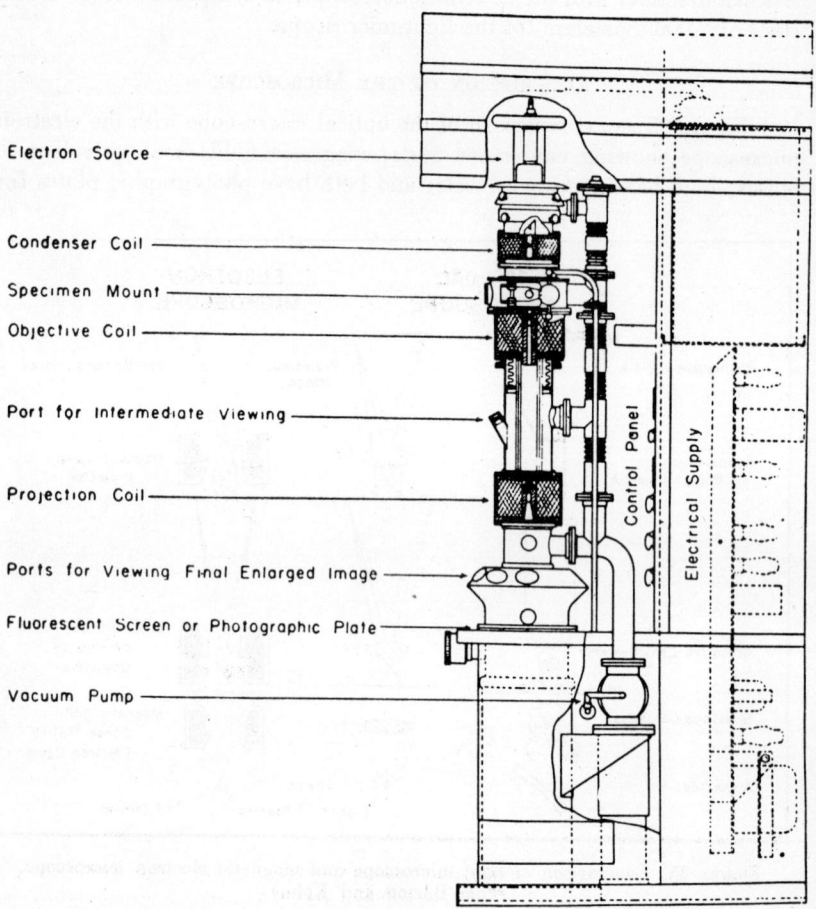

Electron Source

Condenser Coil

Specimen Mount

Objective Coil

Port for Intermediate Viewing

Projection Coil

Ports for Viewing Final Enlarged Image

Fluorescent Screen or Photographic Plate

Vacuum Pump

Control Panel

Electrical Supply

FIGURE 86. Simplified drawing of electron microscope, taken from "Electron Microscope" prepared by R.C.A. Manufacturing Company, Camden, N. J.

plane of the projection-lens coil. By virtue of the relatively low magnification at this point, it is possible to select the most interesting part of the specimen and to move it into position to be magnified further by the projection-lens coil.

Six observation windows enable a number of spectators to view the image simultaneously. With the choice of a selected field of view and the magnification adjusted to the desired value, the fluorescent screen is raised, exposing a photographic plate to the electrons. This plate is carried in a holder in the vacuum system of the microscope. Magnifications of 1,000 to 20,000 diameters are possible, and the definition of the photograph is sufficiently clear to allow further optical enlargement to full useful magnification.

SPECIMEN-MOUNTING TECHNIQUES

It was necessary to devise a special specimen-mounting technique in order to work with the small areas that are enlarged to full magnifications for study. Most specimens are mounted on a 400-mesh screen. This screen is dipped into a solution of collodion, which dries quickly, leaving a strong film approximately one micron (0.001 mm.) thick between the individual wires.

The material for study may be placed on this collodion film in one of several ways: (1) manually, under high-power binoculars; (2) precipitated from solution onto the screen; (3) by passing the screen coated with collodion through a culture of the material; and (4) by placing a drop of material suspended in a liquid onto the screen.

No wet or living tissue can withstand the high vacuum of 10^{-4} to 10^{-5} millimeters of mercury in the electron microscope. However, the microscope is being used extensively in biological studies on materials ranging in size from that of the organs of animals and insects downward through that of the bacteria and of the viruses and even of large molecules. The material in turn must be thin enough to allow the passage of electrons through it. Some materials deteriorate when subjected to the intense electron bombardment, and some materials may heat up during this bombardment. Owing to the high vacuum, this heat cannot be transmitted or conducted away from the subject.

POSSIBLE USES IN CORRELATION WORK

The usefulness of any method of correlation lies in its ability to indicate or prove the existence of equitable or similar ages or environments of deposition between two areas, two wells, or two geologic outcrops. Most methods in geology originally included only the megascopic aspects: for example, similar or identical fossils and equivalent successions of beds, to mention two. With the advances in geologic techniques, more precise correlation has been possible by utilizing microscopic similarities for correlations, as in micropaleontology, sedimentary petrology, and microlithology.

With the electron microscope, it should be possible to achieve the ultimate in utilizing submicroscopic similarities for correlations. Some uses possibly peculiar to this microscope are herein listed and discussed.

Bed Identification

It is possible that many minute similarities exist in beds or formations, which, if they could be seen and studied, could be used to correlate subsurface beds. In studying sandstones, for example, it would not be possible to observe the actual sand grains, but it would be necessary to study the cementing material and any foreign material in the sandstone. Thus, in investigating the clay content versus water conductivity of oil sands, Bates, Gruver, and Yuster [99] isolated mica crystals and photographed them in the electron microscope for study. Sandstones bearing similar mica crystals might be correlative, if other criteria attested to a possible correlation.

Correlations with limestones should involve a different set of conditions. As a general rule limestones are compact or, if porous, contain

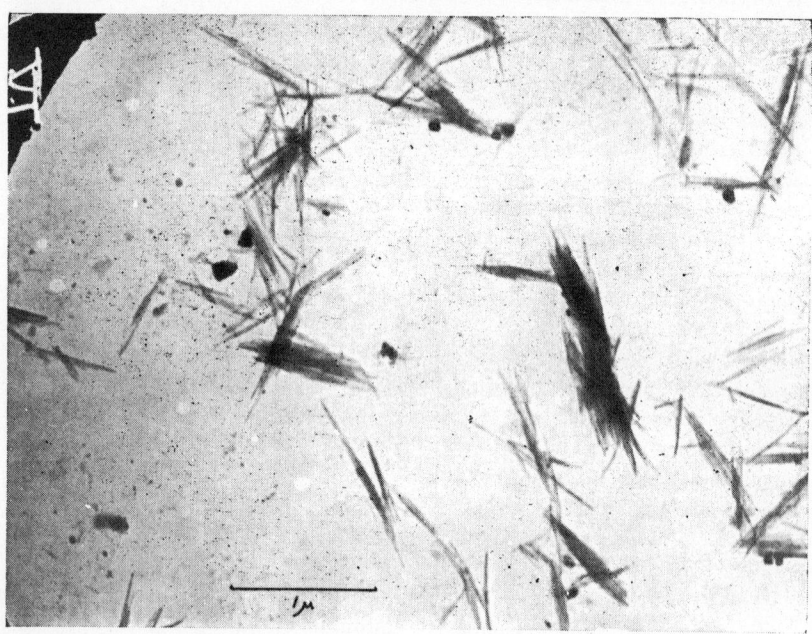

FIGURE 87. Electron-microscope picture of Attapulgus clay ($\times 20,000$). Note minute fibers and bundles of fibers, with very few larger grains. These average less than $\frac{1}{8}\mu$ (= 0.000125 mm.) in diameter and are of colloid size. Courtesy R.C.A. Laboratories and Standard Oil Development Company.)

comparatively large pores and openings. It would be difficult to impossible to grind a thin section of limestone to a thickness allowing the electrons to pass through the specimen and produce an image on the photographic plate. The pores of the limestone are so large as to preclude any precise study of their contour or shape. For these reasons, a possible approach would lie in the study of the residues after the limestone had been dissolved in some suitable solvent. Either the filtrates could be examined for correlatable objects, or the residue, which is often largely clay, might

[99] Bates, T. F., Gruver, R. M., asd Yuster, S. T., *Influence of Clay Content on Water Conductivity of Oil Sands:* Oil Weekly, Oct. 21, 1946.

lend itself to study in the electron microscope. This latter case leads into the problem of the study of shales.

Correlation of Clays

Quoting from Hillier,[100] ". . . particles of various types of clay have probably been subjected to more examination by means of the electron microscope than any other type of material." Some clays are composed of grains of about fifty angstroms in thickness and a few angstroms wide. Studies of the nature and correlation of such minute

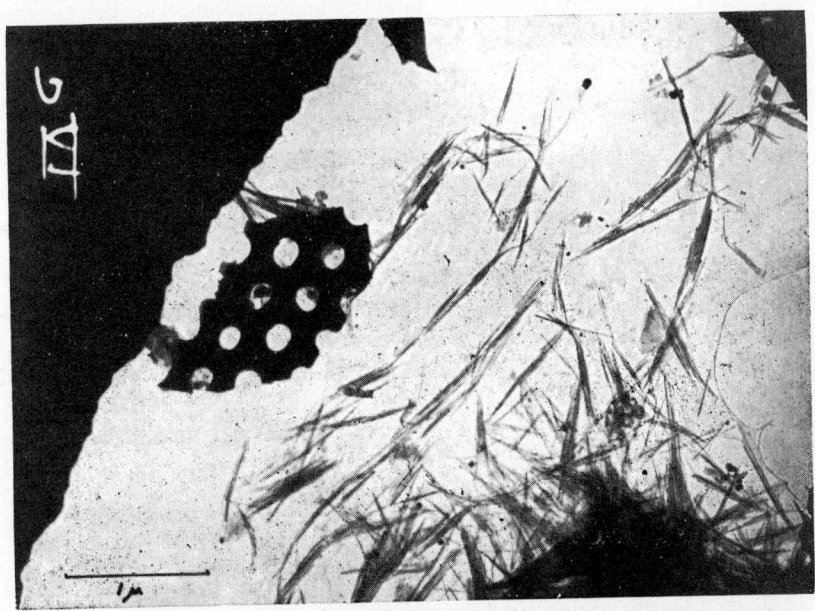

FIGURE 88. Electron-microscope picture of infusorial earth ($\times$ 20,000) sold by Central Scientific Company, Chicago. Note that fibers and bundles of fibers are very similar to those in photograph of Attapulgus clay in figure 87. Diatom fragment near center of photograph is about $1\frac{1}{2}\mu$ in length by 1μ wide. Openings in shell are less than $\frac{1}{5}\mu$ in diameter (0.0002 mm.) and would barely be discernible in the light microscope. (Courtesy R.C.A. Laboratories and Standard Oil Development Company.)

particles in the electron microscope are dependent on characteristic shapes and not on chemical combinations.

One of the clays used extensively in laboratories and refineries for filtering is called "Attapulgus clay," so named for Attapulgus, Georgia. Chemical analyses of this clay show it to be chiefly montmorillonite, a hydrous aluminum silicate, but the individual microcrystalline masses cannot be identified or resolved under the best light microscope. Figure 87 is an electron-microscope picture of this clay, $\times$ 20,000, showing an

[100] Hillier, J., *Electron Microscopy*: Am. Ceramic Soc. Bull., Nov. 1946.

abundance of masses of minute fibers. These fibers are the so-called micro-crystalline masses that cannot be identified or resolved under the polarizing microscope.

Infusorial earth is described as a "siliceous earth made up largely of siliceous fragments of Infusoria, used as fulling material and as a filtering and absorbing agent." Figure 88 is an electron-microscope picture of this material, × 20,000, showing fibers very similar to the Attapulgus clay. The similar shapes of constituent parts of these two materials attest to their similar physical properties.

Clays might lend themselves to study and correlation in the electron microscope in the following ways:

1. The submicroscopic mineralogy and crystallography of clays might be studied. Minute crystals of rutile have been identified in titanium-rich clays. These crystals were too small to be identified under a light microscope. Detailed studies should bring out several similar instances of submicroscopic mineralogy that could be of value in correlation.

2. The presence of submicroscopic organic forms too small to be identified or even noted under a light microscope could provide the means for bed identification. This would involve the development of, shall we say, "electron micropaleontology," wherein organic forms far below the smallest fossil known would be studied.

3. Structural details of clays, pertaining to possible physical and physicochemical properties of the clays, should lend themselves to study in the electron microscope. The importance of this point might be stressed by suggesting that the electron microscope is believed to be capable of resolving giant molecules. At these particle sizes the physical and chemical properties would be dependent one upon the other and should be difficult to separate.

4. In the same general way, clay residues from limestones might be studied. Identification would depend upon the structural details and perhaps the mineralogy. Chemical or spectrographic methods of study would probably be of more value here than would the electron microscope.

5. Physical studies of the response of clays to the high vacuum in the electron microscope and changes due to the electron bombardment, with subsequent heating of the samples, might yield significant similarities and differences of correlative value.

Long-Range Correlation

The foregoing discussion has involved detailed correlations between individual beds. It was pointed out that electron-microscope techniques are not in general use as yet and may not be used except in unusual cases. Long-range correlations, of course, depend upon equivalent criteria being found over long distances. For this reason, long-range correlations with the electron microscope are subject to the same considerations as were the closer, detailed correlations.

Studies of Crude Oil

It will be necessary to develop a technique for studying crude oils in the electron microscope. In one study a specimen of crude oil was mounted in the usual manner on collodion film, and a monotonous gray field was seen, except for one object or group of objects (fig. 89). This was composed of a number of oval bodies; some of these are seen to be solid, while others appear to be breaking up. It is possible that this object was not able to withstand the high vacuum and electron bombardment in the electron microscope, and the photograph caught the material in the process of disintegration.

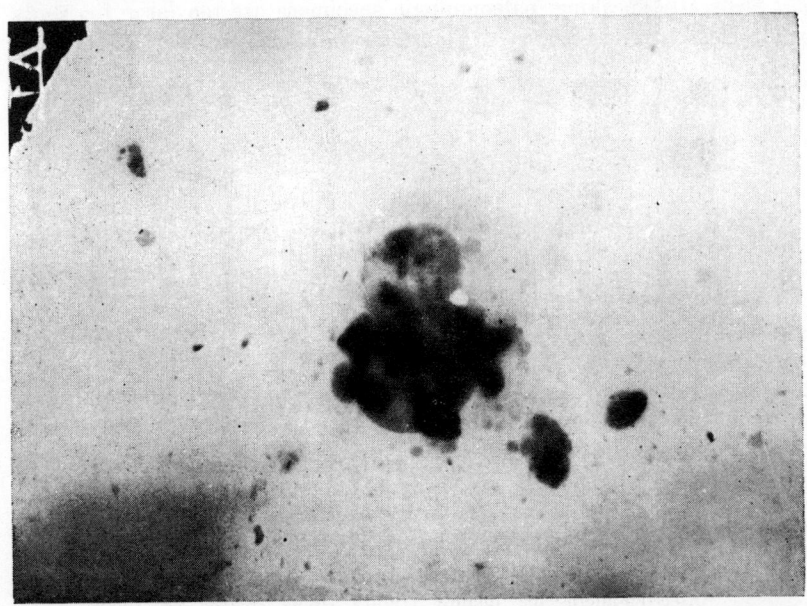

FIGURE 89. Electron-microscope picture ($\times$ 20,000) of object found in sample of crude oil from Athabaska tar sands in Alberta. Note dark, oval bodies associated with two somewhat larger, circular bodies. These oval bodies may be spores or minute protozoan tests. (Courtesy R.C.A. Laboratories and Standard Oil Development Company.)

A possible approach to the study of crude oils in the electron microscope may be as follows:

1. All foreign substances in the oil and in the extracts that are possible to prepare for study in the microscope would be studied.

2. Bacteria in the oils and in the extracts would be observed and identified.

3. The nature of coloring material in some of the darker oils would be studied. Is color due chiefly to the presence of foreign materials, or could it be due to molecular combinations?

4. The behavior of the oil and extracts during preparation would be studied, and the reactions to the high vacuums and to the electron bombardment observed.

5. The R.C.A. engineers and research physicists have photographed what they believe to be giant molecules in the electron microscope. Molecules of crude oil are believed to be disposed in some sort of regular pattern and may be quite large. Perhaps actual molecular differences may be found in crude oils, in the extracts, or in the various fractions that may be used for correlation.

Paleontologic Studies

Generally speaking, paleontologic specimens are too large for study in the electron microscope. It should be valuable, however, in studying

                    A   *B*

FIGURE 90. *A*—Comparison of photograph of diatom shell under light microscope (left) and electron microscope (right) (× 5,000). The light-microscope picture indicates presence of rows of holes, but the electron microscope shows size and arrangement of these rows of holes. Holes are approximately 0.5μ in diameter, separated in the row by a distance of 0.2μ. The rows themselves are approximately 0.9μ apart. *B*—Corresponding photographs of diatom shells as in *A*. Light-microscope picture (left) shows bars in shell and hints at presence of openings in the slots. Electron-microscope picture (right) shows clearly the small holes approximately 0.14μ in diameter. Rows are about 0.2μ apart. (Taken from Burton and Kohl.)

details of fossils too small for study under a light microscope, such as diatoms, spores, algae, and some protozoans. A comparison of diatom shells photographed with a light microscope and the electron microscope (fig. 90) shows that under the light microscope the number and arrangement of the perforations can hardly be determined, while the electron microscope indicates clearly the detail, arrangement, and number of perforations.

CONCLUSION

The electron microscope has opened up a new realm of research and endeavor. It is being adapted to a great number of scientific fields both for research and for industrial purposes. Future developments should increase the resolving power far beyond the best that is available today, but, conversely, this increase in resolving power will be one of the limiting factors of the microscope, because by working with very minute objects it is not possible to mount particular specimens for study. Most geologic techniques do not require these extremely high magnifications, however, and things geologic are usually too large for these magnifications. Further research in strictly geologic fields will have to be limited to particular problems where the microscope can be fully utilized.

X-RAY ANALYSIS

N. CYRIL SCHIELTZ

Today we have developed into the greatest industrial nation in the world with the highest standard of living. This is obviously because, as a nation, we have been able to develop and to accept new scientific methods and tools. Nevertheless, our progress has been greatly retarded because we failed on many occasions to make use of new developments as soon as they were available. This is true particularly of the X-ray-diffraction techniques. Over three and a half decades have elapsed since Laue discovered that X-rays interact with crystalline materials to give diffraction effects; yet a surprisingly great proportion of our scientific and administrative personnel, such as engineers, chemists, and geologists, have so little knowledge concerning it that they are unaware of the possibilities that the method offers, especially as a research tool. As a consequence, many problems have gone unsolved or have required an excessive amount of time and effort before a solution was obtained. Obviously, this regrettable situation exists, at least in part, because, although the technique required to make the X-ray patterns is relatively simple, rather specialized knowledge and considerable experience are essential before one is able to interpret the data properly. Even today many industries fail to appreciate this fact and are attempting to undertake X-ray-diffraction studies with personnel whose training is entirely inadequate to obtain satisfactory results. As a consequence, this otherwise powerful research tool sometimes is soon grossly neglected or abandoned because the returns do not justify the cost of installation and operation. In reality X-ray-diffraction studies have contributed a vast amount of valuable information to industry and research; however, most of it has come from the laboratories of our educational institutions, federal agencies, and a few large industries.

Since this discussion is directed principally toward a reading audience which may have only a limited acquaintance with the method, a brief discussion concerning the mechanism of diffraction appears desirable. All crystalline matter is composed of atoms or molecules arranged in such a manner that they form definite families of planes in various directions through the crystal. By considering primary X-rays to be reflected by these planes in the face of the crystal, the Braggs were able to reduce

Laue's original mathematically complex analysis of this interaction between X-rays and crystalline matter to terms of great simplicity. In figure 91 two such planes *AB* and *CD* represent one of the many families of planes found in a crystal. Two rays *emf* and *gnoph* of the defined X-ray beam are shown to be partly reflected from these planes when striking them with an incident and reflected angle of θ. According to the laws of optics these reflected rays must be in phase to be observed as a reflection. Consequently, ray *gnoph* must be longer than ray *emf* by an integral value of the wave length λ. Inspection reveals that this path difference is the distance *nop* and that $no = d \sin \theta$, and $op = d \sin \theta$; thus $nop = 2d \sin \theta = n\lambda$, which is the statement of Bragg's law.

Although this equation is satisfactory for calculating diffraction effects, it nevertheless reveals little of the actual diffraction mechanism

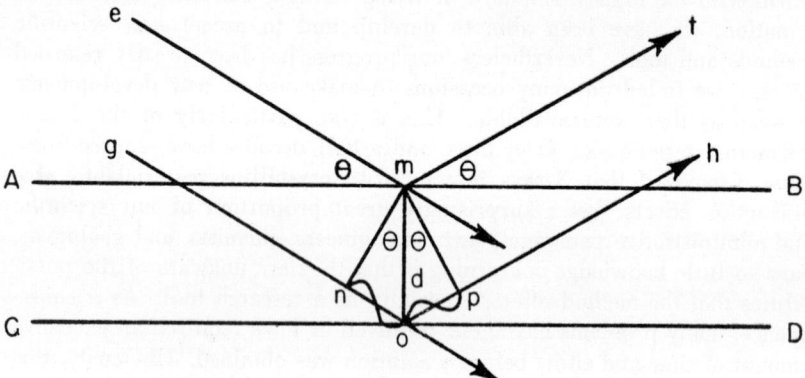

FIGURE 91. Reflection of X-ray beam from planes in face of crystal.

involved. A reasonable understanding of this mechanism can be gained from the familiar two-dimensional analogy of the interaction of waves on water. Figure 92 shows in successive steps (1) the generation of a circular set of waves from a series of parallel wave fronts by a post (or other small object) in a quiet body of water; (2) the interaction of these newly generated circular waves from a row of equally spaced posts produced new diffracted wave fronts; (3) the interaction of these generated circular waves from two rows of posts (two planes) under conditions where Bragg's law is not satisfied; and finally (4) the interaction of these waves where the angle θ has been so chosen that all conditions for the observance of diffraction effects by this particular family of planes have been satisfied. The fact that the diffracted wave fronts from each row of posts are one-quarter of a wave length out of phase with those diffracted by the adjacent rows, under the conditions where Bragg's law is not satisfied, immediately shows that we cannot observe any diffraction from this family of planes under the selected conditions. On the other hand, when the angle θ has been so adjusted that Bragg's law is satisfied, all of these wave fronts coincide: that is, they are in phase and diffraction effects from this particular family of planes are observed. A sketch set into the figure shows how this phenomenon is related to conditions in the X-ray camera.

This simple two-dimensional analogy can be applied to the three-dimensional diffraction of X-rays by crystalline matter if the posts are replaced by a regular assemblage of points (atoms or ions) distributed in space at a distance that is of the same order of magnitude as the wave lengths of X-rays. Spherical waves are created when X-rays, which are electromagnetic waves, cause forced oscillations of the planetary electrons of the atoms which they traverse, the electrons absorbing energy from the X-rays when moving away from the nucleus and radiating energy in all directions when moving toward the nucleus. Inspection reveals that this three-dimensional point system will produce very narrow pencils of rays only in those directions in which these spherical waves are in phase. These reinforced waves are the rays that produce the individual spots in X-ray patterns (Laue, rotation, Weissenberg, etc.) obtained from single crystals. If the single crystal is replaced by a large number of smaller crystals, that is, a powder, the 2θ angle with the undiffracted beam must remain constant since, in Bragg's equation, d for the particular set of planes and the wave length, λ, of the X-rays from a particular target material are fixed. The crystals of the powder with their statistical orientation, unless preferred orientation effects result owing to peculiar crystal shapes, then must produce a whole series of such discrete pencils, so that as a result a continuous diffraction cone with an apex angle of 4θ is obtained. If this cone is now recorded on a photographic film placed perpendicular to the cone axis, the diffraction effect is obtained as a line which is in the form of a ring. A pattern on which the diffraction rings from all families of planes have been recorded is usually referred to as a powder pattern and consists of a series of concentric rings on a flat film, or arcs of rings on a cylindrical strip of film.

Figure 92 reveals that a fixed space arrangement of atoms with definite fixed distances between them must always produce precisely the same X-ray pattern. Furthermore, if the same space arrangement is retained but the distances between atom centers are changed,[1] the X-ray pattern will retain its same general appearance but will either expand or contract. On the other hand, if the space arrangement is altered, the pattern is changed. Consequently, X-ray-diffraction patterns are a sort of fingerprint of crystalline materials. Each individual substance present in a mixture will produce its unique diffraction effects, so that the pattern derived from the mixture is a composite of the patterns of all the materials or compounds in the mixture. Furthermore, the intensities of the lines of the individual patterns are a function of the relative amount of the material present in the mixture, so that the method also has quantitative aspects.

---

[1] Atomic diameters vary from one element to the next; that is, the silicon atom as an ion has a diameter of 0.8 angstroms ($1A = 10^{-8}$ cm.), calcium 2.0 A., potassium 2.66 A., etc.

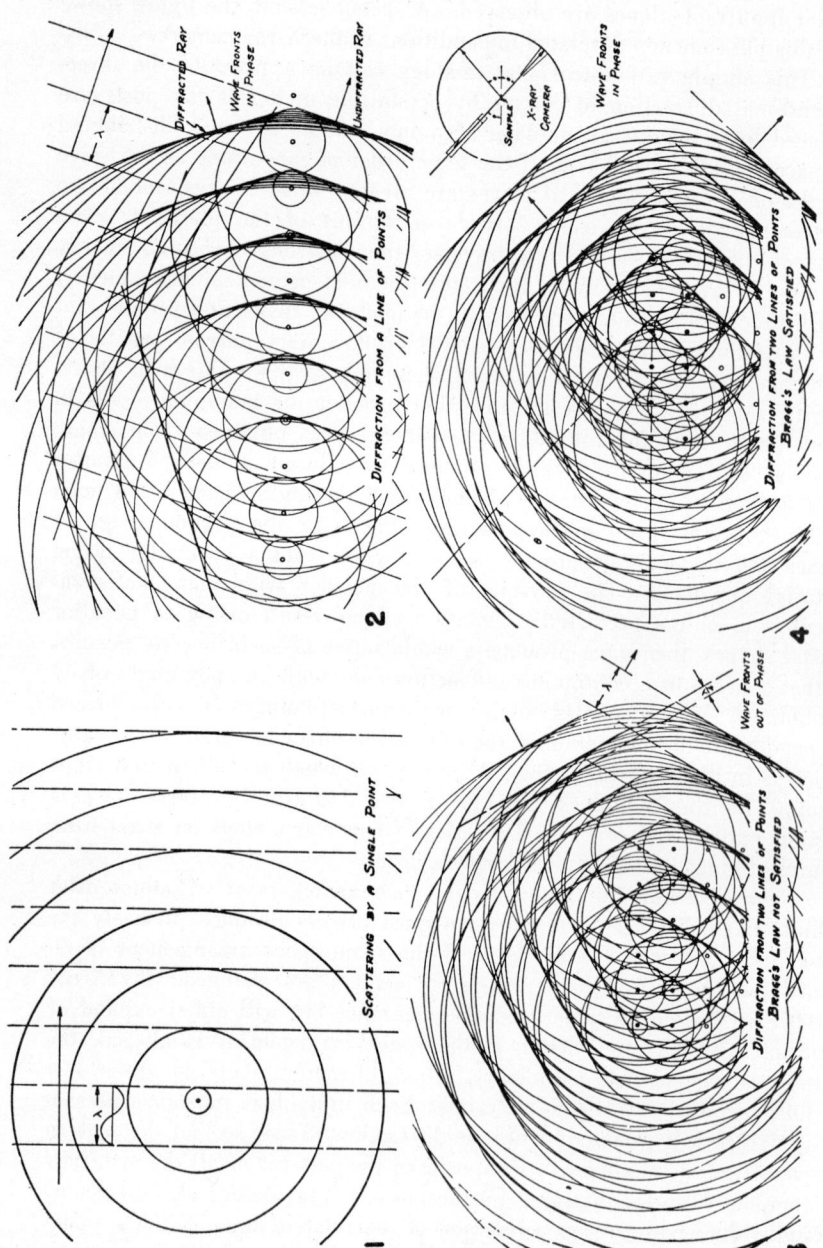

FIGURE 92. Schematic illustration of diffraction mechanism.

Scope

A complete discussion of the various methods of recording diffraction patterns is obviously beyond the scope of this section, and the reader is referred to the original papers and standard texts.[2][3][4][5][6][7] This discussion will be limited to the information required by geologists for the identification of geologic materials. Such information includes the advantages and disadvantages of the X-ray-diffraction method; the fundamentals of recording the data with apparatus employing photographic film or Geiger-counter circuits; the selection and preparation of the material to be investigated; the selection and processing of the films; the conversion of the data into usable form; the interpretation of the data; and when the method can be applied advantageously to geologic problems.

The interpretation of the data requires (1) the conversion of the lines in the X-ray powder pattern to their corresponding interplanar distances so that Hanawalt's [8] method employing the card file of X-ray-diffraction data[9] can be used; or (2) an extended series of standard patterns which are used for direct comparison if complementary data, such as optical measurements, are available to limit the unknown to a workable number of possible materials; or finally (3) the application of the reciprocal lattice to make use of unit-cell data, if powder-diffraction data are lacking and the unit-cell data available.

Advantages and Limitations

The X-ray-diffraction method is advantageous for the analysis of unknown materials, especially mineralogic, because it reveals the state of chemical combination of the constituent elements. Furthermore, the method is nondestructive, and the sample can be used for further studies by other methods. Moreover, satisfactory results can be obtained from very limited amounts of material. Small clusters of powder approximately 0.1 to 0.2 mm. in diameter produce good patterns without necessitating objectionably long exposures. Under extreme conditions, suitable patterns have been obtained from samples that consisted of only a few micrograms of material. Likewise, only limited accuracy in measurements is necessary when making a qualitative analysis. Furthermore, a file of diffraction patterns constitutes a permanent record, which can always be examined and checked by anyone versed in the field.

[2] Hull., A. W., *A New Method of X-ray Crystal Analysis:* Phys. Rev., vol. 10, pp. 661 ff., 1917.
[3] Hull, A. W., *A New Method of Chemical Analysis:* Am. Chem. Soc. Jour., vol. 41, pp. 1168 ff., 1919.
[4] Clark, G. L., *Applied X-Rays*, New York, McGraw-Hill Book Co., Inc., 1940.
[5] Davey, W. P., *Study of Crystal Structure and Its Applications*, New York, McGraw-Hill Book Co., Inc., 1934.
[6] Barrett, C. S., *Structure of Metals*, New York, McGraw-Hill Book Co., Inc., 1943.
[7] Bunn, C. W., *Chemical Crystallography* (Interpretation of Data), New York, Oxford Univ. Press, 1946.
[8] Hanawalt, J. D., Rinne, H. W., and Frevel, L. K., Ind. and Eng. Chemistry, Anal. Ed., vol. 10, pp 457 ff., 1938.
[9] Card file index and first supplement compiled under the joint supervision of the American Society for Testing Materials and the American Society for X-ray and Electron Diffraction. These are available from the American Society for Testing Materials, 260 S. Broad Street, Philadelphia, Pennsylvania.

On the other hand, the X-ray-diffraction method is sometimes considered rather limited as an analytic tool because the relative sensitivity requires that an appreciable amount of a constituent (from one to thirty per cent) [10] must be present in a mixture before its presence can be detected. However, the use of improved techniques will do much to correct this situation. Some materials with patterns having reasonably low background intensities and fairly strong lines can readily be detected in concentrations as low as one-half to one percent, whereas other materials with weaker patterns, such as the montmorillonite-type clays, can be

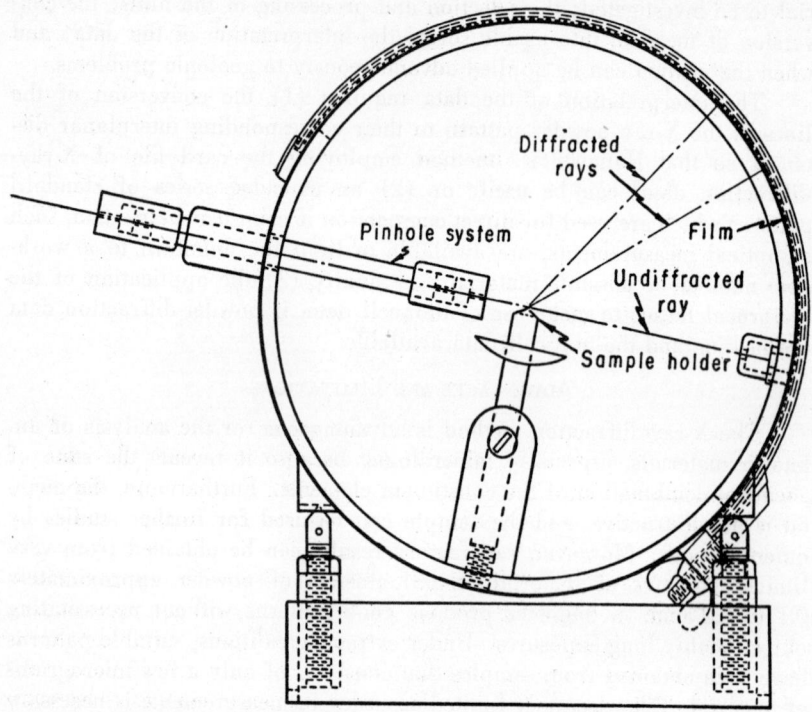

FIGURE 93. Schematic diagram of conventional powder camera.

detected when present in amounts ranging from five to six percent of the sample. It has been reported that special treatment of this clay with glycerol permits detection in amounts as low as one percent.[11] Limitation of the number of detectable constituents in mixtures due to crowding of lines has also been considered a disadvantage by some workers using small-diameter cameras with large pinhole systems.[12] The use of larger camera (10 to 20 cm.) diameters, smaller pinhole systems, and longer

[10] Brosky, S., *P. T. L. News*, Pittsburgh Testing Laboratories, Pittsburgh, Pennsylvania.
[11] Kelley, W. P., *Cation Exchange in Soils:* Am. Chem. Soc. Mon. 109, New York, Reinhold Publishing Corporation, 1948.
[12] Broskey, S., *op. cit.*

radiation wave lengths to spread out the patterns and increase the resolution should increase the number considerably. The most important disadvantages of the method are its inability to detect amorphous phases, such as glasses, when present in only limited amounts, and the fact that solid solutions may not always be observed.

APPARATUS FOR RECORDING THE DIFFRACTION PATTERNS

Two general types of apparatus are commonly used for recording the X-ray-diffraction pattern. Both are essentially the same in regard to the generation of X-rays, being composed of a high-potential (30,000 to

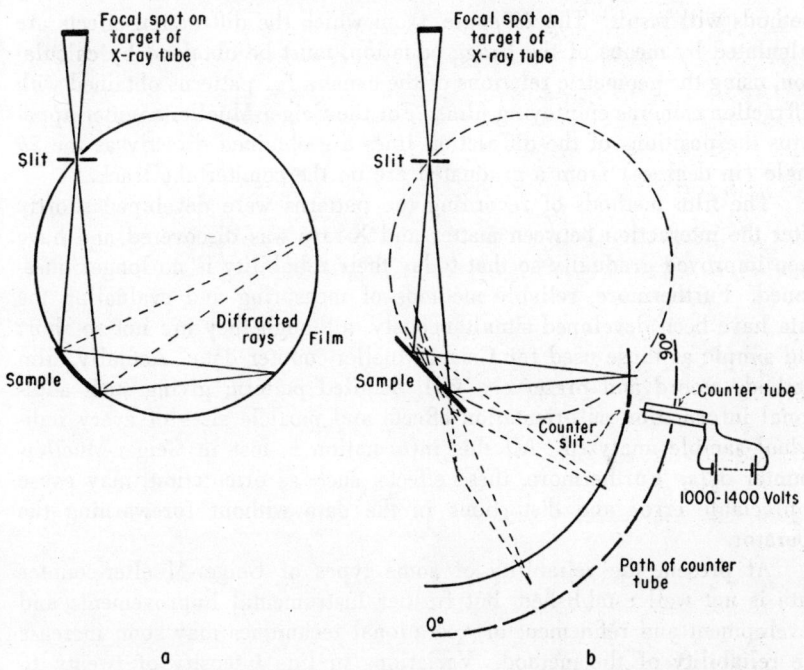

FIGURE 94. (a) Schematic diagram of focusing powder camera. (b) Schematic diagram of relation of focusing camera to Geiger-Mueller-counter apparatus.

50,000 volts) source of current, including a line voltage stabilizer, an auto-transformer to regulate the high potential, the necessary controls, rectifier, and X-ray tubes. The difference in these types of apparatus arises in the manner in which the X-ray-diffraction patterns are recorded; one type uses the conventional diffraction camera with photographic film and the other, a Geiger-Mueller-counter tube with a scaling circuit that may be used to measure the intensity of the diffracted rays, records intensity either by the counting technique or by automatic recording apparatus. The conventionl camera is shown schematically in figure 93.

The Geiger-Mueller apparatus is constructed on the principle of the focusing powder camera (See *a* of fig. 94), although gross deviations from the principle have been permitted in the actual construction of the apparatus (See *b* of fig. 94). In the focusing camera the diffracted lines are focused so that they appear as sharp lines on the film, which lies on the circumference of the circle that passes through the slit. Inspection shows that even though gross deviations from the principles were made in the construction of the apparatus, it nevertheless has relatively good focusing at the countertube slit. On the other hand, even small irregularities in focusing will readily show up when line intensities are based on the number of counts at the peak, and disagreement between the data of the two methods will result. The 2θ angle, from which the diffraction effects are calculated by means of the Bragg equation, must be obtained by calculation, using the geometric relations of the camera for patterns obtained with diffraction cameras employing films. For the Geiger-Mueller-counter apparatus the positions of the diffraction lines are obtained directly as the 2θ angle (in degrees) from a graduated arc on the countertube track.

The film methods of recording the patterns were developed shortly after the interaction between matter and X-rays was discovered and have been improved gradually so that today their reliability is no longer questioned. Furthermore, reliable methods of measuring and evaluating the data have been developed simultaneously, although they are not so short and simple as those used for Geiger-Mueller-counter data. Actually, film methods record and preserve a full, detailed pattern, giving such additional information as orientation effects and particle sizes of every individual sample analyzed. All this information is lost in Geiger-Mueller-counter data. Furthermore, these effects, such as orientation, may cause appreciable error and distortions in the data without forewarning the operator.

At present the reliability of some types of Geiger-Mueller-counter data is not well established, but further instrumental improvements and development and refinement of operational techniques may soon increase the reliability of the method. Variations in line intensity of twenty to thirty percent have been observed by workers [13] studying platy minerals with Geiger-Mueller-counter apparatus. The effect of orientation on the intensity of the strongest line in the pattern has also been studied by the writer. Relatively soft, lathlike organic material was ground for $1\frac{1}{2}$ hours and the powder carefully packed into a special sample holder with a thin-bladed spatula. The sample was then made flush with the surface of the holder with a single pass of the spatula across the holder. The sample was approximately 2 cm. in diameter and $2\frac{1}{2}$ mm. thick. The holder was so constructed that the sample could be rotated in steps about an axis perpendicular to the front surface of the sample. These steps

[13] Beatty, Van D., *X-ray Spectrometer Study of Mica Powders:* Am. Mineralogist, vol. 34, pp. 74 ff., 1949.

were taken at 15-degree intervals and several counts for each step were taken (to obtain an average value) with the counting technique through the peak of the strongest line in the pattern. The resulting intensity variation is roughly shown in figure 95. The maximum variation observed for the strongest line of the pattern was found to be approximately 60 percent when the average mean intensity was used as a basis for the calculation. Naturally, this procedure will give considerably more variation in intensity than will be observed among a group of samples that have all been prepared in essentially the same manner in the conventional type of sample holder. This difficulty of intensity variation might be overcome by

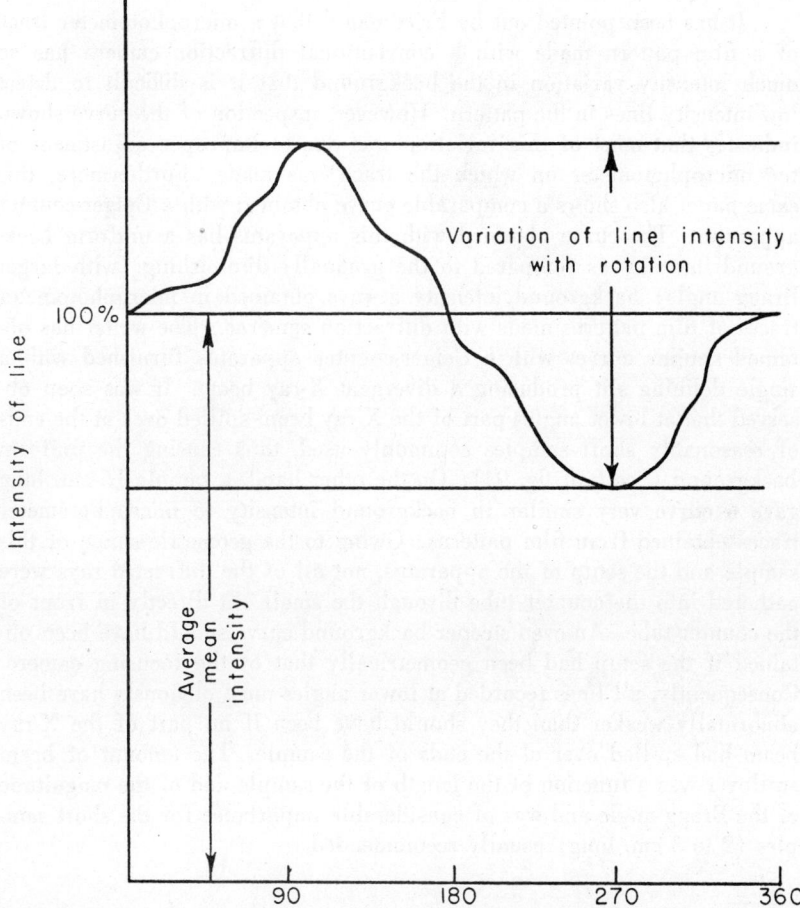

FIGURE 95. Variation of line intensity with rotation of powder sample obtained with a Geiger-counter instrument.

adopting a sample holder capable of rotating the sample at a rate which would average out the orientation effects.

The Geiger-Mueller-counter apparatus is especially recommended by the manufacturers for quantitative analytic work. The observations stated above, however, would introduce some question concerning this application. Furthermore, other research groups have found the apparatus unsuitable for quantitative analytic work, especially when used with recorder apparatus.[14]

Conflicting statements concerning operating characteristics of Geiger-counter apparatus have also been made.[15] [16]

It has been pointed out by Friedman [17] that a microphotometer trace of a film pattern made with a conventional diffraction camera has so much intensity variation in the background that it is difficult to detect low-intensity lines in the pattern. However, inspection of the curve shown indicates that most of this variation was due to improper adjustment of the microphotometer on which the trace was made. Furthermore, this same paper also shows a comparable curve obtained with a Geiger-counter apparatus. The curve obtained with this apparatus has a uniform background intensity as compared to the gradually diminishing (with larger Bragg angle) background intensity always obtained in microphotometer traces of film patterns made with diffraction cameras. The writer has obtained similar curves with a Geiger-counter apparatus furnished with a single defining slit producing a divergent X-ray beam. It was soon observed that at lower angles part of the X-ray beam spilled over at the ends of reasonably short samples commonly used, thus causing the uniform background (See b of fig. 94). On the other hand, a sample 16 cm. long gave a curve very similar in background intensity to microphotometer traces obtained from film patterns. Owing to the geometric shape of this sample and the setup of the apparatus, not all of the diffracted rays were gathered into the counter tube through the small slit directly in front of the counter tube. An even steeper background curve should have been obtained if the setup had been geometrically that of the focusing camera. Consequently, all lines recorded at lower angles must obviously have been abnormally weaker than they should have been if no part of the X-ray beam had spilled over at the ends of the sample. The amount of beam spillover was a function of the length of the sample and of the magnitude of the Bragg angle and was of considerable importance for the short samples (2 to 3 cm. long) usually recommended.

---

[14] Klug, H. P., Alexander, L., and Kummer, Elizabeth, *Quantitative Analysis with the X-ray Spectrometer:* Anal. Chemistry, vol. 20, pp. 607 ff., 1948.

[15] Carl, H. F., *Quantitative Mineral Analysis with a Recording Diffraction Spectrometer:* Am. Mineralogist, vol. 32, pp. 508 ff., 1947.

[16] Lonsdale, Kathleen, *Note on Quantitative Analysis by X-ray Diffraction Methods:* Am. Mineralogist, vol. 33, pp. 90 ff., 1948.

[17] Friedman, H., *Geiger-Counter Spectrometer for Industrial Research:* Electronics, vol. 18, pp. 132 ff., 1945.

PREPARATION AND MOUNTING OF SPECIMEN

Too much emphasis cannot be placed on the selection of the sample for analysis. Because only a very small fraction of the sample placed in the camera is actually exposed to the X-ray beam, it is imperative that all precautions be observed in the choice and preparation of the sample. The sample chosen must be truly representative of the material being investigated as regards composition, structure, or other characteristics for which the sample is examined.

Single Crystals

Single-crystal patterns are seldom made when identification of the material is the only objective, as the powder method is usually considerably simpler. There may be occasions, however, when the sample is limited to a very small, pure, single crystal, insufficient in amount to grind into powder. Under such circumstances, a single crystal ranging from 0.5 mm. to several hundredths of a millimeter in cross section and from several millimeters to about 0.3 to 0.5 mm. in length is mounted on the end of a small glass rod or wire, with one crystallographic axis approximatelly parallel to the axis of the rod so that it can be mounted and adjusted in the goniometer head of the single-crystal camera to turn about this axis. Patterns are recorded successively with alternate rotation about the three crystallographic axes according to procedures found in standard texts.[18] From these patterns unit-cell calculations are made. Under adverse conditions it may be impossible to obtain patterns about all the crystallographic axes, whereupon it may be necessary to calculate the dimensions of the entire cell from a single rotation pattern by means of the reciprocal lattice. A discussion of this concept is beyond the scope of this section, but it may be found in text books on X-ray-diffraction techniques.[19] [20] [21]

Powders

For powder patterns it is usually recommended that several milligrams of representative material be crushed and ground in an agate (or mullite) mortar until the entire specimen will pass a 200-mesh silk bolting cloth or screen. The writer has observed that if the sample is turned or oscillated during exposure to the X-rays it will be sufficient to grind the sample until high lights from individual particles are no longer observed when the powder is examined in a bright light. If the material does not grind readily, it may be filed with a clean single-cut fine-tooth file, using no more pressure than is absolutely essential. If the specimen must be preserved in its original form, the specimen can be mounted in a suitable rotating or oscillating device in such a manner that the sample-to-film distance remains constant.

[18] Buerger, M. J., *X-ray Crystallography*, New York, John Wiley and Sons, Inc., 1942.
[19] Clark, G. L., *op. cit.*
[20] Davey, W. P., *op. cit.*
[21] Bunn, C. W., *op. cit.*

Before the method of mounting the powdered specimen is selected, the optimum thickness of the sample to be used should be determined. The proper thickness can be calculated if sufficient information is available concerning the specimen. Otherwise, the optimum thickness usually can be estimated approximately by an experienced operator from the amount of the undiffracted X-ray beam that penetrates trial specimens, as determined with a fluorescent screen. This thickness can be calculated from the equation:[22]

$$t=\frac{2}{\mu}$$

where μ is the linear absorption coefficient calculated from the mass absorption coefficient according to the relationship:

$$\mu=d\sum p\left(\frac{\mu}{\rho}\right)$$
$$=d\left[p_A\left(\frac{\mu}{\rho}\right)_A+p_B\left(\frac{\mu}{\rho}\right)_B+p_C\left(\frac{\mu}{\rho}\right)_C \cdots \right]$$

d being the density of the material, p the elemental fraction in the compound and $\frac{\mu}{\rho}$ the mass absorption coefficients of the elements for the wave length of the radiation used. The values for $\frac{\mu}{\rho}$ can be found in table form in volume 2 of "*International Tabellen zur Bestimming von Kristallstrukturen*," pages 577 and 578.

For NaCl the optimum thickness for copper radiation is found to be

$$d\sum p\left(\frac{\mu}{\rho}\right)=2.165\ [.396\times30.9+.604\times103.4]=171$$

and

$$t=\frac{2}{171}=.0117\ \text{cm.}$$

This result indicates that the optimum sample thickness is usually considerably less than that generally recommended for capillary mounting.[23]

If too thick a sample is used, a distorted pattern is obtained. Thus, it is obvious that sample thickness becomes important when deciding on a suitable mounting technique. Another very important factor to be considered in connection with the mounting of the specimen is the amount of material available.

For materials of high atomic weight the optimum thickness may be so small as to necessitate dilution of the crystalline material with amorphous diluents such as flour, cornstarch, or gum tragacanth.[24][25] In any

[22] Buerger, M. J., *op. cit.*, p. 182.
[23] *Tentative Recommended Practice for Identification of Crystalline Materials by the Hanawalt X-ray Diffraction Method*: Am. Soc. Testing Materials designation E43-42T, 1942.
[24] Davey, W. P., *op. cit.*
[25] *Tentative Recommended Practice for Identification of Crystalline Materials by the Hanawalt X-ray Diffraction Method*: Am. Soc. Testing Materials designation E43-42T, 1942.

case, however, these diluents should be avoided or kept to a minimum since some of them (e.g., raw cornstarch) produce a crystalline pattern of their own, or an amorphous pattern with very broad lines (halos). These superposed patterns of the diluents often cause a considerably localized background fog, with consequent difficulty in observing lines in the regions of the amorphous bands.

It is recommended that the ground and diluted samples be packed into capillary tubes with an inside diameter of 0.4 to 0.6 mm. and made of plastic materials (materials with amorphous patterns) or glass containing elements of only low atomic weight. The plastic materials are preferred to glass, as measurements on Pyrex tubes with wall thickness just sufficient to permit careful handling show forty- to fifty-percent absorption of the CuK_α radiation. Longer wave lengths are absorbed to an even greater extent. Glass appears to be suitable for MoK_α radiation; however, as will be shown later, Mo radiation is not desirable for use in the identification of components of mixtures.

Another mounting method recommended for long-wave-length studies on materials of low atomic weight consists in mixing the powder of the unknown with about ten percent (by volume) of gum of tragacanth or collodion and extruding it as a rod approximately 0.5 mm. in diameter.

An excellent method of mineral specimen preparation used by some of the most prominent workers in the field, although it is usually not described in standard texts nor recommended in the American Society for Testing Materials procedures,[26] consists in mixing the powder of the unknown with a minimum of Dupont household Duco cement (or other plastic cements) and then rolling the plastic mass between two microscope slides to form a thin rod of the desired thickness. Thickness can be carefully controlled by inserting the microscope slides in a jig which holds them a fixed, predetermined distance apart. The cement acts as binder and diluent, and if kept to a minimum generally will not affect the background of the diffraction pattern. The writer has found this method to be particularly desirable for identifying montmorillonite-type clays, as the Duco cement conditions the clay so that it needs not be specifically treated [27] [28] [29] to be differentiated from other materials, such as muscovite or illite.

Platy or fibrous crystals may become oriented in the cement during rolling of the rod. The lack of random orientation changes the circular lines of the pattern to arcs, especially the lines formed at a small angle to the beam. On film patterns orientation effects usually do not cause any difficulty where qualitative identification is the objective, so long as

[26] *Tentative Recommended Practice for Identification of Crystalline Materials by the Hanawalt X-ray Diffraction Method:* Am. Soc. Testing Materials designation E43-42T, 1942.
[27] Jackson, M. L., and Hellman, N. N., *X-ray Diffraction Procedure for Positive Differentiation of Montmorillonite from Hydrous Mica:* Soil Sci. Soc. Am. Proc., vol. 6, pp. 133 ff., 1941.
[28] Hellman, N. N., Aldrich, D. G., and Jackson, M. L., *Further Note on an X-ray Diffraction Procedure for the Positive Differentiation of Montmorillonite from Hydrous Mica:* Soil Sci. Soc. Am. Proc., vol. 7, p. 194, 1942.
[29] Bradley, W. F., *Diagnostic Criteria for Clay Minerals:* Am. Mineralogist, vol. 30, pp. 704 ff., 1946.

the film is wide enough to include all orientation arcs. On the other hand, these effects frequently are advantageous in that they give some idea concerning the orientation of the crystallographic planes producing these arcs. Moreover, because the orientation arcs are darker than would be the equivalent complete circular line, lower percentages of platy or fibrous minerals can be detected in mixtures than would otherwise be detected.

Furthermore, use of the Duco cement permits preparation of a thinner sample than would the recommended capillaries and consequently makes it possible to obtain a pattern with narrow, sharp lines with maximum resolution. With this type of sample mounting, several lines are frequently obtained at the average position of a single broad line reported in the literature.

Other methods, such as affixing the powder to strings, hair, wire, and glass rods, have been suggested also.[30] These mounts commonly produce abnormal effects and do not appear desirable because of the difficulties involved in obtaining representative samples and the large amount of foreign material (the rod and binder) included in the sample. With glass rods, double lines frequently are obtained in the pattern, a condition that is very undesirable, especially in analysis of mixtures.

If only a very limited amount of material is available, a small lump 0.1 to 0.2 mm. in diameter can be mounted with mucilage or Duco cement on the end of a very thin glass rod. For very fine-grained materials in which the particles are randomly distributed, a powder diffraction is obtained. If the particles are not arranged randomly, the materials first should be crushed with a miscrospatula and the powder worked into a tiny ball with a binder. Such samples require no more than a few micrograms of material and produce satisfactory patterns at approximately double the usual exposure time.

Recently a camera has been developed in the Bureau of Reclamation laboratories to study materials in petrographic thin sections that are not identifiable by microscopic methods. The area on which the pattern is obtained is approximately 0.003 inch in diameter, or about twice the thickness of an average sheet of paper. For this procedure the slide is warmed to soften the mounting medium, and the thin section slid over so that the region to be studied projects over the edge of the glass slide. The cover glass is retracted at the same time. The specimen is mounted in the camera under the petrographic microscope to insure centering of the selected area in the beam. The sample is rotated during the exposure to produce smooth, uniform lines in the pattern. After the pattern has been recorded, the slide is again warmed, and the thin section returned to its original position and covered with the original cover glass.

At present, reliable methods for mounting powder samples for studies

---

[30] *Tentative Recommended Practice for Identification of Crystalline Materials by the Hanawalt X-ray Diffraction Method:* Am. Soc. Testing Materials designation E43-42T, 1942.

with the Geiger-counter apparatus seem to be lacking. H. F. Carl has described a method which he found to yield satisfactory quantitative accuracy.[31] Whatever method is selected, it should be remembered that different materials pack differently into the holder, and the operator should first check his technique on a series of synthetic samples of known composition before attempting to use it quantitatively or on unknown specimens.

POSITION AND TYPE OF FILM

As indicated in figure 92, diffraction lines (rings) can be produced over the entire region from practically 0° to 175° of the 2θ angle. Some materials such as metals and inorganic compounds produce patterns ranging over the entire region from 0° to 175°, whereas organic compounds produce practically an entire pattern at small angles. Again, certain sections of the region from 0° to 175° may be selected for detailed study, as for example in back-reflection work or studies where extreme accuracy is involved, when the region from 130° to 175° is used (See fig. 96). Consequently, the type of camera and film selected depends upon the objective of the investigation. Thus, for example, a camera with the film in the form of a cylinder with the specimen located at the axis is to be much preferred for the identification of rocks, minerals, and soils.

Most, if not all, X-ray-film emulsions available were developed primarily for radiographic work and consequently have a rather high degree of contrast [32] or reveal relatively small differences in absorption by the materials studied. For diffraction work, especially for studying mixtures, a film showing a straight-line function with a moderate slope over a considerable range, when the line density, $\log\left(\dfrac{I_0}{I}\right)$, is plotted against exposure, $(I_0 t)$, is desirable (See fig. 97). At present, such film is not generally available, and one must use the emulsions developed for radiographic work, compromising between exposure time and pattern quality. The fastest emulsions usually show considerable background blackening, whereas slower films producing good, clean backgrounds require considerably more exposure. Thus the choice of film rests on a number of conditions. For rapid and only approximate identifications, the fast films are preferred, whereas slower films are used if all possible information is to be gleaned from the pattern. Films as a rule are duplitized; that is, they have emulsions on both sides. To a slight extent, the double emulsion causes diffuseness in the lines, but rarely sufficiently to justify use of single-layer-emulsion film. All films should be developed according to the time, temperature, and processing conditions recommended by the manufacturer.[33]

Intensifying screens have been used for cutting down exposure time, but this practice is not recommended for mixtures of minerals because the

[31] Carl, H. F., *op. cit.*
[32] *Radiography of Materials*, Rochester, N. Y., Eastman Kodak Co., X-ray Division.
[33] *Radiography of Materials*, Rochester, N. Y., Eastman Kodak Co., X-ray Division.

screens broaden the pattern lines and thereby decrease definition. Patterns so produced have little value for determining the constitution of complex mixtures, for maximum definition with minimum line width is desired to permit measurement of the maximum number of lines.

X-RADIATION

The use of the K_a doublet radiation from molybdenum has been recommended for chemical analysis by the X-ray-diffraction or Hanawalt method.[34] This radiation might be suitable for the identification of pure

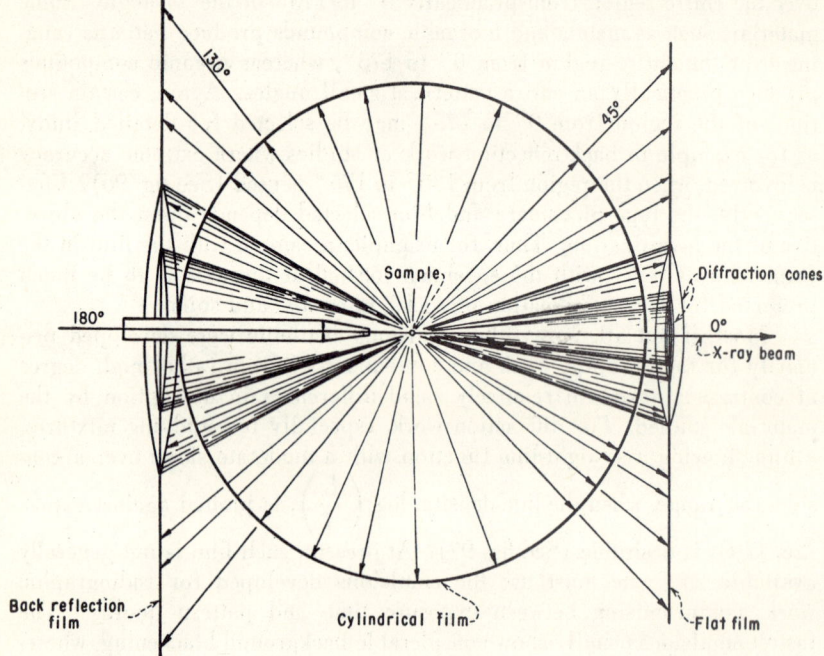

FIGURE 96. Film positions in various cameras used for powder studies.

TABLE 8

ANGULAR RANGE OF CORRESPONDING PATTERNS PRODUCED BY COMMON TARGET MATERIALS COMPARED TO PATTERN RANGE OF CHROMIUM K_a RADIATION

Radiation	Angular range of corresponding patterns (for outer line of pattern with $d = 1.1497$ Å.)
Cr	170° 0′
Fe	114° 48′
Co	102° 10′
Cu	84° 18′
Mo	36° 0′

[34] *Tentative Recommended Practice for Identification of Crystalline Materials by the Hanawalt X-ray Diffraction Method:* Am. Soc. Testing Materials designation E43-42T, 1942.

substances or very simple mixtures, but does not appear to be of much value for complex mixtures such as rocks and soils, for which the patterns should be spread out as much as possible to prevent superposition of lines from the different patterns of the constituents in the mixture. Reference to table 8, which shows the angular range of corresponding patterns

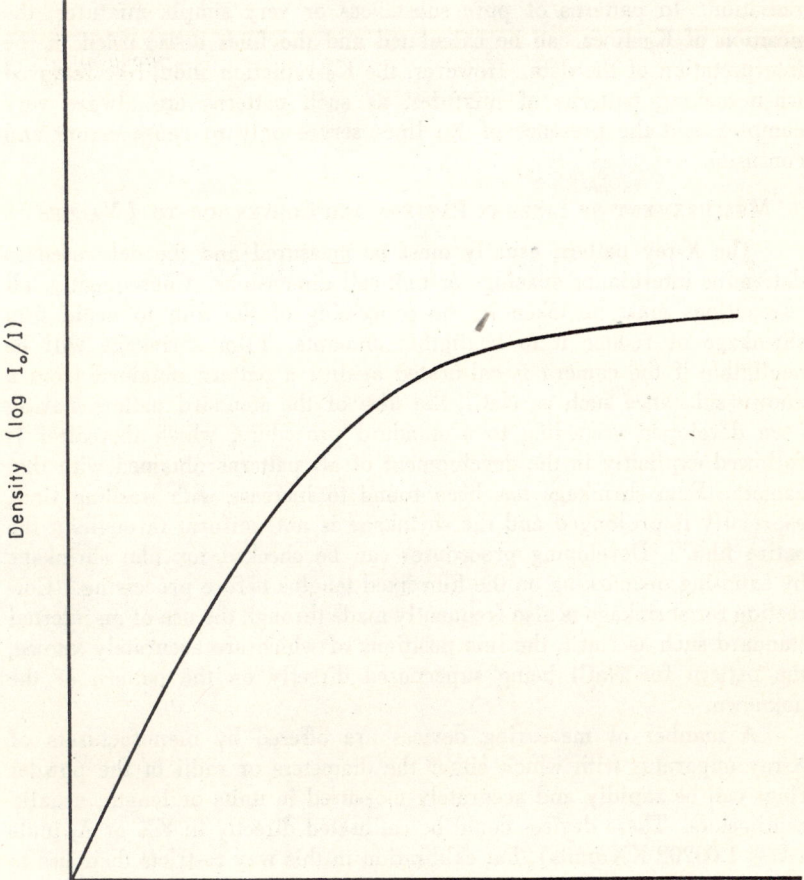

Exposure (intensity x time)

FIGURE 97. Exposure-density curve for typical X-ray film.

produced by the common target materials available as compared to the full pattern range (170°) for chromium K_a radiation, should remove any doubt concerning the foregoing statement. The results given in this table should enable the operator to choose the radiation for his particular needs. Where the operator is restricted to a single type of radiation, the K_a of copper is chosen almost invariably because it favorably combines sample penetration with a reasonably expanded pattern of good quality.

Since the K X-ray spectrum always contains characteristic radiation of several wave lengths, suitable filters [35] [36] or a crystal monochromator should be employed to produce reasonably monochromatic radiation and thus avoid superposition of lines from a second pattern derived from K_β radiation. In patterns of pure substances or very simple mixtures, the position of K_β lines can be calculated and the lines disregarded in the interpretation of the data. However, the K_β radiation should be removed when making patterns of mixtures, as such patterns are always very complex and the presence of K_β lines serves only to cause errors and confusion.

MEASUREMENT OF LINES IN PATTERN AND CONVERSION TO d VALUES

The X-ray pattern usually must be measured and the data used to determine interplanar spacings or unit-cell dimensions. Consequently, all precautions must be taken in the processing of the film to avoid film shrinkage or reduce it to negligible amounts. Film shrinkage will be negligible if the camera is calibrated against a pattern obtained from a known substance such as NaCl, the film of the standard pattern having been developed according to a standard procedure, which thereafter is followed explicitly in the development of all patterns obtained with that camera. Film shrinkage has been found to increase with washing time, especially if prolonged and the shrinkage is not uniform throughout the entire film.[37] Developing procedures can be checked for film shrinkage by exposing or marking on the film fixed lengths before processing. Correction for shrinkage is also frequently made through the use of an internal standard such as NaCl, the line positions of which are accurately known, the pattern for NaCl being superposed directly on the pattern of the unknown.

A number of measuring devices are offered by manufacturers of X-ray apparatus with which either the diameters or radii of the powder rings can be rapidly and accurately measured in units of length, usually centimeters. These devices could be calibrated directly in KX or Å, units (Å. = 1.00202 KX units), but calibration in this way restricts their use to a single type of camera with a fixed radius. Consequently, the measurements in centimeters must be converted into interplanar spacings or unit-cell dimensions by calculation or calibration curves indicating ring diameters or radii (in centimeters or millimeters) as a function of interplanar spacing (in angstrom units). For rapid and approximate measurements, a direct-reading scale on transparent plastic can be prepared from calculated or plotted data, interplanar distances equivalent to each ring being read directly when the scale is superimposed on the pattern.

[35] Clark, G. L., *op. cit.*
[36] Bunn, C. W., *op. cit.*
[37] Claassen, H. H., and Bow, K. E., *Correction of X-ray Powder Diffraction Patterns:* Sci. Inst. Rev., vol. 17, pp. 307 ff., 1946.

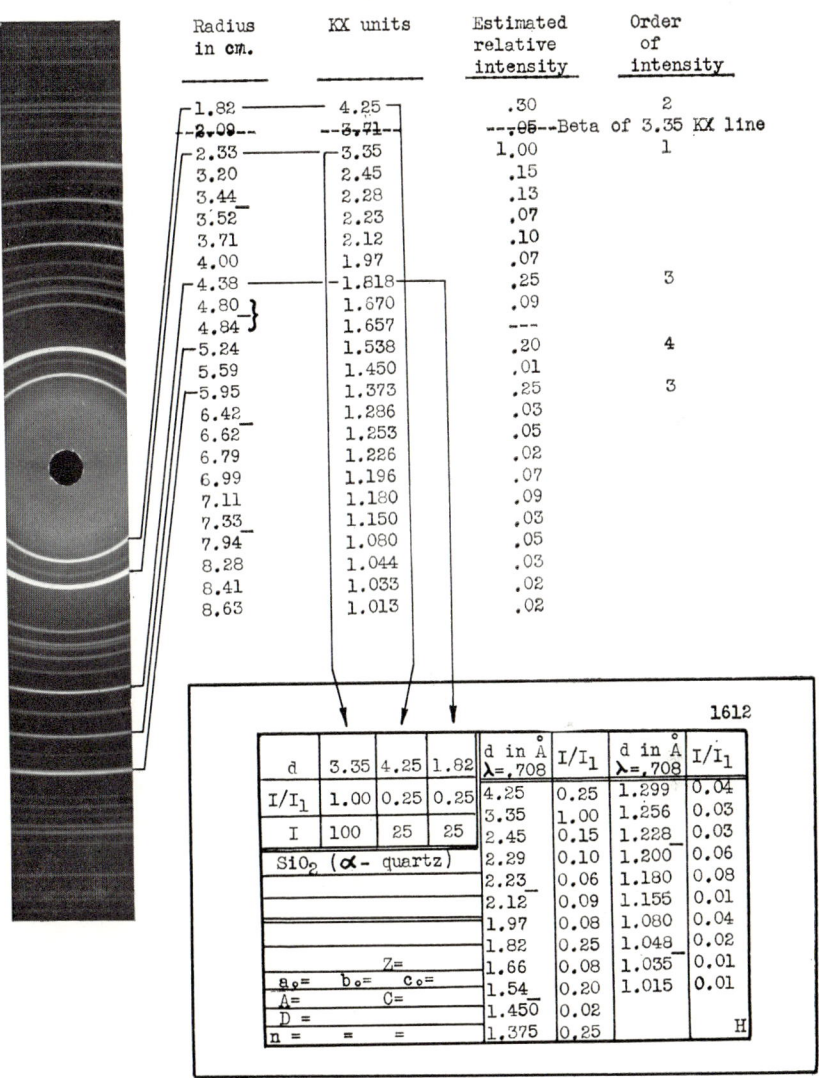

Radius in cm.	KX units	Estimated relative intensity	Order of intensity
1.82	4.25	.30	2
2.09	3.71	.05	Beta of 3.35 KX line
2.33	3.35	1.00	1
3.20	2.45	.15	
3.44	2.28	.13	
3.52	2.23	.07	
3.71	2.12	.10	
4.00	1.97	.07	
4.38	1.818	.25	3
4.80	1.670	.09	
4.84	1.657	---	
5.24	1.538	.20	4
5.59	1.450	.01	
5.95	1.373	.25	3
6.42	1.286	.03	
6.62	1.253	.05	
6.79	1.226	.02	
6.99	1.196	.07	
7.11	1.180	.09	
7.33	1.150	.03	
7.94	1.080	.05	
8.28	1.044	.03	
8.41	1.033	.02	
8.63	1.013	.02	

1612

d	3.35	4.25	1.82
I/I_1	1.00	0.25	0.25
I	100	25	25
SiO$_2$ (α- quartz)			

Z=
a$_0$= b$_0$= c$_0$=
A= C=
D =
n = = =

d in Å λ=.708	I/I_1	d in Å λ=.708	I/I_1
4.25	0.25	1.299	0.04
3.35	1.00	1.256	0.03
2.45	0.15	1.228	0.03
2.29	0.10	1.200	0.06
2.23	0.06	1.180	0.08
2.12	0.09	1.155	0.01
1.97	0.08	1.080	0.04
1.82	0.25	1.048	0.02
1.66	0.08	1.035	0.01
1.54	0.20	1.015	0.01
1.450	0.02		
1.375	0.25		H

PLATE 6. Illustration of the use of the card-index method of identifying an unknown.

If single-crystal-rotation patterns taken perpendicular to each of the three crystallographic axes are available, one dimension of the unit cell can be calculated from each of the three patterns. For cylindrical patterns the angle u_n is calculated from the tangent function of the distance measured on the pattern between the 0 and nth layer line and the film radius. For flat patterns it is calculated from the distance measured between the 0 layer and the apex of the nth-layer-line hyperbola and the sample-to-

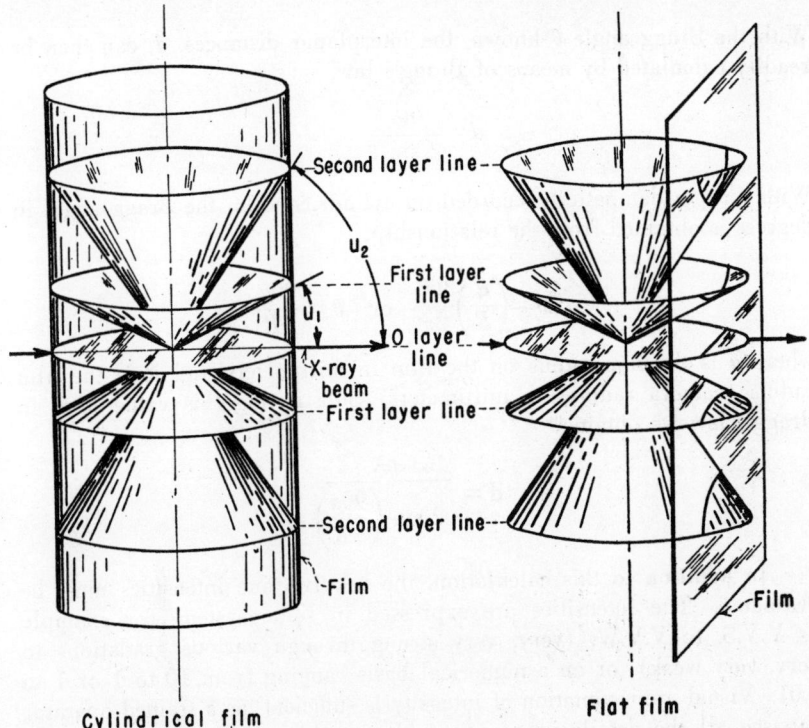

Cylindrical film

Flat film

FIGURE 98. Schematic diagram of single-crystal layer-line positions in cylindrical- and flat-film cameras.

film distance (See fig. 98). This value is then substituted in the equation:

$$I = \frac{n\lambda}{\sin u_n}$$

to obtain the identity period, I, or the distance between planes from one equivalent point to the next along the axis of rotation.[38]

On the other hand, if the diffraction data were obtained by the powder method, a technique considerably simpler than the single-crystal rotation method, the procedures of measurement and calculation are different.

[38] Friedman, H., *op. cit.*, chap. 5.

The process of measuring the powder pattern is the same regardless of how the measurements are to be used. The diameters (or radii) of all lines in the pattern are measured and recorded in centimeters or millimeters. If the X-ray pattern is recorded on flat film, the Bragg angle is obtained from the tangent relationship, namely,

$$\tan 2\theta = \frac{\text{line radius}}{\text{sample to film distance}}$$

With the Bragg angle θ known, the interplanar distances, d, can then be readily calculated by means of Bragg's law,

$$d = \frac{n\lambda}{2 \sin \theta}$$

With the powder pattern recorded on cylindrical film the Bragg angle in degrees is obtained from the relationship,

$$2\theta = \left(\frac{a}{R}\right)\frac{360}{2\pi} \quad \text{or} \quad \theta = \frac{90a}{\pi R}$$

where a is the line radius on the film in millimeters and R is the film radius (camera radius) in millimeters. Substituting this value of θ in Bragg's law we obtain [39]

$$d = \frac{n\lambda}{2 \sin \left(\dfrac{90a}{\pi R}\right)}$$

In addition to this calculation, the relative line intensities must be evaluated. The intensities are expressed in some system, for example as V.V.S. to V.V.W. (very, very strong through various gradations to very, very weak), or on a numerical basis ranging from 10 to 1 or 1 to 0.01. Visual approximation of intensity is sufficient, as a trained operator can see all the details in a pattern that can be detected with a densitometer. As has been suggested previously (See fig. 97), relative intensities of lines in a well-exposed pattern are different from those in an underexposed pattern. Differences in exposures not only result from changes in exposure time from specimen to specimen, but also occur within a single pattern representing a mixture containing both large and small proportions of the several ingredients.[40] ·If a particular constituent is to be determined, the writer has found it advisable to prepare a series of underexposed patterns of the constituent in pure form with exposure times of 1 percent, 2.5 percent, 5 percent, etc., of that used in obtaining the pattern of the mixture. This procedure will demonstrate why rather

[39] Clark, G. L., *op. cit.*, p. 279.
[40] Hellman, N. N., and Jackson, M. L., *Photometric Interpretation of X-ray Diffraction Patterns for Quantitative Estimation of Minerals in Clays:* Soil Sci. Soc. Am. Proc., vol. 8, pp. 135 ff., 1944.

strong lines of the patterns of minor constituents frequently cannot be found in the pattern of the mixture.

Identification of Minerals and Components of Mixtures

Single Crystals

If X-ray-diffraction data have been obtained from single-crystal rotation patterns and unit-cell dimensions calculated, the identity of the compound usually can be determined. However, the determination may not be simple because as yet unit-cell data have not been compiled into tables according to some regular order (decreasing or increasing) of the cell dimensions along the three axes. However, if the unit-cell dimensions of the unknown are found to correspond to those of a previously described compound, the identification can be considered to be reliably established. In fact, unit-cell data are about the most reliable type of X-ray data available for identifying organic materials; and it is most regrettable that no one has undertaken the task of compiling them into some systematic form based upon dimensions. Recently R. W. G. Wyckoff [41] initiated a continuous loose-leaf system for compiling unit-cell and other crystallographic data according to compound classification. This compilation will aid in the identification of compounds, but the difficulties of accomplishing an identification from unit-cell dimensions alone will be manifest.

Powders and Fine-Grained Materials

If the X-ray-diffraction data were obtained from powder patterns, the process of qualitative and semiquantitative identification is considerably simpler than if only single-crystal rotation patterns were available. For identification of a specimen from a powder-diffraction pattern, the radii, or diameters, of all lines in the pattern are measured, and the interplanar spacings calculated. The details of the procedure to be followed depend on the nature of the unknown and on the amount of other data available, such as optical and physical properties and chemical analyses.

If the unknown represents a pure compound or a mixture composed essentially of one constituent with only minor amounts of other ingredients and nothing is known concerning the identity of the compound or the principal ingredient, the Hanawalt method of identification is used.

The Hanawalt method, recommended by the American Society for Testing Materials, is based upon a card-file index system catalogued according to the three strongest lines in the pattern. After the pattern of the unknown has been measured, converted into interplanar spacings, the intensity of the lines estimated, and at least the three strongest lines (more if the three strongest lines are not outstanding) identified, the group of cards representing materials for which the strongest line corresponds to the same interplanar spacing as does the strongest line in the pattern

---

[41] Wyckoff, R. W. G., *Crystal Structures*, New York, Interscience Publishers, Inc., 1948.

is selected from the index. The subgroup for which the second-most-intense line corresponds to the same interplanar spacing as does the second-strongest line in the pattern of the unknown is then examined for correspondence between the third line of the cards and the third-strongest line in the pattern. Finally, the entire pattern of the unknown is checked against the pattern selected from the card index. This procedure is illustrated in plate 6. However, because of differences between the techniques used in obtaining the data for the card index and that used by the operator in obtaining the pattern of the unknown, or because of variations found in the patterns of some types of materials (to be discussed later), the operator should regard correspondence within $\pm$ 0.05 Å. as a satisfactory match for interplanar spacing in comparing his patterns with those recorded in the index. This same possible variation should be allowed in selecting the groups of cards for comparison.

Should the foregoing procedure be unsuccessful or if the specimen to be identified is known to be a mixture of several ingredients all in only small or moderate concentration, a somewhat different method of identification must be used. In mixtures, each of the three strongest lines may belong to patterns of different constituents so that the above procedure (outlined in plate 6) could not be used. For relatively simple mixtures, the procedure above may work if more (ten or twelve) of the strongest lines are used in searching the card index. In general, however, only the strongest line of the pattern can be used as a guide for selecting the group of cards for comparison. All of the lines on each card of the selected group are compared with the pattern of the unknown; bearing in mind, of course, that at least all the strongest lines must be found in the pattern of the unknown, with proper relative intensity. Checking of only a few lines on a card is usually sufficient to indicate whether or not agreement exists. When a card identifies part of the pattern of the mixture, the lines belonging to the pattern of the identified constituent are marked (on the pattern or a corresponding tabulation of data). The procedure is now repeated for the remainder of the pattern, again starting with the strongest remaining line. In this way all the constituents of the mixture can be identified, provided their patterns are catalogued in the index, when fluorescent scattering is small (recognized by light background in the X-ray pattern). The relative amounts of the ingredients present are deduced from the relative intensities of the lines in the pattern, as compared to the intensities of the lines in the pattern of the pure constituents, the exposure times, of course, being the same for all patterns. A series of underexposed patterns (1, 2½, 5, etc. percent of the total exposure time) of the pure constituent in question will be of considerable help in estimating these intensities.

If the absolute proportion of each compound in the mixture is to be determined, a synthetic specimen must be prepared from the identified pure materials in such proportions that the synthetic mixture yields a pattern matching in spacing and intensity all the lines of the original pattern, when both patterns are prepared under identical conditions of exposure and processing. If line shifts, fading of the pattern in general with increasing values of-the 2θ angle, or other differences are observed in the patterns, irregularities of composition, such as solid solutions, are indicated and the compound composition of the specimen must be determined by calculation from a chemical analysis. The chemical analysis frequently is best accomplished by means of standard spectrographic procedures. For thorough study of mixtures of silicates, the methods of X-ray-diffraction analysis [42] [43] [44] are practically indispensable. These methods reveal the various chemical combinations in which the silicon exists, whereas chemical or spectrographic methods alone yield only the total amount of silicon in the unknown, giving no clue as to its mode of combination.

Recently the American Society for Testing Materials has announced the completion and availability in the near future of the new second supplementary set of index cards. The original and first supplementary sets will now be available in the revised form only. Each set includes data for approximately 1,400 compounds.

In the original and first supplementary sets, the values of the d-spacings corresponding to the three strongest lines, together with their corresponding relative intensities, appear in the upper left-hand corner of each card (See pl. 6). There are three cards in the file for each diffraction pattern; the first card has the strongest line of the pattern at the extreme left and also contains the complete pattern data and some crystallographic data where available. The second card has the second strongest line in this position, and the third has the third strongest line in this position. The cards with the second and third strongest lines at the extreme left position were only "follow" cards and did not contain any data other than the d-spacings corresponding to the three strongest lines. The cards are filed in straight numerical order.

The revised original and first supplementary sets and the second supplementary set include only one card for each pattern, so as to reduce the required number of cards. These cards also include the data for the d-spacings corresponding to the three strongest lines of the pattern listed in decreasing order of intensity in the upper left corner of the card. The data for the largest spacing of the pattern are given to the right of the data for the three strongest lines. Wherever available, additional data consisting of the data for the X-ray set-up, crystallographic information, optical information, and information concerning the source, preparation, heat treatment, etc., of the sample are given. In addition, the card con-

---

[42] Clark, G. L., and Reynolds, D. H.. *Quantitative Analysis of Mine Dusts:* Ind. and Eng. Chemistry Anal. Ed., vol. 8, pp. 36 ff., 1936.

[43] Ballard, J. W., Oshry, H. I., and Schrenk, H. H., *Quantitative Analysis by X-ray Diffraction I. Determination of Quartz:* Bur. Mines Rept. Inv. 3520.

[44] Ballard, J. W., and Schrenk, H. H., *Routine Quantitative Analysis by X-ray Diffraction:* U. S. Bur. Mines Rept. 3888.

tains the formulas (chemical and structural for organic compounds), name, and complete pattern data. The cards are arranged into small Hanawalt groups of convenient size for values of the strongest line, and each group is arranged in numerical sequence according to the values of the second strongest line. This difference between the old and the revised-card indices will, of course, alter the above-described procedure somewhat when the revised index is used. With the revised index, the search of the diffraction-data file starts with two lines chosen from the unknown pattern as the strongest and second strongest. If this choice does not locate a corresponding X-ray pattern, it is necessary to reverse the order of the lines and search again.

It may even be necessary to try various other combinations of strong lines in the pattern before the identification can be made. For those who wish to continue the original method of searching the data file, the Society offers additional sets of the revised cards at reduced prices. A numerical index is also supplied with the revised sets of cards. This index has listings arranged in Hanawalt groups with three variations for the three strongest lines in each pattern: namely, first, second, third; second, first, third; and third, first, second.

When considerable investigation is being carried out in a limited field, or if sufficient optical or other data are available so that the possible compounds in unidentified specimens are relatively small, it frequently is advantageous to build up a file of patterns of standard materials. These patterns can then be used for identifying unknowns by direct comparison with their patterns. Plate 7 illustrates this method. However, it is to be strongly emphasized that extreme caution must be observed in selecting the materials for these standard patterns. Errors in identification are found frequently even for specimens obtained from established museum and private mineral collections.

Direct comparison of patterns, when used together with the Hanawalt method described above, is the most satsifactory for identification of materials, both in accuracy and time saved in the analysis. Occasionally, the Hanawalt method fails for mixtures because several strong lines of different ingredients fall in juxtaposition on the pattern and consequently are considered as a single broad line in the interpretation of the data, thus considerably displacing the position of the line in question. If the probable constitution of the mixture can be surmised, direct comparison with standard patterns will immediately disclose such situations, and errors and time-consuming labor are avoided.

Sometimes unit-cell data are available in the literature when powder data are lacking. Unit-cell data for a known material can be used to establish the identity of an unknown material from which a powder-dif-fraction pattern has been obtained. This method is practicable only if some clue suggests the identity of the unknown, and the number of known materials to be compared with the unknown is small. The comparison of

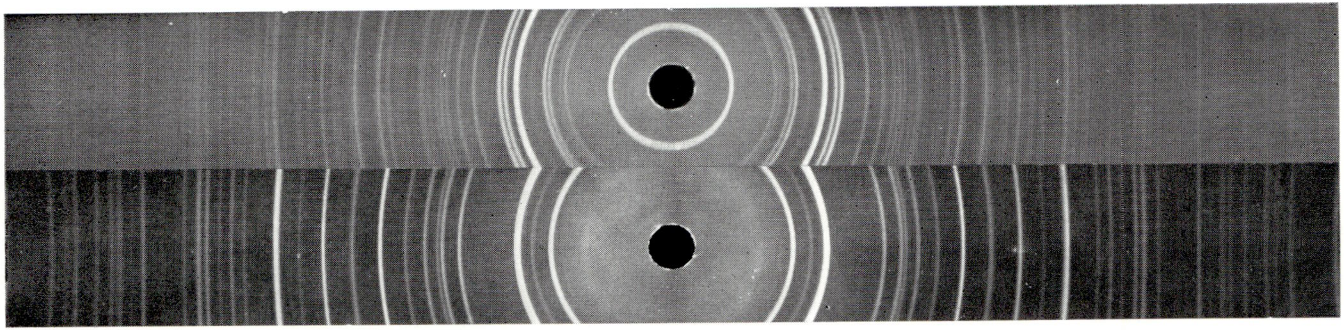

Plate 7. Direct comparison of pattern of unknown with a standard pattern: 1, pattern of unknown mixture; 2, standard pattern of quartz.

the unit-cell data with the powder-diffraction data is accomplished by application of the reciprocal-lattice concept. A complete explanation of this concept is, of course, beyond the scope of this section and the reader is referred to other sources.[45] [46] [47] However, it can be shown that the relationship between the true lattice (real space) and the reciprocal lattice (reciprocal space) can be expressed by the equation,

$$ d = \frac{R\,\lambda}{d^*}, $$

where d is the interplanar distance in the true lattice, d^* the interplanar distance in the reciprocal lattice, λ the wave length of the radiation used, and R a constant called the "magnification factor" applied to convert the dimensions in reciprocal space to such a magnitude that the reciprocal lattice or net can be plotted easily in cm.-units. If the unit-cell dimensions are not much over 10 Å., the value $R = 10$ will produce a reciprocal net of convenient dimensions. If the unit cell has dimensions between 10 and 30 Å., a value of $R = 20$ should be chosen. Briefly, the procedure is the following: the a, b, and c dimensions of the unit cell are converted into reciprocal-cell dimensions by means of the equation above and the resulting three-dimensional net plotted in one plane by folding the vertical planes down into the horizontal plane (See fig. 99). Thereupon, the experimentally determined powder-diffraction data are also converted into reciprocal dimensions by the same equation and the results (rings representing the ends of reciprocal space vectors free to turn about the origin) are superposed on the reciprocal net of the unit cell. If the unit cell fits the experimentally determined powder-diffraction data, there will be a net intersection at the end of each vector; i.e., the rings derived from the·powder data will all pass through one or more intersections of the three-dimensional reciprocal-unit cell net.

A mixture of minerals which are frequently difficult to differentiate by optical examination, especially when examined in the form of a rather fine powder, has been chosen to illustrate this method. Owing to the nature of these minerals, the mixture could be identified as a single homogeneous substance. A diffraction-powder pattern, however, will definitely show it to be a mixture. With the methods described above, one constituent can readily be identified as quartz from these powder data. This identification is further verified by direct comparison with a standard quartz pattern (See pl. 7).

Assuming that powder data are not available for the other constituent of the mixture, it would then be impossible to identify this constituent with the aid of the card index. However, if now through more thorough optical examination, further data can be obtained to limit the number of possible compounds to be checked to a reasonable number, identification

[45] Clark, G. L., *op. cit.*
[46] Davey, W. P., *op. cit.*
[47] Bunn, C W., *op. cit.*

will still be possible if suitable unit-cell data are available. In this case, this would involve careful checking of the refractive indices, obtaining the birefringence, and, if possible, such information as would enable one to classify the constituent as isotropic, uniaxial, or biaxial. Now further checking of the list of possible constituents obtained by the above procedure against the list for which powder data are available would readily

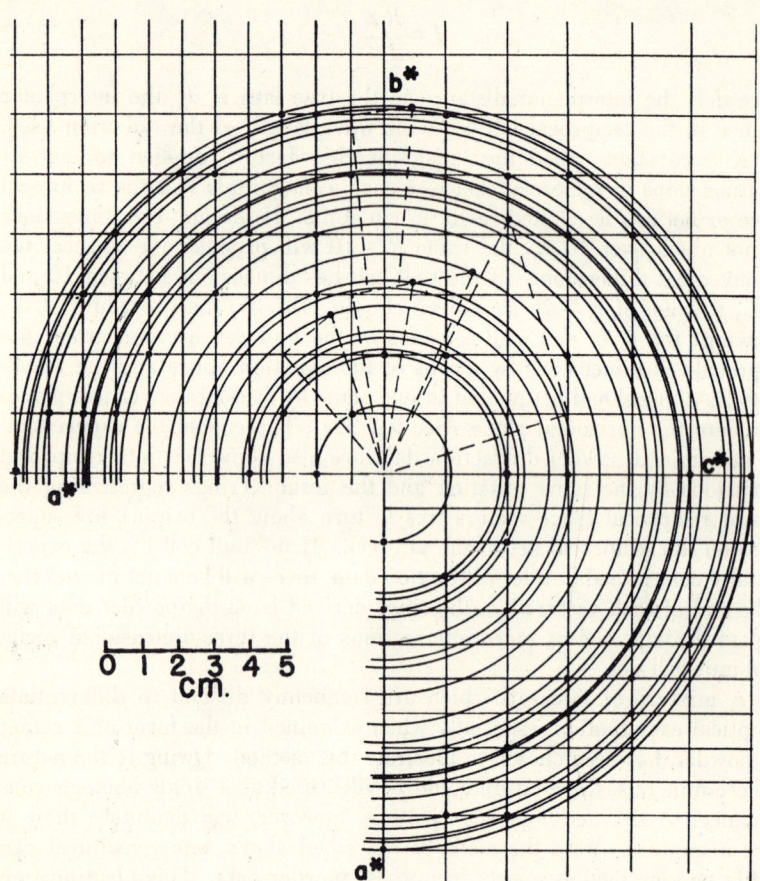

FIGURE 99. Comparison of powder data of unknown with possible
unit-cell data.

reduce the possibilities so that it would become feasible to apply the reciprocal-lattice method shown in figure 99.

All unidentified lines of the pattern of the unknown are converted to lines with reciprocal radii by means of the above equation and then drawn on transparent paper or plastic. The unit-cell dimensions of the possible constituents are then converted to reciprocal dimensions, and these

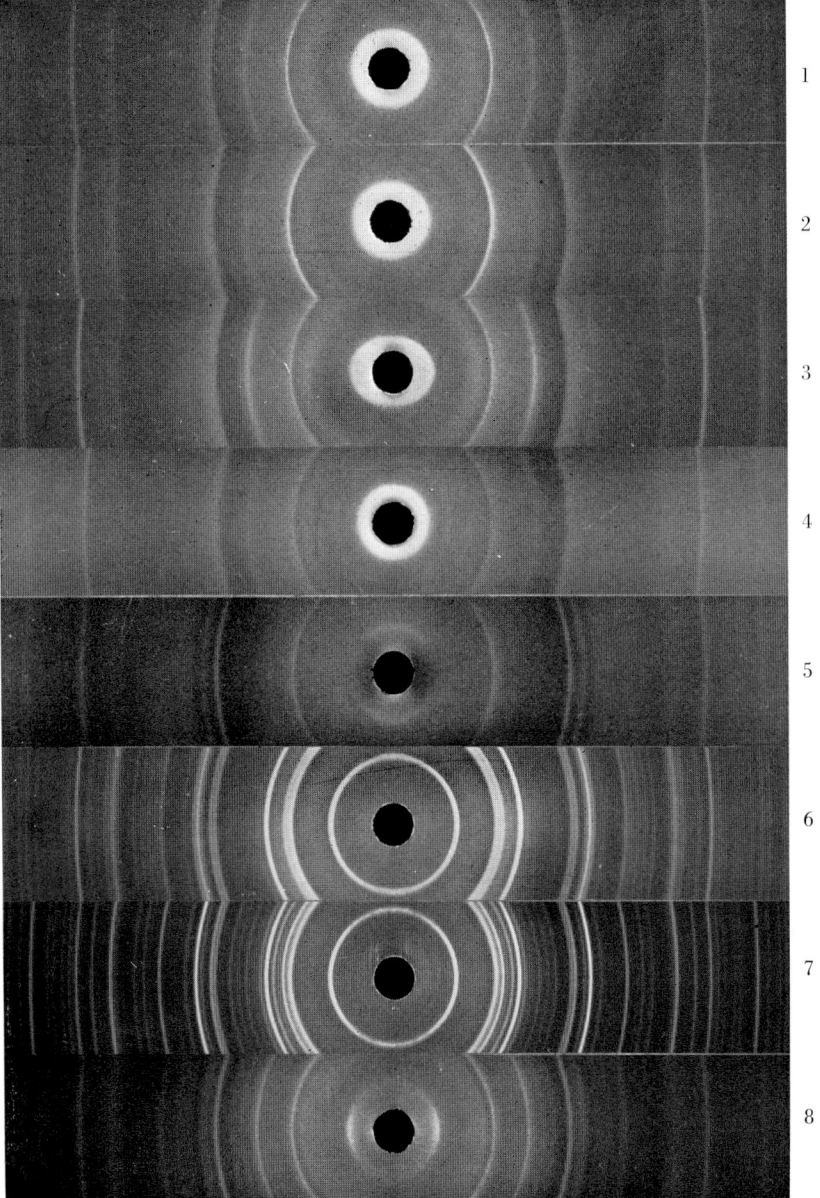

PLATE 8. Typical clay patterns. (1) Wyoming bentonite (montmorillonite), (2) beidellite, (3) hectorite, (4) nontronite, (5) glauconite, (6) kaolinite, (7) dickite, (8) illite.

reciprocal three-dimensional nets are drawn on separate sheets of paper. For figure 99 the unit-cell dimensions are those of cordierite: namely, a = 17.1 Å., b = 9.78 Å., c = 9.33 Å., and the orthorhombic crystal system. The experimental data are then superimposed on these various possible nets and the experimental lines (rings) checked for agreement with the net intersections. If a reasonable number of lines show agreement, then any lines not identified by the net intersections directly as (h00), (k00), (100), (hk0), (h01) and (0kl) are checked for (hkl) agreement (dotted triangles in figure 99 represent coincidence of (hkl) net intersections with experimental data lines). Coincidence of one or more net intersections with every experimental powder-data line identifies the second constituent in the mixture as cordierite.

Special Problems of Identification

The value of the X-ray-diffraction method, especially as a research tool, cannot be questioned. Its greatest effectiveness is derived when the X-ray-diffraction data are supplemented by physical and chemical determinations made by other methods; but the method can be used independently to great advantage in many problems. In some investigations X-ray-diffraction analyses are more rapid and efficient than are alternative methods; in other investigations, X-ray-diffraction analysis only will yield the necessary information. In plates 8 and 9 are examples of materials that are difficult to analyze or identify by methods other than X-ray-diffraction analysis.

Among the clays (pl. 8) some members of the montmorillonite group [48][49] show remarkable similarity, and at present information is insufficient to permit positive differentiation of the several members of the group on the basis of the X-ray-diffraction patterns alone. However, if the clays are calcined at temperatures determined from experimental studies or from thermal-dehydration [50] or differential-thermal [51][52] analyses, their identity can be established definitely from X-ray-diffraction studies. Furthermore, when Dupont household Duco cement is used as a binder for the powdered montmorillonite-type clay samples, the diameter and sharpness of the innermost line in the pattern give some information concerning the identity of the adsorbed cations. Preliminary observations indicate that a broad diffuse line represents a mixture of cations, whereas a sharp narrow line represents a relatively pure single cation. For the potassium ion, the line position corresponds to approximately 11.9 Å., for the sodium ion approximately 12.9 Å., and for the calcium ion approximately 15.5 Å. Likewise, the general degree of expansion or contraction

---

[48] Grim, R. E., *Modern Concepts of Clay Minerals: Jour. Geology*, vol. 50, no. 3, pp. 225 ff., 1942.
[49] Ross, C. S., and Hendricks, S. B., *Minerals of the Montmorillonite Group: U. S. Geol. Survey Prof. Paper 205-B*, 1943
[50] Nutting, P. G., *Some Standard Thermal Dehydration Curves of Minerals: U. S. Dept. Interior Prof. Paper, 197-E*.
[51] Grim, R. E., and Rowland, R. A., *Differential Thermal Analysis of Clay Minerals and Other Hydrous Minerals: Am. Mineralogist*, vol. 27, pp. 746, 801 ff., 1942.
[52] Grim, R. E., *Differential Thermal Curves of Prepared Mixtures of Clay Minerals: Am. Mineralogist*, vol. 32, pp. 493 ff., 1947.

of the over-all pattern is an indication of the chemical nature of the middle, gibbsite or brucite, sheet of the three-layer packet. For an element with a small atomic radius (aluminum) an expanded pattern is obtained, whereas for an element with a large radius (ferrous iron) a contracted pattern results. In general, most clays can readily be recognized and identified without preliminary treatment, as is shown by plate 8. The X-ray-diffraction method is particularly valuable in the analysis of shales, because they are frequently so fine-grained and so heterogeneous in composition as to preclude adequate microscopic analysis. The optimum particle size for X-ray-diffraction studies ranges from about 10^{-3} cm. to 10^{-6} cm., which lies just beyond the limit of the microscope. Plate 9 illustrates differences between various shales.

However, the X-ray-diffraction method is not a panacea for all problems, and its utility is usually considerably enhanced if it is used in connection with other methods, especially microscopic, spectrographic, and chemical procedures. This is particularly true for investigation of certain types of complex minerals or mixtures. There is no doubt that the method becomes more effective and efficient as the mixture becomes simpler or the unknown material purer; and, consequently, it is at times advisable or even necessary to concentrate or purify the constituents for separate study before a mixture can be satisfactorily analyzed. Purification and concentration of ingredients are especially valuable in the investigation of substances whose pattern is not sufficiently distinctive to permit use of merely a few isolated lines. Optical data obtained from microscopic measurements can often reduce time and labor by aiding in the selection of standard patterns to be used in the comparison; of course, positive identification may be accomplished on some materials by microscopy alone. Much time can be saved in a laboratory by using the X-ray method only if satisfactory answers cannot be obtained from microscopic studies.

The fact that nature is not particular, as regards chemical composition, when forming crystals is being recognized by men of science.[53] [54] [55] Very important properties are associated with apparently insignificant changes in chemical composition of minerals. Frequently, when once a geometric space arrangement of a crystal has been started, nature will continue with the building process using indiscriminately any atoms or ions available that are reasonably similar in size as long as the over-all structure is kept electrically neutral. Crystals that have extensive substitution have been referred to as "half-breed" and "stuffed" crystals,[56] depending on the mechanism by which the structure maintains neutrality. As a result of such partial substitutions the refractive indices of some

---

[53] Thompson, J. B., Jr., *Role of Aluminum in Rock-Forming Silicates:* Am. Mineralogist, vol. 33, pp. 209 ff., 1948.
[54] Buerger, M. J., *Crystals Based on the Silica Structures:* Am. Mineralogist, vol. 33, pp. 751 ff., 1948.
[55] Barshad, J., *Vermiculite and Its Relations to Biotite as Revealed by Base Exchange Reaction, X-ray Analysis, Differential Thermal Curves, and Water Content:* Am. Mineralogist, vol 33, pp. 655 ff., 1948.
[56] Buerger, M. J., *op. cit.*

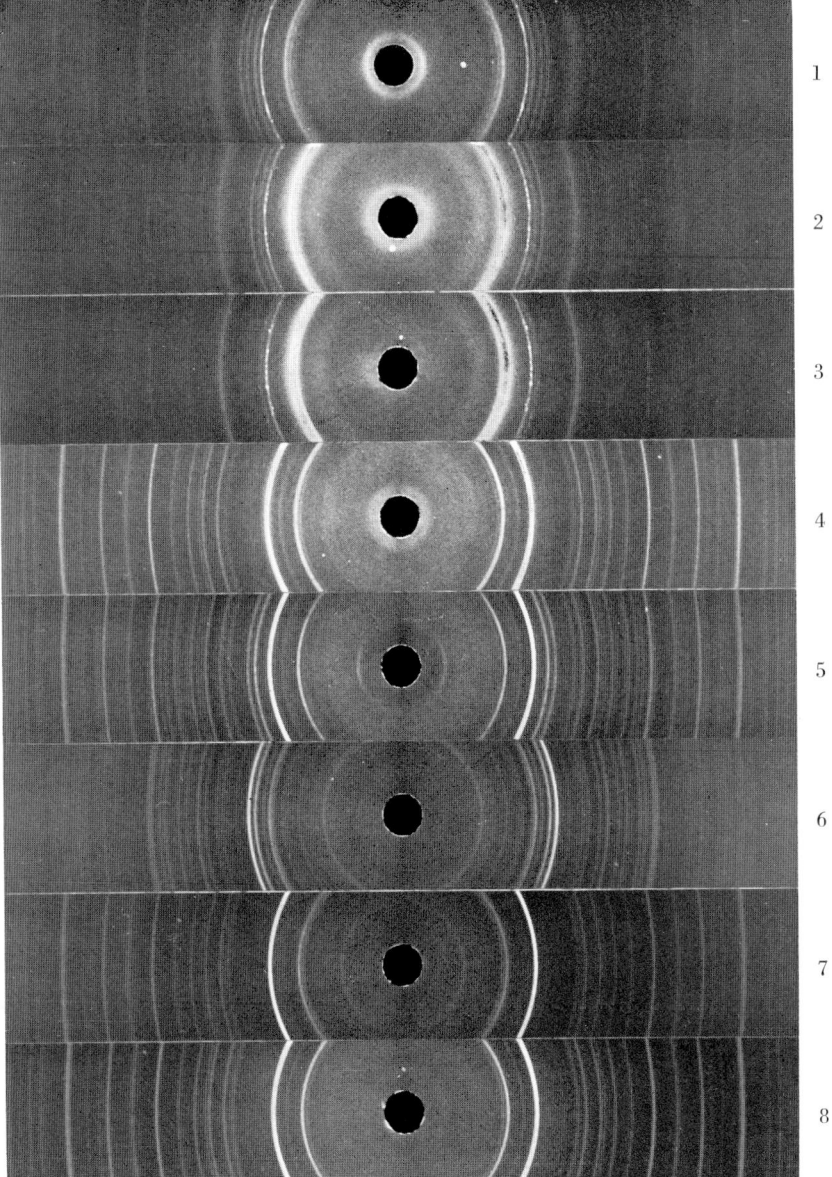

PLATE 9. Typical shale patterns. (1) Puente shale formation, Colton, California; (2) Salinas shale formation, Santa Barbara County, California; (3) Monterey shale formation, Santa Cruz, California; (4) Mowry shale formation, Casper, Wyoming; (5) Black Diamond shale, Metalene Falls, Washington; (6) kerogen (oil shale), Green River formation, Rifle, Colorado; (7) shale from Benton formation, Golden, Colorado; (8) illite-glauconite shale, Granby, Colorado.

materials must be expressed as a range rather than as a definite value. Such variations in chemical composition with resulting changes in lattice dimensions cause differences in line intensities, shifting of lines, appearance or disappearance of lines, and other changes in X-ray-diffraction patterns. Consequently, it is not possible to establish a standard pattern for some minerals, as has been attempted in the card-index system. For proper identification of compounds with variable chemical composition, isomorphism and phase relationships must always be considered. For such compounds, complete knowledge of the identity and structure will be obtained only from simultaneous consideration of chemical composition, crystallography, physical and physical-chemical properties, and X-ray-diffraction data.

Occasionally, evidence of atomic replacement within crystals is hardly detectable in the X-ray-diffraction patterns, especially where the unit-cell parameter is dependent on a certain kind or kinds of atoms or ions which form a rigid geometric-space packing, with the other atoms or ions fitted loosely into the holes of the structure. Substitutions of the latter type of atoms or ions may cause little if any change in the lattice parameters of the crystal.

APPLICATIONS

The X-ray-diffraction methods of analysis of geologic materials can be used in subsurface investigations to supply the geologist with information not otherwise obtainable, to furnish the petroleum engineer with precise knowledge of the composition and certain properties of reservoir rocks, and to trace mineralogic and structural changes of importance in problems of sedimentation and sedimentary petrology.

The precise identification of mineralogic composition made possible by the X-ray-diffraction method will permit the correlation of formations where other data are lacking, or may prevent erroneous correlation based on unreliable information. Identification of the kind and amount of minor constituents in apparently homogeneous, thick formations may subdivide the sequence in such a manner as to demonstrate the stratigraphic relationship to similar formations occurring elsewhere.

The analysis of reservoir rocks by X-ray-diffraction may reveal details of composition otherwise overlooked. In particular, the kind and amount of interstitial clay may critically control effective porosity and permeability of formations by changes in hydration and degree of flocculation, as a consequence of change in the solutions saturating the rock. Flocculation or deflocculation and hydration or dehydration of clays are controlled by their mineralogy as well as by their environmental changes. Hence, the susceptibility of clays to change during the water-flooding or other secondary-recovery programs can be detected by X-ray-diffraction analysis of reservoir rocks.

The geologist and engineer will find that X-ray-diffraction methods increase the reliability of geologic logging. The method supplements petrographic techniques of logging drill core, making possible quick and precise identification of even exceedingly fine-grained types and complex mixtures. In addition, the method can supply basic data on petrography and mineralogy necessary to interpret completely the electric and gamma radiation logs of drill holes. Both engineers and geologists are finding that the X-ray method of analysis is a powerful tool in the identification of potentially unsound materials in foundation strata or construction materials proposed for use in dams, powerhouses, buildings, and other large engineering works.

Finally, X-ray-diffraction analysis, both of geologic materials collected in the field from outcrops and cores and of synthetic materials in the laboratory, will yield detailed knowledge of processes involved in deposition, consolidation, and induration of sediments. The methods of X-ray-diffraction analysis are unsurpassed in effectiveness and efficiency in the tracing of progressive changes in mineralogy and structure of materials. Application of these methods will demonstrate the process of recrystallization during consolidation and induration, such as may occur in unstable minerals like clays, and the formation of new minerals, such as feldspar, mica, and zeolites. Only when these and related processes are understood will the conditions of petroleum formation, migration, accumulation, and production be understood fully.

The versatility and adaptability of X-ray-diffraction methods have justified recognition by the petroleum geologist, engineer, and chemist. For one problem the methods may afford merely a valued supplement to other techniques; for another problem the methods may be indispensable to a successful solution. Consequently, the supervisor of subsurface investigations should be cognizant of the potentialities of the X-ray-diffraction methods so that they will be used when and as required by the nature of the problems to be solved.

MULTIPLE-DIFFERENTIAL THERMAL ANALYSIS
PAUL F. KERR and J. L. KULP

Differential thermal analysis provides a useful technique for the study of specific minerals or mineral groups with distinctive heating curves. The method is suitable for both qualitative and semiquantitative studies of the clay minerals, the hydrous oxides of iron, aluminum and manganese, the carbonates, the zeolites, and a goodly number of other minerals. In general, the method applies to substances that yield characteristic peaks in the differential thermal curves.

In this technique a dual-terminal thermocouple is employed. One terminal is inserted in an inert material which does not undergo exothermic or endothermic reaction through the temperature interval to be studied. The other is placed in the mineral or mixtures of minerals under test. With a constant heating rate a thermal reaction in the sample will be recorded as a deviation from the straight-line plot of temperature difference against temperature. This deviation is dependent upon the nature of the heat change for its direction and amplitude. Peaks may be due to loss of either absorbed or lattice water, decomposition, or changes in crystal structure. They are characteristic for most thermally active minerals. Mixtures show a composite curve of the effects of the individual components in their proper proportion.

Although the original work on thermal analysis was done by Le Chatelier in 1887, it was not until the late 1930's that the method began to be used for semiquantitative study of clay minerals. In recent years studies have been made at the National Bureau of Standards,[57] the Massachusetts Institute of Technology,[58] the United States Geological Survey,[59] the Illinois Geological Survey,[60] the Bureau of Plant Industry,[61] and various United States Bureau of Mines research laboratories.[62]

Publications resulting from these studies emphasize the value of differential thermal analysis as a supplementary method coordinated with chemical, optical, and X-ray methods in studying clay minerals. X-ray data may have certain advantages in indicating a general clay-mineral group. Thermal-analysis curves, on the other hand, may contribute quantitative data on mixtures not readily available from X-ray-diffraction studies. Also, substitution in the clay-mineral lattice is frequently more apparent in the peak shifts of thermal curves than in X-ray patterns that frequently lack suitable definition. In combination, the two methods offer a solution to many complex problems in the study of clays.

The authors wish to acknowledge the helpful criticisms of the manuscript received from R. E. Grim, the Illinois Geological Survey; Ben B. Cox and Duncan McConnell, the Gulf Research and Development Laboratories; M. L. Fuller, T. L. Hurst, L. D. Fetterolf, and D. G. Brubaker, the New Jersey Zinc Company; Robert Rowan and R. H. Sherman, the Creole Petroleum Corporation; and Parke A. Dickey, the Carter Oil Company.

THE APPARATUS

The use of thermal analysis in the study of argillic alteration of a mineralized area or a stratigraphic-correlation problem requires the testing of hundreds of samples. This has involved a tedious laboratory procedure in the forms of apparatus described in the literature,[63][64] where a

---

[57] Insley, H., and Ewell, R. H., *Thermal Behavior of Kaolin Minerals:* Nat. Bur. Standards Jour. Research, vol. 14, pp. 615-627, 1935.
[58] Norton, F. H., *Critical Study of the Differential Thermal Method for the Identification of the Clay Minerals:* Am. Ceram. Soc. Jour., vol. 22, pp. 54-63, 1939.
[59] Alexander, L. T., *et al., Relationship of the Clay Minerals Halloysite and Endellite:* Am. Mineralogist, vol. 28, pp. 1-18. 1943.
[60] Grim, R. E., and Rowland, R. A., *Differential Thermal Analysis of Clay Minerals and Other Hydrous Materials:* Am. Mineralogist, vol. 27, pp. 746-761; 801-818, 1942.
[61] Hendricks, S. B., Goldrich, S. S., and Nelson, R. A., *On a Portable Differential Thermal Outfit:* Econ. Geology, vol. 41, p. 41, 1946.
[62] Speil, Sidney, Berkelhamer, L. H., Pask, J. A., and Davies, Ben, *Differential Thermal Analysis, Its Application to Clays and other Aluminus Minerals:* U. S. Bur. Mines Tech. Paper 664, 81 pp., 1945.
[63] Speil, Sidney, Berkelhamer, L. H., Pask, J. A., and Davies, Ben, *op. cit.*
[64] Norton, F. H., *op. cit.*

single sample is run at a time. Since each run requires several hours including cooling time, a maximum of about three samples a day may be analyzed. To overcome this difficulty, as well as to provide a simultaneous

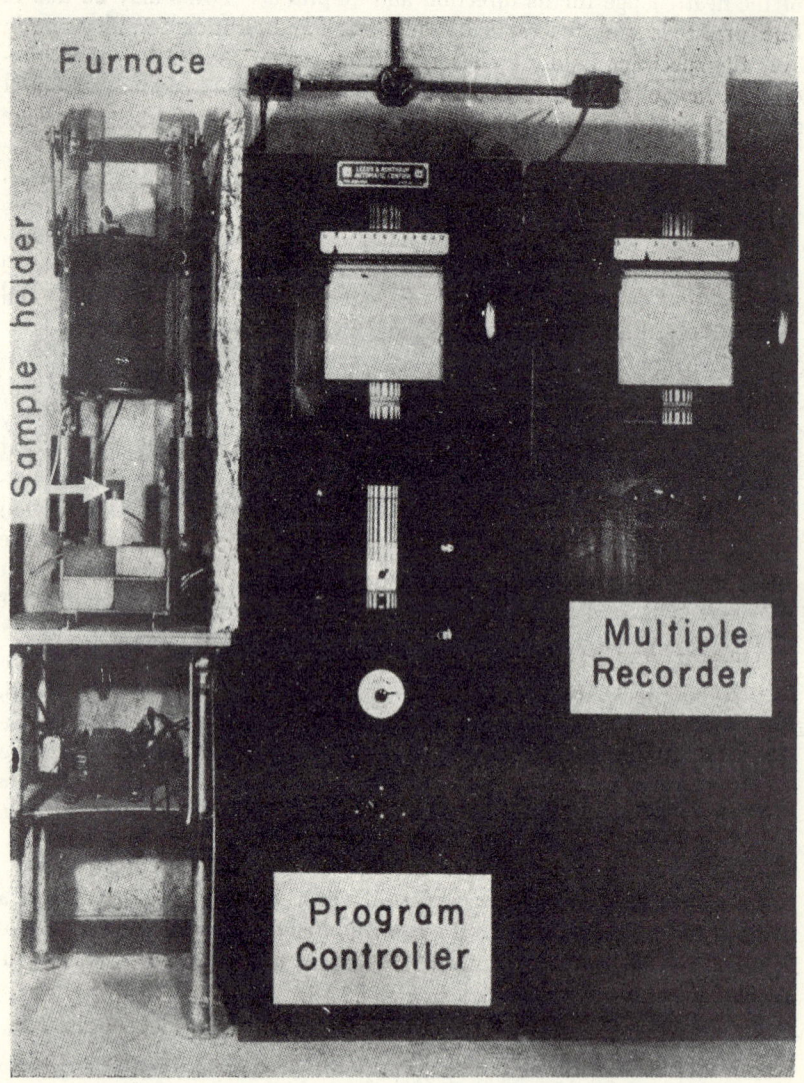

FIGURE 100. Complete multiple-differential thermal-analysis unit.

comparative record, a multiple-thermal-analysis unit was designed.[65] The various parts of the equipment were assembled late in November 1946

[65] Kulp, J. L., and Kerr, P. F., *Multiple Thermal Analyses:* Science, vol. 105, no. 2729, p. 413, 1947.

and were placed in operation about January 1, 1947, and approximately 1,500 samples had been run by August 1, 1947.

Figure 100 shows the apparatus as set up the mineralogical laboratory at Columbia University. For purposes of description, the apparatus may be conveniently divided into four parts: the furnace, the sample holder, the program controller, and the multirecorder.

The furnace is a Hoskins 305 electrical-resistance furnace into which an alundum tube ($1\frac{3}{4}$ inches inside diameter by 12 inches with a three-sixteenths-inch wall) is inserted to diffuse the heat and to insulate the metal specimen holder from the heater coils. The furnace is mounted vertically on a track and can be raised or lowered over the specimen holder by means of counterweights attached to two cables over pulleys.

The specimen holder (fig. 101) is drilled from a cylindrical block of

PLAN VIEW

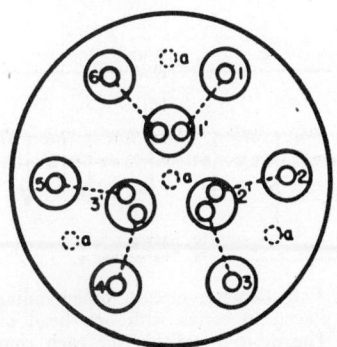

FIGURE 101. Nickel specimen holder.

chrome-nickel steel $1\frac{5}{8}$ inches outside diameter and one inch in height. Both pure nickel and chrome-nickel steel have been used, but the latter has similar heat conductivity and is less subject to scaling. The six samples to be tested are loaded in the outer holes numbered 1 to 6, while the inner holes 1', 2', 3' are used for inert material, which is ordinarily purified alundum manufactured by the Norton Company. The dashed lines indicate the connections between the two terminals of the chrome-alumel differential thermocouples. Thus one hole containing alundum is sufficient for the inert side of two differential couples. Chrome-alumel couples, BXS22, were used for maximum electromotive-force generation and were found to be substantial. The dots a indicate the position of the temperature-recording thermocouples. The terminals of these couples are adjusted to the same height as the differential couples in the samples. The sample and alundum holes are one-fourth inch in diameter and three-eighths inch deep.

The chrome-nickel-steel block is supported by an alundum tube ($1\frac{1}{8}$

inches inside diameter by 6½ inches with a one-fourth-inch wall) and supports a cylindrical cover of solid nickel half an inch thick, placed on the
block to shield the samples from direct radiation. Two complete units of
sample holder and thermocouples were prepared. Thus, if a break occurs
in one thermocouple, the entire unit may be recovered without delay and
a replacement connected. The next run may thus be carried out without
loss of time for repairs.

The program controller is a special Leeds and Northrup "Micro-
max," which is connected to one of the four possible temperature-recording thermocouples by way of a rotary selector switch. This unit is rated
to raise or lower the temperature of the sample at any desired rate from
0° to 50° C. a minute. It will also automatically hold the samples at

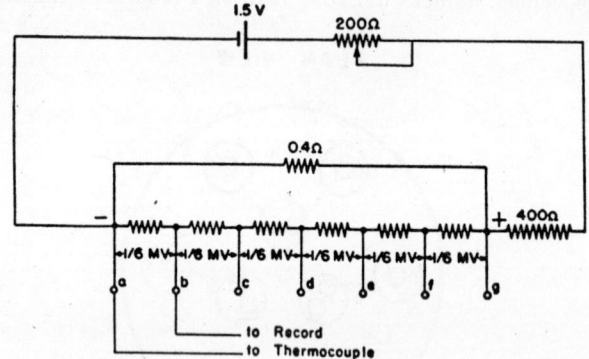

FIGURE 102. Potentiometer circuit for spreading records. The
unit is placed in series with one head of each thermocouple. The desired position for each couple is achieved
by connecting across appropriate terminals from a to g. In
the diagram, connection on a and b would add a constant 1/6 mv. to the base line of the differential thermal
curve.

any desired temperature when that temperature is reached. The pen record
indicates the temperature of the sample. The controller, when properly
adjusted, gives a linear heating curve.

The recorder for the differential thermocouples is a Leeds and Northrup "Speedomax," a six-point, high-speed, high-sensitivity electronic recorder with a maximum range of three millivolts. The chart of this recorder is synchronized with the chart containing the temperature record
on the program controller. This recorder is sensitive to 0.1° C. differential temperature, which, with the present specimen holder, gives a peak
one centimeter in amplitude for the alpha-beta quartz change. Experimentation on increasing sensitivity with accessory devices is in progress.
However, it should be pointed out that, beyond a certain limit of sensitivity, thermal gradients, geometry, thermocouple defects, and other unknown factors cause prohibitive irregularities in the base line. The pres-

ent equipment yields curves that are reproducible to a degree or so in peak temperature and to five percent of the peak amplitude under normal conditions.

Since all the differential thermocouples print at zero millivolts where there is no reaction taking place, it is desirable to spread the six records. This is done by a simple potentiometer circuit (fig. 102), which places the base line of each record about one-sixth of a millivolt from its nearest neighbors. The exact separation desired is achieved by adjusting a 200-ohm resistance in the battery circuit. It is also desirable to have certain sensitivity scales available, since some of the reaction minerals such as alunite, jarosite, kaolinite, and carbonates may extend beyond the chart on high sensitivity. Because this type of recorder measures the electromotive force of the thermocouple, a simple voltage divided with proportionate resistances is efficient for obtaining one-half, one-third, or any other predetermined fraction of the generated electromotive force.

Finally, there are two solenoid pens in series, one attached to the edge of each recorder. By means of a button switch, the solenoids are simultaneously activated, thus marking both records at the same time. Since the temperature at that instant can be read from the program-controller record, the temperature of the six records is also known and can be written on the multirecord chart at the completion of the run.

The advantages of this equipment are worthy of note. One of the greatest is the multiple-record feature, by means of which with three runs eighteen samples may be tested conveniently in an eight-hour day. Also significant is the reduction in the number of potential variables in using six samples under the same heating conditions. This is important when runs of quantitative mixtures are compared. The unit is compact, it does not require a darkened room for operation as in the photographic recording methods, and the results are immediately observable. The chief disadvantage lies in the necessity for applying minor corrections to each curve.

PROCEDURE

The samples to be tested by the differential-thermal-analysis apparatus are passed through a 50-mesh screen and packed to finger tightness around the differential thermocouple. No pretreatment is given for an ordinary run. It has been found by experimentation, as reported by others, that any attempt to attain equilibrium with a specified humidity merely alters the initial absorbed-water peaks ($100°-200°$ C.), the amplitudes of which are usually not used for quantitative analysis. Ordinarily weighing has been found to be unnecessary, and reproducible curves may be obtained for the same pure substance with finger-tight packing with a close-fitting metal plunger to a constant level. In special cases attention must be given to the problems of particle size, weight, and humidity.

After the samples are loaded, the cover is placed on the specimen holder, and the two charts are synchronized, the furnace is started. The heating rate has been standardized at 12° a minute, as this gives sensitive control, produces adequately sharp peaks, and is close to the heating rate used by a number of other workers in this field. The record is made from 100° to 1,050° C. At the beginning and end of the run the button switch activating the solenoid pens is pushed, thus fixing the temperature on the multiple-differential thermocouple record. When 1,050° C. is reached, the furnace is raised from the specimen holder, and

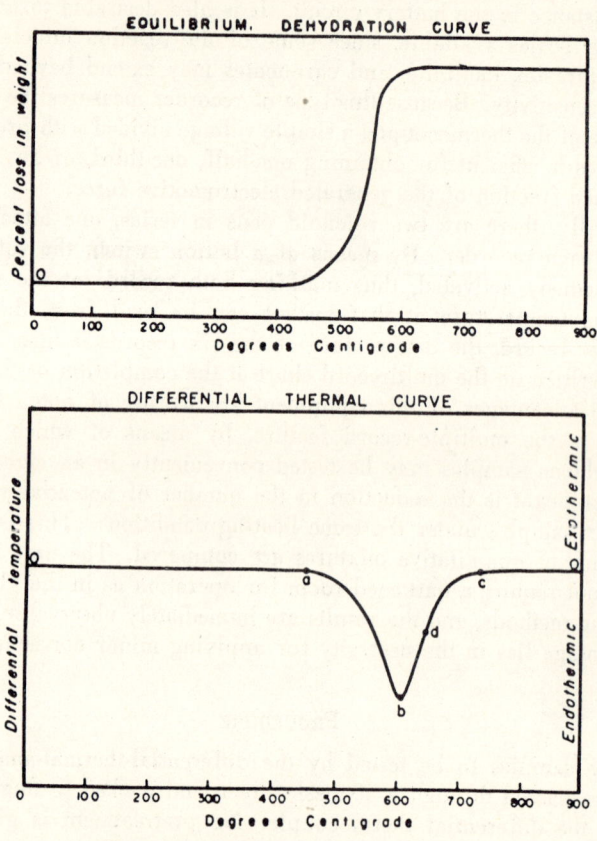

FIGURE 103. Theoretical thermal curves.

the samples are removed by compressed air while they are still hot. This procedure prevents caking, which occurs in certain specimens upon cooling.

The temperature thermocouples are calibrated and recalibrated occasionally with the alpha-beta quartz change. It has been found after trying many thermocouples that the quartz-inversion peak occurs on a differential curve within 5° of 579° C. (This is higher than the equilibrium value.)

The reproducibility has also been observed with the standard Georgia kaolinite endothermic and exothermic peaks. Since this is consistent with data in the literature and since the change in peak temperature of pure hydrous minerals may easily vary 5° C., more precise calibration has been considered unnecessary. Different sample blocks, thermocouples, and furnace windings produce no change in peak temperature greater than 5° C.

Although the thermocouples are made as similar as possible and adjusted to approximately the same heights in the sample holes, the sensitivity varies slightly. Therefore, after replacement of one specimen holder and the corresponding thermocouples by another, the first run is usually made with standard Georgia kaolinite in all sample holes. This indicates the relative sensitivity of the various thermocouples. It has been found that these relative sensitivities remain essentially constant for the life of the thermocouples unless the height of the thermocouple is changed as a result of rough handling.

All curves included in this description are based on the same sensitivity for direct comparison. It has been found convenient to plot the differential-thermal curve so that an exothermic peak is upward, while an endothermic reaction is represented by a deviation downward from the base-line curve.

THEORY

The theory of differential thermal analysis has been presented by Speil.[66] The following account, modified and corrected,[67] is included to aid in introducing the present studies.

Figure 103 compares two methods of dehydrating a clay mineral. The static method produces the equilibrium-dehydration curve, while the dynamic method gives the differential-thermal curve. In the first instance, the sample is held at each successively higher temperature until it has reached equilibrium. In the second, the sample is heated at a constant rate, thus extending the dehydration over a longer temperature range. Since the thermal curve is a differential function, it depends only on those effects that do not occur simultaneously and equally in the specimen and the inert material. Hence, there are only two thermal effects to consider, the differential flow of heat from the block to or from the thermocouple in the center of the sample and the heat of the thermal reaction. The differential-thermal curve of figure 103 represents an endothermic reaction. Below temperature a the heat inflow to both thermocouples, sample and inert material, is the same, and no difference in temperature is recorded. At a the reaction starts absorbing heat from its surroundings, making the sample couple cooler than the alundum couple. This effect increases until at b the rate of heat absorption by the chemical reaction equals the rate

    [66] Speil, Sidney, Berkelhamer, L. H., Pask, J. A., and Davies, Ben, *op. cit.*
    [67] Error in Spiel's derivation pointed out by Dr. D. G. Brubaker, N. J. Zinc Co., Palmerton, Pa.

of differential-heat conductivity into the clay specimen. The rate of heat absorption then continues to decrease more rapidly than the inflow of heat from the block. At this point d between b and c, the reaction ceases. However, since this point cannot be established exactly, a and c are usually chosen as limits.

Under static conditions the heat effect would cause a rise in temperature ΔT_s of a specimen given by:

$$\Delta T_s = \frac{M(\Delta H)}{M_o C} \tag{1}$$

where $M =$ the mass of the reactive mineral,
      $H =$ the specific heat reaction,
      $M_o =$ the total mass of the specimen, and
      $C =$ the mean specific heat of the specimen.
However, the heat flow from the nickel block towards the centers of the two sample cavities must be taken into account.

For any point between a and c, the simplified equation describing the changes in heat content of the thermally active constituent is:

$$M \int_a^x \frac{dH}{dt} dt + gk \int_a^x (T_o - T) \, dt = M_o C (T - T_a) \tag{2}$$

        (A)   (B)   (C)

for the inert sample:

$$g \cdot k' \int_a^x (T_o - T') \, dt = M_o' C' (T' - T_a') \tag{3}$$

               (B)'   (C)'

where $t =$ time,
      $M_o =$ the total mass of the test specimen,
      $M_o' =$ the total mass of the alundum,
      $C' =$ the mean specific heat of the test specimen,
      $C' =$ the mean specific heat of the alundum,
      $k =$ the conductivity of the specimen,
      $k' =$ the conductivity of the alundum,
      $g =$ the geometric-shape constant,
      $T_o =$ the temperature of the nickel block,
      $T_a =$ the temperature at the center of the sample at time $T = a$,
      $T_a' =$ the temperature at the center of the alundum at time. $T = a$,
      $T =$ the temperature at the center of the sample,
      $T' =$ the temperature at the center of the alundum.

Factor A defines the quantity of heat added to or subtracted from the test specimen owing to reaction. In an exothermic reaction $\dfrac{dH}{dt}$ is positive. Factor B [68] defines the quantity of heat absorbed by the specimen [69] $A + B = C$, because at any point x along the differential-thermal curve, the amount of heat used in raising the temperature of the specimen must equal the amount brought in by flow from the metal block plus the amount added or subtracted by the reaction.

In the event sample factor a does not exist, the heat which flows in B' must equal the heat used in raising the temperature of the specimen C'.

Let $$C' = C + \Delta C$$
and $$k' = k + \Delta k$$

Also in the experimental procedure $M_o = M_o'$ within the error of measurement. Subtracting (3) from (2) and rearranging gives:

$$M \int_a^x \frac{dH}{dt} dt + gk \int_a^x (T'-T)\, dt - g\Delta K \int_a^x (T_o-T')\, dt$$
$$= M_o \{ C[(T-T_a) - (T'-T_a')] - \Delta C[T'-T_a'] \}$$
$$= M_o \{ D[(T-T') - T_a - T_a')] - \Delta C[T'-T_a'] \} \tag{4}$$

As $T'-T = T =$ temperature indicated by the differential thermocouple, the equation can be considerably simplified by assuming that the term containing (T_o-T'), C, and K are small in comparison with other terms. By using a and c as integration limits:

$$M \int_a^c \frac{dH}{dt} dt + g \cdot k \int_{a'}^{c} \Delta T dt = M_o C[(T_c-T_c') - (T_a-T_a')] \tag{5}$$

but to a close approximation [69]
$$(T_a-T_a') = (T_c-T_c')$$
and
$$M \int_a^c \frac{dH}{dt} dt = M\Delta H,$$

the total heat of reaction

$$so, \frac{M(\Delta H)}{g \cdot k} = \int_a^c \Delta T dt. \tag{6}$$

The last expression is proportional to the area enclosed by a straight line from a to c and the curve abc, if the deviation from the base line is a linear function of the differential temperature. It is proportional, therefore, to the percentage of reacting material in a given weight of

[68] The temperature gradient in the chrome-nickel steel can be neglected as the thermal conductivity of the metal is so much greater than that of the refractory sample to be tested.

[69] $(Ta-Ta')$ and $(Tc-Tc')$ will equal zero for specimen holders in which the test and alundum holes are symmetrically spaced relative to the heat source. In the present concentric type of spacing $(Ta-Ta') = (Tc-Tc')$ within the error of measurement.

sults obtainable. It is believed that they agree substantially with those sample. This forms the basis for the quantitative use of differential thermal analysis. The linear relationship holds reasonably well. More exact determinations of comparatively simple systems can be made by running known mixtures and preparing a calibration curve of area versus percentage of each component.

The derivation above neglects the differential terms and the temperature gradient in the sample. It shows that the area under the curve is a measure of the total heat effect. The area is also considered independent of the specific heat. This factor, however, actually does affect the shape of the peak and may change the area slightly. For many purposes the approximate relationships are sufficient.

Qualitative Applications

The various clay minerals yield sufficiently different peaks to make the differential-thermal-analysis method particularly useful. When a specimen is relatively pure, preliminary identification by thermal curves is frequently comparatively simple. In addition, two-component mixtures are often resolved and at times even three-component mixtures. If, however, mixtures become too complex, only one or possibly two of the major components may be identified.

The multiple-thermal-analysis apparatus makes possible a rapid, widespread survey of the groups of minerals that can be identified by this procedure. The thermal curves given here are representative of the re- of the other workers on record. Illustrative curves are shown for the kaolinite and montmorillonite groups along with the hydrous oxides of aluminum and iron and some sulphates and carbonates.

The kaolin minerals (figs. 104, 105, and 106) are characterized by a large endothermic peak ranging from 550 to 700° C., owing to the decomposition of the kaolinite lattice into amorphous silica and alumina and a sharp exothermic peak of 980° C. caused by the recrystallization of amorphous alumina to gamma alumina.

Thermal curves of dickite from several localities are shown in figure 104. A number of the samples illustrated have been studied in connection with other investigations to such an extent that they may be considered representative of this clay mineral. The sample from Red Mountain, Colorado, was ground, and curves were run to compare 100–200-mesh, 200–300-mesh, and smaller than 300-mesh material. It is interesting to note that an ordinary specimen from St. Peter's dome yields a curve similar to Red Mountain dickite ground to minus-300 mesh. It is evident that particle size is a factor to be considered.

As the degree of orderliness in the superposition of the kaolin layers decreases from dickite through kaolinite to halloysite, the endothermic peak shifts downward in temperature, indicating less stability of the lattice. The disorder reaches a sufficient extent in the case of halloysite to

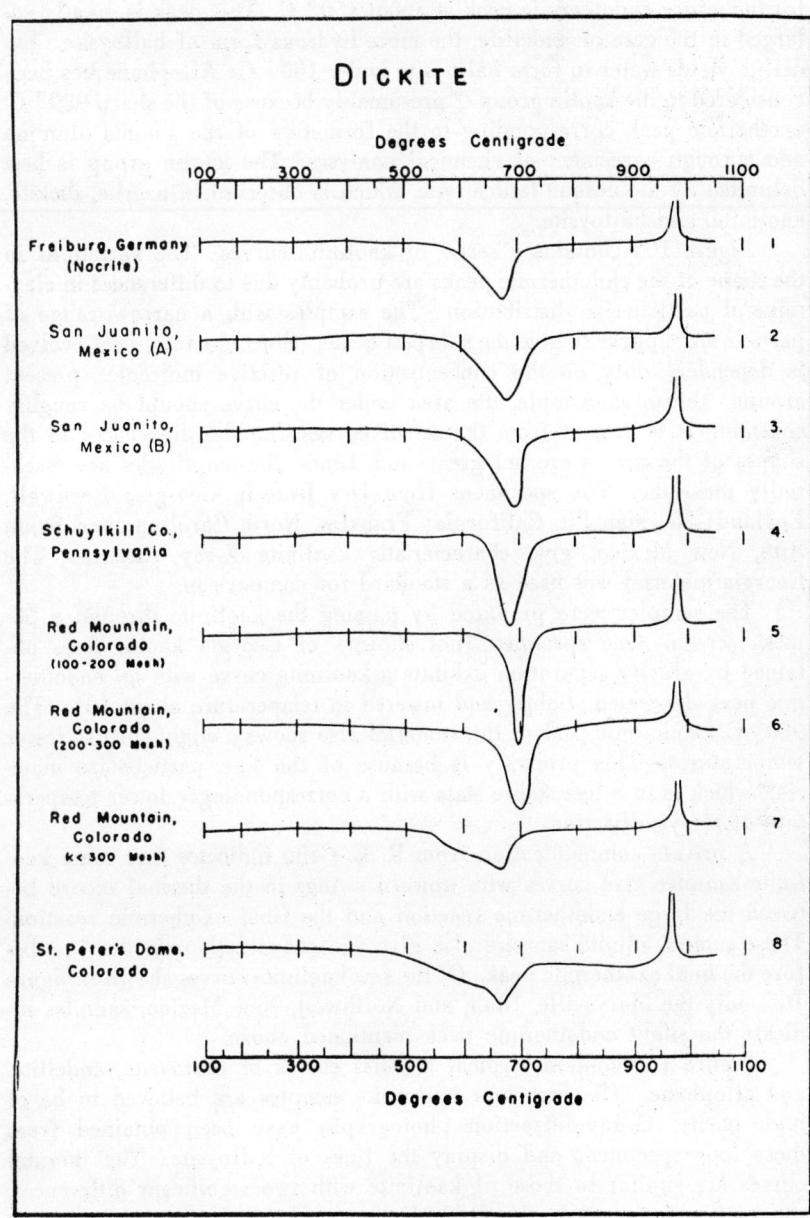

FIGURE 104. Thermal curves of dickite from several localities. Curves 5, 6, and 7 indicate effect of grinding on dickite from Red Mountain, Colorado.

permit absorption of some water between the lattice layers. This accounts for the minor endothermic peak at about 150° C. This peak is greatly enlarged in the case of endellite, the more hydrous form of halloysite. Endellite yields water to form halloysite under 100° C. Allophane has been considered in the kaolin group [70] presumably because of the sharp 980° C. exothermic peak corresponding to the formation of the gamma alumina and a rough agreement of chemical analyses. The kaolin group is best delimited by the unique lattice type, which is observed in nacrite, dickite, kaolinite, and halloysite.

Figure 105 contains a series of kaolinite curves. The variations in the shape of the endothermic peaks are probably due to differences in clay-mineral particle-size distribution. The samples with a narrow range of particle size appear to give the sharpest peaks. Since the total heat evolved is dependent only on the concentration of reactive molecules present around the thermocouple, the area under the curve should be roughly constant. It is evident from the set of curves that the differences in the shapes of the curves are not great, and, hence, the amplitudes are essentially the same. The specimens from Dry Branch, Georgia; Cornwall, England; Newman Pit, California; Franklin, North Carolina; and Santa Rita, New Mexico, give characteristic kaolinite X-ray patterns. The Georgia material was used as a standard for comparison.

The samples were prepared by passing the kaolinite through a 50-mesh screen. One specimen (not shown) of Georgia kaolin fines obtained by gravity separation exhibits a kaolinite curve with an endothermic peak depressed slightly and lowered in temperature about 10°. The 980° C. exothermic peak of this material also shows a slight shift to lower temperatures. This probably is because of the finer-particle-size material, which is in a less stable state with a correspondingly lower temperature of recrystallization.

A private communication from R. E. Grim indicates that some kaolinite samples give curves with upward swings in the thermal record between the large endothermic reaction and the final exothermic reaction. These same kaolinite samples also give a slight endothermic dip just before the final exothermic peak. Of the few kaolinite curves shown in figure 105, only the Marysville, Utah, and Northwest, New Mexico, samples indicate the slight endothermic peak mentioned above.

Figure 106 contains typical thermal curves of halloysite, endellite, and allophane. The first four halloysite samples are believed to be of high purity. X-ray-diffraction photographs have been obtained from these four specimens and display the lines of halloysite. The thermal curves are similar to those of kaolinite with two significant differences, the small endothermic peak at 150° C. due to adsorbed water, and the shift in the main endothermic peak to about 570° C. Grim [71] claims that halloysite does not have a lower temperature for the main endothermic peak than

---

[70] Speil, Sidney, Berkelhamer, L. H., Pask, J. A., and Davies, Ben, *op. cit.*
[71] Grim, R. E., *Modern Concepts of Clay Minerals:* Jour. Geology, vol. 50, pp. 225-275, 1942.

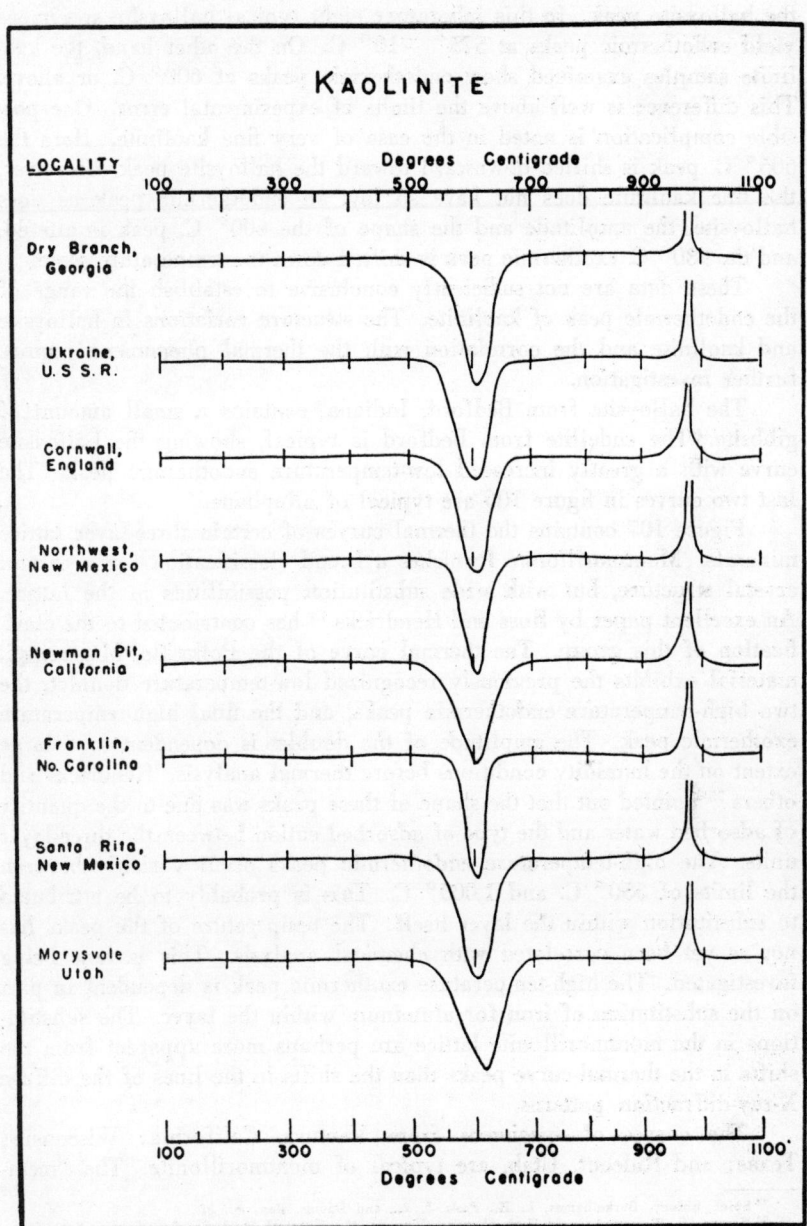

FIGURE 105. Thermal curves of eight specimens of kaolinite from different localities, all similar in character.

kaolinite. Other workers [72] [73] show evidence of the lower temperature of the halloysite peak. In this laboratory eight typical halloysite specimens yield endothermic peaks at 575° ±10° C. On the other hand, the kaolinite samples examined show endothermic peaks at 600° C. or above. This difference is well above the limits of experimental error. One possible complication is noted in the case of very fine kaolinite. Here the 605° C. peak is shifted downward toward the halloysite peak. However, the fine kaolinite does not have so low an endothermic peak as does halloysite, the amplitude and the shape of the 600° C. peak is altered, and the 980° C. exothermic peak is shifted down the temperature scale.

These data are not sufficiently conclusive to establish the range of the endothermic peak of kaolinite. The structure variations in halloysite and kaolinite and the correlation with the thermal phenomena require further investigation.

The halloysite from Bedford, Indiana, contains a small amount of gibbsite. The endellite from Bedford is typical, showing the halloysite curve with a greatly increased low-temperature endothermic peak. The last two curves in figure 106 are typical of allophane.

Figure 107 contains the thermal curves of certain three-layer lattice minerals. Montmorillonite furnishes a broad classification for a certain crystal structure, but with wide substitution possibilities in the lattice. An excellent paper by Ross and Hendricks [74] has contributed to the clarification of this group. The thermal curve of the Polkville, Mississippi, material exhibits the previously recognized low-temperature doublet, the two high-temperature endothermic peaks, and the final high-temperature exothermic peak. The amplitude of the doublet is dependent to a large extent on the humidity conditions before thermal analysis. Hendricks and others [75] pointed out that the shape of these peaks was due to the quantity of adsorbed water and the type of adsorbed cation between the three-layer units. The high-temperature endothermic peaks occur variably between the limits of 550° C. and 1,000° C. This is probably to be attributed to substitution within the layer itself. The temperature of the peaks has not as yet been correlated with chemical analysis. This is now being investigated. The high-temperature exothermic peak is dependent in part on the substitution of iron for aluminum within the layer. The substitutions in the montmorillonite lattice are perhaps more apparent from the shifts in the thermal-curve peaks than the shifts in the lines of the diffuse X-ray-diffraction patterns.

The curves of specimens from Ventura, California; Wisconsin; Texas; and Rideout, Utah, are typical of montmorillonite. The "meta-

[72] Spiel, Sidney, Berkelhamer, L. H., Pask, J. A., and Davies, Ben, *op. cit.*

[73] Norton, F. H., *Analysis of High-Alumina Clays by the Thermal Method:* Am. Ceram. Soc. Jour., vol. 23, pp. 281-282, 1940.

[74] Ross, C. S., and Hendricks, S. B., *Minerals of the Montmorillonite Group:* U. S. Geol. Survey Prof. Paper 205-B, 1945.

[75] Hendricks, S. B., Nelson, R. A., and Alexander, L. T., *Hydration Mechanism of the Clay Mineral Montmorillonite Saturated with Various Cations:* Am. Chem. Soc. Jour., vol. 62, pp. 1457-1464, 1940.

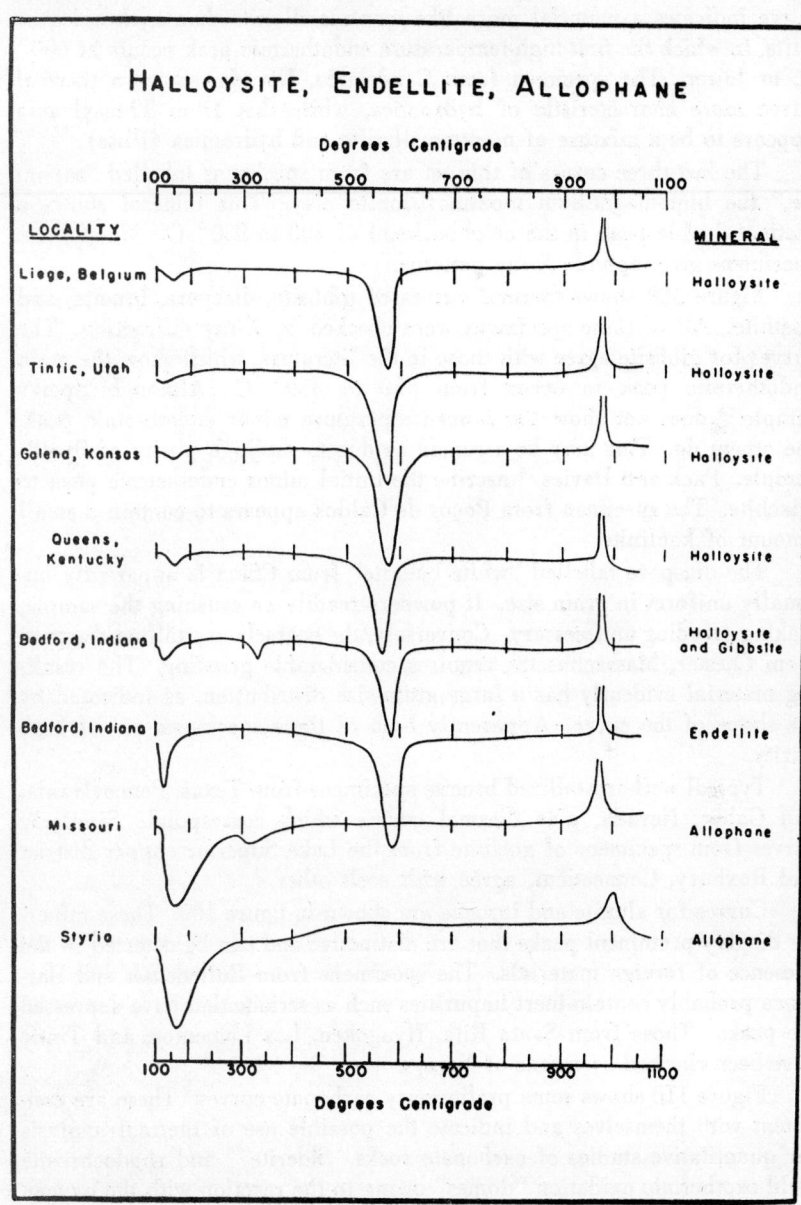

FIGURE 106. Thermal curves of halloysite mixed with some gibbsite, endellite, and allophane.

bentonite" from Highbridge, Kentucky, contains potash, but the thermal curve indicates a material more like montmorillonite than hydromica or illite, in which the first high-temperature endothermic peak occurs at 600° C. or lower. The specimen from Candelaria, Nevada, shows a thermal curve more characteristic of hydromica, while that from Transylvania appears to be a mixture of montmorillonite and hydromica (illite).

The last three curves of this set are from specimens labelled "saponite," the high-magnesium montmorillonite clay. This mineral shows a distinct double peak in the neighborhood of 800 to 850° C. All of these specimens give saponite X-ray patterns.

Figure 108 shows thermal curves of gibbsite, diaspore, brucite, and goethite. All of these specimens were checked by X-ray diffraction. The curves for gibbsite agree with those in the literature, which show the main endothermic peak to occur from 330 to 350° C. Although Speil's sample [76] does not show the lower-temperature minor endothermic peak, the others do. This may be assumed as due to the high purity of Speil's sample. Pack and Davies [76] ascribe the initial minor endothermic peak to cliachite. The specimen from Poços de Caldos appears to contain a small amount of kaolinite.

The diaspore labelled "white bauxite" from China is apparently unusually uniform in grain size. It powders readily on crushing the sample, making grinding unnecessary. Conversely, the coarsely crystalline diaspore from Chester, Massachusetts, requires considerable grinding. The resulting material evidently has a large grain-size distribution, as indicated by the shape of the curve. Apparently both of these specimens are of high purity.

Typical well-crystallized brucite specimens from Texas, Pennsylvania, and Gabbs, Nevada, give thermal curves which correspond. Similarly curves from specimens of goethite from the Lake Superior copper district and Roxbury, Connecticut, agree with each other.

Curves for alunite and jarosite are shown in figure 109. These minerals display prominent peaks that are distinctive and can be detected in the presence of foreign materials. The specimens from Bulledehah and Barranca probably contain inert impurities such as sericite that have depressed the peaks. Those from Santa Rita, Hyagoken, Los Lamentos, and Tintic have been checked by means of X rays.

Figure 110 shows some preliminary carbonate curves. These are consistent with themselves and indicate the possible use of thermal analysis for quantitative studies of carbonate rocks. Siderite [78] and rhodochrosite yield exothermic oxidation "domes" owing to the reaction with the oxygen of the air of the lower-valence oxide produced in the carbonate decom-

[76] Speil, Sidney, Berkelhamer, L. H., Pask, J. A., and Davies, Ben, op. cit.
[77] Speil, Sidney, Berkelhamer, L. H., Pask, J. A., and Davies, Ben, idem.
[78] Kerr, P. F., and Kulp, J. L., Differential Thermal Analysis of Siderite: Am. Mineralogist, vol. 32, p. 678, 1947.

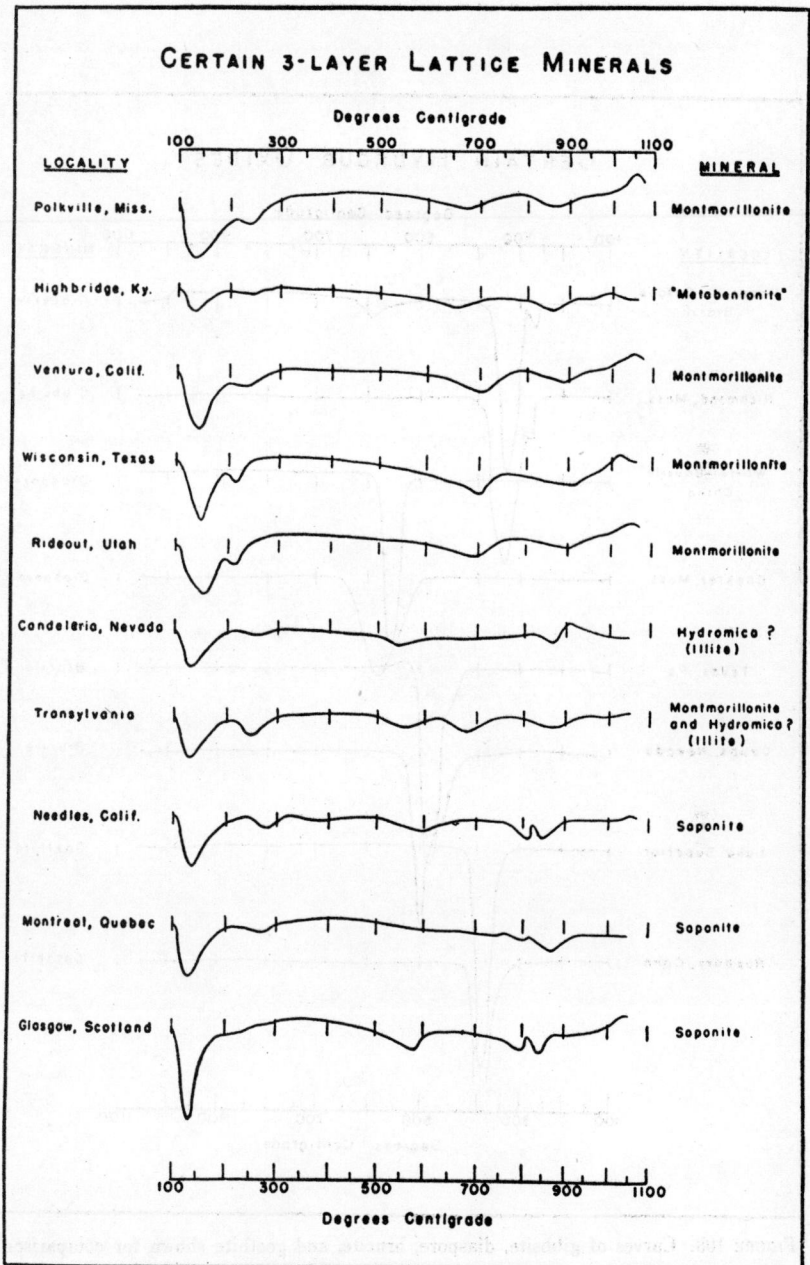

FIGURE 107. Several thermal curves of montmorillonite, saponite, and hydromica (metabentonite and illite) are illustrated.

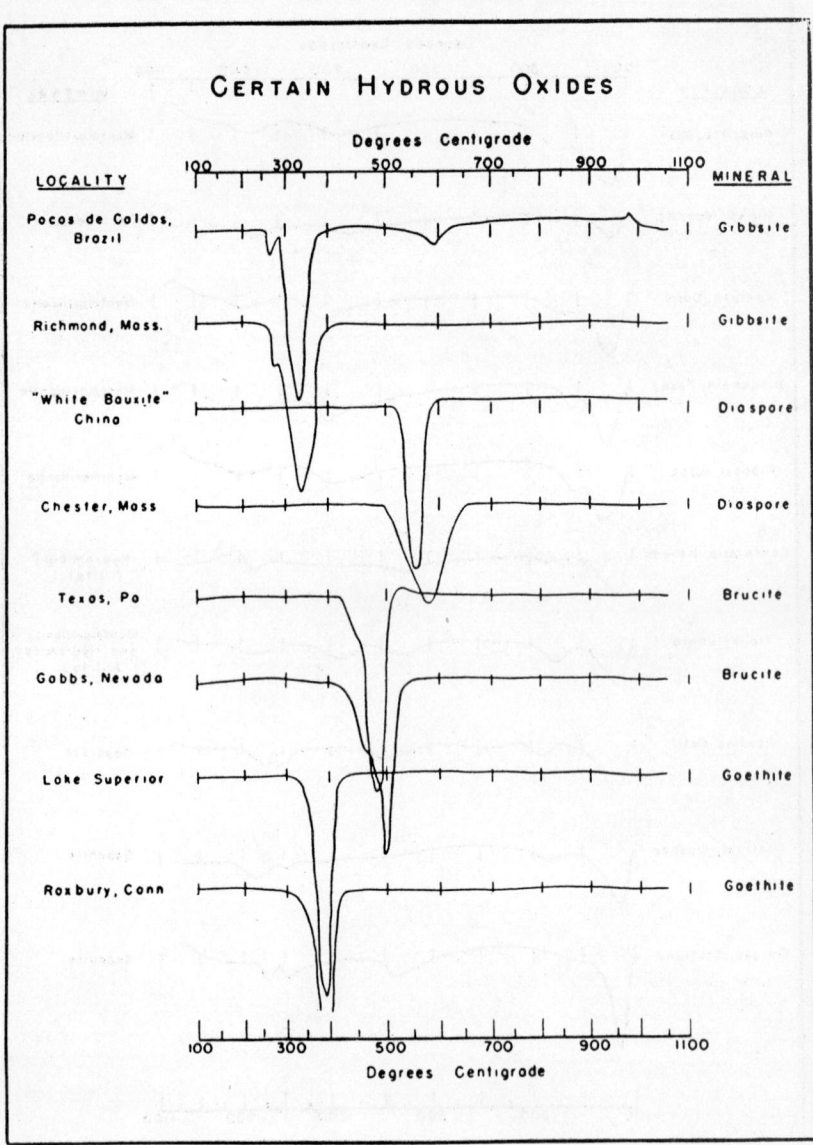

Figure 108. Curves of gibbsite, diaspore, brucite, and goethite shown for comparison.

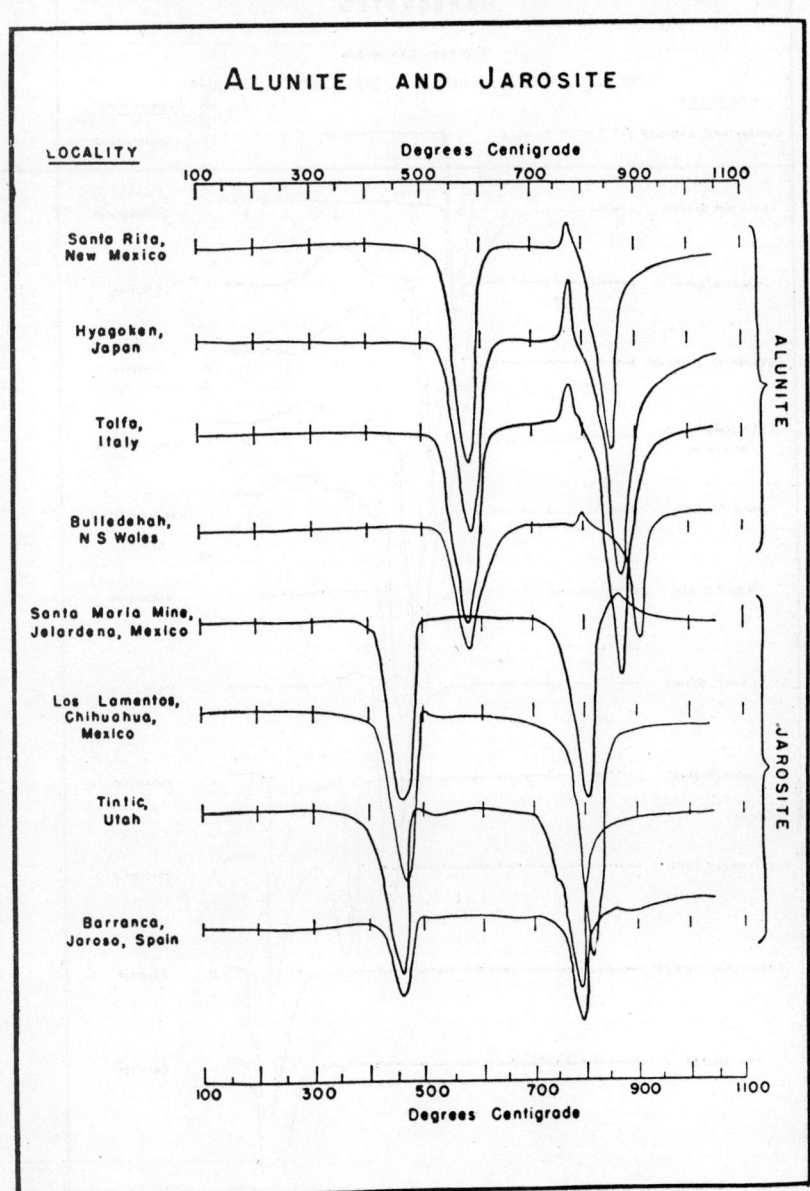

FIGURE 109. Thermal curves for alunite and jarosite.

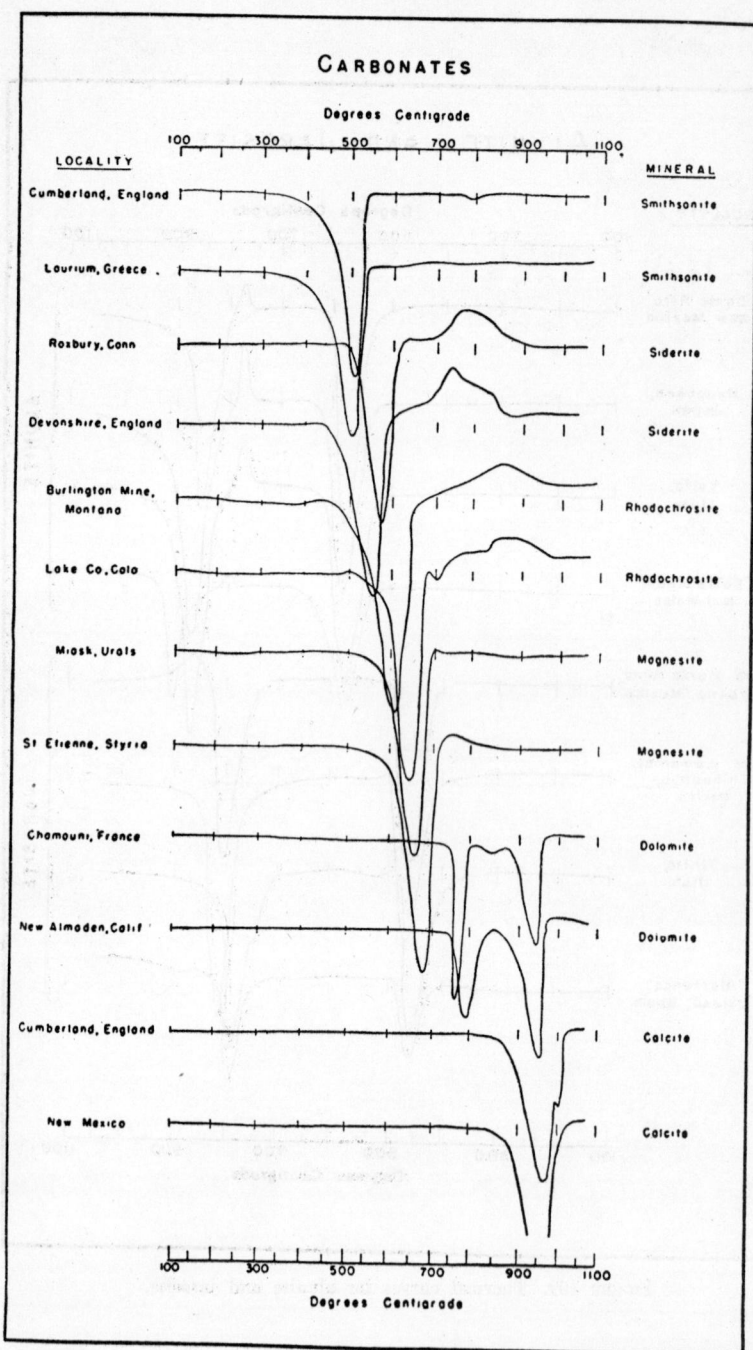

FIGURE 110. Thermal curves for a number of common rhombohedral carbonates.

position. Cuthbert and Rowland [79] published thermal curves of several carbonate minerals.

In these curves the carbonate peaks are low because of the admixture of inert material. The curves from figure 110 closely approximate the carbonate specimens run by other workers.[80][81]

Artificial Mixtures

Figures 111 to 118 show sets of thermal curves of predetermined mixtures ground to 50-mesh.[82] Although theoretically the area under the curve should be proportional to the percentage of the mineral present, this does not strictly hold experimentally. It has been found, however, that for known mixtures the amplitude of the peak plotted against the percentage of the mineral present gives a smooth curve. Moreover, it has been found that this "calibration curve" is not particularly affected by the chemical nature of the other components. Using figures for artificial mixtures containing kaolinite to furnish data, the graph in figure 119 was prepared. The amplitude of the endothermic 605° C. peak for kaolinite is plotted against the percentage of kaolinite in the particular mixture. A different symbol is used for each mixture. The area within the two smooth curves indicates the possible error to be expected from a mixture of kaolinite with an unknown aggregation, as indicated by artificial mixtures. Clay minerals that give such distinctive peaks as kaolinite may be quantitatively estimated with reasonable certainty for simple mixtures within ten or twenty percent. The variation may be due in part to minor differences in the heat conductivity of the foreign constituent.

The necessary assumption to render valid the application of the calibration curve to an unknown mixture is that the clay minerals in the unknown must be in roughly the same physical and chemical condition as in the artificial mixtures. This is probably a good approximation in many cases, particularly in the case of hydrothermal clays formed *in situ*. Grim [83] has already pointed out the need for great caution in making such an assumption for certain sedimentary-clay mixtures.

Figure 111 shows a suite of kaolinite-goethite mixtures. The endothermic decomposition peaks for both minerals are shifted down in temperature with increasing percentage of the other mineral. This shift is to

[79] Cuthbert, F. L., and Rowland, R. A., *Differential Thermal Analysis of Some Carbonate Minerals:* Am. Mineralogist, vol. 32, p. 111, 1947.
[80] Speil, Sidney, Berkelhamer, L. H., Pask, J. A., and Davies, Ben, *op. cit.*
[81] Faust, unpublished. (No reference given in original paper.)
[82] The samples used in these artificial mixtures were essentially uncontaminated materials from well-known localities and were checked both optically and by means of X-ray diffraction. The alunite sample was analyzed chemically by Ledoux and Company.

Mineral	Locality	Mineral	Locality
Alunite	Santa Rita, New Mexico	Quartz	
Jarosite	Santa Maria mine,	Sericite	American Canyon, Nevada
	Jelardena, Durango, Mexico	Dickite	Cusihuirachic, Mexico
Kaolinite	Dry Branch, Georgia	Goethite	Lake Superior
Montmorillonite	Polkville, Mississippi		

[83] Grim, R. E., *Differential Thermal Curves of Prepared Mixtures of Clay Minerals:* Am. Mineralogist, vol. 32, p. 493, 1947.

be attributed to the conductance of the heat away from the particles in the endothermic reaction by the foreign inert neighbors. The 980° C. exothermic peak is not shifted appreciably. This is probably a result of the narrow temperature range of the reaction. Below a certain temperature, under these conditions of molecular structure, amorphous alumina will not change over to gamma alumina. At 980° C., however, crystallization occurs almost instantaneously. Hence, the mixture of 50-mesh inert material with 50-mesh kaolinite does not appreciably shift this peak.

Figure 112 illustrates the effect of mixing quartz with kaolinite. The quartz curve is a straight line aside from a minor peak at the inversion point. The kaolinite curve is depressed by the admixture of quartz, but comparison with figure 105 indicates that otherwise there does not appear to be any substantial change.

Figure 113 represents mixtures of sericite and kaolinite. Sericite shows little noticeable differential effect. On the other hand, even as little as ten percent kaolinite in a mixture with sericite may be detected. Since both minerals are common in zones of hydrothermal alteration, this feature is of interest.

Figure 114 contains curves of kaolinite and alunite, which represent a mixture of two thermally active minerals that may occur together in the same deposits. Both minerals yield sharp and distinctive thermal peaks.

Figure 115 represents a sequence of thermal curves for alunite and jarosite where the samples are artificial mixtures. Both alunite (fig. 109) and kaolinite (fig. 105) are illustrated elsewhere. Where curves show such prominent peaks, mixtures may be studied with reasonable facility. A proportional decrease in the amplitude as well as a downward shift of peak temperatures occurs with an increase in foreign constituents.

A common problem in the study of zones of argillic alteration concerns the estimation of the relative amounts of kaolinite and dickite present in a natural mixture. Figure 116 illustrates a series of artificial mixtures of the two minerals.

Kaolinite-montmorillonite mixtures are illustrated in figure 117. Evidently the apparatus as normally employed is less sensitive for the detection of montmorillonite in a mixture than it is for distinguishing minerals with higher temperatures and more distinctive thermal effects.

Montmorillonite-sericite mixtures are indicated in figure 118. While montmorillonite would be detected in such mixtures, it seems likely that sericite would escape detection. It is evident that the effect of shifting the peaks with the percentage of impurity must be determined for each mineral properly to identify minerals in mixtures. The carbonates appear particularly sensitive to this effect.

The curves above have been used effectively in the semiquantitative determination of the argillic constituents of an altered mineralized area. The application of the technique to this form of problem offers signifi-

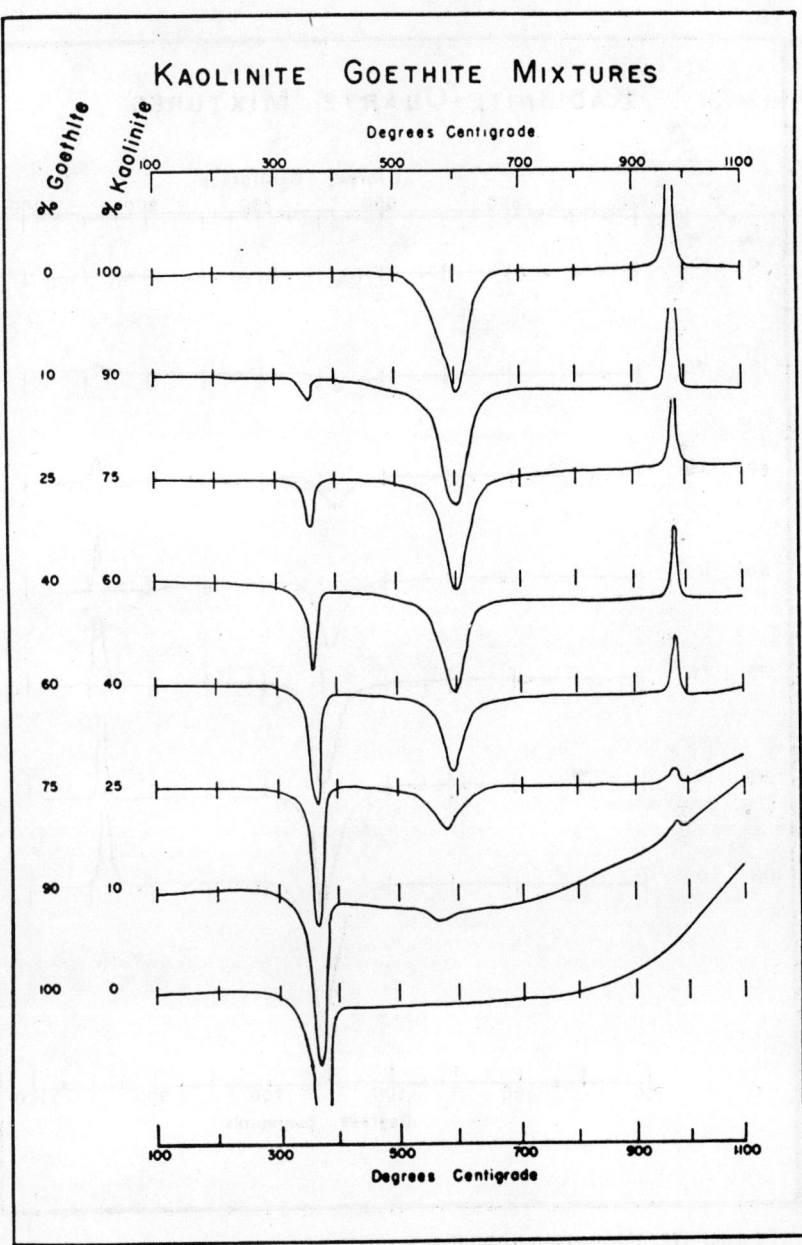

FIGURE 111. Artificial mixtures of kaolinite and goethite arranged to illustrate possible interpretation of natural mixtures.

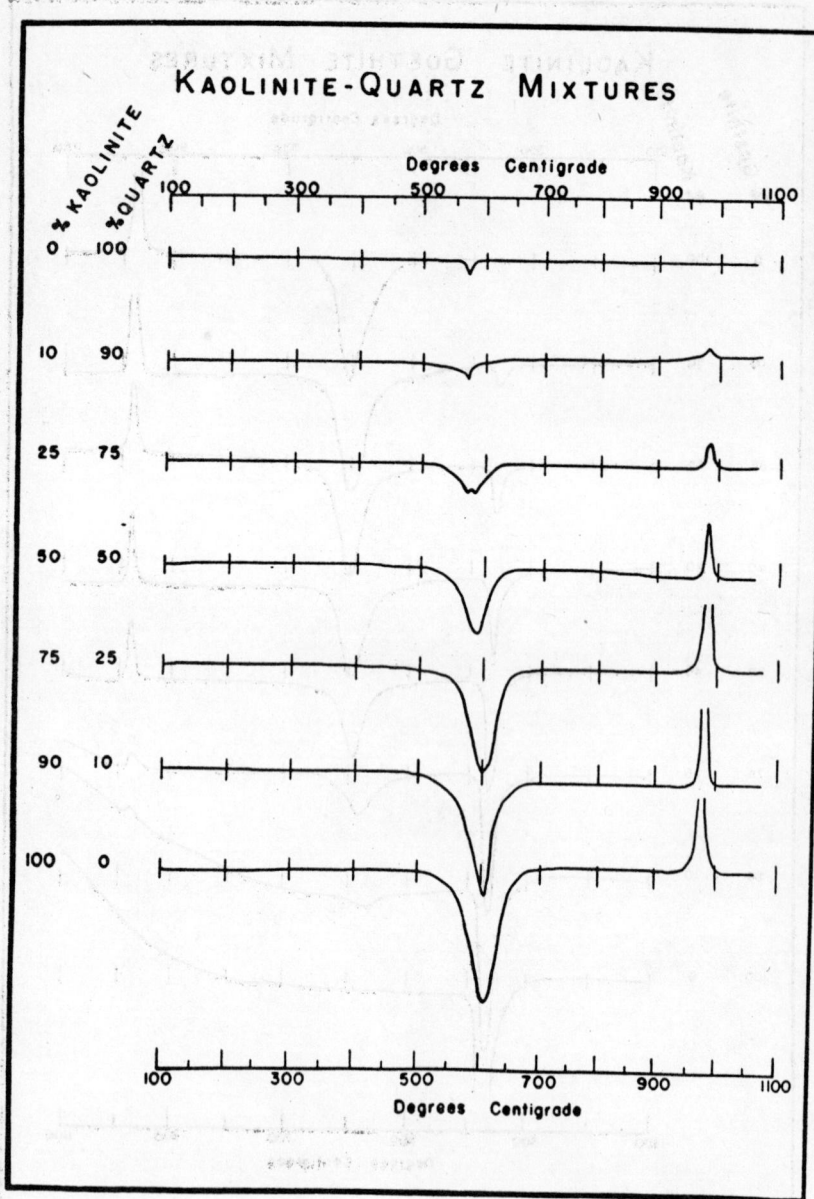

FIGURE 112. Curves indicating effect on kaolinite with quartz as an impurity.

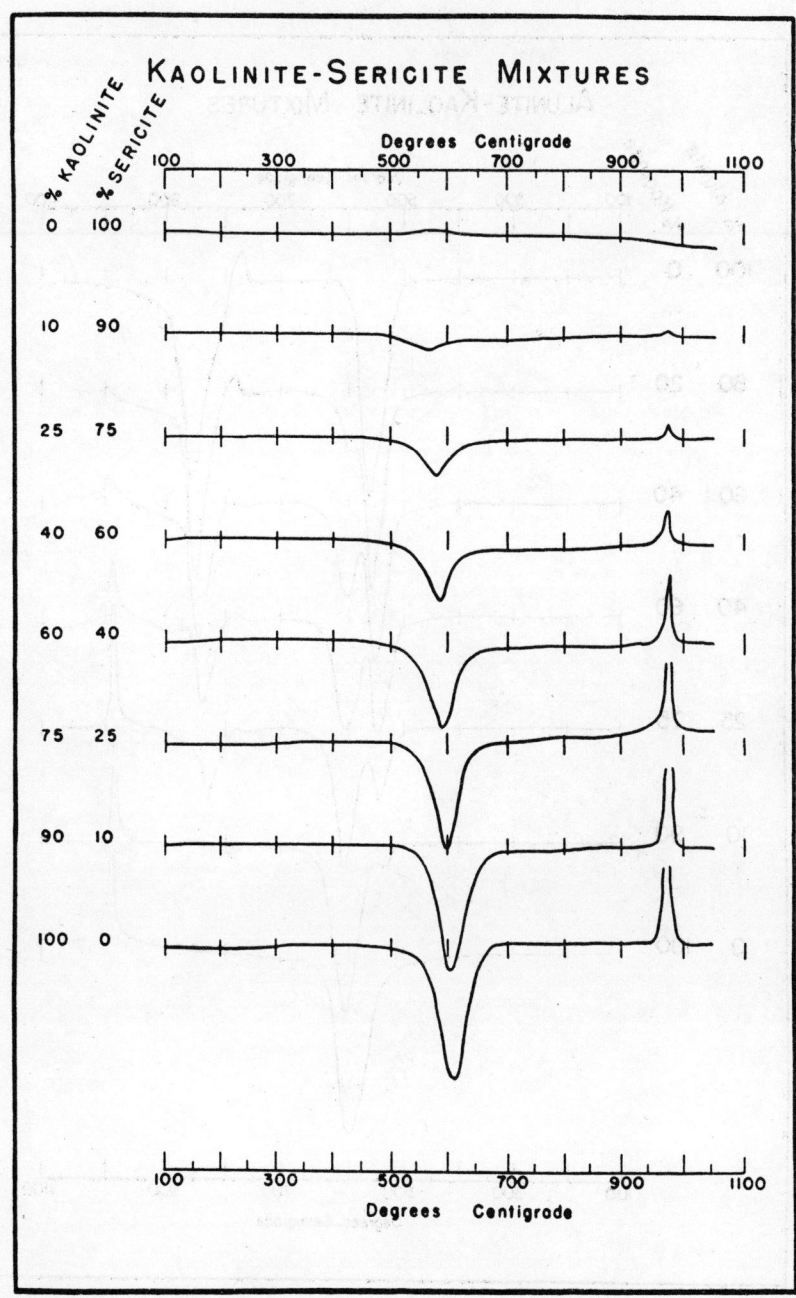

Figure 113. Curves showing mixtures of almost inert sericite and active kaolinite.

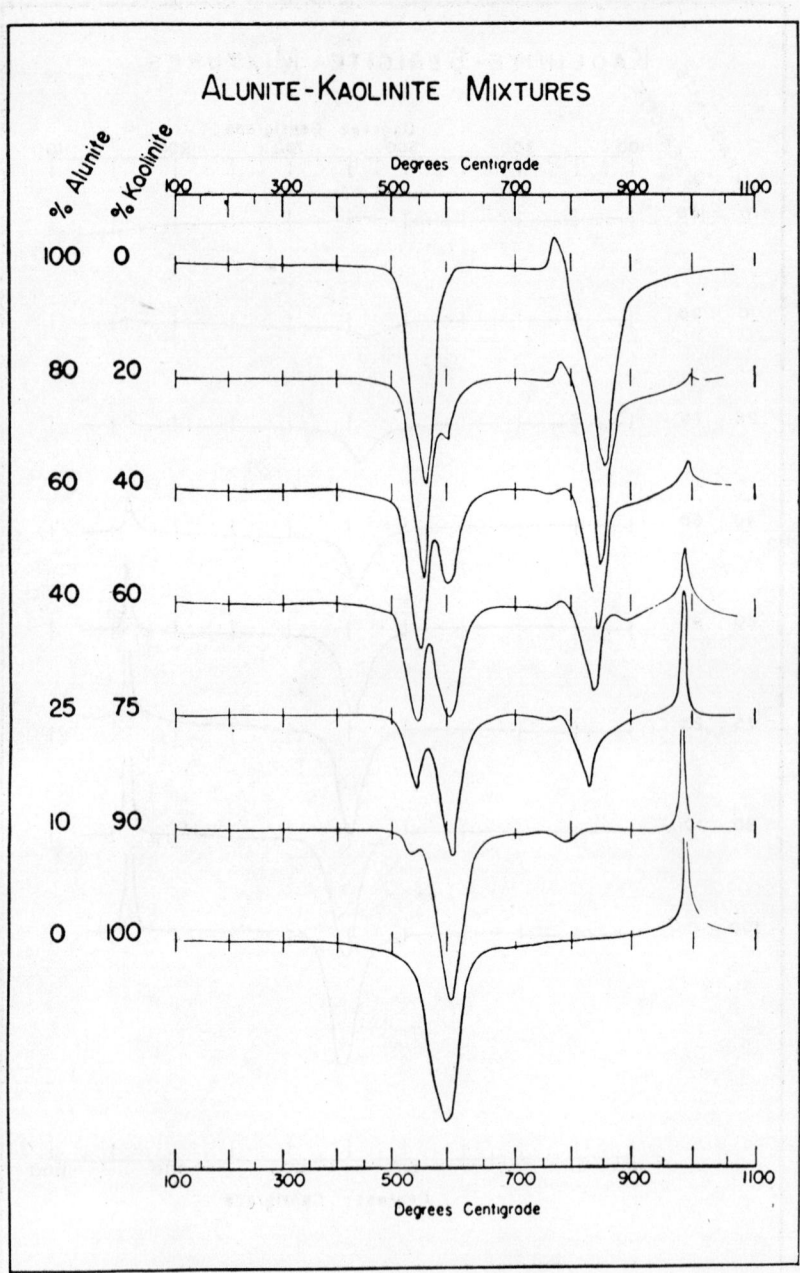

FIGURE 114. Thermally active kaolinite and alunite in artificial mixtures.

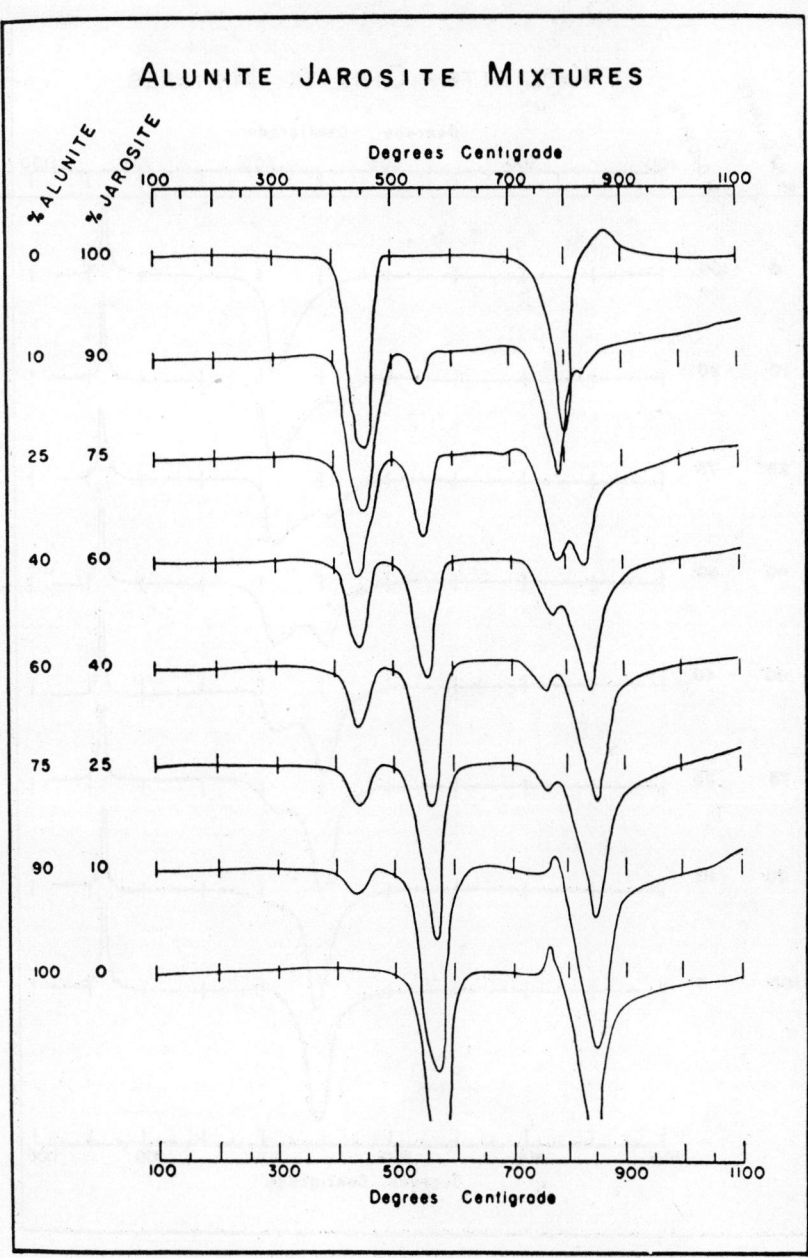

FIGURE 115. Thermal curves of artificial jarosite mixtures.

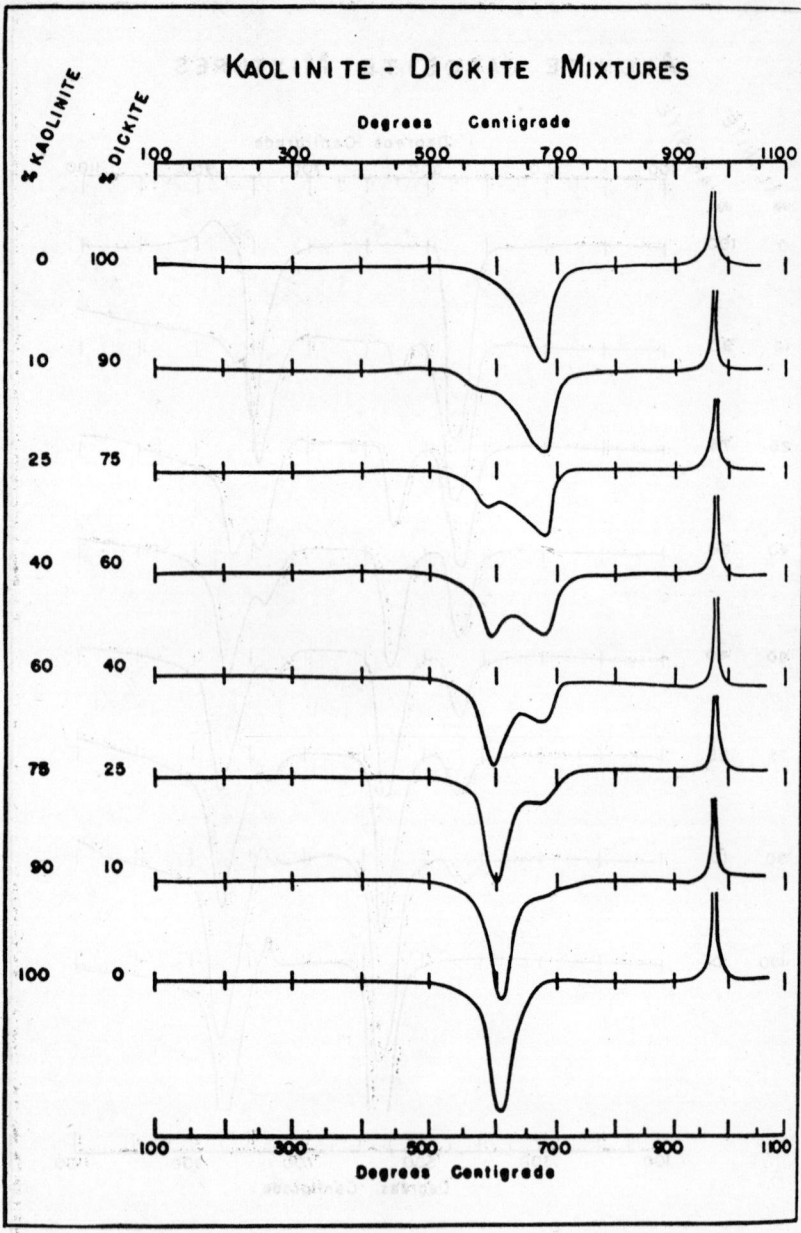

FIGURE 116. Artificial mixtures of kaolinite and dickite showing variation in thermal curves.

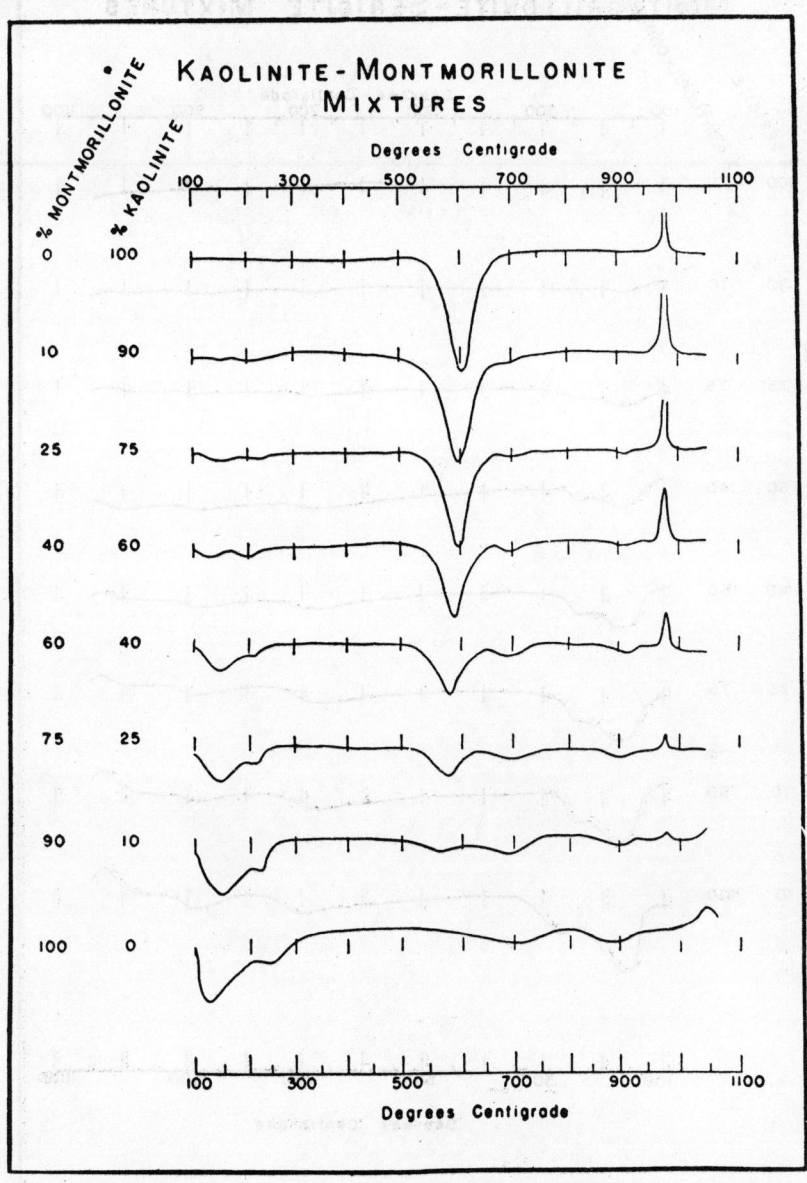

FIGURE 117. Thermal curves of artificial kaolinite-montmorillonite mixtures with a range from 0 to 100 percent.

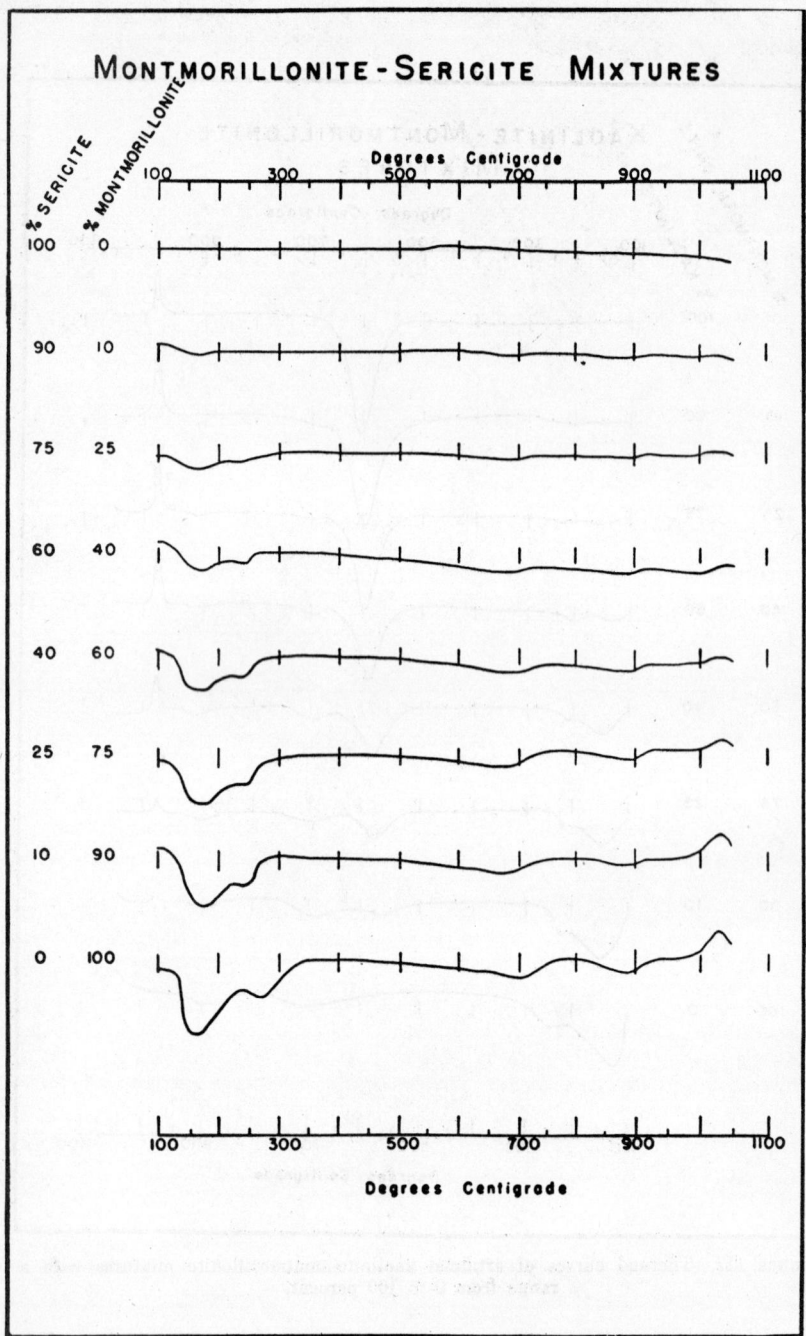

FIGURE 118. Artificial mixtures of almost inert sericite with montmorillonite.

cant possibilities in mapping alteration zones associated with mineral deposits in studies of the type reviewed by Kerr.[84]

The thermal curves thus far obtained in these studies are for the most part consistent with curves recorded in the literature, allowing for the

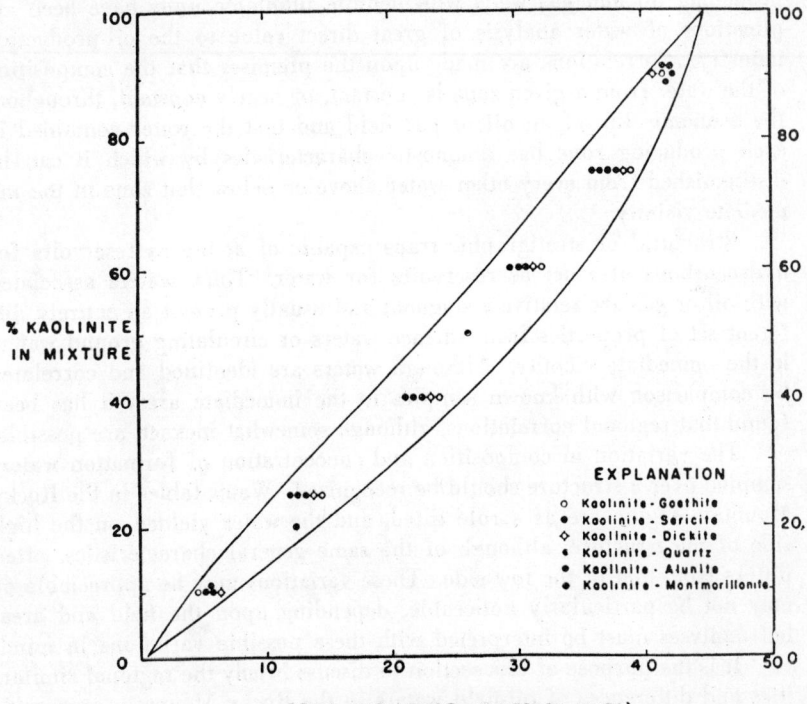

FIGURE 119. Graph showing variation in position and amplitude of the 605° C. peak of kaolinite in various artificial mixtures.

variation in heating rates. The temperatures at which peaks occur have been agreed upon by various observers with different types of apparatus, if the heating rates, the thermocouples, and the size of sample are constant. The amplitude of the peaks for any given concentration of active ingredient is a function of the sensitivity of the individual apparatus.

---

[84] Kerr, P. F., *Alteration Studies:* Am. Mineralogist, vol. 32, p. 158, 1947.

WATER ANALYSIS
(CHARACTERISTICS OF OIL-FIELD WATERS OF THE ROCKY MOUNTAIN REGION)

JAMES G. CRAWFORD

The identification and correlation of waters found in drilling and producing oil and gas wells with definite lithologic units have been applications of water analysis of great direct value to the oil-production industry. Correlations are made upon the premises that the composition of the water from a given zone is constant, or nearly constant, throughout the economic life of an oil or gas field and that the water contained in each producing zone has diagnostic characteristics by which it can be distinguished from every other water above or below that zone in the immediate vicinity.

Structural or stratigraphic traps capable of acting as reservoirs for hydrocarbons also act as reservoirs for water. Thus, waters associated with oil or gas are relatively stagnant and usually present an entirely different set of properties from surface waters or circulating ground waters in the immediate vicinity. Although waters are identified and correlated by comparison with known samples in the immediate area, it has been found that regional correlations, although somewhat inexact, are possible.

The variation in composition and concentration of formation waters sampled over a structure should be recognized. Water tables in the Rocky Mountain region are as a rule tilted, and the water yielded on the high side of the structure, although of the same general characteristics, often differs from that of the low side. These variations may be appreciable or may not be particularly noticeable, depending upon the field and area, but analyses must be interpreted with these possible variations in mind.

It is the purpose of this section to discuss briefly the regional similarities and differences of oil-field waters in the Rocky Mountain area, with particular emphasis on correlation with definite lithologic units.

The writer is indebted to the United States Geological Survey for many of the analyses from which correlations could be made; to the oil companies of the Rocky Mountain region for the many analyses furnished; to H. E. Summerford for his valuable assistance in the preparation of the geologic information; to J. A. Waatti for preparing the illustrations; and to R. M. Larsen for general and specific criticism.

CLASSIFICATION OF WATERS

The Palmer [85] system of water classification emphasizes important differences between waters in geochemical relationship and has been used throughout this paper in the discussion of types of water. The Palmer system groups those radicles that are either chemically similar or geologically associated: Sodium and potassium are grouped as alkalies; cal-

---

[85] Palmer, Chase, *The Geochemical Interpretation of Water Analyses:* U. S. Geol. Survey Bull. 479, 1911.

cium and magnesium are grouped as alkaline earths; sulphates, chlorides, and nitrates are grouped as strong acids; and carbonates, bicarbonates, and sulphides are grouped as weak acids. Thus, according to the reacting value of these four groups, natural waters can be classified into four types, i.e., primary saline, secondary saline, primary alkaline, and secondary alkaline.

An excess of strong acids over weak acids causes salinity. It should be noted that salinity can be due to either the sulphate or chloride radicle or to both. The alkalies in connection with the strong acids cause primary salinity, and an excess of strong acids with an equal value of alkaline earths induces secondary salinity.

An excess of alkalies over the strong acids with an equal value of the weak acids makes up primary alkalinity, and an excess of weak acids combined with an equal value of alkaline earths produces secondary alkalinity.

Secondary salinity and primary alkalinity are incompatible; thus, each natural water will have two or three of the above-mentioned properties but never all four.

Primary salinity is common to all waters, and a primary saline water is essentially a solution of sodium and potassium sulphates and chlorides. A primary alkaline water consists principally of sodium and potassium carbonates and bicarbonates. Calcium and magnesium sulphates and chlorides predominate in a secondary saline water, and the water is permanently hard. Temporary hardness is present in a secondary alkaline water consisting principally of calcium and magnesium bicarbonates.

Surface Waters

Surface waters in the Rocky Mountain region range from the dilute, soft, alkaline waters of igneous terrain to moderately concentrated, hard, rine beds. The usual mountain water derived from melting snow is soft, alkaline, and dilute, but, after it has traversed marine sediments, calcium and magnesium sulphates dominate the chemical system, and the water often takes on a load of salts that makes it unfit to drink.

The North Platte River, for example, rises in a network of mountain streams in North Park, Colorado, and drains the southeastern quarter of Wyoming. Near its source it is a primary alkaline type, but by the time it reaches the Pathfinder reservoir it has been changed to a secondary saline type. The Popo Agie, though dilute, is secondary saline near Lander, Wyoming, whereas Castle Creek, Teapot Creek, and Salt Creek waters are undrinkable because of the alkaline earth salts leached from the Steele shale, as are the waters of many other smaller streams of Wyoming whose drainage does not embrace the higher mountain areas.

The influence of surface waters upon formation waters encountered in drilling wells can be observed in a number of instances in Wyoming. The most striking example is the Shannon sandstone along the western edge of the Powder River Basin, where the influence of primary alkaline waters from the Big Horn Mountains is indicated at Billy Creek, and the influence of secondary saline surface waters can be traced in the Salt Creek area.

In general, the presence of a secondary saline water in Cretaceous and younger sands of Wyoming indicate surface-water infiltration or contamination. Sulphate is absent or negligible in Cretaceous waters, and the presence of this radicle almost invariably indicates drilling-water contamination, particularly as sulphate is the dominant negative ion in most surface waters. The presence of sulphate in persistent and notable quantities usually is not encountered below the ground-water zone until Jurassic beds are reached. Below the Jurassic, however, sulphate is usually the principal negative ion.

TERTIARY

Tertiary beds of unconsolidated variegated shales and sandstones cover most of the plain and basin areas of the Rocky Mountain region with thicknesses up to more than 30,000 feet. Most of these sands are not productive of commercial oil or gas. Small amounts of oil and gas have been produced from the White River formations of Oligocene age at Shawnee, Douglas, and Brenning Basin in central-eastern Wyoming, and oil and gas are being produced commercially from lenticular sand bodies in the Wasatch formation of Eocene age at Hiawatha and Powder Wash, Colorado, and La Barge, Wyoming.

With the exception of the above-mentioned producing fields, the water analyses available from Tertiary sands were sampled for drilling use and do not reflect the stagnant conditions associated with oil-field waters.

Green River Formation

An exceptional type of water has been encountered in a few wells drilled into the lacustrine Green River formation of Eocene age. The formation consists of sandstone, marl, limestone, and sandy shale, with thin beds of oil shale, halite, glauberite, and trona. These salts have influenced the composition of the ground water in this area with the result that sodium carbonate occupies 60 percent of the dissolved salt content of the water, sodium chloride 37 percent, and sodium sulphate 3 percent. Total solids range from 40,000 to 80,000 parts per million, and the water has been used for the production of crude soda ash.

Wasatch Formation

The Wasatch formation at Hiawatha and Powder Wash, Colorado, consists of more than 5,000 feet of shale containing irregular and lenticular fluviatile and lacustrine porous sands. Oil and gas production at Hiawatha comes from three lenticular oil sands between depths of 2,032 and

2,512 feet, and at Powder Wash from zones logged in one well at 3,087 to 3,113 feet and 5,014 to 5,023 feet. The formation contains numerous water-bearing lenticular sand bodies in addition to the oil- and gas-producing zones.

The La Barge field in western Wyoming produces oil from the Wasatch formation. The producing zone, at depths of 650 to 1,100 feet, consists of two to three divisions, the upper a persistent sandstone averaging about twenty feet in thickness, and the lower a series of sandstones separated by interfingering shale beds that vary greatly in thickness.

Wasatch waters are, on the whole, saline, the salinity being due almost entirely to the chloride ion. Concentrations range in total solids from 1,500 to as much as 32,000 parts per million at Hiawatha and Powder Wash, and from 3,000 to 12,000 parts per million at La Barge. The more concentrated and saline water at La Barge occurs in the lower and less-continuous zones. Secondary characteristics are relatively low, though variable, in the Colorado Wasatch waters and are negligible in the La Barge waters.

Lenticularity and lack of continuity are indicated by the erratic and variable nature of the Hiawatha and Powder Wash waters. The uniformity of the La Barge analyses points to the more continuous nature of the sandstones. Representative Wasatch waters are tabulated in table 9.

UPPER CRETACEOUS

Upper Cretaceous beds have yielded a major portion of the oil and gas production in the Rocky Mountain region and, although overshadowed now by pre-Triassic exploration, do and will continue to hold an important place in Rocky Mountain oil production. A basal sandstone of the Mesaverde formation has produced some oil at Simpson Ridge, Wyoming, and is producing oil now at West Poison Spider, Wyoming. The principal Upper Cretaceous oil-producing zones in Wyoming are the Wall Creek and equivalent sandstone beds of the Frontier formation and the Muddy (Newcastle) sandstone member of the Thermopolis shale.

Oil production from Upper Cretaceous sands in Montana has been limited to a few areas, the most important being Cat Creek, but these sands are important gas producers throughout the Great Plains region of the state.

There has been some oil and gas production from Upper Cretaceous beds in Colorado, principally from fractured sandy zones in the Mancos shale, but Lower Cretaceous and older beds are the more prolific horizons.

Montana Group

Montana-group waters are important in oil-field operations in the state of Montana more for identification of intrusive water than for any other purpose. With the exception of gas-producing fields, surface-water

TABLE 9

TYPICAL ANALYSES OF TERTIARY, MONTANA-GROUP AND COLORADO-GROUP WATERS OF ROCKY MOUNTAIN REGION

Field	Name of sand	Parts per million								Type of production
		Total solids	Na	Ca	Mg	SO$_4$	Cl	CO$_3$	HCO$_3$	
Tertiary:										
Hiawatha, Colo.	Wasatch	5,044	1,751	94	40	867	2,009	0	575	Oil
Powder Wash, Colo.	Wasatch	4,984	1,978	26	0	0	2,341	57	1,185	Oil
La Barge, Wyo.	Wasatch (upper)	3,177	1,245	33	20	65	850	0	1,960	Oil
La Barge, Wyo.	Wasatch (lower)	8,558	3,458	0	0	0	3,800	0	2,640	Oil
Montana group:										
Cut Bank, Mont.	Eagle	992	386	0	0	265	14	0	665	Dry
Boxelder, Mont.	Judith River	6,967	2,717	30	0	0	4,095	0	255	Gas
Cedar Creek, Mont.	Judith River	11,143	4,244	127	0	35	6,700	0	75	Gas
Colorado group:										
Cut Bank, Mont.	Blackleaf	14,906	5,706	126	35	0	8,572	0	950	Gas
Bowdoin, Mont.	Blackleaf	8,681	3,263	59	64	0	5,143	0	310	Gas
Cat Creek, Mont.	First Cat Creek	1,861	766	0	0	0	620	0	965	Oil
Shannon sandstone:										
Big Muddy, Wyo.	Shannon	13,050	4,989	77	49	0	7,714	0	450	Oil
Billy Creek, Wyo.	Shannon	2,671	1,100	0	0	0	885	0	1,395	Gas
Cole Creek, Wyo.	Shannon	17,274	6,729	63	24	0	9,900	0	1,135	Oil
Salt Creek, Wyo.	Shannon	2,874	407	126	301	1,527	26	0	990	Dry

characteristics usually predominate, and the available reanalyses are so scattered that generalization is difficult.

The Two Medicine and Eagle sandstones at Cut Bank yield waters of dilute to moderate concentration—from 800 to 8,000 parts per million total solids—consisting principally of sodium sulphate and sodium bicarbonate. Calcium and magnesium are absent or are present in small quantities only, and chloride seldom exceeds 100 parts per million. The Eagle waters are more dilute than the Two Medicine waters and average about 1,200 parts per million total solids. Both waters show ground-water characteristics and are usually shut off with a single string of casing.

In contrast, the Judith River waters encountered in the gas-producing Cedar Creek anticline are saline with concentrations ranging from 8,000 to 15,000 parts per million total solids; chloride varies from 4,000 to 10,000 parts per million; sulphate is absent or is present only in negligible quantities; bicarbonate is low, averaging about 200 parts per million; and calcium and magnesium are present in relatively small but definite quantities. These waters are associated with natural gas, and their higher-than-average concentrations are ascribed to the evaporative effect of the gas. It has been found that as a general rule throughout the Rocky Mountain region waters associated with gas are more concentrated than waters associated with oil.

The extreme difference in composition between the surface-water type at Cut Bank and the chloride-saline type in gas-producing fields is shown in the ionic statements in table 9.

Colorado Group

The Colorado group in Montana includes, besides dark fissile shale, sandy and sandstone members that yield oil or gas at several localities. In Wyoming the Colorado group includes the Shannon sandstone, the important oil-producing Frontier formation, and the Muddy (Newcastle) sandstone.

The Blackleaf sandy member of the Colorado shale yields gas with showings of oil in a number of Montana fields. The waters encountered in this member at Border-Red Coulee, Cut Bank, Kevin-Sunburst, and Bowdoin are essentially solutions of sodium chloride ranging from 6,000 to 16,000 parts per million total solids. Sulphate is negligible in these waters, and secondary characteristics are low but persistent, ranging from 50 to 200 parts per million calcium and a trace to 100 parts per million magnesium. Primary salinity averages ninety percent, primary alkalinity six percent, and secondary alkalinity four percent of the chemical system. It is interesting to note that these waters resemble Judith River water as found in the Cedar Creek anticline; thus the concentrations and salinity can be ascribed to their association with gas.

First Cat Creek sandstone waters at Cat Creek are moderately dilute and balanced, the alkalinity and salinity each averaging about fifty percent of the system. The concentrations vary from about 1,000 to 2,000 parts per million total solids, and the alkaline earths and sulphate are absent or negligible. The Cat Creek field is one of the areas in the Rocky Mountain region in which a dilute to moderately dilute water is found associated with commercial oil production.

Three typical Colorado-group waters are tabulated in table 9.

Shannon Sandstone

Although the Shannon sandstone has been tested in many wells, the only fields now producing from it are Cole Creek and Big Muddy, Wyoming. At Cole Creek the Shannon sandstone is found at a depth of approximately 4,500 feet. A concentrated, saline water averaging about 18,000 parts per million total solids is produced with the oil from edge wells. The water averages about six percent secondary alkalinity, and sulphate is present in quantities of 50 to 500 parts per million. This water, although more concentrated and with little higher alkalinity, resembles equivalent water at Big Muddy.

The top of the Shannon sandstone in the Big Muddy field occurs at a depth of about 900 feet and consists of about thirty feet of alternating lenses of buff to gray sandstone and sandy shale carrying both oil and water. The water is a solution of sodium chloride varying from 9,000 to 15,000 parts per million total solids, with about three percent secondary characteristics; bicarbonate is relatively low, ranging from 450 to 700 parts per million.

It is interesting to note the variation of the Shannon waters along the western edge of the Powder River Basin. The Shannon sandstone in the Billy Creek gas field yielded a balanced water, i.e., a water in which alkalinity and salinity occupy about fifty percent each in the chemical system, with no secondary characteristics or alkaline earths; concentrations ranged from 2,000 to 3,000 parts per million total solids. The sandstone in this field is 900 to 1,300 feet below the surface and is fed by fresh, sulphate-free alkaline water from the nearby Big Horn Mountains.

The influence of secondary-saline surface water can be seen in the Salt Creek area, where the Shannon sandstone forms an escarpment on the east and west sides of the Salt Creek uplift. Here the feed is the gypsum-impregnated waters of Castle Creek, Teapot Creek, and Salt Creek, and the Shannon formation waters encountered during drilling were practically identical to these surface waters.

Saline waters of relatively high concentration are associated with oil production from the Shannon sandstone at Cole Creek and Big Muddy, and it is concluded that they have not been influenced to any extent by surface-water infiltration.

Thus, it is possible in this one area, trending northwest-southeast along the western edge of the Powder River Basin, to observe the effects of surface-water infiltration of two different types upon the connate water originally in the sand, as typified by Cole Creek and Big Muddy waters. Representative analyses of all these waters are given in table 9.

Frontier Formation

The Frontier formation has been one of the most productive oil horizons in the Rocky Mountain region. It is overshadowed now by pre-Triassic production but nevertheless still is an actual and potential oil producer of large capacity. Where it produces oil the formation ranges in thickness from 370 to 1,200 feet and contains from two to nine beds of sandstone. Where the several Wall Creek sands can be separated, the designation of "First Wall Creek," "Second Wall Creek," *et cetera*, are given them; where the separate sands cannot be identified, water samples are designated simply as "Frontier." Frontier sands where identifiable in the Big Horn Basin are termed "Torchlight" and "Peay."

Frontier waters are for the most part solutions of sodium chloride and sodium bicarbonate in varying proportions. Calcium and magnesium occur in small amounts in some waters and are absent in others; in no case is there sufficient calcium or magnesium to give secondary salinity to the water. Sulphate is absent or is present in minor quantities only. Concentrations are quite variable, ranging from about 1,200 parts per million to 50,000 parts per million total solids.

Representative analyses of a number of the more important Frontier waters of Wyoming are tabulated in table 10. The variation in concentration of these waters is due, it is believed, to several causes, among them and most important being the lenticularity of the sands and their permeability development. It is believed that the more dilute and alkaline waters have been modified considerably by meteoric waters, whereas the more concentrated, saline waters are assumed to be connate without any substantial modification by meteoric waters since accumulation.

It is interesting to note from table 10 that the highest concentrated waters are associated with gas in the Baxter Basin fields. These, together with the Montana-group waters cited above, tend to support the contention of Mills and Wells [86] that water can become concentrated at depth by the agency of moving and expanding gas.

Muddy (Newcastle) Sand

Oil production from the Muddy sand had been small and scattered until the devlopment of the Mush Creek and Skull Creek areas along the eastern edge of the Powder River Basin. Here a twelve- to twenty-foot section of medium- to fine-grained, slightly tripolitic sand with coal intercalations interbedded with shales, locally called the "Newcastle sandstone," yields commercial oil production.

The Newcastle water at Mush Creek is a primary-saline water ranging from 10,000 to 15,000 parts per million total solids; secondary char-

[86] Mills, R. Van A., and Wells, R. C., *The Evaporation and Concentration of Waters Associated with Petroleum and Natural Gas:* U. S. Geol. Survey Bull. 693, 1919.

TABLE 10

TYPICAL ANALYSES OF WYOMING FRONTIER WATERS

Field	Name of sand	Total solids	Na	Ca	Mg	SO₄	Cl	CO₃	HCO₃	Type of production
			Parts per million							
Alkali Butte	Frontier	9,756	3,847	27	17	0	4,950	79	1,700	Dry
Badger Basin	Frontier	11,369	4,432	62	0	44	6,384	0	910	Oil
Baxter Basin (N)	Frontier	28,626	10,879	222	125	0	16,667	0	1,490	Gas
Baxter Basin (M)	Frontier	54,100	21,114	170	93	37	30,350	0	4,750	Gas
Baxter Basin (S)	Frontier	21,515	8,790	0	0	0	8,150	1,047	7,175	Gas
Big Muddy	Frontier	7,167	2,967	0	0	0	1,976	0	4,505	Oil
Big Sand Draw	Frontier	5,640	2,252	16	0	26	2,600	0	1,518	Oil
Byron	Frontier	2,632	1,132	0	0	9	129	145	2,475	Gas
Cole Creek	Frontier	17,712	7,033	6	0	63	9,504	0	2,250	Dry
Elk Basin	Frontier	9,784	3,867	24	0	0	5,275	0	1,257	Oil
Garland	Frontier	10,901	4,575	0	0	0	2,305	0	8,180	Gas
Glassner Dome	Second Frontier	3,763	1,570	0	0	141	598	325	2,295	Dry
Golden Eagle	Frontier	4,942	2,051	34	5	198	261	133	4,600	Dry
Little Sand Draw	Fifth Frontier	2,060	887	0	0	0	100	264	1,645	Dry
Lost Soldier	Frontier	13,263	5,376	36	0	0	5,100	0	5,595	Oil
Midway	Second Wall Creek	6,452	2,568	21	0	113	2,850	0	1,830	Oil
Oregon Basin	Frontier	1,559	666	0	0	0	150	79	1,350	Dry
Pilot Butte	Frontier	4,383	1,698	24	57	0	1,395	0	2,472	Dry
Salt Creek	First Wall Creek	4,268	1,822	16	0	0	450	108	3,840	Oil
Salt Creek	Second Wall Creek	10,274	4,104	0	0	0	5,050	0	2,245	Oil
Salt Creek	Third Wall Creek	16,433	6,537	0	0	0	8,900	0	2,025	Oil
Sand Creek	Third Frontier	4,675	1,863	4	0	488	1,225	146	1,930	Oil
Steamboat Butte	Third Frontier	4,246	1,706	0	0	98	1,800	72	1,160	Dry
Stove Creek	Frontier	4,999	1,867	23	50	51	2,900	24	170	Dry
West Poison Spider	Frontier	5,375	1,981	123	10	21	2,989	0	510	Oil
Worland	Frontier	13,871	5,368	73	25	74	7,800	12	1,055	Gas

acteristics and sulphate are negligible; chloride ranges from 3,000 to 7,000 parts per million depending upon the concentration; bicarbonate, though, is more erratic and varies from a low of 1,300 to a high of 6,000 parts per million.

In contrast to Mush Creek, the Newcastle-sandstone waters at Skull Creek are alkaline and vary from 6,000 to 8,000 parts per million total solids. Secondary characteristics and sulphate are negligible, chloride ranges from 800 to 2,000 parts per million, and bicarbonate from 4,000 to 6,000 parts per million. This, the author believes, is evidence that the Skull Creek waters have been modified by meteoric water to a greater extent than the Mush Creek waters.

LOWER CRETACEOUS

Lower Cretaceous beds include the Cloverly formation of Wyoming and Colorado, the Greybull sandstone and Pryor conglomerate of south-central Montana and the Big Horn Basin of Wyoming, and the Kootenai formation of north and north-central Montana. These beds produce oil at various localities in Wyoming and Colorado, and the Kootenai is one of the principal oil- and gas-producing formations of Montana.

In general, Dakota(?) waters of Colorado are relatively dilute, soft, and alkaline, with concentrations ranging from 700 to 3,000 parts per million total solids and with an average of about 1,500 parts per million total solids. Wyoming Dakota(?) waters are quite variable; where associated with oil or gas, concentrations as high as 20,000 parts per million total solids are encountered, but where not associated with hydrocarbon accumulation dilute waters are the rule. Kootenai waters range from about 1,000 parts per million at Cat Creek to as high as 15,000 parts per million total solids at Cut Bank and on the average seem to be more concentrated than equivalent waters in Colorado and Wyoming.

Dakota(?) and Kootenai waters as a rule are more alkaline than Upper Cretaceous waters previously discussed. Alkalinity is the important distinguishing feature between Colorado and Kootenai waters in Montana, the bicarbonate content of Kootenai waters being appreciably higher. Exceptions to this rule in Wyoming, however, are numerous, the North Baxter Basin gas field being an example of high salinity and low alkalinity; the Beaver Creek gas field also violates the general rule, as do Salt Creek and Church Buttes.

The Dakota(?) and Lakota sands more often than any other formation yield potable water in Wyoming. This is particularly noticeable along the eastern edge of the Powder River Basin and in the Poison Spider area of central Wyoming, where the sands are not deeply buried and crop out on nearby uplifts.

Table 11 lists a number of the more important Lower Cretaceous waters of the Rocky Mountain region.

TABLE II

Representative Lower Cretaceous Waters of the Rocky Mountain Region

Field	Name of sand	Total solids	Na	Ca	Mg	SO₄	Cl	CO₃	HCO₃	Type of production
						Parts per million				
Colorado:										
Iles	Dakota (?)	1,034	434	11	0	0	38	0	1,120	Oil
Moffat	Dakota (?)	1,047	447	0	0	0	109	26	945	Dry
Rangely	Dakota (?)	2,260	873	8	0	279	800	0	610	Dry
Douglas Creek	Lakota	80,131	26,756	3,188	839	161	49,000	0	380	Gas
Montana:										
Border-Red Coulee	Cosmos-Vanalta	3,462	1,285	90	52	0	619	0	2,880	Oil
Cat Creek	Second Cat Creek	1,311	512	6	6	17	156	67	1,112	Oil
Cat Creek	Third Cat Creek	930	354	0	0	289	35	13	486	Dry
Cut Bank	Upper Sunburst	10,786	4,132	56	54		6,077	0	950	Gas
Cut Bank	Lower Sunburst	7,631	3,086	0	0	51	3,240	0	2,550	Gas
Cut Bank	Cut Bank	8,465	3,582	12	0	0	1,158	0	7,550	Oil
Kevin-Sunburst	Sunburst	12,074	4,670	66	60	44	5,980	0	2,550	Gas
Wyoming:										
Baxter Basin (N)	Dakota (?)	17,401	6,546	164	81	306	9,800	0	1,025	Gas
Baxter Basin (M)	Dakota (?)	15,818	6,461	0	0	36	5,928	0	6,900	Gas
Bolton Creek	Dakota (?)	579	226	0	0	143	32	18	325	Dry
Bridge Creek	Dakota (?)	821	349	0	0	0	46	72	720	Dry
Church Buttes	Dakota (?)	11,476	4,415	86	24	0	6,500	0	915	Gas
Iron Creek	Dakota (?)	2,201	913	16	13	3	65	126	2,165	Oil
Salt Creek	Dakota (?)	13,009	5,131	38	6	42	6,850	0	1,915	Oil
Steamboat Butte	Dakota (?)	4,567	1,848	0	0	395	1,030	109	2,410	Oil
Alkali Butte	Lakota	6,738	2,672	0	0	446	2,673	79	1,765	Dry
Beaver Creek	Lakota	10,634	4,224	10	0	174	5,247	79	1,830	Gas
Big Muddy	Lakota	3,310	1,318	0	0	485	616	67	1,675	Oil
Cole Creek	Lakota	6,237	2,280	67	0	1,321	2,080	0	995	Oil
Osage	Lakota	780	257	0	0	355	8	0	215	Dry
Salt Creek	Upper Lakota	4,925	2,008	0	0	0	1,928	30	1,950	Oil
Salt Creek	Lower Lakota	2,484	1,035	0	0	0	650	0	1,625	Oil

Jurassic

Marine Jurassic beds include the important Sundance sandstone of Colorado and central and eastern Wyoming, the Nugget of western Wyoming and western Colorado, and the Ellis formation of Montana. The Sundance is an important oil producer in Colorado at Iles, Moffat, and Wilson Creek and in Wyoming at Salt Creek, Lance Creek, Big Medicine Bow, Rock Creek, and other localities. Oil production at Kevin-Sunburst, Montana, comes from the zone at the contact of the Ellis formation and Madison limestone, much of which is from the reworked Madison limestone.

The Morrison formation overlies the Sundance in Colorado and Wyoming and represents the transition zone from Lower Cretaceous to the marine Jurassic. Its waters are included in this section. The Morrison produces oil at Wilson Creek, Colorado, and in extent and productivity outranks the Sundance. Scattered showings of oil and gas have been encountered elsewhere in the Morrison formation.

Morrison waters in Colorado range from soft, alkaline types to saline waters containing appreciable hardness. They vary in concentration from about 3,000 to 15,000 parts per million total solids and usually contain appreciable amounts of sulphate. Most of the Sundance waters of Colorado do not contain sulphate—by which they can be distinguished from Morrison waters—and the alkaline and moderately dilute waters are usually soft. The saline waters generally contain appreciable hardness and in many respects resemble Morrison waters.

The Ellis waters of Montana are quite uniform over the northern portion of the plains and consist principally of sodium chloride and sodium bicarbonate with minor but persistent secondary characteristics. These waters almost invariably contain hydrogen sulphide, and, with the exception of Cosmos-Vanalto waters at Border-Red Coulee, are the youngest waters of the state to carry hydrogen sulphides. They average about 3,500 parts per million total solids.

The Sundance waters of Wyoming are rather variable in concentration and composition. They range from a low of about 1,200 parts per million total solids at Big Medicine Bow to as high as 40,000 parts per million at Steamboat Butte, where the water takes on evaporite characteristics similar to Triassic waters. The Sundance is the youngest formation in Wyoming in which secondary salinity becomes an appreciable and persistent part of the chemical system, yet secondary salinity is present only in a few fields such as Alkali Butte, the third sand at Salt Creek, the basal sand at Lance Creek, and other scattered localities. It is believed that secondary salinity is persistently present in Sundance waters in those beds in which limestone predominates; where sandstone predominates, the water is of the same general character—relatively soft, with little or no sulphate—as Cretaceous waters. Thus, it is inferred that the characteristics and composition of for-

TABLE 12

REPRESENTATIVE JURASSIC WATERS OF THE ROCKY MOUNTAIN REGION

Field	Name of sand	Parts per million								Type of production
		Total solids	Na	Ca	Mg	SO₄	Cl	CO₃	HCO₃	
Colorado:										
Iles	Morrison	3,384	1,382	0	0	424	257	121	2,440	Oil
Maudlin Gulch	Morrison	3,395	1,259	31	11	430	1,400	72	390	Oil
Moffat	Morrison	9,486	3,570	76	30	560	5,000	0	535	Dry
Rangely	Morrison	7,810	3,069	18	0	984	2,092	0	3,350	Gas
Wilson Creek	Morrison	3,666	1,416	26	0	161	1,805	0	525	Oil
Iles	Sundance	3,053	1,307	0	0	0	258	0	3,025	Oil
Maudlin Gulch	Sundance	11,913	4,317	194	27	1,284	5,878	0	434	Oil
Moffat	Sundance	2,532	1,074	6	0	0	272	0	2,400	Oil
Wilson Creek	Sundance	14,898	5,245	374	79	1,044	8,060	0	195	Oil
Montana:										
Border-Red Coulee	Ellis	3,881	1,577	24	13	66	755	0	2,940	Dry
Cat Creek	Morrison	3,364	1,191	17	4	1,309	462	0	775	Oil
Cat Creek	Ellis	2,932	1,101	12	0	598	827	42	715	Oil
Kevin-Sunburst	Ellis	3,481	1,265	88	71	0	643	0	2,875	Oil
Whitlash	Ellis	4,565	1,759	10	0	821	1,154	0	1,670	Gas
Wyoming:										
Alkali Butte	Morrison	12,048	4,629	37	9	989	5,742	0	1,305	Dry
Alkali Butte	Nugget	16,860	5,794	173	65	5,436	5,247	12	270	Dry
Bailey Dome	Sundance	3,240	1,332	20	0	0	2,270	0	2,270	Oil
Baxter Basin (N)	Sundance	12,902	5,288	23	0	151	3,900	0	7,200	Gas
Baxter Basin (S)	Sundance	7,876	3,162	12	0	153	3,250	64	2,510	Gas
Big Medicine Bow	Sundance	1,736	683	0	0	376	146	54	970	Oil
East Allen Lake	Sundance	5,384	2,308	0	0	0	420	370	4,650	Gas
Hatfield	Sundance	2,792	1,194	0	0	30	188	168	2,465	Oil
Lance Creek	Basal Sundance	10,339	3,445	192	35	3,663	2,761	0	495	Oil
Lost Soldier	Sundance	3,992	1,649	15	0	0	996	0	2,710	Oil
Rock Creek	Sundance	2,274	918	0	0	319	227	259	1,120	Oil
Salt Creek	Morrison	11,835	4,580	36	24	334	6,361	0	1,049	Oil
Salt Creek	Third Sundance	11,981	4,009	334	59	2,662	4,568	0	710	Oil
Steamboat Butte	Nugget	30,038	10,012	486	103	10,946	8,316	0	355	Oil
Wertz	Sundance	3,095	1,657							

mation waters depend to some extent at least upon the petrography of the rocks.

Representative Morrison and Sundance waters are tabulated in table 12.

TRIASSIC

Small amounts of oil have been produced from the basal part of the Moenkopi formation of Triassic age in Utah, and oil production has come from stray sandstone beds in the Chugwater formation of Permian and Triassic age at Grass Creek and Hamilton Dome, Wyoming, but for the most part Triassic beds have been dry. Water analyses of Triassic age have been few and scattered, but the Chugwater waters of Wyoming that have been sampled are highly concentrated solutions of sodium sulphate and sodium chloride averaging from 40,000 to 60,000 parts per million total solids; alkalinity is negligible in these waters, the bicarbonate usually averaging about 200 parts per million.

Triassic beds are the dividing line between the post-Triassic primary waters and the pre-Triassic secondary waters of the Rocky Mountain region. Most of the waters in zones younger than Triassic contain few, if any, secondary characteristics; in contrast, secondary characteristics dominate the waters of formations older than Triassic.

PERMIAN

Permian oil production in the Rocky Mountain region is limited principally to the Embar formation in Wyoming, consisting of porous dolomite, limestone, and chert. Much of the exploration work in the Rocky Mountain region since 1940 has been in the Embar and older formations, and these now outrank the post-Triassic beds in productivity.

Embar waters range from 1,800 parts per million total solids at Dallas Dome to 38,000 parts per million total solids at Neiber Dome, but the average concentration ranges from 4,000 to 7,000 parts per million total solids. With but few exceptions these waters are solutions of sodium, calcium, and magnesium sulphates in varying proportions. Sulphate salinity exceeds chloride salinity, and the bicarbonate ion usually is relatively low; most of these waters carry hydrogen sulphide.

The Embar water at Neiber Dome is very unusual in that there is a large amount of alkalinity present as the bicarbonate radicle. This is decidedly out of line in a post-Triassic water, but there have not been sufficient samples from this structure for postulations concerning its source.

Typical Embar waters of Wyoming are tabulated in table 13.

PENNSYLVANIAN

The marine Pennsylvanian beds include the important oil-producing Tensleep sandstone of the Rocky Mountain region, the Amsden formation, and their equivalent of eastern Wyoming, the Minnelusa formation.

TABLE 13

Representative Permian (Embar) Waters of Wyoming

Field	Name of sand	Total solids	Parts per million							Type of production
			Na	Ca	Mg	SO₄	Cl	CO₃	HCO₃	
Alkali Butte	Embar	8,662	2,596	254	59	4,431	1,150	0	350	Dry
Black Mountain	Embar	3,802	427	582	154	2,176	78	0	781	Oil
Dallas Dome	Embar	1,825	502	147	49	263	204	72	1,195	Oil
Derby Dome	Embar	2,001	373	220	43	1,178	12	0	355	Oil
East Warm Springs	Embar	2,995	263	548	135	1,376	235	0	890	Oil
Franks Fork	Embar	2,274	207	435	115	682	360	0	965	Oil
Garland	Embar	6,587	1,417	529	121	4,156	130	12	450	Oil
Gebo	Embar	13,712	4,261	357	80	6,518	1,778	0	1,460	Oil
Grass Creek	Embar	5,821	1,172	543	197	2,584	715	0	1,240	Oil
Hamilton Dome	Embar	12,178	3,913	114	156	5,420	1,930	0	1,310	Oil
Kirby Creek	Embar	4,300	1,437	25	0	2,434	75	0	670	Oil
Little Buffalo Basin	Embar	5,377	777	756	193	2,631	270	0	1,525	Oil
Maverick Springs	Embar	4,563	657	624	136	2,599	284	0	535	Oil
Neiber Dome	Embar	38,583	14,545	322	79	8,402	5,500	835	17,800	Oil
Oregon Basin	Embar	7,674	1,911	459	165	3,929	490	0	1,465	Oil
Pilot Butte	Embar	12,023	3,462	363	76	7,437	525	0	325	Dry
Steamboat Butte	Embar	6,207	1,496	424	85	3,600	371	0	470	Dry
Wagonhound	Embar	4,066	885	393	109	1,556	540	0	1,185	Oil
Worland	Embar	14,987	4,232	450	123	9,691	299	0	390	Oil
Zimmerman Butte	Embar	8,241	2,335	486	111	3,124	562	0	2,490	Oil

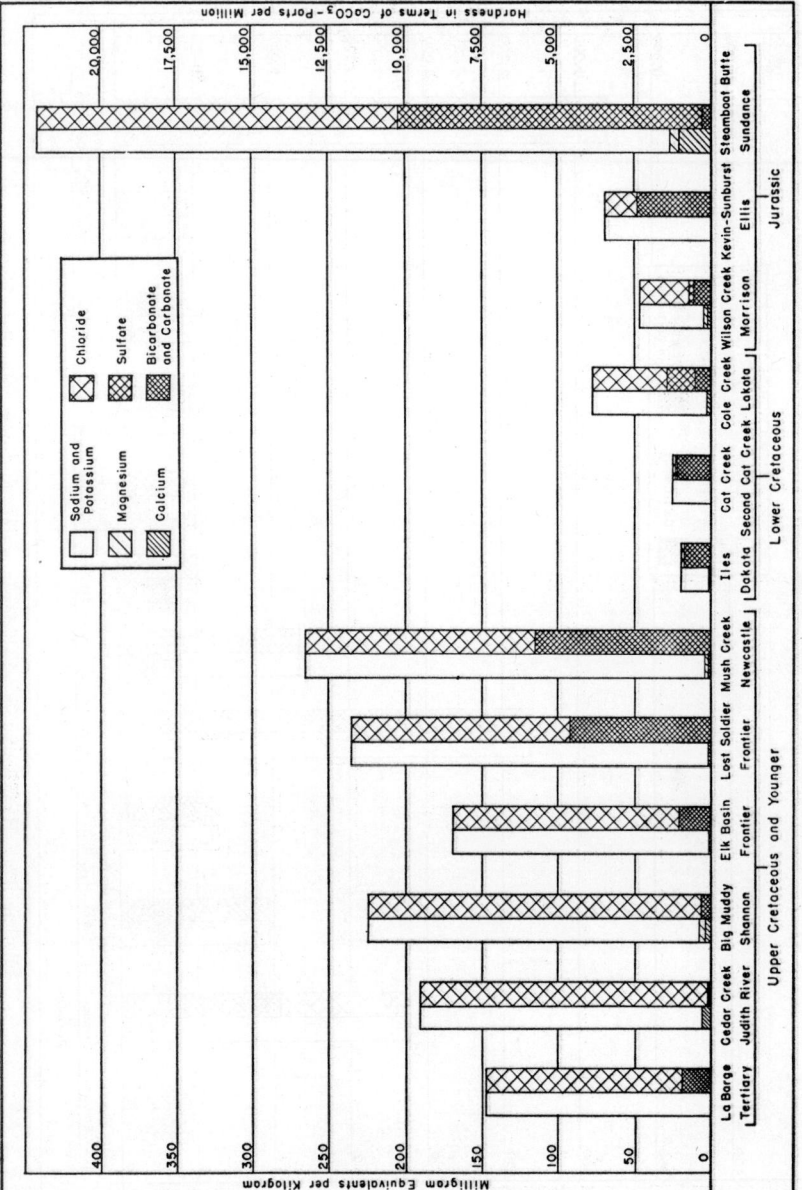

FIGURE 120. Representative post-Triassic waters of Rocky Mountain region.

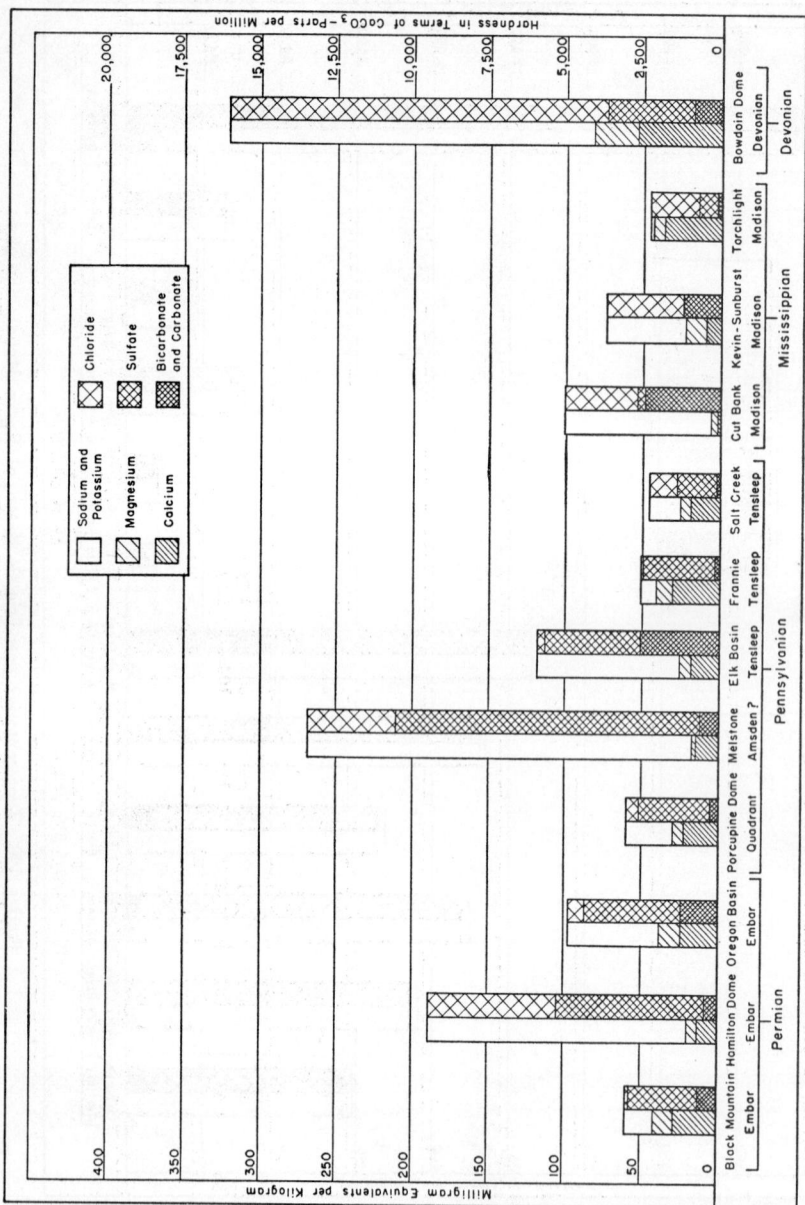

FIGURE 121. Representative pre-Triassic waters of Rocky Mountain region.

These beds also include the Weber sandstone of Colorado and Utah and the Quadrant formation of Montana.

The greater part of the oil production of the Rocky Mountain region now comes from Pennsylvanian beds. Larger and more productive fields producing from Pennsylvanian beds include Lance Creek (Minne-lusa), Elk Basin (Tensleep), Salt Creek (Tensleep), Steamboat Butte (Tensleep), Wertz (Tensleep), Rangely (Weber), and Lost Soldier (Ten-sleep). The Quadrant formation of Montana has yielded a negligible amount of oil in central Montana but for the most part produces only copious quantities of water.

In general, Pennsylvanian waters are saline, with the salinity being due principally to the sulphate ion. Like the Permian waters previously discussed, the Pennsylvanian waters are marked by persistent and appre-ciable secondary characteristics; bicarbonate is usually low. Hydrogen sulphide is commonly present but usually not in any great quantity.

The Weber water in Colorado and Utah is a chloride-saline type with appreciable quantities of sulphate. At Rangely a brine varying from about 100,000 parts per million to 150,000 parts per million total solids is associated with oil, and this is the highest-concentrated oil-field water from a producing field in the Rocky Mountain region. The Rangely brine is principally sodium chloride, the chloride ion ranging from 60,000 to 100,000 parts per million and the sodium from 35,000 to 60,000 parts per million; calcium averages 5,000 parts per million and magnesium about 650 parts per million.

The Quadrant waters of Montana are typical Pennsylvanian waters. They average about 3,000 parts per million total solids and consist prin-cipally of the sulphates of sodium, calcium, and magnesium. Secondary salinity occupies from forty to seventy percent of the chemical system, and traces of hydrogen sulphide are usually found in the fresh water. Artesian flows, usually hot, of 10,000 to 125,000 barrels a day are en-countered in the Quadrant formation in the Montana fields tabulated in table 14, and these waters are the youngest encountered in north-central and central Montana that consistently contain large quantities of sulphate and the alkaline earths.

The area in Montana embracing the new fields of Big Wall, Melstone, and Ragged Point is now controversial as to the age of the producing hori-zon. Water analyses seem to indicate that the three fields are producing from lithologically identical units, but present opinion places Big Wall and Melstone production as Amsden and Ragged Point as Kibby of Mississippian age. The author has placed all analyses in the Pennsyl-vanian table as Amsden(?), but further study of the formation may change this grouping.

Tensleep and Minnelusa waters in Wyoming vary from dilute to mod-erately concentrated (from a low of 200 parts per million at Derby Dome to 13,000 parts per million at Quealy Dome) but on the whole average

TABLE 14

Representative Pennsylvanian Waters of the Rocky Mountain Region

Field	Name of sand	Total solids	Na	Ca	Mg	SO₄	Cl	CO₃	HCO₃	Type of production
					Parts per million					
Colorado:										
Iles	Weber	1,803	741	0	0	114	337	210	815	Dry
Rangely	Weber	108,053	37,725	3,509	568	973	65,000	0	565	Oil
Utah:										
Clay Basin	Weber	18,801	6,962	107	90	2,306	8,500	54	1,590	Dry
Montana:										
Bowdoin	Quadrant	3,441	269	544	177	2,180	200	0	145	Dry
Porcupine	Quadrant	3,995	674	470	104	2,297	288	0	330	Dry
Sixshooter	Quadrant	2,471	331	374	64	1,305	225	0	350	Dry
Big Wall	Amsden (?)	13,642	3,904	427	79	8,645	369	59	305	Oil
Melstone	Amsden (?)	18,021	5,743	312	31	9,576	1,980	96	575	Oil
Ragged Point	Amsden (?)	12,381	3,844	347	79	5,201	2,789	50	145	Oil
Soap Creek	Amsden (?)	2,403	55	537	99	1,575	4	0	270	Oil
Womans Pocket	Amsden	4,055	500	610	137	2,265	350	0	393	Oil
Wyoming:										
Alkali Butte	Tensleep	2,644	671	199	36	982	698	0	118	Dry
Big Hollow	Tensleep	3,428	509	490	53	2,234	58	0	170	Dry
Big Medicine Bow	Tensleep	2,564	838	97	0	945	234	0	915	Oil
Black Mountain	Tensleep	435	74	76	16	44	16	0	425	Oil
Bud Kimball	Tensleep	1,482	278	124	57	878	25	0	244	Dry

TABLE 14 (Continued)

Field	Name of sand	Total solids	Na	Ca	Mg	SO₄	Cl	CO₃	HCO₃	Type of production
					Parts per million					
Byron	Tensleep	2,630	65	576	134	1,528	22	0	620	Oil
Corley Dome	Tensleep	2,172	626	76	24	1,185	74	0	380	Dry
Dallas Dome	Tensleep	481	39	90	39	74	23	0	440	Oil
Dallas Dome	Amsden	321	28	27	23	46	16	0	184	Dry
Derby Dome	Tensleep	212	10	42	24	0	3	0	270	Oil
Elk Basin	Tensleep	12,453	3,940	222	55	6,590	992	0	1,330	Oil
Frannie	Tensleep	3,356	206	639	128	2,230	28	0	255	Oil
Garland	Tensleep	3,057	232	549	143	1,680	157	0	525	Oil
Grass Creek	Tensleep	3,541	556	386	155	1,880	343	0	450	Oil
Hatfield	Tensleep	6,054	1,741	239	56	3,128	508	0	780	Oil
Herrick Dome	Tensleep	8,455	2,027	400	214	5,271	410	0	270	Oil
Lance Creek	Converse	6,673	1,669	454	53	3,873	356	0	545	Oil
Lance Creek	Leo	3,017	632	307	59	1,322	606	0	185	Oil
Lander (Hudson)	Tensleep	249	34	41	17	13	31	7	230	Oil
Little Laramie	Tensleep	6,731	1,693	507	67	2,926	1,500	0	63	Oil
McGill	Tensleep	2,321	155	523	37	1,335	185	39	175	Dry
Mill Creek	Tensleep	3,949	984	261	40	2,327	119	11	365	Dry
North Sage Creek	Tensleep	828	204	36	43	201	282	0	104	Dry
Pilot Butte	Tensleep	2,606	434	306	80	1,414	220	0	310	Oil
Quealy	Tensleep	13,223	4,273	479	77	2,877	5,396	0	245	Oil
Salt Creek	Tensleep	2,836	456	378	82	1,231	604	0	170	Oil
Steamboat Butte	Tensleep	3,225	649	340	78	1,523	244	0	795	Oil
Torchlight	Tensleep	3,516	89	620	251	2,373	14	24	295	Oil

from 3,000 to 4,000 parts per million total solids. Secondary salinity often dominates the chemical system, and even when it does not dominate it is appreciable. Like Embar waters, sulphate is the principal negative radicle, and bicarbonate is usually low.

The dilute Tensleep waters in table 14, Dallas, Derby, and Lander, are artesian even though associated with oil. The Popo Agie River is apparently the source of these waters, and it is surmised that the Tensleep sandstone in these fields is subjected to an active, vigorous natural water drive; it is estimated that about four barrels of water are produced for each barrel of oil. Dilute water also is associated with oil at Black Mountain, Wyoming.

MISSISSIPPIAN

The important Mississippian oil-producing zone is the Madison limestone and its Black Hills equivalent, the Pahasapa. It was the oldest sedimentary formation in the Rocky Mountain region to produce oil until the recent discovery of oil in a Cambrian sandstone at Lost Soldier, Wyoming. The Madison limestone yields oil in Montana and in the Lost Soldier-Wertz area and the Big Horn Basin of Wyoming and has yielded showings of oil in other parts of the region. (See table 15.)

Madison waters of Montana are somewhat variable. Chloride salinity dominates in the fields around Sweetgrass arch and Sweetgrass Hills, and sulphate salinity is predominant in the central and north-central fields. The average concentration is about 5,000 parts per million total solids, and calcium and magnesium are present to some extent in all Madison waters, although less pronounced in the chloride-saline type.

Madison waters from Wyoming fields, with the exception of one or two extremely dilute waters, seem to be more uniform than equivalent Montana waters. As a rule there is more secondary salinity in Madison waters than in Embar or Tensleep waters, and Madison waters usually are more dilute. The average concentration of Madison waters is about 2,500 parts per million; sulphate usually dominates the negative ions, although there are a few waters, such as Torchlight, in which chloride dominates. Bicarbonate usually is low, averaging less than 500 parts per million, and in this respect the Madison waters resemble the average Tensleep water.

Although there is a marked similarity between pre-Triassic waters, it is not difficult usually to distinguish among Embar, Tensleep, and Madison waters in the same field or area. There are sufficient differences in concentration, alkalinity, or sulphate-chloride ratio to correlate each water with its lithologic unit and make identification relatively easy.

DEVONIAN AND OLDER

Showings of oil have been found in pre-Cambrian crystallines, Cambrian strata, Ordovician strata, and Devonian strata at widely separated localities in the Rocky Mountain region. Some geologists believe that the

TABLE 15

Representative Mississippian Waters of the Rocky Mountain Region

Field	Name of sand	Total solids	Parts per million							Type of production
			Na	Ca	Mg	SO₄	Cl	CO₃	HCO₃	
Montana:										
Cut Bank	Madison	5,680	2,187	39	57	232	1,680	84	2,850	Oil
Dunbar	Madison	6,452	2,334	217	64	257	1,177	0	4,825	Dry
Howard Coulee	Madison	4,688	1,021	446	116	1,726	1,116	0	535	Dry
Kevin-Sunburst	Madison	4,057	1,173	200	168	0	1,771	0	1,515	Oil
Pondera	Madison	5,131	1,377	94	427	146	1,150	0	3,940	Oil
Utopia	Madison	5,354	1,551	140	266	264	2,640	0	1,002	Oil
Wyoming:										
Adon	Madison	3,070	611	310	57	1,703	310	0	160	Dry
Alkali Anticline	Madison	2,289	290	339	59	1,503	24	0	150	Dry
Big Sand Draw	Madison	1,387	276	154	26	679	141	0	226	Dry
Black Mountain	Madison	336	5	75	37	15	22	0	370	Oil
Elk Basin	Madison	6,354	1,652	383	105	2,855	1,000	0	730	Oil
Four Bear	Madison	2,385	214	460	100	1,009	110	0	1,000	Oil
Garland	Madison	2,960	290	801	136	1,524	426	0	230	Oil
Sage Creek	Madison	1,240	220	109	68	594	14	22	434	Dry
Steamboat Butte	Madison	2,181	261	369	77	966	183	0	660	Dry
Torchlight	Madison	2,643	33	749	90	570	1,106	0	193	Dry

Bighorn dolomite of Upper Ordovician age yields some of the deepest oil produced at Garland, Wyoming. But for many years, until 1948, the Madison limestone of Mississippian age was the oldest proved commercial oil zone in the Rocky Mountain region. In 1948 a Cambrian sand was proved for commercial oil production at Lost Soldier and Wertz, Wyoming, and the search for oil in pre-Mississippian beds was intensified.

Several wells, particularly in Montana, have penetrated Ordovician and older beds, and a few analyses are available of these older waters, although the number is insufficient to warrant generalizing. The Ordovician waters analyzed appear to be of the same general character as Madison waters in the same well, whereas Devonian waters so far analyzed appear to be a more concentrated, more saline type, marked in particular by high calcium content. One typical Devonian water has a concentration of total solids of 19,000 parts per million and a calcium content of 1,096 parts per million; another Devonian water has a concentration of total solids of 11,000 parts per million and a calcium content of 1,027 parts per million. The chloride-sulphate ratio was 4:1 in the first water and 2:1 in the second.

It is believed, however, that these older waters as a whole will not be substantially different from other pre-Triassic waters, and that limestone characteristics will be the rule and not the exception.

CONCLUSION

The value of analyses as a means of identification of intrusive waters in well bores has been well established in the Rocky Mountain region. This is the primary purpose of water analyses, and the application of such data to engineering and geologic problems and theory is secondary. It is established definitely that there are sufficient differences in concentration and composition to correlate a water with its reservoir zone so that it can be differentiated from all other waters above or below that zone in a particular well or area.

The generally dilute nature of the oil-field waters of the Rocky Mountain region indicates extensive modification and dilution by meteoric waters. Some of the waters encountered seem to indicate little change since deposition, so that it is concluded that for these waters modification and dilution occurred before deposition; other waters seem to indicate extensive modification since deposition. In any respect, the brines commonly associated with oil in other provinces of the world do not occur in the Rocky Mountain region.

It is apparent that the oil-field waters in the Rocky Mountain region have been influenced by the petrography of the rocks in which they occur. Secondary characteristics and sulphate are at a minimum in waters of sandstone reservoirs of Cretaceous age or younger, whereas secondary characteristics and sulphate are prominent in the limestones and limy formations of pre-Triassic age.

CORE ANALYSIS—PREDICTING WELL BEHAVIOR
JOHN G. CARAN

Core analysis has proved a valuable aid in the successful exploration, exploitation, and evaluation of gas and oil reserves. The basic core data make possible the location of fluid contacts and the prediction of the type of production to be expected.

It is readily admitted that core analysis is but one of the useful tools

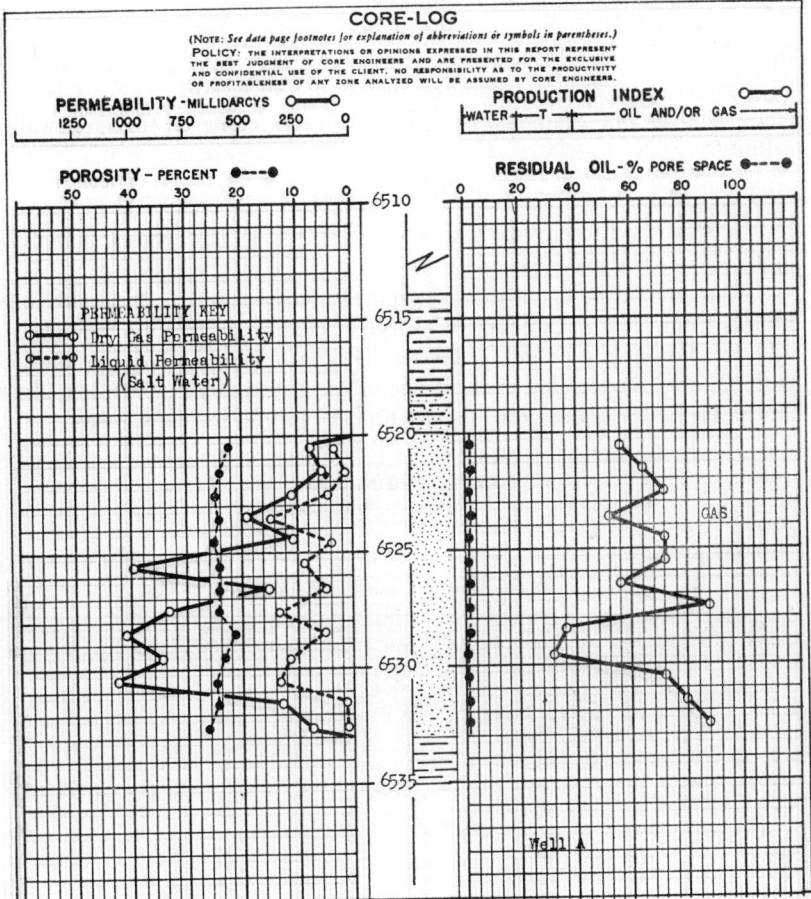

FIGURE 122. Core-log of well A.

available to the modern completion engineer. The use of core data, together with a knowledge of the structural position of the well, the study of electric logs, and comprehensive drill-stem testing, minimizes the possibility of completing a dry hole or missing a productive formation.

No attempt has been made to discuss in detail all of the techniques employed by the various research and commercial core-testing laboratories. Analysis procedures vary for the formations being tested, the depth and pressure of the reservoir, and the type of core sample. Formations usually analyzed include sandstone and limestone. Chalk, serpentine, and conglomerate analyses require specialized techniques. The depth and pressure of the reservoir are reflected in the residual saturation determinations. Cores may be of the conventional and wire-line, full-diameter, core-barrel type, or the smaller sidewall samples.

Core analyses may be defined as the determination and evaluation of the productive characteristics of a formation sample by the measurement of porosity, permeability, and residual fluid saturations. Other tests include acidization, grain size, interstitial water content, and core-water salinity.

CORE SAMPLING

Any core-analysis report is only as reliable as the original sampling and treatment of the core.

Care should be taken to select representative samples, preferably one sample from each foot of core recovery. If changes in the lithology occur, additional samples should be taken. The core should never be washed with water. Samples for analysis should be wiped clean of drilling fluid and immediately sealed from the atmosphere to prevent fluid losses.

CORE ENGINEERS
Reservoir Core Analysts
SAN ANTONIO, TEXAS

COMPANY_____   WELL____A_____

CORE ANALYSIS DATA AND INTERPRETATION
(See footnotes for meaning of symbols or abbreviations)

SAMPLE NUMBER	DEPTH FEET	PERMEABILITY MILLIDARCYS		POROSITY PER CENT	RESIDUAL FLUIDS		SALINITY P.P.M. CHLORIDES	PROBABLE PRODUCTION (1)	REMARKS
		Hor.	Vert.		OIL % VOL. | % PORE	WATER % PORE			
1	6520.5	191	188	22.5	1.9	49.2	5380	Gas	
2	21.5	144	242	24.0	2.2	49.7	4610	Gas	
3	22.5	278	18	25.1	1.7	49.8	3900	Gas	
4	23.5	478	336	24.2	1.9	48.8	4350	Gas	
5	24.5	270	144	25.4	1.3	49.3	4000	Gas	
6	25.5	985	76	24.0	1.5	47.6	4150	Gas	
7	26.5	377	5	24.2	1.6	51.4	5000	Gas	
8	27.5	830	1040	24.3	1.7	43.4	4880	Gas	
9	28.5	1020	79	21.3	1.9	50.6	4880	Gas	
10	29.5	856	96	23.4	0.5	67.5	9750	Gas	High water saturation
11	30.5	1060	872	24.8	0.4	47.4	5070	Gas	
12	31.5	320	288	24.2	0.5	44.5	6350	Gas	
13	32.5	186	240	26.4	0.4	45.1	6280	Gas	

(*) REFER TO REPORT LETTER   (NS) NO SAMPLE,   (NF) NO FLOW,   (NCF) NO COMMERCIAL FLOW,   (LFR) LOW FLOW RATE.   (C) CONDENSATE TYPE RESIDUAL

(1) PREDICTION ASSUMES COMPLETE ZONE ISOLATION,

POLICY: THE INTERPRETATIONS OR OPINIONS EXPRESSED IN THIS REPORT REPRESENT THE BEST JUDGMENT OF CORE ENGINEERS AND ARE PRESENTED FOR THE EXCLUSIVE AND CONFIDENTIAL USE OF THE CLIENT. NO RESPONSIBILITY AS TO THE PRODUCTIVITY OR PROFITABLENESS OF ANY ZONE ANALYZED WILL BE ASSUMED BY CORE ENGINEERS.

FIGURE 123. Tabulated core data and interpretation for well A. Although potential had not been run, it was probable that some oil would be produced as a spray with gas.

The practice of breaking up the core into small pieces to smell and taste for the presence of hydrocarbons should be avoided. The use of an ultraviolet light will detect the presence of liquid hydrocarbons, and a portable gas analyzer will detect even minute quantities of gas. Smelling or tasting cores for the presence of gas is fallible because sweet gases have no apparent odor or taste, and many gas sands have been condemned as water productive because of the inability physically to detect gas.

CORE ENGINEERS
Reservoir Core Analysts

SAN ANTONIO. TEXAS

GAS AND WATER PERMEABILITY RELATIONSHIPS
(SEE REPORT LETTER FOR DISCUSSION OF RESULTS)

COMPANY_____ WELL_ A

SAMPLE DEPTH (FEET)	SPECIFIC PERMEABILITIES PERMEABILITY, MILLIDARCYS (1)			PERMEABILITY RATIOS RATIOS OF SPECIFIC PERMEABILITIES		
	DRY GAS	SALT WATER	FRESH WATER	DRY GAS / SALT WATER	DRY GAS / FRESH WATER	FRESH WATER / SALT WATER
6520.5	191	81	74	2.4	2.6	0.913
21.5	144	40	18	3.6	8.0	0.450
22.5	278	101	74	2.8	3.8	0.733
23.5	478	365	322	1.3	1.5	0.882
24.5	270	103	100	2.6	2.7	0.970
25.5	985	220	163	4.5	6.0	0.742
26.5	377	114	102	3.3	3.7	0.895
27.5	830	340	300	2.4	2.8	0.882
28.5	1020	131	130	7.8	7.8	0.992
29.5	856	298	220	2.9	3.9	0.738
30.5	1060	325	296	3.3	3.6	0.910
31.5	320	46	41	7.0	7.8	0.892
32.5	186	34	14	5.5	13.3	0.412
AVERAGES	538	169	143	3.2	3.8	0.846

Total Capacity (Dry Gas)   • 6995 md.-ft.

Total Capacity (Salt Water) ■ 2198 md.-ft.

Total Capacity (Fresh Water)= 1854 md.-ft.

NOTE: (1) SAMPLE COMPLETELY SATURATED WITH FLUID. (NT) NO TEST. SAMPLE DISINTEGRATED. (°) INFINITE.

POLICY: THE INTERPRETATIONS OR OPINIONS EXPRESSED IN THIS REPORT REPRESENT THE BEST JUDGMENT OF CORE ENGINEERS AND ARE PRESENTED FOR THE EXCLUSIVE AND CONFIDENTIAL USE OF THE CLIENT. NO RESPONSIBILITY AS TO THE PRODUCTIVITY OR PROFITABLENESS OF ANY ZONE ANALYZED WILL BE ASSUMED BY CORE ENGINEERS.

FIGURE 124. Gas- and water-permeability relationships for well A. Results of these tests show that sand is clean and relatively free from hydratable materials. Sand of this type would respond to injection of gas, salt water, or fresh water.

Cores may be analyzed on location or preserved for off-location analyses. Methods of core preservation include quick-freezing and sealing in airtight containers such as small-diameter plastic tubes. These tubes, which may be obtained to fit the core diameters very closely, reduce the void. If the void is sufficiently small, cores may be preserved for indefinite periods in airtight containers. Pressure is created in the container by the evaporation of a very small amount of the fluids in the outer portion of the core; only the center of the core is used for satura-

<hr>

CORE ENGINEERS
Reservoir Core Analysts
SAN ANTONIO. TEXAS

COMPANY _____   WELL __A_____

GAS PRODUCTIVE FORMATION CORE DATA SUMMARY
(SEE FOOT NOTES FOR EXPLANATION OF FIGURES IN PARENTHESES)

FORMATION OR ZONE NO.	1				
DEPTH. FEET	6520 - 6533				
PROBABLE PRODUCTION (1)	Gas				
ANALYZED.—PRODUCTIVE. FEET	13 - 13				
CORED.—RECOVERED. FEET	13 - 13				
CORE RECOVERY. %	100.0				
AVERAGE PERMEABILITY. MILLIDARCYS (2)	538				
CUMULATIVE CAPACITY. MILLIDARCY-FEET (3)	6995				
AVERAGE POROSITY. PERCENT	24.2				
AVERAGE POROSITY. BARRELS PER ACRE FOOT	1878				
AVERAGE RESIDUAL OIL SATURATION. % PORE SPACE	1.3				
AVER. RESID. CONDENSATE SATURATION. % PORE SPACE					
AVERAGE TOTAL WATER SATURATION. % PORE SPACE (4)	49.6				
AVERAGE CONNATE WATER. SATURATION. CALCULATED % PORE SPACE (5)	34				
LIQUID HYDROCARBON GRAVITY °API	39				

GAS RESERVES AND RECOVERABLE GAS
MCF PER ACRE FOOT OF FORMATION

RESERVOIR RESERVE (6)	6.96				
SURFACE RESERVE (7)	1122				
RECOVERABLE GAS (8)	626				

NOTE:

(*) REFER TO REPORT LETTER.
(1) PREDICTION ASSUMES COMPLETE ZONE ISOLATION.
(2) SPECIFIC PERMEABILITY.
(3) CAPACITY OF RECOVERED FORMATION ONLY.
(4) CONNATE PLUS DRILLING WATER.
(5) NON-PRODUCIBLE CAPILLARY WATER.
(6) AT RESERVOIR PRESSURE AND TEMPERATURE.
(7) AT 14.7 PSI AND 60°F.
(8) AT SURFACE CONDITIONS. ESTIMATED RESID. PRESSURE (*).
(9) INSUFFICIENT RESERVOIR DATA FOR CALCULATIONS.

POLICY: THE INTERPRETATIONS OR OPINIONS EXPRESSED IN THIS REPORT REPRESENT THE BEST JUDGMENT OF CORE ENGINEERS AND ARE PRESENTED FOR THE EXCLUSIVE AND CONFIDENTIAL USE OF THE CLIENT. NO RESPONSIBILITY AS TO THE PRODUCTIVITY OR PROFITABLENESS OF ANY ZONE ANALYZED WILL BE ASSUMED BY CORE ENGINEERS.

<hr>

Figure 125. Summary of core data, gas reserve, and recoverable gas for well A. Capacity is sufficiently high for commercial gas production. Recoverable gas volume has been calculated to a residual pressure of 200 p.s.i.

tion tests. Once this pressure is created, further evaporation ceases. The practice of quick-freezing cores apparently has proved feasible. However, the samples selected for saturation measurements should be thawed before analysis because frosting occurs upon their exposure to air. The frosting action picks up water from the atmosphere and alters the core saturation. To avoid contamination from atmospheric water, frozen cores may be thawed in airtight containers. It thus seems logical to place the cores in airtight containers immediately upon sampling at the well site in order to avoid undue water contamination from the atmosphere while thawing and sampling for analysis.

TYPES OF CORE SAMPLES

Normally the conventional and wire-line cores of diameter over 1¼ inches are the best type for reliable analysis.

Cores obtained by a shoveling or scraping action or by percussion bullets often prove of little value because the extent of compaction, fracturing, and abnormal contamination associated with taking samples of these types cannot be adequately evaluated. Microscopic examinations of many side-wall cores has shown varying degrees of mud contamination. The basic fundamentals of core-data interpretation often prove valueless for sidewall cores of small diameter. In general, residual-water saturations of side-wall samples are higher than those for wire-line cores from the same sand. Special techniques have been developed in order to minimize the source of error due to the small size of the samples, but fracturing and compaction alter the permeabilities of small side-wall cores.

PHYSICAL CHARACTERISTICS OF CORE SAMPLES

Porosity

Porosity is the available void or storage capacity of the reservoir and may be expressed as a percentage or in barrels per acre foot.

The effective porosity is of primary interest in the calculation of reserves because it is the ratio of the interconnected pore spaces to the total bulk volume. Porosity is a direct function of the grain size.

There has been confusion in the industry between porosity and permeability. A formation may have a high porosity but low permeability.

Permeability

Whether the fluids will flow from the formation or remain locked in the pores is dependent on the permeability. Conventional core-analysis reports usually present the "specific" dry-gas permeability of cores, for which the unit of measurement is the millidarcy.

Experience has shown that the dry-gas permeability can be a very poor index to the productivity of so-called bentonitic sands because of the swelling action of fresh water on these sands.[87] Water-permeability

[87] Caran, J. G., and Caran, R., *Gas and Water Permeability Relationships of the Navarro Sand in the South Texas Area:* Mines Mag., vol. 38, no. 12, Dec. 1948.

data should be a part of every core-analysis report made on sand samples. Clean, porous limestones usually show the same permeability to water as to dry gas because the measuring fluid does not react with the sample.

Permeability may be measured parallel or vertical to the bedding planes of the formation. Vertical permeability should be measured for all formations having fluid contacts because of the importance of selecting

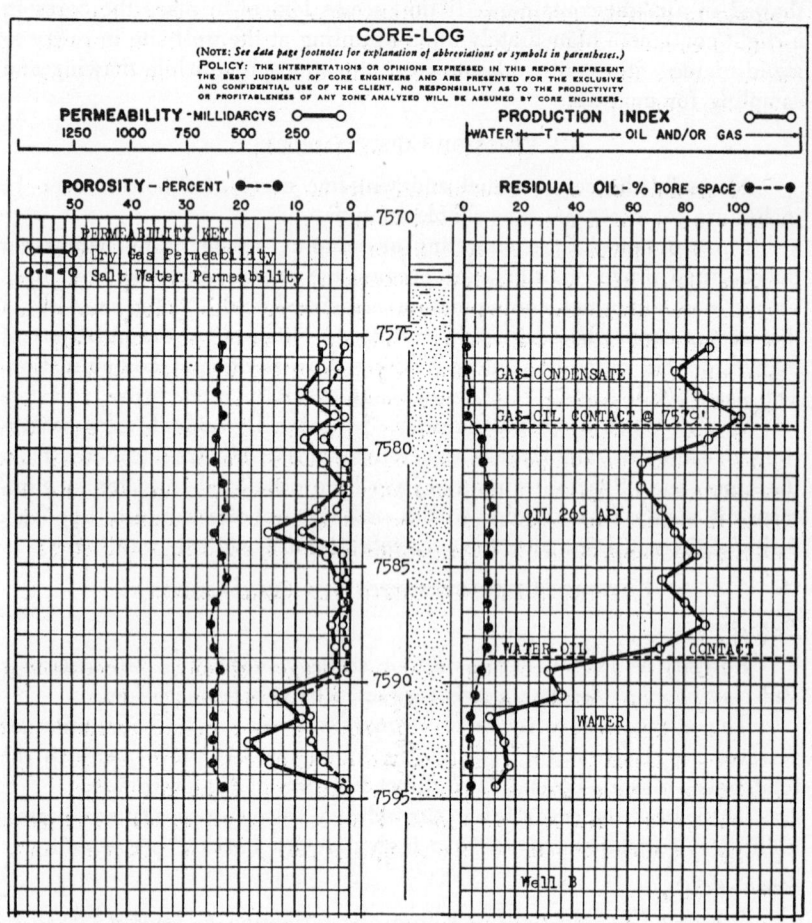

FIGURE 126. Core-log of well B.

the completion zone least likely to be affected by fluid coning. (See figs. 126 and 127). The degree of coning of water or gas is a function of the pressure distribution and the horizontal and vertical permeability ratios and may be calculated for any point in a thick formation if complete permeability data are available.[88] Core-analysis data may be used to cal-

---

[88] Caran, J. G., and Caran, R., *op. cit.*

culate the volume of clean oil that will be produced before bottom water cones into the sand face.[89]

Research has determined that the presence of more than one fluid in the formation affects the flow of the other fluids.[90] The apparent permeability to one particular phase of saturation in a mixture of fluids is called the "effective permeability." Permeability tests using highly saline,

CORE ENGINEERS
Reservoir Core Analysis
SAN ANTONIO. TEXAS

COMPANY_____   WELL____B_____

CORE ANALYSIS DATA AND INTERPRETATION
(See footnotes for meaning of symbols or abbreviations)

SAMPLE NUMBER	DEPTH FEET	PERMEABILITY MILLIDARCYS		POROSITY PER CENT	RESIDUAL FLUIDS			SALINITY P.P.M. CHLORIDES	PROBABLE PRODUCTION (1)	REMARKS
					OIL	WATER				
		Hor.	Vert.		% VOL.	% PORE	% PORE			
1	7575.5	143	67	23.7	0.3	1.3	31.9	3940	Condensate	
2	76.5	149	107	23.9	0.3	1.2	39.7	5000	Condensate	
3	77.5	236	298	24.4	0.8	3.2	34.8	4930	Condensate	
4	78.5	72	26	23.6	0.3	1.4	21.1	9320	Condensate	
	7579.0 GAS-OIL CONTACT									
5	7579.5	206	236	24.1	1.7	7.2	30.1	6780	Oil 26° API	
6	80.5	124	125	24.8	1.8	7.3	36.5	4340	Oil	
7	81.5	49	54	23.1	2.1	9.2	35.3	4070	Oil	
8	82.5	151	205	22.5	2.3	10.1	32.2	4170	Oil	
9	83.5	358	127	24.5	2.4	9.7	32.7	6300	Oil	
10	84.5	103	0.0	23.8	2.2	9.4	27.6	7280	Oil	
11	85.5	51	118	22.6	2.3	10.0	31.0	6250	Oil	
12	86.5	37	22	24.2	2.3	9.5	31.3	7500	Oil	
13	87.5	87	58	25.5	2.4	9.4	30.5	5000	Oil	
14	88.5	68	60	24.6	2.0	8.1	35.2	4830	Oil	
	7589.0 WATER-OIL CONTACT									
15	7589.5	64	128	23.4	1.6	6.8	40.8	7200	Transitional	
16	90.5	296	143	24.4	1.2	4.9	40.1	3030	Transitional	
17	91.5	210	118	24.8	0.9	3.6	46.6	2830	Water	
18	92.5	454	290	24.9	0.8	3.3	44.4	2280	Water	
19	93.5	370	292	24.8	0.8	3.1	39.0	2140	Water	
20	94.5	33	13	23.0	0.8	3.4	42.6	3380	Water	

(*) REFER TO REPORT LETTER   (NS) NO SAMPLE •   (NF) NO FLOW •   (NCF) NO COMMERCIAL FLOW •   (LFR) LOW FLOW RATE ,   (C) CONDENSATE TYPE RESIDUAL ,
(1) PREDICTION ASSUMES COMPLETE ZONE ISOLATION
POLICY: THE INTERPRETATIONS OR OPINIONS EXPRESSED IN THIS REPORT REPRESENT THE BEST JUDGMENT OF CORE ENGINEERS AND ARE PRESENTED FOR THE EXCLUSIVE
AND CONFIDENTIAL USE OF THE CLIENT. NO RESPONSIBILITY AS TO THE PRODUCTIVITY OR PROFITABLENESS OF ANY ZONE ANALYZED WILL BE ASSUMED BY CORE ENGINEERS

FIGURE 127. Tabulated core data and interpretation for well B. The presence of a gas cap and bottom water presented a problem in completion, particularly as vertical permeability exists at both of the fluid contacts. This well was squeezed three times in an attempt to shut off the gas; however, it is still producing with a high gas-oil ratio. The residual-oil saturation is somewhat low for normal-ratio production.

synthetic brines apparently give results which reflect the effective permeability of the formation.[91]

The ratio of the effective permeability to the specific permeability is termed the "relative permeability" and is usually expressed as a percentage

---

[89] Muskat, M., and Wykoff, R. D., Am. Inst. Min. Met. Eng. Trans., 1935.
[90] Muskat, M., *Performance of Bottom-Water Drive Reservoirs:* Am. Inst. Min. Met. Eng. Tech. Pub. 2060. Sept. 1946.
[91] Bolset, H. G., *Flow of Gas-Liquid Mixtures Through Consolidated Sand:* Am. Inst. Min. Met. Eng. Trans., 1940

or a fraction. The relative-permeability characteristic may be used with other data from core analysis and reservoir-fluid analysis to predict the performance of reservoirs under various productive mechanisms.[92] The relative-permeability characteristic. is reflected in produced gas-oil ratios as the gas saturation increases, in the drop of productivity of gas-condensate wells when the sand face becomes saturated with condensate, in the drop in potential of gas wells when water is produced through the completion interval, and in the production of water from zones of high connate-water saturation.

A lower limit for commercial permeability cannot be set because the factors of sand thickness and reservoir pressure directly affect this limit. The extremely low permeabilities of some formations in Colorado, Wyoming, and Texas are compensated for by the great thicknesses and high pressures. Actually it is not the permeability of any one foot of formation that is so critical; it is the total capacity (the permeability times the thickness) together with the reservoir pressure that controls the flow rate.

Residual-Fluid Saturations

The residual-liquid saturations determined by core analyses are the oil or condensate and the total core water. (See figs. 123, 127, and 131.) The gas volume is determined indirectly as the difference between the total pore volume and the liquid volume. These saturations may be determined by extraction or retort methods. One of the most rapid and efficient methods is the electric, water-cooled condenser-fluid stills.

Residual Oil and Condensate

The amount of residual oil remaining in the core sample cut from a high-pressure reservoir is dependent on the following factors.[93]
1. Reservoir-liquid saturation.
2. Formation or reservoir pressure.
3. Mud pressure and water loss.
4. Reservoir-liquid viscosities.
5. Coring time.
6. Vertical and horizontal permeabilities.
7. Pressure-depletion rate while pulling core.
8. Core diameter and type of core.
9. Method of obtaining core sample.
10. Solution gas-oil ratio.

The residual hydrocarbon saturation is dependent on the flushing or contamination by the drilling fluid that takes place as the formation is cored. When coring with water-base mud, flushing takes place ahead of the bit and radially through the core, owing to the difference between

---

[92] Muskat, M., and Taylor, M. O., *Effect of Reservoir Fluid and Rock Characteristics on Production Histories of Gas-Drive Reservoirs:* Am. Inst. Min. Met. Eng. Tech. Pub. 1914, 1945
[93] Caran, J. G., *Core Analysis—Its Interpretation and Application in Reservoir Engineering:* Petroleum Eng., Oct. 1947.

the mud weight and the formation pressure. While the core is being brought to the surface, gas expansion takes place with the reduction in pressure, thus driving out additional oil and gas; by the time the permeable cores reach the surface the pressure depletion is complete.

The residual-condensate saturation is the result of the retrograde-condensation mechanism that takes place as the pressure is reduced.

The residual oil or condensate may be expressed in percentage by

CORE ENGINEERS
Reservoir Core Analysts
SAN ANTONIO. TEXAS

GAS AND WATER PERMEABILITY RELATIONSHIPS
(SEE REPORT LETTER FOR DISCUSSION OF RESULTS)

COMPANY_____ WELL___B___

SAMPLE DEPTH (FEET)	SPECIFIC PERMEABILITIES PERMEABILITY. MILLIDARCYS (1)			PERMEABILITY RATIOS (RATIOS OF SPECIFIC PERMEABILITIES)		
	DRY GAS	SALT WATER	FRESH WATER	DRY GAS / SALT WATER	DRY GAS / FRESH WATER	FRESH WATER / SALT WATER
7575.5	143	42	35	3.4	4.1	0.834
76.5	149	54	46	2.8	3.2	0.852
77.5	236	113	94	2.1	2.5	0.831
78.5	72	29	16	2.5	4.5	0.552
79.5	206	127	111	1.6	1.9	0.875
80.5	124	26	16	4.8	7.8	0.615
81.5	49	10	2	4.9	24.5	0.200
82.5	151	54	42	2.8	3.6	0.778
83.5	358	212	185	1.7	1.9	0.872
84.5	103	39	17	2.6	6.1	0.436
85.5	51	14	9	3.6	5.7	0.643
86.5	37	7	5	5.3	7.4	0.714
87.5	87	33	20	2.6	4.4	0.606
88.5	68	11	9	6.2	7.6	0.818
89.5	64	12	10	5.3	6.4	0.833
90.5	296	220	209	1.3	1.4	0.950
91.5	210	189	175	1.1	1.2	0.925
92.5	454	178	99	2.6	4.6	0.556
93.5	370	129	63	2.9	5.9	0.488
94.5	33	3	2	10.0	16.5	0.667
AVERAGES	163	75	58	2.2	2.8	0.773

NOTE: (1) SAMPLE COMPLETELY SATURATED WITH FLUID. (INT) NO TEST. SAMPLE DISINTEGRATED. (*) INFINITE.

POLICY: THE INTERPRETATIONS OR OPINIONS EXPRESSED IN THIS REPORT REPRESENT THE BEST JUDGMENT OF CORE ENGINEERS AND ARE PRESENTED FOR THE EXCLUSIVE AND CONFIDENTIAL USE OF THE CLIENT. NO RESPONSIBILITY AS TO THE PRODUCTIVITY OR PROFITABLENESS OF ANY ZONE ANALYZED WILL BE ASSUMED BY CORE ENGINEERS.

FIGURE 128. Gas- and water-permeability relationships for well B. The fresh-water-salt-water permeability ratios are somewhat irregular. It is probable that this sand would respond most readily to gas and salt-water injection.

volume, percentage of the pore space, and barrels per acre-foot of formation (figs. 127 and 131).

Residual Core Water

In high-pressure, flush reservoirs cored with water-base mud, the amount of residual or total water in the core at the time of analysis is the sum of the connate plus any drilling water that may have been forced into the pores of the sand while the core was being cut.

If the formation contains minerals characterized by a high chemically bound water content, this water will also be recovered when high-temperature-retort methods are used, and corrections for such water of crystallization should be made. Low-temperature extraction methods seldom recover water of crystallization.

Cores cut with oil-base mud from low-pressure reservoirs normally show total water saturations that may be assumed to be the connate or interstitial water saturation of the formation. Deep, high-pressure reservoirs cored with oil-base muds show total water saturations that may not be the true interstitial-water content because of the high temperatures encountered.

The total core water is usually presented in percentage of the pore space.

Connate Water

Connate water is present in varying degree in all water-wet sand or lime formations. It has been termed both "interstitial" and "connate" water. In the writer's opinion, the term "connate" water is the more desirable and will be used throughout this discussion.

Connate water is defined as the water in the formation at the time the formation is cored. It is immaterial in core-analysis work whether it is the same water that saturated the sand when it was first deposited or water that migrated into the sand during later geologic time. The connate water may be made up of "free" water, and the water may be held in the interstices of the sand by capillarity. The free water may be produced, but the capillary water is not producible. Most oil or gas sands are water-wet and contain some connate water, although water-free fluids may be produced.

A number of methods are available for measuring or calculating the connate-water saturations of sand formations.

1. Capillary pressure versus water saturation by gas-pressure and centrifugal methods.[94]
2. The use of tracers in the drilling fluid.[95]
3. The calculation of saturation based on electric-log resistivities.[96]

[94] McCullough, J. J., Albough, F. W., and Jones, P. H., *Determination of the Interstitial-Water Content of Oil and Gas Sands by Laboratory Tests of Core Samples:* Am. Petroleum Inst. Drilling and Production Practice, 1936.
[95] Pyle, H. C., and Jones, P. H., *Quantitative Determination of Connate-Water of Oil Sands:* Am. Petroleum Inst. Drilling and Production Practice, 1936.
[96] Jones, J. P., *Water Saturation vs. Resistivity:* Petroleum Production, vol. 1, pp. 54-56, 1916.

4. Coring with oil-base mud.[97]
5. Calculations based on salinities of core water.[98]
6. Empirical calculations based on the porosity and residual fluid saturations.[99]

CORE ENGINEERS
Reservoir Core Analysts
SAN ANTONIO, TEXAS

COMPANY_____   WELL___B_____

CORE DATA SUMMARY

ZONE OR DIVISION NO.	Wilcox				
DEPTH, FEET	7579 - 7589				
PROBABLE PRODUCTION (1)	Oil				
ANALYZED—PRODUCTIVE. FEET	10 - 10				
CORED, FEET	10				
CORE RECOVERY %	100.0				
AVERAGE PERMEABILITY, MILLIDARCYS	123				
CUMULATIVE CAPACITY, MILLIDARCY-FEET	1234				
AVERAGE POROSITY PERCENT	24.0				
AVERAGE POROSITY, BARRELS PER ACRE FOOT	1862				
AVERAGE RESIDUAL OIL SATURATION, % PORE SPACE	9.0				
AVERAGE TOTAL WATER SATURATION, % PORE SPACE	32.2				
AVERAGE CONNATE WATER SATURATION, CALCULATED % PORE SPACE	21				
RESIDUAL OIL GRAVITY, °API	26				
FORMATION VOLUME FACTOR, ESTIMATED	1.51				
SOLUTION GAS-OIL RATIO CUBIC FEET PER BARREL (2)	680				
OIL IN PLACE BARRELS PER ACRE FOOT (3)	1471				

RECOVERABLE OIL, STOCK TANK BARRELS PER ACRE FOOT

BY SOLUTION GAS EXPANSION (4)	440				
INCREASE BY EFFECTIVE WATER DRIVE	373				
MAXIMUM RECOVERY, AFTER WATER DRIVE (5)	813				

NOTE:
(*) REFER TO REPORT LETTER.
(1) PREDICTION ASSUMES COMPLETE ZONE ISOLATION.
(2) PRESSURE REDUCTION FROM SATURATION PRESSURE TO ATMOSPHERIC.
(3) OIL VOLUME AT ORIGINAL RESERVOIR CONDITIONS.

(4) AFTER PRESSURE REDUCTION FROM RESERVOIR PRESSURE TO ZERO PSI.
(5) ORIGINAL RESERVOIR PRESSURE MAINTAINED BY WATER DRIVE.
(6) GAS PHASE RESERVOIR; NO ESTIMATE.

POLICY: THE INTERPRETATIONS OR OPINIONS EXPRESSED IN THIS REPORT REPRESENT THE BEST JUDGMENT OF CORE ENGINEERS AND ARE PRESENTED FOR THE EXCLUSIVE AND CONFIDENTIAL USE OF THE CLIENT. NO RESPONSIBILITY AS TO THE PRODUCTIVITY OR PROFITABLENESS OF ANY ZONE ANALYZED WILL BE ASSUMED BY CORE ENGINEERS.

FIGURE 129. Summary of core data, oil reserve, and recoverable oil of well B. Because of the gas cap and the probability of a water drive, it is apparent that the ultimate recovery will be higher than the solution gas recovery indicated.

---

[97] Stuart, R. W., *Use of Oil Base Mud:* Oil Weekly, pp. 41-45, May 27, 1946.
[98] Sage, H. F., and Armstrong, D. M., *Estimation of Connate Water from the Salt Content of the Core:* Am. Petroleum Inst. Prod. Bull. 223, May 1939.
[99] Lewis, J. A., *Core Analysis—An Aid to Increasing the Recovery of Oil:* Am. Inst. Min. Met. Eng. Trans., pp. 68-75, 1942.

Experience has shown that the total water saturations measured for limestone cores from oil fields in Caldwell, Guadalupe, and Milam Counties in south Texas may be considered as the connate-water saturations. Cores cut with diamond-core bits using water-base mud normally show total water saturations between 15 and 25 percent of the pore space when the cores are oil-productive.

The practice of using dextrose in cable-tool coring of partially depleted oil sands in some Pennsylvania fields has proved practical in differentiating between drilling, connate, and flood or extraneous water.[100]

A recent development of the capillary-pressure studies shows possibilities of using the high-pressure mercury pump in measuring that portion of sand-core samples occupied by capillary water.[1]

The use of capillary-pressure measurements for the determination of capillary water is commercially practical. Of particular value are the restored-state techniques in that the core samples need not be fresh. It is highly desirable to know the height above the water table when selecting the portion of the pressure curve applicable to the particular zone being tested. The portion of the curve which represents the irreducible-minimum water saturation may not apply to all reservoirs and is dependent on the closure. In practice, a core more than fifty feet above the water table in the reservoir will probably contain its minimum water saturation.[2]

A series of comparison tests in which the connate-water content of a large number of cores cut with oil-base mud was compared with results obtained by the following methods showed reasonably close agreement; these included the use of the capillary-pressure method, calculations from electric-log resistivities, calculations based on salinities, and distillation measurements.[3]

The salinity of the residual water in cores cut with water-base mud has been used in an attempt to estimate the connate-water saturation. This method has been found to be unreliable for most cores taken from flush, high-pressure reservoirs because of the contamination by the water in the drilling fluid. Zones of very low permeability may show salinities within the range of the formation water, but connate-water content calculated from these salinities may not apply to more permeable flushed zones because of the relationship between permeability and connate water. Comparison tests using electric-log, oil-base, and tracer data have established a logarithmic relationship between permeability, total water, and connate water.[4] Unfortunately, no one relationship will apply to all sand bodies.

[100] Clark, A. P., *A Method for Determining Connate and Drilling Water Saturation for Cable Tool Cores:* Producers Monthly, July 1947.

[1] Purcell, W. R., *Capillary Pressures—Their Measurement Using Mercury and the Calculation of Permeability Therefrom:* Am. Irst. Min. Met. Eng., Fall Meeting, Dallas, Texas, Oct. 1948.

[2] Bruce, W. A., and Welge, R. J., *The Restored State Method for Determination of Oil in Place and Connate Water:* World Oil, Aug. 1947.

[3] Thornton, O. F., and Marshall, D. L., *Estimating Interstitial Water by the Capillary Pressure Method:* Am. Inst. Min. Met. Eng. Tech. Pub. 2126, Jan. 1947.

[4] Earlougher, R. C., *Core-Analysis Problems in the Mid-Continent Area:* Am. Petroleum Inst. Drilling and Production Practice, 1940.

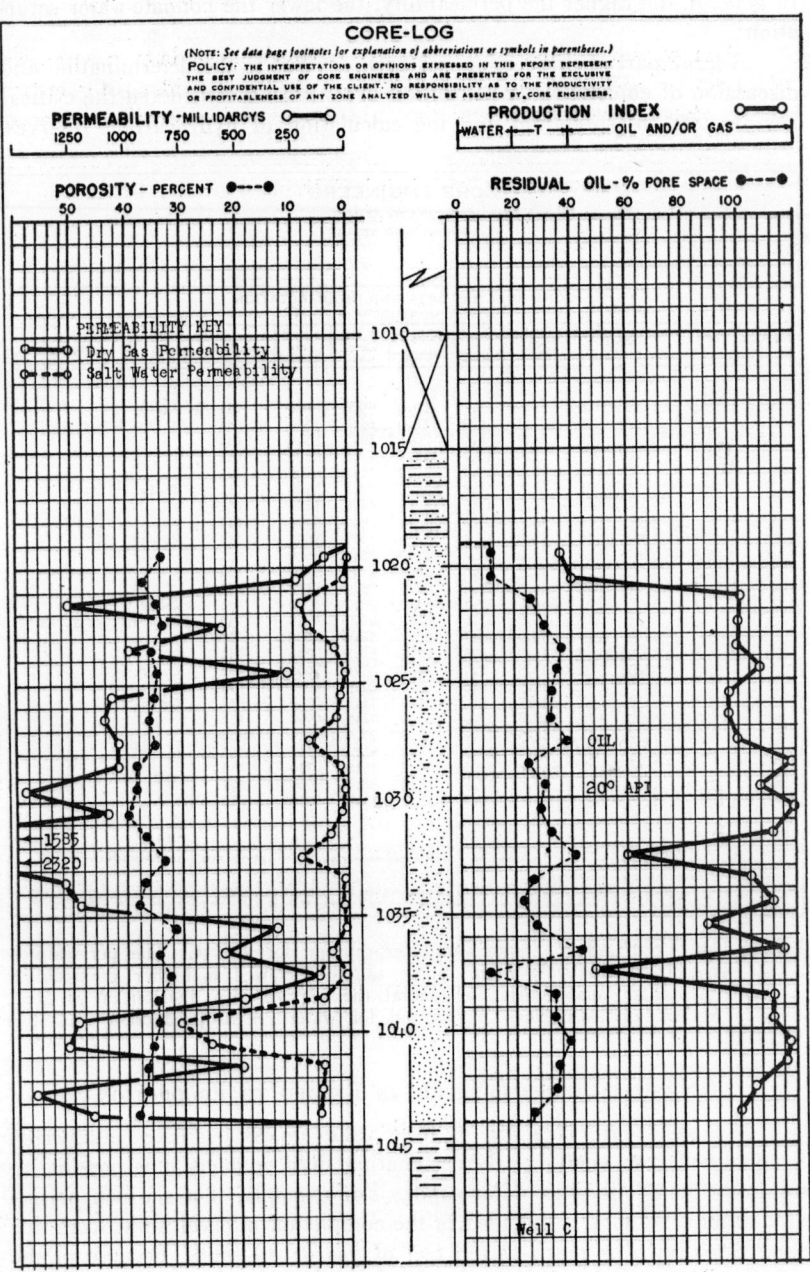

FIGURE 130. Core-log of well C.

In general, the higher the permeability, the lower the connate-water saturation.

A large part of this section has been devoted to the determination and discussion of connate-water saturation because it is considered the critical key in data interpretation and the calculation of hydrocarbon reserves.

CORE ENGINEERS
Reservoir Core Analysis
SAN ANTONIO, TEXAS

COMPANY_____   WELL___C___

CORE ANALYSIS DATA AND INTERPRETATION
(See footnotes for meaning of symbols or abbreviations)

SAMPLE NUMBER	DEPTH FEET	PERMEABILITY MILLIDARCYS Hor.	Vert.	POROSITY PER CENT	RESIDUAL FLUIDS OIL % VOL	OIL % PORE	WATER % PORE	SALINITY P.P.M. CHLORIDES	PROBABLE PRODUCTION (1)	REMARKS
1	1019.5	100	148	33.9	3.6	10.6	59.6	4060	Oil	Shaly, LFR
2	20.5	237	200	37.1	4.2	11.6	61.2	3130	Oil	Shaly, LFR
3	21.5	1260	820	34.7	8.6	26.3	32.7	4440	Oil	
4	22.5	570	855	33.9	10.6	31.1	31.1	4800	Oil	
5	23.5	990	1370	35.9	13.3	36.9	31.9	4200	Oil	
6	24.5	273	91	34.9	12.2	34.9	27.9	4490	Oil	Shaly
7	25.5	1060	1070	35.4	11.5	32.6	33.3	3810	Oil	
8	26.5	1100	910	36.0	12.0	33.4	28.2	4530	Oil	
9	27.5	1090	1400	34.8	13.1	37.5	30.2	3800	Oil	20 deg. API
10	28.5	1085	1110	38.2	9.2	24.0	28.7	3610	Oil	
11	29.5	1470	1360	38.3	11.8	30.8	30.5	4540	Oil	
12	30.5	1085	1020	39.9	11.4	28.5	25.9	4330	Oil	
13	31.5	1585	1200	36.4	12.1	33.2	28.7	3970	Oil	Shale inclusions
14	32.5	2320	1670	33.7	13.6	40.4	40.9	3640	Oil	High wtr. sat., shly.
15	33.5	1290	1295	36.9	9.7	26.2	32.6	4000	Oil	
16	34.5	1220	1240	38.0	8.5	22.4	30.7	4200	Oil	
17	35.5	325	354	31.5	8.4	26.8	43.6	5000	Oil	Shaly
18	36.5	568	412	34.3	14.6	42.6	24.5	5480	Oil	Shaly
19	37.5	132	62	32.7	3.4	10.4	71.8	6260	Oil	Shaly, LFR
20	38.5	480	356	34.8	13.9	33.9	26.9	6960	Oil	
21	39.5	1240	422	34.5	11.9	34.3	28.6	7840	Oil	
22	40.5	1270	950	35.8	13.9	38.8	24.1	6130	Oil	
23	41.5	485	640	36.8	13.2	35.8	35.8	4730	Oil	
24	42.5	1420	700	36.9	12.8	34.6	38.3	5960	Oil	
25	43.5	1170	408	38.5	9.9	25.7	38.5	6120	Oil	

(*) REFER TO REPORT LETTER   (NS) NO SAMPLE   (NF) NO FLOW   (NCF) NO COMMERCIAL FLOW   (LFR) LOW FLOW RATE   (C) CONDENSATE TYPE RESIDUAL

(1) PREDICTION ASSUMES COMPLETE ZONE ISOLATION

POLICY: THE INTERPRETATIONS OR OPINIONS EXPRESSED IN THIS REPORT REPRESENT THE BEST JUDGMENT OF CORE ENGINEERS AND ARE PRESENTED FOR THE EXCLUSIVE AND CONFIDENTIAL USE OF THE CLIENT. NO RESPONSIBILITY AS TO THE PRODUCTIVITY OR PROFITABLENESS OF ANY ZONE ANALYZED WILL BE ASSUMED BY CORE ENGINEERS.

FIGURE 131. Tabulated core data and interpretation for well C. The permeability to dry gas is relatively high. Some samples showed a higher permeability vertically than horizontally. In general the permeability distribution is good. The residual-oil saturations are normal for water-free-oil production from this sand.

CORE-DATA INTERPRETATION IN REGARD TO PRODUCTION

The basic core data are of little value unless correctly interpreted by qualified core analysts. By designating fluid contacts it is possible to minimize completions in water sands. For example, if a six-inch permeable-water-sand zone is included in the completion interval, fluid from this sand can drown as much as ten feet of gas or oil sand. Because of the critical nature of completing oil wells with a gas cap or bottom water, it is highly important that core depth and perforation or completion depths agree. Many times the core analyst is held responsible for poor completions when the fault lies in the disagreement between driller's and electric-log depths. Cores are usually submitted to the core analyst with the driller's depths designated. Electric-log depths may differ by several feet from the driller's depths, and, unless the correction is specifically made, the core-analysis report will be in error as to depths. The writer strongly emphasizes the importance of correct depth designation.

Careful note should be made as to the type of drilling fluid used. Special muds such as low-water-loss, modified-starch drilling fluids are often used in coring sands with low productive capacity. Total water saturations in the same sand may vary with the type of water-base drilling fluid used, and this factor must be taken into consideration in the data interpretation. It is readily apparent that the practice of washing a well with water after drilling or coring with low-water-loss mud defeats the purpose for which these special muds were used. Wells of this type should be washed and cleaned with crude oil. In other words, the completion techniques employed by the operator may nullify accurate work on the part of the core analyst.

Failure properly to cement or squeeze gas zones above an oil sand will result in high gas-oil-ratio production (fig. 127). Sand bodies drilled into bottom water must be squeezed or water will be produced (fig. 127). The use of excessively high pressures to break down the sand while cementing may result in sand fractures which will cause channeling and bad completions. The necessity of squeezing the same sand several times in order to shut off gas or water may permanently damage the sand and result in abandonment of the well.

No definite set of rules or tables can be made for the interpretation of core data. As a matter of fact, the practice of some analysts in making interpretations as to probable production from the data without seeing the cores often leads to bad predictions.

The interpretation of data from flush, deep, high-pressure reservoirs is entirely different from the interpretation of core data of samples cut from shallow, low-pressure, depleted reservoirs. However, there are some characteristics which enable the interpreter to predict correctly the type of production to be expected and to locate the fluid contacts.[5]

GAS SANDS

The following physical characteristics are associated with permeable, high-pressure gas sands:

1. Low water saturation dependent on permeability, formation pressure, and degree of contamination by drilling water.

2. Residual-oil saturation absent or present in varying degree, depending on the gravity of the oil and the height above the gas-oil contact. Residual-oil color will vary from amber to yellow.

3. If thoroughly flushed, the salinity of the core water will approach

---

[5] Caran, J. G., *Core Analysis—An Aid To Profitable Completions:* Mines Mag., vol. 37, no. 2, Feb. 1947.

that of the drilling fluid; if not, the salinity will be normal or comparable to that of cores from the oil column.

4. Characteristic gas odor when fresh cores are broken. It is important to note that some gases have no apparent odor particularly when composed primarily of methane and ethane, so that the lack of odor is not always conclusive.

5. An inspection of the mud sheath on the core may show evidence of gas breaking from the pores of the sand. This is particularly true of sands with low permeability.

6. Connate-water saturations are usually less than forty percent of the pore space.

It is particularly difficult to differentiate between water- and gas-productive sands when the reservoir pressure is very low. Low-pressure gas sands may show total water saturations approaching 80 percent of the pore space and still produce dry gas.

CONDENSATE SANDS

Gas-condensate or distillate sands from high-pressure reservoirs may be recognized by characteristics similar to those for dry-gas sands, but with the following differences:

1. The high gravity of residual liquid hydrocarbons, usually above 50° A.P.I.

2. Characteristic water-white color of the residual hydrocarbons; saturation will vary between two and five percent of the pore space.

Experience is the critical factor in detecting gas-condensate sands, because the fluids retorted from sands with a high-shale content will sometimes show a light-colored meniscus at the top of the water, which may be taken as a condensate residual.

OIL SANDS

Oil sands are not usually flushed by drilling fluids to so great an extent as gas or water sands of comparable permeability, owing to the influence of interfacial forces and the difference in the viscosities of the fluids. Some low-pressure sands with very little solution gas and saturated with oil of very low gravity may show very little flushing by drilling fluid.

The residual-oil saturation of an oil-productive sand will vary with the factors discussed earlier in this paper under the heading "Residual Oil and Condensate."

Table 17 presents a relationship between the residual-oil gravity and the normal saturation for oil sands having a reservoir pressure between 2,000 and 3,500 p.s.i. An inverse relationship is evident from table 17; the higher the oil gravity, the lower the minimum residual oil content for water-free production. For shallower sands with lower formation pressure, the minimum saturation limits will be higher than those shown in the table.

It should be noted that these saturations are based on the assumption that no free gas will be produced from the sand face. The key to high gas-oil-ratio production lies in the residual-oil saturations, assuming water will not be produced.

TABLE 17

RESIDUAL OIL SATURATION OF OIL SANDS

Oil gravity (°A.P.I. at 60/60)	Residual oil (minimum) (percent pore space)
48–40	5–10
40–30	10–15
30–25	15–20
25 or less	20 or more

The connate-water saturations of oil sands vary appreciably with the permeability and structural position of the well. Clean sands with permeabilities above 1,000 millidarcys may have connate-water saturations less than 15 percent of the pore-space.

The maximum connate water that an oil-bearing zone can contain and not produce water varies with the effective permeability to each fluid phase. The limiting or critical water saturation must be determined for each sand in each reservoir.

The critical oil-water ratio may be used as an index to the probable production. This ratio varies for different sands, and no one ratio has been determined that will apply to all formations.

BLEEDING CORES

Since cores were first taken, the geologist has always been wary of so-called bleeding cores or cores that show oil or gas seeping from the pores for some time after the core is taken from the barrel. Many oil-bearing sands have been condemned as probably water-productive because of this bleeding.

In the writer's opinion, cores bleed because the pressure-depletion process is not complete by the time the core is taken from the barrel. Usually bleeding cores are low in permeability, and thus the pressure depletion process takes longer. Formations should not be condemned because of bleeding alone; core analyses can often explain this phenomenon.

OIL-SATURATED LIMESTONE

The analysis of limestone varies somewhat from that of sand or sandstone in that the permeability characteristic is not so critical, particularly if commercial porosity exists. This discussion is limited to porous and does not concern fractured limestone.

Care must be exercised in selecting representative samples of limestone for analysis because of abrupt changes in porosity. Cores from the

Edwards limestone in Caldwell and Guadalupe Counties in south Texas often show as much as 45 percent residual-oil saturation. Where oil productive, this limestone averages 20 percent in total-water saturation. As discussed elsewhere in this paper, it is assumed by core analysts in that area that the total water is the connate-water saturation. Porosity char-

---

CORE ENGINEERS
Reservoir Core Analysts

SAN ANTONIO TEXAS

GAS AND WATER PERMEABILITY RELATIONSHIPS
(SEE REPORT LETTER FOR DISCUSSION OF RESULTS)

COMPANY_____ WELL___C___

SAMPLE DEPTH (FEET)	SPECIFIC PERMEABILITIES PERMEABILITY MILLIDARCYS (1)			PERMEABILITY RATIOS RATIOS OF SPECIFIC PERMEABILITIES		
	DRY GAS	SALT WATER	FRESH WATER	DRY GAS SALT WATER	DRY GAS FRESH WATER	FRESH WATER SALT WATER
1019.5	100	0.4	0.0	250	(*)	0.0
20.5	237	4.0	0.0	59	(*)	0.0
21.5	1260	203.	5.7	6	221	0.028
22.5	570	192.	0.0	3	(*)	0.0
23.5	990	54.	0.0	18	(*)	0.0
24.5	273	3.3	0.3	83	910	0.091
25.5	1060	47.	0.9	23	1178	0.019
26.5	1100	50.	0.0	22	(*)	0.0
27.5	1090	188.	4.8	6	23	0.026
25.5	1085	48.	0.2	23	5420	0.004
29.5	1470	4.3	0.0	342	(*)	0.0
30.5	1085	35.	0.0	31	(*)	0.0
31.5	1585	90.	0.0	18	(*)	0.0
32.5	2320	210.	6.0	11	387	0.029
33.5	1290	8.	0.4	161	3220	0.050
34.5	1220	19.	2.3	64	530	0.121
35.5	325	7.	0.0	47	(*)	0.0
36.5	568	79.	1.1	7	52	0.014
37.5	132	7.	0.0	19	(*)	0.0
38.5	480	110.	1.0	4	480	0.009
39.5	1240	760.	36.	2	35	0.047
40.5	1270	642.	21.	2	61	0.033
41.5	485	110.	0.8	4	606	0.007
42.5	1420	121.	10.	12	142	0.083
43.5	1170	148.	0.1	8	11700	0.001
AVERAGES	954	126.	3.6	8	265	0.029

Total Capacity (Dry Gas)   =   23825 md.-ft.

Total Capacity (Salt Water) =   3140 md.-ft.

Total Capacity (Fresh Water) =   91 md.-ft.

NOTE: (1) SAMPLE COMPLETELY SATURATED WITH FLUID. (NT) NO TEST. SAMPLE DISINTEGRATED. (*) INFINITE.

POLICY: THE INTERPRETATIONS OR OPINIONS EXPRESSED IN THIS REPORT REPRESENT THE BEST JUDGMENT OF CORE ENGINEERS AND ARE PRESENTED FOR THE EXCLUSIVE AND CONFIDENTIAL USE OF THE CLIENT. NO RESPONSIBILITY AS TO THE PRODUCTIVITY OR PROFITABLENESS OF ANY ZONE ANALYZED WILL BE ASSUMED BY CORE ENGINEERS.

FIGURE 132. Gas- and water-permeability relationships for well C. These data emphasize the importance of making water-permeability tests. The sand would favorably respond to gas injection. Some portions of the sand would respond to salt-water injection, but fresh water will cause swelling of the hydratable materials in the sand, and restriction of flow rates will result.

acteristics for the Edwards limestone along the Balcones fault trend in south Texas vary from 15 to 35 percent.

Where the oil exists in fractured limestone, core analyses have very little value.

Chalk and Serpentine Production

Production from fractured chalk or serpentine occurs from the fracture planes and not from the dense, impermeable portions of the formation. Analyses made on the Austin chalk in Frio County, Texas, have proved of little value to the oil operator because the saturation exists in fracture planes. Where the saturation occurs in solution porosity in these formations, the same techniques used in limestone analyses apply.

Conglomerate Production

Analyses made by the writer on conglomerate cores from the north and north-central Texas areas show very high permeabilities. The porosity varies with the sorting and shape of the materials making up the conglomerate. Where the pebbles are of uniform size and cementation is at a minimum, the porosity approaches a maximum. Usually, conglomerate cores are thoroughly flushed by drilling water, and the estimation of the connate-water saturation becomes a problem. Minimum-reserve calculations based on analyses of consolidated conglomerate may be made using total water saturations as the connate water.

Water Sands

Water-productive sands associated with oil sands may be recognized by low and irregular oil saturations in the transitional zones (fig. 127). Hydrocarbons may not be present below the oil-water contact. The oil-water contact of a uniformly permeable sand is defined as the level in the sand column below which water alone will be produced. The total water saturations of water sands are usually higher than those for commercially productive sands and will approach 100 percent of the pore space when hydrocarbons are absent. Sands that produce water contain large amounts of "free" or noncapillary water.

Transitional Zones

Where a gas cap or bottom water is present in a thick sand column, assuming an oil column exists, there are transitional zones from the gas to the oil zones and from the oil to the water-productive zones. These zones vary in thickness within the same reservoir.

The gas-to-oil transitional zone may be recognized by an increase in residual-oil content as the gas-oil contact is approached. It is advisable to confine the completion interval to the oil zone and several feet below the designated gas-oil contact in order to minimize the possibility of gas channeling and high gas-oil-ratio production.

The transitional zone from oil to water is usually characterized by an increase in total water content and a small drop in residual-oil saturation. The oil-water ratio will be smaller than that for the oil-productive zone. This type of transitional zone is of doubtful commercial value; usually, water is produced with the oil upon initial completion, and, as the oil saturation is decreased, the permeability to water rapidly increases to the point where the entire zone is water-productive.

The foregoing discussion emphasizes the importance of making vertical permeability tests. Vertical permeability barriers should be used in selecting the completion interval. Many core-analysis reports contain no vertical permeability data because some core analysts have not recognized the importance of these tests. The writer believes that every possible test should be made on core samples while the cores are available because, once the cores have been discarded, these additional data can only be secured by recoring the formations.

Secondary Migration of Water

There is no way of recognizing oil- or gas-sand zones which have been subjected to the secondary encroachment of water, and many wrong predictions as to probable production may be ascribed to this phenomenon. Cores from these zones can show exactly the same residual hydrocarbon saturations as the cores from zones that have not been flushed by salt water.

The reason these zones cannot be recognized is that there is sufficient gas left in the sand to drive out the excess water and thus make the water saturations similar to those of gas-productive sands. Checks by the use of the gas analyzer would show gas to be present. The only conclusive test is to make a production test through the drill stem. Flooded oil sands can show normal residual-oil saturations and still produce nearly 100 percent salt water.

The inability of the core analyst to recognize these zones solely on the basis of core analysis should be recognized by the oil operators, and all of the data as to the structural position of the well, the probability of faulting, and electric-log correlation should be made available to the interpreter. It is sometimes difficult to understand why the oil operators withhold this information from the core analyst when they are paying for the work. Any and all pertinent data that will help in arriving at the right answer should be used.

Oil in Place

All of the oil in place cannot be recovered by any known means of production. This volume is usually calculated at reservoir conditions and is expressed in barrels per acre-foot of formation. The practice of some engineers in using the residual-core oil as the oil in place may apply for unflushed cores from shallow low-pressure reservoirs but does not apply to cores from high-pressure reservoirs. As indicated in the following equation, if the porosity and the connate-water data are correct, then the oil in place may be readily calculated. $O.I.P. = P_o (1 - S_{cw}) \times 7,758$

$O.I.P.$ = Oil in place, barrels per acre-foot at reservoir conditions.

P_o = Porosity expressed as a decimal.

S_{cw} = Connate-water saturation expressed as a decimal part of the pore space.

$7,758$ = Volume of one acre-foot in barrels.

The oil-in-place equation is based on the assumption that the reservoir pores are completely saturated at reservoir conditions.

MAXIMUM RECOVERABLE OIL

The volume of oil that may be recovered from a high-pressure reservoir by an effective water drive and gas expansion when the reservoir pressure is maintained at or near the original saturation pressure is termed the "maximum recoverable oil."

The principles upon which this recovery calculation is based involve the consideration of the flushing and pressure depletion that takes place while cutting and recovering the core. As discussed elsewhere, the core is subjected to water flood while being cut and to a pressure-depletion process while being raised to the surface. These mechanisms occur in reverse order in the normal productive life of the formation.

The following equation is similar to the oil-in-place equation with the exception of the correction for the residual oil. The residual oil is converted into a volume at reservoir conditions and is subtracted from the oil in place because there is no means of producing this oil.[6]

$$M.R. = \frac{P_o(1 - S_{ro} - S_{cw})}{F.V.F.} \times 7,758$$

$M.R.$ = Maximum recovery, stock-tank barrels per acre-foot of formation.

P_o = Porosity, expressed as a decimal part of the formation volume.

S_{ro} = Residual oil at reservoir conditions, expressed as a decimal part of the pore space.

S_{cw} = Connate-water saturation, expressed as a decimal part of the pore space.

$F.V.F.$ = Formation-volume factor, decimal expression of the volume that one barrel of stock-tank oil occupies at reservoir conditions.

Although this equation is accepted by most core analysts for calculating the maximum recoverable oil from high-pressure reservoirs, the writer believes that the values obtained are sometimes much too high. The critical factor in this equation is the residual oil, and seldom is it

[6] Muskat, M., The Determination of Factors Affecting Reservoir Performance: Am Petroleum Inst. Drilling and Production Practice, 1940.

possible under field conditions to reduce the residual oil to as low a point as that measured in cores in the laboratory.

It should be noted that the permeability factor does not enter in this equation. This factor is definitely reflected in the length of time required to recover the oil. Actually it is the capacity (permeability in millidarcys times thickness in feet) and the pressure that control the flow rate. The higher the capacity, the more rapid the rate of depletion.

SOLUTION-GAS-EXPANSION RECOVERY

Recovery by solution-gas expansion is usually the initial type of recovery from most reservoirs. The gas in solution in the oil expands when a pressure differential is created in the reservoir by producing methods, and oil is forced into the well bore. This type of recovery can occur in water-drive reservoirs when the rate of withdrawal exceeds the rate of water encroachment and a sufficient drop in pressure occurs in the formation surrounding the well bore to release the gas in solution.

Solution-gas-type reservoirs should never be produced at a rate that will cause free gas to break out of solution in the formation. This free gas restricts the flow of oil and results in high gas-oil-ratio production.

Reservoirs operating under a solution-gas drive may be expected to produce from 10 to 40 percent of the original oil in place.[7]

Several methods are available for the calculation of gas-expansion recoveries. One method that employs the basic core-analysis data has been proposed by Johnston [8] and has been used with some effectiveness in fields where it was possible to determine the correlation factor. The correlation factor used for certain California fields was 0.65. The following equation has for its basis the assumption that the gas-filled space in the core at the surface is very nearly the same as that which would result from the normal pressure depletion of a dissolved-gas reservoir.

$$G.E.R. = \frac{P_o(1 - S_{tw} - S_{ro})}{F.V.F.} \times 7,758 \times C.F.$$

$G.E.R.$ = Gas-expansion recovery, stock-tank barrels per acre-foot of formation.

P_o = Porosity, expressed as a decimal.

S_{tw} = Total water saturation, expressed as a decimal part of the pore space.

S_{ro} = Residual oil at reservoir conditions, expressed as a decimal part of the pore space.

$C.F.$ = Correlation factor, expressed as a decimal.

The gas-expansion-recovery equation has been used without the corre-

---

[7] Buckley, S. E., and Craze, R. C., *The Development and Control of Oil Reservoirs:* Am. Petroleum Inst., Drilling and Production Practice, 1943.

[8] Johnston, Norris, *Core Analysis Interpretation:* Am. Petroleum Inst., Drilling and Production Practice, 1941.

lation factor to calculate the maximum recoverable oil from porous limestone.

GAS RESERVES

The dry-gas reserve at reservoir and surface conditions may be calculated from core analysis, reservoir temperature, and pressure data.

$$V_r = P_o \left(1 - S_{cw}\right) \times 7{,}758 \times 0.005615$$

V_r = Gas reserve at reservoir conditions, Mcf. per acre-foot.
P_o = Porosity, expressed as a decimal.

$$V_s = \frac{V_r}{Z} \times \frac{(P_r + 14.7)}{14.7} \times \frac{520}{(460 + T_r)}$$

V_s = Gas reserve at surface conditions, Mcf. per acre-foot at 14.7 p.s.i. and 60° F.
V_r = Gas reserve at reservoir conditions.
P_r = Reservoir temperature, degrees F.
T_r = Reservoir pressure, p.s.i.
Z = Compressibility factor, expressed as a decimal.

RECOVERABLE GAS

The recoverable gas may be calculated to any required residual pressure from the surface gas reserve.

$$R.G. = V_s \times \frac{(P_r - P_f) + 14.7}{P_r + 14.7}$$

$R.G.$ = Maximum recoverable gas to required residual pressure, Mcf. per acre-foot.
V_s = Gas reserve at surface conditions, Mcf. per acre-foot.
P_r = Final residual reservoir pressure, p.s.i.
P_f = Final residual reservoir pressure, p.s.i.

Although this equation is mathematically correct, the calculated recoverable gas estimate is seldom attained. Conservative estimates of the recoverable gas may be determined by multiplying the value obtained from this equation by 0.60.

It is apparent from the discussion of hydrocarbon reserves and recoveries that the calculation of these volumes is not merely the substitution of figures in an equation.

GRAPHIC PRESENTATION OF CORE DATA

The core-log (figs. 122, 126, and 130) is a means of presenting the data for visual comparisons. The gas and water-permeability curves, together with the porosity curve, are of value in the selection of completion intervals and may be used to determine the optimum footage available for gas or water injection in pressure-maintenance or secondary-recovery programs. The water-permeability curve will show the portions of the sand body that will most readily take water in water flooding and will also show where water production will first occur if water encroachment takes place.

The plotting of the impermeable barriers enables the operator to make use of natural shutoffs for the control of high gas-oil-ratio and bottom-water production.

The residual-oil plot shows the oil found in the core and is expressed in percentage of the pore space. Where gas-condensate or distillate is measured, the residual condensate is plotted as oil type (figs. 122 and 126).

The production index is based on a relationship between the analyzed physical characteristics of the cores and is an expression of the type of production to be expected.

The gravity of the residual oil as determined by core analysis or drill-stem tests is presented on the core-log whenever the formation is considered oil-productive.

SUMMARY OF APPLICATIONS OF CORE ANALYSIS

Exploration

The analyses of representative core samples and the intelligent interpretation of the data are aids in the evaluation of the commercial possibilities of wildcat wells, edge wells, and field extensions. Core analyses together with electric and mechanical well logging make possible correlation work.

The determination of the oil gravity and of the productive capacity of oil-bearing formations minimizes the possibility of completing a dry hole or missing a productive formation.

Where wells are drilled to total depth without coring, an electric log can be run, and the zones that exhibit possibilities of hydrocarbon saturation may be cored with any one of a number of sidewall-coring devices. The analysis of sidewall samples is a highly specialized technique, and the interpretation of the core data requires an extensive knowledge of the physical characteristics of each formation as represented by full-diameter-core data.

Exploitation

Core-analysis data may be used with reservoir-fluid-analysis data to predict the productive history of a formation and thus predetermine the required capacity of the field equipment throughout the producing life of the reservoir. Selection of the most effective well-spacing pattern may be guided by core analysis.

Evaluation of oil and gas reservoirs can be made by the use of core data, and the probable recoveries can be calculated.

Core-analysis data are essential for reservoir-performance predictions that form the basis for pressure maintenance, repressuring, and secondary-recovery programs. Unless complete core data are available for all gas- or water-injection wells, it is generally impossible to predict the course of the injected fluids. The permeability profile may be used selectively to perforate or pack water- or gas-injection wells so that zones of low permeability will not be by-passed by the injected fluids.

Core permeability data may be used to plot isoperm maps and thus

CORE ENGINEERS
Reservoir Core Analysis
SAN ANTONIO, TEXAS

COMPANY_____   WELL _C_____

CORE DATA SUMMARY

ZONE OR DIVISION NO.	1				
DEPTH, FEET	1,019 - 10VV\				
PROBABLE PRODUCTION (1)	Oil				
ANALYZED—PRODUCTIVE, FEET	25 - 25				
CORED, FEET	25				
CORE RECOVERY, %	100.0				
AVERAGE PERMEABILITY, MILLIDARCYS	954				
CUMULATIVE CAPACITY, MILLIDARCY-FEET	23,825				
AVERAGE POROSITY, PERCENT	35.7				
AVERAGE POROSITY, BARRELS PER ACRE FOOT	2770				
AVERAGE RESIDUAL OIL SATURATION, % PORE SPACE	32.2				
AVERAGE TOTAL WATER SATURATION, % PORE SPACE	35.5				
AVERAGE CONNATE WATER SATURATION, CALCULATED % PORE SPACE	18				
RESIDUAL OIL GRAVITY, °API	20				
FORMATION VOLUME FACTOR, ESTIMATED	1.05				
SOLUTION GAS-OIL RATIO CUBIC FEET PER BARREL (2)	150				
OIL IN PLACE BARRELS PER ACRE FOOT (3)	2271				

RECOVERABLE OIL, STOCK TANK BARRELS PER ACRE-FOOT

BY SOLUTION GAS EXPANSION (4)	550				
INCREASE BY EFFECTIVE WATER DRIVE	725				
MAXIMUM RECOVERY, AFTER WATER DRIVE (5)	1275				

NOTE:

(*) REFER TO REPORT LETTER.

(1) PREDICTION ASSUMES COMPLETE ZONE ISOLATION.

(2) PRESSURE REDUCTION FROM SATURATION PRESSURE TO ATMOSPHERIC.

(3) OIL VOLUME AT ORIGINAL RESERVOIR CONDITIONS.

(4) AFTER PRESSURE REDUCTION FROM RESERVOIR PRESSURE TO ZERO PSI.

(5) ORIGINAL RESERVOIR PRESSURE MAINTAINED BY WATER DRIVE.

(6) GAS PHASE RESERVOIR; NO ESTIMATE.

POLICY: THE INTERPRETATIONS OR OPINIONS EXPRESSED IN THIS REPORT REPRESENT THE BEST JUDGMENT OF CORE ENGINEERS AND ARE PRESENTED FOR THE EXCLUSIVE AND CONFIDENTIAL USE OF THE CLIENT. NO RESPONSIBILITY AS TO THE PRODUCTIVITY OR PROFITABLENESS OF ANY ZONE ANALYZED WILL BE ASSUMED BY CORE ENGINEERS.

FIGURE 133. Summary of core data, oil reserve, and recoverable oil. This sand has a very high productive capacity as measured by dry-gas flow. The sand should be cored and completed with oil-base fluids. The low connate-water saturation together with the high porosity results in high reserve and recoverable-oil estimates.

make possible the correlation of zones of like permeability in a reservoir. The possibility of gas or water channeling can be minimized by using the permeability data for selective shooting, plugging, acidizing, or packer setting.

Core analysis is a highly specialized phase of petroleum-reservoir engineering. The technique and procedures employed in analyzing cores vary with the formation and type of core samples being tested. These techniques may be considered somewhat mechanical, but the interpretation of core data requires experience. There are many core analysts but few qualified core-data interpreters. The use of one set of rules without common sense or judgment usually results in bad predictions on the part of the interpreter. An attempt has been made to point out the factors beyond the control of the interpreter, and due consideration should be given by the oil operators for these extenuating factors.

The practical value of core analysis to the petroleum industry is evidenced by its ever-increasing use in the exploration and exploitation of gas and oil reservoirs. The cost of this specialized service is small in comparison to the present-day cost of drilling a well. To core a well without analyzing the cores is as obsolete as drilling the well without making an electric-log survey.

Sometimes hundreds of thousands of dollars are spent in order to core but a small fraction of the total depth of the well. Merely smelling, tasting, and blowing through these cores can not tell how much oil or gas is present and how much can be recovered. Such predictions are the work of the qualified core analyst.

Acknowledgments

Appreciation is expressed for the encouragement, valuable criticisms, and suggestions made by the writer's associates, Robert Caran, C. E. Gordon, R. C. Kenney, S. H. Caran, and E. C. Lamon.

All of the well data presented in this paper were taken from the files of Core-Engineers and represent information from actual analyses; the company and well names were withheld because it was felt that these wells were selected merely as examples, and the data were not intended to be used for correlation or exploratory work.

FLUOROANALYSIS IN PETROLEUM EXPLORATION
JACK DE MENT

Luminescence is the emission of light by matter under the influence of energy; such a light emission violates the laws of thermal radiation. Two types of luminescence are generally recognized: one, called "fluorescence," lasts only as long as the matter is under the influence of the exciting energy; the other, called "phosphorescence," persists after the exciting energy has ceased to influence the matter. The fluorescence corresponds to the popular designation "glow"; that of phosphorescence to "afterglow."

Luminescent systems are of two kinds, both of which enter into the theory and application of fluoroanalysis to petroleum detection and exploration. The homogeneous luminescent system consists of matter in which all particles are identical, i.e., possessing no impurities and, in the case of solids, ideal crystallographic and stoichiometric characteristics. In the case of liquids, many highly purified organic compounds very nearly conform to the requirements for a homogeneous luminescent system. The heterogeneous luminescent system, in contrast to the homogeneous, contains impurities and/or structural imperfections, and these may enter into the luminescence. Solid heterogeneous luminescent systems are illustrated by the light-emitting materials called "phosphors," which contain a trace of activating substance, usually a metallic salt, in solid solution in a bulk of relatively inert solvent. Liquid heterogeneous systems are illustrated by crude oil, which contains many different organic substances, one or more giving rise to the luminescent response to ultraviolet light or other forms of exciting energy. Oil in a liquid solvent also comprises a heterogeneous system.

A number of monographs and treatises have been written on luminescence and kindred subjects, and for a more thorough background of fluorochemical methods of geophysical exploration it is recommended that these be referred to.[9] [10] Likewise, extensive literature is available on that branch of analytic science known as fluorochemical analysis, an understanding of which is necessary in petroleum applications.[11] [12] [13] A recent review has been given of fluorescent techniques in petroleum exploration.[14]

PETROLEUM FLUOROCHEMISTRY

The luminescence of petroleum has been recognized since the earliest days of antiquity. This property has, in years gone by, been known from the fact that crude oil may be of a clear color when taken from the earth but is generally greenish in reflected light and claret-red in transmitted light.

Ultraviolet light, filtered free of visible rays, evokes in crude oils a fluorescence that is generally blue, blue-green, or greenish. There may be considerable deviation from these colors, however, and oils from different localities may present typical fluorescent colors.

The fluorescence of crude petroleum, which is excited by visible light, is familiar as "bloom." Ultraviolet light of wave lengths down to 2,000

[9] Pringsheim, P., and Vogel, M., *Luminescence and Its Practical Applications*, New York, Interscience Publishers, 1943.
[10] De Ment, Jack, *Fluorochemistry* (extensive bibliography), New York Chemical Publishing Co., 1945.
[11] Radley, J., and Grant, J., *Fluorescence Analysis in Uultra-Violet Light*, 3d ed., New York, D. Van Nostrand Co., 1939.
[12] Danckwortt, P., *Lumineszenz-Analyse im Filtrierten Ultra-violetten Licht*, 4th. Auff., Leipzeg, Akad. Verlag., 1940.
[13] Haitinger, M., *Die Fluoreszenzanalyse in der Mikrochemie*, Vienna and Leipzig, Emil Haim, 1937.
[14] De Ment, Jack, Geophysics, vol. 12, pp. 72-98, 1947.

angstrom units excites petroleum fluorescence, although the so-called near-ultraviolet, that approximating 3,650 angstrom units, usually excites the brightest response.

Ultraviolet light can be employed on sands, shales, drill cores, and drillings, as well as soil samples, not only to detect petroleum but also to reveal the presence of other formations or for detecting economically important mineral species. More than 200 minerals and gems fluoresce under ultraviolet light and other exciting radiations.[15]

As an activator in solids, little is actually known about the role petroleum plays. In friable soils and earths, it is probable that the oil exists in physical combination, e.g., sorptively bonded, with these materials. Traces of petroleum may be found disseminated throughout a crystalline mineral, producing a fluorescent effect that may be typical of the oil or of the mineral and oil combined. Texas localities yield large crystals of calcite containing oil, and this is responsible for a fluorescence independent of manganese or rare-earths activation.

Oil shales provide interesting fluorescences under ultraviolet light. Common shales ("blaes") appear a very dark brown in the massive state, whereas the color may be absent in the powdered shale. With a kerogen shale, a rich chocolate-brown may be observed in both lump and powdered forms. Torbanite can be distinguished from cannel by the bright yellow streaks on a brown background.[16] [17] [18] Useful indications are obtained as to the origin, preparation, and process of cleaning of Esthonian and Manchurian shale oils; individual fractions can be graded according to boiling points, all by fluorescent response.[19]

There is a good deal of information available on the fluorochemistry of substances related to petroleum.[20] [21] Asphalts, coal tars, bitumens, coals, and sundry organic minerals not only fluoresce, but the fluorescence can often be seen at great dilutions. The origin of such materials can be empirically identified by the response; a trace of coal-tar pitch, for example, shows in asphalt by its greenish-blue emission when present at a ratio of 1:50,000.[22]

Petroleum oils and most refined petroleum products, whether liquid or solid, fluoresce at great dilutions in various liquid solvents. The solvents usually preferred include benzene, hexane, ethyl ether, carbon tetrachloride, various straight-chain hydrocarbons, and the like. The presence of a nitro group or of chlorine in the solvents generally acts to diminish the intensity of the fluorescence.

The detection threshold varies according to the method employed.

[15] De Ment, Jack, Ultra-Violet Prod., Inc., Bull., no. 2, p. 1, 1944; Oil and Gas Jour., vol. 44, pp. 75-80. 1945; *Handbook of Fluorescent Gems and Minerals*, Portland, Ore., Mineralogist Publishing Co., 1949.
[16] Radley, G., and Grant, J., *op. cit.*
[17] Danckwortt, P., *op. cit.*
[18] Haitinger, M., *op. cit.*
[19] Wittich, M., Brennstoff-Chemie, Heft 19, 1927.
[20] Pringsheim, P., and Vogel, M., *op. cit.*
[21] De Ment, Jack, *op. cit.*, 1945.
[22] Teuscher, W. Chem. Fabrik, Band 53, 1930; Band 54, 1930.

For unaided-eye observations, one part of oil in 100,000 parts of carbon tetrachloride can readily be seen. With electronic or photographic methods, the detection threshold may range to the order of parts per hundreds of millions. Heavy oils can be detected at a dilution of one part in 200,000 to 500,000 parts of water, and gasoline only at one part in 20,000 for the same solvent.[23] Dispersed in soils, oil concentrations of a few parts in tens to hundreds of millions can be detected by photographic methods.

METHODS OF FLUOROANALYSIS

Every luminescent substance exhibits certain distinguishing fluorochemical characteristics. Since luminescence is usually visible light, often with varying admixtures of ultraviolet and infrared wave lengths, the methods and means of optics serve for an analytic appraisal of the luminescence.

This means that luminescent light can be subjected to photometric, spectrometric, and spectrophotometric determinations. Likewise, since the luminescent system must absorb light, i.e., exciting energy, before an emission can take place, the absorption characteristics of that system can be studied. With the phenomenologic distinction between fluorescence and phosphorescence, it is evident that the lifetime of the luminescence may be a distinguishing property of light-emitting systems.

These purely physical techniques, directed to the luminescence per se, reinforce a number of chemical and physiochemical methods of analysis which have proved valuable in petroleum work. It is not possible to discuss here in detail the relative merits or limitations of each of the fluoroanalytic methods, since considerable literature is devoted to that subject, but a brief résumé is in order so that their application to the specialized field of petroleum detection and assay can be more adequately appraised.

Fluorometry

The photometry, i.e., the determination of intensity, of fluorescence is called "fluorometry." The intensity or brightness to the eye, being dependent in many instances upon physical and chemical conditions and upon the nature of the exciting radiation and its wave length, provides valuable information about the kind and concentration of a luminescent substance in solution. Fluorescence intensity depends upon several factors, including (1) the concentration of the fluorescent substance in solution, (2) the intensity and wave length of the exciting light, (3) the pH and temperature of the solution, (4) the nature of the solvent, and (5) the effect of interfering materials that may be present.

Electronic fluorometry relies upon the response of a photoelectric cell or, in some cases, upon a photon counter tube, under standardized conditions, for the measurement of fluorescence intensity. A number of these instruments have been designed and constructed for chemical and

[23] Halstrik, J., Archiv für Hygiene und Bakteriologie. Band 128, pp. 155-168, 1942.

biochemical applications, and through their sensitivity fractions of a microgram of certain fluorescent substances may be detected.

Optical photometers have been adapted to fluorometry, such as the Duboscq type.[24]

Fluorography

The appraisal of fluorescent light by photography is called "fluorography." Both black-and-white and color methods can be used. Weak luminescence is often advantageously detected and measured by the long exposures possible with photographic emulsions. An ordinary camera, the lense of which is covered with a filter transparent to visible light but opaque to ultraviolet light, can be employed. In a fluorograph, only the emitted light is recorded and portions of an object that absorb ultraviolet light but do not luminesce appear dark, as is true of ultraviolet reflecting areas, the light from which is intercepted by filter. Through densitometry of the photograph the intensity of the fluorescence may be measured.

The fluorographic method of fluorometry has been adapted to petroleum detection in soil samples [25] and is described in more detail below.

Fluoroadsorption Analysis

The adsorption of fluorescent materials upon an inert, nonfluorescent substance serves both to enhance emission and enable detection of separated substances from oft-complex mixtures.

The fluoroadsorption methods of analysis are either two-dimensional, i.e., involving paper, or, three-dimensional, i.e., involving solid columns, depending upon the adsorbent. In both cases the aim is to isolate oil traces from a large bulk of other material, e.g., soil, rock, or solvent, depending upon the method used.

The sorption phenomenon lays the basis for an extremely delicate physicochemical method of analysis, and the sensitivity is greatly enhanced by examination under ultraviolet light. Molecules that ordinarily do not fluoresce in the condensed phase, as in liquid or solid states, may often emit brightly when dispersed by adsorption.

While the pioneer work on ordinary capillary analysis was done in 1829 by Shonbein, it remained for one of his students, Goppelroeder, just prior to 1901, to apply fluorochemical methods to the technique. Important contributions have been made by numerous investigators;[26] one of the best recent reviews is that of Germann.[27]

When a solid column of adsorbent is used, the solution is drawn therethrough, the method being known as "chromatographic analysis," although erroneously termed "ultrachromatography" when an ultraviolet-light examination is relied upon for an appraisal of the results. Various crude oils in fractions can be separated upon a column of alumina, fol-

[24] Koch, W., Nature (London), vol. 154, p. 239, 1944.
[25] Ferguson, W. B., U.S. Patent 2,356,454, Aug. 22, 1944.
[26] Neugebauer, H., Die Kapillar-Lumineszenzanalyse, Leipzig, Schwabe, 1933.
[27] Germann, F., Colorado Univ. Studies, vol. D, pp. 1-14, 1940.

lowed by elution with different solvents.[28] Qualitative differentiation of the crude oils is made by observing the fluorescence of the different fractions in chloroform solution and the differences in color between the main bulk of the solutions. The capillary-strip method is also employed to distinguish between crude oils of different origins, as crude and refined oils from Digboi and Assam, some artificial crudes, and a chloroform extract of a resin.

Fluorescence Microscopy

The fluorescence microscope is an instrument which permits examination of an object at high or low magnifications under filtered ultraviolet light. Fluorescence microscopy may be of the reflected variety of the transmitted variety. Actually there is little difference between the two, but for opaque specimens the reflected method is generally the best.

Since petrographic methods play an important role in geochemical and geophysical prospecting, fluorescence microscopy would appear to have a promising future in this field as an auxiliary tool. The features of petrographic sections that may not be revealing by bright-field microscopy may well be so in ultraviolet light. The fluorescence microscope detects oil traces in drill cores, earths, muds, cuttings, water residues, and the like.[29]

Quenching Analysis

The fluorescence of petroleum is quenched or reduced in color and brightness by certain substances. The prudent use of such substances makes possible measurement of fluorescence intensity and in certain instances identification. Thus, nitrobenzene has been used to measure the fluorescence, and, by additions of this substance to a sample, the change in response appears to have some value for distinguishing oils.[30] The present writer has found that in complex organic substances, such as essential oils, the application of a given quencher to oil samples of different origin causes the brightness and color of fluorescence to vary greatly among the samples. In addition to nitrobenzene as a quencher, other substances of this character include trinitrotoluene, picric acid, and chlorinated hydrocarbons.

FLUORESCENCE EXPLORATION

The values of fluorescence and ultraviolet light as aids in correlating oil sands were first pointed out in 1936 by Melhase.[31] He stated:

During the past two years the writer [Melhase] has tested oils from many of the California fields and from different sands that may occur in these fields and it was found that no two samples of oil exhibited the same quality and degree of fluorescence unless they were obtained from identical sands. . . . It

[28] Mukherjee, N., and Indra, M., Nature (London), vol. 154, pp. 134-145, 1944; Inst. Petroleum Jour., vol. 31, pp. 173-178, 1945.
[29] De Ment, Jack, Oil Weekly, vol. 103, pp. 17-19, 1941.
[30] Mukherjee, N., and Indra, M., *idem*.
[31] Melhase, J., Mineralogist, vol. 4, p. 9. 1936.

follows, therefore, that when a well is drilled in any particular field and samples from each of the various sands penetrated are tested and classified according to their respective fluorescent qualities, the record thus obtained becomes an index for that field. When subsequent wells are drilled the new samples may be compared with the index and correlated accordingly. It is necessary, of course, that the samples to be tested consist of oil or of sands containing oil.

The observations of Melhase, while classic, have given way to more accurate and standardized, though often empirical, laboratory and field methods. The correlations are not limited to shales, but also include any earth-bearing oil, either of surface or of subsurface origin.

The method, it must be emphasized, is neither foolproof nor perfectly reliable. As many· of the techniques are empirical, developed in the laboratory according to the bents of the technician, there may be considerable latitude in the extent of dependability of results. The inspection of untreated earths with the unaided eye is open to much more error than the study of these materials either with instruments or by extraction and subsequent objective measurement. Trace amounts of highly fluorescent minerals may negate results for an inexperienced worker. Likewise, emphasis should be given the possible untoward effects of air on the sample, since some cores, after aging and exposure to subtropical climate, may lose their fluorescence. The role of hydrocarbonoxidizing bacteria in the destruction of oil traces in rock or earth should not be discounted.

It has been found that well cores and cuttings, even when small, broken, or partly contaminated, fluoresce if they contain liquid hydrocarbons.[32] In general it is found that a producible sand will give uniform fluorescence throughout, but that a sand carrying salt water in addition to oil will show a mottled fluorescence. In local areas increases in oil saturation and productivity appear to increase fluorescence intensity. When properly employed, fluorescence discloses the presence of petroleum undetectable by odor, taste, stain, saturation, analysis, or electric log.

FLUOROLOGGING

Well-log curves prepared from fluorochemical data provide extensive and reliable information.[33] The curves are called "fluorologs" (fig. 134), and the data are obtained from cuttings and core samples as customarily taken for paleontologic analyses. Core samples are not necessary in the procedure which has been developed by Ferguson, but it is desirable to have samples of all cores taken. The cuttings samples are usually taken at an interval of not more than thirty feet, but samples taken at shorter intervals provide more accurate results. Sampling is initiated at the surface and continued at regular intervals to the total depth of the hole.

The assay is made by the photographic method, and fluorescence intensities are plotted against depth. It is not necessary to rely upon the photographic procedure, but it is useful for very weak fluorescences. A

[32] De Ment, Jack, op. cit., 1947.
[33] Ferguson, W. B., and Campbell, O. E., The Fluorographic Method of Petroleum Exploration: 10 pp., Houston, Texas, Fluorographic Exploration Co., Apr. 24, 1945.

log is prepared from composite samples, which are obtained by mixing proportional amounts of the samples from 200 feet of hole. If the fluorescence intensity of a composite sample runs above a given value, a possible pay sand is indicated somewhere in the interval covered by the sample. The depth and the true value of the showing are established by analyzing each of the samples taken within the interval of showing. That part of a log based on analyses of individual samples is referred to as "detailed." The effectiveness of fluorologging for locating commercial pay horizons can be appreciated only from personal experience with the results obtained.

The over-all characteristics of the fluorolog of a well and the fluorolog of a dry hole are markedly different. A dry hole is characterized by low values from the surface to the total depth of the hole. A producer that cuts a fault between the surface and the producing horizon often shows a fluorolog characteristic of a dry hole above the fault, but a fluorolog typical of a producing well below the fault.

On a fluorolog of a producing well relatively high values are shown at and for some distance below the surface, with a progressive increase from the surface to the producing zone. Between pay horizons, in wells having multiple producing beds, the curve shows a continuation of high values, but it shows a decrease below each sand and a buildup as the next producing formation is approached.

Fluorologging is adapted to small samples, such as side-wall cores, which are insufficient in size to be satisfactorily analyzed by core analysis. Moreover, data are obtained on sections of a drill hole lost by stuck pipe, a blowout, and similar difficulties before an electric log is run. Fluorologs supply information concerning the production possibilities of all strata penetrated and afford an independent check on electric logs, core analyses, and other data.

FLUOROGRAPHIC EXPLORATION

Among the recent developments in the fluorochemical method of locating petroleum is the technique of fluorographic exploration, which depends upon the response of soil samples to ultraviolet light; the samples are collected over the area to be explored, and drilling is not required.

Fluorographic exploration was originated several years prior to 1943 by Blau,[34] and recently procedural improvements have been directed to the assay of soil samples.[35][36]

In fluorographic reconnaissance field work, soil samples are taken at stations a quarter of a mile apart on a regular grid pattern. A reconnaissance survey is adequate for most projects, as it establishes approximately the outline of a favorable prospect and develops considerable detail. If greater detail is desired, the samples may be taken at shorter intervals; a

[34] Blau, W., U.S. Patent 2,337,443. Dec. 21, 1943.
[35] Squires, R. M., U.S. Patent 2,451,883. Oct. 19, 1948.
[36] Stevens, N. M., and Squires, R. M., U.S. Patent 2,451,885, Oct. 19, 1948.

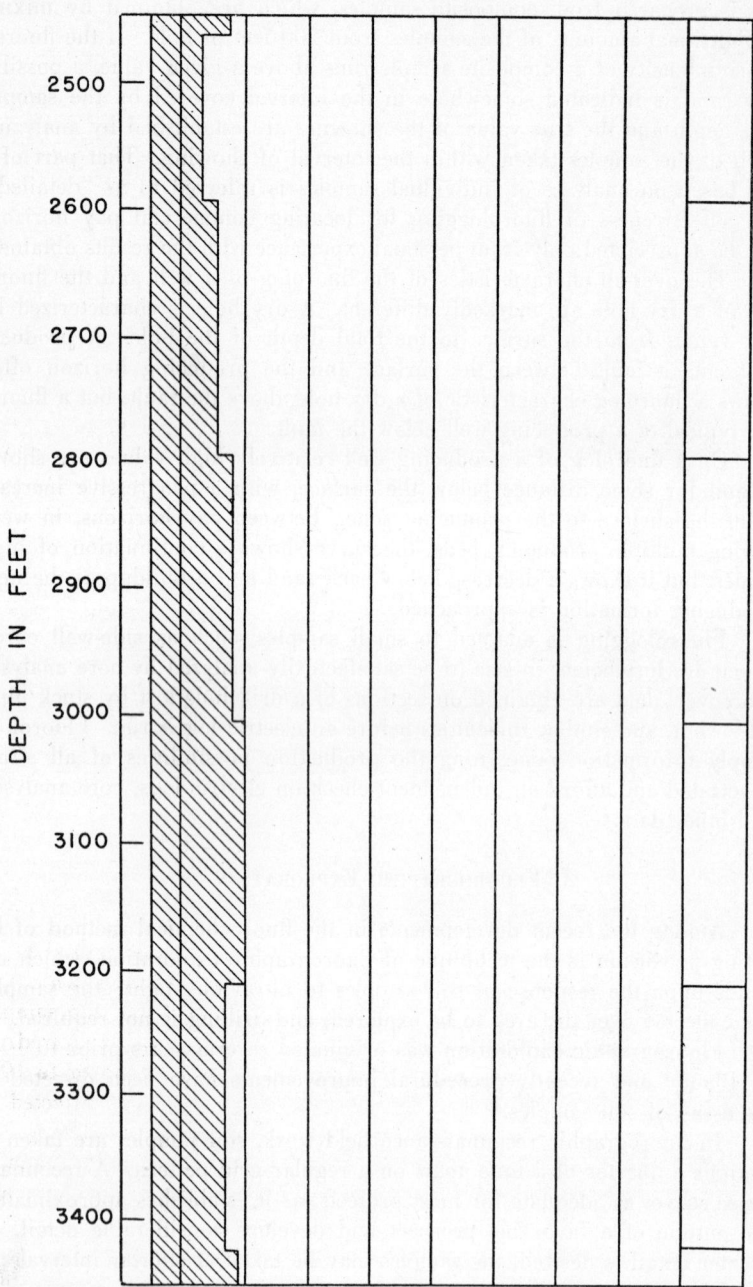

FIGURE 134. Section of a fluorolog, showing manner in which fluorescence intensity of core sample or the like is plotted against depth at which sample was obtained.

minimum area of 25 square miles should be covered in a survey, so that anomalies can be established.

Equally. satisfactory results have been obtained from samples taken in all types of terrain from swamps to sand dunes in all parts of Texas and in parts of Louisiana, Mississippi, Kansas, and New Mexico. Soil samples taken at different times and under different conditions exhibit the same fluorescent characteristics, so that a fluorographic survey can be repeated with equivalent results.

The Blau technique depends upon the fluorescence characteristic of petroleum or its derivatives, the employment of soil standards of known concentration, and the empirical correlation of soil samples with the locality from which they were obtained. The measurement of fluorescence intensity is made by photography, and the interpretation is performed by an experienced oil geologist.

The novel feature of fluorographic exploration is that drilling is not required. The soil samples are collected to a depth of several inches from the surface. Plant material is carefully excluded, so as to avoid contamination. In a given region all samples are taken from uniform depths.

The soil samples as obtained are placed under ultraviolet light and the fluorescence intensity determined by comparison with standards prepared from nonfluorescent soil containing ten, thirty, forty, fifty, and so on parts of oil per million of soil. Visual examination is suitable for samples carrying large amounts of oil.

A fluorographic survey is interpreted by posting the fluorescence-intensity values on the survey grid according to the location from which they were taken and by drawing contour lines, called "isofluors," to correspond to the fluorescence values of the stations. A subsurface accumulation of oil may show isofluoric enclosure. If structure is present, an isofluor map may disclose the geologic pattern of that structure. The isofluor map is effective for locating some faults and minor geologic features, and the details may enable a geologist to differentiate between a structural and a stratigraphic accumulation.

The final interpretation or recommendation is empirical, and valid data are obtained only after a study of a series of maps embracing both the most detailed and the most generalized of information.

SHALE DENSITY ANALYSIS
F. WALKER JOHNSON

Shale density is a criterion of formation evaluation which is sometimes applicable to subsurface geological problems. Shale compaction, a major factor in density variation, has long been recognized by geologists as being an important geological process.[37] Compaction of shale is the result of pressure exerted by the weight of overlying sediments and, in

---

[37] Hedberg, Hollis D., *Gravitational Compaction of Clays and Shales:* Am. Journal of Science, Fifth Series, vol. 31, no. 184, pp. 241-287, 1936.

part, by tectonic movements. Conglomerates, sandstones, limestones, and most chemical precipitates show very little reduction in rock volume as result of gravitational pressures. However, the fine-grained, argillaceous sediments show maximum volume reduction of more than 80 percent. Inasmuch as a large percentage of the sediments of the earth's crust are composed of clays and shales, their compaction is of special interest to the geologist. Reduction of rock volume by compaction is related to the weight of overburden and resultant reduction of porosity, and possibly, to a certain extent, age of the sediments, and tectonics. Compaction therefore results in density increase and porosity decrease.

Investigation by several workers, notably Athy [38] and Hedberg,[39] has shown the relationship of increased density with greater depth of burial.

DEFINITIONS

Several terms used in shale-density analysis work require clarification to form a basis for discussion on this subject. Adopting, in general, the nomenclature of Hedberg,[40] these may be described as follows:

Bulk Density (rock density, lump density) is the density of the thoroughly dry rock, that is, the rock with pore space free of liquids. Hedberg's samples were weighed in air, coated with paraffin and weighed in water. Corrections were made for temperature of water and water content of the sample. The water content was determined from the difference in weight of the powdered sample before and after thorough drying at a temperature of 110 to 120 degrees C. This is also sometimes called "dry density," particularly when applied to naturally dried core samples.

Grain Density (mineral density, absolute density) is the density of the constituent particles of a rock, that is, the rock substance free from pore space.

Natural Density is the density of the rock with all pore space filled with water which is *assumed* to be the usual condition found in nature. A true natural density of a shale sample is, in most cases, essentially impossible to obtain with complete accuracy as there is a certain amount of loss of density due to dissipation of fluid or gases when a core sample is brought to the surface in a well. This is particularly true at the greater depths, where subsurface pressures are very high. However, a result approaching natural density can be obtained with fair accuracy if the sample is weighed immediately after extraction from the core barrel. Some geologists refer to this as "wet density" inasmuch as the true "natural density" can be most nearly obtained from wet core samples while they are still saturated with subsurface fluids.

Apparent Specific Gravity. B. C. Refshauge, in a private report, used the term "apparent specific gravity" as the specific gravity of a porous

[38] Athy, L. F., *Density, Porosity and Compaction of Sedimentary Rocks:* Am. Assoc. Petroleum Geologists Bull., vol. 14, pp. 1-24, 1930.
[39] Hedberg, Hollis D., *op.cit.*, 1936.
[40] Hedberg, Hollis D., *op. cit.*, p. 252, 1936.

body, the pores of which are filled with air. However, the apparent specific gravity of dried core samples, as used in Refshauge's experiments, does not signify oven-dried samples, but those naturally dried in an arid climate. Consequently, they may retain small amounts of water. Density of such samples will, in most cases, average slightly higher than Hedberg's bulk density, but less than his natural density. This term is essentially the same as "dry density," a term commonly applied to naturally dried core samples.

Compaction is expressed as the percentage of reduction of rock volume.

METHOD OF DETERMINATION OF DENSITIES

The most commonly used method of shale-density determination in petroleum work is to weigh the sample in air and then in water. The density is equivalent to the weight in air divided by the difference between the weight in air and the weight in water. The most satisfactory results can be obtained by weighing core samples immediately after extraction from the core barrel. In this manner results approaching the "natural density" of the sediment are obtained. Samples which readily absorb water or show evidence of disintegration from contact with water must be coated before weighing. Ambroid cement or collodion have proved to be satisfactory coatings. However, old X-ray or photographic films dissolved in acetone form an effective coating which may be applied by dipping the samples in the solution. Samples chosen should weigh at least 200 grams in order to obtain the most accurate results. If large-size samples are used, an ordinary laboratory balance may be employed. A very fine wire is attached to one side of the balance for the purpose of suspending the sample in air and water. A Westphall balance can be used for smaller samples.

It is very important that "pure shale" samples be used for density analysis. Every effort should be made to select massive shale free from sand, lignite, secondary mineralization, or other material which might abnormally affect the increase in density brought about by compaction. Experience has shown that very satisfactory results can be obtained if large numbers of samples are carefully selected by simple inspection.

Samples dried in an arid climate and which have been in storage in a laboratory for considerable time will generally require coating before immersion in water. However, successful results have, in many cases, been realized on large, naturally dried samples if the operation is carried out very rapidly. The results obtained are those falling within the range of apparent specific gravity, mentioned above. A series of laboratory checks on samples have shown that they lose their natural fluid content with time so that results obtained by direct weighing of samples in laboratories will approach bulk density. (See fig. 135.)

TIME CURVE OF THE PERCENTAGE CHANGES IN WEIGHT OF CORES FROM WELL W.A.10 ESPERKE

BY DR. W. HAFNER

SAMPLES FROM 418-441 METERS DEPTH GRAYISH BLACK MARLY CLAY FROM THE ALBIAN (LOWER CRETACEOUS)

SAMPLES STORED IN HEATED LABORATORY

———— SAMPLES WITH COATING
-------- SAMPLES WITHOUT COATING

LOSS OF WEIGHT IN PERCENTAGE

SAMPLES GRAY CLAY OR MARLY CLAY

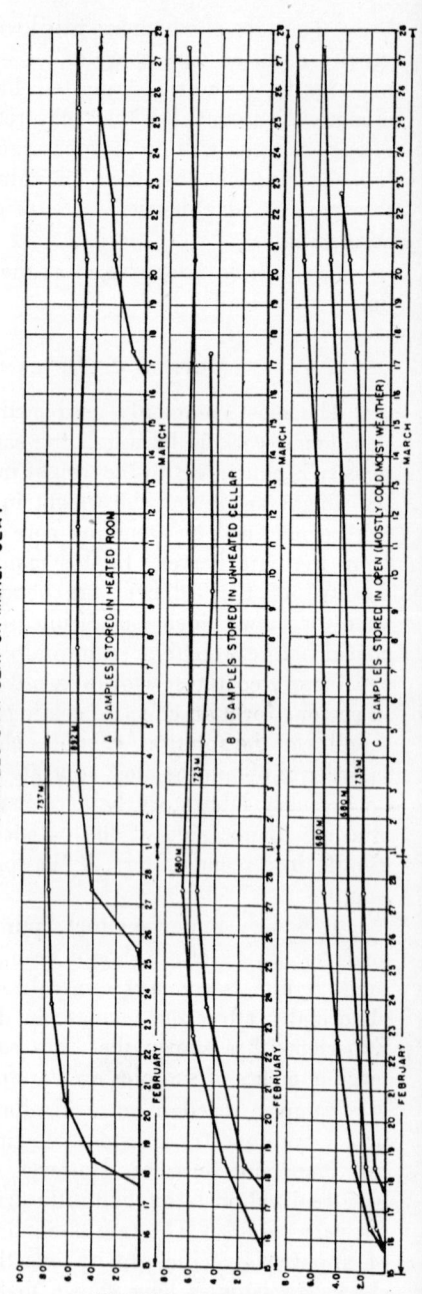

A SAMPLES STORED IN HEATED ROOM

B SAMPLES STORED IN UNHEATED CELLAR

C SAMPLES STORED IN OPEN (MOSTLY COLD MOIST WEATHER)

LOSS OF WEIGHT IN PERCENTAGE

SOURCES OF ERROR

Sources of error in shale-density determinations fall into two categories. These are errors which may be introduced by some inherent conditions of the sample and those acquired during the actual process of weighing the sample.

The inherent conditions of the sample, which result in variations in shale density, may be attributed principally to the effects of weathering and mineralization. Leaching near the surface or at unconformities will result in a decrease in density. Increases in density are often brought

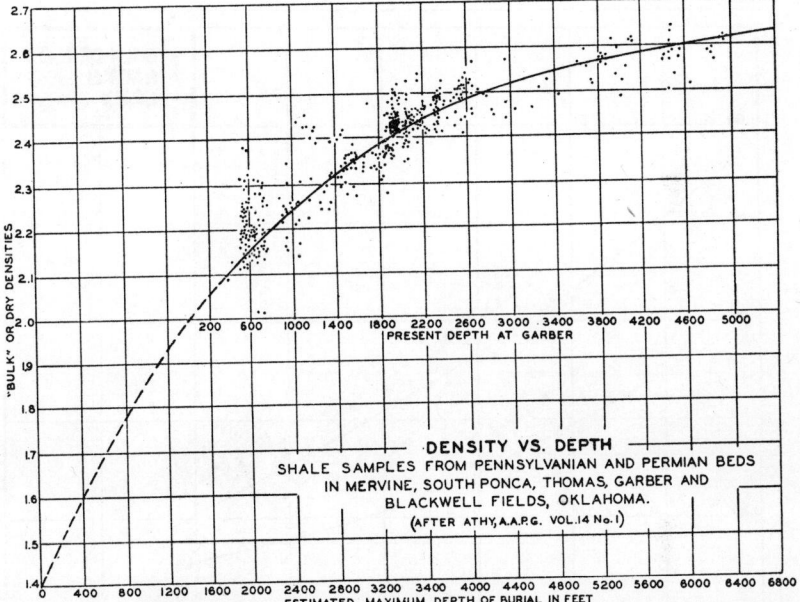

FIGURE 136. Relationship of density to depth of burial. (After Athy. Reproduced permission Am. Assoc. Petroleum Geologists.)

about by secondary mineralization. In some cases pyritization has caused significant increases in rock density.

Several sources of error are associated with the weighing of samples. Size of sample in some cases is an important factor in this respect. Minor errors in weighing of a small sample can have a very important effect upon the specific gravity, whereas the small errors on large samples will have a relatively minor effect upon the ultimate results. Sensitivity of scale used in weighing is another source of error. Water absorption by porous samples can, in some cases, bring about serious error in density determinations.

The effect of the suspending cradle or fine wire and the cellulose coating of the sample may be significant, particularly, in the case of small samples. However, by using samples weighing about 200 grams, Refshauge's work indicates that such errors may be reduced to less than 0.01 of apparent specific gravity in the average sample measured.

APPLICATION OF RESULTS

Shale-density determinations may be applied to the estimation of original depth of burial of sediment and the detection of major unconformities in the section, as evidence of thrust faulting and as indication of

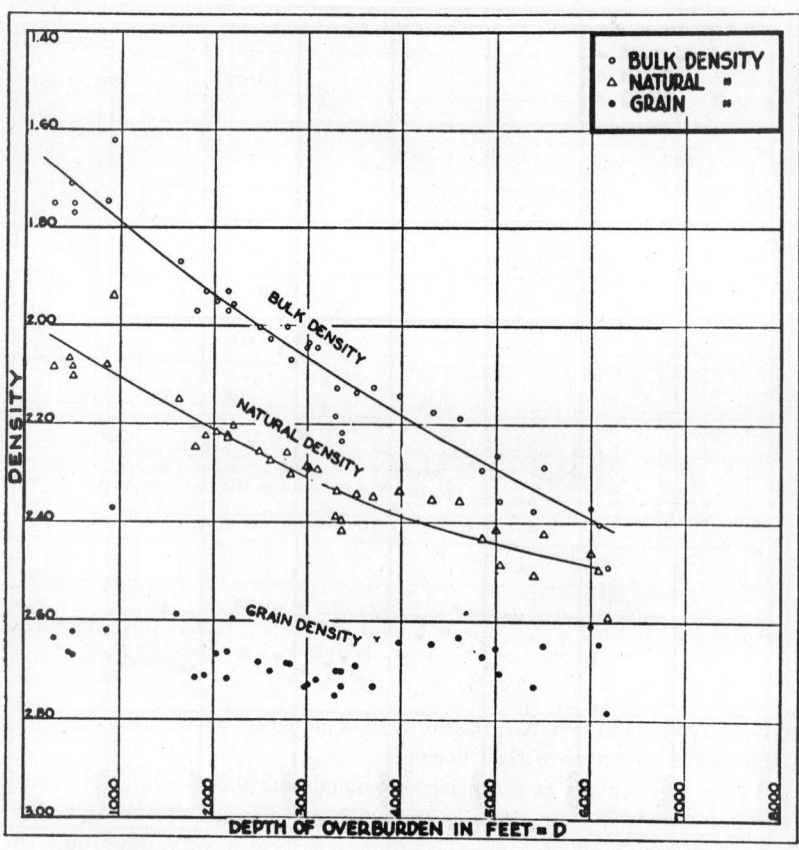

FIGURE 137. Relationship of density to depth of burial in Venezuela.
(From Hedberg, Am. Jour. Sci)

the presence of weathered zones. Such data is also of value in geophysical work, particularly in the interpretation of gravity-meter surveys.

Hedberg has published an excellent summary on gravitational compaction of clays and shales in which he presents charts showing the re-

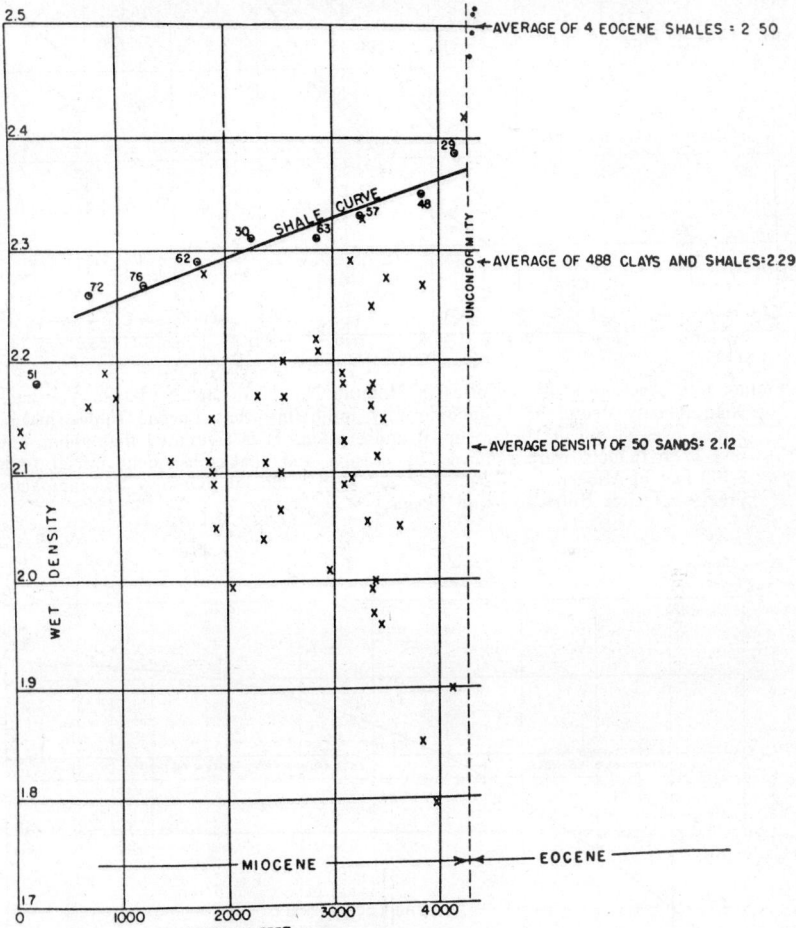

FIGURE 138. Average Miocene and Eocene shale densities in Zamuro No. 1, Falcon, Venezuela. Note density increase at unconformity indicating that Eocene may have been buried much deeper than at present, and that considerable section was removed before deposition of Miocene. (Data by C. C. Fritts, Jr.)

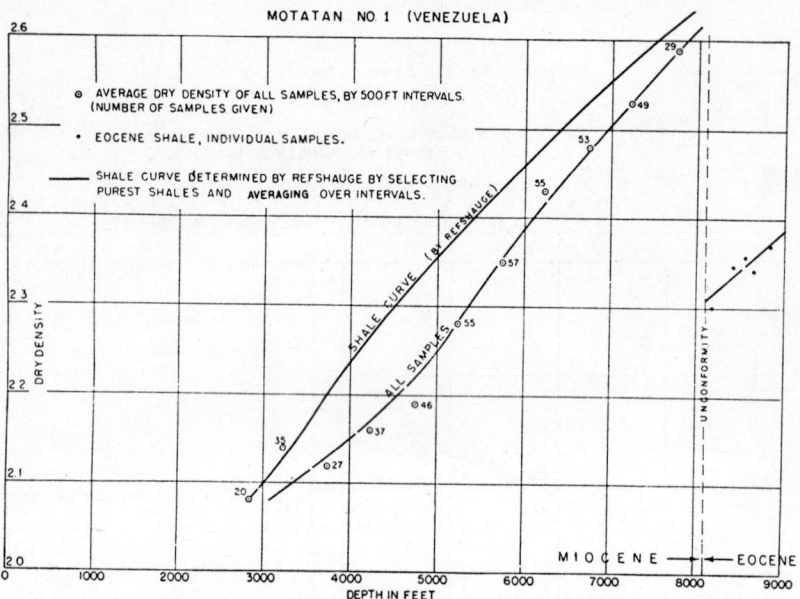

FIGURE 139. Average shale densities in Motatán No. 1, Maracaibo Basin, Venezuela. Note density decrease at unconformity indicating that Eocene shales had attained present density before uplift and erosion. It is estimated that about 4,000 feet of sediments were removed by erosion and that subsequent burial under 8,100 feet of Miocene sediments had not been sufficient to renew the compaction process. (After Refshauge and Skeels.)

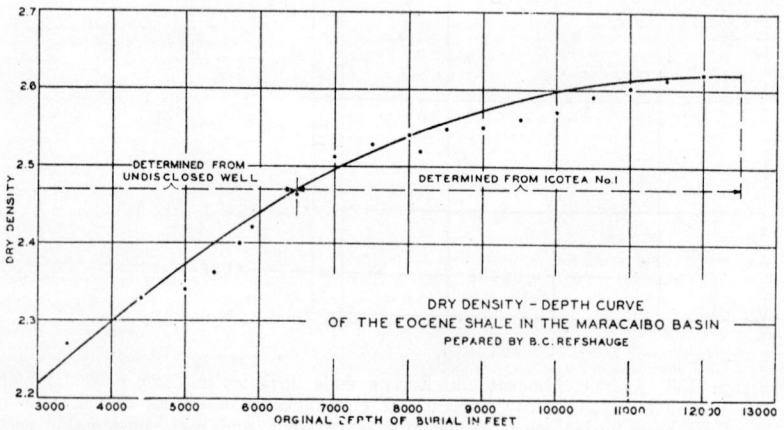

FIGURE 140. Dry densities versus estimated depth of burial in two Eocene wells in Maracaibo Basin, Venezuela. Over 500 samples are represented by these curves. This interpretation conforms to the generalized regional geological data. (After Refshauge and Skeels.)

lationship of shale density and shale porosity to depth of burial.[41] D. C. Skeels, in a private report, has accumulated additional data, some of which is shown on the accompanying figures. However, some geologists believe that age of sediment as well as depth of burial may be a pertinent factor. Therefore, arbitrary determinations of depth of burial from

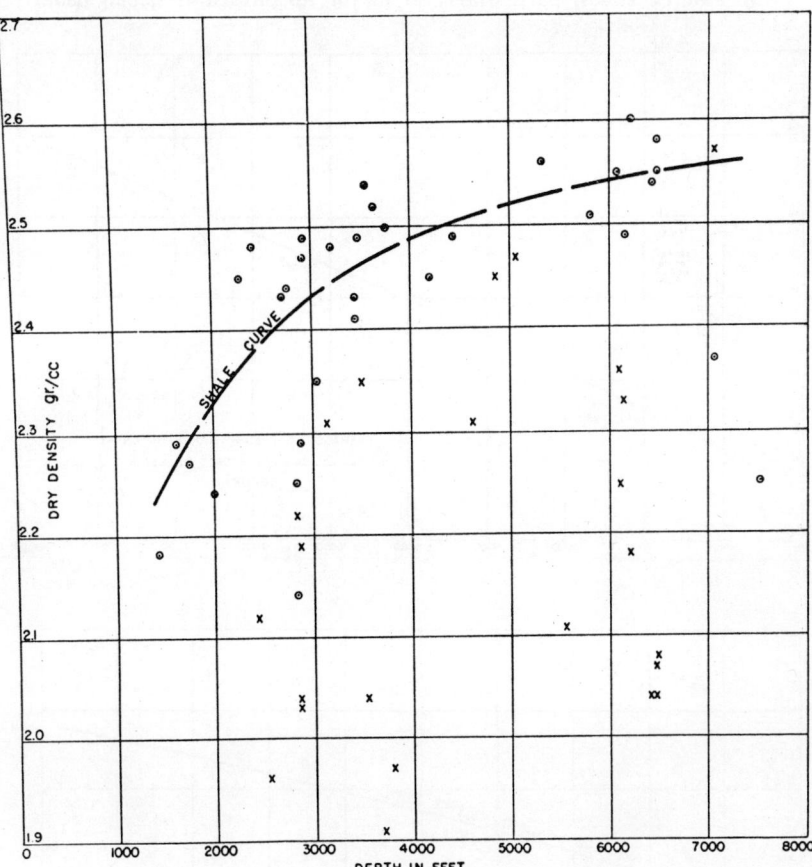

FIGURE 141. Dry densities from well in Colombia. ⊙ = shale samples; ✕ = sands and sandy shales. (Measurements by W. G. Herlithy; reported by C. H. Acheson. After Skeels.)

straight density determinations should be used with care, particularly if the inherent character of the sediment has been effected by secondary mineralization.

Major unconformities can sometimes be distinguished readily by shale-density determination through breaks in density curves. Good examples have been observed in Western Venezuela (See figs. 138 and 139)

---

[41] Hedberg, Hollis D., *op. cit.*

where an abrupt break in the density curve marks the unconformity between the Miocene and Eocene.

Thrust faulting may, in some instances, be detected through careful shale-density studies. This is particularly true in cases where displacement due to thrusting involves several thousand feet of section.

Weathered zones, particularly at major unconformities, can usually

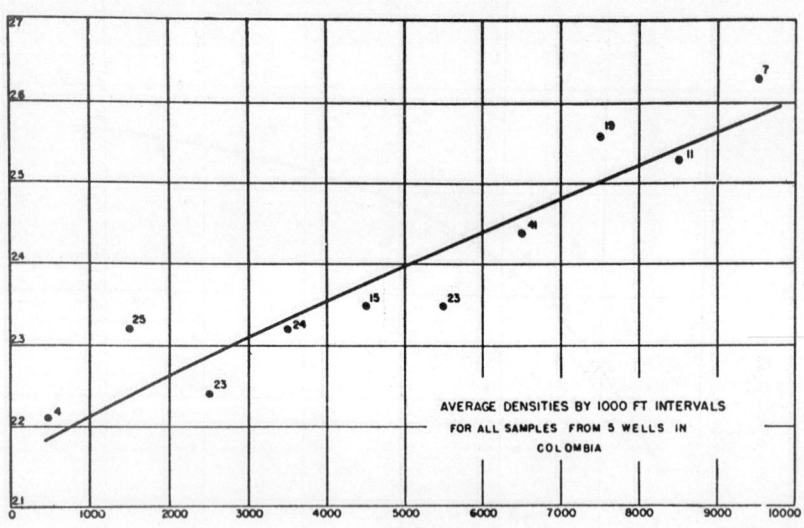

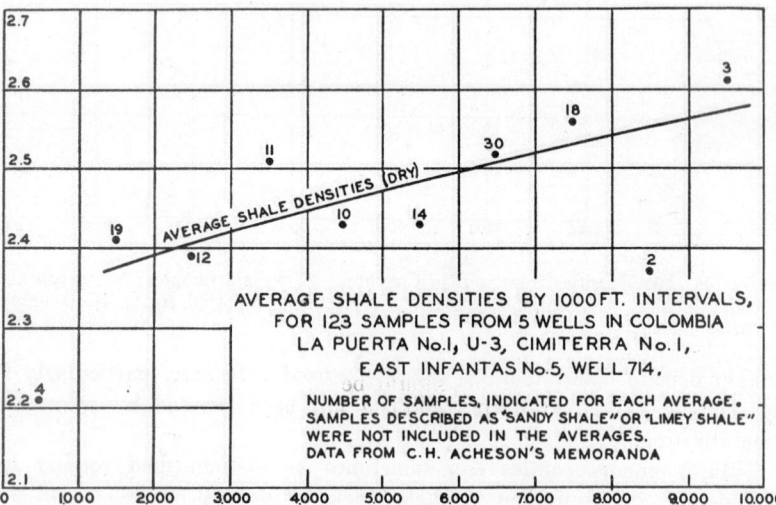

Figure 142. Average shale densities from Colombian wells. Note that densities are higher than averages for all samples. (After Skeels.)

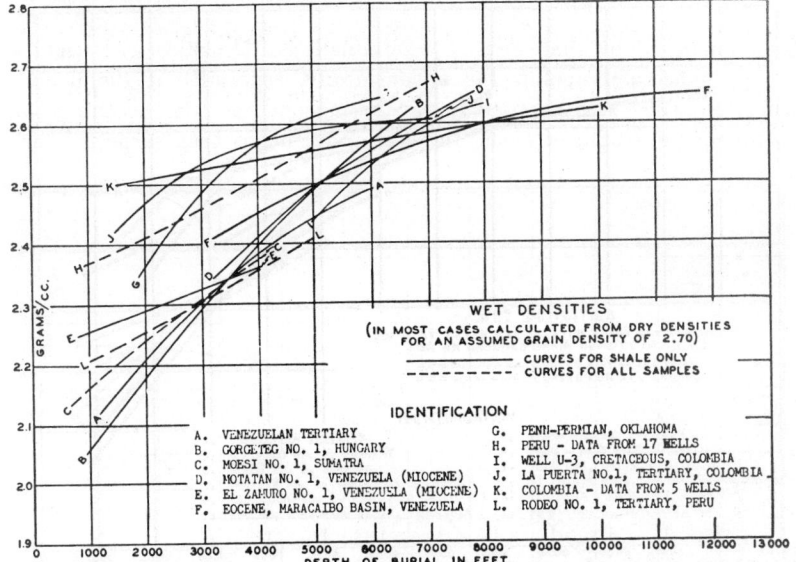

FIGURE 143. Calculated wet density curves. Note that curves *A*, *B*, *C*, *D*, and *I* agree fairly well. Athy's curve (*G*) and La Puerta No. 1 (*J*) fit data much better if curves shifted 2,000 feet and 2,500 feet to right, suggesting greater original depth of burial than now indicated. Curves *K* and *H* fit poorly, probably due to averaging results of several wells. (After Skeels.)

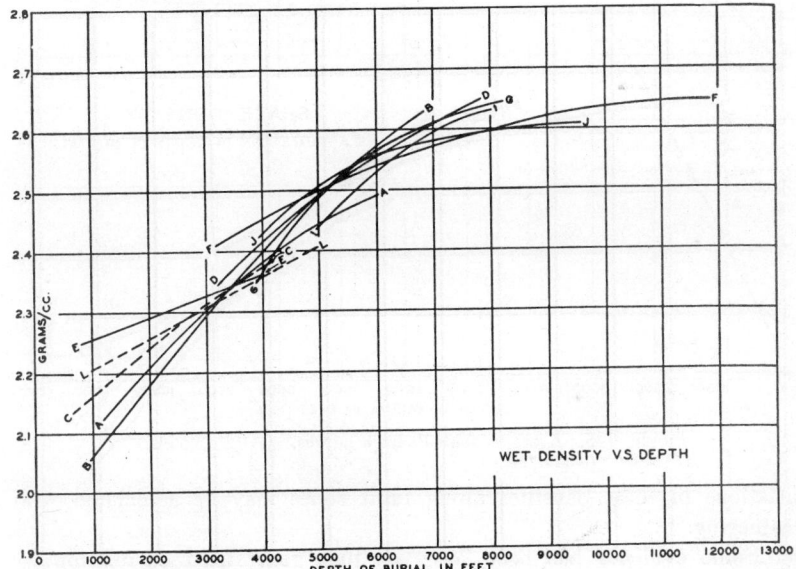

FIGURE 144. Wet density versus depth. Curves *H* and *K* eliminated as non-typical. Curve *G* shifted 2,000 feet and *J* shifted 2,500 feet to right. (After Skeels.)

be distinguished by shale-density determinations. Densities will usually be less immediately below the unconformity, especially if the unconformable surface was exposed to weathering and resulting leaching for an appreciable length of time. In some cases, it has been suspected that

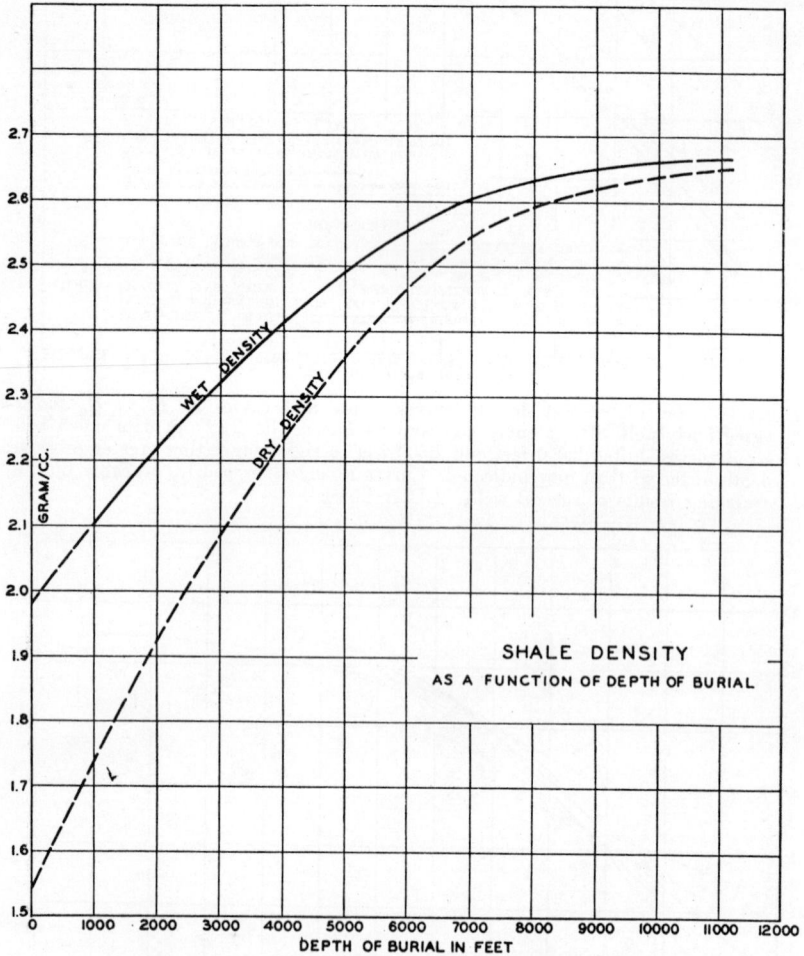

FIGURE 145. Average shale-density curves. (After Skeels.)

variations in shale densities along fault zones may be associated with weathering.

Some evidence has been observed that generalized correlation by shale density may be accomplished in local areas. However, such data must be used with care.

Hedberg [42] has shown the relationship of decrease of porosity with

---

[42] Hedberg, Hollis D., *op. cit.*, p. 263.

increasing pressure as being fundamental in compaction studies. There-fore, density data may be applied to the estimation of porosity in shales (fig. 146).

Shale density adds another valuable tool to geologic investigation, especially when used with other geologic data and interpreted from a broad viewpoint.

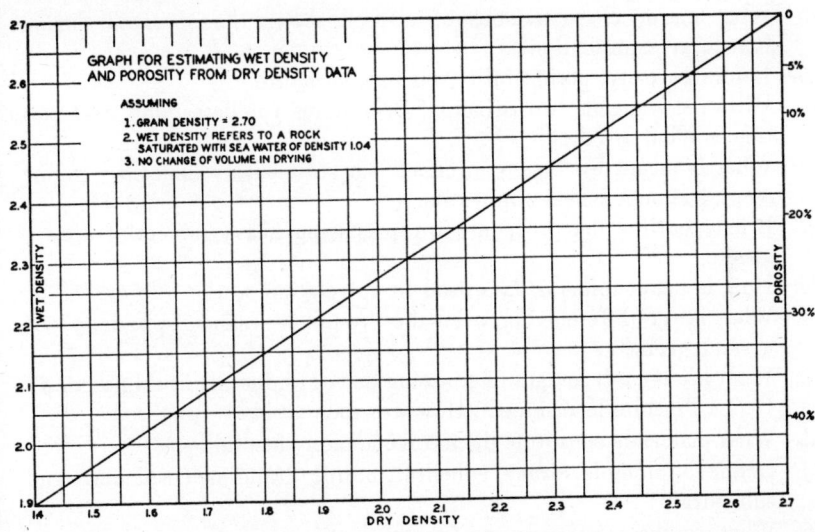

GRAPH FOR ESTIMATING WET DENSITY
AND POROSITY FROM DRY DENSITY DATA

ASSUMING

1. GRAIN DENSITY = 2.70
2. WET DENSITY REFERS TO A ROCK SATURATED WITH SEA WATER OF DENSITY 1.04
3. NO CHANGE OF VOLUME IN DRYING

Figure 146. Density-porosity relationship. Wet density = dry density × .615 + 1.04. This is a poor substitute for the method of obtaining wet densities at the well. (After Skeels, Standard Oil Company of New Jersey.)

Questions

1. In what ways has micropaleontology aided the oil industry?
2. What are Foraminifera, ostracodes, diatoms?
3. The vertical distribution of microfossils and their relative abundance at various stratigraphic positions may be graphically illustrated as shown in figures 37 to 39. How are such charts prepared?
4. What is meant by a "facies log"?
5. May microfaunal assemblages differ along any given time surface? What effect does this have on correlation problems?
6. What morphological features are treated in describing ostracodes?
7. What are calcareous algae, and in what type of environment do they develop?
8. What are conodonts, Radiolaria, and otoliths?
9. Study the flow chart of a medium-sized micropaleontological labora-tory given in figure 50.
10 Should a micropaleontologist in evaluating a stratigraphic section be

342   SUBSURFACE GEOLOGIC METHODS

concerned only with the study of microfaunas? Why?

11. How is detrital mineralogy helpful in establishing subsurface correlations?

12. How are rotary well samples obtained?

13. What is the difference in quality of a rotary and a cable-tool well sample?

14. What are the chief disadvantages of coring?

15. What major characteristics of a rock should be observed during binocular examination?

16. Define porosity, permeability, texture, structure.

17. Hills recommends two principal procedures for describing well cuttings. What are they?

18. What is meant by "heavy minerals" in sedimentary deposits? Name several common heavy minerals.

19. Briefly outline the procedure for preparing heavy mineral fractionates?

20. How is heavy-mineral data used in correlation work?

21. What is the difference between the "roundness" and "sphericity" of a sand grain?

22. What are the advantages of studying detrital minerals in thin section?

23. Define "insoluble residue." How are such residues prepared?

24. What materials constitute the more common "insolubles"?

25. Define: dolomold, drusy, euhedral, oolith, granulated, spicular, and subhedral.

26. What are the primary means for differentiating various cherts?

27. How are insoluble residues used in correlation work?

28. A method of plotting insoluble-residue data is shown in figures 58 and 59. Study.

29. What procedure is followed in preparing a thin section of friable material?

30. Define petrofabrics.

31. Give a map symbol for: plunging syncline, reverse fault, concealed anticline, overturned anticline.

32. What procedure is followed in collecting a sample in the field for petrofabric study?

33. What are the applications of petrofabric studies?

34. How may size analysis be applied in stratigraphic investigations?

35. What is a histogram, simple-frequency curve, and cumulative-frequency curve?

36. How is the coefficient of sorting of a sediment determined, and what is the significance of this value?

37. What staining methods are used for distinguishing calcite from

38. Outline the procedure for preparing a clay sample for stain anlysis.

39. How is kaolinite distinguished from montmorillonite by use of stains?
40. What is the importance of knowing whether a montmorillonite clay or kaolinite clay is involved in construction problems?
41. What factors control the shape of sedimentary grains and fragments?
42. What is the relationship between porosity, grain size and permeability?
43. What magnifications are involved in electron-microscope studies?
44. What are the possible uses of the electron microscope in correlation work?
45. Is there any relationship between subsurface lithologic units and their water characteristics?
46. How are ground waters classified?
47. Study the graphic method of representing ground-water data in figure 120.
48. Compare the composition of Lower Cretaceous waters of the Rocky Mountain Region given on table 11.
49. What are the advantages and limitations of the X-ray diffraction analysis?
50. How is the sample prepared for an X-ray mineral determination?
51. Study the X-ray pattern given on plates 7 and 8.
52. How may X-ray data be applied to stratigraphic investigations?
53. What is the purpose of core analysis?
54. What information is given on a core-log?
55. How should one sample a core from which a detailed analysis is to be made?
56. What factors control the amount of residual oil in a core sample?
57. What is meant by "connate water" and what methods are applied for calculating the connate-water saturation of sand formations?
58. How are core data used for interpreting probable production?
59. What physical characteristics are associated with permeable, high-pressure gas sands?
60. What is meant by "bleeding cores"?
61. What are the applications of core analysis?
62. What is the general principle and procedure of differential-thermal analysis? Could such a technique be applied to certain stratigraphic problems? How?
63. What rock types are best suited for differential thermal studies?
64. What is the value of shale-density data?

CHAPTER 5

SUBSURFACE LOGGING METHODS

SAMPLING AND EXAMINATION OF WELL CUTTINGS

JOHN M. HILLS

Although the techniques described in this paper have been learned during 14 years of subsurface work, the writer does not wish to give the impression that they are original with him. They are rather the results of many years' experience of hundreds of geologists engaged in cuttings sample work, which are summarized especially for the benefit of those entering the profession or setting up subsurface departments in other areas.

Well cuttings are the source of most subsurface data obtained in the Mid-Continent and Permian Basin areas of the United States. The collection and examination of samples of these well cuttings are highly organized and important techniques. In fact, in most holes in this province, well cuttings are the only reliable source of data concerning the formations penetrated. Logs made by experienced cable-tool drillers are very useful, but under present conditions these are rare. It is usual to find the driller's log made by an inexperienced or uninterested rotary driller. Such a log gives only a general idea of the formations drilled, especially in areas where there is a distinct formation change between wells.

Of all the methods of obtaining information concerning rocks cut by the drill, cores are probably the most reliable. In soft unconsolidated formations these cores can be taken rapidly and with comparatively little expense, by means of a wire-line core barrel. However, in the Mid-Continent and Permian Basin areas most of the rocks are well lithified, and many are extremely hard. For this reason wire-line core bits wear out rapidly, and cores are usually obtained by use of conventional core barrels which are capable of cutting a maximum of 20 feet at one time. Naturally this makes complete coring extremely expensive, especially in deep holes. Recently a new technique of coring with diamond-studded core heads has been developed which may lead to much more extensive coring. It is possible that in the future as much as 100 feet of core may be taken at one time by means of a diamond bit without coming out of the hole. This may render practicable coring of complete sections of deep holes.

The writer is especially indebted to W. D. Anderson, who taught him the fundamentals of sample examination; to William Y. Penn for the pictures of porosity; to E. Russell Lloyd, who suggested the writing of this paper and has been generous with help and suggestions; and to John Emery Adams and W. W. West, who kindly read and criticized the manuscript.

ELECTRICAL LOGS AND RADIOACTIVITY LOGS

Electrical-resistivity and self-potential logs have almost supplanted cutting samples and cores in areas where the stratigraphic section is shale and sand and where salt beds do not contaminate the drilling mud. However, in areas where the section is composed of several kinds of rock, especially where limestone and anhydrite are abundant, electrical logs must be used with sample logs since the lithologic variety and high resistance of the rocks introduce many unknowns into the interpretation of the log. Thus, it is impossible to solve the problem without a sample log which will enable one to eliminate many of the possible solutions.

In limestone reservoirs it is very difficult to recognize the fluid content of the reservoir rock by use of electrical logs without additional information from testing the well, because the high resistivity of the limestone and the sulphur water commonly present mask the resistivity effects of the other fluids in the formation. However, electrical logs are useful for correlation in local limestone reservoirs where the section is well controlled by logs from sample cuttings. One fact that should be remembered in using electrical logs is that rock-salt beds cut by the bit cause a salty mud with a very low electrical resistance that results in a featureless self-potential curve that is useless in making correlations.

Radioactivity logs are also useful as an auxiliary to sample logs. Here, also, a wide variety of rock and fluids in the stratigraphic column results in several possible interpretations for the curve, and a sample log is necessary for the correct solution. Radioactivity logs are sometimes very useful in logging old wells that have already been cased off or wells where mechanical difficulties have prevented taking a representative set of cuttings. Radioactivity neutron logs are probably more useful than resistivity logs in indicating a fluid zone, but it is not possible from the neutron log alone to determine whether the fluid content of the formation is oil or water.

Thus, in areas where consolidated formations are encountered, examination of well-cutting samples is generally the easiest and least expensive method of detecting changes in the formation and determining the stratigraphic section penetrated by any well. Much subsurface geologic work depends on the collection of representative samples of the formation penetrated, describing these samples accurately, and plotting the description so that the sections in different wells may be correlated.

CABLE-TOOL SAMPLES

The collection of cuttings samples from cable-tool wells presents comparatively few difficulties. The samples should be collected from the first bailer after each run of the bit, in a bucket hung at the end of the dump box. These samples should be washed enough to carry off all mud. They should be put in cloth sacks, then labeled with the name of the company, name of the farm, number of the well location, and depth of sample. The hole is, of course, bailed clean each run of the bit, and only occasionally are cavings a problem in consolidated formations.

In drilling bentonitic shales, quicksands, or conglomerate, the hole ordinarily caves readily so that the samples are not entirely reliable. When high-pressure gas is encountered, the hole is commonly filled with water or light mud to control the gas. This ordinarily results in samples being finely ground and much material being washed off the upper parts of the hole, an action which contaminates the samples. The same condition results from drilling ahead in a hole full of water coming from the formation. Generally, under these conditions, pipe is run before much hole is made so that the number of poor samples is comparatively small. Where wire line or tools have been lost in the hole and must be drilled up, some samples contain large amounts of iron. This can be removed by means of a magnet and the residue of the sample examined. Surface rock and other materials sometimes are thrown into the hole in attempting to straighten it.

The problem of checking the depth at which samples are taken is not critical with cable tools, since the depth is checked by sand line on the bailer after each run. However, important datum beds in the section should be checked for depth by stringing in the sand line or by running a steel measuring line. Cable-tool samples differ from rotary samples by the general flaky character of the harder formations and by the polishing and rounding of many cuttings from attrition due to turbulence set up by the bit, and the irregular intervals at which they are taken, since the length of bit runs are determined by the character of the formation and mechanical factors.

Rotary Samples

The collection of rotary samples presents many more difficulties than the collection of cable-tool samples and for many years it was considered impossible to obtain reliable samples by this method of drilling. It is still difficult to obtain representative samples from rotary holes in unconsolidated formations, but a technique has been worked out whereby representative samples can be obtained in consolidated formations. These samples are taken from the returning fluid stream at regular intervals. The sample interval is usually 5 or 10 feet, but may be as small as 1 foot or as large as 30 feet. The chief problems in collecting samples of rotary cuttings are to prevent contamination with upper beds, prevent powdering of the sample, prevent loss of the sample, prevent elutriation, obtain correct depth measurement, and wash and properly dry the samples.

Contamination from Upper Beds

Contamination from upper beds is the chief problem of collecting rotary samples. In the early days of rotary drilling before mud building was fully understood, the walls of the hole were poorly plastered and unstable. Thus, the samples contained large amounts of extraneous material. The practice of considering any new material appearing in the samples as composing the entire content of the formation drilled then became general and any other material was believed to be cavings. At present, however, the treatment of rotary muds has advanced so that usually a properly mudded rotary hole caves very little and the percentage content of the sample may be taken as representative of the types of rock in the interval covered. This enables the geologist to follow very closely the lateral gradations in the section.

It should be emphasized that the geologist must work with the drilling superintendent to insure the maintenance of proper mud in the hole not only for the sake of drilling progress but also for the securing of representative cuttings samples. The mud must have a gel strength and viscosity sufficient to bring the cuttings to the surface without recirculation and regrinding. It must be also capable of cushioning the impact of the drill pipe against the walls so as to prevent powdering of the samples. Once on the surface, the mud should be run through pits sufficiently large to insure settling of all the cuttings so that they will not recirculate. A shale shaker will insure complete separation of the larger cuttings and the mud.

POWDERING OF CUTTINGS

The powdering of the sample to a size too small to be examined effectively under the binocular microscope is largely due to poor mud which allows regrinding of the sample by the bit and powdering by whipping of the drill pipe against the walls of the hole. When extraordinarily large cuttings are needed for analysis of porosity and permeability or for other purposes, it is helpful to circulate in reverse of the usual manner, that is, to pump the mud down between the casing and drill pipe and up through the drill pipe. This process results in higher mud velocities returning through the drill pipe which brings larger cuttings from the bottom as soon as they are chipped off by the teeth of the rock bit. Some of these cuttings are $\frac{1}{2}$ to $\frac{3}{4}$ inch across.

Reverse circulation is especially useful when drilling into low-pressure formations, by using oil as drilling fluid, as it prevents any of the fine oil-borne cuttings from plugging the pores of the formation. In regular circulation with oil, the low viscosity of the oil will not carry out the cuttings, and they are commonly reground to a fine putty-like mass which is useless for examination and has a plugging effect on low-pressure pays. Of course, reverse circulation is not necessary in high-pressure gas or oil pays since the high pressures tend to increase the velocity of the circulating fluids and carry the cuttings out of the hole. The advantages of using a water-free drilling fluid in low-pressure oil pays are combined with the advantages of a high-viscosity mud in carrying out the cuttings and cleaning the formation in oil-base muds recently developed. With these muds it is not necessary to use reverse circulation to obtain cuttings

348   SUBSURFACE GEOLOGIC METHODS

large enough for visual examination. However, since methods have been recently developed for determining porosity and permeability from extra-large cuttings obtained by reverse circulation, this method undoubtedly still will be used to some extent in pay sections.

LOSS OF CUTTINGS

Loss of cuttings samples is due to two chief causes: lost circulation, and blow-outs. Both of these are primarily mud problems. Lost circulation is caused either by too high water loss in the mud or by excessively heavy mud which overcomes the formation pressure of a porous or fractured bed so that the mud enters the pores or fractures. The cuttings are then carried into the porous beds and may not ever be recovered unless the well is completed in this zone and the cuttings come out later with the oil. Blow-outs are caused by unexpectedly high formation pressures or by carelessness in handling the mud. The cuttings are blown out of the hole and not recovered.

ELUTRIATION

Elutriation or separation of the coarse from the finer part of the sample by the upward movement of the circulating fluid is due to the use of mud with low viscosity and gel strength. In good mud the cuttings are held in suspension and there is little change in the relative position.

METHODS OF OBTAINING CORRECT DEPTHS

Obtaining proper depth measurement for rotary cuttings is another major problem. The depth of the well must be checked often either by steel measuring line or by measuring the drill pipe under tension. Attention must be given constantly to see that the crew catches the samples at the proper intervals and that they do not anticipate the sample by filling several sacks at one time.

A great help in assuring correct sample depth is to have the driller keep a record of the drilling time on a form, such as table 18. Each interval, 1, 5, or 10 feet, depending on the importance of the section, is shown on this sheet with the time drilling began, the time it ended, the time taken out for mechanical work, and the net time drilling. These intervals are usually marked on the kelly with grease or chalk and checked by the pipe measurement each time the kelly is drilled down to the derrick floor. In addition to assuring the correct depth of the samples, the drilling time gives a very valuable clue to the nature of formations penetrated as the time of drilling has a very close correlation with the lithologic character of the formation. This drilling time also may be taken with a mechanical device.

In pay sections where the exact measurements are very important or in deep holes where the cuttings take a long time to come to the surface, samples should be labelled with the depth at which they are actually cut rather than the depth of the well at the time they come to surface. This measurement is accomplished by placing some easily identifiable substance, such as rice or corn in the drill pipe at the derrick floor when making a connection, and measuring the time required to bring the substance around to the shale shaker or return pipe. If regular circulation is being used, it can then be calculated from the pump pressure and volume handled how long the mud requires to go from the derrick floor through the drill pipe to the bit, and this subtracted from the total return time gives the time necessary for the samples to come from the bit to the surface.

TABLE 18
EXAMPLE OF DRILLING-TIME FORM

Company.................................... No.................................... Farm..
Location............................ County............................ State..
Weight on drill pipe................ Mud weight................ R.P.M................ Mud Viscosity................

| Depth | | Drilling Time | | Actual | |
From	To	Began	Ended	Time	Time Out—Remarks

Actual time is time spent in drilling. Shut-down time, round trips. changing bit, repairs, etc., also condition and type of bit should be noted under "remarks."

Unless it is known that the hole is in extraordinarily good shape with no washed-out places, it is not satisfactory to calculate the return time from the bit by mud volume and velocity, because some eddying and consequent lowering of the mud velocity takes place in all washed-out places. If it is not possible to determine the sample return time from experimental methods of calculation, a rule of thumb is that under ordinary mud pressure with 7- or 8-inch hole, it will take cuttings about 10 minutes per 1,000 feet to return from bottom.

After the return time of the sample is determined or estimated, each sample should be caught that length of time after the appropriate depth mark on the kelly reaches the rotary table. An easy way to do this is to place on the drilling-time sheet the time each sample should be caught. This is done by adding the return time in minutes to the time at which the given mark is down to the rotary table. This method will eliminate lag in the samples and should make the sample log correlate very closely with the drilling time and with the electrical log.

CATCHING SAMPLES

There are many ways to catch representative samples from the returning mud stream. The rotary shale shaker of the Thompson type has become very common in recent years. This is a large cylindrical screen through which the returning mud stream passes. The screen is turned by a water wheel moved by the mud stream. Attached to the large screen is a much smaller screen with fine mesh through which a portion of the main mud stream is diverted, off which comes a small portion of the cuttings which is collected in a box at the end of the screen for visual examination. This type of screen has the advantage that it requires no outside power to operate it, and it catches a representative sample of the cuttings without any further attention from the operator or crew. However, unless a very fine screen is used on the sample catcher, it will not catch all the fine sands. In some formations the mud will not wash out of the cuttings without use of an excessive amount of water which renders it unsuitable for use with low water loss muds.

Another important type of screen commonly used is the vibratory shaker. In this device the mud stream passes across a vibrating screen. The mud passes through the screen into the pits while the cuttings are vibrating off into another receptacle. In consolidated formations good samples can be taken by placing a narrow box under a part of the end of the vibrating screen so that a representative portion will fall into this box. However, this screen is open to the same objection as the Thompson machine in formations consisting of fine sands, since the fine sand tends to pass through the screen and not be caught in the sample.

The simplest and possibly most reliable sample-catching device (fig. 147) consists of a small-diameter nipple welded to the bottom of the mud-return line with a 1½-inch or 2-inch line running to a box 1 foot by 1 foot by 3 feet, with a removable gate about 6 inches high in the end. In this box a representative portion of the cuttings is collected and may be shoveled out at the appropriate time into a bucket and the box cleaned by removing the end gate and letting the mud stream carry out the remaining cuttings. When these are washed away, the gate is replaced and the collection of the next sample is begun. By this means a representative portion of both the fine and coarse parts of the cuttings is obtained. Sometimes such a box is set in the course of the main mud stream, but so many cuttings accumulate that the space behind the gate is filled with cuttings before drilling of the sample interval is completed so that the last part of the interval is not represented by the cuttings in the box.

After the samples are caught, they must be washed properly. This can be done in a bucket by filling it partly full of water and stirring the samples vigorously, letting the clean part settle and decanting the fluid sev-

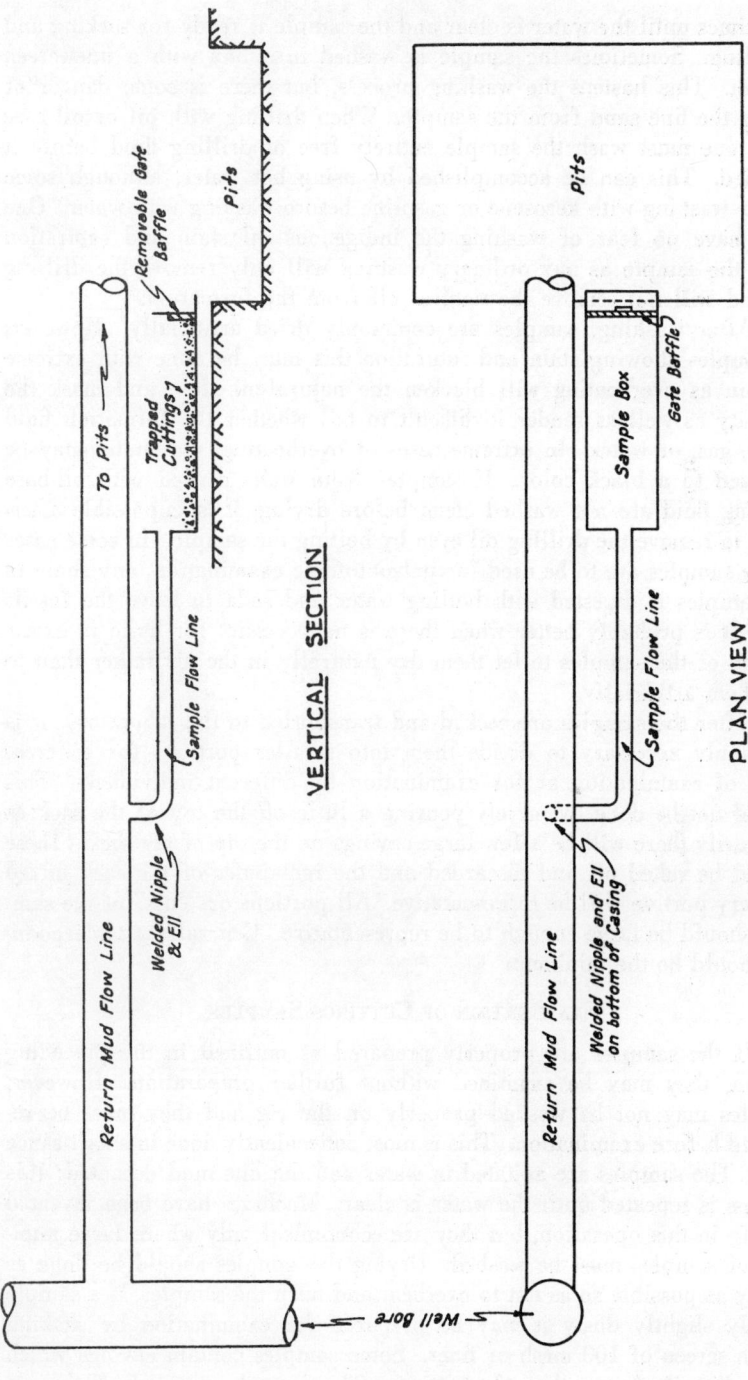

FIGURE 147. Diagram of sample-catching device on a well. Sample-catching arrangements vary considerably. It is also common practice to secure samples from a shaker screen placed in the return mud-flow line.

eral times until the water is clear and the sample is ready for sacking and labelling. Sometimes the sample is washed in a box with a fine-screen bottom. This hastens the washing process, but there is some danger of losing the fine sand from the sample. When drilling with oil or oil-base mud, one must wash the sample entirely free of drilling fluid before it is dried. This can be accomplished by using hot water, although some prefer washing with kerosene or gasoline before washing with water. One need have no fear of washing the indigenous oil stain and saturation from the sample as any ordinary washing will only remove the drilling oil and will not remove the natural oil from the formation.

After washing, samples are commonly dried artificially. However, in samples showing stain and saturation this must be done with extreme caution as overheating will blacken the natural-oil stain and mask the porosity as well as render it difficult to tell whether the formation fluid is oil, gas, or water. In extreme cases of overheating, red shales may be oxidized to a black color. If samples from wells drilled with oil-base drilling fluid are not washed clean before drying, it is impossible afterward to remove the drilling oil even by boiling the sample. In some cases where samples are to be used for paleontologic examination, any shale in the samples is digested with boiling water and soda to leave the fossils free. It is probably better when there is no necessity for haste in examination of the samples to let them dry naturally in the air rather than to dry them artificially.

After the samples are sacked and transported to the laboratory, it is commonly necessary to divide them into smaller portions for different types of examination or for examination by different individuals. This should not be done by merely pouring a little off the top of the sack as ordinarily there will be a few large cavings on the top of the sack. These should be raked off and discarded and the remainder of the sack mixed so every portion will be representative. All portions or "cuts" of the samples should be large enough to be representative. Generally a tablespoonful should be the minimum.

EXAMINATION OF CUTTINGS SAMPLES

If the samples are properly prepared as outlined in the preceding section, they may be examined without further preparation. However, samples may not be washed properly on the rig and they must be rewashed before examination. This is most conveniently done in small sauce pans. The samples are agitated in water and the fine mud decanted; this process is repeated until the water is clear. Machines have been invented to help in this operation, but they are economical only where large numbers of samples must be washed. Drying the samples should be done as slowly as possible so as not to overheat and burn the sample. If a sample is only slightly dusty it may be prepared for examination by shaking over a screen of 100 mesh or finer. Some samples contain cavings which are ordinarily larger than the cuttings. These can be scraped off the top after agitation, and the residue of the sample examined. The same process helps to remove gel flake and other plugging materials added to the mud to control lost circulation. Iron particles in the samples can be removed by a magnet.

Samples may be examined while covered with water. This method brings out by differential refraction some of the qualities of the sample such as oölitic structures and anhydrite crystals which may be overlooked in dry samples. When examining the sample wet, it is not necessary to have them so clean as when examining them dry since a certain amount of washing may be done in the water in which the sample is examined. Dry examination is more convenient since it is not necessary to have a number of petri trays or watch glasses with which to examine the samples in water.

Cuttings samples ordinarily are examined under a binocular microscope of low power. The exact power used varies with the individual geologist and with the nature of the samples. The power should be high enough to see the essential structure and texture of the cuttings. Porosity coarse enough to have a permeability sufficient to make commercial oil production should be seen easily. Oölites, inclusions, and other sedimentary features should be clearly visible also. On the other hand, the microscope power should be low enough to permit examination of large numbers of samples without eye strain and should permit a field of view large enough for estimating percentages of the various constituents accurately with only a few changes in position of the sample. The powers that meet these requirements vary between 12 and 24. Higher powers are used for special purposes such as the description of minor features of microscopic fossils. For general use, however, in the oil industry it is not desirable to have a power higher than 24, since a very high power may give false impressions about the permeability and porosity of cuttings.

The illumination of the microscope field is important. In the past, small incandescent lights which may be focused to give intense illumination have been used widely. However, with the development of the fluorescent lighting tube, it has been found that the white light given by this tube is much superior to the yellowish light of the incandescent bulb. This is especially true when searching for oil stain in the cuttings, since the yellowish incandescent light commonly masks the light-brown oil stain. Usually it is not practicable to focus the fluorescent tube so as to give intense light for high-power work. However, for the usual low-power examination, the fluorescent tube is very satisfactory. In examining samples for oil stain and saturation it may be desirable to use an ultraviolet light.

In examining well cuttings samples the geologist should endeavor to estimate as closely as possible the proportion of each rock type in the sample and to describe the lithologic characteristics of the rock that are essential to the correlation of the beds. In commercial work it is not practicable to give the analytical detail that may be desirable in purely scientific research. The geologist, nevertheless, should endeavor to make the description as complete as possible in the time available. Necessarily, the description should be more detailed on wildcat wells than on field wells and much more detailed in the pay sections than in the upper sections. In field wells it is permissible to omit descriptions of long sections of comparatively insignificant beds between key beds. The geologist should remember that it is much better to have too full a description than too meager a description. Experience has shown that full descriptions have saved much time and money, whereas meager descriptions have resulted in redescription of the cuttings several times. Nevertheless, some redescription can not be avoided; therefore, samples should be carefully preserved and indexed so that they will be available for future study.

There are two principal ways of describing samples. The first of these is the interpretative system, in which the geologist picks out the cuttings which he believes to be representative of the formation penetrated and describes the entire sample as composed of this rock. The rest of the sample is assumed to be cavings. This kind of description brings out formational changes and is of greatest value in areas where the various formations are of wide extent and relatively constant character, as in the Paleozoic of the Mid-Continent region. In areas of rapid lateral gradation in the lithologic character of formations, as in the Permian Basin of west Texas, this method results in masking of lateral variations and misinterpretation of the nature of the stratigraphic column. This, of course, results in miscorrelation of the well logs.

In regions of pronounced lateral gradation it has been found that a second method of sample description is most satisfactory. This is the percentage description, where the geologist describes all material in the sample, disregarding obvious foreign substances and cavings. This system, though making it difficult to determine formational boundaries from the sample log, shows the gradations of the beds and often enables one to trace a horizon through different sedimentary facies.

Percentage of various constituents is estimated by the eye. It has been found that experienced sample examiners, using good samples, will agree very closely on the percentages of constituents. However, with poor samples where some judgment must be exercised in disregarding obvious cavings, the value of the description is dependent on the judgment of the geologist who examines the samples. Mechanical counters or calculators, such as the integrating stage, have not been found to give enough additional accuracy to be worthwhile in ordinary work. Experienced sample examiners can describe 100 to 300 samples in an 8-hour day, according to the nature of the cuttings.[1]

The descriptions may be written out or plotted directly on a log strip. For many purposes, it is desirable to make written descriptions,

---

[1] A full description of sample examination as applied to stratigraphic work is given by L. H. Lukert, Oil and Gas Jour., pp. 49-51, June 1937.

TABLE 19
EXAMPLE OF SAMPLE-DESCRIPTION FORM

Operator............................. No............. Lease.......................... County...............................

Section ... Block ... League ... Survey
Labor........... Twp................ Range........................ Footage..
Elevation................. T.D................ Date......................... Examined by...........................

Depth		Lime or Dolomite	Anhy.				Non Red Shale	Red Shale	Red ss.	Gray ss.		Chert
From	To		%	Lith.			%	Lith.				

since the logs can be plotted in several places at the same time and a record is kept which is not subject to destructive wear as is the plotted log which is used constantly in the field and laboratory. A type of description form is included (table 19). In addition to the constituents shown on this form, others may be specified according to the nature of stratigraphic sections in the area worked. Pyrite and differently colored shales will be the most likely additions. In areas where sand production is important, separate columns for porosity and saturation in sandstone may be desirable.

Color of the rock fragments in cuttings samples is a very important attribute. At present there is a wide variation in the descriptions of color by subsurface geologists. The same rock may be described as tan by one and brown or gray by another. It is desirable in any organization attempting to start sample examination work to standardize some color scheme, possibly that being developed by the inter-society committee.[2]

Size of particles in dolomites and limestones is another matter on which subsurface geologists vary. The following table gives the tentative scale of crystal sizes which was developed by the West Texas Geological Society several years ago. It would be to the advantage of anyone undertaking sample description for the first time to adopt a scale with appropriate abbreviations for the sample-description sheets.[3]

---

[2] DeFord, R. K., *Rock Color Chart for Field Geologists:* Am. Assoc. Petroleum Geologists Bull., vol. 31, no. 10, pp. 1903-1904, 1947.
[3] A more elaborate scale is advocated by Ronald K. DeFord, *Grain Size in Carbonate Rocks:* Am. Assoc. Petroleum Geologists Bull., vol. 30, no. 10, pp. 1921-1928, 1946.

TABLE 20

TENTATIVE SCALE OF CRYSTAL SIZES IN DOLOMITES AND LIMESTONES

Descriptive Adjectives	Crystal Diameter
1. Mat	Invisible
2. Microcrystalline	Less than 10 mm.
a. Cryptocrystalline	Less than 0.02 mm.
b. Finely crystalline	0.02 to 0.1 mm.
c. Mediocrystalline	0.1 to 2.0 mm.
d. Coarsely crystalline	2.0 to 10 mm.
3. Megacrystalline	

DEFINITIONS

Mat.—Compact, exceptionally homogeneous; having a dull but even surface under low binocular microscope; resembling limestones used in lithography; as, a *mat* limestone, a *mat* dolomite.
Microcrystalline.—Having crystals less than 10 mm. long; having crystals small enough to be viewed under low-power binocular microscope.
Cryptocrystalline.—Indistinctly crystalline; showing very small, indistinct crystal faces; composed of crystals too small to be measured under low-power binocular microscope.
Megacrystalline.—Having crystals 10 mm. or more in length; having crystals too large to be readily discernible under low-power binocular microscope.

POROSITY, PERMEABILITY, AND OIL STAIN

Description of the porosity, permeability, and oil stain of cuttings samples is one of the most important parts of the sample examiner's work. In sandstone, the porosity is determined by the size of the grain, the sorting, and the amount of the cement present. The grain size should be described by some standard scale such as Wentworth's [4] or Alling's.[5]

FIGURE 148. Isolated pin-point porosity. Reverse circulated cuttings. x5. Penrose's University 2, sec. 3, blk. 10, University lands, Andrews County, Texas; 4410-4415 feet.

[4] Wentworth, C. K., *A Scale of Grade and Class Terms for Clastic Sediments:* Jour. Geol., vol. 30, pp. 377-392, 1922.
[5] Alling, H. L., *A Metric Grade Scale for Sedimentary Rocks:* Jour. Geol., vol. 51, pp. 259-269, 1943.

FIGURE 149. Granular dolomite with good intermediate porosity. Reverse circulated cuttings. x10. Penrose's University 2, sec. 3, blk. 10, University lands, Andrews County, Texas; 4465-4470 feet.

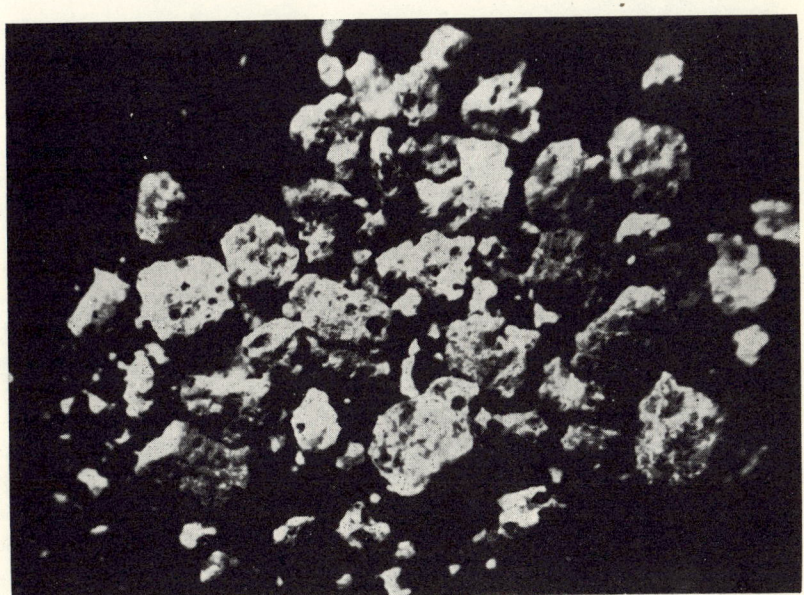

FIGURE 150. Leached oölitic porosity in dolomite. x5. Champlin's University 1-D, sec. 26, blk. 13, University lands, Andrews County, Texas; 8030-8060 feet.

The sorting is dependent on the proportion of fine minerals present, such as silt or shale, or the proportion of large grains, such as large frosted quartz grains commonly found in the Whitehorse section of the Permian Basin. The amount of cement also largely determines the porosity which is at a maximum in free unconsolidated sands and minimum in quartzite. The permeability is largely determined in the same manner as porosity, but here the size of grain and the sorting are most important.

Oil stain and odor are very important and much experience on the part of the sample examiner is required to estimate these accurately. Dry gas will not stain sandstone, but wet gas may show a very light tan stain. Oil is commonly darker, but very high-gravity oil shows no more stain than the gas. Porous sand, containing no stain whatsoever may be suspected of carrying water. In examining samples of oil stain, it is very convenient to have a fluoroscope, which may detect very light stains of high-gravity oil which are not obvious in ordinary light. Best results in detection of oil stains are obtained with mercury vapor lamps emitting light, with wave lengths ranging from 3,300 to 3,800 Angstrom units. Lamps giving light of shorter wave lengths, such as the quartz tube, cause fluorescence of oil, but also cause much mineral fluorescence in the sample, which may confuse the observer.

USE OF CUTTINGS DESCRIPTIONS IN STUDY OF LIMESTÒNE RESERVOIRS

Limestones and dolomites form important oil reservoirs, and cuttings descriptions furnish one of the chief sources of information concerning them. There are two chief types of porosity in these rocks. The first is intergranular porosity which consists of openings between the crystals, oölites, or other discrete particles of the rock which in its geometry is similar to sandstone porosity. The second is fracture porosity (or foramenular porosity of Bulnes and Fitting [6]) which consists of large openings through otherwise solid masses, such as fractures and vugs.

The intergranular porosity is easily observed under the binocular microscope where the tiny openings may commonly be seen connected with each other. The larger openings are not ordinarily visible in their entirety under the miscroscope but are indicated by irregular surfaces lined by crystals which have been formed in a comparatively large cavity. These large openings are much more difficult to detect than the smaller pores and ordinarily they have gone unnoticed unless indicated by the drilling-time or the manner of drilling. Permeability in the fractures or foramenular porosity is almost impossible to determine under the microscope, since there is no means of knowing how far apart the walls of the large openings originally were.

---

[6] Bulnes, A. C., and Fitting, R. U., Jr., *An Introductory Discussion of Reservoir Performance of Limestone Formations:* Petroleum Development and Technology, 1945, Petroleum Division, Trans. Amer. Inst. Min. Met. Eng., vol. 160, pp. 181-201.

Permeability in intergranular porosity can be estimated qualitatively by noticing the size of the pores and their apparent interconnection. In general, the larger the pores the higher the permeability. The converse, however, is not always true. Some dolomites showing very fine porosity are shown by core analysis to be surprisingly permeable. However, one type of porosity is non-permeable almost without exception. This is called pin-point porosity and consists of small isolated holes. Some of these holes contain small amounts of asphaltic material and even may contain live oil and gas, but commercial production is not developed from them. So far, no quantitative results about porosity and permeability have been obtained from ordinary cuttings. However, the coarser reverse-circulation cuttings have been analyzed for porosity and permeability with favorable results.

An interesting method of reproducing and visualizing these pore spaces is presented by Nuss and Whitney,[7] who impregnated limestones with plastic and then dissolved the limestone with acid, leaving a model of the porosity as a residue.

Oil and gas stains are ordinarily readily detectable in limestones and dolomites. The heavy sour oils, as found in the Upper Permian, leave a dark brown stain which is unmistakable. Lighter oils of the Lower Permian rocks show good stains, and the very light oils found in the Lower Paleozoic strata leave an extremely light stain which is difficult to detect under incandescent light, but may be seen in white or ultra-violet light. Since most gases carry a small amount of light oil with them, they will show slight stains in limestones. Gas-oil contacts commonly can be recognized accurately by the darkening of the stain at the top of the oil column. Water may be indicated by lightening of the stain and black asphaltic residues in the samples. Many dolomites have a characteristic sheen on crystal faces within the water zone. However, a well oil-stained section may produce water upon test. This fact may be attributed to later movement of the structure which causes shifting of the water table.

In concluding the discussion of cuttings description, it should be said that no rigid rules can be given for guidance in describing samples. Each geological province and each geological organization have their own problems which must be worked out individually. Since the stratigraphic sections penetrated by wells are as varied as those on the surface, there can be no substitute for experience and judgment on the part of the geologist. Description of samples should never be allowed to degenerate into a mere mechanical process. The better the geological background of the person examining samples, the better will be the description.

---

[7] Nuss, W. F., and Whitney, R. L., *Technique for Reproducing Rock Pore Space:* Am. Assoc. Petroleum Geologists Bull., vol. 31, no. 11, pp. 2044-2049, 1947.

PLOTTING

After the description of the samples is made, it must be plotted in graphic form to make the information easily available for correlation and study. It is usual to plot this material on a narrow strip of heavy paper or cardboard so that a large number of logs can be laid out to compare and correlate the sections. A 3-inch width has been found to be most convenient for this purpose. Upon this 3-inch strip the lithologic characteristics are plotted on a column $\frac{1}{2}$ to $\frac{3}{4}$ inch wide with a vertical scale of 1 inch equals 100 feet. On this column each 10 feet (1/10th inch) is marked so as to facilitate plotting. The rest of the log strip is used for notes on lithologic characteristics that are not easily plotted as symbols. It has been found in plotting the lithologic characteristics that strong contrasting colors facilitate correlation and comparison of the sections. While no standard system of symbols has been adopted, the following colors are widely used.

> Light blue = Calcareous limestone
> Dark blue = Dolomite
> Red = Red shale
> Gray = Gray shale
> Yellow = Sandstone
> Green = Salt
> Purple = Anhydrite

Other colors may be added for local necessities. It has been found very helpful also to plot the drilling time next to the lithologic column on the log strip. In the case of mechanically taken drilling time, a copy of the graph is plotted; with manually taken drilling time, the plotting is done as bar graphs—the common scale being 1 inch equals 100 minutes of drilling time.

Porosity and oil stain in the samples are indicated by symbols on either side of the lithologic column. These symbols may be either in ink or in suitable colors. It is very helpful to make some distinction as to probable gas or water stain even though these fluids can not be differentiated certainly from cutting examination. Sharp changes in lithologic character should be indicated plainly on the plotted log. This indication usually is done best by having the geologist who examined the samples check the plotted log and indicate where he believes the formation boundaries should be, because their exact position is commonly difficult to determine from the written description alone.

All tests and showings should be indicated on the margin of the log strip. These should include all drill-stem tests in rotary wells, and amount and kind of fluid in cable-tool holes. It is also helpful to indicate any swabbing tests taken, the acid used, amount of nitroglycerine used and section shot, and perforations in the casing. Casing seats should be indicated on the margin of the log with a notation of the diameter of the casing and the amount of cement used in setting the casing. Total depth,

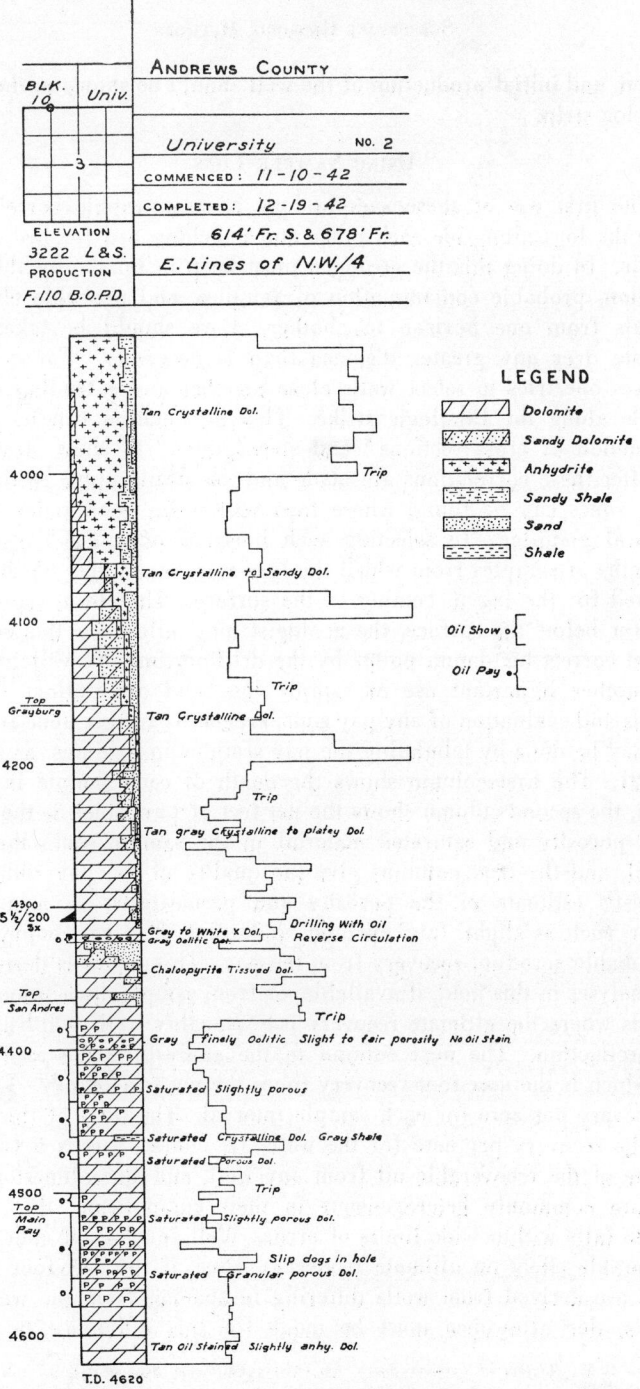

FIGURE 151. Type of sample log of producing section. Lithology is plotted on a percentage basis. Drilling-time data are recorded to right of lithologic column. Note slow penetration rate through anhydrite above 4,000 feet.

location, and initial production of the well should be shown in the heading of the log strip.

Using Sample Logs

The first use of these sample logs is stratigraphic correlation by laying the logs alongside each other and matching bed for bed as far as possible. In doing this the geologist must keep in mind probable lateral gradation, probable contamination of samples, and probable changes in intervals from one horizon to another. Care should be taken not to correlate over any greater distance than is necessary. For correlation purposes one tries to select wells close together and extending as far as possible along the lithologic strike. This information can be used for construction of cross sections [8] and stereograms [9] to show stratigraphy.

After these correlations are made and the stratigraphic section established, zones can be found whose tops will make good index beds for structural mapping. In selecting such horizons one must consider that the depths of samples from which the logs were made may not have been corrected for the lag in coming to the surface. Therefore, especially in zones far below the surface, the geologist must allow for this correction or must correct his datum points by the drilling time, if available.

Another important use of sample logs and descriptions is in the analysis and evaluation of any pay zone, especially in limestone reservoirs. This may be done by tabulating the pay sections in columns, as shown in table 21. The first column shows the depth of each sample in the pay section, the second column shows the net feet of pay which is the percentage of porosity and saturated material in the sample times the sample interval, and the next columns give the quality of the pay which is the geologist's estimate of the porosity and permeability in a qualitative manner, such as slight, fair, medium, and good. The next column shows the probable acre-foot recovery from the pay. This figure is derived from core analyses in this field, if available, or from sample descriptions of pay in fields where the ultimate recovery is reasonably well established from past production. The next column in the analysis is the recovery per acre, which is the acre-foot recovery times the net feet of pay. This gives the recovery per acre for each sample interval. The total of this column gives the recovery per acre for the well. Of course, this is a volumetric estimate of the recoverable oil from any well, and since limestone reservoirs are commonly heterogeneous in their composition, this recovery estimate falls within wide limits of error. Well spacing, of course, has a considerable effect on ultimate recovery. Thus, if the acre-foot recovery figures are derived from wells differing in spacing from the well under analysis, due allowance must be made for this difference in spacing.

---

[8] Hills, J. M., *Rhythm of Permian Seas:* Am. Assoc. Petroleum Geologists Bull., vol. 26, no. 2, pp. 217-255, 1942.
  [9] Lewis, F. E., *Position of San Andres Group, West Texas and New Mexico:* ibid., vol. 25, no. 1, p. 73, footnote 1, 1941.

TABLE 21
DETAILED RECOVERY ESTIMATE, UNIVERSITY WELL NO. 2, ANDREWS COUNTY, TEXAS

Depth in Feet	Remarks	Net Feet Slight Porosity	Net Feet Fair Porosity	Net Feet Medium Porosity	Net Feet Good Porosity	Acre-Foot Recovery	Recover. Per Acre (Barrels)
4380–85	Stained	1				75	75
90	Stained	1				75	75
4400–05	Sat. v. sl. por.	2				50	100
10	Sat. oolitic	3				100	300
15	Sat. isolated por.	3				50	150
20	Sat. gran. v. Sl. por.	4				50	200
25–30	Sat. v. sl. por	4				50	200
35	Sat. v. sl. por	1				50	50
40	Sat. v. sl. por.	2				50	100
45–50	Sat. sl. por.	1				100	100
55–60	Sat. gran.	2				125	250
65–75	Sat.	4				150	300
4500–05	Stained	1				75	75
10	Sat.	1				100	100
15–20	Sat.	5				75	375
25	Sat.	5				75	375
30	Sat. gran.	5				100	500
35	Sat. gran.	3				100	300
40	V. sl. f. por.	5				40	200
45	Sat. gran.	5				100	500
50	Sat. cryst. sl. gran.	3				50	150
55	Sat. cryst.	1	2			200	600
60	Sat. gran.	4				150	600
65	Sat. gran.	1	1			200	200
		65	3			Total	5875

Generally, it is found that this volumetric estimation is somewhat above an estimate made from pressure decline or production decline curves, and allowances should be made. In spite of its imperfections, this method of valuation is the only one available in many fields in which cuttings samples have been kept but no other type of reservoir information is available.

A somewhat similar method may be used for sandstone reservoirs. However, since sandstone reservoirs are much more uniform in their porosity and permeability, it is commonly possible to assign an acre-foot recovery for an entire field and to obtain the ultimate recovery by multiplication of the net pay—sand thickness in any well by the acre-foot recovery, rendering detailed pay analysis unnecessary.

Logs of cuttings samples are also very useful in studying the characteristics of limestone reservoirs. They give some idea of volume and relative permeability of lenticular pay zones. They are a useful supplement to any coring program undertaken as part of a reservoir study. It is often possible to select gas-oil contacts from cutting logs so as to select points for plug-back or for packer settings in remedial work.

SUMMARY

Samples of well cuttings can be taken from both cable-tool and rotary holes in such a manner that the nature of the formations penetrated can be determined accurately. These samples can be described by a geologist and the description plotted so as to give a graphic section of the formations encountered in the well. These sections can be correlated to give a picture of the regional stratigraphy. From the sections, index beds can be selected for use in structural contouring.

From the sample descriptions, volumetric estimates of ultimate well yield can be made, reservoir study facilitated, and well remedial work guided.

ELECTRIC LOGGING

E. F. STRATTON and R. D. FORD

The electric log consists of a spontaneous potential curve and, generally, three resistivity curves. The specific recording practice and the type and number of curves vary from one geologic province to another, depending upon the nature of the formations and the problems to be solved.

SPONTANEOUS POTENTIAL

The spontaneous-potential (SP) log is used to distinguish between permeable and nonpermeable formations, as, for example, sand and shale or permeable and nonpermeable limestone. However, a quantitative relationship between porosity or permeability and spontaneous potential does not exist. Empirical relationships have been found, however, and have been established in specific pools for particular formations.

The spontaneous potential log of a bore hole is a record of the potentials measured in the mud along the hole. In fact, the potentials are measured between an electrode lowered into the hole and another electrode at the surface and are related to an arbitrary constant. As the SP log is generally flat in front of shales and shows positive or negative anomalies opposite permeable beds, it is convenient to take the line obtained in front of shales as the base line.

Spontaneous potential anomalies in a bore hole are due primarily to the electromotive forces generated by two different electrical phenomena. The first of these and the more important is the electrochemical cell formed between the drilling fluid, the fluid in the permeable zone, and the shale surrounding the permeable section. This may be expressed as:

$$E = K \ \log \frac{R_2}{R_1}$$

(1)

where E=electromotive force of spontaneous potential in millivolts.

R_2=resistivity of drilling fluid in ohmmeters.

R_1=resistivity of the fluid in the permeable zone in ohmmeters.

K=factor dependent upon the chemical composition of the two fluids and upon the *character of the shale adjacent the permeable bed.*

The second of these electromotive forces may result from the filtra-

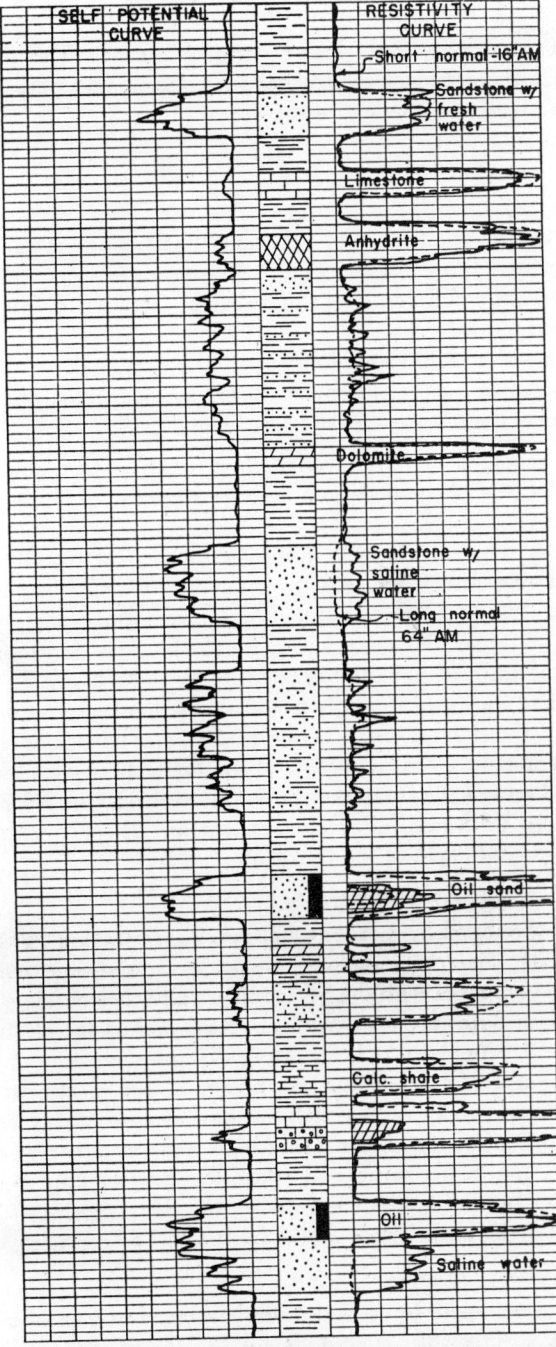

FIGURE 152. Schematic electrical log showing relationship between electrical characteristics and various
lithologic types.

tion (2) of the drilling fluid into a permeable zone. The principle is a recognized phenomenon of electrochemistry (streaming potential) and, if effective in a well, may be expressed as:

$$E = \frac{MR_2P}{V} \tag{2}$$

where E=electromotive force or spontaneous potential in millivolts.
   R_2=resistivity of drilling fluid.
   P=pressure differential (atmospheres) between drilling fluid and formation.
   V=viscosity of filtering fluid.
   M=complex factor dependent upon the nature of the permeable zone, the filtrate, and the filter (mud cake).

There may be other factors effective in generating bore-hole potentials, but, at present, the phenomena just described appear to be those of major importance.

Whatever their origin may be (electrochemical or electrokinetic), the electromotive forces give rise to a current, which flows through the permeable layers, then spreads into the adjacent impervious formations, and returns through the mud filling the hole. The SP anomalies correspond to the drop of potential created by the circulation of the current in the hole, and thus measure only a part of the total electromotive forces. Consequently, the characteristics of the SP log, and particularly the amplitude of the anomalies, are a function of several factors, such as the salinity of the mud and of the formation fluid, the resistivity of the surrounding formations, bed thickness, hole diameter, amount of shaly material in the permeable bed, and depth of mud invasion.

Fresh-Water-Bearing Formations

The SP developed by a fresh-water-bearing formation is usually very small (fig. 153), frequently nonexistent (fig. 154), and sometimes reversed (fig. 155), as compared with the SP across a salt-water-bearing formation. As most drilling fluids are comparatively fresh and as the electrochemical effect has been recognized as being generally preponderant, it is obvious from the formula (1) describing the action of the electrochemical cell that, when the resistivity of the drilling fluid is appreciably higher than that of the formation water, the SP is high and negative; when the two are the same, the SP is zero; and when the drilling-fluid resistivity is lower than the formation-water resistivity, the SP is positive. Therefore, in the case of fresh-water sands and a fresh mud, SP's usually are small.

Salt-Water-Bearing Formations

The SP developed by a salt-water-bearing formation is generally sharp, has an appreciable magnitude up to 100 or 200 millivolts, and is negative with respect to the surrounding shale or nonpermeable forma-

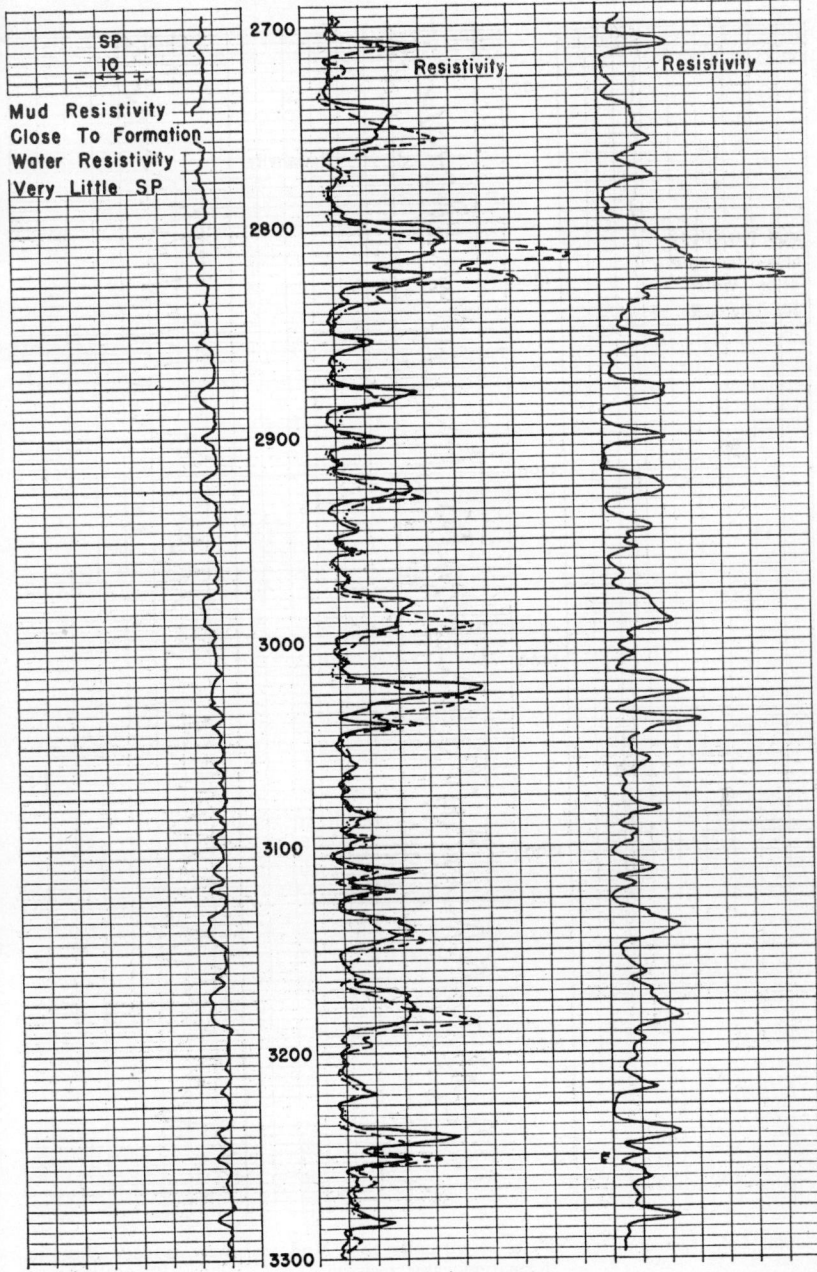

SP
10
– ← → +

Mud Resistivity
Close To Formation
Water Resistivity
Very Little SP

2700

Resistivity

Resistivity

2800

2900

3000

3100

3200

3300

FIGURE 153. Log showing SP to be very small. Mud resistivity is close to formation-water resistivity.

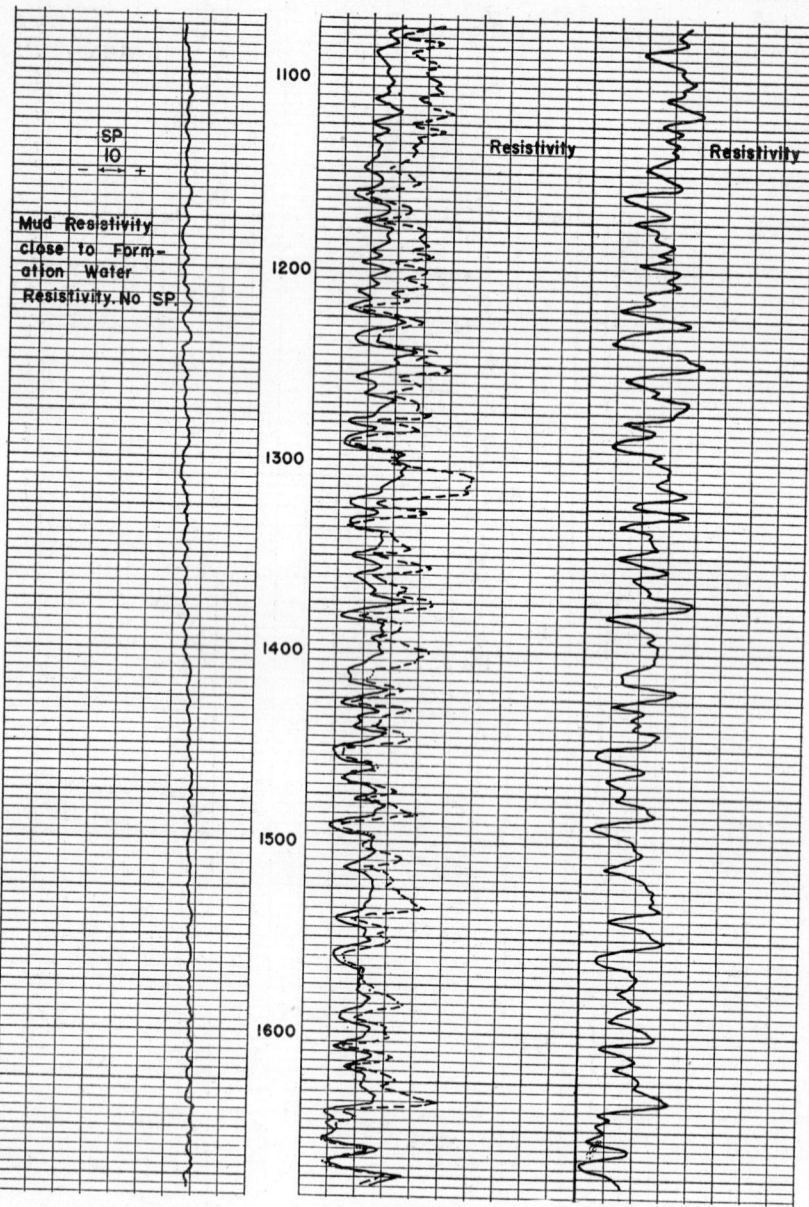

FIGURE 154. Log showing SP to be virtually nil. Mud resistivity is very close to formation-water resistivity.

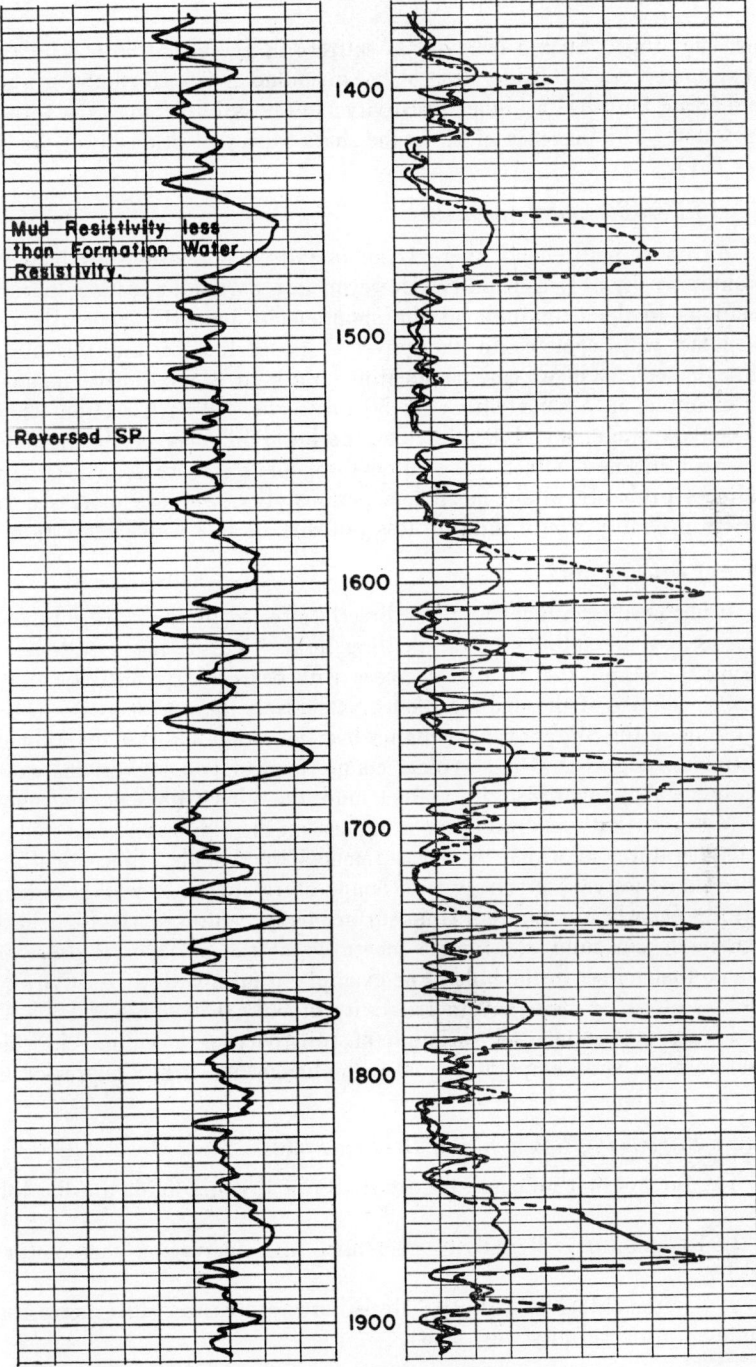

FIGURE 155. Log showing polarity of SP reversed. Mud resistivity is less than formation-water resistivity.

tions (fig. 156). Such a behavior is entirely in agreement with results to be expected from a consideration of the formulae as described above when the drilling fluid has a higher resistivity (lower salinity) than the formation water. The presence of oil in the shaly sand will often lower the SP (fig. 157).

Effect of Porosity and Permeability

As noted beforehand, there is no quantitative relationship between SP and porosity or permeability. However, in a particular section, marked variations in the magnitude of the spontaneous potential generally are associated with changes in the physical properties of the formation, although such changes can be identified only in a qualitative manner. For example, in a sand with only 50 millivolts, SP in a section where the sands average, say, 100 millivolts, the lower SP may be the resultant of a bed-thickness effect, of shaliness, or of an increased resistivity. Shaliness probably would mean less permeability, whereas increased resistivity probably would indicate less porosity or less water saturation.[10]

Effect of Drilling Mud

It has been seen that the SP is directly affected by the resistivity and, therefore, by the salinity of the drilling fluid. Considering a salt-water-bearing formation, the SP will decrease with decreasing resistivity or increasing salinity of the drilling mud. Not only is there a decrease in the magnitude of the SP, but the anomalies lose definition and the log becomes featureless. Figure 158 is a typical comparison of two logs in the same well, one with a low-resistivity (salty) mud, the other with a normal mud.

Mud-resistivity measurements are given on the log headings as well as the temperatures at which the measurements were made. Since drilling-fluid resistivities vary inversely with temperature and since well-bore temperatures usually are different from surface temperatures, corrections have to be made for mud resistivities measured at the surface in order to evaluate their effect in the hole. For example, a mud with a resistivity of 2.0 ohms at 64° F. will have a resistivity of only 0.70 ohms at 200° F.

Figure 159 shows the variation of resistivity of a sodium-chloride (salt) solution with temperature and with changes in salinity expressed in parts per million.

Factors Influencing Resistivity of Drilling Fluids

The factors that influence the resistivity of drilling fluids are the following:

1. Temperature. Resistivity decreases with increasing temperature (fig. 159).

2. Sodium-chloride salinity. Resistivity decreases with increasing sodium-chloride salinity (fig. 159).

---

[10] Doll, H. G., *The S. P. Log: Theoretical Analysis and Principles of Interpretation:* Am. Inst. Min. Met. Eng. Petroleum Technology, Sept. 1948.

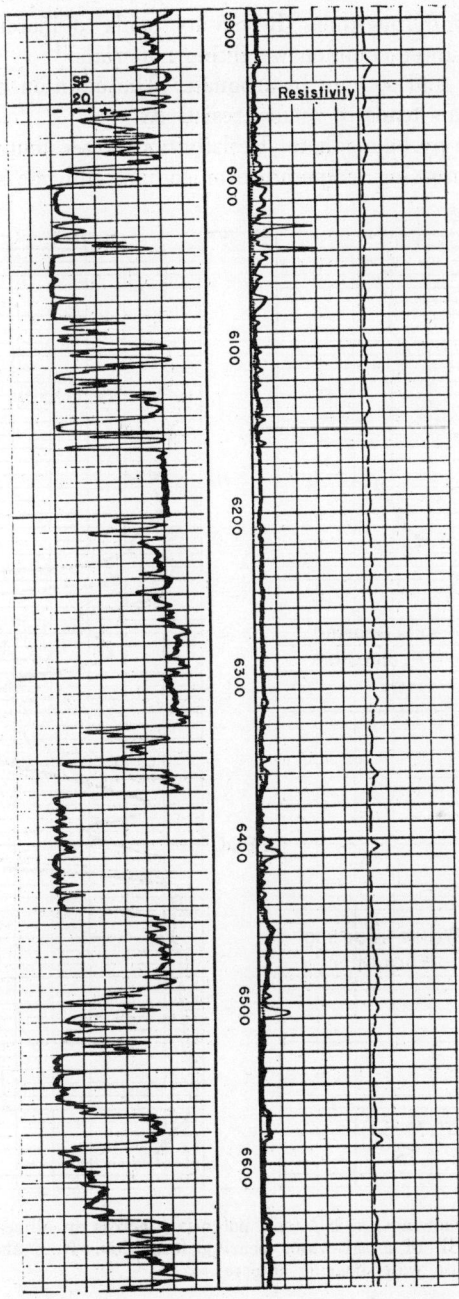

FIGURE 156. Negative SP in salt-water-bearing formations (sands). Mud resistivity is greater than formation-water resistivity.

3. Barite and limestone. Resistivity tends to increase slightly with the addition of these common weighting materials.[11]

4. Cement and sodium bicarbonate. The addition of either or both of these materials tends to reduce resistivity.

5. Sodium pyrophosphate. Resistivity decreases but not uniformly as it does with increasing temperature or sodium chloride salinity.

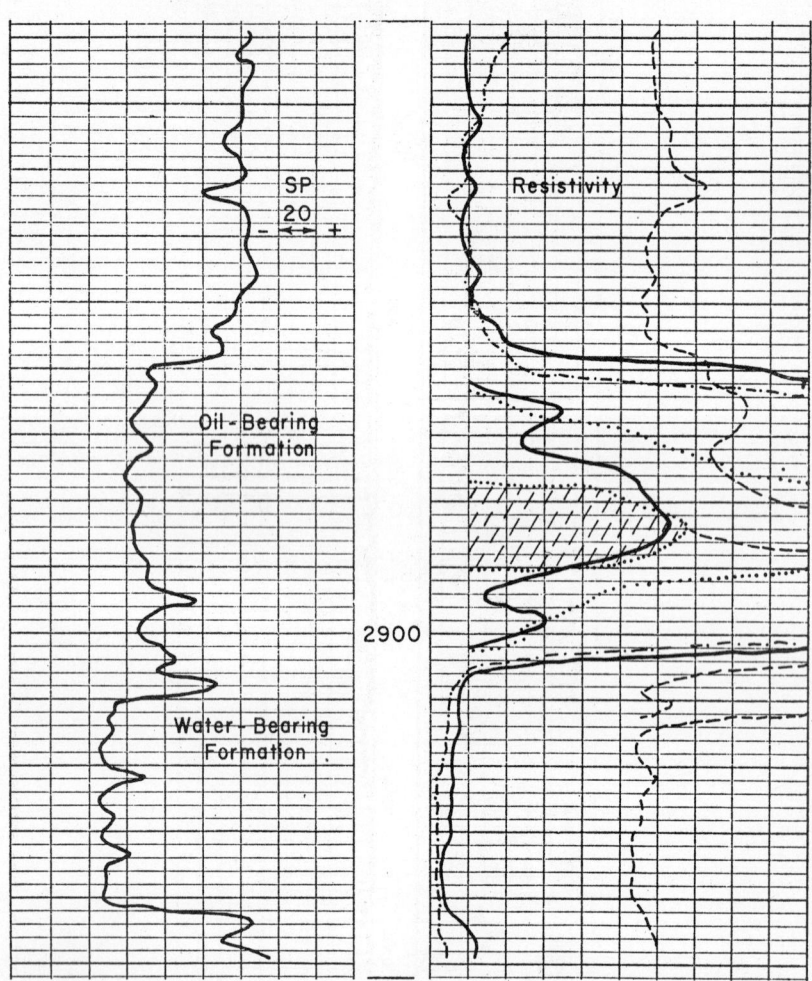

FIGURE 157. Log showing less SP (self potential) across an oil-bearing formation as compared to SP of a salt-water bearing formation. Note also decrease in resistivity opposite water-bearing interval.

6. Quebracho. The effect on resistivity of quantities generally used is negligible.

7. Starch (i.e. "Impermex"). The resistivity is unaffected except as the preservatives may change the resistivity.

8. Caustic soda. The resistivity is decreased with the addition of caustic soda.

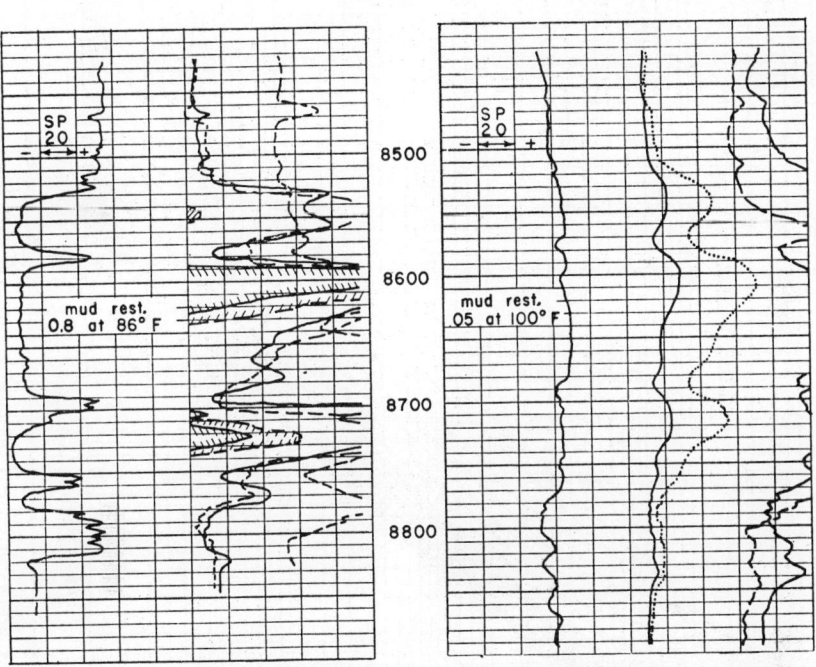

FIGURE 158. Comparison of two logs in the same well, one with a low resistive (salty) mud, the other with a normal mud.

Effect of Bed Thickness

Research has indicated that, all other factors remaining the same, when the bed is less than four times the hole diameter in thickness, the SP will decrease with decreasing bed thickness.

A section composed of interbedded thin shales and permeable sands may have much less SP than the sand would have if the thin shale beds were not present. This is an important factor in evaluating thin sand zones and explains the good production obtained sometimes from sandy zones with low SP.[12]

Effect of Special Muds

The normal constituents of drilling muds, natural clay, "aquagel," "baroid," starch, and other additives have no effect on the SP log except

---

[12] Doll, H. C., *op. cit.*

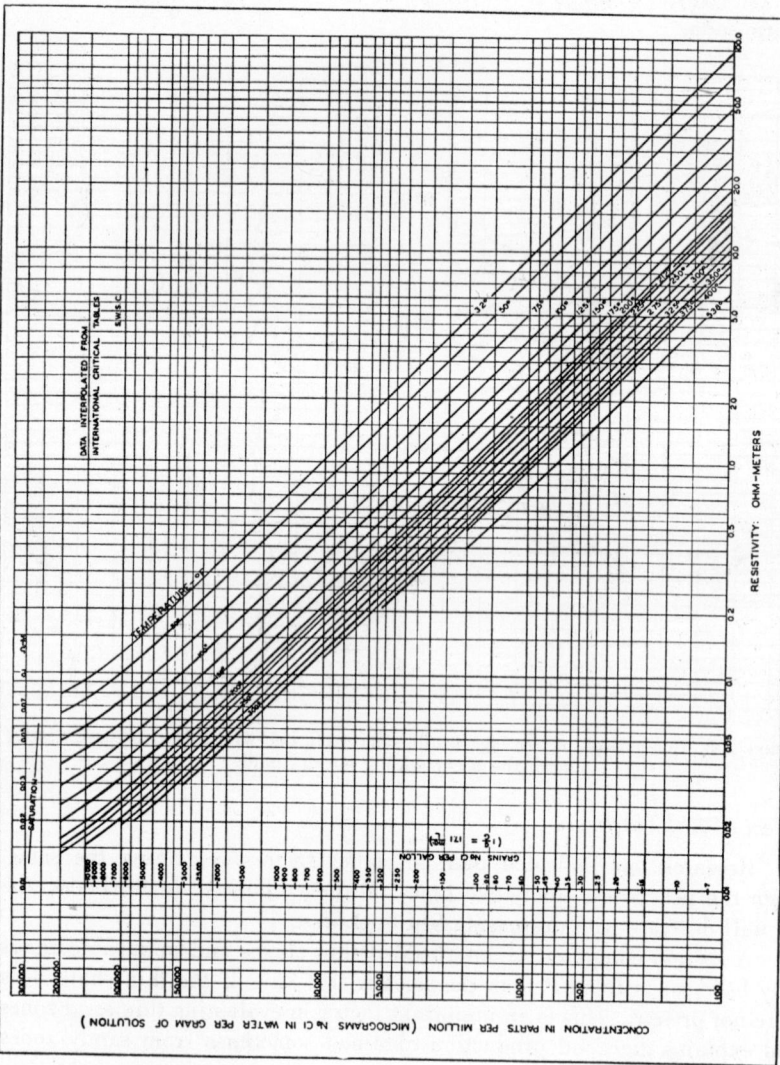

FIGURE 159. Resistivity graph for salinity and temperature of NaCl solution.

as they change the resistivity of the mud. On the other hand, two muds used occasionally, silicate and oil-base, affect the SP in a manner unrelated to resistivity.

A positive SP generally is recorded in a silicate mud opposite a permeable sand, although the mud resistivity is higher than the resistivity

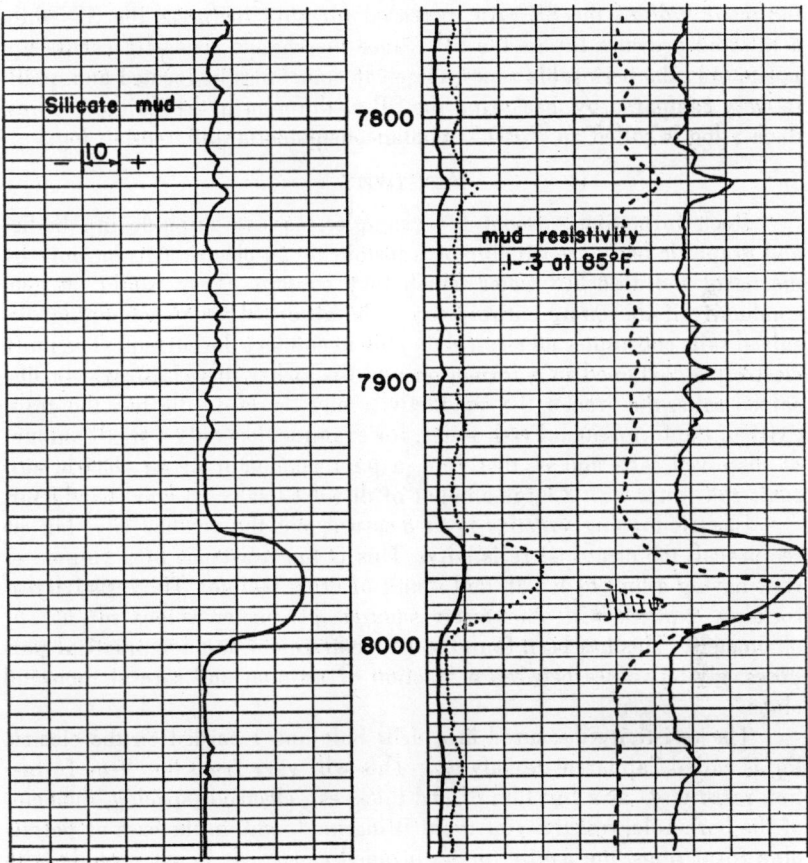

Figure 160. Example of a log made in sodium-silicate mud showing reversal of SP curve.

of the formation water (fig. 160). The anomalies are small but sharp and are easily seen when the proper SP scale is used.

SP logs in true oil-base muds, those with a water content of less than five percent and a resistivity of several thousand ohms, usually are not satisfactory for distinguishing between permeable and nonpermeable formations or for locating formation boundaries. On the other hand, normal SP logs are obtained in oil-emulsion muds. Such muds are a mixture of water-base and oil-base mud, and the resistivity generally has a value of from two to ten ohms m²/m.

Effect of Variation in Hydrostatic Pressure

It has been demonstrated many times that a change in the hydrostatic pressure exerted by the mud column on a permeable formation will change the magnitude of the SP. An increased pressure increases the SP while a lowered pressure lowers the SP. Since the change generally occurs opposite only the permeable zones, these can sometimes be located and qualitatively compared by measuring the SP at different hydrostatic pressures. Such a log is called an "SPD," spontaneous-potential-differential, log.

RESISTIVITY

Rock formations, except for example, massive sulphide ore bodies and graphitic beds, are capable of transmitting an electric current only by means of the absorbed water which they contain. They would be nonconductive if they were entirely dry. The absorbed water containing dissolved salts constitutes an electrolyte able to conduct the current. The more electrolyte contained in a formation and the richer this electrolyte in dissolved salts, the greater the conductivity and therefore the less the resistivity of the formation. Fresh water, for example, has only a small amount of dissolved salts and is, therefore, a poor conductor of an electric current; salt water with a large amount of dissolved salt is a good conductor.

Electric-logging practice is to measure, not the conductivity, but its reciprocal, the electrical resistivity. This is the resistance of a volume of rock having a unit of length and a unit of cross section. The resistivity of rocks is expressed in ohm meter squared per meter (ohms m²/m) or ohmmeters. This has been found to be a convenient unit for practical purposes, giving values between a fraction of an ohm and several thousand ohms.

The resistivity measured in a drill hole and recorded on the electric log is called "apparent resistivity." This will vary from the "true formation resistivity" as a function of bed thickness, electrode spacing, diameter of the bore hole, resistivity of the drilling mud, and, in the case of permeable formations, the nature of the invaded zone.[13] It is not feasible with present measuring procedures to eliminate these effects; however, the influence of bed thickness on apparent resistivity is negligible if the formation is many times as thick as the AM spacing of the "normal" curve or several times as thick as the AO spacing of the "lateral." The other factors noted above can be taken into account and true resistivity determined from the apparent resistivity by the use of resistivity-departure curves.[14]

The electric-log resistivity or apparent resistivity is satisfactory for all problems except the determination of the fluid content of permeable

    [13] Doll, H. G., Legrand, J. C., and Stratton, E. F., *True Resistivity Determination from the Electric Log—Its Application to Log Analysis:* Am. Petroleum Inst. Drilling and Production Practice, p. 215, 1947.
    [14] *Resistivity Departure Curves,* Schlumberger Well Surveying Corporation, Sept. 1947.

zones. The true resistivity required for this latter work may be obtained from the departure curves or sometimes directly from the long normal or lateral curve.

Experience and research have proved the general over-all utility of the multi-electrode method for making resistivity measurements, as it minimizes the effect of the drilling fluid and the well bore, and it makes possible a direct comparison of the several recorded resistivity curves. Multi-electrode recording, as distinguished from single- or point-electrode measurements, is made with a system of four electrodes; two of these are current emitting and two are for potential measurement. The curves recorded are termed "normal" or "lateral," depending upon the electrode arrangement.

Terminology

Electrodes—The current electrodes are designated "A" and "B," the measuring electrodes "M" and "N." Common practice is to have the two current electrodes, A and B, and one potential-measuring electrode, M, in the hole with the other potential electrode, N, at the surface. Some of the electrodes in the hole are mounted on a mandrel, called a "sonde," which serves as a guide and weight for the cable.

Normal Curve—A normal curve is a resistivity log recorded with the four-electrode system where the distance between one current and one potential-measuring electrode, AM, is of primary importance. The position of the other current electrode, B, is relatively unimportant as long as the distance, AM, is small as compared to AB (fig. 161).

Amplified Normal Curve—A resistivity log recorded with the normal-curve-electrode arrangement but using an amplified or exaggerated scale is an amplified normal curve. For example, the normal curve might be recorded on a 20-ohm scale, the amplified normal with a 4-ohm scale.

Long Normal Curve—A long normal curve is a resistivity log recorded with the same electrode arrangement as the normal but with the distance, AM, several times as great as the normal.

Lateral Curve—A resistivity log recorded with the four-electrode system, where the distance between one potential-measuring electrode and a point midway between the two current electrodes, AB, is of primary importance, is a lateral curve. The distance AB is small as compared with the distance AM (fig. 162).

Long Lateral Curve—A resistivity log recorded with the same electrode arrangement as the lateral but with the distance between M and the midpoint of AB longer than that of the regular lateral is a long lateral curve.

Electrode Spacing

Normal Curves—The spacing is considered as the distance, AM, between the current electrode, A, and the potential-measuring electrode, M. Depending upon the geologic province, this spacing varies between eight inches and eighteen inches for the normal curve and between five and six feet for the long normal curve and is indicated on the Schlumberger log heading as "AM."

Lateral Curves—The spacing is considered as the distance between the potential-measuring electrode, M, and a point midway between the two current electrodes, AB. This spacing is always large as compared with the distance, AB, and is indicated on the Schlumberger log heading by "AO" (fig. 162).

Depth of Investigation—The depth of investigation is an indefinite matter since, for both the normal and lateral arrangement, it varies with many factors. It can be stated, however, that in general the greater the AM or AO spacing the greater the depth of investigation.

Characteristics of Resistivity Curves

A resistivity log has primarily a twofold purpose: one, to locate and determine the boundaries of all resistive formations; the other, to determine the fluid content, both qualitatively and quantitatively, of permeable

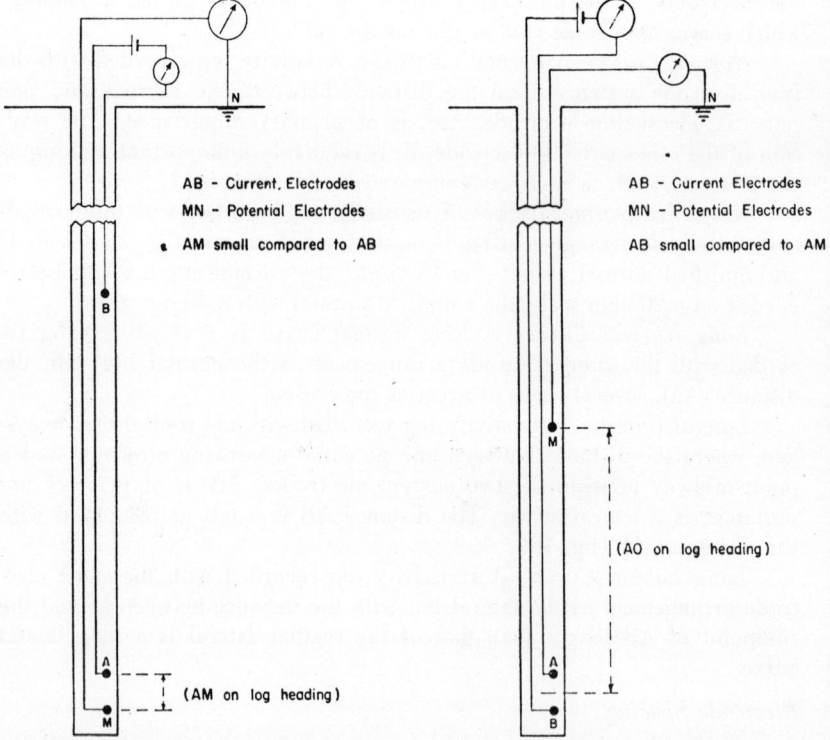

FIGURE 161. Schematic diagram showing arrangement of electrodes for recording normal resistivity curves.

FIGURE 162. Schematic diagram of electrode arrangement for recording lateral curves.

formations. The first condition, with a normal curve, is achieved best by a short electrode spacing, AM; the second, by using a longer electrode spacing in order to minimize the effects of the drilling fluid, the diameter of the hole, and the invaded zone. As a result two or more resistivity curves, one with a short spacing and others with a somewhat longer spacing, are commonly recorded and presented on each log.

The particular behavior of normal curves in resistive beds of various thicknesses is illustrated in figure 163. It is to be noted especially that a resistive bed equal to the electrode spacing gives no indication; a resistive bed where thickness is less than the electrode spacing is shown as an inverted anomaly.

The lateral-type curve, with the spacings commonly employed, is usually adequate to minimize the effect of the invaded zone and, at the same time, to indicate the position of resistive zones. Figure 164 indicates the behavior of the lateral curve with resistive beds of different thicknesses. Note that it shows beds of all thicknesses, but that the top boundary of formations whose thickness is greater than the AO spacing is indefinite, and true values are shielded out for a distance equal to the AO spacing. The actual thickness of beds less than AB is exaggerated by an amount equal to the distance between the current electrodes. Below thin resistive beds an abnormally low resistivity is measured for a distance equal to the AO spacing regardless of the nature of the formation opposite the section (fig. 165).

APPLICATION OF THE ELECTRIC LOG

Correlation—The utility of the electric log in detailed structural pool studies or in general stratigraphic investigations is well known. Figure 166 is a typical correlation study in the Midcontinent area.

Distinction between Porous and Permeable Formations and Nonporous and Nonpermeable Formations—The spontaneous potential curve usually indicates permeable formations containing saline interstitial water by a marked negative anomaly. This characteristic is common for sands as well as limestones or dolomites and for shallow as well as deep formations. Formations containing fresh interstitial water, on the other hand, are usually indicated by their lack of SP anomaly or by a positive anomaly.

An electric-log analysis through a limestone or cemented-sandstone section, where permeable zones occur interbedded with otherwise nonpermeable beds, needs particular mention. The permeable zones, whether oil- or water-bearing, are usually more conductive (less resistive) than the surrounding nonpermeable formations because of the saline interstitial water. The resistivity log exhibits, therefore, a lower value across the permeable zones than across the nonpermeable ones (fig. 167). The anomalies on the SP log spread above and below the permeable beds to such an extent that permeable-zone boundaries are not determined easily by a cursory ex-

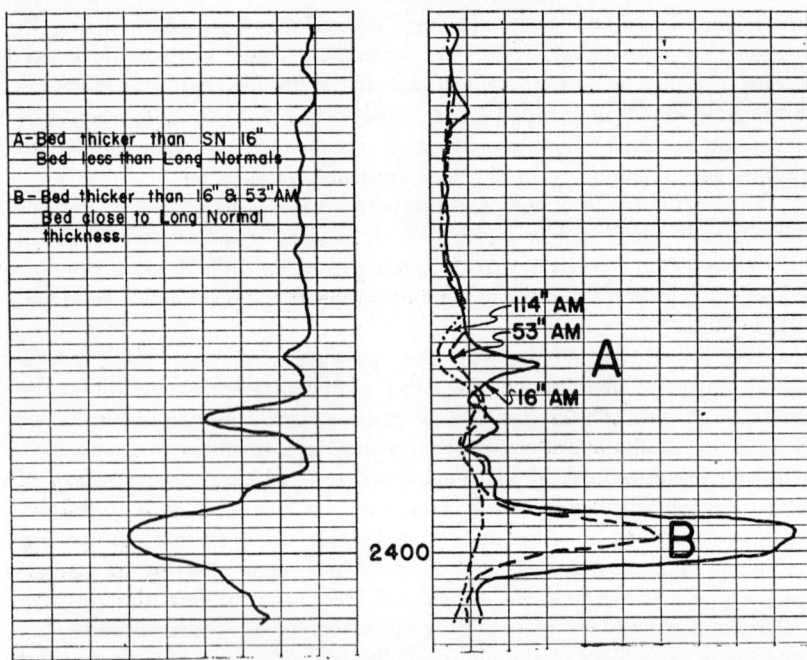

FIGURE 163. Log illustrating effect of resistive beds of various thicknesses on normal curves.

amination. The latter problem has been discussed in detail by Doll,[15] and reference to his paper will indicate that permeable-zone boundaries in an otherwise resistive and nonpermeable formation can be determined with accuracy by a careful study of the SP log in conjunction with the resistivity diagram.

The Location and Exact Depth of All Formations—A stratigraphic log is obtained indicating the presence and depth position of all formations. This practically eliminates the possibility of passing up a potentially productive oil- or gas-bearing formation.

Sand Studies—Changes in the physical characteristics of reservoirs can be studied, aiding the solution of many exploration and production problems.

Determination of Thicknesses—The thicknesses of all formations can be determined, and the net producing thickness of sand reservoirs can be calculated.

Fluid-Content Determinations—It is possible in most cases to distinguish between an oil or gas reservoir and a water-bearing formation; in many sand reservoirs a quantitative determination of the percentage of void space containing oil or gas or, conversely, the percentage of void space containing interstitial water can be made.

---

[15] Doll, H. G., *op. cit.*

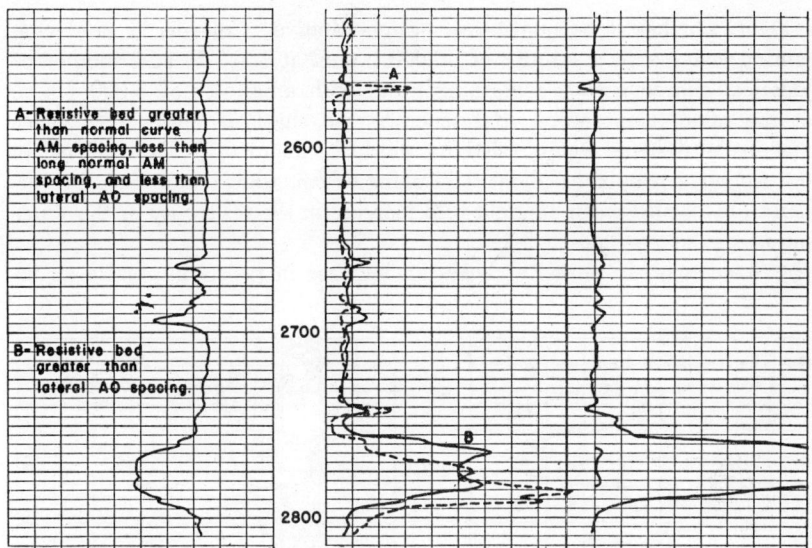

FIGURE 164. Log illustrating behavior of lateral curve with resistive beds of different thicknesses. Comments on curves to right of depth column: dashed line represents the lateral AO $= 15'$; solid line represents the normal AM $= 16''$; solid line curve at extreme right represents the normal AM $= 63''$.

It has been proved experimentally that the resistivity of a water-saturated permeable formation is related to the resistivity of the water, to the amount of void space, and to the size, shape, and distribution of this space. This relationship may be expressed as:[16]

$$R_o = F R_w \qquad (1)$$

where R_o = resistivity of formation 100-percent water-saturated.

F = formation-resistivity factor.

R_w = resistivity of water saturating the formation.

The factor F has been shown, in general, to be equal to:

$$F = \frac{1}{p^m} \qquad (2)$$

where p = fractional porosity.

m = proportionality factor.

The factor m is related to size, shape, and distribution of the void space. Experimental data have shown that it varies from 1.3 to somewhat over 2.0. The lower values are generally found in unconsolidated sands, the higher in consolidated sands.

A knowledge of the value of R_o, the resistivity of a formation 100-percent water-saturated, is fundamental to any interpretation for fluid

[16] Archie, G. E., The Electrical Resistivity Log as an Aid in Determining Some Reservoir Characteristics: Am. Inst. Min. Met. Eng. Tech. Pub. 1422, Jan. 1942.

content, whether it be qualitative or quantitative. Consider a sand containing water with a salinity of 50,000 p.p.m. at 100° F. and another of identical characteristics containing water with a salinity of 5,000 p.p.m. at the same temperature. Reference to the chart of figure 159 shows that in the first the water resistivity would be about 0.10 ohms m²/m; in the second, it would be about 0.85 ohms m²/m, which, applying the formula (1), would mean an eight-fold increase in the resistivity of the water sand.

A decrease in porosity causes an increase in resistivity, while an in-

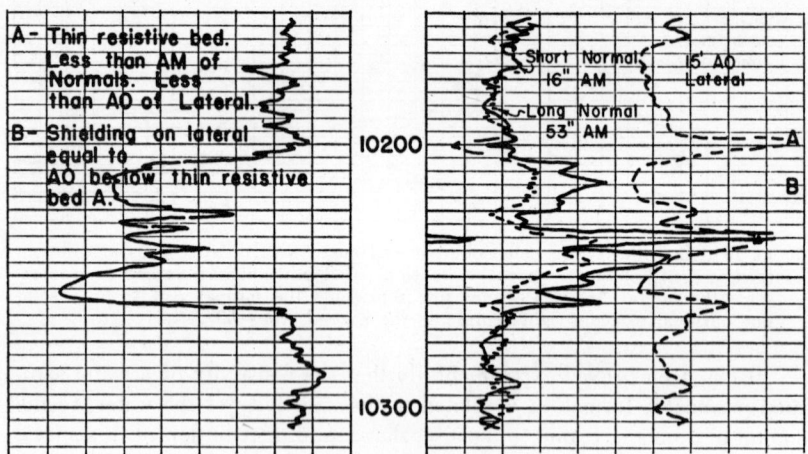

FIGURE 165. Portion of a log showing shielding effect of thin resistive beds on lateral curve.

crease in porosity causes a decrease in resistivity. This factor is not especially important for small changes in porosity—less than five percent —but must be given major consideration when there are marked changes in percentage porosity.

The evidence indicates clearly that fluid-content interpretation can be made if the resistivity of the sand is known when it is 100-percent water-saturated. These data may be obtained in one of two ways, either (1) direct from the electric log, or (2) by laboratory resistivity measurement on a core.

The second method requires a measurement of the resistivity of a core sample 100-percent water-saturated. This information permits a determination of the formation-resistivity factor; and, thereby, knowing the salinity of the water in the formation, one can calculate the resistivity of the formation when it is completely water-saturated.

It has been shown by a number of investigators that the resistivity of an oil reservoir is, among other things, related to the percentage of water

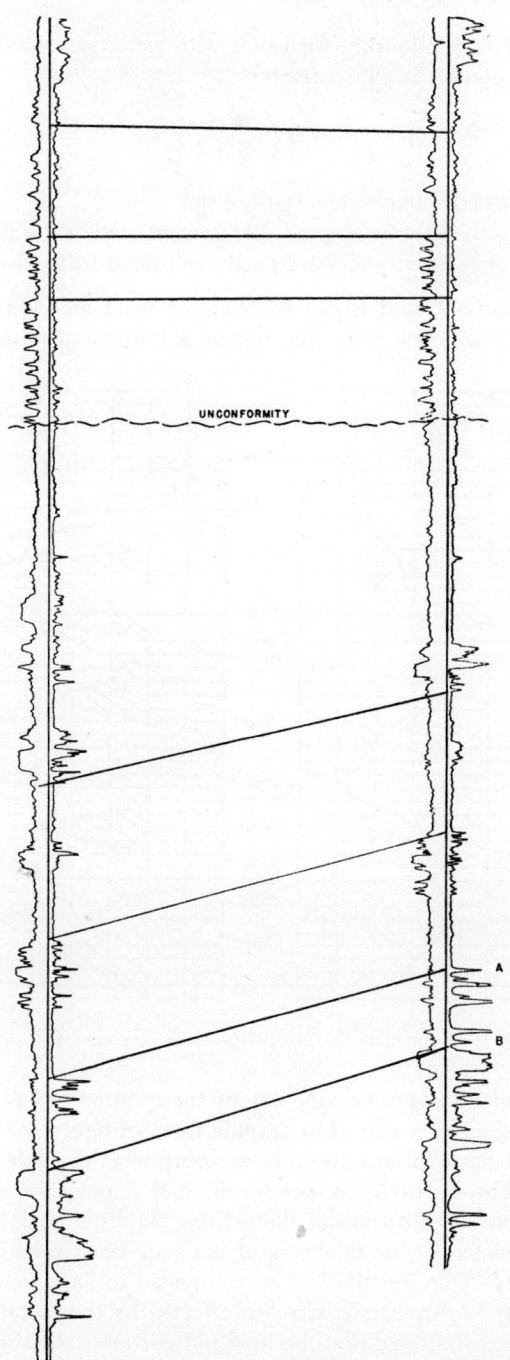

FIGURE 166. Use of electric log for correlation in Midcontinent area.

it contains. The relationship for sands with water saturations higher than about fifteen percent is approximately

$$S = \left(\frac{R_o}{R}\right)^{\frac{1}{2}}$$

where S = fractional water saturation.

R_o = resistivity of sand 100-percent water-saturated.

R = resistivity of sand partly saturated with oil or gas.

Since the factors R_o and R can be obtained from the electric log, a procedure is available for using the log as a tool in quantitative reservoir

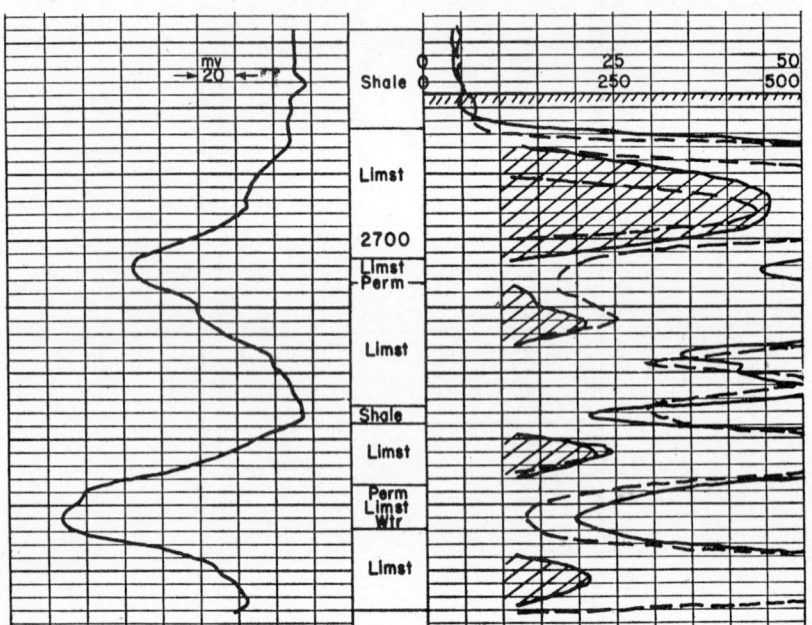

Figure 167. Common characteristics of an electric log in limestone.

study. In order to facilitate the use of the relationship expressed above, the equation has been placed in graphic form in figure 168. Actual quantitative log interpretations are only as accurate as the data on which they are based. The resistivity values for R_o and R must be obtained practically unaffected by invasion of the drilling fluid, the salinity of the fluid, the hole diameter, or the thickness of the bed. Such resistivities are commonly termed "true resistivities" as contrasted to "apparent resistivities," a value related to true resistivity but affected by the factors noted above. Experience has indicated that the long normal and lateral curves are generally adequate to minimize the effect of drilling-fluid invasion and hole

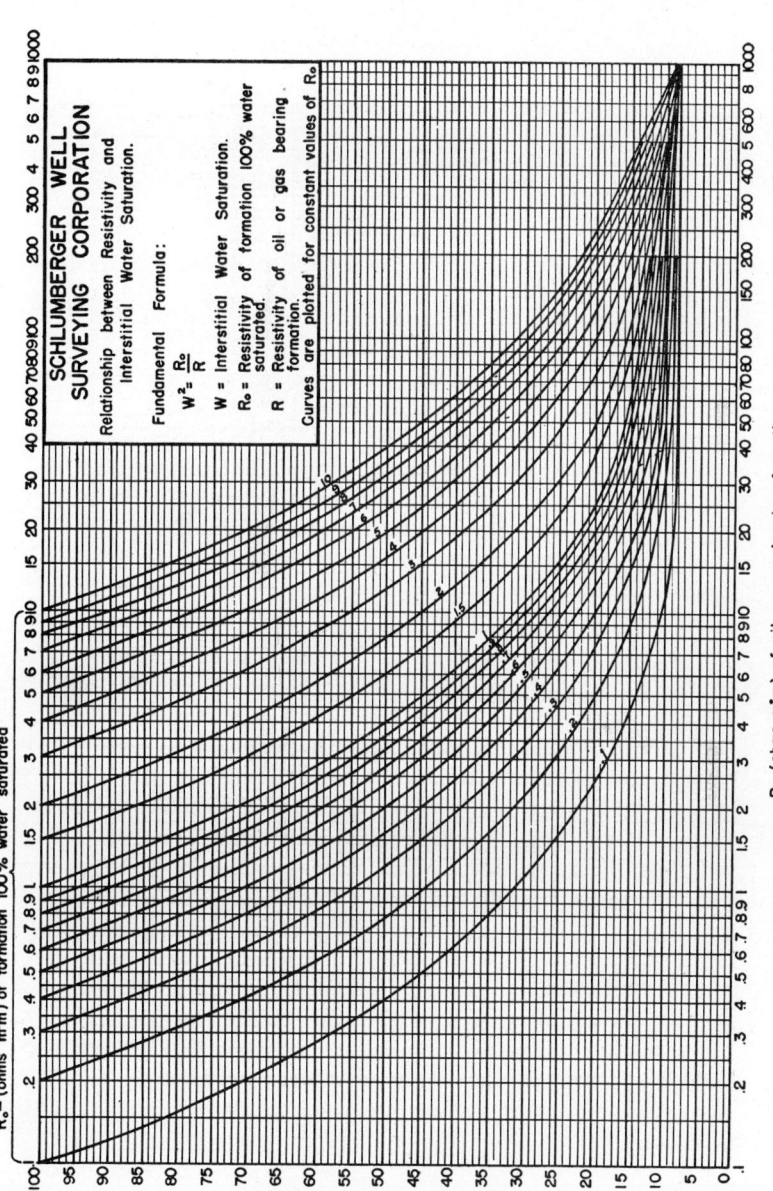

FIGURE 168. Graph showing relationship between resistivity and interstitial-water saturation.

SCHLUMBERGER WELL SURVEYING CORPORATION

Relationship between Resistivity and Interstitial Water Saturation.

Fundamental Formula:

$$W^2 = \frac{R_o}{R}$$

W = Interstitial Water Saturation.

R_o = Resistivity of formation 100% water saturated.

R = Resistivity of oil or gas bearing formation.

Curves are plotted for constant values of R_o

7 — 1938

R_o — (ohms m²m) of formation 100% water saturated

R — (ohms m²m) of oil or gas bearing formation

W — Interstitial water % of pores space

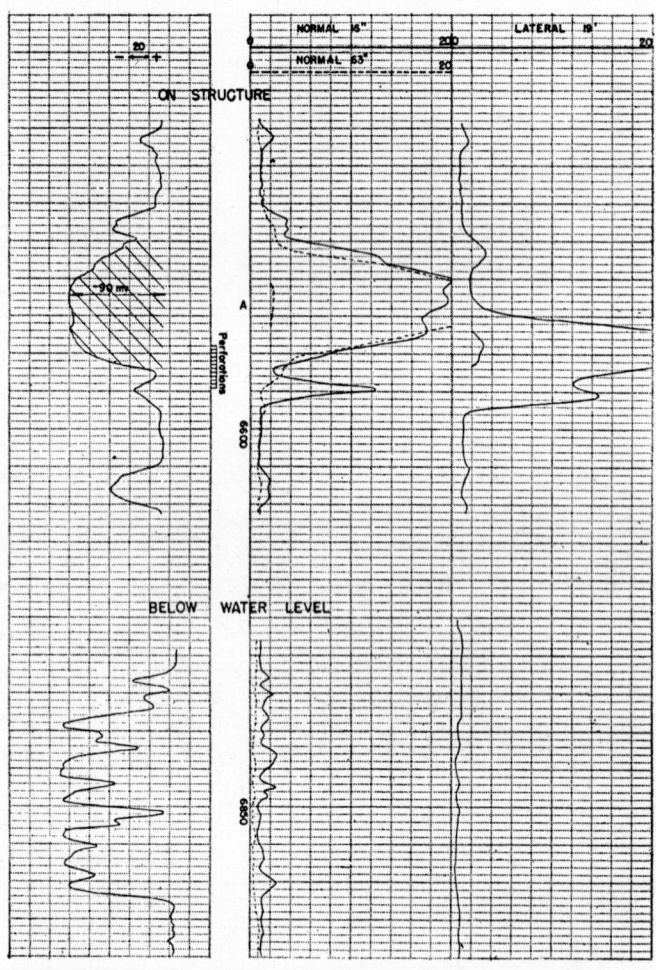

Figure 169. Combination log and core analysis and production results.

Explanation of figure 169

LOG ANALYSIS

1) Water saturation S

At water level $R_o = 0.3$ ohms

At level A, $S = \sqrt{\frac{.3}{21}} = 12\%$ approx.

2) Formation water resistivity R_w

$SP = -70 \log_{10} \frac{R_m}{R_w} = -90$ mv.

thus, $R_w = 0.05$ ohm at BHT

3) Porosity p

$$F = \frac{R_o}{R_w} = \frac{0.3}{0.05} = 6$$

$$p = \sqrt[m]{\frac{1}{F}} = 30\% \text{ for } m = 1.5$$

CORE ANALYSIS

Average Porosity   30%

Average Permeability   1000 md.

Connate water determined by restored state methods 10-12%

PRODUCTION RESULTS

Perforated: 6581-6594

Initial Production: 97 bbls. per day 30° API

Gas oil ratio: 4400/1

Gas increasing with time.

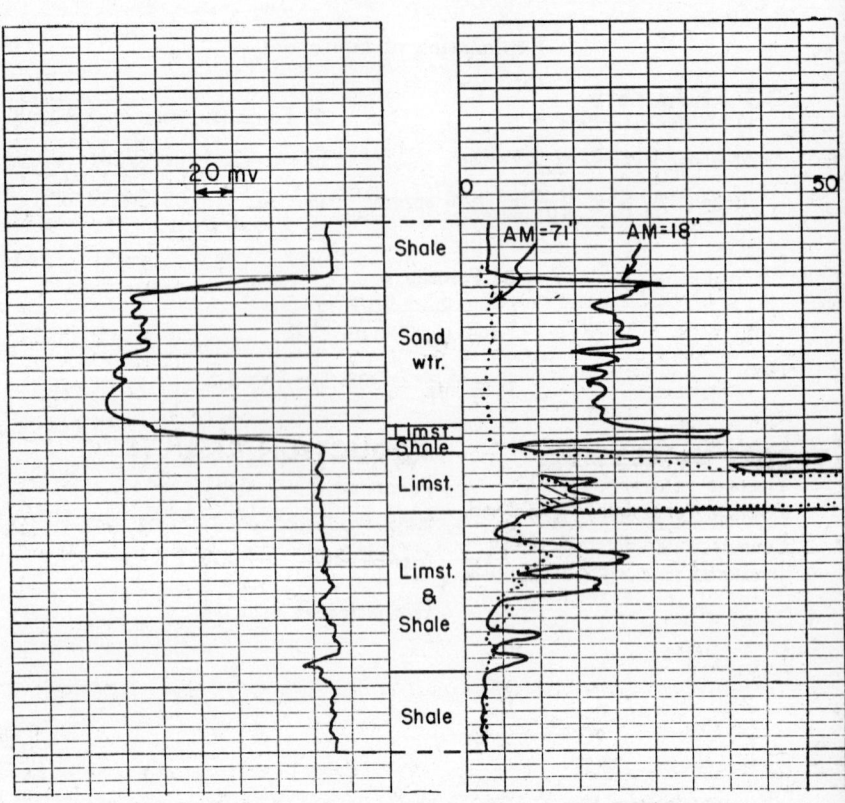

FIGURE 170. Portion of electrical log, showing typical profile pattern reflecting interbedded limestone, shale, and a salt-water-bearing sandstone.

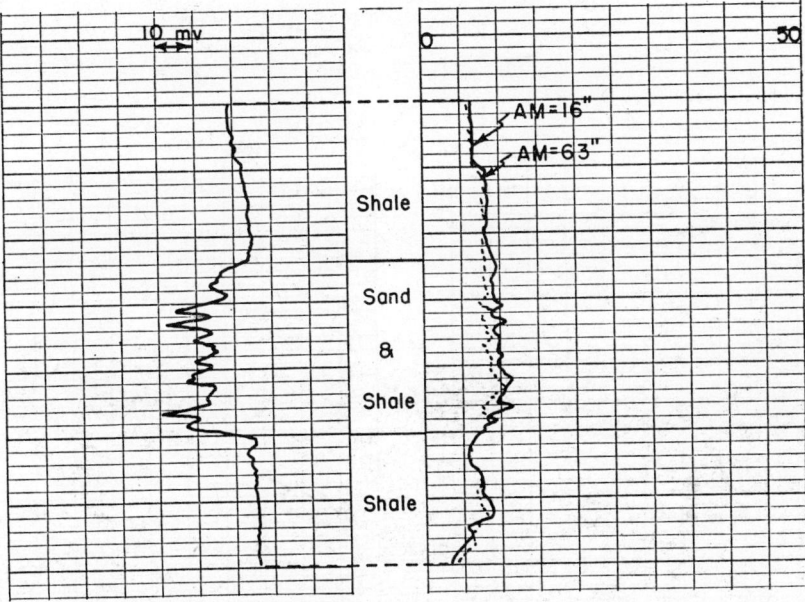

FIGURE 171. Portion of electrical log showing typical profile pattern of a section consisting of thin interbedded sandstones and shale (Cretaceous of Wyoming).

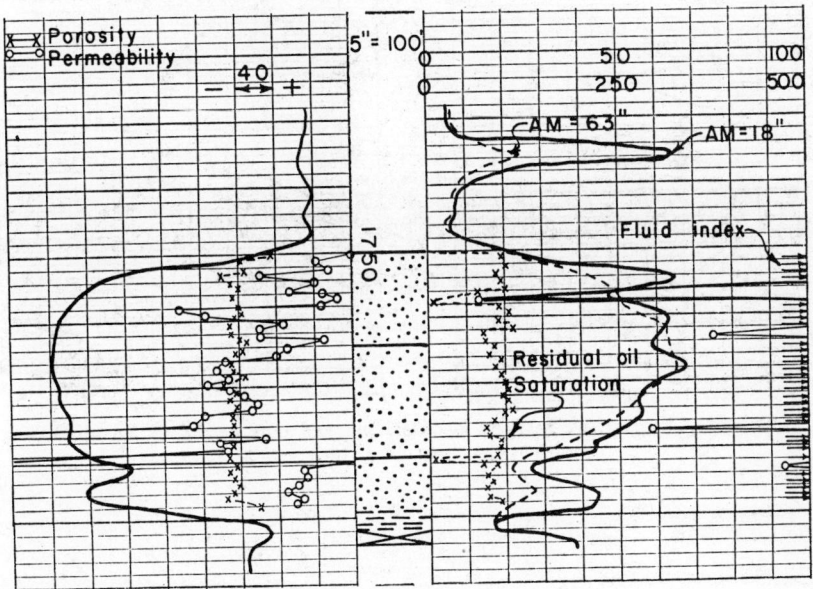

FIGURE 172. Portion of electrical log showing typical profiles produced by an oil sand, and a comparison with the core record and core analysis (Mississippian of Illinois).

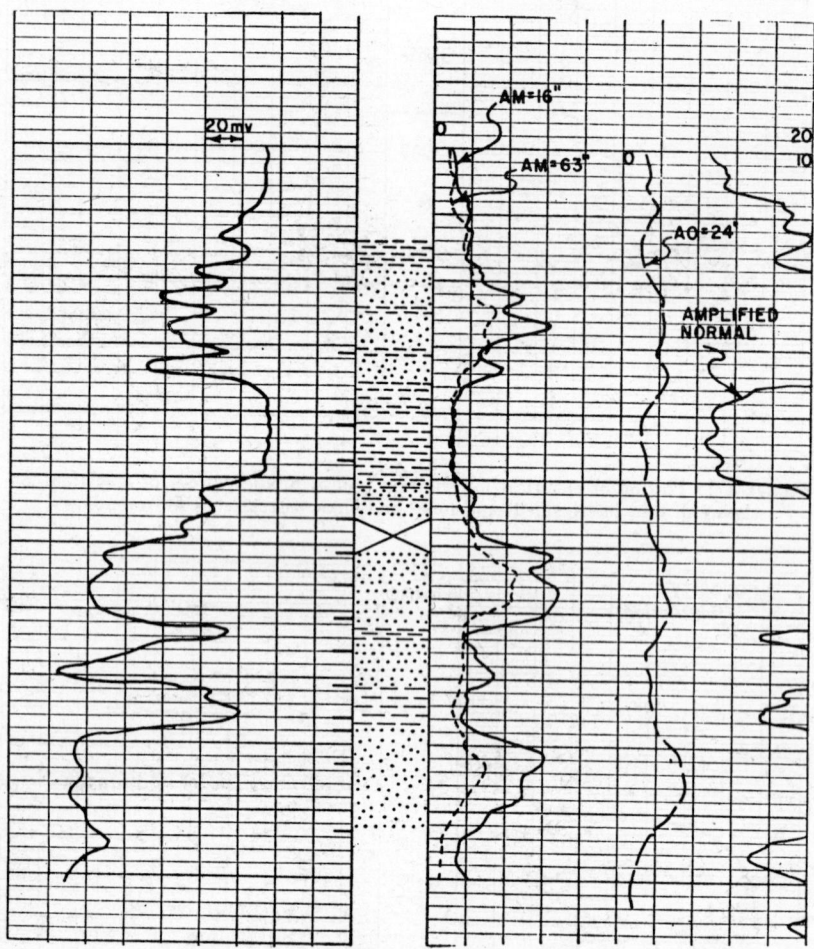

FIGURE 173. Comparison between electrical log and core record in an alternating sandstone-shale section (Texas Tertiary).

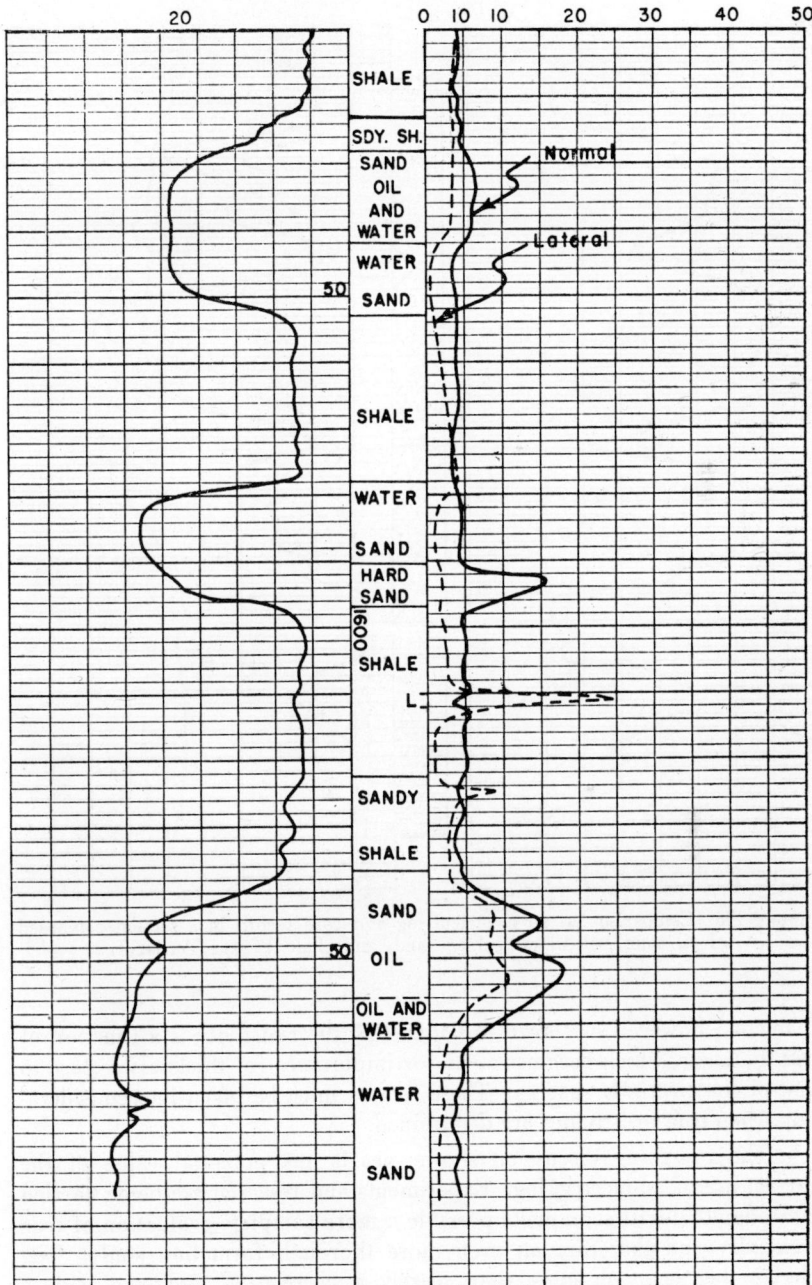

FIGURE 174. Portion of electrical log through a section composed essentially of shale, thin limestone, oil sands, and salt-water-bearing sands (Texas Tertiary).

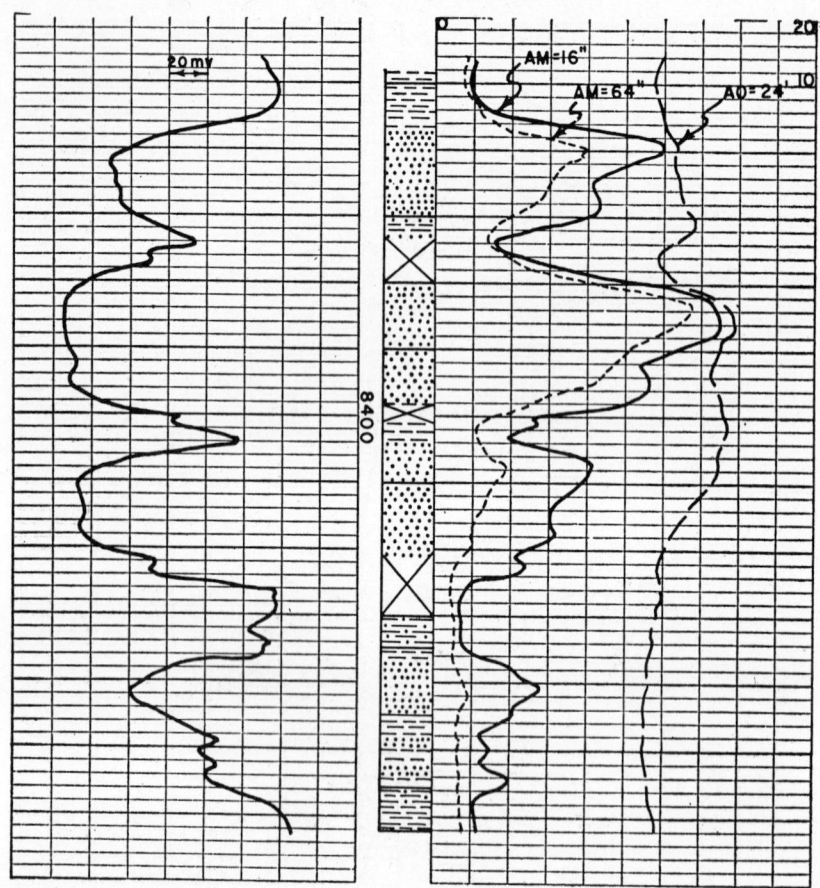

FIGURE 175. Comparison of an electrical log and core record in a section composed of oil sand, salt-water-bearing sand, and shale (Texas Tertiary).

diameter; however, as shown in figure 163, the resistivity is a function of bed thickness, which becomes of major importance for thicknesses close to that of the electrode spacing. These factors must be taken into consideration when true resistivities are determined.

The foregoing discussion shows clearly the potential value of the correct use of an electric log. Experiments and research are being carried out today that will soon make possible a more complete evaluation of true resistivity and therefore an even more thorough formation study. The electric log is much more than merely a correlation tool, and with a complete understanding of the factors influencing the log much valuable information may be obtained.

INDUCTION LOGGING AND ITS APPLICATION TO LOGGING OF WELLS DRILLED WITH OIL-BASE MUD

H. G. DOLL

The measurement of the resistivity of the formations traversed by drill holes has become standard practice in oil-well drilling during the last twenty years. The technique used requires that direct contact be made with the mud filling the bore hole by means of electrodes connected to the insulated conductors of the supporting cable. A current of constant intensity is generally made to flow in the surrounding medium through one or two of these electrodes called "power electrodes." It produces in the surrounding medium, by ohmic effect, potential differences which are proportional to its average resistivity. These potential differences are picked up by one or more measuring electrodes and are recorded continuously at the surface of the ground, giving the resistivity log.

There are cases, however, where a direct contact between the electrodes and the drilling mud is not possible; for instance, in holes drilled with cable tools, which are generally dry, or in holes where nonconductive oil-base mud is used in rotary drilling. The conventional electric-logging method then requires scratcher electrodes, which are forced by springs on the wall of the hole to make direct contact with the formations. In some cases the results are fairly satisfactory, but sometimes, particularly in wells drilled through hard formations, the measurements are not reliable because of poor contacts with the formations. It is particularly for that reason that a new method of electric logging, known as "induction logging," has been introduced for resistivity measurements in oil-base mud.

The induction-logging system does not require any direct contact with the mud or with the ground. As indicated by the name of the method, the formations surrounding the logging apparatus are energized by induction. To that effect, alternating current of appropriate frequency is made to flow through a coil, referred to as the "transmitter," which is supported by an insulating mandrel. The alternating magnetic field thus created generates eddy currents, which follow circular paths coaxial with the hole and the coil system, in the formations surrounding the hole. These eddy currents create a secondary magnetic field, which induces an electromotive force in a second coil, referred to as the "receiver," mounted on the same nonconductive mandrel at a certain distance called "spacing" from the transmitter.

If the transmitter current is maintained at a constant value, the intensity of the eddy currents is proportional to the conductivity of the ground. Thereby, the conductivity of the ground determines the secondary field created by the eddy currents and the signal generated in the receiver.

As in regular logging with electrodes, the signal is recorded continuously at the surface of the ground while the apparatus is moved along the hole. The record thus produced, which is frequently called an "induction log" because of the way in which it is obtained, shows the variations of the ground conductivity—and, consequently, of its inverse, the ground resistivity—with respect to depth. It is, therefore, equivalent to the resistivity log obtained by the conventional method of electric logging with electrodes in water-base mud.

The advantages of the method are more immediate, especially in view of the difficulties encountered by the conventional method of electric logging. This does not mean that the induction-logging method will not work in water-base mud; on the contrary, it is believed that in that case also the method will have important advantages. Experience in water-base mud is, however, still very limited, not only because the available instruments were all applied to oil-base-mud operations, where they were badly needed, but also because certain improvements, which are still being studied, are to be introduced for best operation in water-base mud. This is why it is felt advisable not to discuss the use of induction logging in water-base mud at this time and to wait for the results of field tests that will be made for that case.

In conventional logging practice, the resistivity unit is the ohm-meter. Conductivities are expressed in mhos per meter. It is preferred, however, to use units of millimhos per meter for induction logging in order to get a range of values that does not require an extensive use of decimal figures. Accordingly,

$$C \text{ mmhos/m} = \frac{1,000}{R \text{ ohm/m}}$$

Thereby, a bed with a resistivity of 100 ohm/m has a conductivity of 10 mmhos/m.

RESISTIVITY MEASUREMENTS BY INDUCTION LOGGING

The apparatus used for induction logging is shown schematically in figure 176; it is in fact a mutual-impedance bridge. It comprises essentially a transmitter coil T, fed with alternating current by an oscillator, and a receiver coil R connected through an amplifier to the recording galvanometer. In the absence of any conductive medium around the apparatus, as, for example, when it is suspended in the air from a wood frame high enough above ground, the coupling between the transmitter and receiver coils is fully balanced, so that the measuring apparatus reads "zero." When the apparatus is in a drill hole, the alternating field set up by the transmitter coil produces in the surrounding medium, i.e., in the ground, induced currents, generally known as "eddy currents," which are proportional to the conductivity of the ground. The electromotive force induced in the receiver coil by the eddy currents, referred to hereafter as the "signal," and designated by E, is proportional to the conductivity of the ground. If, therefore, the apparatus is properly calibrated,

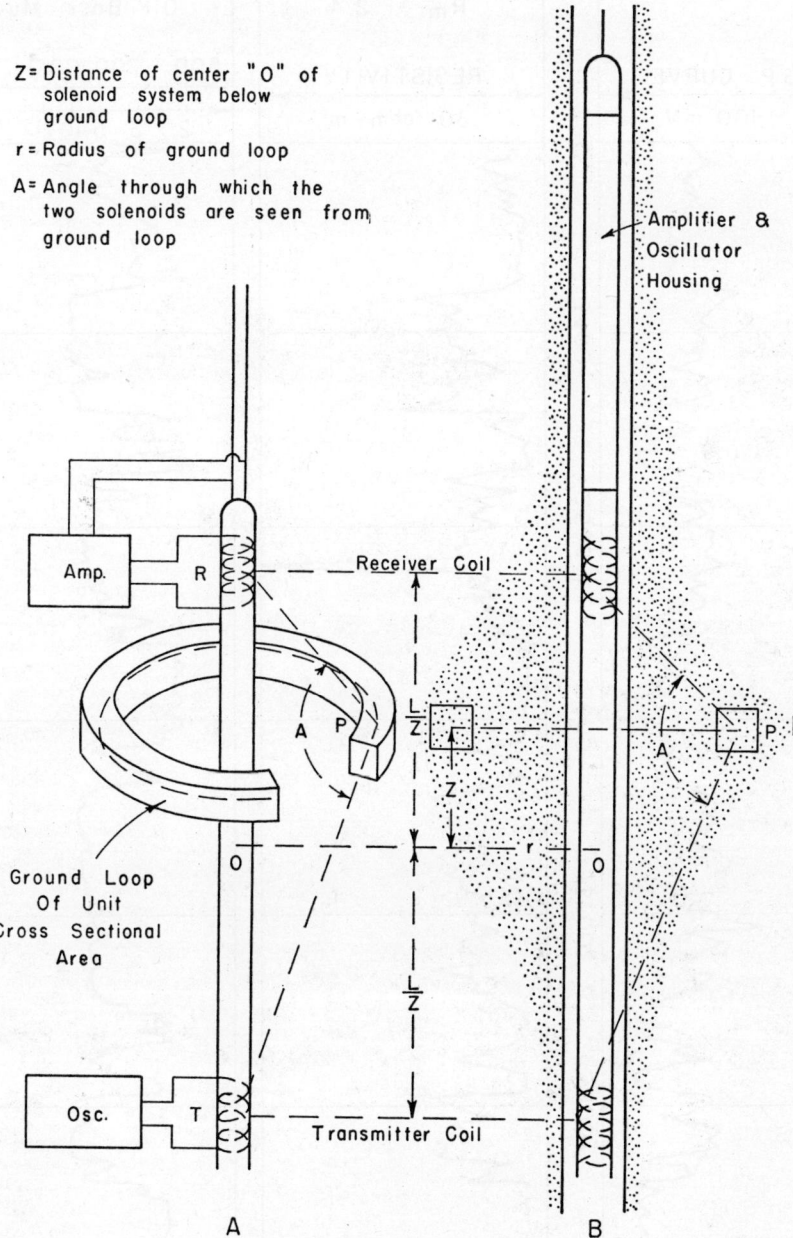

Z = Distance of center "O" of solenoid system below ground loop

r = Radius of ground loop

A = Angle through which the two solenoids are seen from ground loop

Amplifier & Oscillator Housing

Amp.

R

Receiver Coil

A P

$\frac{L}{2}$

Z

r

O

O

Ground Loop Of Unit Cross Sectional Area

A P

$\frac{L}{2}$

Osc.

T

Transmitter Coil

A   B

FIGURE 176. Elecetrical principle (A) and apparatus (B) used for induction logging. (From *Oil and Gas Jour.*)

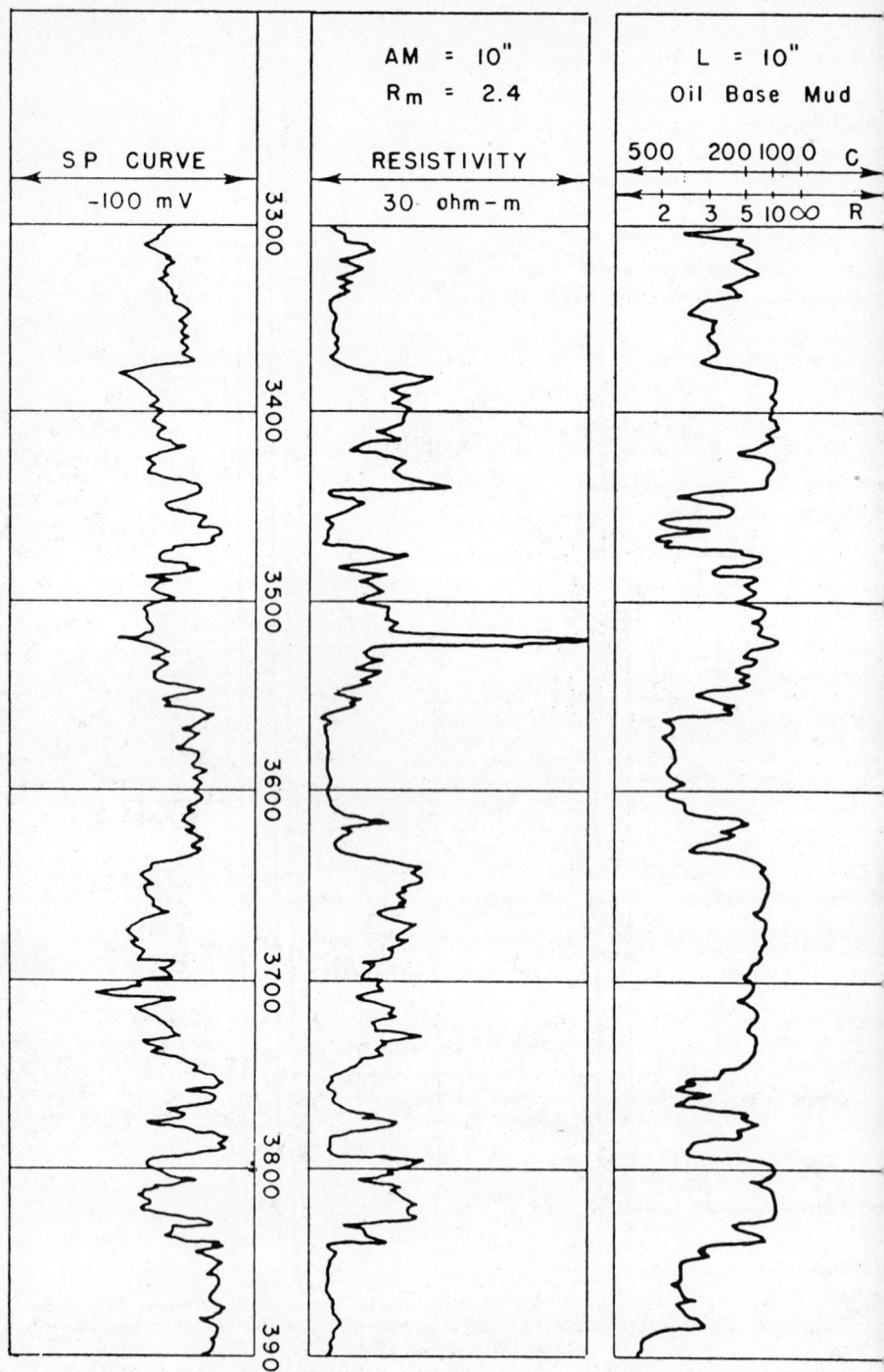

FIGURE 177. Induction log (right) recorded in oil-base mud, alongside a conventional electric log (left) of same well recorded later in water-base mud.

a measure of the signal constitutes a quantitative determination of the conductivity of the ground.

The signal is amplified and rectified into direct current for transmission in the cable to the surface where it is automatically recorded. A remote-controlled test signal is provided in the apparatus to check the calibration.

The oscillator and the amplifier are contained in a pressure-proof housing, called the "electronic cartridge," on top of the coil assembly. The subsurface instrument is represented schematically in B of figure 176. An induction log recorded in oil-base mud by this equipment is given in figure 177, alongside the conventional electric log of the same well recorded later in water-base mud for comparison.

When the ground surrounding the coil system is homogeneous, as is practically the case for a thick bed which is not appreciably invaded by the mud fluid, the conductivity, as measured by the apparatus, is equal to the true conductivity of the ground. When, however, the ground around the coil system is not homogeneous, as, for example, in the case of a thin bed surrounded by formations of appreciably different conductivity, the conductivity, as measured by the apparatus, represents a combination of the conductivities of the different media surrounding the coil system and is referred to as the "apparent conductivity." This is similar to what happens for electric logging with electrodes, where the apparatus also measures an apparent resistivity. In both cases, a better approximation of the true conductivity can be obtained by applying corrections deduced from the departure curves [17] or correction charts.

An important advantage of the induction-logging system is that the measured values, even without corrections, are already nearer to the true values; furthermore, the corrections themselves are much easier to compute than in the case of logging with electrodes, particularly when influence of bed thickness is to be taken into consideration.

Geometry of Induction Logging

In the logging method using electrodes for the determination of the ground resistivity the flow of current is of the radial type, and it is not possible to study separately the influence of the different regions of ground surrounding the electrode system. The reason is that the lines of current flow cross the boundaries between the different media, such as, for example, the boundary between a given bed and the bed next to it or the boundary between the mud and a bed. If the resistivity of any given medium is changed, this affects the lines of current flow even in their path through the other media. This is the reason that the mathematical com-

[17] Departure curves for electric logging with electrodes have been published earlier in a booklet entitled "Resistivity Departure Curves," 1947, by Schlumberger Well Surveying Corporation, Houston, Texas. The application of the curves was discussed in a paper on "True Resistivity Determination from the Electric Log—Its Application to Log Analysis," by H. G. Doll, J. C. Legrand, and E. F. Stratton, presented at the 1947 spring meeting of the Pacific Coast District, Division of Production, American Petroleum Institute, at Los Angeles, California.

putation of departure curves is rather complicated and can only lead to a fair approximation when the beds become thinner and more homogeneous.

In induction logging the situation is entirely different. If the hole is vertical, as will be assumed to simplify the discussion, the lines of current flow are horizontal circumferences having their centers on the axis of the hole. Since there is generally a symmetry of revolution of the ground around the axis of the drill hole, each line of current flow remains in the same medium all along its path and never crosses a boundary between media of different conductivities. On the other hand, if the frequency is not extremely high, the reaction of the different circular currents on one another can be neglected. In this condition, the action of the different regions of ground, which individually have a symmetry of revolution around the hole, can be considered separately, and the measured signal is simply the sum of the individual signals given by the different regions. The consequence is that the theoretical computation of charts or of typical logs corresponding to any distribution of ground conductivities is always possible, if, of course, there be a symmetry of revolution, as is usually the case.

CONCLUSION

The conductivity or the resistivity of formations traversed by a drill hole can be determined by the induction-logging method. This new technique is particularly useful at present for logging dry holes and holes filled with oil-base mud, in which direct contact with the formations is difficult to establish.

The method has great flexibility and is quite promising. Coil systems can be designed to give a focusing effect in order to obtain directly on the log a more accurate value for the conductivity of beds of finite thickness.

Since the different regions of ground generally have a symmetry of revolution around the axis of the hole and since, therefore, the induced currents have circular paths around that axis, the currents never cross the boundary from one region to the other. In these conditions the contributions of the different regions to the measured signal can be considered independently. For that reason it is relatively easy to compute typical logs and correction charts, which should greatly improve the possibilities of quantitative interpretation.

THE MICROLOG

H. G. DOLL

In conventional electrical logging, the spontaneous potential (SP) log is used to delineate the permeable beds, and the resistivity logs are used primarily to provide indications concerning the fluid content of the beds.

When the formations are much more resistive than the mud, as happens, for example, in limestone fields, the SP currents are short-circuited by the more conductive mud column, with the result that the SP log is quite rounded. In that case, the SP log generally gives the approximate location of the permeable formations, but it cannot be used for an accurate determination of the boundaries of each permeable bed.[18]

Solutions for the problem of obtaining a better determination of the permeable beds in limestone fields were developed from two angles. One approach consisted in improvements of the logging of the SP, as given by Selective SP logging and Static SP logging.[19] These new methods, which have been described in an earlier paper, give good results when the mud is not too salty; but they are still in a somewhat experimental stage, mostly because the development efforts have lately been concentrated on another approach to the problem, i.e. the microlog.

The microlog, which is the subject of the present paper, has been developed primarily as a means for the accurate determination of the permeable beds, where the SP log alone does not give a satisfactory answer. For that reason, this new development has found its first field of application in limestone areas, where the usefulness of micrologging is most obvious. The microlog is, however, also of importance in sand and shale formations, if only for a more precise determination of the boundaries between successive beds, and for a better evaluation of the sand count.

It is emphasized that the present paper is not intended to give an exhaustive and definitive description of the subject. In fact, the application of the new method has had a partly experimental character up to the present time, and several features of the corresponding technique are still being improved. It is possible that some of the improvements now under way will modify, to a certain extent, the response of the micrologs and the procedure of interpretation. These differences, however, should not bring about any fundamental changes in the principle of this method, and should rather make its application easier and more reliable.

Principle of Micrologging

A microlog is a resistivity log recorded with electrodes spaced at short distances from each other in an insulating pad which is pressed

---

[18] Doll, H. G., *The SP Log: Theoretical Analysis and Principles of Interpretation:* Petroleum Technology, vol. II, or Transactions AIME, Petroleum Branch, vol. 179, p. 146, Sept. 1948.
[19] Doll, H. G., *Selective SP Logging:* Paper presented at the AIME Columbus, Ohio Meeting Sept. 25-28, 1949, and at the San Antonio, Texas meeting, Oct. 5-7, 1949.

against the wall of the drill hole. Under those conditions, the system measures the average resistivity of the small volume of material—hereinafter referred to as a "microvolume"—which is located under the pad, and which is, therefore, electrically shielded against the short-circuiting action of the mud. Two different electrode systems, with different depths of investigation, are generally used in combination to provide two logs that are recorded simultaneously. For both electrode systems, the spacings are very small—usually one inch or two inches. In the discussion, the systems which have the smallest and the largest depth of investigation are respectively referred to as the "short spacing" and the "long spacing."

When the pad is applied to a permeable bed, the mud cake represents a substantial porportion of the microvolume. Inasmuch as the mud cake has a resistivity R_{mc} which can be estimated to be only about twice the resistivity R_m of the mud, the resistivities recorded through micrologging—hereinafter referred to as micro-resistivities—are never very high opposite permeable beds, and are appreciably related to the resistivity of the mud. The other part of the microvolume is constituted by a fraction of the solid structure of the permeable bed whose pores are almost completely filled by the mud filtrate. The resistivity of that part of the microvolume is, therefore, not much different from the value $F \times R_m$ [20] which corresponds to complete mud filtrate saturation, so that it is also directly related to the resistivity of the mud.

It can easily be deduced from these considerations that the microresistivities measured opposite a permeable bed cannot generally be higher than a certain number of times the resistivity of the mud, unless the mud cake is very thin, and unless, simultaneously, the formation factor F is very high. A corresponding limit R_{lim} can be set at about 20 or 30 times the resistivity of the mud for the average case; therefore one of the rules of interpretation for the microlog is to classify as most probably impervious all formations for which the microresistivities are higher than a certain limit R_{lim} directly related to the resistivity of the mud.

Because of a smaller depth of investigation, the short spacing is more influenced by the mud cake, and, therefore, generally gives a smaller apparent resistivity than the long spacing. This difference between the microresistivities recorded using two different depths of investigation is called "departure," and is said to be positive when the longer spacing gives the larger resistivity. When there is a large percentage of "positive departure," the formation can almost certainly be interpreted as permeable.

When the pad is applied to an impervious bed of low resistivity, both spacings measure substantially the same resistivity, which is that of the formation, and there is no appreciable departure between the microresistivity curves. If the resistivity of the impervious bed is very high, the microresistivities can differ appreciably from the formation resistivity,

[20] Archie, G. E., *The Electrical Resistivity Log as an Aid in Determining Some Reservoir Characteristics:* Petroleum Technology, vol. J, 1942. F is the formation factor.

but they are both higher than the limit R_{lim} above which beds should be classified as impervious. For intermediate values of the resistivity, and because of the limited dimensions of the pad, a departure between the

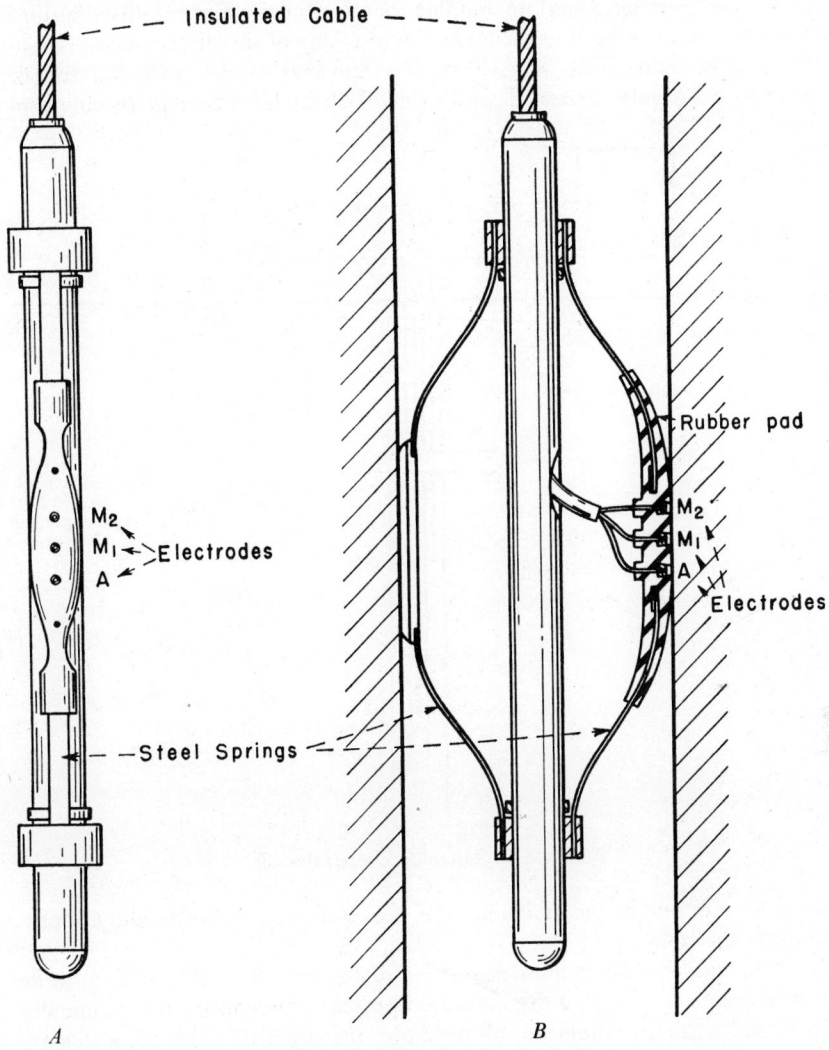

FIGURE 178. Micrologging apparatus.

microresistivity curves is sometimes observed on impervious beds; but, in that case, the departure is negative; i.e. the longer spacing gives the smaller value of the apparent resistivity, so that there can be no confusion in the interpretation.

These different features of the interpretation will be discussed in more detail in a later section.

DESCRIPTION OF THE EQUIPMENT

The micrologging apparatus consists essentially of a rubber pad, which is pressed against the wall of the drill hole, and in the face of which are inserted a certain number of electrodes. Several distributions of electrodes have been experimented with. One of the distributions is represented in figure 178. The electrodes are nearly flush with the rubber surface, or slightly recessed, and each of them is connected by an insu-

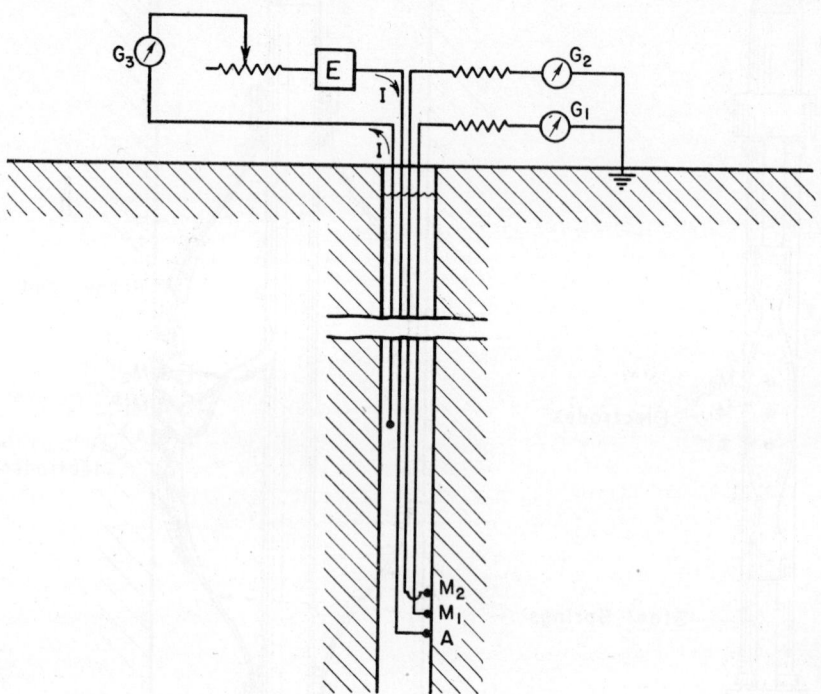

FIGURE 179. Microlog electrical setup.

lated wire to one of the conductors of the cable used to lower the apparatus into the hole.

The rubber pad is molded on one of the branches of a spring guide whose design is such that the pressure applied to the pad is approximately independent of the diameter of the hole, provided that this diameter remains between certain limits which, for one of the guides presently in use, are respectively $4\frac{1}{2}$ inches and 16 inches. The rubber pad fits the wall of the drill hole over a substantial area surrounding the electrodes because of its shape and the pressure exerted upon it. The pad also shields the electrodes from the mud column, while the electrodes themselves are in direct electrical contact either with the formation, or with the mud cake between the pad and the formation.

The resistivity of the small volume of ground surrounding the electrodes can be measured, for example, by sending a current of known intensity I through electrode A (fig. 179), and by measuring with galvanometer G_1, the potential difference created by that current between electrode M_1 and the reference electrode N.

Similarly, a slightly larger volume of ground can be taken into account by measuring, with galvanometer G_2, the potential produced at electrode M_2 by the same current. Because the spacing AM_2 is twice as long as the spacing AM_1, the corresponding investigation is also twice as deep. This is an important feature of the system, it is possible not only to measure the average resistivity of the ground under the pad, but also to determine whether or not the resistivity varies with depth from the pad. This determination is of significance, for a resistivity variation with depth from the pad usually occurs when there is a mud cake.

Electrode combinations other than AM_1 and AM_2 can, of course, be used. When a current is sent through electrode A, it is also possible to measure the potential difference between electrodes M_1 and M_2. The three-electrode system AM_1M_2 is more influenced by the mud cake than the two-electrode system AM_1, so that a combination of the device AM_1M_2 and the device AM_2 would generally give a larger departure opposite permeable beds than a combination of the two devices AM_1 and AM_2.

To simplify the wording, the two electrode and three electrode systems are called "normal" and "lateral" devices, respectively. It should be pointed out, however, that micrologging devices, because they have much smaller depth of investigation and because the electrodes are shielded from the mud column, are different from the normal and lateral devices used in conventional logging.[21]

On most of the pads presently in use, the three electrodes are placed on a vertical line, in the middle of the pad, with a spacing of one inch between the successive electrodes. Three electrode combinations which are thus obtained are listed below for reference.

Measuring device
AM_1 normal
AM^2 normal
AM_1M_2 lateral
    Reference
    $1''$ normal (or $1''$)
    $2''$ normal (or $2''$)
    $1'' \times 1''$ lateral (or $1'' \times 1''$)

The combination of the $1'' \times 1''$ lateral and the $2''$ normal is preferred at the present time.

The electric circuit used for the recording of the microresistivity curves is somewhat similar to that used for conventional logging. Changes

[21] Doll, H. G., Legrand, J. C., Stratton, E. F., True Resistivity Determination from the Electric Log— Its Application to Log Anaylsis: Oil and Gas Jour., vol. 46, no. 20, Sept. 20, p. 297 ff., 1947.

have been introduced, however, to compensate for the high resistance to ground of the small electrodes.

INTERPRETATION OF MICROLOGS

A discussion of the interpretation of the micrologs will be aided by placing the different cases usually encountered in practice into categories;

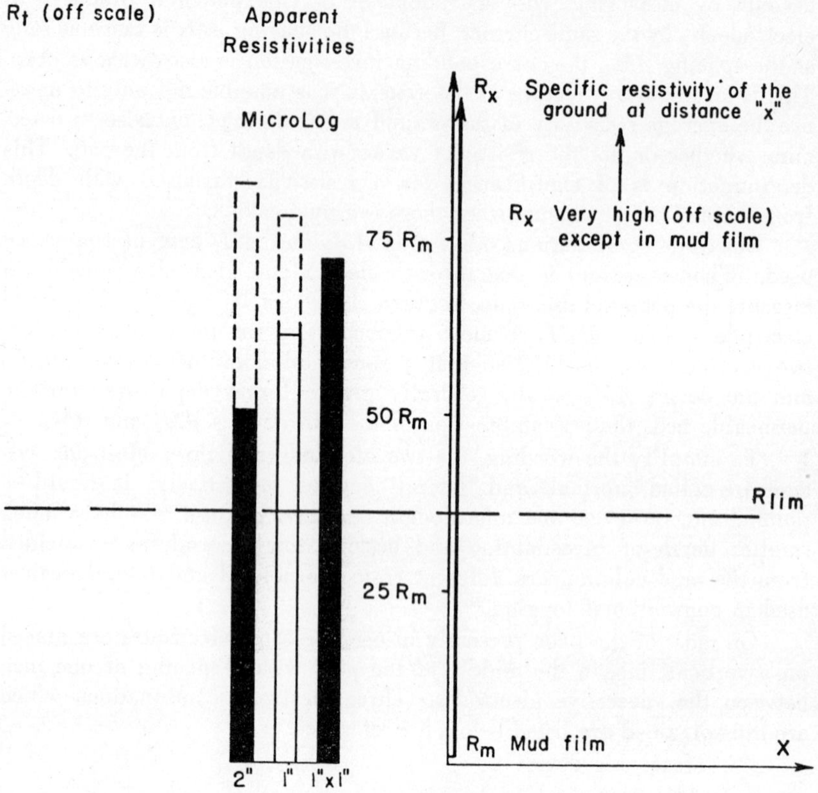

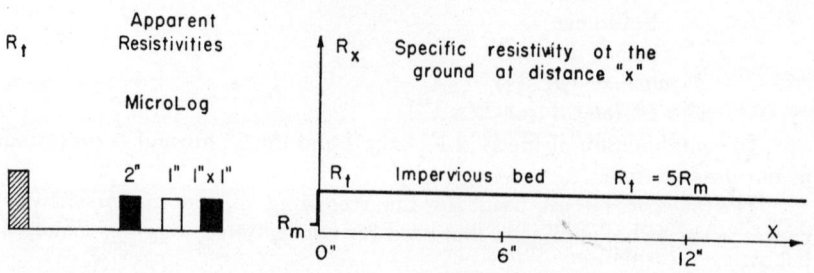

FIGURE 180. Impervious formation of high resistivity with a mud film of 1/64-inch thickness.

FIGURE 181. Impervious bed of low-resistivity shale (category 1₃).

which will be illustrated by schematic drawings (figs. 180, 181, 182, 183).
Each category will be given a reference letter, namely, I_1, I_2, I_3 (Impervious beds), P_1, P_2 (Permeable beds). These letters will be placed on the field examples exhibited (figs. 184 through 189) at the levels of each section of bore hole which can be considered as belonging to the corresponding category.

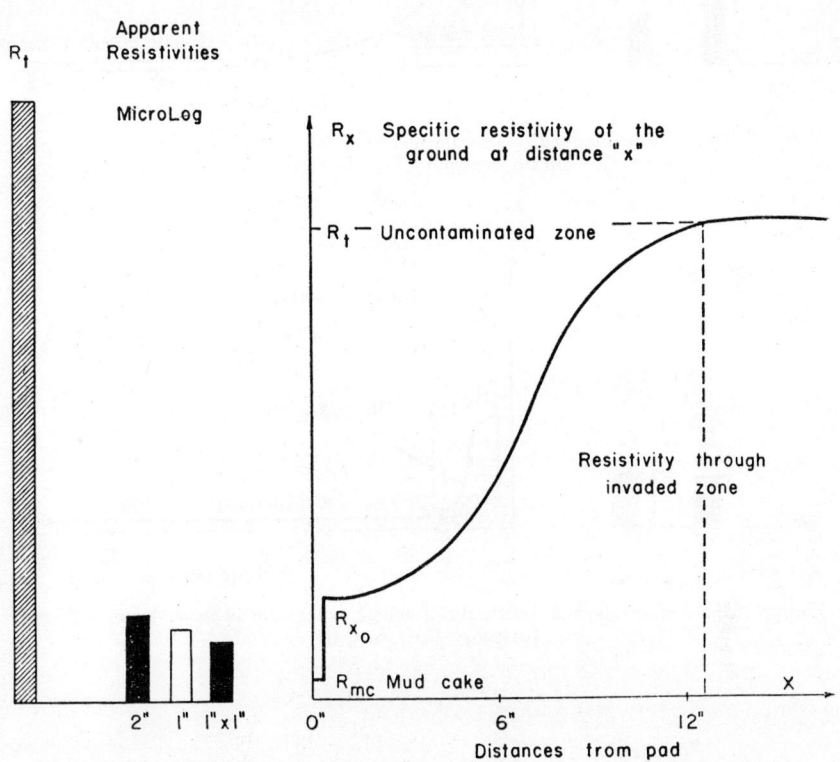

FIGURE 182. Permeable bed (oil-bearing) invaded by mud filtrate (category P_2).
$$R_t < R_{x_0}.$$

Highly Resistive Impervious Formations (Category I_1)

In impervious formations of high resistivity—i.e. whose resistivity is, for example, more than 50 times that of the mud—all the microresistivities are high. Because the wall of the drilled hole is generally somewhat rugose, the rubber pad cannot fit perfectly against it, and a sort of mud film may remain between the two. For that reason, and also because of the limited dimension of the rubber pad, the apparent resistivity recorded on the micrologs can be substantially lower than the true resistivity for that type of formation. When the formation resistivity is very high, the rugosity of the wall, which causes a mud film to remain under the pad,

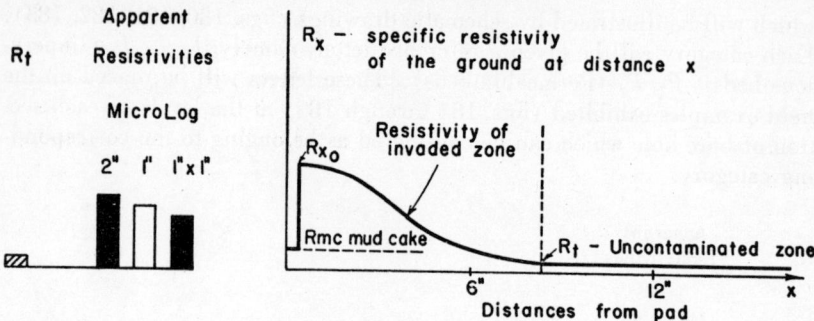

FIGURE 183a. Permeable bed (salt-water-bearing) with moderate invasion by mud filtrate (category P_2). $R_t < R_{x_0}$.

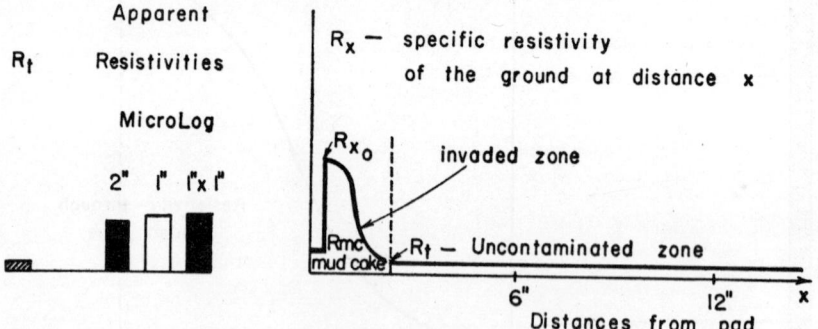

FIGURE 183b. Permeable bed (salt-water-bearing) with little invasion by mud filtrate (category P_1). $R_t < R_{x_0}$.

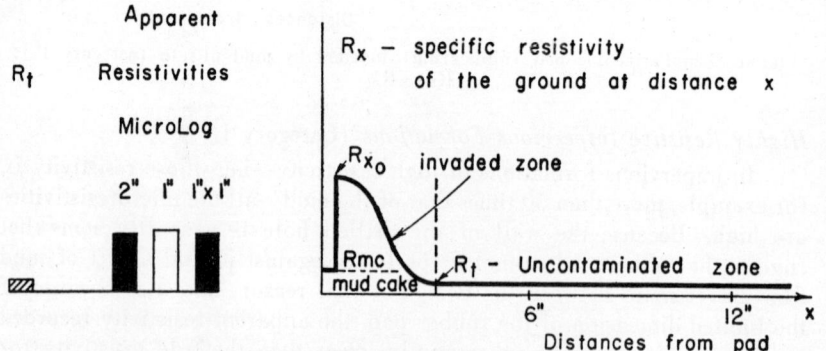

FIGURE 183c. Permeable bed (salt-water-bearing) with very little invasion by mud filtrate (category P_1). $R_t < R_{x_0}$.

almost entirely controls the microresistivities. Too much importance should, therefore, not be attributed to the absolute values determined for the microresistivities, nor to the amount and type of departure between the two curves.

The diagnostic is based in this category on the fact that all microresistivities are superior to a certain limit R_{lim} which, as said before, can be taken equal to about 20 or 30 times the resistivity of the mud in the average cases. Such high microresistivities on both micrologs could not normally be recorded for a permeable formation. Both the invaded zone of the permeable formation and the mud cake would contribute to lower

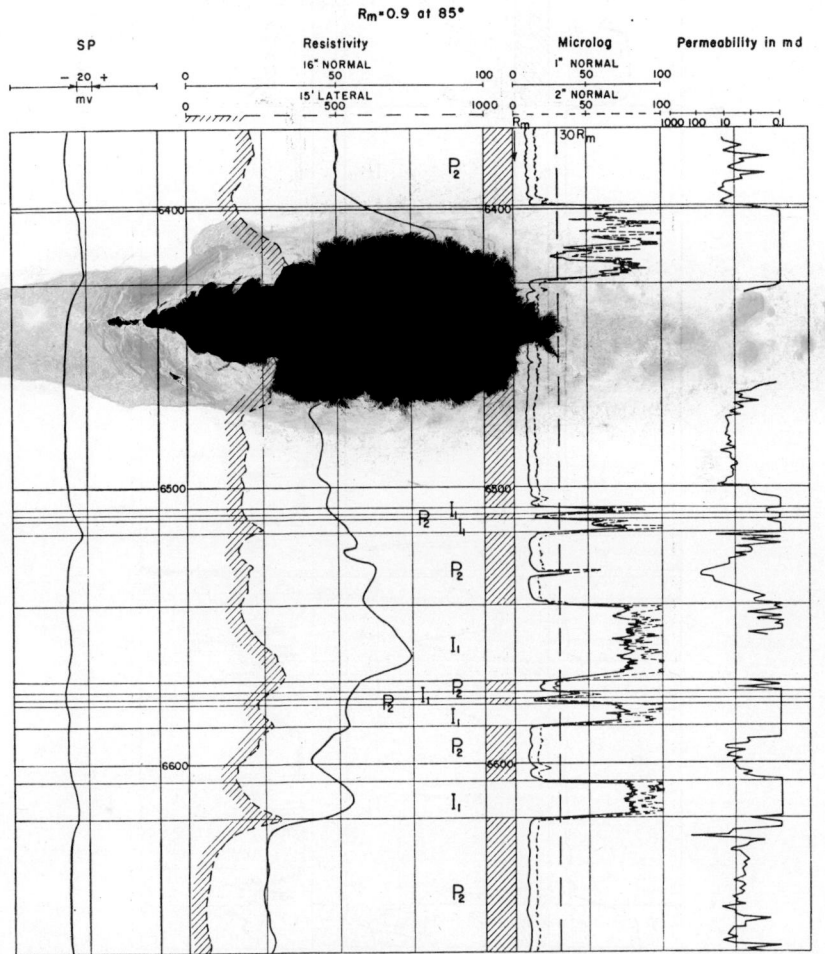

FIGURE 184. Example of microlog.

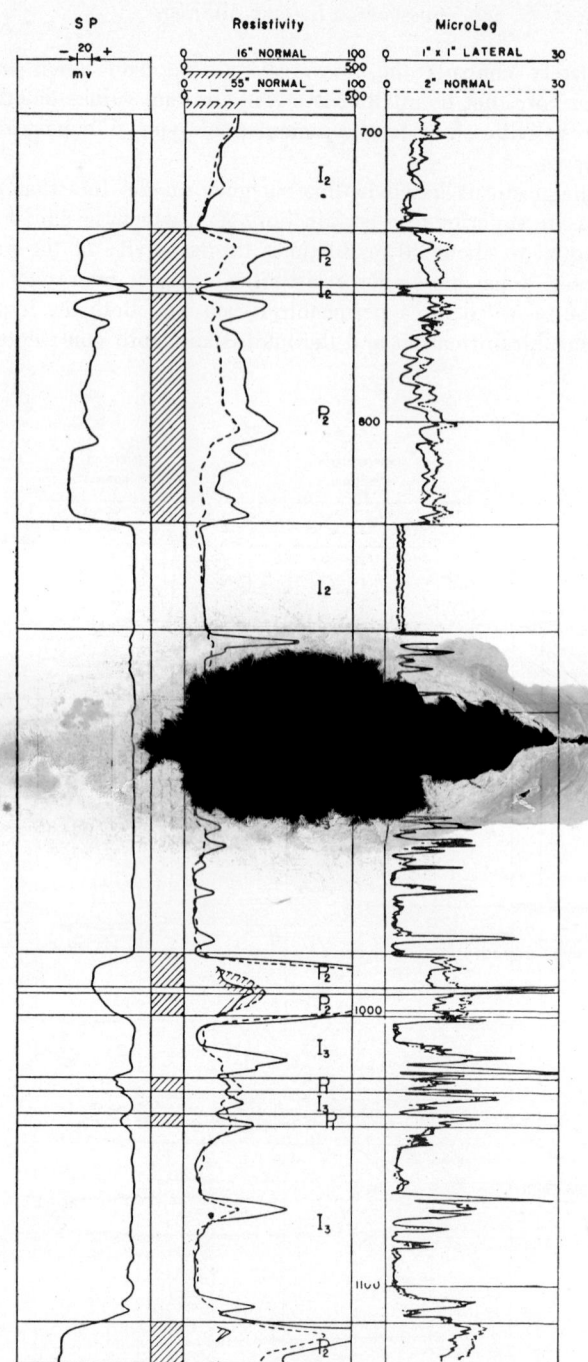

FIGURE 185. Example of microlog.

the apparent resistivity on the micrologs, and keep at least one of them, the 1 x 1-in. lateral or the 1-in. normal, under the limit R_{1im}.

The case of a compact bed of high resistivity is illustrated in figure 180, where it is supposed that the formation is more than 200 times as re-

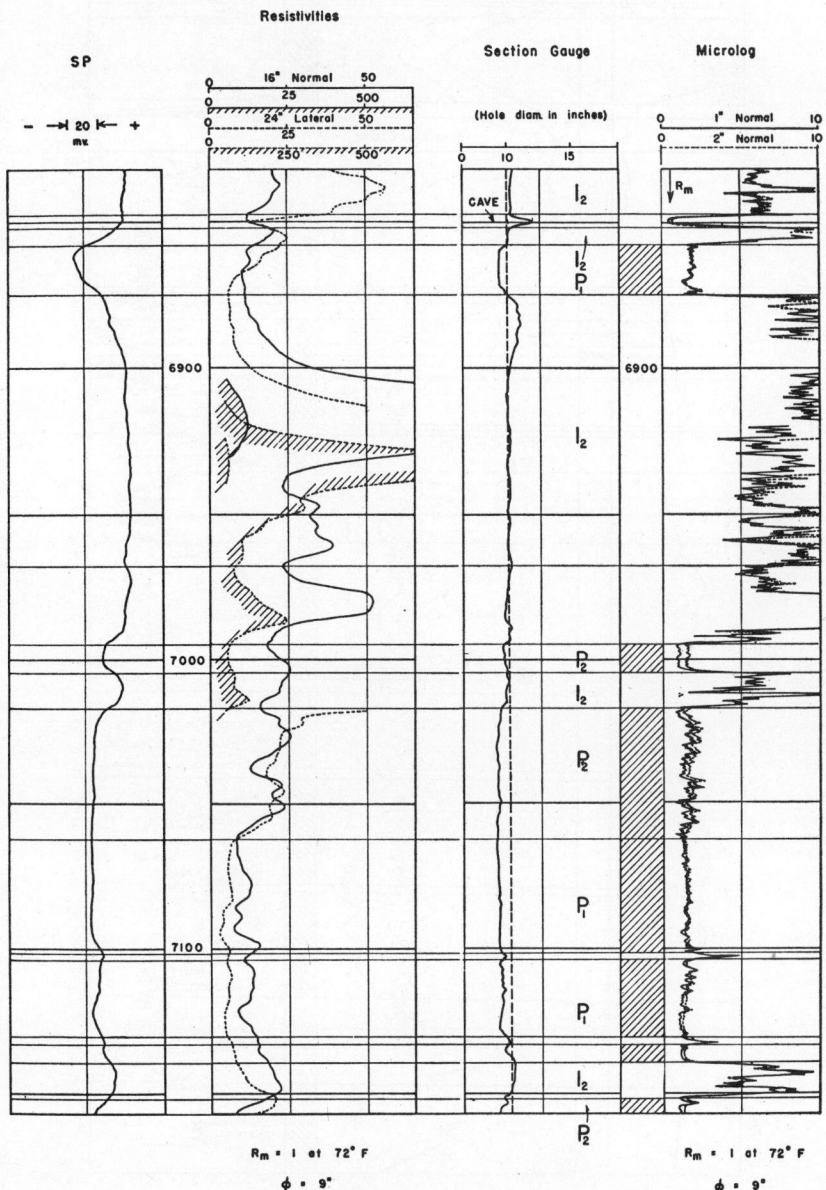

FIGURE 186. Example of microlog.

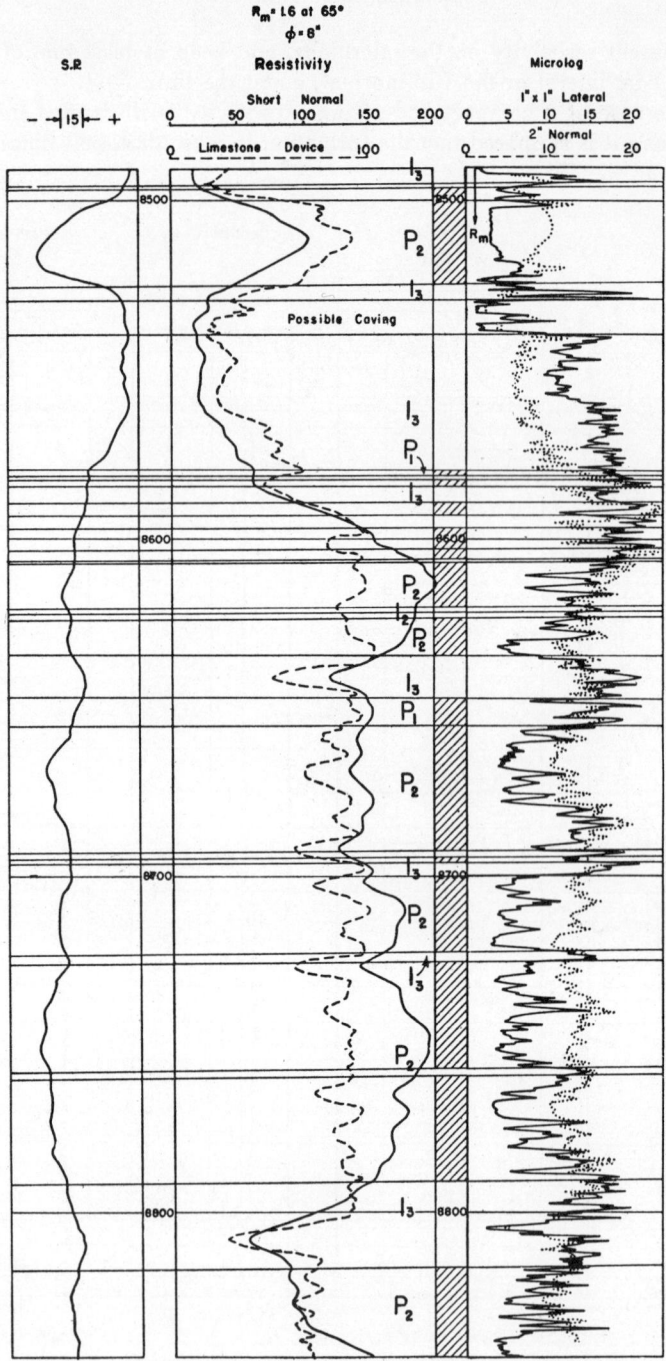

FIGURE 187. Example of microlog in vuggular limestone (Ellenburger).

sistive as the mud, and that there is, between the pad and the formation, a uniform mud film 1/64 inch thick. The order of magnitude of the different microresistivities is about the same, but there can be a slight departure either positive or negative, depending on the shape of the pad, the rugosity of the wall, the ratio of formation resistivity to mud resistivity, etc.

Impervious Beds of Low Resistivity (Categories I_2 and I_3)

Figure 181 represents an impervious bed whose resistivity has been

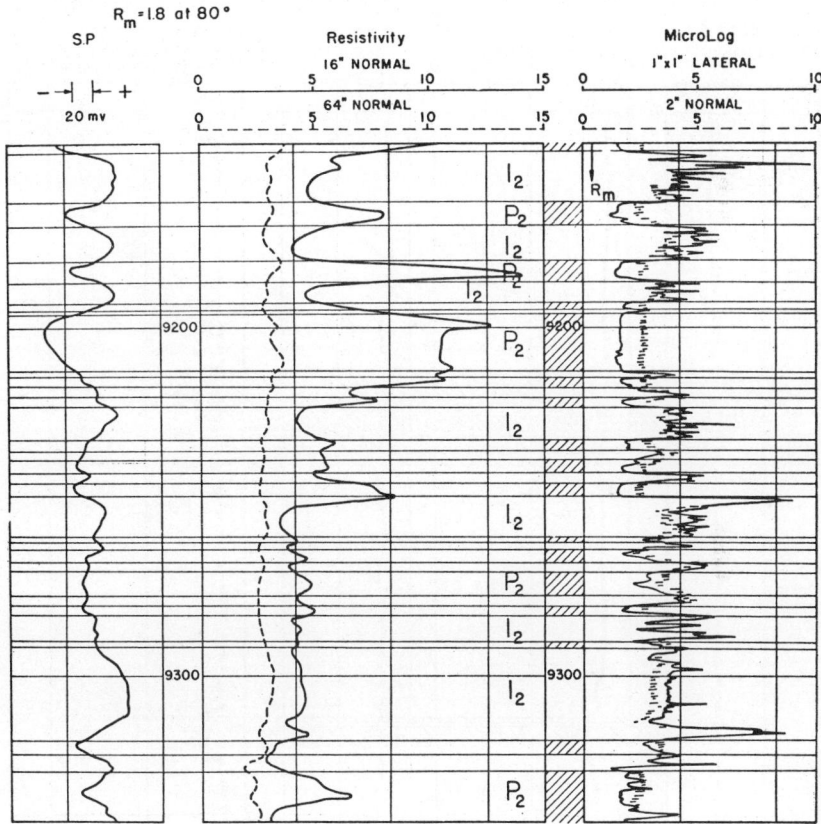

FIGURE 188. Example of microlog in shaly sandstone.

assumed to be five times the resistivity R_m of the mud. The resistivity diagram on the right of the figure illustrates the fact that the resistivity is uniform and does not vary with the distance from the pad, as would occur for a permeable bed. This diagram also shows the resistivity scale in relation to the resistivity R_m of the mud.

The rectangles at the left of figure 181 represent, at the same scale, the respective values of the different microresistivities that would be obtained in that case and the formation resistivity R_t. The apparent resistivities measured on the different microresistivity logs are slightly lower than the true resistivity R_t of the formation because of the limited dimensions of the pad.

When the resistivity of the impervious bed is not much different

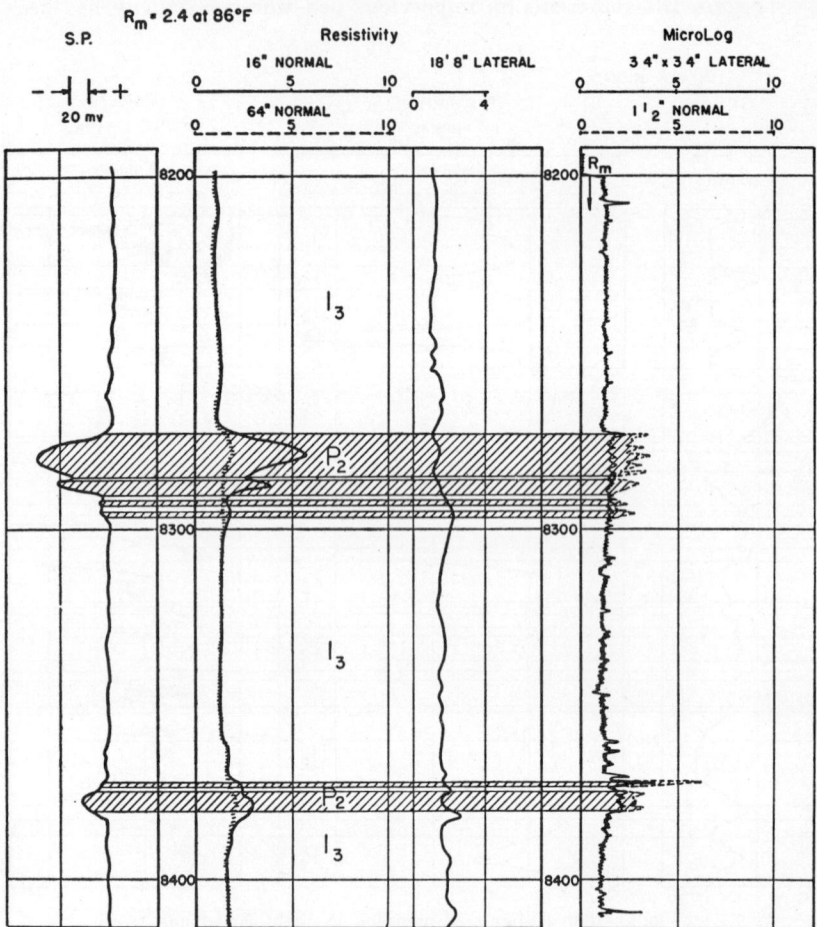

FIGURE 189. Example of microlog in shaly sandstone.

from that of the mud, there is practically no departure between the micro-resistivity curves (Category I_3). When, on the contrary, the resistivity of the impervious bed is appreciably higher than that of the mud, there is generally, at least with the pads presently used, a substantial negative departure which is characteristic of an impervious bed (Category I_2).

Permeable beds (Categories P_1 and P_2)

At the level of a permeable formation, the rubber pad slides over the mud cake against which it is applied, and the mud cake itself is separated from the uncontaminated part of the formation by the invaded zone, wherein the original fluid has progressively been replaced by mud filtrate.

Two cases must here be considered, depending on whether the invaded zone is less or more resistive than the uncontaminated zone.

The first case, which is the simplest as far as interpretation of the micrologs is concerned, is represented in figure 182, which is similar to figure 181 already discussed, except that now the resistivity varies in the formation with the distance from the pad. Here again, the abscissae on the diagram represent the depths from the pad, while the ordinates show the resistivity of the ground at the corresponding depths.

The resistivity R_{mc} of the mud cake has been assumed to be twice the resistivity R_m of the mud. The resistivity R_{x_0} immediately behind the mud cake, where the permeable bed should be practically saturated by mud filtrate, has been taken equal to 10 times R_m; this would correspond to a value of 10 for the formation factor, if the saturation by mud filtrate is complete, and if the pores are reasonably free of conductive solids.[22]

On the left of figure 182 are represented, at the same scale, the approximate values of the microresistivities that would be obtained in that particular case from the microlog. As can be seen, the departure is positive and quite substantial. This result is general when R_t is larger than R_{x_0}. A large positive departure between microcurves is characteristic of a permeable bed, provided, however, that the resistivity measured by the 1x1-in. lateral, or by the 1-in. normal, be lower than approximately 30 times that of the mud (Category P_2).

The interpretation is less definite for beds, such as salt-water-bearing beds, where the resistivity of the invaded zone is larger than that of the uncontaminated zone. This is particularly true when the mud is of the low water-loss type, with the consequence that the mud cake is thin and the formation is invaded by the mud filtrate to only a short distance from the wall.

When the mud cake has an appreciable thickness, and when, simultaneously, the depth of invasion is large enough, the microresistivities measured with the different electrode combinations show good positive departures (Category P_2). This effect is represented on figure 183a. For a smaller penetration of the mud filtrate into the permeable bed, the departure would disappear (Category P_1), as represented in figure 183b. For still less invasion, the departure might even be slightly negative, as represented in figure 183c, but this negative departure never exceeds 20 percent (Category P_1).

On the left-hand side of figures 183a, b, c are represented the ap-

---

[22] Panode, H. W., and Wyllie, M. R. J., *The Presence of Conductive Solid in Reservoir Rocks as a Factor in Electric Log Interpretation:* AIEE Meeting, San Antonio, Oct. 5-7, 1949.

proximate resistivities that would be measured on the microlog in the different cases, and for the different electrodes combinations with the assumption that R_{xo} is about five times as high as the resistivity of the mud cake, which again corresponds to a formation factor of about 10. With them are also represented the formation resistivities. When the departures between the microresistivities are small, nil, or slightly reversed (within 20 percent), the interpretation can be aided by the fact that these microresistivities are larger than the formation resistivity, contrary to what would normally happen for an impervious bed of low resistivity, as discussed in connection with figure 181 above. It is simpler, however, to refer to the SP log to resolve the ambiguity in this case.

In all the cases represented in figures 180, 181, 182, and 183, the microresistivity corresponding to the 1x1-in. lateral could be computed with reasonable accuracy. The microresistivities corresponding to the 1 and 2-in. normals are only approximate, because the effect of the limited side of the pad cannot be accounted for accurately with these devices.

It is interesting to notice that the microresistivity for the 1x1-in. lateral is the same for the different cases represented on figures 180, 181, 182, and 183, which cases differ only by the value of true resistivity R_t and the depth of invasion by the mud filtrate. This illustrates the fact that the microresistivities, when measured with an electrode combination having a very small depth of investigation, are essentially responsive to the resistivity R_{xo} of the invaded zone and to the thickness and resistivity of the mud cake.

SUMMARIZED RULES OF INTERPRETATION

From the above discussion, it is possible to derive a certain number of simple rules of interpretation for the microresistivity logs of formations, whereby two electrode combinations of different depths of investigation have been run simultaneously. These rules, which will apply in the great majority of cases, are schematically represented on chart 1.

Case I—The two microresistivities are higher than R_{lim}, that is, higher than about 20 or 30 times the mud resistivity. The formation then under study is a compact one, and should be interpreted as impervious, regardless of the departure (Category I_1).

Case II—The microresistivity determined by the shorter spacing, generally 1x1-in. lateral, is smaller than the limit R_{lim}. In that case, the sign and magnitude of departure should be examined, and, in case of doubt, the ambiguity resolved by reference to the SP log. This gives the following interpretation categories:

(a)   Large negative departure (more than 20 percent)—the formation is impervious (Category I_2).

(b)   The departure is not definite enough (less than 20 percent)— the microlog alone cannot determine, in general, whether the formation is permeable or not in this case, except that, when the resistivities of all

CHART 1

				Category
$R'_{1"x1"} > R_{lim}$			Impervious	I_1
	$R_2" < R'_{1"x1"}$ large negative departure		Impervious	$= I_2$
$R'_{1"x1"} < R_{lim}$	$R_2" \cong R'_{1"x1"}$ departure nil or small	S P trend positive	Impervious	$= I_3$
		S P trend negative	Permeable	$= P_1$
	$R_2" > R'_{1"x1"}$ large positive departure		Permeable	$= P_2$

$R_{lim} =$ to about $\begin{cases} \text{10-15 } R_m \text{ for fresh mud.} \\ \text{20-30 } R_m \text{ for average mud.} \\ \text{40-50 } R_m \text{ for very salty mud.} \end{cases}$

SP trend positive if convexity SP log toward positive, or if SP log on positive limit.
SP trend negative if convexity SP log toward negative, or if SP log on negative limit.

formations are known to be much higher than that of the mud, the fact that the microresistivities are comparatively low gives a good probability that this is due to mud infiltration in a permeable bed. The ambiguity, if any, can be resolved by noting the trend of the SP curve. If that trend is positive, i.e. if the convexity is toward the positive side, or if the SP is on the positive limit (shale line), then the bed must be interpreted as impervious and the microlog gives its exact boundaries (Category I_3). If, on the contrary, the trend of the SP log is negative, the bed is permeable, and again the microlog gives the accurate boundaries (Category P_1).

(c) Large positive departure (more than 20 percent)—the formation must be permeable (Category P_2).

Remarks

(1) It is indeed wise to check the trend of the SP log in any case; unless this log is completely flat, it should always be possible to determine the trend of the SP.

When the mud is saturated with salt, the SP curve is completely flat. In that case, however, there is generally little doubt about the interpretation of the microlog, because the permeable beds, which are invaded by the saturated mud filtrate, are the only ones to give microresistivities lower than the limit R_{lim}, which itself is lower than the true resistivity of all impervious beds. The interpretation can then be based on the observation of the lows in the microlog, while a good positive departure, if it exists, brings a useful confirmation.

(2) The interpretation is also facilitated by the consideration of the conventional resistivity log, insofar as this log makes it possible to evaluate the true formation resistivity R_t.

When the value of the microresistivity R_{micro} is less than R_{lim}, it can reasonably be assumed that a bed cannot be impervious unless R_t is comprised between values respectively equal to about R_{micro} and $2\,R_{micro}$. When R_t is between the two values, the SP log will generally give a definite anomaly that will make the interpretation safe. When R_t is not between the above limits, there is a strong probability that the bed is permeable.

(3) In the particular case of a highly resistive permeable bed, with a comparatively conductive invaded zone, and interbedded with shales and highly resistive compact formations, the SP log may show no appreciable deflection with respect to the shale line. But, in that case, a good correct departure should be obtained from the microlog.

Exceptions to the rules

There are few cases which do not fall within the simple set of rules discussed in the previous section. Fortunately, these cases are rare, and the ambiguity can generally be resolved if the remarks made in the previous section are taken into account.

A first exception, which is obvious, corresponds to cavings whose diameters are larger than the maximum expansion of the spring system. If, in such a case, the pad remains a few inches from the wall, the two microresistivities will be equal to the resistivity R_m of the mud, irrespective of what the formation is. If, on the contrary, the pad happens to be near to the wall, the 1x1-in. lateral will measure approximately the resistivity of the mud, whereas the 2-in. normal will already be substantially affected by the formation resistivity. Inasmuch as the formation resistivity is usually higher than that of the mud, even for a conductive shale, there will be a positive departure between the micrologs which could be falsely interpreted as indicating a permeable bed. These two cases can generally be detected by observing the abnormally low value given by the short spacing (1x1-in. lateral) for which the microresistivity is frequently close to R_m in that case, and they would not fail to be recognized if a section-gauge log were run, as is highly recommended when large cavings might exist. In those cases, the microlog indications should be disregarded, and the interpretation based on the conventional log and, in particular, on the SP log, as far as the determination of the permeable beds is concerned.

The interpretation rules could also be at fault if the mud cake, instead of being built on the wall, and within the hole, were built within the pores of the permeable bed. This case has not been encountered yet, but it seems that it could occur in coarse-grain sands. If that did happen, the part of the permeable bed where the pores are filled with the mud deposit would be a little more resistive than the invaded zone behind, with the result that the 1x1-in. lateral would measure an apparent resistivity slightly higher than that measured by the 2-in. normal. This would result in a negative departure, and the bed might, therefore, on the basis of the microlog, be falsely classified as impervious. Here again, the SP log would generally settle the question.

Other exceptions, which have not yet been observed, may exist and will reveal themselves when more examples are acquired. For that reason, it will always be useful to confront the indications given by the microlog with those obtained from the SP log, and also from the conventional resistivity logs, supplemented by the section-gauge log when large cavings are to be expected.

FIELD EXAMPLES

Figure 184 shows an example of a microlog recorded in a sequence of shales and limestone. In this instance, most of the compact beds give rise to microresistivities which are definitely higher than 30 R_m, so that the discrimination between permeable and impermeable formations is particularly easy. The permeability record, provided by core analysis, was available in this hole and is reproduced in the figure as a check on the indications of the electrical logs.

In figures 185 and 186, sequences of sands, sandstones, limestones, and shales are exhibited. At the upper part of figure 186, a sharp depression in the microlog brings the microresistivity down to approximately the resistivity R_m of the mud, a fact which is almost a certain indication of caving. This interpretation is confirmed by the section-gauge log.

Figure 187 shows the behavior of the microlog in the case of a thick limestone formation (Ellenburger), composed of compact zones and of fissured zones with vugular porosity.

Figures 188 and 189 illustrate the features of the microlog in sequences of non-consolidated sands and shales, such as those commonly encountered in the Gulf Coast or in analogous geological provinces. In these regions, the conventional SP and resistivity logs differentiate very well the different sections where sands or shales are respectively predominant, but they are not capable of delineating each separate sand or shale streak, if these are thin. A very detailed record of the individual permeable and impervious beds is obtained from the microlog, as shown on the figures. This result is of interest for an accurate determination of the proportion of shale and sand in shaly sands, or, in colloquial terms, for the evaluation of the sand count.

POSSIBILITIES OF QUANTITATIVE INTERPRETATION

To date, the microlog has been used primarily for a qualitative determination of the permeable beds and an accurate determination of their boundaries. It must be kept in mind, however, that the microresistivities measured for permeable beds are largely dependent on the resistivity and thickness of the mud cake, and on the resistivity R_{x_0} of the formation immediately behind the mud cake. It is possible that the developments presently under way could bring the micrologging equipment to a point where R_{x_0} and the thickness of the mud cake could be determined quantitatively.

When the first few inches of the permeable bed immediately behind the mud cake are practically saturated with mud filtrate, and when the mud cake is entirely built outside of the formation, and not partly within its pores, R_{x_0} is equal to $F R_{mf}$, and, therefore, gives a direct measure of the formation factor F if the resistivity R_{mf} of the mud filtrate at the corresponding temperature is known. The possibility of determining the formation factor in situ and continuously would obviously be of great interest because of the close relationship of that factor with the porosity, at least when the permeable material is reasonably free from conductive solids such as clay.

It is more difficult to predict whether useful information could be derived from the thickness of the mud cake. Since laboratory experiments [23] have shown that, for a given mud, the thickness of the cake and

[23] Byck. H. T., *The Effect of Formation Permeability on the Plastering Behavior of Mud Fluids:* API Drilling and Production Practice, p. 40, 1940.

the water loss are constants which do not depend on the nature of the permeable formation, and particularly on its permeability, it is logical to expect that all the mud cakes in a given hole will be found to have the same thickness. It is, however, possible that, even if the thickness of all mud cakes should be the same in a given well, a determination of that quantity would give valuable information about the behavior of the mud itself in the drill-hole conditions. If further experience happens to show that the mud cake thickness varies from bed to bed, these variations could likely be related to some particular properties of the permeable formations, which might be of interest.

Conclusion

A new electrical logging method, called micrologging, has been discussed. This method makes use of electrodes applied to the wall of the drill hole under a nonconductive pad which shields them from the mud column. Two different microresistivity logs are recorded simultaneously, with two different electrode systems, both of which correspond to very small electrode spacings.

Permeable beds are, in general, clearly indicated on the micrologs by a positive departure between the two microresistivity curves. Even when the departure is not definite, the interpretation is easy, thanks to simple rules based on the magnitude of the microresistivities and on the behavior of the SP log. In all cases, the boundaries of the permeable beds are determined with great accuracy.

The microlog is, therefore, an important addition to the conventional electrical log, and should contribute to a better and more accurate determination of the permeable beds, particularly in limestone territories. Because of its accuracy in the determination of boundaries, it is, however, probable that the microlog will also render important services in sand and shale formations, where it could appreciably increase the accuracy of the sand count.

Acknowledgments

The writer is indebted to the many engineers of the Schlumberger Well Surveying Corporation who, at headquarters and in the field, cooperated in the development of the method described in this paper. Acknowledgments are also due to the oil companies for their courtesy in making examples available.

RADIOACTIVITY WELL LOGGING
V. J. MERCIER

Radioactivity well logging as practiced commercially in the United States and South America is at the present time composed of two curves, the gamma-ray curve and the neutron curve.

The gamma-ray curve is a relative measurement of the natural radioactivity occurring in the strata of the earth. Minute quantities of radioactive materials in one form or another are universally distributed. Measurable quantities are found in all kinds of igneous, metamorphic, and sedimentary rocks. From laboratory measurements and the experience gained in logging more than 15,000 wells, certain conclusions can be drawn concerning the relative intensity of radioactivity in different kinds of sedimentary rocks. Figure 190 is a chart that demonstrates the relative radioactivity values of various formations encountered in well logging. Anhydrite, salt, and coal are very low in radioactivity, while

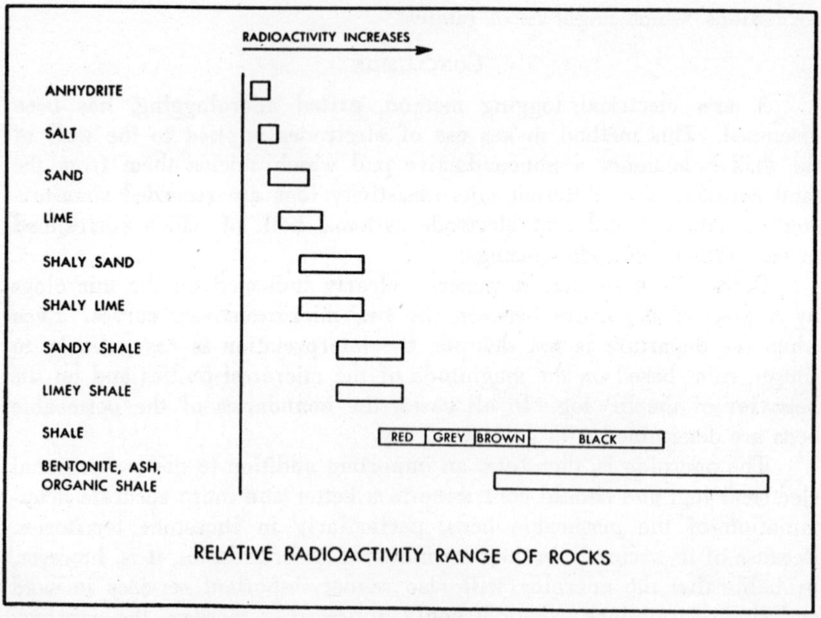

RADIOACTIVITY INCREASES →

ANHYDRITE	□
SALT	□
SAND	▭
LIME	▭
SHALY SAND	▭
SHALY LIME	▭
SANDY SHALE	▭
LIMEY SHALE	▭
SHALE	RED | GREY |BROWN| BLACK
BENTONITE, ASH, ORGANIC SHALE	▭

RELATIVE RADIOACTIVITY RANGE OF ROCKS

FIGURE 190. Drawing indicating radioactive values of various formations encountered in well logging.

shale, bentonite, ash, and organic shale have the highest values of radioactivity encountered. Producing formations such as sand, limestone, and dolomite are relatively low in radioactivity.

The neutron curve might well be referred to as a "fluid content or hydrogen curve," since hydrogen is the controlling factor on the action behavior of the curve in well logging. Neutron well logging is the process of bombarding the strata with a strong source of fast-moving neutrons and recording the secondary gamma rays that have been excited by the neutron bombardment. Where hydrogen is present in the strata, the neutrons are slowed down or stopped, and this in turn gives a low value on the curve. Where there is no, or very little, hydrogen present, the response is quite high in value and is shown as a throw to the right.

Instrumentation

The object of radioactivity well logging is to measure the radiations emitted by radioactive substances in the rock formations adjacent to the walls of the drill hole and to plot the intensity of the radiations in the

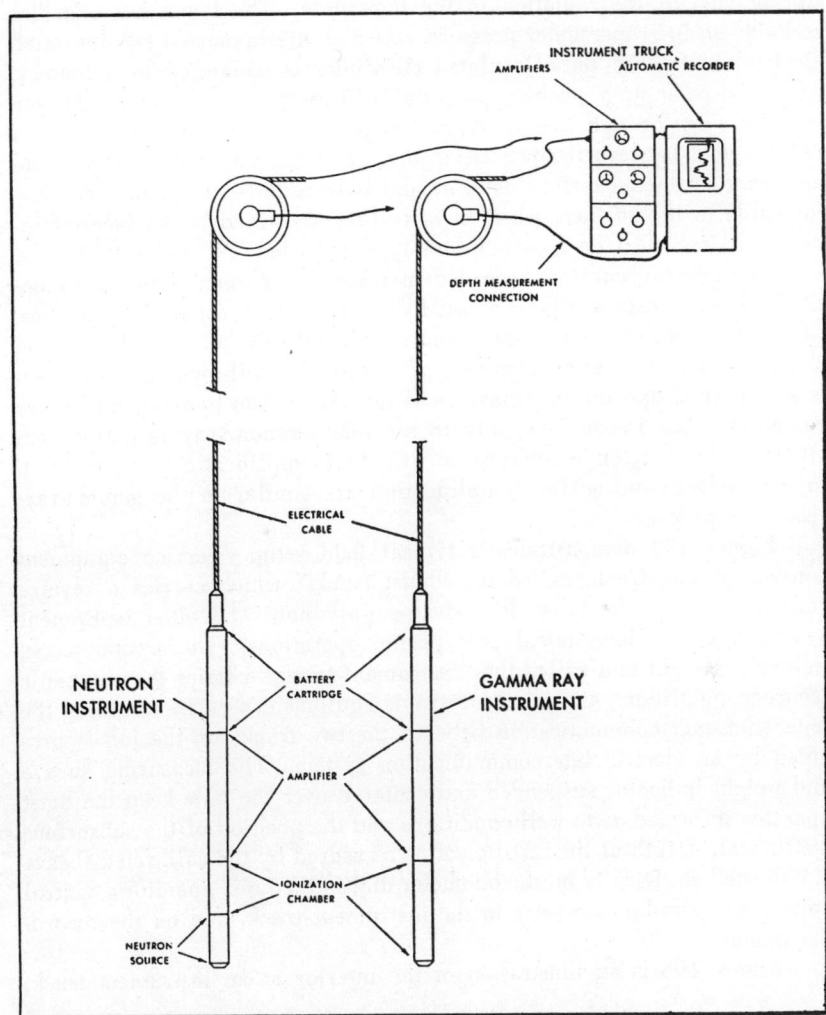

Figure 191. Radioactive logging apparatus. (Lane-Wells.)

form of a graph versus depth. The order of magnitude of the radioactivity to be expected in everyday practice is exceedingly small. Average midcontinent sedimentary rocks contain from two to twenty micromicrograms of radium per gram of rock. This concentration of radium is so small that one thousand metric tons would be required to extract a few milligrams of the radium element. The well-logging instrument employs an ionization chamber to observe the radiations. The thick wall of the ionization chamber serves two purposes, which are (1) to resist the pressure of the fluid head in the well, and (2) to select the most penetrating part of the radiation in the bore hole. The ionization chamber contains an inert gas under pressure, in which are immersed two insulated electrodes. One of these insulated electrodes is connected to a battery, which keeps it at a positive potential with respect to ground. Gamma rays passing through this inert gas partly ionize the gas, permitting a current flow between the two electrodes. This current is amplified by an amplifier in the subsurface detector and is transmitted through a conductor cable to the surface, where it is further amplified by surface equipment and recorded on a pen-and-ink-type recorder.

The neutron curve is recorded in a similar manner with the exception that the strata are bombarded by a very strong source of neutrons, which are contained in a neutron source immediately below the ionization chamber. The ionization chamber of a neutron well-logging instrument is so designed that the instrument will not respond to natural radioactive emanations but is sensitive only to secondary-gamma-ray radiations excited by the neutron bombardment. Current amplification, transmission to the surface, and surface amplification are similar to the gamma-ray-recording process.

Figure 192 demonstrates a typical field setup. Service equipment consists of one truck, called the "hoist truck," which carries a reverse concentric cable, the hoist, the power-supply unit, and other equipment necessary to the mechanical part of the operations. The second truck, lighter in weight and called the "instrument truck," carries the automatic recorder, amplifiers, and other electronic equipment used in recording the logs. Constant communication between the two trucks on the job is provided by an electric intercommunication system. The measuring sheave and weight indicator suspended and centered over the hole keep the hoist operator informed as to well conditions and the position of the subsurface instrument. Depth of the instrument as measured by the calibrated sheave is indicated electrically on the odometer dial in the hoist operator's control panel, on a similar odometer in the instrument truck, and on the recording paper.

Figure 193 is an illustration of the interior of an instrument truck.

INTERPRETATION

Any well log must be interpreted in terms of geology and stratigraphy before its utility can be realized. Because the proper interpretation of a radioactivity well log will identify the various formations represented on the log and determine their characteristics and extent, such interpretation necessarily involves a wide variety of both geology and bore-hole

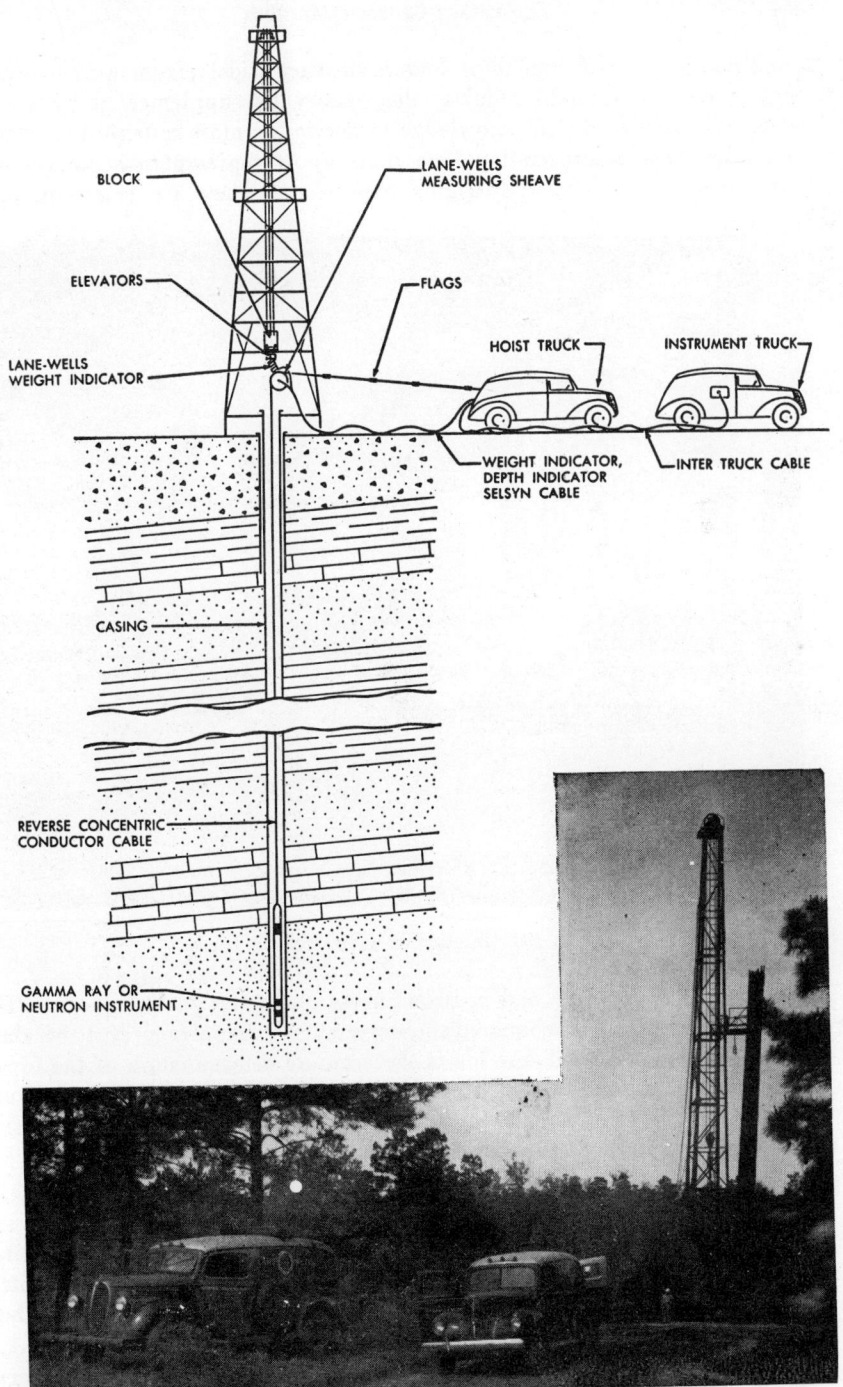

BLOCK

LANE-WELLS
MEASURING SHEAVE

ELEVATORS

FLAGS

LANE-WELLS
WEIGHT INDICATOR

HOIST TRUCK

INSTRUMENT TRUCK

WEIGHT INDICATOR,
DEPTH INDICATOR
SELSYN CABLE

INTER TRUCK CABLE

CASING

REVERSE CONCENTRIC
CONDUCTOR CABLE

GAMMA RAY OR
NEUTRON INSTRUMENT

FIGURE 192. Typical field setup for radioactive well logging.

conditions. The fact that these factors do vary widely from well to well and from field to field requires that theory be supplemented by wide experience and empirical knowledge to derive the most reliable log interpretation. The interpretation data that follow represent a summary of extensive radioactivity-well-logging experience. They are based on the

FIGURE 193. Interior of an instrument truck.

behavior of gamma-ray and neutron curves under the geologic and bore-hole conditions most commonly observed. An important part of the correct interpretation of any log is the accurate determination of the tops and bottoms of formations. This procedure requires particular attention for radioactivity well logs because of the characteristically sloping transitions, which represent formation contacts.

Figure 194, which is a portion of a gamma-ray curve, represents the correct point for determining the top and bottom of a zone. In all cases, regardless of the magnitude of the break, the midpoint or center of a minimum-maximum intensity value on the curve will most reliably indicate the actual formation contact. Shown previously in figure 190 is the relative radioactivity range of the rocks encountered in oil-well logging. A working knowledge of the local stratigraphy is necessary for the correct geologic interpretation of the gamma-ray curve.

Table 22 is a laboratory analysis of several hundred rock samples showing the average radioactivity in radium equivalents per gram of rock. This table is taken from "The Total Gamma Ray Activity of Sedimentary Rocks as Indicated by Geiger Counter Determinations" by Russell,[24] who has pointed out that an increase in radioactivity is directly proportional to an increase in shale or silt in the strata. Furthermore, the shade or color of the rock has a relationship to the amount of radio-

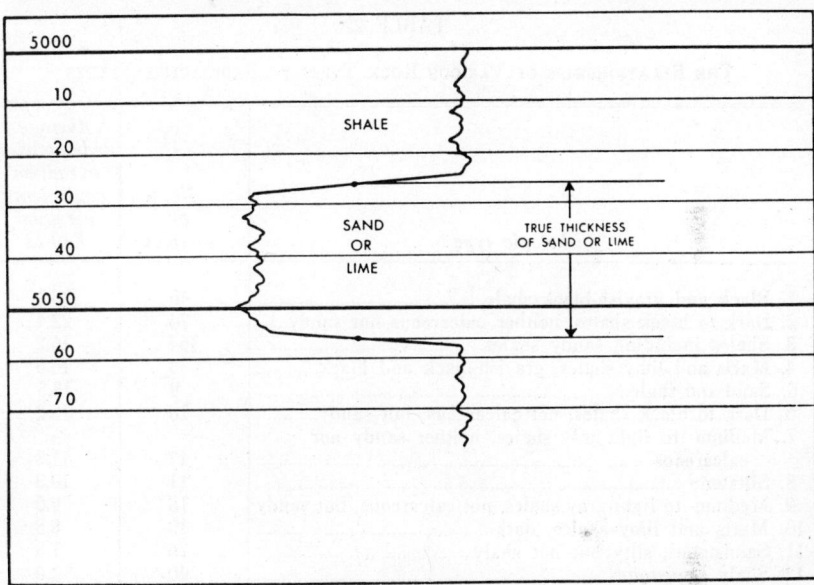

FIGURE 194. Showing method of determining true thickness of sandstone or limestone using midpoint of transition for top and bottom of zone.

activity. The darker the shade or color, the higher the radioactivity, except for rocks stained by oil or asphalt. One notable exception is coal, which is very low in radioactivity. For general interpretation purposes, we can conclude that formations encountered during a radioactivity survey will be of the value shown in table 22.[25] Two things must be kept in mind, however. The first is that a knowledge of the local stratigraphy is imperative for the correct interpretation because sandstone, limestone, and dolomite have so nearly the same value that they cannot be accurately differentiated by gamma-ray evaluation alone. The second is that in some areas of complex geology some sandstone and/or limestone have been encountered that are as high in radioactivity material as the usual shale. This phenomenon has occurred only in some Gulf Coast and California areas; none has been logged in Midcontinent fields.

[24] Russell, W. L., *The Total Gamma Ray Activity of Sedimentary Rocks as Indicated by Geiger Counter Determinations:* Geophysics, vol. 9, no. 2, Apr. 1944.
[25] Russell, W. L., *op. cit.*

Neutron-curve interpretation is a matter entirely different from gamma-ray interpretation. It has been stated previously that the response of the instrument is controlled mostly by hydrogen. Response of the instrument opposite hydrogen is very low. It matters not in what form the hydrogen occurs, whether in gas, oil, water, or shale, the curve is characterized by a low reading on the graph or to the left of the chart. For this reason a neutron curve used alone cannot be interpreted further

TABLE 22

THE RELATIONSHIPS OF VARIOUS ROCK TYPES TO RADIOACTIVE VALUES

Lithologic type	No. of samples	Average radioactivity in radium equivalents per gram X10-12
1. Black and grayish-black shale	40	26.1
2. Dark to black shales, neither calcareous nor sandy	74	22.4
3. Shales including sandy shales	164	16.2
4. Marls and limy shales, grayish-black and black	3	16.5
5. Sand and shale	9	13.5
6. Dark to black shales, not calcareous, but sandy	16	13.2
7. Medium- to light-gray shales, neither sandy nor calcareous	17	11.3
8. Siltstone	11	10.3
9. Medium- to light-gray shales, not calcareous, but sandy	18	9.0
10. Marls and limy shales, dark	10	8.8
11. Sandstones, silty but not shaly	26	7.3
12. Shaly sandstones	40	7.0
13. Marls and limy shales of light shades	16	6.8
14. All sandstones, including shaly sandstones	131	5.3
15. All sandstones, excluding shaly sandstones, but including silty types	105	4.0
16. Sandstones free from silt and shale	76	4.1
17. Shale-free limestones and dolomites	64	4.1
18. Microcrystalline to earthy limestones and dolomites of medium to light shade	28	4.0
19. Medium- to light-shade, shale-free limestone	33	3.8
20. Finely to coarsely crystalline limestones and dolomites of medium to light shade	24	3.1
21. Medium to light shade, shale-free dolomite	21	3.1
22. Effect of shade in shale-free limestone and dolomite:		
A. Light gray to white	30	3.1
B. Medium shade	22	4.1
C. Dark to black	10	6.1
23. Estimated original permeability of sandstone before cementation:		
Very high	35	2.9
High	37	5.1
Low	40	6.6
Very low	24	7.5

than to show the presence or absence of hydrogen. However, the neutron curve used in combination with the gamma-ray curve can be interpreted to locate possible porous zones in producing strata very accurately.

In figure 195 a portion of a combination radioactivity log is shown with the gamma-ray curve to the left, indicating a productive limestone of considerable extent. The tops and the bottoms of the formations have

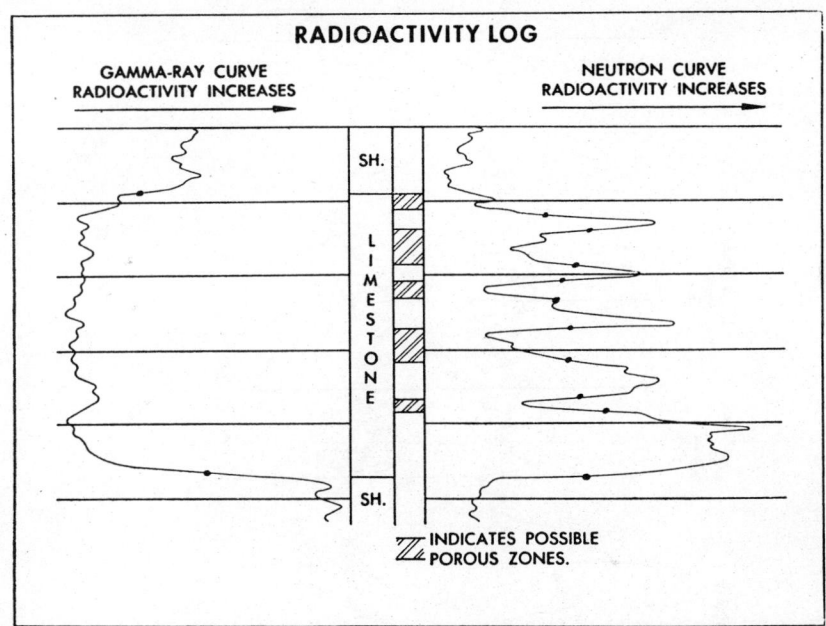

FIGURE 195. Combination radioactivity log indicating productive limestone of considerable extent.

been determined as previously explained by selecting the midpoint of the transition. The neutron curve shown on the right is also interpreted by selecting midpoints on the transitions. "Throws" to the right are considered barren or impervious strata, while "throws" to the left indicating the presence of hydrogen are considered to have fluid in place and are therefore interpreted as possible porous zones.

APPLICATIONS

The radioactivity well log was originally developed with the thought that some direct relationship might exist between natural radioactivity and the presence of petroleum. Experience to date indicates the absence of any such relationship. However, the indirect applications of the radioactivity well log to the location of petroleum reserves are numerous and varied.

The first advantage of the radioactivity well log over any other type

Radioactivity Log   **Electrical Log**

Gamma Ray   Resistivity

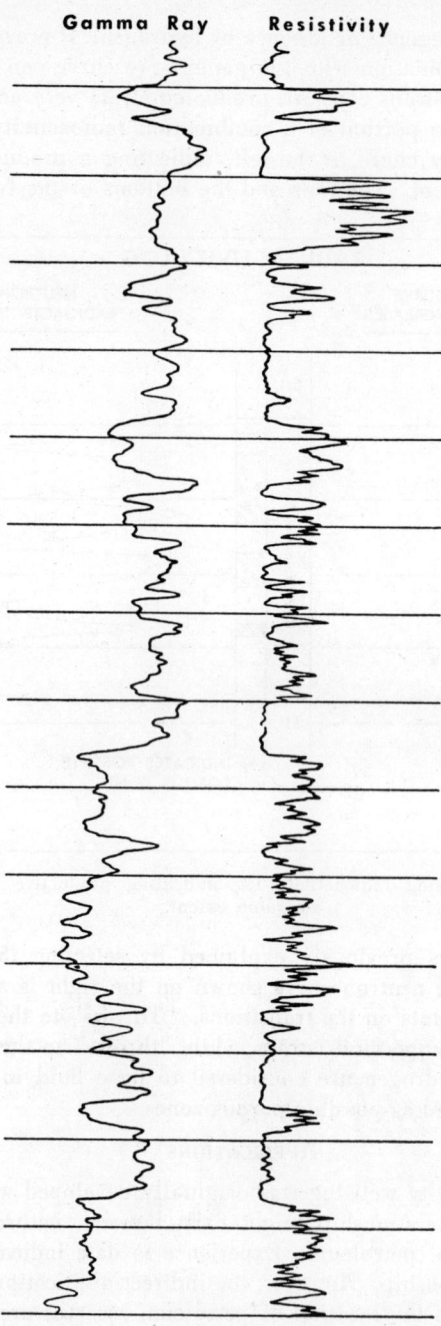

FIGURE 196. Comparison between a gamma-ray curve and an electrical-resistivity profile taken four years previously in same well, Rincon area, California. (Lane-Wells.)

of well-logging method is its ability to log through steel casing. This permits the stratigraphic study of old wells drilled prior to the development of geophysical well-logging methods now in common use. The location of upper cased-off potential producing zones has presented a logical field for this type of logging, particularly where no information was available or where the available information was doubtful.

Variations of this primary application are apparent. The location

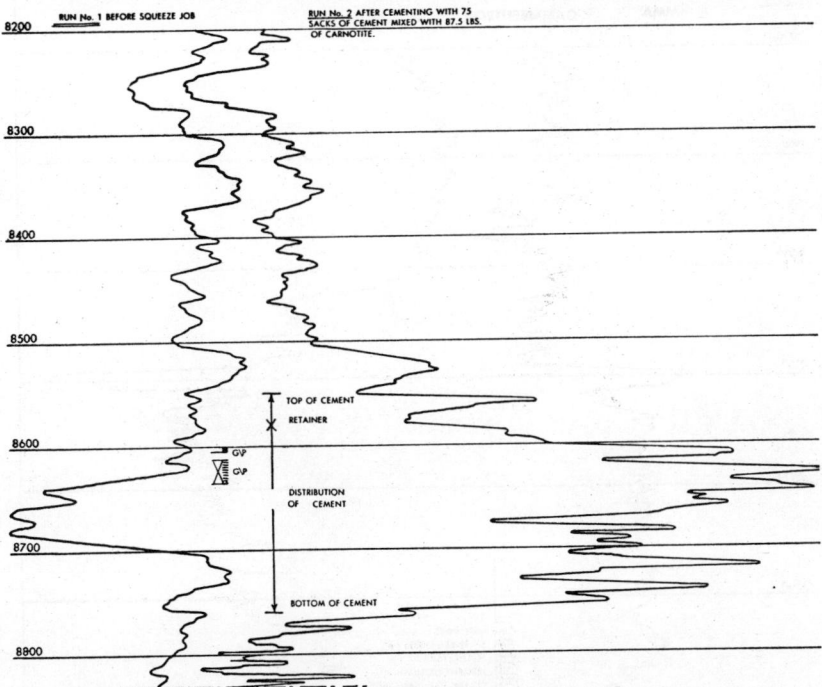

FIGURE 197. Location of carnotite (radioactive) cement with gamma-ray curve.
(Lane-Wells.)

of the top of the producing zone for bottom-water shutoff, the correction of some of the earlier drillers' logs, the location of upper potential fresh-water sands for salt-water disposal, the supplying of additional information where cores were not completely recovered or were lost, and the location of the top and bottom of an oil-producing zone for gas-oil-ratio control are all applications of this type.

Many logs are run for measurement checks of various kinds. Drill-pipe measure, casing measure, and electric-log measure often disagree. On many of the older wells the original zero point has been lost. On other wells, both old and new, a new bottom has been established. Many of the earlier sample logs did not take into consideration the time lag between the point of origin of the sample and the depth of the well at the time the sample was caught. Because of the tested accuracy of the measuring system employed and the ability of the instruments to indicate the changes in formations behind casing, radioactivity logs have become a popular means of resolving all manner of measuring discrepancies.

An additional feature of radioactivity well logging is the collar log,

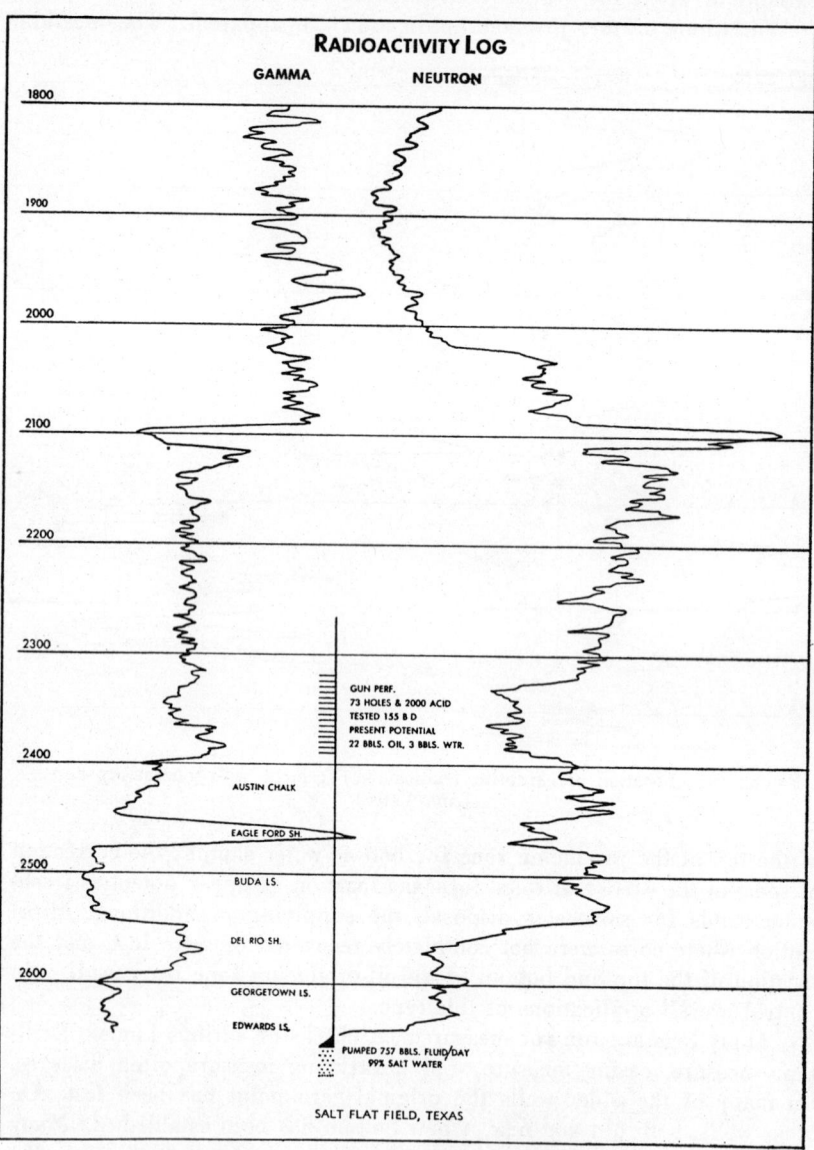

FIGURE 198. Radioactive curves in Branyon field, Caldwell County, Texas. Radioactivity increases from left to right. (Lane-Wells.)

which is a record of the exact location of each casing collar. It is recorded electrically by means of a collar locator which is a component part of the gamma-ray instrument. Thus, the collar log is recorded simultaneously and on the same chart with the gamma-ray curve. This provides a permanent record of the fixed relationship of collars to formations.

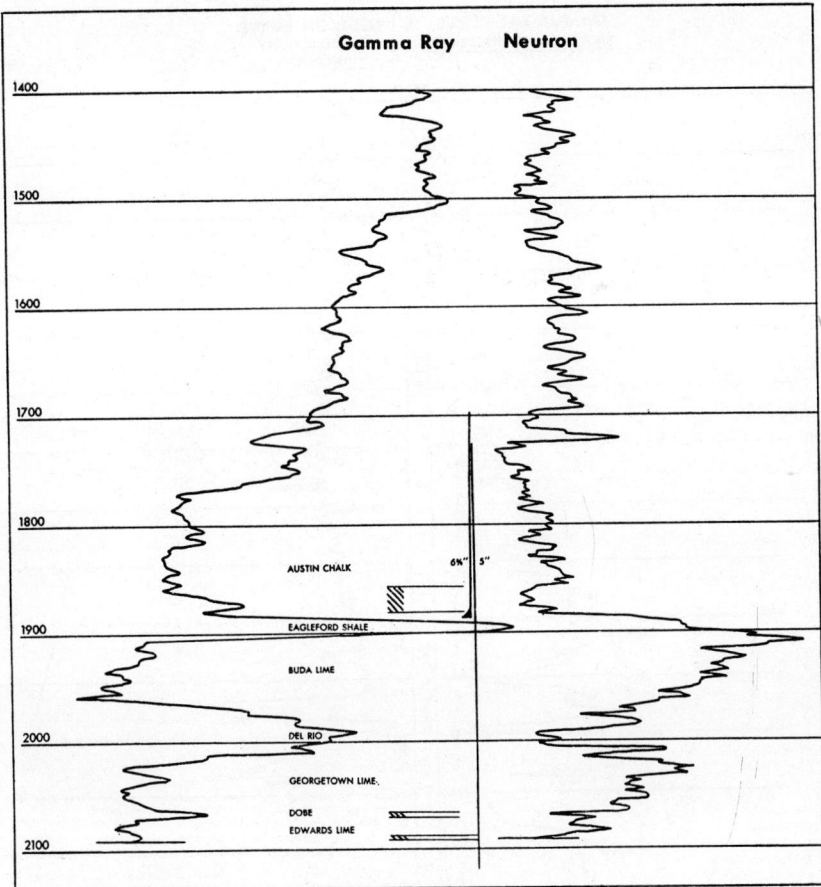

FIGURE 199. Radioactivity curves in Salt Flat field, Texas. Radioactivity increases from left to right. Note maximum point on gamma curve opposite Eagle Ford shale. (Lane-Wells.)

Formation tops and bottoms are accurately established in relation to the nearest casing collar. This combination eliminates the many measuring discrepancies that may occur in various means of well measuring.

The second advantage of radioactivity well logging is its ability to log in contaminated well fluids. In areas such as central Kansas where salt beds are encountered in drilling, the resistivity of the drilling fluid is lowered to a point where it is impossible to get a good electric log. The high resistivity of fresh-water muds presents a similar problem in attempts to log fresh-water sands electrically. Oil-base mud is used in some areas to eliminate the infiltration where the productive zones are partly depleted or where bottom-hole pressures are low. Since oil is

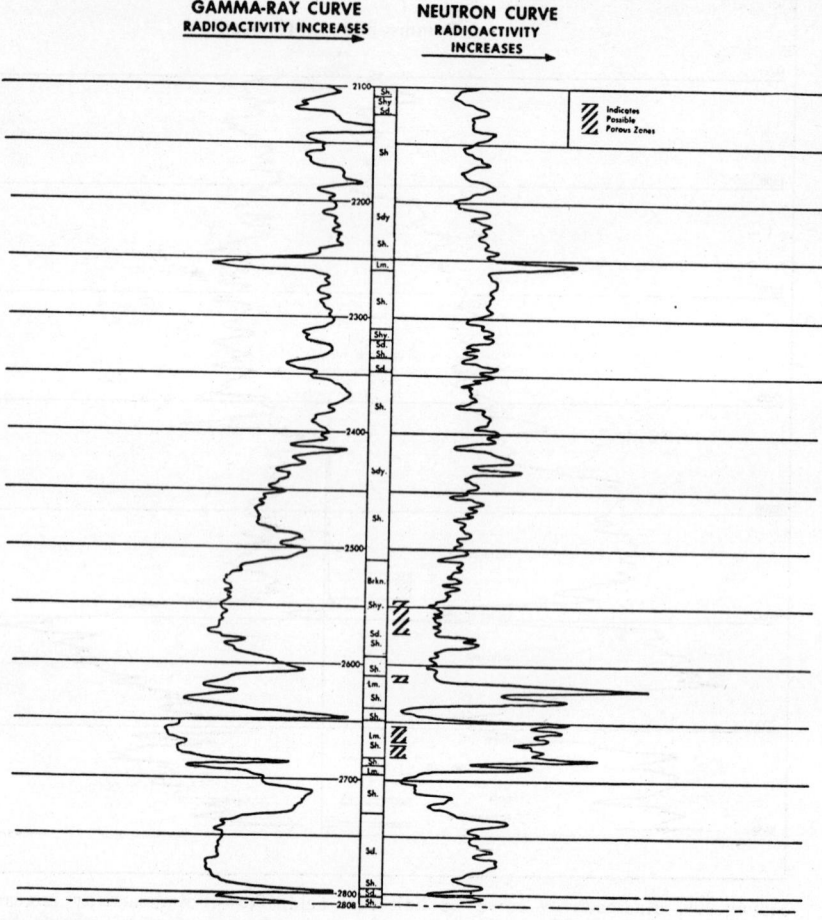

FIGURE 200. Characteristics of gamma and neutron curves in Cushing-Drumright field, Oklahoma. This well is 20 years old. (Lane-Wells.)

a dielectric substance, it shows a high electric resistance, which makes it difficult to obtain satisfactory electric logs. Electric logs have been run in oil-base mud by using electrodes that make a sliding contact to the walls of the well. However, washed-out places and irregularities in the well bore tend to defeat the purpose of the contact electrode and cause a curve to be produced that is sometimes difficult to interpret. Since the radioactivity well log is a measurement of radioactive emanations, it is not affected by salt-water mud, fresh-water mud, oil-base mud, or other contaminated fluids.

Radioactivity well logs are easily correlated with other types of geophysical well logs and such information obtained from well bores as sample logs and core analysis. Radioactivity logs and the types of electric surveys in common use today are easily correlated. Although there is no relationship between the two types of surveys and they are

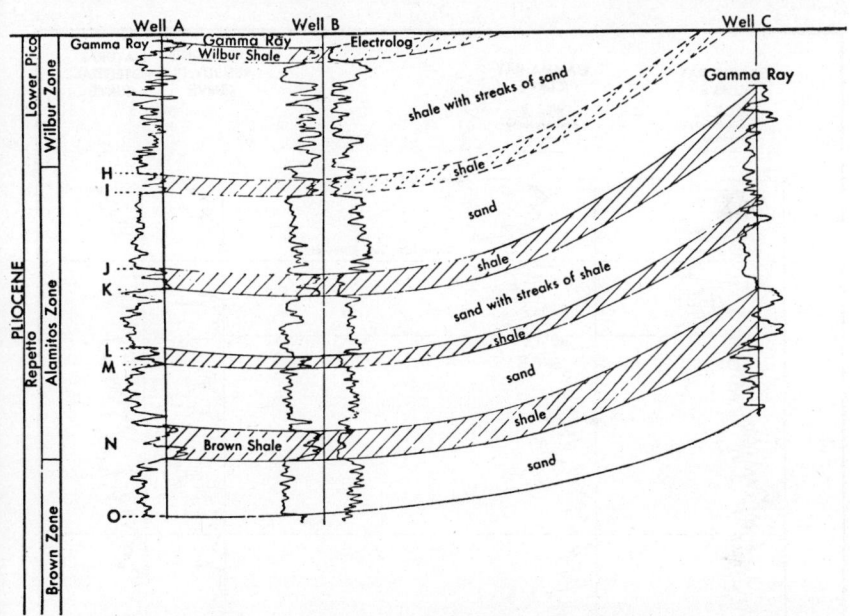

FIGURE 201. Correlation of radioactivity logs made in cased wells along a two-mile strike section, Long Beach field, California. (Lane-Wells.)

often surveyed under different well conditions or years apart, there still exists a similarity in response that permits accurate correlation. This is most fortunate, for it is simple logic to comprehend that a multiconductor cable such as is employed in electric logging used under varying fluid and mud conditions cannot be expected to measure with complete accuracy.

Figure 203 illustrates the typical response of radioactivity and electric logs to the usual formations encountered in oil-well drilling. These generalized curves are shown to illustrate the most typical response of radioactivity and electric logs to the various types of formations. These typical curves should not be used as a criterion for analysis of any particular log since in practice a wide range of response may exist. In making correlations between radioactivity and electric logs, the gamma-ray curve is correlated with the self-potential curve, while the neutron curve is compared with the shallow-resistivity curve.

Radioactivity well logs are also employed for a variety of specialized applications. One of these is the logging of added radioactive material. In the study of squeeze cement jobs on the Gulf Coast, it has been found desirable to know the vertical travel of the cement and its mass distribution behind the casing string. By means of a radioactive tracer such as carnotite mixed with cement, it is possible to determine the direction and extent of the cement travel. The usual procedure is to record a gamma-ray curve of the well under normal conditions. The mixture of

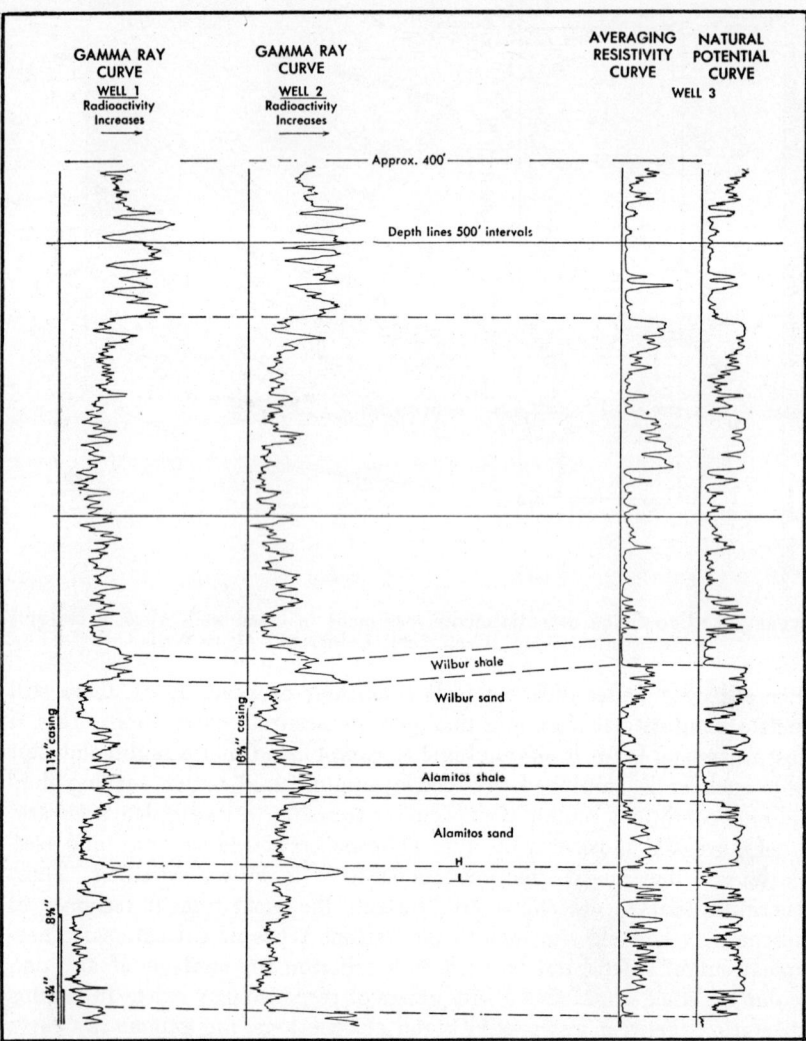

FIGURE 202. Gamma ray-electrical log correlation of section in Long Beach field, California. (Lane-Wells.)

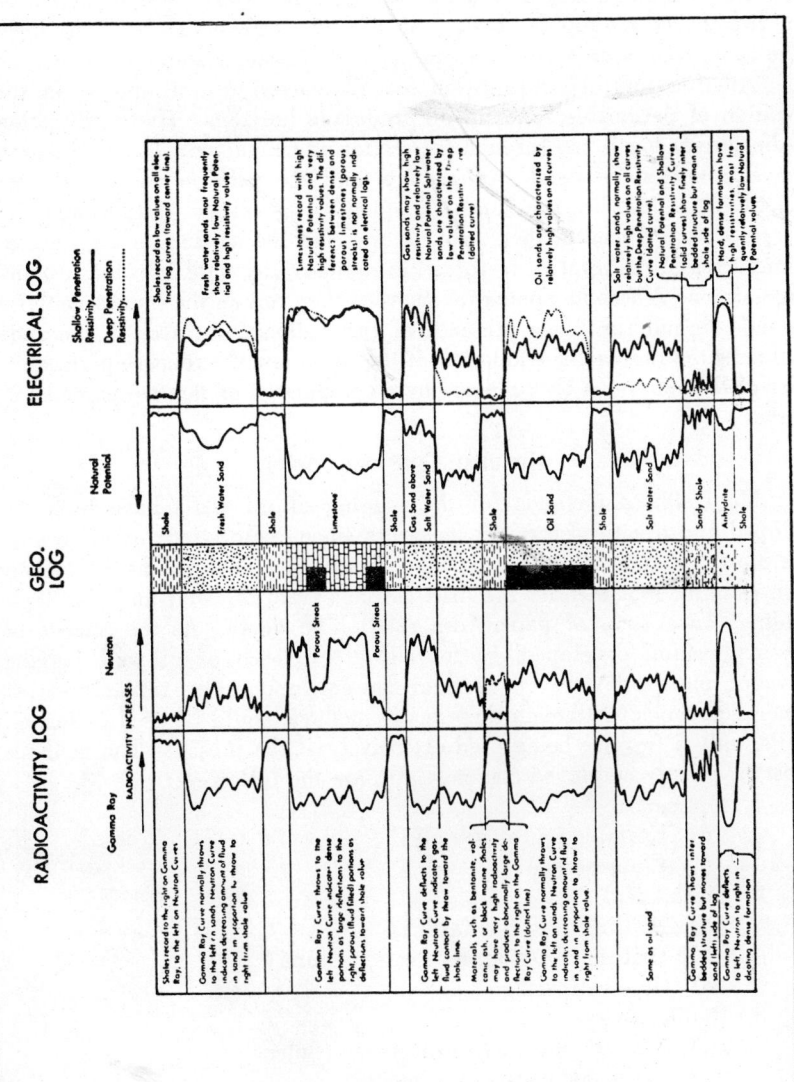

Figure 203. Showing response of radioactivity and electric logs in usual formations encountered in oil-well drilling.

carnotite and cement is then squeezed. After the cement has set and the plug has been drilled, a second gamma-ray curve is run on the same sensitivity as the first. The difference in gamma-ray intensity between the first and second runs is caused by the carnotite. Thus, by comparison of the two curves, the top and bottom of the cement travel can be determined. The same technique applies to squeezing of plastic material.

Another unusual application for radioactivity well logs is in the location of permeable zones in oil-producing horizons. The combination radioactivity log is run through the section to be studied. The gamma-ray curve provides the local stratigraphy, and the neutron curve indicates the possible fluid-bearing zones. A quantity of oil or water containing a radioactive tracer is then spotted opposite the section to be studied. Pump pressure is applied to force the radioactivity fluid into the producing horizon. A second gamma-ray curve is then run on the same sensitivity as the original run. The presence of the radioactive tracer in the zone increases the radioactive valuation of the zones and the relative permeabilities of the zones can be estimated by a comparison of the two gamma-ray curves.

GENERAL CONSIDERATIONS

Many varied methods for the logging of oil wells have been developed and practiced through the years since exploration for oil began, the most common being the use of a sample log obtained by geologists, who examine and identify the drill cuttings from the well and plot their findings on a strip of paper with relation to depth. As the science of oil exploration developed, so too did the science of oil-well logging advance, until today the oil operator can call upon approximately eleven general methods. Although these many methods could be used to log oil wells, only a few are being used extensively at the present. The methods most frequently employed commercially are the following:

1. Optical
   (a) Examination of samples
   (b) Fluorescence
2. Mechanical
   (a) Drilling rate for hardness
   (b) Drilling reaction, hardness, texture
   (c) Hole calipering
3. Radioactivity
   (a) Natural, radioactive-substance content
   (b) Artificial, hydrogen content
4. Electric
   (a) Natural potential
   (b) Formation conductivity
   (c) Mud conductivity

In the selection of the best method for the logging of an oil well, consideration must be given to the problems to be solved, the methods available, and the adaptability of the methods and their cost. The method selected should, therefore, have as many as possible of the following properties:

1. Significant relationship to the lithology of the rocks.
2. Detection of oil, gas, or certain minerals penetrated by the bore.
3. Many vertical variations, ability to differentiate strata.
4. Persistence of vertical variations laterally, to provide good subsurface correlations.
5. Ability to work under almost any bore-hole condition.
   (a) Cased holes, dry or filled with fluid
   (b) Open holes, dry
   (c) Open holes filled with mud or water
   (d) Open holes filled with oil or gas
6. Location of porous strata.
7. Economy of operation.
8. Simplicity of interpretation.

Radioactivity well logs fit nearly all of these requirements. Coring and core analysis constitute the only positive means of lithologic identification, but owing to excessive cost the method is not very widely employed. The relationship of the radioactive content of the strata to lithology has already been pointed out, and a suitable lithologic identification can be executed by the gamma-ray curve.

No positive means of the detection of either oil or gas has yet been determined other than fluorescent and chemical examination of samples and cores. At present the combination radioactivity log makes no claim to the direct detection of oil or gas, but the indirect use of the log by correlation or structural determination has been the means of locating many new pay horizons. It is therefore stated that radioactivity logs are as capable in the detection of oil or gas as other logging methods, with the exception of visual examination of conventional cores. Research and future development may well provide a sure means of oil detection, either through the development of an additional curve or through the better interpretation of analysis. The requirements 3 and 4 given above, vertical variations for differentiating strata and vertical variations laterally of ample magnitude to provide subsurface correlations, are adequately filled by the radioactivity well log. Abnormally high radioactive shales in most areas serve as base markers for the long-range correlation of gamma-ray curves. Within confined areas it has been found that the neutron curve correlates readily, even to the extent that the evaluation of potential reserves is possible.

In considering requirement 5, the advantages and the flexibility of the radioactivity log are most obvious. The radioactivity log can be obtained in cased holes, dry or containing bore-hole fluid of any type, and in open

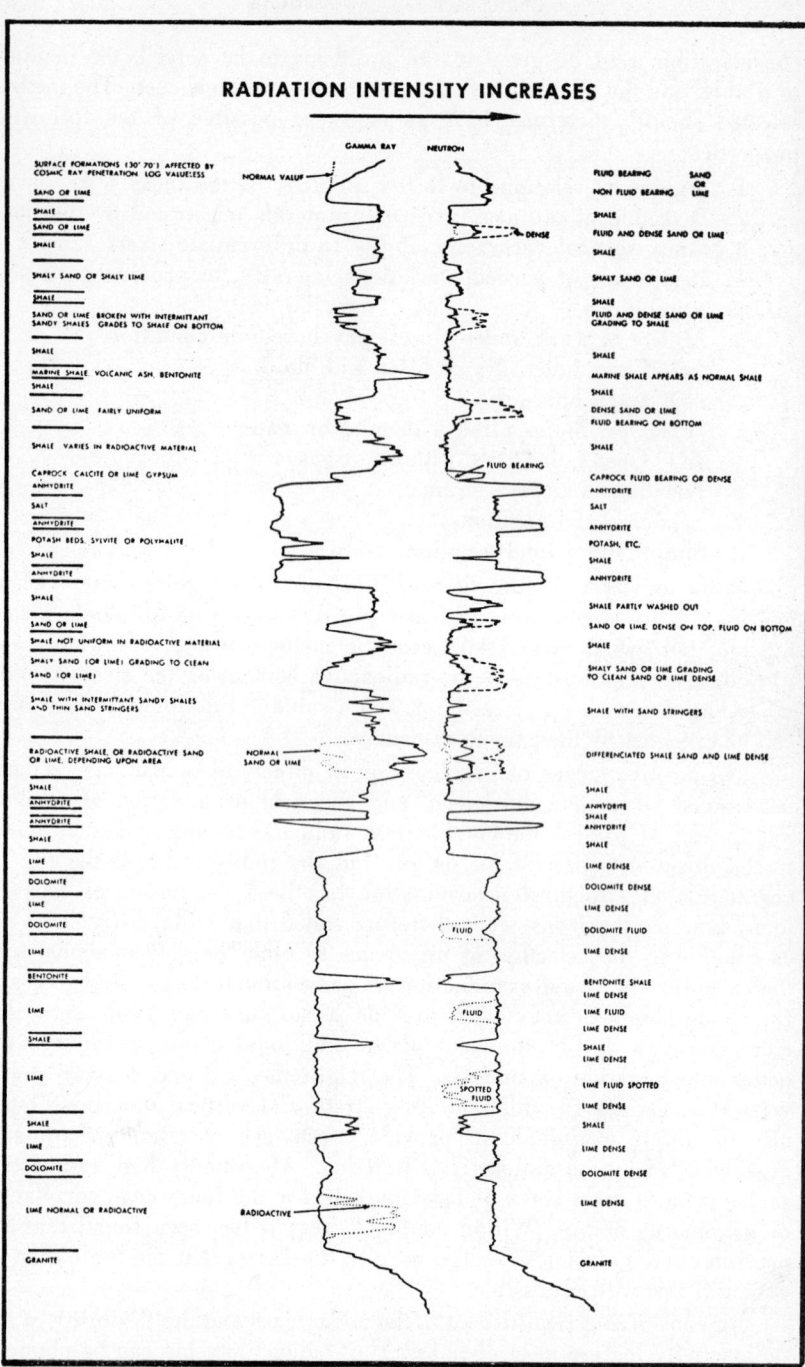

FIGURE 204. Showing correlation of radioactivity-log profile with various rock types.

holes, dry, producing gas, containing salty mud, conditioned mud, fresh water, oil, oil-base muds, or any mixture of these fluids.

The location of porous strata by the neutron curve, especially when it is used simultaneously with the gamma-ray curve, has been described and explained. The ability to determine porous strata by radioactivity means has been proved beyond all doubt by the hundreds of successful completions of oil wells in which radioactivity logs were employed.

Because of the flexibility of the radioactivity log it becomes also the most economical survey available. No special arrangements or hole-conditioning procedures are necessary.

Simplicity of interpretation is one of the primary advantages of logging by radioactivity means. It is not necessary for the oil operator, engineer, or geologist also to become a nuclear physicist. A very basic understanding of the principle upon which a radioactivity log operates, the action behavior of the recorded picture obtained, and a knowledge of the stratigraphy of the region are all that are necessary for accurate interpretation.

CALIPER AND TEMPERATURE LOGGING
WILFRED TAPPER

The uses of caliper and temperature records are sometimes so inter-related that a discussion of one log presupposes a discussion of the other. Here, however, for purposes of clarity, the records and the tools used to obtain them are treated separately.

It is the purpose of this section to give an outline of the history, development, construction, and uses of caliper and temperature electrodes. A knowledge of the physical construction of both types of electrodes results in a clearer understanding of the data obtained and more efficient utilization of the logs.

The few anomalies cited as examples do not pretend to be comprehensive. The uses for the various logs listed surely do not exhaust present or future possibilities in the oil industry or in other fields.

The writer is indebted to Mr. H. K. McArthur, Mr. J. K. Reynolds, and Mr. W. D. Owsley, of the Halliburton Oil Well Cementing Company, for advice and criticism in the preparation of this section.

CALIPER LOGGING

Even in the early days of cable-tool drilling, oil men were well aware that drill holes did not stand true to gauge. As holes were drilled deeper and exploration moved southward to younger sediments, this fact became painfully obvious.

One of the factors influencing the development and use of the rotary drill was its ability to cut through young, unconsolidated sediments and

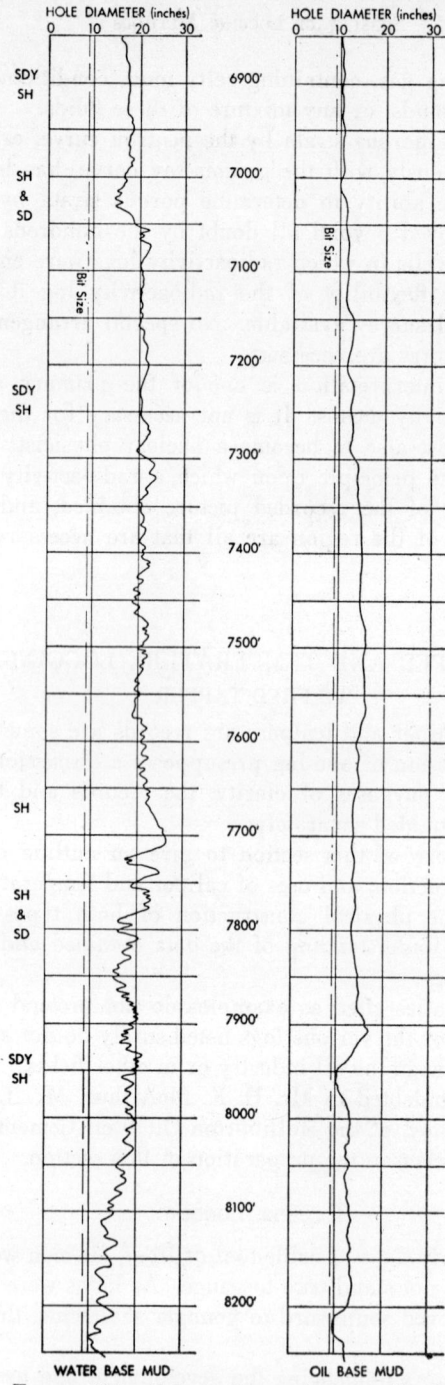

FIGURE 205. Caliper logs of two wells in
Mercy field, Texas, showing influence of
type of mud on caving.

still have the bore hole tend to stand up. Nevertheless, numerous problems connected with rotary drilling and oil producing were a direct result of hole caving in rotary holes. The exact nature and extent of this caving was unknown. Every driller and every geologist had a theory, but there was no exact knowledge. Subsurface bore-hole caving was likened to surface erosion, but owing to the different character of the sediments and the greatly accelerated subsurface erosional forces, this parallel could not be drawn too far.

Present knowledge indicates three major reasons for differential hole size in a hole drilled with rotary tools: (1) action of the drilling fluid, (2) action of the bit, and (3) action of the drill pipe. Of these, it is believed that the action of the drilling fluid is most important.

The hole change caused by mud or any conventional drilling fluid is due either to a chemical effect (hydration) or a mechanical effect (attrition or dissolution). Of these two, it is believed that the former is the more important cause.

Water-base muds must have a tendency to cause many shales to swell and heave or to disintegrate. Many shales disintegrate beyond the 32-inch range of the modern caliper tool. Interestingly enough, a mud cake may be built up on the face of a formation, causing a hole to caliper smaller than bit size.

In order to reduce hole change, many muds other than water-base muds are used. Oil-base, oil-emulsion, silicate-base, and salt-base muds are all commonly used to prevent the caving of shales or soluble formations. It is not the purpose of this section to discuss the merits of various types of muds. The effect of these muds on hole size has been clearly shown to the industry by caliper logs. Figure 205 shows caliper logs on offset wells, one drilled with oil-base mud and one with water-base mud.

In 1932 M. M. Kinley made a successful attempt to caliper a hole. A few years later R. B. Bossler measured the enlargement of a hole caused by shooting. These early tools measured a short section of the well and were used mostly in shallow areas. The original Kinley caliper had four separately actuated arms, each arm giving a separate record. A stylus-recorded strip chart was obtained.

The present modern caliper used by the oil industry was developed from the original M. M. Kinley tool by the Halliburton Oil Well Cementing Company in 1940.

The present caliper (figs. 206 and 207) is a chrome-plated tool three inches in diameter and approximately five feet long. It consists of an oil-filled chamber containing the electrical components, the four caliper arms, and the releasing mechanism. The tool itself is standard equipment on an electrical-service truck and is run in a well on a five-sixteenths-inch logging line.

The four spring-actuated arms of the tool contact the walls of the bore hole when they are released. The motion of these arms is transmitted to a rheostat inside the oil-filled chamber by means of a flexible bronze cable-and-pulley system in such a manner that the change in resistance of the rheostat is always proportional to the change in average diameter as measured by the four arms. Owing to the spring tension in the arms, the tool will be approximately centered in the well, unless the hole is considerably off vertical.

The arms are held in a closed position by a steel band when going in the hole. This band can be broken at will, either by firing a brass projectile located underneath or by spudding on bottom.

The chamber of the tool is filled with oil and kept hydrostatically

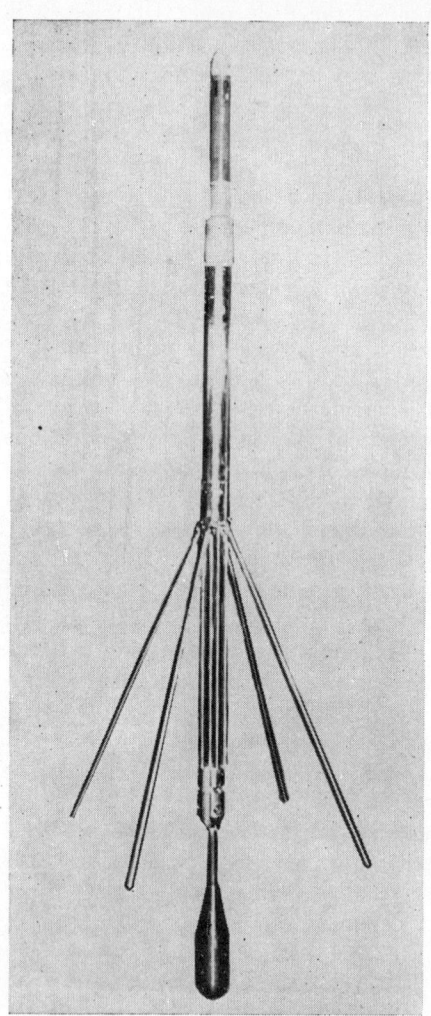

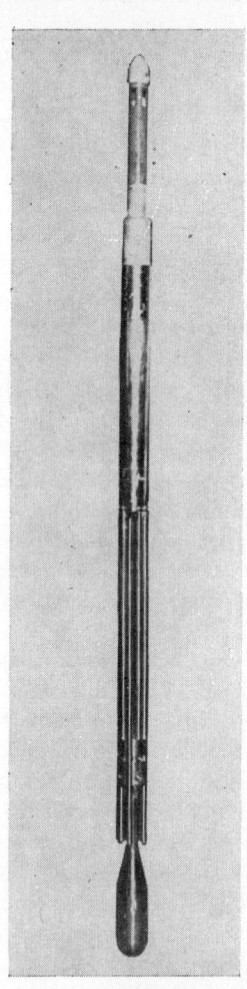

FIGURE 207. Caliper in closed position.

FIGURE 206. Caliper in open position.

balanced by means of a large rubber diaphragm, which acts as a volume equalizer between the oil and the mud, compensating for the difference caused by the motion of the push rods and by changes in temperature.

A constant direct current is supplied to the rheostat in the tool, and the resultant potential drop across it is measured and recorded as a caliper log. These logs take the form of a continuous galvanometer trace recorded on film showing the average diameter of the bore hole, recorded as a function of depth.

A study of numerous caliper logs soon leads one to the conclusion that caving patterns exist. Certain generalizations may be made as to the relative ability of rocks to stand up to bit size. These generalizations are shown in graphic form in table 23.

TABLE 23

ABILITY OF ROCKS TO STAND UP TO BIT SIZE

Rock	Poor	Medium	Good
Sand	....	x	x
Shale	x	....	....
Chalk	....	....	x
Limestone	....	....	x
Dolomite	....	....	x
Anhydrite	....	....	x
Salt	x	....	....

Although similar drilling conditions and similar muds tend to standardize caliper logs, some astonishing long-range correlations may be made by using them, even for wells drilled under entirely different circumstances. In the west Texas area, a number of horizons may be recognized on caliper logs, from Upton County, Texas, to the Hobbs field, Lea County, New Mexico, a distance of 120 miles. Here caliper-log correlations are more trustworthy than those made with electric logs.

The most obvious use for the caliper log in the oil industry is as a tool to calculate the proper amount of cement necessary to fill up the annular space between the casing and the open hole to a desired point. The actual amount of cement necessary for a desired fill is often two or three times the amount one would use from theoretical calculations. Figure 208 is an actual caliper log of a well in Smith County, Texas. The hole was drilled with an $8\frac{5}{8}$-inch bit, with $5\frac{1}{2}$-inch casing set on bottom. Theoretical fill-up is 22 sacks of cement per 100 feet of hole. For more than 1000 feet of section 220 sacks of cement is the theoretical amount necessary to fill back to 9500 feet. The actual amount of cement needed is 544.4 sacks, or more than twice the theoretical quantity.

Another problem encountered in the successful completion of an oil well is the location and reaming of tight spots in the hole, so that casing can be set and successfully cemented without channeling and bridging. The value of a caliper log in locating these zones has been demonstrated many times.

Modern drilling practices presuppose the use of many scratchers, centralizers, and guides welded to the casing to assist in obtaining a better cement job. All these tools have proved extremely useful in obtaining better cementing jobs and eliminating costly squeeze jobs, but they are useless unless properly positioned in the hole with the aid of a caliper log.

In plugging back with cement, plastics, or gravel packing, a knowledge of hole size is invaluable. In remedial work of this kind, well records are usually inaccurate or nonexistent, and it is only through an application of the knowledge gained from a caliper log that a successful job may be performed.

In the field of drill-stem testing, a knowledge of caving conditions has saved oil operators an untold amount of money. Figure 209 is a caliper log of a well in Smith County, Texas, where the operator wished to find a packer seat to drill-stem test the Paluxey sand at 7140 feet. A glance at the log will show that, using trial-and-error methods, the chances of successfully setting a packer are relatively small. By applying the information to be gained from the use of a caliper log, the proper point to set the packer and the proper size of packer are easily determined.

Many other uses have been found for caliper logs. The successful completion of a fishing job may depend on a knowledge of the size of the hole above the junk. A log made after a fishing job is completed will provide the necessary data successfully to resume drilling operations.

A knowledge of hole size is essential in evaluating an acidizing job, picking a zone to side-wall-core, evaluating the results of shooting with nitroglycerine, and finding a proper zone to gun-perforate. The present caliper log has fulfilled all these functions.

TEMPERATURE LOGGING

As early as 1869 Lord Kelvin conducted experiments in measuring earth temperatures at a depth of 350 feet in the ground. Since then, geologists have speculated on the geothermal gradient in the earth's crust. Even with such an early start, little has been done in the way of quantitative work with earth-thermal measurements.

At the present time thermal measurements in either a cased or open hole are usually obtained by means of a continuous-recording, extremely accurate, electronic thermometer. Such a tool is standard equipment on an electric-logging truck and is run on a five-sixteenth-inch conductor cable.

The temperature electrode is a three-inch rubber-covered tool about six feet long. In a groove in the rubber coating of the electrode is a twenty-inch length of platinum wire, which is exposed to the mud column.

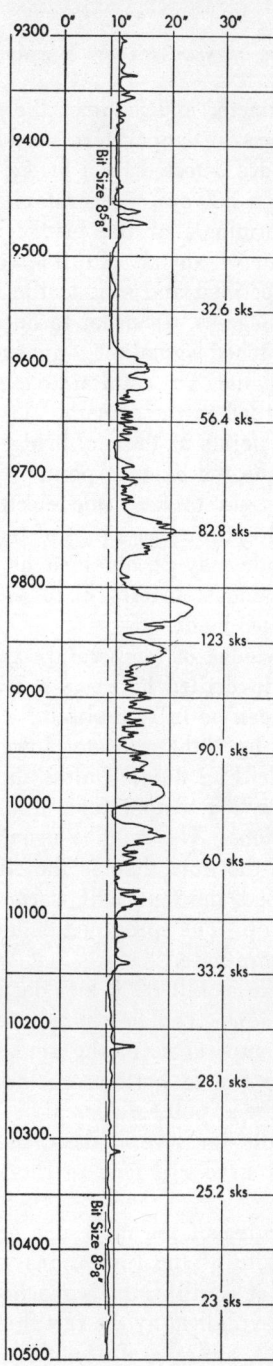

FIGURE 208. Caliper log of well in Smith County, Texas, showing effect of hole size on cement calculations.

This wire is small in diameter and assumes the temperature of the fluid around it rapidly. Changes in temperature produce changes in the resistance of the wire, which are detected by a bridge circuit in the electrode. Alternating current from a 500-cycle generator is supplied to the bridge terminals. The signal terminals of the bridge are transformer-coupled to the grid-cathode circuit of an a.c. amplifier circuit in the tool. The amplified a.c. signal is rectified and sent to the surface as a d.c. signal, where it is calibrated in degrees. In order to buck out the static value of this signal, a known matched signal of opposite polarity is placed in series with the electrode signal. The resultant d.c. signal is amplified in the instrument tray and recorded in the camera. A switch is provided in the tool for changing signal points at the a.c. bridge, so that the bridge can always be operated close to the balance point. This is necessary for two reasons, to reduce noise and to keep the electrode system from being saturated. The resultant log comes as a plot of temperature versus depth.

The standard electrode may be obtained in two temperature ranges, from 20° F. to 280° F. and from 60° F. to 340° F. Tools capable of being run in tubing are also made.

A fundamental knowledge of temperature gradients and temperature anomalies, as expressed in drilled holes, is necessary before the myriad uses of temperature logs can be fully realized.

Measurements made by a thermometer lowered in a drill hole give the temperature of the drilling fluid. Unless the hole has not been circulated for several weeks, the temperature of the mud is very different from that of the formations. The mud is usually colder at the bottom and hotter at the top of the hole than is the surrounding strata. Thus, when circulation is stopped, the mud will warm up in the lower part of the hole and cool at the top. The speed of this heat exchange will depend on the lithology of the bore hole.

To illustrate this, take a well 8000 feet deep, as illustrated in figure 210. The geothermal gradient can be represented by a straight line, as shown in curve 1. The temperature of the mud, when circulation ceases, is almost the same from top to bottom, as indicated by curve 2. The temperature gradients cross at point A.

If the well is left idle for several days, the temperatures will tend to equalize and the mud curve will tend to rotate from 2 to 1 about the axis A.

The cooling or warming of the mud at a certain depth will depend on the thermal conductivity of the formations and the size of the hole. Experience has shown that equilibrium is reached sooner opposite sands than shales. This can be explained by the fact that (1) hole size is smaller opposite sands than shales, and, therefore, the volume of the mud is less, and (2) the thermal conductivity of sands is greater than that of shale.

Thus it will be seen that during thermal evolution, sands will exhibit a lesser temperature than adjacent shales in the top part of the hole and a

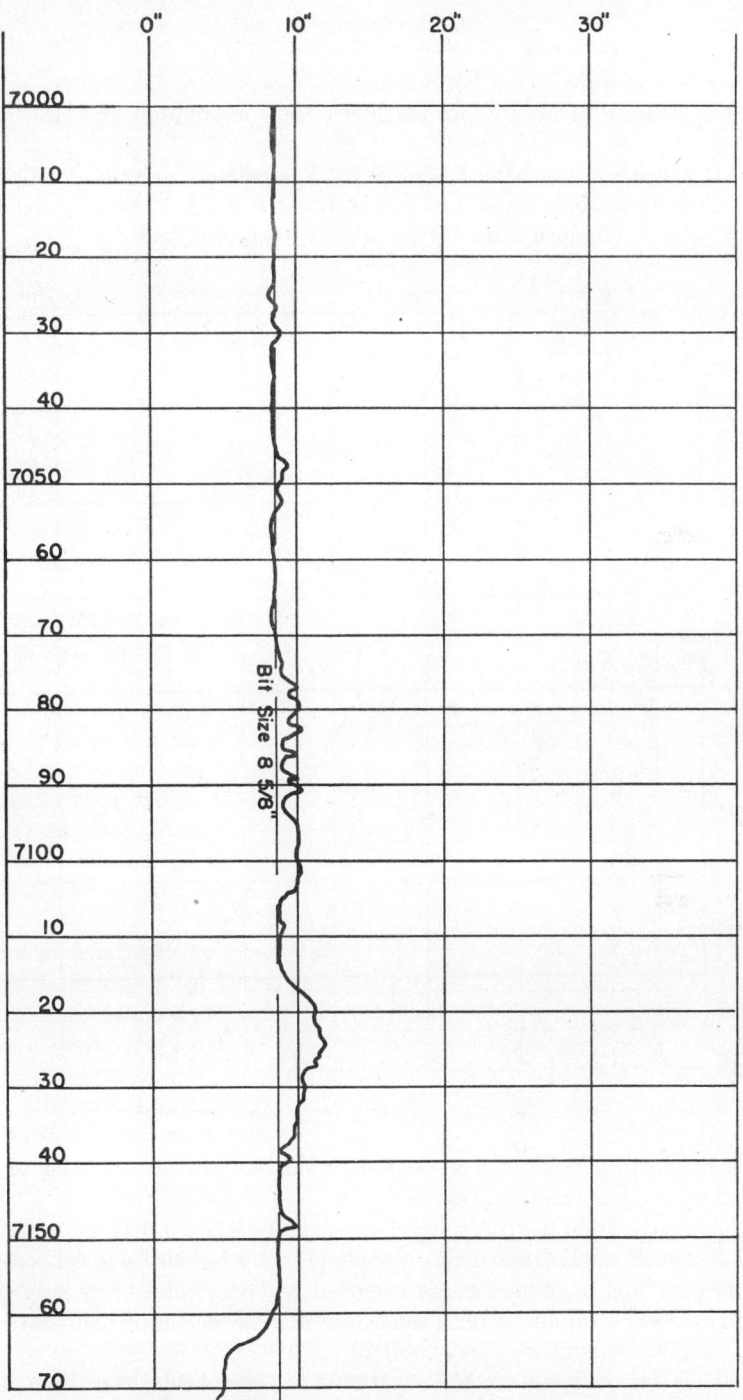

FIGURE 209. Caliper log in Smith County, Texas, used to find packer seat.

higher temperature in the bottom part. Curve 3 of figure 210 represents the temperature of mud about ten hours after circulation and illustrates this point.

If a temperature record is obtained immediately after circulation, a flat curve resembling curve 2 of figure 210 will result. The same type of curve will be obtained after several days have passed and equilbrium has

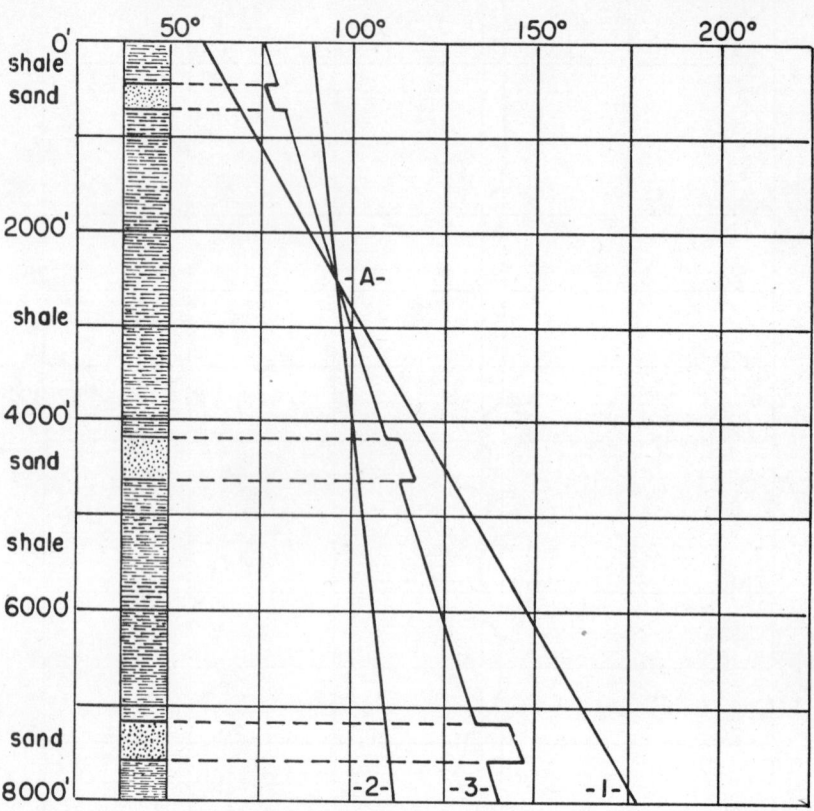

FIGURE 210. Chart showing temperature gradients.

been reached. The most pronounced anomalies are obtained 24 to 36 hours after circulation has ceased.

Although from the foregoing discussion one can see that it is possible to distinguish sands from shales, the temperature log will not yet replace the electric log. Factors such as chemical reaction, changes in hole size, the movement of fluids between sands, and the movement of hydrocarbons can alter the temperature in a well.

Thus far only open-hole-temperature measurements have been considered, but the presence of a string of casing does not disturb the thermic state of a bore hole. Therefore, all of this previous discussion applies to cased holes when they have been cemented, provided the temperature log is made four or five days after cementing.

Cement generates considerable heat as it sets up, and this factor has resulted in the principal application of temperature logs: The determination of a cement top behind casing by means of thermal measurements. The magnitude of this temperature increase varies with the time elapsed since cementing and the quantity of cement used.

Most of the heat is generated a few hours after the cement job has been completed. After this interval it is quite possible that the formation will absorb heat faster than it is generated by the setting cement, thereby cooling the mud. As a rule the best time to run a temperature survey is from four to eighteen hours after the cementing plug has hit bottom. The exact interval depends on a number of factors. Most operators prefer not to release the pipe pressure until after the initial set of the cement, which is a function of the type of cement, the type of water, the temperature, the pressure, and other variables.

The quantity of cement also affects the magnitude of a temperature anomaly. The amount of deflection tends to vary as does the amount of cement, which is a function of hole size. The joint use of a caliper log when trying to interpret a temperature log is often useful in this respect.

Several precautions should be taken when obtaining temperature logs in a cased hole. Circulation after cementing has been completed results in the heat evolved by setting cement being dissipated, and a trustworthy record is not obtained.

Temperature surveys in their present form will tell how high cement is in the annular space and, comparatively, how much cement is behind pipe. However, the survey does not indicate where the cement is. It is, for all practical purposes, impossible to detect channeling on a cement job by means of a temperature survey.

Numerous other uses have been made of temperature records in both cased and uncased holes. As gas enters a well, either through a hole in the casing or into a bore hole on an uncased well, it expands and cools. This characteristic has been used to find oil-gas contacts or a hole in the casing. However, any use of a temperature log presupposes that the temperature anomalies being measured are of a greater magnitude than four or five degrees, these being the order of formation thermal differences.

WELL LOGGING BY DRILLING-MUD AND CUTTINGS ANALYSIS

ARTHUR LANGTON

Today the determination of the fluid content of the porous formations penetrated by the bit so that no oil- or gas-bearing formations will be overlooked constitutes one of the most important problems in drilling for oil and gas. Modern mud-analysis, well-logging equipment, and improved technique allow for the detection of the most minute quantities of oil and gas and the exact placing of these shows at the proper depth.

The Baroid well-logging service has been developed using the following facts as a basis: In drilling a well the bit disintegrates a cylindrical section of the formation. If the pore spaces of this cylinder contain oil or gas, part of these fluids will be entrained by the drilling mud and part will be retained on or in the cuttings. If the drilling fluid and the cuttings are continuously tested on their return to the surface and the results of these tests correlated with the depth of origin, the presence or absence of oil or gas in specific formations can be determined.

In the application of the method continuous tests are made on the

FIGURE 211. Baroid mud-analysis well-logging unit on location.

mud and cuttings returning to the surface. The test for gas in the mud is made by diverting a portion of the circulating mud from the flow line to a separator or gas trap, where the mud is thoroughly mixed with air and a portion of the gas in the mud is removed. A stream of air is drawn countercurrent to the flow of mud in the gas trap, thus materially assisting the separation of the gas from the mud. The air-gas mixture is then drawn into a "hot-wire" gas-detector instrument, where the percentage of combustible gas is determined. The amount of methane gas present in the total is determined by controlling the temperature of the filament in the gas detector.

Thus, the gas readings in the mud are recorded as total gas, which includes all the combustible gases and methane gas. This "breakdown" of gas readings is made because recent studies have indicated that, with few exceptions, all productive horizons logged by the mud-analysis method have shown a definite increase of methane gas. Therefore, inclusion of the methane curve on the new log is important to operators, since, normally, those zones that do not show that an increase of methane may be condemned.

The test for gas in the cuttings is made by placing a small sample of

FIGURE 212. Operating equipment in Baroid mud-analysis well-logging unit.

the cuttings with a measured amount of water in a closed-container-type, high-speed grinder. After cuttings are ground, the air-gas mixture in the container is examined for gas in the same manner as described in the preceding paragraph. Here, too, the gas is reported as total gas and methane gas.

The presence of oil in the drilling fluid is detected by a physical examination of the drilling mud under ultraviolet light. A sample of the mud is treated to reduce the surface tension and gel strength, after which it is placed in a viewing box. This box is so constructed that all external light is excluded; thus the sample may be subjected selectively to either ultraviolet or white light. The magnitude of the oil shows in fluid-logging units is based on the amounts of observed flourescence of the crude oil.

Oil in the cuttings is determined by using the same ultraviolet-light arrangement. Freshly washed cuttings are placed under the light and observed for fluorescence. The cuttings are then treated with various leaching agents to see if they will give a "cut." Certain minerals fluoresce when subjected to ultraviolet light. This fluorescence can be distinguished from crude-oil fluorescence by the fact that mineral fluorescence will not give a "cut" when treated with a leaching agent.

FIGURE 213. Control panel of Baroid mud-analysis well-logging unit.

Further, the cuttings are subjected to a detailed examination under the microscope, in which the operator records the estimated percentage of limestone, sandstone, shale, and anhydrite. Note is also made of indications of porosity and permeability.

A depth meter, pump-stroke counter, and pump-rate meter provide the necessary data for obtaining a rate-of-penetration curve and for properly correlating with depth any shows of oil and gas found in the mud and cutting samples at the surface. The log is prepared as a plot of the magnitude of the oil and gas shows versus depth. Also given is a sand- or lime-index curve showing the relative amounts of either sand or lime in the cuttings. This curve is intended to be used as a supplement to the drilling-rate curve. The drilling-rate curve and the sand-index curve are used as indications of porosity and permeability values and of changing formation. They are also used for correlation between wells in the same general area because of characteristic variations in formation hardnesses and sand or lime content.

The well-logging service is used as wells are being drilled, and results on any formation are available as soon as the mud used in drilling the formation reaches the surface. The system does not interfere with the drilling operations in any way, and no changes in ordinary drilling equip-

FIGURE 214. Special filter press designed to test efficiency of fibrous materials for restoring lost circulation. "Core" consists of sized rock.

ment or procedure are required to make it workable. The logging units are in reality mobile field laboratories. In addition to the logging instruments, each unit is equipped with complete mud-testing equipment. Tests for mud weight, viscosity, water loss, cake thickness, salinity, and other characteristics can be made so that a satisfactory drilling mud may be maintained.

The principal application has been in the drilling of wildcat or exploratory wells, where its usefulness is apparent in many phases of the well program. Coring can be reduced to a minimum by using mud-analysis logging to choose only those sections to core which contain oil or gas. The procedure is to drill ahead until a significant increase in drilling rate, "a drilling break" indicating a change to a softer, more-easily-drilled formation, is encountered. After two to four feet of the soft formation are penetrated, further drilling is suspended until the mud, which was exposed to this new zone, and the corresponding cuttings are pumped to the surface and samples reach the logging unit for analysis. If oil or gas shows are obtained, cores of the formation are taken. If nothing of interest to the operator is indicated, drilling is resumed until another drilling break is encountered; whereupon, the procedure is repeated. Such a program has an operational advantage in that the information is available a relatively short time after the zone is penetrated. This program allows the operator to core, or, if desired, to make drill-stem tests while the effect of contamination by the drilling fluid is at a minimum.

Baroid Well-Logging Units also contain complete core-analysis equipment by means of which the operator may make determinations of porosity, permeability, salinity, and oil and water saturations.

The type of log produced is especially useful in those cases where conditions exist which make it difficult to get good electrical logs. These conditions may apply to an entire area such as the Permian Basin of West Texas. Here it is almost impossible, because of the properties of the lime formation and the high-saline content of the drilling muds, to obtain good electric logs; whereas, the Baroid log gives the operator accurate information on the oil and gas content of the formations. Similar conditions exist when the temperature of the drilling fluid becomes exceedingly high, as in some of the deep Louisiana wells, or when the chemical treatment of the mud is such that it will interfere with the electrical log. Also to be considered are isolated formations which are difficult to interpret by the electric log. In this category are such formations as the Eocene Wilcox, the Cotton Valley, the Travis Peak, and numerous others. The electric logs taken through these sections do not readily indicate the fluid content of the formation. With the Baroid log, a positive identification of the fluid content is made, thus reducing testing to a minimum, with a resultant decrease in time and money expended on the completion of the well.

In certain sections of the country some exploratory wells are drilled without taking any cores. After the electrical log is run, sidewall samples are taken to check the possible indications shown on the electric log. In this manner, the wells are drilled with a minimum output of time and expense, but ever-present is the possibility that a productive sand or lime will be missed. This possibility is removed by using the Baroid Well Logging service during drilling. This type of logging can be depended upon to pick up indications of all possibly productive formations, and the fluid content of such formations can be further checked by side-wall cores.

This exploratory method is particularly useful in areas of abnormally high pressure, where sloughing shale is encountered, and, in general, where coring is difficult or impossible. In cases on record, hole trouble developed so seriously that it was impossible to make other surveys, and the mud and cuttings-analysis log was the principal, if not the only, source of information on the lower section of the hole. An operator desiring to quit a well because of severe hole trouble, such as junk, lost circulation, or high pressures, may spend tens of thousands of dollars conditioning the hole just to run the final electric log over perhaps only a few hundred feet drilled since the last log was run. If a Baroid log had been obtained, it would have provided reasonable assurance that nothing had been passed, and the operator could have plugged and abandoned the hole as economically as possible.

The method does not present a complete subsurface picture. It does not give quantitative determinations of the amount of oil and gas occurring in the formation, nor does it furnish quantitative information on the productivity of the oil and gas horizons. A quantitative estimate is prevented by the numerous factors which affect the concentration of oil and gas in the mud and the cuttings. Some of these factors are the ratio of the volume of formation drilled to the volume of mud used to drill it; the flushing action of the drilling fluid, which in itself is affected by the mud-filtration characteristics, drilling rate, speed of rotation of the bit, differential pressure, and effective porosity and permeability; and the amount of oil and gas which is recirculated, which depends on the viscosity and gel properties of the drilling mud. However, the method does give reliable qualitative information on the occurrence of oil and gas, and the interpretation of the log should be made in the light of all available information on that section of the hole logged. It is extremely valuable in minimizing coring, as outlined before, in removing the ever-present possibility that a production zone will be missed and in rounding out the picture presented by other formation data.

DRILLING-TIME LOGGING
G. FREDERICK SHEPHERD

Rate of penetration is considered here as the time required to rotary-drill a linear unit of depth of a geologic formation of the earth's crust. It is believed that drilling-time characteristics constitute a diagnostic property of a rock resulting from its composition, mode of deposition, degree of compaction, and other known physical features by definition of which the rock is described. At present, drilling-time properties, measurable in terms of rate of penetration, are qualitative in scope and relative in their interpretation. That they may become quantitative and definable in fixed values that may be significant in the determination of lithologic types is anticipated.

Rate of penetration may be measured in terms of drilling time and drilling rate, each of which has its own specific uses and limitations. The relationship between the two and their differences are discussed. Drilling-time data may be applied to many engineering and drilling problems and have proved of considerable value to contractors, drilling crews, and operators. The multiple uses to the geologist involve general correlation problems and detailed studies of lithology. The geologic application of drilling-time data is an additional technique of particular value in the determination of potentially productive sections of a well bore and in the calculation of recoverable reserves.

Many methods have been used to measure and record drilling time. A technique used by the author is described and illustrated in which mechanical recording of depth in reference to elapsed time is translated into graphs or logs, which may be used in the solution of a great many geologic and engineering problems. The amount of section so logged and the scale employed are determined by the requirements of the individual problem. Detailed instructions are included whereby a geologist not experienced in the use of this technique may learn how to prepare and interpret drilling-time logs in any area in which he may be interested.

DEFINITIONS

Early methods of determining rate of penetration were crude and approximate and generally consisted in recording the time required to drill a certain number of feet of hole or the number of feet drilled per hour. Data thus acquired may be adequate for some purposes, but as the technique of using drilling time became more widely employed, certain advantages were observed in determining the specific net amount of time required to drill each foot. Before discussing the application of drilling-time data, the distinction should be understood between drilling rate and drilling time.

By definition, rate of penetration is a fixed lithologic property, even though it may be diagnostic only when used as a relative term. Drilling time is the duration of time required in the actual drilling of a unit of depth. Drilling rate is the number of units of depth drilled in a unit of time.

The foregoing may be illustrated by comparison to an automobile speedometer. When a car travels a mile in a certain number of minutes and seconds, it is a measure of speed comparable to the time occupied in the drilling of one foot of formation, which we have defined as drilling time. When a speedometer indicator points to 45, it indicates that the car is traveling at the rate of 45 miles per hour. This is comparable to rate of penetration measured in the number of feet drilled per hour, which is the definition given for drilling rate. The distinction is more than academic and should be clearly understood, because the interpretation of rate of penetration is strongly influenced by the method of recording.

Drilling Time: Diagnostic of Lithology

The relationship between rate of penetration and lithology has been understood for many years. The application of drilling time was recognized at least seventy years ago,[26] and the interpretative value has been appreciated for more than fifteen years. The use of drilling-time data has widened continuously as mechanical devices for their recording have become available and new applications of the data have been found.

Drillers probably were the first to learn that a change in drilling rate meant a change in the type of rock penetrated by the bit. Limestone, cap rock, shale, or sandstone would be recognized prior to confirmation by examination of the cuttings. Until recently the application of drilling-time data was essentially one of qualitative significance, and methods of observing rate of penetration were formerly far from exact.

Regardless of the method of observation or the crudeness of technique in measuring drilling time, one fact remains unchanged and should be emphasized at the outset. Drilling-time characteristics are but one of many diagnostic properties of a rock; therefore, the use of rate-of-penetration data must be considered as corroborative of other techniques by which lithologic properties are recognized. Sample examination, coring, electric and radioactivity logging, temperature and caliper surveying, drilling-time logging, and other means of geologic observation must go hand in hand to meet today's demand for more scientific methods for finding oil and gas reserves.

If, when drilling under uniform conditions, the bit's penetration changes from a slow rate to a faster rate or vice versa, it is an indication that a new type of lithology has been encountered. The obvious examples are readily recognizable and well known. A driller could hardly fail to know when he has encountered cap rock or a sandstone by "the way the bit acts." Certainly a change from crystalline limestone to dense dolomite or from hard shale to limestone is more difficult to observe, but any change in the characteristics of lithology should cause a change in the rate of penetration, provided all contributing factors remain constant.

One might question whether rate of penetration is scientifically a property of a rock because, at the present time at least, it is not capable of being catalogued in quantitative terms. This is a weakness in technical procedures, or the fault may be the inability to evaluate contributing factors, but this does not alter the fact that rate of penetration is a petrographic property. If there were no means of determining the identity of mineral constituents of a rock, it would not be wrong to state that mineral composition is a diagnostic property by means of which a specific lithologic type could be identified.

It can not be claimed that a certain sandstone having ninety percent quartz, six percent feldspar, three percent femags, and one percent auxiliary minerals, for example, will drill at a rate of one foot in three minutes

---

[26] Carll, J. F., *Discussion of Drilling Time:* Second Geological Survey of Pennsylvania, vol. 3, 1880. By personal correspondence with Dr. J. V. Howell.

and fifteen seconds; nor, conversely, that any rock which drills at that rate is necessarily that particular type of sandstone. Under one set of conditions it may drill in exactly that amount of time, and with other drilling conditions it may require much less or much more time. Nevertheless, we are defining rate of penetration as a fixed lithologic property, comparable to electric, radioactive, or mineralogic properties, and the hypothetical fixed time required to drill one foot of sandstone such as that described above could be defined as diagnostic of that rock.

It is hoped that some procedure for quantitatively evaluating contributing factors will enable drilling-time properties to be understood as fixed characteristics after allowing for the amounts of time required in the drilling of a unit of depth that are not attributed to the inherent lithology of the rock. Among the contributing factors referred to are the size of the hole, the type of bit, the drilling weight employed, the rotary speed, torque and friction, and the condition of the mud. This subject is worthy of study as a research project in order to determine the net-drilling-time value of a formation having uniform characteristics over an area large enough that adequate drilling-time data could be accumulated and studied.

The qualitative interpretative value of drilling-time data, however, is not impaired by the absence of quantitative calculations. Observations by the author in the drilling of hundreds of thousands of feet of hole have shown that drilling conditions are insignificant in comparison to lithology in determining the rate of penetration of a rock formation. Obvious exceptions have been noted, but the foregoing observation holds true. It has been shown in many instances that a change in formation will be reflected by a change in drilling time even when very dull bits have been in use or where other conditions would be expected to obliterate any evidences of change in drilling time. Perhaps the condition that affects drilling time more than any other is holding up on drilling weight as when straightening a hole tending to deviate. Under such disadvantageous conditions, there may be no pronounced change in the actual time required to drill a unit of depth when passing from one lithology into another, but the pattern of the curve plotted from drilling-time data seldom fails to reflect the change in lithology. Adverse drilling conditions do require more careful interpretation than favorable drilling conditions, but the effect of changes in lithology is seldom completely obscured.

Because drilling time is a qualitative property of a rock, it is important that correct identification of lithologies be based on the observation of relative values. A foot of hole that is drilled in five minutes at one depth may be interpreted as a sandstone, and another foot drilled under conditions differing from the first-mentioned foot requiring also five minutes for drilling may be interpreted as a shale. In each case the interpretation is based on the relative time in comparison to previously drilled feet. The value of the application of drilling-time data to geologic and engineering problems lies in the recognition of this relative interpretation.

DRILLING TIME AND DRILLING RATE

The two means of measuring rate of penetration have been defined; and, as shown, drilling time is a specific value for each foot, and drilling

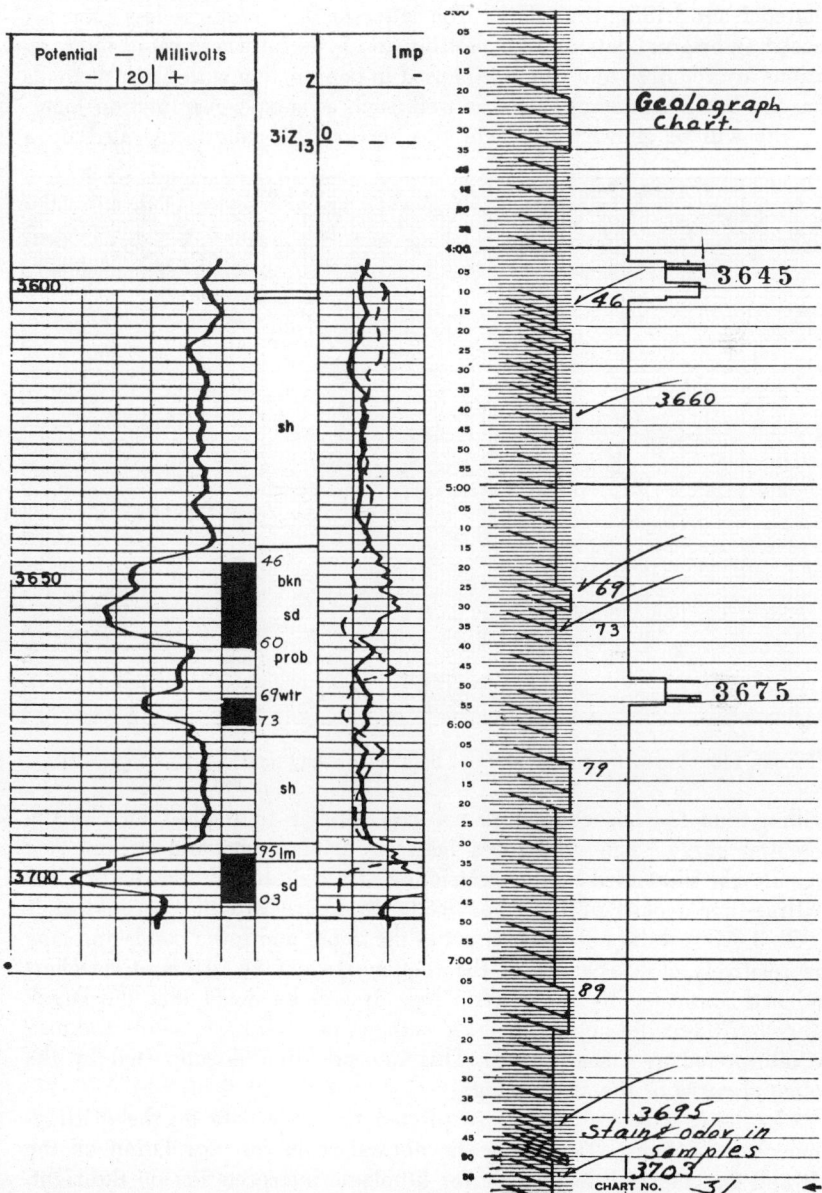

FIGURE 215. Relationship between electric log and drilling-time data. Black areas on electrical log indicate intervals of rapid penetration.

rate is an average value involving the drilling of several feet. The former is more exact and is useful where detailed lithologic information is required and where difficult correlation problems may be solved only in the study of minor features that are best disclosed in the pattern of a log plotted from drilling-time data. The latter method requires less time for recording original data and for plotting and is useful where rapid interpretations are required and where it is used in conjunction with other methods that are based on average data as well, such as sample-examination logs.

As will be shown further in this section the pattern revealed by a

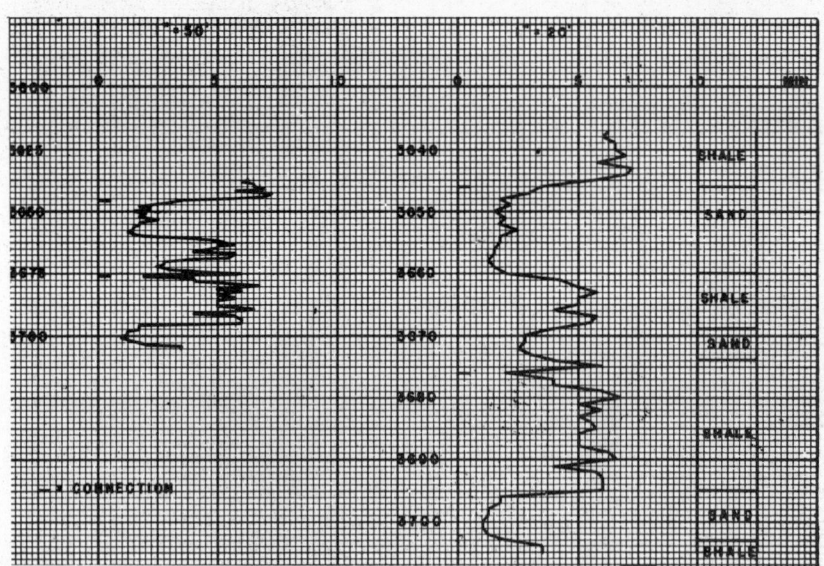

Figure 216. Two curves plotted from data on drilling-time chart of figure 215.

drilling-time log has characteristics very similar to that of an electric-potential curve. The differences between drilling-time and drilling-rate logs may be illustrated by the electric log of a well drilled and the original drilling-time record of the same well. In figure 215 the porous sands 3,669–3,673 feet did not drill as fast as the upper and lower sands and may the relatively close spacing of the foot marks on the drilling-time chart and are confirmed by the electric log. It will be noted that the streak 3,669–3,673 feet did not drill as fast as the upper and lower sands and may be interpreted as a shaly sand. This interpretation is supported by the potential curve of the electric log.

In figure 216 are two curves plotted from the data on the drilling-time chart of figure 215, using the normal scale for correlation on the left and the enlarged detail scale for lithologic interpretation on the right. A close comparison of the detailed curve with the potential curve in figure 215 reveals not only the corresponding sand sections but also a close resemblance between the patterns of the two curves. Note, for example, such minor features as the slightly sandy shale streaks at 3,666 feet and 3,691 feet, which are represented as slight bulges in the potential curve. The characteristics of electric and drilling-time curves are determined by changes in lithology. The drilling-time break at 3,676 feet is interpreted as anomalous to lithology because a connection was made at this depth and part of this foot was drilled with the clutch out, causing an erroneously timed foot to be registered. By marking on the log where interruptions in drilling occur, such features may be recognized with ease and incorrect interpretations prevented.

The curves of figure 217, which were plotted from the same data as those of figure 216, show by solid lines the loss of detail in using five-foot intervals instead of one-foot intervals and by dotted lines the effect of averaging when using drilling-rate values. In the upper solid curve on the correlation scale the total time for five feet was used and plotted as a bar curve according to the manner generally practiced on sample logs. In the lower solid curve on the detailed scale, the average time per foot was plotted as a point-to-point curve. In both of these curves the presence of two sands and one shaly sand is observed, but the exact depths at which they occur, their net thickness, and minor lithologic breaks are absent. Obviously, it requires more time to plot the curves in figure 216 than in figure 217, and the information to be gained is disproportionate to the time saved. There is some value in large-interval drilling time and in drilling-rate curves to be sure, but their use is restricted to problems where only general impressions are needed either for correlation or lithologic interpretation. In plotting sample logs on the basis of percentage of lithologic types present in each sample, the position of major breaks may be determined by plotting a drilling-rate curve similar to the upper dotted curve in figure 217. Where difficult full-length correlation problems are encountered, however, the drilling-time curves of figure 216 will be found far more reliable and useful. Drilling-rate logs are useful in sample examination work in determining sample lag, but here again the information is only exact within the limits of accuracy of the method used. Further discussion of the application of drilling time and drilling rate follows at the end of the next section.

Several devices purport to record changes in rate of penetration in terms of feet per hour, and mechanical instruments have been marketed that provide drilling-time logs or data from which these logs may be plotted manually. One of the drawbacks to the use of drilling-time logs has been the time required to plot curves of the type illustrated in figure 216. There is no machine available that will reliably record or plot drilling-time logs of this type and eliminate human errors. Considerable experimental work along this line has been done, and the need for such a device is great. Among the best known to the author is the extensive research, laboratory testing, and field demonstrations conducted by Mr.

Thomas A. Banning, Jr., of Chicago, Illinois. Under the title "Measuring and Recording Various Well Drilling Operations" Mr. Banning has filed application for Letters Patent in the United States Patent Office. This application reviews thoroughly the history of the art of drilling-time recording and investigates means by which data measurable during the drilling of a well can be recorded and plotted to produce logs such as are described in this article. Field tests are on record in which remarkable accuracy in depth measuring and time coordination were achieved. Such equipment, when available, will increase the application of drilling-time data as here discussed, and undoubtedly will open up new channels of research into lithologic properties hitherto inadequately understood.

APPLICATION OF DRILLING-TIME LOGS

Many of the uses for drilling-time logs have been cited and illustrated in the literature. The value to drillers, contractors, and operators is well known. The chief concern is the drilling of the greatest amount of hole in the shortest time possible consistent with good safety practices. At the same time anyone connected with the drilling of an oil well knows that the purpose of drilling a hole is to gain information, the use of which may lead to production of oil or gas. The contractor may find drilling-time records most useful in the analysis of operations and the study of down time as well as pay time. The performance of different types of bits in various formations can be observed directly on drilling-time charts which reflect the types of formations being penetrated. The driller finds drilling-time charts of value as a record of his tour showing exactly how much he drilled, what type of formations were encountered and their depths, and a record of time down for sundry purposes. The guess work is eliminated. But the driller, like the operator, is concerned with finding oil, and he knows that reservoir beds are porous media overlain by hard or impervious layers. If he knows he is drilling in a section where potentially productive formations may occur and has not been given precise instructions in respect to coring, logging, or testing, the driller uses his best judgment in the interest of the operator. When he observes a drilling-time break, he may stop drilling and circulate samples for examination and wait for orders before drilling past any formation that might carry oil or gas. The drilling-time charts provide an indisputable record of where the top of the break was encountered and how many feet have been drilled in it.

The geologic importance of drilling-time logs is evident primarily in the fact that foot-by-foot information is available for correlation. As has been shown in figures 215 and 216, a drilling-time log, plotted on a time scale such that the amplitude between fast and slow peaks generally corresponds to the range between the shale base line and the maximum peak of an electric-potential curve, provides the geologist with data that may be used, within reasonable limits, for much the same purposes as an electric log itself. It is obvious, therefore, that if a drilling-time log of a well corresponds in pattern to an electric-potential log of that well after the hole has been drilled, the drilling-time log made during the drilling could be used for the purposes served by the electric log. This has proved true particularly in the Gulf Coast area where long-range or

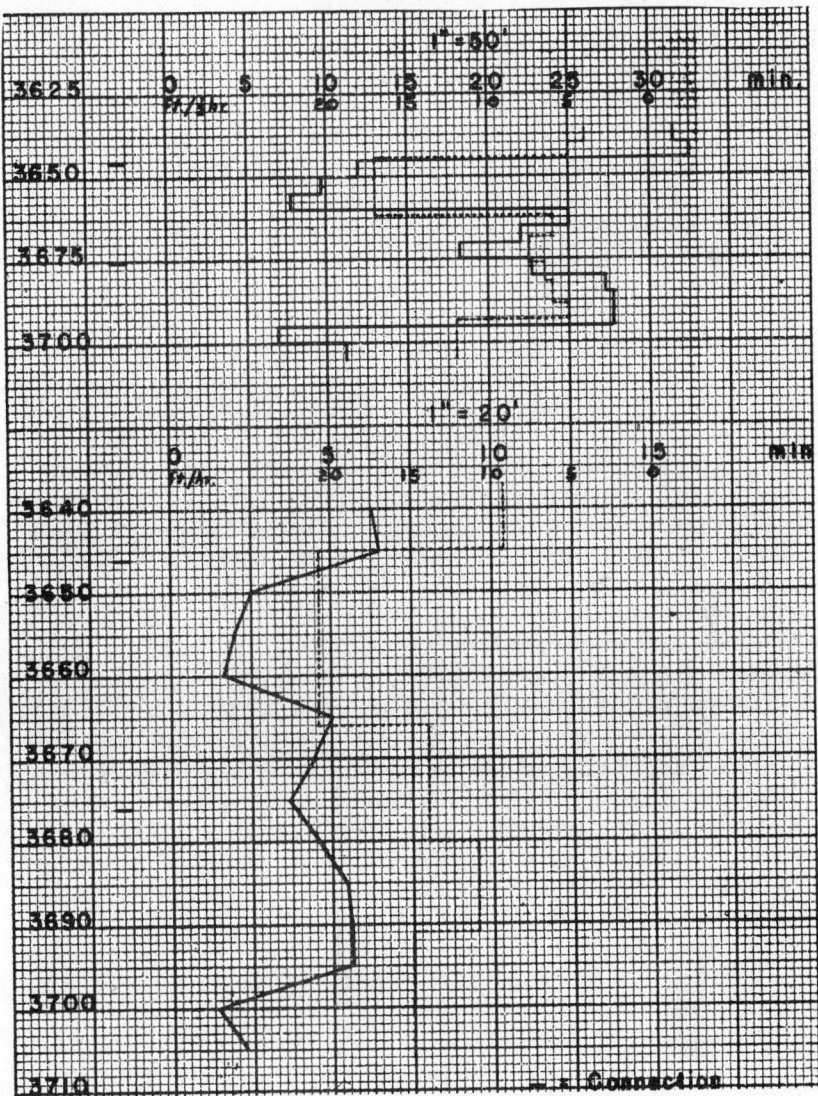

FIGURE 217. Curves plotted from same data as those of figure 216. Solid line represents a plotted five-foot interval. Dotted line reflects an average plot.

local correlation is based on the succession of a series of beds predominantly shale and sandstone. The stratigraphic position of any portion of a drilling well can be established in advance of electric logging by observing the sequence of beds penetrated as revealed by a drilling-time log.

It would be easy, for example, for the sequence of beds illustrated in figure 216 to correspond to a similar sequence of beds above or below the portion of the well shown. It would be difficult, if not beyond the realm of possibility, for the full length of section drilled and logged, below that depth at which drilling time becomes diagnostic of lithology, to correspond and be correlated erroneously with the same stratigraphic section of another well where such correlation could otherwise be established.

The economic and geologic value of this use of drilling-time logs is apparent. In the writer's experience many preliminary correlation runs of electric logs have been unnecessary because the purpose for which they would have been run was adequately served by drilling-time logs plotted as the drilling took place.

The most widely recognized geologic use of drilling time is in connection with coring operations. Careful recording aids in obtaining accuracy in well measurements, particularly where continuous coring is done over a long section. Drilling time often makes it possible to interpret the lithology of missing portions of cores recovered and identifies the portion of section cored from which the recovered core came.

Unless 100 percent of the core is recovered, it may be difficult to determine the net thickness of productive formations, even with the aid of an electric or radioactivity log. In areas where limestone streaks are interbedded with saturated sandstone, as in the Oligocene formation of south-central Louisiana, it is nearly impossible to interpret an electric log correctly without corroborative data. Greatest accuracy may be obtained in such problems by the use of electric logs, drilling-time logs, and cores combined.

Another important use of drilling-time data is their aid in the interpretation of electric logs. It is a common practice in drilling wells in the Gulf Coast area to rely on sidewall cores to check questionable shows of saturation in beds not cored during the drilling. Some of these questionable shows are thin calcareous beds which produce resistivity kicks on electric logs that are not unlike those that might be caused by saturated sandstones. The detailed examination of an electric log in conjunction with an accurate drilling-time log may reveal information on these questionable beds sufficient to identify them as calcareous or arenaceous. This use of drilling-time logs reduces the cost of side-wall coring and effects a further saving of rig time.

The use of drilling-time logs as an aid in the interpretation of electric logs may be applied to the problem of reserves estimate. Estimates of ultimate reserves of oil and gas often fail to represent the actual amount of eventual recovery. Although great progress has been made in understanding physical and chemical reservoir conditions and factors relating to the recovery of petroleum resources, the amount of reserves in place is given only as an estimate. Some even discount the value of making estimates of this character because of the lack of knowledge or possession of empirical data necessary to arrive at reliable conclusions. Efforts are continuously being made to increase the accuracy of reserves estimates. The oil or gas content of a reservoir bed is generally given in barrels of oil or mcf. of gas per acre foot. Lack of adequate knowledge pertaining to the reservoir conditions limits the accuracy of the estimate of the formation's content per unit volume, but the factor given is the best available in light of present-day scientific understanding. An estimate of the areal extent of a reservoir bed is also subject to considerable latitude because of the lack of knowledge concerning migration channels and underground drainage conditions. The estimate again is made on the best information that can be supplied by the subsurface geologist after mapping the structure in which the producing horizon is found.

In establishing the net effective thickness of the reservoir bed, there is a greater means of eliminating the necessity of an estimate, provided adequate data are available. In a section where reservoir beds are very uniform in respect to lithology, porosity, and permeability, the thickness may be determined by the driller's record, an electric log, a radioactivity log, or other reliable methods of logging or observing a formation. In those reservoir beds where the lithology is not uniform all available means may be required to determine the net effective thickness of that bed. Core information and analyses, electric and radioactivity logs, and drilling-time logs contribute to the best possible answer. As shown in figure 215, the less-permeable character of the bed from 3,669 feet to 3,673 feet was indicated both by drilling-time and electric logs.

There are many cases where thin shale breaks in a sandstone reservoir or tight calcareous streaks interbedded with saturated sandstone are not indicated on electric or radioactivity logs. If recorded on a short enough depth interval, however, drilling time will seldom fail to disclose the presence and thickness of such breaks. The writer has used drilling time recorded at intervals of one-tenth of a foot in a productive section where many and very thin impervious streaks were present and onl by this means was able to determine the exact net effective thickness of the reservoir. Therefore, if positive information can be gained as to the thickness of a formation, one of the three essential data making up a reserves estimate can be assigned a fixed value, and the accuracy of the final answer is increased.

Perhaps the greatest argument for the use of drilling-time logs is their value as insurance against the loss of geologic information in the event that other types of logging are precluded because of well conditions or in case of a junked hole. Because a drilling-time log provides essential data corresponding to an electric-potential log, it can be used for correlation to determine the stratigraphic and structural position of a well which might otherwise remain unknown. It may also reveal the presence of probably porous beds that may be saturated because of their structural position, and therefore the economic risk of drilling a new hole or abandoning a location may be substantially reduced.

Since the publication of the first edition of this symposium, the writer has received communication from J. A. Simons, geologist with Creole Petroleum Corporation, regarding the use of drilling-time and drilling-rate curves in Venezuela. He states, "This technique (drilling-time logging) is extremely useful in this area (Pedernales District) and is used as a method of bottoming a field well at the base of a known productive sand lense and to prevent the penetration of the next lense, known to be salt-water bearing . . . It developed that the plotting of drilling time in minutes per foot had been tried and poor results were obtained. A different method of plotting—feet per hour per one foot—was tried and has been adopted as a standard practice."

Simons states further, "If it requires a certain number of minutes to drill one foot, that is the drilling time for that foot. But that foot was also drilled at a certain *rate* which can be expressed in units of depth per units of time . . . When drilling time is plotted in feet per hour for one-foot intervals, a semi-logarithmic form of curve is obtained which dampens the effect of very slow feet caused by harder streaks or by the inattention of the driller, and which conversely exaggerates the effect of fast drilling that cannot be caused by anything but the bit entering a zone of easier digging. The scale is chosen to fit the fastest drilling observed, and fluctuations in the hardness of shale do not cause a widely varying curve, and the S.P. log in the shale section is more closely approximated."

The above discussion is illustrated in figure 218 and is introduced in this paper as an illustration of individual adoption of variations in selection of scale and method of plotting. It should be pointed out, however, that drilling-rate values cannot be determined without first measuring drilling-time values and it would appear that calculating the reciprocal values is an unnecessary step. The use of the zero base line at the right and the plotting of reciprocal values, even on a straight arithmetic basis, might have advantages in the analysis of special problems.

Commenting on the above, Mr. T. A. Banning, Jr., Chicago, Illinois, writes*: "The answer to the foregoing question (drilling time or drilling rate—which?) must depend on the use to which the information given is to be put. For some purposes *time* will be the most important factor of the ratio—for other purposes *rate* will be the important factor. In any case drilling time and drilling rate are mathematically interchangeable, and under the conditions presented in the case of well drilling these data are in all cases sufficient to permit presentation of the ratio either way."

---

* Personal communication.

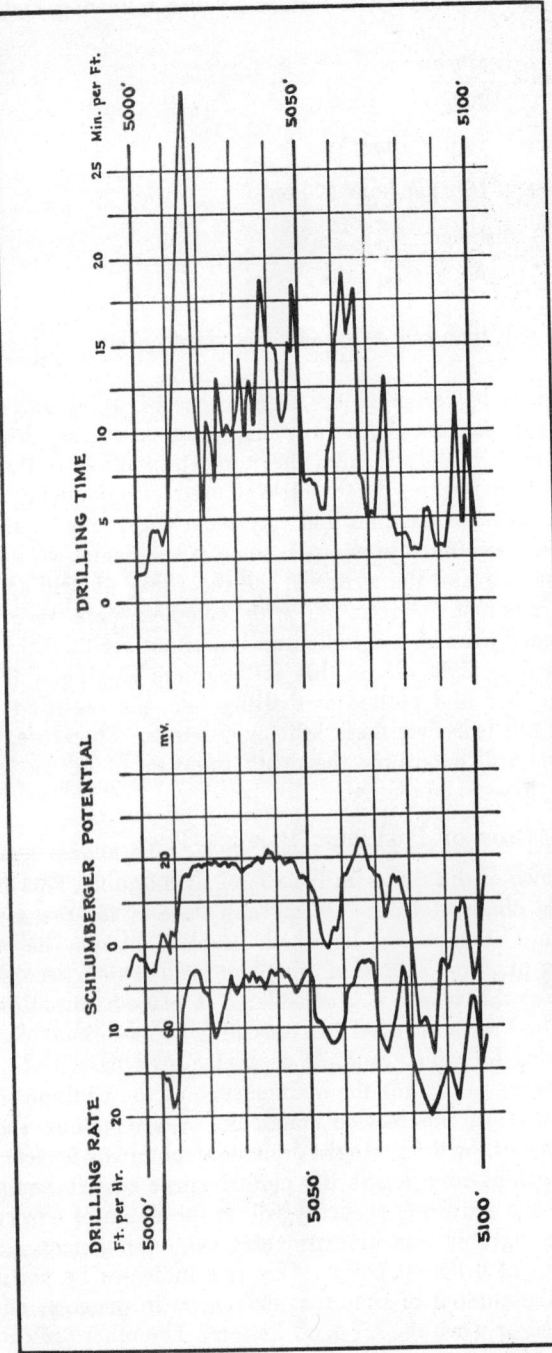

FIGURE 218. Correlation of the electrical (SP) curve (center) with a drilling-time log (right) and its reciprocal log in semilogarithmic form, drilling-rate log (left). Data from Creole Petroleum Corporation's No. 2-PSX, Pedernales District, Venezuela. (After J. A. Simons.)

This fact may be illustrated further by the following mathematical relationships:

$$\text{rate} = \frac{\text{distance}}{\text{time}}$$

$$\text{distance} = \text{rate} \times \text{time; or}$$

$$\text{distance} = \text{feet per hour} \times \text{hours}$$

$$\text{time} = \frac{\text{distance}}{\text{rate}}$$

If distance $= 1$, then rate $= \dfrac{1}{\text{time}}$ or, conversely, time $\dfrac{1}{\text{rate}}$

It is important to recognize the significance of using unit-depth intervals in these calculations. If, by any mathematical means, drilling rate is calculated from drilling-time data, the rate represents only the average speed of penetration for the depth interval used. Consequently, drilling rate obscures any minor variations that may occur within the depth interval used. As has been mentioned previously, some experimental work has been done by Mr. Banning and the writer in the recording of drilling time on one-tenth foot intervals. Correlation with complete core recoveries of sections thus logged showed a definite and important relationship between lithology and drilling time. If, in this drilling-time logging, full foot intervals had been used and plotted as drilling rate, the resulting average-value log would fail to reflect these lithologic details. Therefore, the purpose to be served will determine the depth intervals to be used and the character of the curve to be plotted.

METHOD OF PREPARING DRILLING-TIME LOGS

The experience of the writer in the use of drilling-time data has been based on records obtained from "Geolograph" charts for the most part. Although there are other means by which usable data may be collected, perhaps the most practical source of complete drilling-time records is the "Geolograph." For this reason it is considered in place here to describe in detail the technique recommended in translating the original record into the drilling-time log for which multiple uses have been described.

It is unnecessary to include the maintenance of the equipment, which is the responsibility of the service company. A few points should be kept in mind, however, by the geologist desiring to obtain as perfect records as possible. A drilling-time log is the plotted curve of two components, time and depth, each requiring accuracy within the limits of observational errors. The "Geolograph" machine provides two inking pens, generally supplied with inks of different colors. One pen indicates by vertical and horizontal lines the amount of time that drilling is in progress and when it has been stopped or when the bit is off bottom. The other indicates by a

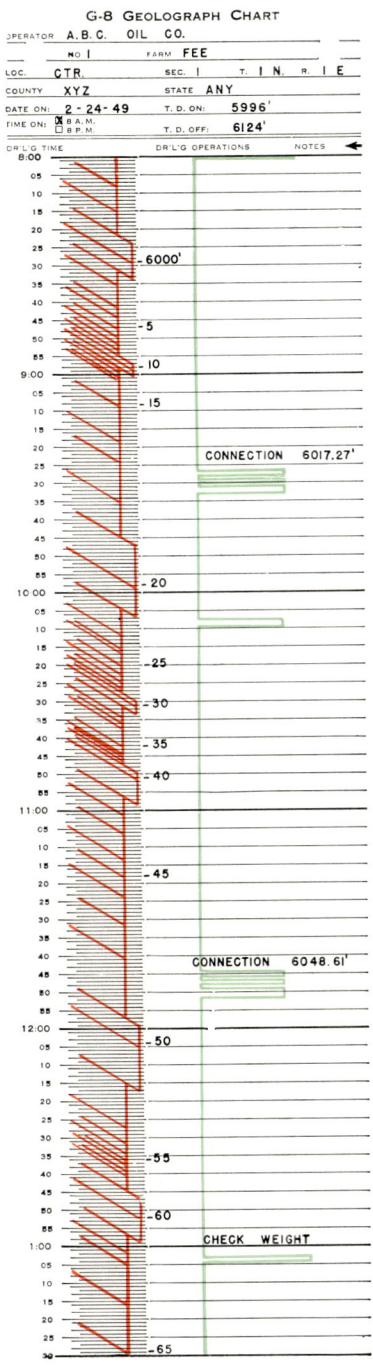

PLATE 10. "Geolograph" chart showing essential data used in preparing a drilling-time log. Displacement of red line to right marks every ten feet of section penetrated.

slanting stroke the completion of exactly one foot drilled, or two feet if set for recording on two-foot intervals. Therefore, the exact depth marked by each stroke of the pen must be known and the length of time occupied in the drilling must be determined. Plate 10 is a "Geolograph" chart showing essential data used in making a drilling-time log.

It is strongly recommended that the geologist using complete drilling-time data keep his own pipe tally. Practices differ among contractors in keeping pipe tallies, but inasmuch as the object here is to know the depth indicated by each mark on the chart, the geologist's tally must show all pipe in the hole at the time the mark is made. This becomes most important when changes in the drilling string are made, as for coring. The tally should show first the bit, subs, and drill collars in the string, followed by each joint of drill pipe added. The kelly should be measured accurately and its length recorded. The geologist should also observe whether it is the practice of the driller to drill the kelly down or make his connections with some length up on the kelly. The "Geolograph" automatically shows when a new joint of pipe is added, and the geologist should write on the chart the kelly-down depth of each connection, accurate to the nearest hundredth of a foot, as shown on his pipe tally. (See connections at 6,017.27 feet and 6,048.61 feet in pl. 10).

The charts are generally changed twice a day and when received for translating will show the date, time of chart change, depth of beginning and end of each chart, and correct depth at each connection or beginning of a round trip. The depth of each mark or every fifth mark between connections should then be written on the chart, as 6,000 feet, 6,005 feet, 6,010 feet, et seq., plate 10. If the drillers have been very careful in throwing in the clutch at the proper time, the number of marks should be identical with the number of feet drilled. Often this is not the case, and observation of drilling operations and experience with how such discrepancies occur will show the geologist where errors take place. The most common error of this type is made by the driller in failing to throw in the clutch after making a connection at exactly the same depth as when it was thrown out prior to making the connection. Changes in drilling weight caused by settling out of rock cuttings may cause what might be called "false drilling" or the redrilling of depth without drilling a new formation. Or, if the driller fails to throw in the clutch as soon as the bit reaches bottom with the same weight as when it left bottom, some new hole may be drilled without being recorded. (See 3,676 feet in figs. 215, 216.) Errors of these kinds are readily understood when one realizes that the drilling crew is most busily occupied when connections are made. Often the correction for depth will be made immediately after a connection, but the geologist must use his best judgment in making all corrections so that when a depth is assigned to a mark it will be correct.

Where no "extra" marks have been made and no "skips" noticed, there may accumulate fractional-foot errors, which may be designated as "creep." Over long-continued drilling, involving several connections or even round trips, a correction may be required, and it is difficult to know where it should be made. Having written the kelly-down depth at each connection and the correct depth at a round trip, accurate in each case to the nearest hundredth of a foot, it will be obvious what depth should be assigned just before or just after making a connection. For example, if a connection has been made at 6,079.04 feet in figure 219 and a foot mark is shown immediately after the connection, it would be reasonable to indicate the depth of the mark prior to the connection as 6,078 feet, as the total number of feet drilled corresponds to the number of foot marks on the chart. Had the connection been made at 6,078.97 feet, however, and a foot mark shown just prior to making the connection, with a normal drilling interval following the connection before the next foot mark was recorded, it would be reasonable to assign a depth of 6,079 feet to the foot mark prior to the connection. Apparent errors of this type would be the result of "creep."

The foregoing is only suggestive of some generalizations that may be made in preparing the chart for time determinations. Frequent use of drilling-time charts will increase the speed and accuracy in obtaining proper depth designations. The chart reproduced in plate 10 provides 24 inches for recording twelve hours of time, or two inches per hour. The hour is divided into twelve divisions of five minutes each, and these are further divided into one-minute divisions. For high speed, when the requirements demand it at the sacrifice of accuracy, the eye can measure the distance between foot marks within an accuracy of approximately twenty to thirty percent. Many uses of drilling-time logs require greater accuracy than this, however, particularly when minor breaks in the over-all pattern are used to correlate with minor breaks on an electric-potential log. To obtain greater accuracy, with an error of less than one percent, measurements may be made with an engineer's scale, using the thirty-divisions-per-inch scale for the purpose. This scale, placed on the chart having two inches per hour, will provide sixty divisions per hour or one per minute.

The use of a printed form to tabulate the time readings may be considered as an extra step, and some may suggest plotting directly from observation of the chart. It has been found that this extra step not only avoids many errors that might otherwise be made but actually saves time. In addition one often wishes to plot the same data on more than one scale, in which case the time saving is considerable. The same person may make the readings and write the tabulation, but if one person keeps his eye on the scale and chart and another writes down the time on the form, it will save two-thirds of the time required for this operation. Additional time can be saved in the following step, plotting the log from the tabulation, by having one person read the units off as another plots the coordinates.

Figure 219 illustrates a type of form on which drilling-time data may be tabulated.

Determination of the time factor from the chart, using the engineer's scale, is made by placing the scale parallel with the long dimension of the chart and at the left of the base line from which the slanting stroke is made. The zero of the scale is then placed at the top side of the depth stroke and the position of the top side of the succeeding depth stroke noted on the scale. The observer can read the scale to the nearest quarter of a division or fifteen seconds of time, and this reading is the time factor for the foot being measured. Because of the type of coordinate paper recommended for plotting the log, it will facilitate recording and plotting these time factors if decimals instead of fractions are used, as in figure 219.

No difficulty will be encountered in thus reading and tabulating the time where there have been no interruptions in drilling. If drilling has been stopped at other than at the completion of an even foot, as at 6,021–6,022 feet in plate 10, the time out must be subtracted from the total time to indicate only the net time consumed in the drilling of the foot. The position of the 6,022-foot stroke on the scale is shifted to the top of the bottom horizontal line, indicating the end of the time out. The position on the scale of the top side of the top horizontal time-out line will be the net total time required in the drilling of this foot. If more than one interruption has been made in the drilling of one foot, this process is repeated for each pair of horizontal lines indicating interruptions. All interruptions should be recorded on the tabulation, as in figure 219, and by conventional symbols shown on the plotted log, as in figure 220, to prevent false interpretations. Care should be given to the pens so that they are in exact horizontal registry in respect to each other at the top of each chart; otherwise, errors in calculating time-out intervals may be made.

After the chart has been prepared with the proper depths indicated, any instructed person may make the time determinations and plot the log. Office clerical assistance is used often by geologists supervising the work on several wells being drilled at the same time. When this is done, it is imperative that the original charts be received with the correct depth designations shown.

Plotting the log starts with the tabulation of data as in figure 219. The selection of the time scale is important and has been discussed and illustrated in previous papers by the writer. The log strip found to be most satisfactory for interpretative work consists of coordinate paper with twenty divisions per inch and cut six to eight inches in width. The heading at the top should show the name and location of the well, datum elevations, and the vertical scale used. The name of the observer or plotter may be useful for the record. Referring to figure 220, the depths are given at the left edge of the log at intervals of 10, 25, 50, or 100 feet as determined by the vertical scale used. The time scale is shown across the top of the log in terms of minutes per foot.

Sometimes the upper part of a well may be drilled at such a rate that two-foot timing is desirable. In this case the horizontal or time scale will be one-half that for one-foot timing so that the relative amplitude of the curves will remain the same. The time scale should be given on the

Operator___A. B. C. Oil Co.___
Well_____No. 1 Fee___
Field_____Wildcat___
Parish_____XYZ County, Any State___

#	Value	Note	#	Value	Note
60.01	4.50		51	12.50	
02	4.25		52	12.00	
03	6.00		53	4.50	
04	2.75		54	2.50	
05	3.00		55	1.50	
06	2.25		56	1.00	
07	1.75		57	2.00	
08	2.00		58	.75	
09	1.00		59	4.25	
10	1.25		60	5.25	
11	1.50		61	6.50	
12	1.25		62	5.50	
13	1.00		63	2.00	Interruption
14	1.00		64	11.00	
15	7.50		65	13.25	
16	10.00		66	10.00	
17	5.50		67	7.75	
18	4.50	Connection 6017.27'	68	5.00	
19	9.50		69	3.00	
20	11.50		70	1.00	
21	7.75		71	1.75	
22	5.25	Interruption	72	1.25	
23	3.00		73	2.25	
24	1.00		74	2.00	
25	3.50		75	2.50	
26	2.75		76	3.50	
27	1.75		77	2.75	
28	2.00		78	3.00	
29	.75		79	3.50	Connection 6079.04'
30	1.50		80	3.75	
31	2.50		81	8.25	
32	3.00		82	4.00	
33	1.50		83	7.25	
34	1.50		84	3.00	
35	5.00		85	3.25	
36	1.50		86	3.00	
37	1.00		87	3.50	
38	.75		88	4.25	
39	1.25		89	7.00	
40	1.50		90	9.25	
41	7.25		91	10.50	
42	5.25		92	9.25	
43	5.00		93	12.75	
44	6.50		94	6.00	
45	5.50		95	6.00	
46	5.50		96	6.25	
47	8.25		97	5.25	
48	9.00		98	7.75	
49	8.50	Connection 6048.61'	99	10.50	
50	6.00		61.00	14.50	

Date Record___Feb. 24, 1949___
Drilling Notes___Bit No. 12   Type ....   Mud Weight 9.6 lbs.   Viscosity   38"___

Remarks___Depths off 1 ft. 6063-6080' due to pick up and creep.___

FIGURE 219. Drilling-time record.

log when changing from two-foot to one-foot recording. It has been found by experience that a time scale of five minutes per inch on one-foot recording is most acceptable to yield an amplitude corresponding to electric-potential curves. On this scale each unit represents a quarter of a minute, which is the base unit tabulated in figure 219. If conditions in an area are such that drilling is much faster or much slower than normal, the time scale may be changed to meet the requirements of that area. It may be preferable, where drilling rates are very slow, to shift the base line instead of changing the time scale. In either manner of plotting the log, exceptionally slow feet will be encountered which will exceed the visible scale on the log. Such off-scale footage may be plotted on scale using the thirty-minute time as the zero base and cross-hatching the off-scale portion for convenience. The resulting log will appear similar to off-scale electric-resistivity logs.

The question may arise as to what to do with apparent errors in time values. If during continuous drilling, the brake is set while the driller is busy elsewhere, the weight may drill off the bit before the drill stem is lowered. This often occurs in drilling soft formations, and the "Geolograph" will fail to record the true net time required in the drilling. It is recommended that the values be plotted as they actually are recorded even when they are known to be fictional. The reason for this is that, if an effort is made to make arbitrary corrections, the value selected may have as little relation to the correct value as the recorded value. Therefore, if all values are plotted as recorded, the correction can be made in the interpretative phase of the work. One very fast foot in the midst of normally slow drilling, or one very slow foot in the midst of normally fast drilling would present no problem in its interpretation. The writer has followed this practice consistently because sometimes an apparent time error may not be an error at all, but rather may represent a very abrupt lithologic change over a short distance. A thin streak of shale in a soft sandstone, an ironstone bed located in soft shaly sand, or a thin streak of soft sandstone interbedded with a very hard limestone would appear as false recordings such as are indicated above.

As a suggestion, there are many advantages to be gained from plotting the log with black ink rather than with pencil. The tabulated values are plotted as coordinates on the log with respect to depth and time. A sharp, medium-hard pencil is ideal for this purpose. These points may then be connected by an ink line, using a ruling pen and metal-edged ruler. The inked curve is much easier to examine than a penciled curve, particularly under poor lighting conditions. An inked curve is preferable to a penciled curve also because it facilitates reproduction by direct printing or photostating.

In conclusion, it may be suggested that the free exchange of drilling-time logs would prove of considerable mutual benefit to geologists in much the same way that the interchange of electric and radioactivity logs

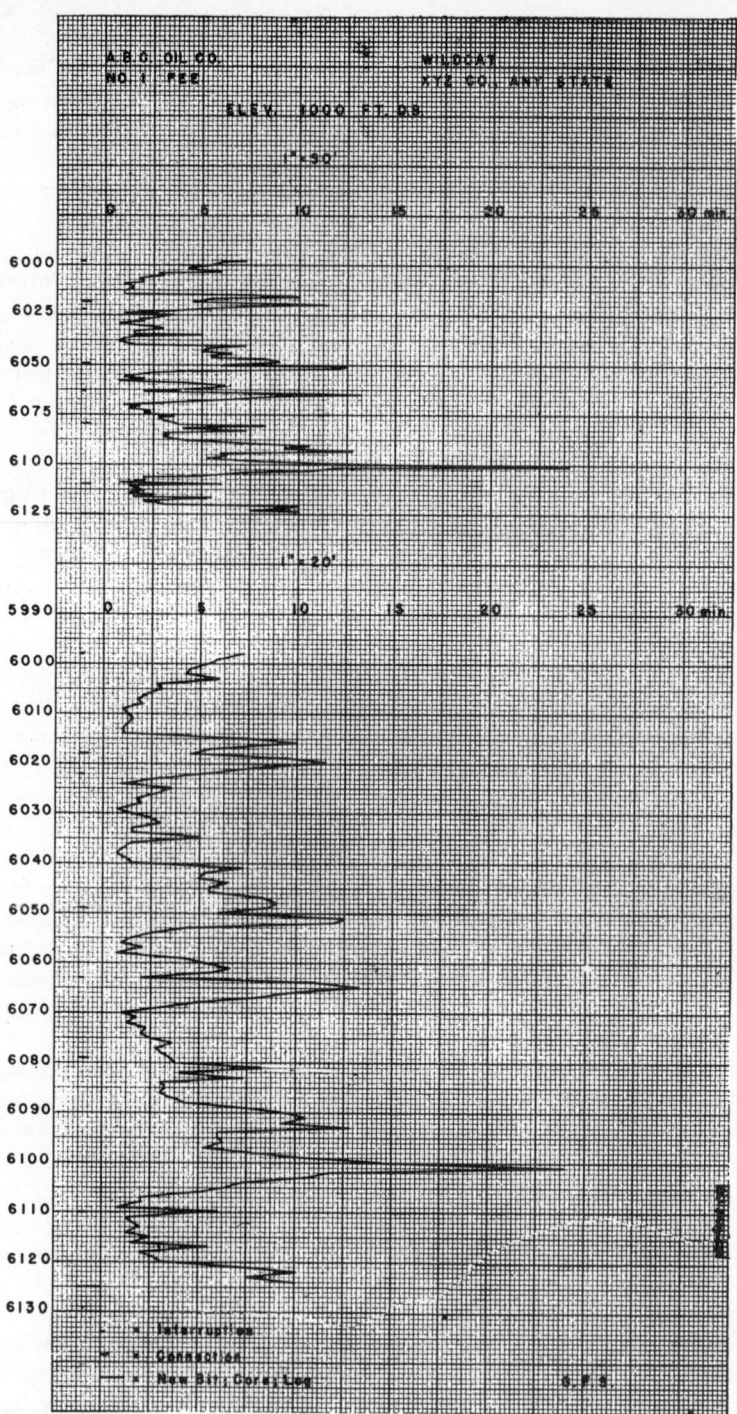

FIGURE 220. Drilling-time log strip.

is at the present time. Many companies publish catalogs of logs available for distribution. If drilling-time logs were added to these lists, they would benefit geologists whose responsibility it is to interpret properly the records made in the drilling of a test for oil or gas.

DRILLER'S LOGGING
L. W. LeROY

The driller and his assistants assume major responsibility for the successful operation and completion of their assigned well. Frequently these men fail to receive full credit for their role in exploration and exploitation programs. Too often have their opinions and suggestions been neglected or considered unworthy by "technical" personnel.

One of the many duties of the driller during drilling operations is to record and tabulate to the best of his ability the character of the penetrated strata. Frequently his lithologic "calls" are unreliable; many of them, however, are exceptionally accurate, particularly if he is familiar with the drilled section. There are numerous examples where a driller's recordings of formational "tops" are as exact as those determined by the geologist.

Drillers' logs would be greatly improved if geologists would take the time and effort to create a lithologic interest among drillers. Since oil companies are constantly demanding more accuracy in well logging, the driller can be of great assistance in this phase, if he has an understanding of the rock types through which he is drilling. It would not be out of line if these men were given, prior to drilling, an introduction to the various rock varieties to be expected in their wells.

DRILLERS' TERMINOLOGY

Examination of drillers' logs (See table 24) often reveals such terms as "lime," "sand," "shell," "gumbo," "broken sand," "broken lime," "dry sand," "quicksand," "hard rock," "pebble shale," "muck," "sand rock," "loam," "caving shale," and "niggerheads." Possible interpretations of these terms are as follows: caving shale, muck, gumbo, loam, the equivalent of soft shale; sand, broken lime, hard rock, shell, the equivalent of limestone or sandstone; lime, broken sand, hard rock, shell, the equivalent of limestone.

"Shale with niggerheads" may represent a shale containing thin, calcareous streaks or concretionary structures. "Quicksand" could be interpreted as soft, easily penetrated sand. "Caving shale" may suggest to the geologist the presence of bentonite. "Lime" recordings often incorporate such rock types as anhydrite, highly calcareous sandstone, light-colored tuff, or dolomite.

The color of rock types many times is inadequately recorded by the

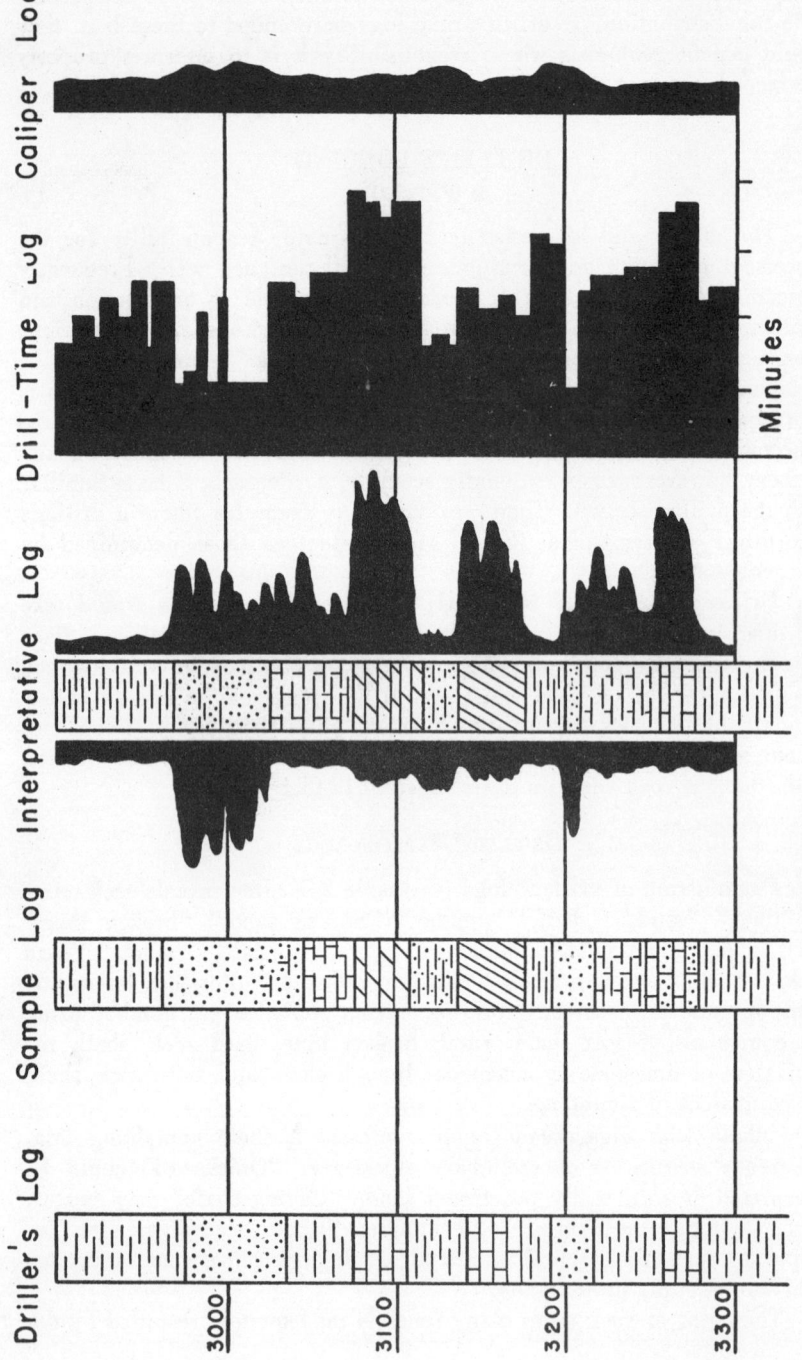

FIGURE 221. Showing relationship of driller's log to geologist's interpretative log.

driller. His "blue" shale may upon closer examination actually be green or greenish-gray; "red" shale may range from deep maroon to light pink. The term "red" may be continued on the tour sheet through a gray-colored section if mud contamination is excessive and samples are not thoroughly cleaned.

INTERPRETATION OF DRILLER'S LOG

Even though a driller's interpretation of lithology may be incomplete from a geologist's or engineer's viewpoint, his log is generally a result of careful and sincere observation and should under no circumstances be disregarded. Many early drillers' logs are impossible to interpret because of inconsistencies and erroneous lithic recordings; many have been found to aid materially in re-evaluation of stratigraphic sections, however, particularly when supplemented by more recently and carefully analyzed stratal data. Interpretation of these logs involves the understanding of routine procedures of the driller and the significance to him of their results.

It is now standard practice in rotary drilling to adhere to constant weight on the bit and rotational speed of the drill pipe. As these opera-

TABLE 24
A TYPICAL DRILLER'S RECORDING OF LITHOLOGY

Formation	Top (ft.)	Formation	Top (ft.)
Lime rock	24	Green shale	718
Sand	30	Chocolate shale	719
Blue shale	234	Gray shale	720
Blue lime	264	Green shale	721
Black shale	454	Hard lime	724
Sand rock; top of *Dakota*	456	Chocolate shale	730
White gumbo	458	Gray water sand; water in top 4 ft.	745
Wind rock	460	White sand	775
White gumbo	462	Brown sand	787
Sand rock	468	Hard red shale; top of *Morrison*	844
Water sand	470	Hard red and green shale	902
Sand rock	485	Brown sand, oil showing	906
Fine clay	487	Chocolate shale and green shale	944
Sand rock	498	Brown and green sand; oil	950
Hard sand rock	532	Green shale	956
Lime rock	544	Green sand; oil	962
Fine clay	554	Gumbo	963
Sand rock	574	Green sand; oil	969
Fine clay	584	Green sandy shale	986
Sand rock	620	Green lime	1007
Shale	668	Hard, brown lime	1010
White lime	687	Brown clay	1013
Red clay	689	Green, limy shale	1022
Lime rock	692	Green and blue limy shale	1050
Red clay	694	Black or bluish lime, weathers	
Chocolate shale	706	gray. Total depth	1065
Gray shale	715		

tions are now controlled by instruments, the driller has a better opportunity to record the relative hardness of formations by keeping a record of footage drilled per unit of time.

In the event that a geologist is assigned to watch or "sit on" a well, it becomes his responsibility to adjust the driller's lithologic "call" and suggest necessary modifications on the tour report before it is submitted to the field or home office. A driller's log in conjunction with other log data is useful in making final interpretative logs (fig. 221).

DRILLING-TIME LOGGING

P. B. NICHOLS

Subsurface lithologic units, which may be identified by means of sample-examination, electric-logging, or radioactivity methods, may also be recognized and delineated by drilling-time logging. In rotary drilling under normal operating conditions, major changes in the rate of penetration are due to changes in the lithology or texture of the formation being drilled. In other words, lithology governs the rate of penetration. Mechanical factors such as weight on the bit, mud-pump volume and pressure, and rotary r.p.m. are important in making hole, but they are not responsible for the type of drilling-time "breaks" that stand out prominently on mechanical well logs.

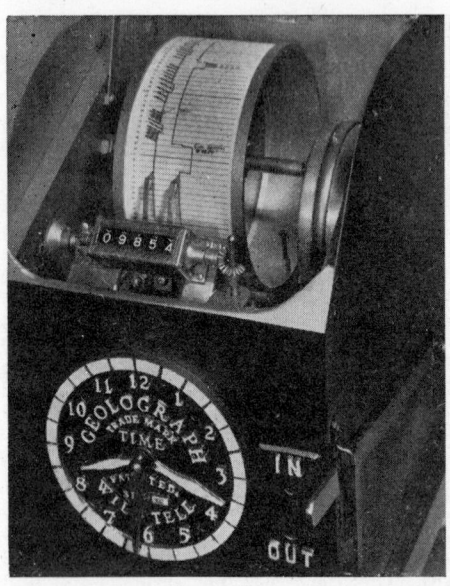

FIGURE 222. "Geolograph" recording unit showing footage pen (left) and down-time pen (right); depth meter and revolving indicator are on front. Actuating cable passes through guide pipes in back.

Formerly the drilling time was obtained by stripping the kelly in ten-, five-, or one-foot segments, depending upon the degree of control desired, and recording the time as each mark reached the kelly bushing. Within the last decade mechanical recorders have been developed which not only are much more accurate in timing but are designed to show additional relevant data such as the depth of the hole and the time consumed in making connections, round trips for bit changes, and down time for repairs.

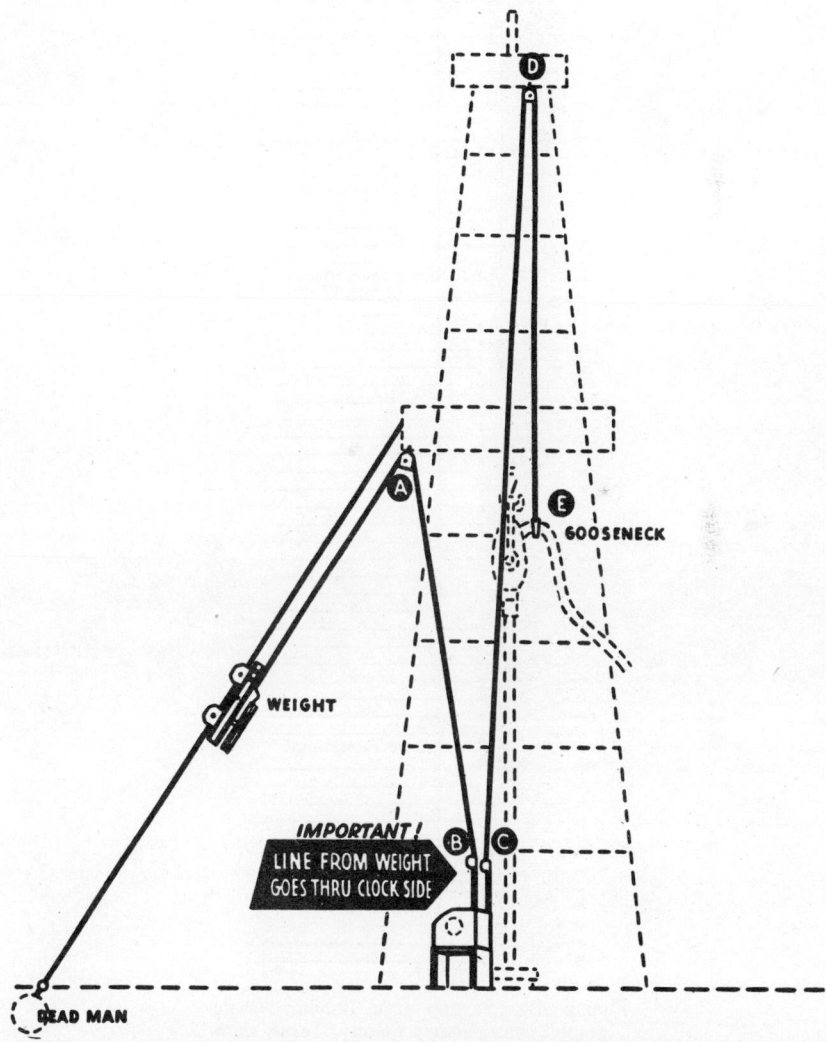

FIGURE 223. Schematic drawing of "Geolograph hook-up."

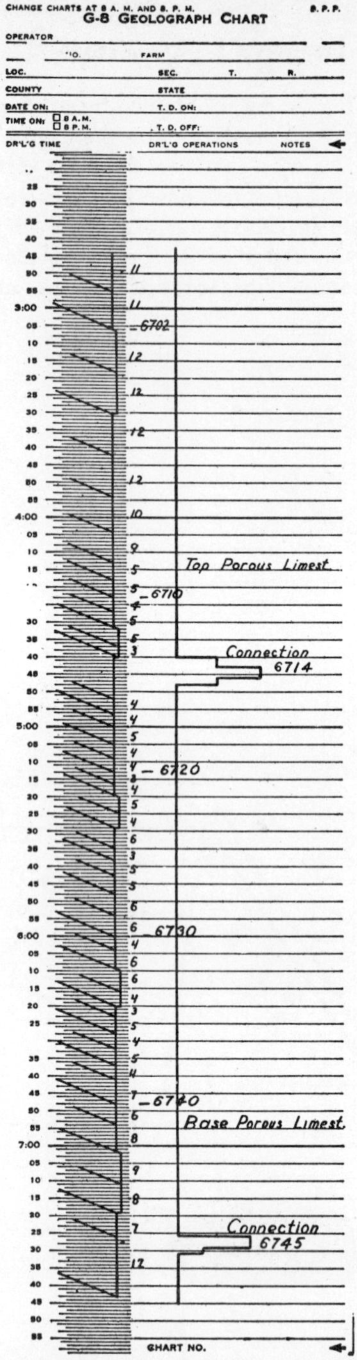

FIGURE 224. Portion of a 12-hour "Geolograph" chart, Scurry County, Texas, showing drilling change at top of reef-limestone
producing horizon at 6,708 feet with break
in the pay at 6,741 feet. Pipe was added
to string at 6,714 feet and at 6,745 feet.

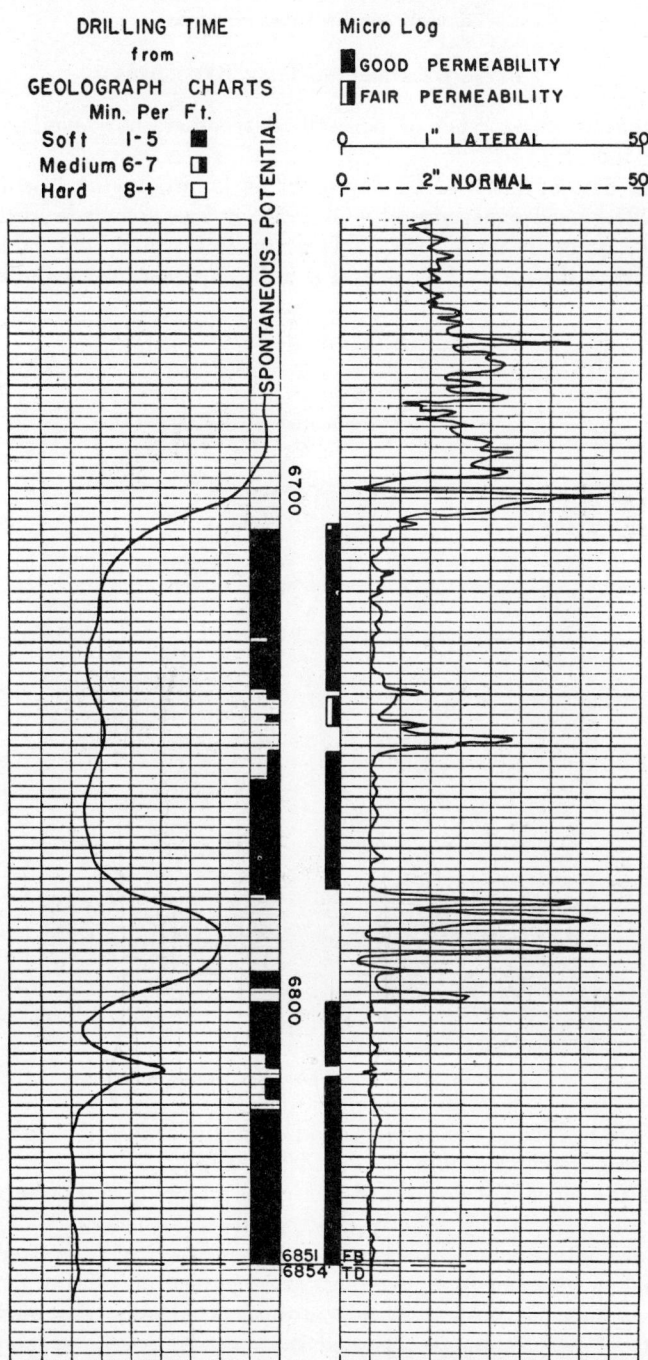

FIGURE 225. Drilling breaks from "Geolograph" charts may be transferred to electric log for direct comparison. Refer to figure 224 for actual drill-time data from upper part of Scurry reef limestone.

TYPES OF DRILLING-TIME RECORDERS

There are three types of penetration recorders now available to the drilling industry.

The first of these to be developed is known as the "Geolograph" (fig. 222). Used experimentally in 1937, it was not made available to the trade generally until 1943. This device is actuated by a cable which, made fast to the rotary hose swivel coupling, passes over a pulley at the

FIGURE 226. "Log-O-Graf" recording unit.

crown, down through the recorder, and up in the derrick through a second pulley to a weight movable on a guideline (fig. 223).

As the bit moves downward, the cable passes through the recorder, where each foot is individually checked off on a time chart. This chart is divided into minute, five-minute, and hour periods. The recorder is designed to give one-foot, two-foot, and ten-foot increments of drilling time. As time is constant in this method of logging, any variation in the rate of penetration is indicated by a change in the spacing of the foot marks on the chart. Logged in this manner, a great amount of detail is made available at the time such information is most needed by those

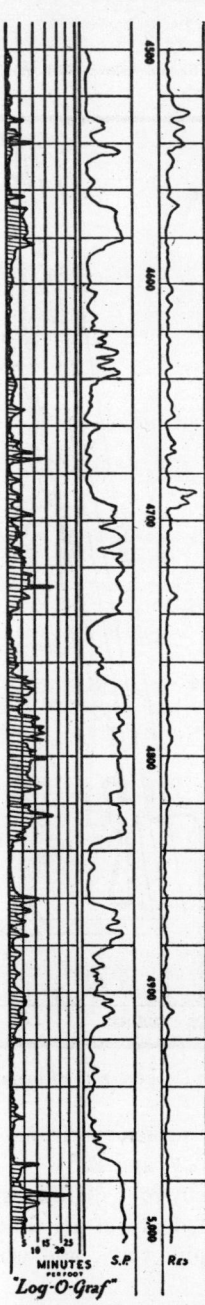

FIGURE 227. "Log-O-Graf" chart in juxtaposition with electric log showing its rate-of-penetration curve. Graph is formed as footage is drilled.

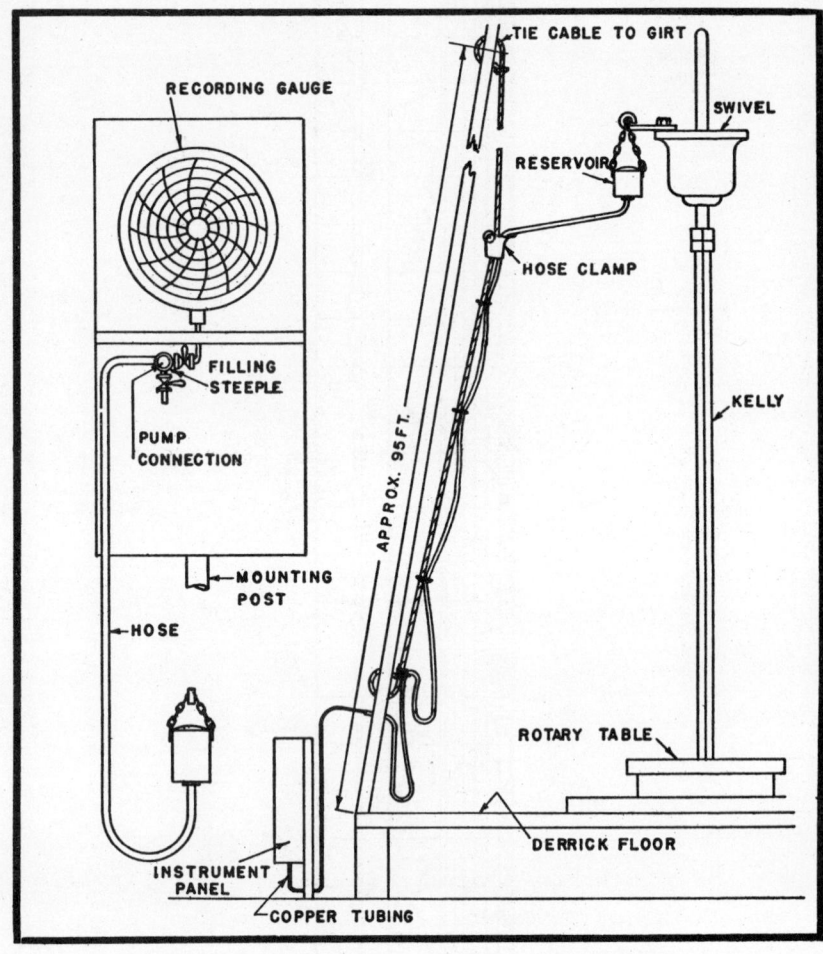

FIGURE 228. Martin-Decker, rate-of-penetration recorder.

directly responsible for the proper decisions that must be made during the drilling operation (figs. 224 and 225).

In addition to the foot-by-foot drilling-time record, a meter is provided which shows the total depth at all times, which thus provides for a continuous check on the pipe tally, and which eliminates many errors in depth.

The front of the unit is provided with a dial on which an indicator revolves so that the driller can tell to the inch where he is on any particular foot of hole.

Parallel to and contemporaneous with the drilling-time column of the "Geolograph" chart, a second pen records the down time. When the bit is raised off bottom, this pen moves to the right, automatically marking on the chart the time drilling ceased. When drilling is resumed, the pen returns to its normal drilling position.

The operations record shows to the minute how much time is consumed in making trips, connections, repairs, straight-hole tests, and like operations. This record not only aids the contractor in spotting the equipment requiring repairs that are becoming so costly that replace-

Figure 229. Martin-Decker recorder showing fluid reservoir, hose, and recording gauge with chart.

ment would be profitable, but also is being used in the development and testing of new drilling techniques and equipment.

A number of mud-logging companies have included "Geolograph" units in their logging equipment.

As the accuracy of mud logging is dependent on the ability to correlate the shows of oil and gas with the source beds, it is important that enough detail be obtained from the time log to enable the operator definitely to check the show in the mud stream back to its point of origin.

A second type of instrument, called the "Log-O-Graf, graphs the time horizontally and the depth vertically. The chief advantage claimed for this type of instrument is that the resulting graph, when made on the same vertical scale as the electric log, may be directly correlated with it without redrafting (figs. 226 and 227).

A third instrument known as the "Martin-Decker rate-of-penetration recorder" is based on the hydraulic principle. A fluid container on the kelly is connected by a hose to the recording mechanism (fig. 228). Lowering the kelly changes the hydrostatic pressure, which is recorded on a polar chart (fig. 229).

Two transparent templates inscribed with drilling-rate curves are provided. To reduce the drilling-rate curve on the chart to feet per hour, the template is placed over the chart and rotated until a drilling-rate curve coincides with the curve on the template, which may then be read directly in feet per hour (fig. 230).

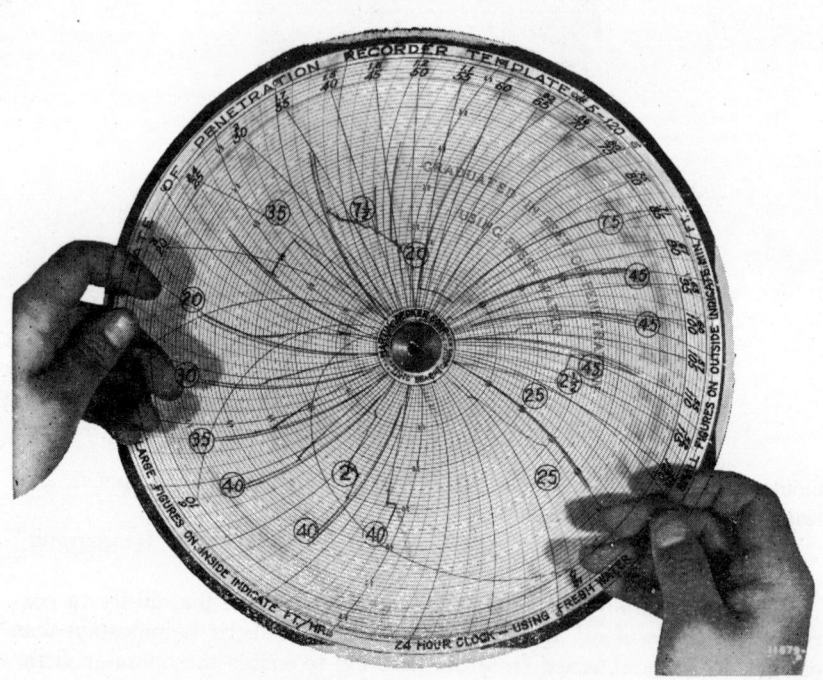

FIGURE 230. Martin-Decker recorder.

SPECTROCHEMICAL SAMPLE LOGGING

L. L. SLOSS AND S. R. B. COOKE

Subsurface stratigraphers have long encountered difficulties in differentiating and correlating thick stratigraphic sections involving monotonous successions of similar lithology which lack lithologic or paleontologic markers or identifiable zones. In some areas and in some parts of the rock column, heavy-mineral analyses, insoluble residues, and electrical or radioactive character, singly or in combination, have served effectively in the solution of such problems. In other cases, notably where thick shale or carbonate sequences are concerned, these methods have failed to achieve a satisfactory basis for detailed structural and stratigraphic interpretations. In these cases differentiation of stratigraphic units and their correlation have, perforce, been based on tedious microscopic examination of samples. Microscopic sample study has added greatly to our understanding of sedimentary petrography, but as a correlation technique it often fails in the very cases that demand its application—that is, in rock sequences not adapted to "in-the-hole" techniques or analyses of special mineralogic attributes.

These commonly encountered circumstances require an approach to stratigraphic differentiation and correlation based on characteristics other than gross lithology, observable mineralogy, or the electrical or radioactive constants of the rock or its contained fluids. Further, the applicability of any new technique would be increased if easily reproducible data were involved and if the impact of the personal equation were held at a minimum.

While the writers were associated at the Montana School of Mines they became involved in correlation problems that arose from detailed studies of Mississippian limestones in the subsurface and outcrop areas of Montana and southern Alberta. In this area the Mississippian rocks do not contain suitable microfossils in sufficient numbers to permit the establishment of paleontologic zones. Lithologic aspect varies from place to place in response to tectonic influences, which range from the border of a geosyncline, across a stable shelf area, and into an active intracratonic basin. Correlation is sufficiently difficult within any single tectonic province and becomes increasingly complex as provincial boundaries are traversed.

It occurred to the writers that since the entire area was submerged by a continuous seaway any changes in the chemical composition of the marine waters would be transmitted throughout the area and be reflected more or less simultaneously in the concurrent sedimentation. To test this hypothesis qualitative spectrochemical analyses of the trace and minor-element constituents of successive samples were run. The results proved negative in that all samples contained about the same suite of elements, and no data of significance to correlation were obtained.

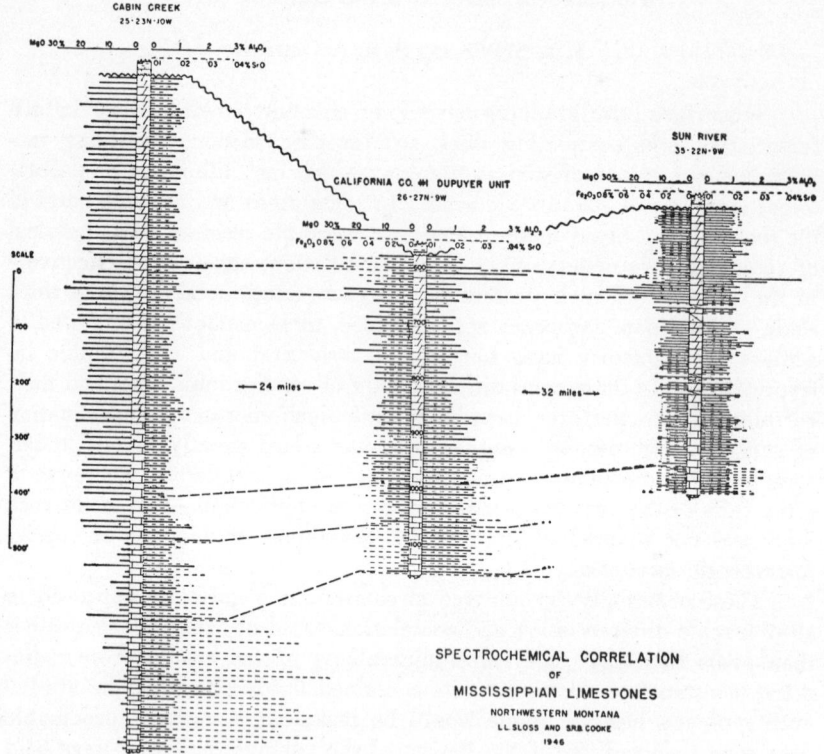

FIGURE 231. Graphic method of recording spectrochemical data. (Reproduced permission Am. Assoc. Petroleum Geologists.)

As it became apparent that qualitative analysis was not the answer, efforts were turned toward quantitative analysis by spectrochemical methods, and, after some difficulties, a satisfactory technique was evolved and useful data were obtained. The results of this investigation and the procedure followed were presented in 1946.[27] Figure 231 is reprinted from the original paper; it illustrates a number of zones, determined by quantitative spectrochemical character and traceable from section to section.

The writers, in the intervening period, have not pursued the matter further at a publication level. Subsequent development, to the writers' knowledge, has taken place in the realm of industrial research, but the results are not available for publication. This note is included in the present symposium in the interest of completeness and is intended to bring up to date certain questions of technique, application, and interpretation. Interested readers are referred to the original paper for details not repeated here.

---

[27] Sloss, L. L., and Cooke, S. R. B., *Spectrochemical Sample Logging of Limestones:* Am. Assoc. Petroleum Geologists Bull., vol. 30, no. 11, pp. 1888-1898, Nov. 1946.

SAMPLING

In considering the interpretation of a spectrochemical sample log it is important to keep in mind the fact that analyses are based on five-milligram samples, an amount that may be placed under the little fingernail with slight discomfort. Therefore, every precaution must be taken to insure that the tiny sample is as representative as possible of the ten-foot (or smaller) interval from which it is taken. Where cores are available and core recovery is high, proper sampling is achieved simply by knocking off closely spaced chips, lumping together the chips from each five- or ten-foot interval. Good cable-tool cuttings yield adequate samples, with rapid inspection serving to eliminate obvious extraneous materials. As in other samples studied, rotary cuttings pose the most difficult problems. However, by judicious use of corollary data from electric logs and drilling time it is usually possible to assemble a sample containing a high percentage of material representative of the indicated interval. In most cases a cursory examination of the sample under binocular microscope and picking of chips with tweezers are unavoidable. In other cases where the history of the drilling procedure is well-known and confidence can be placed in the purity of the cuttings, a portion for analysis can be drawn directly from the sample sack. The writers have used both techniques and have derived interpretable results from unpicked samples. However, in view of the postsampling investment in time and effort, it seems more prudent to keep the sampling method as rigid as is practical, at least in the exploratory stages of any investigation.

SAMPLE TREATMENT

In the writers' first paper on spectrochemical well logging, the methods employed were limited by the equipment available and by the difficulty of applying quantitative spectrum analysis to relatively refractory and quite nonconducting rock powders. In conventional metallurgical spectrum analysis both arc- and spark-excitation methods may be used, because the material under investigation usually consists of regularly shaped cast or machined pieces of metal or alloy. The regular shape permits reproducibility of arc length, and the high conductivity of the metal permits the specimens to form the electrodes themselves. Further, the reasonably uniform composition of metallic alloys permits more or less constant volatilization of the contained constituents once the arc has been established.

On the other hand, quantitative analysis of minerals or of rock powders by the spectrograph presents some unusual problems. In the first place the sample must be representative of the particular mineral or of the total amount of the specimen gathered. In the case of opaque minerals the specimen may be polished and samples taken by using a micro-drill under a microscope, insuring that only the mineral under investigation is sampled. For heterogeneous rocks, the entire sample must be reduced to a reasonably fine size by careful crushing; mixed, sampled, and this sample reduced further in particle size; then remixed and resampled, the process being repeated until the operator is satisfied that the few milligrams taken for analysis are representative of the entire sample. For relatively homogeneous rocks such as compact limestones the sampling requirements are perhaps less rigid than those for rocks containing a variety of minerals of different crushing hardnesses and different sizes of liberation, but the operation is nevertheless tedious.

The only satisfactory method for use in analyzing nonconducting powders consists in placing the powder within a cup bored in an electrode made of some conducting material. In this case the arc (the high-tension spark cannot be used) is struck between the electrode holding the powder and another electrode of the same material. The high temperature reached by the cupped electrode (the anode or positive electrode) gradually volatilizes the constituents of the powder, and the spectrum lines of the constituents of the powder are excited in the arc column. High-purity silver and copper have been used as electrode materials, but the most convenient and easily obtained substance is graphite. This is available in regular shapes, usually as rods of various dimensions, at reasonable cost, and of very low impurity content. It has the further advantage of easy machinability.

For quantitative analysis all powder samples must be weighed. This is tedious work, and any method that reduces time spent in weighing without materially sacrificing accuracy is to be commended. The Roller-Smith spring balance has been found most useful in this respect, for it is extremely rapid in operation and is quite consistent in behavior. Samples weighing five milligrams can be weighed with an accuracy of 0.2 percent. Specimens can be loaded into the cupped electrodes directly from the balance pan through a small glass funnel or by means of a shaped assay-silver or platinum launder. Even with these aids, however, the time consumed in weighing is unduly great. It should be possible to expedite weighing and loading by using a machine that produces pellets of the rock powder. As the accuracy of spectrochemical analysis is not of a very high order, limited accuracy in reproducibility of pellet weight would be acceptable. Care would be necessary that pellets were not contaminated by dust from preceding specimens.

Arc analysis of limestone powders presents an unusual problem. The rapid heating in the arc gives rise to copious evolution of carbon dioxide, which, if the lower electrode is poorly designed, tends to blow part or all of the sample from the cup. This difficulty has been overcome by using a relatively deep cup [28] so that the carbonates are calcined at low temperature before volatilization of the constituent elements commences. The increased exposure time thus enforced adds considerably to gen-

---

[28] Sloss, L. L., and Cooke, S. R. B., *op. cit.*

eral background fog and to the intensities of the carbon and cyanogen bands from the electrodes, but there seems to be no way to eliminate this difficulty, which becomes exaggerated when a wide slit is used.

EXCITATION

The direct-current arc is the simplest form of excitation that is used, as it requires very little equipment, and, as it operates satisfactorily at voltages between 100 and 250, it is comparatively safe. It has the reputation of being less stable than other types of arc owing to wandering of the cathode "spot" and, consequently, of lack of reproducibility. Hampton and Campbell,[29] however, have recently demonstrated that the direct-current arc is capable of high reproducibility under carefully controlled conditions.

The high-voltage alternating-current arc was introduced by Duffendack and Thompson [30] to avoid the irregularities of the direct-current arc. It has acquired a good reputation in spectrochemical work, but the writers have had no experience with it. It requires more equipment and safety precautions than the low-voltage arc.

In all types of arc excitation care must be taken that the entire specimen is volatilized, otherwise serious errors will be introduced. For instance, although there is some overlap, the order of volatilization in a limestone is sodium, calcium, silicon, and last titanium. Interruption of the arc before the titanium has boiled off obviously leads to error in determination of the titanium content of the limestone. The arc draws a heavy current, as is shown by an ammeter in the circuit, during the period of volatilization of the major elements in the sample, and when this is complete for those elements the current abruptly drops to the steady value of the graphite arc. Unfortunately there is no significant corresponding drop to indicate the completion of volatilization of what are usually minor quantities of titanium or of other refractory elements. It seems best under these conditions to operate the arc for some time after the current drops so as to insure complete removal of the sample and to design the shape of the lower electrode so that the walls are thin and burn evenly down to the bottom of the cup.

INSTRUMENTATION

To utilize the spectrochemical method to its greatest advantage a spectrograph of high dispersion, yet of relatively high speed,, should be available. Large quartz spectrographs of the Littrow type are advantageous and should be equipped with glass prisms to obtain adequate dispersion in the visible region. The Gaertner, Bausch and Lomb, and Hilger instruments are typical of this class. In recent years the grating

[29] Hampton, R. R., and Campbell, H. N., *Measurement of Spectral Intensities in the Direct Current Arc:* Optic. Soc. America Jour., vol. 34, no. 1, p. 12, 1944.

[30] Duffendack, O. S., and Thompson, K. B., *Developments in the Quantitative Analysis of Solutions by Spectroscopic Means:* Am. Soc. Testing Materials Proc., vol. 36, no. 2, p. 301, 1936.

spectrograph has become widely adopted, and, with the advent of specially ruled gratings capable of throwing much of the incident energy into a specified order, it is competing successfully with the prism instrument. Where the design of grating spectrograph is such that dispersion is approximately linear, identification of spectrum lines is facilitated by using linear interpolation rather than the cumbersome Hartman formula required for prism spectrograms. In the United States grating spectrographs are available from the Applied Research Laboratories, the Baird Associates, and the Jarrell-Ash Company.

STANDARDS

The writers used Slavin's "total-energy method" [31] as being best suited to the problem under investigation. This method utilizes an external standard, which in this case consisted of a mixture of equal quantities of United States Bureau of Standards standard samples No. 1a (argillaceous limestone) and No. 88 (dolomite) of known composition. This mixture gave consistent results and performed in the arc similarly to the materials under investigation. Initially synthetic standards were used, consisting of carefully purified calcium carbonate containing known quantities of added barium sulphate, strontium sulphate, silica, and ferric, aluminum, magnesium, and titanium oxides. These, however, gave erratic results, primarily owing to a proved inability to obtain thorough mixing of the constituents and also to the fluffy nature of the precipitated calcium carbonate, which led to improper volatilization in the arc.

The employment of the internal-standard method is theoretically sounder than that of any method using external standards, but the additional loss of time required in weighing out the standard and the very difficult problem of obtaining thorough mixing of each unknown with the standard make the method seem puite unattractive. Kvalheim [32] has described a useful internal-standard method for determining sodium, potassium, aluminum, calcium, magnesium, iron, and manganese in minerals, rocks, and slags, employing strontium carbonate as the standard material. When silicon is to be determined, a mixture of beryllium carbonate and sodium chloride is introduced as the internal standard.

A most interesting method of spectrochemical analysis based on measurement of line widths has been described recently by Coheur. [33] This procedure would seem to have the advantage of rapidity and furthermore is applicable to elemental concentrations higher than those usually measured in such work.

In using Slavin's or any other external-standard method it is necessary to photograph at least six spectra of the standard on the same film

---

[31] Slavin, M., *Quantitative Analysis Based on Spectral Energy:* Ind. and Eng. Chemistry, vol. 10, p. 410, 1938.

[32] Kvalheim, A., *Spectrochemical Determination of the Major Constituents of Minerals and Rocks:* Optic. Soc. America Jour., vol. 37, p. 585, 1947.

[33] Coheur, P., *A Method of Quantitative Spectrochemical Analysis Based on Line Widths:* Optic Soc. America Jour., vol. 36, p. 498, 1946.

or plate on which the spectra of the unknown samples are photographed, in order to obtain the working curve. To avoid unnecessary duplication and to maintain constant development conditions, it is best, therefore, to use a spectrograph that will take the maximum possible number of spectrograms on one plate.

DENSITOMETRY

A Gaertner microdensitometer was used by Sloss and Cooke. This is a subjective type of instrument, in which the density of any given line is matched against the density of a calibrated neutral-density wedge. It is rapid in operation, but, because a definite area of lime is required for matching, the spectrum lines must be broadened by operating the spectrograph with a wide slit. Such line broadening leads to two difficulties: first, the resolving power of the spectrograph is greatly reduced, requiring that the lines chosen for analysis be free from proximity to neighboring lines, which may overlap when the slit is opened; and second, the background fog is greatly increased, which leads to the embarrassing question of just how the background correction should be applied. Undoubtedly, the photoelectric microphotometer, of which many types are available, is best suited for measuring line density. Instruments of this type are available from Hilger, Bausch & Lomb, the Applied Research Laboratories, and Leeds & Northrup.

A comparatively new method of analysis [34] [35] [36] eliminates the photographic plate as an integrating device. By using combination of slits in the focal plane of the spectrograph, lines of those elements the quantities of which are under investigation are isolated, and the respective light energies are integrated by multiplier phototubes. The method seems to have been singularly successful where it has been applied, and it would seem applicable to the present problem.

INTERPRETATION

Limited experience with spectrochemical logging indicates that, once the necessary data are accumulated, interpretation of the results in terms of correlation is fairly straightforward. Graphic presentation of data is necessary before they are intelligible. The writers have used multiple-bar diagrams, as illustrated in figure 231, and curves simulating electric logs. The latter are not entirely justified, since a sloping line connecting the values for adjacent samples suggests a transitional series of values from sample to sample; nevertheless, such a representation seems to be the most workable, perhaps because stratigraphers are accustomed to handling similar curves. In any case, when the data are plotted for a series of wells

[34] Boettner, E. A., and Brewington, G. P., *The Application of Multiplier Photo-Tubes to Quantitative Spectrochemical Analysis:* Optic. Soc. America Jour., vol. 34, p. 6, 1944.
[35] Saunderson, J. L., Caldecourt, V. J., and Peterson, E. W., *A Direct-Reading Spectrochemical Installation,* The Dow Chemical Co., 1945.
[36] Nashtoll, G. A., and Bryan, F. R., *An Application of Multiplier Photo-Tubes to the Spectrochemical Analysis of Magnesium Alloy:* Optic. Soc. America Jour., vol. 35, p. 646, 1945.

and the resulting diagrams or curves compared, correlations are indicated by similarity of pattern.

In their original investigation the writers used iron, magnesium, strontium, and aluminum as critical elements. Of these, aluminum may be considered as being universally applicable to problems in which the clay content of limestones is considered a significant attribute, but this element is of no value in highly argillaceous strata. Magnesium proved quite useful in the area investigated but would, of course, have no correlation significance in areas where dolomitization is known to transgress stratigraphic boundaries. The writers encountered no difficulties in using strontium, but the element may be suspect in highly porous rocks, since it is usually associated intimately with the barium in drilling muds. In other words, there is insufficient documented experience to permit a proper evaluation of significant elements for analysis. Each new investigation must be considered an independent problem, and a number of elements must be determined before a working choice is made. It is reasonable to expect, however, that in the average case two or three elements will draw the required distinctions, additional elements merely repeating and supporting the pattern. Very little work has been done on shales, but preliminary results indicate that consistent and interpretable data can be gained. It is to be hoped that further investigations will be forthcoming.

Aside from the application of spectrochemical logging to questions of correlation, the technique is capable of yielding data useful in other fundamental stratigraphic problems. It would be enlightening, for instance, to have a volume of spectrochemical data on sediments of known depositional environments, from restricted evaporite basins, brackish lagoons, reefs, and so forth. It is probable that each of the environments had an effect on the composition of sea water and a corresponding response in the trace-element chemistry of the sediments. If such environmental responses could be established it might be possible to aid in the determination of the depositional environment of less well understood strata.

CONCLUSION

The spectrochemical method has a definite usefulness in application to difficult problems of correlation and is capable of yielding a mass of data for aiding our understanding of fundamental geologic questions. Industrial and academic acceptance of the method is inhibited by the relatively high initial expenditure required for laboratory equipment and the relatively high per-sample cost of analysis. The first item is justifiable if the method is applicable to a large volume of samples. It can not be practical to employ the method, however, if the costs per sample are excessive. Therefore, it is necessary to trim laboratory running expenses wherever possible. This can best be accomplished in two ways. First, mechanization of sample grinding, weighing, and electrode loading can potentially make an enormous reduction in the most time-consuming operations of the procedure. Second, the elimination of steps leading to precise quantification of absolute values will reduce much of the laborious control of standards, excitation, and energy measurement. In the majority of cases figures on variations in the relative amounts of various elements from sample to sample are as useful as more precise absolute data. It is the pattern formed by changes in percentage with well depth that is more important than the exact amount of each element in each sample.

COMPOSITE-CUTTINGS-ANALYSIS LOGGING

R. J. GILL

Information obtained simultaneous with the drilling of a wildcat well is of extreme importance, and this knowledge can be acquired at no other time. The establishment of a 24-hour-a-day logging service is based on this premise.

In order to log a well adequately, the cuttings must be examined and tested for oil and gas concurrent with the drilling, owing to the rapid dissipation of the lighter hydrocarbons that may be brought to the surface in the cuttings. Recommendations concerning coring or drill-stem testing may be quickly and accurately made at a time when it is most important. Conversely, much rig time can be saved by eliminating many unnecessary cores and tests.

It has been proved that cuttings from a well brought to the surface by the mud stream will retain a small fraction of oil or gas, if an oil- or gas-bearing stratum has been penetrated. However, this oil or gas may be present in the cuttings in such minute quantities that it cannot be visually or nasally detected. Mechanical and electrical instruments have been developed to detect and relatively measure obscure oil and gas showings in the samples that might otherwise be overlooked.

This continuous logging process is carried on in a trailer-laboratory, which is located at the well as near to the mud-discharge line as possible. A portion of the mud from the flow line is diverted into the laboratory, where it goes through a miniature shale shaker, which separates the cuttings from the mud, allowing the mud to return to the pits and providing a continuous flow of samples available for examination. The availability of the cuttings within the laboratory in a steady flow permits examination more frequently and without contamination by previously drilled sediments.

A cable is run from the laboratory over a sheave near the top of the derrick and down to the swivel. Thus, as the kelly is lowered, the drilling-time mechanism in the laboratory in turn operates and records the time for each foot of hole drilled. The importance of accurate drilling time has

been proved, and this chart is closely watched during the process of drilling
for any drilling breaks that might occur. A drilling break usually indicates a porous zone which is a potential oil or gas reservoir. When a
drilling break is encountered, drilling is suspended, the cuttings are circulated up from that zone, and a determination is made of any oil or gas
content. The suspension of drilling near the top of any oil or gas zone is
of great importance in the event of a relatively thin pay section. A re-

FIGURE 232. Laboratory located at well where steel mud tanks were used above
ground. Note high flow line and connections to laboratory.

liable drill-stem test cannot be made if the bit has penetrated the water
zone, if one is present, beneath the oil or gas.

Routine tests are made each two feet of hole drilled from cuttings
collected at a predetermined interval. The length of time necessary for
cuttings to reach the surface from the bottom of the hole varies with the
depth and condition of the hole as well as the capacity of the pump. This
amount of sample lag is determined frequently to allow sampling for the
correct depth. The following method is a reliable means of determining
the lag: Some material that can be easily detected, preferably about the
size and density of the actual cuttings, is placed in the drill pipe when a
connection is being made. The length of time required for the material to
reappear at the surface is measured. The time required for the pump to

replace the fluid in the drill pipe is calculated, and this time subtracted from the round-trip time will give the sample-lag time at that particular depth. Two clocks are mounted in a panel, one clock showing the correct time, and the other set back to allow for the sample lag. When the hands on the second clock show the time that a certain depth was drilled, cuttings are collected for that interval. Thus all tests and examinations are made from the true depth.

FIGURE 233. View of a sample separator within laboratory. Geologist has collected uncontaminated cuttings from predetermined depth preparatory to examination and testing for oil or gas.

until they are clean. A measured amount of cuttings is put into a container, which is filled half full with fresh water. The container is constructed with blades in the bottom, which revolve at a very high speed, pulverizing the cuttings. This has the effect of exposing the pore spaces within each chip. Any gas that might be contained in the pores or might adhere to the sample is released by the agitation and brought to the sur-

The cuttings for each two feet are washed and decanted several times face of the water. A measured amount of cuttings is put into a container, which is filled half full with fresh water. The container is constructed with blades in the bottom, which revolve at a very high speed, pulverizing the cuttings. This has the effect of exposing the pore spaces within each chip. Any gas that might be contained in the pores or might adhere to the sample is released by the agitation and brought to the surface of the water. By applying a vacuum to the container, the air-gas mixture is drawn over a hot platinum filament of a balanced electrical circuit. If any combustible gases are present in the mixture, they will increase the temperature of the platinum filament and consequently throw the system out of balance. This differential is recorded on a galvanometer. Methane and ethane ignite at a temperature considerably higher than propane, butane, etc., and shale gas. Thus it is possible to distinguish readily between the higher- and lower-burning-point gases. The temperature of the platinum filament is ad-

FIGURE 234. Interior view showing a sample agitator with gas analyzer above. Infrared sample drier can be seen at left and mud-testing equipment at right.

justed to ignite all hydrocarbon gases present and is then lowered to a point at which the methane and ethane will not ignite. The difference between the two readings will thus be the reading of the methane and ethane present. As this is a qualitative test only, results must be correlated with the lithologic characteristics of the section, for porosity and permeability control the magnitude of the readings obtained.

Petroleum oils show a fluorescence under an ultraviolet light. The maximum fluorescence occurs with a lamp having a wave length of between 3,200 and 3,800 angstroms. A 230-volt mercury-vapor lamp with suitable filters is considered most effective for detecting residual oil in the cuttings. The lamp is enclosed in a light-tight box with a viewer that has a magnification of approximately $3\frac{1}{2}$ times. Samples are examined under the fluoroscope every two feet for any oil content. As many minerals will give off fluorescence of various colors, it is necessary to distinguish between mineral and oil fluorescence. This is done by putting a portion of the fluorescent material into a color-reaction plate and applying a petroleum solvent to it. If oil is present, a "cut" will be obtained; that is, some of the oil will be dissolved out and will come to the surface of the cutting agent, causing the fluid to "bloom" or fluoresce. If no cut is obtained, it can be assumed that the fluorescence is from a mineral source. Examination under the microscope will probably verify the presence of the mineral. Refined rig greases and oils that may contaminate the cuttings may easily

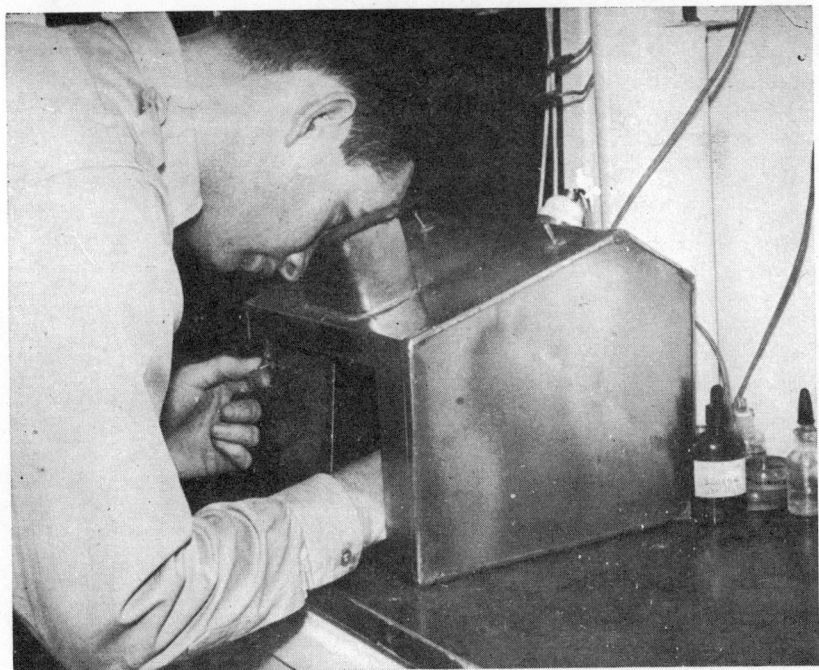

FIGURE 235. Examining cuttings under ultraviolet light for oil content.

be detected by experimentation. The color of the fluorescence obtained from crude petroleum may range from brown to yellow and through various tints of green and blue to near-white, dependent upon the gravity or impurities. The color may be important in distinguishing separate zones in the same well where cavings from above may be present, as well as in correlating the same zone from well to well.

A microscopic examination of all cuttings is made by the geologist assigned to the laboratory as the well progresses. A detailed interpretative lithologic log is made from this examination. An attempt is made before and during the logging of a well in an unfamiliar area to obtain as much available geologic knowledge as possible. These studies, together with the interpretative geologic log, make it possible to pick important formations tops. The geologist is present when all cores are removed, and these can, if desired, be sealed preparatory to core analysis.

Equipment is maintained for the testing of drilling-mud characteristics; tests are made as often as it is desired for weight, water loss, viscosity, filter cake, salinity, or any other properties to insure a satisfactory drilling mud.

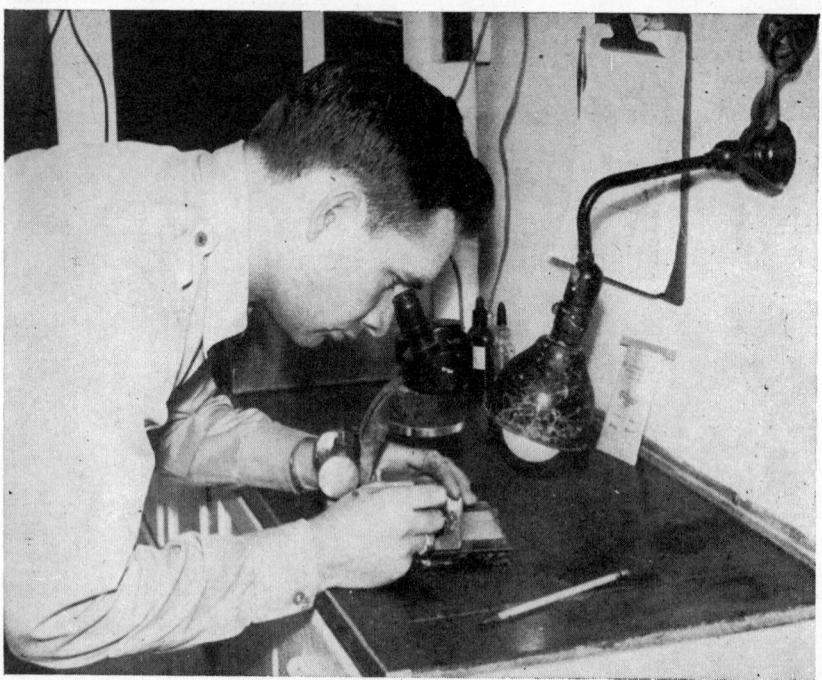

FIGURE 236. A detailed microscopic examination by a competent geologist is very necessary on all exploratory wells.

The results of all tests and examinations, as well as the drilling-time record, are plotted graphically on a log having a scale of five inches to one hundred feet to permit the showing of as much detail as possible. The first column on the final composite log denotes the characteristics of the mud at the depth each check was made, as well as symbols indicating new bits and circulating and coring points. The drilling-time curve in the next three columns is plotted in minutes per foot, and the horizontal scale is amplified to show minor variations in the drilling rate. The center two columns of the log show the depth and graphic lithologic log. To the right are shown the results of oil and gas tests. Since these tests are purely qualitative, an arbitrary scale of 100 percent has been established to show comparative readings. Oil is shown as a solid line; the "high" gas reading is plotted with a dotted line, and the "low" gas reading with a dashed line. The oil and methane shows are colored green and red, respectively, on the final log for quick noting. The last column of the log is devoted to a detailed escription of the lithology. Formation tops and results of coring and drill-stem tests, as well as any other pertinent data, are suitably located on the log. A modified daily log covering the previous 24-hour period is sent each day to operators upon request.

FIGURE 237. Plotting composite daily log, which includes drilling time, oil and gas shows, and detailed lithologic descriptions.

The cuttings-analysis-logging method is a direct means of determining the presence of oil- or gas-bearing formations. Oil and gas in the cuttings can be detected by the instruments maintained in quantities too minute to be observed in any other way. Accurate determinations can be made of odorless gases as well as high-gravity oils or distillate. These may be too elusive to detect with a microscope and the human senses. With a geologist or technician on duty 24 hours a day, an accurate valuation of the section may be made consistent with the drilling. Experience has shown that not all productive zones are drilled at a faster rate even though porosity is present, an example being the Simpson sand of southern Kansas and Oklahoma. However, since cuttings are analyzed each two feet regardless of the rate of penetration, a zone of this type would be detected from any oil or gas showing. As the cuttings from the well are used for all tests, any mud that might carry combustible gas due either to chemical action or recirculation does not affect the results.

The development of mechanical and electric well-logging instruments has greatly eliminated the chance of human error or misjudgment. It is of no less importance, however, that qualified personnel correlate and interpret the data in order that the greatest benefit be obtained.

Statistics show new oil fields progressively becoming harder to find. Many new areas are being explored and new zones in old areas being tested. It is becoming necessary to employ the best available man power and methods in the search for new petroleum reserves. After a location has been made, it is of the utmost importance that the well be accurately evaluated for oil or gas, or the preliminary work will have been done in vain. The cuttings-analysis-logging method should leave the operator without question regarding the section logged.

QUESTIONS

1. How are cable-tool samples collected?
2. Discuss contamination of ditch samples from rotary wells.
3. What methods are used for obtaining true, ditch-sample depths?
4. What is the general procedure in examining cuttings samples?
5. Discuss the plotting of well samples.
6. For what purposes are well-sample logs used?
7. What two basic curves are obtained in electrical logging?
8. What are the characteristics of the resistivity curve opposite fresh-water and salt-water-bearing strata?
9. What effect does mud resistivity have on the self-potential profile?
10. What factors influence the resistivity of drilling fluids?
11. Does hydrostatic pressure exerted by the mud column affect the character of the self-potential curve?
12. How is the resistivity curve obtained and what does it actually represent?
13. What applications does the electrical log have?
14. What would be the characteristics of the self-potential and resistivity curve opposite the following lithologies: dense limestone, sandy shale, sandstone with fresh water, sandstone with salt water, anhydrite, calcareous shales, quartzite.
15. Is it advantageous to know the basic lithologies of a penetrated section before making a lithic determination based only on the electrical profile?
16. May an electrical-log survey be made in cased holes?
17. What is meant by "induction logging"?
18. Under what conditions can the induction log be used?
19. Does the induction-logging system require any direct contact with the mud or with the ground?

20. What advantages does induction logging have over electric logging?
21. The radioactive log involves the gamma and neutron curve. Upon what principles are these curves based?
22. What rock types are most radioactive? Which are least radioactive?
23. Why is the neutron curve sometimes referred to as the "hydrogen curve"?
24. Which radioactive curve is used for defining probable porous zones in a well section?
25. What are the applications of the radioactive well log?
26. Carefully study figure 203 which shows the response of radioactivity and electric logs in usual formations.
27. What does the caliper-log profile show, and for what purpose is it used?
28. For what purposes are thermal surveys of drill holes made?
29. The cooling or warming of a drilling mud at any given depth depends on what factors?
30. Why it is necessary to circulate the drilling fluid before running a temperature survey?
31. How long should the drilling fluid remain stationary before a temperature survey is made?
32. In the section "Well Logging by Drilling—Mud and Cutting Analysis," the term "sand or lime-index curve" is used. What is the value of these curves?
33. Periodic testing of drilling-mud qualities is required for successful completion of a well. Why?

34. What size sample is used in the spectrochemical logging method?
35. What is the standard method of graphically representing spectrochemical data?
36. What is meant by "drill-time logging"?
37. Study figure 218 which correlates the electrical profile with the penetration-rate log.
38. What are the applications of a drill-time log?
39. How are drill-time logs prepared?
40. What factors control penetration rate?
41. Compare the relationship of the driller's log to other log data given in figure 221.
42. Upon what principle is composite-cuttings-analysis logging based?

CHAPTER 6

MISCELLANEOUS SUBSURFACE METHODS

CONTROLLED DIRECTIONAL DRILLING

J. B. MURDOCH, JR.

Probably the first directional-drilling work ever done was to side-track tools lost in a well. When the driller considered that the churn-drill bit was irretrievably lost, he went into the nearby forest and cut down a tree of slightly smaller diameter than the hole that was being drilled. Starting at the top of this log he cut a long, tapered face on one side. This crude whipstock was forced down the hole until it lodged on top of the fish. He ran the tools back into the hole and drilled carefully, until he had sidetracked the lost equipment. It is interesting to note that this same hand-hewn whipstock is used today by water-well drillers in many parts of the country. The art of controlled directional drilling has advanced greatly from this first primitive sidetracking operation. Today, with the assistance of service companies specializing in this work, operators are able to do what would have been considered impossible twenty years ago. One exploratory well has been forced to deviate at drift angles as great as 80 degrees and to bottom over 9,000 feet (nearly two miles) horizontally from its surface location.

To a certain extent it might be considered that directional drilling is a logical outgrowth of well surveying. Operators had discovered that many wells deviated great distances underground by natural means. They considered that it might be possible to control this deviation to their advantage, forcing the well to bottom at almost any predetermined point at a desired vertical depth. Certain ingenious individuals set about building experimental tools and devising practical methods for carrying on such operations.

A number of different types and sizes of deflecting tools have been constructed and are in general use wherever directional drilling is required. Primarily, they must be able to alter the course of the hole either in drift or direction or in a combination of both. This deflection must be within practical limits of angle so that the well has no "dog legs" in its course toward the objective. The design of these tools must be such that they can be used in conjunction with standard rotary-drilling equipment. They must be safe to run in the well. They should not require that cement or other foreign material remain in the well after they are used. Owing to the increased cost of drilling wells, tools of this type must occasion the loss of as little rig time as possible. Simplicity in design of such tools is desirable, as it is with any other oil tool used on the drilling string. The deflecting tools described below have been designed and constructed with these considerations in mind.

The Eastman removable whipstock is used more than any other type of deflecting tool. Run in open hole, it is a safe and easy tool to operate, and can be depended upon for accurate and reliable results. Illustrated in figure 238A, the removable whipstock is a cylindrical steel casting five to twelve feet long, the length varying with the diameter. It is made with a chisel point on the bottom, which, when imbedded in the formation, prevents it from turning. It is cast with a ring at the top, by means of which it may be lowered into and withdrawn from the well. The back of the ring is provided with a tapped hole into which a shear bolt can be screwed. A concave, inclined groove is formed on one side from the bottom of the ring to the chisel point. Special spiral-type drag bits are designed to be used with each size whipstock. The diameter of the bit is too great to pass through the ring ·of the whipstock, and spiral fins on the bit cause it to drill smoothly along the face. A tapped hole in the shank of the whipstock bit accommodates the end of the shear pin. From an inspection of table 25 it will be noted that removable whipstocks are cast in diameters ranging from four inches to thirteen inches. They are run in holes from one to two inches larger in diameter than the whipstock, according to the type of formation, the quality of the drilling mud, and the depth of the well. Some of the larger tools are fluted or ribbed to reduce their weights (fig. 238C and fig. 238D).

The procedure for running a whipstock is outlined below. The tool should be run after the use of a flat-bottomed bit. The removable whipstock is held vertically against the rotary table by the cat line, with its face toward the hole, and the drill pipe is lowered through the ring from above. The whipstock bit is made up tight on the end of the drill pipe. The cat line is removed and the drill pipe picked up until the bit takes the weight of the whipstock. It is lowered part of the way into the hole, and the slips are set. The shear bolt is screwed through the back of the whipstock and into the tapped hole provided in the whipstock bit. Thus, the drill pipe and the tool turn as a unit as they are run into the well. The slips are removed and the whipstock is lowered into the well by adding the drill pipe stand by stand. Care should be used in running in with a whipstock, since if the hole is bridged or under gauge, the whipstock bolt is liable to shear before bottom is reached. When the whipstock is near bottom, it is faced in the desired direction, and the drill pipe is spudded lightly to force the chisel point into the formation. Part of the weight of the drill pipe is then applied to the shear bolt. After the bolt has sheared, the bit is turned slowly and very little weight is applied as it drills down the tapered face of the whipstock. As shown in figure 239, the bit enlarges one side of the original hole until it reaches the bottom of the whipstock. At this point it starts making a new, small-gauge hole. More weight is carried on the bit after it penetrates past the bottom of the whipstock. Approximately twenty feet of this small "rathole" is drilled.

The drill pipe is hoisted until the bit engages the whipstock ring.

TABLE 25
Eastman Removable Whipstock

Size whipstock	4-in.	4⅝-in.	5½-in.	6¼-in.	7¼-in.	8½-in.	10-in.	10½-in.	13-in.
Type	Plain	Plain	Plain	Plain	Plain	Fluted	Ribbed	Plain	Ribbed
Actual O.D.	4-in.	4⅝-in.	5½-in.	6¼-in.	7¼-in.	8⅝-in.	10-in.	10-in.	12¼-in.
Ring. I.D.	2½-in.	3½-in.	4-in.	4½-in.	5-in.	6½-in.	6½x6⅞-in.	7½-in.	8-in.
Ring length	4½-in.	6-in.	6-in.	7-in.	9-in.	8-in.	10-in.	9-in.	10-in.
Whipstock length	4 ft.-6½ in.	8 ft.-7 in.	9 ft.-2 in.	9 ft.-8 in.	10 ft.-10 in.	11 ft.-0 in.	12 ft.-3 in.	11 ft.-7 in.	12 ft.-1 in.
Degree of angle	4°00'	2°25'	3°00'	3°00'	5°10'	3°20'	3°00'	3°00'	3°40'
Weight	118	252	480	550	830	1096	1450	1750	2317
Weight—crated	175	375	560	700	1020	1300	1800	2050	2670
Size drill pipe	1½-in.	2-in.	2½-in.	3½-in.	3½-in.	4½-in.	4½-in.	4½-in.	4½-in.
Size drag bit	2⅞-in.	3⅞-in.	4¼-in.	5⅝-in.	5⅝–6-in.	7¼–7⅝-in.	7¼–7⅝-in.	8½-in.	8¾-in.
Size rock bit	2⅞-in.	3¾-in.	4¾-in.	5⅝-in.	5⅝–6¼-in.	7½–7⅝-in.	7½–7⅝-in.	8½–8⅝-in.	8½–8⅝-in.
Approx. size hole	4¾–5-in.	6-in.	6⅝-in.	7⅞–8½-in.	8½–9⅝-in.	10⅝-in.	11¾–12¼-in.	12¼-in.	15–17-in.

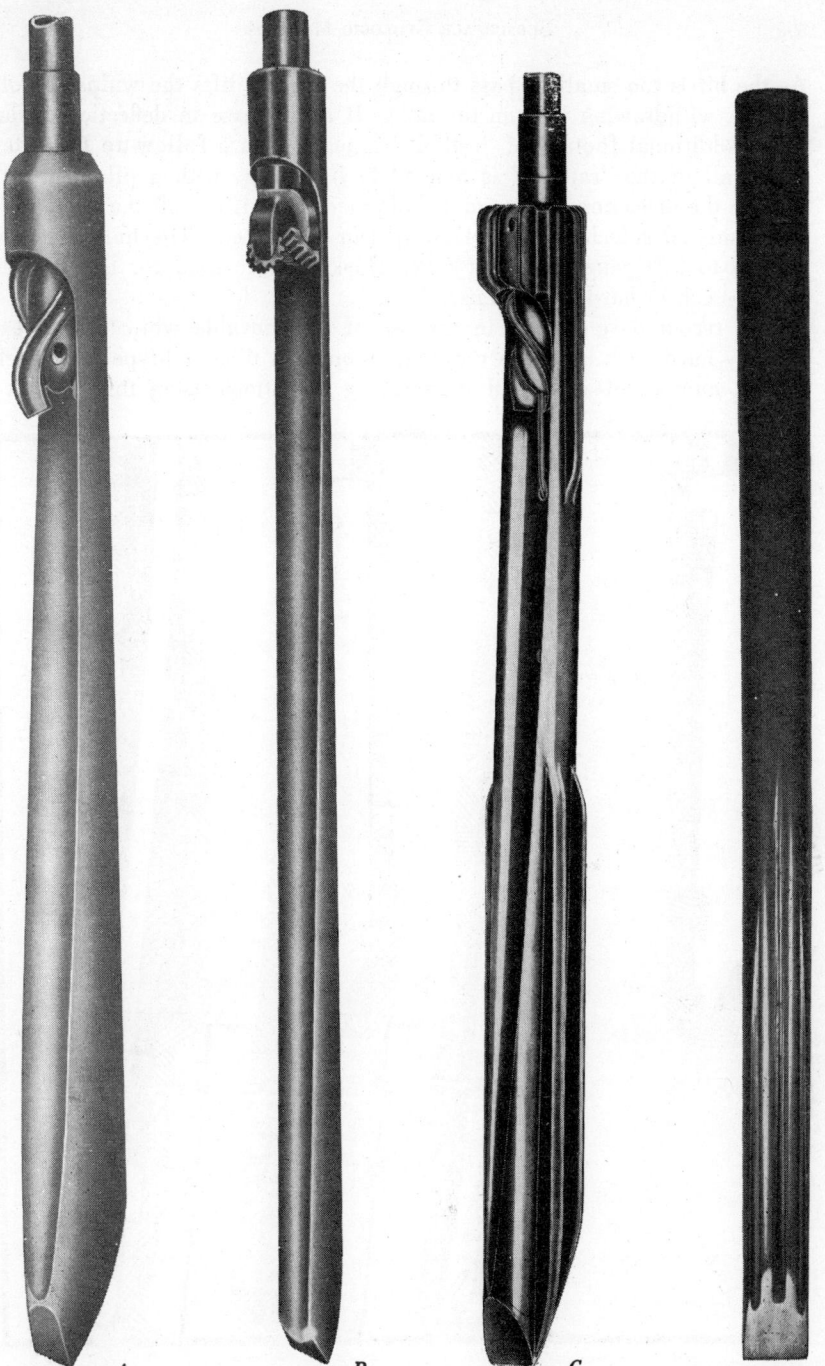

FIGURE 238. *A*—Regular removable whipstock. *B*—Full-gauge removable whipstock. *C*—Ribbed removable whipstock. *D*—Fluted removable whipstock.

As the bit is too small to pass through the ring, it lifts the whipstock off bottom, withdrawing it from the hole. If an increase in deflection is desired, additional footage of "rathole" is made with a follow-up bit, after which all of the "rathole" is reamed to full gauge with a pilot reamer. In case the deflection of the small hole made in drilling off the whipstock is considered sufficient, the follow-up run is omitted. The hole must be opened to full gauge in either case. Rock bits are used for drilling off a whipstock in hard formations.

A recent development in the use of a removable whipstock in extremely hard rock has been the adaptation of a diamond-type whipstock bit. A number of successful sidetracking operations using this bit have

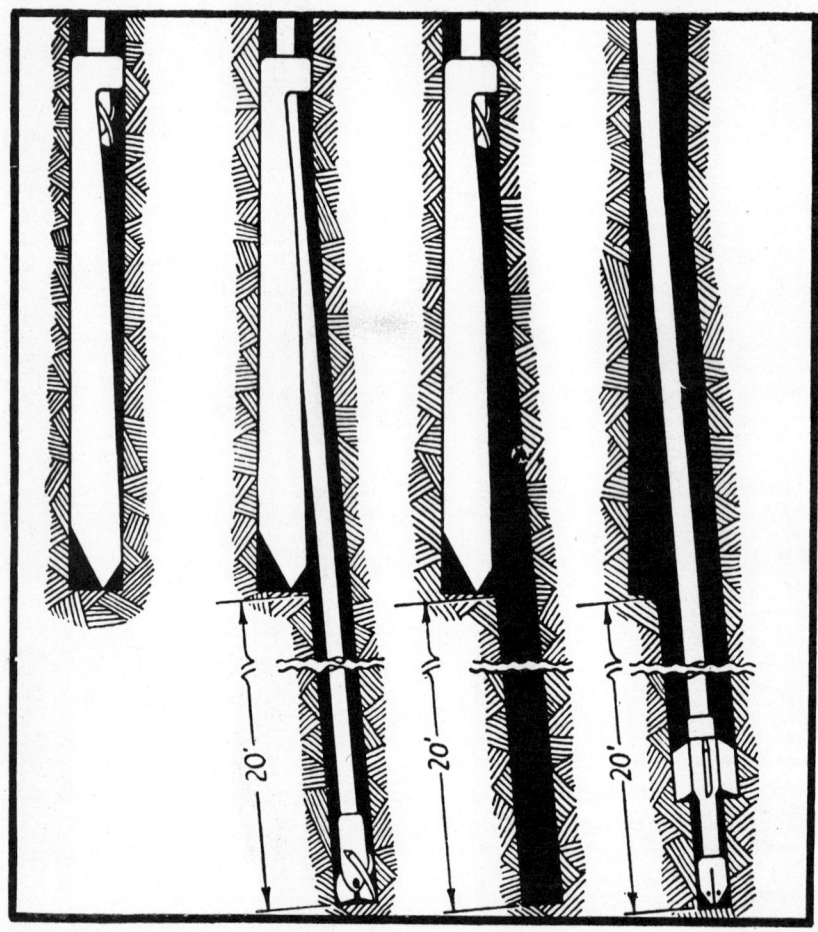

FIGURE 239. Operation of whipstock. Left to right, (1) on bottom in oriented position; (2) pilot hole made; (3) picking up whipstock; (4) reaming pilot hole to full gauge.

been done in the Rocky Mountain area with 5½- and 7½-inch whipstocks. The diamond-core bit, because of its superior cutting qualities in very hard formations, has justified its greater cost over that of a standard oil-well rock bit. Figure 240 illustrates one of these special bits.

The Eastman full-gauge removable whipstock was especially designed

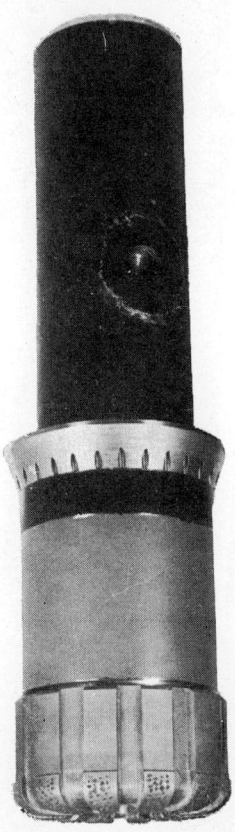

FIGURE 240. Diamond
whipstock bit.

to use a bit larger than that used with the conventional removable whip-stock. The new hole made below the whipstock is slightly smaller in diameter than the hole in which the whipstock is set. Figure 238B shows a full-gauge whipstock equipped with a rock bit. This tool has a much thin-ner section than the conventional whipstock. Data on size, length, etc., are shown on table 26.

Figure 241 illustrates the operation of an Eastman full-gauge whip-stock, which is almost the same as that of the conventional tool, except that the "rathole" need not be reamed. After the whipstock has been withdrawn, a full-gauge bit is used to proceed with further drilling. This type of whipstock was developed to save rig time. The time and expense of making a round trip to ream the "rathole" is considerable in deep wells.

The Eastman knuckle joint is a mechanical deflecting tool of simple

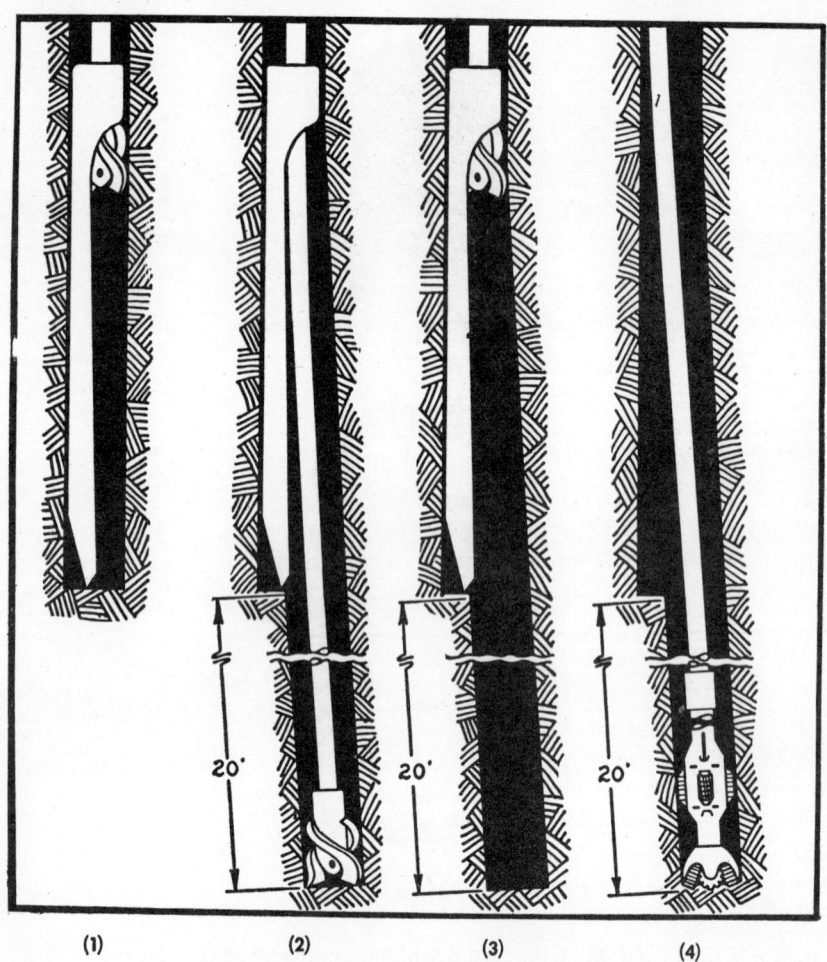

FIGURE 241. Operation of Eastman full-gauge whipstock.

construction, which has been used with very successful results for a number of years. It has certain advantages over the removable whipstock. As provision is made to circulate through the tool, sand bridges in the well may be removed when necessary. No pin is present in the knuckle joint, which may shear prematurely. The danger of sticking a removable whipstock in the hole from cuttings settling out of the mud is much greater than with the knuckle joint. In operation the latter tool is rotated continuously. As illustrated in figure 243, the knuckle joint is essentially a universal ball-and-socket joint, with the lower drill collar held at a fixed angle by a spring-actuated cam. A short drill collar carries a pointed

TABLE 26

EASTMAN FULL-GAUGE WHIPSTOCK

Size whipstock	7⅝ in.	8⅞ in.	10 in.
Ring O.D.	7 in. x 7⅝ in.	8¼ in. x 8⅞ in.	9¾ in. x 10¼ in.
Ring I.D.	5 in. x 5⅝ in.	6 in. x 6⅝ in.	7 in. x 7½ in.
Length	10 ft. 0 in.	11 ft. 8 in.	12 ft. 6 in.
Weight, lbs.	350	738	870
Weight, lbs.—Crated	480	875	1050
Approximate size hole	8⅜ in.–9⅝ in.	10⅝ in.	11¾ in.–12¼ in.
Bit Size	6⅝ in.	7⅞ in.	9 in.

pilot bit at the bottom end and a reamer on the upper end. Either drag- or rock-type bits and reamers may be used on a knuckle joint. Table 27 shows the three diameters in which knuckle joints are made and the recommended hole sizes in which they are used.

The procedure for running a knuckle joint is outlined below. (Also see figure 244.) The tool should be run after the use of a flat-bottomed bit. The knuckle joint is attached directly to the drill pipe (without a drill collar) to obtain maximum flexibility. The tool is lowered to the bottom of the hole and faced so that the bit is in the desired direction. Mud circulation is started and the pumps are allowed to run slowly. The tool is spudded lightly a number of times so that the pilot bit will form a pocket at one side of the hole. The tool is set down on bottom and drilling is commenced by slow rotation of the drill pipe. Light but constant pressure is kept on the bit. At no time should the knuckle joint be allowed to "drill off" or rotate without weight on bottom. After the tool has drilled enough deflected "rathole" for the body to enter the deflected hole, the ball-and-socket action is restricted, and the entire assembly drills as a unit. About twenty feet of "rathole" is drilled. The amount of weight applied to the bit and the type of formation in which the drilling is done determine the amount of deflection made in the well. After the "rathole" is made, the tool is withdrawn from the hole, which is opened to full gauge by means of a pilot reamer.

The Eastman full-gauge knuckle joint is very similar to the regular knuckle joint except that it is designed to drill a nearly full-gauge hole. The differences in the construction of the two tools can be seen from comparing figures 242 and 243. Since the reamer is almost equal in size to the hole in which the tool is run, no reaming of the deflected hole is necessary after the full-gauge knuckle joint has been used. Thus, the operator is able to go back in the hole with a full-gauge drilling bit and proceed without making a special round trip to ream. In deep wells this

TABLE 27

EASTMAN KNUCKLE JOINT

Size (in.)	*Length*	*Tool joint A.P.I. (in.)*	*Weight (lbs.)*	*Drill pipe run on*	*Size hole to run in*	*Drill collar O.D. (in.)*	*Pilot-bit size (in.)*	*Reamer O.D. (in.)*
5¼	6 ft. 1 in.	3½	227	3½ in. double	6⅝ in.– 8⅝ in.	3¼	4	5¼
6⅛	7 ft. 6 in.	3½	295	3½ in. double	7⅝ in.–10⅝ in.	4¼	4¼	6⅛
8	10 ft.	4½	......	4½ in. double	10⅝ in.–12⅝ in.	4½	5¼	8

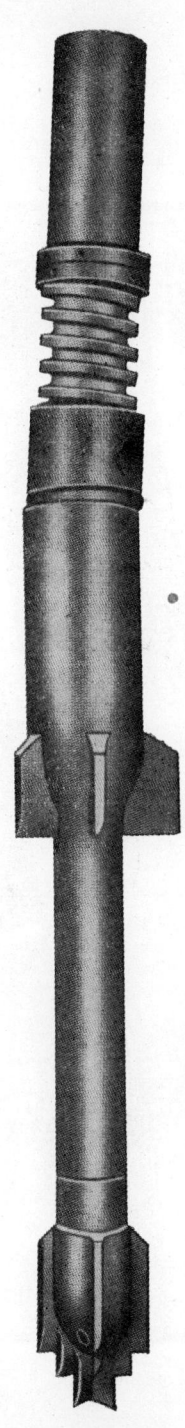

FIGURE 242. East-
man full-gauge
knuckle joint.

FIGURE 243. East-
man knuckle
joint.

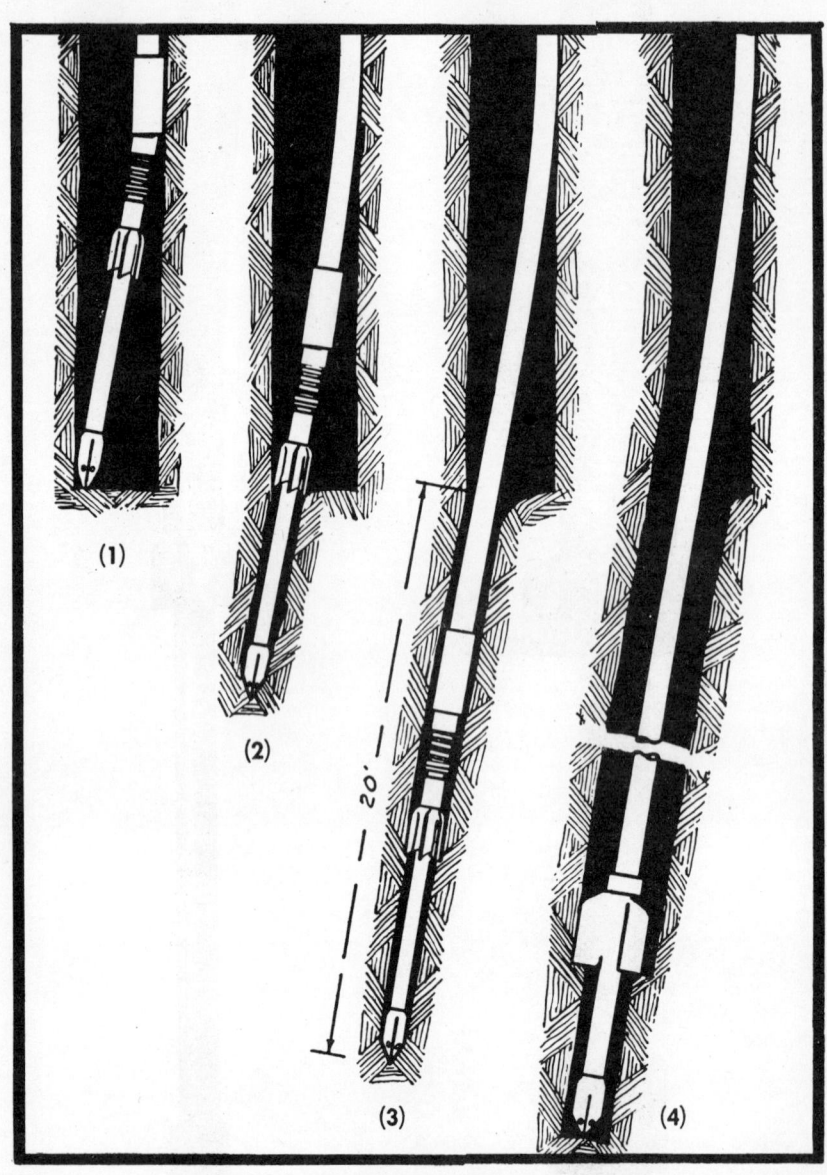

FIGURE 244. Operation of Eastman knuckle joint.

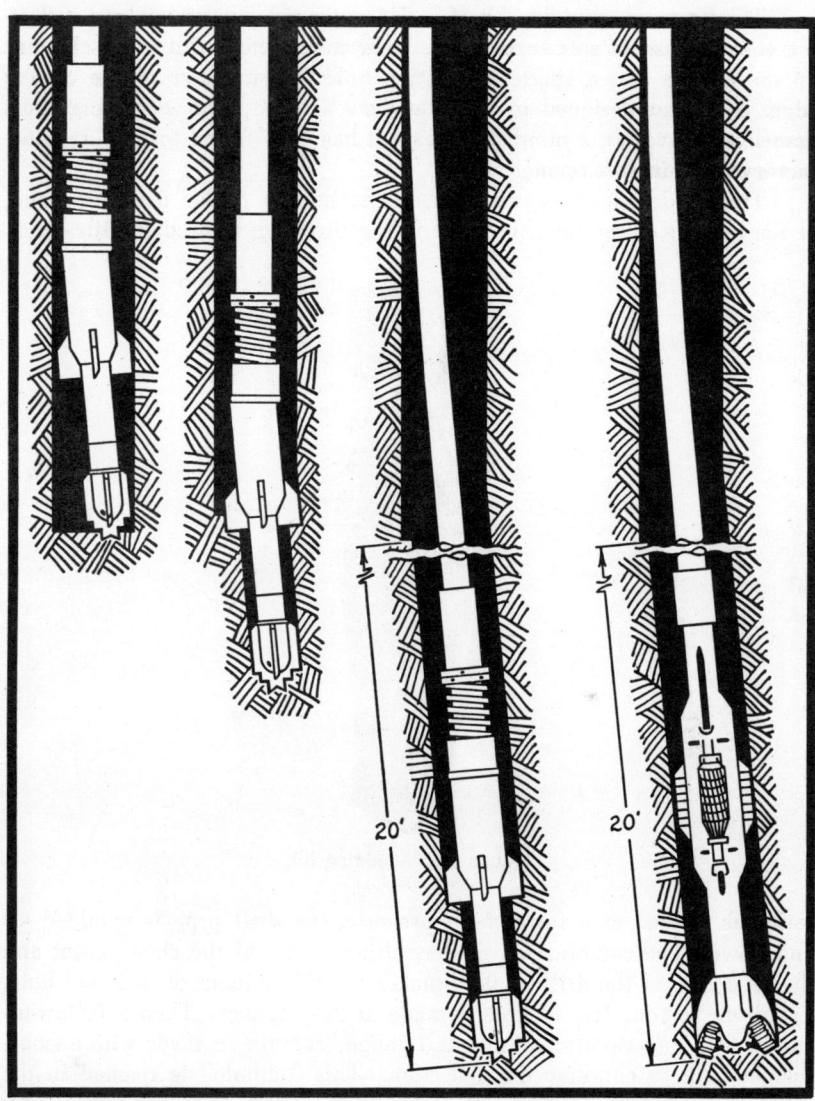

FIGURE 245. Operation of Eastman full-gauge knuckle joint. Left to right, (1) on
bottom in oriented position; (2) starting deflected hole; (3) completing deflected
hole; (4) drilling ahead with full-gauge bit.

occasions a considerable saving in rig time. Rock-type bits and reamers are used in hard formations.

The Eastman spudding bit (fig. 246) is a semipercussion-type deflecting tool for use in soft formations. It is constructed with a chisel point on the bottom and a single circulating hole in the center of the cutting edge. It is not designed to be rotated as are the other deflecting tools described above. In a number of cases it has been found to do very satisfactory work in soft formations.

The spudding bit is made up on the bottom of the drill pipe and, at the bottom of the hole, is faced in the direction desired. While circu-

FIGURE 246. Spudding bit.

lation is carried at a fairly high pressure, the drill pipe is spudded up and down. The combination of the cutting action of the chisel point and the jet action of the drilling fluid makes a small amount of deflected hole. From one to four feet of hole is made in this manner. Then a follow-up run of fifteen to twenty feet of small gauge "rathole" is made with a small diameter bit to continue the deflection. This "rathole" is reamed in the same manner as described for the removable whipstock and knuckle joint. The amount of hole made with the spudding bit and the weight carried on the follow-up bit have a very great influence on the amount of deflection made in the well. This tool often is used to assist in sidetracking a cement plug in soft formations. A spudding bit is run and six inches to one foot of hole made. Then a removable whipstock is faced so that the chisel point of the whipstock sets down into the wedge-shaped hole made by the

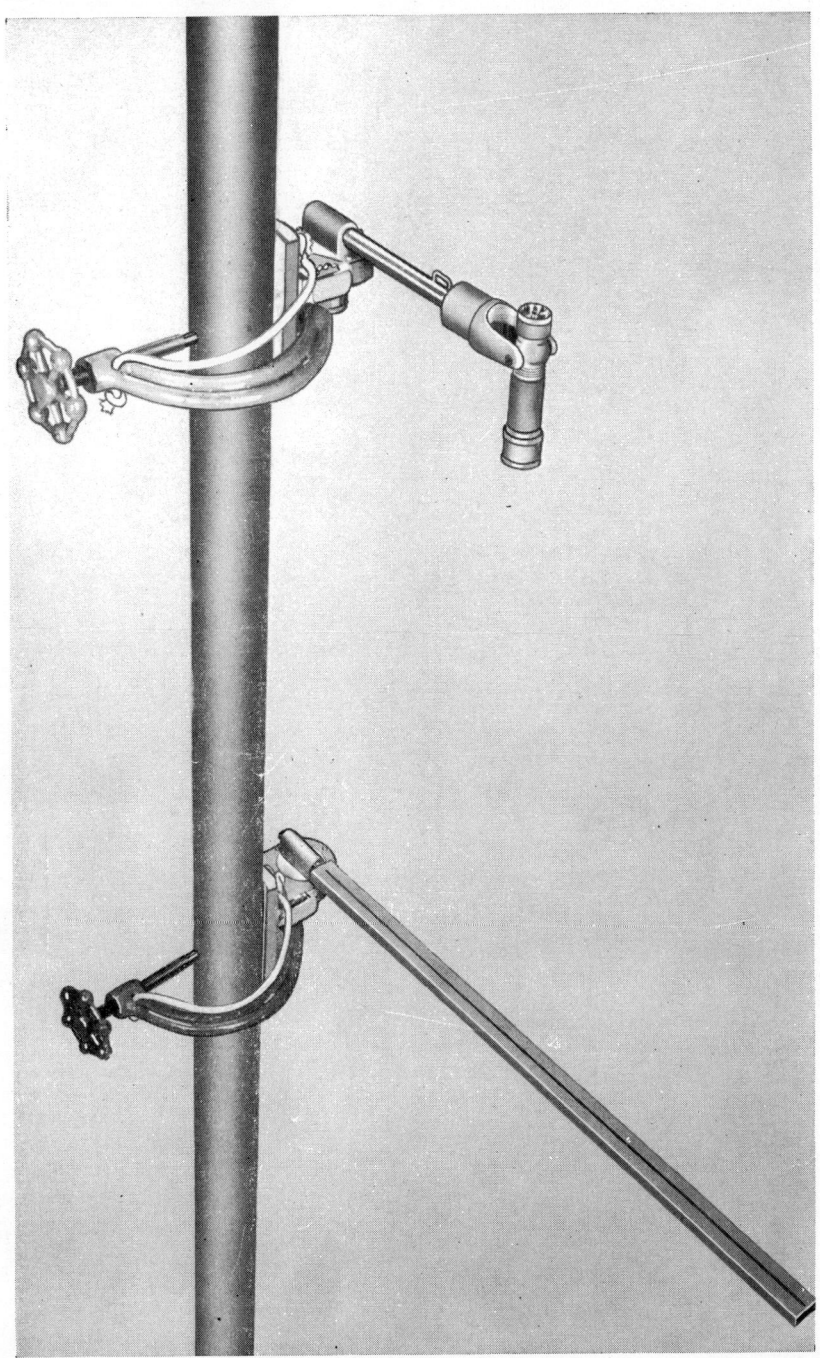

FIGURE 247. Stokenbury equipment.

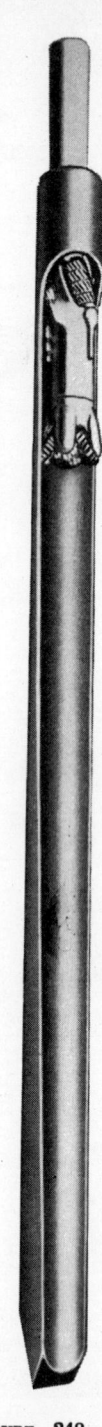

FIGURE 248. Roller
reamer used with
full-gauge remov-
able whipstock

spudding bit. The resulting increase in angle derived from the whipstock run usually will sidetrack the plug. A new technique has been used recently in running whipstocks when a maximum deflection is desired. A roller reamer is made up above the whipstock bit when a full-gauge whipstock is used (fig. 248). This set-up prevents the tendency of the bit to decrease deflection as it drills by the bottom of the whipstock. This set-up has been used in a number of cases to sidetrack plugs when ordinary methods failed.

It is good practice to check the results of a deflecting-tool run by taking a reading in the "rathole." When a small-diameter "rathole" is made, a small-type single shot should be used to eliminate the danger of sticking the single shot in the well. Some operators prefer to make a drift and direction determination after the "rathole" has been reamed. If it is found that the tool run has been unsuccessful, another tool may be set immediately for correction, or the deflected hole sidetracked by drilling with a sharp bit used below a long drill collar.

Orientation of a deflecting tool consists in making the tool face in the direction desired by the operator as it is landed at the bottom of the hole. Two methods are in general use for accomplishing this orientation.

The drill-pipe alignment method is used whenever a tool is set in a well which is nearly vertical at the setting point. It also is used for orientation at shallow depths, as it is more rapid than the bottom-hole orientation method. The system of orientation requires a crew of two engineers. They use drill-pipe clamps, a sighting telescope, and a sighting bar (fig. 247). After the tool has been made up on the drill pipe, it is faced in the direction in which it is desired to land it at the bottom of the well. A sighting bar is placed in an orienting clamp, which is affixed to the drill pipe just above the deflecting tool and sighted at one of the derrick legs. The telescope is placed in a drill-pipe clamp affixed to the upper end of the drill-pipe stand. The telescope is sighted downward along the drill pipe. This upper clamp is adjusted so that the telescope cross hair is aligned with the sighting bar. Then the telescope is removed from the upper clamp. The lower clamp is removed, the pipe is lowered into the well, and the sighting bar is placed in the same clamp from which the telescope was removed. The bar is sighted upon from above in the same manner as before. This process is repeated until the deflecting tool is near bottom. The sighting bar is placed in the last clamp and the drill pipe turned until the bar again is sighted at the same derrick leg. Thus, the tool is faced on bottom in the same direction as it was at the surface. Briefly stated, a line of sight is projected upward in a vertical plane to the top of each succeeding stand of pipe as the tool is lowered into the well. In practice the orientation by this method is very rapid and highly accurate (fig. 249).

The other orientation method extensively used is called the "bottom-hole-orientation system." It is used in wells that have more than two degrees of angle at the point at which a tool is to be set. It requires less time than other orientation systems when tools are run in deep wells. The drift and direction of the well at the proposed setting point must be determined by a single-shot picture or from a multiple-shot survey. A special substitute (fig. 250) containing two opposite-pole magnets is made up immediately above the whipstock bit or the knuckle joint (fig. 251). This orienting substitute has a tapped hole in its side, which receives the end of the shear pin screwed through the back of the whipstock. The magnets

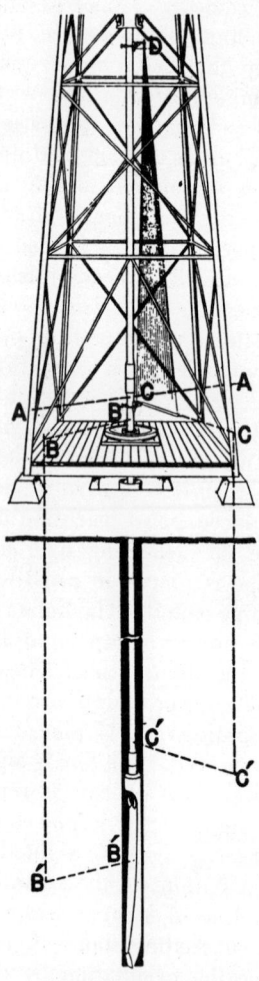

FIGURE 249. Diagram of Stokenbury system of drill-pipe orientation.

are placed in the substitute so that they are in exact alignment with the face of the whipstock. The tool is run into the well on the drill pipe. Just before it reaches the bottom of the hole, a special bottom-hole orientation instrument is run down inside the drill pipe and positioned between the magnets of the substitute. This special instrument consists of an Eastman type "M" drift-indicator body, onto the bottom of which is screwed a magnetic orienting unit (fig. 252). This orienting unit contains a disc

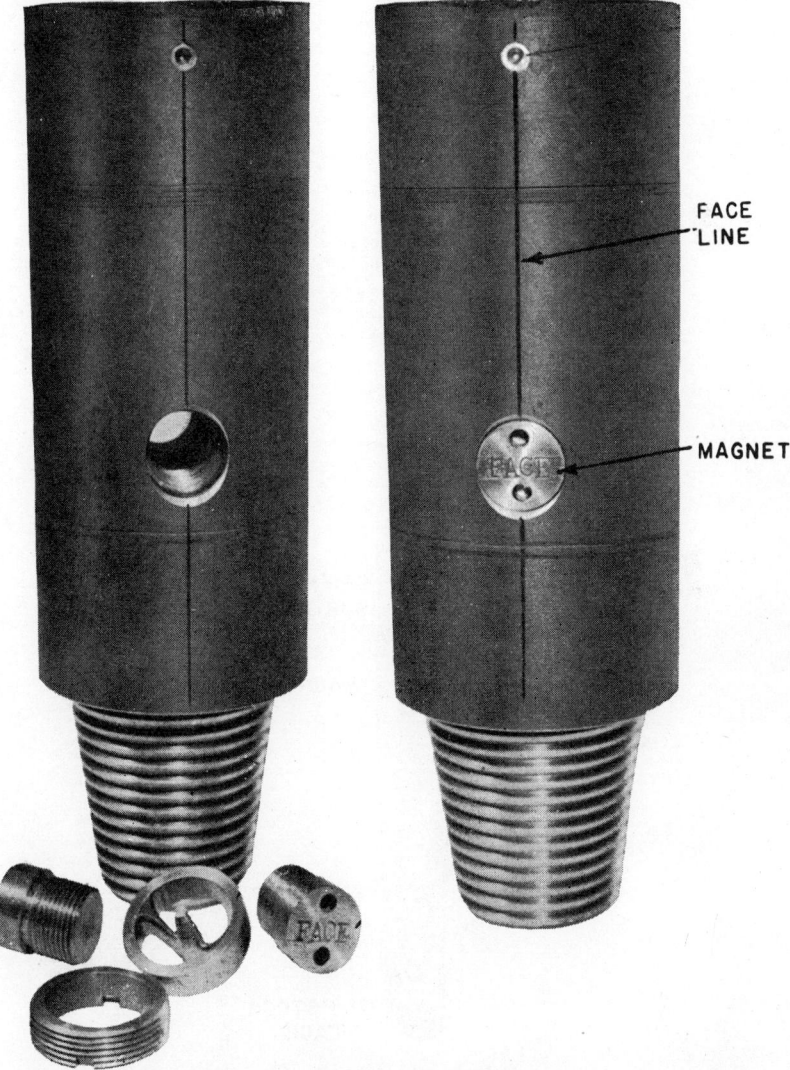

FACE
LINE

MAGNET

FIGURE 250. Bottom-hole-orientation substitute. Whipstock shear-pin hole is in back opposite face magnet.

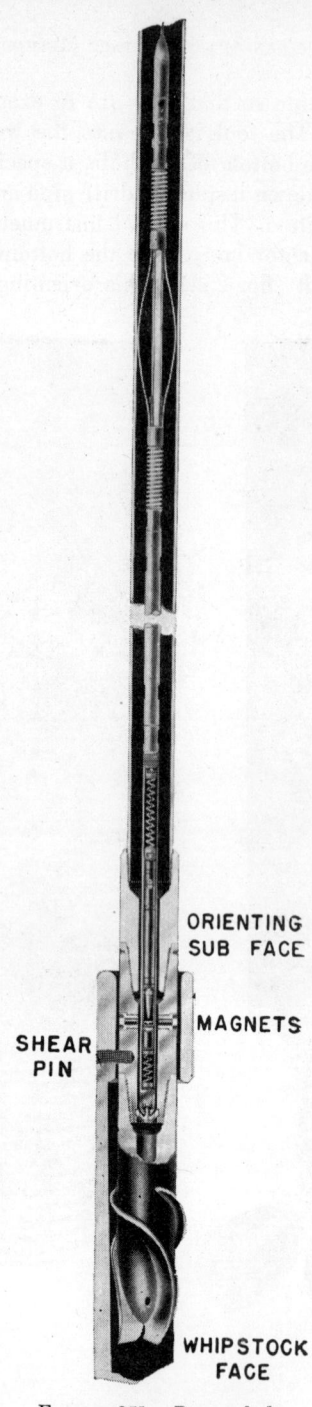

ORIENTING
SUB FACE

MAGNETS

SHEAR
PIN

WHIPSTOCK
FACE

FIGURE 251. Bottom-hole-
orientation equipment
with instrument in
place.

cup similar to that used with the drift indicator, except that it is free to rotate. A small magnet is attached to the bottom of the cup. When the orientation instrument is positioned so that the magnetic disc cup is between the magnets of the substitute, a picture is taken with the instrument. The spot on the disc indicates the low side of the hole. The instrument and barrel are withdrawn from the hole and the unit, with the disc in place, is unscrewed from the instrument and placed in a reader (fig. 253). This reader is constructed with two magnets in exactly the same relation that they have in the drill-pipe substitute. The angular relation between the face of the whipstock and the low side of the hole is obtained by means of the reader. The drill pipe is turned the required amount, either right

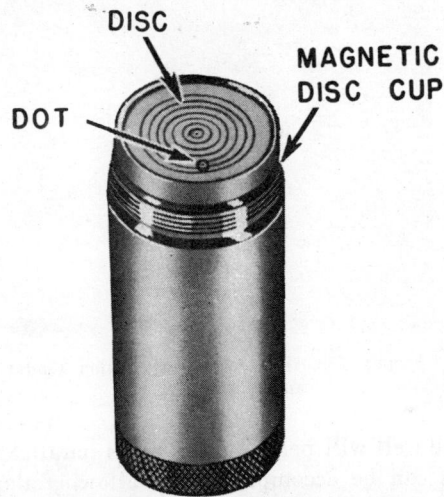

FIGURE 252. Magnetic orienting unit.

or left, to correct for this difference. In this way the tool is faced in the desired direction. A check on the orientation should be made by running the instrument a second time, after the pipe has been turned. If a check reading is taken, this method is considerably slower than the alignment method of orientation when used at shallow depths.

Typical deflection tools and the methods of orienting them into the well have been considered. The art of directional drilling consists not only in the employment of such tools, but in directing a well to an objective throughout its whole course by a knowledge of directional-drilling practices. Some directionally drilled wells are very shallow; others have been drilled to great depth. The carrying out of a successful directional-drilling job calls for careful planning and consideration of the great number of factors involved.

Subsurface geologic conditions in the area where the work is to be carried out are very important. The engineer must have an idea of the productivity of the formation, as well as laws and regulations controlling bottom-hole spacing. Directional wells should be planned to take advantage of multiple-zone completions, possible further deepening to lower zones, or the perforating of casing at a shallower depth. For instance, when geologic conditions are such that a directional well must penetrate a number of oil sands all lying close to a salt dome, a proposal can be

FIGURE 253.   Bottom-hole-orientation reader and disc in place.

drawn so that the well will penetrate the zones parallel to the side of the salt dome. This can be accomplished by allowing the well to become vertical before entering the first zone, or by directing the course of the well in a plane parallel to the side of the salt dome. The dip and strike of the formation and its physical properties affect the drilling of a directional well. Usually wells drilled in a direction normal to the strike will be the easiest to control, while those which parallel the strike will cause the greatest difficuty.

The proper starting point for the deflection depends upon the formation characteristics, the maximum drift angle selected, and the rate of increasing the drift angle. The foregoing factors also are governed by the total vertical depth to the oil sand and the horizontal deflection required. The well proposal must be laid out so that the rate of increase of angle will be feasible in the formation to be drilled. An average rate used in many instances is $2°$ $30'$ to $3°$ $00'$ per 100 feet of hole drilled. Some directional-drilling work has required an increase in the drift angle of from $6°$ to $8°$ per 100 feet until a maximum drift angle from $60°$ to $80°$ has been attained. The best practice is to increase the angle at a lower rate and drill a little more directional hole to reach the objective. Maintaining a very low drift angle or an unnecessarily high angle is not recommended. Drift angles from 15° to 45° have been successfully maintained for thousands of feet. It is within this range that most economical directional drilling is accomplished. Very low angles definitely increase the cost of a well because of the greater amount of directional hole which has to be drilled and the difficulty of controlling the direction of a low-angle hole during drilling. Extremely high-angle wells present special difficulties in logging, surveying, and running casing. The depth at which the deflection work is started materially affects the cost. If a reasonably high angle is used, the directional work can be started deeper, and more straight hole may be drilled first. The formations to be drilled and their rates of penetration should be studied in order to select the best zone in which to do directional drilling. Whenever possible deflecting tools should be set in formations which are not badly fractured. The starting point, the bottom depth, the desired deflection, and the drift angle chosen all bear a definite relationship to one another.

A directionally drilled well is not a "crooked hole" but is a directed well wherein all bends are controlled to stay within safe limits. Frequent causes of mechanical trouble in both vertical and directed wells are excessive "dog-legs" and unnecessary wandering of the well course. A "dog-leg" may be either a change in drift or a change in direction or a combination of both. Calculated graphically, "dog-legs" are expressed in degrees for a certain specific length of well. Different formations vary in their tendency to keyseat. It would not be practical to attempt to determine the maximum permissible "dog-leg" for each formation; therefore, a safe angle is chosen which will suit the requirements for efficient drilling and production. Recommended practice is to limit the average increase in drift to 2° 30′ to 3° 00′ per 100 feet drilled, and to limit the maximum "dog-leg" caused by deflection tools to 3° 00′ in any 50-foot section and 5° 00′ in any 100-foot section of well bore.

Directional wells are of two general types. In one, the drift angle is increased at a uniform rate to the desired maximum deflection angle, which is maintained until the oil zone is reached. In the second type, the angle is increased at a uniform rate and maintained until the desired deflection is obtained, at which point the well is brought back to vertical at a uniform rate of decrease in angle. These two basic types of wells are shown in figure 254. The first type of well generally is easier and less expensive to drill. It also has the advantage that the oil zone is penetrated at an angle, increasing the exposure of sand in the producing zone. The second type of well is used in situations where deeper horizons are to be explored or in drilling when a well is to be bottomed very near a lease boundary or along salt domes. Owing to the fact that the straightening process in the deeper portion of the well is often a slow process, this type of well generally is more costly.

The casing program for a directional well will vary only slightly

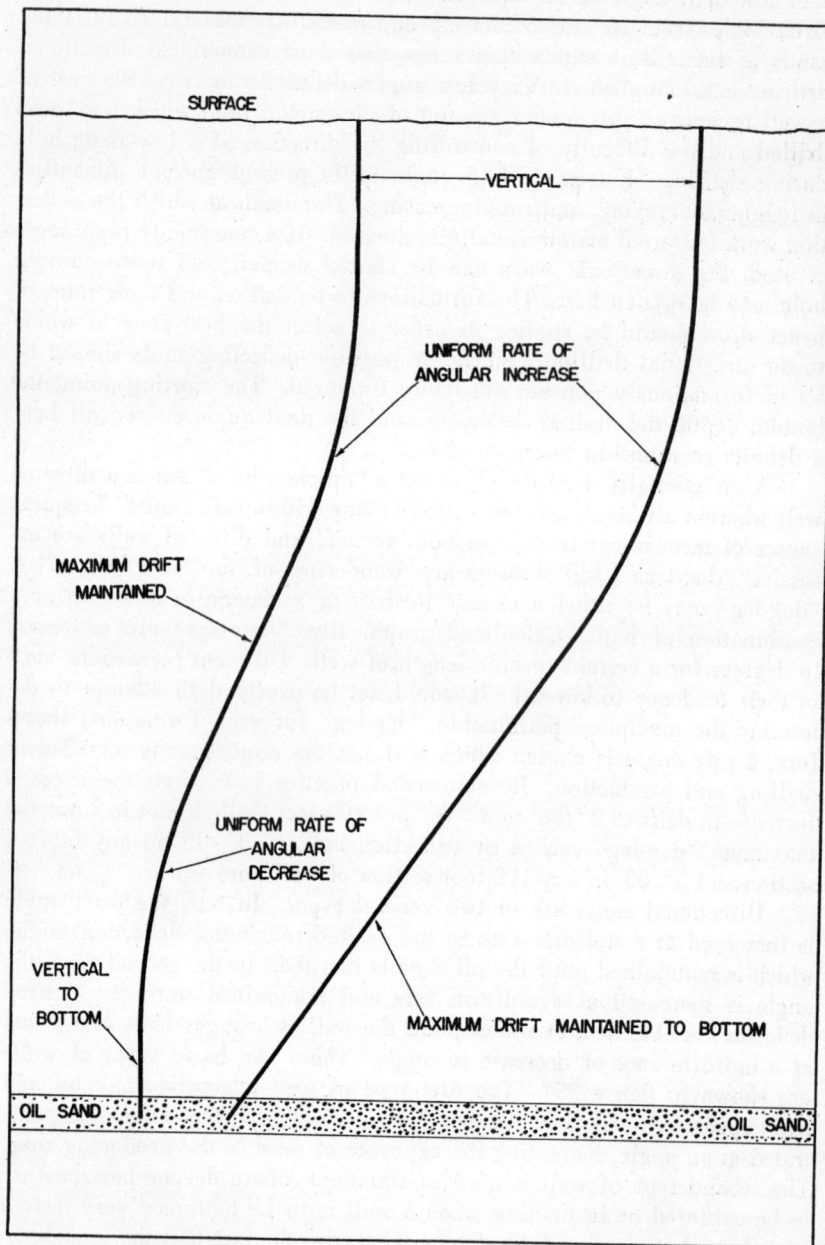

FIGURE 254. Two basic types of directionally drilled wells.

from that which would be chosen for a straight hole. Bending stresses set up in the casing by curvature of the hole are small; however, it is good practice to increase the thickness of the casing throughout the curved portion of the well bore to resist wear by the drill pipe. Some operators use one grade better casing on a directional hole than they would in a comparable straight hole. On very deep or difficult directional wells it may be necessary to use a protective string of casing to safeguard the upper portion of the hole. The constant use of wall reamers in directional drilling keeps the well bore in good condition, and usually casing is run without difficulty. Flush-joint casing is seldom used in slant holes, as it has a tendency to stick. Some flush-joint liners have been run successfully when drilling in the oil zone was done with oil as the circulating fluid.

Having considered a number of varying factors in planning a directional well, definite decisions must be made and a proposal for operations drafted. Tables and charts similar to the one illustrated in figure 255 are used by engineers for this purpose. The chart shown gives the vertical depth and deviation for a hole, the angle of which increases at the uniform rate of 2° 30′ per 100 feet of hole drilled. A short section of vertical hole should be drilled from the surface to prevent future difficulty in pumping the well. Whenever possible it is advisable to set surface casing in vertical hole and start deflecting the well 100 feet or so below the shoe. In some cases, where maximum deviation was desired at a shallow depth, deflection has been started at 100 feet below the surface. After a tentative starting point has been chosen, the average deviation from this point to the objective is computed by considering the vertical depth to the producing zone. This gives a rough approximation of the maximum angle of the well. As the maximum angle will be attained by increasing the angle at a uniform rate, the actual maximum angle in the well will be somewhat greater than the average. It is advisable to assume two or three logical starting points and figure the maximum angle for each. Other factors affecting the drilling of the well may be evaluated in relation to each of these starting points. The theoretical course of the well is computed by means of special charts and plane trigonometry after the starting point has been chosen. The measured depth of the well course and depths of different formation markers are obtained in this way. In fields where a considerable amount of directional drilling has been done, it will be necessary to plot wells already drilled and those proposed on the same plan and section to give assurance that the course of the wells will not interfere or intersect.

After the calculations have been completed and consideration has been given to other wells in the vicinity, a drilling proposal is drawn to scale (fig. 256). A vertical section of the well is made, as well as a plan view showing the course of the well from the surface to the objective. The proposal is drawn on cross-section paper, so that a single-shot survey of the well made as it is drilled may be plotted upon both the plan and

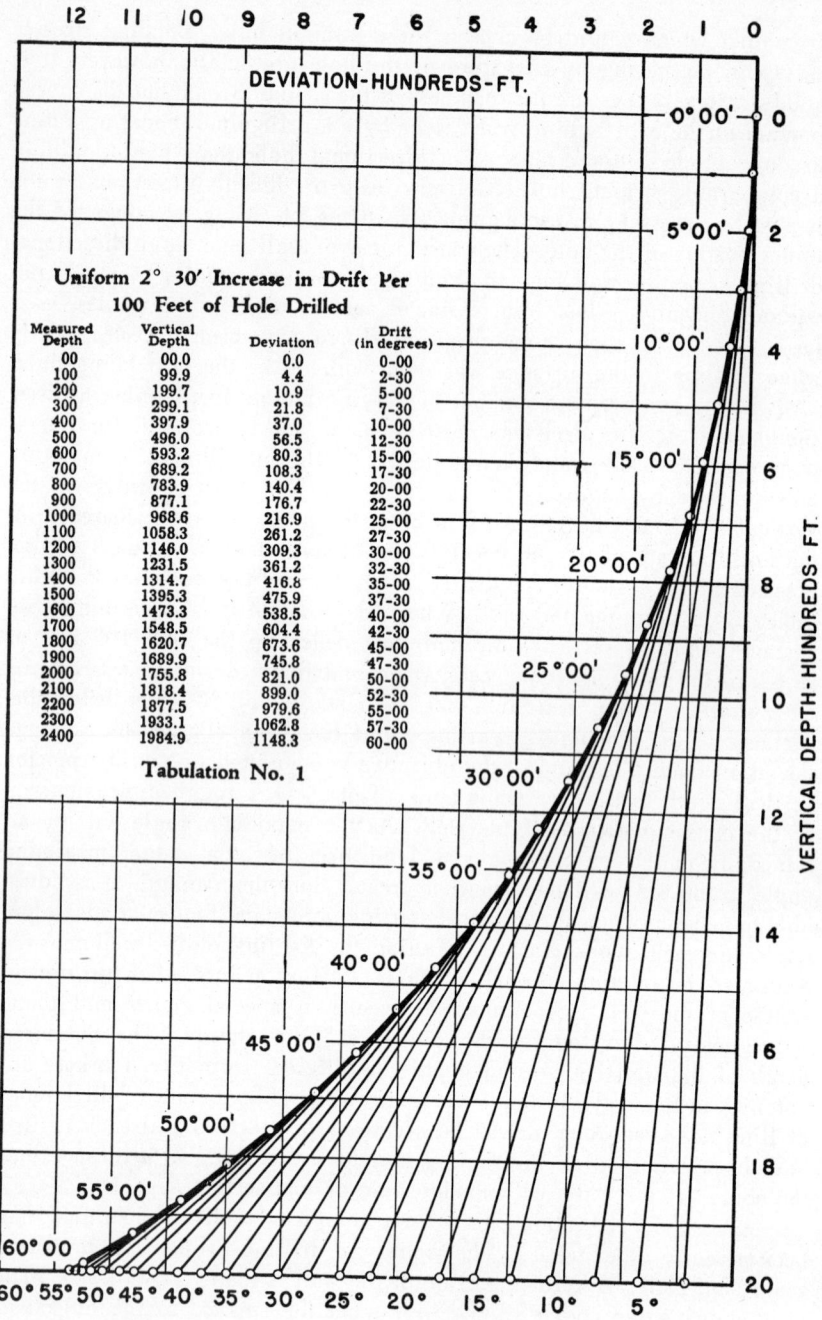

FIGURE 255. Estimating chart for planning directional wells.

Uniform 2° 30' Increase in Drift Per 100 Feet of Hole Drilled

Measured Depth	Vertical Depth	Deviation	Drift (in degrees)
00	00.0	0.0	0-00
100	99.9	4.4	2-30
200	199.7	10.9	5-00
300	299.1	21.8	7-30
400	397.9	37.0	10-00
500	496.0	56.5	12-30
600	593.2	80.3	15-00
700	689.2	108.3	17-30
800	783.9	140.4	20-00
900	877.1	176.7	22-30
1000	968.6	216.9	25-00
1100	1058.3	261.2	27-30
1200	1146.0	309.3	30-00
1300	1231.5	361.2	32-30
1400	1314.7	416.8	35-00
1500	1395.3	475.9	37-30
1600	1473.3	538.5	40-00
1700	1548.5	604.4	42-30
1800	1620.7	673.6	45-00
1900	1689.9	745.8	47-30
2000	1755.8	821.0	50-00
2100	1818.4	899.0	52-30
2200	1877.5	979.6	55-00
2300	1933.1	1062.8	57-30
2400	1984.9	1148.3	60-00

Tabulation No. 1

section. A number of copies of this proposal are reproduced. One is used by the directional-drilling engineer, who has charge of carrying out the plans illustrated. The service company whose engineers are doing the directional drilling will keep a copy up to date in their local office, so that they may be able to work with the operator. The operator's engineer or geologist definitely will need a copy to keep in touch with the work as it proceeds. Each of these copies should be kept up to date each day from the survey data.

Cylinder drilling is a technique that resulted from the great number of directional wells drilled at Huntington Beach, California. Since 1938 the prevalence of drilling wells within theoretical cylinders has increased steadily. To a certain extent this system insures the drilling of mechanically correct holes by limiting the tolerance of the drift and direction of the course of the well. An imaginary cylinder is described about the proposed well course as a center. The radii of these cylinders usually are 50 to 100 feet, but it has been necessary to make some of them much smaller. On the proposal (fig. 256) a 25-foot-radius cylinder was specified from the surface to the 1,800-foot depth. As the deflection work started at this point, the radius of the cylinder was increased to fifty feet. The operator desired to bottom the well within a 25-foot-radius target; therefore, the cylinder was tapered at the bottom to this size. This proposal shows a cross section of the two different-size cylinders in its upper right-hand corner. The survey data taken as the well is drilled is first plotted on the plan and section and then is transferred to the section of the cylinder. This indicates graphically to the engineer whether it is necessary to increase or decrease drift or to turn the well right or left.

To illustrate some of the techniques used, a procedure is outlined briefly for drilling a well similar to that illustrated in the proposal.

The well would be drilled to the 1,800-foot depth, taking the usual precautions to keep the hole straight. The drilling crew would take single-shot pictures at intervals of not over 100 feet as hole was made. This information would then be converted into rectangular coordinates and vertical depth for each point at which a reading was taken. This survey data would be plotted on the plan and section of the proposal. When the 1,800-foot-depth point had been reached, a directional-drilling crew would be called to assist in starting the deviated hole. A deflecting tool would be set at this point and would be faced in the correct direction to start the well on its proposed course.

After the deflecting-tool run, a single-shot survey would be made of the results to determine whether they were satisfactory. The engineers would attempt to increase the angle of the hole at the rate specified. Usually this deflection can be accomplished by using different drilling setups and varying the weight on the bit, speed of rotation of the pipe, and amount of mud circulation. About 100 feet of hole would be drilled below the deflected hole with a short drill collar and a roller reamer immediately

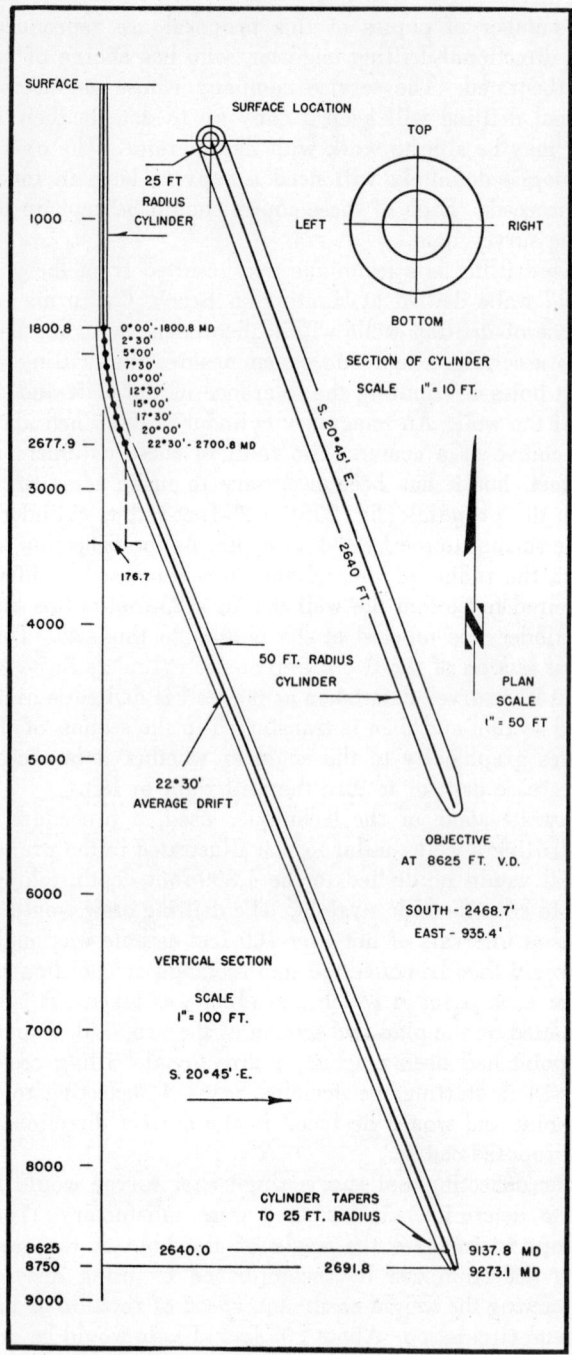

Figure 256. Directional-drilling proposal.

above the bit, applying weight to the drill pipe and rotating the pipe about sixty r.p.m. with moderate pump speed. After drilling this distance, a single-shot reading would be taken either through a trigger bit, in a non-magnetic drill collar, or with an open-hole single shot to determine whether the drift and direction were in accordance with the plan. The location of the well at this point would be plotted on the plan and section. If it were found that the course of the well was not as desired, another deflecting tool would be set to keep the well on course. The maximum angle and correct direction are attained by using different drilling set-ups and setting deflecting tools. Then it is probable that long drill collars would be used above the bit to maintain the course of the hole in a straight line. Changes in any of the factors which affect the directional work would not be made unless a single-shot picture was taken to check the effect of the last setup used. Techniques in drilling directional wells and maintaining them on their true courses vary greatly with different formations. Types of drilling equipment, bits, reamers, and drill collars and methods of surveying, as well as the relation of the course of the hole to the structure, are factors to be considered.

Few ironclad rules can be followed, as directional-drilling techniques are learned only by experience. Certain basic methods influence the increase or decrease of drift. An increase in the angle of drilling is accompanied by the application of weight to the drilling bit, moderate rotation speeds, and moderate pump pressure. Drift generally is gained by using a short drill collar and a roller reamer directly above the bit. If it is necessary to increase drift quite rapidly, the drill collar may be omitted, but the well possibly may wander from its course. Reduction of drift generally is accomplished by the use of sharp bits, light drilling weight, fast rotation, and high pump pressure. Decrease in drift can be assisted by using a long drill collar above the bit with a roller reamer at the top of the drill collar. After a start has been made in reducing drift, the application of weight often will increase the rate of straightening the hole.

Little definite evidence that any type of bit will consistently turn a well to the right or to the left has ever been presented. Many factors influence the course of a directional well. Most directional drilling engineers base their usage of bits upon experience in the field in which they are working or in fields having similar drilling conditions. Some bits have been designed especially for use in directional drilling work. Most of the deflecting tools set, after the drift has been increased to its maximum, are to correct the directional course of the hole. As the drilling proceeds, the engineers plot the location of the well on the proposal as single-shot readings are taken. The location of the well within the cylinders is the basis for their recommendations as to methods to be used in drilling. Aside from noting the effect of different drilling setups and methods which they recommend, the control men try to anticipate and

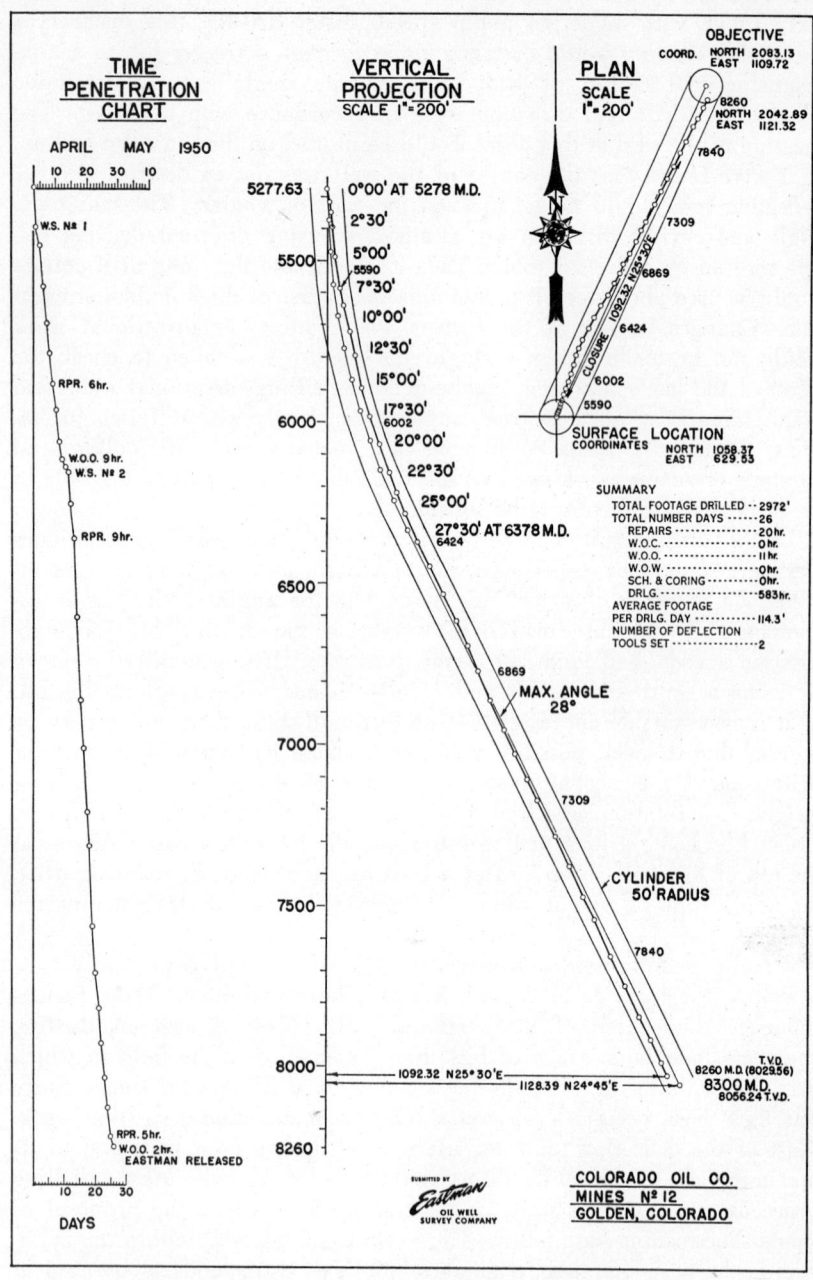

FIGURE 257. Typical directional-drilling completion report.

provide for eventualities that might occur as the well is deepened. Thus, they set deflecting tools in such a manner that if an unexpectedly large "dog leg" results, it will be of benefit in the long range program. In a way it might be said that the successful directional-drilling engineer is planning 300 to 400 feet ahead of the bit.

When the directional drilling has been finished, a completion report is submitted to the operator for which the work was done. This report shows a plan and vertical section of the well, together with a time-penetration curve as illustrated in figure 257. The course of the well as it was actually drilled within the cylinder is plotted on the plan and section. The time-penetration curve is drawn alongside the vertical section. The horizontal axis represents days on the job, and the vertical axis represents depth. Time lost due to repairs, waiting on orders, waiting on cement, logging, weather, etc., are noted at the appropriate depths as they occurred. A summary tabulation gives the number of hours lost due to the causes mentioned above as well as the number of hours drilling. From this information, an average footage of hole drilled per day is derived. The information on this completion report has been found to be of considerable value to the operator concerned, especially in estimating subsequent directional-drilling operations. In a number of cases where comparisons could be made, it has been found that a directional well was drilled at a faster rate than a straight hole could have been probably because of the fact that, after a directional well has attained maximum drift and correct direction, more weight is applied to the bit than in a straight well. Therefore, the rate of penetration is increased.

APPLICATIONS

The applications of directional drilling are numerous and varied. Its first use was to explore productive zones off the California coast by drilling wells from shore to the subsea formations. Since that time many applications of directional drilling have been made throughout domestic oil fields as well as in foreign countries. The discovery of oil-bearing formations in the Gulf of Mexico recently has given added impetus to directional-drilling operations.

Undoubtedly the greatest amount of directional-drilling work is done in deflecting wells from convenient and accessible surface locations to bottom directly beneath a location where it is impossible or expensive to set up derricks. Much oil has been produced under such conditions that never could have been obtained without directional drilling.

Since 1933 and to the present time, an offshore structure in the tideland area seaward from Huntington Beach, California (fig. 262), has been explored and produced by wells drilled from shore. The old field at this location saw the first extensive application of directed drilling. Some 60 or 80 slant holes were drilled from surface locations in the townlot field bordering the shore to subsea bottom locations nearly half a mile

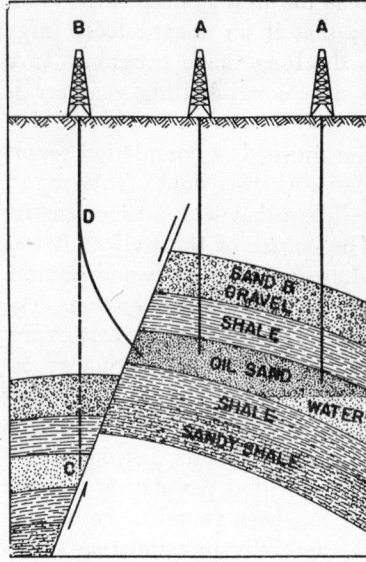

FIGURE 258. Controlled directional
drilling for crossing fault lines.

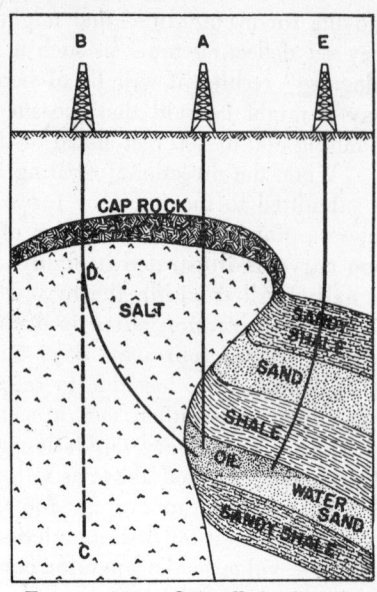

FIGURE 259. Controlled directional
drilling for overhanging salt domes.

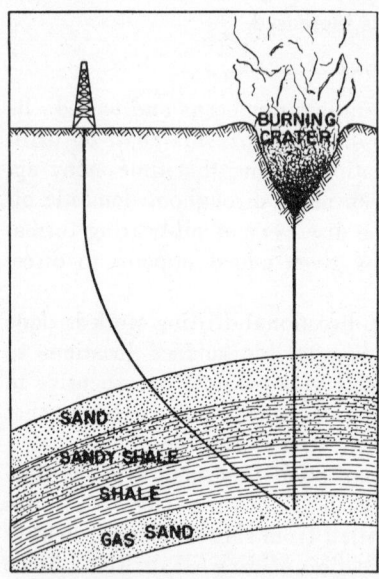

FIGURE 260. Controlled directional
drilling for relief wells.

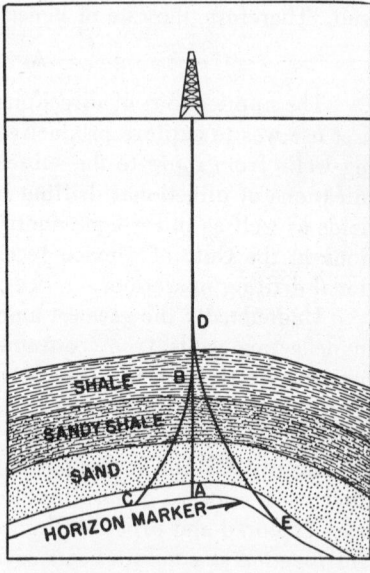

FIGURE 261. Controlled directional
drilling for geologic exploration.

offshore. Some of these wells attained deflection angles greater than 60 degrees. Many of these holes were very poorly drilled directionally, but much of the experience gained from these wells developed the art of directional drilling to a standard well practice as used today. Numerous wells were badly "dog-legged" with abrupt changes in drift and direction. Some wells were unintentionally sidetracked while drilling; others collided. Producers sometimes began pumping mud from a nearby drilling well and found

FIGURE 262. Huntington Beach, California, town-lot field about 1935. The first directional drilling was done in this field.

their pumping wells were ruined. These angled wells presented new problems in well surveying, problems which accelerated the development of better instruments and equipment.

Since 1938 one company has been developing a state lease lying under the Pacific Ocean west of Huntington Beach. Wells have been drilled continuously since then; a number are in process of drilling; and an extensive program is planned for the future. Paralleling the shore, wells are surface-spaced 27 feet apart so that a single steel derrick mounted on wheels can serve a group of five pumping wells (fig. 263). These wells are drilled in directional cylinders, in many cases threading around wells already producing, to bottom from 700 feet to a mile from their surface locations at vertical depths of 4,000 to 5,000 feet. Wells drain an area

FIGURE 263. Directional wells at Huntington Beach, California. Surface locations spaced so that a single production derrick mounted on wheels can service a group of wells.

FIGURE 264. Crater at Conroe, Texas, and relief well. Note stub derricks between crater and well erected to permit completion of relief operations if oil in crater should become ignited.

of five to ten acres each. When the wells cease flowing, electric-motor-driven pumping units are used. About 230 directional wells have been drilled in the twelve years since the program was started. Plans are being prepared for drilling a new off-shore lease recently granted to the company. It will be necessary to drill a number of very high-angled wells to exploit this submarine-oil formation since the lease to be developed lies between one and two miles off shore from the proposed surface locations. Records kept by the company on production costs show that wear on pumping equipment is not greater than it would be in average vertical wells because of the care used in executing the directional drilling.

Development of offshore structures from littoral locations has been practiced in the Wilmington, Long Beach, and Elwood fields, California. Much work of this same type has been completed successfully along the Gulf Coast of Texas and Louisiana.

The drilling of relief wells is a spectacular and well-known application of controlled directional drilling to the problem of bringing wild wells under control. The first work of this kind was done by Eastman engineers in the Conroe field in Texas in 1933 (fig. 264). A well cratered and started uncontrolled production of oil at the rate of 6,500 barrels a day. This well produced over 1,400,000 barrels of oil before being brought under control, drawing heavily upon the producing formation of the field. A relief well was spudded at a location 412 feet from the center of the crater, and the hole was directed to very near the bottom of the wild producer at a depth of about 5,100 feet. By pumping water into the producing formation near the bottom of the cratered well, it was possible to shut off the flow of oil completely. Elaborate precautions were taken during the course of drilling the relief well to prevent the nearby crater filled with oil from becoming ignited. The relief well later was plugged at the point where the deflection started and redrilled as a vertical well. It is producing today.

When an uncontrolled well has not damaged the casing badly, it is possible for competent fire-fighting crews to extinguish the fire and cap the well. Once a well is badly cratered, a relief well is the only economical means for placing the well under control. Since 1933 a number of relief wells have been drilled, all successfully. They include wells at Laguna Madre, Falfurrias, Premont, and Silsbee, in Texas; Arcadia and St. Martin Parishes in Louisiana; Cement field in Oklahoma; Leduc in Canada; and a burning gasser in Medias, Roumania.

Skill in directional drilling, together with very accurate well surveying, is demonstrated in the drilling of successful relief wells. A typical relief well is shown in section in figure 260.

Different types of drilling to produce accumulations of oil about salt domes are illustrated in figure 259. Several direction-drilling applications are shown. B is a well that continued in the salt mass to C without breaking through the overhang because of its placement in relation to the

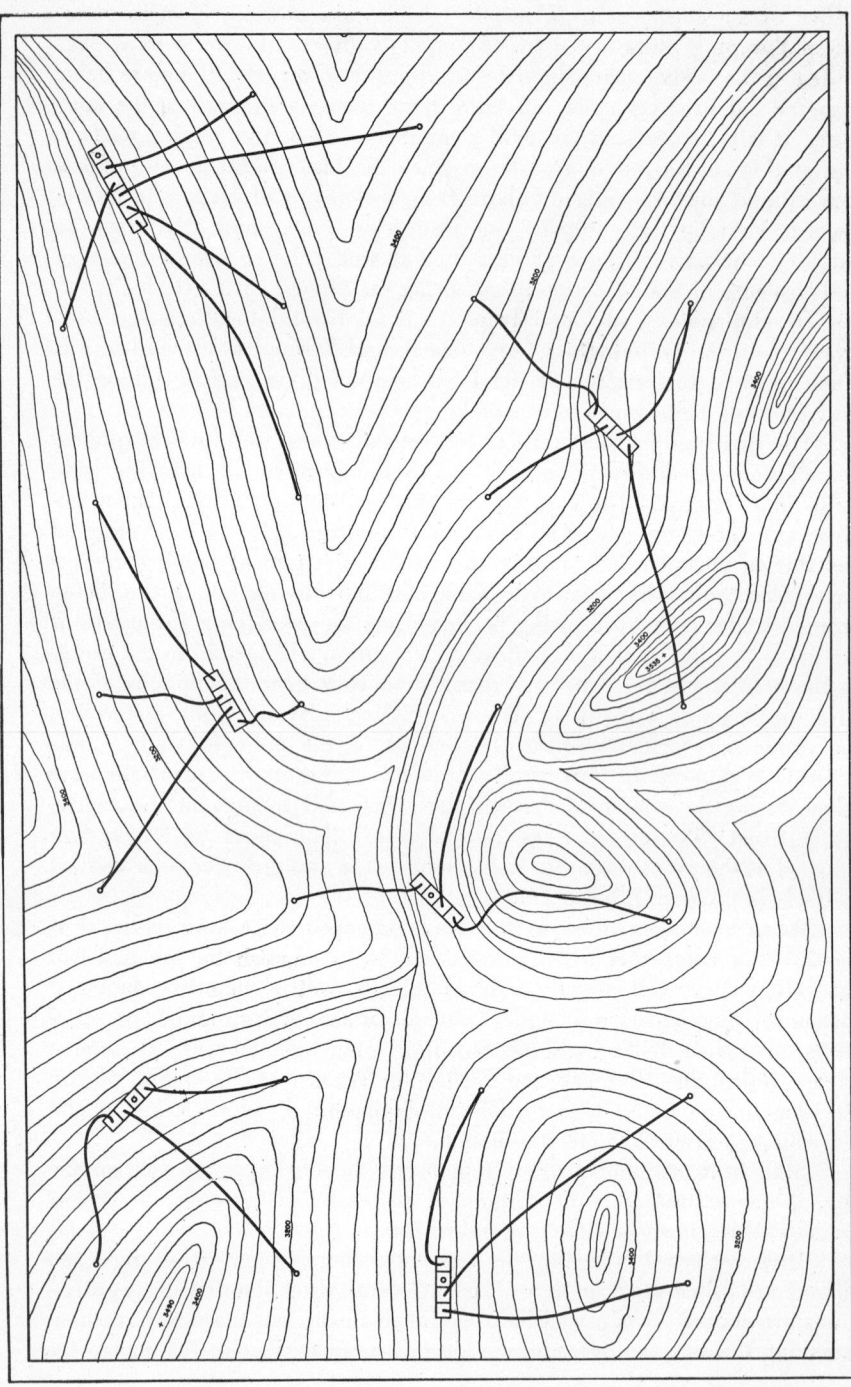

FIGURE 265. Development of mountainous lease by drilling of directional wells from surface valley locations.

dome. Slant-drilled from D, the well was converted from a dry hole to a producer. Well A is a successful but expensive vertical well. The cap rock and salt or mineralized sand, through which it was necessary to drill, presented costly difficulties. Some cap rock is very hard and makes the drilling of such a hole very costly owing to the time and bits consumed in penetrating this formation. The mineralized sand is very porous and full of cavities, which often results in loss of returns from circulation. Location of well E permits the well to be drilled in better formation and directed under the overhang. When geologic conditions are such that a directional well must penetrate a number of oil zones all lying close to a salt dome, it can be directed so that it will penetrate the zones parallel to the side of the salt dome. It may be allowed to become vertical before entering the first zone, or its course can be made parallel to the side of the dome. The last method permits penetration of a number of horizons at points equidistant from the face of the salt dome without the necessity of changing the drift of the directed well.

Since the subsurface geology may be very complex near salt domes, it is not uncommon to redrill a well three or four times in exploring for producing formations. A well will be drilled, taking into consideration the best information available. If it penetrates salt, it is plugged back and directed to a new location, possibly again striking the salt. If this occurs, the well is redirected to a point thought to be more favorable from the geologic information previously obtained. Much exploratory work of this kind has been done in the Gulf Coast area with the assistance of directional drilling.

Adverse surface topography has caused the adoption of directed drilling. An operator in developing a mountainous lease must balance the cost of access roads for wells drilled vertically against the cost of segregating surface locations at conventient locations and resorting to directional drilling to space their bottoms correctly.

It was found that, on a number of leases in Ventura and Santa Barbara counties in California, the cost of building and maintaining roads for the life of the wells (say 25 years), together with the expense of laying oil lines to each of them over the rough terrain, was much greater than development by slant drilling. Figure 265 is a typical example. Wells have been grouped in valley locations for ease of access and convenience in drilling and servicing. The courses of the wells were directed to correct bottom-hole spacing in the oil-bearing sand.

The development of oil resources under tidal swamps and bayous presents situations in which it is impossible, or at best very expensive, to drill vertical wells. The cost of piling and mat foundations is great, and they are sometimes unsatisfactory. Directional wells as drilled on the Gulf coast of Louisiana and Texas offer the solution to this problem.

Directional wells permit the tapping of oil reservoirs beneath wide rivers, lakes, and similar bodies of water. Wells directed under the Mississippi and the Cimarron rivers at various points are examples. An oil structure beneath Lake Centralia in the Salem, Illinois, field was developed with wells spaced at bottom by slant drilling. This is an outstanding example of directional drilling in hard-rock country. The average deflection of 250 feet was attained without difficulty.

Other works of man present similar opportunities for the use of slant drilling. Deep-water ship channels, dredged turning basins, wharves, and transit sheds, as well as industrial sites, such as steam-generating plants, ship yards, dry docks, and manufacturing establishments, would have pre-

FIGURE 266. Wells drilled to produce oil beneath harbor area of Long Beach, California. Many wells shown in this figure are directional.

vented development of the multiple-zoned oil reservoir beneath Long Beach harbor, California (fig. 266). By carefully planned directional wells drilled in cylinders the whole harbor area is being drained of the subsurface oil without disturbing the expensive surface installations. Production of oil from beneath the large Navy establishment at Roosevelt Base has not been hampered by the repair yards and graving dock at the surface. Oil is produced through directed wells, which honeycomb the underground formations.

A spectacular example of this type of work is offered by the production of oil from under the capitol and other state buildings in the heart of Oklahoma City. Directional wells are extracting the oil from beneath the buildings without disturbing the buildings or impairing the beauty of the grounds.

Western business and residential areas of Long Beach, California, are underlain at shallow depths with oil-bearing sands. Zoning prohibiting the erection of derricks has prevented the development of these resources by vertical wells. One group of operators has leased a strip of land 27 feet wide and 6,200 feet long adjoining the area restricted by zoning. Using this land as surface locations for directionally drilled wells, the oil beneath the zoned area is being removed without disturbance to the surface improvements. These wells, spaced in line eighteen feet apart, are deflected at very shallow depths at rates of increase in drift as great as eight degrees per hundred feet of drilled hole. Some wells are bottomed as distant as eight or ten city blocks from their surface locations to produce from a shallow sand at about 3,000 feet. The drift angle of many of these wells exceeds 60 degrees. All wells must be drilled within cylinders to prevent collision between wells. This orderly development of subsurface community leases is still being carried on.

An example of producing from a small lease that is inaccessible because of zoning restrictions is shown in figure 267. The subsurface rights to a small, irregularly shaped parcel of land were obtained by the operators. The lease was completely covered by buildings, and the area was zoned against the use of drilling rigs. Plans called for two wells to be drilled to a shallow oil zone of about a 3,000-foot depth. Surface rights were obtained for rig locations about 1,100 feet from the subsurface lease in a district where city zoning restrictions permitted oil operations. An easement was obtained to permit the passage of the wells under property between the surface and bottom locations. The known presence of lower oil sands made it advisable to straighten the wells on bottom, as in future redrilling operations the liners could be pulled and the wells deepened vertically to keep them within the lease boundaries. Proposals were drawn by the survey engineers and checked by the petroleum engineer and operator.

The drawing shows plans and vertical sections of both wells as proposed (light lines) and the single-shot survey of the wells as they were directionally drilled (heavy lines). The proposal specified that one well was to be deflected at a 200-foot depth, increasing the drift at the rate of 4° 00′ per 100 feet of hole drilled to a maximum drift of 32° 36′. After about 1,250 feet of hole was drilled at this angle, the drift was to be reduced at 3° 00′ per 100 feet of hole drilled to vertical at bottom. The other well was drawn on a similar plan except that deflection started at a 400-foot depth. The increase and decrease in drift necessary to deviate the wells an average of 1,150 feet at a 3,000-foot vertical depth with the bottom hole vertical were not considered too difficult. However, the operators were unable to obtain an easement across certain adjoining property,

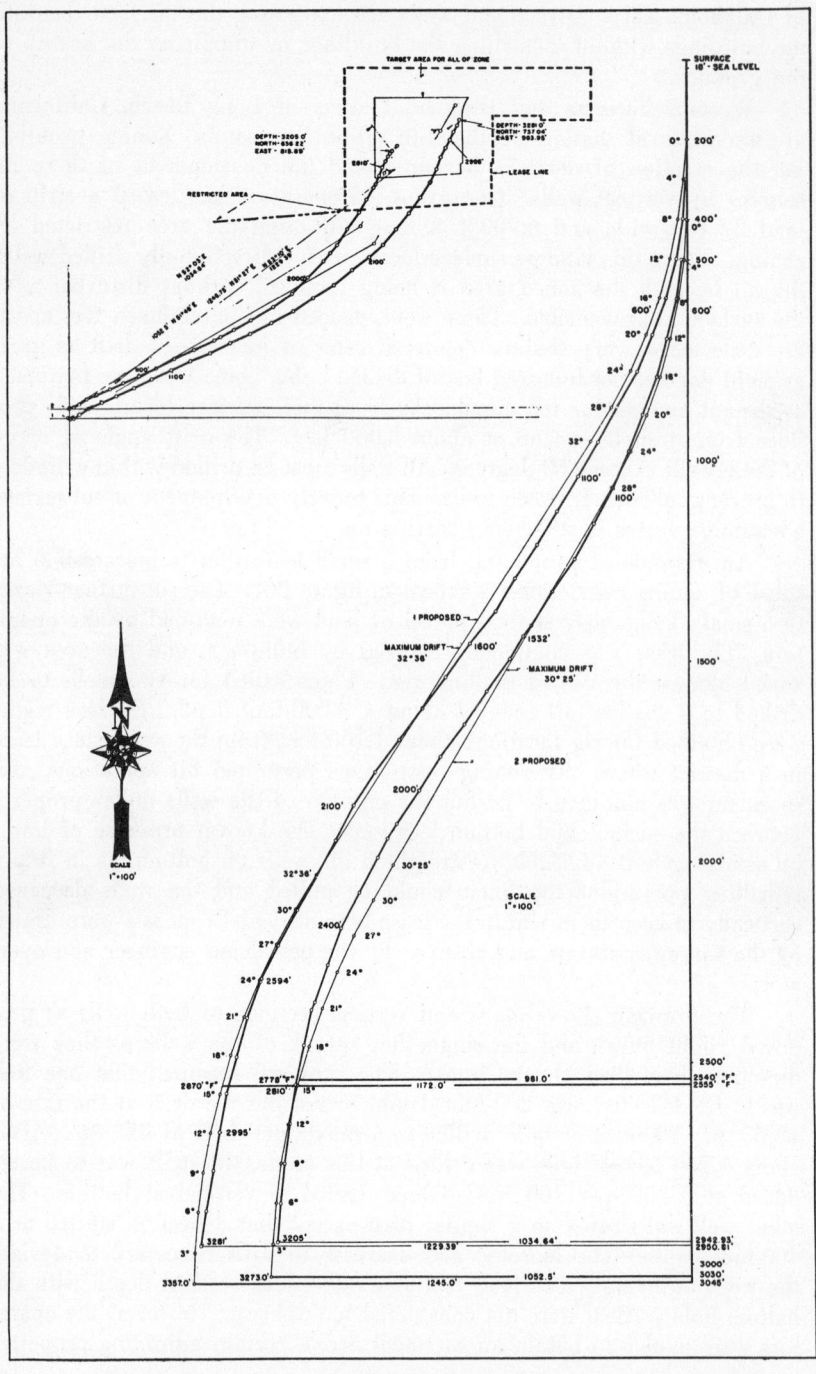

FIGURE 267. Two wells directionally drilled to small specified target.

so it was necessary to turn the courses of both wells in that part of their courses where the drift was greatest. Specialized directional crews supervised the operations on both wells, bottoming one well 20 feet and the other 40 feet from their exact objectives. Considering the difficulty of the work the record made was very good. The wells averaged 19 days each in drilling, or 170 feet of hole made each day. Seventeen deflecting tools were set for an average of 8¼ tools to the well. Casing was run without difficulty and both wells now are producing. Obviously the oil from this lease never could have been produced without directional drilling.

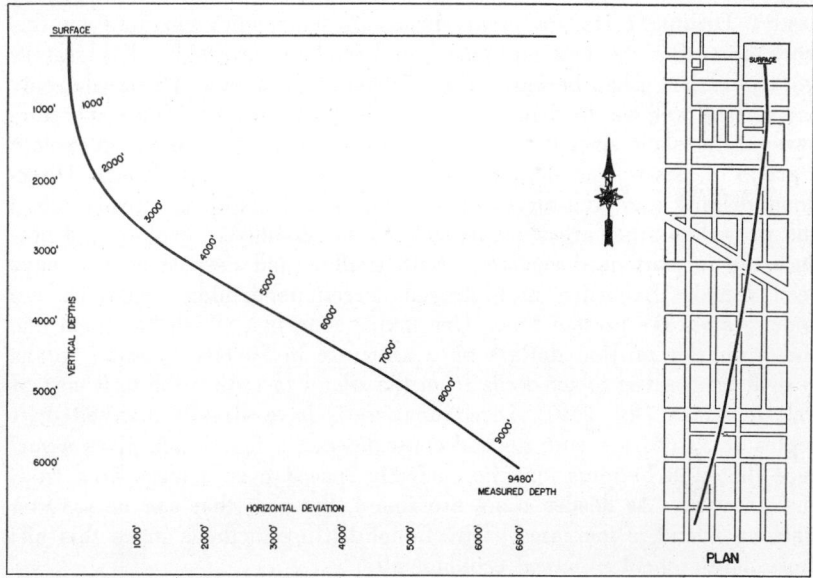

FIGURE 268. Exploratory directional well drilled under residential district of a city.

An interesting example of exploratory directed-drilling operations is illustrated in figure 268. It was suspected that an oil structure existed at about 6,000 feet below the residential district of a city. It was impossible to set up a derrick and drill vertically to explore the formation. A surface location was chosen about fifteen city blocks from the proposed bottom of the well. Leases were secured from each individual lot owner before any actual drilling could be started. Since the well was located near a city hospital, it was necessary to completely sound proof the rig. Deflection was started at about a 600-foot depth by setting an Eastman removable whipstock. A number of different types of drilling bits were used at rotary speeds varying from 50 to 100 r.p.m. and with drilling weights from 3 to 14 points. Drift and direction readings were taken by regular single shots run in open hole on sand line and by small single shots run in nonmagnetic drill collars and out through trigger-type surveying bits. Magnetic multiple-shot surveys were made in the deeper portions of the well. This well deviated horizontally 6,738 feet at a vertical depth of 5,949 feet. A drift angle varying between 52 and 62 degrees was maintained in the well below 2,500-foot measured depth.

Offshore leases in the Gulf of Mexico granted by the states of Texas and Louisiana to major oil companies in the last three years have confronted engineers with new problems in constructing artificial drilling locations for exploratory wells. As some of the drilling programs already under way are as far as 20 miles offshore and are in water 50 feet deep, the construction and maintenance of such man made islands is very costly. Drilling costs are great; hence the companies carrying on this work are using the best equipment and methods available. Exploratory work must be rushed because of provisions of the leases. Obviously, only one vertical well can be drilled from an island of this type. Some operators have set the island structure on the corner of four leases in order to explore a group of leases from the one point by drilling directional wells. Directional-drilling and well-surveying methods and techniques already tested and proved in other areas are being used successfully to explore and produce the formations discovered. Both shallow and deep structures have been found. Extensive multiple-well directional-drilling programs are under way at the present time. One major company, which has spent one and a quarter million dollars on a structure in 50 feet of water, plans to drill from seven to ten wells from the island to reduce the unit cost of well locations (fig. 269). Directional wells have already attained drift angles of 57 degrees with no bad consequences, a fact which gives assurance that well bottoms may be correctly spaced over a large area from one structure. As deeper zones are found, the area that can be covered from an island is increased. Directional-drilling methods make this offshore development practical economically.

A question which often is asked when the drilling of a directional well is under consideration is whether there will be any serious pumping difficulties encountered in producing the well. A survey of this problem was made in Los Angeles Basin fields, since a large number of high-angle wells have been pumping there for years. A number of major oil operators and representatives of rod and tubing manufacturers were asked for their actual experience in this matter. Their opinion was that there was little or no difference in tubing and rod wear between straight wells and those drilled at high angles by modern directional-drilling methods. In fact, it was found that several straight wells which had tight spirals or "dog legs" in their courses gave more production trouble than nearby directional wells. One company had made a private investigation of this kind on several hundred straight and deviated wells of their own. Sand content of the oil, drift angle in the well, sizes of rods and tubing, types of pumps used, and other pertinent factors were taken into consideration.

Their conclusion based on this study was that wear of rods and tubing was no more severe in the deviated wells than it was in straight wells.

Directional work often is done in drilling vertical wells where the limiting drift angle may be three or five degrees as specified in contracts. Deflecting tools faced in the opposite direction from the angle of drift of a well will straighten its course. In many cases the contractor profits through straightening the well by this method, owing to the time saved in comparison with bringing the well back to vertical by conventional straight-hole methods.

FIGURE 269. Offshore marine structure in the Gulf of Mexico. Exploration work is being done by means of directionally drilled wells.

Vertically drilled wells located near lease lines have been kept within correct boundaries by turning their courses with deflecting tools. Strip and very small locations have been exploited by careful drilling, consistent surveying, and an occasional deflection-tool run to correct the course of the well.

The occasional loss of tools in the well are hazards of well drilling. Expensive fishing operations ensue, sometimes to no avail. In cases of this sort the hole is bridged with cement and deviated away from the "fish" by directed drilling, effectually sidetracking the lost equipment. A few very skillfully executed jobs have been successfully done when casing or drill pipe becomes lodged at shallow depths in deep holes. By side-tracking around the "fish" and drilling back into the original hole below, much valuable drilled hole has been salvaged. An example of this occurred at Huntington Beach when a sharp bit was being run into a directional well. Before the driller realized it, the hole had been unintentionally side-tracked. A cement plug was used to bridge the hole, and a whipstock faced in the same direction as the course of the lost well forced the bit to break through into the original hole.

Drilling of wells in close proximity to faults may present difficulties. Figure 258 illustrates three wells drilled near a fault. Wells on the foot-wall side of the fault plane at *A* will produce from a reservoir accumulated against the fault zone. Well *B*, which originally bottomed at *C*, was not a producer but was redrilled from a cement plug set at *D* to penetrate the gouge zone and encounter oil on the other side.

Holes drilled through faults should have courses as nearly normal to the fault plane as possible and should enter the zone with a high drift angle. Wells bordering the south side of Sunnyside Cemetery at Long Beach were deflected to reach oil accumulated alongside a fault beneath the cemetery, obviating the necessity of removing bodies and abandoning the area as a burial ground.

Wells have drilled into the gouge zones of faults and followed the plane of the fault for the remainder of their course. Such wells have been plugged back and redrilled to make producers.

Edge wells have been converted into productive wells by directed drilling. A large investment can be saved by plugging the well and deflecting the bottom portion upstructure away from edge water. Edge property, partly productive and partly barren, presents this hazard. Wells drilled on such a lease are directed to bottom on the productive portion of the lease or are redrilled directionally to gain production and avoid short-lived edge wells.

Some operators have considered the directional drilling of wells to increase the exposure of the oil-bearing formation. Any angled hole will increase the amount of sand penetrated if drilled in nearly flat structures, parallel to the strike, or down the dip of the structure. This increased exposure of productive sand is true of practically all directional wells.

Figure 261 illustrates an exploratory operation to define a horizon marker. Multiple holes from one location were drilled by plugging the well and deviating the hole twice after establishing the depth of the marker with the original vertical well. Drilled hole and time are saved by such methods. Recontouring of producing horizons from data collected in locating the markers by actual drilling causes many contour anomalies to disappear. Consideration of the surveys of a number of wells on a lease, together with the logs of these same wells, generally clarifies the subsurface geologic picture. In exploring near salt domes, it sometimes is necessary to plug and redrill wells that strike the salt initially in order to locate productive zones.

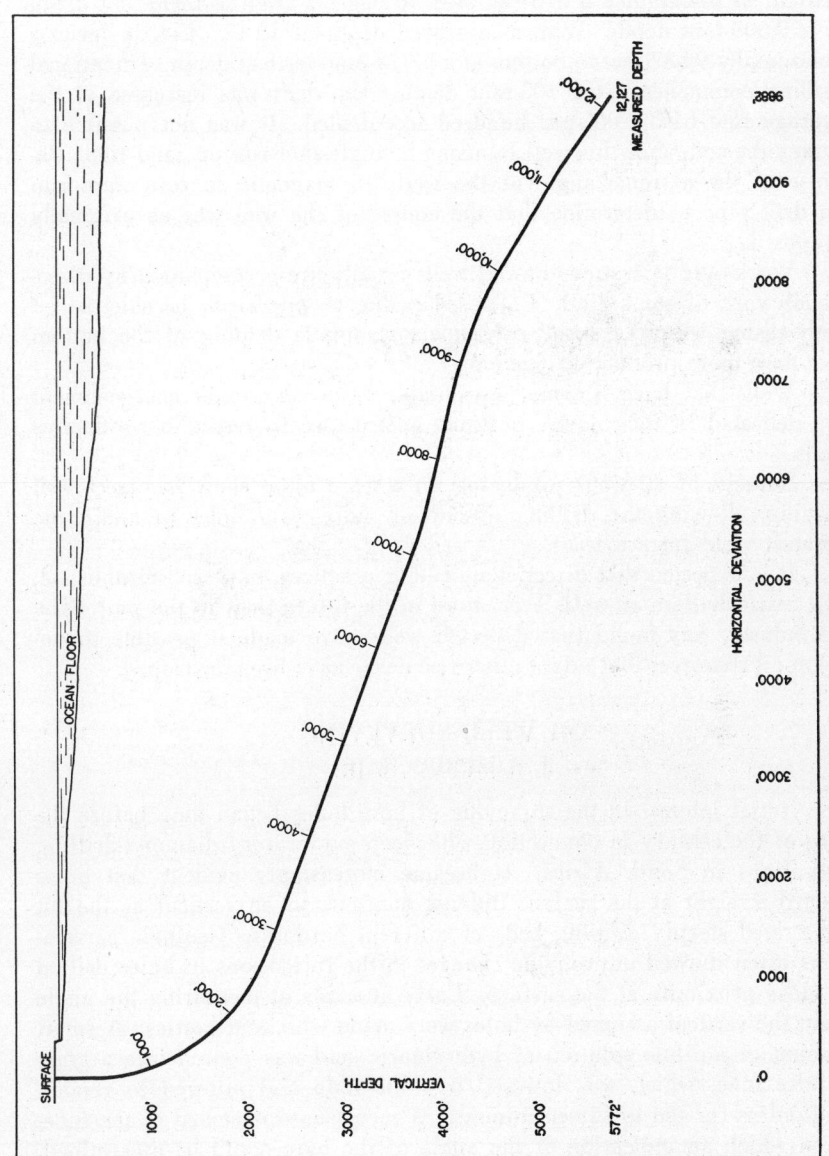

FIGURE 270. Exploratory well directionally drilled under the ocean.

A section of a well drilled for subsurface exploration from a shore-line location to bottom under the Pacific Ocean is shown in figure 270. As far as is known, this well deviates the greatest distance of any well ever drilled. It maintained a drift of over 70 degrees from a 3,000-foot depth to a 9,500-foot depth. With a measured depth of 12,127 feet, it deviates horizontally 9,882 feet to bottom at a 5,772-foot vertical depth. Directional drilling commenced at a 400-foot depth. The drift was increased at the average rate of 3° 00' per hundred feet drilled. It was not possible to survey the course of this well by using a single-shot run on sand line. Because of the extreme angle of the well, 15 magnetic surveys were run on drill pipe to determine that the course of the well was as originally proposed.

Work-over operations on old wells usually are accompanied by directional work of some kind. Collapsed casing or liners can be sidetracked with casing whipstocks and subsequent slant-hole drilling of the bottom to a new, more productive location.

Wells that have become unprofitable from edge-water encroachment are deflected in their lower portions upstructure to result in productive wells.

Surveys of all wells producing on a lease often show incorrect well spacing, allowing the drilling of one or more slant holes to attain the greatest yield from a lease.

It is expected that directional-drilling practices, now well established, will assist drillers of wells even more in the future than in the past. The oil industry has found that deflected wells have made it possible to develop oil resources that might otherwise never have been developed.

OIL-WELL SURVEYING
J. B. MURDOCH, JR.

Initial interest in the surveying of bore holes began long before the turn of the century in connection with deep exploratory diamond-drilling operations in South Africa. It became increasingly evident that holes started straight at the surface did not continue to be vertical as the bit penetrated steeply dipping beds of different hardness. Geologic correlations often showed improbable changes in the formations in holes drilled in close proximity at the surface. Early attempts at measuring the angle from the vertical assumed by holes were made with acid bottles. A small amount of a dilute solution of hydrofluoric acid was poured into a glass culture tube, which was lowered into the hole and allowed to remain motionless for ten to fifteen minutes. A meniscus was etched in the tube, from which an indication of the angle of the bore could be ascertained. Another method was to float a magnetic needle in fluid gelatin and allow it to remain at rest for a time. The drift (inclination from the vertical) and the direction were obtained. When the gelatin hardened, the compass was held in place and indicated magnetic north. The drift of the hole was read from the level of the gelatin.

When oil-well drillers felt the need for such survey information, the development of devices for this purpose was accelerated and resulted in a number of types and makes of well-surveying instruments now widely used.

The three basic types of instruments in general use are drift indicators, single-shot instruments, and multiple-shot instruments.

Drift indicators record the angle of inclination from the vertical (drift) of the well bore but do not indicate the direction of this inclination. Different methods are used to determine this information. Mechanical, photographic, electro-chemical and fluid-operated devices are used to record the drift in the well. These instruments give a permanent record for each run. One of the mechanical types makes a double record to assure the operator that the instrument was motionless at the time the determination was made.

Single-shot instruments record the drift and direction of the well bore at the depth at which they are run. Most successful instruments of this type are photographic. Mechanical survey instruments are being developed because of the increasing depth and consequent rise of temperature in wells. One reading or "shot," as it is generally called, is taken on a recording disc which can be preserved as a permanent record.

Multiple-shot instruments take a number of readings of the drift and direction of the bore as the machine is lowered into the well. This information usually is recorded upon photographic film. Pictures are taken at predetermined time intervals by the making and breaking of electrical contacts controlled by constant-running motors or clockwork within the instrument itself. A complete survey of any well may be made in one run, records being taken at intervals as desired. One type of multiple-shot machine is run on conductor cable so that it may be controlled from the surface. A contact on the surface-control panel regulates the photographing of the angle unit from which the reading is obtained.

The direction of drift is often obtained with a magnetic compass when the machine is used in open or uncased hole. Various methods are used to determine the direction within the casing where the compass is ineffective. Some instruments contain a gyroscope for this purpose. An accepted, long-used method is to start the machine into the well faced in a predetermined direction, after which the rotation of the drill pipe is measured as stands are added. These pipe-rotation angles are added algebraically, and this correction is applied to the survey film as it is read. This drill-pipe-orientation method has the advantage of allowing the surveyor to compare the oriented and magnetic readings when a machine with a compass is used in open hole.

The drift indicator probably is the most extensively used well-surveying instrument available today. As stated before, it records the drift of the well at the particular depth at which it is run. Many drilling con-

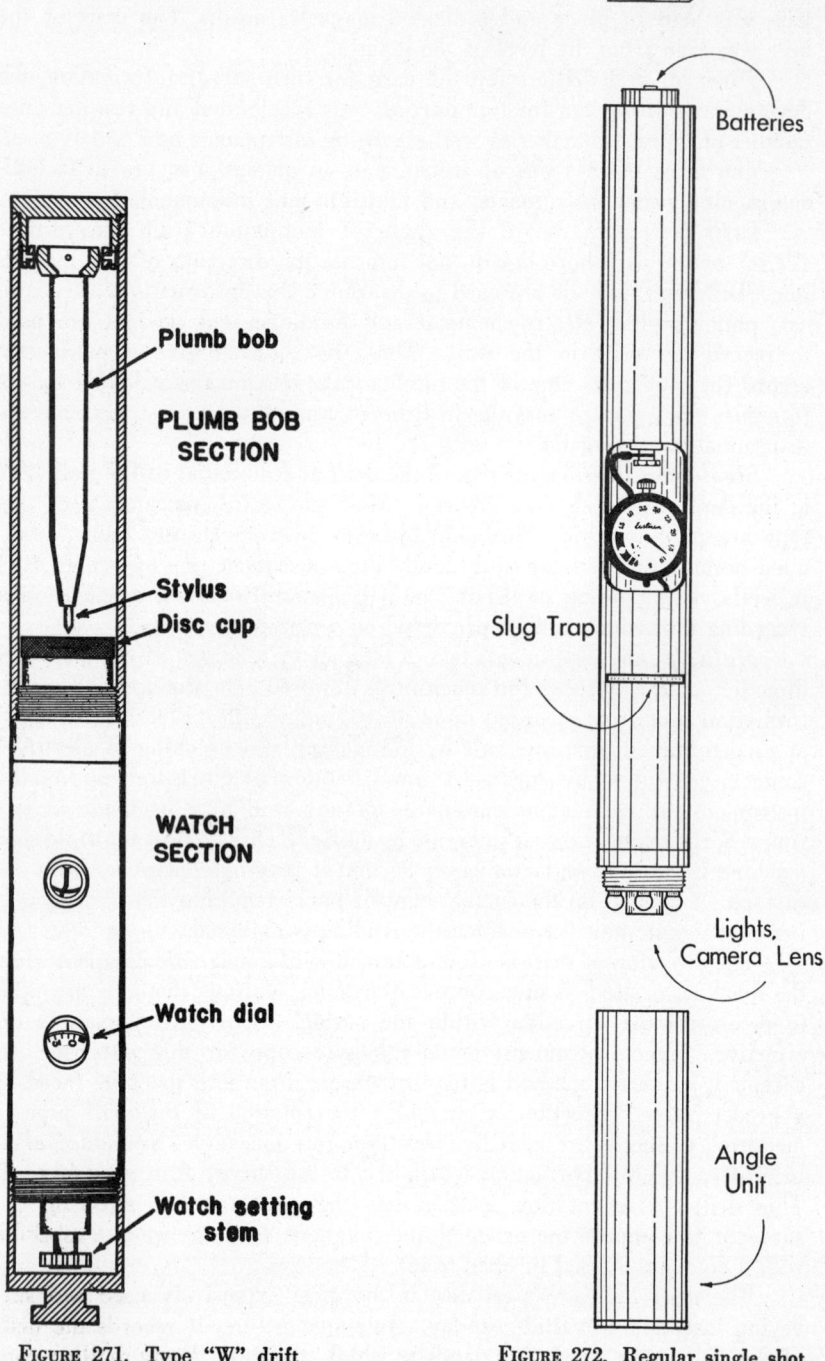

Plumb bob

PLUMB BOB
SECTION

Stylus
Disc cup

WATCH
SECTION

Watch dial

Watch setting
stem

Batteries

Slug Trap

Lights,
Camera Lens

Angle
Unit

FIGURE 271. Type "W" drift
indicator.

FIGURE 272. Regular single shot.

tracts specify that the angle of the well will at no point exceed 3 or 5 degrees of drift. This limitation helps to assure the well owner that no sharp kinks or "dog legs" will render difficult the running of casing and that mechanical difficulties will not be encountered in producing the well. On a large lease where wells are spaced far apart, some assurance of correct bottom-hole spacing is obtained. Where productive formations are not dipping steeply and hence the bottoming of the well is not especially critical, the drift indicator gives as much information as is necessary. Consistent use of the instrument at intervals of not over 100 feet warns the driller in order that the well may be straightened before it has exceeded the contract limit.

The Eastman Oil Well Survey Company manufactures two types of

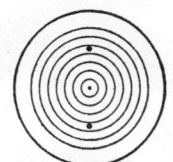

FIGURE 273. Type "W"
drift-indicator disc.

drift-indicator instruments. The most popular kind is the type "W" self-checking instrument (fig. 271). It is a mechanical machine which uses no bulbs, batteries, or sensitized discs.

A contact-timing watch is mounted in the bottom of the drift indicator. It controls the movement of the disc cup containing the record disc (fig. 273) by means of an ingenious arrangement of cams and levers. A plumb bob suspended in a universal-joint mounting houses a pointed marking stylus buffered by a spring. The plumb bob points directly downward when it is motionless.

To operate it, one unscrews the plumb-bob section from the watch section. A soft paper disc, on which concentric rings representing degrees of drift have been printed, is pressed into the disc cup. The knurled disc cup is rotated manually as far as possible in the direction of the arrow and held in this position. A contact-timing watch of special construction is both set and wound in one operation by turning the watch-setting stem. The amount of time set is observed through a window in the case. Functioning of the clockwork timing mechanism is checked by inspection through a second window. Enough time is set so that the instrument will reach the bottom of the well before the time indicates zero. A surface watch (fig. 274) is set at the same time for reference. The drift indicator is screwed together, slid into the barrel, and run into the well. When the time set has elapsed, the disc cup is lifted and turned 180 degrees in azimuth by a cam device. A tiny punch mark is made on the disc by the plumb-bob stylus. This turning action causes the plumb bob to swing violently. Forty seconds later (after the plumb bob has come to rest) the disc and cup again are raised and turned by the cam until a second punch mark is made. The disc-cup assembly retracts immediately, after which the instrument may be removed from the well. If the instrument was at rest at the bottom of the well when the records were made, the punch marks are on opposite sides of the disc (180 degrees in azimuth) and at approximately the same drift as indicated by the concentric rings on the disc. If the marks on the disc are not in this relationship, the instrument was moving when one or both of the records were

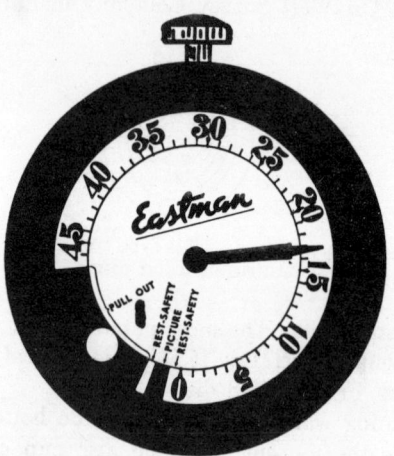

FIGURE 274. Instrument-contact watch
(surface watch identical).

made. This double-marking feature guarantees that an incorrect reading will not be accepted.

High bottom-hole temperatures do not affect the operation of this mechanical instrument or its record disc. The disc need not be protected from light since it is not printed on sensitized paper.

The instrument is dropped or run on a piano-wire line inside the drill pipe. Two types of steel barrels are used (fig. 275). Both types are similar in that they are equipped with spring-type bottom shock absorbers, and a mud tight closure is insured by use of an O-ring. The instrument is attached to a rubber shock absorber inside the barrel to cushion it. The drop or "go devil" barrel (fig. 275A) is equipped with a feather-head at the top to center the barrel in the drill pipe and to retard the fall of the barrel by friction. When the instrument is lowered into the drill pipe on a piano wire or sand line, the barrel shown in B of figure 275, equipped with a rope socket, is used. A bridger, or baffle plate, is placed above the bit before the drill pipe is run into the hole. The

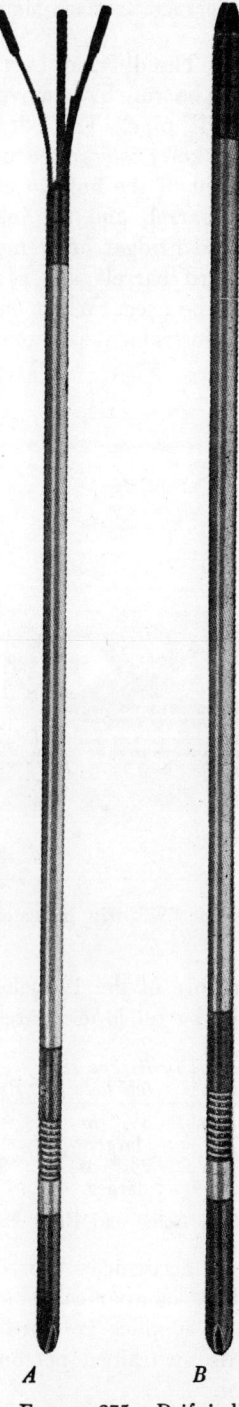

A   B

FIGURE 275. Drift-indi-
cator barrels.

barrel lands on the bridger. The different barrel tops are interchangeable so that the instrument may be run by removing the kelly and lowering the instrument into the drill pipe. The barrel is withdrawn after the reading has been taken, the kelly is screwed on, and drilling is resumed. If it is necessary to come out of the hole to change bits, the featherhead may be screwed onto the barrel, and the instrument dropped into the drill pipe. When the bit and bridger are removed, the barrel is slid out of the drill collar. Standard barrels are $1\frac{5}{8}$ inches in diameter but a smaller $1\frac{3}{8}$-inch barrel may be used. When the larger barrel is used, the seven-eighths-inch-diameter instrument is screwed into an inner barrel to

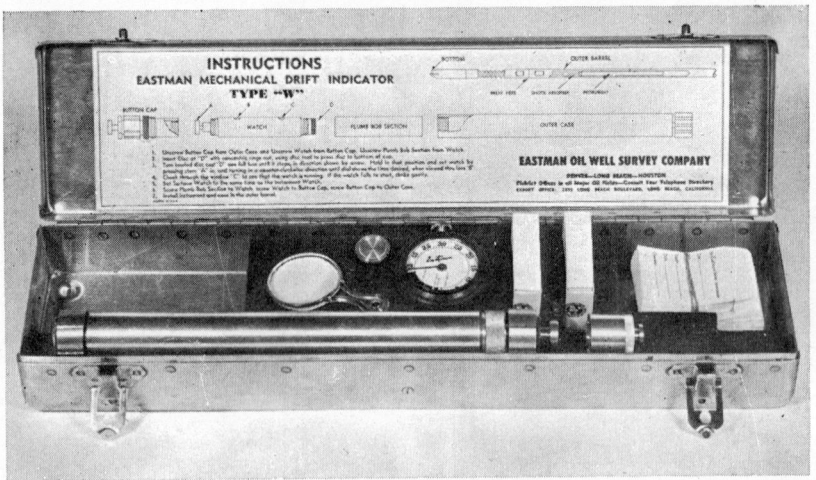

FIGURE 276. Type "W" drift indicator and supplies.

compensate for the larger bore of the $1\frac{5}{8}$-inch barrel. The smallest sizes of drill pipe in which each barrel is used are shown below.

Barrel diameter	Internal flush	Drill pipe size full hole	Regular	External flush
$1\frac{5}{8}''$	$2\frac{3}{8}''$ or larger	$3\frac{1}{2}''$ or larger	$4''$	$4''$
$1\frac{3}{8}''$	$2\frac{3}{8}''$ or larger	$2\frac{3}{8}''$ or larger	$3\frac{1}{2}''$	$3\frac{1}{2}''*$

* Except Hughes External Flush Acme and Reed External Flush Slim Hole Type.

The instrument and all accessories are rented on a daily, monthly, or term-lease basis to drilling companies. They are operated by the well crews. Kit, instrument, and supplies are furnished (fig. 276). Frequent inspection and service calls by trained personnel insure the satisfactory operation of the equipment.

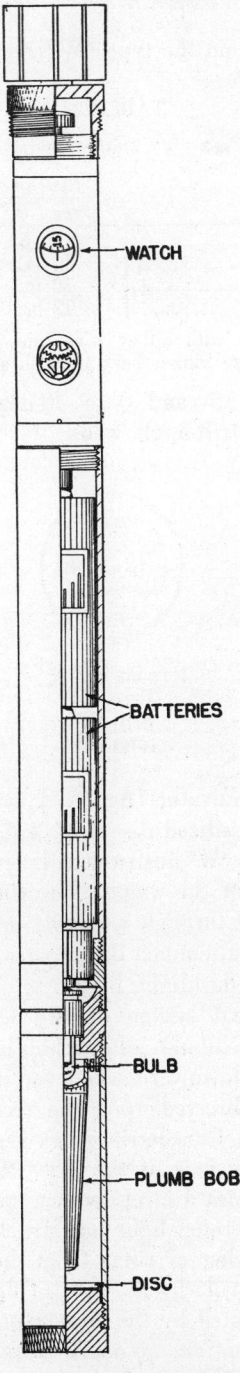

WATCH

BATTERIES

BULB

PLUMB BOB

DISC

FIGURE 277. Type "M"
drift indicator.

Table 28 shows data on the type "W" instrument and barrels.

TABLE 28

TYPE "W" DRIFT INDICATOR

Data table	Instrument type "W"	Barrels* Standard	Barrels* Small
Angle range	0 to 6°	—	—
Diameter	⅞ in.	1⅝ in.	1⅜ in.
Length	10 in.	5 ft. 10 in.	5 ft. 8 in.
Weight	12 oz.	28 lbs.	22 lbs.

* Barrels are interchangeable with either wire line, sand line, or "go devil" connections. Sinker bars are also available.

A few special 0 to 15- and 0 to 30-degree-range instruments are available for use in high-drift-angle wells.

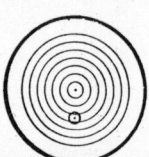

FIGURE 278.
Type "M"
drift-indi-
cator disc.

The type "M" drift indicator (fig. 277) is photographic and makes a permanent record on a sensitized disc (fig. 278). A contact-timing watch similar to that used in the "W" instrument is set and wound in one operation by turning the stem of the watch. The time setting and functioning of the watch are observed through windows in the case. When the time set has elapsed, an electrical contact illuminates the bulb in the instrument with current from small flashlight batteries. This light is concentrated into a beam by an optical system of two lenses mounted within the plumb-bob tube. Since the tubular plumb bob is suspended in a universal-type mounting, it points downward whenever it is motionless. This spot of concentrated light is directed upon the record disc, which has been centered in the disc cup. Concentric rings representing each degree of drift are printed upon the face of the disc. Sixty to ninety seconds of exposure is sufficient to print a circle with a small dot at its center upon the disc where the concentrated light has struck. The amount of drift is read by counting the number of rings from the center and interpolating between rings to the dot made by the spot of light. The acceptance of an incorrect reading is prevented by the slow exposure qualities of the disc. If the plumb bob is in motion, no mark is made upon the disc. Discs are not developed or fixed, but are paper- or foil-wrapped and kept in envelopes before and after use. They may be subjected to weak light for some time before the disc is discolored.

The type "M" drift indicator is run in the same barrels and in the same manner as was described for the type "W" instrument. Similar rental agreements and service arrangements are made.

Table 28a shows data on type "M" instruments and barrels.

A few 0 to 15-degree-range instruments are available for use in high drift-angle wells.

TABLE 28a

TYPE "M" DRIFT INDICATOR

Data table	Instrument type "M"	Barrels* Standard	Barrels* Small
Angle range	0 to 6°	—	—
Diameter	⅞ in.	1⅝ in.	1⅜"
Length	13½ in.	5 ft. 9¾ in.	5 ft. 8 in.
Weight	12 oz.	28 lbs.	22 lbs.

* Barrels are interchangeable with either wire line, or "go devil" connections. Sinker bars are also available.

Drift-indicator readings usually are taken at about 100-foot intervals as a well is drilled. By simple computations, and applying the drift to the correct segments of the hole surveyed, one can obtain a maximum possible deviation. By inspection of the different readings taken as drilling proceeds, remedial measures can be taken to assure that the well will not be kinked or "dog-legged," or that the hole does not assume an angle higher than the contract limit specified.

All single-shot instruments in general use are the photographic type, the data being recorded upon a sensitized disc by a simple camera which photographs an angle unit and compass. These instruments indicate both the drift and the direction of the bore at any point in a drilling well where the hole is open or uncased. The single-shot machine is the most convenient instrument for surveying when directional-drilling techniques are being used. Its use is essential in vertical-hole drilling near lease lines and where correct bottom-hole spacing is vital. Since, like the drift indicator, it is run as the drilling proceeds, it enables the driller to take steps to change the course of the well if it is not satisfactory. Detailed advantages of the use of a single-shot instrument will be discussed later in this section.

There are two kinds of Eastman single-shot instruments, basically similar in construction and operation. The regular or open-hole single shot is 2¼ inches in diameter while the small type or inside single shot is 1½ inches.

The Eastman regular single shot (fig. 272) contains a contact-timing watch which controls the time at which a record is taken by turning on the lights. Large flashlight cells supply the current for the light bulbs which illuminate an angle unit. A simple camera lens focuses the image of the angle unit upon a metal-edged disc of bromide photographic paper. The angle unit consists of a skeleton-type plumb bob suspended in a universal mounting with a ring carrying two cross hairs forming its base. Immediately below is a glass, on which are etched concentric rings representing degrees of drift. This glass also serves as the top closure of the fluid-filled compass chamber. Within this chamber a compass card marked in bearings is free to turn on a spring-mounted pivot needle. The compass mechanism is lubricated and buffered from extreme shock by compass fluid. A sylphon, which closes the bottom of the compass cup, permits expansion of the fluid when the instrument is used in high-temperature wells.

A loader and an unloader (fig. 279) make possible the insertion

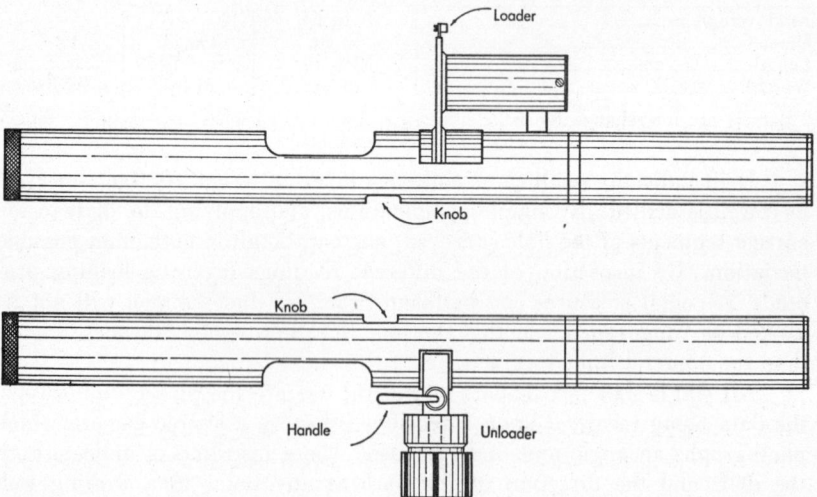

Figure 279. Loading disc in single-shot instrument (top). Unloading disc from single-shot instrument (bottom).

and removal of the sensitized disc from the machine in daylight. The loader fits snugly over and engages a slot in the machine. By pressing a knob in the instrument the slug trap is opened. The loader slide is raised and lowered to force a disc into the machine, after which the trap is closed by releasing the knob. Developing of the disc is accomplished by filling the base of the unloader with developer-fixer solution. The instrument is placed on the unloader so that the slots coincide, and the unloader handle is turned to a horizontal position. When the slug trap is opened by pressure on the knob in the instrument, the disc falls into the unloader and is immersed in the solution. After the handle on the unloader is closed, the single-shot machine is removed. Four minutes is sufficient time for developing and fixing the disc, after which it may be washed in clean water and read. An illustration of a reading from the 10-

degree angle-unit instrument is shown in figure 280. Drift is read by counting the rings from the center and interpolating when necessary. The direction bearing is obtained by prolonging a line from the center dot through the intersection of the plumb-bob cross hairs to the edge of the compass card.

The compass cards of all single-shot instruments used in California

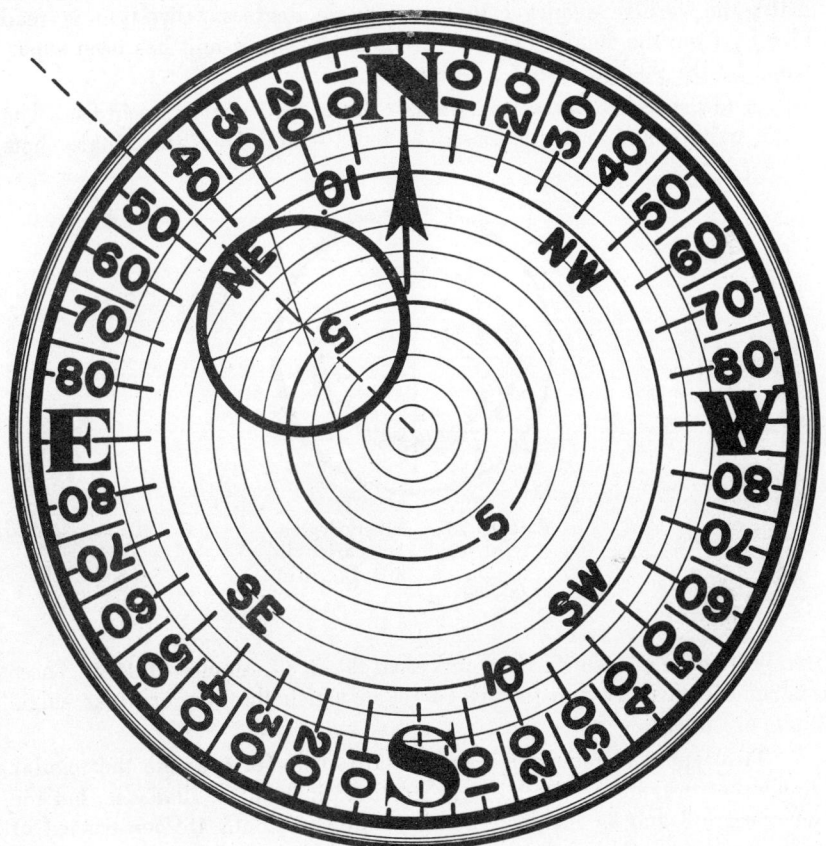

FIGURE 280. Single-shot disc, direction N. 45° E., drift 6° 00'.

are corrected for local magnetic declination. Instruments rented in other parts of the United States have compasses which indicate magnetic bearings. The true bearings are obtained from the latter readings by applying a correction for local magnetic declination.

Angle units are interchangeable in maximum drift values of 10, 18, and 90 degrees. The first two units are constructed as described above. The 90-degree unit, a disc from which is shown in figure 281, is a much more intricate mechanism completely immersed in compass fluid (fig.

282). A yoke which is weighted on one side is mounted on a vertical shaft. This shaft is free to turn on small ball bearings. Therefore, the weighted side of the yoke seeks the low side of the hole. Two vertical quadrants and a compass card are cradled in the upper portion of the yoke. This sub-assembly is weighted on the bottom and is free to move about a horizontal axis in the yoke. Wire-index lines are used to read the drift and direction. One quadrant is graduated at 10-degree intervals of drift; the vernier quadrant indicates single degrees. Direction is read directly from the compass card. The 30-degree-angle unit has been superseded by the 90-degree unit in most cases.

A few special 5-degree maximum-reading instruments are in use. The angle unit is similar to that used in the 10- and 18-degree single shots

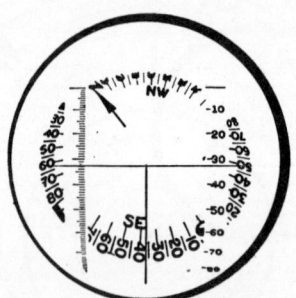

FIGURE 281. Ninety-degree-type single-shot disc—direction S. 36° E., drift 33° 00′.

except that the concentric rings represent 20 or 30 minutes of drift. These instruments may be run in any barrel in which the other regular single shots are used.

Two types of nonmagnetic barrels are made for use with the regular single-shot instrument. Both are bored to 2¼-inch inside diameter, but the barrels are 3 and 3½ inches in outside diameter. Both are constructed of nonmagnetic duralumin and bronze and have a bull plug using an O-ring-type pressure closure. A shock-absorber plunger contained in the bull plug is spring and rubber buffered to protect the instrument from severe vertical shock.

The 3½-inch barrel (fig. 283) is used for running the instrument in open hole on sand line after the drill pipe has been removed from the well. It is equipped with a tubing collar at the top so that it may be made up on two joints of tubing. The tubing is attached to the sand line with a rope socket or bailer top. A tool-joint connection is provided on the top of the barrel if it is necessary to run the instrument on drill pipe. A rubber mud shield slipped over the barrel keeps the single shot clean as it is unloaded from the barrel.

The three-inch barrel is provided with top connections to fit large diameter wire-line coring equipment when the instrument is run down inside the drill pipe and out through the bit. (See description of the small-type single shot below.) It is similar to the larger barrel.

The regular single shot generally is run after the drill pipe has been

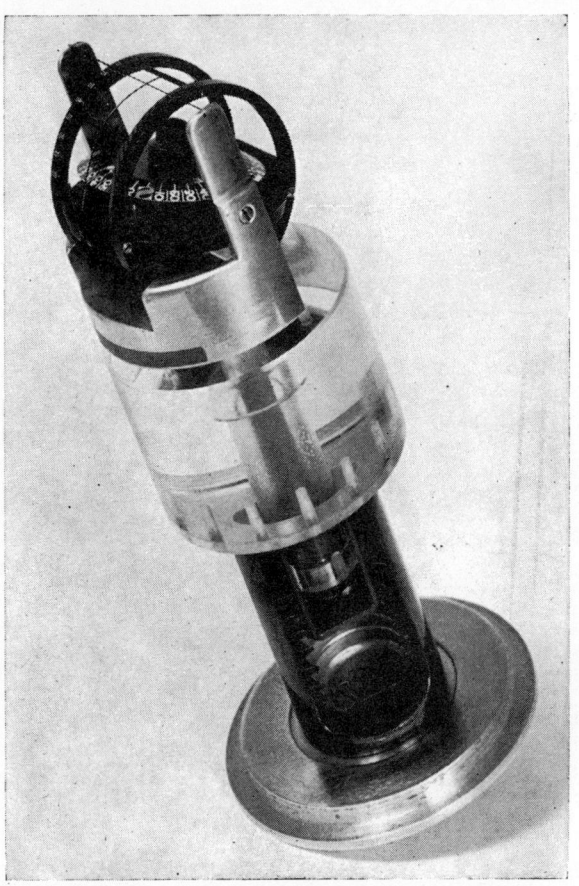

FIGURE 282. Cutaway model of 90-degree-angle unit.

removed from the well (fig. 286). The angle unit is unscrewed from the instrument and the lights are checked to insure that they are operating. After the angle unit is replaced, a disk is loaded into the machine with the loader, and the contact-timing watch is set for the amount of time it will take to run the instrument to the required depth in the well. The surface watch (fig. 274) is set at exactly the same time for reference. The

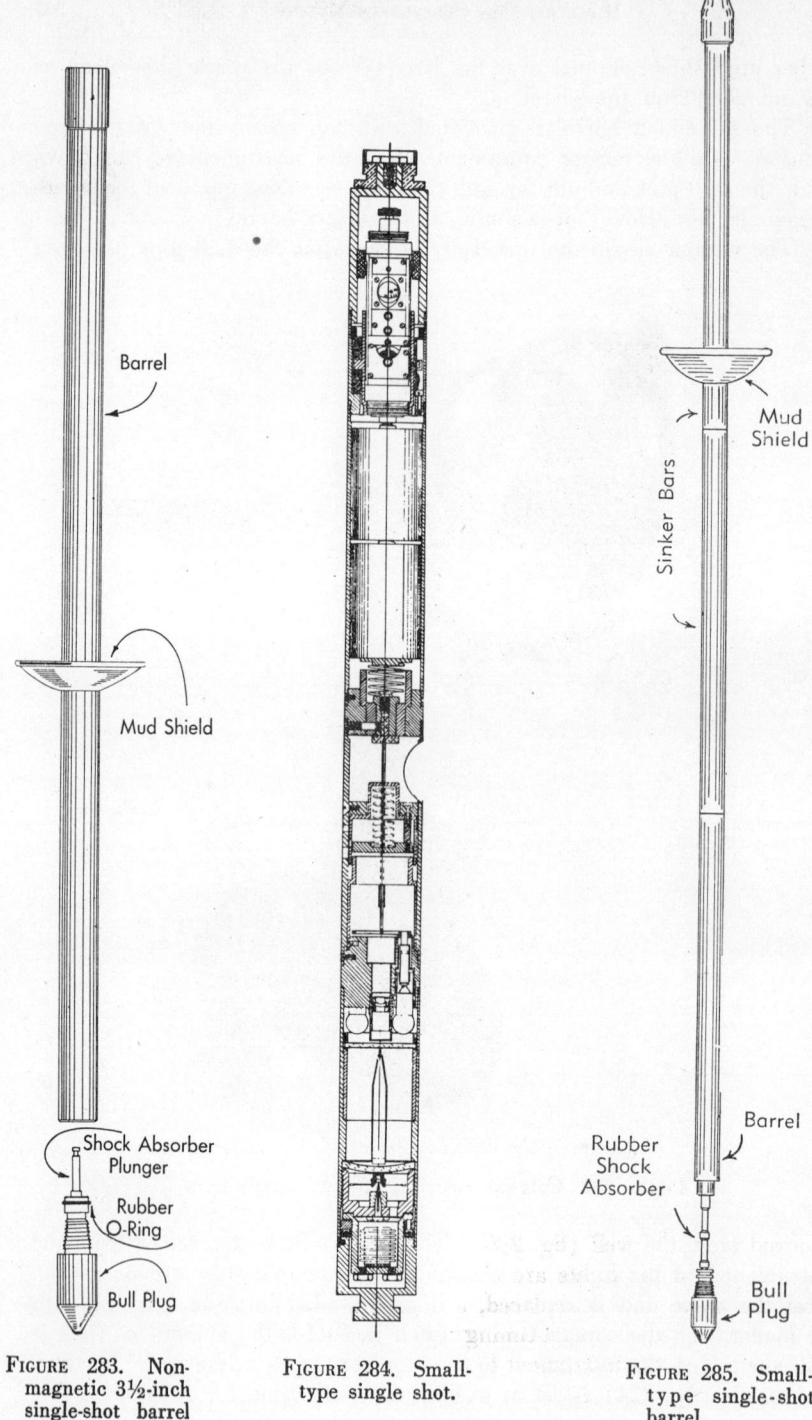

Barrel

Mud Shield

Shock Absorber
Plunger

Rubber
O-Ring

Bull Plug

FIGURE 283. Non-
magnetic 3½-inch
single-shot barrel
assembly.

FIGURE 284. Small-
type single shot.

Sinker Bars

Mud
Shield

Rubber
Shock
Absorber

Barrel

Bull
Plug

FIGURE 285. Small-
type single-shot
barrel.

instrument is loaded into the barrel with the bull-plug shock-absorber plunger engaging the bottom of the machine. The bull plug is made up tight. Then the assembly is lowered into the well and held motionless at the correct depth. When the surface watch indicates that the picture has

Figure 286. Preparing to run regular single shot in open hole on sand line.

been taken (approximately one-minute exposure) the equipment is hoisted to the surface. The instrument is removed from the barrel, and the film is developed and fixed in the unloader. The reading of the disc indicates the drift and direction of the well at the depth at which the picture was taken.

The single shot, barrel, and accessories are rented to operators on a daily or monthly rental basis. All supplies are furnished and service calls are made at the well to check and adjust the instrument (fig. 287). Most drilling crews are well acquainted with the operation of this instrument and can use it whenever necessary.

Table 29 shows data on the regular single shot and barrels.

The Eastman small-type single shot (fig. 284) is similar in construction and operation to the regular machine except that it is 1½ inches in

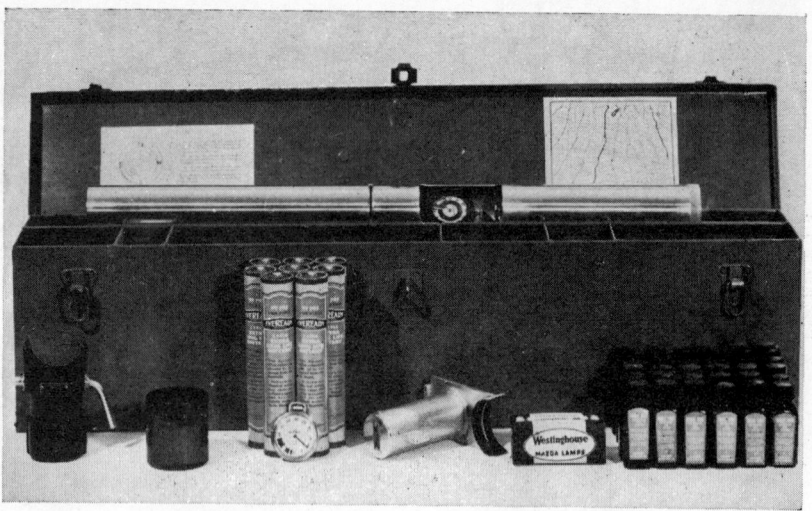

FIGURE 287. Regular single shot and supplies.

TABLE 29

REGULAR TYPE SINGLE SHOT

	Diameter	Length	Weight (lbs.)
Instrument (5° to 30° units)	2¼ in. O.D.	26 in.	5
Instrument (90° unit)	2¼ in. O.D.	29¾ in.	6
Open hole barrel*	3¼ in. O.D.	7 ft.–8 ft.	112
Inside Barrel**	3 in. O.D.	9 ft.–17 ft.	145
Kit box and instrument	7 in. x 7 in.	30¾ in.	29½

* Collar threaded 2½ in. A.P.I. external upset 10-thread tubing.

** Inside type barrel available with wire line coring equipment connections as required. diameter instead of 2¼ inches. The contact timing watch is a duplicate of that used on the type "M" drift indicator. Except for size, angle units are identical in range and construction with those for the regular single shot. No 5-degree, small, single-shot angle units are made. The disc is smaller, as are the loader and unloader. This surveying instrument was developed for a special purpose. With the increasing use of retractable wire-line coring equipment, a demand was created for an instrument that could be run inside the drill pipe and allowed to protrude from the end of the drilling bit into open hole to obtain a reading. Originally the small single shot was developed to meet this need. Drillers realize a considerable saving in drilling time by making an accurate single-shot survey through the bit without the necessity of removing the drill pipe from the hole.

The $2\frac{1}{8}$-inch bronze instrument barrel illustrated in figure 285 is similar to the open-hole barrel used for the regular single shot except that sinker bars are added as needed for weight. A rubber shock absorber has been substituted for the shock-absorber plunger in the large barrel. The upper portion is made with a landing shoe which seats in the bit when the instrument is in the correct position for taking a reading. In taking a picture the retractable core barrel is retrieved from the bit with an overshot run down through the drill pipe on a wire line. The drill pipe is hoisted about 10 to 20 feet off bottom. A spearhead on top of the small single-shot barrel is engaged by the overshot. Then the instrument and barrel are lowered down the drill pipe until the barrel landing shoe fits into the bit. Various lengths of nonmagnetic sinker bars are used (according to the drift of the well) to position the instrument-angle unit from 6 to 17 feet beyond the magnetic influence of the bit and drill pipe. After the picture has been taken, the barrel is hoisted out of the drill pipe, the core barrel is dropped back down the pipe, and ordinary drilling is resumed. Figure 288 shows the instrument barrel in position in a drilling bit.

The development of special surveying bits, among them the "trigger bit," has increased the use of small single shots. These types of bits, for digging both soft and hard formations, are constructed with a hole of sufficient size to accommodate the $2\frac{1}{8}$-inch barrel. A spring-actuated trigger mounted in the bottom of the bit breaks the core when the bit is drilling. When a survey reading is taken, the drill pipe is hoisted off bottom so that the single-shot barrel may protrude beyond the bit. A reading is taken with the small single shot by lowering the barrel on wire line through the drill pipe. The trigger, which is forced back by the barrel, resumes its normal position when the survey barrel is withdrawn. Figure 290 shows one type of trigger bit.

Non-magnetic drill collars used in the drilling string immediately above the bit have reduced the cost and delay in surveying wells. K-Monel collars from 10 to 23 feet long are used, according to the drift angle of the well being drilled. The survey barrel is lowered through the drill pipe on wire line and comes to rest on a landing ring in the top of the collar. Some barrels are made with nonmagnetic spacer bars below the bull plug, and are a length that positions the instrument correctly in the drill collar. By either method the barrel is always positioned so that the instrument-angle unit is in the center of the length of the nonmagnetic collar. Below 18- or 20-degree drift the magnetic influence of the bit and drill pipe does not affect the magnetic compass in the single shot. It has

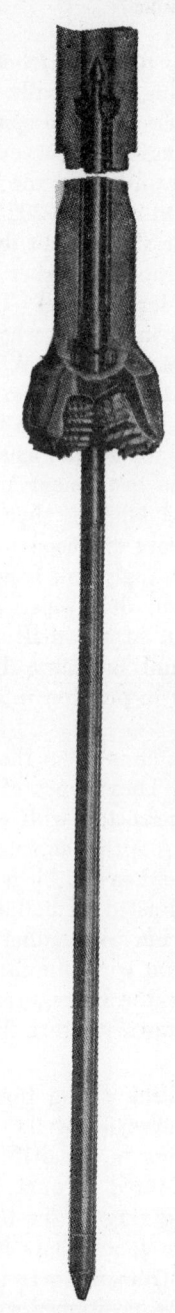

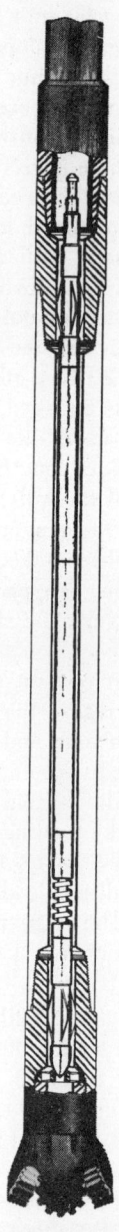

FIGURE 288. Small
single shot bar-
rel in position in
retractable coring
bit.

FIGURE 289. Single
shot positioned in
nonmagnetic drill
collar.

been found that corrections above 20-degree drift must be made to the bearings as the magnetic influence becomes apparent. A survey barrel stopped in position on the landing ring in the top of a nonmagnetic drill collar is illustrated in figure 289.

A drop or "go devil" type of barrel has been developed recently for

FIGURE 290. Trigger-type surveying bit.

use with nonmagnetic drill collars. This barrel, equipped with exterior spring and rubber shock absorbers and with the instrument suspended from a rubber shock absorber, is being used successfully. A spearhead at the top permits it to be retrieved from the drill pipe with an overshot run on piano-wire line. If the drill pipe is to be removed from the hole, the barrel is dropped and later recovered when the bit is unscrewed from the drill collar, as is done with a drift indicator. Figure 289 illustrates a "go devil" barrel in a K-Monel drill collar.

At times a small single shot is used for surveying in open-hole run on a sand line in the same manner as the regular single shot. In small-diameter holes, such as the "rathole" made in drilling off a whipstock, some operators consider it safer to use the small instrument. In exceptionally hot wells an inside single shot is sometimes run in a regular (3½-inch) single-shot barrel. Spacer rings position the smaller instrument in the barrel. The air in the annular space between the single shot and the barrel acts as an insulator. Thus, pictures have been taken when otherwise it would have been impossible. Discs are similar to those used by the regular single shot except that they are 1⅛ inches in diameter.

The small single shot is a rental machine serviced at regular intervals and run by the drilling crews themselves.

Table 30 gives pertinent information on the small instrument, barrel, and accessories.

Most operators take single-shot readings at depth intervals not greater than 100 feet. At critical points in drilling, in checking deflecting-tool runs, and whenever the situation requires, pictures are taken at shorter distances. After reading the discs for drift and direction, computations

TABLE 30

Small Type Single Shot

	Diameter	Length	Weight (lbs.)
Instrument (12° and 20° units)	1½ in. O.D.	23 in.	3¼
Instrument (90° unit)	1½ in. O.D.	25¼ in.	4
Survey barrel	2⅛ in. O.D.	8 ft.–17 ft.	80–170
Kit and supplies	7 in. x 7 in.	30¾ in.	28

are made for the vertical depth and deviation of each course. Rectangular coordinates for each depth at which a reading was taken are derived by use of the directions read from the discs. After vertical depths for each measured depth have been obtained, a plan and vertical sections of the course of the well may be plotted on cross-section paper. The advantage of a single-shot survey is that the foregoing data can be obtained quickly. If the course of the well is not correct, measures can be taken immediately to deflect or straighten the well. The position of the well as it is being drilled is information essential to the conduct of controlled directional-drilling operations. Plans for future action affecting the course of the well can be made from this information. Since "dog legs" in a well are usually a combination of both change in drift and change in direction, the amount of "dog leg" can be ascertained quickly when the single-shot readings are available.

A type "A" single-shot instrument is in process of manufacture at the present time. It is a further development in which the most desirable

EXPANSION HEAD

BATTERIES

MOTOR

FILM BOX

CAMERA LENS

LIGHTS

FIGURE 291. Eastman DX-type multishot machine.

features of both the regular and small machines have been preserved. Complete redesign indicated by field experience has been done on all parts of the instrument. Through this redesign a machine will be produced which will be easier and cheaper to manufacture and maintain. The instrument will contain fewer and simpler parts which to the operator means longer and more trouble-free operation.

The type "A" single-shot instrument will be $1\frac{1}{2}$ inches in diameter, but will use a disc $1\frac{5}{16}$ inches in size. This disc should be easier to read.

A telltale device on the slug-trap operating lever will indicate whether a disc is loaded in the machine or not. A larger sylphon and a more sensative shock-mounted compass needle in the angle unit will better provide for high-temperature surveys and make the readings more accurate. Newly designed loaders and unloaders are being manufactured for this modern instrument.

It is probable that in time the type "A" instrument will displace both the regular and small single-shot machines now in use.

Multiple-shot surveying instruments, as the name implies, are capable of making a number of readings in one run. They are used principally to survey wells that were drilled and put on production before the importance of surveying was generally accepted, to learn the course of a well after casing has been set, or to locate the course of a well on which a drift indicator has been used during drilling. Multiple-shot machines generally effecting a saving in rig time over the single-shot method by surveying both cased and uncased wells of any depth drilled to date in a single run. However, the information usually is obtained too late to change the course of the well without plugging and sidetracking the original hole. Photographic means for recording data are used in modern multiple-shot machines.

The Eastman DX-type multiple-shot machine is illustrated in figure 291. A small motor, powered by batteries in the machine and controlled by a main switch, runs throughout the survey. This motor turns a gear train, which alternately lights the bulbs for an exposure interval of six seconds, and moves the special 16-mm. recording film ahead. Film is metered from the new film spool to the takeup spool over a sprocket wheel. The lights illuminate an angle unit almost identical with that used on the regular single shot. A camera lens focuses the image of the angle unit on the film. Simultaneously, the face of a timing clock is photographed on the film as a reference. Enough heat-resistant film is spooled in the film box to enable the machine to take records for 17 hours at 90-second intervals. The angle unit has an off-center, tapered groove cut in its bottom so that it will fit into the barrel in exactly the same position each time. An orienting lug, exactly parallel to the groove, is located in the angle unit where it is photographed. The upper end of the machine is fitted with an expansion head so that the instrument may be tightly clamped in the barrel.

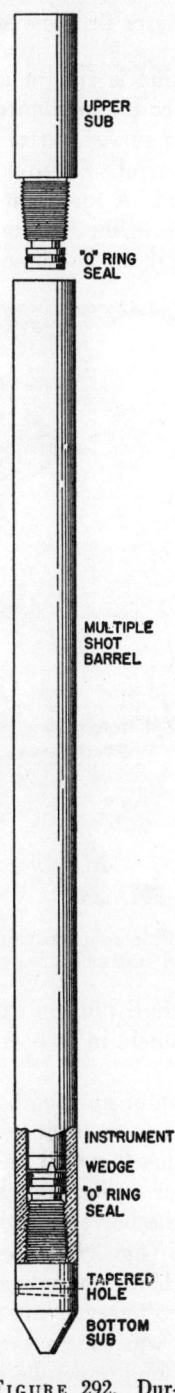

FIGURE 292. Dur-
aluminum mul-
tiple-shot barrel
(3½-inch diame-
ter) for DX in-
strument.

The multiple-shot machine is run in a protective nonmagnetic case of bronze or duralumin. Since the instrument is $2\frac{1}{4}$ inches in diameter, as is the regular single shot, the survey barrel is $3\frac{1}{2}$ inches. The machine is loaded into the top of the barrel. An upper substitute closes the barrel mud-tight with an O-ring seal. A tool-joint box is cut in the top of this upper substitute. The groove in the bottom of the angle unit fits over a corresponding tapered wedge in the bottom of the survey barrel. A tapered

FIGURE 293. Starting multiple-shot machine; synchronizing surface and instrument watches.

hole is bored in the bottom bull plug in exact relationship to the wedge. The multiple-shot machine is held in its barrel by tightening the expansion head (fig. 292).

Magnetic surveys in open or uncased hole are run on drill pipe, sand line, or conductor cable. The instrument, loaded with film and with fresh batteries, is started. The watch that will be photographed in the machine is synchronized with a similar surface watch (fig. 293). After the instrument has been tightened in the barrel and the closure made, the assembly is made up on the drill pipe (fig. 294). As the pipe is run into the well stand by stand, the pipe is held motionless for two minutes whenever a record is desired. These time intervals are noted on a field sheet so that the correct pictures are read when the survey is developed. When a magnetic survey is run on sand line in open hole, two joints of tubing usually are used as a sinker bar. The machine is run to bottom and readings are taken (usually at 100-foot intervals) as the line is being drawn out of the well. Conductor-cable runs are made with a survey instrument similar to that described, except that neither batteries nor an instrument watch is used. When a reading is required, a button that is pushed on the surface-control panel puts the machine in operation for one cycle, takes a photograph of the angle unit, and winds the film (fig. 295). This method of running a magnetic survey is very rapid.

FIGURE 294. Making up multiple-shot barrel on drill pipe preparatory to running survey.

Oriented surveys are made in a cased hole, where lost tools have been forced into the wall of the hole, or in sidetracking operations near casing. The magnetic compass in the instrument-angle unit will not give correct direction readings because of the magnetic properties of the casing. Oriented surveys are run on drill pipe or tubing. A few surveys have been oriented on sucker rods inside two-inch upset tubing by a special survey machine with fairly successful results.

When an oriented survey is made, the rotation of the drill pipe as it is run into the well is measured and recorded for each stand. Measurement is made during the time in which the pipe is held motionless when the instrument is taking a picture. The instrument and barrel are faced initially in a known direction by a sighting device that fits into the tapered hole bored in the bottom bull plug of the survey barrel. A two-man crew with drill-pipe clamps, a derrick telescope, and a special surveyor's transit measures the amount and direction of the turn of each length of pipe as it is run into the well stand by stand. The derrick man sights through the telescope and aligns the clamp upon a distant target (fig. 296). The floor man measures the amount the length of pipe has turned in relation to the target by use of the transit (fig. 297). These pipe-rotation angles are added algebraically to the azimuth of the direction in which the instru-

FIGURE 295. Running magnetic multiple-shot survey with conductor cable on electronic truck.

ment was originally faced. Thus the direction of the orienting lug photographed in the angle unit is ascertained for each reading. The equipment used for orienting a survey is shown in figure 298.

Combination surveys are made when a hole is partly cased. The drill pipe is oriented to obtain directions in the casing, and magnetic bearings are read from the compass when the machine is in open hole below the casing.

After the survey has been taken, the film is developed, fixed, and dried (fig. 299). It is placed in a special projector (fig. 300), and an enlarged image of each picture ten inches in diameter is projected on a special protractor. The protractor is constructed so that the compass card can be rotated in relation to the base, and an orienting lug similar to that shown on the film is mounted on the protractor base. In reading the film the time interval noted on the field sheet at which the instrument was held motionless in the well indicates the correct picture to be projected. Magnetic surveys are read in the same way that a group of single-shot pictures would be solved for drift and direction. An oriented survey is read by aligning the orienting lug photographed in the angle unit with the lug on the protractor. The compass card of the protractor is turned

FIGURE 296. Derrick man aligning drill-pipe clamp and telescope upon distant target.

until the correct azimuth resulting from summing the pipe-rotation angles for that particular reading is directly under the lug. Then the direction of the well is determined by superimposing a line from the center of the picture through the intersection of the plumb-bob cross hairs to the edge of the compass on the protractor. Drift determinations are made by counting the concentric rings from the center of the picture and interpolating between rings.

Drift and direction readings are derived in the same manner for each point at which the drill pipe was allowed to remain motionless during the running of the field survey. The multiple-shot instrument exposes the film every 90 seconds. As the drill pipe is being run into the well, many pictures are taken while the plumb bob and compass in the angle unit are moving. These pictures are blurred and are not read since they are not points on the survey.

The same data that were available on a single-shot survey have been derived from the survey film. Single-shot and multiple-shot survey computations are similar. A plan and vertical sections are drawn of the course of the well and made into a comprehensive report which is sent to the operator. Data included in a typical report of this kind are shown in figures 301, 302, and 303. The report consists of a title sheet giving the

FIGURE 297. Orienting transit and drill-pipe clamp. Engineer is preparing to read pipe-rotation angle.

company name, the well name, the location of the well. and the type of survey made. Record of survey sheets shows the measured and vertical depths and rectangular coordinates of each point at which a reading was taken on the subsurface survey. A horizontal plan depicts the course of the well in a horizontal plane. Vertical sections in two planes show the course of the well as if it were viewed from two sides. These drawings and data sheets, all made on tracing paper or cloth so that they may be reproduced, are bound in a survey folder and thus may be filed easily. Affidavits are furnished if requested.

FIGURE 298. Equipment used in orienting survey.

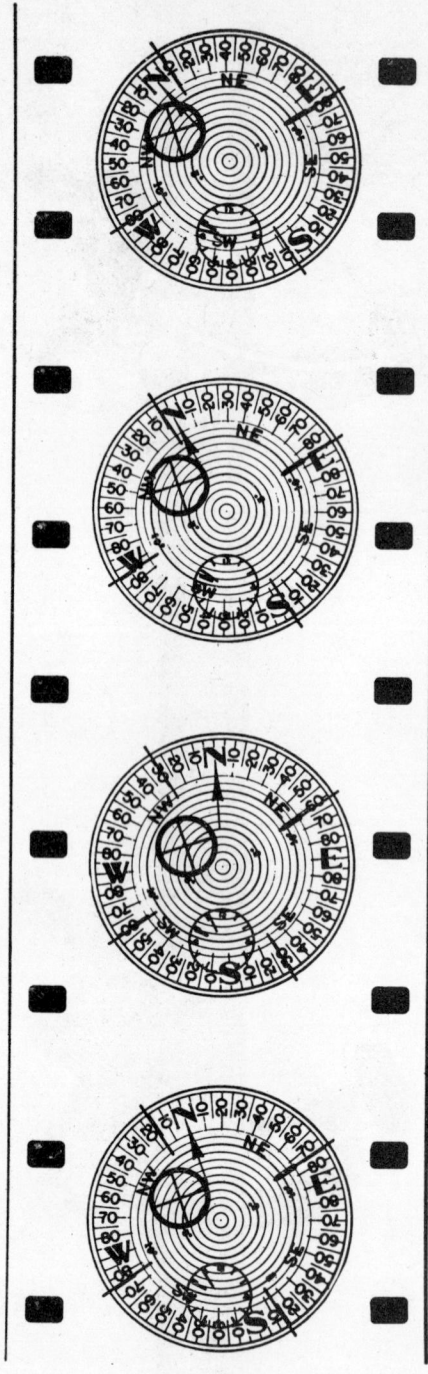

FIGURE 299. Enlarged section of multiple-shot film.

Multiple-shot-angle units are similar to those used on the regular single shot except that a tapered groove is cut in the bottom of the unit in exact alignment with an orienting lug which is photographed in each picture taken by the instrument. Angle units are constructed in maximum drift values of 10 degrees, 18 degrees and 90 degrees. A few units are built as combinations. An 18-degree plumb bob is mounted over a 90-degree angle unit in such a way that both are photographed at once. In this way a well which starts vertically and increases drift to 90 degrees can be surveyed accurately.

FIGURE 300. Projector and protractor used for reading multiple-shot film.

At the present time a special, multiple-shot machine is being developed (fig. 304). It is designed to save operators rig time in the surveying of their wells. The instrument is $1\frac{1}{4}$ inches in diameter and has made test runs in a $1\frac{3}{4}$-inch diameter bronze barrel. It photographs an angle unit similar to that used with the small single shot except that the plumb bob as well as the compass is housed in a fluid-filled chamber. The control mechanism is a sturdy clock which makes a light contact for illuminating and photographing the angle unit. Also it starts and stops an intermittent-running motor which winds the film. Various operation cycles may be used by changing contact wheels to accommodate the speed of the

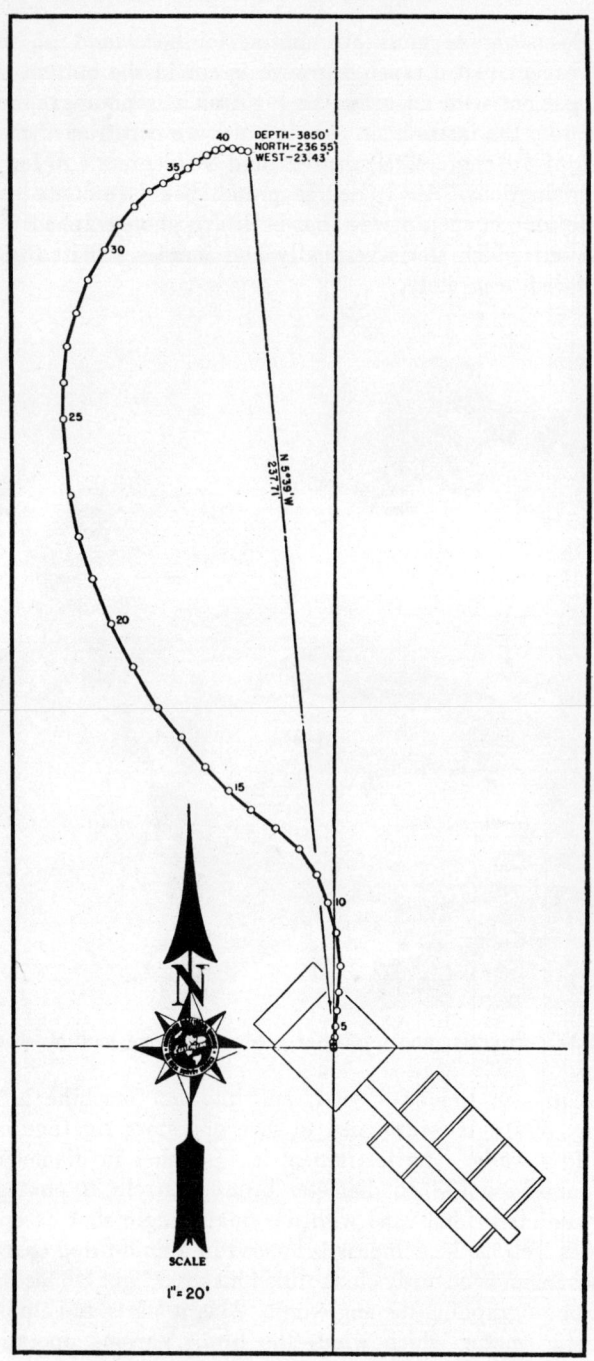

DEPTH—3850'
NORTH—236.55'
WEST—23.43'

40

35

30

25

20

15

10

5

N 5°39'W
237.71'

N

SCALE

1" = 20'

FIGURE 301. Plan view of oil-well survey.

RECORD OF SURVEY

JOB NO. W-7647   DATE July 23, 1947

	MEASURED DEPTH	DRIFT ANGLE	TRUE VERTICAL DEPTH	COURSE DEVIATION	DRIFT DIRECTION	RECTANGULAR COORDINATES				REMARKS
						NORTH	SOUTH	EAST	WEST	
1	70 40	0°10'	70 40	20	N 14°00' W	19			05	
2	160 48	0°15'	160 48	40	N 2°00' W	59			06	
3	249 93	0°30'	249 98	78	N 2°00' E	1 37			03	
4	339 73	0°55'	339 72	1 44	N 5°00' E	2 80		10		
5	430 89	1°45'	430 83	2 78	N 7°00' E	5 56		44		
6	523 31	2°35'	523 16	4 17	N 6°30' E	9 70		1 45		
7	613 06	3°20'	612 76	5 21	N 6°00' E	14 88		1 91		
8	703 06	4°10'	702 53	6 54	N 4°00' E	21 40		60		
9	795 14	5°30'	794 19	8 82	N 8°00' W	30 13				
10	886 39	5°15'	885 06	8 35	N 15°15' W	38 19			1 52	
11	976 39	5°00'	974 72	7 85	N 23°30' W	45 39			4 65	
12	1066 06	5°15'	1064 01	8 20	N 34°00' W	52 19			9 24	
13	1156 39	5°30'	1153 92	8 65	N 49°15' W	57 84			15 79	
14	1247 39	5°10'	1244 55	8 20	N 52°45' W	62 60			22 32	
15	1336 31	5°00'	1333 13	7 75	N 50°00' W	67 78			28 26	
16	1425 81	5°50'	1422 16	9 09	N 47°00' W	73 98			34 91	
17	1515 81	6°15'	1511 63	9 80	N 39°30' W	81 54			41 14	
18	1606 23	7°00'	1601 37	11 02	N 37°00' W	90 34			47 77	
19	1695 90	7°35'	1690 26	11 94	N 34°00' W	100 16			54 39	
20	1785 15	8°15'	1778 59	12 81	N 27°45' W	111 50			60 35	
21	1876 57	8°00'	1869 13	12 73	N 23°15' W	123 20			65 37	
22	1966 65	7°30'	1958 43	11 76	N 18°00' W	134 38			69 00	
23	2056 40	7°05'	2047 50	11 07	N 11°45' W	145 22			71 25	
24	2146 32	6°50'	2136 70	10 70	N 8°00' W	155 82			72 74	
25	2236 74	6°15'	2226 67	9 85	N 2°45' W	165 66			73 21	

FORM D-1113 COLD PHOTO-LITHO

RECORD OF SURVEY

JOB NO. W-7647   DATE July 23, 1947

	MEASURED DEPTH	DRIFT ANGLE	TRUE VERTICAL DEPTH	COURSE DEVIATION	DRIFT DIRECTION	RECTANGULAR COORDINATES				REMARKS
						NORTH	SOUTH	EAST	WEST	
26	2326 90	6°00'	2316 33	9 42	NORTH	175 09			73 21	
27	2418 15	6°05'	2407 07	9 67	N 5°15' E	184 71			72 33	
28	2508 90	5°50'	2497 35	9 22	N 15°00' E	193 62			69 94	
29	2599 57	5°55'	2586 54	9 24	N 20°15' E	202 29			66 74	
30	2688 57	5°40'	2676 10	8 88	N 29°00' E	210 06			62 43	
31	2778 41	5°05'	2765 59	7 96	N 31°45' E	216 83			58 24	
32	2868 16	4°00'	2855 12	6 26	N 40°00' E	221 63			54 22	
33	2958 49	3°40'	2945 27	5 78	N 45°30' E	225 68			50 10	
34	3048 57	2°45'	3035 24	4 32	N 60°00' E	227 84			46 36	
35	3137 73	2°35'	3124 31	4 02	N 58°15' E	229 96			42 94	
36	3227 81	2°15'	3214 32	3 54	N 55°00' E	231 99			40 04	
37	3319 23	1°55'	3305 69	3 05	N 57°00' E	233 65			37 48	
38	3409 56	2°00'	3395 97	3 15	N 58°15' E	235 31			34 40	
39	3500 56	1°45'	3486 92	2 78	N 62°30' E	236 59			32 33	
40	3590 23	1°25'	3576 56	2 21	N 75°00' E	237 16			30 20	
41	3680 56	1°30'	3666 86	2 37	EAST	237 16			27 83	
42	3770 81	1°35'	3757 07	2 49	S 84°30' E	236 92			25 35	
43	3850 00	1°25'	3836 23	1 96	S 79°00' E	236 55			23 43	
			CLOSURE:	237.71'	N 5°39' W					

FORM D-304 COLD PHOTO-LITHO

FIGURE 302. Tabulated survey data used in making oil-well survey.

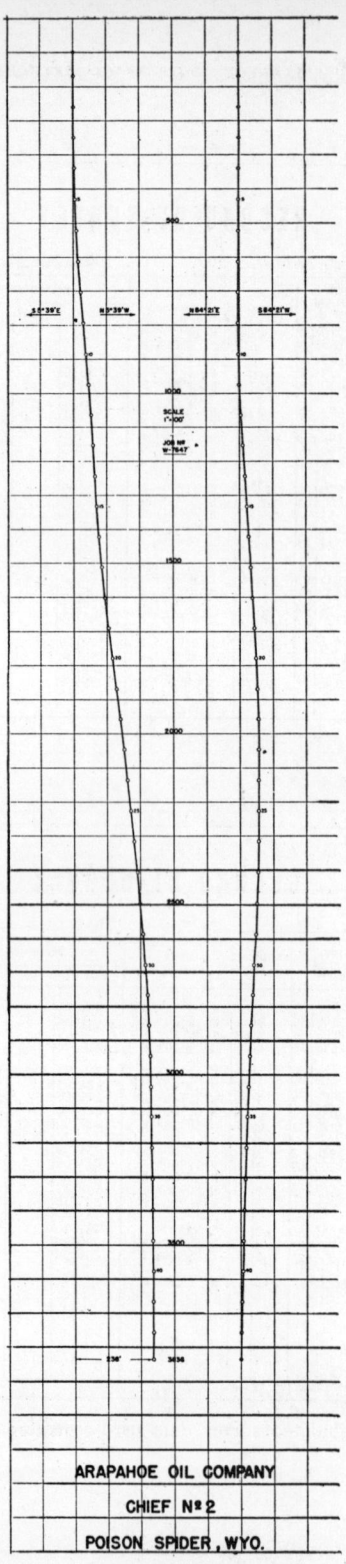

FIGURE 303. Vertical sections of
oil-well survey.

instrument to the well on which it is run. No instrument watch is photographed by the lights which are supplied current from flashlight batteries. An exceptionally accurate, synchronized surface clock is used to number the pictures taken. The instrument is mounted in its barrel between springs and rubber shock absorbers to protect it from vertical shock.

In use, the machine is started in exact synchronization with the surface clock and placed in its barrel. The barrel is lowered on a wire line or dropped into the drill pipe when the driller is ready to take the pipe from the hole to change bits. The instrument barrel is of correct length

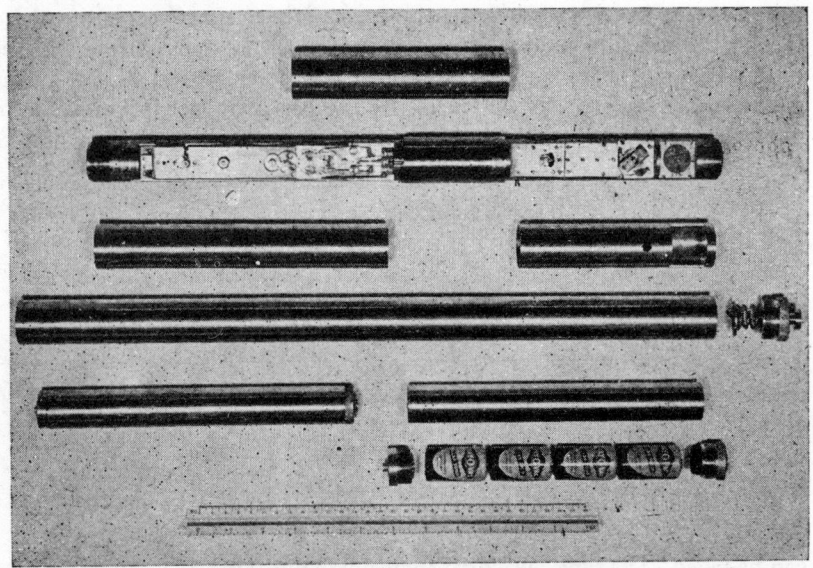

FIGURE 304. Experimental 1¼-inch multiple-shot machine disassembled.

to position the compass in the angle unit in the center of a length of K-Monel drill collar, which is in the drilling string immediately above the bit. After the survey equipment has reached bottom, the pipe is withdrawn from the well. A magnetic survey is made by taking a reading as each stand of drill pipe is unscrewed and set back. Usually an extra minute of time is taken for each stand over and above the normal pull-out time. No special round trip of the drill pipe is made in order to survey the well.

This same machine was used to make an oriented survey on sucker rods inside of two-inch upset tubing in a well in which the tubing was stuck.

It is contemplated that further development will produce an instrument by which an oriented survey can be made as the pipe is withdrawn from the well in much the same manner as described above for running magnetic surveys. Perhaps the equipment will be perfected to be lowered out through a "trigger" type bit in those cases where the operator is not using a nonmagnetic drill collar.

Multiple-shot machines are not rented to operators. Surveys with them are made by service company employees who are especially trained to perform this type of engineering work. Charges vary with the amount of hole surveyed and the type of survey run.

Table 31 gives data on DX-type multiple-shot machines.

Throughout the detailed discussion of well-surveying instruments above it should be noted that these machines have been designed to indicate an erroneous reading when it is taken. The drift and direction indicating means in each of these machines are free to move at all times; no locking devices are used. The instrument must be motionless, and the plumb bob and compass must be allowed to come to rest before a true and accurate reading can be taken. By the use of long exposure time in photographic machines, movement is detected readily and another picture

TABLE 31

MULTIPLE SHOT SURVEY INSTRUMENT

	Standard multiple shot		
	Instrument assembled with 10° unit	Instrument with 90° combination unit	Barrel Standard bronze
O. D. instrument	2¼ in.	2¼ in.	....
Length instrument	46 7/16 in.	50 3/16 in.	....
Weight instrument	22 7/16 lb.	22¾ lb.	....
Max. O. D. barrel	....	....	3½ in.
Length barrel	....	....	6 ft. 7 in.
Weight barrel	....	....	87 lbs.
Connection up	....	....	2⅞ in. A.P.I.
Type connection	....	....	Box

can be taken. The type "W" drift indicator, by making two determinations forty seconds apart, assures a correct reading.

The importance of surveying wells has been recognized for some time, and the use of well-surveying machines is increasing rapidly. Operators find that many times well surveys are among the most valuable records to which they refer for information.

Reasons for surveying wells are many and varied. Much could be written dealing with individual instances in which surveys have made large savings in the cost of drilling new wells, in which unforeseen difficulties have been minimized by knowledge of the course of the bore, and in which work-over operations have been conducted on a scientific basis from survey data.

APPLICATIONS

Some of the more common uses of surveys of wells made as they are in process of drilling are noted in the following paragraphs.

To the petroleum geologist an accurate survey of a well is of inestimable value. When depths measured in the well are considered as vertical depths, erroneous conceptions are formed of geologic horizons, contours of subsurface structures, the thickness of formations, and the depth and location of faults, salt domes, and similar features. Not only the vertical depth, but the location of the well bore throughout, is essential to the correlation of wells to depict the actual geologic conditions underground. Many of the assumptions made by geologists in older fields where drilling was done before the general use of surveying are incorrect. Someone has referred to the crooked hole as an "alibi for geologists." The intelligent employment of electric-coring data is dependent upon knowledge of the course of the hole throughout its length.

In many wildcat-well operations oriented cores are taken at a considerable expenditure of time and money. These cores are of little value unless the dip and strike indicated are corrected for the angle and direction of the hole itself.

A number of mining companies doing exploratory work by diamond-drill coring methods have realized the value of surveying small-diameter core holes made in rock formations. A special $1\frac{7}{8}$-inch diameter barrel has been used in surveys made in shallow AX size holes drilled at Leadville, Colorado, and in Missouri. Larger diameter holes were surveyed with a regular single shot at Rifle and Aspen, Colorado. Special length barrels were manufactured for this purpose so that two instruments could be used at once. Time necessary to make the surveys, run on drill rods or on heavy wire line, was reduced by half by use of these long barrels. No magnetic attraction was detected in the strata penetrated by the diamond drills.

The exact location of the producing portion of a well on a known structure is of considerable economic importance. The petroleum geologist attempts to obtain the largest production with the greatest ultimate yield consistent with sound economic development of the lease as a whole. Surface spacing of wells may be made consistent with good practice in the field concerned; however, if the wells are not drilled vertically, the entire plan is defeated. As the surveying of new wells becomes standard practice, it becomes evident that bottom-hole spacing is of far greater importance than the orderly surface arrangement of wells. In fact directional drilling has made possible the surface spacing of wells at any convenient location, with wells bottomed in almost as exact a pattern as if they were at the surface. It is common knowledge that edge wells are not very desirable because they are short-lived producers. Many wells have been spudded high on the structure but bottomed in edge locations or even too low on the structure to be worth producing. Consistent checking of the well for drift and direction would have insured a better well or obviated the extra expense of redrilling a portion of the well to get production.

When wells are drilled to bottom near faults or are caused to cross them to productive formations or in exploring for oil near salt domes, complete knowledge of the course of the well is essential. The more particular the type of formation encountered, the more necessary accurate data of this kind become. Surveys of old wells have shown that wells intersecting faults at acute angles have a tendency to continue in the plane of the fault. Some were productive; others were not. In any event, they usually wandered far from the bottom-hole location originally intended by the geologist.

If a single shot is used, the wandering course of the hole is known as the well is drilled. Inspection of the survey plot gives ample time to correct the tendency of a well to wander from the course intended. If it is found impossible or impractical because of cost to straighten the course of the well by natural means, a deflecting tool may be set to correct it. Some operators will permit the well to wander (provided no "dog legs" are in the hole) within limits until it nears the producing zone, when a directional-drilling crew is called and the bottom section of the well is deflected to the point desired. This has often been found to effect a considerable saving of rig time as compared to drilling slowly to keep the hole very nearly vertical. A well rarely is bottomed directly under the derrick unless it is deliberately completed there by controlled directional drilling. Unless wells are surveyed to determine their positive underground location, it would appear that geologic and engineering conclusions based on surface locations and spacings will be in error in some ratio similar to the deviation of the wells.

An acceptable bottom-hole-spacing plan sometimes can be made, even after some of the wells on a lease have been drilled and are producing. By surveying the producers and plotting their bottom locations, new wells may be drilled to provide correct drainage. This can be done by directional drilling or by drilling nearly vertical wells at the appropriate surface locations, in which case the surface appearance of the lease will not be uniform. The pattern of wells at the oil sand will be as desired, however, which is the important consideration.

The collision of a drilling well with a neighboring well on the same lease or on a nearby lease has sometimes occurred. The bit has at times cut the casing of a producing well and caused expensive repairs and loss of production while the work was in progress. Delay in drilling, the running of a cement plug, and the sidetracking of the new hole materially increased the cost of the newer well. At times a drilling well has mudded-off another well, damaging the older well and necessitating the redrilling of a considerable portion of the new well to make a satisfactory producer. Production of wells in low-gas-pressure sands has been impaired permanently in this way, and a few of the wells have been abandoned. Surveys of both wells concerned would have prevented such costly errors.

From the standpoint of the geologist, whose work it is to take the information derived from many wells and integrate the whole into an orderly and plausible picture, the knowledge of the extent and direction of the deviation in the various wells is of the greatest value.

If the elimination of difficulties in drilling is considered, well surveying of some sort is of paramount importance. As stated before, most contracts for vertical holes specify that the deviation at no point in the hole shall be greater than 3 or 5 degrees. Aside from the fact that some measure of bottom spacing is thereby specified, a mechanically correct hole is assured. Wells that are allowed to become crooked with "dog legs" or sharp kinks in them are harder to drill, more liable to cause mechanical failures in the drilling equipment, and more difficult to case and cement. Production problems are simplified in a straight well as wear on the pump, abrasion of sucker rods on tubing, and excessive use of power required for pumping are minimized. Down-time with its consequent loss of production is reduced. Many correctly drilled directional wells with great deviations actually are cheaper to produce than some so-called straight wells which deviate very short distances but which are crooked.

A continuous survey of a well as it is drilled is often of greatest value when least expected. Occasionally a well will get out of control, exhausting gas and oil from the formation and sometimes igniting to form a great fire hazard. Relief wells to choke the flow of such wild wells are drilled much more easily if the original hole has been surveyed and the point from which the well is producing is located accurately.

A twist-off or a stuck drilling string is not an uncommon occurrence in drilling. Fishing jobs are often long and very expensive operations. Many times it is less troublesome and costly to set a cement plug on top of the "fish" and sidetrack and redrill around that portion of the hole filled with lost tools. Surveys made as drilling proceeds pay off in a situation of this kind; the new hole may be drilled in such a direction that there is positive assurance that the "fish" will not be encountered again.

Directional holes have been sidetracked unintentionally in soft formations while running in the hole with a sharp bit or during reaming operations. Because of the excellent survey that must be kept on a directional well, it has often been possible by skillful deflection of the bit to break back into the original hole. Thus thousands of feet of drilled hole have been saved that ordinarily might have been lost. In a very few instances casing or tools permanently lodged part of the way up the hole have been cemented to hold them in position; then the well has been sidetracked around the "fish" and back into the original hole below. In deep holes the bottom footage is very costly. An attempt of this kind is very worth while, provided the original hole was well surveyed so that the sidetracked hole can be directed very accurately to intersect the old hole below the "fish." Needless to say, this is a delicate operation depending upon a good survey.

Single-shot surveys are made at frequent intervals on all directional-drilling operations. The exact course of the hole must be known at all times as drilling proceeds, since plans for the use of deflecting tools, a change in drilling setups, and the like are dependent on knowledge of the present position of the well and its relation to the objective. In high-angle-deflection work, the difference between the measured and vertical depths of a well is often hundreds or thousands of feet.

Legal aspects of drilling are such that wells near the boundary of a lease or on very small leases or strips must be surveyed to assure the adjoining property owner that none of the course of the hole trespasses upon his property. Costly lawsuits, court-order surveys, the loss of production, and the possible redrilling of a well to keep it on the correct property are eliminated by surveying as the hole is made. Under existing laws some wells on narrow strips and very small leases never could have been drilled without benefit of surveying.

Some operators prefer to use a drift indicator while drilling, running a magnetic multiple-shot survey of the drilled hole just previous to setting casing. Often an oriented survey is made after the casing is cemented. Such surveying consumes less rig time than if a single shot is used as the drilling proceeds and is, therefore, less costly to the operator. However, the information is available after the hole has been made; hence, any remedial operations taken will be costly. Such surveys serve as a permanent record for future geologic and engineering reference.

Legally one of the advantages of a multiple-shot survey is that the information on the whole well is obtained in one run. Since machines used for this purpose are run by service-company employees only, they are able to attest to the authenticity of the survey record. Single-shot surveys necessarily are made by the driller and crew on tour at the time the hole was drilled. No one man could give an affidavit as to the accuracy of the survey. Both oriented and magnetic multiple-shot surveys have been accepted by courts dealing with subsurface trespass cases.

Positive assurance that the subsurface portion of the well is on the lease is given if a well is checked with a drift indicator while being drilled and surveyed before being put on production.

Many wells drilled and producing for years are being surveyed whenever the necessity of redrilling is indicated by loss of production or water encroachment. Operators often are surprised by the crookedness of older wells; some have been surveyed which attained drift angles over 45 degrees. The reason some wells are water cut and others always have been poor producers often is self-evident from the surveys. At times two well have been found that have been bottomed within twenty or thirty feet of each other. The survey, plotted on a map of the subsurface structure along with the surveys of other wells on the lease, often reveals the desirability of drilling new wells to drain the oil sand properly. Some major companies make it a practice to survey all wells on a lease before considering the redrilling of any one well. Then an over-all plan for the whole lease is drawn up from the subsurface conditions discovered. The best point at which to set a casing whipstock to commence sidetracking can be determined from the survey. Often it is found that a well that could be deepened to produce from a lower zone has wandered from under the lease. In order to prevent possible legal action by the adjoining landowner, the well must be redrilled from some depth shallower than where it crossed the lease line. Surveys of all wells drilled on a lease give a sound basis for proper determinations of recovery and per-acre yield, since interference between wells can be evaluated. A subsurface map of the wells on a lease is an aid in the selection of input wells in repressuring and flooding operations.

When it is intended to perforate casing to produce an upper zone it is vital that the measured depth in the well corresponding to the vertical depth of the oil sand be known accurately. The location of the well on the structure also should be considered. A well survey supplies all of this information.

Operators have been very much pleased that they had surveyed a well on those occasions when casing collapsed or liners became stuck permanently. They were able to deflect the sidetracked hole to a new, more productive location under the lease.

An unusual application of well surveying has been made in an oil field on the Pacific Coast. Oil operators noticed that the cement was beginning to break away from the casing in well cellars. A network of surface levels run across the area of the field indicated that the whole area was sinking. This subsidence has continued to where several wells in the field have been shut down during the last two years. It was suspected that the movement was occurring at subsurface slippage planes. The operators decided to run especially accurate, multiple-shot surveys of the wells affected. By making surveys of the same wells at monthly intervals the depth, location, speed, and direction of movement at the slippage planes was detected.

The Eastman Oil Well Survey Company cooperated with the operators in making special equipment for running these surveys. It was considered advisable to make the survey assembly as small in diameter as possible in order to enable the assembly to pass through the partially collapsed casing at the planes of earth movement. A standard DX multiple-shot machine was shortened by removal of the outer battery case. The shortest protective barrel possible was made to house the instrument. Basket stabilizers, which were free to turn on the outside of the barrel, were made to center the barrel accurately in the casing. These sets of spring stabilizers were made to fit two sizes of casing and also to run through chokes on the wells. The surveying assembly as run into the well consisted of the short survey barrel, stabilized by spring baskets, together with two universal joints separated by a short torque tube. This unit was screwed to the drill pipe or tubing by an upper connection. The articulated assembly followed the course of the bent casing accurately. Oriented surveys were made by taking readings at intervals of six inches or one foot through the fault planes. Subsequent surveys plotted over the originals indicated the direction and amount of movement taking place in the wells. The operators concerned have been able to predict the speed

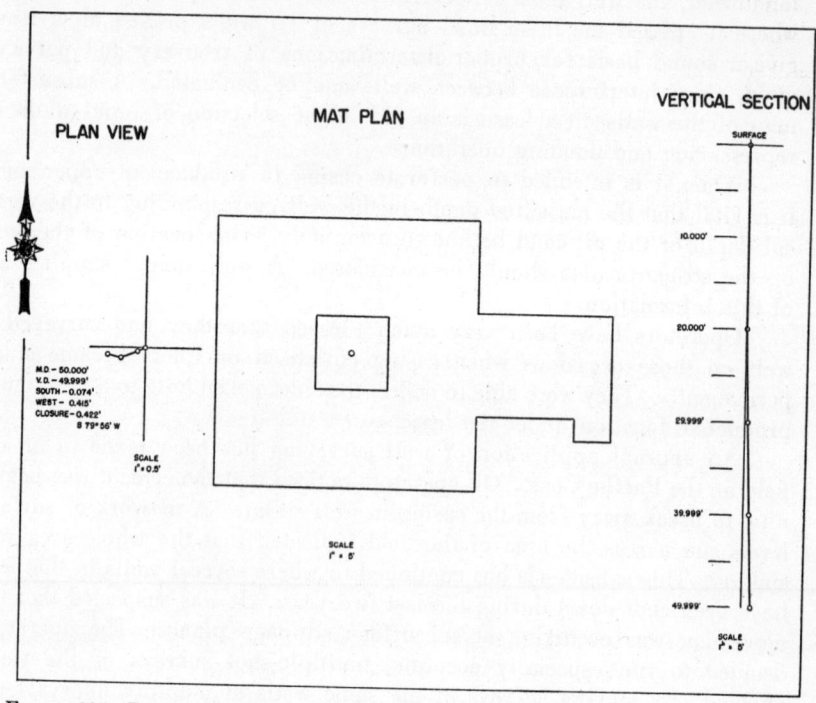

FIGURE 305. Report of sucker-rod survey made on producing well for pump alignment.

of the movement from this survey information. Successful remedial measures have been devised to prolong the life of the producing wells. New methods to combat the effect of this earth movement have been used in drilling new wells.

A major company has surveyed all the wells on one of its large leases in order that a study could be made of the difficulties experienced in pumping some of the wells. Vertical sections on a distorted scale were made for each well so that "dog legs" could be detected more readily. It was found that, by relocating pumps in some wells and by protecting the rods at other points where sharp kinks were evident, the difficulty and attendant expense of pumping were reduced materially.

A few surveys have been oriented on sucker rods inside 2½-inch up-set tubing to shallow depths in producing wells. A small-diameter, multiple-shot survey instrument in a steel barrel was used. These surveys, on which readings at 10-foot intervals to depths of 50 to 100 feet were taken, were run to determine the position of the tubing in the well. Since the inside diameter of large casing provides ample room for the tubing to move, the exact position of the tubing is unknown unless surveyed. It is not impossible for the tubing to deviate as much as 18 inches from vertical in the upper 50 feet of casing. This misalignment results in severe wear on the polish rod and difficulty with stuffing boxes. A plot and section of the course of the tubing (fig. 305) gives the operator adequate information for adjusting the level of the pump slab being poured, or for shimming the pump base so that the polish rod may be correctly aligned. This type of oriented survey may be made when the well is down and the sucker rods have been removed from the tubing.

Most drilling companies find it false economy to save the small amount of time and expense involved in making a directional survey of each well drilled because of the advantages gained by having such a record.

ORIENTED CORES

KIRK CARLSTEN

Core orientation is a method by which a sample cut by an oil-well coring bit can be removed from the well bore so that it may be oriented in space in the same position that it occupied in the formation from which it was taken. Presuming that the core contains bedding planes, the dip and strike of the formation can be measured. The importance of taking oriented cores in drilling wildcat wells is obvious. The dip and strike of the new structures encountered are necessarily of vital concern to the subsurface geologist. Other methods are used for ascertaining this geologic information, but actual cores of the formation itself are the best evidence available.

The Eastman Oil Well Survey Company offers a service for and constructs specialized equipment capable of obtaining oriented cores cut by either conventional core barrels or by retractable wire-line coring equipment.

The equipment for one method applicable to conventional core barrels is illustrated in A of figure 306. When a core is cut, it is scratched at one point on its circumference by a scriber attached to the inner-core barrel. The inner barrel is prevented from rotating by the friction of the core catcher against the core, as well as the scriber, which is imbedded in the core. Thus, the outer barrel containing the diamond cutter cuts the formation, and the core is forced up into the inner barrel. It can be seen from A of figure 306 that a shaft extending from the top of the inner bar-

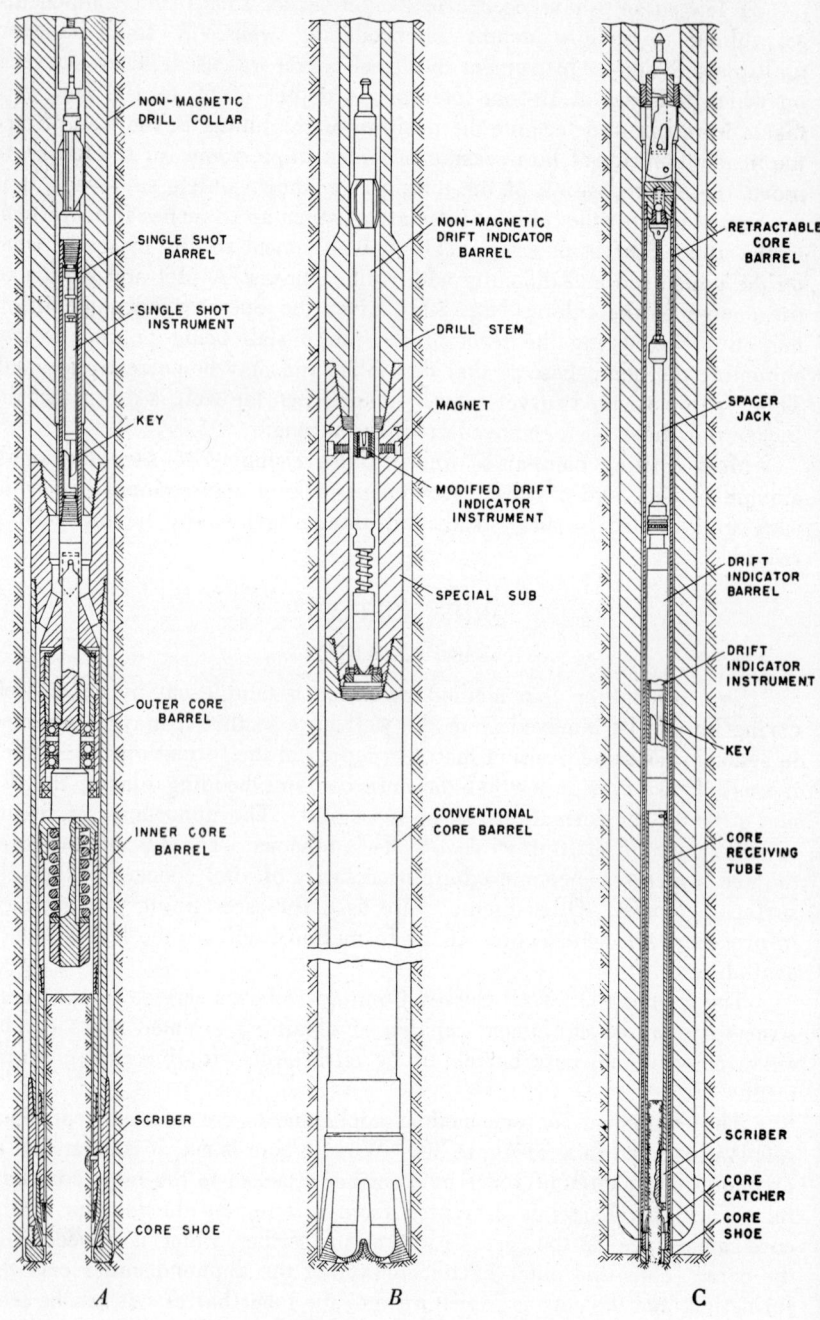

FIGURE 306. Core-orientation equipment.

rel is surmounted by a horizontal pin. A nonmagnetic drill collar is used directly above the core barrel, in order that a single-shot instrument may be used to ascertain directions. An Eastman single shot is used in a special barrel, the bottom of which is equipped with a mule-shoe device, which fits accurately over the end of the inner barrel shaft and engages the pin. The surveying instrument is slidably mounted within its barrel, so that it may be shock absorbed without changing position with respect to its longi-

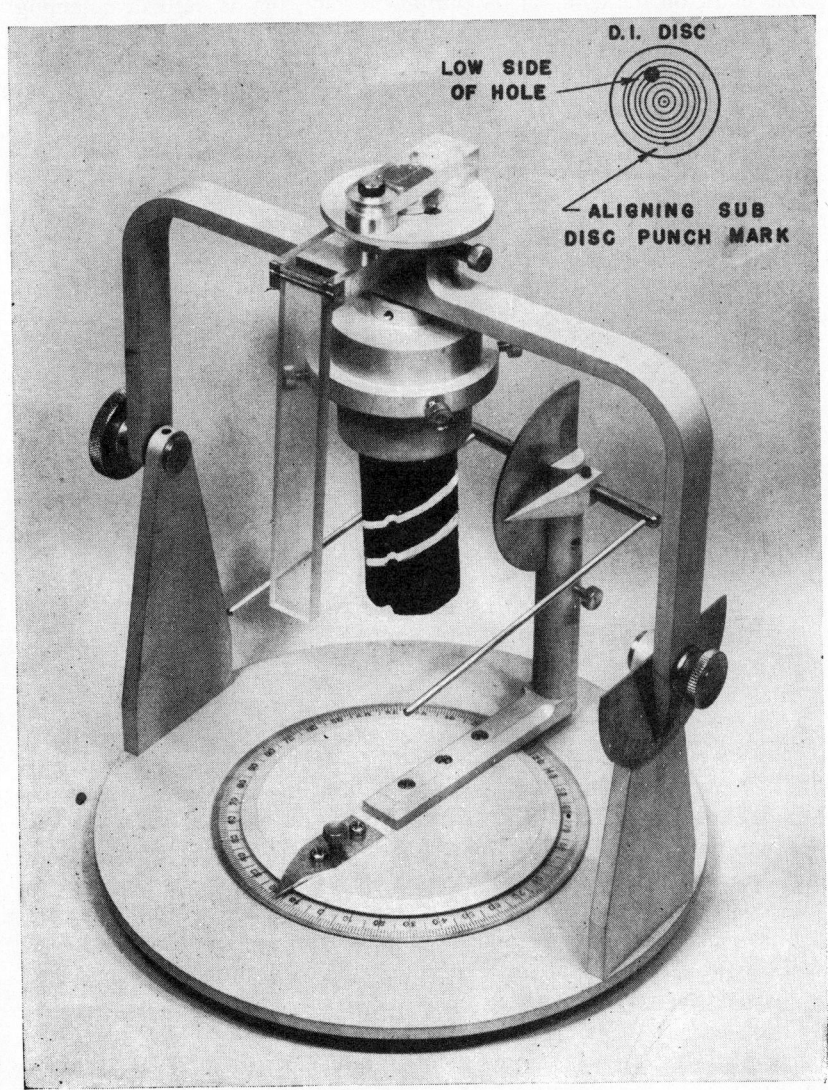

FIGURE 307. Core-orientation reader.

tudinal axis. A lug is mounted in the angle unit of the single shot, so that it is in perfect alignment with the scriber of the inner core barrel. The lug and compass card are both photographed by the single shot machine. The survey instrument and special barrel are run into the hole and engage the inner-core-barrel shaft just previous to breaking off the core from the formation. The reading taken by the instrument indicates the drift and direction of the well, and the special lug shows the true direction of the scribe mark on the core. This equipment generally is used in obtaining oriented cores from holes that are very nearly vertical.

In figure 306, *B* illustrates the equipment for taking oriented cores

FIGURE 308. Oil-well diamond-core bits.

with a conventional core barrel when the hole is not vertical. No nonmagnetic drill collar is used. The drift and direction of the point where the core is to be cut are found by running a single-shot picture in open hole, either before or after the core is cut. A special substitute containing two opposite-pole magnets is made up directly above the conventional core barrel. A bridger is mounted in the inner part of the substitute so that a special type "M" Eastman drift indicator will be positioned with its disc cup directly between the two magnets. A special disc cup, which is free to rotate and contains a small magnet, is used. A small index pointer contained in the disc cup marks the drift-indicator disc and thus records the direction of the magnetic field of the disc cup. After the core is cut and before it is broken off, the special drift indicator is run into the well and a picture is taken. The dot made on the disc by the point of light from the plumb bob indicates the low side of the hole. Thus, the position of one of the magnets in azimuth in respect to the low side of the hole can be ascertained. After the core barrel has been removed from the well bore, a line is extended optically from the magnet along the core barrel to the

FIGURE 309. Oil-well cores.

edge of the core. A mark made at this point on the core is used as a basis for orientation. It will be noted that no scriber need be used in this method.

The "Corex" equipment is illustrated in C of figure 306. This method is applicable only to cores taken with wire-line retractable coring equipment. A special type "M" drift indicator is used for orientation. The retractable core-receiving tube contains a scriber and is connected directly to the drift-indicator barrel mounted above it. A special disk cup has a base which is positioned by a cylindrical key, so that an index pointer in the disk cup is in accurate alignment with the scriber in the receiving tube. The whole assembly illustrated, including the drift indicator, is dropped or run into the drill pipe on wire line and seated. Previously, the amount of time it will take to cut a core has been estimated and set on the watch in the drift indicator. When this time has been consumed in cutting the core, the drill pipe is held motionless while the picture is taken by the drift indicator. The retractable assembly is removed from the well. Ordinary drilling can be resumed after the core breaker is dropped into the bit. The core thus cut is oriented, since the drift-indicator disk shows the relation of the scribe mark on the core to the low side of the hole, as in the second method described above.

It should be noted that the last two methods are dependent upon a drift angle in the well of $2°$ or greater for the orientation of cores.

The core obtained by any of the three methods described above is placed in the reader (fig. 307). It is adjusted so that it is positioned in space exactly as it was in the subsurface formation from which it was cut. The center section of the reader is turned, and the arms are aligned parallel to the dip of the bedding planes of the core. The true dip of the formation is read directly from the vertical protractor, and the direction of the dip is obtained from the horizontal protractor.

Whenever possible, the data obtained are checked by cutting a second oriented core from the formation immediately below the first one.

Typical diamond-core bits used in oil-well coring are illustrated in figure 308. Figure 309 is a photograph of two very hard sandstone cores cut with different sizes of diamond-core bits from subsurface formations in Wyoming.

Design and material changes are being constantly made to simplify and improve the orienting of cores by the "Corex" methods. Determining the dips and strikes of formations by actually measuring them from cores has proved to be the most accurate and reliable method yet devised.

MAGNETIC CORE ORIENTATION
M. G. FREY

Many attempts have been made to develop a method by which cores from oil-well borings can be accurately oriented. Normally it is impossible to determine the original orientation of a core when it reaches the surface and is released from the core barrel. Mechanical, electrical, and magnetic methods have been used with various degrees of success in efforts to orient cores. The purpose of this section is briefly to review the magnetic method and discuss its limitations. For a full description of the orienting apparatus constructed and developed by the California Research Corporation, the reader is referred to the articles by Edward D. Lynton.[1][2][3] No attempt will be made in this section to discuss laboratory equipment and technique other than to give the following brief summary.

The magnetic orienting apparatus used by the California Research Corporation is essentially an astatic magnetic balance consisting of two mutually opposed bar magnets suspended by a thin wire to which is attached a small mirror. The core is introduced beneath the magnetic balance. A soft-iron core is then lowered about the balance and core and shields the magnets from the earth's field and ambient magnetic fields. When the core is slowly rotated, the magnetic field associated with the core causes a deflection of the suspended magnets. This deflection is amplified and recorded on photographic paper by means of a beam of light focused on the mirror and reflected to a photographic drum. The resulting record is in the form of a sine curve from which the direction of polarity of the core can be determined. The assumption is made that the direction of polarity of the core is parallel to that of the earth's field at the location of the well. Once the magnetic directions of the core are established and marked thereon, it is a simple matter to determine the direction of strike and dip.

THEORY

The basic observation that prompted a study of magnetic orientation was made some years ago and was of considerable significance. The magnetism of a piece of basic igneous rock was found to have polarity which agreed essentially in direction with the magnetic field of the earth at that locality. Was this not a reliable way to orient a core in which directions other than vertical are wholly unknown? With this in mind, a number of outcrops were visited, where strike-and-dip measurements of rocks in place were made. Rock samples were carefully marked and collected from these outcrops and sent to the laboratory, where they were oriented magnetically. The directions of strike and dip determined magnetically checked surprisingly well with the field measurements. Directional differences of but 5° to 10° were obtained. A maximum difference of 20° was found.

To obtain a better understanding of magnetic orientation some of the theoretical aspects that have a direct bearing on this problem are briefly considered. All diamagnetic and paramagnetic substances must be eliminated, for the latter possess magnetism only when placed in a magnetic field. Only ferromagnetic materials (of iron, nickel, and cobalt)

---

[1] Lynton, E. D., *Laboratory Orientation of Well Cores by Their Magnetic Polarity:* Am. Assoc. Petroleum Geologists Bull., vol. 21, pp. 580-615, 1937.
[2] Lynton, E. D., *Recent Developments in Laboratory Orientation of Cores by Their Magnetic Polarity:* Geophysics, vol. 3, pp. 122-129, 1938.
[3] Lynton, E. D., *The Mechanics of the Upside Down Core:* Geophysics, vol. 5, pp. 393-401, 1940.

have residual or permanent magnetism that can be measured when the core is effectively shielded from all magnetic fields.[4]

To be ferromagnetic a material must:

1.   Be composed of atoms having permanent magnetic moments. The magnetic moment of an atom is caused by uncompensated spin of certain of its electrons. This theory is diagrammatically illustrated in figure 310.

2.   Have strong interatomic (exchange) forces that maintain the magnetic moments of many atoms parallel to one another.

The atoms of only a few related elements meet these rigid requirements. In addition to the foregoing, a ferromagnetic substance must have a favorable ratio of R (the distance of atomic separation) over r (the

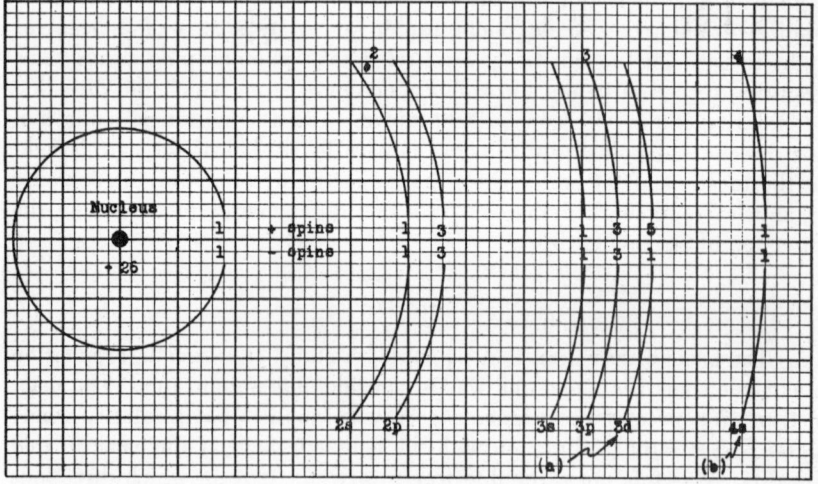

FIGURE 310. Iron (Fe) atom, where in *a* the 3*d* shell is the uncompensated spin 5+ to 1−) giving rise to ferromagnetism; and *b* represents free electrons of the metal. Atomic number = 26; protons to nucleus = 26; electrons in shells = 26. (After Bozarth, "Reviews of Modern Physics," 1947.)

diameter of the shell of uncompensated spin electrons). To illustrate this, manganese, which is normally not ferromagnetic, may become so when the atoms are separated by abnormally large distances, as they are in some compounds and alloys. Figure 311 shows this relationship.

To elaborate somewhat on requirement 1 above we refer to figure 310, which represents a highly magnified structure of an atom of iron. True, modern theory regards the shells not as precise orbits as shown, but rather as zones where the expectancy of finding rapidly moving electrons is greater than in adjacent regions. However, the diagram is regarded as essentially correct and is ideal for explaining such phenomena as emission spectra and to a lesser extent ferromagnetism.

---

[4] Newer instrumentation eliminates the susceptibility effects caused by the presence of the astatic magnetic balance and thus measures the polarity of the ferromagnetic minerals more accurately.

In the study of metals and alloys the next-larger unit of ferromagnetism is the domain. A domain is of small size (approximately 10^{-6} cubic millimeters) and contains many ferromagnetic atoms, all of which are aligned parallel so that their magnetic moments are additive. Adjacent domains have different orientations. The real existence of domains has been shown by powder patterns as well as by the Barkhausen effect, which is caused by a boundary shift of unit domains (the size of one domain increases at the expense of its neighbors) in response to an increasing directionally applied magnetic field.

Numerous magnetic measurements of both intrusive and extrusive igneous rocks indicate that a considerable amount of residual magnetism is acquired upon crystallization and cooling and thereafter tenaciously retained by the ferromagnetic minerals. The same is true of those ferromagnetic minerals that are deposited from water solutions.

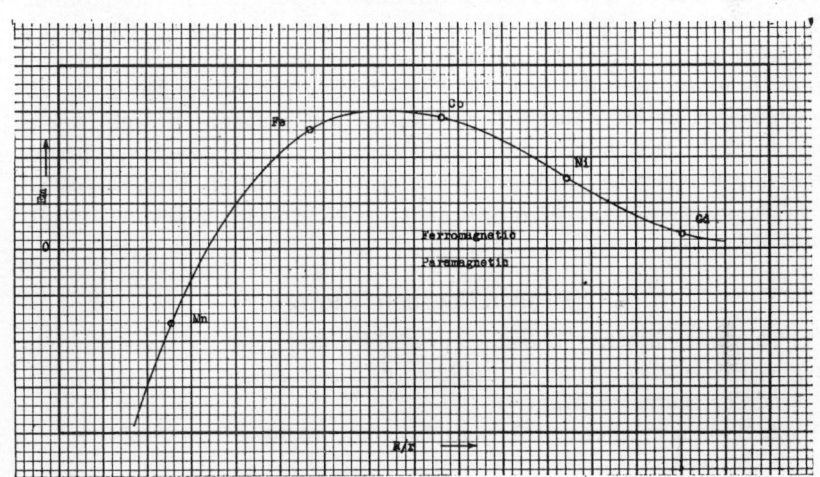

FIGURE 311. Bethe's curves, where Em = energy magnetization; R = atomic separation; r = diameter of uncompensated electron shell ($3d$) of the atom. (After Bozarth, "Reviews of Modern Physics," 1947.)

Crystals display the property of anisotropy, that is, they exhibit preferred crystallographic directions of magnetism. Thus, when a preferred or easy direction of magnetism in an incipient crystal approaches or parallels the direction of the earth's field we may expect the growing crystal to acquire strong directional ferromagnetism. The very powerful interatomic "exchange" forces hold the atoms in the positions of crystallization, and it is only when high temperature or applied magnetic fields of appreciable strength (much greater than the earth's field) are introduced that this directional ferromagnetism may be altered or destroyed.

Residual or permanent magnetism should not be confused with induced magnetism. The relationship is evident in the formula expressing the total intensity of magnetization (polarization) of a rock:

$$I=kH+Ip$$

where I = total polarization

     k = susceptibility factor

     H = earth's magnetic field strength (horizontal component)

     Ip = permanent polarization of a rock

The few measurements that have been made on sedimentary rocks show rather inconclusively that the permanent polarization Ip is on the average but a fraction (one-fifth to four-fifths) of the magnitude of the induced polarization kH. However, in some basic igneous rocks Ip has been found to be several to many times as great as kH. The direction of Ip is fixed at the time the rock forms and is rigidly held in that position. The direction of the vector kH always agrees with that of the earth's magnetic field, whereas Ip, although agreeing initially, may be and generally is moved out of coincidence by geologic processes. Moreover kH is dependent on H, whereas Ip is entirely independent. Because kH is dependent it completely vanishes when effectively shielded and symmetrically shaped (cylinder), where there is no axis of susceptibility, so that Ip can then be accurately determined. This is the basis of magnetic core orientation.

Variable Geologic Factors

The more important geologic variables that must be considered in magnetic core orientation are listed and discussed below:

1. Coarseness of sediment.
2. Secular variation of the earth's magnetic poles.
3. Secondary residual magnetism.
4. Distribution of residual magnetic material in a core.
5. Rock movements.
  (*a*) Tilting.
  (*b*) Stress and strain.
  (*c*) Temperature.
6. Magnetic field of core barrel and drill pipe.
7. Cores from deflected holes.

Coarseness of Sediment

It is an elementary geologic principle that coarse sediments are normally associated with strong currents, finer sediments with weaker currents. We may also state that all clastic particles possessing residual magnetism have a measurable magnetic force that strives to align the particle magnetically with the earth's field during the process of sedimentation. In the deposition of coarse clastic sediments the forces associated with current action and gravitation normally far overbalance those of magnetic attraction. As a result a minimum degree of alignment of residual magnetic particles with the earth's magnetic field is accomplished. In isolated cases good alignment may be attained, but normally it is not to be expected in coarse sediments. In fine shales the ability of the earth's magnetic field in regard to rotating and orienting small residual magnetic particles may be substantial as compared to all opposing forces. As the fine particles slowly settle in relatively quiet water the magnetic forces have ample time to play an important role in sedimentary alignment. The relatively short time available in the more rapid deposition of coarser sediments does not permit the magnetic forces to be so effective. Thus, we may confidently expect the finer clastic sediments (shales and sandstones) to show stronger and more consistent polarization than their coarser equivalents (grits and conglomerates).

Secular Variation of the Earth's Magnetic Poles (Refer to figure 312).

Information is available to show that in historic time the magnetic poles of the earth have migrated an appreciable amount. Records of the magnetic declination have been compiled from as early as 1540 in London, England, at which time a 7° E. declination was registered there. In 1580 the magnetic declination at London had reached a maximum easterly declination of 11° E. From that date the declination moved progressively westward and reached an extreme of 24° W. in 1810, a movement of 35° in 230 years. Since 1810 the declination has again gradually shifted back eastward, and it was therefore assumed that the magnetic poles had a period of rotation of approximately 460 years (1580 to 1810 being half a cycle). Recent information elsewhere tends to disprove any simple and regular period of rotation of the magnetic poles and instead indicates a more complicated phenomenon.

Declination records in the United States date back to 1620 at Eastport, Maine, where a 19° W. declination was measured then. A spread of but 10° has been observed at Eastport, the extremes being 12° W. in 1760, when a reversal in trend occurred, to more than 22° W. at present. At San Francisco the earliest documented information is a 12° E. declination in 1780. By 1940 the declination had increased to 18° E. and it has since continued to move eastward. From the foregoing data the following generalizations may be made:

1. The declination of the magnetic poles changes appreciably with time.

2. Neither the amount nor rate of this change is universally uniform but is dependent upon geographic location and time.

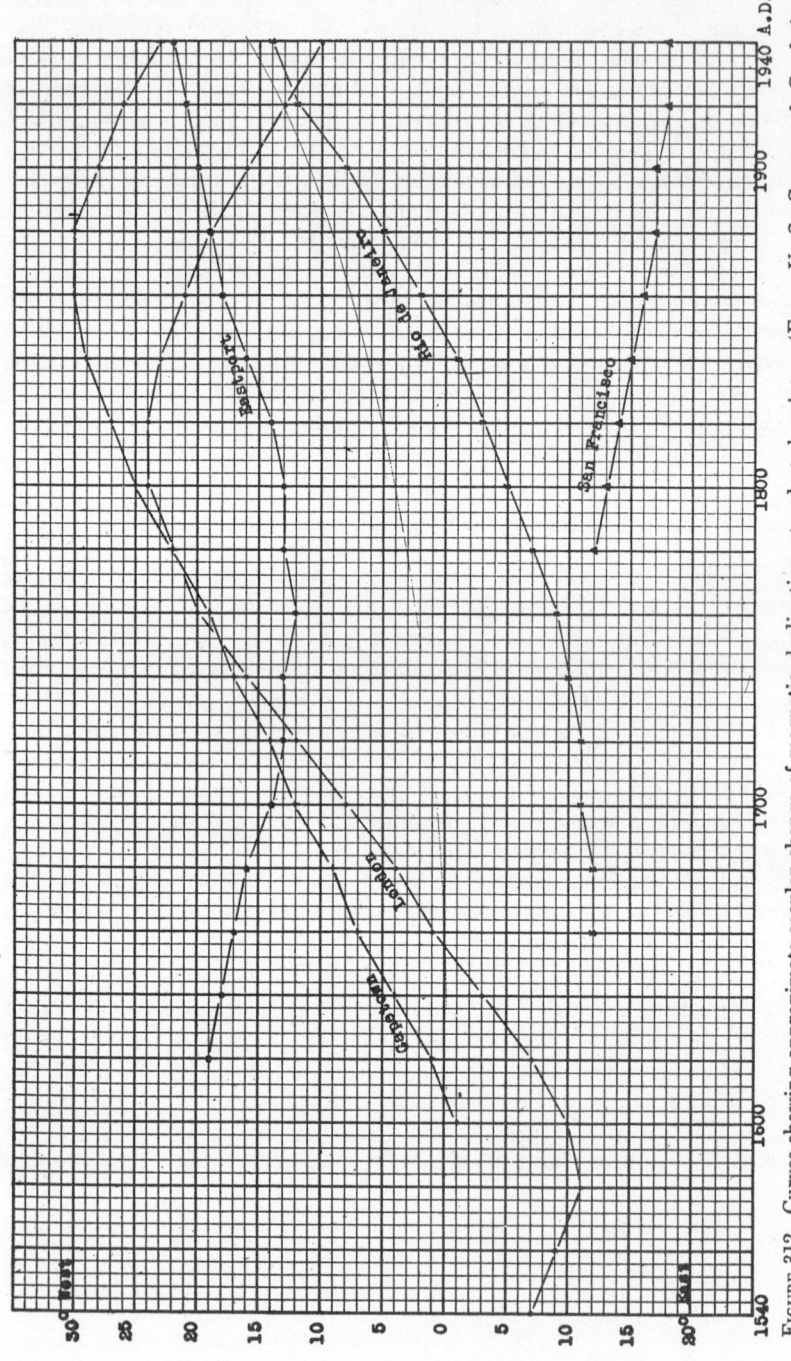

FIGURE 312. Curves showing approximate secular change of magnetic declination at selected points. (From U. S. Coast and Geodetic Survey Ser. 663, 1945.)

3. There is probably no simple and regular period of the rotation of the magnetic poles, although measurements at London indicate a 460-year cycle. Measurements of the magnetic declination of varved clays and lava flows have been reported as demonstrating this cyclic action.

We may thus expect considerable error to be introduced by assuming coincidence between the present magnetic declination and that existent at the time the sediment was deposited.

Secondary Residual Magnetism

Occasionally cements such as vivianite (iron phosphate) are deposited in the pore space of sedimentary rocks. Vivianite, which normally contains some Fe_2O_3, is ferromagnetic and has been found as the principal ferromagnetic mineral in cores showing strong directional magnetism.

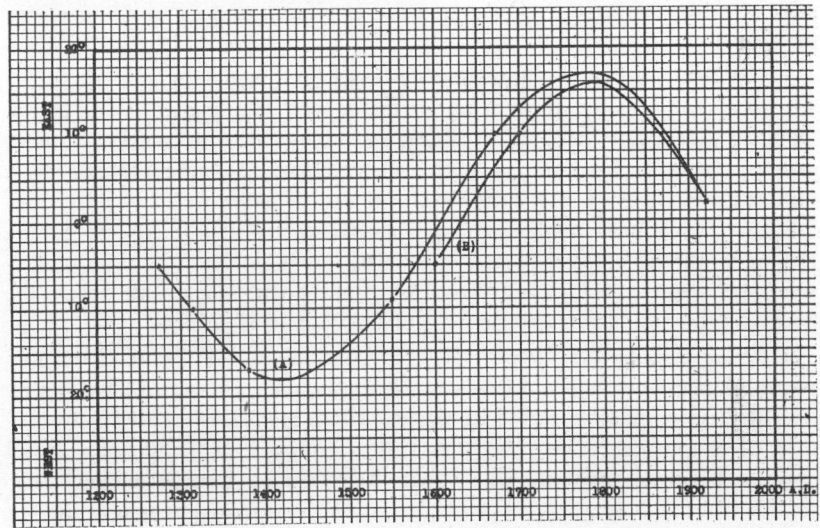

FIGURE 313. Graph showing declination of residual magnetism in lava flows of different ages from Mount Etna (A) and historically observed magnetic declinations at same locality (B). (After Koenigsberger, "Terrestrial Magnetism and Atmospheric Electricity," 1938.)

Lacking experimental data, we must assume that such secondary ferromagnetic minerals succeed in aligning their magnetic moments with that of the earth's field during their period of growth in the sediment. Thus if the residual magnetic properties of a core are found to result substantially from the ferromagnetic cementing material, it may be possible to make rather reliable orientation determinations on (1) coarse clastic sediments that normally possess little directional alignment, (2) sediments in which the matrix is nonmagnetic, and (3) sediments that were highly tilted or folded prior to cementation.

Distribution of Residual Magnetic Minerals in a Core

Under ideal conditions we assume the ferromagnetic minerals to be regularly distributed throughout the core and thus to give rise to a uniform magnetic field. It is possible, however, that isolated crystals having residual magnetism may be present near the surface of the core. Because of this advantageous position they may exert an abnormal effect on the magnetic field of the entire core. The operator may be able to detect this abnormality of position, however, and correct it empirically.

Rock Movements

A history of the movements undergone by the sedimentary rocks should be taken into consideration in magnetic orientation. Rock movements that most concern magnetic orientation of cores are (1) tilting of rock after deposition, (2) stress and strain effects, and (3) high-temperature effects.

Tilting—Tilting of sedimentary rocks is the most obvious and widespread geologic factor of error under the general heading of "Rock Movements." A certain amount of tilting is associated with every oil field, ranging from a maximum in highly folded and faulted structures to a minimum in types of stratigraphic traps. In all cases we must remember that any directional magnetism that was imparted to the matrix of the sediment was accomplished at the time of deposition, unless the polarity was later altered by high temperatures. Subsequent movements rotate the rock and the residual magnetic field of the rock as a unit.

Stress and Strain Effects—We may assume the effect of stress and strain on ferromagnetic minerals to be somewhat similar to that on crystals of ferromagnetic metals and alloys. These materials display the property of magnetostriction, i.e., they either contract or lengthen a small amount when magnetized. The effects of strain on magnetism and magnetism on strain are interdependent. In other words, the magnetism of some materials (those that lengthen when magnetized) is increased by tension; that of others (those that contract when magnetized) is decreased by tensional strain. Likewise, in weak magnetic fields some materials have positive magnetostriction (lengthen) and in strong magnetic fields negative magnetostriction (contract).

In highly folded rocks, where large external forces are active, the strength and possibly the direction of magnetism may be altered in a fashion to comply with strain-ellipsoid relationships. However, we believe this consideration to be of more theoretical and less practical importance as measured by core-orientation standards. The normal association of high temperature with high pressure may make it difficult to single out the effects of one in the presence of the other.

Temperature—High temperature destroys residual magnetism. The temperature above which residual magnetism ceases to exist is called the

"Curie point." The Curie point for nickel is 358° C., for iron 770° C., and for cobalt 1,120° C. At temperatures above the Curie point ferromagnetic minerals become paramagnetic. Conversely, as the temperature is lowered residual magnetism is again acquired. For pure magnetite, Fe_3O_4, the Curie point is 585° C.; for pyrrhotite, FeS, 350° C. Much magnetite, however, is a mixture of Fe_3O_4 and FeO. The Curie point of this contaminated magnetite depends upon the amount of FeO present and decreases to approximately 0° C. for mixtures containing large amounts of FeO. On the other hand, the addition of TiO_2 to the ferromagnetic minerals increases the coercive force and raises the Curie point accordingly.

The high temperatures brought about by deep burial, by neighboring igneous activity, or by other means may raise the temperature of a rock above the Curie point of the ferromagnetic minerals that it contains and thus destroy their residual magnetism. Upon subsequent cooling the ferromagnetic minerals again acquire permanent magnetism, which is essentially coincident in direction with the earth's magnetic field at that locality. This new residual magnetism may form a considerable angle with the direction of the earlier-generation magnetism that was destroyed by the high temperature. In this manner the change of direction of residual magnetism by periodic heating tends to establish alignment with the direction of the earth's magnetic field. On the other hand, the negative geologic factors, which destroy coincidence such as tilting and secular variation, are constantly in action and normally overbalance the intermittent positive effects of temperature. This lag of temperature in keeping pace is often measured in geologic periods or eras, so that divergence between the directions of residual magnetism of a rock and that of the earth's field is the rule and not the exception.

Magnetic Field of Core Barrel and Drill Pipe

The suggestion has been advanced that a strong magnetic field, preferably parallel to the earth's field, be applied at the bottom of a well bore prior to coring, the intent being to create or strengthen the residual magnetism in the rock. Theoretically, this should permit the cores to be more reliably oriented and thus allow the vital strike and dip determinations to be made with greater precision.

As the angle of inclination of the earth's magnetic field approaches the vertical and parallelism with the long axis of the drill pipe, the distortion of the earth's field about the end of the pipe decreases to a minimum value. In the United States, where the angle of inclination is large, the distortion of the magnetic field about the pipe is small, so that the direction of the earth's field may be considered essentially unaltered by the presence of the pipe. The magnitude of the field, however, is materially increased, the strength being a function of the permeability of the steel pipe as well as of its dimensions. The effect of this field on the core, however, is of rather short duration, as upon entering the core barrel the core is shielded to a considerable extent from all magnetic fields until it is finally released at the surface.

Any effect on the core caused by the magnetic field associated with the string of drill pipe and core barrel should be favorable for orientation. The tendency will be to reinforce the magnetic field of the core in a direction parallel to that of the earth's field. Thus, little benefit may be expected by applying special magnetic fields in the bottom of the well bore, when in reality a strong magnetic field is already present through the medium of the drill pipe.

Cores from Deflected Wells

The fact that bore holes always deviate a greater or lesser amount from a vertical position must be borne in mind when making magnetic orientation determinations. Strike-and-dip measurements are always made with the assumption that the core was cut vertically. Corrections should be applied for any appreciable divergence between the vertical position and the true position of the core.

Bore-hole deviations fall into two natural classifications, accidental and intentional. The first type is normally of small magnitude, and corrections for such deviations are unnecessary. The usual drilling contracts specify an allowable deviation up to several degrees from the vertical, and thus most holes are drilled within this permissible cone of error. In the second category, however, the deviation is generally large and corrections are mandatory when the direction and amount of deviation are known.

Several methods may be used to correct for vertical deviation. Instruments have been constructed which duplicate the geometric relations existing in deflected wells. Such instrumental solutions are desirable in that they are rapid as well as visual in three dimensions. Mathematical and graphic solutions have also been developed. A recent paper [5] adequately sums up and evaluates the various methods of correction for hole deviation at the cored horizon.

Experimental Results

Little experimental work has been conducted on the significance of magnetism of rocks. Much remains to be done. Investigations of the "cyclic" secular change have been made as a method of geologic dating. On the basis of observation and experimental evidence largely concerned with investigations of ancient pottery and bricks, Koenigsberger [6] reached the following conclusions which are worthy of note. The application to the residual magnetism of rocks is apparent.

(a) It was found that the magnetism of well-baked bricks and pottery is

[5] McCellan, Hugh, Core Orientation by Graphical and Mathematical Methods: Am. Assoc. Petroleum Geologists Bull., vol. 32, pp. 262-282, 1948.

[6] Koenigsberger, J. G., Residual Magnetism and Measurement of Geologic Time: 16° Cong. géol. internat. Compte rendu, pp. 225-230, 1933.

so strong as to resist at 20° C. the effect of a magnetic field of opposite sign and ten times larger than that of the earth. Such a field weakens the residual magnetism by only five to fifteen percent, according to the conditions of baking. In bricks or pottery that had been well baked at about 500° C., an earth field of 0.45 oersted of direction opposite to the magnetism of the material has only a slight effect when the temperature is raised to 150° C.

(b) If any material has remained about twenty minutes in a magnetic field, a prolongation of the time at the same temperature and in the same field causes no change larger than ten percent. The time effect is asymtotic after a day; the change, with exposure for weeks or months, is less than one percent.

(c) Strong percussions change the magnetism slightly during the first 1,000 percussions. But 10^4 or 10^5 percussions cause no further appreciable decrease.

(d) The apparent coercive force of the thermoresidual magnetism of highly baked bricks, etc., at 20° C. is about 70 oerstads—therefore higher than any natural magnetic field on the earth has ever been, except in close proximity to lightning.

(e) The intensity of the residual magnetism is less if the cooling goes on very rapidly—for example, from 600° C. to 200° C. in ten minutes—than if the cooling is slower; that is in accord with the observations of Melloni. If the time of cooling from 600° to 200° C. is more than an hour, there is no appreciable change for the longer period. In the baking of pottery and in the natural cooling of rocks the time was mostly longer than an hour.

This all adds up to the fact that the residual magnetism of rocks is a rather indestructible feature as affected by the normal forces of nature. On the contrary, one other conclusion reached by Koenigsberger should be mentioned: "The older the rock the lower the residual magnetism. It is not known whether the magnetic field in former geologic periods was weaker or whether some unknown effect has weakened the residual magnetism."

It has been found that the residual magnetism of Tertiary granites is somewhat greater than that of Cretaceous granites and considerably greater than that of Paleozoic granites. This gradual weakening with age has exceptions but is general for any particular kind of rock, such as granite, diorite, or gabbro. Rock type, however, is predominate over age; for example, a Paleozoic gabbro or basalt normally has more residual magnetization than a Tertiary granite or rhyolite. Upon exposure to the chemical-physical agents of weathering, residual magnetism is rather rapidly destroyed.

CONCLUSIONS

Limestones, shales, and sandstones normally have sufficient residual polarity to permit accurate magnetic orientation. However, cores should be carefully selected in the field so as to eliminate as far as possible those which possess little bedding as well as those which are strongly crossbedded. Little pertinent strike and dip information can be expected by properly orienting such cores. Likewise, cores should not be stored near strong magnetic fields nor be exposed to excessive heat. The top of the core should always be so marked.

By properly evaluating the many variables involved, a better understanding of the limitations of the magnetic method of orienting cores is possible. Further, a knowledge of these variables should be of assistance in interpreting the orientation results.

Because of the magnitude of the variables it is obvious that too much weight should not be attached to any individual determination. Instead a statistical approach seems essential in that an average value of a number of determinations is of much greater significance than any particular determination. Likewise, an average direction signifying a dip in a certain quadrant, as NE, NW, SE, SW, should be of the right order of magnitude. Thus, whenever a magnetic dip is to be determined four or five pieces of the core should be submitted. Each piece should be at least two to three inches in length. By orienting each piece and averaging the results an average vector specifying the direction and amount of dip of the cored interval is obtained. The individual determinations may be made to the degree, and perhaps the average as well, as long as it is understood that the average vector is to be interpreted as specifying a general and not a precise direction.

Magnetic orientation is by no means a "cure all." In fact, determinations should be carefully weighed in regard to the magnetic variables that were operative during the history of the sediment. Next, the weighed opinion should be considered in the light of all other available geologic and geophysical evidence. Only in this manner can the results of magnetic determinations of strike and dip be used to advantage.

Practical Applications

Normally, cores are oriented for the purpose of increasing the subsurface control and thus aid in the drilling program of the well as well as in the development of a field. The recognition of subsurface faults is facilitated by an oriented-core program. The direction of dip found magnetically may show that the well has cut a fault. The beds above the fault may dip in one direction, while those below, perhaps of the same formation, may be found to dip an equal amount but in an entirely different direction. In this case, without the information as to the sudden change in the direction of dip with increasing depth, there may be little reason to suspect the presence of a fault, the simplest explanation being merely an assumption that there existed an entirely conformable relationship, perhaps with some thickening and no faulting. The passage of a well from one limb of a steep fold to the other may be similarly detected by oriented cores.

In addition to such practical problems, others, such as those involving rates of sedimentation and short-range correlations, may prove worthy of serious consideration.

CORING TECHNIQUES AND APPLICATIONS

H. L. LANDUA

Since the early days of the oil industry, efforts have been exerted to find better methods of obtaining information about the subsurface formations penetrated during drilling operations. Early techniques included examining (1) the cuttings that resulted from bit action on the formation, (2) a solid piece of the formation obtained by coring, and (3) the circulating fluid as it returned from the well bore for possible oil and gas shows. Subsequently numerous methods of formation logging were made available to the industry, such as electric logging, radioactivity logging, permeability-profile logging, and mud-analysis logging. It is the purpose of this section to discuss (1) the various types of coring techniques now being used by the industry, (2) the application of coring to geologic problems, (3) coring in relation to production work, (4) correlation between coring and electric logging, and (5) the limitations of coring techniques and applications to geologic problems.

TYPES OF CORING

Two types of general drilling techniques are currently being used in oil-field development. One is rotary drilling, in which a bit is attached to pipe and lowered to the formation to be drilled. Bit cutting action is obtained by turning the string of pipe to which the bit is attached. A suitable circulating system is employed to pump fluid down the drill pipe, through the bit, and back to the surface. Formation cuttings are continuously removed from the hole by circulation, and the bit is kept cooled. The other technique is cable-tool drilling, in which the bit is lowered to the formation on a line of some kind, and drilling action is obtained by raising the bit a short distance off bottom and dropping it against the formation to be drilled. This up-and-down action is carried on continuously and at numerous intervals each minute, usually until too many cuttings accumulate on the bottom to allow further drilling. Then a bailer is run on a line and the cuttings are bailed out, after which drilling is resumed.

When it becomes necessary to core in rotary drilling, five general types of coring are used: namely, conventional, diamond, wire-line, reverse-circulation, and side-wall. Conventional-cut cores usually range in diameter from $2\frac{3}{8}$ to $3\frac{9}{16}$ inches; diamond cores, $2\frac{7}{8}$ to $4\frac{7}{8}$ inches; wire-line cores, 1 to $2\frac{3}{16}$ inches; side-wall cores, half an inch to $1\frac{1}{2}$ inches; and reverse-circulation cores usually are in the same range as conventional and diamond cores, depending upon the size of the drilling string. The maximum length of core may vary from the small $1\frac{1}{4}$-inch length obtained in some side-wall coring to ninety-foot lengths that may be obtained in some diamond coring.

Conventional Coring

For conventional coring, only the surface equipment that is used for routine drilling is necessary. A core barrel, shown in figure 314, is

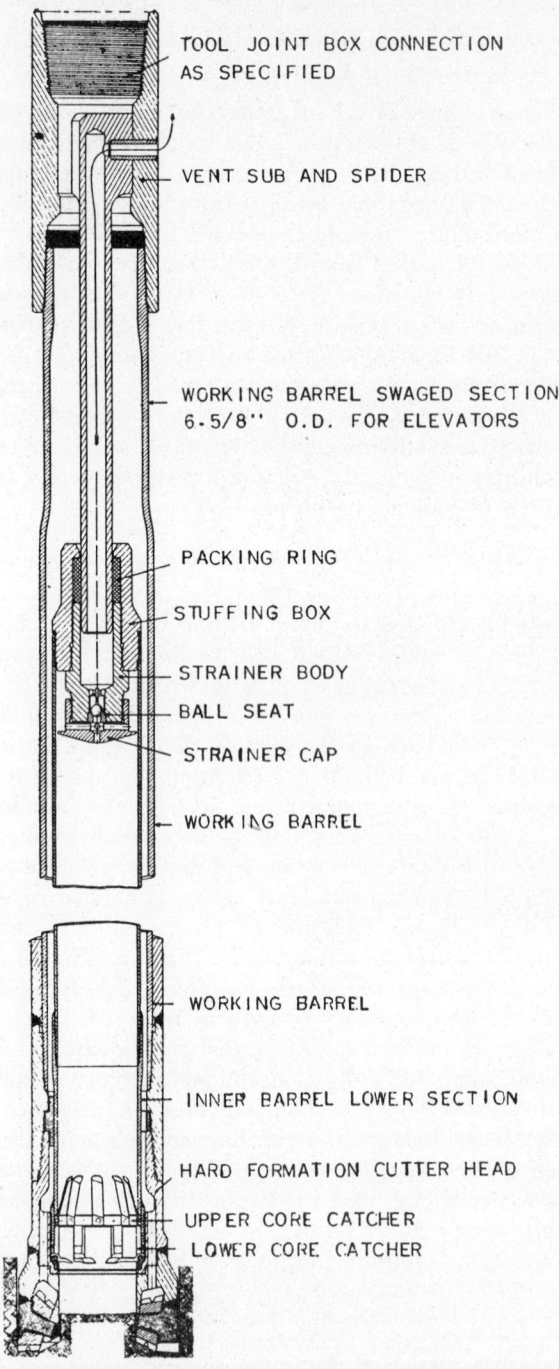

TOOL JOINT BOX CONNECTION
AS SPECIFIED

VENT SUB AND SPIDER

WORKING BARREL SWAGED SECTION
6·5/8" O.D. FOR ELEVATORS

PACKING RING

STUFFING BOX

STRAINER BODY

BALL SEAT

STRAINER CAP

WORKING BARREL

WORKING BARREL

INNER BARREL LOWER SECTION

HARD FORMATION CUTTER HEAD

UPPER CORE CATCHER

LOWER CORE CATCHER

FIGURE 314. Conventional rotary core barrel.

needed in addition to most other routine surface drilling equipment. The complete core-barrel assembly, which is made up on the bottom of the drilling string, consists essentially of a cutter head, an outer barrel, a floating inner barrel, and a finger-type "catcher," which retains the core in the barrel when the assembly is raised from the bottom of the hole. Mud circulates from the drill pipe between the two barrels to the cutter head. Either drag-type or roller-type cutters may be used, depending on the formation to be cored. Ordinarily, the conventional core barrel will accommodate a twenty-foot core, but sometimes cores of much shorter

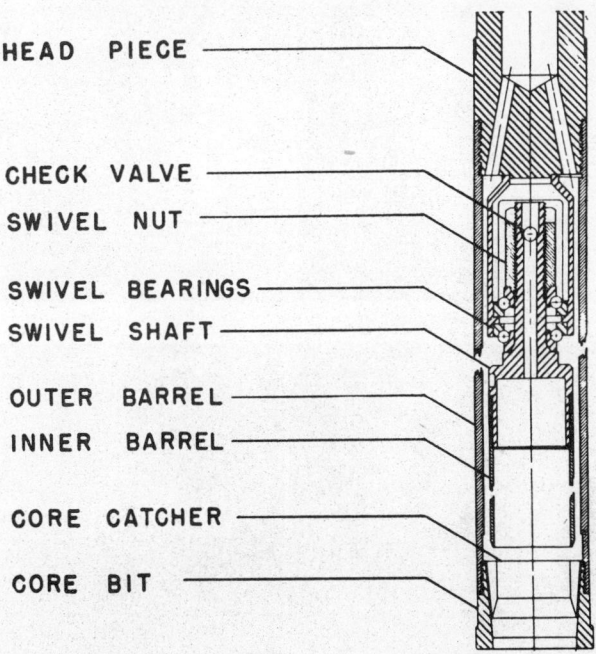

HEAD PIECE

CHECK VALVE
SWIVEL NUT

SWIVEL BEARINGS
SWIVEL SHAFT

OUTER BARREL
INNER BARREL

CORE CATCHER

CORE BIT

FIGURE 315. Core barrel used for diamond coring.

lengths are cut because of local conditions or requirements. Weight on the bit, rotary speed, and circulation rate are dependent upon local conditions, but usually they are varied appreciably from routine drilling techniques. To recover a core cut while using the conventional barrel, the entire drilling string must be hoisted. Advantages of this type of coring include the following: (1) the obtaining of a large-diameter core for a given hole size, (2) the usual allowance of a maximum percentage of recovery of the formation cored, (3) its adaptability to all except the most abrasive types of formations, and (4) the usual requirement of no additional surface drilling equipment Disadvantages include (1) the limitation of cutting only a twenty-foot core during each run and (2) the necessity of pulling the drill pipe from the hole after each core has been cut, to recover the core.

Diamond Coring

Diamond-coring equipment has been used in the mining industry for many years to core small-diameter holes. In the past few years experi-

FIGURE 316. Four designs of diamond-core-heads.

mental work with full-size diamond bits and core bits in the oil industry has led to the adoption of diamond coring for oil-field use in some cases. Experience indicated in general that the largest-diameter core that could be cut in a given-size hole gave a maximum penetration rate at a minimum cost. Usually no additional surface drilling equipment is necessary to do diamond coring. The core barrel, shown in figure 315, is a two-tube assembly similar to the conventional barrels used in the conventional-coring technique, which was discussed previously. The inner tube is supported on two open ball bearings so that it will remain stationary while the core is being cut. Improvements in the design of the barrel are being made as more experience is obtained. Diamond-core bits are all basically of the same construction. The diamonds are placed in a powdered metal, usually tungsten alloy, in a mold of desired size and shape. The matrix under heat and pressure forms a solid mass with the diamonds exposed at the surface. This section, known as the "crown of the bit," is then fastened to a steel bit hub by brazing or mechanical pinning. Four general designs of diamond-core bits are shown in figure 316. The techniques employed in diamond coring involve the use of little weight on the bit, low pump pres-

FIGURE 317. Wire-line coring reel.

sure, and normal to high rotating speeds. Diamonds drill by abrasion, and, while increased weight on the bit may increase drilling rates slightly, high weights tend to fracture the diamonds. Pump pressures and fluid velocity are kept low to prevent washing of the core and to reduce erosion of the bit-matrix metal. Since the cutting rate of a diamond should increase with its linear velocity, high rotating speeds should be best for diamond coring. In actual practice, however, rotating speeds are limited by the drilling equipment available. Experience has also indicated that the removal of all junk iron from the well bore before beginning to core is absolutely essential. When using the proper design of barrel, cores up to ninety feet in length can be cut during each run, if no mechanical difficulty develops.

Advantages of diamond coring include (1) the fact that in areas where the formations are abrasive and hard enough, coring is sometimes more economical than drilling; (2) the usually longer bit life; (3) the possibility of cutting up to ninety feet of core at one run; and (4) the high percentage of recovery. Disadvantages of diamond coring include (1) the high initial expense for the barrel and bits, (2) the requirement of proper operating conditions, (3) the necessity for strict supervision by a person who specializes in diamond-coring operations, and (4) the generally higher costs than in other types of coring.

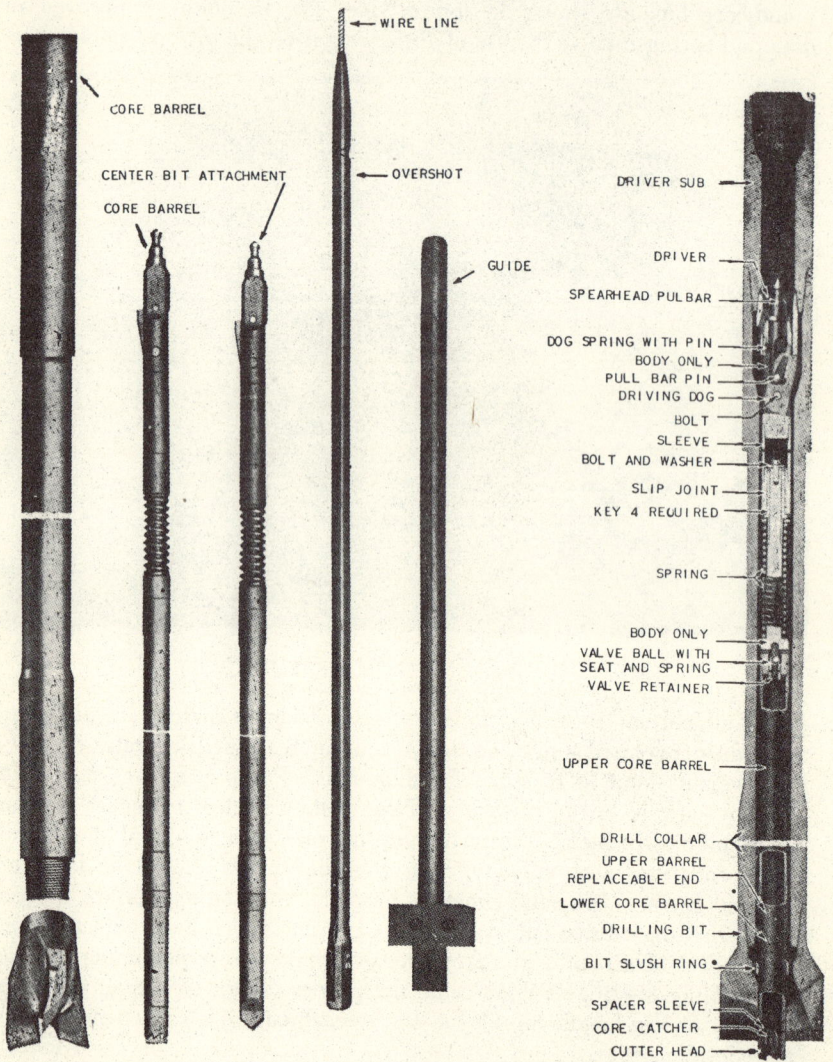

FIGURE 318. Wire-line rotary core barrel and other subsurface equipment.

Wire-Line Coring

Advances in the development of coring apparatus eventually resulted in obtaining a wire-line core barrel, primarily to eliminate the necessity of pulling the drill pipe from the hole to recover each core. In wire-line coring, a suitable hoisting assembly, shown in figure 317, including a wire-line reel, a wire line, a prime mover, and a sheave, is needed in addition to the usual surface drilling equipment. Additional subsurface equipment includes a special core-drill collar and bit, a core barrel and bit, and a wire-line guide and overshot, all of which are shown in figure 318. The core-drill collar and bit are run into the well on the bottom of the drill pipe. Routine drilling may be done by dropping a bit plug inside the drill pipe to shut off the main core-barrel passage through the bit and drilling the formation in the center of the hole. When it is desired to take a core of the formation, the bit plug is removed by means of an overshot that is run inside the drill pipe on a wire line. The core barrel, with cutter head and core-catcher assembly on the bottom, is then dropped inside the drill pipe and automatically latched into place in the drill collar. After the core has been cut in lengths up to twenty feet, depending upon local conditions and requirements, the core barrel, with core inside, is removed with the same overshot and wire line used to recover the bit plug. Consecutive cores may be cut until the core bit is dulled. Weight on bit, rotary speed, and circulation rate are varied somewhat from those used during routine drilling. The advantages of wire-line coring include these: (1) that consecutive cores may be cut until the core bit has been dulled, without the necessity of pulling the drill pipe from the hole to recover each core; (2) that coring and drilling may be done intermittently, until the bit is dulled, without the necessity of making a round trip with the drill pipe; and (3) that a lower coring cost usually is realized. The disadvantages include (1) the requirement of an appreciable amount of additional surface equipment, (2) the limit of the use to coring relatively soft formations, (3) the use of cores comparatively smaller than conventional and diamond-cut cores, and (4) the usually lower core recovery than in conventional and diamond coring.

Reverse-Circulation Coring

Reverse-circulation coring is a very specialized type of coring and can be used only in limited cases. Usually casing must be set at or near the top of the section to be cored, and a special drilling-head assembly and swivel arrangement must be used along with regular surface drilling equipment. The subsurface equipment generally includes a regular conventional or diamond-core bit, which is attached to a special interval-flush-drilling string. Mud circulates down the drill-pipe-casing annulus through the core bit, up the drill pipe, and back to the surface. Cores are continuously circulated up the drill pipe to the surface and caught in a screen basket, through which all returns from the well bore are directed.

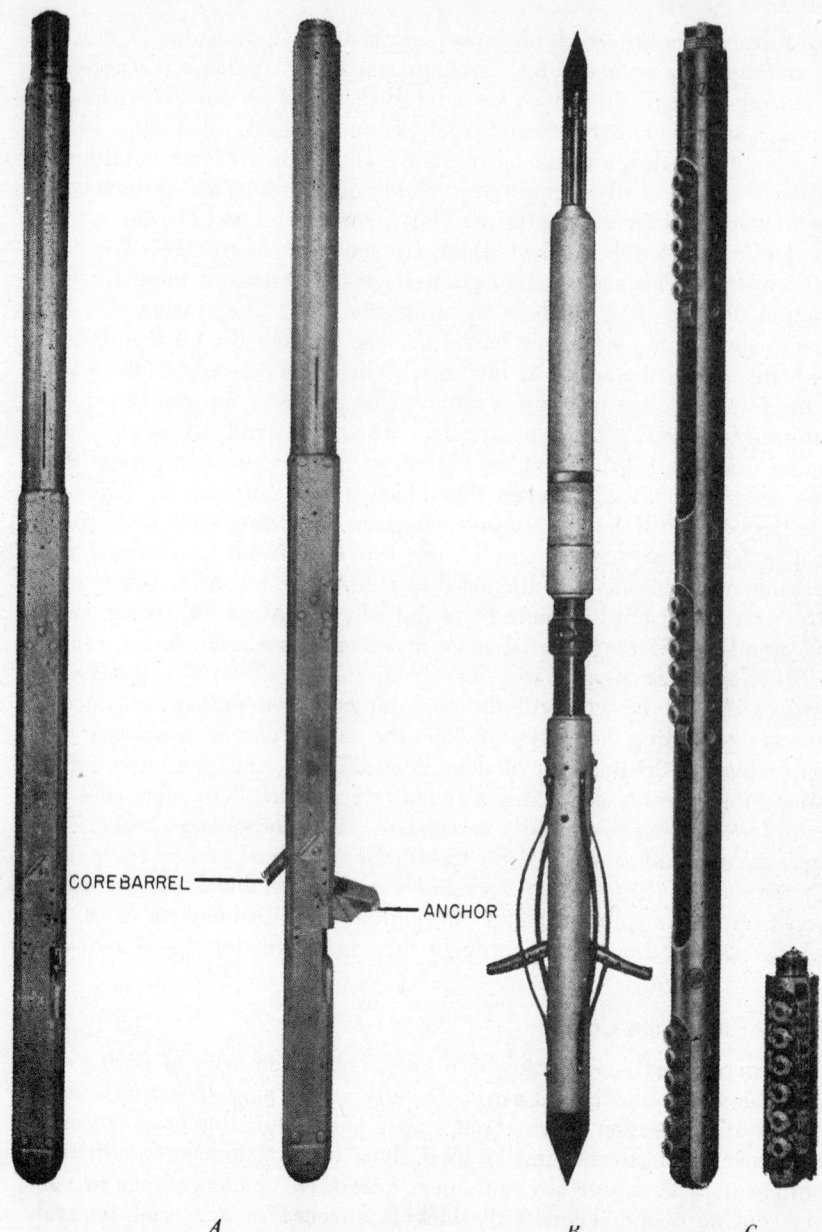

FIGURE 319. *A*—Hard-formation electric-motor-driven side-wall coring device. *B*, *C*— Side-wall coring tools in which core tubes are driven into formation by an electrically detonated charge.

The major advantage of the method is excellent recovery. Some disadvantages are that (1) it usually requires more rig time, (2) it requires an appreciable amount of special equipment, (3) it is generally adaptable only to formations that are consolidated enough to stay intact as they are washed to the surface, (4) it may increase the circulation-loss problem, and (5) it usually requires installation of casing at or near the top of the formation to be cored.

Side-Wall Coring

Side-wall coring techniques were developed to obtain a sample from the wall of a previously drilled hole at desired intervals. In general no surface equipment is required in addition to that used in some of the other coring methods, except possibly with the electrically motivated devices, which require a regular logging truck. Figure 319A illustrates a hard-formation, electric-motor-driven side-wall coring device. In the same figure B and C show two general types of side-wall-coring tools, in which the core tubes are driven into the formation by an electrically detonated charge. The electrically driven devices are lowered into the hole on an electric-logging line, and these devices are usually run into the open-hole section without requiring installation of drill pipe. The core tubes are generally driven into the formation either directly or indirectly by an electrically detonated powder charge or an electrically driven motor and are retrieved by hoisting the assemblies.

Wire-line side-wall samplers, as shown in B of figure 320, are run on an ordinary wire line, and the core tubes are deflected to the formation by suitable assemblies, which were previously attached to the drill pipe. The core is obtained usually by lowering the drill pipe until the core tube has penetrated the formation and then retrieving the core barrel, using the wire line and a suitable overshot assembly. The bit of a one-wire side-wall-coring assembly, shown in A of figure 320, is drilled into the side-wall formation by rotation of the drill pipe. A third type of side-wall-coring device, shown in figure 321, is run on drill pipe, and the core tubes are projected into the formation by mud pressure. These cores, obtained by the hydraulic method, must be retrieved by pulling the drill pipe.

The general advantages of the side-wall-sampling techniques are (1) that a sample can be obtained from the wall of a previously drilled hole, generally at any desired interval, and (2) that the method can be a valuable aid in confirming electric-log interpretations. Disadvantages are (1) that samples are usually too small for ordinary laboratory core analysis, and (2) that samples usually have been subjected to considerable flushing action of filtrate from the drilling mud.

In rotary coring it is very important to have a circulating fluid in the hole that has very good physical characteristics in order to reduce to a minimum the flushing by mud filtrate of the section to be cored. The treatment of water-base muds to filtration-rate values in the very low ranges

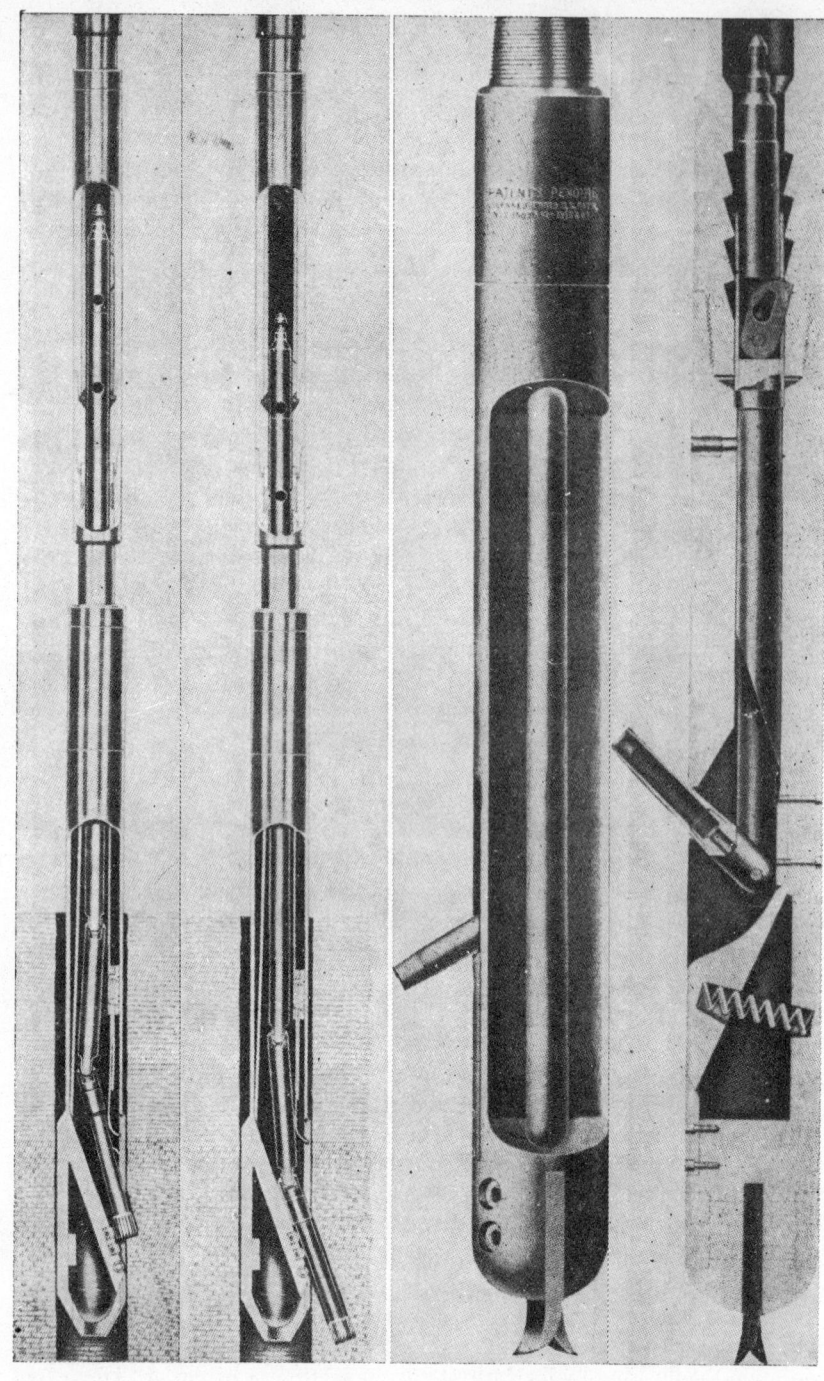

FIGURE 320. A—Rotating-type wire-line side-wall core barrel.
B—Wire-line side-wall core barrel.

and the use of oil-base muds have contributed appreciably to increasing the quality of data obtained from core analysis and have probably resulted sometimes in obtaining increased core recovery. Tracers placed in the circulating fluid may also ultimately help determine the extent of flushing of cores by mud filtrate.

In cable-tool drilling, when it becomes necessary to obtain a core, a barrel such as is shown in figures 322 and 323 is used. It consists primarily of an outer drilling barrel and an inner core-retaining tube, which does not move upward after coring is started until the tool is pulled from the hole. The outer drilling barrel slides on the core-retaining tube and cuts away the formation around it. The core tube follows down over the core of the formation, and the core catcher or trap ring traps the core as the barrel is pulled upward. The core is brought to the surface merely by pulling the barrel, which is usually attached to a wire line. Usually no additional surface equipment is needed for this operation.

APPLICATION OF CORING TO GEOLOGIC AND DEVELOPMENT PROBLEMS

In the past it has been found that coring is an aid in the solution of some of the problems in the geology and development of oil fields that may arise when subsurface formations are prospected. Those problems are usually general, or they may be related directly to either exploratory drilling or field or proved-area drilling. The geologist must evaluate the formations penetrated by the drilling, and to do this adequately he usually utilizes the best tools and methods available.

General geologic problems usually result from a well condition in which a desired tool or method cannot be used. For example, frequently it may be desired to obtain an electric log of a well, but the circulating fluid in the well may be of such a nature that the logging instrument cannot function properly, or, because of sloughing formations or other hindrances, the condition of the hole may be such that the instrument cannot be lowered to the desired depth. Also, in many areas, the geologist prepares a formation-sample log from data obtained by examination of formation cuttings. At times, however, these cuttings may not be satisfactory because (1) mud-circulation rates may be too low to get them to the surface without their being reground; (2) the presence of caving or sloughing formations may mask the drilled cuttings; or (3) the formation cuttings may be so soft that disintegration occurs before they reach the surface. In such circumstances coring, even though possibly less desirable, would provide a means of obtaining substitute data for geologic use.

In exploratory geologic work the geologist may use coring and core-analysis data ta (1) obtain a detailed formation description of the beds that are penetrated, (2) determine the dip and direction of dip of formation beds, and (3) determine the probable fluid content of prospective pay sections. After a core has been obtained and removed from the well, it is usually laid out in the same linear position it held in the core barrel so

FIGURE 321. Hydraulic-type side-wall core barrel.

that the recovery amount and the formation depth can be recorded. By visual examination and measurement, a detailed description of the core is made, the formation composition, the texture, the probable geologic age, the dip of formation beds, and the probable fluid content being observed. If indications are that the core may contain oil or gas, further field and laboratory tests are usually made. Should a core contain oil or gas, the visual examination can usually determine only what section may be a potential producing zone. A special coring method must be used to determine the direction of the dip of formation beds, and obviously such a determination may be very valuable in locating the probable direction in which a structure may be located.

Field or proved-area geologic problems usually resemble those encountered in exploratory work to some extent. The principal difference is that proved-area geologic work generally is directed toward obtaining data that may be used to evaluate a known pay zone or locate a formation marker. When an exploratory well encounters an oil- or gas-bearing zone, the geologic problems then resemble proved-area problems. In proved areas, the geologist may use coring and core data to aid in (1) determining the amount of pay section present and (2) estimating the amount of oil and gas in place.

Coring in Relation to Production Work

Frequently coring and core data can be used to an appreciable advantage in production operations and petroleum-engineering work, in addition to helping the geologist with his various problems. Probably the greatest use of coring in production work is for determining zones that should be formation- or production-tested and for determining, if possible, the gas-oil and water-oil contacts in those zones. Also, at times in production operations, it may be desirable to make an open-hole completion; that is, one in which the oil-string casing is set above the pay section to be tested and produced. Coring can be used to aid in determining the presence of undesirable upper sloughing shales and water-bearing zones immediately above the pay zone. The casing seat may then be picked at a point that would shut off the undesirable formations and yet allow the entire desired section to be tested. Information about the texture of the formation in a prospective pay zone usually aids in picking the most desirable type of completion method and helps determine the possibility of sand production problems. At times diamond coring can aid in operations in areas where very hard and abrasive formations are encountered, because it may be found that coring is more economical than drilling.

Certain petroleum-engineering work is aided greatly by core-analysis data. An understanding of the formation characteristics and composition may lead to the location of (1) oil zones that subsequently might be overlooked and (2) impermeable zones that may aid greatly in workover operations to shut off undesirable water or free gas. Core data almost always help the engineer evaluate subsequent work-over possibilities. One of the earliest engineering uses of core-analysis data was for evaluating and planning secondary oil recovery by water flooding. Today, in addition to that same use, it also aids in evaluating gas cycling and pressure maintenance by gas-injection projects. Probably the latest and perhaps one of the most valuable uses of modern core-analysis data for an engineer is to provide basic data for reservoir-analysis studies.

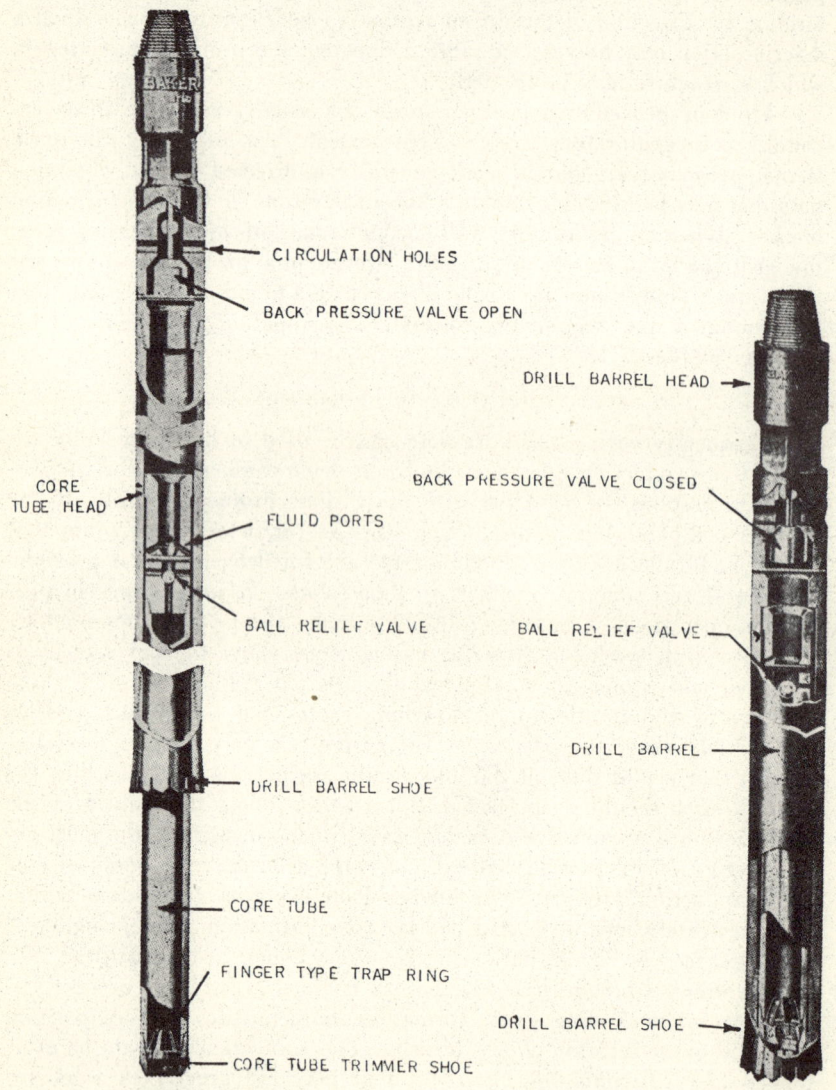

FIGURE 322. Cable-tool core barrel in upstroke position.

FIGURE 323. Cable-tool core barrel in downstroke position.

CORRELATION BETWEEN CORING AND ELECTRIC LOGGING

In considering the value of core data, it must be remembered that the examination of formation cores, either in the field or in the laboratory, provides the only direct information concerning the physical properties of the formation that the drill penetrates and that these core data are the basis of electric and other log interpretations. In figure 324 are illustrated graphic means of presenting electric-log and core-analysis data. An examination of formation cuttings and the use of other logging devices may substantiate core-analysis data, but individually those methods are usually subject to variable factors and broad interpretations. When the geologist uses core data for correlation, he is generally certain that they are accurate; and once the reaction of an electric log in a particular formation in a certain area has been established or substantiated by core-data interpretations, the log becomes a very useful tool. Likewise, formation tests are necessary to determine electric-log characteristics in regard to the probable type of fluid that a formation will produce. It has been found that an interpretation of electric logs of certain holes is often very misleading; the logs are very valuable, however, in determining the tops and thicknesses of certain sections when they are used in conjunction with core data, especially when core recovery has been poor.

Since electric logs are influenced sometimes by the type of drilling mud and the mineral content of the section logged, core logs are often the only means of formation interpretation. Sometimes a correlation between an electric log and a core log on producing formations results in finding measurement errors, which may cause subsequent difficulty. In using electric logs for purposes other than correlations, limitations must be recognized, such as failure to register the presence of sands or other zones in a producing horizon and the reverse reaction on producing sands in certain areas.

LIMITATION OF CORING TECHNIQUES AND APPLICATIONS TO GEOLOGIC PROBLEMS

Even though coring is the best tool available for the correlation and interpretation of geologic formations, it too has certain limitations that must be recognized. Perhaps the greatest limitation of coring is that samples of the complete section cored are rarely obtained for examination. The amount of cored section recovered is often as low as sixty percent, and sometimes there is no recovery. When full recovery is not obtained, it is usually difficult to place the recovered section in the correct position in the formation log and to assign physical values to the entire section from those obtained by analysis of the recovered section. Then too, after a core has been recovered, it is not possible to analyze the entire recovered section; thus more limitations are introduced, because the core has to be sampled and tests obtained only on the sampled portions. Work on gas-oil and water-oil contacts from core-analysis data has certain limitations.

Usually oil percentages, determined from core analysis, are much lower above the gas-oil contact than below it, but this is not always true, especially when the oil has a rather high gravity. Frequently a recovered core near the water-oil contact appears to contain more oil than those higher in the oil column. Past experience has indicated that cores usually are subjected to considerable flushing by filtrates from drilling muds. When water-base muds are used, oil .is generally flushed from the core and oil-saturation values are determined from core analyses may be too low. Likewise, when oil-base mud is used, the water-saturation values determined on some cores, especially those near or below a water-oil contact, may be erratic. It has also been found that some oil sections contain argillaceous materials, which may have various types of nonproducible

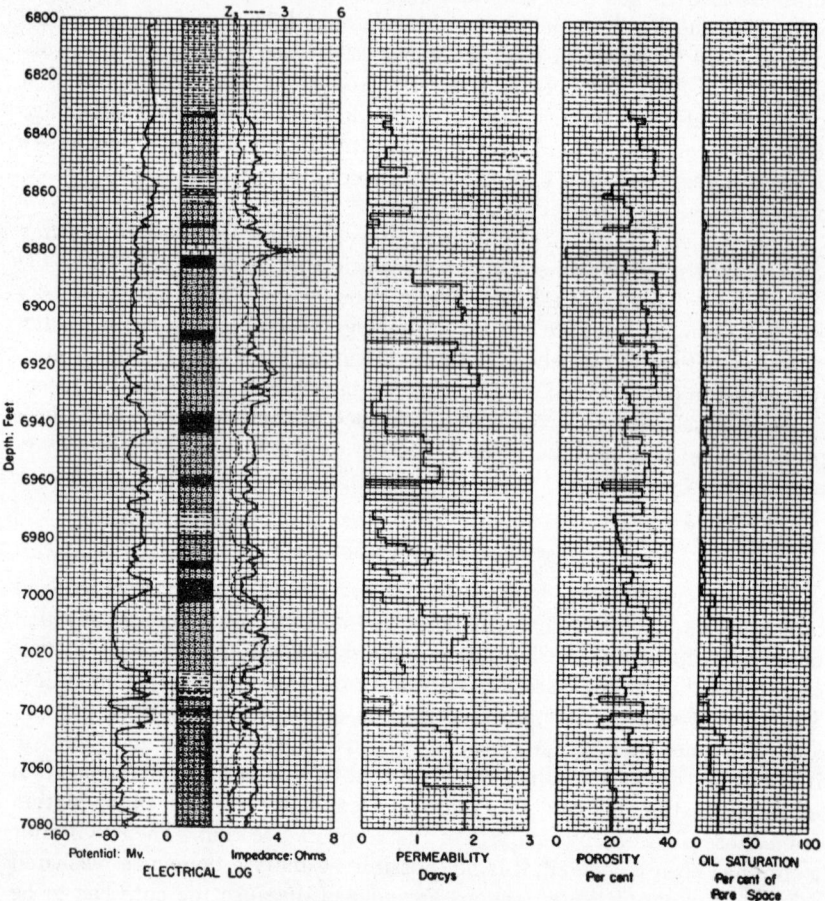

FIGURE 324. Core-analysis data from cores obtained with low-filtration drilling fluid in hole.

bound waters, and these in turn may introduce a considerable error in core-saturation determinations.

Another appreciable limitation of coring results from the fact that it is usually very expensive and causes marked increases in over-all well costs when it must be used. The development of more economical methods to evaluate subsurface formations accurately as well as to find ways to reduce coring costs would be a major contribution to the oil industry.

APPLICATION OF DIPMETER SURVEYS

E. F. STRATTON AND R. G. HAMILTON

The SP (spontaneous-potential) dipmeter was described [7] several years ago as a means for determining the direction and magnitude of formation dip *in situ*. This instrument was designed to record simultaneously three SP curves of known orientation, 120° apart along three generatrices of a well bore. Each curve fixes thus one point on a bedding surface and the position of the surface can be determined (fig. 325) by the displacement between the curves.

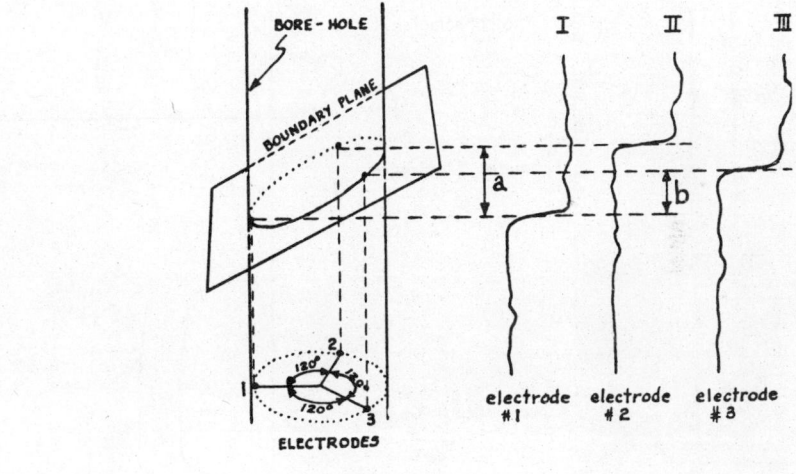

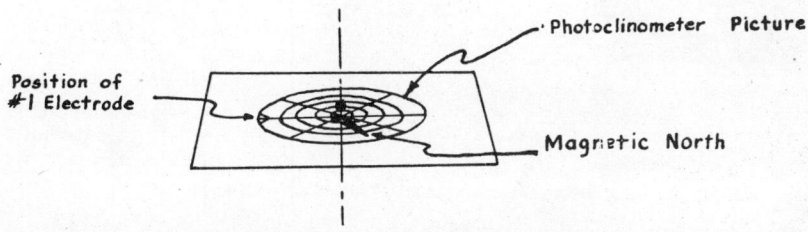

FIGURE 325. Basic principles of dipmeter.

[7] Doll, H. G., *The SP Dipmeter:* Am. Inst. Min. Met. Eng. Tech. Pub. 1547, Jan. 1943.

It was thought desirable in some geologic territories to record three resistivity curves instead of three SP curves. Accordingly, the design and development of a resistivity dipmeter was undertaken in our research department, particularly by H. G. Doll and W. B. Steward. The availability of a resistivity dipmeter, in addition to the SP dipmeter, has extended appreciably the application of the art. Some six or seven thou-

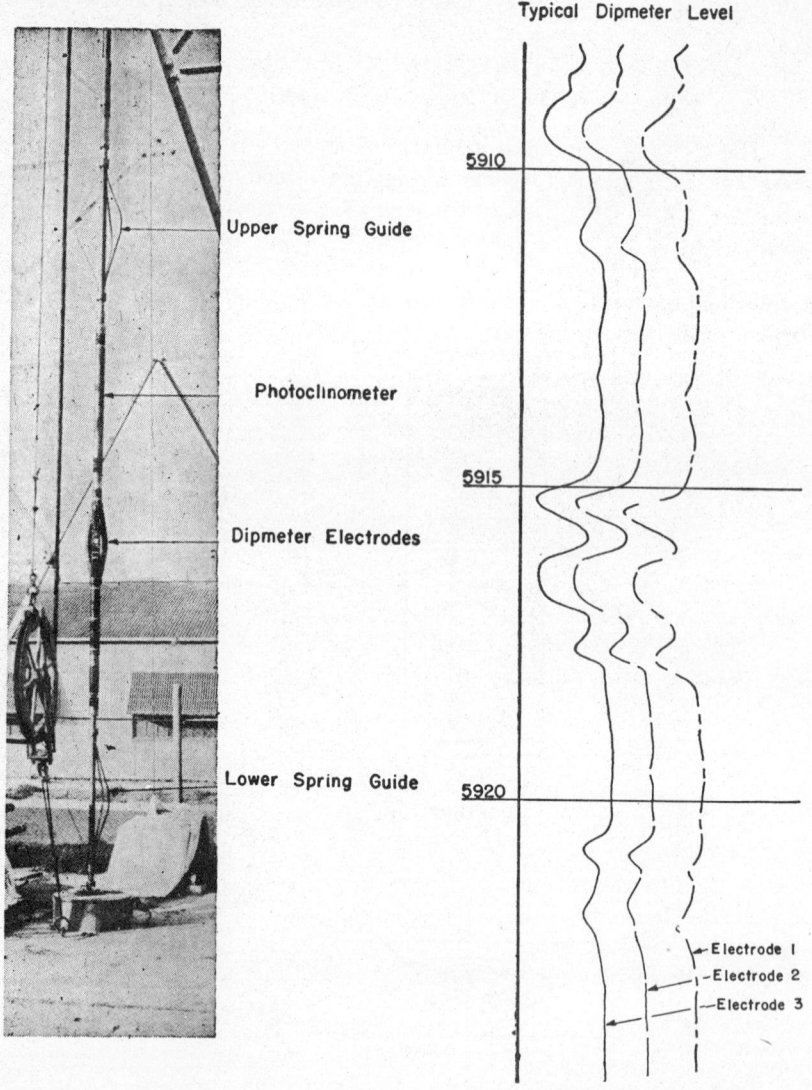

FIGURE 326. Dipmeter hole assemblage.

FIGURE 327. Record of typical dipmeter level.

sand dipmeter levels, SP and resistivity, now have been recorded in wells throughout most of the major oil provinces in the country and in many foreign fields. It seems advisable to analyze these data and to describe their application to some of the problems that have been encountered in exploration and development work.

The hole assembly, about 25 feet long, is shown in figure 326. It consists of a mandril to which are attached three hard-rubber arms spaced 120° apart; in the center of each arm and positioned on the same plane at right angles to the axis of the instrument is one of the three recording electrodes. Attached to the electrode unit is a photoclinometer, which determines photographically the orientation of each of the three SP or resistivity curves and gives the drift and azimuth of the well bore. Spring guides above and below the photoclinometer-electrode assembly serve to keep the device centered in the hole and to prevent it from turning.

The curves are recorded photographically at the surface on the standard electric-logging recorder. Figure 327 is a record of a typical dipmeter level.

AREAS OF APPLICATION

Those regions where the geologic section is primarily sand and shale, i. e., California or the Gulf Coast, are best suited for the use of the SP dipmeter. The spontaneous potential here in general shows sharp, well-defined anomalies at formation boundaries, which give definite dip determinations. Likewise in these areas a series of bedding surfaces between sand and shale can usually be found at any depth in a well where a dip determination is needed.

The resistivity dipmeter, on the other hand, has proved of utility in such areas as west Texas, the Mid-continent, and the Rocky Mountains. The numerous resistivity anomalies between, for example, shale and limestone found in most wells in these areas provide satisfactory levels for dip determinations at almost any position in a well.

FIELD PROCEDURE

Dipmeter surveys are made with the same cable and, as noted above, surface-recording equipment used for electric logging. Common practice is to make the dipmeter run immediately after the electric logging.

The dipmeter levels must be chosen, as described later, from the electric log. The assembly—electrodes, photoclinometer, and spring guides—is lowered to the base of the shallowest level and a photoclinometer picture taken. This picture determines the orientation of the electrodes and the curves, as well as the drift and azimuth of the well bore at the base of the level. The curves are recorded to the top of the level, and a second photoclinometer picture is taken. The latter gives the orientation of the electrodes and the curves, as well as the drift and azimuth of the well bore at the top of the level. After this step, the assembly is lowered to the base of the second-shallowest level, and the procedure noted above is repeated. Other levels are recorded in the same manner.

After the deepest level has been recorded, it is resurveyed with a photoclinometer picture taken in the middle of the level as well as at the top and base. Each of the higher levels, similarly, is repeated as the equipment is withdrawn from the hole.

INTERPRETATION

All dipmeter surveys are analyzed by a staff specializing in such work, and two independent interpretations are made for each operation. Likewise, the data from the original and check runs on each level, generally recorded with different electrode orientations, must agree, or the results are discarded. After the orientation of the curves, their displace-

STA-TION	Depth Interval	From Magnetic North			Displacement of Curves in reference to I.		DIP			OBSERVATIONS
		Drift Azimuth	Drift Angle	Orient. No. I	II	III	Dip Angle	Direction from Mag. b.	True North	
AA1	1645 1659 to 1653	36	1 00	92	Uncertain					
AA2	1673	36	1 00	97	Uncertain					
BB1	2000 2020 to 2004	342	0 45	77	U 0.4	U 1.2	9	120	S 50 E	
BB2	2024	342	0 45	77	U 0.4	U 1.2	9	120	S 50 E	
CC1	2856 to	234	2 00	229	D 0.6	D 2.2	17	91	S 79 E	
CC2	2875	234	2 00	16	D 0.8	U 1.2	15	93	S 77 E	
DD1	3135 to	241	1 00	119	U 1.8	U 0.6	13	75	N 85 E	
DD2	3158	241	1 00	299	D 1.9	D 0.9	13	81	S 89 E	
EE1	3520	237	2 00	214	-0-	D 1.8	15	88	S 82 E	
EE2	3535	237	2 00	75	U 1.2	U 2.0	15	94	S 76 E	
FF1	3608	238	2 15	292	D 0.8	D 2.2	14	140	S 30 E	
FF2	3622	238	2 15	292	D 0.8	D 2.2	14	140	S 30 E	
GG1	3844 to	242	2 00	265	D 2.0	D 1.2	15	60	N 70 E	
GG2	3870	242	2 00	355	D 0.5	U 1.6	16	66	N 76 E	
HH1	3990 to	242	2 15	312	D 2.0	-0-	17	66	N 76 E	
HH2	4000	242	2 15	195	-0-	D 2.2	17	72	N 82 E	
II1	4155 4183 to	234	1 30	343	D 2.3	D 2.3	26	74	N 84 E	
II2	4191 4173	234	1 30	343	D 2.3	U 1.6	26	74	N 84 E	

Smith Petroleum Company, T. Henry #3, Run #1, September 1, 1947

FIGURE 328. Typical dipmeter computation sheet.

ment, and the drift and azimuth of the well bore have been determined over the interval of a level, the magnitude and direction of the dip are obtained quickly by mechanical means.

A typical computation sheet is shown in figure 328. The first column designates the level; A1, for example, is the original run, A2 the check run, etc. The second column gives the depth interval of the level. The third column shows the azimuth of the well in degrees from magnetic north; the fourth column the drift angle. The fifth column is the position of the No. 1 recording electrode in degrees from magnetic north. The sixth and seventh columns indicate the displacement in inches of curves recorded on electrodes two and three with respect to the No. 1 curve. The last three columns give the amount and direction of the dip computed from the previous data.

SELECTION OF LEVELS

Widespread experience indicates that the accuracy of a dipmeter survey depends principally upon a careful selection of the zones in a well over which the measurements are made. It is obvious, too, that numerous dip determinations made at relatively close intervals in a well will more clearly define structural and stratigraphic conditions than a few randomly spaced levels.

After it has been decided at what depth positions dip determinations are needed, a zone or level from 25 to 50 feet in length is chosen nearby for recording the three curves. Each level of such length thus provides a number of bedding surfaces on which a dip determination is made; if the dip measurement is made on but one contact surface, a freak dip direction due to minor bedding irregularities might be considered as representing the true formational dip.

It has been found that the most satisfactory zones for a dipmeter level are those consisting of relatively thin beds, 2 or 3 to 10 feet thick, having sharp contacts with adjacent formations. Such zones for a resistivity-dipmeter level are thin limestones or resistive sandstones interbedded with shale, for an SP-dipmeter level thin sands or sandstones interbedded in shale.

Thin shale or sandy shale beds, on the other hand, within thick sand sections and thin shale beds or minor resistivity variations within a massive limestone frequently give erratic dip determinations. Sometimes, too, dipmeter measurements at the contact between thick sandstones or limestones and the adjacent shales show abnormal results, although reliable values frequently have been obtained at such contacts. The two composite logs in figure 329 illustrate the selection of dipmeter levels.

APPLICATION

Dipmeter surveys provide data assisting in the solution of many structural and stratigraphic problems encountered in exploratory and pool development wells. The correct location of offset wells after one has been drilled is a common problem. If the initial well is a wildcat and a dry hole, it is necessary to know, first, whether the sediments are flat or whether there is some evidence of structure, and, second, what the direction and amount of dip are. Similarly a thin oil reservoir may be found overlain by unwanted gas or underlain by water. It is obvious that in the first case additional wells should be downdip, whereas in the second other wells should be updip.

Well 2 in figure 330, drilled in the Gulf Coast, had noncommercial oil shows in the top of the Frio at 6,222 feet. A dipmeter survey indicated that the dip through the section was 31° S. 73° E. A second, well 1, was found productive in the Frio topped at 5,925 feet. According to the dipmeter data, well 1 should have been 288 feet higher than well 2, whereas it was found to be 298 feet higher.

A similar problem may arise in the steeply dipping beds around piercement-type salt dome. One well may find all sands water-bearing, but laterally a short distance away they may be structurally higher and some of them oil-bearing. The original dry hole may be sidetracked and production obtained if the direction and magnitude of the dip are known.

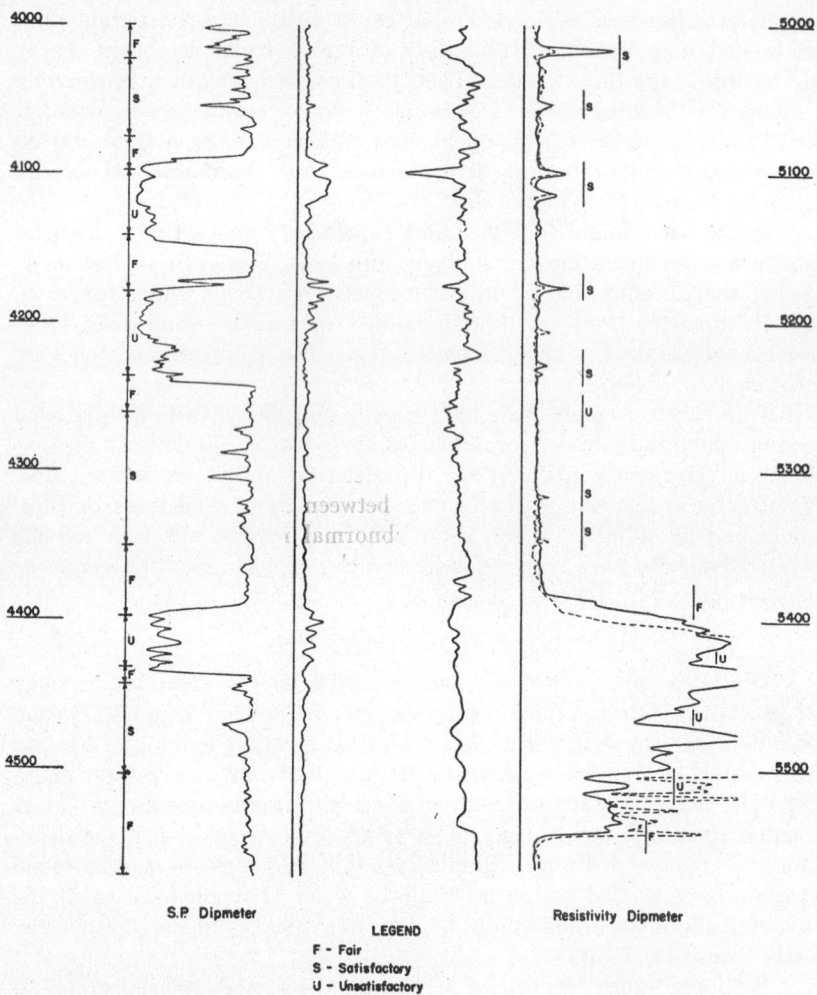

S.P. Dipmeter   Resistivity Dipmeter

LEGEND
F - Fair
S - Satisfactory
U - Unsatisfactory

FIGURE 329. Composite electric logs illustrating level selection for dipmeter surveys.

Figure 331 is a typical example of the use of dipmeter data in such work. The original well, No. 1, was dry in all sands to a depth of about 7,500 feet; some slight saturation, however, was found in the top of the sand reservoir at 6,500 feet. The dipmeter survey showed a dip of 67° S. 33° E, on the top of the sand and a dip of 76° S. 36° E, near the base.

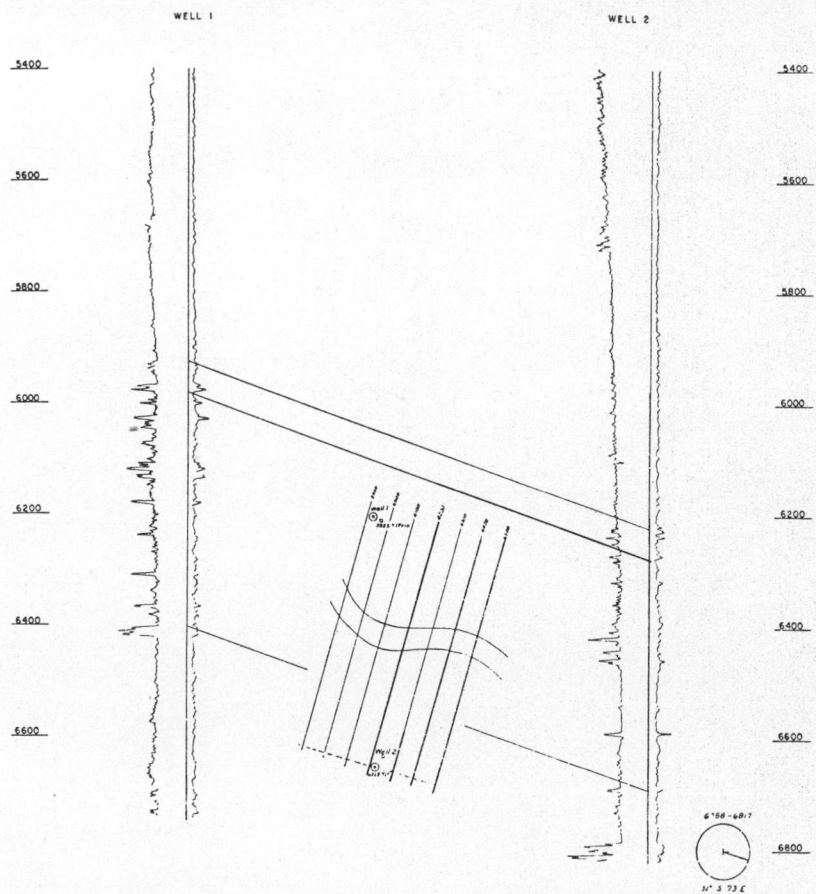

FIGURE 330. Dipmeter-survey results; useful in determining structural relationships of wells and may also assist in determining contour trends.

The well was sidetracked northwest on the basis of this data, and the sand was found to be productive at about 6,000 feet, some 500 feet higher than in the original hole.

The sand topped at 6,500 feet, *A*, has an apparent thickness of 410 feet; the actual thickness, however, using an average dip of 70°, is only

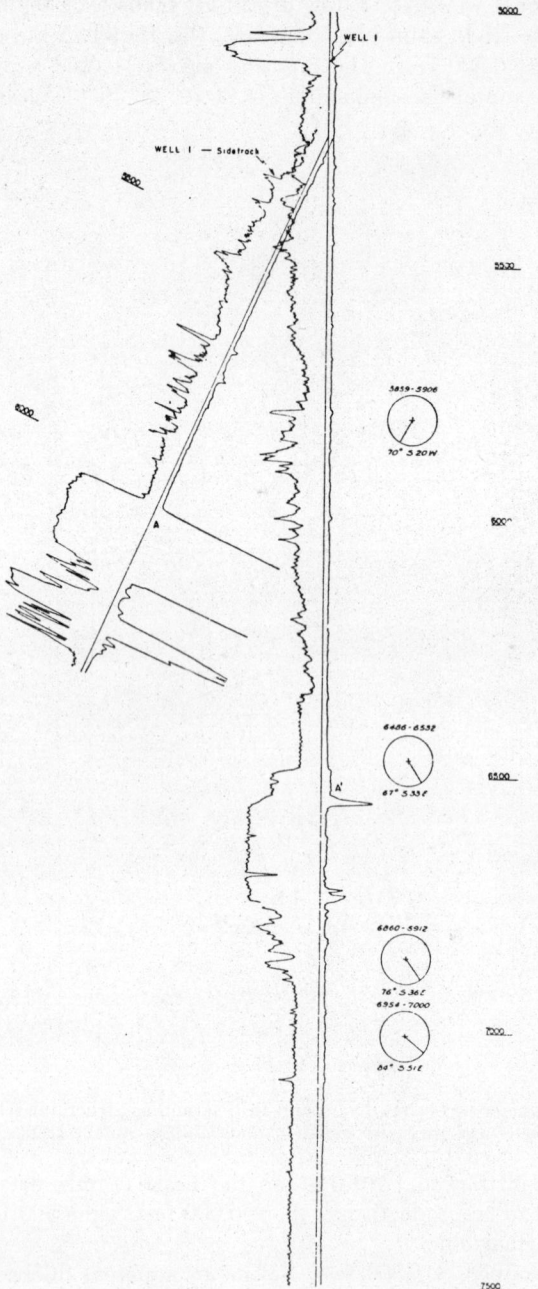

FIGURE 331. Dipmeter-survey results; useful in evaluating structural control in directional-drilling problems.

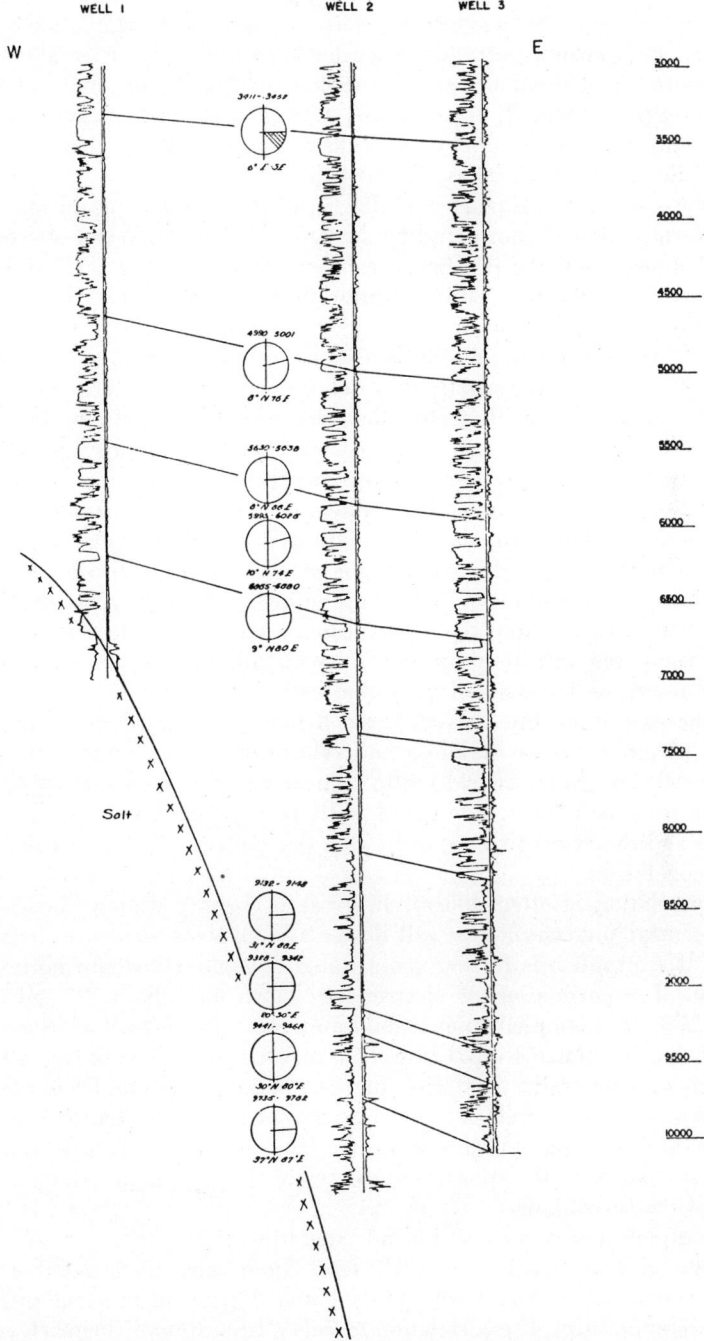

WELL 1 WELL 2 WELL 3

W   E

3000
3500
4000
4500
5000
5500
6000
6500
7000
7500
8000
8500
9000
9500
10000

Salt

3911–3952
6° S 3E

4990 5001
6° N 76 E

5630 5638
6° N 66 E
5993 6018
10° N 74 E
6055 4880
9° N80 E

9138 – 9148
3° N 60 E
9318 – 9342
10° 30° E
9441 – 9468
30° N 80° E
9935 – 9762
37° N 87° E

FIGURE 332. Use of dipmeter data in solving structural conditions
near a piercement salt dome.

140 feet. The dip in the original well increases with depth, indicating that it is approaching salt; it is expected then that dips in the sidetrack, also approaching the dome, may be higher than in the original hole and that the sand section, *A*, is practically vertical. The net thickness of the sand in the sidetrack on this basis is the same as in the original well, although the apparent thickness is 330 feet.

The use of the dipmeter in the solution of other problems near piercement domes is shown further in figure 332. A dipmeter survey on well 2 showed that the dip increases from 6° or less east at 3,400 feet to 37° east at 9,700 feet. This condition indicates a thinning section westward with the possibility of pinchout traps or accumulation against the dome. This is verified by the short sections in well 1 compared with those in well 2 and by the long sections in well 3 compared with those in well 2.

The dip increase with depth is believed to indicate again an approach to the salt; the dip thus may be expected to flatten laterally as well as vertically away from the dome, or conversely to steepen near the dome. Correlation of well 2 with well 1, which encountered salt at about 6,800 feet, shows a steeper dip than that indicated by the dipmeter survey in well 2. The dip obtained by correlation of well 2 with well 3 is less than that shown by the dipmeter. The lower dip obtained by correlation between wells 2 and 3 and the higher dip by correlation between wells 1 and 2 compared with those on the dipmeter are normal and could have been foreseen on the basis of the dipmeter record in well 2.

The steepening dip westward and flattening eastward are important too in planning further development. The productive reservoir at 9,700 feet in well 2 might be considered to be near water, if the steep dip shown thereon was continuous; additional wells to the east, however, could be drilled and the reservoir extended if the dip flattens. Well 3 verifies this latter conclusion.

Correlation is often difficult in areas of steeply dipping beds such as those near piercement-type salt domes and on steep structures in other areas. The problem is further complicated if angular unconformities are present. The correlation of electric logs taken in wells 1, 2, and 3 in figure 333, for example, is not readily apparent. The formation *B* in well 2 might be correlated with *D* in well 3, or even with *F*, assuming a fault between the two wells. Likewise, the correlation of *B* and *D* in well 2 with similar beds in well 1 is not obvious, owing to the lengthened section in well 1 below the unconformity. The problem is further complicated by changes in the appearance of the electric log due to lateral variations of the formations.

A dipmeter survey in well 3 indicated dips of 30° to 37° S. 67° W. at depths of 1,035, 1,242, and 1,540 feet. These data made possible correct correlations of well 3 with wells 1 and 2 lying to the southwest in the direction of dip. The correlation reveals a truncation of Pennsylvanian beds against the overlying Permian. The dip value obtained by correlation

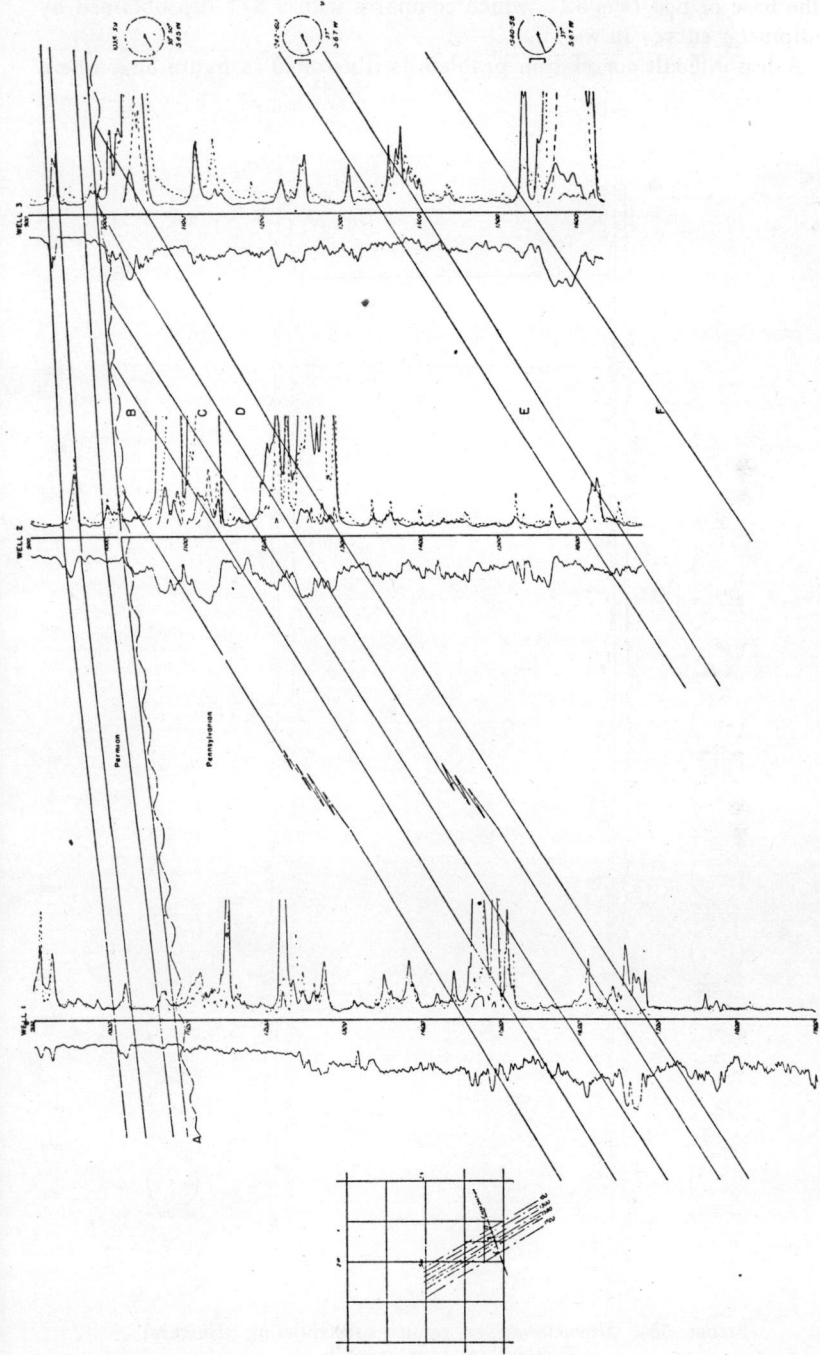

FIGURE 333. Application of dipmeter surveys in identifying and evaluating an angular unconformity.

of bed *D* is 32°, which compares with a 37° dip obtained by
~~er~~ survey in well 3.

~~s-~~difficult correlation problem is illustrated in figure 334, where

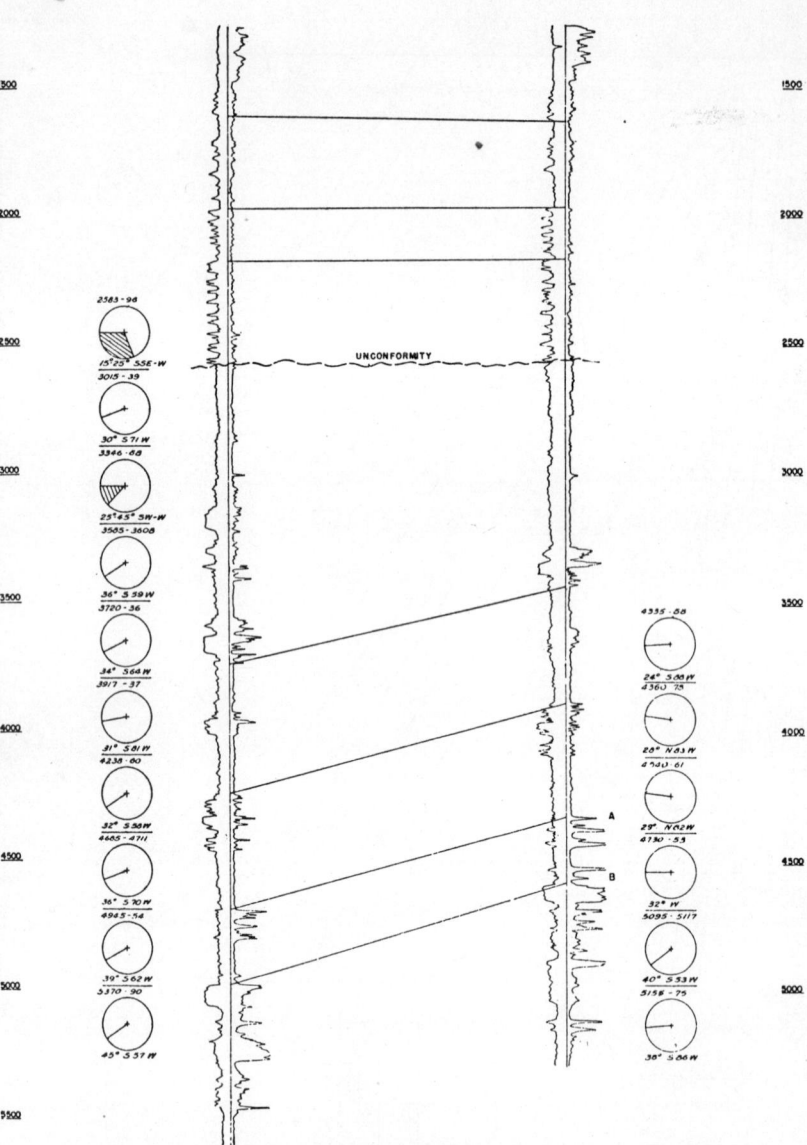

FIGURE 334. Dipmeter-survey results substantiating structural relationships between wells.

relatively flat beds overlie steeply dipping beds. The dipmeter survey in well 2 indicates at *A* dips of 24° and 28° almost due west. The dip value obtained by correlation with well 1 to the west is slightly over 29°. Likewise, the dipmeter results indicate a dip of 29° almost due west at *B*, while the dip obtained by correlation with well 1 is 30°. Dipmeter results on beds *A* and *B* in well 1 show steeper dips than in well 2. The steeper dips appear to be substantiated by the fact that correlative formations in well 1 have greater apparent thickness than in well 2. In view of the close agreement of dipmeter results in well 2 with correlation dip values, it is believed that the formation dips steepen appreciably in the

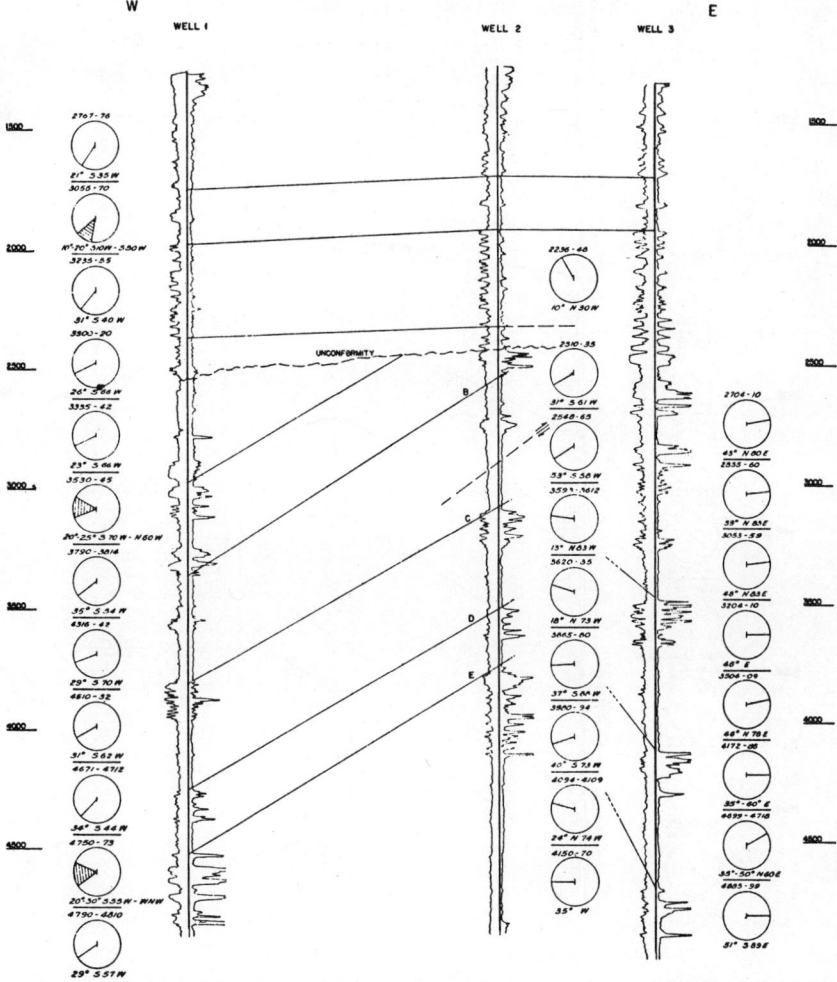

FIGURE 335. Structural reversal on basis of dipmeter and electric-log control.

vicinity of well 1 and that the dipmeter results correctly represent the subsurface condition.

The use of dipmeter results in conjunction with other data to solve problems of correlation in complex structures is further illustrated in

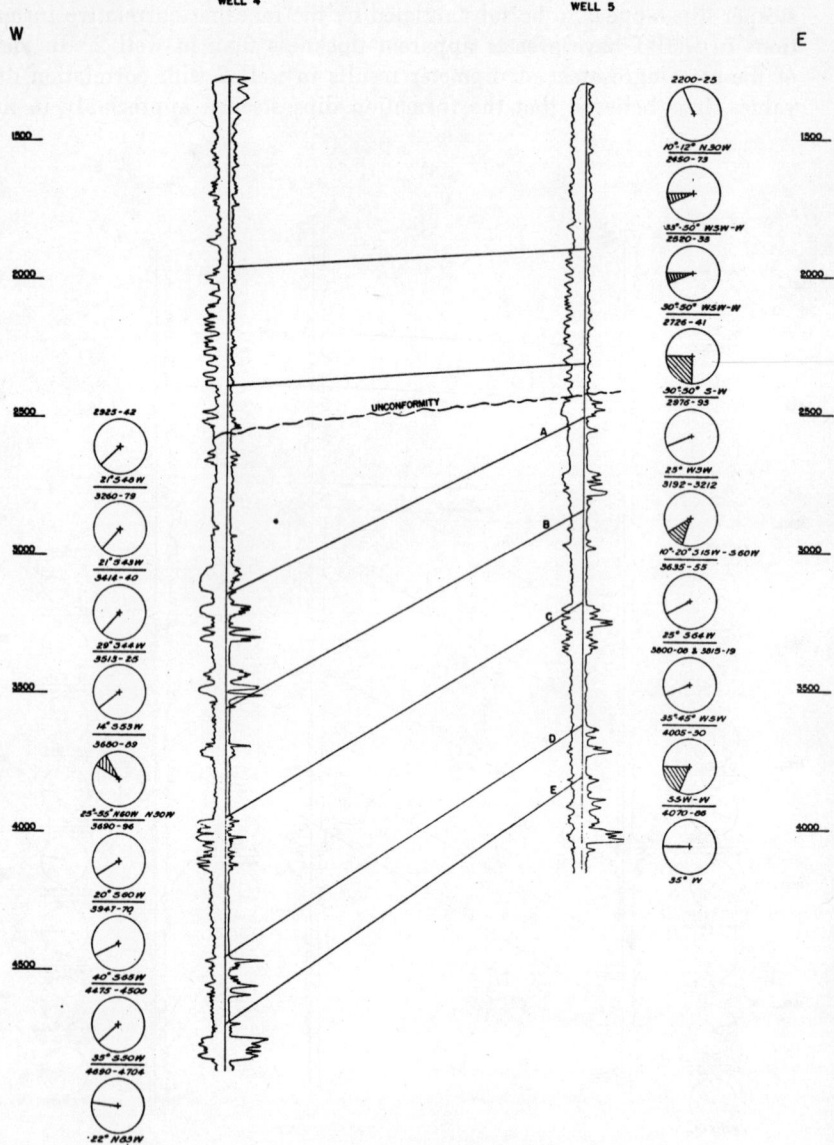

FIGURE 336. Structural interpretation between wells as based on dipmeter and electric-log data.

figures 335 and 336, where a dipmeter survey was made in each of five offset wells. Wells 4 and 5 are direct offsets to wells 1 and 2, respectively.
The correlation of well 1 with well 4 is fairly obviously from the logs;
likewise, the correlation of well 2 with well 5 is apparent. However, owing
to thicker sections, the correlation of well 1 with well 2 and well 4 with
well 5 is more difficult. Dipmeter results in wells 1, 2, 4, and 5 indicate
steep dips to the west and southwest on beds A to E inclusive, thus ex-

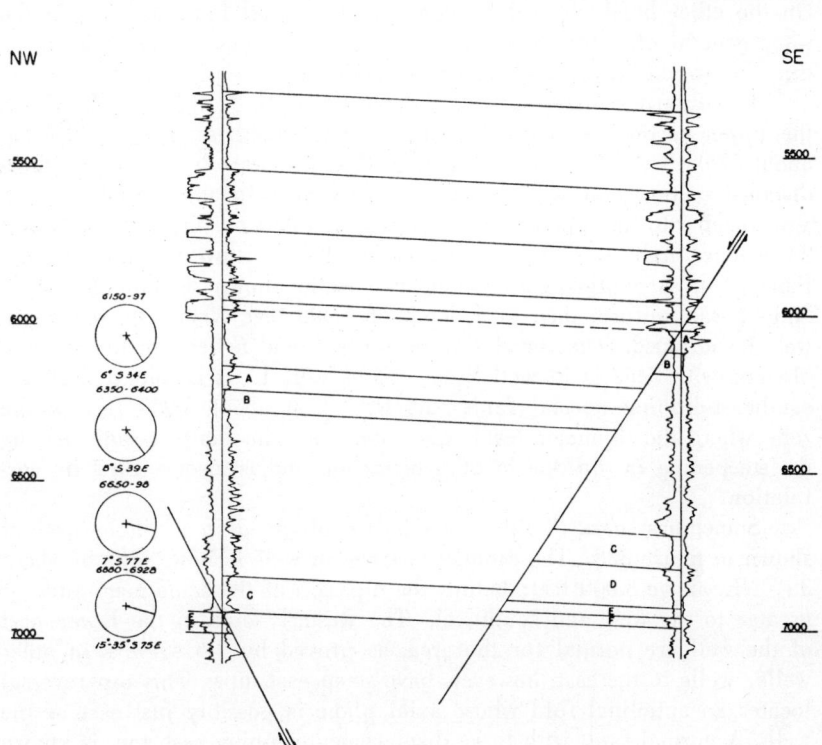

FIGURE 337. Dipmeter data as used in evaluating fault problems
encountered in wells.

plaining the thicker sections in wells 1 and 4. A dip of about 10° N.
30° W. in the upper beds contrasts with much steeper and westerly dips
in the lower sections, and thereby identifies the angular unconformity. It is
of interest to note that the dip in wells 1 and 4 increases with depth. Likewise, the apparent thickness of the formations is greater in well 1 than in
well 2. The dipmeter determinations, thus, in general reflect a thickening
of the section downdip and off-structure from east to west.

Dipmeter results in well 3 indicate dips as steep as 50° to 60° due
east, a sharp reversal from the west dip of well 2. The recognizable forma-

tions also are of much greater apparent thickness. This abrupt reversal from west in wells 1 and 2 to east in well 3 represents a sharp asymmetric fold or drag near a normal fault of large displacement.

An interesting sidelight is the abnormally thick shale section *BC* in well 2. This thickening represents duplication of section by a thrust fault of small displacement, a phenomenon not uncommon in the area.

Drag folding adjacent to a normal fault dipping in a direction opposite to formation dip may result in sharp dip reversals as noted above. On the other hand, drag distortion near a normal fault dipping in the same general direction as the bedding will cause only steepening of the dip. Figure 337 illustrates the latter condition.

The dipmeter survey in well 1 showed a 6° to 8° SE. to ESE. dip in the upper section of the well, with a sharp increase to 15° ESE. at about 6,900 feet. The correlation of well 1 with well 2, the latter a short distance southeast of well 1, gave a small SE. dip above 6,000 feet in agreement with the dipmeter results. Below this depth the section in well 2 shortened, indicating that the well was cut by a northwest-dipping normal fault. The correlation between the two wells, shown in intervals *A, B,* and *C,* was uniform then to a depth of 6,900 feet, where the section in well 1 shortened. The correlation of zones *E* and *F* between the wells is obvious with zone *D* in well 2 cut out of well 1. It is apparent that a southeast-dipping normal fault cuts well 1 at about 6,900 feet in the zone where the dipmeter level was recorded. The fault is indicated by the steepening in dip due to drag distortion and is also verified by correlation.

Sometimes marked dip reversals are observed in a single well as shown in figure 338. The dipmeter survey in well 1 shows dips of about 35° NE. above 3,000 feet; below, the dips are of the same magnitude or greater to the west and southwest. The westerly dips in the lower part of the well are normal for the area, as proved by dip surveys in other wells; wells to the east, however, have steep east dips. This dip reversal locates an anticlinal fold whose axial plane is possibly just east of the well. A normal fault with large displacement dipping east, too, is known to lie just east of the well. The east dip in the upper part of the well represents possibly an effect of the folding accentuated by drag near the fault.

It is interesting to note that bed *C,* by correlation with a type section shown in well 2, appears to be a reptition of *C.* This situation again indicates a small thrust fault between *C* and *C'.*

Dipmeter results, in conjunction with seismic and geologic data, have been utilized to determine a thrust fault of much greater displacement. The electric log in figure 339 has a normal succession of beds to approximately 5,800 feet. Formations below this depth are identified as young beds definitely out of place for a normal stratigraphic sequence. Bed *A* at about 7,200 feet is identified as a repetition of the same bed at about 3,700 feet. This fact indicates thrust faulting. Other information too

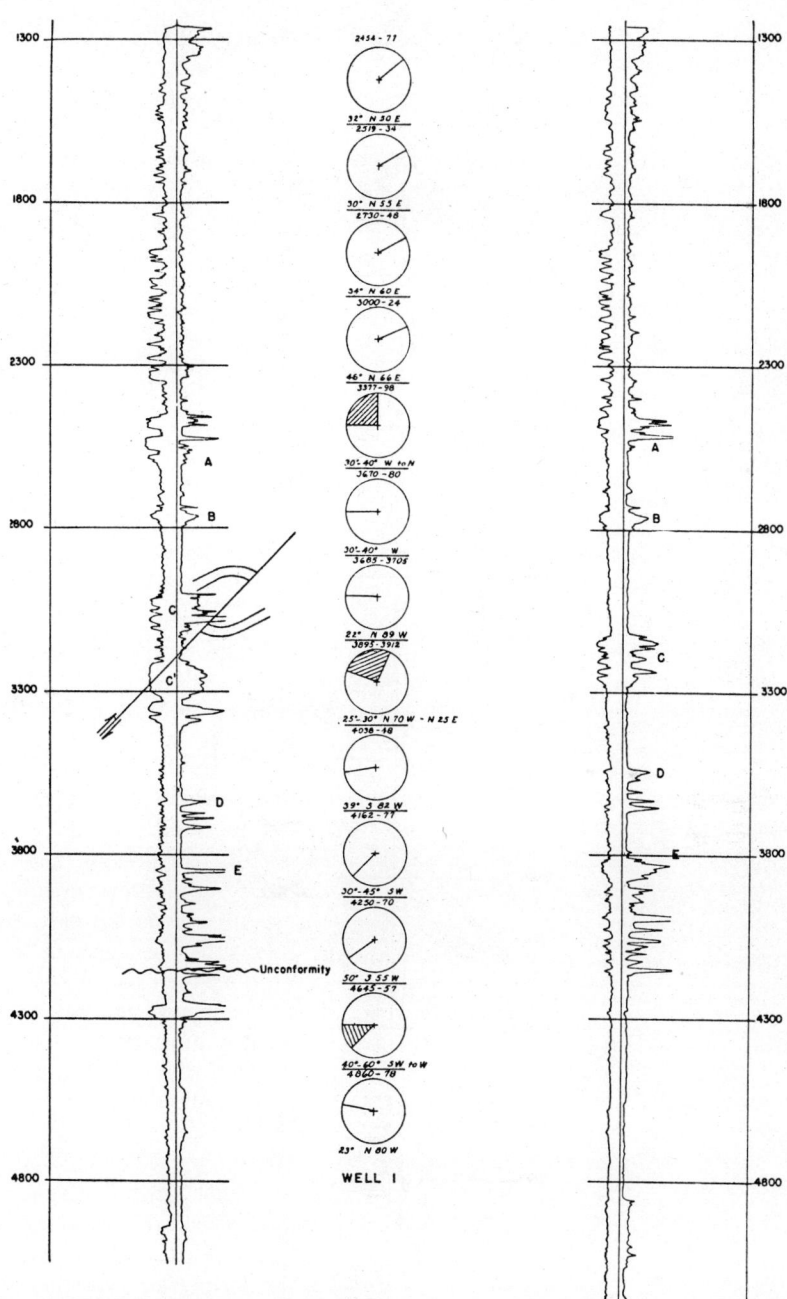

WELL 1   WELL 2

FIGURE 338. Marked dip reversal shown in well No. 1. Dips above 3,000 feet are to northeast; below 3,000 feet they are in west and southwest direction, thus indicating presence of a fault. Well No. 2 is unfaulted.

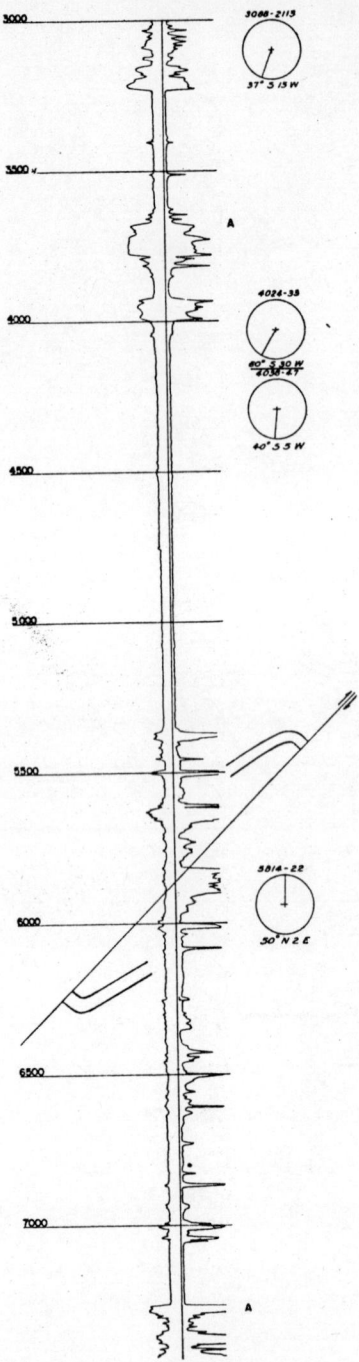

FIGURE 339. Use of dipmeter for evaluating fault conditions in a well.

indicates that a fault cuts the well at approximately 5,814 feet. Dipmeter determinations at 3,088, 4,024 and 4,038 feet show dip values of 37° to 60° almost due south. A determination at 5,814 feet gives a due-north dip of 50°, a sharp dip reversal. The reversal may be interpreted as an unconformity, a fold, or a fault. In view of the information available, it is logical to interpret the reversal as indicating the presence of a thrust fault, the reverse dip occurring in the drag of beds near the fault. Additional dip levels both above and below the fault would have substantiated this interpretation and further localized the fault zone. The apparent thickness of bed *A* at 7,200 feet is less than at 3,700 feet. Additional dip levels might have shown a lower dip at 7,200 feet.

CONCLUSION

Dipmeter data can be used to aid the solution of many structural and stratigraphic problems. Some of these have been described, others only made obvious—for example the location of potential trap areas. Additional applications will undoubtedly develop.

DESIGN AND APPLICATION OF ROCK BITS

L. L. PAYNE

The strength of formations encountered in drilling a hole into the earth's crust varies over an extreme range. For instance, the soft, unconsolidated shales and sands of the Cenozoic are relatively easy to drill as compared to the high-strength lime, dolomite, and chert of the Paleozoic. The rock bits designed for economical performance in drilling formations of the Cenozoic must be altered to provide the types of rock bits developed for the optimum performance in the older, high-strength formations of the Paleozoic. The manner in which the arrangement of the cutting teeth and their action on bottom are altered to provide the types of. bits available for drilling the various formations encountered is discussed under the basic geometry of the tricone bit and the nomenclature that identifies each type of bit.

All Hughes tricone rock bits have the same basic geometry; three-cone construction with chisel-tooth arrangement and action on bottom especially designed to drill the multitude of formations encountered efficiently. This basic geometry has several desirable features. The cone-shaped cutters enclose the antifriction bearings and bearing supports, thereby eliminating any portion of the bit head near the bottom of the hole to invite balling up in sticky formations. The interfitting of rows of teeth on the inner conic portion of each cone into grooves on mating cones provides more space which permits the use of deeper teeth and larger bearings without sacrificing the strength of the cone body. The interfitting of teeth into grooves provides mechanical cleaning of the teeth and grooves, as the cutters are rotated on the formation. This relationship is beneficial in helping the drilling fluid clean the bit.

The geometry of the cone-shaped cutters and the position of the bearing pins supporting the cutters are altered to provide more or less twisting-tearing-scraping action of the teeth as they penetrate the formation. When extremely hard and abrasive formations such as tightly cemented sandstone, sandy dolomite, quartzite, chat, chert, and similar rock types, are being drilled, rock bits perform most efficiently with cutters that have an approximately true rolling motion which provides a chipping-crushing action of the teeth on the formation. A minimum of

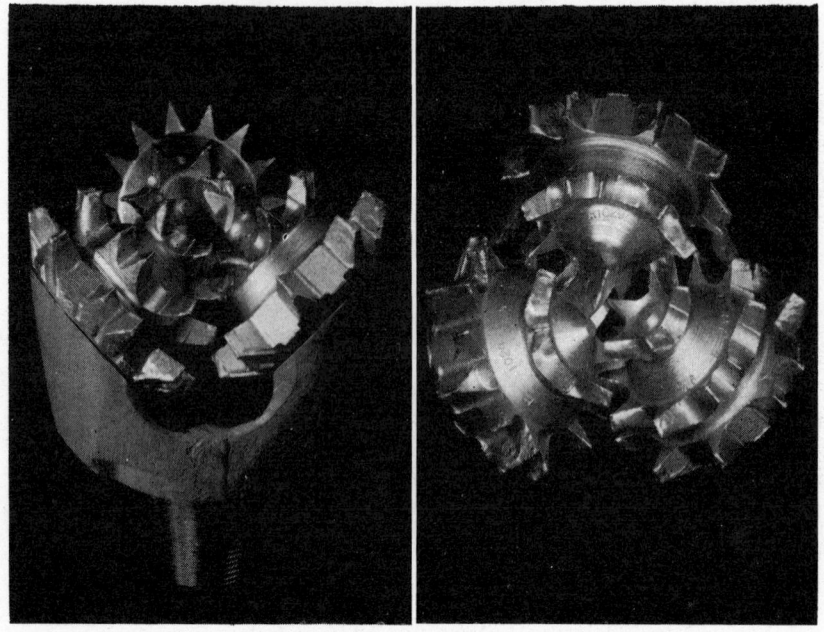

Figure 340. Hughes 9⅞-inch type OSC-3 tricone bit.

abrading or scraping action of the cutter teeth must be embodied in the hard-formation-rock bit design; otherwise, accelerated wear of the teeth will result in a short bit life.

The rock bit designed to drill efficiently soft, unconsolidated, low-strength formations retains the desirable chipping-crushing action and also embodies the twisting-tearing-scraping action on bottom, which is essential for fast rates of penetration.

The different types of tricone bits, developed for drilling the different formations, embody chipping-crushing action. However, the twisting-tearing-scraping action varies in the different types, which, along with general operating practice, are discussed in the following pages.

Soft Formation Bits

The OSC-3 tricone bit (fig. 340) is designed for drilling unconsolidated formations, such as soft shales, clays, redbeds, salt, and broken shale, which have a low compressive strength and high drillability.

The teeth on each cone of the OSC-3-type tricone are more widely spaced and deeply cut than those on the OSC-1 or the OSC types. The very widely spaced and deeply cut teeth permit effective penetration into the low-strength formations and result in the excavation of large cuttings at a rapid rate. The wide spacing of the deeply cut teeth provides more space to handle cuttings extruded from the formation and facilitates cleaning of the cutting surface. Cleaning is of the utmost importance in the softer range of drilling, because of the characteristic sticky properties of strata frequently penetrated.

In addition to the "ventilation" obtained through the use of widely spaced and deep teeth, the full interfitting of the inner rows of teeth into grooves of mating cutters provides mechanical cleaning of the grooves between rows of teeth on each cone. In addition to mechanical cleaning, interfitting provides necessary space for the long penetrating chisel teeth, which are essential for the not uncommon penetration rates of 50 to more than 100 feet an hour.

In general, the soft, unconsolidated formations do not wear the teeth rapidly, and a minimum number may be used. The use of relatively few teeth in each row on the OSC-3 tricone means less tooth contact on bottom at any instant, deeper penetration into the formation, and a faster drilling rate. The teeth are strong but are made as slim as modern metallurgy will permit, thereby prolonging the fast drilling rate. Further to assure a slow rate of dulling and maximum footage per bit, Hughesite hard facing is applied to surfaces of all teeth as well as the gage surface.

Although the teeth are widely spaced, the teeth on successive revolutions do not track in previous impressions made on bottom. Instead, the outermost rows generate a rock-tooth pattern on bottom having crests spaced approximately one-half the spacing of the cutter teeth. The inner rows of teeth also generate a finer-spaced, rock-tooth pattern on bottom. Thus, though teeth are widely separated, particles cut from bottom are relatively small and cannot readily wedge between cutter teeth but are easily washed away by the drilling fluid being ejected from the water-course nozzles.

To increase the cutting speed of the OSC-3 tricone further, especially when light weight must be used to keep the hole straight, the cone axes are offset forwardly off-center to provide a twisting-tearing action on bottom as the teeth penetrate. This feature is effective especially in shale formations having a tough structure not readily chipped.

The OSC-1 Tricone

The OSC-1 design (fig. 342) differs from the OSC-3 in that the teeth are not so widely spaced or so deeply cut. The arrangement of teeth re-

FIGURE 341. Hughes 9-inch type OSC tricone bit.

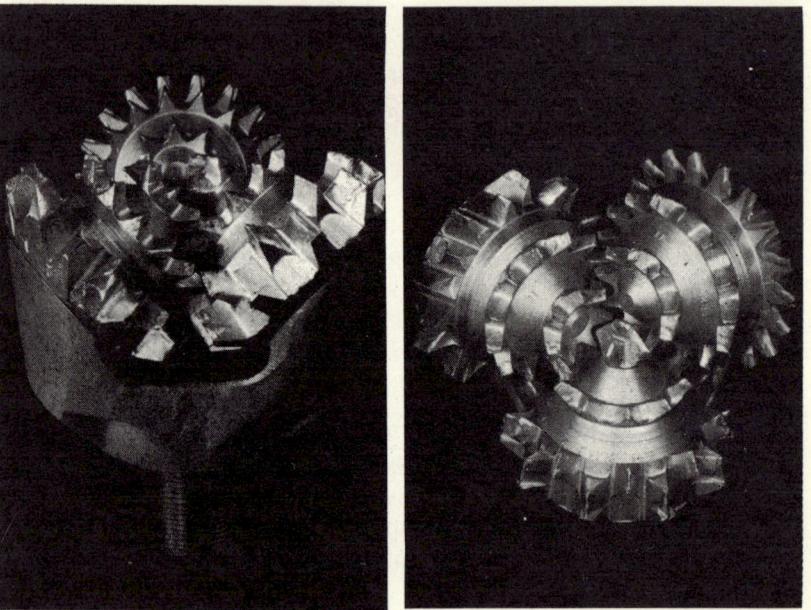

FIGURE 342. Hughes 11-inch type OSC-1 tricone bit.

tains the previously discussed merits of providing ample space for the handling of cuttings and ease of cleaning by the mud stream ejected from the water-course nozzles. The mechanical cleaning obtained from inter-fitting the rows of teeth into grooves on mating cones and the twisting-tearing-scraping action of the teeth as they penetrate the formation are fully retained in the OSC-1 tricone. The design differences which do exist are essential for good rock-bit performance in formations slightly hard for the OSC-3 bit.

FIGURE 343. Hughes 9-inch type W7R tricone bit.

The OSC Tricone

The OSC tricone bit (fig. 341) embodies the same basic principles of design noted in the discussion of OSC-3 and OSC-1 types. The teeth are not so widely spaced or so deeply cut. Although the OSC type is extensively used for drilling unconsolidated shales, redbeds, salt, gypsum, anhydrite, soft lime, etc., it has been used successfully in many oil fields in drilling formations formerly drilled with harder-formation bits such as the OSQ-2 (fig. 344) or OWS (fig. 346) tricone. The footage obtained with the OSC in many instances has been comparable to that obtained with the OSQ-2 or OWS, but the rate of penetration was increased with the OSC design.

The OSC is more versatile than the OSC-3 in regard to the range of formations which can be drilled. For this reason the OSC is available in a greater range of sizes.

The interfitting teeth previously discussed; the twisting-tearing and chipping-crushing action of long, penetrating, chisel-shaped teeth; a safe thickness of cone shell; a strong bearing; all are carefully balanced to provide a long-life bit and a fast rate of penetration.

MEDIUM-HARD FORMATION BITS

The OSQ-2 Tricone

Although most soft to medium-hard formations may be drilled economically with one of the OSC series of tricone bits, there are strata encountered having hardness and abrasive characteristics that require a rock-

FIGURE 344. Hughes 9-inch type OSQ-2 tricone bit.

bit design to give a longer cutting life without a marked reduction in drilling speed.

The OSQ-2 rock bit (fig. 344) is of a special design for drilling formations on the harder side of medium such as packed sands, anhydrite, medium-hard lime, and the harder shales, which respond best to twisting, tearing, chipping, and crushing actions as the cutter teeth penetrate. The OSQ-2 bit has these actions on the formation. The cone axes, as in the OSC series, are offset to provide additional twisting-tearing action. The cutter teeth are spaced apart comparable to the OSC, but no portion of any tooth is removed, thereby providing more cutting teeth to insure longer life and greater footage per bit in harder formations. The cutting structure and action on bottom have been properly adjusted for a good pene-

FIGURE 345. Hughes type OSQ-2 tricone bit and pattern it generates while drilling on bottom. Note that outermost or heel rows of teeth form relatively coarse-pitch rock teeth, which are joined to bottom and wall. Inner rows of cutter teeth form relatively fine-pitch rock teeth, which are readily broken off during rotation of bit.

tration rate and maximum footage in drilling medium-soft formations referred to above.

The interfitting of teeth into grooves of mating cones is embodied in the OSQ-2 rock bit to retain the desirable features of mechanical cleaning: long chisel-shaped teeth; a safe thickness of cone-shell; and a strong bearing. As in the OSC series, the teeth are hard-faced with tungsten carbide to prolong their cutting life, and the same arrangement of water-course nozzles insures an effective flushing action of the drilling fluid. The main difference in the types OSC and OSQ-2 is the increased amount of cutting surface on the OSQ-2 bit.

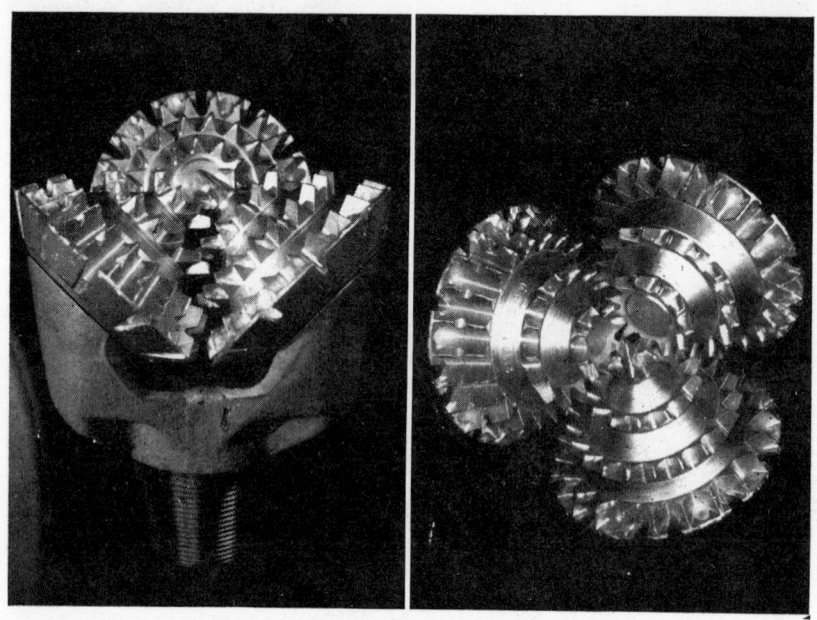

FIGURE 346. Hughes 9-inch type OWS tricone bit.

MEDIUM-HARD TO HARD, NONABRASIVE FORMATION BITS

The OWS Tricone

The OWS rock bit (fig. 346) is designed to drill formations having a compressive strength of a magnitude somewhat greater than that noted in medium-hard formations, which are economically drilled with the OSQ-2. Formations classed as medium-hard to hard, nonabrasive include hard rock interbedded with waxy, tough shales; hard lime; hard anhydrite; dolomitic lime; hard shale; slate; and dolomite. These formations respond to a drilling action combining twisting-tearing and chipping-crushing by the cutter teeth acting on bottom.

The OWS tricone provides teeth with a larger included angle to resist breaking and battering forces encountered in drilling dolomite, hard lime, and hard shale. The chipping and crushing action is obtained from the sturdy, closer-spaced teeth that support the load applied on the bit.

The twisting-tearing action is retained but to a lesser degree than in the OSC series and the OSQ-2 design. This twisting-tearing action of the teeth penetrating the formation is desirable when softer, tough shales, gypsum, and the like are interbedded with hard rock and when factors such as deviation control prevent the application of weight required to establish the most effective chipping and crushing action.

Oftentimes layers of soft, sticky shale are encountered in medium-to-

FIGURE 347. Hughes 9-inch type W7 tricone bit.

hard-rock drilling. The interfitting rows of teeth of the OWS bit mechanically clean the cutting surfaces of sticky formations, help prevent balling up, and retain a good penetration rate. Also, the interfitting teeth provide the additional space needed for effective depth of teeth, thickness of cone shell, and a strong bearing.

Modifications of the OWS, Type OWC

In an area where hard rock such as dolomite or hard limestone predominates, there may be a tendency for the gage point to break down or round off. The OWC rock bit has a T-shaped structure milled where the crest of the heel teeth and the gage surface intersect to fortify this strategic point against wear or breakdown. Otherwise, the OWC bit is identical to the OWS.

HARD AND HARD, ABRASIVE FORMATION BITS

The W7 and W7R Tricones

For hard sandy or chert-bearing limestone and dolomite, quartzite, novaculite, chat, chert, granite, and other hard, abrasive rocks the W7 (fig. 347) and W7R (figs. 343 and 348) tricone bits are favored. The W7R differs from the W7 in its gage design. The W7 has the conventional tri-angular-gage surface design, and adjacent heel teeth are spaced apart at gage. The W7R tricone bit has a gage design fortified by a circumferential cutting edge at gage, formed by a web, which joins adjacent heel teeth in pairs.

The web-type gage surface developed for the W7R has several distinctive features. A gage surface is formed on which a much larger volume of tungsten carbide is applied to resist the severe abrasion encountered in drilling hard and abrasive formations. The conventional triangular-gage teeth form a rock tooth on the formation, which is supported by its cementation with the wall of the hole and the bottom of the hole. This support has an angular connection of approximately 270° and offers a great resistance to the removal of the rock teeth by the heel teeth on the cutters. The W7R bit, with its webbed-gage surface design, cuts a smooth wall, eliminates the junction of the rock teeth with the wall, and readily prevents the buildup of strong rock teeth on bottom.

In drilling hard, abrasive formations that have a high compressive strength, a minimum of twisting-tearing-scraping action is a prerequisite for the design of a well-engineered rock bit. Any twisting-tearing action results in a fast rate of wear on the cutting teeth with a corresponding reduction in footage. The cones on W7R and W7 tricones are designed to minimize the twisting-tearing-scraping action and provide the essential chipping-crushing action. The cone-bearing pins are located on center for approximately true-rolling action to prolong the gage life and cutting structure. The included angle of teeth is greater to provide the strength required in preventing broken or battered teeth. The teeth are more closely spaced and are arranged to prevent tracking and insure a long-cutting life.

The rows of teeth interfitting into grooves in mating cones are retained, as in other Hughes rock bits, to obtain mechanical cleaning of the cutting structure, a suitable depth of teeth, an essential thickness of cone shell, and a strong bearing.

The selection of the W7 or W7R is contingent upon the particular characteristics of the formation being drilled. When gage wear is not serious on the conventional triangular-gage surface and only moderate weight on the rock bit is required to establish an effective chipping-crushing action by the cone teeth contacting the formation, the W7 may be the preferred bit. When gage wear is a factor or when heavy weight must be applied on the bit to establish the desired chipping-crushing action on bottom, the W7R is the most suitable design of rock bit.

SOFT TO MEDIUM-SOFT FORMATION BITS

Weight Applied

In drilling top-hole shales, unconsolidated and broken formations of shale, gypsum, salt, chalk, medium-soft anhydrite, and the like, the

FIGURE 348. Hughes type W7R tricone bit and pattern it generates while drilling on bottom. Note that circumferential gage cutting teeth prevent relatively coarse-pitch rock teeth from joining wall, and that resulting rock teeth are more easily removed. The fine-pitch rock teeth generated by inner rows of cutter teeth are readily removed during rotation of bit.

weight applied on an OSC-series tricone bit need not be of a high magnitude to establish a fast rate of penetration. These formations have a low compressive strength, and good penetration by the rock-bit teeth is obtained with a moderate load on the bit. An initial weight of 500 to 1,000 pounds to the inch of rock-bit diameter and a rotary speed of 150 to 175 r.p.m. are considered good practice. As the teeth on the rock bit wear and gradually dull, the applied weight should be increased to maintain a satisfactory penetration rate.

Rotary Speed

Usually an increase in weight should be accompanied by a reduction in rotary speed to compensate for the increased load on the bearings and permit them to function properly during the normal life of the cone teeth. Common practice, used in drilling low-strength formations, involves a maximum applied load of 2,000 to 2,500 pounds to the inch of bit diameter with the rotary-table speed decreasing from 175 r.p.m. to 75 r.p.m. as additional weight is applied.

In certain areas rotary speeds in the range of 250 r.p.m. to 350 r.p.m. and applied loads in the range discussed above are accepted practice. The formations in these areas are predominantly shale and sands and are relatively free from any hard rocks, which cause severe shock forces that would result in abnormal stress on the rock bit and the drill string.

Volume of Drilling Fluid Circulated

The volume of drilling mud circulated when drilling formations having low strength and high drillability should be sufficient to provide a return velocity up the annulus equal to at least three feet a second. This volume will establish the required velocity through the rock-bit water-course nozzles for good cleaning of the cutting structure and bottom of the hole. Furthermore, the cuttings will reach the surface at a faster rate, and the danger from a high density of cuttings in the annulus sticking the drill string will be less. Since formation characteristics vary widely, a definite pump capacity suitable for one area may be inadequate for a different area.

BITS FOR FORMATIONS ON THE HARDER SIDE OF MEDIUM-SOFT

Weight Applied and Rotary Speed

In drilling formations on the harder side of medium-soft, which include packed sands, anhydrite, medium-hard lime, and medium-hard shales, moderate weight is applied on the OSQ-2-series tricone bit. Many of these formations have a low strength, but harder zones usually present warrant the application of more weight than is normally required in drilling the relatively soft, low-strength formations. As a general rule the OSQ-2 tricone, which is designed to drill these formations, will give satisfactory performance with drilling weights ranging from 775 to 1,000 pounds to the inch of the diameter of the bit in its early life to a maximum of 3,000 pounds to the inch of the diameter of the bit when it is well worn, with rotary speeds decreasing from 150 to 75 r.p.m. as the drilling weight is increased.

Volume of Drilling Fluid Circulated

The pump capacity should be equivalent to that suggested for drilling soft formations. The average penetration rate is less, but the softer, sticky zones are apt to be present in formations classed as "on the harder side of medium-soft," and high drilling-fluid velocity through the rock-bit nozzles is good insurance against the cutting structure balling up. To a certain extent drilling is "blind," and, as formations are not uniformly free of sticky breaks, the pumps should be designed and operated to maintain a clean, free-cutting bit at all times.

BITS FOR MEDIUM-HARD TO HARD, NONABRASIVE FORMATIONS

Weight Applied and Rotary Speed

Medium-hard to hard, nonabrasive formations, like all others, can be drilled with relatively light weight on the rock bit. The rate of penetration with light weight will be slow, particularly when the cutter teeth are partly dulled from their initial sharpness. When drilling conditions permit, it is better practice to apply sufficient weight on the OWS tricone bit to establish effective chipping-crushing action of the cutter teeth working on the formation. This action, in combination with the twisting-tearing action always present in the OWS tricone bit, will materially increase the rate of penetration and footage obtained from the bit.

During the early life of the bit, it is good practice to drill with moderate weight, in the range of 1,000 to 1,500 pounds to the inch of diameter of the bit, with rotary speed in the range of 85 to 125 r.p.m. The cutter-teeth "pattern bottom" and breakage of sharp teeth are minimized. As the teeth dull, the weight applied on the OWS tricone bit should be increased and the rotary speed decreased to retain a fast penetration rate. It is common practice to apply weight in the range of 2,500 to 3,500 pounds to the inch of diameter of the bit, with a rotary-table speed in the range of 60 to 70 r.p.m.

In some areas where other factors such as limited hours on bottom, smooth drilling characteristics of formation, and flat-formation bedding planes will permit, weights in the range of 3,500 to 5,000 pounds to the inch of diameter of the bit are accepted practice. With the OWS tricone bit carrying this heavy weight, the rotary-table speed should not exceed 60 r.p.m.

Volume of Drilling Fluid Circulated

The presence of sticky shales, gypsum, and the like is to be considered in the selection of pump capacity. In general, harder formations are encountered below the soft, unconsolidated formations having high drillability previously discussed, and proper pump capacity for drilling these formations should insure that an adequate volume of drilling fluid is being circulated when drilling with OWS tricone bits.

DRILLING HARD TO HARD, ABRASIVE FORMATIONS

Weight Applied and Rotary Speed

The application of heavy weight on the W7R tricone bit will materially improve performance in drilling hard to hard, abrasive formations, which generally are firmly cemented, have a high compressive strength, and are the most expensive to drill. This class of formations include chert, chat, pyrite, granite, quartzite, dolomitic sandstone, cherty limestone, quartzitic sandstone, and the like.

During the first few minutes of operation a moderate weight in the range or 1,000 to 1,500 pounds to the inch of diameter of the bit is good practice. The cutter teeth will "pattern bottom," and all rows of teeth will function properly. Having "patterned bottom," the weight may be increased into the range of 3,000 to 4,000 pounds to the inch of diameter of the bit to establish an economical rate of penetration. Under heavy-weight operating practice the rotary-table speed is usually within the range of 45 to 60 r.p.m. The desired chipping-crushing action can be realized only when the load supported by the rock-bit teeth is in excess of the compressive strength of the formation.

In extremely hard, abrasive formations the application of weight in the range of 4,000 to 6,000 pounds to the inch of diameter of the bit has been successfully used. With these extreme weights applied on the bit, it is advisable to limit the rotary speed from 40 to 50 r.p.m.

Volume of Drilling Fluid Circulated

In general, balling up of the cutting structure is not liable to occur when drilling in hard to hard, abrasive formations. However, shale breaks are sometimes encountered, and a strong jetting action may be needed to remove any formation from the cutter teeth readily. In addition a return velocity of the drilling fluid in the range of two to three feet a second will give a low density of cuttings in the mud stream and lessen the danger of sticking the drill pipe.

WEIGHT FROM TANDEM DRILL COLLARS

The use of drill collars for providing the weight carried on rock bits offers several advantages. The weight is concentrated closer to bottom and is more effective in forcing the cutter teeth into the formation. The drill pipe is subject principally to tensile loading only, and fatigue failure resulting from reversal of stress is materially reduced.

Long strings of 10 to 20 drill collars, each 30 feet long, are becoming increasingly popular in areas where hard rocks having a high compressive strength are encountered. These hard rocks such as hard dolomite, chert, and limestone can be drilled at faster rates of penetration when all factors involved permit loading the rock bit in the heavyweight range. (See discussion under OWS and W7R tricones.) The weight of the drill collars used should exceed the load applied on the rock bit. In general, from 25 to 50 percent additional weight of drill collars is considered good practice. In any case it is not good practice for the neutral zone (change from compression to tension loading) to occur near the upper end of the top drill collar, as under this condition the joint of drill pipe attached to the top drill collar is not restrained laterally and is subjected to severe bending stresses, and fatigue failure is almost certain to occur in a relatively short time. The weight applied on the bit and the number of drill collars used should be balanced so that the neutral zone is well within the drill-collar string or in the drill pipe several joints above the top drill collar.

Rock-bit failures have been known to occur under lightweight loading when long drill-collar strings were in use. It was found that the excessive drill-collar weight, some 15,000 to 18,000 pounds, in conjunction with rotary speed, rough-running characteristics of the formation, and lightload application on the bit, sets up a harmonic vibration in the drill string of a magnitude sufficient to cause the drill collars to act as a hammer. The force transmitted to the rock bit exceeded the strength of the materials, and fatigue failure occurred. A change in applied weight on the rock bit or in rotary speed will eliminate this condition.

The principal difficulty encountered with long tandem drill-collar strings is failure of the threaded connections. Proper inspection of the threads and shoulders on each trip is good practice. The connections between breaks should be checked for tightness; and, if makeup is noted, the connections should be broken apart, cleaned, inspected, and reconditioned, if necessary, to remove fins and slivers from the threads and galls from the shoulders. The connection should be redoped and tonged tight.

When only a few drill collars (one to three) are used, most of the applied weight is obtained by placing the lower portion of the drill pipe in compression. As a result, the relatively limber drill pipe is deflected off center until restrained by the wall of the hole; bending stresses invite fatigue failures through the drill pipe; the pipe is subjected to O.D. wear, which reduces the wall thickness and may cause failure; a portion of the applied load is lost in friction established by the drill pipe and tool joints forced against the wall of the drilled hole; tool-joint life is materially reduced, particularly in abrasive formations; and the average load on the rock bit is less owing to wall friction and the spring characteristics in a long string of drill pipe.

The compressive strength and drillability of formations vary widely and prevent adoption of a universal drilling practice. With the proper design and type of tricone on bottom, the weight on the bit, the speed of rotation, and the volume of drilling fluid are all important factors in obtaining a maximum penetration rate for efficient rotary drilling.

The graphs reproduced in figures 349 to 354 show the compressive strengths of various formations, the drilling rates in various formations with different loads applied on the rock bit, and common drilling practice in balancing rotary-table speed with the weight applied on different sizes of rocks bits.

All rotary rock bits require weight to make them cut. Up to the foundering point Hughes tricones will generally drill faster and make more hole the greater the weight applied. With weights greater than the foundering load, particularly in soft, low-strength formations, the teeth become fully embedded and the cutter shell bears on bottom and slows

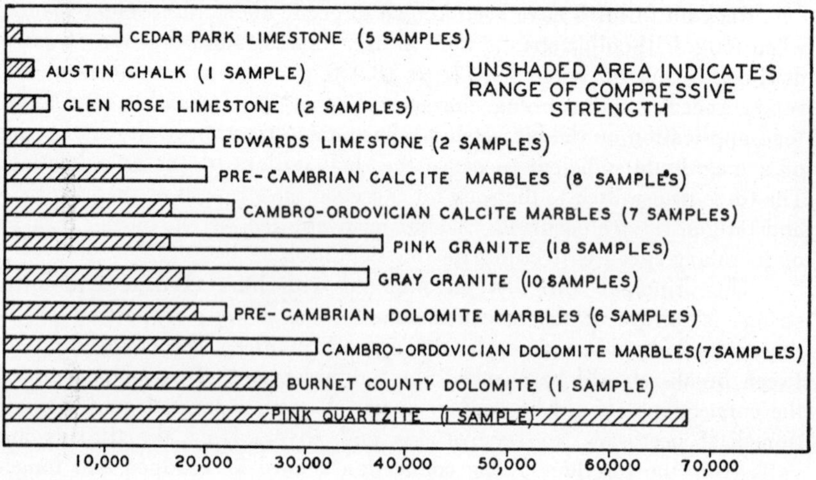

FIGURE 349. Chart showing compressive strength (pounds per square inch) of various rock types.

up progress. The determination of optimum weight, as well as other factors affecting drilling practice, must be the result of actual field testing and the experience thus acquired.

The efficiency of a rock bit requires an adequate volume of fluid circulation, particularly when formations in the softer range are being drilled. The harder range of formations may not require the large volumes used in soft formations. The presence of sticky-shale breaks, which are often encountered, however, warrants the selection of pumps designed to deliver an adequate volume of drilling fluid. The volume of drilling fluid circulated can be established when the pumps are purchased.

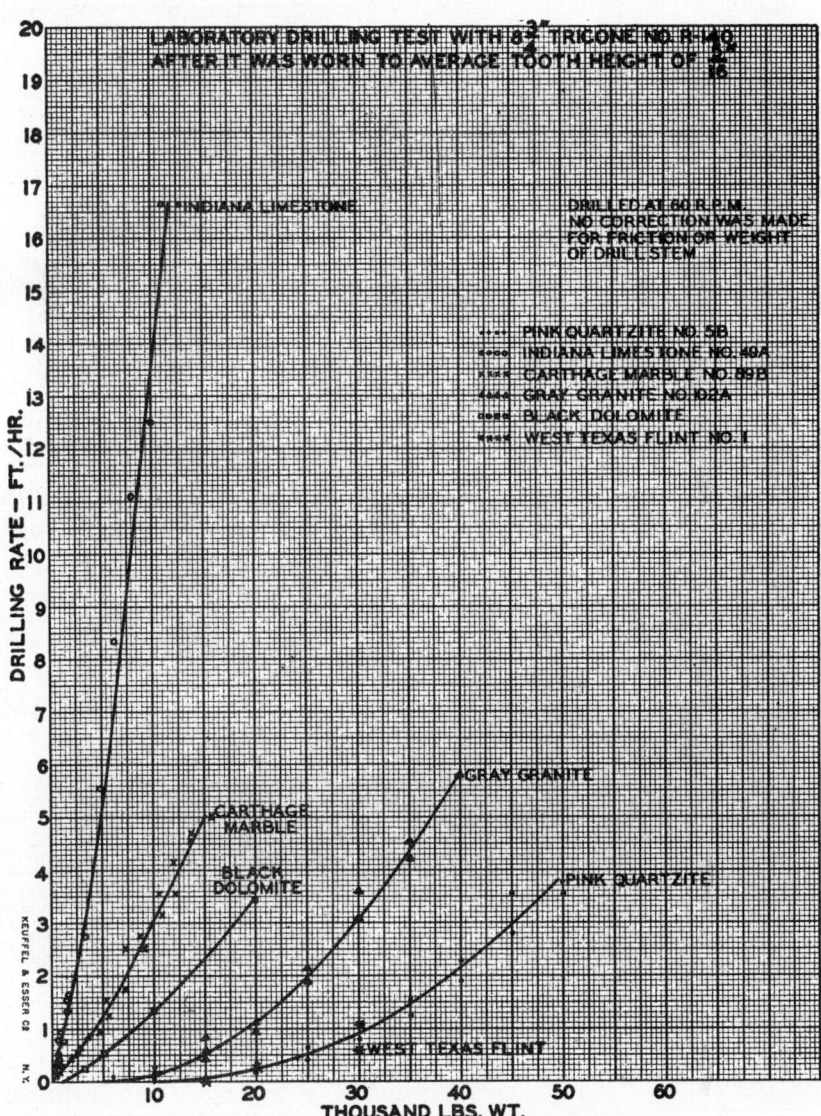

FIGURE 350. Graph of drillability of formations.

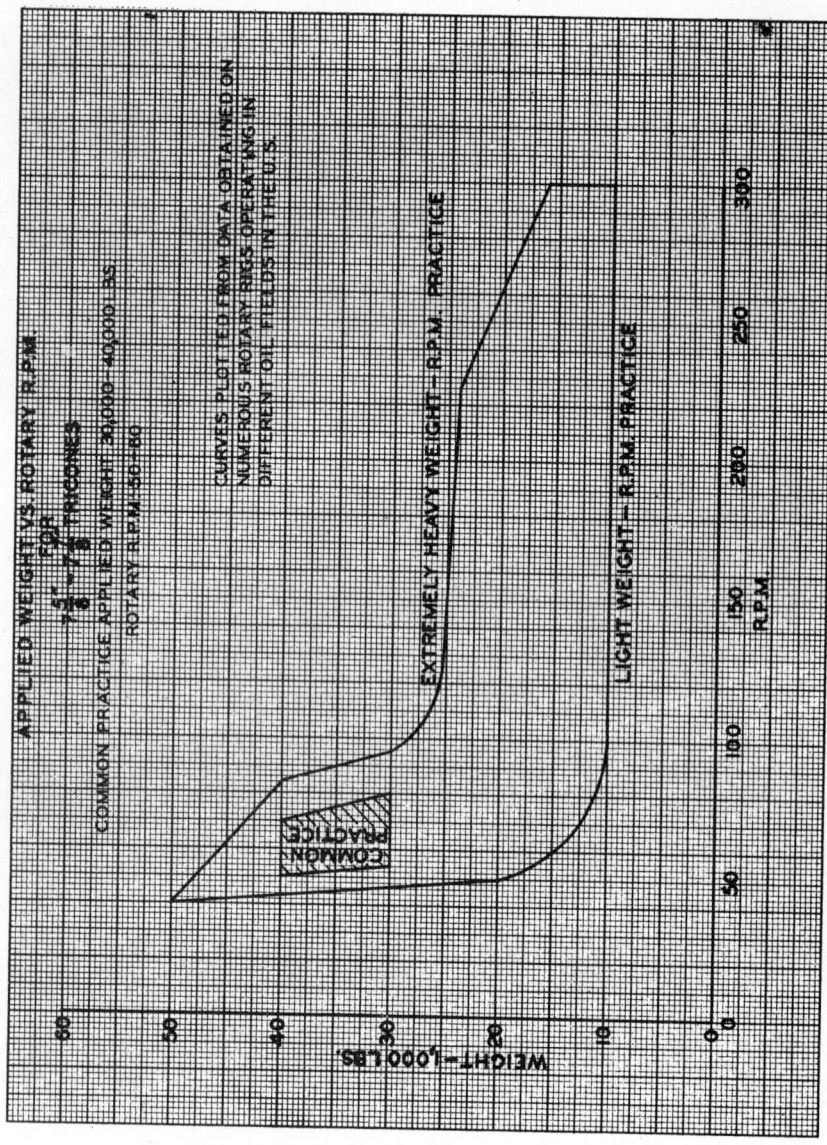

FIGURE 351. Graph showing balancing of rotary-table speed with weight for 7⅞-inch to 7⅞-inch tricones.

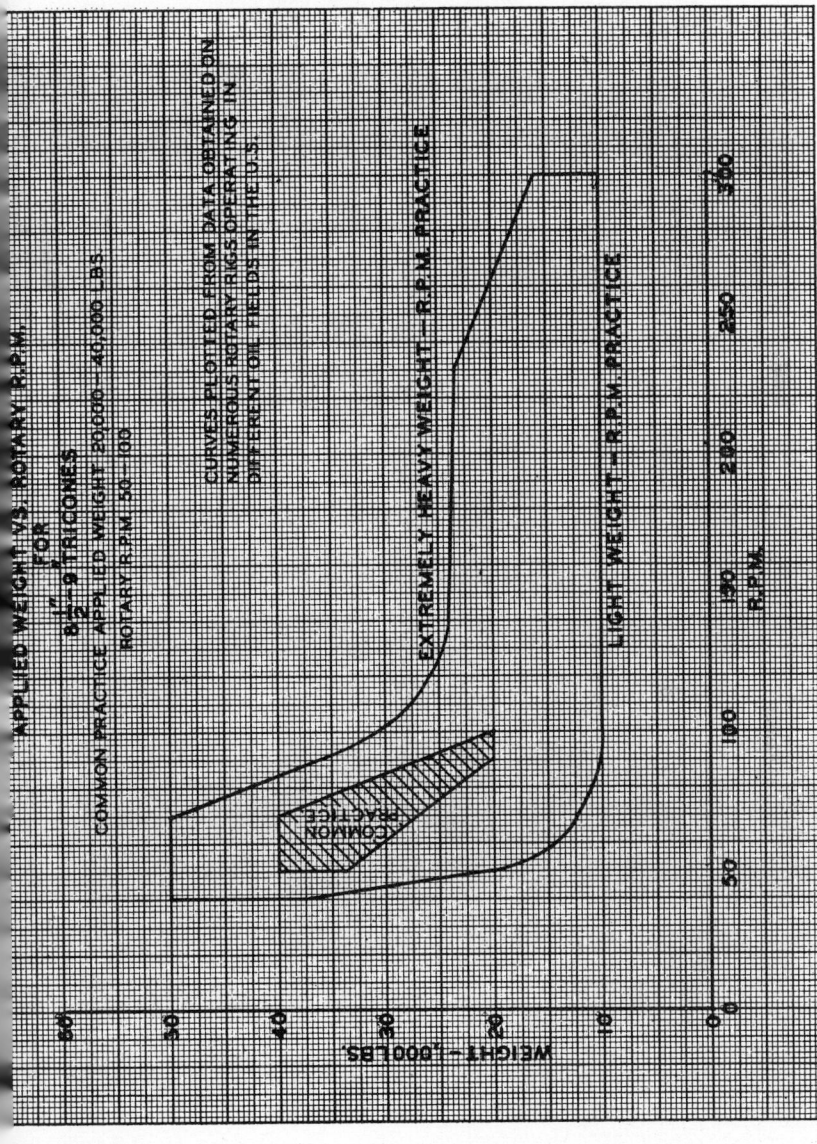

FIGURE 352. Graph showing balancing of rotary-table speed with weight for 8½-inch to 9-inch tricones.

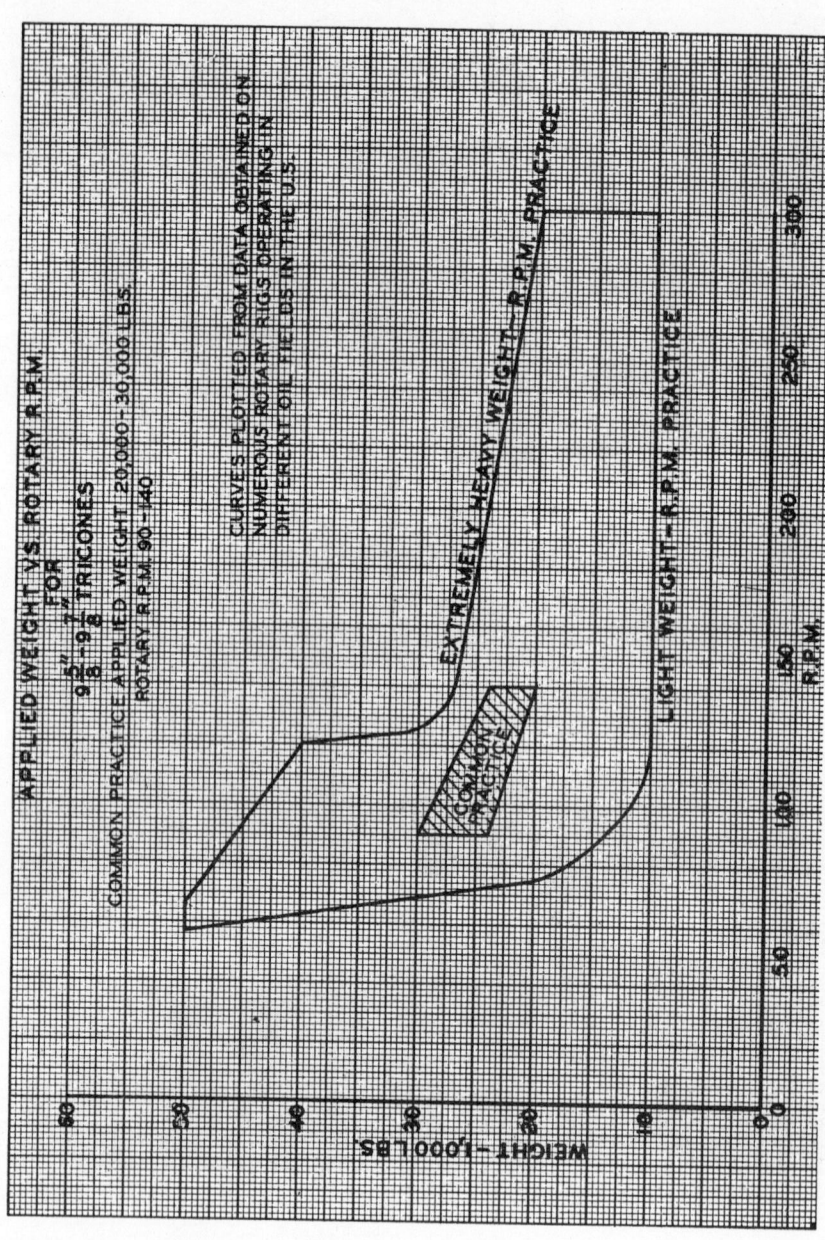

FIGURE 353. Graph showing balancing of rotary-table speed with weight for 9⅝-inch to 9⅞-inch tricones.

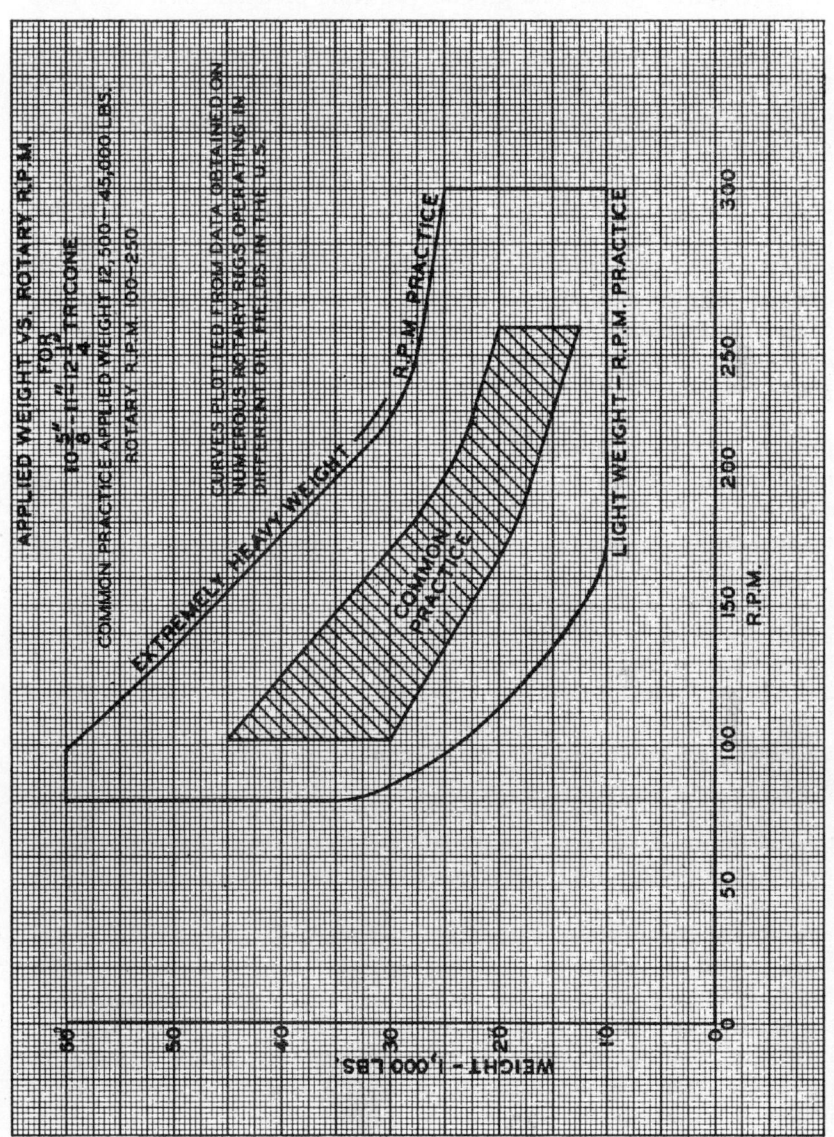

FIGURE 354. Graph showing balancing of rotary-table speed with weight for 10⅝-inch to 12¼-inch tricones.

DEEP-WELL CAMERA

O. E. BARSTOW AND C. M. BRYANT

The optical system and general principle of operation of the deep-well camera are shown schematically in figure 355. For clarity, the dimensional proportions of the figure have been somewhat distorted. In the figure are a cut-away drawing more nearly to scale, giving a better idea of how some of the parts are constructed and arranged, and a photograph of the exterior of the assembled apparatus.

The camera is run into the well on an electric cable, which serves to transmit the power required by the apparatus and to provide remote control of the camera from the surface. As shown in figure 355, the apparatus consists of two main parts, a camera chamber and a water chamber. The camera chamber contains air at atmospheric pressure and houses the camera, including the lens, the film, and the film-drive mechanism. It must be strong enough to withstand the external pressure of several thousand pounds to the square inch encountered in deep wells.

The water chamber, which is filled with clear water, is used merely to provide a bath of clear liquid between the lens and the rock formation being photographed, an idea originally proposed by Reinhold. The water chamber houses a light to illuminate the subject and an inclined mirror, which permits the camera to view the wall of the well horizontally through the cylindrical picture windows in the side of the water chamber. The bellows shown in figure 355 serves to equalize the internal and external pressures on the water chamber, so that the picture window can be relatively large without encountering difficulties due to high pressure. In the bulkhead at the top of the water chamber is a pressure window immediately below the camera lens. The window must withstand full well pressure, but this is a comparatively simple matter because the window diameter can be small.

Attached to the outside of the case is a spring which pushes the picture window against the well wall, thus minimizing the thickness of turbid well fluid through which the picture must be taken. This spring also assists in keeping the camera stationary during exposures. The importance of pushing the camera close against the formation is illustrated in figure 355. Here, one end of a shallow glass baking dish has been blocked up about two inches above the table, and the dish filled with water made turbid by the addition of a little aquagel. A scale on the bottom of the dish is nearly invisible at one end but becomes progressively more distinct as the water depth decreases from two inches to zero. The scale figures denote actual immersion depth. In this case, water which would be murky enough to blot out the picture completely at a distance of two inches from the picture window would permit reasonably clear photographs to be made at distances of one-fourth or one-half inch. By pushing the picture windows against the formation, satisfactory pictures have actually been made in water so turbid

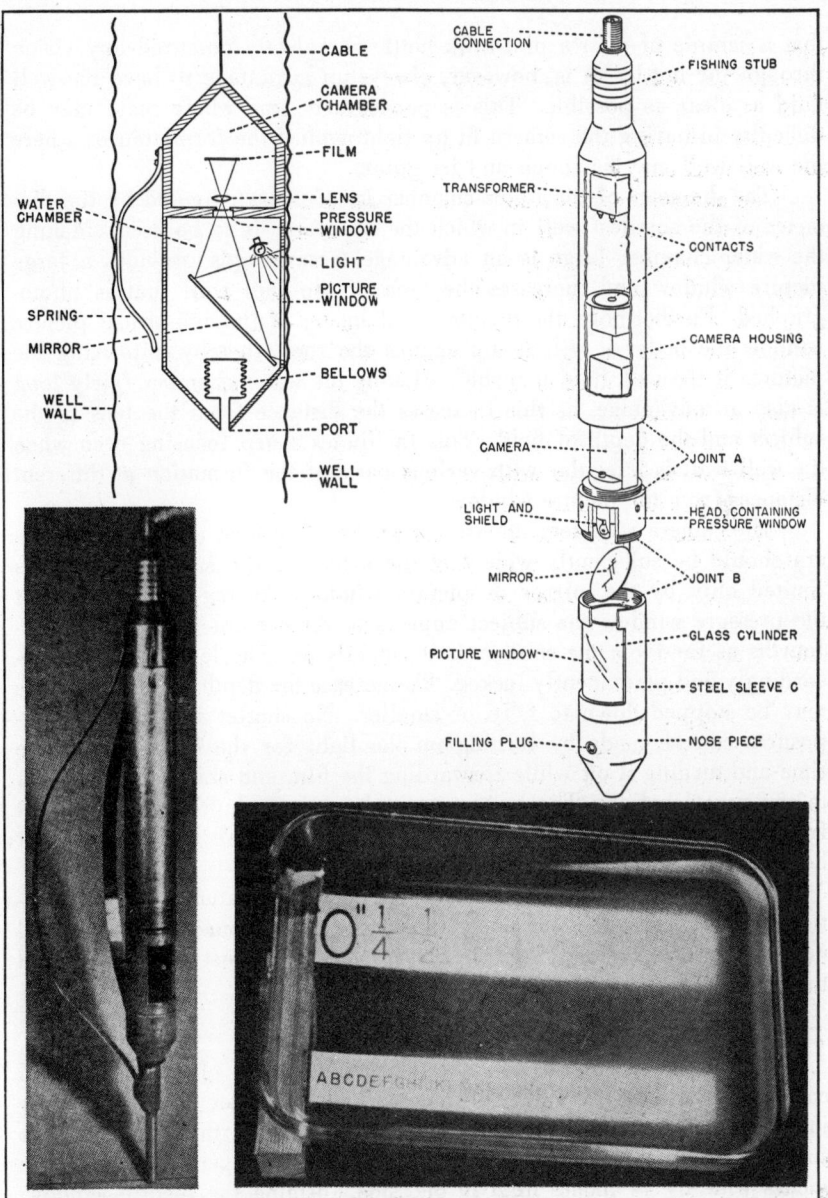

FIGURE 355. General principles of operation of deep-well camera (upper left). General construction and arrangement of parts (upper right). Camera assembled and ready to go into well (lower left); this view shows "push-over" springs that hold picture window of camera close to wall of bore hole. Effect of depth of turbid well fluid on sharpness of picture (lower right); figures on inclined scale represent inches of immersion of scale in murky water.

that a sample of it in a pint milk bottle completely obscured any vision through the bottle. It is, however, always an advantage to have the well fluid as clear as possible. This is particularly true where there may be difficulty in having the camera fit up tight against the formation or where the well wall may be rough and irregular.

The diameter of the water chamber is, of course, limited by the diameter of the smallest well in which the apparatus is to be used. Making the water chamber large is an advantage, because this provides a large picture window and increases the area of the well wall that is photographed. Furthermore, the greater the diameter of the cylindrical picture window the better it will fit up against the rock, thereby improving the pictures if the well fluid is turbid. Making the water chamber fairly long is also an advantage, as this increases the distance from the lens to the subject and the depth of field. This facilitates sharp focusing even when the well wall is irregular with various parts of the formation at different distances from the picture window.

The camera requirements are not severe. The field angle of the camera should be sufficiently wide that the extent of the subject covered is limited only by the mirror or picture window. Owing to refraction at the pressure window the subject appears to the camera to be only three-fourths as far from the camera as it actually is. The lens is focused accordingly and permanently locked. To increase the depth of field, the lens may be stopped down to f/16 or smaller. No shutter is needed, as exposures can be made by turning on the light for the desired exposure time and turning it off while forwarding the film and shifting the camera to the next location. The camera must have enough film capacity that frequent reloading is not necessary, and the film-drive mechanism must be controlled electrically.

The light or lights must be located so as to give satisfactory illumination of the subject without undesirable reflections from the window and mirror, must be shielded from the camera lens, and must be able to stand the well pressure.

The Equipment

Figure 355 shows the actual construction of one of these deep-well cameras. The glass cylinder forming part of the water chamber is four inches in outside diameter by $4\frac{1}{2}$ inches long by one-fourth inch thick. It is surrounded by a steel sleeve having a rectangular opening about three inches wide by $4\frac{1}{2}$ inches high in one side, forming the picture window. The ground ends of the glass are sealed by means of "neoprene" gaskets against the steel head and steel nosepiece, which are screwed into the steel sleeve. These joints are subjected to only small differential pressures because of the equalizing action of the bellows in the nosepiece. Attached to the head is the glass mirror making an angle with the vertical somewhat less than 45°. The mirror is shaped elliptically to fit inside the glass cylinder. The back surface is silvered. In the top of the head is the glass pressure window sealed with a thin rubber gasket to withstand the high pressure in the water chamber. This window is half an inch thick by $1\frac{1}{8}$ inches in diameter with an unsupported diameter of three-fourths inch.

The head carries two General Electric No. 1129 21-candle-power automotive-type lamps, which are operated on six-volt, sixty-cycle alternating current. These lamps were found to withstand the high external pressures satisfactorily. Using distilled water in the chamber with lamp contacts and connections insulated from the water, the leakage current was found to be negligible, and no bubbles due to electrolytic decomposition of the water could be observed. The exact position of the lights and the shape of the reflectors and shields were chosen to give good illumination of the subject while avoiding reflections in the mirror and glass cylinder, which might otherwise show up as bright spots in the picture. Tipping the mirror slightly steeper than 45° also helped to eliminate these reflections.

The camera is screwed to the top of the water-chamber head with its lens directly above the pressure window. The camera housing slides down over the camera making a pressure-tight threaded connection with the watertight chamber head. This joint is the one that is taken apart whenever the camera is loaded or unloaded. The camera housing, which is $4\frac{3}{16}$ inches outside diameter with a three-sixteenth-inch wall, was tested with an external pressure of 2,500 pounds to the square inch. The camera housing and water chamber are zinc-plated inside and out for corrosion resistance.

THE CAMERA

The camera itself, shown in figure 356, is a modified Bell and Howell model 151 magazine-loading 16-millimeter movie camera with a 7.4-millimeter by 10.3-millimeter frame size. It has a built-in feature permitting it to take single exposures in the same manner as with any still camera. The control button has been coupled to a solenoid for remote electrical control. The power for driving the film still comes from the spring motor of the camera, which will give 450 exposures on one winding. A lens having 15-millimeter focal length is used, and provision has been made for stopping the lens down as small as f/32 if desired for greater depth of field. At f/32 an exposure time of eight seconds is satisfactory when negative panchromatic safety film is used.

The upper end of the camera housing is threaded to make a water-tight connection to the socket on the end of the electric cable. Below this connection is a "fishing stub" to facilitate retrieving the camera from a well in case of accident. A transformer fixed in the upper end of the housing provides the six volts required by the lights. When the housing is in place, two spring contacts below the transformer engage contacts on the upper end of the camera to provide connections to the camera and lights.

Two springs (see fig. 355) with their upper ends fixed to the camera housing and their lower ends attached to a sliding block hold the camera against the formation in wells up to twelve inches in diameter. By removing the springs the camera, which is $4\frac{3}{16}$ inches in diameter, can be used in a hole as small as $4\frac{1}{2}$ inches in diameter.

For comparison it might be noted here that the camera model preceding the one described above used an Argus model AF 35-millimeter

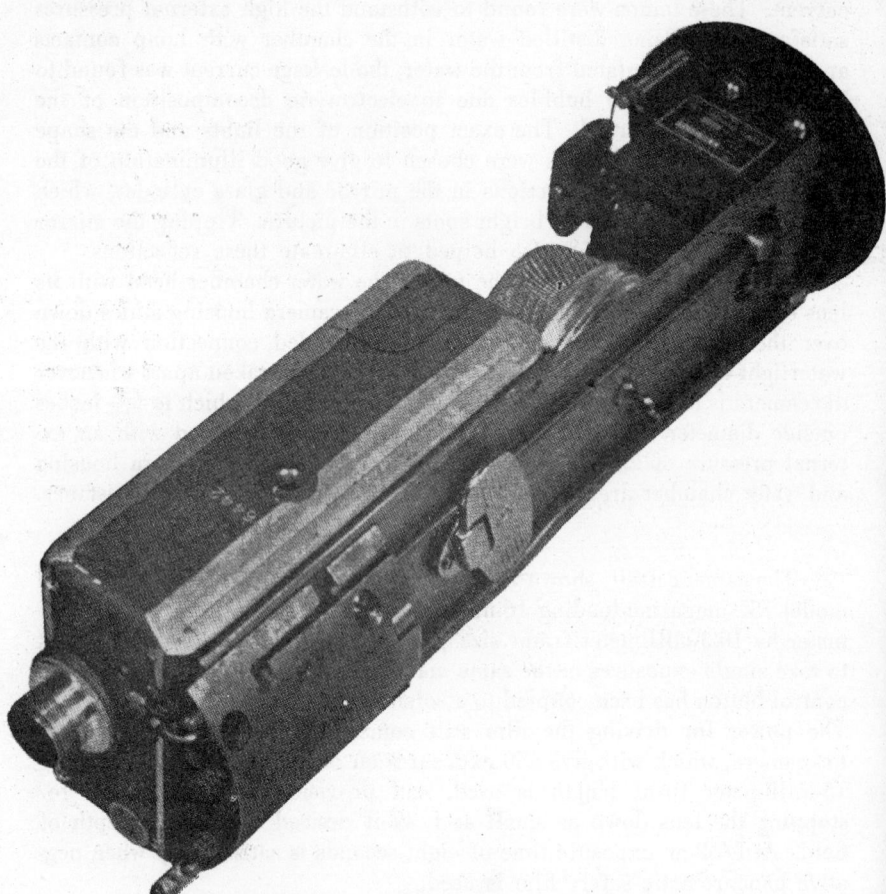

FIGURE 356. Bell and Howell 16-millimeter camera fitted with a solenoid release for use as a deep-well camera.

camera with electric-motor drive requiring a case with an outside diameter of six inches. Use of this camera was limited to large-diameter wells. A further handicap was the fact that only 36 exposures could be made without reloading.

When a well is to be photographed, it is first necessary to determine the type of fluid covering the section of formation where the pictures are to be taken. Most wells that produce formation water will have reasonably clear fluid covering the formation. If there is some doubt as to the type of fluid in the hole, a sample can be obtained by running a thief or bottom-hole sampler. If it develops that the hole is filled with oil or water too turbid for good pictures, it will be necessary to replace the well fluid with a relatively clear fluid. This is easily accomplished by spotting the clear fluid, usually salt water, in the well bore through tubing. The tubing is then removed from the well in order to run the camera.

If the well is flowing, it should be killed before an attempt is made

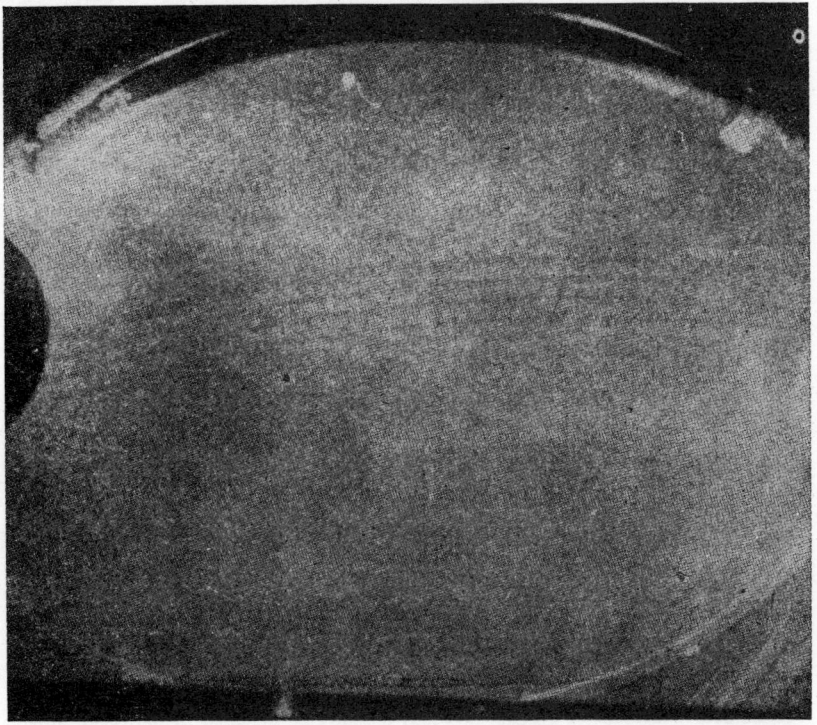

FIGURE 357. Section of casing at 2,361 feet, 8 inches, which contains a collar.

to operate the camera. By using a clear fluid to kill the well, it is possible in one step both to accomplish this and to condition the bore hole so that optimum conditions exist for taking pictures.

In wells that have a layer of oil on top of the clear fluid covering the section to be photographed it is necessary to cover the window of the camera with a matèrial that is very soluble in water but insoluble in oil. This is done so that any drops of oil that might cling to the window of the camera will be released when the material dissolves in the water. Thus the window will be clean when the camera reaches the section of the well to be photographed.

Operating Equipment

The equipment used to operate the camera includes a special non-rotating stranded steel cable, the core of which is a rubber-covered electrical conductor. The metal sheath of the cable serves as the path for the return circuit. This cable is attached to a hoisting winch that has commutator rings enabling the cable to be raised and lowered in the well bore

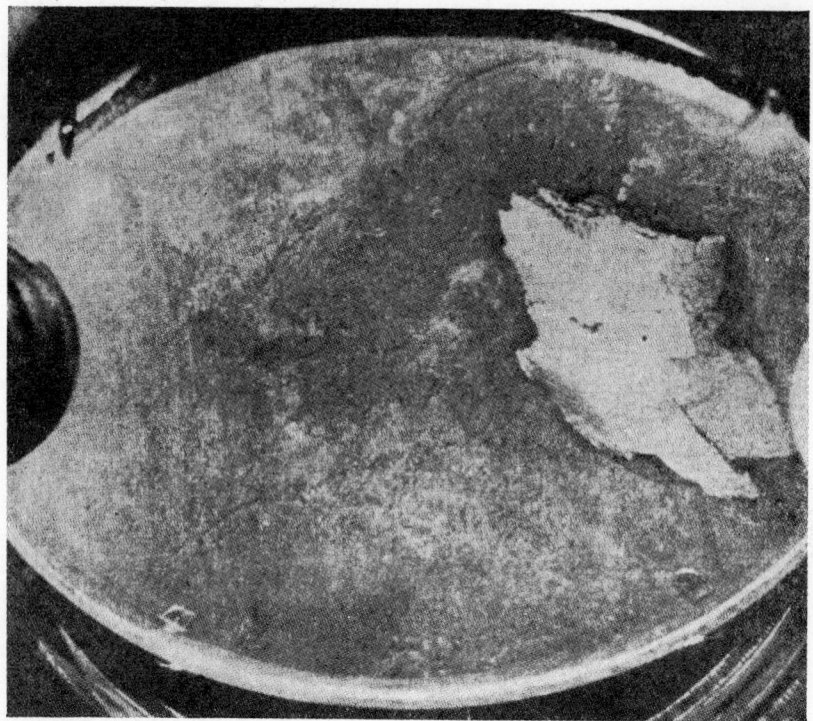

FIGURE 358. Fragment of gypsum that may actually be included in the rock, although shadows around the edge would seem to indicate a possibility that fragment is merely adhering to wall of well (2,429 feet).

without breaking the electrical circuit. A portable 110-volt alternating-circuit generator and an instrument panel containing a voltmeter and an ammeter with suitable control switches for operating the camera are also provided. This equipment is mounted in a large panel-bodied truck so arranged that the truck motor drives the hoisting winch. The depth of the camera in the well bore is determined at all times by running the cable over a measuring wheel, which is connected by selsyn motors to a depth counter on the operating panel of the truck.

After the well-bore conditions have been determined and corrected if necessary, the camera is loaded with the film and attached to the cable. Each roll of film is identified by taking a photograph of a card containing the roll number, date, well name, and other pertinent data. This is done before the camera is lowered into the well and serves also as a check on the electrical circuit and the operation of the camera and lights.

The camera is lowered into the well at a rate of approximately 50 feet a minute. When the camera reaches the well section to be photo-

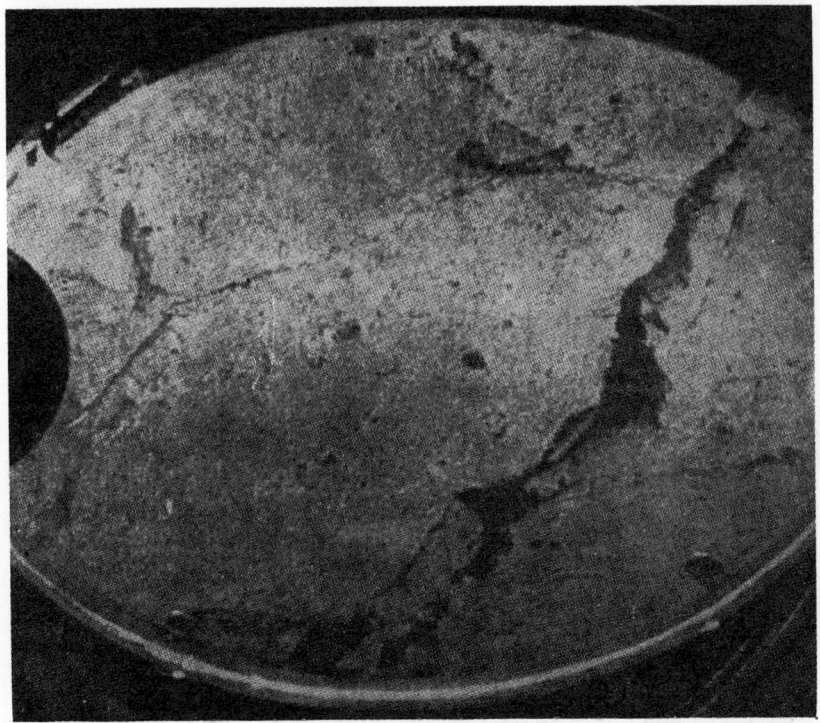

FIGURE 359. Well-developed porous zone on right side of picture. This cavity may be due to enlargement of a fracture by leaching. Entire section shows considerable amount of porosity as shown by small openings appearing all through picture. There is one small stylolite traversing picture, beginning in upper part on right and extending across to join another small fracture on left (2,528 feet).

graphed, it is stopped and an exposure made by operating the surface controls. The correct exposure time depends on the lens-stop setting and the type of film being used and generally amounts to a few seconds. With the present cameras the exposure time is controlled from the truck by the length of time that voltage is applied to the cable. This can be done with a manually operated switch or with an automatic time switch.

Two types of picture surveys are made: one a continuous strip photograph of the well bore showing every inch of the formation, and the other a series of single-shot pictures spaced at intervals of one foot, five feet, or any other desired interval. Usually the one-foot interval is used. The continuous-strip picture is made by lowering the camera only two inches between exposures and then using only the center section of each

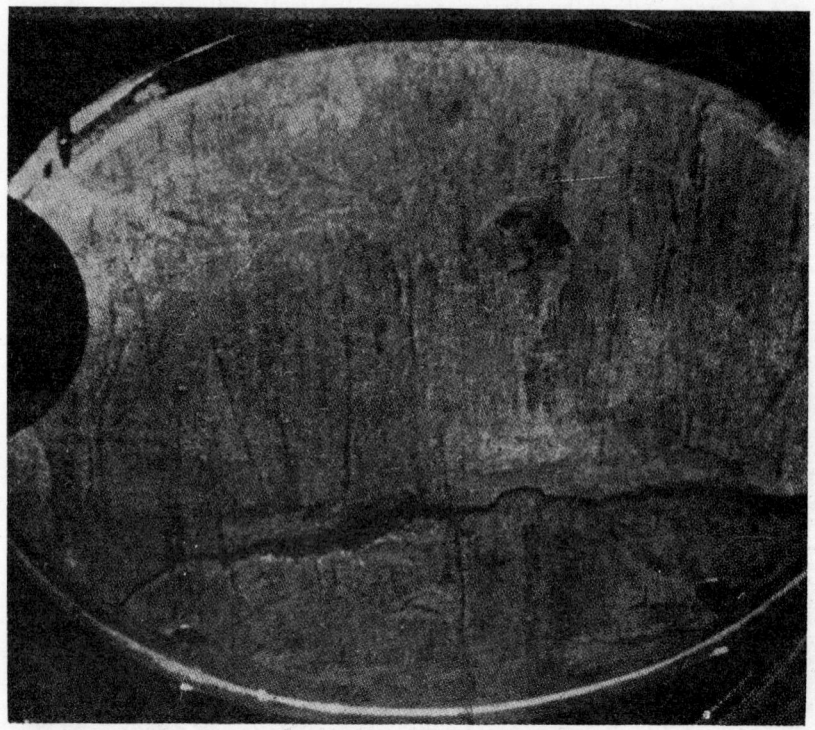

FIGURE 360. Well-developed stylolitic zone that has been leached out by action of subsurface fluids. Vertical striae may be due to presence of stylolites running transversely to the one appearing in bottom of picture. The phenomenon of a leached stylolitic zone is interesting because in some limestones these zones serve as reservoirs of commercial deposits of oil (2,559 feet).

print. The prints are fitted together to make a continuous picture of the well bore in much the same manner that aerial photographs are fitted together to make photomaps.

Figures 357 to 363 are examples of photographs taken in a well in west Texas. The formation is the San Andreas section of the Permian lime in the Hendricks pool. More than 400 pictures were taken in this well at depths of from 2,300 to 2,725 feet. When the pictures were taken, the well fluid was within 200 feet of the top of the well.

Future Developments

In some gas wells the collection of dust on the picture window during the descent of the camera may present a difficulty not encountered in liquid-filled wells. To overcome this difficulty a camera has been built with an extra steel sleeve, which slides over the water chamber and camera housing covering the picture window. When the camera reaches the point

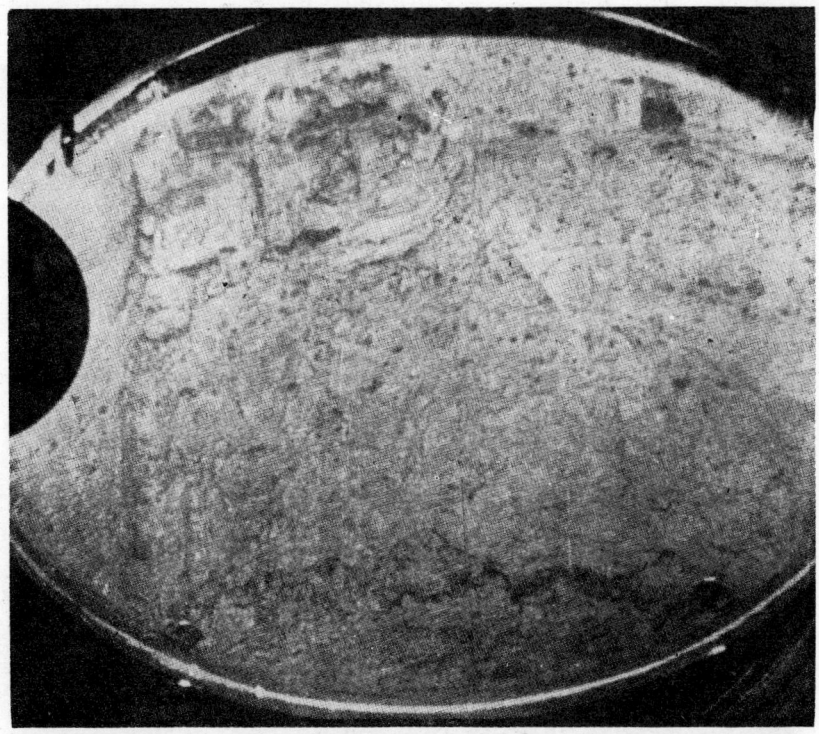

Figure 361. Some very fine porosity in upper half of picture and a leached stylolitic zone at bottom of picture. There seems to be a distorted zone near top of picture which may have been caused by movements of sediments while in a plastic state (2,676 feet).

in the well where the pictures are to be taken, the sleeve is released electrically and slides down far enough to bring a hole in the sleeve into register with the picture window. This device appears to be very workable but has not been field-tested.

A new, very compact camera has been built which is electrically driven, uses 35-millimeter film, and takes 500 pictures 24 millimeters by 36 millimeters with one loading. These dimensions are more than three times those of the pictures taken with the 16-millimeter camera. The new camera fits into the same $4\frac{3}{16}$-inch-diameter case as the 16-millimeter camera. It also has not been field-tested.

In the discussion of the general principles involved in the design of these deep-well cameras, it was pointed out that the use of as large a water chamber as the well bore would permit was an advantage, because the size of the area photographed at each exposure could thereby be increased, and because a better fit between picture window and well bore could be attained. A logical step in this direction might be to build a

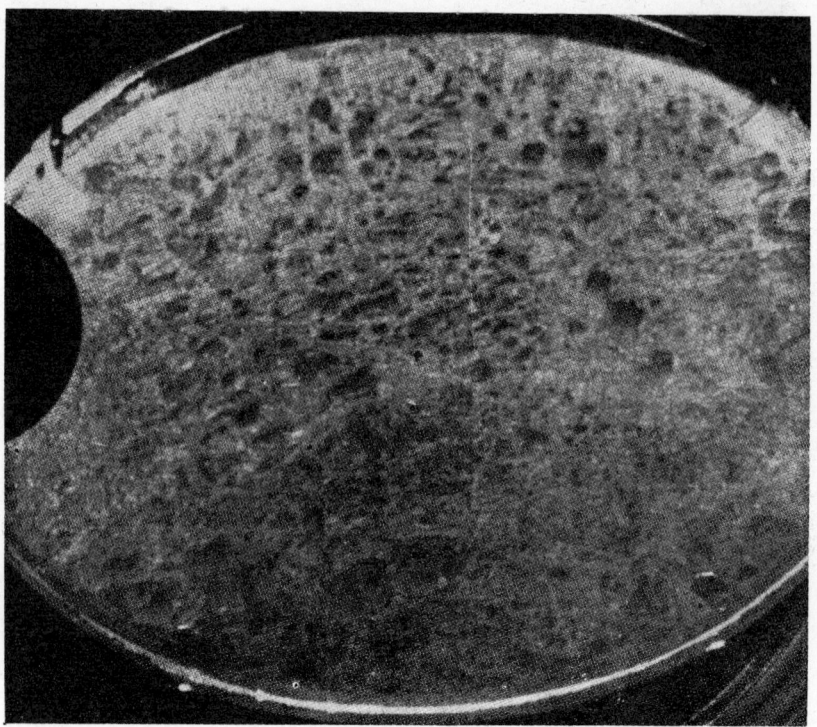

FIGURE 362. Well-developed primary porosity which may occur in all types of limestones; generally the pores are not connected, and rocks of this type do not generally make good reservoirs (2,678 feet).

series of water chambers of different diameters with the upper heads designed to connect to the same standard small-diameter camera housing.

Reinhold [8] ran his camera into shallow wells on a string of tubing and was thereby able to orient the camera to take pictures in any desired direction, this direction being indicated on the film by a compass located

[8] Thomas Reinhold, chief geologist of the Geological Survey Department of Holland. See Haddock, *Deep Bore Hole Surveys and Problems*, p. 179, New York, McGraw-Hill Publishing Co., Inc., 1931; British patent 226,079 and U.S. patents 1,658,537 and 1,790,678; Colliery Engineering, p. 371, Aug. 1926; and Mining and Metallurgy, Feb. 1932.

in the field of view of the camera. To orient the apparatus when it is supported by a cable is somewhat more difficult. Even with a slight deviation of the axis of the bore hole from the vertical there is a strong tendency for the camera to twist around so that it consistently faces the low side of the hole with the "push-over" springs facing the high side. Pictures taken in the same well on different trips usually show the same views because of this tendency of the camera to orient itself. Many schemes have been suggested for taking pictures in all radial directions in the bore hole. One

FIGURE 363. Pay-zone section of well. Also shown are some very large openings together with well-developed smaller openings, most of which are confined to lower one-third of picture. Perhaps some of this porosity may have been formed by leaching out of remains of organisms such as corals. Some of the apparent distortion around large openings may have been caused by growth of reef structure (2,719 feet).

that looks promising is that for using a cylindrical or bowl-shaped picture window extending completely around the circumference of the water chamber, omitting the mirror, and using a very wide-angle camera lens. Another method would be to substitute for the plane mirror a cone-shaped mirror or a submarine-periscope-type mirror with its axis vertical. The distortion produced by such a mirror could be cancelled out by using the same optical system in reverse for projection printing or projection viewing of the developed film. One rather simple method of taking pictures in different directions would be to use a plane 45° mirror as in the present camera but to provide a means, perhaps a motor or a solenoid, for rotating the mirror progressively a slight distance around a vertical axis each time the film is advanced.

STEREOSCOPIC PAIRS

Related to the problem discussed above is the one of taking pictures in stereoscopic pairs for stereoscopic viewing. The sense of perspective offered in this way should be useful in studying the porosity, cleavages, and general structure of the formation. The usual method of taking the views from slightly different lens positions would be applicable here. In this case rotating the mirror is equivalent to shifting the lens position. Thus for steroscopic pairs if would be necessary only to take two separate pictures at each camera location, rotating the lens through a small fixed angle between the two exposures.

Another idea offering real possibilities is the one of replacing the photographic camera with a television camera and transmitting to the surface, where the subject viewed by the camera would be immediately and continuously visible on a screen. This subject has been considered by many. Wired-television techniques now being employed industrially for the observation of remote locations should be applicable here. The Diamond Power Specialty Corporation of Detroit is now engaged in developing a deep-well television device. It is believed that with present television tubes and techniques such a development is entirely feasible. The deep-well camera has been developed to the point at which it can be of very practical value in examining bore holes. Developments in the near future should make it even more valuable.

THE ELECTRIC PILOT IN SELECTIVE ACIDIZING, PERMEABILITY DETERMINATIONS, AND WATER LOCATING

P. N. HARDIN

The three major uses for the electric pilot in oil-field completions and remedial work are in selective acidizing, permeability surveys, and water-locating surveys. Selective acidizing is employed to control the injection of acid into specific sections of the well bore during the acidizing of a well, thus increasing the permeability of these sections without affecting the permeability of other exposed zones. Permeability surveys are made to determine the thickness of the various permeable sections of the bore hole, the vertical position of these zones at the bore hole, and the relative capacities of the individual zones. Water-locating surveys are conducted to determine the points of entry of formation waters into a well.

THE ELECTRIC PILOT

The electric pilot is an apparatus through which the interface between two liquids with dissimilar electric conductivities can be accurately located in a well. Primarily the electric-pilot unit is made up of a source of electricity, a circuit consisting of an insulated single conductor cable that can be run into a well, and a special electrode assembly that can be run into the well on the cable. A schematic diagram of the unit is shown in figure 364. The cable is carried on a power-driven reel, which together

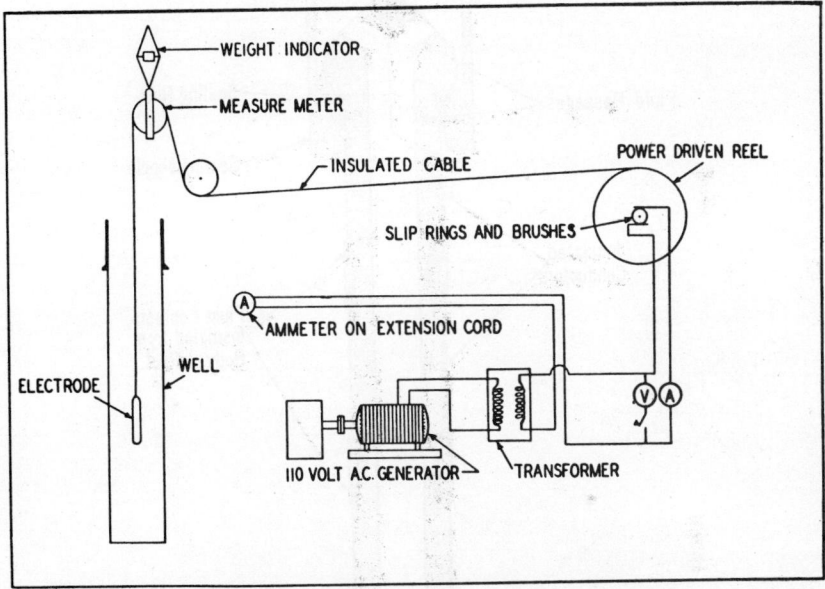

FIGURE 364. Diagram of electric-pilot equipment.

with the generator and all other auxiliary equipment is truck-mounted as a compact unit. Slip rings and brushes permit completing the electric circuit from the cable mounted on the reel to the generator and recording instruments.

A diagrammatic view of the electrode assembly is given in figure 365. The instrument consists essentially of a tube several feet long, on the outside of which are mounted two electrodes. The electrodes are completely insulated from the tube. The insulated conductor wire from the cable is connected to the upper electrode, and the upper and lower electrodes are connected by means of an insulated wire having a resistance of several hundred ohms. In order to complete the electric circuit in the well, it is, therefore, necessary to establish electric connection between one or both of the electrodes and the body of the assembly.

The actual operation of the electric pilot is seen by referring to fig-

ure 366. In the three illustrations of this figure, the electrode assembly
is shown completely immersed in oil, immersed half in acid and half in
oil, and completely immersed in acid. Each of the illustrations shows the
electrode assembly connected with a battery (*B*) and an ammeter (*A*) and

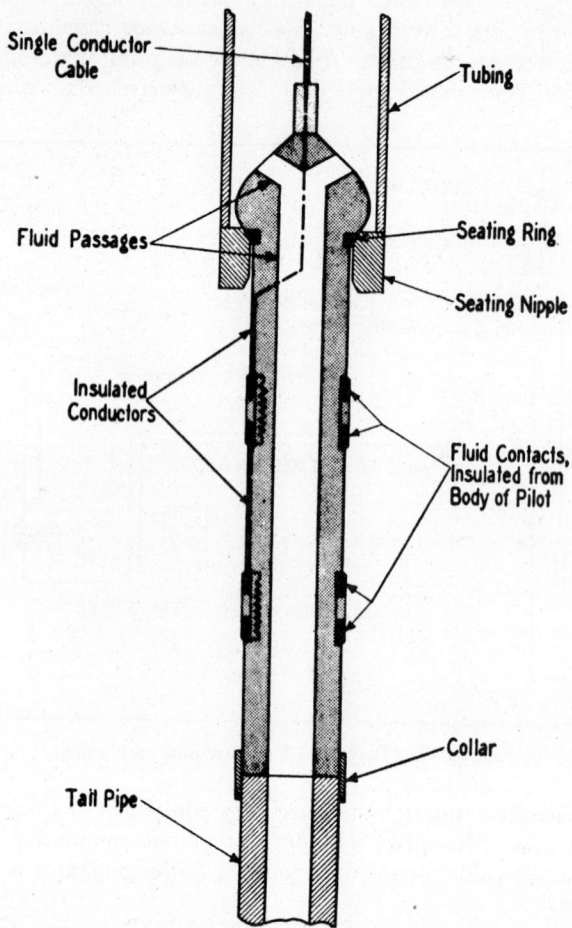

FIGURE 365. Diagram of electrode.

the ammeter readings that can be expected under each of the three conditions that are possible at the bottom of the hole during acid treatment.

In the left-hand illustration of figure 366, only oil is in contact with
the assembly, and, as oil is a nonconductor, the electric circuit between
the electrode and the body of the assembly is not completed. No current
flows in the circuit and the ammeter registers zero.

In the center illustration of figure 366, acid is in contact with the lower electrode and with the lower half of the body of the assembly, and oil is in contact with the upper electrode and the upper part of the body. As acid is a conductor of electricity, the electric circuit is completed between the lower electrode and the body assembly. Current now flows in the circuit from the battery to the top electrode, through the resistance to the bottom electrode, through the acid to the body of the assembly, to the ammeter, and back to the battery. A definite amount of current is now flowing in the circuit, and the ammeter will show about a one-half-scale deflection.

In the right-hand drawing of figure 366, only acid is in contact with the assembly. In this case the electric circuit is completed between both the lower and the upper electrodes and the body of the assembly. The current now flows in the circuit from the battery to the top electrode, through the acid to the body of the assembly, to the ammeter, and back to the battery. As the resistance between the two electrodes now has been shorted out, considerably more current will flow than was possible with the circuit as shown in the center illustration. The ammeter now will indicate practically a full-scale deflection.

The illustrations in figure 366 are analogous to what actually happens during a well treatment. The engineer observes an electrical instrument during the treatment, and when this instrument reads zero, the engineer knows that the acid is below the bottom electrode. When the instrument gives approximately a half-scale deflection, the engineer knows that the acid level is between the bottom and top electrodes. When the instrument gives a full-scale deflection, the engineer knows that the acid is at or above the top electrode. The relative acid- and oil-pumping rates employed during the treatment are then determined by the readings on the electrical instrument. If the acid level is too high, as indicated by a full-scale deflection on the instrument, either the acid pump will be slowed down or the oil pump speeded up or both. If the acid level is too low, as indicated by a zero reading on the electrical instrument, the acid pump will be speeded up or the oil pump slowed down or both.

SELECTIVE ACIDIZING

In the past many methods such as the use of packers, plugging-back, and two-pump treatments have been developed for use in acidizing processes where conventional treating was not applicable. While these methods improved upon the conventional technique, they still left much to be desired, inasmuch as there was no definite check means to indicate where the acid was actually going during the treatment. With the electric pilot the engineer has knowledge of the location of the acid and of what portions of the formation are taking the acid during the entire treatment.

The following outline briefly points out some of the reservoir and well conditions under which selective acidizing should be considered.

1. New wells in which a change in lithology is present in the open

hole or pay section. Such changes may involve sand, lime, dolomite, or primary or secondary anhydrite.

2. New wells in which saturated zones of various permeabilities are present.

3. Wells showing excessive gas-oil ratios, but with the possibility of additional available oil present in tighter saturated sections.

4. Wells in which it is desirable to make additional gas available.

5. Wells with intermediate water zones.

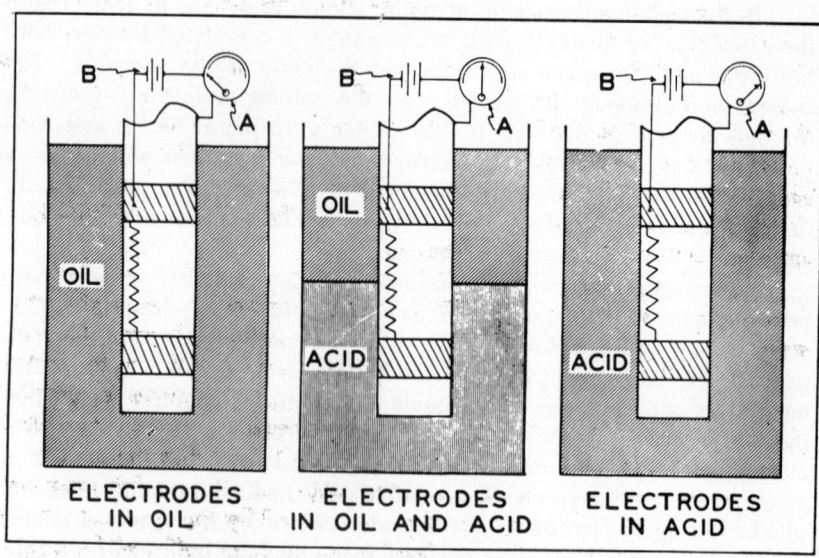

FIGURE 366. Operation of an interface locator.

6. Wells deepened to new pay zones.

7. Old wells with depleted zones but with oil remaining to be recovered from the tighter sections.

8. Wells in which it is desirable to hold acid away from plug-back or other repair work, including "Securaloy" lines, plastic or cement squeeze, and redrill jobs.

9. Wells in which conditions are such that it is important to keep acid away from the casing seat.

10. Injection wells in which it is desired to have uniform distribution of the gas or water being returned to the formation or in which it is desirable to alter existing injection characteristics.

In general, selective acidizing is applicable to wells producing through perforations in casing set through the pay section, together with those wells producing from open hole with the casing set above or on top of the producing zone. In some instances, however, there may be conditions present that will cause complications in wells completed in either of these ways. Channeling behind the pipe and vertical fissures or fractures in the producing formation are two such conditions that might be mentioned. Oftentimes control can still be effected in some instances under these conditions, however, through the proper application of materials that tend to bridge and build up an effective filter cake, thus blocking the movement of fluid into the channels or fractures. In other instances, the conditions mentioned may be too diffcult to permit effective control through the use of these materials.

It might be well to mention a few items that, if given thought and study in specific cases, may go a long way toward assuring satisfactory results from the selective type of treatment. All available bottom-hole data should be plotted and correlated. Permeability surveys have been found to be of primary importance for comparison and correlation with geologic information, and, under certain conditions, temperature surveys are quite valuable. Other factors for determination and consideration are the possibility that emulsion troubles may be encountered, the gravity of the oil to be used, and the solubility of the formation to be acidized After considering all the available information, including past experience with conventional and selective acid work in the area, where available, the proper procedure for the individual well can be determined.

Figure 367 illustrates several typical selective-acidizing applications.

PERMEABILITY SURVEYS

In determining the permeability profile of a section of well bore with the electric pilot, the results obtained are in terms of relative permeability, i.e., the permeability of any given zone is expressed in a percentage relative to the permeability of the entire open section. The relative permeability of an exposed section is determined by measuring the fluid-injection rate into all parts of the entire exposed formation. Various techniques have been used for these relative-permeability surveys, but the one that seems to give the best results in most of the wells consists in introducing sufficient salt water into the bottom of the well completely to cover all the section to be surveyed and then forcing this salt water into the formation by the introduction of oil into the well. The rate of fall of the salt-water-oil interface is determined by means of the electric pilot during the injection of salt water into the formation.

In permeability-profile work the hookup for the electric pilot is different from that used in selective acidizing in that it is necessary to follow the downward-moving two-fluid interface with the electrode assembly. In this procedure the electrode assembly is not seated at the bottom of the tubing, no tubing being required in the well for this type of job. In permeability-survey work, furthermore, the pumping rate at the surface must remain constant throughout the survey. Figures 368 and 369 illus-

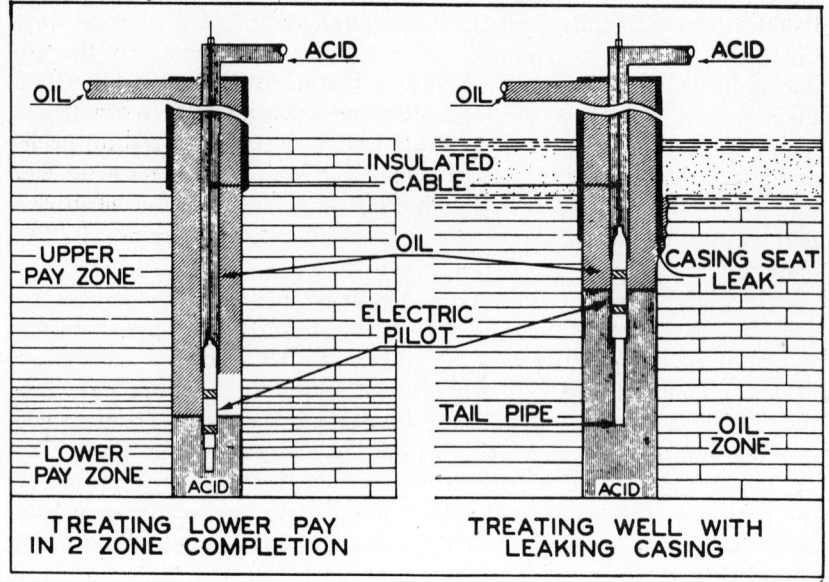

FIGURE 367. Application of electric pilot.

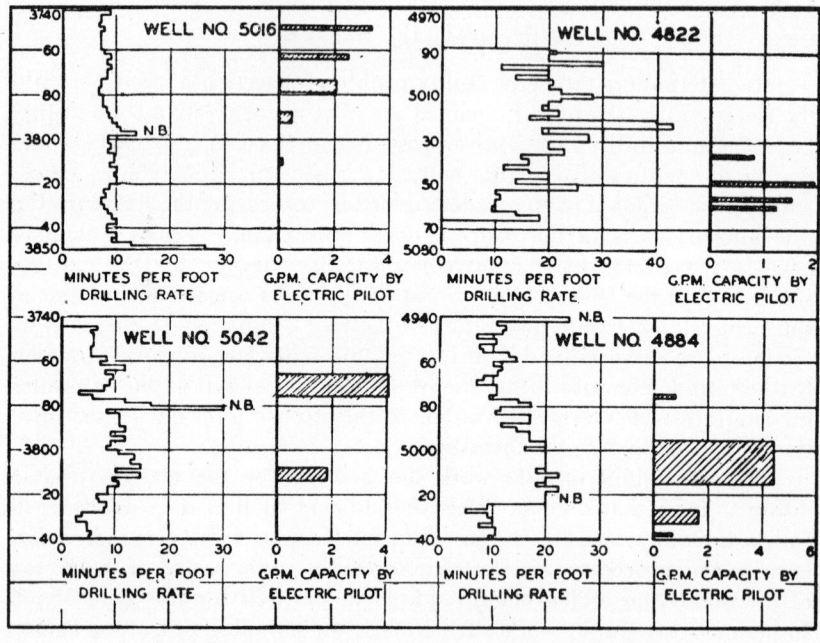

FIGURE 368. Record of a pilot survey and drilling time.

trate the permeability surveys compared with drilling-time and core-analysis data.

In those wells where it is undesirable to use two fluids for the injected-fluid-type survey or in those wells where it is desired to measure the rate of produced rather than injected fluids, either liquids or gas, an instrument known as the "spinner" is used instead of the electrode assembly. The spinner consists essentially of a housing in which a freely rotating propeller is mounted on pivots. As the propeller is rotated by the flow of fluids past the instrument assembly, each rotation is recorded at the surface by means of an electric counting mechanism contained in the spinner

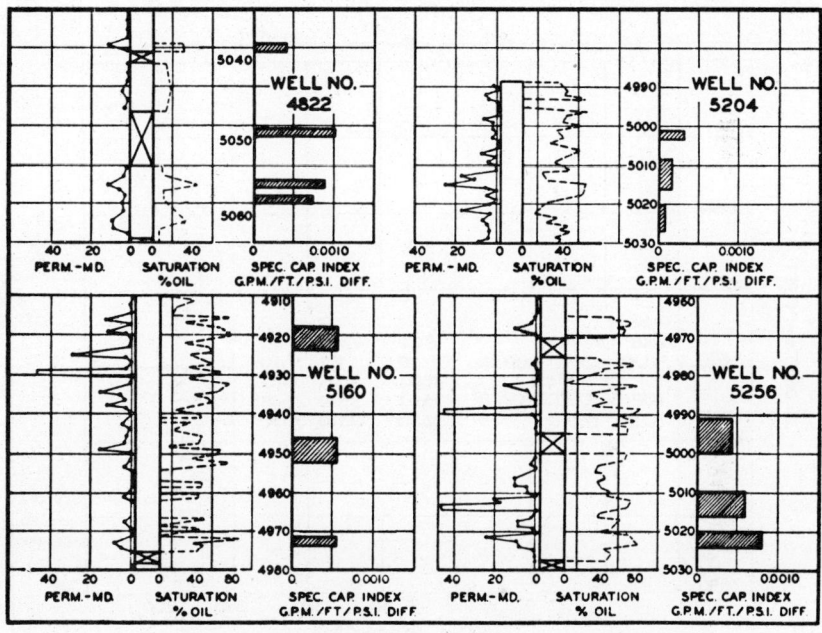

FIGURE 369. Record of a pilot survey and core analysis.

housing. The electric-pilot surface equipment and cable are used, the only difference being in the data-recording meters.

It is seen that knowledge of the variations in bore-hole size is essential to the accuracy of the spinner survey. This knowledge can best be obtained through use of a caliper, another instrument run in conjunction with the electric-pilot unit. This caliper is highly accurate, automatically recording on a chart at the surface the irregularities in the bore hole. Uses for the caliper other than to provide data for permeability and other types of surveys are the following: (1) to locate packer seats in correcting high gas-oil ratios or controlling water intrusion, (2) to gain information for plastic plug-back work, (3) to determine the size of shot holes,

(4) to determine the size of pipe in the hole, (5) to gain information for setting liners, (6) to locate casing seats, and (7) to aid in calculating cement volumes before setting pipe in open holes.

WATER-LOCATING SURVEYS

The accurate location of points of water entry into a well bore is valuable in the planning of reconditioning or work-over programs and is the first step in remedial operations to halt intrusion of the waters. The water-locating survey is one of the several services that employ the electric-pilot unit. For water-locating work the unit is used in conjunction with an electrode consisting of two dissimilar metals, which, when immersed

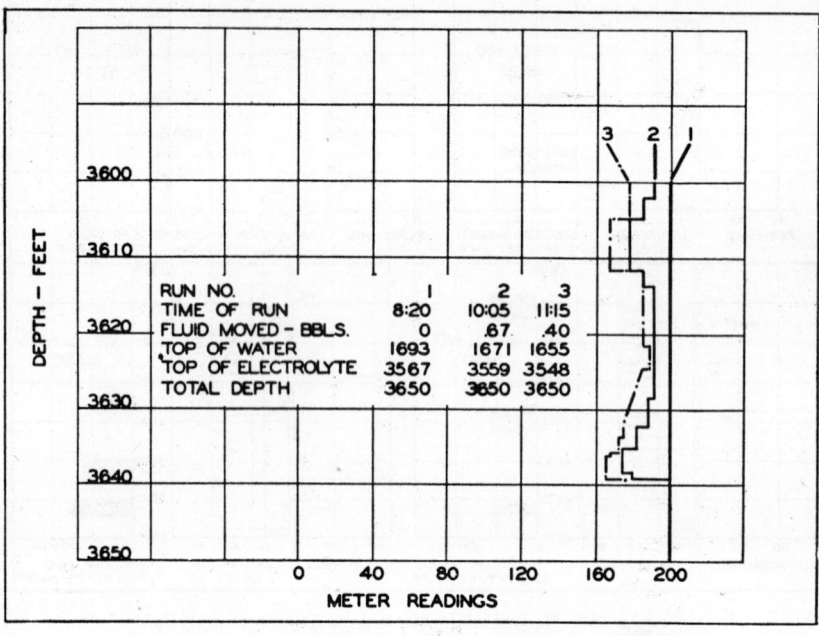

FIGURE 370. Chart of a typical water-locating survey.

in electrolyte solution, act as a primary cell, generating electrical current. This current, generated at the electrode, is measured at the surface, the meters being capable of detecting very slight voltage fluctuations at the electrode.

In operation it is necessary to condition water in the well with a solution of electrolyte prior to conducting the survey. Then, as the survey is made, any local weakening of concentration of the electrolyte caused by infiltrating formation waters is immediately detected by a voltage drop at the electrode as it reaches that point. After the survey the meter readings plotted against depth indicate points where water is entering the well. Figure 370 illustrates a typical water-locating survey, comprising two traverses of the formation after the check run to determine conditioning of the fluid in the well. Two points of water entry are indicated. A subsequent plug-back to 3,626 feet reduced the water seven percent, the amount entering the well at 3,613 feet.

A clean well, free of cavings and other material that might interfere with the operation of the water-locator equipment, is an important prerequisite to a successful survey. If the well has been killed with water, or water has set on the formation over an extended time, the well should be produced and then allowed to come to a static level before the survey.

The fluid electrolyte is distributed throughout the water in the well by means of a special bailer. In a flowing well it is necessary to lubricate the survey to be run immediately after the electrolyte has been dumped.

After a test run to verify proper conditioning of the well fluid, the well is produced, either by flowing, swabbing, or bailing, to allow formational fluids to enter the well. A traverse then is made with the electrode to determine where dilution of the electrolyte is occurring. The number of traverses that may be necessary usually depends upon the rate at which the well produces water.

THE POROSITY AND PERMEABILITY OF CLASTIC SEDIMENTS AND ROCKS

GEORGE H. FANCHER

The occurrence of petroleum depends fundamentally upon the ability of fluids to move through strata of the earth's crust. From a physical point of view, all problems relating to the origin and accumulation of oil, natural gas, and water in, and their production from, the depths of the earth's crust involve the flow of fluids through porous media. The porous media of consequence are sedimentary in origin in the main and include such clastic sediments as silts and soils; unconsolidated sands and consolidated rocks, such as sandstone and limestone, both oölitic and massive; and other porous rocks of lesser economic significance in the oil industry.

SOME CHARACTERISTICS OF SEDIMENTS

The characteristics of the solid constituents of sediments important to evaluation of those physical properties which determine the behavior of fluids in the sediments may be expressed in terms of at least six measureable parameters. These are

(1) size
(2) shape (sphericity)

Note: The illustrations in this section were prepared by William E. Fickert, Department of Petroleum Engineering, University of Texas.

    (3)   roundness (or angularity)
    (4)   texture of the surface
    (5)   orientation
    (6)   mineralogical composition

Each of these characteristics may be defined in several ways, as for example:

Size as a width of an opening in a screen, a diameter calculated from either the velocity at which a particle settles or from the measured volume of the particle, or a dimension observed and estimated visually from microscopic examination.

Sphericity in terms of a shape factor or the ratio of the surface of a sphere equivalent in volume to that of the particle to the surface of the particle.

Roundness in terms of the ratio of the average distance to the corners and edges of a particle to the radius of a circle inscribed within a projection of the particle.

Surface texture in terms of relative smoothness or in measurements of pitting or striations on the surface of the particle.

Orientation as the direction of the principal axes of a particle referred to suitable coordinates in space.

Mineralogical composition as interpretation of factors of hardness, density, wettability, swelling upon hydration, shrinking upon dehydration, color and the like on some systematic basis.

Sediments are composed of multitudes of solid particles differing in size, shape, and other attributes so that each of the characteristics of a particle which have been enumerated for a particular sediment can be described as a frequency distribution which has as parameters a *mean, a standard deviation, a uniformity coefficient, probability attributes,* and other qualities. Indeed, the characteristics of sediments as a problem in geometry are expressible in terms of at least a dozen statistical parameters, depending upon which are considered to be germane to a particular argument. Furthermore, in addition to the unique and specific characteristics of the component particles, sedimentary materials possesses aggregate or mass physical properties such as, for example, porosity, permeability, tensile strength, and capillary properties. The mass properties logically must depend greatly upon the resultant physical properties of the component particles as well as upon geo-historical phenomena, such as sorting, packing, weight of overburden with resultant stress and strain, and contact with fluids. The mass properties are determined best by experiment, but the theory of the relationship of the properties of the component particles to those of the aggregate not only is of great interest and value to the student but also is a challenge to the intellect as a problem in statistical geometry for which a solution ultimately should be possible.

The solution of this problem is a goal toward which petroleum geologists and engineers are striving to arrive analytically and experimentally.

DEGREE OF CONSOLIDATION

The sedimentary materials found in petroleum reservoirs range in degree of consolidation from loose, unconsolidated materials to hard, competent, and even very dense sandstones, quartzites, dolomites, and limestones (oölitic and massive), the extreme conditions occuring less commonly—dolomite and limestone excepted. Generally, some 90 percent or more of the granular material in a true sand or sandstone is quartz; and, consequently, the physical properties of quartz contribute greatly to the resultant. Usually something less than five percent of the remaining material is composed of feldspar, heavy minerals in considerable variety, and micaceous minerals. Clay minerals, shale, silt, and calcareous material usually make up the remaining five or more percent which usually also serve as a cement which binds together the component particles of a consolidated sedimentary rock. For example, the common bonding agent in the Gulf Coast reservoir rocks is clay and, more specifically, usually is the clay mineral, illite.[9] [10] Those sandstones, containing five percent or somewhat less of clay minerals or shale and little or no calcareous matter, also may be consolidated and can be clean in appearance. The conclusion may be drawn that something less than five percent of clay or shale is the minimum amount required for consolidation. Generally more than five percent of these materials occur in sandstones, the amount being large in so-called "dirty" sandstones. Loose sands and relatively unconsolidated sandstones frequently contain more than five percent of these bonding agents, the amount being larger in "dirty" sandstone.

Unconsolidation must be attributed to some difference in geo-history as well as to some difference in physical properties of the materials. The cementing material in a rock is one of the more important factors which affect the behavior of fluids therein.

STRUCTURAL CONFIGURATION

The shape and size of sand grains range between wide limits. The component particles usually possess the approximate form of *parallelepipeds* having various axial dimensions with rounded sides and corners. The particles range in size from small pebbles to that characteristic of fine-grained silt. A preponderance of either extreme in a sand or rock is unusual. Screen-analysis data for various sands may be found in the literature.[11] [12] [13] The diversity and magnitude of sizes reported in the literature

[9] Fancher, G. H., and Oliphant, S. C., AIME, Trans., vol. 151, pp. 221-232, 1943.
[10] Fancher, G. H., Lewis, J. A., and Barnes, K. B., The Pennsylvania State College, Mineral Industries Experiment Station Bull. 12, pp. 65-171, 1933.
[11] Halbouty, M. T., Oil Weekly, vol. 83, no. 11, pp. 21-24; no. 12, pp. 22-26; no. 13, pp. 36-48, 1936; vol. 84, no. 1, pp. 34-50; no. 2, pp. 36-46; no. 3, pp. 36-42, 1936.
[12] Muskat, Morris, The Flow of Fluids through Porous Media, McGraw-Hill Book Company, 1933.
[13] Muskat, Morris, Physical Principles of Oil Production, McGraw-Hill Book Company, 1949.

are typical of the majority of petroleum reservoir strata, although the distribution may differ markedly from sample to sample in the same reservoir.

The configuration of the grains in a sediment determine the nature of the pore spaces in the sediment. Obviously, the component grains may be packed in any manner, but only an exceedingly irregularly spaced and shaped pore always results no matter how regular and systematic the packing may be. Although the pores of sandstone are irregular with respect to both size and shape, the pore pattern of massive limestone and dolomite is even more irregular and complicated than that of sandstone, being determined largely, not by conditions of sedimentation, but by secondary processes of weathering, leaching, partial solution, and secondary deposition of minerals within fluid passageways.[14] The behavior of fluids is much easier to study in sandstones and oölitic limestones than in other rocks for this reason. Although the individual pores of a rock may be highly irregular in shape and size, the large number of pores imparts relatively consistent mass properties to the rock as a whole, and the mass physical properties do reflect the statistical distribution of size and shape of the pores and do permit mathematical analysis of the gross behavior of the sedimentary body with fluids as an entity.

Porosity

The porosity of a rock is a measure of the capacity of that rock to store or hold fluids. In this sense porosity is a static quality completely determined by simple geometry. Porosity is expressed quantitatively as the ratio of the pore volume of the rock on either a fractional or percentage basis. For example:

$$f = \frac{\text{volume of pores}}{\text{bulk volume}} = \frac{\text{volume of pores}}{\text{volume of pores} + \text{rock}} \qquad (1)$$

and

$$\text{percentage porosity} = 100f$$
in which f is the fractional porosity $\qquad (2)$

According to this definition, the porosity of porous materials could have any value up to 100 percent, but the porosity of sedimentary rocks usually is considerably less than 100 percent. If, for example, spheres of equal diameter are arranged systematically, so that each sphere touches another, the wide-packed or cubic system (fig. 371) has a porosity of 47.6 percent, and the close-packed or rhombohedral system has a porosity of 26.0 percent. Clearly the porosity for such a system is independent of the diameter of the sphere and depends solely upon the arrangement of the spheres. If smaller spheres are mixed among the spheres of such a system or, if small enough, are placed within the pores, clearly the ratio of voids to solids becomes less and porosity is reduced. Likewise, if

---

[14] Bulnes, A. C., and Fitting, R. V., Jr., AIME, Trans., vol. 160, pp. 179-201, 1945.

non-spherical particles such as rods or plates are mixed with the spheres, pores can either be partially filled or bridging can result in the formation of even larger pores with the result that porosity would tend to be either decreased or increased, depending upon which phenomenon predominates. The resultant effect of both factors usually is such that the porosity of oil-bearing clastic sedimentary rocks is much lower than the maximum of 47.6 percent for wide-packed spheres of equal diameter because of the effectiveness of plugging and the ability of forces of compaction to minimize bridging.

The clastic sediments are differentiated by names somewhat according to particle size and include the gravels, sandstones, shaly sandstones,

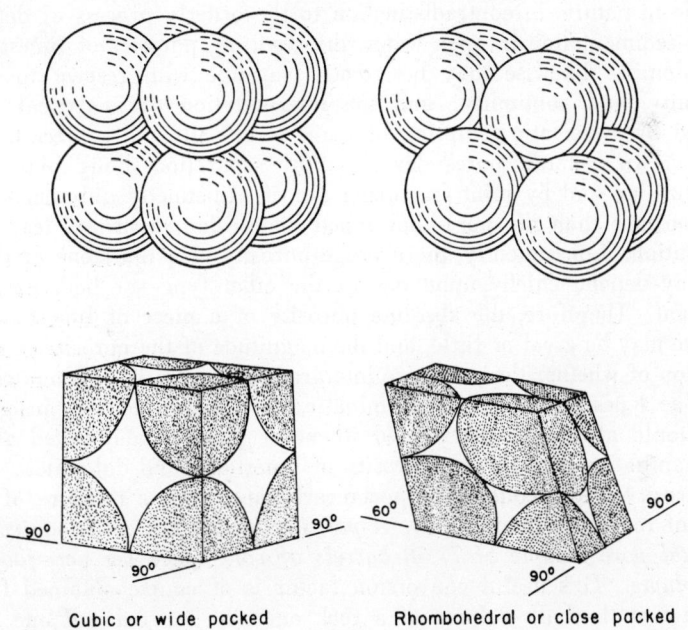

Cubic or wide packed   Rhombohedral or close packed

FIGURE 371. Unit cells and groups of uniform spheres for cubic and rhombohedral packing. (After Graton and Fraser, *Jour. Geology.*)

and clays. The porosity of the finest grained sediments, such as the muds and oozes which eventually become clays and shales, originally was high (40 to 85 percent) because of the great degree of hydration of these tiny particles when deposited. Prolonged compaction under the great weight of overlying sediments reduces porosity and decreases permeability of these rocks, sometimes almost to the vanishing point. The porosity of shale, for example, decreases exponentially with depth, according to Athy.[15] The coarser grained sediments, sands, sandstones, sandy shales, and oölitic limestones, were better able to resist forces of compaction;

---

[15] Athy, L. F., Am. Assoc. Petrol Geologists Bull., vol. 14, no. 1, pp. 25-35, 1930.

and the porosity of these rocks generally is greater than 26 percent and less than 47.6 percent unless plugging of pores also has taken place during geologic time.

The porosity of massive limestones and dolomites depends chiefly upon two geologic processes: namely, a fracturing, jointing, or cracking of the rock matrix resulting from relief of stresses caused by movement of portions of the earth's crust and a development of such porosity as may exist in the matrix by weathering, erosion, solution, and leaching by the circulation of waters. Reduction of porosity by crystallization and deposition of minerals from solution in the pores of limestones and dolomites also is a common geologic phenomenon. Inasmuch as these processes are sporadic in nature, in contradistinction to the orderly process of deposition by sedimentation in open water, the resulting porosity of limestones and dolomites likewise may be greatly variable with respect to size, uniformity, and continuity; and the interpretation of analytical data obtained by systematic sampling of cores is difficult and subject to uncertainty. In general, pores developed by separation along planes of crystallization and by joint fracturing are of a distinctly different order of dimensions than the pores developed by erosion, fissuring, leaching, and solution. Consequently, the average porosity of a limestone or dolomite may depend chiefly upon one or the other type—or both may be significant. Therefore, the absolute porosity of a piece of limestone or dolomite may be great or little, and the magnitude of the porosity is little indication of whether the pores are intergranular or chiefly solution channels. The types of porosity predominating, or those which are insignificant, should always be determined whenever possible and stated along with quantitative data for the porosity of limestones and dolomites.

Porosity is of economic importance because it is a measure of the ability of rocks to contain fluids. *A porosity of one percent is equivalent to a total pore volume of 77.58 barrels of void space per acre-foot of bulk volume.* This useful conversion factor is of course obtained from the fact that the bulk volume of a rock one acre in area and one foot thick is 43,560 cubic feet. Only one percent of this volume is pore space, and a barrel is equivalent to 5.615 cubic feet, so that total pore volume

of the rock $= \dfrac{(43,560)\ (0.01)}{5.615}$ barrels per acre-foot per percent porosity.

This conversion factor makes possible the writing of the general formula

$$\text{Total pore volume in barrels} = 77.58 \text{ at } (100f) \text{ in which} \qquad (3)$$

A is area, acres

t is thickness of formation in feet, and

f is the fractional porosity

The relative extent to which the pores of a rock are filled with specific fluids, i.e., oil, water, and gas, is expressed best as percentage saturation of the pores. If the pores are filled with only one fluid, for example oil, the saturation with respect to that fluid, oil, is 100 percent; if the pores are only half filled with oil, the oil saturation is only 50 percent. Consequently, the total barrels of oil in place at reservoir pressure and temperature may be computed from the relation

$$\text{Total oil content} = 77.58 \text{ at } (100f) \ (0.01S_o) \tag{4}$$

in which S_o is the average-percentage-oil saturation of the rock. Obviously, if the total oil content is desired in barrels of stock-tank oil, the oil-content at reservoir conditions obtained from equation (4) should be multiplied by an appropriate shrinkage factor based upon the particular change in pressure and temperature appropriate to the situation.

Equation (4) may be used to obtain the total water content at reservoir conditions for a particular bulk volume of rock by substitution for S_o of the numerical value of S_w, the percentage-water saturation of the rock appropriate to the particular situation.

The free-gas content in cubic feet at reservoir pressure and temperature may be estimated from the relation

$$\text{Total free-gas content} = 43,560 \ atf \ [1.00 - 0.01 \ (S_o + S_w)] \tag{5}$$

The total free-gas content at reservoir condition from equation (5) can be converted to cubic feet of gas at standard conditions by use of the ideal gas laws corrected for deviation therefrom. An apropriate multiplier accomplishing this objective is

$$\frac{p_r T_b}{p_b T_r Z}$$

in which

p_r is the reservoir pressure, psia
p_b is the base pressure, psia
T_r is the reservoir temperature, $°F$ absolute
T_b is the base temperature, $°F$ absolute, and
Z is the compressibility factor at T_r and p_r, no unit.

A commercial oil-bearing rock must contain sufficient petroleum to warrant development and exploitation. Consequently, the minimum economic porosity of sands and sandstones is 8 to 10 percent (621-779 barrels per acre-foot), and the maximum porosity seldom exceeds an average of 35 percent (2715 barrels per acre-foot). The porosity of commercial oil-bearing massive limestones and dolomites extends over a much larger range of values than that of sandstones. Some limestones and dolomites are so cavernous and solution chasms and vugs therein so large that the porosity becomes so great locally that statistical implications of porosity become meaningless. If, however, the porosity of oil and gas-bearing limestones and dolomites results chiefly from inter-granular interstices, the porosity can be as little as four to six percent and still have commercial value under favorable conditions. Obviously, the pores of rocks containing oil and gas must be interconnected sufficiently well to permit fluids to pass from one pore to another for an oil-and gas-bearing rock to have commercial value. Some pores or interstices may be isolated; and, consequently, sometimes the terms total and effective porosity are employed where differentiation between the porosity based upon the interconnected pores and those that are not is necessary. Total porosity obviously refers to all pores, both interconnected and isolated, whereas effective porosity obviously refers to only interconnected pores, these only being effective in fluid flow. The isolated pores seldom amount to more than ten percent of the total volume of pores in any sedimentary rock.

MEASUREMENT OF POROSITY

The measurement of the porosity of a rock is simple in principle, but can be complicated in practice. The definition of porosity clearly indicates the quantities that must be evaluated in measurement: namely, the pore volume and the bulk volume. Any method of measurement that will enable the reliable determination of these quantities with the degree of accuracy desired may be employed. The bulk volume may be determined by direct measurement for simple geometric forms or by the accurately observed displacement of a suitable liquid. The pore volume may be measured by filling or emptying the pores of a sample with a suitable fluid and measuring the amount required. If the fluid employed is a gas, the accurately observed relations between pressure and volume in isothermal expansion and contraction of the gas within a closed system, part of which includes the pore space to be evaluated, makes possible the computation of the pore volume of the sample. The porosity also is calculable by comparison of the apparent specific gravity of the sample: namely, the ratio of the weight of the sample to its bulk volume, with the true specific gravity of the rock.

Details of the exact procedure required in the laboratory measurement of porosity are available in the literature.[16][17][18][19][20][21][22][23][24][25]

If the pores are so small that penetration by a fluid such as mercury is slight, the bulk volume is readily observed by direct displacement in a suitable volumeter or is calculated from the loss of weight obtained upon submersion of the sample in a liquid. Penetration of the sample by a fluid can be prevented by coating the surface thinly with a lacquer, collodion, or wax. The protected sample can be immersed for determination of the bulk volume. The sample can be saturated with a suitable fluid such as water, tetrachlorethane, or oil, and then the displacement or loss in weight resulting from immersion can be observed. The familar Russel volumeter is useful for this purpose [26] (fig. 372).

[16] Barnes, K. B., The Pennsylvania State College, Mineral Industries Experiment Station Bull. 10, 1931.
[17] Coberly, C. J., and Stevens, A. B., AIME, Trans., vol. 103, pp. 261-269, 1930.
[18] King, F. H., USGS, 10th Annual Report, pt. II, pp. 50-294, 1897-1898.
[19] Melcher, A. F., AIME, Trans., vol. 65, pp. 469-497, 1921.
[20] Plummer, F. B., and Tapp, P. F., Am. Assoc. Petroleum Geologists Bull., vol. 27, pp. 64-84, 1943.
[21] Ritter, H. L., and Drake, L. C., Ind. Chem. Anal. Ed., vol. 17, pp. 782,786, 1945.
[22] Russell, W. L., Am. Assoc. Petroleum Geologists Bull., vol. 16, pp. 231-254, 1932.
[23] Taliaferro, D. B., Johnson, T. W., and Dewees, E. J., U. S. Bur. Mines, R. I. 3352, 1937.
[24] Washburn, E. W., and Bunting, J., Am. Ceramic Soc., vol. 5, pp. 48-56, 112-129, 1922.
[25] Westbrook, M. A., and Redmond, J. F., AIME, Trans., no. 165, pp. 219-222, 1946.
[26] Russell, W. L., op. cit.

The sand-grain volume likewise can be obtained by displacement of a fluid. The Washburn-Bunting porosimeter [27] (fig. 373) is used widely in obtaining the pore volume by direct observation.

The porosity of a zone or stratum composing part or all of a petroleum reservoir must be ascertained from laboratory data obtained from small core samples coming from wells. A single average quantity is

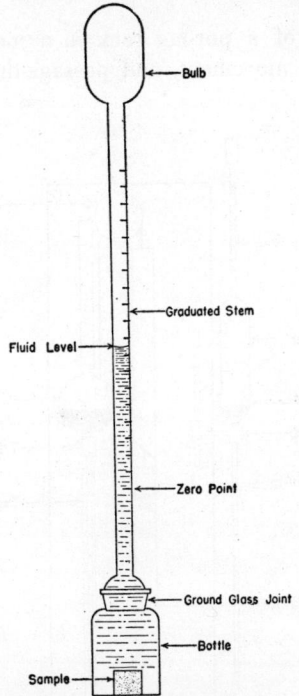

FIGURE 372. Russell volumeter.

desired for many quantitative uses. The accuracy of the figure employed depends not only upon the accuracy of measurement but also upon the adequacy of sampling. The distribution of porosity horizontally and vertically within a stratum is not uniform. The porosity of a reservoir, then, is a statistical quantity, the numerical value of which depends not only upon the porosity of the individual sample of rock but also upon the thickness and areal extent to which that porosity is assigned.

Average porosity, therefore, may be expressed by the relationship

$$f_{av} = \frac{\sum_{1}^{n} (aft)}{\sum_{1}^{n} (at)} \tag{6}$$

[27] Washburn, E. W., and Bunting, J., *op. cit.*

in which a is the areal extent assigned to a sample
  f is the fractional porosity of a sample
  t is the thickness of the interval assigned to a sample, and
  n is the total number of samples

PERMEABILITY

The permeability of a porous rock is a measure of its ability to permit the penetration, movement, and passage-through of fluids. Quan-

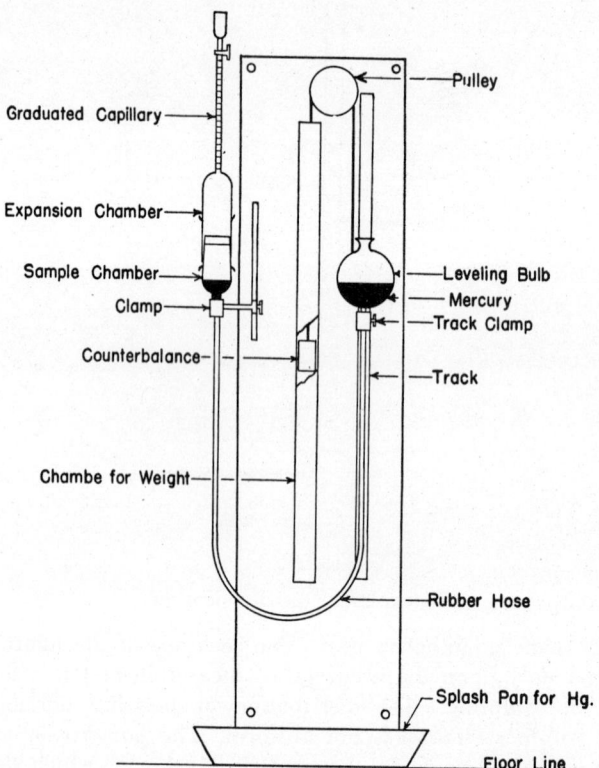

FIGURE 373. Washburn-Bunting type porosimeter used
at University of Texas.

titatively the definition of permeability is based upon the empirical relation between pertinent variables which is known as Darcy's law.[28] Darcy's law, based upon experimentation, states that the velocity of a homogeneous fluid through a porous medium is directly proportional

---

[28] Darcy, Henri, *Les Fountains publiques de la ville de Dijon*, Victor Dalmont, Paris, 1856.

to the hydraulic gradient and inversely proportional to the viscosity of the fluid. Conseqeuently,

$$V \propto (\frac{1}{\mu}) \ (\frac{dp}{dl})$$
(7)

which, if k be a coefficient of proportionality, becomes

$$V = (\frac{k}{\mu}) \ (\frac{dp}{dl})$$
(8)

and for rectalinear flow of a simple fluid (figure 374), becomes upon integration,

$$\frac{Q}{\theta A} = \left[\frac{k}{\mu} \right] \left[\frac{\Delta p}{L} \right]$$
(9)

in which

Q is the volume of a fluid in cubic centimeters.

θ is the time required for Q, seconds.

A is the cross-sectional area perpendicular to the rate of flow, square centimeters.

μ is the absolute viscosity of the fluid, centipoises.

Δp is the pressure difference, atmospheres.

L is the length of the path of flow, centimeters.

k is the coefficient of proportionality or the *permeability* of the porous medium darcys.

Clearly by this definition, the permeability of a porous rock to a simple, homogeneous fluid is dependent entirely on the geometry of the porous system; the permeability, i.e. the quantity k, has the dimensions of an area (L^2). The qualitative relation of permeability to porosity likewise is clear, a rock being permeable by virtue of porosity. Moreover, a rock may be porous but impermeable, but cannot be both permeable and nonporous. The exact quantitative relation between the porosity and permeability has not been determined because of the complexity of the problem. More specifically, the nature of the constant, k, has occupied the attention of many investigators whose findings and experience have been reported extensively in the literature. Although much of value has been learned as a result of these efforts, a general solution to the problem has not been obtained, despite the fact that all investigators agree that the coefficient, k, is an expression of those characteristics of a porous body which require that the relations implied by equation (8) be satisfied and, consequently, that k is substantially independent of the nature of the fluid and, therefore, is determined solely by the geometry of the system.

A rock has a permeability of 1.0 darcy, if the permeability be such as to allow a flow of 1.0 cubic centimeter per second per square centimeter of cross-sectional area perpendicular to the path of flow of a fluid which has a viscosity of 1.0 centipoise at the temperature of flow in response to a pressure gradient of 1.0 atmosphere per centimeter.[29] A millidarcy is one-thousandth of a darcy. Accordingly, a rock with a permeability of two darcys is twice as permeable as one with a permeability of one darcy: i.e., twice as much of a given fluid will pass through the first rock as through the second under equal pressure gradients, or just as much of a given fluid will pass through the first as through the second under half the pressure gradient, or under equal pressure gradients just as much fluid will pass through the first rock as through the second if the viscosity of the fluid in the first rock is twice that of the fluid in the second.

Geologists and engineers in the oil industry are interested in permeability chiefly from the standpoint of measurement of permeabiliy and use of the data thereby obtained for practical purposes. Accurate measurement of permeability and precise use depend basically upon the validity

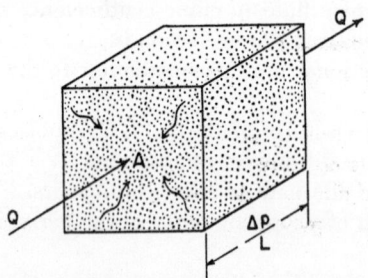

FIGURE 374. Sand model for rectalinear
flow of fluids.

of Darcy's law and analysis of the conditions to which it can be applied with reasonable accuracy. Fortunately, research has established rather clearly the limitations as well as the scope of Darcy's law.

For example, an obvious limitation is that the porous medium must be saturated with a homegeneous fluid for equations (8) or (9) to hold. Implicit in this limitation, however, is the implication that the coefficient of equation (8) is established solely by the geometry of the system and is independent of the nature or kind of fluid. This implication can be rationalized and put in accord with the fact that some fluids "react" with some porous media through hydration, dehydration, or some other phenomenon by means of the concept of a new medium or a change in geometry of the medium resulting from these effects.

Another restriction is the requirement for linear relation of pressure gradient and rate of flow in equation (8). Obviously, k, the coefficient of proportionality (permeability) can be constant only if this linear relation obtains. By analogy to the flow of fluids in pipes, and not in a

[29] American Petroleum Institute, *Standard Procedure for Determining Permeability of Porous Media,* API Code No. 27, 2d ed., April, 1942.

rigorous sense, the term "viscous-flow" is applied to this necessary condition. Likewise by similar analogy, the conditions under which Darcy's law does not hold is designated "turbulent flow." The change in flow from one type of flow to the other can be correlated with the familar Reynold's number.[30]

$$R_e = \left[\frac{d\mu p}{Z} \right]$$

Equation (8), which properly applies to one pore in a rock, must be integrated for the particular geometry of many pores, which is determined by the boundary conditions that prevail in a given situation and for the physical properties of the fluids. For example, in the radial

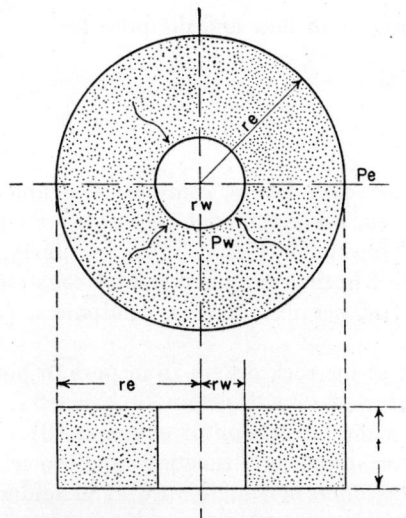

FIGURE 375. Sand model for radial flow
of fluids to central well bore.

isothermal flow of an homogenous fluid from a circular boundary to a central well bore (fig. 375), upon integration, equation (8) becomes

$$\frac{Q}{\theta} = q \frac{2\pi k t (p_e - p_w)}{\mu ln \dfrac{r_e}{r_w}} \tag{10}$$

in which the symbols undefined so far have the following meaning:
    t is the thickness of the producing zone, centimeters
    ln is the natural, or Naperian logarithim
    r is the radius of the sample, any units

---

[30] Fancher, G. H., and Lewis, J. A., Ind. and Eng. Chem., vol. 25, pp. 1139-1147, 1933.

the subscript e refers to the extreme outward radius of drainage and the pressure at that point, and

the subscript w designates the radius of the well bore and the pressure at that point.

Likewise equation (8) becomes, for the inward radial isothermal flow of a perfect gas to a central well-bore, the following

$$\frac{Q}{\theta} = \frac{2\pi kt(p_e - p_w)}{\mu ln \dfrac{r_e}{r_w}} \tag{11}$$

in which $\dfrac{Q}{\theta}$ is the volume rate of flow of gas in cubic centimeters per second at the temperature of flow and the pressure

$$\left[\frac{p_e + p_w}{2} \right]$$

If units of barrels per day, darcys, centipoises, pounds per square inch, and feet are employed, the right-hand side of equation (10) should be multiplied by the conversion factor 7.07. Similarly, insertion of the conversion factor, 39.7 in the right-hand side of equation (11) is required for units of cubic feet per day, darcys, centipoises, pounds per square inch, and feet.

If the structure of the rock differs from pore to pore, k is a variable which must be expressed as a function of geometry, and the function must be employed in the integration of equation (8). The expression of the variation in permeability as a function of distance or of geometry is required for the integration of equation (8) and seldom can be accomplished from a practical point of view with any degree of accuracy, although qualitatively it can be seen to depend upon the unhomogeneity of the rock. Furthermore, this requirement can be fulfilled only to the extent that the student of geology can decipher the intricate process of sedimentation and subsequent lithification for a given rock body and translate this story into the necessary mathematical relationships. Systematic variation of permeability can be expressed mathematically, but haphazard variation virtually defies analysis. Sometimes a sufficient number of core samples are available so that statistical data can be obtained from measurements of permeability in the laboratory, which will enable the student to work out an expression for systematic variations of permeability with some parameter. For example, the coarser grained components of a sandstone deposited offshore should be found closest to the shore line, and the finest grained shales and slates farthest from the shore. Consequently, sometimes under these conditions a linear variation of grain size with distance results, and core-analysis data may make possible determination of the constants in the following equation:

$$k = Cr + b \qquad (12)$$

Equation (8), integrated for the isothermal inward radial flow of a homogeneous liquid from a circular boundary through a zone in which permeability varies according to equation (12), becomes

$$\frac{Q}{\theta} = \frac{2\pi bt(p_e - p_w)}{\mu ln \left[\dfrac{r_e(Cr_w + b)}{r_w(Cr_e + b)} \right]} \qquad (13)$$

MEASUREMENT OF PERMEABILITY

The basis for the measurement of permeability from core samples has been provided in the preceding discussion. The rate of flow of a fluid of definite viscosity in response to an observed pressure gradient through a sample of known or easily determined dimensions provide the essential data.[31] Practically, a core holder [32] of convenient form, which can be combined with reliable and accurate auxillary devices for observation of rate of flow and pressure gradients, is necessary, the assemblage of equipment constituting a permeameter (fig. 377). Air usually is the fluid used in a permeameter because of availability and convenience from every point of view. Liquids such as oil-field salt water, or synthetic brine may be preferred in the testing of samples containing clay and other hydratable minerals. American Petroleum Institute Code 27 provides a convenient and complete guide and reference for the measurement of permeability, and the details of procedure in measurement need not be repeated. Samples for testing should be cut or shaped into convenient and simple geometrical forms so that the permeability can be measured either perpendicular or parallel to the planes of bedding.

RELATIVE PERMEABILITY

The concept of permeability was formalized, the unit defined, and its measurement standardized on the basis that the permeability of a rock or porous body constituting a reservoir for fluids was constant in magnitude and uniquely determined by the geometry and mineral composition of the rock when the pores contained only a simple, homogeneous fluid. Seldom, however, does an oil-bearing rock contain only a homogeneous fluid because of the presence of connate water. Usually two or three fluids (oil, gas, water) compete for space. Under these practical conditions of multifluid saturation, the movement of a particular fluid, although obviously dependent in a qualitative sense upon permeability, is impeded by the other fluids, the degree of interference depending upon their distribution and the relative amounts present. In other words, the

[31] Clough, K. H., Oil Weekly, vol. 83, no. 3, pp. 33-34; no. 4, pp. 27-34; no. 5, pp. 54-58, no. 6, pp. 46-54; no. 7, pp. 42-50; no. 8, pp. 39-44, 1936.
[32] Fancher, G. H., Lewis, J. A., and Barnes, K. B., op. cit.

permeability of the rock to a particular fluid phase becomes not only a
function of the geometry and mineral composition of the rock and the
pertinent physical properties of the respective fluid but also a function
of the fluid saturation. Nevertheless, with appropriate modification, the
concept of permeability may be extended to those porous systems containing two or more fluid phases. The ability of a porous rock or body

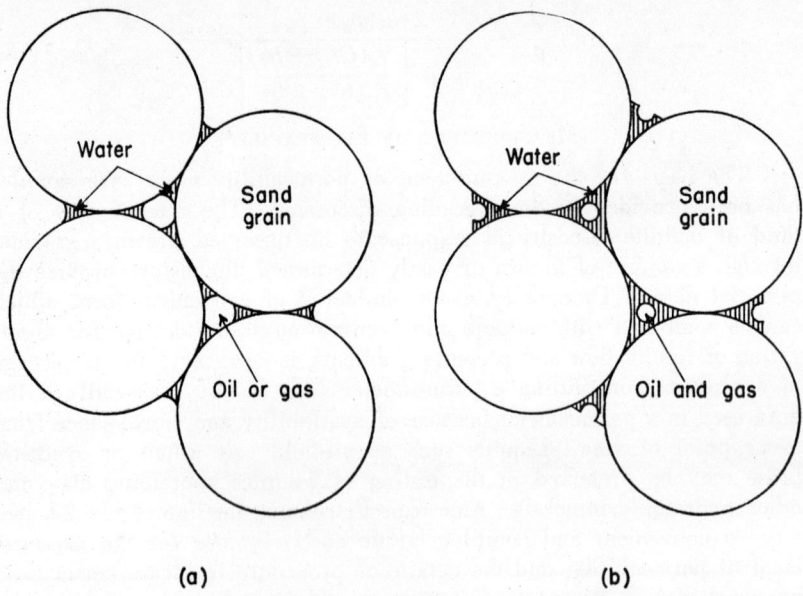

(a)   (b)

Figure 376. Idealized representation of distribution of wetting and nonwetting fluid
    phase about intergrain contacts of spheres. (a) Pendular-ring distribution; (b)
    funicular distribution. (After Muskat.)

to allow passage of a particular fluid is designated *effective permeability*
when expressed in darcys or millidarcys or is termed *relative permeability* when expressed as a ratio of the effective permeability to the preme-
ability of the rock to a homogeneous fluid completely filling the pores.
Consequently, the numerical value of the effective permeability of a rock
will be between the limits of zero and the permeability, k, of the rock in
conventional units of darcys or millidarcys and for that of relative permeability between the corresponding limits of zero and one. Between these
limits the numerical values of effective and of relative permeability are
a function of fluid saturation. Standard terminology requires statement
of the fluid saturation and consequently

$$k_0 (65,20) = 560 \text{ md} \tag{14}$$

is read as the effective permeability of a particular rock to oil at a fluid
saturation of 65 percent oil, 20 percent water, and 15 percent gas (650

millidarcys). If the permeability of this same rock should be 850 milli-darcys, then

$$k_{r0}\ (65,20) = \frac{k_0}{k} = \frac{560}{850} = 0.66 \tag{15}$$

which states that the relative permeability of the rock to oil at a fluid saturation of 65 percent oil, 20 percent water, and 15 percent gas is 0.66—or that at this condition of saturation, the effective permeability of the rock to oil is only about two-thirds or 66 percent of the permeability—so great is the interference of the water and gas also occupying space in the pores to the movement of the oil.

Although only relatively few studies of the relationships between

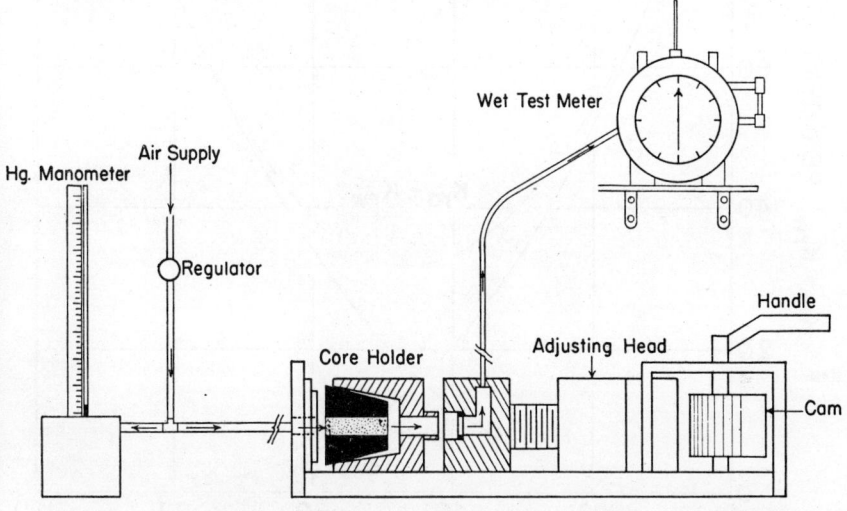

FIGURE 377. Permeameter used at University of Texas. Fancher-type core holder.

effective or relative permeability and saturation have been published, the data which are in the literature have some common characteristics. Whether a fluid wets the minerals composing a rock is important in all permeability-saturation relationships because that fluid phase which wets the rock will separate the nonwetting phase from the walls of any pore passageway; and, consequently, wettability determines the spacial distribu-tion of fluids in a porous rock (fig. 376). Furthermore, wettability is a relative term—immiscible fluids differ not only in wettability but also in the fact that the fluid with greater wettability can displace the one of lesser wettability if the opportunity to do so comes about. The amount and distribution of each fluid phase determines the relative and effective permeability of the rock to fluids. Research data available in the literature demonstrate that:

(1) The permeability of a rock to the wetting-fluid phase decreases rapidly from unity at 100 percent saturation for small decreases in the saturation of the wetting phase and correspond-

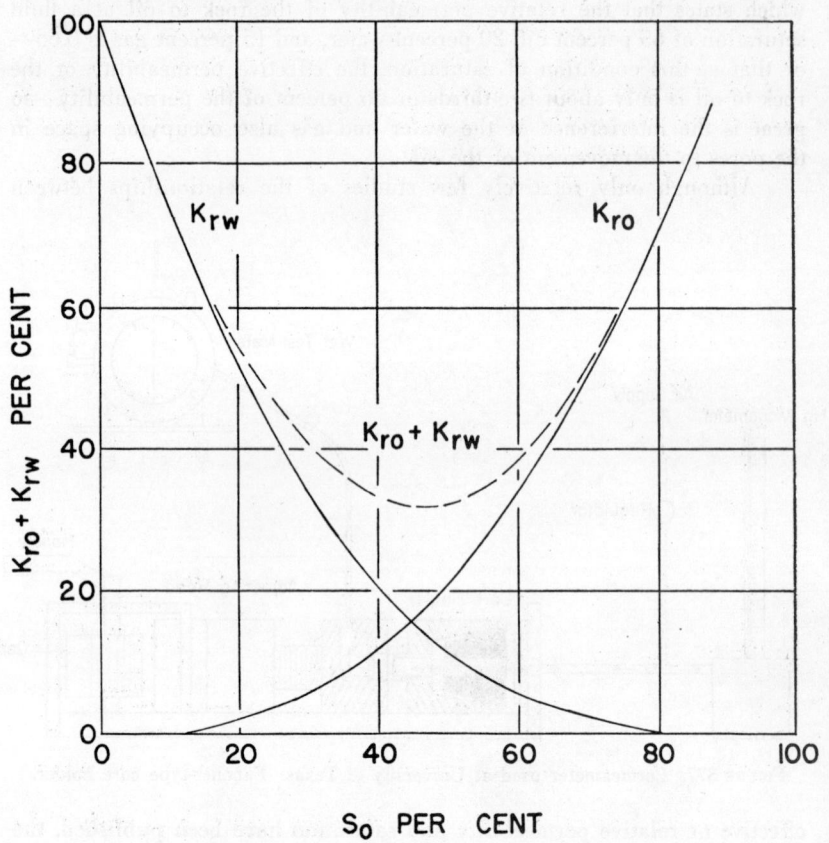

FIGURE 378. Variation of relative permeability of unconsolidated sands to oil and to water with oil saturation. K = 1.04 − 6.8 darcys. (After Leverett.)

ingly for small increases in the saturation of the nonwetting phase (fig. 378). Moreover, the permeability to the wetting phase decreases virtually to zero at a saturation of as great as 15 to 40 percent, depending upon the structure and composition of the rock. Furthermore, the permeability to the wetting-fluid phase is independent of the nature or number of nonwetting-fluid phases.

(2) The permeability of a rock to the nonwetting-fluid phase (figs. 378, 379, 380, 381, 382, and 383) has a real, consistent, and experimentally reproducible value greater than zero only after its saturation has increased from zero to a minimum value of from 5 to 20 percent, depending upon the structure and com-

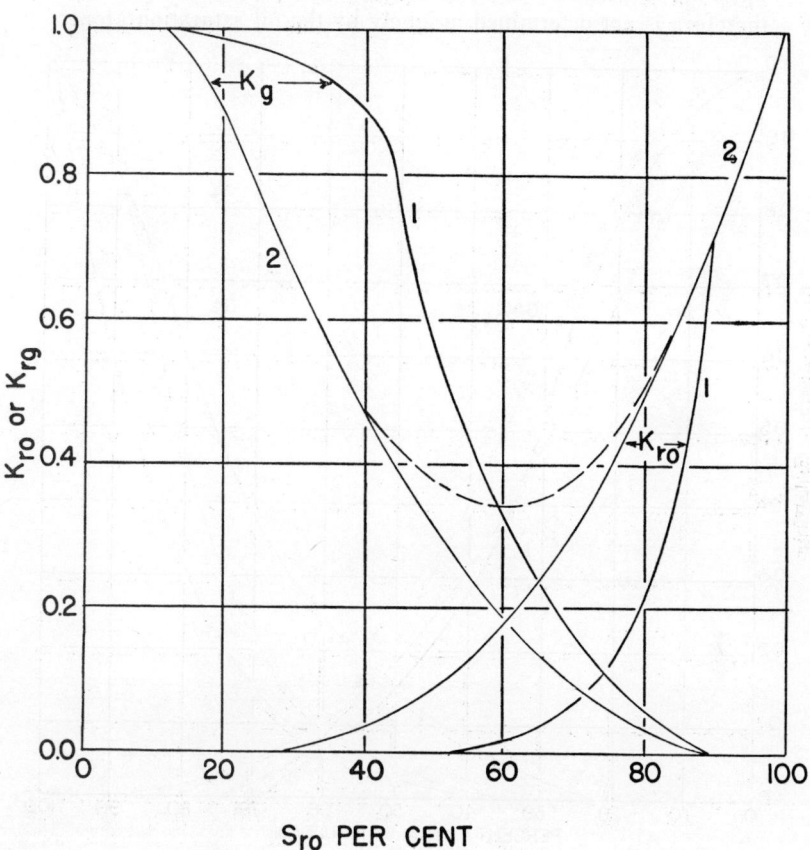

FIGURE 379. Variation of relative permeability of sands to oil and to gas with oil saturation. 1—Consolidated sands; 2—unconsolidated sands.

position of the rock. Thereafter, as the saturation to the non-wetting phase increases slowly at first and then more and more rapidly until at a saturation of from 60 to 90 percent, the exact percentage depending on the structure and composition of the rock, the permeability to the non-wetting phase (figs. 378, 379) virtually becomes equal to the permeability obtained at 100 percent saturation. Furthermore, the permeability to gas (a nonwetting phase) is determined solely by the amount of gas present—i.e., the gas saturation. However, the permeability to oil, usually a nonwetting phase in sedimentary rocks, does depend upon the distribution of other fluids, such as water and gas, and therefore is not determined uniquely by the oil saturation alone.

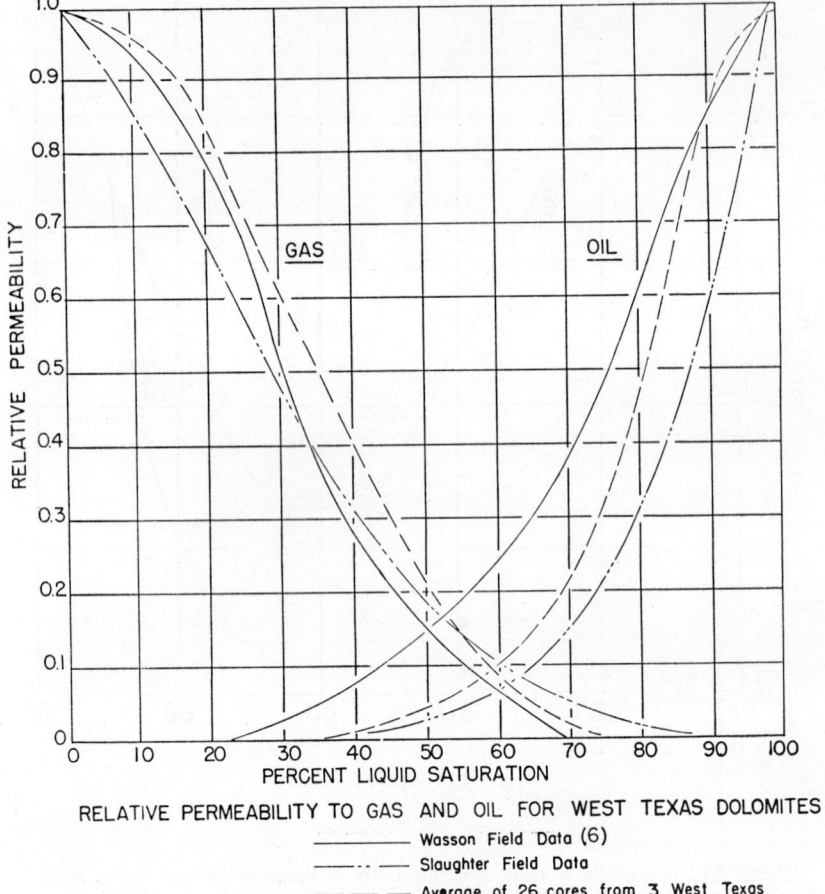

RELATIVE PERMEABILITY TO GAS AND OIL FOR WEST TEXAS DOLOMITES

——————— Wasson Field Data (6)

———··——— Slaughter Field Data

——— ——— Average of 26 cores from 3 West Texas
Permian Dolomites (6)

Figure 380. Variation of relative permeability to oil and to gas of some dolomites.

The mineral composition no doubt has some effect on the permeability-saturation relationships exhibited by a particular rock—but less than one would suspect—the general form of the relation being quite similar in all respects for rocks as dissimilar as the Woodbine sand, the Big Lake dolomite, and the San Andres dolomite (figs. 381, 382, and 383).

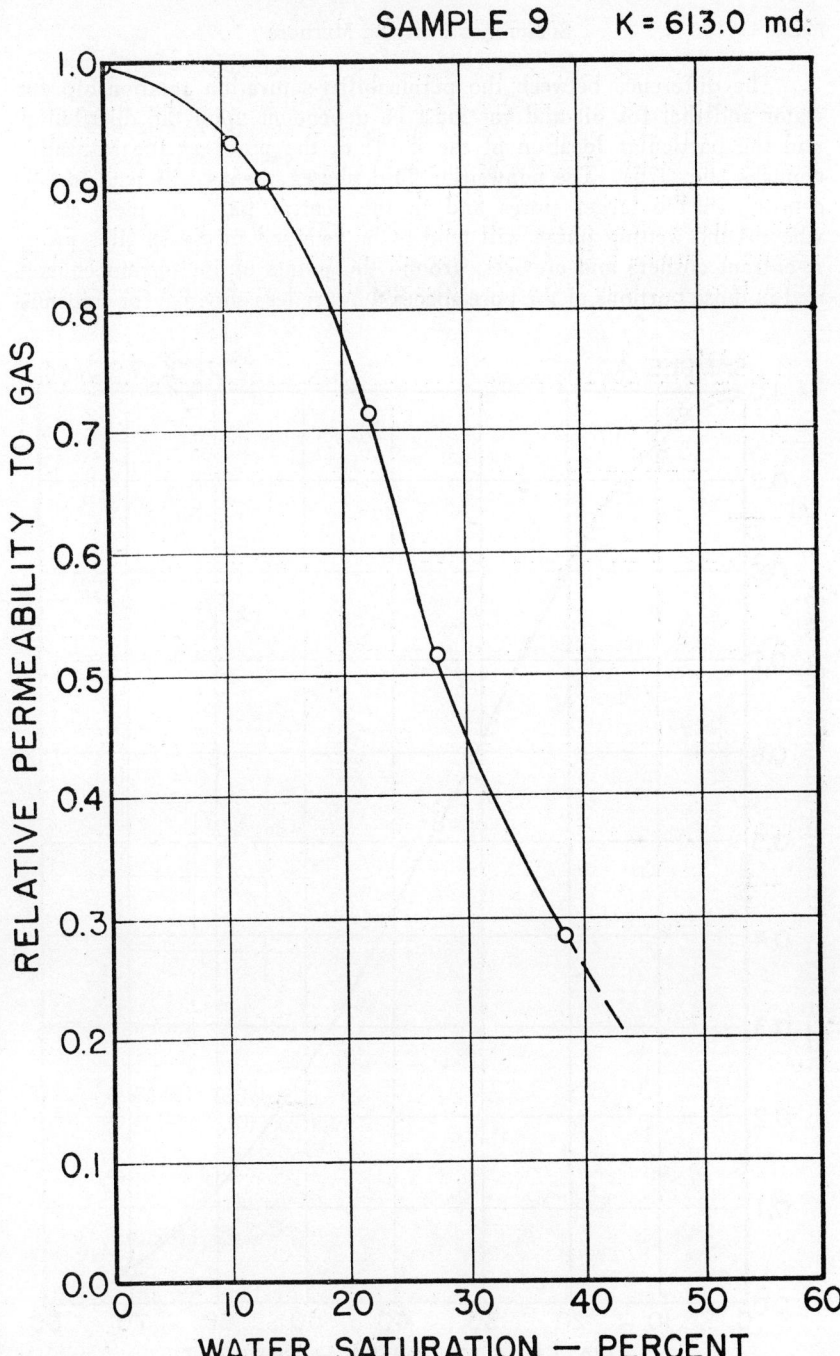

SAMPLE 9   K = 613.0 md.

FIGURE 381. Variation of relative permeability to gas of Woodbine sand with connate-water saturation, Cayuga field, Texas.

The difference between the permeability-saturation relationship for water and that for oil and gas must be dependent upon the distribution and the particular location of the fluids in the pores as the saturation changes (fig. 376). The nonwetting-fluid phases always will tend to concentrate in the larger pores and in the central parts of these pores, whereas the wetting phase will tend to be retained in the smaller pores, re-entrant corners and crevices, around the points of inter-grain contact, and in those portions of the pore spaces that are less effective for transmis-

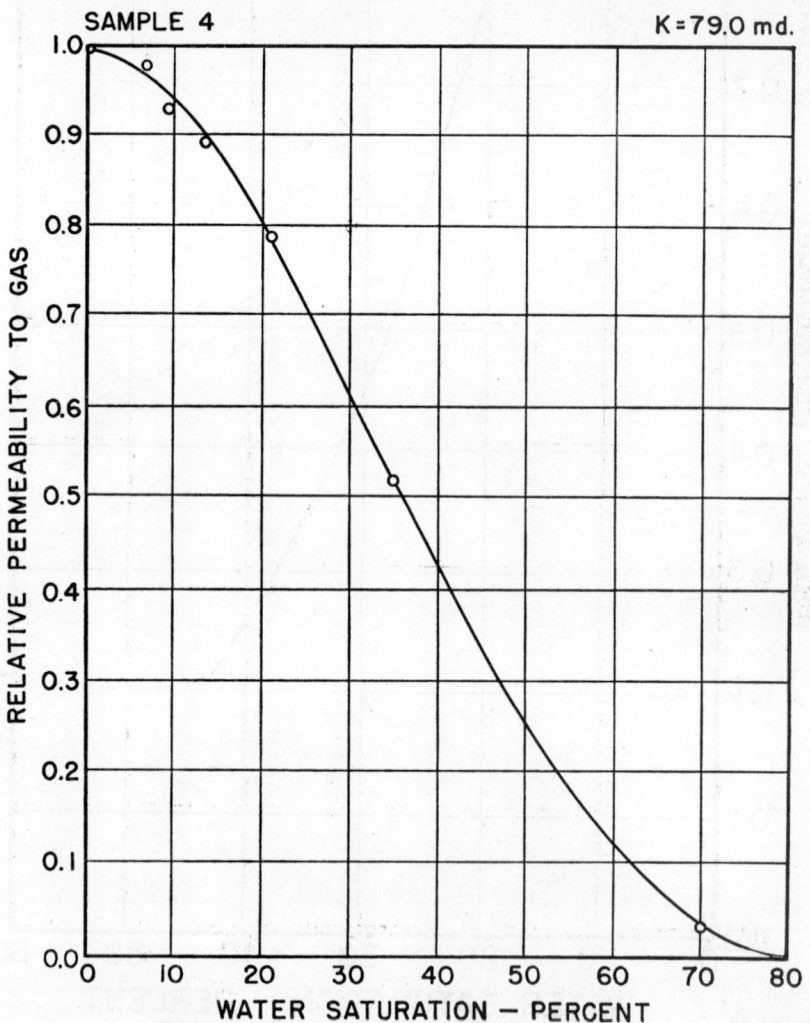

FIGURE 382. Variation of relative permeability to gas of Big Lake dolomite with connate-water saturation, McElroy field, Texas.

sion of fluids. Furthermore, whenever the amount of wetting fluid is too small to allow interconnection of the wetting phase from pore to pore, the wetting phase assumes a position and distribution within the pore system known as *pendular* because of the fact that the wetting fluid is held by surface forces at the solid-liquid interface in the form of pendular rings

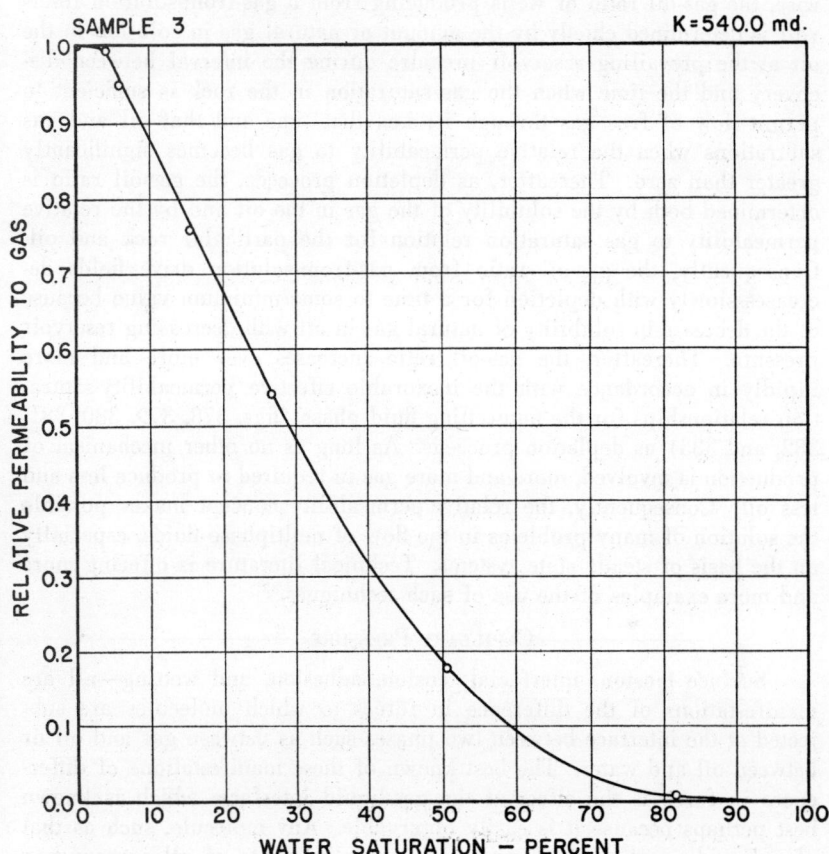

FIGURE 383. Variation of relative permeability to gas of San Andres dolomite with connate-water saturation, Goldsmith field, Texas.

about the points of inter-grain contact figure 376a. If the amount of wetting fluid within the pore system is increased sufficiently, the pendular rings of fluid in adjacent pores coalesce and unite, resulting in the formation of a continuum of the wetting-fluid phase throughout the pore system in what is designated as the funicular arrangement or distribution (fig. 376). Permeability of a rock to a particular fluid phase and maintenance of the steady state of flow of that fluid depend on and require the funicular distribution of the fluid to obtain in the rock. Many of the broad features of migration and accumulation of oil, of reservoir, and of well behavior are explainable and understandable as a result of these permeability-saturation relations. For example, oil free of water can be produced from reservoir rocks containing considerable amounts of connate water when the effective permeability of the rock to water is trivial at the particular saturation with respect to the wetting phase, water (fig. 378). Likewise, the gas-oil ratio of wells producing from a gas-from-solution reservoir is determined chiefly by the amount of natural gas in solution in the oil at the prevailing reservoir pressure during the interval between discovery and the time when the gas saturation in the rock is sufficient to permit flow of free gas through it: i.e., that time and that oil and gas saturations when the relative permeability to gas becomes significantly greater than zero. Thereafter, as depletion proceeds, the gas-oil ratio is determined both by the solubility of the gas in the oil and by the relative permeability to gas saturation relation for the particular rock and oil. Consequently, the gas-oil ratio from gas-from-solution drive fields decreases slowly with depletion for a time to some minimum value because of the decrease in solubility of natural gas in oil with decreasing reservoir pressure. Thereafter, the gas-oil ratio increases ever more and more rapidly in accordance with the inexorable effective permeability-saturation relationships for the nonwetting fluid phase (figs. 378, 379, 380, 381, 382, and 383) as depletion proceeds. As long as no other mechanism of production is involved, more and more gas in required to produce less and less oil. Consequently, the relative-permeability concept makes possible the solution of many problems in the flow of multiphase fluids, especially on the basis of steady state systems. Technical literature is offering more and more examples of the use of such techniques.[33]

CAPILLARY PRESSURE

Surface tension, interfacial tension, adhesion, and wetting—all are manifestations of the difference in forces to which molecules are subjected at the interface between two phases such as between gas and oil or between oil and water. The best known of these manifestations of differences in force is the effect at the gas-liquid interface, which is known best perhaps because it is easily observable. Any molecule, such as that of a liquid, is always subject to the attractive force of all surrounding molecules—those in the liquid and those in the gas. The molecule at the interface between a liquid and a gas is pulled on one side by the liquid and on the other by the gas. Usually the attractive forces differ in magnitude, and the result is a difference in stress. The difference in stress causes the molecules in the interfacial layer to behave something like an elastic membrane in tension. The stress in the surface layer is measurable and is known as the surface tension. Because of interfacial tension, a pressure differential always obtains through the interface, the pressure

[33] Russell, W. L., op. cit.

being greater on the concave side—i.e., the nonwetting side—by an amount in accordance with the following equation:

$$\Delta p_c = \gamma \left(\frac{1}{r_1} + \frac{1}{r_2} \right) \tag{16}$$

in which the terms have the following significance:

Δp_c = difference in pressure, psi

γ = interfacial (surface) tension, pounds per inch

r_1 and r_2 = radii of curvature in any two mutually perpendicular directions. This difference in pressure is inversely proportional to the mean radius of curvature of the interface and may be either negative or positive. The algebraic sign simply indicates that fluid phase in which the pressure is lower, this phase always being that which wets the surface of the rock. Furthermore, every student of science knows that if one end of a glass capillary tube is under water and the other is open to the air, water rises in the tube to a height above the free-water level so that at constant temperature

$$\Delta p_c = (p_w - p_a) \, h = \gamma \left(\frac{1}{r_1} + \frac{1}{r_2} \right) = 2 \frac{\gamma}{r} \tag{17}$$

in which the symbols have the following meaning:

Δp_c = capillary pressure, psi

γ = interfacial tension between fluids, lbs. per foot

p_w = density of the water, pounds per cubic foot

p_a = density of the air, pounds per cubic foot

r_1 and r_2 = radii of curvature of two mutually perpendicular planes

h = height of water column in the tube, feet

r = mean radius of curvature

The pressure difference across the interface in a capillary system frequently is called "capillary pressure." The capillary pressure caused by the surface tension of pure water in a glass capillary tube 1.0×10^{-3} inches in diameter is about 17 psi, which is equivalent to a capillary rise of about 40 feet. Similar effects are observed with systems composed of liquids and porous media modified only in degree, and not in kind, by the increased complexity of the geometry of such systems.

When water is in contact with another immiscible liquid—for example, oil—capillary pressure is developed as a result of the water striving to maintain a surface of minimum area, just as water does in the presence of a gas. The variation of grain size, bonding material, and mineral composition, particularly in the smaller grains such as clay and shale that may be in the pores of sedimentary rocks and between larger grains, affects the magnitude of the capillary pressure in a particular rock. Furthermore, the capillary pressure of a particular rock at any horizontal plane varies with the distance above the free-water

surface and with the fluid saturation of the rock. The conventional capillary-pressure curve (figs. 384 and 385) depicts the relationship between
saturation and the capillary pressure above the free-water level and
illustrates the fact that the equilibrium between capillary pressure and
gravitational forces required by equation 17 results in a vertical saturation gradient in a petroleum reservoir, which is known as the transition
zone.

The concept of capillary pressure demands that, at equilibrium, the
capillary pressure be everywhere the same at the same horizontal level.
Inasmuch as the geometry of a homogeneous rock and, hence, the radius
of curvature of fluid interfaces are constant, the fluid saturation must be
constant at equilibrium in accordance with equation (17). Consequently,
the amount of connate water in a rock is that which can be held in the
rock by capillary forces. Experimental capillary-pressure data provide
a means of determining the connate water saturation, this being the so-
called irreducible water saturation on the curve obtained when saturation
is plotted against capillary pressure (figs. 384 and 385). If the geometry
and mineral composition of a rock were similar and uniform in all respects and in all directions, the permeability and connate water saturation should be constant and uniform throughout the rock. The connate
water saturation of actual sedimentary rocks, formed by the deposition
of particles graded to size or by some other natural process, would be
expected to vary with permeability, increasing as permeability decreases.
Differences in mineral composition resulting in different degrees of wettability likewise result in differences in connate-water saturation. The
Woodbine sand represented by samples 4, 5, and 6 (fig. 383) is an even-
grained, friable, almost pure quartz sandstone containing a small amount
of argillaceous cementing material, according to Puls.[34] The porosity
and permeability of these samples are of the same order of magnitude:
namely, 21.1 percent and 401 millidarcys, 26.6 percent and 113 millidarcys, and 27.1 percent and 641 milldiarcys respectively. The assortment of grains of different sizes is that to be expected of a well sorted
sandstone; and, hence, the capillary pressure-saturation curve is regular
and smooth. Nevertheless, the minimum connate-water saturation of the
samples is 9, 26, and 10 percent respectively. The high connate-water
saturation of sample 5 is due to a greater amount of cementing material,
as well as, perhaps, to a streak of lignitic and feruginous material in the
sample. The Grayburg formation (samples 1, 2, and 3, figure 385) is a
uniformly fine-grained, gray dolomite—in some places sandy and in
others oölitic—the porosity of which is chiefly intergranular, although
samples exhibit scattered, very small solution channels. Consequently,
both the porosity and permeability are low, being 8.5 percent and 8.2
millidarcys, 9.8 percent and 1.4 millidarcys, and 7.9 percent and 1.3
millidarcys respectively. The great difference between the average dimen-

---

[34] Puls, W. L., Thesis for M.S., Degree in Petroleum Engineering, Unpublished, 1950.

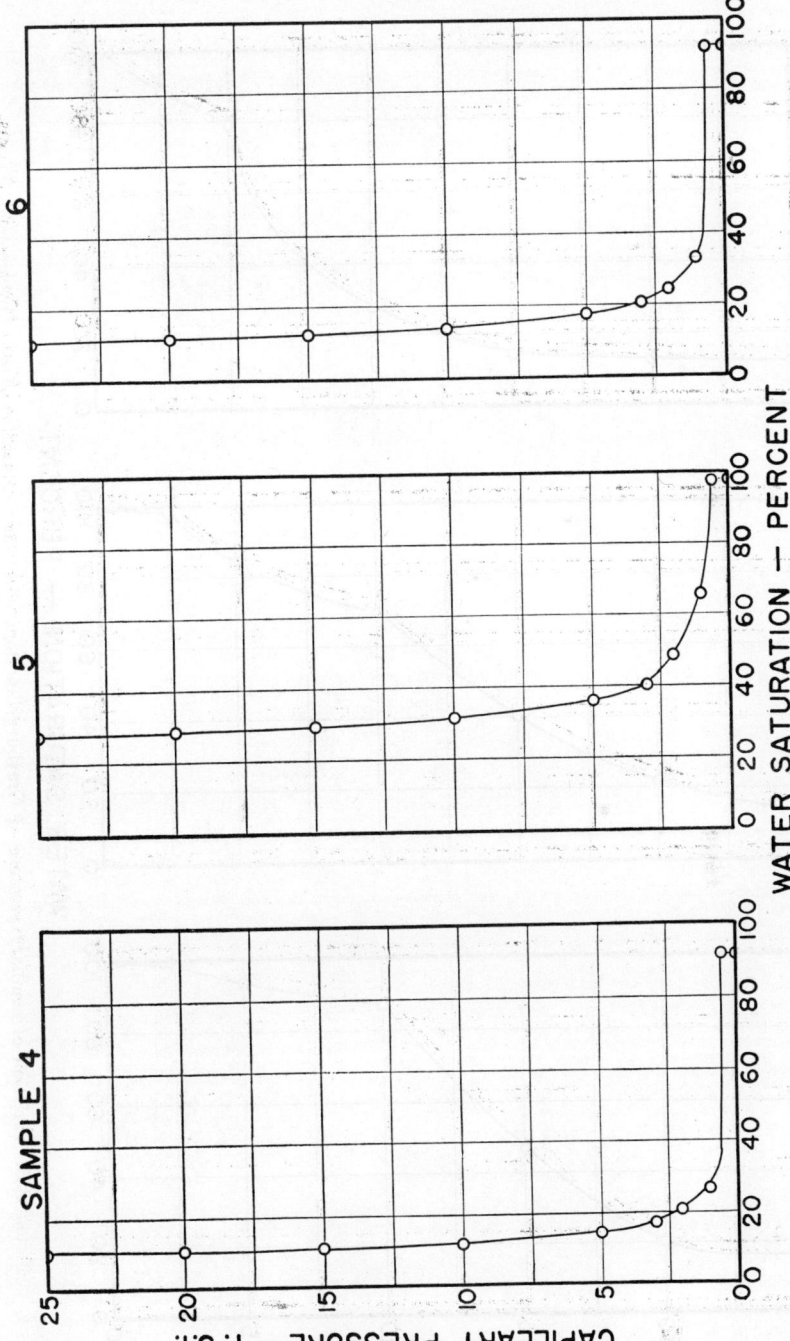

FIGURE 384. Variation of capillary pressure of Woodbine sand with water saturation, Cayuga field, Texas.

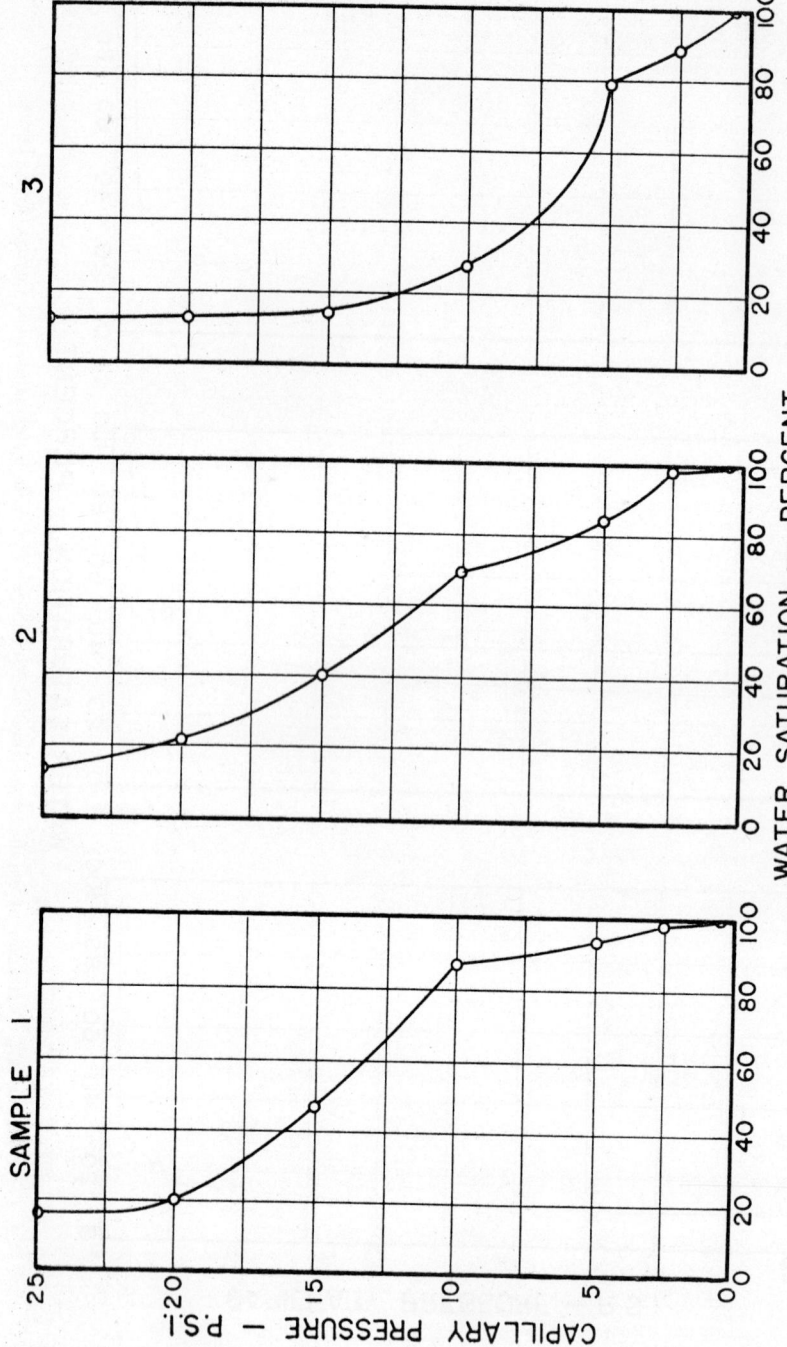

FIGURE 385. Variation of capillary pressure of Grayburg limestone with water saturation, North Cowden field, Texas.

sions of the intergranular pores and those of the solution channels is reflected very well by the abrupt changes of slopes in the capillary-pressure versus water-saturation curves of figure 385. Nevertheless, the connate-water saturation of all the samples is about the same (17, 16, and 17 percent respectively), the ability to retain water being controlled chiefly by the dimensions of the intergranular pores and the mineral make-up of the rock, both of which are essentially constant in these samples.

The net result of all these factors is that connate-water saturation may be expected to differ from place to place in sedimentary rocks but that capillary pressure at the same horizontal datum plane in a rock that is permeable vertically as well as horizontally is single valued. The porosity and permeability of rocks are intimately related to capillary pressure. Although the relationship is intricate, the problem of the relationship, leaving aside the effect of the properties of fluids, is one of stereogeometry, which in sedimentary rocks or clastic sediments is fundamentally one of geology. The concepts of capillary phenomena aid the geologist who is interested in the problem of origin, accumulation, and discovery of oil and gas. These concepts also aid the petroleum engineer who is confronted with production problems, the solution for which must be based upon the behavior of fluids in porous media. Quantitative data are made available from capillary-pressure tests of core samples, and these data can be used advantageously by both geologist and engineer. Capillary data, combined with porosity, permeability, and other physical properties, can be applied practically to increase the efficiency of the finding and the production of oil and gas, which is the ultimate objective of the production branch of the petroleum industry.

DRILLING FLUID CHEMISTRY

H. F. SUTTER

The properties of the drilling fluid used in rotary drilling of oil wells should be such as to promote safe and speedy drilling and completion of the well with maximum productive capacity. This fluid under the propulsion of powerful pumps is forced down through the drill pipe and out into the well through holes in the bit. Jetted against the bottom of the well with high velocity, the circulating fluid is deflected upward and flows back to the surface between the drill pipe and the walls of the well, carrying in the ascending stream the cuttings formed by the drill. At the surface, the drill cuttings are segregated by screening or gravity settling, and the fluid is discharged into a mud pit, later to be again circulated through the well.

The drilling fluid performs a variety of different functions. Prompt and continuous removal of the material loosened by the drill prevents its accumulation in the well with the possibility of freezing the drill pipe. The fluid must possess thixotropic properties so that, in the event of unexpected interruption in circulation, it will gel and prevent settling of drill cuttings to the bottom of the well. The drilling fluid must absorb heat generated in the drill pipe and bit by frictional contact on the walls and bottom of the well. To prevent caving and to lubricate the drill pipe, the fluid must deposit a thin sheath of clay on the walls of the well. The clay, thus deposited, assists along with the density of the drilling fluid in withholding high-pressure gas or water, and also seals unusually permeable low-pressure formations through which the mud might be drained away in sufficient quantity to cause loss of circulation. The drilling fluid must have sufficient density to be capable of providing ample hydrostatic pressure to prevent high-pressure gas, oil, or water from entering the well and causing a destructive blow-out.

To satisfy these requirements, the drilling fluid must have satisfactory viscosity and density and must have well-developed colloidal properties. It must be free of sand and dissolved substances that might cause rapid flocculation, coagulation, or settling of clay particles. The density and viscosity must permit separation of drill cuttings and entrained gas in the facilities provided for such separation; and at the same time, the drilling fluid must not be so viscous that undue pump pressure will be necessary to force it through the system.

The drilling fluid is usually a clay-laden fluid and is aptly termed "drilling mud." Generally, drilling muds may be classified in two systems: water-base and oil-base. Water-base muds are used much more extensively than the oil-base types as the latter are primarily restricted to special-purpose drilling.

PROPERTIES OF DRILLING MUDS

While drilling a well, one may encounter many subsurface formations and drilling conditions which affect the mud adversely or require special mud properties for satisfactory handling. Since the basic system of drilling muds is the water-clay type, it is a highly reactive type of fluid, easily subject to contamination and development of objectionable properties. In some cases fluid systems are especially compounded for the particular purpose of eliminating one or more of the objectionable properties of water-clay muds.

Among the properties of drilling fluids that are important in judging their performance are viscosity, density, colloidity, gel or shear strength, sand, and salt content. The viscosity of the mud influences its fluidity and its resistance to flow through the circulating system. The density is significant in lifting drill cuttings and offsetting high formation pressures. Colloidal properties of the drilling fluid are important in determining ability to form a mud sheath on the wall of the well, lubricate the drill pipe, prevent loss of fluid to the formation, and restrict admission of formation fluids to the well. The gel or shear strength is also an expression of colloidity and measures the resistance offered to the settling of drill cuttings when circulation stops. Sand content is undesirable in a drilling fluid because of the abrasion of the sand on metal surfaces during circulation. Salt content may have an adverse effect on the colloidal properties of a drilling fluid and render it unsuitable for use under certain conditions.

Viscosity

The viscosity of a drilling fluid depends upon the amount and character of the suspended solids. In general, the greater the percentage of suspended solids, the greater will be the viscosity; and plastic clays develop higher viscosities than noncolloidal substances. Bentonite has well-developed colloidal and thixotropic properties. When the clay used in a drilling operation is deficient in colloidal material and lacks thixotropic properties, controlled amounts of finely ground bentonite may be added to the drilling fluid to improve its quality. Viscosities of drilling fluids may fluctuate over a wide range from but a few centipoises to more than 300 centipoises.

Several instruments are available for measuring drilling fluid viscosity; some are primarily for laboratory use, others have been designed for field use. The most commonly employed instruments for measuring relative viscosities of drilling fluids on location are the Marsh funnel and the Stormer viscosimeter. The Marsh funnel is a cone-shaped funnel with an accurately bored discharge tube at the bottom. The funnel is filled with the fluid to be tested and the time in seconds necessary to discharge a measured volume of the fluid is an indication of its relative viscosity.

The Stormer viscosimeter is adapted for determining drilling fluid viscosity and gel strength. The instrument consists principally of a spindle which is rotated in a test cup by a pair of gears driven by a weight. The fluid in the test cup is agitated and then the weight is adjusted until the spindle is made to revolve at a rate of 600 r.p.m. By means of a calibration chart furnished with each instrument, the weight required to achieve this rate may be converted into centipoises.

Density

The amount and the specific gravity of the suspended solids determine the density of the drilling fluid. Many wells are drilled with fluid densities not over 9 lb. per gal. or 67 lb. per cu. ft. In other cases where abnormal pressures are encountered, fluid densities as high as 20 lb. per gal. or 150 lb. per cu. ft. may be prepared that are not too viscous to be handled by the pumps. Fluids weighing more than about 11.5 lb. per gal. or 86 lb. per cu. ft., if prepared exclusively with clay and water, are liable to be too viscous to be readily handled by the pumps, and gas and sand will not readily separate out. If heavier fluids are needed, finely ground heavy minerals may be used to contribute high density without a corresponding increase in viscosity. The most commonly used weighting material, barite, has a specific gravity of 4.3 and, by adding it in a finely ground condition to at least an equal volume of water along with a small amount of colloidal material to keep the solids in suspension, a mud having a density of 20 lb. per gal. or 150 lb. per cu. ft. may be produced without excessive viscosity.

Several instruments are available for measuring fluid density. One such device is the Mudwate hydrometer. The hydrometer consists of a float spindle with calibrated stem, a detachable bakelite cup, and a metal carrying case. The cup is filled with the fluid to be tested and replaced on the spindle. The carrying case is filled with water and the spindle floated therein. The density of the fluid determines the depth to which the spindle sinks, and the stem is calibrated to indicate fluid density in suitable units.

Another convenient and entirely dependable instrument for measuring fluid density is the Baroid Mud Balance. It consists essentially of a base and graduated beam with cup, lid, knife-edge, rider, and counterweight. The cup is filled with the fluid to be tested, and the lid is placed in position to insure constant volume. The balance arm is placed on its base with the knife-edge resting on the fulcrum. The rider, adjusted on the beam until a balance is obtained, indicates the fluid density.

Colloidity

The colloidal properties of a drilling fluid determine its ability to form a suitable cake on the wall of the well, to seal the pores of the wall formations, and to lubricate the drill pipe. The wall-building properties of the drilling fluid are of great importance. As the clay-laden fluid is circulated over the walls of the hole, there is a tendency for the fluid either to enter the pores of the formation or, if the pores are too small, to permit the solid particles to enter. There is also a tendency for the liquid phase of the mud to be squeezed from the fluid into the surrounding formation and to leave the solid matter deposited as a cake. The thickness and permeability of the cake may exercise considerable influence over the drilling of the well. If the formation and cake permeabilities are both high, a thick wall cake will be developed. This cake may become so thick that it will interfere seriously with movement of the drill pipe and may even result in sticking the pipe.

The fluid which enters the formation may encounter shales and clays susceptible to hydration, and the resulting swelling may cause heaving or slipping of the shale or clay into the hole. The hydration or swelling of clays in producing sands may impair production through the loss of permeability.

There is no unit of colloidity, and the colloidal value of a drilling fluid cannot be measured quantitatively. Relative colloidal values and their effect on the properties of drilling fluids are indicated by a wall-building tester. This instrument provides an observation of the ability of the fluid to form a cake under conditions similar to those deep in the well. The wall-building tester consists principally of a filter cell and a device for applying pressure to the fluid enclosed within it. The fluid is forced through a screen-supported filter paper on which the filter cake is formed. The thickness of the filter cake and the filtration rate of the base fluid through it are considered a measurement of the effectiveness of the quantity and type of colloidal materials in the fluid.

Gel Strength

Whereas viscosity is a measure of the resistance of fluids to infinitesimal uniform shear, the gel strength of plastic fluids, such as drilling fluids, is a measure of the minimum shearing stress necessary to produce slipwise movement. Such a gel strength is zero for true fluids, no matter how viscous, but it is appreciable for suspensions such as drilling fluids. Gel strength in drilling fluids is responsible not only in part for the ability of the fluid to transport cuttings upward, but is wholly responsible for the ability of the fluid to hold cuttings and weight material in suspension when circulation is suspended. Non-colloidal particles will eventually sink in a true fluid having zero gel strength, regardless of its viscosity.

Many fluids exhibit an increase of gel strength upon standing quiescent for some time and can be restored to fluidity upon subsequent agitation. Such fluids are termed thixotropic and are tested for initial gel strength immediately upon cessation of agitation. A measure of the thixotropy of a fluid is given by the difference in gel strengths observed immediately after agitation and after some period of quiescence. An adequate gel strength is a valuable characteristic when shutdowns occur, as even fairly large cuttings do not sink very far before the fluid gels sufficiently to support them.

With the Stormer viscosimeter described previously, gel strength determinations may be made over a wide working range. This wide working range is essential in the close of control of the gel-strength property, which is required in meeting certain drilling conditions and also in evaluating treating chemicals designed to reduce the gel strength of drilling fluids.

The fluid to be tested is thoroughly agitated and immediately poured into the test cup. The initial gel-strength measurement is made by determining the minimum weight required to effect a movement of approximately one-quarter revolution of the spindle after the brake is slowly released. The total weight in grams required to obtain this rotation is reported as the initial gel strength.

A second gel-strength determination is made after the fluid in the cup has been allowed to remain quiescent for ten minutes following the initial gel-strength observation. Additional driving weights are added to the line and the brake is slowly released. If no rotation of the spindle occurs, the brake is set again, more weight is added to the line, and the brake is released again. This process is repeated until the minimum weight required to cause rotation is found. This weight is reported as the ten-minute gel-strength reading.

Sand Content

The sand content of a drilling fluid may be extremely detrimental. Sand in the drilling fluid is abrasive and may result in rapid wear of pump liners, drill pipe, casing, and other metallic equipment with which it comes in contact. Sand-laden fluid also has poor wall-building properties, and develops a thick filter cake. If present in sufficient amount, it may freeze the drill pipe to the walls of the well and, perhaps, cause a twist-off.

All material coarser than 200 mesh may be regarded as sand, and good practice requires that the sand content of a drilling fluid be maintained below five percent. To determine whether or not the fluid meets this requirement, occasional tests for sand content should be made. Several test methods are employed, including elutriation, centrifuging, dilution and gravity settling, and sieve analysis.

CHEMICAL TREATMENT OF DRILLING FLUIDS

Chemical treatment of drilling fluids has been widely accepted because it saves money and improves fluid quality. Savings in costs often more than justify chemical treatment. More important, when viscosity, weight, and filter loss of the drilling fluid are controlled to suit the conditions, wells are drilled faster, depreciation of equipment is lessened, and insurance against blowouts and stuck drill pipe is provided.

Chemical treatment of drilling fluids is generally for the purpose of decreasing the viscosity and gel strengths, and reducing the filter rate. Treatment for the reduction of viscosity and gel strengths is commonly called "thinning," and chemicals used for this purpose are often termed "thinners." These chemicals are deflocculating agents which cause a higher degree of dispersion of the colloidal particles of the fluid. More complete dispersion of the particles, in turn, produces a thinner filter cake, lower filter loss, and lower viscosity and gels.

Chemicals should be selected on the basis of safety, simplicity, and economy. When chemicals are tested in any fluid, the effect of repeated treatment with the same chemical should be considered. Some chemicals that work well at first may later hinder treatment when it is urgently needed.

During the drilling of oil wells, the quality of the fluid in circulation changes constantly. Drilling fluids may be adversely affected by:

(1) Cement, anhydrite, gypsum, etc.
(2) Salt water
(3) Excess bentonitic shale
(4) Gas
(5) Heat

When tests show poor fluid conditions, particularly when viscosity or filter loss are too high, small quantities of chemicals, properly applied, will often quickly restore satisfactory fluid quality.

There are two classes of compounds that are particularly effective for treating drilling fluids. The complex phosphates are good for lowering viscosity and have a beneficial effect on filter loss. The tannins are very effective for lowering water loss and frequently have a beneficial effect on viscosity. Under certain conditions it is possible to use only complex phosphates or tannin compounds for maintaining satisfactory properties. In other cases complex phosphates and tannin compounds may be used together. When first applied, these chemicals are very effective in nearly every case, and only small quantities are required. With the repeated treatment that is required from time to time during the drilling of the well, the further addition of chemicals does not always have the expected effect. This is one of the problems in the successful chemical treatment of drilling fluids.

Only complex phosphates, such as tetrasodium pyrophosphate, sodium acid pyrophosphate, sodium tetraphosphate, and sodium hexametaphosphate, are useful for treating drilling fluids. The orthophosphates are not used because they are ineffective as thinners.

Tannic acid or tannin is a complex organic compound occurring in certain trees and plants. Pure tannic acids can be extracted from these and other sources, but because of the expense involved, the less costly crude extracts are used for treating drilling fluids. The tannin extracts such as quebracho extract are commonly used primarily to reduce viscosity and filter loss and to condition the filter cake. The proper use of tannin extracts will almost always yield a fluid which loses very little water to underground formations and will form a thin, tough cake that protects the walls of the open hole.

One of the most common troubles in drilling fluids is cement contamination. Usually no "symptoms" are required to locate this trouble, because it may be expected whenever operations require that cement be drilled, and particularly when the cement has not set thoroughly. Calcium hydroxide is present in cement slurries and is mostly responsible for the observed changes in the fluid properties. The calcium ion replaces the sodium on the clay particles and tends to convert them to calcium clays. The calcium clays are not so highly ionized as the sodium clays. Consequently, the degree of hydration and dispersion of the clay colloids may be reduced. This fact results in an increase in filter loss and a flocculation of the clay colloids. Flocculation results in high viscosities and gel strengths.

Commercial materials have been devised to recondition cement-cut drilling fluids. These materials remove the calcium ions as an insoluble precipitate, reduce the pH of the fluid to the approximate value before contamination, and redisperse the clay aggregates. Frequently, as a result of the treatment of the fluid for cement contamination, the fluid displays better performance characteristics than it did originally. This is because some of the commercial products include substances which counteract the calcium ion.

When it is known in advance that cement contamination will take place, pretreatment of the fluid is advisable if a considerable amount of cement is to be drilled. This pretreatment will reduce flocculation greatly during cement drilling by precipitating most of the undesirable constituents of the cement before serious damage has occurred.

As mentioned in the section on cement contamination, any soluble calcium as calcium phosphate or calcium carbonate and leave the sulfate drilling, sections of lime or limy shale, anhydrite, gypsum, etc., are encountered. These calcium salts are sufficiently soluble to cause trouble from clay flocculation.

When calcium contamination takes place, the fluid will have a fast gel rate and a weak gel strength that prevents the proper settling of sand; the viscosity will be abnormally low; and the water loss will usually be moderately high.

The primary aim in the chemical treatment for contamination by soluble calcium salts is to precipitate the calcium as an insoluble compound. This may be accomplished by use of disodium phosphate, soda ash, or barium carbonate. Disodium phosphate and soda ash as chemical precipitants for the calcium ions are similar in many respects. In the treatment for calcium sulfate, for instance, these chemicals precipitate the calcium salt will cause the flocculation of colloidal clays. In normal ions behind as soluble sodium sulfate. Continued accumulation of such soluble salts acts to increase the gel strength of the fluid. Barium carbonate, on the other hand, is superior to soda ash or disodium phosphate for the removal of anhydrite or gypsum. Complete precipitation of both the calcium and the sulfate ions as well as the barium and the carbonate results, leaving no soluble salts in solution.

When rock salt or salt water contacts a fresh-water drilling fluid, there is initially an increase in apparent viscosity, gel strength, and water loss. Later, as flocculation and dehydration of the colloids takes place, the viscosity falls below normal, and the water loss rises in rough proportion to the salt concentration.

The treatment for salt contamination is not an easy one. In the case of a salt-water flow, the first step in treatment is to increase the fluid weight to the point required to stop the flow.

For moderate salt concentration, disodium phosphate, soda ash, or barium carbonate can be used to precipitate soluble calcium salts present in the salt water and help reduce the gel strength of the fluid. The addition of large quantities of tannin extracts can be used for reducing the high gel strength produced by high salt concentrations.

Where a water-clay system is being used and the salt contamination does not exceed one percent, common practice is usually to continue to use this type of fluid and to keep the physical properties in hand as much as possible. The most extensive contamination of the fluid with salt comes from the drilling of salt beds and domes. In these cases depths of several thousand feet of salt may be penetrated and the drilling fluid soon becomes saturated. Before entering such salt beds, a change to a salt inert-type fluid is necessary.

With certain saline types of drilling fluids, the filter loss and gel strength cannot be maintained by the use of bentonite. In this case it is necessary to use special salt-resisting clays which will furnish suspending properties. These clays generally do not possess good wall-building characteristics, and colloidal additives like pregelatinized starch are used to provide these properties.

Various organic colloids and particularly the starches have been found useful in controlling the wall-building properties of salt-water fluids. These organic colloids are subject to bacterial attack, which reduces their effectiveness and produces objectionable odors in the fluid. Chemical preservatives, high salt content, or high pH of the fluid must then be resorted to for protection against the decomposition.

The highly colloidal, heaving shales are similar in many respects to clay colloids of the type used in drilling fluids. Therefore, any chemical treatment designed to prevent the swelling of those shales is likely to be detrimental to the fluid. The presence of such "bentonitic" shale is evidenced by a sharp increase in viscosity and, usually, an appreciable reduction in water loss of the fluid. In most cases it is possible to control bentonitic shale by conditioning the drilling fluid to minimize the colloidal tendencies of the clay. Special drilling fluids which prevent hydration of the bentonitic shale by maintaining an excess of calcium ions in the fluid have been successfully used. These fluids are amenable to weighting to high densities and to treatments to obtain low filter losses.

Gas cutting is largely a mechanical rather than a chemical problem in drilling-fluid control. The hydrostatic head of the fluid column must be slightly greater than the formation pressure at every point in the hole in order to prevent the flow of gas into the fluid system. Gas in the fluid indicates that the hydrostatic head of the fluid is too low to meet this requirement, and the obvious solution is to increase the fluid weight by the addition of commercial weighting materials.

TEMPERATURE EFFECTS ON DRILLING FLUIDS

In very deep wells, or in drilling near shallow salt domes, abnormally high temperatures are encountered. Such elevated temperatures may have a detrimental effect upon the drilling fluid. The viscosity of most liquids lecreases with temperature increase. The rate of fluid filtration through a filter cake is inversely proportional to the viscosity of the filtrate. Thus the wall cake will tend to be thicker in holes of high bottom-hole temperatures.

One effect of increased fluid temperatures is to revert the complex phosphates to the orthophosphate form, in which condition they are ineffective as thinners. This requires constant additions of large amounts of the complex phosphates to maintain a satisfactory viscosity.

Increased fluid temperatures also increase the chemical reactive properties of the fluid. The reaction rates of cement, gypsum, and salt with the fluid thus become more pronounced at the higher temperatures.

LOST CIRCULATION

One trouble encountered frequently in drilling is the loss of fluid in porous low-pressure sand or gravel beds. It is usually possible in this case to regain circulation by the use of various mechanical plastering agents. Among those most commonly used are shredded cellophane, sugar-cane fibers, and mica flakes. These materials are added to the fluid in the pit and are pumped down opposite the zone to which the fluid is being lost. Such materials must be removed from the fluid after circulation has been regained.

TYPES OF DRILLING FLUIDS

As stated before, the water-clay type is the basic, most widely used fluid system. This usage is the natural result of the availability and normally satisfactory functioning of water as the fluid vehicle. Other fluid systems, however, have been developed to overcome drilling conditions which water-clay fluids have difficulty in handling. These special systems are usually more costly to build and maintain than the water-clay type, but their use for the situation for which they are best fitted is often the difference between completing or abandoning a hole. Some of these special-purpose fluids and some of the conditions for which they are useful are

(1) Water-starch-high pH fluid.

   Primarily used for salt contamination over 1 percent and less than 15 or 20 percent, for drilling gypsum, anhydrite, heaving shale, and for drilling into the production zone.

(2) Salt-water-starch fluid.

   Used primarily where excessive salt contamination results in saturation of the fluid. May also be used for drilling-in or completing wells in formations subject to contamination by water or for wells with low formation pressures.

(3) Sodium-silicate fluid.

   Used for drilling heaving-shale formations or high-pressure salt-water flows.

(4) Oil-base fluid.

   Used primarily for drilling-in or recompleting wells in formations subject to contamination by water or for wells with low formation pressures.

(5) Oil-emulsion fluid.

   Used where low fluid-loss, very thin cake, and good lubrication of the drill pipe are of primary importance, such as in directional drilling and drilling into the production zone.

GENERAL COMMENT ON DRILLING FLUIDS

The history of drilling fluids is so recent that advances in the technology have largely consisted in improving testing techniques, determining the desirable properties of drilling fluids, and making studies of the commercial fluid-making materials. The art of compounding and controlling the properties of drilling fluids has passed the first experimental stage, and the direction of future development seems to be toward the use of standardized commercial materials and the increasing use of fluid systems having specific properties for overcoming special difficulties in the drilling and completion of wells.

HYDRAFRAC* TREATMENT
W. E. HASSEBROEK

The Hydrafrac treatment increases the production of a well to which it is applied by generating new and greater effective permeability within the zone of the well being treated. It was originally intended as a means of rejuvenating wells, especially in sand, but has found excellent application as a new well-completion method.

Hydrafrac treatment produces new permeability by fracturing or splitting the formation to which it is applied, as illustrated in figure 386.

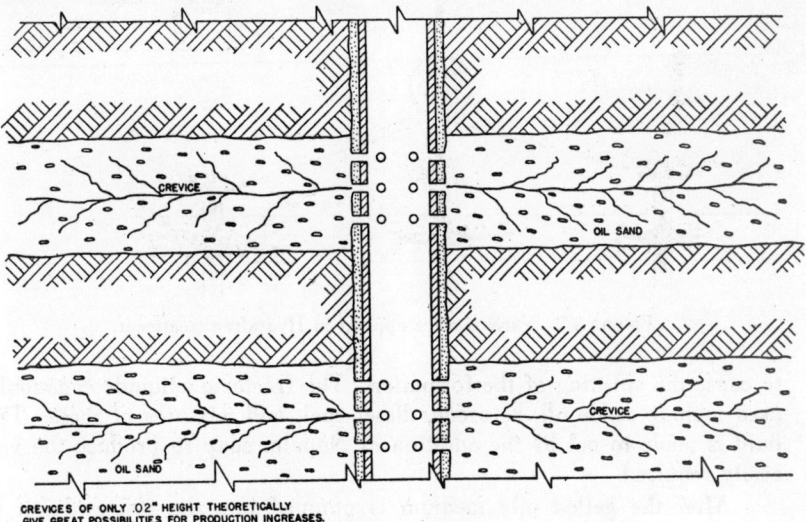

CREVICES OF ONLY .02" HEIGHT THEORETICALLY
GIVE GREAT POSSIBILITIES FOR PRODUCTION INCREASES.

FIGURE 386. Fracture pattern developed by hydraulic pressure transmitted with a high-viscosity liquid pumped into zone to be treated.

* Hydrafrac is a service mark owned by Stanolind Oil & Gas Company.

It is carried out in three steps:
1. Fracturing the formation.
2. Breaking the viscosity of the gel.
3. Placing the well on production.

Fracturing is accomplished (fig. 387, step I) by hydraulic pressure transmitted with high-viscosity liquid pumped into the zone to be treated. It is necessary that the viscous medium, referred to as gel, be of such viscosity that entry into the formation at a given rate of flow can produce pressure in excess of that attributable to the overburden in order

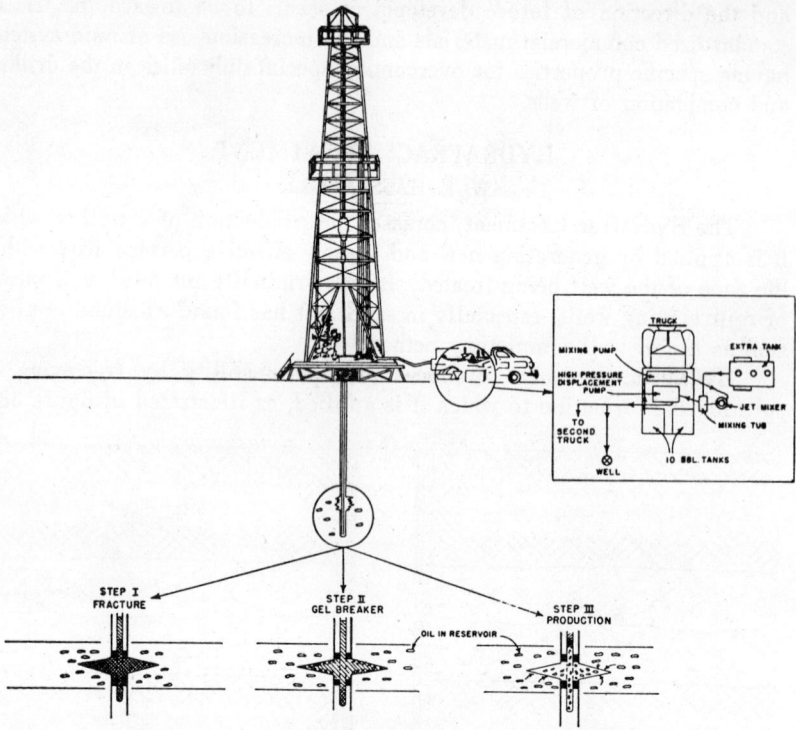

FIGURE 387. Various steps applied in Hydrafrac treatment.

to cause the splitting of the formation. The fracturing liquids commonly used include crude oil, kerosene, diesel fuel, and mixtures of these. The fluid is made to gel by the addition of Napalm soap to produce the viscosity required.

After the gelled oily medium is pumped into the formation, it is caused, at a later time, to revert to approximately its original viscosity by following it down the well with oil carrying certain chemicals (fig. 387, step II), which causes the gel to break down. The gel itself, because it contains a small percentage of water, will usually revert to the viscosity of the original fluid within 24 to 48 hours; however, it is customary to use the gel-breaker chemicals to assist breakdown of the gel.

At the conclusion of the treatment the well is allowed to remain shut-in, under the final pressure remaining when displacing the breaker fluid, for a period of approximately 24 hours, after which it may be placed back on production (fig. 387, step III) without any clean-out, as is necessary after shooting with nitroglycerine. The breaker fluid, the gel medium which has reverted to its original viscosity, and the new oil may all be conveyed to the tank battery without further treatment.

The gel used in the Hydrafrac process carries into the well, quantities of graded sand of approximately 0.02 inches diameter, which are added when the gel is prepared. The sand acts as a propping agent to hold open the fractures produced by the application of hydraulic pressure; thus when it is released and the fractures tend to close because of the overburden, they are held at least partially open by the sand.

It is important when considering the application of Hydrafrac to any well that several factors be thoroughly studied to arrive at a decision as to whether the application of the process is practical and economical. These factors are principally:

1. *State of depletion of the well.* There must be potentially producible oil in place, with its movement to the well bore restricted by low effective permeability, but with sufficient bottom-hole pressure to cause flow of oil into the well bore if permeability can be generated.

2. *Formation permeability.* Although Hydrafrac procedures can be applied to almost any formation permeability, they are usually applied to wells of rather low permeability. The permeability of the zone being treated is one of the factors that controls the viscosity of the gel to be used and the rate of input to produce the pressures necessary to fracture the formation. The procedure has been applied to wells with permeability of almost zero and on other wells with permeabilities of larger values, as for example, certain parts of the east Texas sands wherein permeabilities may run above 1,000 millidarcys. As a general rule, the commercial application of the process has been on zones of permeability of less than 300 millidarcys.

3. *Formation thickness.* With producing zones of great thickness it is often necessary to isolate and treat portions of the zone one at a time in order to gain maximum benefit from the process. With thin zones, if closely associated with water or gas, complications may be introduced, for it is quite possible that a fracture may be produced in the water- or gas-bearing section rather than in that which will produce oil. Further, the thickness of the formation, coupled with permeability, controls the viscosity of the gel to

be used and the rate of input necessary to secure high pressures.

4. *Prior workovers.* The effects of all prior workovers should be carefully considered, especially jobs that may have resulted in some fracturing of the formation, as, for example, shooting. If these fractures have occurred opposite gas or water zones, then it may be necessary to seal or pack them off; for extending them would simply produce more gas or water where oil was desired. Furthermore, prior workovers such as acidizing or shooting may have altered the nature of the bore in open hole in the well and made it impossible to use segregation packers for the isolation of sections within the zone which is to be treated.

5. *Isolation of formation.* The ability to isolate the zone to be treated, whether it be behind casing and contacted by means of perforations, or in open hole, is a most important factor to be considered. Obviously packing off in open hole is more difficult than in casing, especially when enlarged sections of open hole occur in the zone to be treated. In open hole it is desirable to have a caliper log of the section to be treated before attempting to determine whether or not packers, if necessary, can be used. In wells produced through perforations, the original cementing job behind the casing should be studied to determine if it is satisfactory to hold pressures at the points desired and not allow the fluids to migrate into other zones along and behind the casing.

6. *Condition of well equipment.* It is necessary to check the condition of casing and tubing in place in the well since high pressures will be encountered. Pressures to be applied will seldom exceed one pound per foot of depth as the total sub-surface pressure at the face of the formation. In a few cases the theoretical maximum of one pound per foot of depth has been exceeded; however, these are rare and are usually attributable to zones of local high rigidity.

7. *Production of a well to be treated.* A thorough study of the production history of the well to be treated is highly desirable since much information from such study can be had for comparison of results after the treatment. Such items as size of the pump installed in the well, length of the stroke, condition of all the equipment and their effects on the production of the well will influence the comparison of the old and new results of a Hydrafrac treatment.

Hydrafrac is no cure-all and will not produce oil where there is no producible oil present, yet it has resulted in vast improvement in a large number of wells where proper planning was undertaken and the well conditions were evaluated as being good for the application of the procedure. It is believed that where Hydrafrac improves production of any given

well, the improvement is not necessarily due to a mere increase in the
rate of production, but due to a greater effective permeability produced
as a result of the process; the ultimate recovery from the well so treated
will also be increased.

Pressure and volume of injection changes during the course of Hydrafrac treatment are shown in figures 388, 389, and 390. Well *A* in figure
388, completed in the Springer sand in the Sholem-Alechem field of Oklahoma, shows a typical pressure curve with a declining pressure during

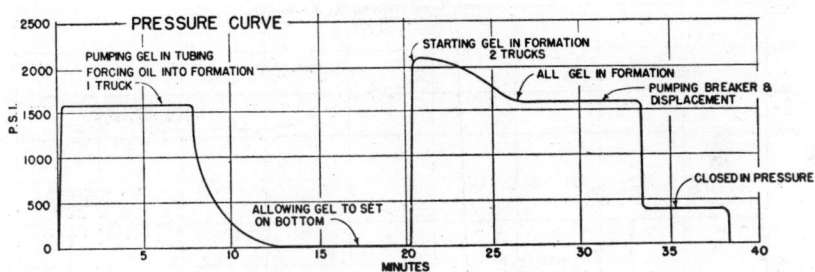

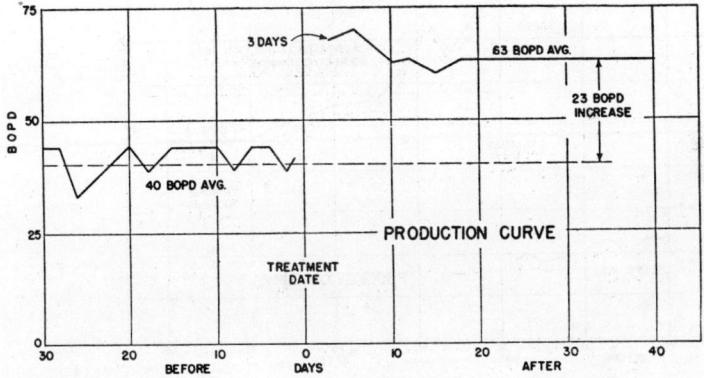

FIGURE 388. Pressure and volume of injection changes during Hydrafrac treatment.

the injection period of gel but with no sharp pressure break. This well was
treated through perforations from 5345 to 5358 feet. Pressure while displacing the gel with two cementing trucks handling the pumping, after
the gel was all in the formation, is about the same as with one truck
pumping oil into the formation before treatment. Obviously the fracture
produced had resulted in considerable increase in the effective permeability of the hole. The production curve after treatment shows a typical
flush-production period lasting a few days.

During this period the well had a peak production of approximately 70 BOPD, whereas the production immediately before the Hydrafrac treatment was averaging 40 BOPD. After about the 17th day of production following treatment, this well had settled down to a 63 BOPD average; thus an increase of 23 BOPD resulted from the Hydrafrac treatment.

Well *C* in figure 389, also a Springer sand well in the Sholem-Alechem field of Oklahoma, shows an entirely different type of pressure break during the course of the Hydrafrac treatment. In this case the sharp break

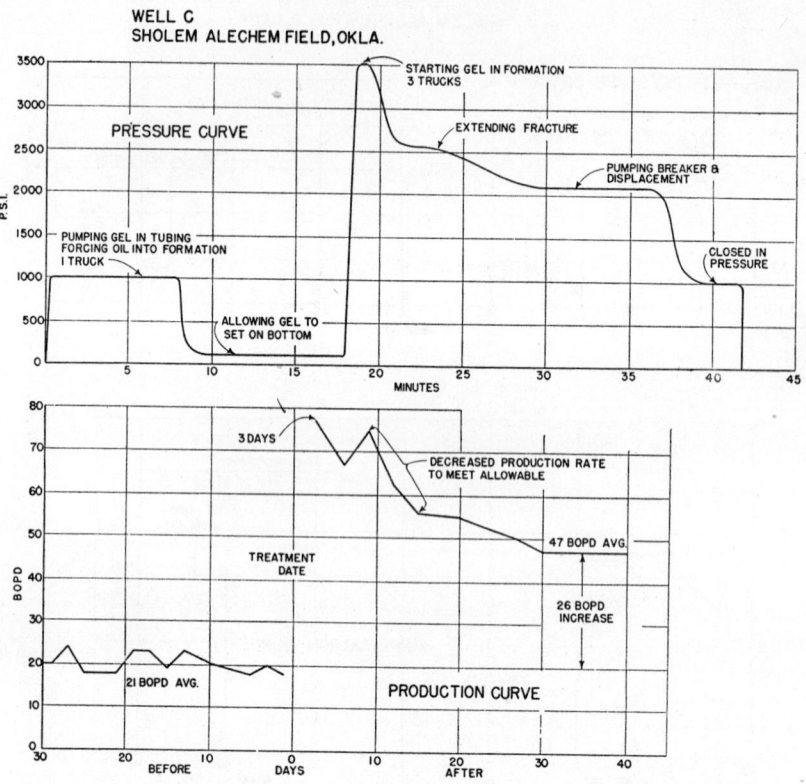

FIGURE 389. Pressure and volume of injection changes during Hydrafrac treatment.

of pressure probably occurred at the initial fracture. The declining curve indicates a rather typical extension of the fracture. Treatment in this case was through perforations of 5286 to 5310 feet. The production curve after treatment shows a flush production peak of about 75 BOPD, with a decline for a few days and a net increase of 26 BOPD resulting after the treatment. In this particular instance the 47 BOPD was the result of voluntarily lowering the production of the well to meet the allowable rate of the field. The well would produce more oil if desired, the exact amount of which is unknown.

Wells *E* and *F* in figure 390, being respectively in the Strawn sand in the Wichita Falls, Texas, area and in the Pennsylvanian sand in the New Hope field of Oklahoma, both give further information with regard to the shape of the pressure curves during the course of treatment. The break in pressure as in well *E* may often show no increase in production. This fact is probably due to breaking into a weak interval, such as a shale zone or

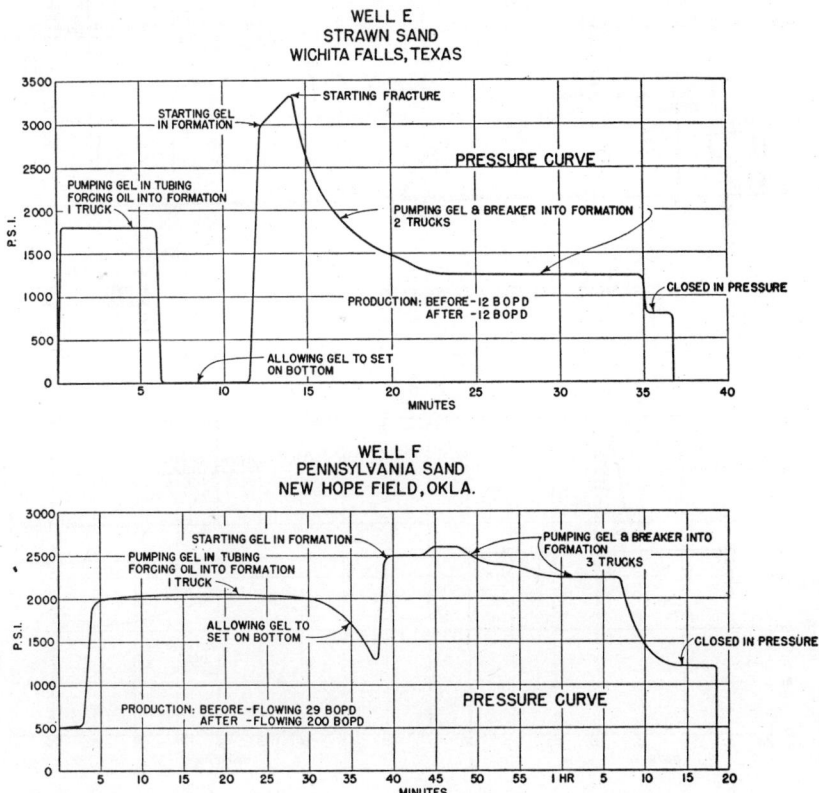

FIGURE 390. Pressure and volume of injection changes during Hydrafrac treatment.

some other nonproductive part of the section being treated. In the case of well *E*, the production before and after treatment was 12 BOPD in each case. Well *F*, on the other hand, shows a typical curve of fracture extension with even a slight rise in pressure occurring after the gel was in the formation for a short distance, and then a gradual tapering off, in spite of injection at high volumes. Curves of this type have appeared on a large percent of the treated wells which were helped.

Well *F* is also interesting because a large volume of a special treating oil had been pumped into the formation before the Hydrafrac treatment. The well was then tested and all the treating oil recovered, but there was no increase in production. The increase in production after the Hydrafrac treatment resulted in bringing the production from 29 BOPD up to 200 BOPD.

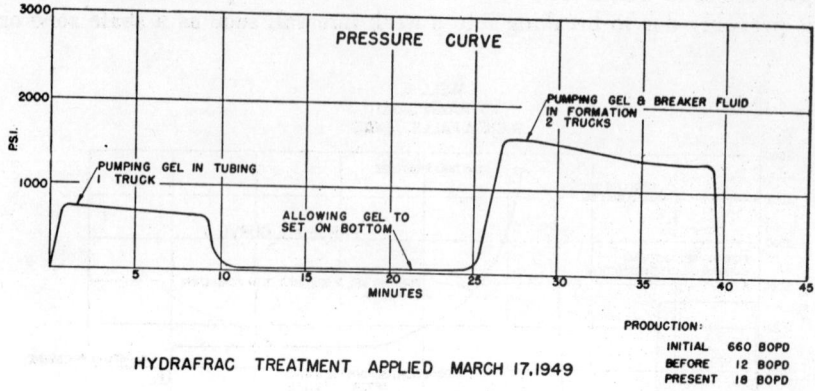

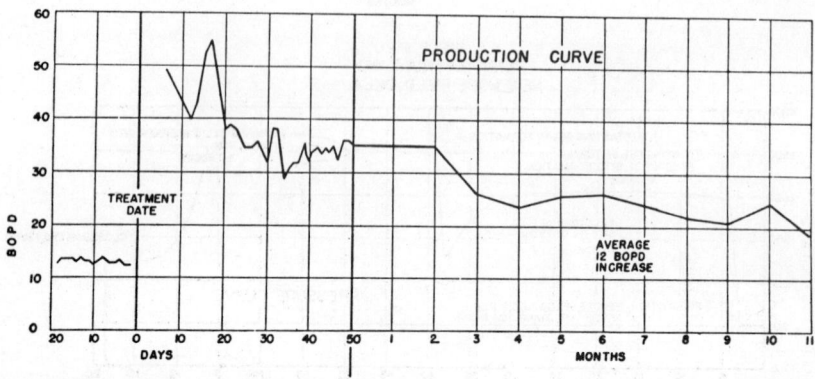

FIGURE 391. Pressure and production curves of a Hydrafrac operation in Daume sand, Strawn formation, Duame field, southwest of Holliday, Texas. Shot on completion with 40 quarts of nitroglycerine, 1945.

Normally, production increases from a Hydrafrac treatment are maintained for a long period of time. Figure 391 shows the pressure curve recorded during the Hydrafrac job on a well in the Strawn sand located southwest of Holliday, Texas, and the production for one year after treatment. This well was approximately four years old and had been shot on completion. At the time of treatment it was producing 12 BOPD. Immediately after treatment the well produced 48 BOPD and gradually decreased to 35 BOPD at the end of one month. This level was maintained for two more months, after which the production gradually decreased to where, at the end of one year, it was producing 18 BOPD. Average increase during the last eight months was 12 BOPD.

A survey of one twelve-month period showed 330 Hydrafrac treatments were performed on 285 wells in the United States with a production increase obtained in 70 percent of the wells treated. Approximately 40 percent of these wells were in Oklahoma and were producing from 28 different formations. Another 25 percent of this number were located in the north central Texas area. The remainder were in various sections extending from the Rocky Mountain region to the Louisiana Gulf coast.

The extremely dense sands of Colorado and Wyoming have made treatments effective in that region. Long sections of producing formations in firm open holes have allowed the use of special packers permitting several treatments per well. Certain sands and dolomite formations in Kansas, where acid has not been effective, have responded and resulted in good production increases. Low-pressure formations in the region including Illinois, Indiana, and Kentucky have caused limitations; however, successful treatments have been accomplished in that area. The long sections of slotted liners in California has made the factor of zone isolation most difficult, but successful treatments have been performed there on wells completed with Hydrafrac in mind.

Generally speaking, the sands of east Texas and the Texas and Louisiana Gulf coast are highly permeable, a fact which limits the need for a treatment of this nature. In the lower coastal region they are even unconsolidated. There are some sections throughout this region in which favorable results have been obtained, but not to the extent found in the mid-continent and northern areas.

FORMATION TESTING
W. A. WALLACE

Formation testing, sometimes designated as drill stem testing is designed to produce economically samples in sufficient quantities, of fluids at the surface from subsurface formations, the samples being accompanied by recorded subsurface formation pressures to determine a satisfactory well-completion program. This method of testing presents more direct evidence of formation-fluid content in both quantity and in quality than any method other than actual production of a completed well. The present technique of this type of testing, developed from the original unique ideas, now is accepted as conforming to conventional practices of oil-well drilling. Originally the procedures were simple, but increased demands for more information has resulted in an assembly of highly specialized equipment qualified to produce desired specimens from shallow to extreme-depth wells. These tools and techniques have made possible the

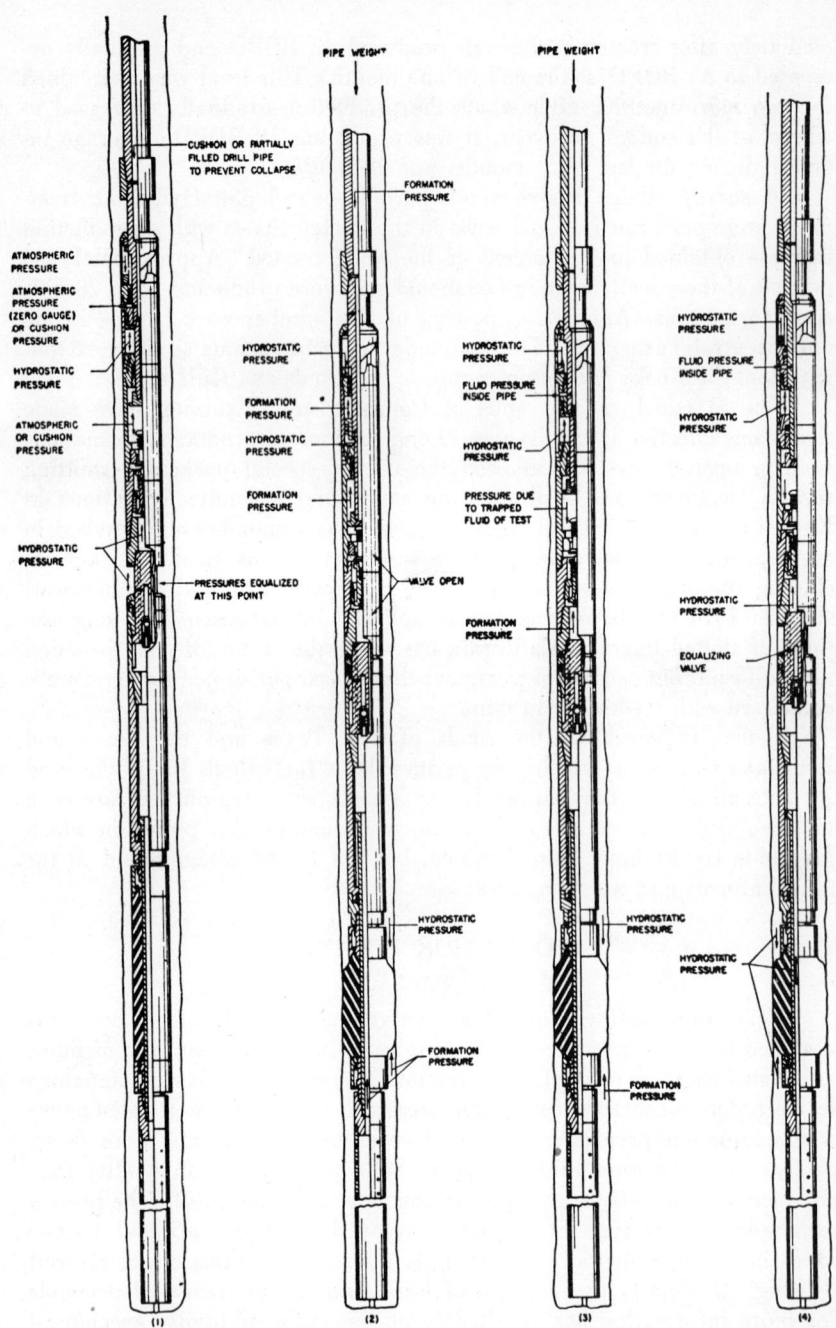

FIGURE 392. Operational sequence of a formation tester: (1) going in hole; (2) packer set and tool open; (3) packer set and tool closed; (4) packer being unseated.

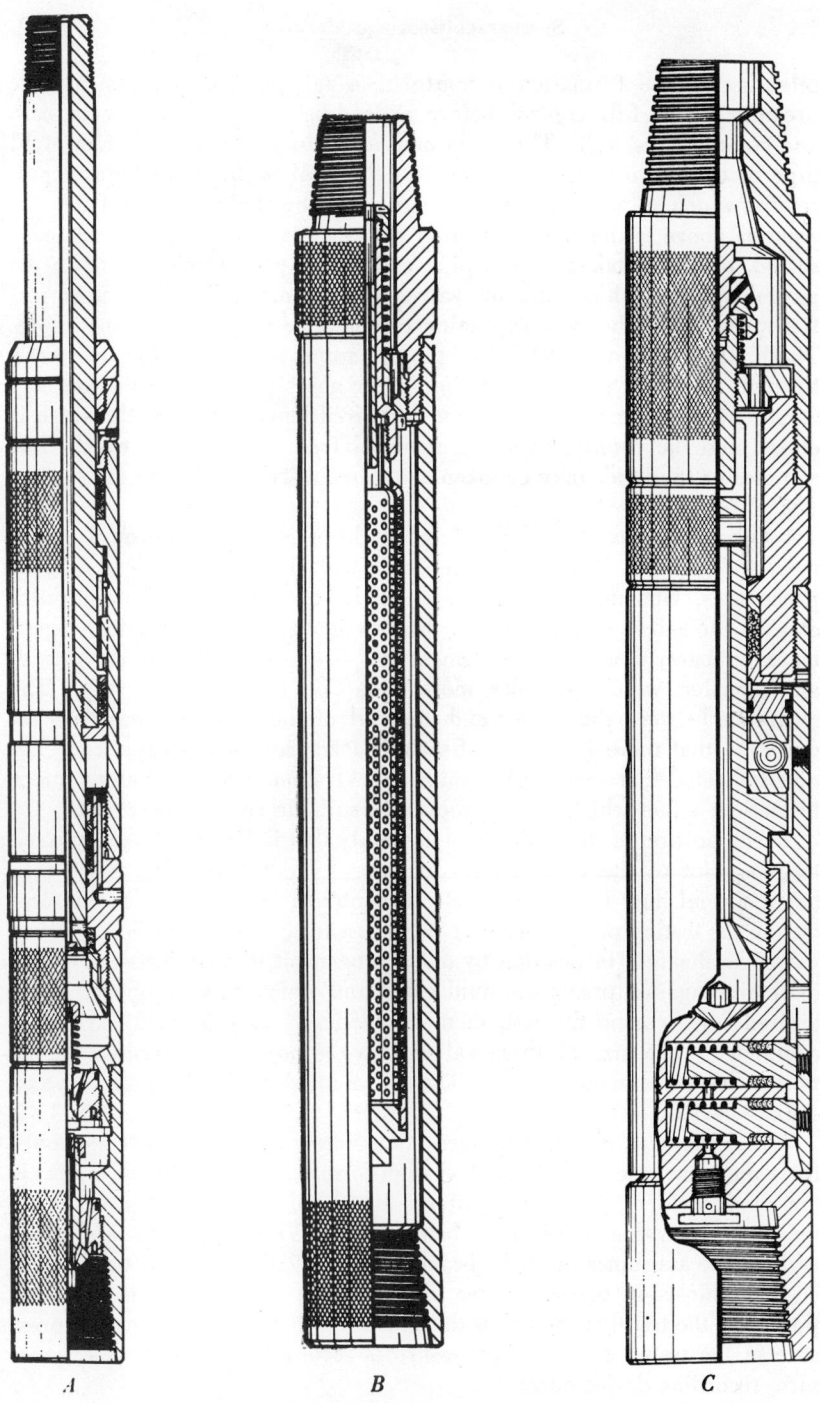

FIGURE 393. *A*—BJ tester. *B*—Long by-pass choke.
C—J-20 closed-in pressure valve.

releasing of the formation pressure of a selected zone to atmospheric pressure under full control before expending time and money to case and complete the well. This is accomplished by setting a packer on drill pipe or tubing to effect a seal against the wall of the bore hole, a procedure which eliminates communication of the drilling fluid from the annulus above to the isolated section below. After the packer is properly sealed, torque is taken on the pipe and added weight released upon the assembly forces which open the valves of the tools and permit the formation pressure of the isolated section to enter the empty drill pipe above. The drilling fluid trapped below the collapsed packer necessarily passes through the tester valves before liquids or gases from the formation may enter. After the test has progressed a desired length of time, the valve is closed and the running-in string removed from the well, at which time samples of the fluids may be taken as the recovery is raised to the surface (fig. 392).

Constant research and development have provided improved equipment and methods of testing to keep pace with the fast changes in drilling procedures. New mechanical adaptations have been added which have increased the safety of the process. Knowledge gained by practical experience indicates that the time element has been extended without hazard, an extension which provides more time for increased recovery. The clearance between the packer and the wall of the bore hole has been reduced so that there is a more effective packer seal. This experience has also indicated that shales erode more quickly than sand formations under bit action, a fact which renders them less suitable for a packer seat.

The portion of the testing tool assembly that is lowered to the selected testing point of the well consists of several individual units. Each unit has a special duty to perform, yet all combined are operated as one unit during the testing procedure. The valve assembly usually consists of three valves held safely in position by a *J* arrangement that protects the empty drill pipe against premature fluid entry and which may be opened when desired for securing the test, then automatically closed when the test is completed. The size of these valves may be adjusted according to the type of test to be made. Figure 393A shows in section one popular type of testing device.

A choke bean is usually placed immediately below the valve assembly for a dual purpose: (1) to control high-formation pressures and decrease the shock load at the opening of the tool, and (2) to determine better the potential value of the zone being tested. Very few wells are tested without the assistance of choke beans. Figure 393B shows this unit.

The closed-in pressure valve, figure 393C, was designed to close off posively the testing zone below the packer to obtain static-formation pressure in the tested zone. This pressure is recorded on a subsurface-pressure, recording-device chart.

The by-pass, or equalizing valve, is usually assembled below the

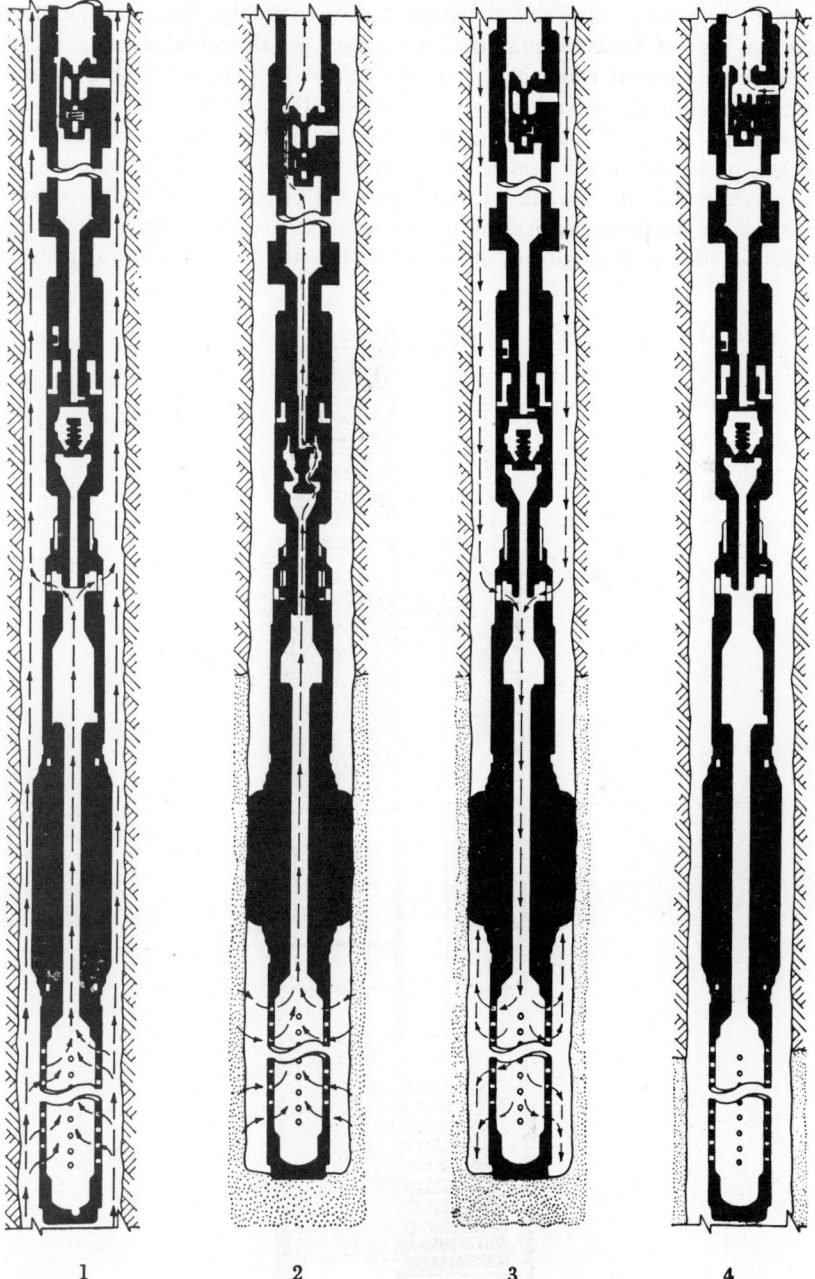

FIGURE 394. Effect of equalizing (by-pass) valve on packer; (1) going in hole;
(2) making test; (3) equalizing; (4) reversing.

closed-in pressure valve and above the packer assembly. This unit permits the free flow of fluids through the packer, as well as around the outside of the packer element while it is run into the well. It equalizes the hydrostatic pressure above and below the packer after the test has been completed. Releasing of the mud weight above the packer permits the removal of the packer from the seat without lifting the weight of the column of mud retained in the annulus, leaving only the friction hold of the packer against the wall to be broken loose. This tool also reduces swabbing effect when it is brought out of the hole after the test. The usual packer

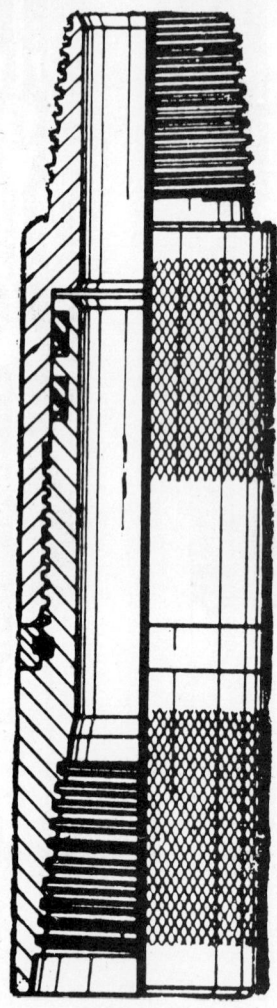

FIGURE 395. Safety joint.

damage caused when releasing the packer from the seat is relieved by this unit. (figs. 392 and 394).

Since the area of contact is greatest at the packer seat, a safety joint is usually run immediately above. This permits the removal of all equipment above that point should a fishing job occur. Several types of safety joints are available which may be run at any desired point in the string. It is a suggested practice to run safety joints on formation tests; however, it is optional with the well operator. One type is shown in figure 395.

Because of varied drilling programs that have incorporated formation testing, it has been necessary to develop testing packers for all well conditions, some of which, although infrequently used, nevertheless remain available.

(1) The cone packer, so named because of its conical shape, was designed to effect a seal on the shoulder created when coring at reduced size below the drill hole. This type of packer is now very seldom used .

(2) The wall packer has almost completely replaced the cone type of packer. This packer was developed to be used in wells where a core shoulder was not available or where testing zones had been penetrated with the drill bit before coring. Its results are more effective because of its ability to be used in the regular drill hole as well as in the smaller core holes. (fig. 392).

(3) The practice of running a wall-over-cone packer combination has been eliminated. Although it was popular several years ago, improved testing techniques and packer construction has made possible the completion of open-hole tests more satisfactorily without the added cone packer.

(4) Double-wall packer testing utilizing two wall packers, one above the other, is effective many times when a single packer has failed. This is especially true when attempting to set the packer opposite a thin lens or in a broken-sand and shale formation.

(5) Testing between two packers is common in many areas. This procedure is usually conducted by setting one packer at the top of a given sand and one at the bottom with a perforated section of pipe between the packers. This setup provides a means of testing a zone that has known productive horizons above and below. Although this procedure is practical in many areas, it is not a suggested practice for an area such as the Gulf Coast or upper Gulf Coast regions where very soft and cavy formations are found.

(6) That the casing has been set and cemented does not necessarily mean the well is ready for production. Certain information is frequently desired before final completion. These data can only be obtained by formation testing or production testing. Formation testing frequently offers as much information in much less

time and at considerably less cost than production testing. It is for this reason that casing or hook-wall packer testing has become so prominent. This type of testing is made possible by attaching the testing equipment to a hook-wall packer that may be

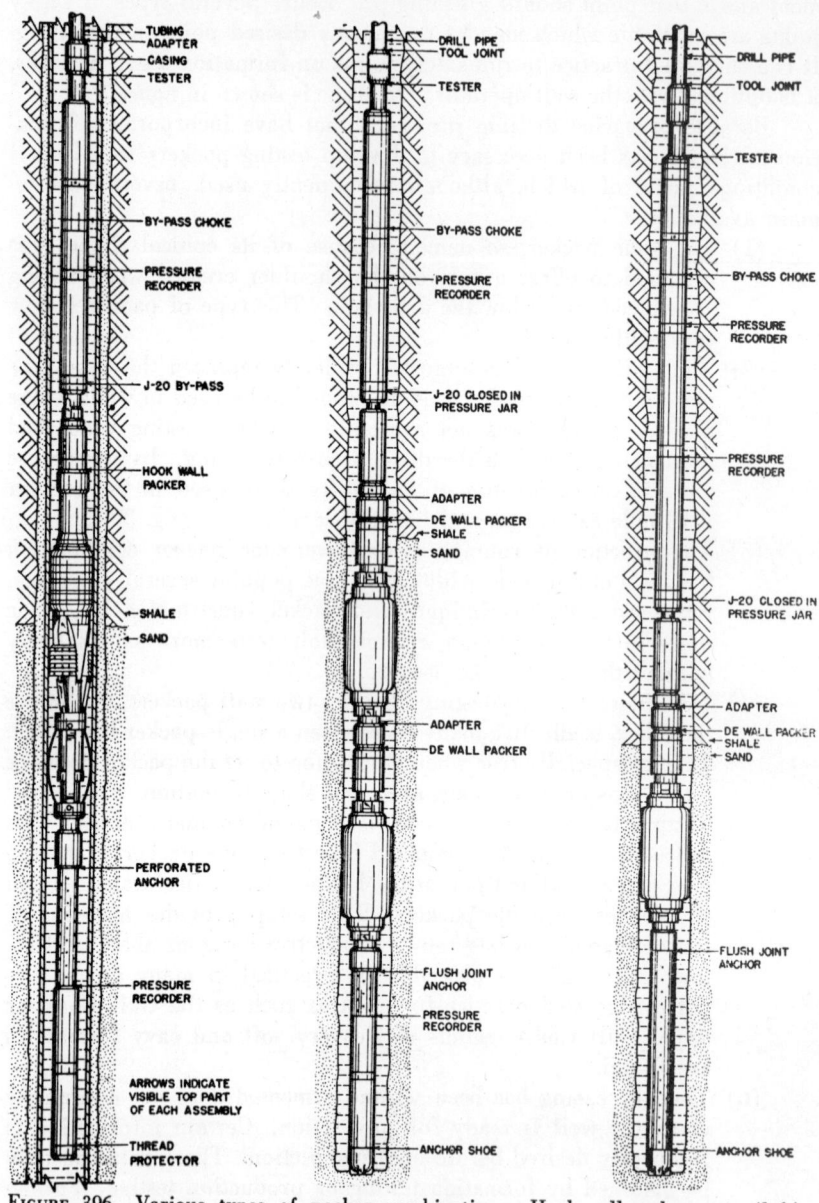

FIGURE 396. Various types of packer combinations. Hook-wall-packer test (left); double-wall-packer test (middle); wall-packer test with type AP pressure-recording devices (right).

set at any desired point in the casing by means of slips that, when released, bind between the slip body and casing and act as an anchor on which to set the packer and to operate the tool mechanism.

There are several applications for which this type can be used:

(A) Productivity through perforations in the casing.

(B) Productivity from the open hole below a cemented casing.

(C) Testing for water shut-off after cementing.

(D) Testing for holes in casing.

Figure 396 shows examples of various packer combinations.

Debris in the drilling fluid, at the time of testing, should be removed and the mud properly strained before entrance is made into the choke area of the testing assembly. Two sets of strainers are provided for this purpose. The anchor, as used for open-hole tests, serves as one strainer, and built within the anchor and mounted over the choke is a secondary strainer. This strainer is smaller in diameter and length, but contains many more holes of smaller diameter to catch debris that has passed through the outer screen. In like manner, double screens are used in casing or hook-wall packer testing (figs. 392, 393B, 394, and 396).

Great strength is required in the anchor pipe for open-hole testing to withstand the weight of the mud column in the annulus above, plus the limited weight of the drill pipe used in collapsing the packer. Flush pipe with tool joints has replaced collared pipe as an anchor. In many areas, drill collars are used to withstand the severe loads, especially when testing thick zones at great depth.

A safety precaution employed during a test is the running of a circulating valve. This valve is usually placed in the string several joints above the tester assembly. Provision is made for free flow of the recovery upward through the valve during the test, yet permitting circulation downward through the running-in string and upward in the annulus between pipe and well bore if necessary. This is made possible by a back-pressure valve opposing fluid entry from outside the drill pipe. This device has proved invaluable in preventing blowouts, controlling wells, and conditioning drilling mud after the test.

For testing in cased wells, a reverse circulating valve has been developed. This unit permits the pumping of fluids, either down the running-in string or down the annulus between the casing and running-in string. This arrangement facilitates the removal of the test recovery while the running-in pipe is still in the well and the recovery of a dry string when coming out of the hole with the tools. Its use is quite popular when drilling town-site leases or on wells over water where the recovery must be controlled at all times. (figs. 392 and 394).

Surface control is an essential phase of a test during the operational procedure. Although it is known that in some areas surface control is of no importance, it is necessary in many others to place special emphasis on control heads and blow-out preventors for protection. For this reason high-pressure control heads have been developed to handle safely the ever-increasing pressures produced on formation tests. The surface control head consists of a heavy-metal valve case containing two valves with

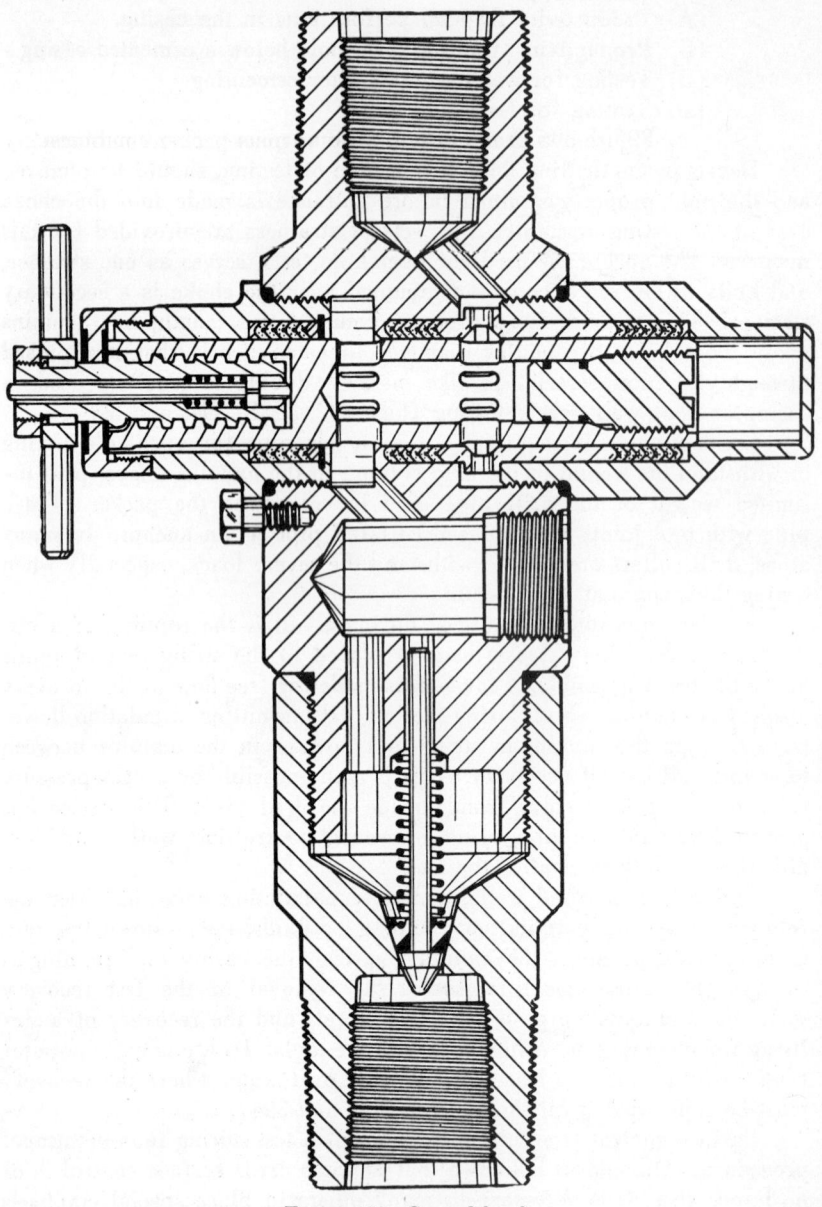

Figure 397. Control head.

an outlet on the side. One valve serves as a master cut-off below the outlet, whereas the other, at the top, when closed may be used to divert the flow through the side port or upward through the kelly joint when it is open. The usual procedure, while testing, is to divert the flow through the side port and attached choke units, then through a flowing hose to a separator and tanks prepared for the purpose. The strength of this type of head is sufficient to withstand heavy pull when the pipe is removed from the well (fig. 397). A multiple choke unit may be attached to the side outlet to provide for the use of various sizes of surface chokes that may be changed at will to produce pressure and flow changes that are invaluable in preparing completion programs.

Improved methods of oil-well-drilling procedures have had a tremendous effect on the condition of the bore hole as well as on the drilling fluid. Practices used for maintaining the wall of the bore hole for drilling or coring are of great aid to formation testing. For this reason, to a large degree, permissible testing depth has increased to any drilling depth desired. Theory, followed by experience, has shown that formation testing is more effective in small drill holes than large ones. Drilling programs, that have adopted smaller holes and smaller casing sizes, have enhanced formation testing success to a large degree. Although many tests are made in wells with a bore-hole diameter of eleven inches or greater, the most desirable diameters are less than $9\frac{5}{8}$ inches.

For any given well depth and mud weight, the smaller the hole bore the less total differential load the packer must withstand and the less circumferential distance that must be sealed off. Coupled with the above facts and of great importance is that the less the annular clearance between packer bottom shoe and hole bore, the better are the chances for effective packer sealing. These things explain why smaller, rather than larger, holes are desired for formation testing.

Condition of the drilling fluid and its effect on testing should not be overlooked. The quality of the mud must be such that it will remove cuttings from the well bottom to provide a solid bottom on which to anchor the testing equipment. Shales, debris, and foreign material of any kind that would impair fluid passage through the screen section should be removed before attempting a formation test. The weight of the drilling mud as used to control the well during the progress of drilling is ordinarily sufficient for protection while making a drill-stem test. However, it is frequently found that the pressure as recorded on a subsurface gage indicates a mere two or three hundred pounds over the formation pressure of the tested zone. This indication should demand and receive immediate attention. Low water-loss muds of all types greatly assist success in obtaining correct data from a test since in such cases the formation has less foreign fluid content to produce before giving up its contained natural fluids.

The selection of the packer seat, when testing an open hole, is of major importance. Only by proper sealing can a successful test be completed. Practical experience has indicated that the most effective seats are in the top of the formation to be tested. In recent years, electrical and caliper logs have borne out this accepted practice, many times locating where least expected packer seats that otherwise would have been overlooked. This is especially true where the well depth has been carried several hundred feet below a desired testing zone. Sand, lime, chalk, and dolomite usually accurately hold to bit diameter, a fact which makes them desirable for packer seats.

Increased depth and mud weight often impose greater pressure on the outside of the empty drill pipe than its strength will withstand. To overcome this condition a fluid cushion is placed in the otherwise empty drill pipe in sufficient quantity to secure the pipe against collapse. The quantity of fluid required for the cushion varies with well depth and mud weight, and should be placed in the pipe while going into the well to eliminate air pockets. Fluid cushion also reduces the shock load at the opening of the tool valves. This is an important factor in formation testing and should always be given consideration.

Subsurface-pressure-recording devices have removed any doubt concerning pressure changes during the testing procedure. Pressures that were previously theoretical have been proved a reality by a continuous recording of all pressure changes that develop while making a formation test. Information obtained from pressure-versus-time recorded charts has induced continued development in this phase of testing such that at the present time the instruments are highly accurate. These recorded pressures are now invaluable in compiling information by geological, engineering, or production departments, in checking the operation of the subsurface equipment, in calculating productivity, in reservoir analysis, and in selecting production equipment. Efforts are being extended to accumulate sufficient material whereby formulas can be set up on a method for evaluation of productivity of wells from drill-stem tests. The basic material for this project necessarily involves pertinent information obtained from pressure-recording-device charts.

Operational procedures that produce predominant pressure changes are (1) lowering the tools into the well, (2) seating the packer, (3) opening the tool valves, (4) flow from formation, (5) closed-in pressure valve closed, (6) opening of fluid by-pass, and (7) removing the tester assembly from the well.

Adverse conditions that are readily recorded by these units include plugged choke, plugged tool, plugged perforations, sloughing of unconsolidated formations below the packer, and irregular diameters of the walls of the drill hole. Formation-testing procedures usually involve the use of at least one pressure recorder. More recently it has been found that the use of two gauges is more practical. If one only is used, it should be blanked off from the flow stream. When two are used, the lower recorder is blanked off while the other is placed in the flow stream. This arrangement permits detection of the part of the equipment that becomes plugged, should such occur, and also indicates sloughing around the recorders and perforations.

If the well conditions permit, pressure-recording devices are placed below the packer. If not, above-packer recording devices are placed above the packer and below the choke. This is true whether one or two recorders are desired. Pressure values remain the same regardless of the location of the recorders, as long as they remain below the bottom hole choke. If placed above the choke, recorded pressures will indicate only the weight of the fluid recovery and gas pressure trapped within the running-in string.

Requirements for testing between packers (whether in open hole or in casing) include the use of one pressure-recording device below the bottom packer and one or two recorders in the testing section between the packers. The recorder below the bottom packer is used to indicate the sealing results of the lower packer, and the two in the test zone record the pressure changes during the operational procedure of the test. In many cases it has been found where highly permeable formations have been sealed below the bottom packer, dehydration of the drilling fluids below the packer into the permeable formation have indicated the static pressure of the formation sealed off. This is not true in all formations. Less permeable formations below the packer that lend no avenue of escape by dehydration maintain the original hydrostatic pressure during the course of the test, and this pressure is released only when the bottom one of the two packers has been removed from its seat. The addition of more than one packer above a test zone has no material effect on the pressure recording.

Various types of subsurface pressure-recording devices have been developed for the purpose of securing recorded pressures during the progress of drill-stem tests. Of these the spring and piston type and the Bourdon tube type are the ones most frequently used. The design of these devices was originally patterned after wire-line, bottom-hole-pressure instruments used for taking bottom-hole pressures of wells on production. For formation testing the spring and piston type of devices were first used. This type of gauge permitted the development of a rugged and sturdy unit that would withstand the rough treatment so common in formation testing. Although rugged in construction its calibrated springs and highly polished pistons operate to a degree of accuracy sufficient for standard use on formation tests. Periodical calibrations of these units by dead-weight testers and corrected to temperatures have continued their value. Units of this type are usually referred to as standard pressure-recording devices. (fig. 398).

Developed in more recent years is the Bourdon tube type of recording device whose duties in formation testing remain comparable to that of the standard pressure-recording devices. The mechanical reaction of this unit

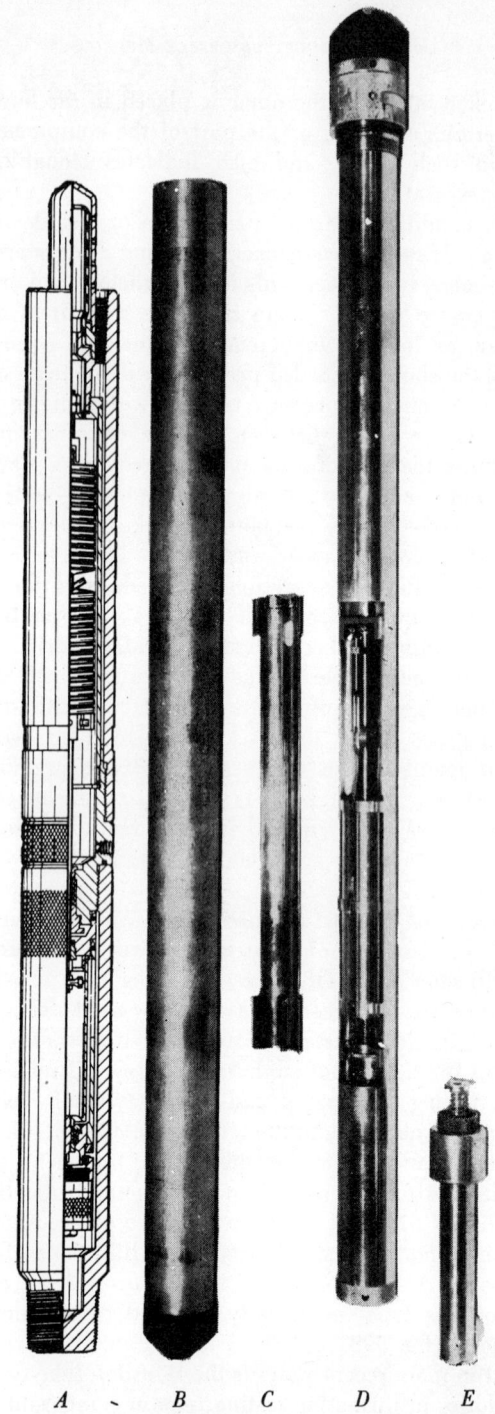

A   B   C   D   E

FIGURE 398. *A*—Howco standard pressure-recording device. *B-E*—Bourdon pressure-tube-type recording device; *B*—outer case; *C*—inner door; *D*—"BT" prd with chart and stylus; *E*—clock.

comes from the expansion and contraction of a Spiral Bourdon tube from which it gets its name. (fig. 398). Although delicate in construction, by using certain precautions it is possible to obtain very accurate recordings without destruction. Because it is highly sensitive to reaction, slight variation of pressure changes are recorded which to a certain degree have im-

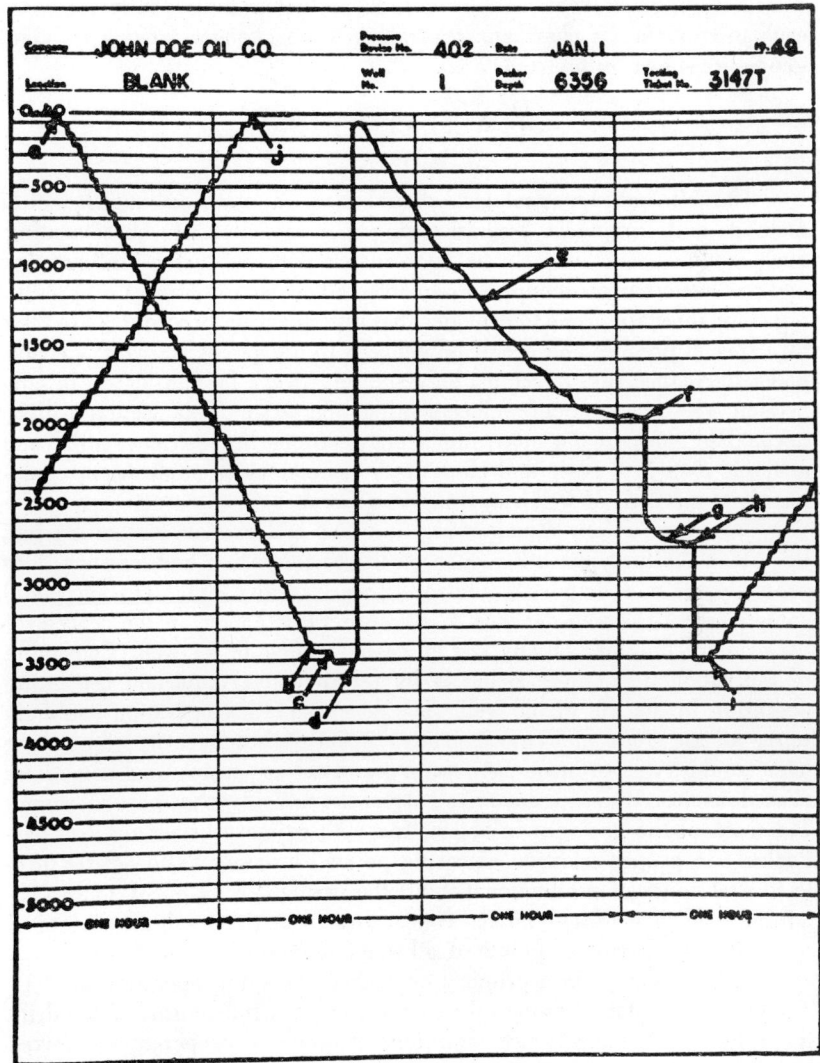

FIGURE 399. Chart from standard pressure-recording device: (a) test started; (b) reached packer seat; (c) packer seated; (d) tester opened; (e) flowing pressure; (f) tested closed; (g) closed-in formation pressure; (h) packer unseated; (i) started out of hole; (j) reached surface.

proved the accuracy of the unit. This type of recorder is also periodically calibrated by a dead-weight tester and to temperature, as are units of other types. The mechanical setup of the Bourdon gauge is such that it may be adapted in the testing assembly as other recorders. Unlike the standard recorder, these instruments record upon a blank chart. The recordings that are made during a test are measured in inches of deflection and converted to pressure from a calibration curve that is individually made for each instrument. A chart and typical pressure changes from a standard gauge are shown in figure 399.

OIL-WELL CEMENTING
W. D. OWSLEY

The oil industry geologist should have a working knowledge of methods and equipment used in casing and remedial oil-well cementing. Such methods must usually be coordinated with subsurface information and various other procedures. Frequently other departments of oil-producing companies are prone to consider that all cementing procedures are their own special premise and are of no concern to the geologist. This is a mistaken idea since subsurface geology should always be considered in connection with any kind of well-cementing operation. This section will be devoted to some of the more important relations between cementing and geologic data in a basic and elementary manner as to application, equipment and methods of well cementing in general.

Cementing of any casing string in a well is a primary cementing job, this being the first cementing done on the casing string and, as such, it is probably the most important cementing work of any kind. The successful completion of the well depends to a considerable extent on the success of the primary job done on any and all casing strings in the well. Cementing of the production casing string is by far the most important cementing operation in the life of the well. The success of this operation can fully repay all the study and care expended in its performance by eliminating later troubles due to incomplete sealing and the expensive, and sometimes hazardous procedures necessary to correct the difficulty thus brought about. The work of the geologist in connection with primary cementing lies particularly in proper selection of casing points, decisions as to desired protection of zones of production behind the casing, exclusion of zones containing corrosive waters, advice concerning formations subject to loss of returns, and selection of points of oil-water or gas-oil contact.

The usual steps in a primary or casing cementing operation are (1) The hole is circulated through the casing to obtain full returns of the drilling fluid; (2) Cement to suit conditions of time and temperature is mixed and pumped into the casing above the bottom plug separating the slurry from the mud; and (3) A top plug is released and pumped down to force the liquid cement slurry to the bottom and up into the annulus between the bore and casing.

For a successful operation of this type, some of the results sought are: (1) sealing of the producing zone from encroachment of water from other zones penetrated; (2) prevention of migration of fluids between any zones covered by the cement; (3) support and protection of the casing against collapse, corrosion, and other damage; and (4) prevention of blow out on the outside of the casing.

From a mechanical standpoint and without regard to any geological consideration, the following points are vital to any good casing cementing job:

(1) A properly drilled and conditioned hole, which is as true a bore as possible for the area with regard to both vertical drift and diametral trueness to gauge. Free circulation with as low a fluid loss as can be obtained. Mud of low water loss, low viscosity and sufficient weight to hold back subsurface fluids. A final clean up of the bore to rid it and the mud system of as much suspended or deposited cuttings, debris, and gelatinized mud as possible.

(2) A proper string of casing to suit existing well conditions, equipped with a guide shoe and at least one float collar, all correctly tallied and checked.

(3) Cementing equipment of high reliability to suit the conditions of the area of operation and well depth as to pump pressures and volumes desired. Strong, reliable, and speedy casing-head connections. Two plugs, a bottom and a top one, of the best possible design to separate mud from cement slurry, wipe the casing walls, and produce a positive shut-off when the slurry is landed in place.

(4) Capable, well-trained, alert, and reliable cementing crew and drilling-crew personnel.

(5) A type of cement, as portland, slo-set, of high early strength to suit time and temperature conditions of the hole being cased.

(6) Ample supplies of water and mud to effect mixing of the cement and pumping it into place.

The above items are basic; without them no casing-cementing job can be successful; yet, even with all of them, many such jobs are unsuccessful because water or gas is not properly sealed in place. Remedial measures are necessary later to correct the original failure. Several things may be done to better insure success on any casing job in practically any area of operation, and it is with these added features that the geologist should be especially concerned.

(1) A study of the various well logs in regard to oil and/or gas zones, water zones, water-oil contacts, and gas-oil contacts. Zones of weakness which might cause or have already caused lost circulation. Size of the final bore at various depths and these depth relations to formations which must be properly covered with cement. Temperature existing in the well.

(2) From a study as in (1) above, select the best position for the

casing shoe, giving careful consideration to gas-oil contact and nearest upper water. Determine minimum height of fill-up of cement required. Assist in selecting proper cement for temperature and time requirements and resistance to corrosive waters.

At this point several modern practices in casing cementing should be considered, selected, and applied as the prior log study may indicate.

In the past it was believed that cement slurry exerted a powerful scavenging action on the walls of the bore, an action which would free it from all mud cake and allow the cement to bond directly to the face of the exposed formations. It is now recognized that this is not unusually true and that, in fact, generally the flow of cement will not entirely rid the formation face of mud accumulations. This fact led to the adoption of mechanical abrading devices (called wall cleaners or scratchers) and centralizers to assist in completely surrounding the casing with cement and allowing it to contact and adhere to the formation face.

Two types of wall-cleaning devices are in general use. One requires reciprocation of the casing in the bore to remove the mud cake. The other cleans sections of the hole by rotation of the casing after it reaches the selected shoe position. Both utilize spring-wire wickers in support members set around the casing for the reciprocation type and axially along the casing for the rotation type.

Both the above wall cleaners have certain advantages and disadvantages in regard to operation and initial cost. Only a study of each should guide the prospective user; however, of one fact he can be assured: Mechanical wall cleaning at proper positions in the well bore, coupled with centralizers, will vastly improve the possible success of the casing-cementing job.

Location points of wall cleaners in relation to sections of hole to be cleaned up, and therefore best protected, is a definite geological premise. It is neither necessary nor economically practical to place wall cleaners and centralizers indiscriminately on the full length of a casing string. Place them where they have an opportunity to accomplish their intended work and obtain the best results—for example across oil-water contacts and oil-gas contacts; at the top and bottom of known productive or possible later productive formations; between closely spaced zones which must be acid treated, Hydrafrac treated, or separately produced; and contacts of and across water zones which might cause excessive corrosion or infiltrate into another zone.

It is unnecessary and definitely uneconomical to apply abrading and centralizing devices in thick nonproductive zones, or in sections having excessively large hole diameters due to wash-out during *drilling*.

Certain admixtures to the usual water-cement slurry offer added assurance to cementing success. Bentonite added to the cement is advantageous because it prevents settling of the cement particles which cause water-filled cavities to form behind the casing. Some increase in volume of the set cement is obtained. A better flow characteristic results. A lighter weight slurry which is desirable in areas subject to loss of returns can be produced. Less shattering of the set cement may occur during gun perforating.

Various materials may be added to cement slurry to prevent loss of circulation. Chopped cellophane flake is frequently used for this purpose. Caution must be exercised with certain other lost circulation materials in cement slurry as some organic fibers will effect the pumpability and setting times of the cement.

Remedial cementing includes all type of jobs, other than casing cementing, to correct difficulties encountered in drilling and completing wells. Of greatest importance among such operations are squeeze cementing, plug backs, lost returns, and liner jobs. These have as their more frequent purposes the shutting off of water or excess gas, correction of lost circulation, changing of hole depth, and lengthening of the cased section of the hole. These features may be sought after either separately or in various combinations at any one time. Geological data is always involved in connection with all such work.

Exclusion of bottom water which has come into the well by rise of the water table is a comparatively simple procedure. In open hole, usually a plug-back job is involved. This procedure of setting a plug of cement slurry from bottom upward across the oil-water contact. Generally, this is accomplished by pumping a quantity of cement to bottom through tubing or drill pipe and spotting it in place by measured displacement with mud or water. In soft formations the ordinary plug back does not always have a very long life because water soon comes in around the cement plug. This restricted longevity gave rise to the practice of putting pressure in excess of the hydrostatic head of the fluid in the well on the liquid slurry to force the slurry against the face of the formation and to dehydrate the slurry against the formation, a procedure which developes an improved bond. This practice was enlarged upon and is known as squeeze cementing —the forcing of slurry out into, and into contact with, a formation under high pump pressure.

It is incorrect to assume that the solid particles of cement in a slurry actually penetrate the interstitial spaces in a sand body. Actually these particles will not enter the pores of even a coarse sand for more than a fraction of an inch. In spite of this fact a large amount of cement can be forced out into a sand formation. This injection is due to the splitting of the formation by subsurface pressure which exceeds the overburden load. It is also true that a considerable portion of the mixing water is forced out of the slurry into the permeable media with which it is in contact, and action which creates a tight cement-to-formation bond and leaves a dense mass of cement within the well bore or casing.

Squeeze cementing is the most effective means now known for excluding water from production either from open hole or from production through perforations. It is also frequently applied to cut off excess gas issuing from the gas-cap section of a producing zone.

It is always important when attempting the exclusion of water or gas that the squeeze job or plug-back cover the offending section and overlap the contact point. When reperforation is initiated, the new holes should not be allowed to enter the sections which were cemented to correct production.

The brief and elementary details which have been given are intended merely to bring attention to certain items which should be kept in mind by the subsurface geologist during oil-well cementing practices.

WELL ACIDIZATION
L. W. LeROY

Well acidizing is defined "as the process of introducing inhibited hydrochloric acid into predominantly limestone formations to enlarge the pores by removing obstacles and constrictions from them and extending them into new drainage areas, thus lowering the resistance offered to the flow of oil and gas through the oil-bearing formations."[35] Acidizing of a well results in the modification of the physical aspects of the reservoir strata so that pressure differentials across a portion of it are reduced, thereby permitting a more efficient utilization of the available energy.[36]

As a result of well acidization, new oil reserves have been discovered, oil-field developments have been improved, and production has been increased. Acidizing has become an adopted procedure in completing and reworking of wells which produce from limestone and/or dolomite rocks. In certain cases, acidization of sandstone-producing intervals has promoted greater and more efficient recovery.

Limestone and Dolomite

The carbonate rocks are represented by limestone and dolomite and mixtures of both types. Some limestones are true clastics, whereas others are chemical or biochemical precipitates. Diagenetic changes frequently produce extreme modification of the original deposit. Limestones and dolomites vary widely in composition and texture. They include such foreign constituents as quartz grains, clay, anhydrite, gypsum, iron oxides, and chert. These impurities frequently play an important role in acidizing programs. The texture of limestones and dolomites may range from very fine to coarse crystalline. Dolomites more frequently exhibit coarser crystallinity than limestones. The chemical composition of the rocks is perhaps the most basic of all geologic factors which control the results of acidizing. However, the texture and structural fabric of the rock are also important factors.

[35] Love, W. W., and Fitzgerald, P. E., *Importance of Geological Data in Acidizing of Wells:* Am. Assoc. of Petroleum Geologists Bull., vol. 21, no. 5, p. 616, 1937.

[36] Fitzgerald, P. E., James, J. R., and Austin, R. L., *Laboratory and Field Observations of Effect of Acidizing Oil Reservoirs Composed of Sands:* Am. Assoc. Petroleum Geologists Bull., vol. 25, no. 5, p. 850, 1941.

The character, continuity, thickness, and extent of porous and permeable intervals within a carbonate section are of primary concern to the operator. Rock characteristics may change drastically both vertically and horizontally.

Two types of porosity are recognized in carbonate rocks: (1) *primary porosity*—the percentage of pore space present at or just subsequent to deposition, and (2) *secondary porosity*—the pore space developed subsequent to or during lithification. In the latter case, the original openings have been enlarged or new pore space developed by ground-water percolation. Development of secondary porosity and permeability gives rise to several types of solution patterns.[37] (1) *equi-solution* type, which involves a carbonate rock of uniform composition; (2) *channel pattern*, developed in an alternating limestone and dolomite section where differential solution is evident or in fractured carbonate rocks; and (3) *cellular* type which is typified by shallow vugs formed by solution on exposed surfaces.

Carbonate rocks possessing a vuggular pattern do not necessarily indicate continuous porosity. However, this type of solution pattern coupled with the equi-solution and channel types tend to develop favorable permeability. Acid treatment of the rock fosters interconnection of these solution patterns and thus improves the flow of fluids and gases.

Sandstone

Acidization of sandstone reservoirs has in some cases resulted in increased production; however, a large number of such treatments have been failures.[38] Acid testing of core samples prior to formation treatment serves to determine whether increased production may be expected. Fitzgerald, James, and Austin carried on a core-acidizing study in which 332 cores from 30 different producing formations were involved. Of this total 271 (81 percent) showed an increase in permeability after acidizing. The average permeability increase was 337 percent. Several examples of increased porosity and permeability in sandstones are given below.

Formation	Average
Bartlesville (Oklahoma)	
Porosity, % (before)	18.4
Porosity, % (after)	19.2
Permeability, M'd (before)	70.0
Permeability, M'd (after)	81.0
"Wilcox" (Oklahoma)	
Porosity, % (before)	12.6
Porosity, % (after)	12.4
Permeability, M'd (before)	83.3
Permeability, M'd (after)	479.2

---

[37] Howard, W. V., and David, M. W., *Development of Porosity in Limestones*: Am. Assoc. Petroleum Geologists Bull., vol. 20, no. 11, p. 1402, 1936.

[38] Fitzgerald, P. E., James, J. R., and Austin, R. L., *op. cit.*, p. 851.

UNTREATED   TREATED

FIGURE 400. Kansas Arbuckle core. Dense core with permeable streak running obliquely. Appearance of core and treating response indicate uniform structure within permeable section. (From Chamberlain, *Oil Weekly*.)

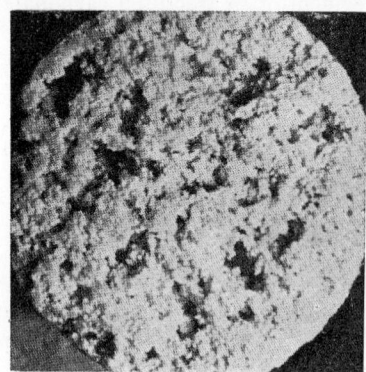

UNTREATED   TREATED

FIGURE 401. Niagara dolomite treated with 15-percent hydrochloric acid containing a surface-tension-lowering agent. Uniform structure, moderate channeling. (From Chamberlain, *Oil Weekly*.)

UNTREATED   TREATED

FIGURE 402. Viola limestone treated with 15-percent hydrochloric acid plus an aromatic penetrating agent. High resistance to acid until acid channels through core. Small crystals dislodge from walls and plug pores. (From Chamberlain, *Oil Weekly*.)

Structural, Stratigraphic, and Lithologic Influence on Acidizing

To perform properly an acid treatment of a well, the acid engineer should have a concept of the rock structure and the type and relationships of the rocks involved. The subsurface geologist should be consulted on the problem and should assist in the treatment procedure. Before acidizing the attitude of the beds should be known in order to predict the effect of acidization along bedding surfaces. The type and amount of fracturing should be evaluated in order to avoid water entry or increased gas-oil ratios following the treatment. The opening of a fracture by acidization, a fracture in which water is confined under high pressure, may influence serious production difficulties.

The stratigraphy of a well section should be properly determined before the initial acid treatment. In some cases several porous zones may be present; one may be more permeable than another; consequently, it will absorb the greater portion of the acid, and leave the less-permeable interval unacidized. A permeable sandstone bed in a carbonate section may give similar results. Sections of this type require stage acidation.

Rarely are carbonate-producing sections lithologically homogeneous. Limestones and dolomites commonly contain impurities which affect acidization treatment. Love and Fitzgerald [39] comment that sandy limestones are always treacherous to acidize and on acidizing may occasionally result in production decrease. This decrease is attributed to partial plugging of the pores by sand grains freed by the reaction of the solvent. Many carbonate rocks contain variable amounts of the clay minerals. Upon acidization the clay fraction is released and transported into the minute pores, where it restricts the flow of the acid and retards the outflow of the contained oil and gas. Iron and aluminum compounds occur in carbonate deposits. The reaction between these compounds and the injected acid tend sometimes to create colloidal precipitates which act as plugging agents.

Examples of Acidization

The results of acidization depend upon the character (composition, texture, and structure) of the rock and the applied method of treatment. Several examples of acidization are given.

(1) In the Buckeye field, Gladwin County, Michigan,[40] production is obtained from along modified bedding surfaces and fractures in the Rogers City limestone and from lenticular porous zones near the top of the Dundee formation (Devonian). Acidization is initiated immediately after completion of the well. It is customary to inject from 2,000 to 3,000 gallons opposite the producing intervals. Occasionally as many as 10,000 gallons have been used.

(2) In the Cunningham field, Kingman and Pratt Counties, Kansas,[41]

[39] Love, W. W., and Fitzgerald, P. E., *op. cit.*, pp. 625-626.
[40] Addison, C. C., *Buckeye Oil Field, Gladwin County, Michigan:* Am. Assoc. Petroleum Geologists Bull., vol. 24, no. 11, p. 1981, 1940.
[41] Rutledge, R. B., and Bryant, H. S., *Cunningham Field, Kingman and Pratt Counties, Kansas:* Am. Assoc. Petroleum Geologists Bull., vol. 21, no. 4, p. 518, 1937.

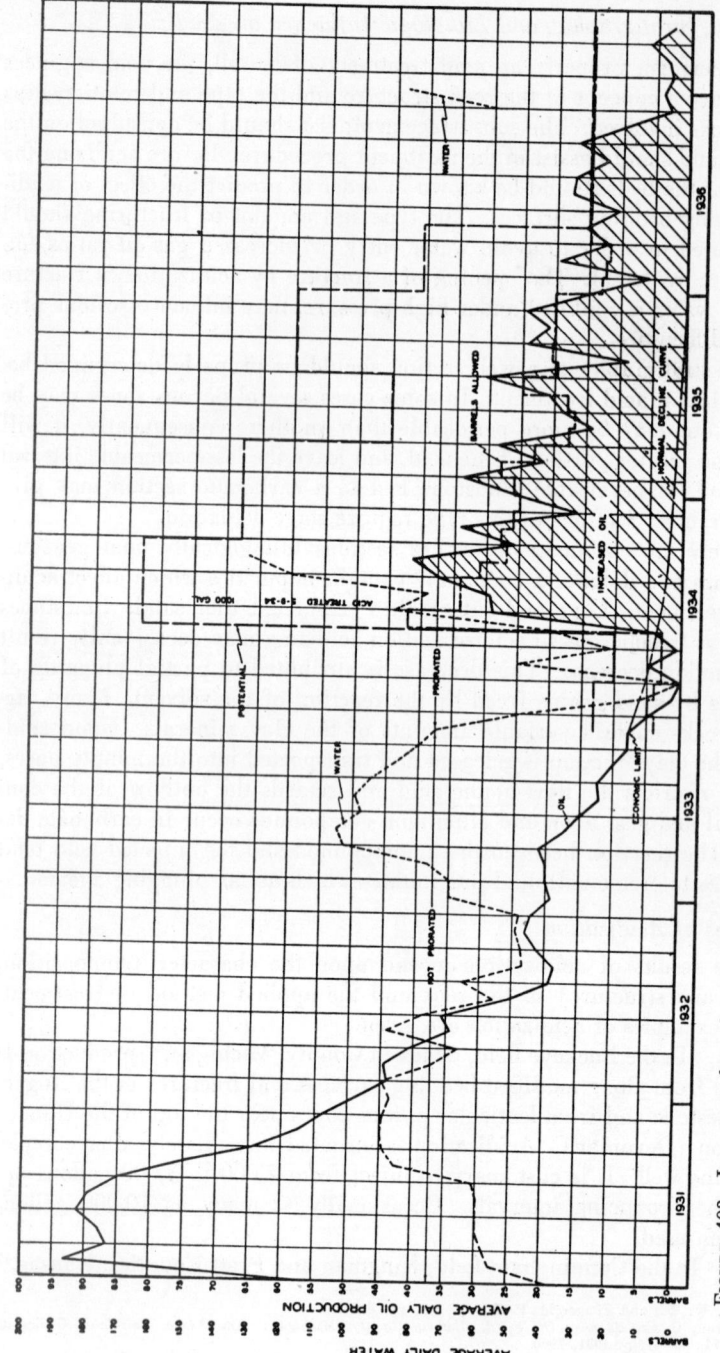

FIGURE 403. Increase of production by acidization, Viola limestone, P. F. Lygrisse "D" No. 1. Greenwich field, Kansas. (From Bunte. Reproduced permission Am. Assoc. Petroleum Geologists.)

the Lansing limestone (Pennsylvanian) is treated just after the first short production test with 2,000 to 5,000 gallons of acid under pressure. The initial production is increased up to 320 percent. The limestone is coarsely crystalline, fossiliferous, and oölitic and contains streaks of chert and shale.

(3) Production from the Viola limestone of the Greenwich field of Sedgwick County, Kansas,[42] is substantially increased by adding 1,000 gallons of acid before potential tests and toward the end of the well's economic limit (fig. 403).

(4) In the Hugoton gas field in southwestern Kansas,[43] the dolomitic-limestones of the Big Blue series (Permian) are generally treated with 8,000 gallons of acid applied in two or three stages. Most wells increase in open-flow capacity of about 200 percent. Nine hundred percent increase has been the maximum.

(5) In the Wasson field, Yoakum and Gaines Counties, Texas,[44] the common acid treatment consists of three stages involving a total of 8,700 gallons. The acidization is carried out in the San Andreas porous dolomite of Permian age. Six to fourteen days are required for the treatment. The chemical used includes inhibitors, demulsifiers, activators, and other physical and chemical modifiers to improve the treatment process.

Methods of Locating Porous Zones

(1) Diamond Coring: Because of high core recovery by diamond coring, porous and fractured intervals in carbonate rocks may be more clearly delimited.

(2) Well Cuttings: Examination of carefully controlled well cuttings frequently afford a means of locating permeable intervals. The details of the permeable intervals, however, cannot be accurately evaluated by this method.

(3) Lost Circulation: Interruptions in the mud circulating system serve as an index to the position of permeable zones. Lost circulation, however, does not indicate a petroliferous interval; water may be the prevalent fluid.

(4) Thermal Data: Thermal-log surveys have frequently indicated hot and cool points in the mud column, a fact which suggests flow of either fluid or gas or both.

(5) Electric Pilot Survey: This method involves covering the exposed well section to be evaluated with an electrical conducting medium (salt water). The fluid is then forced into the formation at a uniform rate by an oil column. The rate of depression of the oil-water interface is then recorded with a current-contact electrode. These data are recorded and the permeable intervals determined.

[42] Bunte, A. S., Subsurface Study of Greenwich Pool, Sedgwick County, Kansas: Am. Assoc. Petroleum Geologists Bull., vol. 23, no. 5, p. 657, 1939.
[43] Barth, G. G., and Smith, R. M., Relative Porosity and Permeability of Producing Formations of Hugoton Field as Indicated by Gas Withdrawals and Pressure Decline: Am. Assoc. Petroleum Geologists Bull., vol. 24, no. 10, p. 1803, 1940.
[44] Schneider, W. T., Geology of the Wasson Field, Yoakum and Gaines Counties, Texas: Am. Assoc. Petroleum Geologists Bull., vol. 27, no. 4, p. 521, 1943.

(6) Spinner Surveys: The spinner survey is based on the measurement of variations in the velocity of fluid being displaced into an exposed formation section. Recordings are controlled by a small propeller-like spinner, which is placed in the fluid column. The spinner increases in r.p.m. when permeable zones are reached as a result of fluid entry into the formation.

(7) Drill-Time Data: Changes in penetration rates frequently reflect intervals of permeability. Close attention is given to sections wherein penetration rates rapidly increase. The introduction of the mechanically operated Geolograph has permitted better definition of porous phases in both carbonate and shale-sandstone sections.

(8) Formation Testing: Systematic testing of the penetrated section by standard formation-testing procedure is by far the most conclusive method for evaluating reservoir fluid characteristics. In addition to yield-

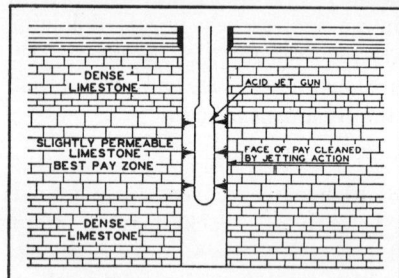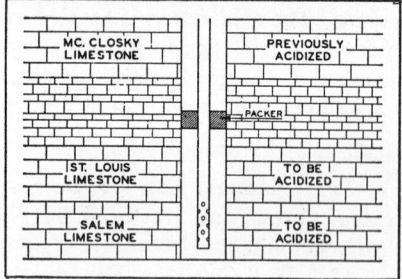

FIGURE 404. Acidizing through the jet gun permits more effective treatments because the acid can be directed into all important pays by setting gun opposite these zones (left). Acidizing below packers is necessary to secure maximum results in treatment of certain types of wells which have been deepened to new pay horizons (right). (From Love, Dowell Incorporated.)

ing relative permeability data, the method also affords a measure of the saturation values of the formation.

Acidizing Procedure

Deeper drilling, selection of thin producing intervals, and increased costs have called for improved efficiency in acidizing procedure. Greater attention is being given the volume of acid to be introduced into the well and to whether or not the acid should be added in one stage or in several stages. Study is being given to the length of.time of acidization and to the reaction and distribution of the acid in various carbonate rocks. Other problems being considered are the plugging effects of the insolubles in low-pressure formations and the point at which the spent acid should be displaced from the hole.

The general technique and operation of well-acidizing follows: The drilling fluid opposite the exposed section should first be displaced with water. The water is then replaced with oil. It is good practice to evaluate the exposed section by formation testing prior to acidizing. This procedure gives an index to the initial productive characteristics of the well, permits formational flushing, and leaves an oil column opposite the exposed section. Formation testing at this stage permits definition of oil-gas and oil-water interfaces which should be established before acidization. Initial flushing also permits removal of drilling fluid present in cavities and solution channels. Following this stage a spinner-type survey may be made. The next step involves acid washing of the hole wall. This stage may be eliminated if bottom-hole pressures are sufficiently high.

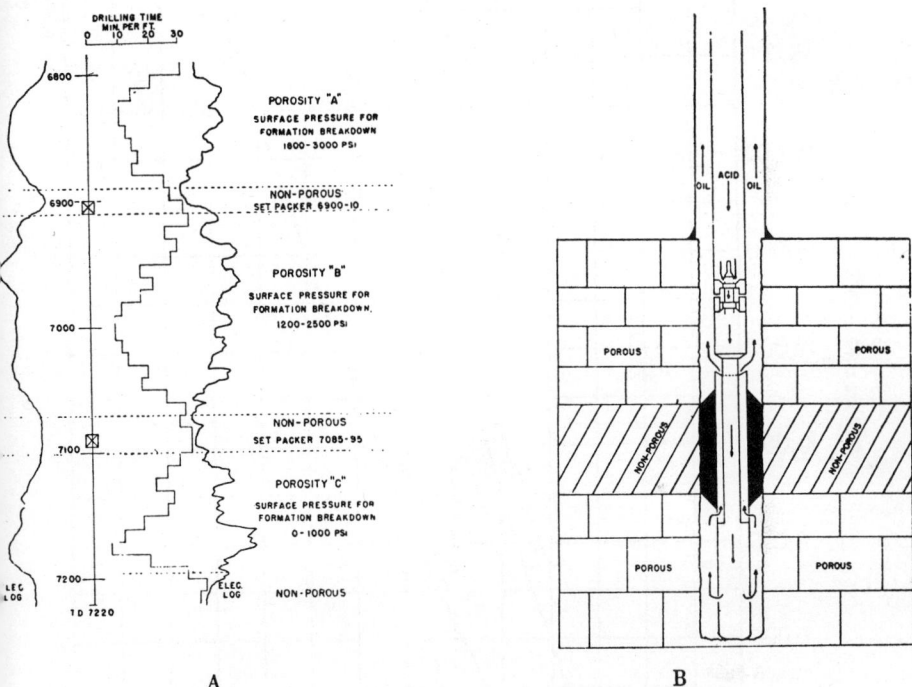

A   B

FIGURE 405. A—Typical producing interval. B—Circulating acid in place opposite formation. (From Moore and Adams, *World Oil.*)

Following the foregoing preparation procedure, the well is then ready for selective acidizing. There are several methods for isolating the individual zones for treatment. One method follows the two-pump procedure of controlling the displacement and injection of the acid. A second method makes use of a packer or packers to control the acid placement and injection. The packer method has the advantage of more positive control of the acid under high injection rates and pressures. Other methods involve acidization through selective gun perforations through solid cemented casing and by use of temporary plastic and gel seals.

Shooting

In certain cases if acidization of a well does not increase production, shooting is initiated. Some operators do not favor this procedure because of the possibility of the many major problems which may arise as a result. Other operators recognize the procedure as routine. Actual examples have occurred where a reduction in production resulted from shooting.

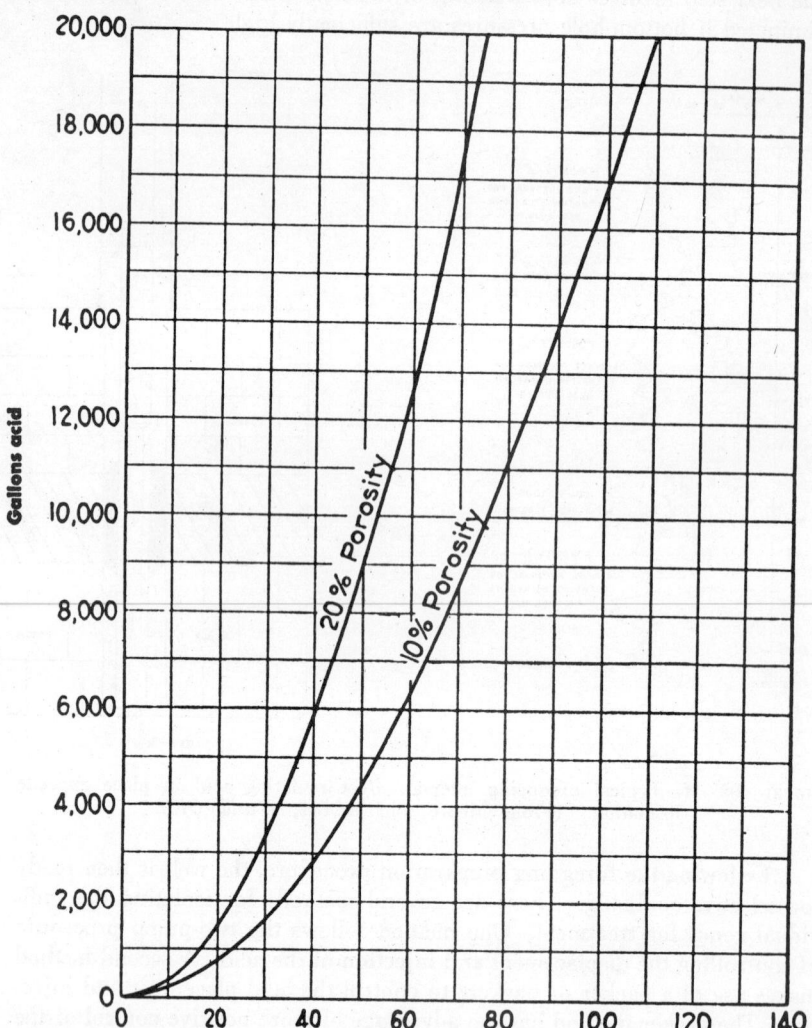

FIGURE 406. Relationship between radius of penetration and quantity of acid used for pays of 10-percent and 20-percent porosity. Radius of penetration in inches for a 100-foot thickness of uniformly permeable pay. (From Moore and Adams, *World Oil.*)

Best shooting results are obtained in sandy limestones, dolomitic limestones, anhydritic limestones, and cherty limestones.

Once a well is shot, there is little opportunity of subsequent remedial work being effective because of the fractured condition of the formation.

In some wells, production is increased after shooting for a short period—then declines rapidly.

Several facts should be considered in shooting of carbonate sections: (1) It is not good practice to shoot closer than 20 to 30 feet from the casing shoe; (2) Formations exhibiting less than 10 percent porosity are

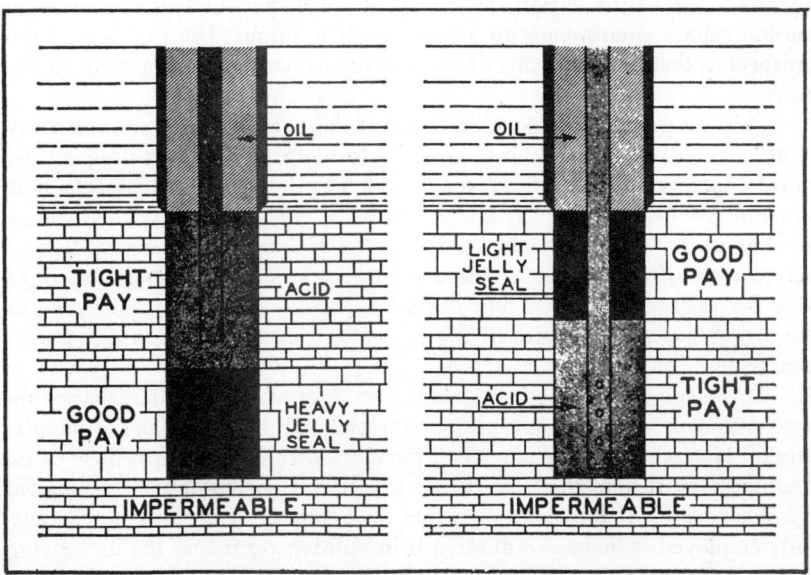

FIGURE 407. Use of jelly seal to permit blocking off a permeable section for acidizing a tighter zone and to direct acid into more productive section. (From Love, Dowell Incorporated.)

rarely shot; (3) The average shooting intervals vary between 150 to 225 feet; (4) Solidified nitroglycerine is considered safer than the liquid form in shooting procedures; (5) Normal loadings range from $2\frac{1}{2}$ to 5 quarts of nitroglycerine per foot of section. Frequently 7 to 10 percent gun cotton is added to increase the shooting power by several percent.

In the North Cowden Field, Texas, "only 24 wells, or approximately 11 percent of the field total, would be commercial producers without the use of nitroglycerine."[45] Most of the clean-up work in this field is accomplished by the reverse-circulation method in which oil is circulated down the casing and upward and out through the tubing to give a high-carrying velocity. In the Turner Valley field of Alberta, Canada, shooting results

[45] Giesey, Sam C., and Fulk, Frank F., *North Cowden Field, Ector County, Texas:* Am. Assoc. Petroleum Geologists Bull., vol. 25, no. 4, pp. 593-629, 1941.

have been disappointing. It is believed that the porous zones are probably too soft to be artifically fractured.[46]

GEOCHEMICAL METHODS
R. MAURICE TRIPP

The use of geochemical techniques and methods to aid in determining subsurface conditions as a means of exploration for petroleum depends upon acceptance of the following philosophy: The occurrence of any substance in the earth's crust in unusual concentrations tends to shift the equilibrium value of the various chemical, physical, and biological factors in the immediate environment to an abnormal position. The problem of the prospector, then, becomes one of recognizing deviations from normal conditions.

It is not practical or possible to detect these deviations from normality in all properties. Some lend themselves to more precise measurement because of more highly refined analytical techniques or instruments now available or because the variations are on a grander scale. A surprising number of the early oil-field discoveries are credited to the recognition of macro-seeps of oil and gas. It is a natural consequence of logic, then, to look for micro-seeps of oil and gas which are not normally detectable to the eye or nose, but require highly specialized instrumentation and analytical techniques.

It is important that one does not lose sight of the fact that petroleum exploration is first, last, and always a geological problem. The accumulation of data, whether it be chemical, physical, or biological, has little or no usefulness until a method has been defined for expressing its geological significance. This does not mean that an exploration aid can not be usefully employed if there is a difference in opinion regarding the underlying theory any more than one cannot drive a car because the origin of oil is not fully understood.

Historical Development

In 1930, V. A. Sokolov, a young Russian nuclear physicist, and his co-worker, M. G. Gurevitch, devised an apparatus for measuring the minute quantities of gas liberated during the radioactive disintegration of certain elements. It occurred to Sokolov that there was a possibility that micro-gas seeps might exist over oil reservoirs, and they made micro-tests for such gases at Grozny and Baku with the appartus designed for nuclear studies.

At about the same time the same idea occurred to G. L. Hassler in this country and to G. Laubmeyer in Germany. These investigators were all concerned with analyzing the free-soil gas for hydrocarbon constituents.

About 1937, E. E. Rossaire and L. Horvitz introduced the technique of taking soil samples in the field rather than gas samples and then, removing the intrained occluded and adsorbed gas in the laboratory.

---

[46] Mackenzie, W. D. C., *Paleozoic Limestone of Turner Valley, Alberta, Canada:* Am. Assoc. Petroleum Geologists Bull., vol. 29, no. 9, pp. 1620-1640, 1940.

It was further reasoned by American investigators that either some of the liquid hydrocarbons might also have migrated to the surface or that some of the gases had been converted by oxidation and by the influence of sunlight into liquid and solid hydrocarbons through a series of condensation, polymerization, and oxidation reactions. Further thinking and field experimentation along this line suggested that there might also be a vertical circulation of formational fluids which would bring higher contents of mineralized waters to the surface.

The anomalous concentration distribution of certain trace-elements and ions which have been observed over oil fields has also been explained as a result of evaporation of the soil solution by passage of the gaseous hydrocarbons.

GENERAL THEORY

The underlying theory for all the geochemical methods and techniques require that some of the hydrocarbon components in the oil and gas reservoir migrate toward the surface by a process of diffusion, effusion, and/or permeation. The actual process of migration appears to be a rather complex one which is not very clearly understood. The occurrence of oil or gas at depth is recognized by the unusual concentration of these once-migrating hydocarbon constituents or by disturbances in other chemical, physical, or biological components at the surface brought about by the action of these migrating hydrocarbons.

The application of a modified D'arcy's equation to determine the rate of flow of gas through the semi-permeable rocks of the sedimentary section yields the interesting result that significant quantities of the gaseous hydrocarbons are migrating toward the surface from a subsurface reservoir. The equation is as follows:

$$Q = \frac{KA}{\mu \, (l+m) \, L_n} \left[p_l^{l+m} - p_n^{l+m} \right]$$

where $Q =$ flow of gas in cubic centimeters per second
$K =$ permeability in D'arcy's.
$A =$ cross-sectional area of flow in square centimeters.
$\mu =$ viscosity of gas in poises.
$m =$ the thermodynamic character of expansion of the gas: for isothermal $m = 1$, for adiabatic $m = C_v/C_p$.
$L_n =$ distance from reservoir to surface in centimeters.
$p_l =$ reservoir pressure in atmospheres.
$p_n =$ partial pressures of gas at surface in atmospheres.

If the permeability be taken equal to 10^{-7} D'arcy's, the value of η for methane at $20°$ C. equals 120×10^{-6} poises; m equals approximately 0.86; L_n, the depth of the reservoir, is assumed to equal 4500' at which the hydrostatic pressure will be equal to 138.5 atmospheres and p_n will be something less than one atmosphere. Then, Q becomes 376 cubic feet of methane per square yard per year at the mean temperature of 53° F. It should be pointed out that the other gaseous hydrocarbons would show a faster rate of flow because of their lower viscosity. It is rather paradoxical that methane has the highest viscosity of all the gaseous hydrocarbons and, consequently, will migrate slower under the assumed conditions. It is interesting to observe that, perhaps, this paradoxical condition concerning the viscosity of the gaseous hydrocarbons has had something to do with determining the high methane content of most natural gases. Since methane is much slower to migrate under the same conditions than the other gases, it would be retained longer and, consequently, would form the major portion of the residual gas in any given reservoir.

A thermodynamic consideration of this migrating gas leads one to some interesting conclusions, as follows. If one assumes that one cubic foot of methane in a reservoir 4500' below the surface at a temperature of 205° F. and a pressure of 138.5 atmosphere migrates toward the surface along a pressure gradient of 0.0306 atmospheres per foot and a temperature gradient of 0.0337° F. where the mean temperature is 53° F. and the mean pressure 0.8 atmospheres, then the condition of state at any depth may be represented by the following equation:

$$PV = nzRT$$

where P = pressure
      V = volume of gas
      n = mol fraction
      z = compressibility factor
      R = universal gas constant
      T = absolute temperature

In figure 408 is plotted the water-vapor content of that same volume of methane, assuming that it is saturated at all depths along its migratory path.

Attention is drawn to the fact that water vapor is being condensed from the gas during the lower three thousand feet. More than 80 percent of the total water which is evaporated from the formations comes from the twenty-five feet immediately below the surface. It is possible that evaporation of these waters might cause the concentration of dissolved mineral matter to such an extent that substances with a very small solubility product constant which occur in the soil solution at saturation levels might be precipitated in the interstices of the rock and soil. This is one reason for determining the variation in concentration of certain minerals and substances of low solubility and more or less universal occurrence in the soil solution. It has been reasoned that substances of this sort may be precipitated and thereby stabilized or fixed in the path of the migrating gases; and, if sufficiently sensitive techniques are devised for their detection, they become indices of subsurface conditions. In some cases, it is probable that they influence the subsequent developments of micro-flora and fauna which may, in turn, become the observable index.

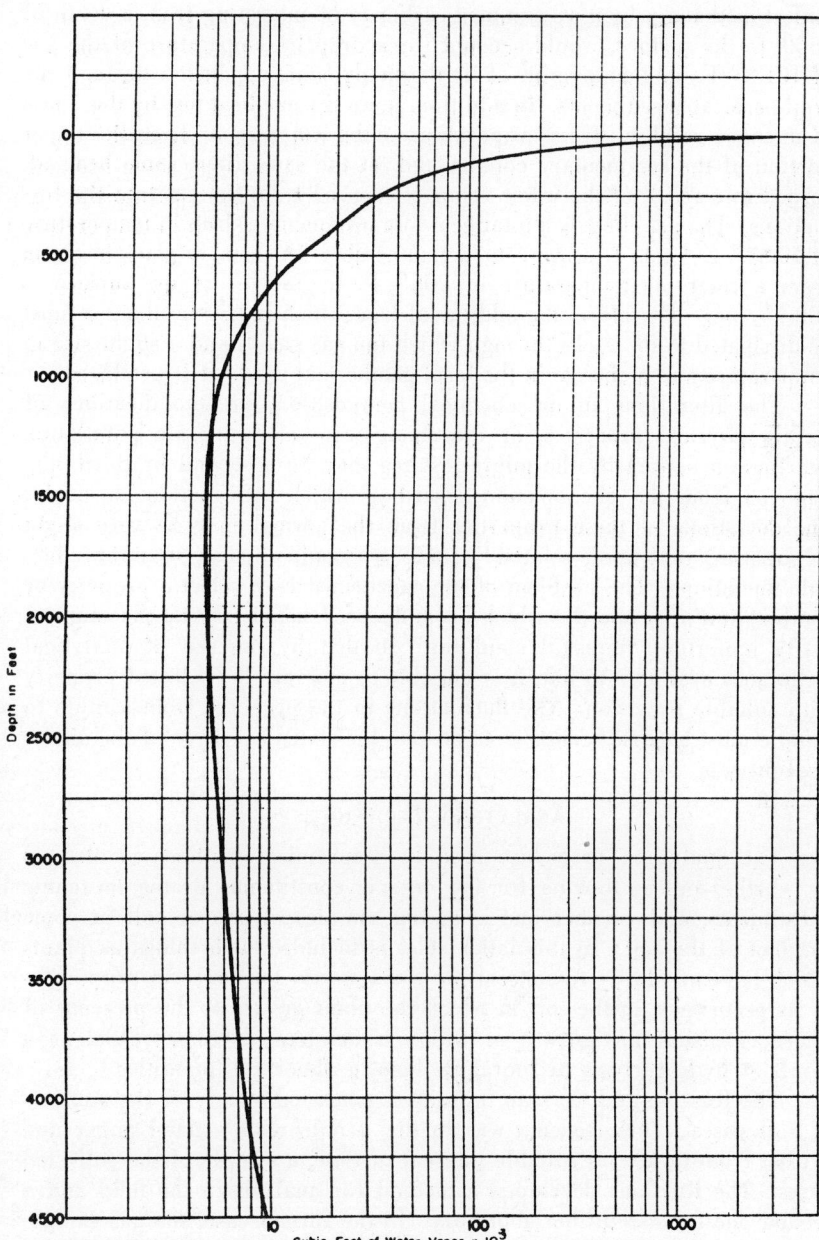

FIGURE 408. Water-vapor content of 0.3045 mols (1 cubic foot at reservoir conditions) of saturated methane under the temperature and pressure conditions at the respective depths.

It is also of theoretical interest to point out that the heat lost by the 1 mol of methane, which we assumed earlier to be migrating from a depth of 4500' to the surface, would account for a drop in temperature of the gas of 101.8° F./mol as a result of its thermodynamic expansion through the semi-permeable sediments. In addition, there is some heat lost by the 1 mol of methane resulting from evaporation of the water vapor from the upper portion of the sedimentary column and, at the same time, some heat adsorped as a result of the water vapor condensed from the gas into the formations. The net effect is a total heat loss producing a drop in temperature of 102.3° F./mol. However, the actual cooling which takes place in going from a reservoir temperature of 205° F. to 53° F. at the surface is 152.4°/mol. Therefore, the additional heat which is lost by the gas must be dissipated to the rocks through which the gas passes and a slight rise in temperature and a change in the local geothermal gradient is produced.

The alterations in the chemical composition, the modifications of certain physical processes, or the changes in micro-organic population distribution caused by the migrating gas may be a second or third step removed from the phenomenon regarding which information is sought. The deviations of these properties from the normal may be very slight or superimposed upon relatively large seasonal, diurnal, or nonpredictable fluctations. The problem of the geochemist is to select a property or combination of properties which are influenced sufficiently by the economically important "unusual condition" so that by the use of analytical techniques available to him it is possible to measure the selected property with suitable precision. The fluctuations in the property from sample to sample must be significantly greater than the statistical error of the analytical process.

ANALYTICAL TECHNIQUES

The analytical procedure used in geochemical methods is dictated by whether one is looking for the organic constituents or the inorganic constituents, and whether one is taking samples of gas or soil or some product of the soil. In this latter class is included such things as plants which have an ability to concentrate a diagnostic substance in proportion to its occurrence in the soil in which the plant grows, or the presence of microorganisms, the growth of which is accelerated or depressed as a result of hydrocarbons or inorganic ions in abnormal quantities.

The Russian and German techniques consisted mostly of the analysis of soil gases. Their scheme was to dig a hole with a hand auger and extract from it, after a suitable interval of time, a sample of the collected gases. The Russians developed a method for analysis in the field and a second one for use in the laboratory. In the former case, the gas sample was passed through a caustic solution to remove the carbon dioxide and then into a combustion chamber. The resulting carbon dioxide was allowed to bubble through a barium-hydroxide solution until the first turbidity was observed. The volume of gas required to do this is compared with that required when a known concentration of carbon dioxide is bubbled through the same apparatus. This comparison is a measure of the total hydrocarbon content of the sample without any indication of the type of hydrocarbon present.

The laboratory technique was a little more refined in that after freeing the gas from carbon dioxide, water vapor, and ammonia, it was circulated through a trap immersed in liquid air. All the constituents other than methane and ethane were condensed in the trap and the two lightest constituents were oxidized. The resulting carbon dioxide was collected and its volume determined with a McLeod gauge. This is referred to as the light fraction. The heavy fraction which is a mixture of complex hydrocarbons and their derivatives with a boiling point above the temperature of liquid air was allowed to volatilize and its total volume determined directly with a McLeod gauge.

The German technique consisted of digging a shallow-bore hole and suitably sealing it against the entry of gas other than soil gas for a suitable period to establish equilibrium, usually from twenty-four to forty-eight hours. The gas was then pumped out of this hole across a heated platinum filament in the presence of oxygen. The oxidation of the hydrocarbons raised the temperature of the filament and thereby, increased its resistance. The change in the resistance was used as a measure of the hydrocarbon concentration.

A modification of the gas analysis scheme was proposed and used by S. J. Pirson. This process endeavors to measure the absolute rate of emanation per unit area of the earth's surface for one or more of the diagnostic gases. The process consists of placing over a specified, confined surface of the sample a suitable adsorbant for the soil gas or gases. This period is usually twenty-four hours. After equilibrium is attained, the adsorber tubes are sealed and taken to the laboratory for degasing, fractionation, distillation, identification, and quantitative measurement.

The soil analysis techniques which were instigated largely by Rossaire and Horvitz consist of taking samples of the soil at different depths, depending upon the kind of analysis to be performed. The usual practice requires samples from a depth of 0 to 6 inches, from the interval between one foot to just below the water table, and from well cores and cuttings. The samples are analyzed for hydrogen, methane, hydrocarbon gases higher than methane, pseudo hydrocarbon liquids and solids, and inorganic ions such as sulphates, chlorides, and carbonates.

The organic constituents are removed by suitable solvents, by low-temperature degasing or by high-temperature degasing. The selection of a solvent to be used for the extraction of the organic constituents depends somewhat upon the nature and composition of the critical constituent. Carbon tetrachloride or other chlorinated solvents are suitable if the soil does not contain appreciable quantities of waxy material derived from plants such as higher alcohols, aldehydes, and ketones. Similarly, benzene and related aromatic solvents can be used in certain instances. Some of the aliphatic solvents have a relatively high solvent action toward the mineral waxes and a low desolving ability toward the vegetable waxes. In most instances it is necessary to prepare a synthetic solvent compounded from a mixture of polar and nonpolar solvents so designed as to have a high selective solvation for a specific compound or group of compounds.

The degasing of the samples consists of placing the sample in a flask and reducing the pressure to about 60 mm. of mercury. The sample is treated with phosphoric acid to decompose the carbonates, and the carbon dioxide resulting therefrom is removed. The flask is heated at 100° C. and the evolving gas is carefully scrubbed and introduced into the analytical apparatus at 10^{-5} mm. of mercury. The analysis consists of condensing the gases in a liquid nitrogen trap at $-196°$ C. The substances which are still gaseous are drawn off, and then the temperature is raised to $-145°$ C. and the resulting volatile constituents removed. A third fraction is made up of the constituents which are liquid at $-145°$ C. but which are gaseous at atmospheric conditions.

The first group consists mainly of air, methane, and hydrogen. The second group contains any ethane, propane, and butane which might be present; and the third fraction is made up of pentane and heavier hydrocarbons. Analysis for the methane consists of burning the gas over a glowing platinum wire, and measuring the volume of the resulting carbon dioxide. The quantities of the other fractions are determined by measuring the volumes with a McLeod gauge before and after combustion. The weights of the different fractions are calculated from the gas laws and the final results expressed in parts per billion.

Several other methods have been used for ultra-sensitive detection and quantitative determination of the hydrocarbon constituents. Important among these are infra-red absorption spectroscopy, fluorophotometry, polarography, and mass spectroscopy.

The fluorescence of certain of the hydrocarbons when activated by ultraviolet light has offered an extremely sensitive method for detecting vanishingly small amounts of these substances. Also, the presence of some of the inorganic, secondary minerals is quickly and readily determined by fluorescence. The details of the fluorescence technique differ considerably with different users. In some instances the sample of soil is exposed to ultraviolet light and the fluorescence determined either by observation with the eye, with a photo-cell, or in the extreme cases by long exposure to a photographic plate. Increasing fluorescence is related to increasing concentration of the sought-after constituent in the sample. Another variation of this technique is concerned with the ultraviolet examination of the solvent extract from the samples. In still other instances, an effort is made to enhance or depress the fluorescence of certain constituents with respect to others in the sample and to compare their relative quantities as well as certain ratios.

Attempts to use the infrared absorption spectrometer for soil analysis has not been too successful when applied to the soil samples proper. Some success has been attained by analyzing qualitatively and quantitatively the solvent extracts and the gaseous components.

The mass spectrometer has been used mostly for the analysis of the soil gases for micro quantities of hydrocarbon constituents.

The polarograph has been adapted to the qualitative and quantitative determination of some of the metallic ions as well as organic constituents in solutions taken from the samples. The techniques vary considerably in the use of this instrument for these types of analyses. Its greatest usefulness appears to be in the determination of the metallic trace-ions.

In many instances, the biological effects caused by the presence of minor or trace amounts of either hydrocarbons or inorganic substances have been relied upon as a basis for the analytical technique. The presence of vanishingly small quantities of certain metallic ions in the soil solution often inhibits or accelerates the growth of some indicator plants. Very often these effects are magnified to such a point that they are easily visible to even an untrained observer. In other instances, the effects are much more subtle and changes in the metabolic rate or cell development of certain tissues in specified plants can be observed only under a microscope. In still other instances, a diagnostic ion may be concentrated in the plant structure or in specified tissues of the plant in proportion to its occurrence in the soil solution, but at much higher concentration values than in the soil.

Sometimes the content of a second element in the plant is caused to vary over a fairly broad range when the primary element concentration varies only a small amount in the nutrient solutions in the soil. Occasionally the variation in the ratio of concentration of certain diagnostic substances is the most indicative.

The macroorganisms are sometimes quite sensitive to changes in the composition of their environment both with respect to metallic ions and the presence of hydrocarbons in the soil gas.

Several methods have been used with varying degrees of success for determining the population of the microorganisms in the soil. One scheme involves the determination of the total organic matter which is oxidized by hydrogen peroxide. Modifications of this method are concerned with the strength of the oxidant used in making the determination for organic matter. The strength of the oxidant determines the class of organic matter included in the reaction.

In some areas it has been practical to determine the population density of certain hydrocarbon-consuming bacteria such as bacillus methaniscus and to compare their occurrence from point to point over the prospect. The theory is that the number of colonies and the size of the colonies of a hydrocarbon-consuming bacteria will be at a maximum in the presence of an abundant food supply. The well-known paraffin dirt of the Gulf Coast is reputed to be the by-product of metabolic activity of certain hydrocarbon-loving microorganisms. The process sometimes works in the reverse direction also. The presence of sulfur or chlorine compounds in abnormal abundance as a result of the secondary-mineralization effects mentioned earlier, or perhaps the presence of sulfur compounds from some crudes, creates a toxic condition that inhibits the growth and development of certain microorganisms.

The preparation of a geochemical well log consists in taking samples of well cuttings and combining them into a composite sample. The vertical extent of the resulting sample will be determined by the amount of detail desired. It may vary from 5 to 100 feet. The usual interval is about a 30-foot section. For most of the procedures which involve the taking of soil samples or formation samples, a 100-gram sample is considered essential. Since the quantities being measured occur in parts per million for the liquids and solids and parts per billion by weight in the case of the gases, the larger the sample that is practical to collect the better and more reliable the analytical routine will be.

CORRECTION FACTORS

It can readily be seen that variations in the inherent characteristics of various sediments and soils such as the crystal structure, the free-surface energy, the interfacial tension, the surface work function, etc., may very well influence the amount of adbsorbed, intrained and occluded gaseous, liquid and solid hydrocarbons, as well as the fixation or stabilization of certain inorganic ions present in the environment. Variations in the particle size and physical condition of the surface of the mineral grains plus an almost infinite range in the combinations of the different minerals composing the various soils and sediments complicates the problem further. Superimposed upon these inherent characteristics are the transient effects of changing barometric pressure, wind velocity and solar radiation, the annual amount of precipitation and the seasonal distribution of that precipitation, infiltration of meteoric water, surface evaporation and movement of the ground waters by capillarity, the slope of the land, the rate of erosion, and the season of the year when the erosion is most intensive, the length of time which the soil has been in its present position as it influences the establishment of equilibrium conditions, and the amount and type of vegetational cover.

Various means are used for minimizing these differences or evaluating their effect so that subsequent measurements can be reduced to a common base for relative comparisons. Without these corrections the geochemical data usually has a rather restricted significance because the quantities in question may vary only slightly above the background.

RESULTS AND INTERPRETATIONS

Horvitz has pointed out that the analytical data collected from a large number of geochemical well logs have indicated the existence of definite relationships between the measured constituents and occurrence of petroleum at greater depth. Wells centrally located over an oil accumulation have low values of hydrocarbon content in the cuttings from the upper portion of the well. At a considerable distance above the accumulation the values begin increasing, gradually reaching a maximum at the reservoir.

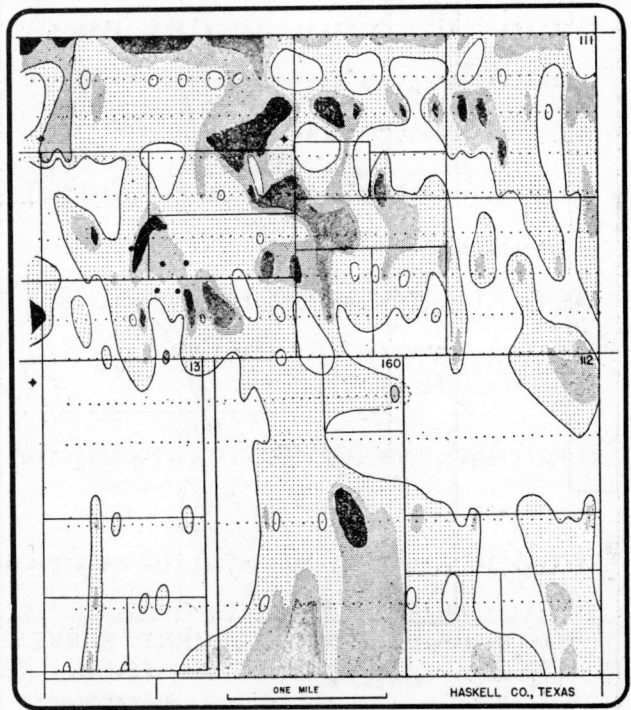

FIGURE 409. Carbon-anomaly map showing distribution pattern of some organic compounds thought to have a unique relationship to petroleum. White area is low concentration, below 15 parts per million. Light stippled area ranges from 15 to 25 ppm; medium stipled area ranges from 25 to 30 ppm; dark area is above 30 ppm. Note that production discovered to date appears to fall on west edge of a halo pattern.

If multiple-producing zones are present, the values decrease abruptly after passing the upper zone and then build up gradually again as the next lower production is approached. The higher the gravity of the oil in the reservoir, the greater the distance above the deposit the increased concentration is observed.

Approximately equal quantities of ethane-propane-butane and pentane plus heavier hydrocarbons is generally indicative of gas-condensate accumulations. The pentane plus heavier hydrocarbons predominate over oil reservoirs.

The hydrogen content reaches a maximum a considerable distance above an accumulation. The amplitude of the hydrogen curve is highest, and the verticle extent of the high values is a minimum near the central

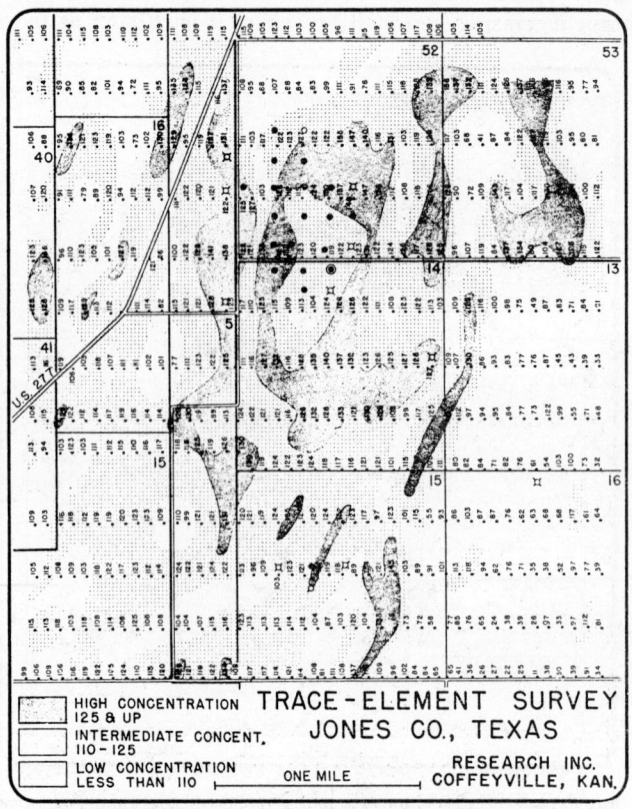

FIGURE 410. Concentration distribution of a group of trace elements in top soil from an area in Jones County, Texas.

part of the accumulation. Near the edge of the accumulation the hydrogen values are lower, but are distributed over a much greater verticle range.

The logs of nonproductive wells which are only a short distance latterly removed from known fields show extremely low hydocarbon concentrations throughout the section.

Analyses for the inorganic constituents of well cuttings have been found to be most useful as an aid in lithologic correlation. Small amounts of chlorides and sulfates often influence the electrical-log interpretation more than their percentage composition would suggest.

The early results of soil-gas analysis obtained by the Russians indicated a high value over the accumulation. American geochemists have observed, except for a few minor exceptions, a ring or halo of high values surrounding commercial production.

Figure 409 shows the results of a survey in which a group of organic constituents were removed from the soil samples by a highly selective sol-

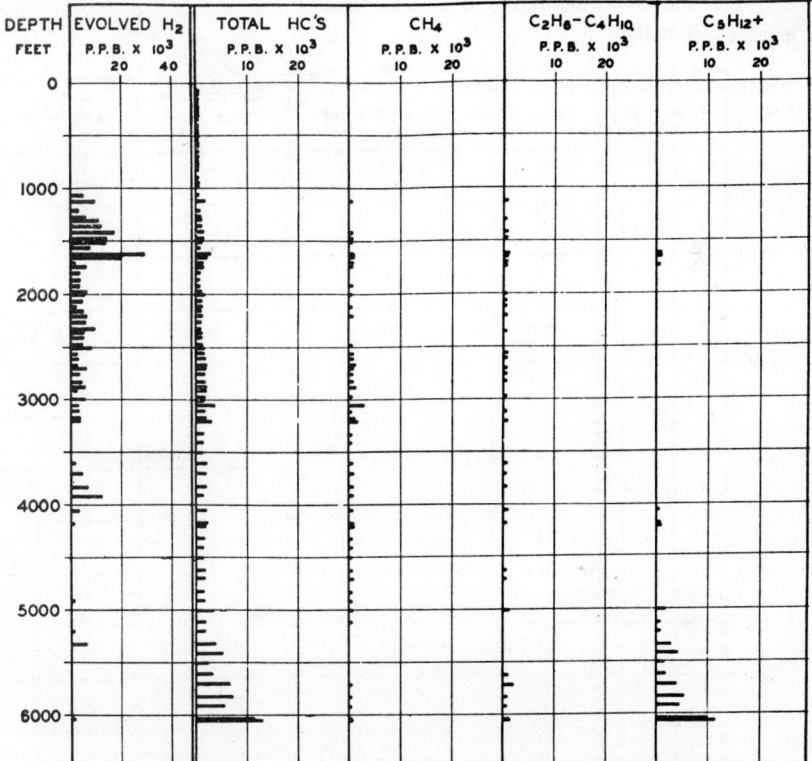

FIGURE 411. Geochemical well log of Godwin B-5 well, Friendswood field, Harris County, Texas. (After Horvitz.)

vent thought to have a unique origin in the surface as a result of subsurface petroleum. The anomalous area has since been drilled with the discovery of a Pennsylvanian reef trap. Seismograph and gravity work was also done across the area, and the combined data was instrumental in the drilling of the exploratory well.

Figure 410 is the result of a survey in Jones County, Texas, in which determinations were made of a group of trace-elements in the soil. Two rather well-defined concentric halos were found in the distribution of the diagnostic elements. This condition was predicted to be the result of multiple producing zones. Subsequent drilling of the anomaly resulted in the discovery indicated. This field is reported to be producing from the Canyon sand at a depth of between 3500 and 3600 feet. The well in the central portion of section 15 has only a small production from the Palo Pinto line at 2640 feet.

Figure 411 is taken from a geochemical well log of a well in the Friendswood Oil Field, Harris County, Texas, from data supplied by L. Horvitz. The oil is approximately 40° Baume' gravity, and produces from a depth of 6040 feet.

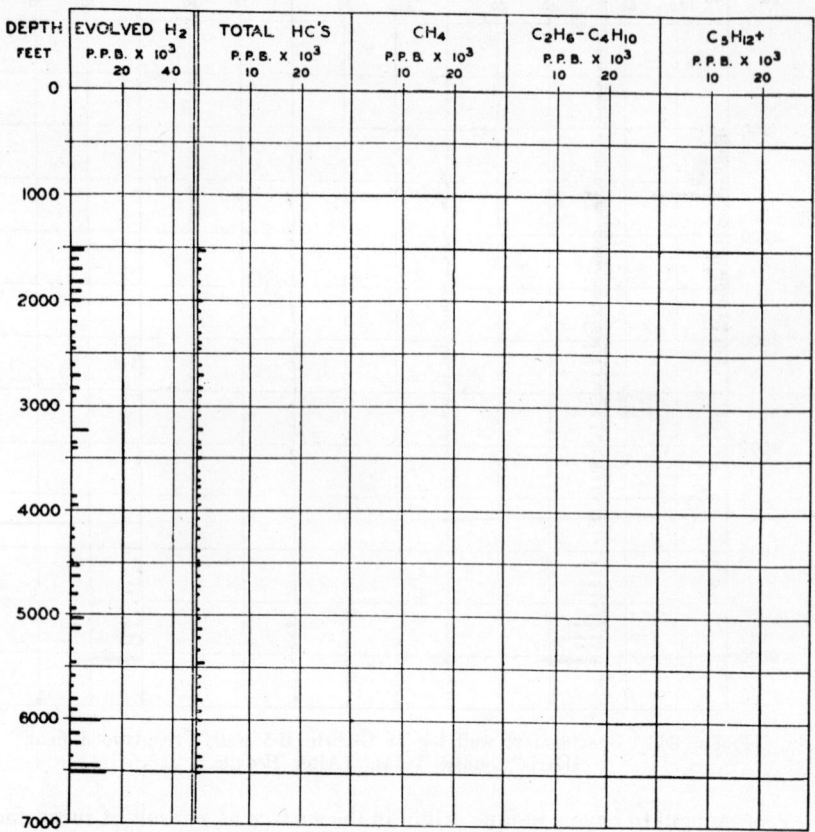

FIGURE 412. Geochemical well log of dry hole, Alemeda area, Harris County, Texas. (After Horvitz.)

FUTURE POSSIBILITIES

Results from basic research in the field of geochemistry and the trend in thinking among geologists who have been using geochemistry as an aid in petroleum exploration indicate that it might be possible in the future to evaluate the petroliferous possibilities of an entire sedimentary basin or province. In other words, by studying the fundamental geochemical characteristics of an area, it would be possible to determine the probability of finding commercial oil by drilling in any of the structural or stratigraphic traps in that particular province. This would be an enormously important step in the direction of focusing exploration efforts on those areas most likely to produce the reserves of the future.

QUESTIONS

1. What is meant by "controlled directional drilling"?
2. What is a whipstock and what is the purpose of running the device in a drill hole?
3. Refer to figure 239 ff. for operation of the whipstock.
4. What is the recommended drift in a directionally drilled well?
5. Directional wells are of what two types?
6. What are the applications of directional drilling?
7. In well surveying what three basic types of instruments are used? What are the principles of each?
8. For what reasons are wells surveyed?
9. What is the basic principle of magnetic-core orientation?
10. What geologic variables must be considered in a magnetic-core orientation?
11. Discuss the practical applications of magnetic-core orientation.
12 How are oriented cores taken with the conventional-core barrel? With the wire-line retractable-coring equipment?
13. Upon what principle are dipmeter surveys based?
14. How are dipmeter levels selected in a well?
15. Give several applications of dipmeter surveys.
16. What are the five general types of coring procedures?
17. What are the advantages of conventional coring? Disadvantages?
18. What are the advantages of diamond coring? Disadvantages?
19. What are the advantages of wire-line coring? Disadvantages?
20. What are the advantages of reverse-circulation coring? Disadvantages?
21. Give the advantages and disadvantages of side-wall coring.
22. How may the geologist use coring data to advantage?
23. How does information obtained from cores assist in the evaluation of the electric-log profile?
24. What is the basic geometry of tricone bits and what is the purpose of such construction?
25. What are the advantages of using several drill collars during drilling?
26. What is the optical system and general principle of operation of the deep-well camera?
27. Two types of deep-well-camera surveys are made. Discuss each.
28. The electric pilot is used for what three purposes in oil-field completions and remedial work?
29. What are some of the reservoir and well conditions under which selective acidizing should be considered?

30. How are water-locating surveys made and for what purposes?
31. What are the steps followed in applying a Hydrafrac treatment to a formation?
32. What factors should be considered before applying the Hydrafrac treatment to a formation?
33. In what areas has the Hydrafrac treatment proved successful?
34. What is meant by well "acidization"?
35. Define primary, secondary porosity?
36. What methods are used in locating porous intervals in carbonate sections?
37. Briefly outline two methods used in acidizing a well.
38. What is the purpose of shooting a well? Why should precautions be taken in this procedure?
39. What are the functions of a drilling fluid?
40. What are the major properties of drilling fluids?
41. Why is sand in drilling fluids undesirable?
42. The viscosity of a drilling fluid depends on what factors?
43. What is the average fluid density of a drilling fluid? What mineral is commonly added to drilling fluid to increase its weight?
44. What is meant by "wall cake"? What property of a drilling fluid controls the development of this cake?
45. Drilling fluids may be adversely affected by what factors?
46. Comment on the term "thinners" as applied in chemical treatment of a drilling fluid.
47. What are the two classes of compounds which are effective in chemically treating a drilling fluid?
48. Does cement unbalance a drilling fluid? Why?
49. How does temperature effect the properties of drilling fluids?
50. List five types of drilling fluids and give reasons for their particular uses.
51. What is meant by "formation testing"?
52. What is the purpose of a "choke bean" in a formation tester?
53. What is a "cone packer," "wall packer," "double wall packer"?
54. What applications does hook-wall packer testing have?
55. What is the purpose of a high-pressure control head in making a formation test?
56. What is the most desirable hole diameter for formation testing?
57. What are the usual steps followed in cementing casing?
58. From a mechanical view point what points are of vital concern for a good casing cementing job?
59. What features should a geologist be concerned with during a casing cementing operation?
60. What materials are added to a cement slurry to prevent loss of circulation?
61. What are two types of wall-cleaning devices and for what purpose are they used?

CHAPTER 7

SECONDARY RECOVERY OF PETROLEUM

PAUL D. TORREY

The following definition of secondary recovery by Johnson and van Wingen [1] is used for the purpose of this paper:

Recovery by any method (natural flow or artificial lift) of that petroleum which enters a well as a result of augmentation of the remaining native reservoir energy (as by fluid injection) after a reservoir has approached its economic production limit by primary recovery methods.

It is appreciated that there is a growing tendency to include all water- and gas-injection operations, regardless of the time during the life of a reservoir in which they are commenced, within the scope of secondary recovery. Some change in definition, therefore, may be adopted generally in the future and, in fact, is recognized by certain authorities at present.

From the foregoing definition it will be evident that there are three fundamental requirements for the application of secondary methods: (1) an injection well, (2) a producing well or wells drilled into, (3) a common porous and permeable oil-bearing formation, through which liquids or gases are forced under artificial pressure.

This paper considers the history of secondary-recovery operations in the United States, the various methods that have been employed to increase recovery from oil fields in which the primary reserve has become depleted, secondary oil reserves in the United States, the susceptibility of oil fields to the application of secondary methods, and the costs of development and operation of secondary projects and some of the results that have been obtained. A brief review of secondary operations in the Rocky Mountain states is presented.

The oil that may be produced by secondary methods does not require discovery, and the application of these methods serves to make available more of that which has already been found. For this reason, the additional oil which is so obtained does not have to return exploration and leasing expenses to yield a profit. The economic opportunities for secondary recovery, consequently, are sometimes more attractive than in the development of primary production. However, for both primary and secondary production, the amount of oil that can be obtained and the development and operating expenses will determine the success or failure of a project. Improvements in oil-recovery technology and increased prices for crude oil are just as important in secondary operations as they are in primary

[1] Johnson, Norris, and van Wingen, Nicco, *Glossary of Terms and Definitions Pertaining to Secondary Recovery Operations*, submitted to Standing Subcommittee on Secondary Recovery Methods, Am. Petroleum Inst., Oct. 20, 1948.

production and may enable the working of inferior reservoirs at a profit; in some instances, a successive reworking of a field is made possible. In this respect, secondary-recovery operations resemble the mining of ores, where an improvement in production methods or an increase in price will permit profitable recovery of lean or more difficult deposits.

Secondary-recovery operations, although yielding only a small part of the oil production of the United States, nevertheless are an important factor in the business of oil production in several states. To a large extent they are confined to the older fields, where their use has maintained the production of oil far beyond the time when the natural decline of the wells would have enforced abandonment. Therefore, the application of secondary methods may be regarded as a true conservation measure, resulting in an increased recovery of oil that otherwise could not be obtained profitably and in many cases in the preservation of natural-gas reserves which might be dissipated.

HISTORY OF SECONDARY-RECOVERY OPERATIONS IN THE UNITED STATES

The use of compressed air or air-gas mixtures, gas, steam, water, and other suitable fluids to increase the recovery of oil is about as old as the art of removing oil from the earth. Shortly after oil was discovered in Venango County, Pennsylvania, patents were issued covering mechanical devices and techniques designed to stimulate the extraction of oil from underground reservoirs. Many of these patents were based on inventions involving the creation of a vacuum in producing wells or the injection of fluids into the reservoir by means of input wells equipped especially for this purpose.

The first recorded use of vacuum was in the Triumph pool, Pennsylvania, in 1869. It has been well-established that an attempt to apply a combination of vacuum and gas repressuring was made in Clarion County, Pennsylvania, in 1895. This project was unsuccessful on account of the tightness of the producing formation. However, shortly thereafter, combined vacuum and repressuring were employed successfully in Venango County, Pennsylvania.

The first known intentional injection of gas into oil-bearing rocks to increase production was accomplished by James D. Dinsmoor on the Benton farm, Venango County, Pennsylvania, in 1890. In this operation gas from a lower formation was introduced into the Third Venango oil sand at a pressure of about 100 pounds per square inch. The production of oil was more than doubled. Subsequently, repressuring operations, which were sometimes combined with application of vacuum, were carried on by Dinsmoor in the vicinity of St. Marys, West Virginia, and in other parts of Venango County.

In 1911 I. L. Dunn commenced the historic air- and gas-repressuring operations in the Chesterhill field of southeastern Ohio, which have been described so ably by Lewis.[2] Only a few years later, in 1917, repressuring operations were started in the Midcontinent region (Nowata County, Oklahoma), and subsequently the injection of air and gas has been employed to varying extent in practically all of the important oil-producing regions of the United States.

Until about 1935, water-flooding operations were far more restricted geographically in the United States than air- and gas-repressuring projects. However, as a source of crude-oil supply water flooding has been of great importance in certain of the Eastern states, where a greater part of the production of oil is derived from the application of this method. It is reasonably certain that intentional water flooding was initiated in the Bradford field, McKean County, Pennsylvania, and Cattaraugus County, New York. Effects of increased production from this source were first noted in 1907, although it is believed that floods were being operated secretly prior to that time. Because of the clandestine nature of most of the early water-flooding operations in northern Pennsylvania, little detailed information has been preserved on the results that were obtained, such as is available on the early air- and gas-injection operations in Pennsylvania, West Virginia, and Ohio.

Unsystematic water floods were commenced in Nowata County, Oklahoma, in 1931, which were followed, in 1934, by a systematic operation developed similarly to the methods employed in the Bradford field. The latter project established the effectiveness of water flooding in the Bartlesville sand, and in succeeding years it has been followed by ever-expanding secondary-recovery activity in the Midcontinent and Southwestern regions.

Mining for petroleum is the most ancient known production method. The early mining operations consisted of the enlargement at the surface of natural seepages, and of shallow pits or short drifts into the outcrop of oil-bearing sands, from which the accumulated oil could be removed by bailing. More recently, actual underground mining for petroleum has been practiced in France, Germany, and Japan, in fields where primary production has declined to a low level.

There have been several attempts to mine petroleum in the United States, the most recent of which are projects in Miami County, Kansas, at Richards, Missouri, and at Rocky Grove, Venango County, Pennsylvania. These operations differ from the European mines in that no drifts or cross cuts have been dug into the oil-bearing formation, but rather the sand has been penetrated by a series of horizontal holes drilled in a cartwheel pattern back away from the central shaft. None of these recent operations in the United States has been economically successful, but the Pennsylvania mine did provide a great deal of valuable scientific and technical information.

[2] Lewis, J. O., *Methods for Increasing the Recovery from Oil Sands:* U. S. Bur. Mines Bull. **148**. Oct. 1917.

SECONDARY METHODS OF OIL RECOVERY

Vacuum

Although the application of vacuum is not regarded as a secondary method, according to the definition accepted in this paper, brief mention of vacuum is justified because it was one of the first methods to be applied purposefully to wells to increase the production of oil. Vacuum consists of creating a pressure differential in the annular space in an oil or gas well by applying suction, thereby causing a movement of liquid and gaseous hydrocarbons toward the well bore. The amount of pressure differential that can be created by the application of vacuum is, of course, very limited in comparison to the injection of gas under pressure into the reservoir. It was soon recognized that production benefits resulting from the use of vacuum were not permanent, but it was maintained on many properties because of the increased richness of casing-head-gasoline production.

Air and Gas Injection

Gas repressuring, in contrast to pressure maintenance, usually is applied in a field when the point of depletion by primary methods of production has been reached and oil no longer can be produced profitably. For this reason, in the older fields, many repressuring projects have a very humble beginning; and the engineer who may be called upon to supervise the secondary operation frequently will be confronted with a dismaying collection of antiquated and worn-out equipment, junked holes, wells in bad condition, and poor records.

Unitization of the field for air- and gas-injection operations generally is desirable and will result in lower development costs, maximum economy of operation, and better control of reservoir performance.

In northwestern Pennsylvania, where air- and gas-injection operations have been conducted on an extensive scale for many years, it is common practice to use old holes for producing wells and to drill and core new holes for injection operations. Wide variations in permeability, which many times are present, are controlled by segregation of the sand body into two to five sections. This segregation is accomplished by setting packers in such position that the air or gas can be injected into each sand section separately and under different pressures.

Usually an attempt is made to locate the intake wells so as to form as symmetrical a well-spacing pattern as possible in relation to the producing wells. However, more frequently the well patterns are irregular and the well spacing is variable. In some of the more recent, intensively developed projects in Venango County, Pennsylvania, the intake wells are located in the center of a hexagon formed by six producing wells at the corners. The common distance between the intake and producing wells is from 150 to 250 feet.

As a secondary-recovery operation, the injection of gas into a partly depleted oil reservoir is essentially a continuous circulation of the gas into and through the producing formation. Since the gas is a nonwetting phase, it will pass through pore channels already opened by the previous removal of fluids from the reservoir and will tend to move the remaining oil by viscous drag rather than by direct displacement, such as takes place by the action of an expanding gas cap.

The rate of movement of a gas drive will be proportional to the pressure gradient established in the reservoir. The movement of the gas through the reservoir serves continually to reduce the oil saturation and results in a very rapid increase in permeability to gas, producing what may soon become a prohibitively high gas-oil ratio. This effect is responsible for the relatively low efficiency of the gas-drive recovery process, and limits definitely the amount of oil which can be obtained by gas injection.

The chief justification for the injection of gas is that it will make available additional oil at a commercial rate that probably would not be obtained otherwise because of the low rate of production which prevails toward the end of the primary-production phase. In other words, gas injection serves to accelerate recovery during the late life of a field by retarding the normal rate of decline.

The history of air and gas repressuring in many parts of the United States is very similar. When primary production of oil by conventional pumping methods is no longer economically justifiable, the operator must make some change in practice or abandon his property or field. Since a minimum investment in new wells and production equipment is generally required for the injection of air or gas, repressuring by these fluids has been found to be the cheapest method for maintaining or increasing the production of oil. As a result, profitable production can be maintained, and the ownership of the working interest in wells, properties, and possibly entire fields preserved to the producers until conditions may become more opportune for the application of some other secondary method, such as water flooding.

Water Flooding

The process of water flooding consists in applying water under pressure to an oil-bearing formation by means of specially equipped intake wells. It has been most successful in fine-grained, tightly-cemented sands, which are frequently characterized by high residual oil content after the primary phase of production. In the initial stage of water flooding an oil bank is formed ahead of the advancing water if the mobility [3] of the

---

[3] "Mobility" is defined as the effective permeability of a reservoir rock to the fluid phase divided by the reservoir viscosity of that phase.

oil concurrently is greater than that of the water. The initial stage is followed by a viscous drag stage, at which time the permeability of the reservoir surfaces to oil is greatly reduced and the permeability to water is greatly increased, resulting in high produced water-oil ratios.

Most water-flooding operations are developed on what is known as the "five-spot" pattern, with the producing oil well located in the center of a square formed by water-input wells at the four corners. Other patterns have been employed to a lesser extent, but the theoretical flooding efficiencies of the various patterns are so close that the convenience of the "five-spot" usually encourages its use.

The spacing between water-input and producing wells has been determined in the past largely by long experience. In some of the earlier "five-spot" developments in northern Pennsylvania the distance between input and producing wells was as low as 150 feet. More recently, distances of from 225 to 250 feet have become common practice, which results in a considerable reduction in development expense. Experience in the Bradford field (Pennsylvania) has shown that there has been no appreciable decrease in the oil recovery obtained from the wider-spaced floods, although a longer period of time has been required to obtain the total recovery. However, in many fields the lack of continuity of individual beds of the reservoir rock might cause the trapping of considerable oil if much wider spacing should be employed, and for that reason there seems to be rather definite practical limitations to further expansion of well-spacing patterns.

The time required to deplete the wider-spaced flooding projects, of course, can be reduced by the use of higher injection pressures. However, there is a limit to the pressure that can be applied because of the tendency of the rocks to break or rupture under excessive pressures. It has been found that pressure-parting of the rock can be avoided if the bottom-hole injection pressure does not exceed about 1.25 pounds per square inch per foot of depth. Variations from this breakdown pressure can be attributed to differences in the strength and rigidity of the producing formation and the overburden.

Delayed drilling of producing wells for a predetermined period after the injection of water into the reservoir has been commenced has resulted in a much improved recovery of oil where a wide range of permeability exists. However, the water must be introduced into the sand at a balanced rate through each intake well so as to prevent an off-center concentration of oil within the pattern.

When the producing wells of some of the first delayed floods were drilled, it was found that they would flow on account of a buildup of pressure in the reservoir. This discovery immediately suggested that complete secondary recovery might be obtained by flowing if some back-pressure could be maintained on the producing wells. Flowing secondary production has been practiced successfully on a few properties in Pennsylvania and on several properties in Oklahoma and Kansas where permeability and viscosity relations are favorable for the injection of substantial quantities of water. The advantages of flowing over conventional production by pumping are found in the economy of cost and in the economy and simplicity of operation. Experience with flowing projects indicates that there is no loss in ultimate recovery under comparable conditions.

Mining

Except for the operation at Rocky Grove, Pennsylvania, attempts at oil mining in the United States have been on such a small scale that no standard technique has been developed similar to some of the foreign operations. The information that has been obtained from the United States mines, therefore, has not been adequate to determine whether mining operations might be economically feasible in this country. Data on the Rocky Grove operation have not been released, but it may be assumed that it was a failure because of cessation of the development program.

No attempt was made at Rocky Grove to excavate the oil-bearing rock as has been done in various European operations, where a recovery of 40 percent of the original oil content of the reservoir has been reported. Such recovery is definitely superior to most of the secondary-recovery operations in the United States, and encourages the belief that the feasibility of mining for petroleum in this country clearly justifies further investigation.

SECONDARY OIL RESERVES

The estimation of secondary oil reserves in the United States that can be made at the present time can be regarded only as a first approximation, and, undoubtedly, will be revised from time to time just as estimates of the total proved primary reserves are revised. In the aggregate, figures which can be developed currently on secondary reserves are probably conservative, but, specifically, they may be somewhat in error, because it is not yet possible to break down and classify all estimations of secondary reserves as proved, probable, and possible. The secondary oil reserves of the United States are estimated currently to be in excess of 7 billion barrels, but the author is willing to venture the guess that the physically recoverable secondary oil reserve of the nation may be as much as twice this figure.

Undoubtedly, important secondary reserves exist in other oil-producing countries. So far as the author has been able to determine, no systematic effort has been made to evaluate the magnitude of foreign secondary reserves.

SUSCEPTIBILITY OF OIL FIELDS TO SECONDARY METHODS

From the time of the earlier secondary operations in the Eastern states it has been recognized that all oil fields are not adapted to the application of secondary-recovery methods. Consequently, it is just as important to provide information which will assist in avoiding failures as it is to encourage the use of secondary methods in fields where the probability of success is high. Some of the things that will control the success or failure of secondary-recovery operations, and some of the factors which should be considered in the formulation of plans for a secondary recovery project are discussed briefly hereafter.

It is notable that most of the successful secondary-recovery operations in the United States are restricted largely to fields where primary recovery of oil has been obtained by the action of dissolved-gas drive. Fields in which active natural water encroachment has been effective are generally not so well adapted to secondary-recovery operations after the primary-recovery phase, on account of low residual-oil content. It is also pertinent to emphasize that better reservoir performance efficiency in certain of the more recently discovered dissolved-gas-drive fields should eliminate, in part, the necessity for future application of secondary methods. If oil had been produced efficiently in many of the older fields, it is doubtful whether there would be such a high present activity in secondary-recovery operations. Therefore, secondary recovery, in a certain sense, is a salvaging operation which may be partly avoided in the future by the more efficient development and operation of oil fields. In this, studies of primary pressure control, undoubtedly, will play a very important part.

It will be obvious that geologic factors will have an important bearing on the adaptability of an oil field to secondary-recovery operations. In order to evaluate these factors properly, all available subsurface geologic information and production data should be plotted on a structure map, showing the configuration of the top of the oil-producing formation being investigated, and on an isopachous map of the pay zone. A number of graphic cross-sections of the pay zone, both parallel and at right angles to the long axis of the field, will assist greatly in the interpretation of the subsurface geology, and will permit taking better advantage of all favorable geologic features which may be utilized for the improvement of oil recovery. Homogeneity and continuity of the reservoir rock will promote uniform movement of injected fluids, and the depth and thickness of the oil-bearing formation will have an important influence on the economics of the recovery operation. The distribution of gas, oil, and water in the reservoir rock is most important, and must receive careful consideration. Faults, which may seal off segments of the field, may seriously obstruct uniform fluid movement. Where the rocks are horizontal or where the rate of dip is low, the effects of structure may be disregarded in the design of a secondary-recovery operation. However, where oil and gas accumulation has been controlled by a steeply dipping anticline or is associated in any way with steeply dipping beds, the injection of fluids with reference to structural position is most important. Because of differences in gravity,

Homogeneity and continuity of the reservoir rock will promote uniform movement of injected fluids, and the depth and thickness of the oil-bearing formation will have an important influence on the economics of the recovery operation. The distribution of gas, oil, and water in the reservoir rock is most important, and must receive careful consideration. Faults, which may seal off segments of the field, may seriously obstruct uniform fluid movement. Where the rocks are horizontal or where the rate of dip is low, the effects of structure may be disregarded in the design of a secondary-recovery operation. However, where oil and gas accumulation has been controlled by a steeply dipping anticline or is associated in any way with steeply dipping beds, the injection of fluids with reference to structural position is most important. Because of differences in gravity, it is preferable to inject gases at high structural position and liquids at low structural position in the reservoir. The shape and geologic pattern of the reservoir, developed by the conditions controlling the deposition of the sediments involved, and the presence of shale partings or beds of low or negligible permeability, may have very definite influence on the locations best suited for producing and injection wells.

The permeability of the reservoir rock will influence the distance between wells, the well-spacing pattern, and the pressures which must be exerted to promote the effective movement of fluids through the reservoir. The principal use of permeability determinations is for the prediction of rates of flow through the reservoir. In secondary-recovery operations fluids are injected into the reservoir and other fluids are withdrawn from the reservoir. The control of the amount and rate of fluid injected and produced is very important in order to insure the maximum economic recovery of oil, for in order to produce oil from a reservoir in which the original energy has been dissipated, it must be displaced and moved by extraneous force. Control of injection rate, therefore, is one of the factors which help to control recovery.

Although uniformity of the permeability profile is, admittedly, a most desirable condition, secondary-recovery operations have been conducted successfully in formations having a wide permeability range. Methods have been developed which enable the plugging off of zones of extraordinary permeability, and a concentration of oil around the periphery of producing wells can be induced by a delay in their drilling until the voids of the producing formation approach complete fluid saturation.

It should be evident that the amount of oil that may be recovered from a partly depleted sand by secondary methods, using either gas or water injection, is dependent upon the amount of oil remaining in the sand. Of almost equal importance is the amount of gas or the amount of water which will have to be injected in order to obtain a given amount of oil. Observations in both the laboratory and the field have demonstrated that when a sand contains less than 20 percent oil saturation, practically no oil will flow through it. It is, therefore, virtually impossible to obtain oil by known secondary methods from sands possessing a residual oil content of 20 percent or less of the effective pore space, and for this reason secondary-recovery methods have not been successful in some fields, even though the total oil content of the reservoir may amount to many thousands of barrels per acre. As a practical matter, the recoverable oil may be estimated by subtracting the minimum saturation determined by laboratory tests from the total saturation, with a correction allowance to account for lenticular-sand conditions, heterogeneous permeability, or other factors which may affect the uniform displacement of the remaining oil.

Experience derived from studies of fluid saturations of oil-bearing rocks has shown that the higher the oil saturation, the larger the percentage of oil in place that is recoverable, and this is a factor that would encourage early injection of gas or water into the reservoir to insure the maximum economic recovery.

One of the chief physical quantities entering into the process of recovery of oil by secondary methods is the viscosity of the crude oil. Unfortunately, all of the effects that a change in viscosity may have on the rate of flow of fluids from an intake to a producing well cannot be predicted, for the change may not be uniform throughout that part of the reservoir in which a differential in pressure has been created. However, at any particular point in the reservoir the rate of flow of a fluid phase is inversely proportional to its viscosity, if the pressure gradient and permeability to the same phase remain constant.

The time required for any given recovery process, therefore, will depend on the viscosity of the crude oil, and an increase in viscosity may result in a corresponding diminution in total recovery. The viscosity of crude oils which have responded successfully to secondary-recovery operations has ranged from two to thirty times the viscosity of water.

The phenomena of capillarity and surface tension are important factors in determining the efficiency of the recovery processes of gas and water injection into oil-bearing formations. Oil is held within the pores of the reservoir by the action of capillary forces and it is also adsorbed on the reservoir surface as a film. Much remains to be learned about the effects of these phenomena on secondary-recovery operations, but some conclusions are rather generally accepted.

Displacement of the oil from the reservoir involves the disturbance of a state of equilibrium that exists between the interfacial tensions of four phases: gas, oil, water, and solid minerals; and the fluid with the greater adhesion tension with reference to the solid phase will tend to displace the fluid with the lesser adhesion tension.

In most petroleum reservoirs it is believed that both oil-wet and water-wet surfaces exist, with the latter predominant except in a few fields. It is likely that hydrophobic sands will not respond successfully to water flooding.

Capillary forces, undoubtedly, affect the permeability of the oil-bearing formation to a fluid phase and, as such, are important in determining the most advantageous rate of advance of the injection medium.

Artificial control of interfacial tension between water and oil by use of surface-tension depressants in the water has so far failed in practical application, because the surface-active compounds tend to be absorbed on the solid surfaces. The advancing water front, therefore, is depleted of these agents before any beneficial effects can be observed.

Many of the factors which will control the success or failure of secondary-recovery projects can be determined in advance by laboratory analysis of cores and from studies of reservoir conditions. Indicated recoveries, as obtained from laboratory flooding tests, have proved to be remarkably accurate if reliable relative permeability data are available. Gas and water saturations are particularly important in ascertaining the ratio of injected fluid to oil produced. If the initial mobility of the water is greater than that of oil, water flooding will probably not be economically feasible on account of excessive water-oil ratios.

Information of great value in the formulation of plans for the development and operation of secondary-recovery projects can be obtained from small-scale pilot-plant tests of gas- or water-injection capacities and pressures. Such tests can be made usually at moderate expense with portable compressors and pumps and using the existing production facilities. The information so gained provides a useful check on laboratory determinations.

SOME COSTS AND RESULTS OF SECONDARY-RECOVERY OPERATIONS

Development and operating costs in secondary-recovery operations are so variable that any figures presented must be specific in order to avoid misinterpretation.

Present-day costs (1949) for the complete development of water-flooding projects in the Bradford field of northern Pennsylvania will range from $4,000 to $5,000 an acre, depending on the drilling depth.

The two earliest water floods in north Texas have had an over-all development and operating expense, adjusted to prices prevailing during the early part of 1948, of $1.40 and $1.47 a barrel respectively. Similarly adjusted average costs, which include land cost or over-riding royalty, on 17 water-flooding operations in Oklahoma are $1.55 a barrel, and on 16 water-flooding operations in Kansas are $1.32 a barrel. These figures do not include any allowance for interest on the investment or for federal taxes. They were compiled from records of the earlier secondary operations in both states, all of which are now depleted or are in the final stages of depletion and where drilling depths have ranged from 700 to 1,500 feet. The cost of development and operation in deeper fields, where any considerable amount of redrilling will be required will, of course, be subtantially higher.

Some of the results obtained from secondary-recovery operations are most impressive. Certain of them are cited briefly as follows:

Water flooding has increased the production of oil in New York by 500 percent over what it was in 1920, and the ultimate recovery of oil in the state will at least be doubled.

In Pennsylvania about 80 percent of the current oil production comes from secondary-recovery operations, and the recovery from the great Bradford field has been more than doubled by the application of water flooding. Elsewhere in Pennsylvania it has been shown that an average increased recovery of 42 barrels per acre-foot can be obtained by systematic injection of air and gas into the Venango sand series. The reserve of oil which may be produced in Pennsylvania by secondary methods is believed to be adequate to maintain the current rate of production for about 50 years.

In Illinois information obtained on the results from 128 secondary projects indicates that oil recovery can be increased by about 370 million barrels, which is somewhat more than the present proved primary reserve of the state.

Approximately 140 million barrels of oil have been produced by the application of secondary methods in Oklahoma, and the proved secondary reserve is estimated to be at least 1 billion barrels of oil.

In Texas, it has been estimated that the water-injection program in the East Texas field will result in an increased recovery of more than 600 million barrels of oil. This project cannot be classified strictly as a secondary-recovery operation, in accordance with the definition proposed previously, but it does represent an outstanding example of the application of secondary techniques.

Recent studies by the Arkansas Oil and Gas Commission [4] have shown that the secondary reserves of Arkansas, which include all additional oil that may be obtained by the injection of gas or water, are almost equal to the existing proved primary reserve. This determination is of significant importance on account of the low current rate of discovery of new fields in the state.

SECONDARY-RECOVERY OPERATIONS IN ROCKY MOUNTAIN STATES

The Rocky Mountain region, which embraces the Rocky Mountain system and subordinate folded areas, the Colorado Plateau, and the western part of the Great Plains, extends from central New Mexico to the Canadian border. Oil of highly different physical and chemical characteristics is produced from rocks ranging in age from the Cambrian to the

---

[4] Fancher, G. H., and Mackay, D. K., *Secondary Recovery of Petroleum in Arkansas*, El Dorado, Ark., Arkansas Oil and Gas Comm., 1944.

Oligocene, and under a multitude of structural conditions. Classical examples of anticlinal accumulations, such as Salt Creek and Lost Soldier in Wyoming, and Rangely in Colorado are found in this region. The Cut Bank oil and gas field, in northern Montana, is one of the largest fields ever discovered where accumulation has been controlled by a stratigraphic trap. Many of the Rocky Mountain fields are highly faulted and produce from more than one formation, and a wide variety of reservoir conditions and reservoir performance prevails throughout the region. It is believed that there are many opportunities for effective application of secondary methods in reservoirs which will not have an efficient primary recovery.

Montana

Secondary-recovery operations in Montana are restricted to experimental gas injection on several properties in the Cut Bank field of Glacier County. Primary recovery from the large Cut Bank field will be inferior, and it is estimated that some 318 million barrels of oil will remain unrecovered by the present methods of production in the 53,000 acres of proved oil land in the field. It is believed that there is a distinct possibility that the primary recovery of the Cut Bank field can be doubled by the application of secondary methods.

Wyoming

Thirteen separate secondary-recovery projects currently are in operation in Wyoming. Several of these projects are in different sands in the same field. Ten projects employ gas injection, two are experimental water floods, and one is a unique sand-heating operation. Three earlier gas-injection projects have been abandoned.

Gas injection was first commenced in the Elk Basin and Salt Creek fields in 1926, and has been continued with little interruption to the present time with a substantial increase in recovery resulting. Injection of gas in the Grass Creek field has been primarily for gas storage, although some increase in oil recovery is indicated. Recent gas injection into the Tensleep reservoir of the Elk Basin field and the injection projects in the Lance Creek, Lost Soldier, and Wertz fields are classified as pressure maintenance operations.

Experimental water floods are in operation in the Shannon sand of the Cole Creek field and in the Second Wall Creek sand of the Salt Creek field.

The deposition of excessive amounts of paraffin over the exposed surface of the Newcastle sand in wells of the Osage field, Weston County, was a serious operating problem until portable electric heaters were devised to be lowered into the wells for the purpose of melting the paraffin, which could then be pumped to the surface before it had a chance to solidify. This heating operation must be repeated periodically, but it has proved to be quite successful, and undoubtedly has been a contributing factor to the maintenance of production in the Osage field during recent years.

Colorado

Former gas-injection projects in the Wellington and Fort Collins fields of Larimer County, Colorado, have been abandoned, and there are no secondary-recovery projects in operation in the state at the present time.

Gas injection into the Morrison sand reservoir of the Wilson Creek field of Rio Blanco County, is a recent pressure-maintenance project. The same may be said for the injection program in the North McCallum field, of Jackson County, which has been in operation since 1944. The North McCallum pressure-maintenance project is unique because the gas which is injected is about 94 percent carbon dioxide.

Consideration has been given to a pressure-maintenance program in the Rangely field of Rio Blanco County. Unaided primary recovery from the Rangely field probably will be very inferior on account of the low permeability of the producing formation and because of excessive waste of gas from the gas-cap. The original oil content of the Weber sand reservoir of the Rangely field was about 1.5 billion barrels, most of which will not be recovered by the present method of operation. Therefore, an unusual opportunity exists for the application of some form of improved recovery technology.

SOME LIMITING FACTORS OF SECONDARY OPERATIONS

The production of the large secondary reserve of the United States is limited, just as primary production is limited, by the ability to produce oil at a profit. Profitable secondary production can be assured in many fields by the ability to utilize existing production facilities and by the establishment of unit and cooperative projects.

The ability to utilize existing production facilities, thereby reducing development costs to a minimum, is going to have a great bearing on the secondary possibilities in deeper fields and in fields possessing thin pay sections, even though the productive horizons are otherwise suitable for gas or water injection; for it may not be profitable to redrill the field for secondary-recovery operations. Obviously, this is a problem which merits careful consideration by operators who desire to continue in the business of oil production in fields which are approaching or have reached the point of primary depletion. Except for mining operations, all known and proposed methods for increasing oil recovery require the utilization of wells for the injection and production of fluids into and from oil-bearing rocks. For this reason, it should be evident that no well should be abandoned until consideration is given to the effects that such abandonment might have on future attempts to increase recovery. Also, the completion of all wells should be planned so that they may be utilized to the greatest advantage as soon as the need for pressure maintenance or for secondary-recovery operations becomes evident.

In order to apply secondary recovery or pressure maintenance effectively and economically, the unitization or cooperative development and operation of oil fields is usually necessary. This objective cannot be attained as long as state laws permitting the unitization of oil and gas fields are either inadequate or lacking and as long as operators are liable for violations of antitrust laws if they join a unit or cooperative project.

Nothing can be gained by advocating the application of methods for increasing oil recovery as long as they may place the operator in legal jeopardy, and it is certainly desirable that the executives and legislatures of the various oil-producing states should be acquainted with this fact. Opposition to laws permitting unit operations is based usually on either ignorance or avarice, for so far as the author has been able to determine there are no unsuccessful unitization agreements where commercial oil production has been developed. This being the case, the facts concerning the general success of unitized operations should receive wide publicity so that the objections to unitization resulting from ignorance may be eliminated. Likewise, it seems clear that the oil industry and the states must cooperate to remove the nullifying effects that may be imposed by a selfish minority, who will purposely obstruct a program of unquestioned conservation designed for the common good of a field solely for the nuisance value it can create.

Although the production of synthetic liquid fuels may have some limiting influence on future secondary-recovery activity, it seems more probable that the cost of finding and producing oil by primary methods will continue for a considerable period to be the more important factor in controlling the extent of application of secondary methods.

CONCLUSIONS

The magnitude of the unrecovered oil in the United States has never received the attention which it deserves and, therefore, is an unknown quantity. Some 64 billion barrels of oil have been discovered in this country since the completion of the first commercial well in Venango County, Pennsylvania, in 1859. Consequently, the conclusion cannot be escaped that the United States has been supplied abundantly with oil, and it seems equally certain, from the author's studies and observations, that a great deal of oil will not be recovered by common production methods that have been in use for a period of time which now is approaching one century. However, it should be recognized that effective oil-recovery techniques, just like the conservation of soil and forest resources, have not always been known, and their development has taken place over a period of many years. Likewise, for years the price received for crude oil has been in no way commensurate with the value of the product and has not permitted, in many instances, a normal replacement of produced reserves. As a consequence, in many producing areas it has not been profitable in the past to apply methods for increasing oil recovery, even though the techniques for so doing have been known and have been applied successfully in certain parts of the country for over half a century.

Questions

1. Define "secondary recovery."

2. Why has secondary-recovery procedure become more important in the oil industry during the past few years?

3. Summarize the early history of secondary-recovery operations.

4. What methods are used in secondary-oil-recovery procedure?

5. What type of field is most adaptable for secondary-recovery methods?

6. What geologic factors have an important bearing on the results of secondary recovery?

7. Give a general statement on the costs of secondary-recovery operations.

8. What are some of the limiting factors on secondary operations?

CHAPTER 8

VALUATION AND SUBSURFACE GEOLOGY
JOHN D. TODD

The art of valuing producing petroleum properties (and it is more of an art than a science) has grown up in the oil business contemporaneously with the maturing of subsurface geological knowledge. Of course, both started in Colonel Drake's time from zero, and during the past ninety years, valuation has at times out-distanced geology. In other decades geological knowledge has grown faster than knowledge of valuation.

When Professor Wright investigated the oil business at the close of the Civil War, he found that subsurface geology was a maze of superstitions and that valuation attempts were dominated by promotion. As thousands of wells were drilled and millions of barrels of oil were yearly produced, many things were learned about oil reservoirs—both in the drilling of new wells and in the performance of the old wells. Wright found that the average life of a well was eighteen months in 1865, whereas shortly thereafter wells were being sold on twice that payout period.

One of the earliest standards of valuation was the "days of payout," on which much Allegheny production was sold. It was common custom in early Pennsylvanian days to value net daily barrels of settled-oil production at one thousand times the price of oil. For instance, a property producing 1,000 net barrels per day would be worth $3,000,000 if oil was selling at $3.00 per barrel. Although this method gave no consideration to subsurface geology (and properly so, since almost nothing was known about it), it represented a composite of oil men's opinions on operating costs, probable rate of decline, and expected profits. Even today, the net daily barrels produced is used as a yardstick in evaluating some properties.

As the oil industry grew, some systematic operators began to keep charts on the wells and to draw curves on their production. As every well was always produced at its maximum rate, these charts recorded not only the decline of daily production but also the decline of daily *capacity*. As none of the wells ever increased, these curves became known as "decline curves." From the study of many of these curves, it became apparent that in many fields, the decline followed a certain pattern. Observers found that they could draw an average curve of all the wells, and could fit any well curve into a portion of a master curve—and that, properly placed, the future performance of that well followed the master curve. From this fact evolved the Law of Equal Expectations expressed as: "If two wells under similar conditions produce equal amounts during any given year, the amounts they will produce thereafter, on the average, will be approximately equal regardless of their relative ages." In other words, a fifty-barrel-a-day well that is ten years old will produce about as much oil in the future as a fifty-barrel well that is one year old.

Of course, decline curves had little to do with subsurface geology. Our knowledge of subsurface geology was expanding each decade, but figuring reserves was a matter of daily production figures and graph paper. Family curves, composite curves, running averages, comparative ratios, cumulative-productive curves—all had their sponsors and special uses. Plotting the curves on logarithmic or semi-logarithmic papers made a straight line out of the curves and aided in their projection.

The matter of estimating how much oil a well would produce in the future from facts on its past production, and the past production of other wells, attained a high degree of proficiency with an exacting standard of accuracy. The future production of a well could be closely predicted, with or without a knowledge of subsurface geology. It is still true that when decline curves are available (as in the later life of wells, when they will no longer make their allowable, and are thus being produced at capacity) a proper projection of the curve gives the best possible estimate as to both the ultimate production of a well and the rate of that production.

PRORATION

One of the really great changes in the oil industry came in about 1931 with the advent of over-production. The oil industry has always been plagued by over-production and periodic price declines. As early as 1926 legal efforts were made to curtail production equitably (always claimed in the cause of conservation, but actually to prevent price decline). The development of the giant East Texas Field in a financial depression paved the way to permanent proration by statute. Proration meant that wells would henceforth produce not at capacity, but at their MER (Maximum Efficient Rate), a percentage thereof, or some other alleged equitable system of restriction. This also meant payouts in terms of years instead of weeks, and the need of much more capital—the first well no longer would finance the drilling of all the later wells on a large tract; money had to be borrowed.

As wells no longer produced at capacity, there were now no "decline curves;" so all of the curve liturgy had to be replaced by some other method by which the ultimate production of a lease could be predicted early in its life. Loans were to be arranged, mergers made, unitizations accomplished, and properties to be bought and sold—the oil business had to go on in spite of proration, and some new scheme of valuation had to be adopted.

Fortunately, subsurface geology had grown greatly during the "decline curve" days. Much had been learned about structure, reservoir rocks, connate water, explusive mechanisms, and a host of other important items. In arriving at an estimate of recoverable oil (usually referred to as reserves) the estimator could use all of these data. With all of this modern information, some of the speculation about the subsurface geology of the reservoir was eliminated, and the reserve estimate was rendered progressively safer.

While proration was still an infant, the electric log came to the aid of subsurface geologists. It is hard to overestimate the utility of electrical logs in any phase of subsurface geology, but valuation as we know it today in most areas could not exist without electrical logs. In fact, reliable reserve estimates are often made with no other information except a map and a set of electrical logs.

MARKET VALUE

Almost all valuations fall into two classes (1) market valuation or (2) the engineering or analytical valuation. Little has been written about the market value of a petroleum property except in the law books. The fair market value is said to be "the amount that would be paid on a certain day by an informed buyer, able and willing to buy, and accepted by an informed seller, willing but not forced to sell." This theoretical trade made by two theoretical parties is the courts' attempt to determine what a property is *really* worth at a specific time. For most legal purposes, it is this "fair market value" in which the judge and jury are interested, and which they are trying to determine. For some tax and estate purposes, this "fair market value" is the appraisal that is needed, and in such cases no amount of counting acre-feet and calculating will do. One must attempt to reconstruct what would have been a fair trade at the stated time.

The best evidence of fair market value is an actual sale of that particular property, or some part of it, at that time. But, of course, that never happens; disputes do not occur when the answer is known. Litigation and tax disputes arise in the area of uncertainty, and usually are due to a conflict between optimists and pessimists. If there were a sale, a fair sale, of that particular property at that time, one could establish fair market value without knowing any subsurface geology, or for that matter, without knowing anything about the oil business.

But fair market value must always be proved by comparison, inference, and analogy. The next best evidence is records of sale from county deed records of adjoining properties or of similar properties elsewhere. Oil companies keep records of transactions submitted to them, and these tend to show fair market value. Also, traders in that area frequently know of transactions or near-transactions on similar properties, which are admissible evidence in some proceedings.

As soon as one begins to study adjoining or similar properties, he is dealing in subsurface geology. An adjoining lease may raise such questions as: Is it up or down structure? Is it a gas or water-drive reservoir? Does the producing section thicken or thin in that direction? Does the reservoir increase or decrease in porosity and permeability in that direction? And, when similar properties are considered, it is almost entirely a study of subsurface geology—and the compared property is always similar in some respects and dissimilar in many others. And the larger the area covered, the more subsurface geology needed.

Most disputes about market value end in a compromise, somewhere between the optimists and the pessimists. But, in those cases which proceed to final legal adjudication, victory usually comes to the appraiser who works the hardest and knows the subject best—there is so much subsurface geology to know, and so much evidence on both sides, that either side can win.

ENGINEERING VALUATIONS

In contrast to fair market value, an engineering valuation is an analytical study of every known fact about a property to arrive at a theoretical value (as of now or any other time). Such valuation assumes (1) that the property will produce a certain amount of money—the computed reserve at some assumed price (2) that it will cost a certain amount of money to develop and operate the property (except in the case of a royalty) (3) that the difference between gross income and expenses is the future profit—which when discounted at some assumed rate of interest gives a present value of the property.

Such engineering valuations are usually the basis of mergers and unitizations; they are required by banks and insurance companies in the making of loans; and they form a trading background for most sales of properties. All major companies make and constantly revise such valuations on their own properties as a kind of perpetual inventory. The largest oil companies keep up such reserve estimates on all producing properties (those of all of their competitors as well as their own) to stay posted on their relative reserve position, and to help their pipe-line subsidiaries in their competition for control of the available oil.

In spite of the fact that these valuations are in daily use by every part of the oil and gas business, and although they are computed on large, expensive mechanical calculators that carry the answer out to four decimal places—still they are all based on assumptions and are far from exact. Every engineering valuation assumes (1) some continuity of reservoir rock and porosity, (2) some amount of connate water, (3) a recovery factor, (4) a future rate of production, (5) a future price for oil or gas, (6) future costs, and (7) some discount factor or factors. In addition to the above seven estimates, at least some of the following factors are estimated in every valuation (usually one half of them are estimates): (8) productive acreage, (9) thickness of productive zone including gas-oil and oil-water contacts, and the amount of gross section that is porous, (10) percentage of porosity and how permeable, and (11) shrinkage factor.

As mentioned above, quite reliable valuations can be made by a geologist who is well acquainted with the general area if he uses only a map and a set of electrical logs; and very many of the valuations are based only on that data. Rather than bemoaning the mistakes in valuations, we should marvel that they are so near right and have formed a basis on which the oil and gas business could grow to where its present annual production exceeds the annual national production of coal, iron, copper, gold, silver, and all other metals combined.

FIGURING THE RESERVE

In actual practice the figuring of the reserve by the engineering (often called the volumetric) method involves (1) delineating the reservoir area and thickness, and calculating the acre-feet of gross rock, (2) figuring the amount of pore space in that reservoir, (3) deducting the part of space occupied by water, (4) figuring the oil or gas in place, (5) deducting the shrinkage it will suffer on coming to atmospheric pressure and temperature, (6) deducting the part of oil or gas that will not be recoverable by present methods and will therefore be left in the ground, and (7) deducting the past cumulative production—what has already been produced.

For oil, these steps are usually expressed in the following formula:
$$R = 7758 \times A \times T \times P \times (1-I) \times S \times F$$
in which R is the recoverable reserve.

The *7758* is the number of barrels of tank-stock oil needed to fill one acre, one foot deep.

The A is the number of acres in the reservoir. This factor requires the most careful and detailed kind of subsurface geology. Some important information is almost always lacking, a fact which requires the most thoughtful consideration of what is known and the drawing of the most reasonable inferences. It has been said by some of our most eminent valuation experts that *properly dimensioning the reservoir is the greatest single source of error in reserve estimates.*

The T is the thickness of the producing measure. This factor is ordinarily arrived at from electrical logs, supplemented by such coring or cuttings and core analysis as is obtained from that property or any related property. This procedure always involves the picking of the top of the formation, frequently involves the location of a gas-oil contact, and usually requires the selection of an oil-water contact. Experience in the general area of the property is of commanding importance in the making of these various judgments. There is much to know that is not written in books. An opportunity to see many wells cored and logged, to produce them, and to be intimately acquainted with their entire history—this type of experience, to which geologists with major oil companies are daily exposed, is the best possible teacher in such matters.

The P in the formula is the porosity. Frequently no information is available from the particular wells—then an estimate for normal porosity in that formation at that depth is used. Often a little core analysis at one or two points in the producing horizon is available—this fact is carefully placed on the log and an effort made to realize whether the data represent normal porosity or whether the cores were from a good streak, or perhaps a tight part of the horizon. Often only porosity data from other scattered wells in the field are available—in such cases an educated guess is made, and porosity is assumed to be the same under the entire property.

Although a rock may be porous, it also must be permeable to be a producible petroleum reservoir. If sufficient permeability (the continuity of the pore spaces, which permits passage of fluids through the pores) does not exist, the fluids in the porous reservoir cannot be produced—or they can be produced only so slowly that they are not commercial.

Fracturing may contribute to the porosity of limestones. It is notoriously hard to recover cores from lime, and usually cores give no information as to total amount of fracturing. The amount of fracturing is sometimes estimated by computing what the capacity of the well should be from its total section, bore hole, porosity, and permeability—then attributing any excess capacity to fractures. (Engineers can compute anything if you let them make a few innocuous "assumptions" while they get their slide rule out of the case.) Cores taken with a diamond bit quite often result in the recovery of fractured cores intact, so that estimates of fracture porosity can be made.

Measurement of porosity in the now popular reef, conglomerate, or other "heterogeneous" reservoirs introduces new problems. The microlog shows promise of enabling one to make a good estimate of the amount of permeable zone in such a section, although the percentage porosity still must be obtained by coring or estimation. If one has micrologs nearby, he can pretend that the section will be the same elsewhere. If one has only electrical logs, he just does his best, remembering that assuming 50 percent of the reef to be porous seems to be about the average used.

Porosity in producing fields varies from 2 to 40 percent, with the average being between 20 and 25 percent. It is a great variable. The porosity between wells is not known, even though one may have measured it in cores from the producing section of the wells. It is always assumed that the well bores are representative of the reservoir. Performance of the wells themselves is often some help—good, high-capacity wells are not made in thin, nonporous sections.

Connate or interstitial water (denoted by I in the formula) is nonproducible water contained in the reservoir rock along with the oil and/or gas. It is thought that the formations, when laid down, contained seawater in the available pore spaces and that the reservoir rocks retain this seawater (sometimes called fossil water) until it was driven out by the inmigrating gas and oil. The gas or oil displace most of the sea water, but not all of it—some of it is held by capillary forces and remains as a coating of water around each sand grain. Producing formations nearly always contain some interstitial water, which occupies part of the pore space and reduces the oil or gas reserve.

The amount of connate water is extremely difficult to determine. Cores are often hard to secure, and are usually contaminated by drilling fluids. Only rarely are cores taken with oil-base mud in order to prevent flushing of the core by water from drilling mud. The "restored state" method gives close approximations by determining how much water a core should retain in the face of gas or oil flushing. In this method, capillary equilibrium similar to that existing in the reservoir is established in the laboratory within the core. Connate water can often be estimated from a study of electrical log resistivities. Connate water varies between 0 and 60 percent; it will average 20 percent in good porous sands and 30 percent in tight, impermeable sands. Since it is a reducing factor, it is dealt with as $1 - I$; thus, 20 percent connate water is $1-20$ percent, or a multiplier of 0.80.

The shrinkage factor is denoted by S in the formula. A barrel of reservoir oil shrinks when brought to the lower temperature and pressure at the surface—a barrel of reservoir oil being equal to only a fraction of a barrel of stock-tank oil. This reduction in volume is due almost entirely to the gas coming out of solution when the barrel of oil is brought to the surface—a barrel of oil with hundreds of feet of gas in solution in it occupies more space than that same barrel of oil with the gas removed. The more gas in solution, the greater is the shrinkage; and the deeper the reservoir, the greater is the amount of gas in solution—so that, in general, the deeper the well the greater the shrinkage factor. The shrinkage is usually measured in the laboratory or estimated from the gas-oil ratio of the wells and the gravity of the oil. This shrinkage is a substantial reducing factor where ratios in excess of 2000:1 of solution gas are encountered; the factor may be 50 percent or less. At great depths it takes over two barrels of such reservoir oil to make one barrel of stock-tank oil.

The recovery factor is designated as F in the formula. Knowing how much oil or gas is in the reservoir, how much of this is going to be producible—how much will be left in the reservoir when the field is abandoned? *This is always an estimate; no one can know.* The recovery is influenced by many things: by the amount of gas in solution, the viscosity of the oil, the rate of production, the primary expulsive energy, the price of the oil, and a number of other things. Recovery factors vary widely, but generally are within the following ranges: 20 to 40 percent of the oil in place for dissolved-gas-drive fields, 30 to 70 percent for expanding gas cap plus gravity, and 50 to 80 percent for fully effective water drive.

A HYPOTHETICAL CASE

An example always best illustrates the actual workings of such a formula, and affords an opportunity to point out the dependence of a valuation on subsurface geology. Assume that one is to evaluate a conventional seven-eighths lease on a 160-acre tract with eight wells which are located midway down the flank of a rather steep Gulf Coast anticline and which are producing oil by effective water drive from a sand at −5,000 feet.

A map is secured showing the location of all the wells on the structure, plus electrical logs of all the wells in the field. A structure map is drawn on the top of the producing formation—where several sands produce, a structure map should be drawn on each sand. This procedure involves carefully picking the sand "top" on each well in the field, which is checked against all obtainable core data to verify that what appears to be top of sand is the first productive sand encountered. After the structure map is drawn, an isopachous map of the producing sand is prepared. This map is based on "effective sand"—from a minute examination of the detailed section (100' = 5") of the electrical log, plus all known coring, core analysis, caliper log, gamma-ray log, drilling-time chart, and any other data obtainable. Shale and/or hard streaks are excluded in order to isopach only the actual producing section.

After the isopach is drawn, each isopach interval is planimetered to give the actual number of acres situated in that zone of thickness. Thus, the area between the 50- and 45-ft. isopach lines is planimetered and assumed to have a sand thickness of 47½ ft. Only a single tract of 160 acres is to be valued, and it is found that 48 acres is underlaid by 40 ft. of effective sand, 68 acres by 36 ft. of sand, and 44 acres by 32 ft. of sand. Thus, in valuing the 160 acres, three valuations are made; or, all three segments are combined to get a composite of acre-feet of sand. To composite these figures, multiply 48 x 40 to get 1920 acre feet, 68 x 36 to get 2448 acre feet, and 44 x 32 to get 1408—or an aggregate of 5776 acre feet of producing sand under the 160 acres (the product of A, acreage, and the T, thickness, of the formula). It is already known that oil fields in this trend can be expected to produce 400 barrels per acre foot. An experienced valuator in the trend thus could estimate around 2,310,400 barrels (5776 × 400) of ultimate recovery.

It is found that when the eight wells on the lease were drilled, no cores were obtained. It is learned that core laboratories ran cores from a well on an adjoining lease; average porosities of 21, 23, and 24 percent were obtained from three cores. The Gulf Company has run porosities, which average 23 percent on most of its wells. Permeabilities averaged 550 millidarcys. It is concluded that 22 percent is about the average porosity for the lease being valued, and that permeabilities are satisfactory. This conclusion appears to be in line with known porosities in nearby fields producing from the same formation and on strike.

The next factor in the formula is I (interstitial or connate water). In an older field, the amount of connate water will probably involve an estimate—as in the early days there were no connate-water determinations. In fact, many then thought connate water did not exist. If no connate-water determinations are available, estimates can be made based on known determinations in similar formations in on-strike fields. Connate water ranges from 10 to 50 percent of the pore space; and, in estimating, around 20 percent in good porous reservoirs and 30 percent in mediocre reservoirs are fairly safe averages.

In this field, the development has been more recent—all of the wells have electrical logs. Cores from one nearby well were analyzed by the restored-state method in the Texas Company's laboratory; some companies do systematic core determinations on all of their own wells and sometimes on adjacent wells. Three pieces of cores were analyzed on this well and showed connate water content of 35, 32 and 36 percent. Comparison with the electrical log indicates that the higher values came from the normal parts of the sand and that the 32 percent was from the most porous part of the sand.

It is learned that the Shell Company has made connate water determinations in their own laboratory on all the cores taken from their wells on this structure. The valuator has a geologist friend at Shell, and while he cannot see their core analysis reports, he is able to "swab" the information that they are using 33 percent for connate water. Putting all of this information together, he concludes that 34 percent is probably a safe factor for the interstitial water.

Fortunately, a nearby well logged the producing sand below the water contact, so that the resistivity of this sand saturated with water (R_o factor) is known. Thus, the R_o factor is known, and from the electrical log the probable connate water content is computed as 35 percent—an amount which appears to confirm the other data.

With a porosity of 22 percent and connate water of 34 percent, reference to figure 413 shows that there are 1126 barrels of available pore space per acre foot in the reservoir. All kinds of charts, graphs, and short cuts are used by various reservoir engineers. This combination of porosity and connate water is one of the more useful ones and saves much routine calculation in both oil and gas problems. In the present instance, it enables one to substitute "1126 barrels of pore space" for the $P \times (1 - I)$ part of the formula. Therefore, 5776 acre feet $\times$1126 barrels of available pore space equals 6,503,776 barrels of space under the lease—or, in this instance, 6,503,776 barrels of reservoir oil in place.

The S or shrinkage factor is largely dependent on the amount of dissolved gas and the gravity of the oil. Gas-oil ratios are regularly determined by various state regulatory bodies and are almost always available. The wells on this lease produce oil of 35° gravity with an average gas-oil ratio of 510:1. Referring to published charts and curves, one finds that the shrinkage factor (sometimes called formation volume factor) for a 510:1 ratio and 35° gravity oil will be 78 percent; or that a barrel of oil in the reservoir is only equal to 78/100 of a barrel when it reaches the stock tank and is ready to be sold. This is always a reducing factor but

PERCENT POROSITY

CONNATE WATER PERCENTAGE	10	11	12	13	14	15	16	17	18	19	20	21	22	23	24
10	698 / 3,920	768 / 4,312	838 / 4,704	908 / 5,097	978 / 5,489	1,047 / 5,881	1,117 / 6,273	1,187 / 6,665	1,257 / 7,057	1,327 / 7,449	1,396 / 7,841	1,466 / 8,233	1,536 / 8,625	1,606 / 9,017	1,676 / 9,409
11	690 / 3,877	760 / 4,265	829 / 4,652	898 / 5,040	967 / 5,428	1,036 / 5,815	1,105 / 6,203	1,174 / 6,591	1,243 / 6,978	1,312 / 7,366	1,381 / 7,754	1,450 / 8,141	1,519 / 8,529	1,588 / 8,917	1,657 / 9,304
12	683 / 3,833	751 / 4,217	819 / 4,600	888 / 4,983	956 / 5,367	1,024 / 5,750	1,092 / 6,133	1,161 / 6,517	1,289 / 6,900	1,297 / 7,283	1,365 / 7,667	1,434 / 8,050	1,502 / 8,433	1,570 / 8,817	1,638 / 9,200
13	675 / 3,790	742 / 4,169	810 / 4,548	877 / 4,927	945 / 5,306	1,012 / 5,685	1,080 / 6,064	1,147 / 6,443	1,215 / 6,822	1,282 / 7,200	1,350 / 7,579	1,417 / 7,958	1,485 / 8,337	1,552 / 8,716	1,620 / 9,095
14	667 / 3,746	734 / 4,121	801 / 4,495	867 / 4,870	934 / 5,245	1,001 / 5,619	1,068 / 5,994	1,134 / 6,368	1,201 / 6,743	1,268 / 7,118	1,334 / 7,492	1,401 / 7,867	1,468 / 8,242	1,535 / 8,616	1,601 / 8,991
15	659 / 3,703	725 / 4,073	791 / 4,443	857 / 4,813	923 / 5,184	989 / 5,554	1,055 / 5,924	1,121 / 6,294	1,187 / 6,665	1,253 / 7,035	1,319 / 7,405	1,385 / 7,775	1,451 / 8,146	1,517 / 8,516	1,583 / 8,886
16	652 / 3,659	717 / 4,025	782 / 4,391	847 / 4,757	914 / 5,123	978 / 5,489	1,043 / 5,854	1,108 / 6,220	1,173 / 6,586	1,238 / 6,952	1,303 / 7,318	1,369 / 7,684	1,434 / 8,050	1,499 / 8,416	1,564 / 8,782
17	644 / 3,615	708 / 3,977	773 / 4,339	837 / 4,700	901 / 5,062	966 / 5,423	1,030 / 5,785	1,095 / 6,146	1,159 / 6,508	1,223 / 6,869	1,288 / 7,231	1,352 / 7,593	1,417 / 7,954	1,481 / 8,316	1,545 / 8,677
18	636 / 3,572	700 / 3,929	763 / 4,286	827 / 4,644	891 / 5,001	954 / 5,358	1,018 / 5,715	1,081 / 6,072	1,145 / 6,429	1,209 / 6,787	1,272 / 7,144	1,336 / 7,501	1,400 / 7,858	1,463 / 8,215	1,527 / 8,573
19	628 / 3,528	691 / 3,881	754 / 4,234	817 / 4,587	880 / 4,940	943 / 5,293	1,005 / 5,645	1,068 / 5,998	1,131 / 6,351	1,194 / 6,704	1,257 / 7,057	1,320 / 7,410	1,382 / 7,762	1,445 / 8,115	1,508 / 8,468
20	621 / 3,485	683 / 3,833	745 / 4,182	807 / 4,530	869 / 4,879	931 / 5,227	993 / 5,576	1,055 / 5,924	1,117 / 6,273	1,179 / 6,621	1,241 / 6,970	1,303 / 7,318	1,365 / 7,667	1,427 / 8,015	1,490 / 8,364
21	613 / 3,441	674 / 3,785	735 / 4,129	797 / 4,474	858 / 4,818	919 / 5,162	981 / 5,506	1,042 / 5,850	1,103 / 6,194	1,164 / 6,538	1,226 / 6,882	1,287 / 7,227	1,348 / 7,571	1,410 / 7,915	1,471 / 8,259
22	605 / 3,398	666 / 3,737	726 / 4,077	787 / 4,417	847 / 4,757	908 / 5,097	968 / 5,436	1,029 / 5,776	1,089 / 6,116	1,150 / 6,456	1,210 / 6,795	1,271 / 7,135	1,331 / 7,475	1,392 / 7,815	1,452 / 8,154
23	597 / 3,354	657 / 3,690	717 / 4,025	777 / 4,360	836 / 4,696	896 / 5,031	956 / 5,367	1,016 / 5,702	1,075 / 6,037	1,135 / 6,373	1,195 / 6,708	1,254 / 7,044	1,314 / 7,379	1,374 / 7,714	1,434 / 8,050
24	590 / 3,311	649 / 3,642	708 / 3,973	766 / 4,304	825 / 4,635	884 / 4,966	943 / 5,297	1,002 / 5,628	1,061 / 5,959	1,120 / 6,290	1,179 / 6,621	1,238 / 6,952	1,297 / 7,283	1,356 / 7,614	1,415 / 7,945
25	582 / 3,267	640 / 3,594	698 / 3,920	756 / 4,247	815 / 4,574	873 / 4,901	931 / 5,227	989 / 5,554	1,047 / 5,881	1,106 / 6,207	1,164 / 6,534	1,222 / 6,861	1,280 / 7,187	1,338 / 7,514	1,396 / 7,841
26	574 / 3,223	632 / 3,546	689 / 3,868	746 / 4,190	804 / 4,513	861 / 4,835	919 / 5,158	976 / 5,480	1,033 / 5,802	1,091 / 6,125	1,148 / 6,447	1,206 / 6,769	1,263 / 7,092	1,320 / 7,414	1,378 / 7,736
27	566 / 3,180	623 / 3,498	680 / 3,816	735 / 4,134	793 / 4,452	850 / 4,770	906 / 5,088	963 / 5,406	1,019 / 5,724	1,076 / 6,042	1,133 / 6,360	1,189 / 6,678	1,246 / 6,996	1,303 / 7,314	1,359 / 7,632
28	559 / 3,136	614 / 3,450	670 / 3,764	726 / 4,077	782 / 4,391	838 / 4,704	894 / 5,018	950 / 5,332	1,005 / 5,645	1,061 / 5,959	1,117 / 6,273	1,173 / 6,586	1,229 / 6,900	1,285 / 7,214	1,341 / 7,527
29	551 / 3,093	606 / 3,402	661 / 3,711	716 / 4,021	771 / 4,330	826 / 4,639	881 / 4,948	936 / 5,258	991 / 5,567	1,047 / 5,876	1,102 / 6,186	1,157 / 6,495	1,212 / 6,804	1,267 / 7,113	1,322 / 7,423
30	543 / 3,049	597 / 3,354	652 / 3,659	706 / 3,964	760 / 4,269	815 / 4,574	869 / 4,879	923 / 5,184	978 / 5,489	1,032 / 5,793	1,086 / 6,098	1,140 / 6,403	1,195 / 6,708	1,249 / 7,013	1,303 / 7,318
31	535 / 3,006	589 / 3,306	642 / 3,607	696 / 3,907	749 / 4,208	803 / 4,508	856 / 4,809	910 / 5,110	964 / 5,411	1,017 / 5,711	1,071 / 6,011	1,124 / 6,312	1,178 / 6,612	1,231 / 6,913	1,285 / 7,214
32	528 / 2,962	580 / 3,258	633 / 3,555	686 / 3,851	739 / 4,147	791 / 4,443	844 / 4,739	897 / 5,036	950 / 5,332	1,002 / 5,628	1,055 / 5,924	1,108 / 6,220	1,161 / 6,517	1,213 / 6,813	1,266 / 7,109
33	520 / 2,919	572 / 3,210	624 / 3,502	676 / 3,794	728 / 4,086	780 / 4,378	832 / 4,670	884 / 4,961	936 / 5,253	988 / 5,545	1,040 / 5,837	1,092 / 6,129	1,144 / 6,421	1,195 / 6,713	1,247 / 7,004
34	512 / 2,875	563 / 3,162	614 / 3,450	666 / 3,737	717 / 4,025	768 / 4,312	819 / 4,600	870 / 4,887	922 / 5,175	973 / 5,462	1,024 / 5,750	1,075 / 6,037	1,126 / 6,325	1,178 / 6,612	1,229 / 6,900
35	504 / 2,831	555 / 3,115	605 / 3,398	656 / 3,681	706 / 3,964	756 / 4,247	807 / 4,530	857 / 4,813	908 / 5,097	958 / 5,380	1,009 / 5,663	1,059 / 5,946	1,109 / 6,229	1,160 / 6,512	1,210 / 6,795
36	497 / 2,788	546 / 3,067	596 / 3,345	645 / 3,624	695 / 3,903	745 / 4,182	794 / 4,461	844 / 4,739	894 / 5,018	943 / 5,297	993 / 5,576	1,043 / 5,854	1,092 / 6,133	1,142 / 6,412	1,192 / 6,691
37	489 / 2,744	538 / 3,019	587 / 3,293	635 / 3,568	684 / 3,842	733 / 4,116	782 / 4,391	830 / 4,665	880 / 4,940	929 / 5,214	978 / 5,489	1,026 / 5,763	1,073 / 6,037	1,124 / 6,312	1,173 / 6,586
38	481 / 2,701	529 / 2,971	577 / 3,241	625 / 3,511	673 / 3,781	721 / 4,051	770 / 4,321	818 / 4,591	866 / 4,861	914 / 5,131	962 / 5,401	1,010 / 5,672	1,058 / 5,942	1,106 / 6,212	1,154 / 6,482
39	473 / 2,657	521 / 2,923	568 / 3,189	615 / 3,454	663 / 3,720	710 / 3,986	757 / 4,251	805 / 4,517	852 / 4,783	899 / 5,049	946 / 5,314	994 / 5,580	1,041 / 5,846	1,088 / 6,111	1,136 / 6,377
40	465 / 2,614	512 / 2,875	559 / 3,136	605 / 3,398	652 / 3,659	698 / 3,920	745 / 4,182	791 / 4,443	838 / 4,704	884 / 4,966	931 / 5,227	978 / 5,489	1,024 / 5,750	1,071 / 6,011	1,117 / 6,273
41	458 / 2,570	503 / 2,827	549 / 3,084	595 / 3,341	641 / 3,598	687 / 3,855	732 / 4,112	778 / 4,369	824 / 4,626	870 / 4,883	915 / 5,140	961 / 5,397	1,007 / 5,654	1,053 / 5,911	1,099 / 6,168
42	450 / 2,526	495 / 2,779	540 / 3,032	585 / 3,284	630 / 3,537	675 / 3,790	720 / 4,042	765 / 4,295	810 / 4,548	855 / 4,800	900 / 5,053	945 / 5,306	990 / 5,558	1,035 / 5,811	1,080 / 6,064
43	442 / 2,483	486 / 2,731	531 / 2,980	575 / 3,228	619 / 3,476	663 / 3,724	708 / 3,973	752 / 4,221	796 / 4,469	840 / 4,718	884 / 4,966	929 / 5,214	973 / 5,462	1,017 / 5,711	1,061 / 5,959
44	434 / 2,439	478 / 2,683	521 / 2,927	565 / 3,171	608 / 3,415	652 / 3,659	695 / 3,903	739 / 4,147	782 / 4,391	825 / 4,635	869 / 4,879	912 / 5,123	956 / 5,367	999 / 5,611	1,043 / 5,854
45	427 / 2,396	469 / 2,635	512 / 2,875	555 / 3,115	597 / 3,354	640 / 3,594	683 / 3,833	725 / 4,073	768 / 4,312	811 / 4,552	853 / 4,792	896 / 5,031	939 / 5,271	981 / 5,510	1,024 / 5,750

FIGURE 413. Barrels of void space in one acre foot. Cubic feet of void space in one acre foot; with percent porosity. One acre foot = 7,758 bbls.; one acre foot = 43,560 cu. ft.

at shallower depths and with under-saturated crudes, it is not too large—at great depth with large amounts of dissolved gas, the neglect of this factor gives greatly exaggerated reserves.

The last item in the formula is F for recovery factor, which is the percentage of the inplace oil that the estimator believes is recoverable. It is always an estimate, and all too often just an off-hand guess. It is one of the greatest sources of errors in estimates. Many, many things go into a recovery factor, and lots of them are not found in the science books.

Chief among the scientific factors affecting recovery is the type of explosive energy that is bringing the oil to the bore hole, and perhaps removing the oil from other leases. If there is a fully effective water drive, the lease high on the structure will have a very high recovery factor—usually recovering more oil than could be calculated to be under it. This fact is due to the Law of Capture; the oil belongs not to the lease owner under whose property it is discovered, but to the well owner who produces it (reduces it to capture). Courts are making some feeble efforts to get away from the Rule of Capture, but it is still very strong in most jurisdictions and must always be considered in every estimate of recovery. The workings of the Rule of Capture are intimately related to the subsurface geology of the property—and afford the geologically minded operator many opportunities to take advantage of, or to protect himself from, natural drainage.

Just as the top lease in a water-drive field produces all of its own oil and some of all the oil in the down-dip lease—conversely, the edge lease produces only part of its own oil and none of anyone else's oil. How much oil an edge-lease owner will produce depends on the allowable production compared with those of up-dip leases. The down-dip lease always shares its oil with the up-dip leases, but a larger allowable sometimes results in the down-dip lease getting a fairer share. Water encroachment is not always regular; frequently water will intrude through the more porous permeable zones to reach the higher wells—this is particularly true if the up-dip wells are pulled hard. In such cases, the edge lease is left to produce its oil long beyond its computed time. Many edge leases which should have gone to water years ago (if the water level evenly moved up proportionately to the oil withdrawn from the reservoir) are yet producing, and still make their allowable every day.

Big recoveries from the leases high on structure are matters of common knowledge. But enormous recoveries from a lease low on the structure are just as likely if drive is from gas and the dip is steep. Gas drive is only gravity drainage under high pressure—and the edge lease will produce all of its oil and all of the oil that runs down into it, whereas the top lease will produce only part of its oil and then go to gas.

The study of "who gets the oil" is a fascinating combination of subsurface geology, some engineering, and an analysis of various statutes and regulations of oil and gas boards. A change in field rules or in the method of computing allowables can move the recoverable reserve from one lease to other leases. By the rules of oil and gas boards, the Law of Capture can be modified or annulled, and the to-be-produced oil can be moved from one lease to another—just as the President transfers funds from the Justice Department to the Labor Department. Such transfers are not spelled out in the engineering terms of the rule change—but they are there in between the lines, and they become apparent when one analyzes the value of various leases. This matter does not come up when an entire field is considered—but one company seldom owns an entire field.

Other scientific factors which contribute to the recovery factor are the gravity of the oil, its viscosity, the amount of dissolved gas (the viscosity goes up as you lose the gas), the porosity, and the relative permeability of the reservoir rock. The rate at which the field will be produced has a bearing on the recovery factor.

Aside from the engineering angles, the political implications of the situation have some bearing on the recovery factor. An operator's political connections or antipathies may influence his allowables. The brother-in-law of a member of the Oil and Gas Board seldom gets an unfair allowable—that is a positive factor in the recovery under his lease; it could be a negative factor if you are valuing the adjoining lease.

It is ascertained that the bottom-hole pressures do not decline in the 160-acre lease and that water encroaches into the edge wells. It is concluded that there is a fully effective water drive. The producing formation covers a vast area and logically might contain an expanding-water blanket. Other operators in this field consider it to be water drive also.

The oil is of 35° gravity, and the 500 m.d. permeability is fair. One would expect water drive to be of the lower order of effectiveness and tentatively estimate a recovery factor of 50 percent. The lease is half way up the structure, so it will surely draw some of the oil out of the down-dip leases; some of the oil will be lost to the up-dip leases. Consideration of the amount of oil the reservoir has produced, how far the water has advanced, and how much farther it must come to reach the lease being valued leads one to the conclusion that about as much oil will be gained as will be lost by drainage. Therefore, the recovery factor is left at 50 percent.

Using a shrinkage factor of 78 percent and a recovery factor of 50 percent, 6,503,776 (reservoir oil in place) × 78 percent (shrinkage) × 50 percent (recovery factor) = 2,536,472 barrels of original recoverable oil under the lease. An examination of production records shows that the lease has already produced a total of 289,336 barrels; this deduction leaves 2,247,136 barrels of remaining reserve under the lease. There is also contained in this oil 1146 million cubic feet of gas (2,247,136 × 510); but, as there is no present means of recompressing this gas (to raise it to pipe-line pressure) and no market for the low-pressure gas, no

804 SUBSURFACE GEOLOGIC METHODS

value will be included for the gas. As the operators has a full seven-eighths lease, he owns 1,966,224 barrels of oil and 1003 mmcf of gas.

Now that it has been concluded that 1,966,244 barrels of oil can be recovered by primary recovery means, how many dollars will this oil sell for? Will the price of oil go down, or will it go up? The usual procedure is to assume that the price of oil will remain the same (of course it never has, but that is an easy way out). The operating costs are likely to go down and the value of dollars to increase if oil declines in price; and if oil goes up in price, operating costs will probably increase and the purchasing power of the dollar will decline. The present price of 35° oil in the area of this lease is $2.78 per barrel (March 1, 1950). So, multiplying 1,966,244 barrels x $2.78, one finds that the anticipated sale price of the operator's oil from this lease is $5,466,158.32. It will be left to the executive or operator who receives the report to predict the future price of oil (since that is really a part of management's duties), and the report assumes that all this oil will be sold at $2.78 a barrel.

When one attempts to estimate the future price of oil, he has left the field of geology and entered economics. When he attempts to estimate the operating costs of bringing this oil to the surface, he is in the field of accounting. But, obviously, it will cost money to produce any oil (unless it is royalty oil) and those costs must be deducted from receipts before the profit will be known. Again the usual procedure is to assume that the costs of labor, materials, and other items will remain the same. Examination of the books of the operator indicates that it now costs $0.28 per barrel to produce and handle the oil, including advalorem and severance taxes. The wells are all flowing and with good water drive will flow all their oil—but as they grow older reworks will come oftener. So, for the future an increased cost is assumed to be $0.31 per barrel. Deducting this, one finds that the future net profit from oil will be $4,856,622.68. The salvage value at abandonment, estimated to be $40,000.00, may be added to the value of the oil.

Actually, it will cost more to produce the oil if the producing rates are small, as more years of operation are required. Operating costs, aside from taxes, are fairly uniform for a given well regardless of producing rate. Therefore, the per-barrel cost becomes higher as the rate of production becomes lower. What kind of allowables will these wells have in the future? Will they be permitted to produce more, or will they be required to produce less? Much depends on economic trends. But again, in this case, it is assumed the status quo will continue. The lease now has an allowable of 65 barrels of oil per day per well, which amounts to 15,600 barrels per month. At that rate the lease will produce for 144 months, or 12 years, and will produce a monthly income of equal monthly payments for that period.

Now, what is such a future income worth at the present time? What is money worth? Money is loaned at different rates, depending on who the borrower is and of what his security consists. Normal discount rates used for oil evaluations vary from three to six percent. The $4,856,622.68 discounted at five percent per year yields a present value of $3,587,101.51. To this amount is added the present worth of the $40,000 salvage 12 years hence, which is $22,289.48. This gives a total present worth of $3,609,-390.99. In some valuations, no discount is figured. In such cases, the recipient of the valuation report is left to make his own estimate of what that future income is worth now.

A valuation report on oil should also mention the amount of gas to be produced—as some market may be developed for this gas. Also, the report should mention any geology pointing to deeper possibilities under the lease, but no value should be accorded unless the deeper sands have been drilled to and tested.

In conclusion a word of warning: The latest production data on the lease should always be obtained—an actual trip to the property should be made by the appraiser just before he signs the report. The gauger on the lease not only knows how each well performed last month or last year; he knows how it performed yesterday afternoon and this morning. The production of each of the wells should be carefully checked—gas-oil ratios, pressures, and water percentages should be measured. Despite all the formulas, maps, logs, etc., the value of the property is the ability of some holes in the ground to produce material that will sell. If those wells quit producing material that sells, there is nothing much to value. The writer knows of wells that have declined as much as 75 percent while a valuation was being prepared—if a last minute check-up had not been made, the decline never would have been suspected. When valuing an oil or gas well, always remember Mark Twain's definition of a mine, "A hole in the ground owned by a liar."

GAS PROPERTIES

The geology of the structure and the intimate study of the reservoir rock itself is the same for the valuation of gas property as for oil. But, because gas is dealt with as a vapor and is so highly compressible, some other considerations enter into gas valuations.

In oil properties, the deeper the well, the greater the shrinkage factor: 1,000 barrels of reservoir oil at 5,000 feet is usually more stock-tank oil than 1,000 barrels at 10,000 feet—or, to put it another way, 20 feet of sand usually contains more recoverable oil at 5,000 feet than at 10,000 feet. But just the opposite is true for gas—20 feet of sand at 10,000 feet contains nearly twice as much recoverable gas as the same 20 feet of sand would contain at 5,000 feet. Oil at greater depth is worth less because of the excessive cost of drilling deep wells—gas values are affected less because gas wells are always so widely spaced, usually on 320- or 640-acre units. By and large, oil decreases in value with depth, but gas increases in value with depth.

Assume that the same 160 acres overlies gas instead of oil. In gas valuation, the formula begins with 43,560 as the number of cubic feet in one acre (one always deals with cubic feet in gas—mcf, thousand cubic feet—or mmcf, million cubic feet). In order to obtain cubic feet of reservoir space it would be necessary to multiply by porosity and connate water factors—in this case again refer to figure 413 and find that for 22 percent porosity and 34 percent connate water, there are 6,512 cubic feet of gas space in each acre foot. Again, using 5,576 acre feet of reservoir, one finds that there are 36,310,912 cubic feet of storage space in the reservoir. At atmospheric pressure and temperature, this reservoir would contain 36,310,912 cubic feet of air, and would hold that much gas.

But this reservoir is not at atmospheric conditions; it is under very high pressure and is always much hotter than the atmosphere. According to Boyle's Law, as the pressure is increased, the amount of gas a reservoir contains also increases. If the pressure is doubled (temperature remaining the same) there is twice as much gas—so that each time an atmosphere of pressure (14.7 lbs.) is added, a reservoir full of gas, or 36,310,912 cubic feet, is added.

If bottom-hole pressures have been taken on the gas wells, exactly the pressure existing in the reservoir will be known. Or, if the well-head pressures, fluid levels, and the weight of the gas are known, the bottom-hole pressure can be computed. Often the original, bottom-hole pressure can be estimated from the depth of the formation. Normal pressure in post-Eocene sediments on the Gulf Coast increases $46\frac{1}{2}$ lbs. per square inch with each 100 feet of depth (There are a few low-pressure and a few high-pressure reservoirs, but not many which vary from the 0.465 rule—which incidentally, is the weight of a column of sea water). When depth in feet is multiplied by 0.465, the approximate bottom-hole pressure is obtained. In this case, 5,000 × 0.465 gives 2,325 lbs. as the bottom-hole pressure. This 0.465 gradient applies well along the Gulf Coast on reservoirs of Oligocene and younger age—in other areas factors as high as 0.760 and as low as 0.420 are used. These values have been published and are usually well known to experienced geologists.

A reservoir pressure of 2,325 lbs. amounts to 158 atmospheres; (2,325 ÷ 14.7) so there is 158 times as much gas in the reservoir as it would hold at atmospheric pressure. This statement, however, is not quite true because Boyle's Law does not work exactly at higher pressures—it was worked out at lower pressures and later research has indicated variance at high pressures. This amount is known as supercompressibility, and the factor is called the "Z factor." If the composition or density of the gas is known, one can figure the Z factor for different pressures and temperatures. In the present case, the Z factor could be estimated for normal Gulf Coast gas at that depth to give a multiplier of 1.225. The Z factor results in an additive multiplier down to depths of around 10,000 feet, and with normal pressures is greatest around 4,000 feet.

Gas reserves are figured to atmospheric pressure of 14.7 lbs.; but sometimes the gas is already dedicated under contract of sale at a higher pressure, such as 16.4 lbs. In such a case, the 16.4 is divided into the reservoir pressure, and a smaller quotient results. It is obvious that it is to the advantage of the gas pipe line to buy the gas on as high a pressure base as possible—on a 29.4-lb. base only half as many cubic feet are paid for as on a 14.7-lb. Gas pipe lines nearly always sell gas on a 14.7-lb. basis.

In addition to correcting for pressure and supercompressibility, correction must be made for the reservoir temperature. Charles' Law states that as the temperature is increased, volume decreases in proportion to absolute temperatures. If bottom-hole temperature is known, this reduction in volume can be figured by direct application of Charles' Law to reduce the gas to the volume it would have at 60° F., which is the standard temperature. If reservoir temperature is not known, it can be estimated from temperature-gradient curves for the area—such temperature gradients vary grossly from one province to another, but in any particular area are quite regular and are well known. For the example lease, from curves available on this area, it is found that the reservoir temperature should be 170° F., and by application of Charles' Law, 0.825 is arrived at as the temperature-correction factor.

Only the recovery factor (F in the formula) remains to complete the calculation. Many gas fields have been produced to exhaustion—some in recent enough times for complete production figures to be available. Recoveries run very high, and with proper location of wells, one could recover about all the gas in a water-drive reservoir. But, because of improper structural location of wells, lenticularity, and other irregularities, recoveries seldom run over 90 percent. On the Gulf Coast in water-drive fields, the factor of 85 percent is commonly used at this time.

In confined reservoirs where only gas expansion drive is available, it is usual to fix some abandonment pressure. Trunk-line carriers do not desire gas at a pressure of less than 500 to 750 pounds per square inch—as the saying goes, "it will not buck the line" at lower pressures. Gas at less than line pressure must be compressed before it can be sold, and compressing requires equipment and costs money. Depending on the outlet to which the gas is going, an abandonment pressure of 500 lbs. or more is allowed for in fields where the pressure will decline. If one neglects supercompressibility change with pressure, the recovery factor is equal to the original-minus-abandonment-pressure difference divided by the original pressure.

For the example lease, a water-drive recovery factor of 85 percent is used. The calculation therefore is 36,310,912 × 158 × 1.225 × .825 × .85, a total of 4,928,368,853 feet of recoverable gas. As gas is sold in one-thousand-cubic-foot units, this amount would be written as 4,928,368 mcf—and as reserves are usually estimated in millions of cubic feet, this amount would be written as 4,928 mmcf. From production data it is found that 227 mmcf have been produced, and that a reserve of 4,701 mmcf is left.

The converting of gas reserves into dollars is the same as for oil, except possibly that it is more of a certainty. Gas is sold under long-term contracts, usually 10 to 20 years, at stipulated prices; for this reason the price does not fluctuate as greatly or as rapidly as the price of oil. It is easier to foresee what the gas price is to be in years ahead, and it is much more likely to remain at the present level than is oil. Also, gas deliveries are scheduled for years ahead, so it is easier to estimate over how long a period gas production will be extended.

If the future price and future deliveries of the well are not covered by contract, it is assumed that the rate of production and price will remain the same. A schedule of future income year by year is set up, with expected future expenses deducted. The costs of operating gas wells are usually much lower than costs for oil wells. The present value of each year's income is discounted back to present value; final salvage value is discounted; and these are all added together to give the total present worth of the property.

Many gas wells, particularly those in high-pressure reservoirs, produce a liquid called "condensate." This liquid was in the sand, but it was there as a vapor. It condenses into a liquid when brought to the lower pressure and temperature at the surface. The amount varies from less than 10 to more than 100 barrels per million cubic feet. If the reservoir is under water drive, the yield of condensate will remain substantially constant; but, if it is under gas-expansion drive, the yield will decline as the pressure drops.

In the example field, it is assumed that the yield is five bbls. per mmcf of 54° API condensate. Since there is a water-drive, it is further assumed that the yield will remain the same, and that the price will remain at $2.90.

As in the case of oil, the future sales price of this 21,690 barrels of distillate (4,338 × 5) is discounted back to present value by using a five-percent discount factor. This factor is added to the present value of the gas. The distillate is worth more if the gas is to be produced quickly than if it takes a long time, as time brings on a heavier discount factor.

CHECKING THE VALUE

The final result of the valuation study should be checked by the geologist by any and every means available to him. As stated above, he should know the average acre-foot recoveries from similar recoveries. If his reserve estimate is far out of line, he should know why.

He should know the average selling price of oil in the ground and of gas in the ground. This price he can constantly know by analyzing the sales of producing properties. For instance, oil has sold recently from $.80 to $1.05 a barrel in the reservoir, and gas at 1 to $2\frac{1}{2}$ cents per thousand. These figures often enable one to check the present-worth estimate because sales are often 60 to 70 percent of present worth.

Whenever decline curves can be constructed, they should always be used. The solution should be worked out by projecting the curve. The curve should come out in the proximity of the computed estimate—if it does not, another check should be made. If any material-balance equations can be solved, they should always be used—even though they give data only on the entire field, they often confirm or deny the conclusions.

The evaluator should always go on the property, physically inspect everything he can see, quiz the gauger who handles the lease, and get the latest production figures he has. In setting up a reserve estimate, the valuator says by implication that he expects the wells to produce for years and years—he must be certain they are not already beginning to fail.

CONCLUSION

Accurate valuation of oil and gas properties now depends entirely on a proper understanding of the conditions existing in the producing reservoir. Since one cannot go down in the reservoir, as one does in a mine, all conclusions must be based on inference, comparison, and experience.

Although an understanding of the fundamentals of production engineering is needed, it serves only as a background. The geologist who is familiar with all of the producing structures in the trend, and who knows the geologic section in the wildcats is best equipped to analyze the production data from any one field. There is much more to outguessing Mother Nature than being able to reduce well performance to exact figures, or to analyze an electrical log into precise millivolts and ohms.

Geology, being a study of the earth itself, involves its students in a consideration of the source of sediments, conditions of deposition, up- and down-dip facies changes, lateral variations in thickness or character of reservoir rocks, and countless other kindred problems of sedimentology. Because of his training in these matters and his continued interest in them, the well-grounded subsurface geologist makes the most reliable evaluator.

QUESTIONS

1. What is meant by "decline curves"?
2. What is the Law of Equal Expectations?
3. Discuss "proration."
4. Every engineering valuation of an oil property assumes what?
5. In figuring reserves what factors must be considered?
6. What is the significance of the number "7758"?
7. Analyze and discuss the formula $R = 7758 \times A \times T \times P \times (1 - I) \times S \times F$.
8. What is the significance of figure 413?
9. How and why should valuation results be checked?
10. Are gas reserves figured to atmospheric pressure?
11. Is the statement "the deeper the well, the greater the shrinkage factor" correct?
12. What is meant by the "Z factor," "law of capture," and "recovery factor"?

CHAPTER 9

DUTIES AND REPORTS OF A SUBSURFACE GEOLOGIST

GEORGE W. JOHNSON

The work of a subsurface geologist as he "sits on" a well may be divided into two general categories: collecting and assembling data, and reporting. Such work will be rather generally described first; then specific examples of field practices will be given.

DUTIES OF A SUBSURFACE GEOLOGIST ON AN EXPLORATORY WELL

It is common practice during exploratory drilling for a geologist to "sit on" the well. His responsibilities are to assimilate and evaluate all subsurface geologic and frequently engineering data during penetration. These responsibilities include (1) collecting, preparing, examining, describing, and shipping ditch, core, gas, and fluid samples; (2) keeping the log of the well up to date; (3) recommending coring, formation tests, and logging surveys; (4) evaluating all oil and gas shows; (5) witnessing or supervising formation tests; (6) following drilling operations in contiguous areas; (7) selecting proper casing points; (8) forecasting possible strata that may cause drilling difficulties; (9) running salinity and water-loss tests on mud; (10) submitting geologic progress and summary reports; and (11) cooperating with the petroleum- and production-engineering personnel and with the subsurface laboratory.

Well Log

The well-site geologist is called upon to keep a current log of the well at all times. All pertinent data such as lithology, cored intervals, electric profile, casing records, tested and perforated intervals, hole deviation, the formation-test summary, and oil and gas shows should be plotted on tracing cloth, thus permitting periodic progress copies of the logs to be submitted to the home office.

Ditch Sampling

Ditch samples should be taken at ten-foot intervals down to the proximity of important markers and at five-foot or lesser intervals thereafter until the desired point is established. Fresh ditch samples should be examined by ultraviolet light through all possible producing intervals while drilling is in progress.

There is always an interval between the time at which a formation is drilled and that of the return of its fragments to the surface. This time interval depends on such factors as the size, rate, and efficiency of the circulation pumps, the mud viscosity, and the diameter and depth of the hole. When these variables are known, the approximate time lag can be calculated and adjustments of sample depth made. The adjustments can be verified by adding such material as corn, rice, or wheat to the circulation system and then making a record of the time required for the material to reappear in the surface overflow. From these data a graph may be constructed for estimating the time lag for various depths. Comparison of lithologic breaks with the electrical profile and drill-time data also permits correction of sample lag.

Continuous coring does not necessarily require retainment of ditch samples; it is advantageous, however, to collect them if core recovery is inadequate.

The quality and reliability of ditch samples decrease with depth and penetration-rate increase. In soft formations where penetration is rapid, sample contamination is excessive, and it is impossible to determine the actual depth position of the cuttings. When penetration is slow, samples are more representative, and their stratigraphic positions can be more accurately determined.

Samples should be washed thoroughly of drilling fluid, dried, carefully examined by hand lens or microscope. The ratio of the various lithologies is noted, described and plotted, tagged, and submitted to the subsurface laboratory for refined examination.

Samples of the drilling fluid should be continuously examined while fresh under the ultraviolet light at frequent intervals during penetration. A gas detector is recommended to record gas shows in the drilling fluid.

Drilling Rate

A careful record of the penetration rate should be kept continuously. Extreme care should be exercised to tabulate any abrupt increase or decrease in the penetration rate while drilling is in progress. These breaks should then be evaluated in terms of the stratigraphy.

Coring

Coring programs vary with the problem involved and with the company policies. Several companies developing the same area may follow entirely different avenues of coring procedure. There cannot be too many cores from an exploration drill hole.

Cores are generally extracted and washed on the derrick floor and placed in core boxes or trays. It is important that the cores be carefully marked so that correct depth sequence can be followed.

The label of each core should include the interval cored and the footage recovered. If more than one tray of the same core is involved, a label should be placed on each tray; in addition, each tray should be labeled as, for example, "Tray No. 1 of 3 trays." Labels should be placed on the bottom end of the tray. Tray No. 1 is commonly assumed to include the top three feet of the cored section.

Cores should be taken at the discretion of the well-site geologist in collaboration with the district geologist when important changes of formations are noted. Coring is undertaken to determine the position of marker beds, contacts, oil and gas shows and whenever required for additional paleontologic or geologic information. The thickness of the cored section should be reduced to a minimum yet be sufficient to satisfy requirements.

The interval between cores in exploratory holes should not exceed 50 to 100 feet within possible productive strata. Coring may begin at the first bit change below the surface casing; thereafter, cores should be taken at intervals of no more than approximately 500 feet down to the top of a potentially productive zone.

When drilling to an important stratigraphic marker, coring at approximately 50-foot intervals should be started about 300 feet above the anticipated depth. When establishing marker positions it is sometimes preferable to drill alternately with a wire-line core bit and a drilling bit.

If a decided change in the penetration rate indicates that a porous phase has been encountered while a potentially productive formation is being drilled, two to five feet of section should be penetrated and drilling suspended until circulation cuttings from the bottom have reached the surface and have been checked lithologically and by the ultraviolet light for oil showings. When oil or gas showings are encountered in this or any other manner, coring should begin. If no indications of oil or gas are obtained, drilling may be resumed through the porous body and returns continuously analyzed by ultraviolet light. A core should be taken regardless of shows after drilling 15 feet in any continuous porous phase within a potentially productive part of the section.

Ditch samples and cores frequently give oil shows from thin sand laminae in shales that are not potential reservoirs. If the first core or cores taken after encountering such shows should indicate that the formation is not potentially productive, drilling may be resumed; but extreme care must be taken to define drilling breaks in petroliferous sections, at which coring should again commence.

If ditch samples should show an increase in sand content of a petroliferous section being drilled according to the foregoing procedure, indicating that a potential reservoir stratum is being penetrated, coring should be started to determine the nature of the formation.

If limestones or dolomites are being drilled, check cores should be taken when the ditch-sample fragments show evidence of oil staining, coarse crystallinity, vuggulation, or fracturing.

On each exploratory well, core samples of prospective oil or gas sands should be immediately preserved by the well-site geologist according to laboratory instructions for general core analysis (porosity, permeability, and fluid determinations).

On wells drilled subsequent to the first in a field, coring should be reduced as a result of information already available, although care should be exercised to refine or confirm important stratigraphic data by selective coring.

Side-wall cores should be taken in uncored sands that appear promising on the electric log and also at points where samples may be required for special geologic information. All sand samples from sidewall cores should be tested in ultraviolet light and by other methods immediately after their extraction.

Preparation of Core Samples for Analysis and Shipment

A short description of the procedure in wrapping cores for shipment follows. It is not necessary to ship excessively large cores for testing, the ideal length being about eight inches. A different technique is employed in shipping cores, when fluid saturations are to be determined, from that when permeabilities and porosities are desired.

When only the permeability, porosity, or grain size is required, precautions against fluid loss need not be considered. Generally, it is better to leave the mud cake intact rather than to remove it, as the cake offers some support to the core. The core sample should be labeled with the name or number of the well, the depth of the core, and, where applicable, the zone name. The core can then be packed in waste, sawdust, excelsior, or shredded newspaper in an individual container. The most satisfactory containers are, in order of preference, sealed metal cans, cylindrical cardboard ice-cream containers, and glass fruit jars. The containers should be packed in suitable wooden boxes and packing material added to minimize breakage.

When the fluid saturations are desired, that is, the oil and water content of the core, rapidity in preparing the core for shipment is essential in order to prevent evaporation losses. As soon as possible after the core is removed from the core container, it is tightly wrapped in at least two layers of lead foil. After the core is wrapped, the label is applied, and the core is dipped in melted parafin wax for proper sealing. The core can then be packed in the individual container. Care should be taken to avoid breaking the wax coating during packing and shipping.

Core Descriptions

The geologist should examine the core immediately upon its extraction and prior to cleaning it in order to observe any oil or gas indications that may appear on or within the mud cake. However, the core should be thoroughly cleaned before a description of the lithology or sampling is attempted. Certain types of cores are best cleaned immediately after they are removed from the core barrel; others may be cleaned after the mud cake has almost dried.

The first data recorded when describing cores are the cored interval and the footage recovered. Core dips, fracturing, and striae direction should be noted and carefully rated.

Cores are described individually from the top down. If the entire core is of homogeneous material, it may be described as a unit. If it is lithologically heterogeneous, each lithic unit requires description.

An example for describing a core of diversified lithology is given below:

Core No.	Cored interval	Feet recovered		Lithology
321	5,919–29	8	1.5	*Mudstone:* dark-gray with occasional irregular streaks and lentils of light-gray, slightly calcareous siltstone.
			2.0	*Sandstone:* fine-grained, somewhat argillaceous, micaceous, friable, gray-white; contains several two-inch layers of shale in lowermost part; dip 15°.
			2.0	*Limestone:* fine- to medium-crystalline, light-gray; minutely vuggulated in middle part; fossil fragments conspicuous throughout; trace of glauconite in lower six inches; few minute fractures extending in all directions; few minor *traces of residual oil* along fracture planes; upper boundary gradational into overlying sandstone; lower contact sharp and distinct.
			1.5	*Dolomite:* coarse-crystalline, light-tan, highly vuggulated, massive; containing irregular *oil staining* throughout; no fracturing evident; vugs range in diameter from one-eighth to one-half inch and are uniformly distributed.
			1.0	*Shale:* black, carbonaceous, well-laminated; fish scales common on parting surfaces; dip 20°.

Comments: Drill-time data indicated at least four feet of sandstone between the depths 5,920.5 and 5,924.5. The corrected lithologic subdivision follows: mudstone, 5,919–5,920.5, sandstone, 5,920.5–5,924.5, limestone, 5,924.5–5,926.5, dolomite, 5,926.5–5,928, shale, 5,928–5,929. Minor amounts of gas appeared in the porous dolomite unit; formation test is recommended.

Core Sampling

Core-sample intervals will vary in different wells according to the rock types and subsequent evaluation methods to be applied. During sampling the stratigraphic position of the sample and the interval of which it is typical should be noted. Even though a composite sample of several small fragments is taken, it is advisable to include sections that exhibit dip, porosity, or other special features that may be of significance in evaluating the cored interval.

Samples should be dried under atmospheric conditions before being submitted to the laboratory. A tag showing the well number and the type and depth of the sample should be attached on the outside of the sacked sample. If it is a core sample, the cored interval, the recovery, the location of the sample in the core, and the interval of which it is typical should be noted.

Fossils should be wrapped separately to prevent breakage during shipment.

Core sections to be analyzed for magnetic polarity should be carefully selected, labeled, and packed. Immediately after removing the core from the inner barrel, the geologist should clearly mark the top and bottom of the core. The greatest source of error in this method is the incorrect labeling of the tops and bottoms. Core sections should be at least eight inches in length and should exhibit sufficient bedding to provide dip data. Dips exceeding 10° give the most satisfactory results. Interbedded sandstone and shale cores are most desirable. Rarely do massive limestones and dolomites give favorable orientation results. When core samples are sent for analysis, the following information should accompany them: the type of material, the age and stratigraphic position, and the inclination and direction of the hole at the point at which the sample was taken. All cores should be removed from stray magnetic fields, wrapped wet in cellophane, and sent as soon as possible to the laboratory.

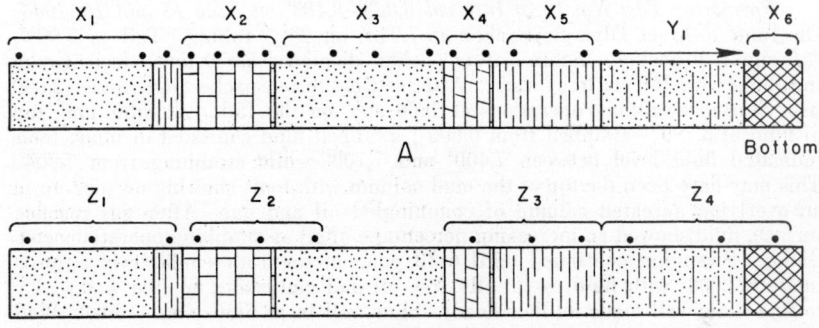

FIGURE 414. Correct *(A)* and incorrect *(B)* manner of sampling a well core for analysis. Spot samples (●) may be taken at any desired interval and composited (X_1, X_2, etc.). Groove sampling (Y_1) is sometimes followed. Samples composited as shown by the Z series should be used only for general purposes, since such samples represent mixed lithologies.

Formation Testing

Information given on a drill-stem or formation test depends on the nature of the equipment used. Some of the points that should be considered are the name and type of the tester, the depth of the hole, the depth at which the packer it set, length of pipe below the packer (whether blank or perforated), and choke data. The time the packer is set should be noted. The hole should be full of mud, and the geologist or engineer should note at the time of packer setting whether the fluid level in the hole drops, and, if so, the rate of the drop. After the tester is opened, results should be carefully observed and tabulated. Pressure data should be recorded at regular intervals and all variations recorded. The time the tester is released should be recorded. When the drill pipe is extracted, the number of stands of recovered fluid should be listed. From the volume and type of fluid recovered and the time duration of the test, the rate of flow into the pipe can then be calculated. All fluids and gases should be sampled and described.

When depth and other drilling conditions make it practical, depending upon the nature of the formation encountered, formation tests in wildcat wells should be made on the upper five to ten (or less) feet of any permeable interval where production possibilities appear favorable. When a productive interval is encountered, subsequent tests should be made on overlapping intervals to locate gas-oil and oil-water contacts. On wells subsequent to the first on a structure, intervals to be tested can be more closely planned from results of the first well. Provision must be made to record volumetric measurement of all oil, gas, or water recovered.

In exploratory work the well-site geologist may be responsible for recording formation-test data. A recording form commonly followed is given below:

Formation Test No. 12 of Interval 8,028'–8,183' on June 15 and 16, 1945: On June 15th set Olympic packer at 7,948' on 2⅜" tubing. Tail to 7,999', ¾" choke in packer. Before setting packer, displaced 5,670 lineal feet of mud in tubing with 160 cu. ft. of water. After setting, tested packer; OK under 2,000 psi. Started swabbing operations at 1:00 a.m. on the 15th. At 8:00 a.m., top of fluid at 5,790'—swabbed from 6,600'; swabbed fluid consisted of mud; tools indicated fluid level between 7,400' and 7,700' while swabbing from 7,904'. This may have been the top of the mud column with tools showing no pick-up in an overlying aëreated column of commingled oil and gas. After gas reached surface, fluid showed an increasing percentage of oil as swabbing operations continued. Swabbing continued until 6:30 p.m. on the 15th—fluid level approximately 7,000' with fluid essentially free oil and emulsion; no free water. Up to this time, well made no flow heads. Shut well in at 6:30 p.m. (15th). Well remained shut in from 6:30 p.m. to 12:30 p.m. with shut-in surface pressures as follows:

$$7:30 \ p. \ m. — \ \ 575 \ psi$$
$$8:30 \ p. \ m. — \ \ 825 \ psi$$
$$9:30 \ p. \ m. — 1,125 \ psi$$
$$10:30 \ p. \ m. — 1,350 \ psi$$
$$11:30 \ p. \ m. — 1,500 \ psi$$
$$12:30 \ a. \ m. — 1,650 \ psi$$

Opened well at 12:30 a.m. through a 8/64" bean. After releasing pressure, flow was continuous—fluid oil with some emulsion. No free water After equilibrium flow conditions had been established, tubing-head flow pressure approximated 225 psi.

During a two-hour gauge ending at 12:45 p.m. (16th) production totaled 8.61 bbls. (4.3 B/H or 103.2 B/D) through a 8/64" bean. Total cut 3.8%— water testing 192 G/G salt. Without changing the bean setting, measured gas rates between 1:20 and 2:25 PH as follows:

Time 24-hour rate		Time 24-hour rate	
1:20 p. m.	360 Mcf	1:55 p. m.	330 Mcf
1:25 p. m.	350 Mcf	2:00 p. m.	348 Mcf
1:30 p. m.	348 Mcf	2:05 p. m.	340 Mcf
1:35 p. m.	352 Mcf	2:10 p. m.	314 Mcf

1:40 p. m.	330 Mcf	2:15 p. m.	306 Mcf
1:45 p. m.	340 Mcf	2:20 p. m.	319 Mcf
1:50 p. m.	296 Mcf	2:25 p. m.	314 Mcf

At 4:30 p. m. (16th), increased flow bean setting from 8 to 12/64″. While on gauge from 5:25 to 7:45 p.m. (2 hrs. and 20 min.), well produced a total of 12.59 bbls. of fluid (5.4 B/H or 129.6 B/D rates). The tubing head flow pressure varied between 90 and 110 psi. Trap pressure 50–60 psi. No cut was taken during this gauge.

At 8:00 p. m. on June 16th, pumped 193 cu. ft. of 82-lb. mud into tubing forcing hole fluid in tubing back into formation. Pulled packer loose at 9:00 p. m.

Gas Samples

Where newly discovered oil- or gas-bearing intervals are evaluated by means of a formation test or a conventional production test, representative samples of the oil and gas should be collected in suitable containers to be subjected to complete analysis later.

Tests for Oil

All sands concerning which there is any question of their possible productivity should be tested for oil and gas. Such tests can be made in various ways. In the case of a core, the sample tested should be taken from the interior to avoid exterior contamination. Chloroform, ether, and carbon bisulphide as solvents of petroleum are used for making these tests. In the procedure a few cubic centimeters of the crushed sample is placed in a test tube, enough solvent is added to cover the sample, and the test tube is shaken. If petroleum is present, the liquid will show various shades of brown, the shade depending on the amount and type of petroleum present. A doubtful or faint test can sometimes be confirmed by slowly dropping the supernatant liquid from the test tube on a piece of filter paper. If the test is positive, a dark ring will form around the periphery of the expanded drop when the paper has dried.

In the event very light-fraction oils are involved, a more delicate test is made with acetone. This test is carried out by placing about 10 cc. of the sample in a test tube and adding 20 to 30 cc. of acetone. After the mixture is shaken, 20 to 30 cc. of distilled water is added. The presence of oil turns the liquid milky. As this is a delicate test, an oily thumb held over the test tube during the shaking will frequently give a positive test. A positive acetone test cannot be considered as absolute proof of the presence of oil, but a negative one will prove its absence.

Salinity Test

A salinity test should be made on the drilling fluid periodically, preferably not less frequently than every 24 hours, and recorded on the daily drilling report. If water is recovered from formation tests, a salinity determination should be made and a properly marked sample retained for possible subsequent analysis. Salinity tests should also be made on core samples of sands.

Electric Logging

Electric logs should be periodically run in order to check the penetrated section. Anomalies should be evaluated as soon as possible by sidewall cores, by re-examination of cuttings or by formation testing. All lithologic adjustments should be made on the progress log.

Radioactive Logging and Caliper Logging

Radioactive logging should be used to determine porous intervals in limestone and as an aid to accurate selective perforating. Before the casing is set, a caliper profile should be obtained in order that proper cement computations can be made.

Through the courtesy of the Richmond Exploration Company the following procedure for sampling, logging, and testing of wildcat wells in Venezuela is given.

Ditch Samples

Cuttings—These must be taken at ten-foot intervals down to the vicinity of the first important marker and at five-foot intervals below that point to total depth, always including intervals cored. Ditch samples must be examined by ultraviolet light during drilling through a possible producing section, as soon as possible after the cuttings reach the surface. Samples showing fluorescence must be tested for "cut." Samples must be preserved in sacks and tags properly marked for identification. Ditch samples should be used for general logging purposes.

Drilling Fluid—Samples of drilling fluid must be examined while fresh under ultraviolet light at frequent intervals during drilling through all possible producing sections.

Coring

General—Cores should, in general, be taken when changes of formation are noted, and to determine location of marker beds, at contacts, at oil shows and also whenever required for essential paleontologic or geologic information. Thickness of section cored should be kept at a minimum to satisfy requirements.

Maximum Interval—The maximum interval between cores in any exploratory hole should not exceed 500 feet. This interval may vary according to desired results.

Interval—Important Markers: In any case where drilling depth to an important stratigraphic marker is not predictable within fifty feet with a reasonable degree of certainty, spot coring at 50-foot intervals should begin 300 feet above the anticipated depth to the marker.

Porous Zones—When a decided change in penetration rate indicates that a porous sand or other porous zone may have been encountered, two feet to five feet of penetration should be taken into this zone and drilling suspended until circulation samples from bottom have reached the surface and have been checked for oil showings by ultraviolet light. When oil or gas showings are encountered in this or any other manner, the full vertical extent of the zone should be ascertained by continuous coring; this coring should be continued to a minimum of 20 feet below the last-observed oil show. If no indications of oil or gas are obtained as first noted above, drilling should be resumed through the porous zone and continuous analysis of returns by ultraviolet light made until the porous zone is penetrated. A core should be taken regardless of shows after drilling 30 feet in any continuous sand body or other porous zone.

Core-Analysis Samples—In each exploratory well, core samples of prospective oil and/or gas sands should be carefully selected and properly preserved according to laboratory instructions immediately after extraction for general core analysis, particularly for porosity and permeability, and also for fluid saturation whenever occasion arises. Representative samples from among these should be forwarded to the main office in the event that local facilities for analysis are inadequate.

Step-Out Wells—On wells subsequent to the first on a given structure, coring should be reduced as a result of information already at hand.

Side-Wall Cores—Side-wall cores should be taken in sands not cored originally, which look promising from electric or radioactive log data and at points where samples may be required for geologic information. All sand samples from side-wall cores should be tested in ultraviolet light immediately after extraction.

Core Recovery—Coring operations must be continuous until satisfactory recovery is obtained or until the objective of coring has been satisfied. On certain occasions, a small percentage recovery may be considered as "satisfactory recovery," as in many limestone sections in which experience has proved that normal recovery is low.

Formation Testing

When depth and other drilling conditions make it practical, formation tests in a wildcat well should be made on any sand or other permeable zone where productive possibilities are indicated. Ordinarily, when a productive zone is encountered, formation testing should be continuous in stages up to 100 feet throughout the vertical extent of the productive material to locate gas-oil and oil-water interfaces, if present, as well as to evaluate productivity within reasonable limits. Overlap intervals between tests should be minimized to preclude obtaining misrepresentative information. On well subsequent to the first on a given structure, zones to be tested can be more closely planned from results of the first well. Provision must be made to record volumetric measurements of all oil, gas, and/or water recovered.

Mud and Water Tests

Salinity Tests—Drilling-fluid salinity and water-loss tests should be made periodically, preferably not in excess of 24 hours, and should be reported on daily drilling reports. In case water is recovered from formation tests, a salinity test should be made and a properly marked sample retained for possible subsequent geochemical analysis.

Mud Tests, General—Other mud tests and analyses conforming to accepted current practice must be made and reported as a routine procedure. Special mud tests should be made as called for.

Electric and Mechanical Logging

Electric Logs—Electric logs should be taken prior to landing all casing, except surface casing, on completion of hole to final depth, and also at other points as required by special circumstances in which specific information is required regarding a portion of open hole during the process of drilling. Recognition of any zone, the possibilities of which have not been satisfactorily indicated by other means, should constitute such a special circumstance. Electric-log data should not be relied upon as final evidence regarding productivity in the early stages of development, but should be carefully checked with other evidence obtained during the drilling of the well.

The maximum interval during which the hole should be exposed to drilling fluid prior to electric logging should be kept under careful consideration, taking into account all known factors which will affect logging results, such as lithology and mud type and quality.

Gas Detector—A gas detector should be used to log gas indications.

Dipmeter Surveys—Dipmeter surveys should be used wherever practicable. Coring should be reduced to whatever extent dipmeter surveys can be safely relied upon to substitute for structural-control data otherwise obtainable only from cores.

Penetration Logging—Penetration rates must be accurately logged. A mechanical device which has proved satisfactory for this purpose by field use should be utilized, the log from which should be checked against drillers' observations.

Radioactive Logs—Radioactive logs should be run opposite all carbonate sections. Use of these methods opposite sand and shale sections should depend upon estimated value of resulting information.

Caliper Logging—Caliper logging should be utilized wherever available. and when prospects of useful resulting information are estimated to justify costs of the service.

Oil and Gas Samples

Where newly discovered oil- or gas-bearing measures are tested, either by means of a formation test or a conventional production test, representative samples of the oil and gas should be collected in suitable containers so that they may be shipped to the United States for analysis, if the local management deems the productive possibilities to be of sufficient magnitude to justify this expense.

Facilities for determination of gravity of oil samples should be available at the well, and determination of gravity should be made on all samples immediately upon recovery of these. Observation of the general character of oil and gas should also be made immediately upon recovery from tests, particularly for presence of sulphur.

An example of the form used in recording data from a wildcat well, as followed by the Richmond Exploration Company, is given on page 821. The use of the form is explained below:

General

Purpose—The purpose of "Exploratory Well Summary" is to provide a routine channel for a concise summary, arbitrarily limited to one page, of the most widely used geologic and related well information for the use of all persons properly concerned.

Preparation Schedule—The form should be completed for each significant exploratory well, whether active, completed, or abandoned. It should be revised for any given well whenever new or better information becomes available for such well.

Uniformity—Uniformity of usage adhering to the following explanation must be maintained in order to assure continued clear understanding by all readers of data presented. Suggestions for major changes in usage of the form are in order at any time, but these must be agreed on with the head office before implementation.

EXPLORATORY WELL SUMMARY
ASCENCION NO. 1
(Pacific Exploration Company)
(Well)

| PARAGUAY |
| (Country) |
| PARANA |
| (Region) |

Date: Nov. 17, 1941

LOCATION: Ascencion structure, Santa Rosa department. Coord:
N. 176, 261 E. 163,644. (Origin: Cruz Nevada N. 200,000 E.
200,000).

ELEVATION: 148.6' (DF) 139' (Gd).
SPUDDED: Sept. 12, 1940   DEPTH: 12,680' T.D. (7-25-41).
OPERATIONAL STATUS (and Date): Suspended. (7-26-41).
STRUCTURE: Faulted anticline.   OBJECTIVE: Braden sand.

CASING	
SIZE	SHOE
16"	638'
8⅝"	10,454'

PRODUCTION TESTS AND SHOWS

		TESTS		SHOWS
Interval	Test	Rise	Recovery	
7160-7250'	JFT	2500'	20% oil; 60% mud.	7170--7268' Minor oil and gas shows.
8235-8425'	JFT	5000'	Muddy water.	8220-8460' Good oil and gas shows.

FORMATION	INTERVAL	DEPTH
FINAL		
Pleistocene		
Santa Maria	900'	0- 900'
Miocene		
Rosa	1,150'	900- 2,050'
Upper Dalmar	2,550'	2,050- 4,600'
Lower Dalmar	1,945'	4,600- 6,545'
Oligocene		
Concepcion	615'	6,545- 7,160'
Eocene		
San Pablo	3,520'	7,160-10,680'
(Upper Salina)	(58')	(7,182- 7,240')
(Lower Salina)	(155')	(7,600- 7,755')
(Castro sand)	(172')	(8.243- 8,415')
Rocha	869'	10,680-11,549'
(Braden sand)	(350')	(10,680-11,030')
TENTATIVE		
Paleocene		
Cesana	576'	11,549-12,125'
Cretaceous		
Upper Winther	555' Penetrated	12,125-12,680' TD
ESTIMATED		
Cretaceous		
Winther	860'	12,680-13,540'
Taube limestone		13,540-

BY: HBM/dc DISTRIBUTION: RML HFT NFE WRS AMP PCC LOR WELL: Ascension
No. 1.

Confidentiality—Particularly considering that this form summarizes much
of our most significant drilling data, confidential status of compiled forms pertaining to company wells must be carefully preserved. At the same time, the
usefulness of the form will be measured by the extent to which these data may
be fully utilized by all company personnel properly concerned with this information. Distribution must be correspondingly inclusive.

Location

Enter the following data under this heading in the following order:

Designation—The designation of the prospect or area, such as "Gualanday No. 1 anticline," "Totumal anticline," "Berreberre anticline"; the general concession, such as "Dubasa"; and the state or province, such as "Choco."

Location Details—Enter coordinates and coordinate reference point where known; otherwise enter shot point or reference to other generally known survey point.

Elevation

Enter elevation in feet referred to local standard elevation datum. Specify "derrick floor" (DF) or "ground" (Gd). Enter both if available.

Spudded

Enter spud date.

Depth

Enter last available depth. Indicate total depth by "T.D." Add date on which depth was reached.

Operational Status (and Date)

Enter current operational status as of date, such as "gravelling road," "rigging up," "drilling," "coring," "suspended," and "abandoned." Add date on which current status was reached if different from current date, as: "Completed (Aug. 17, 1946)."

Casing

First enter size and shoe depth for strings already set. Immediately underneath last string list, enter the word "programmed" and add casing program for remainder of hole.

Structure

Indicate type of structure and closure and method of exploration used to determine these factors, such as "fold closure, fault, seismic."

Objective

Indicate the principal stratigraphic objective zone.

Production Tests and Shows

Present the best summary of tests and shows to current date, which can be shown within the space allotted on the form for this purpose.

Stratigraphy

Outline stratigraphy as indicated below, including all potential producing formations within reach of the drill (about 15,000 feet for a standard deep-drilling rig), regardless of scheduled drilling depth.

Formation—Refer to the example given. List geologic epochs, formational units, and important members (sand and/or others) of formations in outline order under "Formation." As main headings in this outline, use terms qualifying the degree of certainty of the interval and depth figures given, such as "final," "tentative," and "estimated." Insert these main-heading terms wherever they properly apply to the data. For instance, after drilling into the Eocene, all intervals and depth data down through the Oligocene-Eocene contact may be considered as final, whereas all figures below the top of the Eocene would remain in the status of estimates.

Enter important members (sands and/or others) of formations in outline form and in stratigraphic order under their respective formation headings. Place member names and figures in parentheses to emphasize that depth ranges of these units occur as insertions in the "Depth" column, and do not follow consecutively in the regular consecutive sequence of formation contact depth figures.

After the main heading "Estimated" add in parentheses an estimated figure for allowed error in depth figures as "±300'." Alternatively, refer by asterisk to a brief footnote concerning amount of error allowed in estimated depth figures.

Explain other major existing discrepancies by footnote.

Interval—Enter net drilling interval of each separate stratigraphic unit.

Depth—Enter depth range of each separate stratigraphic unit.

By—Indicate author and typist by initials, as in ordinary correspondence.

Distribution—Indicate distribution of copies.

OBJECTIVE REPORT WRITING [a]

A report from a professional geologist is the tangible presentation of his work and opinion, for which his clients pay a fee.

Objective report, as here defined, is one that gets down to cases at once to the satisfaction of a client in industry. It is assumed in the following suggestions that the geologist is retained to solve a problem that the client cannot solve for himself. It is not implied that all professional, consulting problems can be flexed to the following matrix, but a very great number of them may be.

Preliminary Procedure

Diagnose the problem.

Make adequate investigation (field and reference) for all data that might be pertinent to the solution of the problem.

Include in the assembled data information which seems reliable and pertinent, but which cannot be verified on account of time or physical limitations. (For example, an inaccessible, back-filled stope in a mine, or a log of a long-abandoned well.

Integrate these data into an answer to the problem, or into a pattern of recommended operations that will contribute to that end.

Writing the Report

Write the story completely. Do not stint on words, pages, or time.

Revise and correct the sentences. Good English may not be complimented, but bad English is "felt" even if the reader himself does not quite know why it is bad. Good English inscribes in the mind of the reader, a client, "Here is a man who cares."

Select an assortment of colored pencils. Then go through the composition and underline, in color, "pertinent" data and "incidental" data.

[a] This section, "Objective Report Writing," is by Paul H. Keating, Department of Geology, Colorado School of Mines.

"Pertinent" data are data essential to your conclusions and recommendations. "Incidental" data are information that is, or may be, of interest to those who are to investigate further, or to a reader who seeks more comprehensive information.

Reread the composition, one color at a time. Try to assume that you have never heard of the problem involved. Ask yourself, "Would I understand it?" Let your judgment be your guide on deletions or additions.

Assemble the composition in semi-final form, as indicated by the colored underlinings. At all times keep in mind first the problem, and second the data pertinent to the solution of the problem.

Write the final draft and use, of course, suitable headings and introductory summaries so that the reader will be properly oriented at all times. The following suggestions for the writing of the final draft are paramount.

1. State the problem as you have diagnosed it.
2. Use maps and diagrams that are well titled, scaled, and oriented in preference to long-winded descriptions.
3. State conclusions and recommendations. If there are qualifying, or unknown and unpredictable factors to the conclusions, tabulate such factors immediately below the conclusions.
4. Tabulate or discuss the pertinent evidence used to formulate conclusions and recommendations. For example, point out parallel situations in your experience where the drawing of similar conclusions was to the advantage of your client. Also point out differences in such situations.
5. Compact into a sort of epiloque all loosely relevant data: history, scenery, weather and road conditions, temperature, personnel contacts, labor troubles, and all other such trivia for the use of those who want minutely detailed information. But do not allow trivia to obscure your work and opinion.

Summary

The function of a professional geologist is to solve a problem. When called upon by industry, you must diagnose, investigate, and report. Your professional integrity and standing will grow in proportion to the competency of your diagnosis, the adequacy and accuracy of your recorded data, and the soundness of your conclusions and recommendations.

If you are sure, say so. If subsequent evidence proves you to be wrong, you are still free from guilt, in that your conclusions were honest in the light of the facts available at the time the conclusions were made. If you do not know, or are not sure, say so. Many a banker, doctor, lawyer, editor, and prelate has suffered oblivion because he was afraid to admit he did not know the answer. If you do not know the answer, say so. Honesty has the curious property of retaining respect under all circumstances.

Remember, an objective report is the tangible presentation of your work submitted to a client who has a problem he cannot solve. Give him the best answer you can, the best thinking you can contribute, the most data you can collect, the clearest English you can write—all in the fewest words.

PUBLICATION STYLE

If a report is written for a definite publication, the style of that pubpublication should be ascertained and followed; if the report is not for publication, the writer should select a style and follow it. Abbreviation, capitalization, compounding, use of numerals, punctuation, tabular form, and references must be correct, but they must also be consistent. Commercial publishers and learned societies [1] usually have style books, and it is suggested that the prospective writer obtain one of these.

Foremost style manuals to be recommended are the *United States Government Printing Office Style Manual*,[2] *A Manual of Style*,[3] *Words into Type*,[4] and *Suggestions to Authors of Papers Submitted for Publications by the United States Geological Survey*.[5] Further information on abbreviation is available in *Abbreviations for Scientific and Engineering Terms*.[6]

The preparation and reproduction of illustrations cannot receive full consideration here. Again the reader is referred to handbooks on the subject. *The Preparation of Illustrations for Reports of the United States Geological Survey*,[7] includes also brief descriptions of processes of reproduction. *Times-Series Charts*,[8] a manual of design and construction, has been issued as American Standard Z15.2—1938, reaffirmed 1947.

CREOLE WELL REPORTS

Through the courtesy of the Creole Petroleum Corporation, Venezuela, the following procedures are given for making weekly chronological well reports, weekly well-sample reports, completion log reports, and sand data reports. This material was prepared by the geological staff of the Creole Petroleum Corporation.

Weekly Chronological Well Report

The weekly chronological well report, as the name implies, is a detailed chronological account of activities of the well from the time rigging

[1] For example, *Preparation of Manuscripts*, Am. Assoc. Petroleum Geologists Bull., vol. 32, no. 2, pp. 307-311, 1948.

[2] *United States Government Printing Office Style Manual*, Superintendent of Documents, U. S. Government Printing Office, Washington 25, D. C., 1945. (An *Abridged Style Manual*, containing material in the unabridged edition except the part that is of interest solely to printers, is obtainable from the same source.)

[3] *A Manual of Style*, University of Chicago Press, Chicago, Ill., 1937.

[4] Skillin, Marjorie E., and Gay, Robert M., *Words into Type*, Appleton-Century-Crofts, Inc., New York, N. Y., 1948.

[5] Wood, G. McL., *Suggestions to Authors of Papers Submitted for Publication by the United States Geological Survey*, 4th ed., Superintendent of Documents, U. S. Government Printing Office, Washington, 25, D.C., 1935.

[6] *Abbreviations for Scientific and Engineering Terms:* Am. Soc. of Mechanical Engineers, 29 West 39th St., New York, N. Y., 1941.

[7] Ridgeway, J. L., *The Preparation of Illustrations for Reports of the United States Geological Survey*, Superintendent of Documents, U. S. Government Printing Office, Washington, D. C., 1920.

[8] *Time-Series Charts:* Am. Soc. of Mechanical Engineers, New York, N. Y., 1938, reaffirmed 1947.

up begins until its completion. When properly submitted, this report becomes a composite file of all drilling and testing activities, geological information, and production data from which subsequent records and pertinent facts can be accurately obtained. Experience has shown that this
procedure of recording well data is the most satisfactory because it is presented in sequence, an arrangement which facilitates reference and allows
for complete recording of data in a compact form for filing. Too often
valuable well information is lost because there is no standardized procedure
for recording the data, or more often because important facts are registered
only in the mind of an individual. Consequently, at a later date certain
operations, previously performed, will have to be repeated. It cannot be
over-emphasized that no matter how insignificant a detail, it should be
recorded in the weekly chronological well report, for a triviality may become an important fact at a future date. The report should be edited daily
from the daily drilling report. This procedure offers two distinct advantages: (1) All the prevailing information is recorded immediately in order
to preclude the omission of small but possibly pertinent details; and (2)
when the report is due, only final typing of the master copy is necessary.

The weekly chronological well report is to be presented in duplicate
form. Elite type is preferable. The report terminates each Monday at
7:00 a.m., or at some other convenient time, depending on the district
operations and prior agreement with the petroleum engineering and geological offices in the division covering the activities of the well for the
preceding week. The district geological, drilling, and petroleum engineering groups should become thoroughly familiar with this form and fully
cooperate in its preparation. One group is usually assigned the duty of
preparing this report. The contents should be checked by other interested
groups before they are typed in final form.

A description of the report is included on the following pages with
a typed example illustrating the form which is to be followed. It should
be noted that these selected examples are not from a single well. All districts must use the same patterns, as illustrated by the examples, to avoid
confusion.

Information Required in Weekly Chronological Well Report

Example 1—Rigging Up—The progress of rigging up should be reported from the time the rig is moved to the location until rigging-up
operations are completed. The first report on rigging-up operations should
state from where the rig was moved. See figure 415.

Example 2—Spudding—The time of spudding, date, size of bit, and
casing details, if any, should be given the day the well is spudded.

At this time, a brief resume of the drilling program should be presented, giving the expected depths of the most important markers and/or
sand tops, as well as a separate paragraph stating the location of the well,
ground and rotary elevation, drilling A.F.E., objective, concession, classification (Lahees), etc. See figure 416.

6-196 3-46

INFORME SEMANAL DE PERFORACION DE POZO

DISTRITO_____   SEMANA QUE TERMINA 7.00 A. M._____

POZO No_____   A.F.E. No_____   TALADRO No_____   GABARRA No_____

FECHA	PROFUNDIDAD TOTAL	PROGRESO	DIAS EN PERFORACION	OBSERVACIONES
10-22	-	-	1	Moving in rig from MP-34 to E2-I. Loc. 15½ r.u.

FIGURE 415

FECHA	PROFUNDIDAD TOTAL	PROGRESO	DIAS EN PERFORACION	OBSERVACIONES
10-24	530'	530	3	Fin. r.u. and spd. at 11:00 P.M., Oct. 23rd. Drld. from 0-530' w/12-1/4" Brewster bit. Ran and cmt'd. 16 jts. (506') of 9-5/8" csg. at 517' w/250 sx. of Atlas cmt. Job compl'td. at 3:00 AM/

The prog. calls for drlg. to the top of the La Pica fm. where a Sch. corr. surv. will be made. Depending upon the well's struct. position, the J-6 sd. will be cored and the well drld. 200' into the Carapita fm. after which Sch. and dip meter survs. will be made. S.W.C. will be secured of unusual sds. Est. markers are:

La Pica fm.	4500'	(-4000')
Textularia zone	6100'	(-5600')
J-6 sd.	6500'	(-5650')
Carapita fm.	6800'	(-6300')

Well will be compl'td. dually with the J-6 to J-12 sds. prod. thru the csg. and the J-17 to J-21 sds. prod. thru the tbg.

Well formerly designated as grid E2-I; loc. at coord. N-188,708.86 E-152,535.92. Grd. Elev. 490'; R.T. Elev. 500', AFE-J-2050 Conc. 391-2; Field Well.

FIGURE 416

FECHA	PROFUNDIDAD TOTAL	PROGRESO	DIAS EN PERFORACION	OBSERVACIONES
10-26	1700'	1000	5	Drld. from 700-950' w/8-5/8" Zublin bit. Pulled out, chg'd. bits and drld. from 950-1700' w/8-5/8" Hughes 08Q-2 bit w/3-1/2 pts. wgt. In hole drlg. Mud wgt. 10.7 lbs/gal. Visc. 45. Sypho at 1000' showed 1/2"dev.

FIGURE 417

Example 3—Drilling—Drilling operations should include footage drilled with each bit, size and type of bit, bit changes, points of weight, mud weight, viscosity, and deviation surveys. See figure 417.

Example 4—Cores—When core data is submitted, the first paragraph should embody the usual well activities. The second paragraph should contain the interval cored, recovery, and general description of each core. Examples are also given for submitting side-wall cores. Detailed description should be included in the weekly core and sample-description reports. See figure 418.

Example 5—Electrical Surveys—Before recording the information obtained from the electrical surveys, the first paragraph should state in

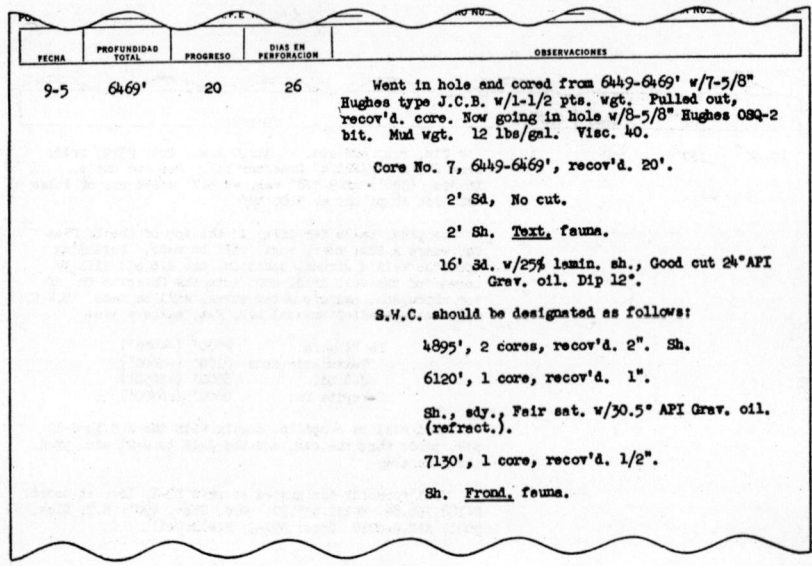

FECHA	PROFUNDIDAD TOTAL	PROGRESO	DIAS EN PERFORACION	OBSERVACIONES
9-5	6469'	20	26	Went in hole and cored from 6449-6469' w/7-5/8" Hughes type J.C.B. w/1-1/2 pts. wgt. Pulled out, recov'd. core. Now going in hole w/8-5/8" Hughes OSQ-2 bit. Mud wgt. 12 lbs/gal. Visc. 40.
				Core No. 7, 6449-6469', recov'd. 20'.
				2' Sd, No cut.
				2' Sh. Text, fauna.
				16' Sd. w/25% lamin. sh., Good cut 24°API Grav. oil. Dip 12°.
				S.W.C. should be designated as follows:
				4895', 2 cores, recov'd. 2". Sh.
				6120', 1 core, recov'd. 1".
				Sh., sdy., Fair sat. w/30.5° API Grav. oil. (refract.).
				7130', 1 core, recov'd. 1/2".
				Sh. Frond, fauna.

FIGURE 418

chronological order the activities of the well during the past 24 hours. The second paragraph should include the depth to which the log was run, total depth, formation and zone tops based on the electrical log, cores and ditch samples, correlation of the structural position of the well with nearby wells, and other geological data of interest or significance. Then a listing of the sand data should follow. This listing should describe the sands penetrated, intervals, and net sand content. Below these data a general interpretation should be made of the sands. The interpretations should be based on the electrical log, cores, ditch samples, and structural position.[9] See figures 419 and 420.

Example 6—Testing Program—The entire testing program should be outlined and recorded below the information set forth in example 5.

---

[9] These data are sometimes deleted from the chronological report on wildcat wells or other wells as directed by the Division or Caracas offices.

Any alteration in the program at a later date must be presented and reported under the following caption: *CHANGE IN TESTING PROGRAM*. See figure 421.

Example 7—Casing and Liner—All casing and liner records must be tabulated in detail and must state the intervals, size, number of joints, and weight of pipe. See figure 422.

Example 8—Isolation—In squeeze-cementing operations, the amount of cement, pressure required to break down the formation, maximum- and final-squeeze pressures, and the time cementing operations were completed should be recorded. Pressure at which the squeeze job is tested after drilling out should always be stated. See figure 423.

Example 9—Perforating—The time perforating was initiated and

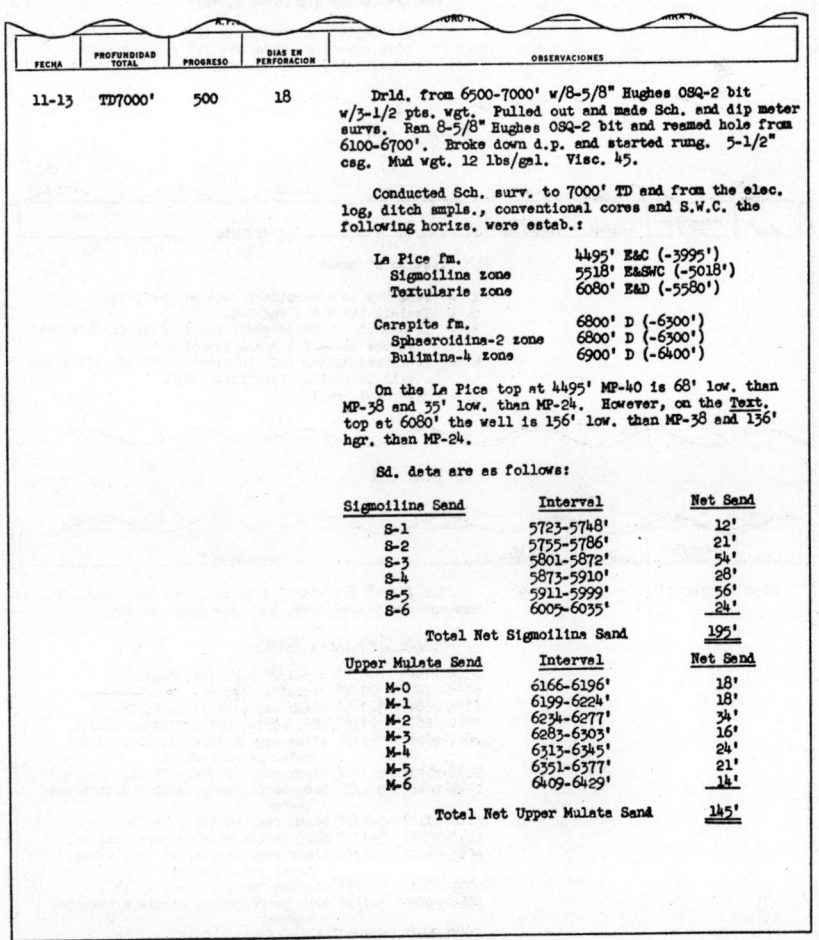

FIGURE 419

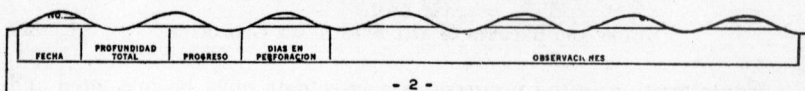

FECHA	PROFUNDIDAD TOTAL	PROGRESO	DIAS EN PERFORACION	OBSERVACIONES

- 2 -

Lower Mulata Sand	Interval	Net Sand
M-7	6483-6492'	5'
M-8	6509-6513'	3'
M-9	6524-6549'	22'
M-10	6562-6566'	3'

Total Net Lower Mulata Sand   33'

The S sd. development is comparable to that of MP-38 w/the reappear. of S-1. The resist. features of all the S sds. are the same as that of MP-38 w/the usual low third curve values in S-3 and S-4.

The only change from MP-38 in the M sd. development is the disappear. of 11' of sd. from the base of M-0. M-5 is compl'ty cut by the third curve and consid. wet which confirms the core data on this sd. M-3 and M-4 have gnrl. similar curves to those of MP-38 but are quest. at this struct. position.

Results of the dip meter survey:

6905-6915', S 13° W, dip 14°
7223-7230', N 3° W, dip 14° (calculated)
8110-8120', N 68° E, dip 10°

FIGURE 420

FECHA	PROFUNDIDAD TOTAL	PROGRESO	DIAS EN PERFORACION	OBSERVACIONES

Testing Program:

1 - Test M-9 to determine if oil or gas prod.
2 - Isolate btw'n M-1 and M-2.
3 - If S-4 sd. in MP-39 prod. W., isolate in this well at the base of S-3 and top of S-6.
4 - If corresponding int. is clean in MP-39, all S sds. will be perf'd. for final compl.
5 - Make dual compl.

FIGURE 421

FECHA	PROFUNDIDAD TOTAL	PROGRESO	DIAS EN PERFORACION	OBSERVACIONES
6-10	TD3001'	-	49	Ran 7-3/4" Brewster bit to bot. and cond. mud. Pulled out and ran 5-1/2" comb. lnr. Now prep. to cmt.

5-1/2" Comb. Lnr. Record.

2185-2190'	8-5/8" x 5-1/2" B.R. Lnr. hngr.
2190-2193'	5-1/2" circlt'n nipple
2193-2664'	5-1/2" blank csg., 15 jts., 15 lb.
2664-2667'	5-1/2" Bkr. Series Whirler cmtg. collar.
2667-2678'	5-1/2" blank csg. nipple, 15 lb., w/reg. metal petal basket.
2678-2741'	5-1/2" blank csg., 2 jts., 15 lb.
2741-2746'	5-1/2" Bkr. perfd. cmtg. nipple w/inverted basket.
2746-2910'	5-1/2" blank csg., 5 jts., 15 lb.
2910-2913'	5-1/2" Bkr. Series Whirler cmtg. collar.
2913-2921'	5-1/2" blank csg. nipple, 15 lb., w/reg. metal petal basket.
2921-2981'	5-1/2" Slotted lnr., 2 jts., 15 lb.
2981-2986'	5-1/2" Bkr. perfd. cmtg. nipple w/inverted basket.
2986-2998'	5-1/2" blank nipple, 15 lb.
2998-3000'	5-1/2" Hal'b. comb. float and guide shoe.

FIGURE 422

completed, and the total number and size of shots should be stated. A separate listing should be made giving the name of the sand perforated, over-all interval or intervals perforated in each sand, and the amount of net sand perforated per producing zone. See figure 424.

Example 10—Production Tests—The production test should include

FECHA	PROFUNDIDAD TOTAL	PROGRESO	DIAS EN PERFORACION	OBSERVACIONES
10-10	TD8260' PB6844'	-	48	Sch. perfd. 5823-5825' w/8 8.5 mm. shots. Set Yowell tool at 5810' and sqzd. 70 sx. of Lehigh cmt. thru above perfs. Break down press. 2500 psi. Max. and final press. 4400 psi. Job compl'td. at 8:00 AM. Started drlg. out cmt. at 11:00 PM. Cleaned out from 5811-5826'. Tstd. cmt. job of int. 5823-5825' w/1500 psi which held OK. Cleaned out to 5880'. Cond. mud 11.6 lbs/gal. Visc. 45. Sch. perfd. 5848-5853' and 5860-5875' w/37 8.5 mm. shots. Now going in hole w/Hal'b. tstr. to test above perfd. ints.

FIGURE 423

FECHA	PROFUNDIDAD TOTAL	PROGRESO	DIAS EN PERFORACION	OBSERVACIONES
5-4	TD6936'	-	28	Sch. fin. perfg. for final compl. at 12:00 AM making a total of 526 8.5 mm. shots. Rmd. perfs. from 5750-6200'. Now prep. to set ret. for dual compl.

Sigmoilina Sands

Sand	Interval	Net Perfd. Sd.
S-2	5754-5780' 5785-5790'	31'
S-3	5800-5827' 5830-5855'	48'
S-4	5874-5916'	34'
S-5	5916-5925' 5930-5940' 5944-5972'	41'
S-6	6010-6035' 6038-6048'	32'
	Total Net Perfd. Sigmoilina Sand	186'

Upper Mulata Sands

Sand	Interval	Net Perfd. Sd.
M-0	6190-6200' 6207-6220'	20'
M-1	6223-6248'	18'
	Total Net Perfd. Upper Mulata Sand	38'

Lower Mulata Sands

Sand	Interval	Net Perfd. Sd.
M-9	6605-6620'	13'
	Total Net Perfd. Lower Mulata Sand	13'

FIGURE 424

the name of the sand or sands tested, the perforated interval, hours, rate, choke, G.O.R., pressure, gravity, and cuts. See figure 425.

Example 11—Completion—When the official initial production has been obtained, it should be stated as such. A second paragraph should include the date the I.P. was established, type of completion, name of producing sands, and net perforated sands. See figure 426.

Weekly Well-Sample Report

This form will be used by subsurface geologists to record all sample descriptions: i.e. ditch, core, A-1 cores, and sidewall cores obtained on drilling wells. See figure 427.

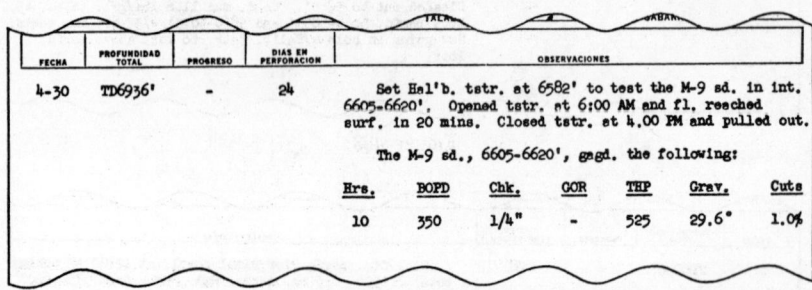

FECHA	PROFUNDIDAD TOTAL	PROGRESO	DIAS EN PERFORACION	OBSERVACIONES
4-30	TD6936'	-	24	Set Hal'b. tstr. at 6582' to test the M-9 sd. in int. 6605-6620'. Opened tstr. at 6:00 AM and fl. reached surf. in 20 mins. Closed tstr. at 4.00 PM and pulled out.

The M-9 sd., 6605-6620', gagd. the following:

Hrs.	BOPD	Chk.	GOR	THP	Grav.	Cuts
10	350	1/4"	-	525	29.6°	1.0%

FIGURE 425

FECHA	PROFUNDIDAD TOTAL	PROGRESO	DIAS EN PERFORACION	OBSERVACIONES
12-24	TD6936'	-	31	Set ret. as packer at 6074'. Started wshg. well at 11:00 AM and csg. came in at 2:15 PM and tbg. at 3.00 PM. Turned well to station and gagd. the following official I.P.:

Hrs.	BOPD	Chk.	GOR	THP	CHP	Grav.	Cuts
24	606	1/4" C	587	-	1000	33.9°	Clean
24	358	1/4" T	839	500	-	30.9°	0.2%W

The well was compl'td. as a dual prod. and the I.P. established Dec. 22, 1943 with the S-2 thru S-6 sds. (Net perfd. sd. 186') prod. thru the csg. and the M-0, M-1 and M-9 sds. (net perfd. sd. 51') prod. thru the tbg. LAST REPORT.

FIGURE 426

Entries to be made in a weekly well-sample report are
1. Number of weekly well sample report for the particular well. These numbers should be in chronological order.
2. First day of week covered by report.
3. Last day of week covered by report.
4. Well number.
5. Page number of weekly well-sample report for the well in reference.

6. Depth in feet of the interval sampled.

7. Under "Type" indicate whether "Ditch," "Core" (give number of core), "A-1 Core" or "S.W.C." The report should have the descriptions in depth sequence. Even if an interval is cored and described, the ditch sample corresponding to the cored interval should also be described. (See description for ditch samples and core No. 57, interval 8965-75.) Sidewall cores and/or A-1 cores can be covered in a separate weekly well-sample report and, if at all possible, should be described in a descending-depth order.

8. All descriptions should follow a definite order when possible.

WEEKLY WELL SAMPLE REPORT NO: 30 ①

FROM August 9 ②	TO August 15, 1948 ③		WELL NO: ④ JGE-000	PAGE NO: ⑤ 37
INTERVAL SAMPLED	TYPE	LITHOLOGY	DIP	REMARKS
8950-8965 ⑥	Ditch	SHALE - gr. to dk. gr., med. to hd. fiss., some sltly. mic. and lig., vitreous lustre, w. tr. SANDSTONE, lt. gr., f. compt. sltly. calc., barren. ⑧		⑪ 5% ss.
8955-8965 ⑦	Core No. 56	Rec. 6'8" 3'8" SHALE, gr. to dk. gr., fairly hd., fiss., v. sltly. lig. 3'0" SHALE, as above, somewhat fract'd. & slicks'd., lam'td. w. small amt. barren SANDSTONE, contains numerous macrofossils.	⑩ 35° 40°	15% ss. Fossils
8965-8975	Ditch	SANDSTONE - lt. gr., f.g., firm to hd., carb. & mic. por., well sat. w. oil, tr. SHALE.		27.6°API Susp 28.3°-28.9°RI
8965-8975	Core No. 57	Rec. 10' 0" 10' 0" SANDSTONE, lt. gr., f.g., hd., carb., mic., well sat. w. oil, contains occ. lam. SHALE.		Top Col. "A" 8965 95% ss. 28.2 API susp. 28.9° refract.
	⑨	Use of abbreviations in core and sample descriptions is optional. If the geologist describing the cores prefers to write out the full descriptive terms, he may do so with the following results:		
8955-8965	Core No. 56	Rec. 6'8" 3'8" - SHALE, gray to dark gray, fairly hard, fissile, very slightly lignitic. 3'0" - SHALE, as above, somewhat fractured and slickensided, laminated with a small amount of barren SANDSTONE, contains numerous macrofossils.	35° 40°	15% ss. Fossils

FIGURE 427

It is particularly desirable that the rock name appear near the beginning of the description so that it does not become lost in the mass of less important detail. In the following outline probably only half or less than half of the features listed need be noted for any one sample. Items which are of greatest importance and which should be particularly checked in each description are shown in capitals in the outline. In the description it is desirable that the rock name be written in capitals and that remarks concerning special features such as "fossils" and "oil and gas" be underlined.

ROCK NAME—texture—COLOR—luster—HARDNESS—cohesion—fracture feel—BEDDING—original structure—secondary structures—concretions mineral and mechanical composition—special features—FOSSILS—OIL and GAS—DIP.

When a core is described (see example given for core No. 56), the total recovery is shown first, followed by the description of each lithologic unit *from top to bottom*.

9. Use of abbreviations in core and sample descriptions is optional. If the geologist describing cores prefers to write out the full descriptive terms, he may do so.

10. Dips as measured from cores should be indicated in a column titled "Dip."

11. A "Remarks" column is used for clarifying any data shown in the lithologic description.
   a. Estimated percentage of minor constituent of rock described should be given: i.e., 5% ss.
   b. Presence of fossils should be noted.
   c. Determination of oil gravities should be given, including the method used, whether "Suspension" or "Refractive Index."
   d. Any formation top, faunal zone, sand top, etc., should be indicated, together with their respective depths. These data should be underlined to make them stand out.

Method of Preparation—All entries must be typewritten. Elite type is preferred. A sheet of carbon paper is reversed in order that the print appears on both sides. Reproductions of this report are made by the Ozalid process.

Copies—The original copy is retained in the district geological group files for additional reproduction for possible well-exchange trades, particularly on wildcat wells.

Normal distribution (Creole 100% acreage—operated by Creole) requires one copy for each well file.

Completion Log Report

It has become a common practice among geologists in many areas to prepare a completion log report upon termination of drilling a well.

All pertinent geological and mechanical data are drafted on the log, which is then reproduced and distributed among parties interested in the well. The following set of instructions has been included in the Geological Procedure Guide of the Creole Petroleum Corporation and is submitted as being one method of log preparation which has proved very useful in Venezuelan operations. Completion logs are usually drafted on the 1:500-scale electrical log. This scale is commonly adopted by all companies operating in that country.

.01   Heading

.011 Well Data

a. Wrico lettering guides are used in the drafting of well number, field, name of operator and/or owner, and scales.

b. All other data on heading exclusive of map is prepared in capital letters with a typewriter having a good ribbon. A piece of carbon is placed behind the log heading in such a manner that the letters are typed on both sides. An Ozalid opaque typewriter ribbon may be used.

c. "COORDINATES," "ORIGIN," "GRID," "STATE," "DISTRICT," and "CONCESSION" are otained from the district's "Well Number Assignment" letter, or the well's location plat.

d. "SPUDDED" and "I.P." established are the dates, expressed in month, day and year (in that order) in which these events occurred. This date should coincide with the date I.P. was established, as stated at the foot of the log. Month should be expressed by abbreviation rather than numeral.

e. "GD. ELEV." is the elevation in feet above sea level of the location as obtained from the Engineering Department.

f. "R. T. ELEV." is the height in feet above sea level of the rotary table, as determined by the type of foundation on lake wells—obtained from the engineering deparment for land operations. The district's "Well Number Assignment" letter is also consulted for rotary-table elevation.

g. "BRADEN HD. ELEV." is height in feet above sea level of the Braden head.

h. "T. D." is the depth of the hole in feet upon completion of the well, as measured from the rotary table.

i. "P. B." is the depth, in feet, of the top of the uppermost plug back or bridge plug. When there is no plug in a well and, therefore, no entry to be made after "P.B.," a dash is entered to indicate that nothing has been omitted.

j. "TYPE OF COMPLETION" indicates the type of completion, normal, dual, or triple zone, effected in the well.

k. "PRODUCING UNIT" is the stratigraphic name of the section open to production. The unit or units producing through the casing are listed under "CASING," whereas those open through the tubing are listed under "TUBING." Care is taken, and abbreviation of unit names made, to assure that the names appear in the proper column. In normal completions when there is no entry to be made under "CASING," dashes are entered to indicate that nothing has been omitted.

l. Self-potential and resistivity scales are drafted on the completion log heading using Wrico lettering guide No. 90, Pen No. 7. When ever the recorded curves run off the negative, they commence again at the zero-scale line and are usually recorded at an altered scale. When this is the case, altered scales should be included in the log heading. The second galvanometer portion of the normal curve, however, is shaded for clarification. If a scale unlike that of the normal scale is used for the fourth curve, the altered scale is included in the heading immediately below the normal resistivity scale.

.02 Location Map

a. Map is normally traced from a convenient scale company map.

b. A 1:20,000 scale is preferred for exploitation wells. Wildcats or outpost wells sometimes require smaller-scale maps.

c. All wells are represented by a 2-mm. dot on a 1:20,000-scale location map, by a 1½-mm. dot on 1:50,000-scale location map, and by a 1-mm. dot on all location maps 1:100,000 or smaller. The scale of completion log heading-location maps is always indicated in lower right-hand corner. The surveyed well to which the log pertains is identified by a circle 5 mm. in diameter drawn around the dot representing the well.

d. The surveyed well is placed as close as possible to the center of the map, yet not so close that such a position prevents the inclusion, normally, of at least three nearby, previously drilled wells.

e. All well numbers are drafted with Wrico Lettering Guide 90, Pen No. 7.

f. One north-south and one east-west, finely drawn, coordinate lines are included on the map. Even thousand coordinate lines are preferable. Coordinate numbers are drafted free-hand, as inconspicuously as possible, parallel to and to the left of the north-south line as one reads from bottom to top, and parallel to and above the east-west line as one reads from left to right.

g. Concession lines are drafted somewhat heavier than coordinate lines.

h. The title of the concession in which a well occurs is shown on the location plat by using Wrico Lettering Guide 90, Pen No. 7.

.03 Miscellaneous

a. Whenever an entry is not made in the heading, a dash indicates nothing has been omitted.

b. The heading transparent is not attached to log negative but is left separate in order to permit individual reproduction of heading and log, which normally are of varying densities and require different velocity and light to reproduce. Prints of heading and log are later joined together.

.04 Log

.041 Geological Data

a. All geological period, formation, member, zone, and sand names are drafted in accordance with the draftsman's guide. Variations from the standard guide are made only when the characteristics of the log are such as to require change. Only faunal zone tops will be shown. Well depth and zone are shown by using inclined Wrico Guide 90, Pen No. 7.

b. Conformable formation contacts are represented by a solid line, and unconformable formation contacts by a wavy line which passes completely across the log. These lines do not interrupt the graphic representation of cores or perforations if they are included in the log.

c. Sand tops and bottoms are represented by a solid line passing from the S. P. curve through the center of the log to the normal resistivity log (without interrupting the graphic representation of core or perforations).

d. Depths of formation tops, sand tops, and base of sands are drafted in vertical numbers with Wrico Lettering Guide 90, Pen No. 7. Subsea depths are indicated in parenthesis opposite the well depth with a minus sign opposite the depth.

e. When the thicknesses of sand bodies permit, sand names are written vertically from bottom to top on one line, unabbreviated in accordance with the draftsman's guide. When necessary, the sand names are abbreviated and/or written on two lines. Occasionally, when necessary, sand and formation names are written horizontally to fit within their limits.

f. Net sand thicknesses determined by log and/or cores are drafted freehand. They are slightly inclined opposite the sand body on the resistivity side of the log in accordance with the draftsman's guide. Net oil sand open to production may be shown in parenthesis to the right of the net-sand figure.

g. To facilitate the addition of net oil-sand figures (or, oil sandstone) in any one sand or formation containing numerous sand bodies, the sum of all net oil sands (or oil sandstones) is drafted at the base of the sand, group, zone or formation on the resistivity side of the log in accordance with the draftsman's guide. Whenever the total net-oil sand for any sand or formation can be readily determined without considerable addition, the sum is omitted.

h. Whenever a sand, such as the Lower Lagunillas sand, is broken down into zones, each zone is bracketed and the zone title is drafted on the left side of log in accordance with the draftsman's guide on one line. Abbreviations are used when necessary. Wrico Guide 90, Pen No. 7 are recommended.

i. Fossil zones or subzones are indicated in accordance with the draftsman's guide. The name of the zone and the depths are written with an inclined Wrico Guide 90, Pen No. 7.

j. A recognized or accepted fault encountered in a well is represented as shown on the log. Postulated faults are represented by similarly located, dashed lines. Throw (vertical displacement) of fault or fault zone, when known, is included in accordance with the draftsman's guide on the resistivity side of the log.

k. Dipmeter readings and bearings are represented by the conventional dipmeter symbol—an 8-mm., four-quadrant circle with the direction of dip indicated therein—drafted on the resistivity side of the log at the depth where measured. The type of instrument used is represented by the letters "EM" (electro-magnetic), "SP" (self potential) or "RES" (resistivity) drafted above the circle. Angle of dip is drafted below the symbol in accordance with the draftsman's guide. Doubtful dipmeter results are indicated by a question mark in parenthesis after the dip.

l. Wire-line or conventional coring is represented by a symbol 2 mm. wide adjacent to the zero line on the resistivity side of the log. A line 3 mm. long marks the top and bottom of the cored interval. The amount recovered is filled in solid black opposite the section from where recovery is believed to have been obtained; whereas the unrecovered section is left blank between the two vertical lines which are 2 mm. apart. When Schlumberger negatives, with depths recorded down the center of the log, are used, cored intervals are limited to the 10-mm. strip in the middle of log. Placement of such graphic data will vary with the type of log used. In all other respects the representation is identical. The core number is indicated to the right of the symbol for cored section.

m. Graphic representation of cored sections is in accordance with the legend of lithologic symbols. Ditch sample and core data may be represented on a percentage basis.

n. Dips measured in cores are represented in degrees and drafted to the immediate right of the core-recovery symbol of the log at the depth where measured.

o. Side-wall cores are represented in accordance with the lithologic-symbol legend in a box approximately 3 x 6 mm., conveniently located to the right of the zero line of the resistivity curve.

p. Ditch samples are not normally represented graphically on completion logs unless they can materially aid in the log's interpretation. When designated, they appear as approximate percentage of each lithology represented. Clastics are plotted on left side and chemical sediments on right. When an oil cut is obtained from ditch samples, it is represented by a 3-mm. circle located in the center of the graphic log.

q. Any show, which is considered significant by the geologist, of oil, gas, or water, is represented by the letters *So*, *Sg*, or *Sw* respectively in the center of the log at the depth where encountered. The letter *S* is colored by red, yellow, or blue, depending on the type of show.

r. Whenever a well blows out, the sand from which the blowout originated is indicated by an *X* in the center of the log. The word *Blowout* and depth at which it occurred are recorded to the left of the *X* symbol in accordance with the draftsman's guide.

.05 Mechanical Data

.051 Casing and C.O.S.

a. Blank casing is represented in the conventional manner by a solid, thin, vertical line.

b. "C.O.S." is represented in the conventional manner by a solid line (blank pipe), and a dashed (preperforated pipe) vertical line.

c. The size of the blank casing and depth at which cemented are represented numerically to the left of the casing shoe and graphically by a solid casing shoe pointing to the left. The size of the preperforated casing and depth where set are represented numerically to the left of the "C.O.S." and graphically by an open shoe pointing to the left.

d. Tubing is represented graphically only for dually completed wells or where a packer is run on the tubing.

.06 Perforations

a. In "C.O.S." a solid line representing the blank casing is drafted to the point where the preperforated pipe begins. The graphic representation of the preperforated pipe always commences with a gap, then a dash and gap, etc. The bottom of a preperforated interval, when located between two pieces of blank pipe, likewise always begins and ends in a gap.

b. Gun-perforated intervals open to production are represented by a series of horizontal lines 2 mm. long that intersect the solid casing line. The top and bottom of a gun-perforated interval are indicated by a 4-mm. horizontal line.

c. Gun-perforated intervals not open to production and not squeezed off are identical to those open to production. A bridge plug or reason for the section not being open is indicated.

d. Gun-perforated intervals for squeeze-cementing operations are represented by top and bottom lines and in-between lines, none of which intersect the casing.

e. Intervals gun-perforated for production test and squeezed off are represented by a series of 1-mm. lines on the left side of the casing line. The top and bottom of the gun-perforated interval are indicated by 4-mm. horizontal lines.

.07 Plugs
  a. Cement plugs are represented by the conventional hour-glass and cement symbols.
  b. Bridge plugs are represented by a condensed hour-glass symbol, the top of which is drafted at the recorded depth of the top of the plug.

.08 Surveys
  a. Electrode spacing of each resistivity curve made is recorded in a place where it will not interfere with the interpretation of the log. If more than one run is made, each run used for the completion log should have the electrode spacings recorded on the film. Likewise the temperature (degrees Fahrenheit), and the resistivity of the mud are recorded on the film for each run.
  b. Mud resistivity is the resistivity of the mud as measured by the loggers at the time the survey was made. Entered with the resistivity is the temperature in degrees Fahrenheit at which it was measured. On completion logs, where two or more runs make up a composite log, the mud resistivity and temperature for each run and the interval surveyed on each run are plotted at the top of the log of the respective runs. Hole size and run number should also be recorded. This data is usually entered on the preliminary log.
  c. "SCHLUMBERGER DEPTH," "DRILLER'S DEPTH," and the final "TOTAL DEPTH" of the well are recorded with a Wrico Guide CVC, CVL, CVN-185, Pen No. 7 at the foot of the log in the space indicated by the draftsman's guide. "FIRST READING" is no longer recorded at the bottom of the completion log unless it varies appreciably from the logger's total depth.
  d. Whenever a well is carried deeper after an electrical survey has been run and no additional surveys have been made, a notation explaining the discrepancy between "TOTAL DEPTH" and "SCHL. DEPTH" appears at the foot of the log.

.09 Well Number and Scale
  a. The well number and scale, having been drafted on the negative prior to the reproduction of the 1:500 preliminary log, remain unchanged for single wells in the preparation of the completion log.

.10 Production Data
  a. Intervals tested are embraced by brackets drafted on the resistivity side of the log. Each bracket is labeled with the test number in a circle. Results of tests, including test number, sand or sands tested, interval, what obtained on test, indication as to type of testing method used and length of test, size of choke (if

bottom-hole choke used also include), gas-oil ratio, pressures, and gravity and cuts are recorded, in the order given, at foot of log under title "TESTS." Bottom-hole pressures and differential data should be included, if available, as well as the method of inducing production. When selected intervals are tested, only the uppermost and lowermost depths are listed, followed by "Sel. Perfd."

b. The initial production of the well is recorded at the foot of the log under the original date when the I.P. was established. Immediately below the date I.P. was established is recorded in parenthesis the type of completion. On the next line is listed the producing sand or sands, depths of producing interval, and, in parentheses, "GUN PERFD.," if blank casing has been perforated, or by "C.O.S.," if a combination oil string of blank and preperforated pipe was used. If selective intervals are open to production, only the uppermost and lowermost depths are listed after the producing sand or sands. A note "SEL. PERFD." or "C.O.S. SEL. PERFD." in parenthesis refers the observer to the log for actual depth of the selected intervals. The initial production always commences on a new line, although the sequence of reporting the production is "BOPD," "CHK," "GOR," "THP," "CHP," "GRAV," "CUT." The cut is broken down to show the components of which it is made. If bottom-hole pressure and pressure differentials are available, they are entered after the cut.

c. Original date when I.P. was established and initial production at foot of log should agree with the dates and figures listed in the completion report on this well.

d. Data at the foot of the log, if short, is drafted on the negative. If considerable data are included at the foot of the log, it is typewritten on transparent paper with a good typewriter ribbon. A piece of carbon paper is placed behind the transparent paper in such a manner that the letters are typed on both sides of the paper. An Ozalid opaque typewriter ribbon may be used. Ozalid copies are then attached to the log.

.11 Miscellaneous

a. Normal curves recorded by the second galvanometer are shaded in accordance with the draftsman's guide by means of oblique lines 3 mm. apart, running from upper right to lower left through the altered curve.

b. API gravities determined from oil-saturated cores or ditch samples are drafted freehand at the end of the bar graph to represent gravity on the resistivity side of the log at the depths where the saturation was encountered. Refractometer gravities are used when available. When suspension bead or hydrometer methods are employed, it is so noted in parenthesis after the gravity.

A horizontal line, 1 mm. long for each degree of the oil's API gravity, is drafted between the zero-resistivity line of the log and the numerical representation of each gravity, a procedure which creates a bar graph of oil gravities.

.12   Coloring

     a.   When the completion log negative has been completed and 1:500 prints have been made, those copies to be distributed within the company are painted with water colors to indicate the district's interpretation of the log. Both the SP and normal curve of sand bodies untested but believed to be petroliferous, either by correlation, core analysis, or electrical-log interpretation, are outlined in red. Sand bodies, either producing oil or tested and found to be capable of producing oil, are filled in solidly with red from the center line of the log to the normal curve on the resistivity side of the log and to the limits of the SP curve on the SP side. Sands believed to be, or found to be, gas- or water-bearing are similarly designated by the appropriate color, yellow or blue. Cores are appropriately colored also.

     b.   Tests are distinctively colored in accordance with the legend of color symbols used in tests.

     c.   Formation limits are designated by distinctive colors, in accordance with the formation color chart used in the area.

     d.   When circles, representing petroliferous ditch samples have been employed, they are colored also. Similarly, when used, the blowout symbol is appropriately colored.

.13   Trimming and Folding

     a.   When final prints are prepared, they are trimmed to a width of $6\frac{1}{2}$ in., with $\frac{3}{4}$ in. left at the bottom of the log below the last line of printed material, and an equal distance, $\frac{3}{4}$ in., above the topmost margin of the log heading to permit perforating for filing. When the log is wider, effort should be made to hold the width to $8\frac{1}{2}$ in. to facilitate filing.

     b.   Completion logs, prior to distribution, are folded in a manner most practical for filing purposes. The first fold is made 13 in. below the top of the trimmed print. The next fold is made immediately below the well number on the heading approximately $10\frac{1}{4}$ in. above the first fold. All subsequent folds are made identical to the first at the bottom, or to the second fold immediately below the well number. The last fold or folds are made wherever practical to permit the bottom of the print to fall in coincidence with the first fold so that the scale, well number, initial production, etc., at the bottom of the log are visible when filed.

.14   Distribution

     a.   The district geological department preparing the completion log keeps one copy, complete with color, for filing in the respective well file and makes distribution in accordance with the current distribution chart.

.15 Preservation of Original Negatives and Headings

    a. The original log negatives and production data attachments at bottom of log should be carefully rolled and wrapped in paper, clearly marked with well number, and filed in numerical order in

SAND DATA SHEET

AGE	SAND OR FORMATION	INTERVAL	TOP SUBSEA	NET SAND	NET OIL SAND	NET PROD. SAND	EVALU-ATION	REMARKS
							O	
	Lagunillas Fm.	1600 – 3282	–1585				O	
M I O C E N E	Bachaquero Sd.	1600 – 2940	–1585	178	46	33	◉	Perf'd Sd.12' Int 1610–22' BOPD 722 Gr. 16.2°
	Lwr. Lag. Sd.	2940 – 3282	–2925	98	73	–	◉	BOPD 850 Gr. 22.1°
	La Rosa Fm.	3282 – 3552	–3267	–	–		O	
	La Rosa Sd.	3297 – 3335	–3282	22	22		◉	
	Intermed. Sd.	3360 – 3387	–3345	24	27		☼	950 MCF Gas
	Sta.Barb. Sd.	3490 – 3552	–3475	18	18		◉	
OLIG	Icotea Fm.	3552 – 3561	–3537	6	–		O	
	Ambrosio Fm.	3561 – 3682	–3546	–	–		O O	Shale
	Las Flores Fm.	3682 – 4041	–3667	26	26		◉	Dips 5–25°
M I O C E N E	Potreritos	4041 – 7880	–4026				O	
	B-6-72 SD.	4145 – 4360	–4330	142	142		◉	BOPD 2500 Gr. 29.3°
	B-7-44 Sd.	4372 – 4580	–4357	163	102		◉	
	B-8-43 Sd.	4592 – 4676	–4577	18	–		O	Wet
	Misoa Fm.	8100 – 9846	–8085				◉	
PAL	Guasare Fm.	9846 – 10275	–9831				◉	BOPD 260 Gr. 36.2° Dip Avg. 8° S. 45° E
	Mito Juan Fm.	10275 – 10447	–10260				O	Shale
	Colon Sh.	10447 – 11571	–10432				O	"
U S C R E T A C E O U S	Socuy Fm.	11571 – 11633	–11556				O	Massive Ls.
	La Luna Fm.	11633 – 11870	–11618				◉	
	Capacho Fm.	11870 – TD	–11855				◉	Strong Flow Salt Water
							O	
							O	
							O	

ELECTRICAL SURVEY 962 – 12,106 (Schl.)	COORDINATES S. 42,962.00 E. 16,481.03		TOTAL DEPTH 12,106	PLUG BACK 1630
OTHER SURVEYS Dipmeter, Temperature, Caliper.	GRID W-29	CONC Agua-468	COMPLETION Normal	COMPLETION DATE June 1, 1949
DATE PREPARED June 15, 1949	CORES 1600 = 11870	SWC Yes	RTE 15	WELL NO. XJ-982
PREPARED BY J. Doe	MAXIMUM DEVIATION & DEPTH 3° at 4590	AREA OR FIELD Lagunillas		

TRIM LINE 8½"

FIGURE 428

a fireproof location. The heading should be filed in a folder in numerical order so they may be readily available. It is emphasized that the original logs are to be carefully preserved.

Sand Data Report

The sand data report is used by subsurface geologists to record in a readily available manner certain formation and test data obtained during the drilling of a well. See figure 428.

Heading Entries—Heading entries to be made in sand data report are
1. General
    a. A sand data sheet, which contains in a convenient form the name of sand or formation, interval, subsea depths, net sand, net oil sand, net producing sand, evaluation, and remarks, is prepared upon the completion of each well by the district geological group.
    b. Data and symbols are entered on a blank sand data sheet by a geologist. Any necessary drafting of symbols on a transparent original is then accomplished and the well data is typed in. A good typewriter ribbon is used. A piece of carbon paper is placed behind the original in such a manner that the letters are typed on both sides of the transparent original. Upon completion, a sufficient number of copies are reproduced by the Ozalid method to provide distribution in accordance with the current distribution chart.
    c. The symbols used on the sand data sheets are distinctive and require no coloring.
2. Preparation of Sand Data Sheet
    a. *Age*—In the column entitled "Age" are placed the correct age designations for the sands or formations within the over-all determination (i.e. Miocene, Eocene, Paleocene, Cretaceous).
    b. *Sand or Formations*—In the column entitled "Sand or Formation" are listed, in stratigraphic order, the important sands or formations normally encountered in the area where the well is situated. Wherever the stratigraphic section is somewhat questionable, blank spaces are to be provided to permit the recording of the sands or formations at a later date when the encountered sequence has been definitely established.
    c. *Interval*—Insert the top and bottom electrical-log depths of each sand and/or formation encountered. Whenever a sand or formation, normally encountered in the area, is found to be absent in the well, the abbreviation "Abs" is entered in the "Interval" column. When drilling ceases before a sand or formation is completely penetrated, no bottom depth is indicated; and the letters "TD" are substituted.
    d. *Top Subsea*—This calls for the subsea depth, preceded by a minus sign, of the tops of the horizons encountered.

e. *Net Sand*—This is the actual net-sand thickness, in feet, found within the sand or formation as determined from electrical-log interpretation or cores. The net sand figure entered for any sand or formation not completely penetrated should be followed by a plus sign.

f. *Net Oilsand*—This is the number of feet of net saturated-oil sand assigned to the sand or formation from electrical-log interpretation, core valuation, or actual tests. The net oilsand figure entered for any sand or formation not completely penetrated should be followed by a plus sign.

g. *Net Producing Sand*—This is the number of feet of oil sand which can be effectively drained with the present perforations open. For example, a sand body may contain 23 ft. of net producing oil sand but may have only a 10-ft. section perforated (i.e. in homogeneous sand bodies, vertical permeability may be assumed up to and down to the nearest shale break).

h. *Evaluation*—This column shows graphically, where possible, the interpretation of the sand as determined by tests, ditch samples, or cores. Production-test results are indicated by modifying, where necessary, the 2-mm. circle appearing on the transparent original in accordance with well symbols for use on reservoir isopach maps. Results of drill-stem tests are similarly represented with the addition of the letters "DST" to the right of the circle. Evaluation of ditch samples or cored intervals are presented in accordance with the symbols. The unit left open to production is designated in accordance with the appropriate symbol. See figure 429.

i. *Remarks*—This column will list pertinent information which will further explain or clarify data on the sand or formation under discussion. The "Remarks" column will show "Net Perfd. Oil Sand Open," which is defined as *only* that number of feet of oil sand opposite the perforated interval (i.e. if 1610–1622 ft. were perforated in a sand body containing 46 ft. of net oilsand, *only* 12 ft. would be shown for this "Net Perfd. Sd." figure in the "Remarks" column). This fact would be indicated by "Perfd. Sd." followed by the number of feet. Also in the "Remarks" column will be shown initial production in BOPD and API gravity of the oil. If producing rates are determined from drill stem or production tests, they are also indicated in the "Remarks" column.

j. *Electrical Survey*—Under "Electrical Survey" the uppermost and lowermost depths logged should be indicated. The type of equipment used for making such a survey should be shown in parenthesis: i.e. (Schl.) for Schlumberger or (B.G.) for Blau Gemmer. If some other logging unit is used, it should be indicated accordingly.

k. *Other Surveys*—Indicate whether dip-meter, temperature, caliper, or other type of survey has been made on the well.

l. *Other Information*—Coordinates, total depth, plug back, grid, concession, type of completion, completion date, date prepared, etc. should be shown as indicated in figure 428. The depth of the top of the first core and the depth of the bottom of the last core should be shown under "Cores." If the section were cored continuously, a single dash should be shown between the top and bottom depths. If it were cored intermittently, a double dash

WELL SYMBOLS FOR USE ON RESERVOIR ISOPACH MAPS
SIMBOLOS USADOS EN LOS POZOS EN MAPAS DE "RESERVOIR" ISOPACOS

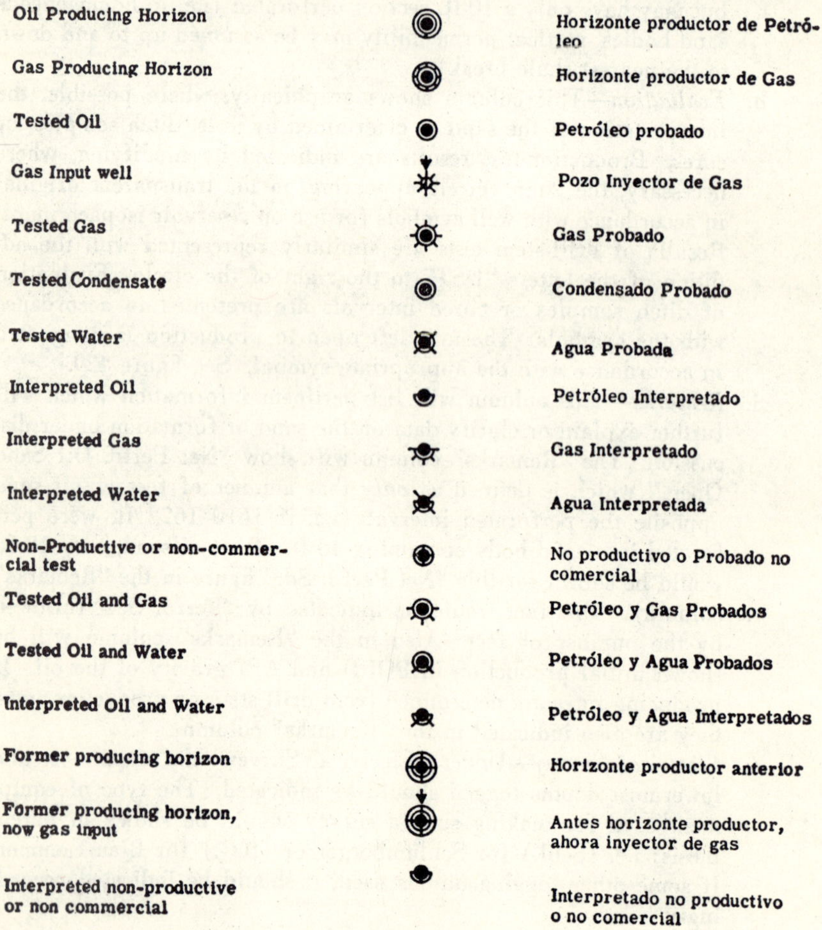

Oil Producing Horizon	◉	Horizonte productor de Petróleo
Gas Producing Horizon	◉	Horizonte productor de Gas
Tested Oil	◉	Petróleo probado
Gas Input well	✳	Pozo Inyector de Gas
Tested Gas	☀	Gas Probado
Tested Condensate	◉	Condensado Probado
Tested Water	✖	Agua Probada
Interpreted Oil	●	Petróleo Interpretado
Interpreted Gas	☀	Gas Interpretado
Interpreted Water	✖	Agua Interpretada
Non-Productive or non-commercial test	◉	No productivo o Probado no comercial
Tested Oil and Gas	☀	Petróleo y Gas Probados
Tested Oil and Water	◉	Petróleo y Agua Probados
Interpreted Oil and Water	✖	Petróleo y Agua Interpretados
Former producing horizon	◎	Horizonte productor anterior
Former producing horizon, now gas input	◎	Antes horizonte productor, ahora inyector de gas
Interpreted non-productive or non commercial	●	Interpretado no productivo o no comercial

FIGURE 429

should be shown between the top and bottom depths. Under "SWC," if side-wall cores are taken, "Yes" should be inserted. If no side-wall cores are taken, "No" should be inserted. Rotary table elevation, name of recorder, whom prepared by, maximum deviation and depth of same, and area or field should all be indicated. The well number should be drafted in with Wrico Template No. 175.

3. Trimming—Trim lines 8½ x 11- or 12-in. lengths are indicated on the sand-data-sheet forms. Copies distributed to well files should be trimmed to an 11-in. length, whereas those distributed for insertion in books of sand-data sheets should be trimmed to a 12-in. length. The 12-in.-length prints will be filed in a special folder which will be prepared for the accumulation of sand-data sheets.

Date Due—The sand-data report is due as soon as possible after completion, suspension, or abandonment of the well.

Method of Preparation—Material is typewritten (elite type preferred) with carbon paper reversed to print on back also. Symbols are drawn in ink by the draftsman.

Copies—Copies of the sand-data report are in the following:
Creole well files (5) Trimmed 8½ in. x 11 in.)
Sand Data Sheet Book (5) Trimmed 8½ in. x 12 in.).

EXAMPLES OF OUTLINES OF REPORTS ON SUBSURFACE GEOLOGY

The organization of reports on subsurface geology varies according to the subject. Thus, reports of regional scope differ from those devoted to more local assignments.

Examples of outlines of various types of subsurface reports follow:

Report on Exploratory (Wildcat) Well

I. Introduction
Name of well and company, location of well, elevation (derrick floor and ground), date commenced, date of completion or abandonment, date of final depth, casing record (perforated intervals, size, cement data), total depth, type of drilling equipment, summary of well history (hole deviation, coring program, logging surveys, mechanical difficulties, etc.)
II. Purpose of drilling well
III. Summary
    1. Age, thickness (with corresponding depths), relationships, and lithologic summary of formations.
    2. Statement on the regional and local structure.
    3. Comments on oil and gas shows; age of host sediment.
IV. Conclusions and recommendations
V. Geology
    1. Stratigraphy (general statement)
        (a) Regional: variation in facies, thickness, unconformities, etc.
        (b) Local: (Based primarily on penetrated section) ; relationship to regional stratigraphy; paleontologic and mineralogic summary (contributed by stratigraphic laboratory).

2. Structure (general statement)
   (a) Regional: discussion of broad structural grain (fault and fold pattern).
   (b) Local: areal geologic map, structure map, cross sections; description of local structure and its relationship to the regional structural fabric.

VI. Discussion of logging data (electric, radioactive, thermal, drill-time, core-analysis, etc.)

VII. Accompanying enclosures
   1. Index map (local and regional).
   2. Geologic, structural, isopachous, and lithofacies maps.
   3. Local and regional cross sections.
   4. Well log (lithic and electric, showing cored intervals, casing points, tested intervals, etc.).
   5. Chronologic chart.
   6. Correlation chart.
   7. Photographs.

VIII. Miscellaneous data (summary)
   1. Mud variations (salinity, viscosity, and temperature).
   2. Condition of hole (caving, deviation).
   3. Bottom-hole pressures.
   4. Gas recording.
   5. Bit, core, and casing problems.
   6. Formation tests.
   7. Perforation program.

IX. Detailed description of ditch and core samples

Interval (ft.)	Description
2,110–25	Coarse, gray, unconsolidated sand.
2,125–35	Red, slightly mottled claystone.

———————————Top Cramer formation (Oligocene)———————————

2,135–90	Medium-grained, friable sandstone with 10 percent red claystone.
2,190–2,210	(Core 1) (15 feet recovered). Dark-gray shale with thin streaks of fine-grained, slightly oil-stained sandstone (medium acetone cut, strong fluorescence); shows few high-angled fracture planes with faint striations at 60° to the axis of core; average dip about 16°; locally 12° and 22°; hole deviation 2° off vertical.

Report on Well in Proved Field

STRATIGRAPHIC AND PALEONTOLOGIC ANALYSIS OF JONES 36-1,
INDEPENDENT OIL COMPANY, COLORADO

Well: Jones 36-1.
Company: Independent.
Location: Center SW¼SW¼, sec. 32, T. 3 N., R. 46 W.
Date commenced: March 15, 1946.
Date final depth reached: September 18, 1946.
Casing record: 16-inch cemented 1,200 ft.; 8⅝-inch cemented 6,634 ft.
Total depth: 6,960 ft. (bottomed in Morrison formation (Jurassic).
Status: Abandoned as dry hole.

Stratigraphic-Faunal-Mineralogic Summary

(a) Formations penetrated:

	Thickness (feet)
Upper Cretaceous:	
Pierre	0–5,790
Niobrara	5,790–6,080
Benton	6,080–6,660
Dakota group	6,660–6,950

————————————Unconformity————————————

Jurassic:	
Morrison	6,950–6,960

(b) Lithology, mineralogy, and paleontology
Upper Cretaceous:
Pierre (0–5,790 ft.)—Dark-gray shale and mudstone with a few thin interbeds
of fine-grained, gray, extremely tight quartzose sandstone containing
scattered nondiagnostic ilmenite, zircon, and tourmaline grains; some
small Foraminifera, ostracodes, mollusc fragments, and *Inoceramus*
prisms throughout.

Niobrara (5,790–6,080 ft.)—Essentially dark-gray-black, highly calcareous,
foraminiferal shale; base marked by a fine-crystalline, gray limestone
(probably Timpas).

Benton (6,080–6,660 ft.)—Gray-black shale containing thin bluish-white bentonite streaks throughout; few scattered Foraminifera; mineralogy undetermined.

Dakota group (6,660–6,950 ft.)—Consists of three units, upper sandstone,
middle shale, and lower sandstone; sandstone fine- to medium-grained,
gray; shale black to dark gray, carbonaceous, and siltaceous; some mollusc fragments noted; detrital heavy minerals essentially zircon and
tourmaline, and sparse garnet; a few minor oil and gas shows in upper
part of upper sandstone member.

Jurassic:
Morrison (6,950–6,960 ft.)—Light-gray, slightly calcareous mudstone.

Detail Stratigraphy

Upper Cretaceous:
Pierre (0–5,790 ft.)

Lithology: Dark-gray to grayish-black shale and mudstone with many thin,
irregular streaks and laminae of light-gray siltstone; a few layers of fine-
grained, medium-gray sandstone in upper and middle part; fine carbonaceous
material disseminated through the sandstone.

Stratigraphic relationship: Conformable but sharply distinguishable from the
underlying Niobrara. The Pierre is essentially noncalcareous, whereas the
Niobrara contains considerable lime. This boundary reflects a major environmental anomaly.

Paleontology: Foraminifera *(Cibicides, Anomalina,* and *Gumbelina)* are
found sporadically throughout; a few ostracodes and molluscan fragments
noted. The paleontologic aspects of the Pierre do not permit faunal zonation.

Mineralogy: Zircon, ilmenite, and tourmaline present in most arenaceous
phases. The homogeneity of this assemblage does not permit mineralogic
subdivision of the section.

Electrical characteristics: As the Pierre is represented essentially by shales,
the electrical profile is exceptionally monotonous, although the more silty and
arenaceous phases are moderately well-defined, particularly in the intervals
1,885–2,350 and 3,285–3,445 feet. The resistivity profile shows a conspicuous
increase at the Pierre-Niobrara boundary.

Correlation: The Pierre is widely distributed in eastern Colorado. The silty
phases in the upper part may be correlated moderately well from the electric log over considerable distances. The base of the Pierre is correlative
over wide areas from lithology and electrical characteristics.

Productive character and possibilities: Sandstones within the Pierre are
extremely fine-grained; permeability determinations from eighteen cores
showed average values of 35 millidarcys; porosity values are from eight
to twelve percent. No oil or gas shows were observed in the formation.
Some sandstones may prove productive, although on the whole they would
not be expected to contain commercial quantities.

Depositional environment: The Pierre shales accumulated under marine
conditions as indicated by the foraminiferal faunas. The sparseness of the
benthos microfaunas, however, suggests unfavorable bottom conditions.

Succeeding formations should be treated as given above. Information pertaining to core descriptions, cored intervals, formation tests,
mechanical troubles, and hole deviation should also be included.

Enclosures should consist of a combined graphic, electric, and drill-time log, including all pertinent stratigraphic, paleontologic, and operating data, a chronologic chart, a correlation chart, and a map showing the well location with reference to local and regional geologic conditions. Any other information relating to the drilling and completion of the well should be presented.

Report on Oil Field

HAWKINS FIELD, WOOD COUNTY, TEXAS [10]

E. A. WENDLANT, T. H. SHELBY, JR., AND J. S. BELL, HUMBLE OIL AND REFINING COMPANY

Abstract
Introduction
Location
Topography and drainage
History
Stratigraphy
Structure
    Surface
    Subsurface
Prospects for deeper production
Relation of accumulation to structure
Reservoir characteristics
Production
Crude-oil characteristics
Production statistics
Plants and pipe-line outlets
Figure 1.   Index map.
FIGURE 2.   Surface geology and structure map.
Figure 3.   Core-test map; contoured on top of Cretaceous.
Figure 4.   Isopach map showing effective Woodbine oil-sand thickness.
Figure 5.   Isopach map showing effective Woodbine gas-sand thickness.
Figure 6.   Isobaric map.

OREGON BASIN FIELD, PARK COUNTY, WYOMING [11]

PAUL T. WALTON, PACIFIC WESTERN OIL CORPORATION

Abstract
Introduction
History
Stratigraphy
    Surface formations
    Subsurface formations
Structure
    Surface
    Subsurface
Relation of structure to accumulation
Producing formations
Production
    Drilling methods
    Completion practice
    Crude-oil characteristics
    Production curves
    Pipe lines and outlets
    Future development
Figure 1.   Index map, Big Horn Basin. Wyoming.
Figure 2.   Map of Big Horn Basin showing location of Oregon Basin field.
Figure 3.   Oregon Basin oil field (air photograph).
Figure 4.   Composite lithologic and electric log.

[10] Am. Assoc. Petroleum Geologists Bull., vol. 30, no. 11, pp. 1830-56, Nov. 1946.
[11] Am. Assoc. Petroleum Geologists Bull., vol. 31, no. 8, pp. 1431-58, Aug. 1947.

Figure 5. South stratigraphic cross section: Dinwoody, Embar, Tensleep, Amsden zones.

Figure 6. Structure map of Oregon Basin, Wyoming; contours on Embar limestone.

Figure 7. West-east structural cross section; south dome Oregon Basin.

MENE GRANDE OIL COMPANY

Oil Fields of Greater Oficina Area, Central Anzoategui, Venezuela [12]

H. D. HEDBERG, GULF OIL CORPORATION, AND L. C. SASS AND H. J. FUNKHOUSER.

Abstract
Introduction
Major features
Acreage ownership and nomenclature of districts, fields, and wells
Discovery
History of exploration and development
Surface features
    Physiography
    Geography
    Areal geology and vegetation
    Surface indications of oil and gas
Stratigraphy
    Regional
        Mesaverde formation
        Sacacual group
        Freites formation
        Oficina formation
        Temblador formation
        Basement
Correlation
Structure
Geologic history
Oil and gas reservoirs
    Reservoir pressures
    Reservoir temperatures
    Reservoir fluids
        Oils
        Formation waters
Notes on origin, migration, and accumulation of oil
Drilling and completion practice
Production
    Production practices
    Production data
    Production processes
    Pressure maintenance and secondary recovery
    Outlet
Summary of principal features of individual fields

Regional Reports

Cambrian and Ordovician Rocks in Michigan Basin and Adjoining Areas [13]

GEORGE V. COHEE, UNITED STATES GEOLOGICAL SURVEY

Abstract
Introduction
Pre-Cambrian rocks
Cambrian rocks
Lower Ordovician rocks
Middle Ordovician rocks
Upper Ordovician rocks

[12] Am. Assoc. Petroleum Geologists Bull., vol. 31, no. 12, pp. 2089-2169, Dec. 1947.
[13] Am. Assoc. Petroleum Geologists Bull., vol. 32, no. 8, pp. 1417-1448, Aug. 1948.

Structure
Showings and oil and gas production from "sub-Trenton" rocks
Oil and gas production from Middle Ordovician rocks
Other areas of Trenton production
Oil and gas possibilities
Figure 1. Outcrop areas of Cambrian and Ordovician rocks, and major structural features around the Michigan Basin. Areas of oil and gas production from the Trenton limestone and older rocks are also shown.
Figure 2. Thickness of Upper Cambrian and Lower Ordovician rocks in the Michigan Basin.
Figure 3. Subsurface section from southeastern Michigan to northeastern Ohio showing lithologic characteristics and thickness of Cambrian, Lower and Middle Ordovician rocks.
Figure 4. Subsurface section from northeastern Indiana to northeastern Ohio showing lithologic characteristics and thickness of Cambrian, Lower and Middle Ordovician rocks.
Figure 5. Lithofacies map of Middle Ordovician rocks.
Figure 6. Thickness of Middle Ordovician rocks in the Michigan Basin.
Figure 7. Thickness of Upper Ordovician rocks in the Michigan Basin.
Figure 8. Contours on top of pre-Cambrian.
Figure 9. Contours on top of Trenton limestone.
Table 1. Wells, with depths to top of Ordovician and older rocks determined from mounted drill cuttings (depths and elevations in feet.)
Table 2. Wells with oil and gas production below Middle Ordovician rocks.

SUBSURFACE TRENTON AND SUB-TRENTON ROCKS IN OHIO, NEW YORK, PENNSYLVANIA, AND WEST VIRGINIA [14]

CHARLES R. FETTKE, CARNEGIE INSTITUTE OF TECHNOLOGY

Abstract
Introduction
Stratigraphy
Outcrops
Subsurface sections
Stratigraphic sections
Correlation
    Upper Cambrian or Croixian series
    Lower Ordovician or Canadian series
    Middle Ordovician or Champlainian series
Unconformity at the base of the Middle Ordovician
Structure
Oil and gas possibilities
    Trenton
    Sub-Trenton
Figure 1. Regional structural map contoured on the top of the Trenton limestone.
Figure 2. Correlation chart of the Middle and Lower Ordovician and Cambrian formations around the northern rim of the Appalachian Basin.
Figure 3. Stratigraphic section across the northern part of the Appalachian Basin from northern Ohio to south-central New York.
Figure 4. Stratigraphic section along the 79th meridian of longitude.
Figure 5. Stratigraphic section along the 82nd meridian and across central Ohio.
Figure 6. Thickness map of Middle Ordovician limestones in the northern part of the Appalachian Basin.
Figure 7. Thickness map of Upper Ordovician series in the northern part of the Appalachian Basin.

SUBSURFACE LOWER CRETACEOUS FORMATIONS OF SOUTH TEXAS [15]

RALPH W. IMLAY, UNITED STATES GEOLOGICAL SURVEY

Abstract
Introduction

[14] Am. Assoc. Petroleum Geologists Bull., vol. 32, no. 8, pp. 1457-92, Aug. 1948.
[15] Am. Assoc. Petroleum Geologists Bull., vol. 29, no. 10, pp. 1416-1469, Oct. 1945.

Acknowledgements
Stratigraphic summary
    Hosston formation
        Definition
        Distribution and thickness
        Stratigraphic and lithologic features
        Correlation
    Sligo formation
        (treated as above)
References
Table 1. Correlation of Lower Cretaceous formations of the coastal plain of Arkansas, Louisiana, and Texas.
Table 2. Formation tops in Comanche and older rocks in south Texas and Mexico.
Table 3. Known range in thickness of the Cretaceous formation of pre-Gulf age in south Texas.
Figure 1. Index map of south Texas and northern Mexico showing the location of wells.
Figure 2. Columnar sections from central mineral region to Frio County, Texas (graphic and electric log).
Figure 3. Columnar section from Limestone County to Atascosa County, Texas (graphic and electric log).

Report on a Petroliferous Province[16]

I. Introduction.
    1. Importance today and in earlier history.
    2. Location and boundaries. Illustrate with map showing location of productive areas and names of more important fields. Give states involved and relative importance of each.
    3. Subprovinces if any.
    4. Date of discovery and history of development. Most active areas at present.
    5. Unusual characteristics in the geology and in the occurrence of oil and gas.
    6. Surface indications of oil and gas.
II. Geomorphology and general geology.
    General statement regarding physiography, range in age of rocks, thickness of strata, regional and local structure and their influence on topography, igneous activity, extent of exposures, explanation of techniques best adapted to area.
III. Stratigraphy.
    1. Thickness, character, age, and distribution of the rocks.
    2. Stratigraphic table or, preferably, columnar section or sections.
    3. Description of surface and subsurface formations by systems. Correlation chart.
    4. Lateral variations in thickness and character (facies changes), sources of sediments, shifting of axis of geosyncline of deposition, old shore lines.
    5. Unconformities. Wedge areas. Overlap and offlap relations. Buried topographic features, etc.
    6. Key horizons.
    7. Methods of subsurface correlation.
    8. Producing horizons, continuity, lithologic character.
IV. Structure.
    1. Description of surface and subsurface regional structure and relation to other major tectonic features. Rate of dip, etc. (Use structure contour map or maps and cross sections). Age and origin.
    2. Influence on distribution of formations.

---

[16] By F. M. Van Tuyl, department of geology, Colorado School of Mines.

   3. Modifying structural features (including faults and buried hills and ridges). Nature, size, and amount of closure (if any). Trends.
   4. Changes in structure with depth. Times of subsidiary folding and faulting and modification of earlier structures by later deformation. Isopach maps.
   5. Brief statement regarding relation of production to regional and local structures.
 V. Paleogeology.
   1. Geologic history of area. Influence on sedimentation. Evolution of present structure. Subsurface areal and structure maps below unconformities contrasted with surface-geology maps.
   2. Times of igneous activity. Types of intrusion and extrusion and their distribution.
   3. Degree of metamorphism of various horizons.
 VI. Occurrence of oil and gas. (If several subprovinces, consider each separately.)
   1. Types of traps. Describe each and illustrate by producing fields. Relative importance of each. Relations to structure.
   2. Methods of prospecting.
   3. Barren structures and possible reasons therefor.
   4. Reservoir horizons. Number, age, relative importance, character, thickness, continuity, porosity, permeability, kinds of cement, thickness, and degree of saturation of each and amount and character of occluded water present.
   5. Water horizons.
   6. Reservoir pressures.
   7. Composition of edge water.
   8. Possible sources of oil and gas and probable time or times of accumulation.
   9. Possibility of applying carbon-ratio theory.
   10. Possibility of extending producing areas or discovering new producing horizons. Indicate most promising areas and ages of prospective horizons.
 VII. Description of several typical pools illustrating each important mode of occurrence.
   1. Location, date of discovery, exploration methods employed, etc.
   2. Geography and physiography.
   3. Surface and subsurface geology. Contour maps and cross sections.
   4. Type of trap.
   5. Producing horizons. Character, extent, and depths of each.
   6. Grade of oil.
   7. Methods of drilling.
   8. Depth of drilling.
   9. Completion and production techniques.
   10. Water drive, gas-cap drive or dissolved-gas drive.
   11. Secondary-recovery operations.
   12. Conservation practices.
   13. Ultimate recovery per acre anticipated for each horizon.
 VIII. Production statistics and reserves, using latest data available.
 IX. Bibliography.

QUESTIONS

1. What are some of the responsibilities of the subsurface geologist?
2. Discuss ditch sampling, coring, and preparation of core samples for analysis and shipment.
3. Give an example of organizing a core description.
4. What is meant by formation testing and how is the information recorded?
5. What solvents are used for testing the presence of oil in a sample?

Which test is most sensitive?
6. Carefully read the section "Objective Report Writing" and discuss.
7. What sources may be followed in establishing a publication style?
8. What data should be included on a weekly chronological well report?
9. What information should be included in a weekly well-sample report?
10. A completion log report should include what information?
11. What is a sand data report and why is such important?
12. Discuss the organization of a report of an exploratory (wildcat) well.
13. Discuss the organization of a report on a well in a proven field.
14. Carefully study the report outline on a petroliferous province and discuss.

CHAPTER 10

GRAPHIC REPRESENTATIONS

L. W. LeRoy

Carefully selected and drafted illustrations of any report permit minimization of the text material and enable the reader to obtain a more concise understanding of the subject treated. Concepts and interpretations should be displayed graphically whenever possible. Each drawing should be adequately captioned and correlated with the text.

WELL LOGS

Data to be included on final well logs are governed by company policy. Some logs contain all information relating to the well, whereas others may contain only the minimum facts. Well-log scales vary considerably (1 inch = 50 feet, 1 inch = 100 feet, 1 inch = 200 feet) and are determined by the amount of detail to be shown. Logs are commonly prepared on standard commercial strips (1 inch = 100 feet (fig. 430).

Log headings should include the following information: name of well and company; location (state, county, area, section, township, and range); elevation (ground, derrick floor); date well commenced, completed, or abandoned; total depth; type of drilling equipment; casing record, and by whom logged. The graphic log should include the lithologic column with lithology represented by graphic symbols (fig. 431), colors (pl. 11), or a combination of both; formation description (abbreviated); cored intervals; the position of oil and gas shows (symbols); core dips and fractured phases; casing points (depth, size, and cement data); formational and age boundaries; mineralogic and paleontologic data; marker beds; tested intervals, including a brief summary of results; perforated and plugged intervals; points of loss of circulation or fluid entrance; hole deviation; mud and temperature data; drilling progress (day, week, month); and points of mechanical trouble and bit changes.

Electric-, radioactive- and drill-time-log profiles should be added whenever possible, and the condition of the mud at the time these survey runs were made should also be given.

Plotted lithic data obtained from ditch cuttings frequently do not correspond exactly with other logging data (electric, drill-time, etc.), consequently reinterpretations and adjustments are required.

Some subsurface geologists prefer plotting lithology on a percentage basis, if the penetrated section is not well-known. Straight lithic calls are common practice after formational units have been adequately established.

Colors symbolizing various lithologies are widely used although the color pattern varies among companies. Generally, yellow indicates sand; blue, limestone; and hues of such colors as green, red, gray, and tan, represent shales. Advantages in the use of color symbols are rapid plotting

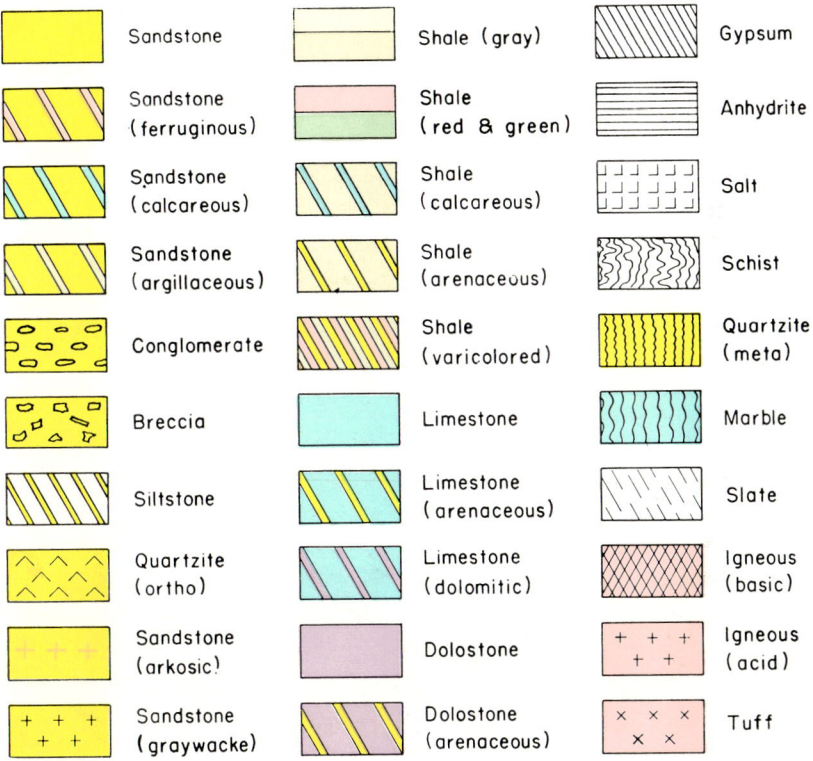

PLATE 11. Colored well-log symbols.

and visual clarity of the lithologies. Plastic templates are frequently employed during plotting. Graphic symbols are favored in the event multiple copies of the log are required.

It is now possible to duplicate colored logs directly from the original by photographic methods. This procedure eliminates checking for errors and permits wide distribution of log strips and the establishment of complete log files. The photographic method of reproducing log strips will be gratefully received by sample-log services and companies.

A unique method of combining lithologic and electric data is shown in figure 433, wherein the electric-log-profile characteristics serve as boundaries for the lithologic plot.

Abbreviations commonly used in log description work are the following:

abundant—A or abd.
agglomerate—aggl.
angular—ang.
arenaceous—aren.
argillaceous—argill.
arkosic—ark.
anhydrite—anhyd. or anhy.
bentonite—bent.
biotite—bio.
bituminous—bit. or bitum.
black—blk.
blue—bl.
bottom—bot. or bott.
brown—brn.
calcareous—calc.
carbonaceous—carb.
cement—cmt.
chert—cht. or ch.
clay—cl.
coarse—cse.
common—C
compact—cpt.
composition—comp.
concretion—conc.
crystalline—xln.
dark—dk.
dense—ds. or d/
distributed—dist. or distrib.

dolomite—dol. or dolo.
ferruginous—ferr.
fibrous—fib.
fine—fn.
fissile—fss.
flint—fl.
formation—fm.
frosted—fstd.
Foraminifera—Forams. or F.
fossil (iferous) —foss.
fracture—fract.
glauconite—glauc. or g/
grain—grn.
granite—grt.
granular—grnl.
gray—gy.
green—gn.
gypsum—gyp.
hard—hd.
igneous—ign.
indurated—indur.
laminated—lam.
limestone—ls.
loose—l/
massive—mass.
marl—ml.
material—matl.
medium—med.

micaceous—mic.
mollusc—moll. (M.F.)
mottled—mot.
oölith—oöl.
pyrite—pyr.
rare—R.
regular—reg.
residue—res.
rock—rk.
rounded—rdd.
sand—sd.
sandstone—ss.
sandy—sdy.
scarce—S
shale—sh.
siderite—sid.
slightly—sl/
siliceous—sil. or silic.
some—s/
streak—stk. or strk.
the—t/
thin-bedded—t.b.
tuffaceous—tuff.
variable—var.
very—v/
with—w/
yellow—yl.

CORRELATION CHARTS

Correlation charts are variously arranged and drafted. Scales used are governed by the problem. Stratigraphic sections may be shown by single lines or by graphic or colored columns. Sufficient data should be included in order to clarify fundamental lithologic and paleontologic characteristics of each formation. If correlations are well established, lines connecting correlative units should be solid; if inferred or questionable correlations are involved, the lines should be dashed or dotted. If a series of well columns are involved, it is required that the sea-level datum and correlative depth and formation names and ages be shown. A scaled index map showing the location of sections and a graphic vertical

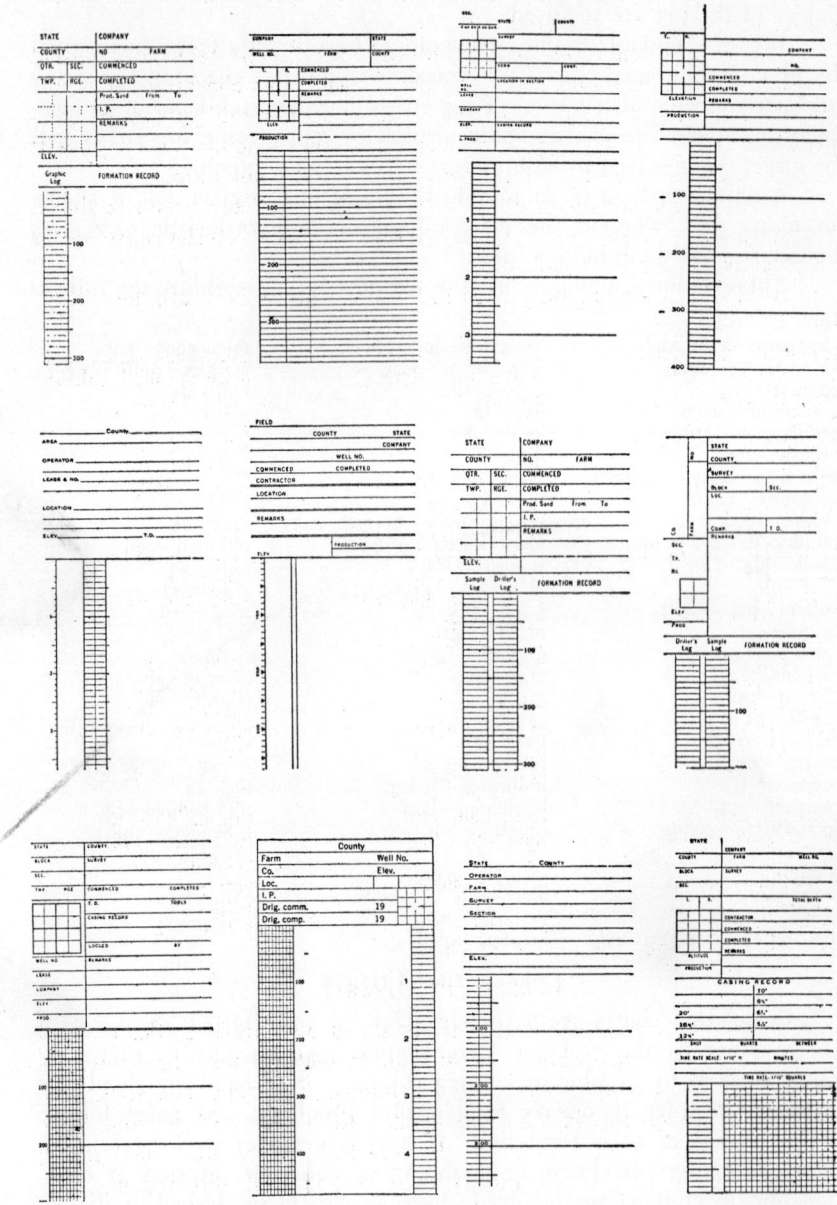

FIGURE 430. Types of standard well-log strips.

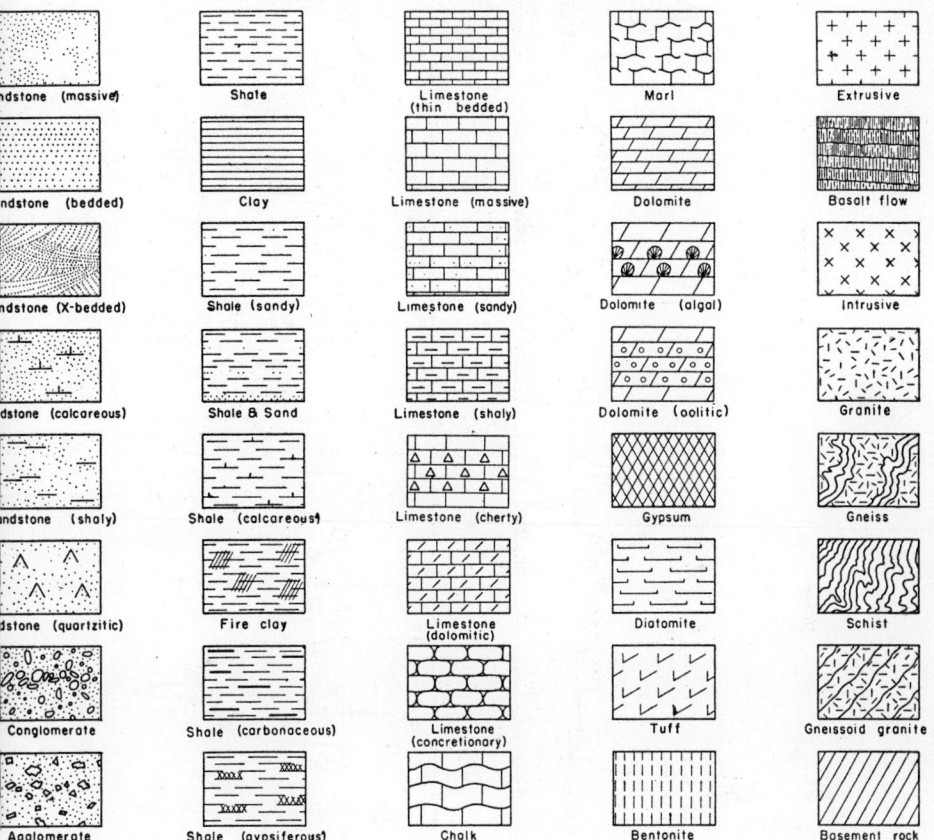

FIGURE 431. Lithologic graphic symbols used in plotting well and surface sections.

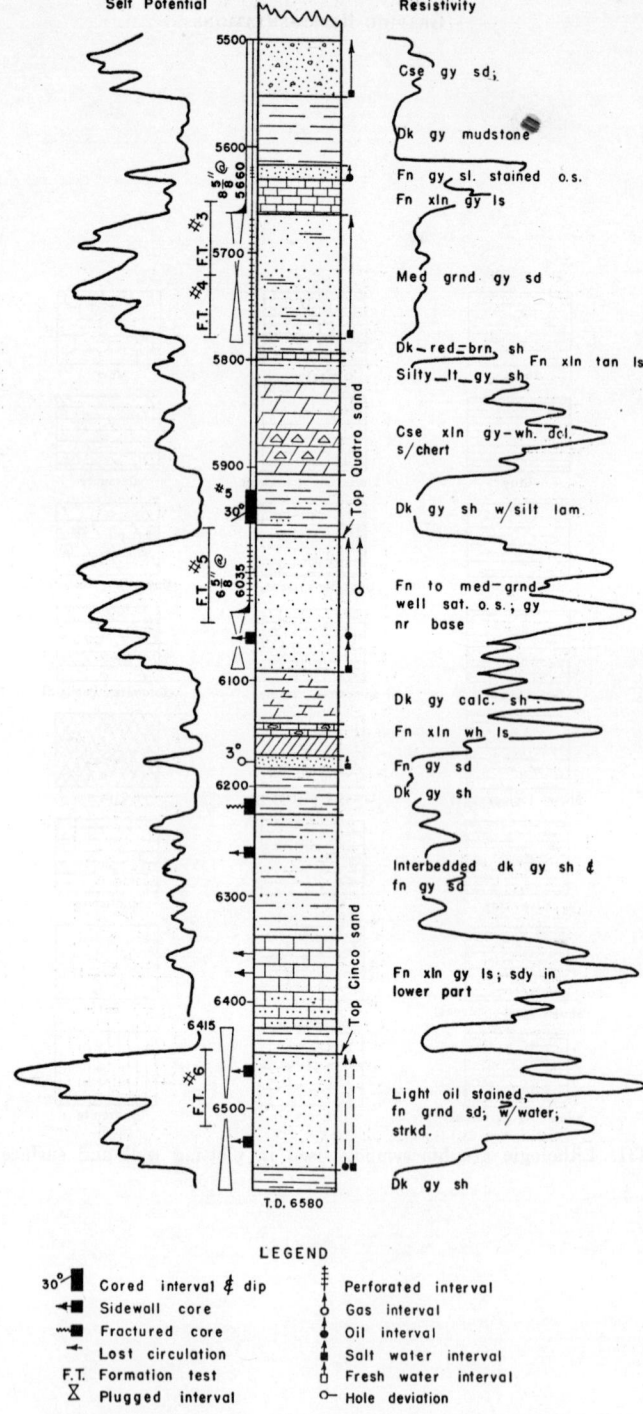

Self Potential

Resistivity

5500

Cse gy sd;

5600

Dk gy mudstone

85" @ 5660

Fn gy, sl. stained o.s.

Fn xln gy ls

F.T. #3

5700

F.T. #4

Med grnd gy sd

5800

Dk-red-brn sh   Fn xln tan ls

Silty lt gy sh

Cse xln gy-wh. dol. s/chert

5900

#5

3°

Dk gy sh w/silt lam.

Top Quatro sand

F.T. #5

65" @ 6035

Fn to med-grnd well sat. o.s.; gy nr base

6100

Dk gy calc. sh

Fn xln wh ls

3°

Fn gy sd

Dk gy sh

6200

Interbedded dk gy sh & fn gy sd

6300

Top Cinco sand

Fn xln gy ls; sdy in lower part

6400

6415

F.T. #6

Light oil stained; fn grnd sd; w/water; strkd.

6500

Dk gy sh

T.D. 6580

LEGEND

30° ▮ Cored interval & dip
◄■ Sidewall core
〜■ Fractured core
◄- Lost circulation
F.T. Formation test
⋈ Plugged interval

╪ Perforated interval
○ Gas interval
● Oil interval
▮ Salt water interval
□ Fresh water interval
○- Hole deviation

FIGURE 432. Data to be included on a final well log.

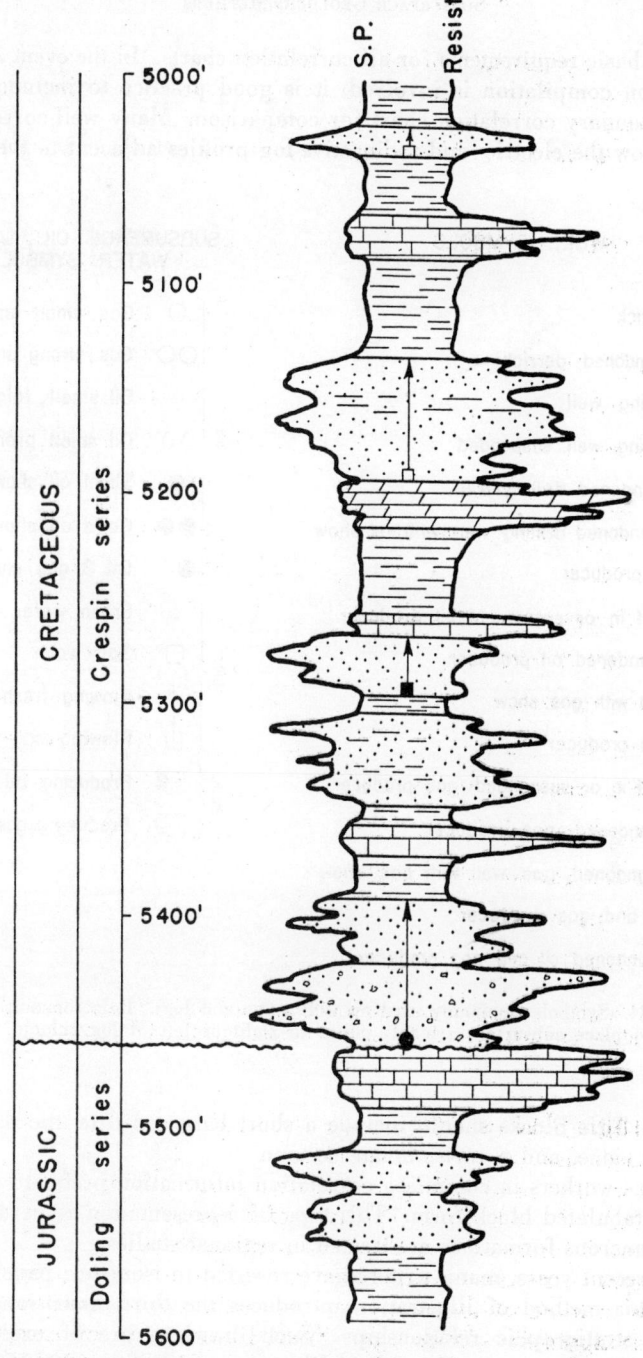

FIGURE 433. An application of electric profiles for plotting
lithologic logs.

scale are basic requirements for all correlation charts. In the event a local correlation compilation is involved, it is good practice to include a regional summary correlation chart for comparison. Many well-correlation charts show the electric- and radioactive-log profiles adjacent to the lithic

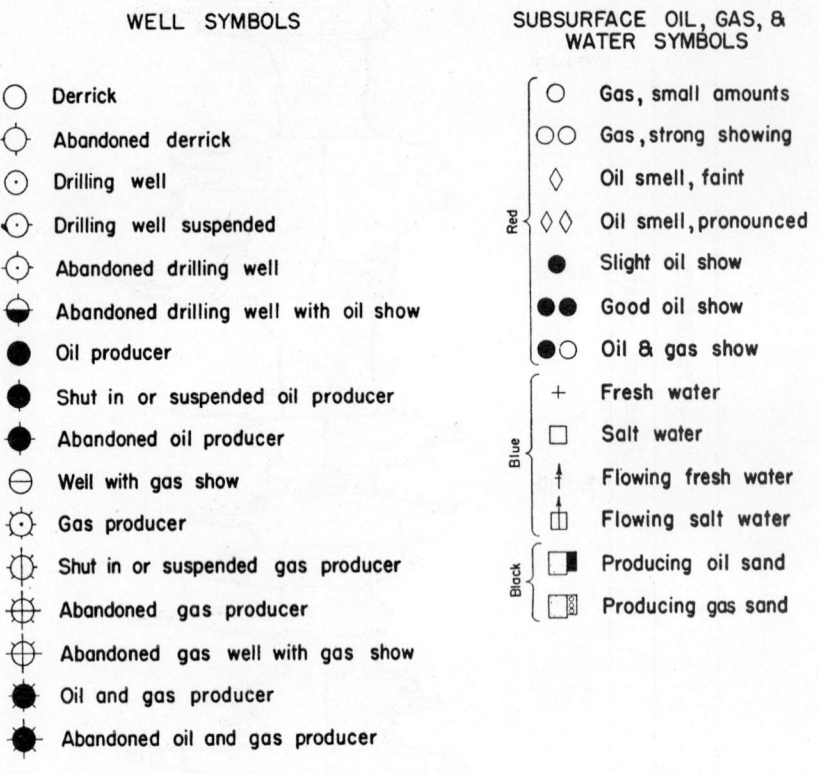

WELL SYMBOLS

SUBSURFACE OIL, GAS, & WATER SYMBOLS

○ Derrick

◔ Abandoned derrick

⊙ Drilling well

◒ Drilling well suspended

◔ Abandoned drilling well

◓ Abandoned drilling well with oil show

● Oil producer

● Shut in or suspended oil producer

◀ Abandoned oil producer

⊖ Well with gas show

☼ Gas producer

☼ Shut in or suspended gas producer

⊕ Abandoned gas producer

⊕ Abandoned gas well with gas show

✸ Oil and gas producer

✸ Abandoned oil and gas producer

Red {
○ Gas, small amounts

○○ Gas, strong showing

◇ Oil smell, faint

◇◇ Oil smell, pronounced

● Slight oil show

●● Good oil show

●○ Oil & gas show
}

Blue {
+ Fresh water

□ Salt water

↑ Flowing fresh water

⊤ Flowing salt water
}

Black {
▣ Producing oil sand

▨ Producing gas sand
}

FIGURE 434. Symbols commonly used on well maps and logs. It is common practice to place subsurface symbols either to right or left of log column.

column. Title blocks should include a short balanced title, the date, the author's name, and a revision-date column.

Some workers in compiling correlation information prefer to present data in tabulated block form. This type of representation is permissible when numerous formations are treated in regional studies.

In recent years, many writers have reverted to isometric panel drawings. This method of illustration introduces the third-dimensional concept of stratigraphic relationships. Such drawings are of exceptional value in that they are easily and readily interpreted (figs. 462, 463).

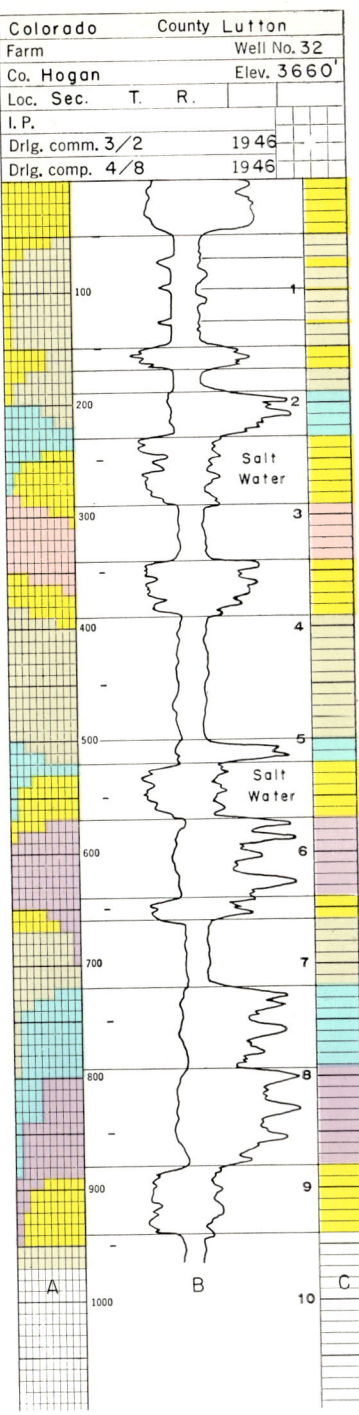

Colorado	County Lutton
Farm	Well No. 32
Co. Hogan	Elev. 3660'
Loc. Sec. T. R.	
I. P.	
Drlg. comm. 3/2	1946
Drlg. comp. 4/8	1946

PLATE 12. *A*, combination lithologic percentage; *B*, electrical; and *C*, interpretative well logs. The interpretative log is based on data secured from the percentage and electrical logs. Note the readjustment of lithologic boundaries as shown on the interpretative log. Considerable interpretative information is lost if a percentage log is not plotted.

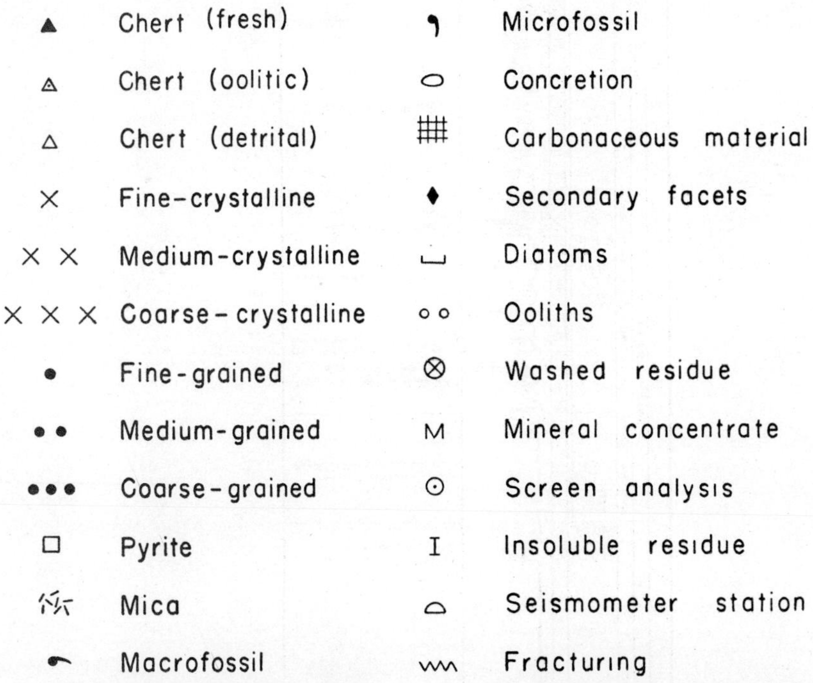

▲   Chert (fresh)   ꟼ   Microfossil

△ (oolitic) Chert (oolitic)   ◠   Concretion

△   Chert (detrital)   ▦   Carbonaceous material

×   Fine–crystalline   ◆   Secondary facets

× ×   Medium–crystalline   ⌴   Diatoms

× × ×   Coarse–crystalline   ∘∘   Ooliths

●   Fine-grained   ⊗   Washed residue

●●   Medium-grained   M   Mineral concentrate

●●●   Coarse-grained   ⊙   Screen analysis

□   Pyrite   I   Insoluble residue

𛀀   Mica   ◠   Seismometer station

⌒   Macrofossil   ⌇   Fracturing

FIGURE 435. Miscellaneous symbols used on well logs. Symbols are generally placed to left of lithologic column. Purpose of these symbols is to minimize written text on log.

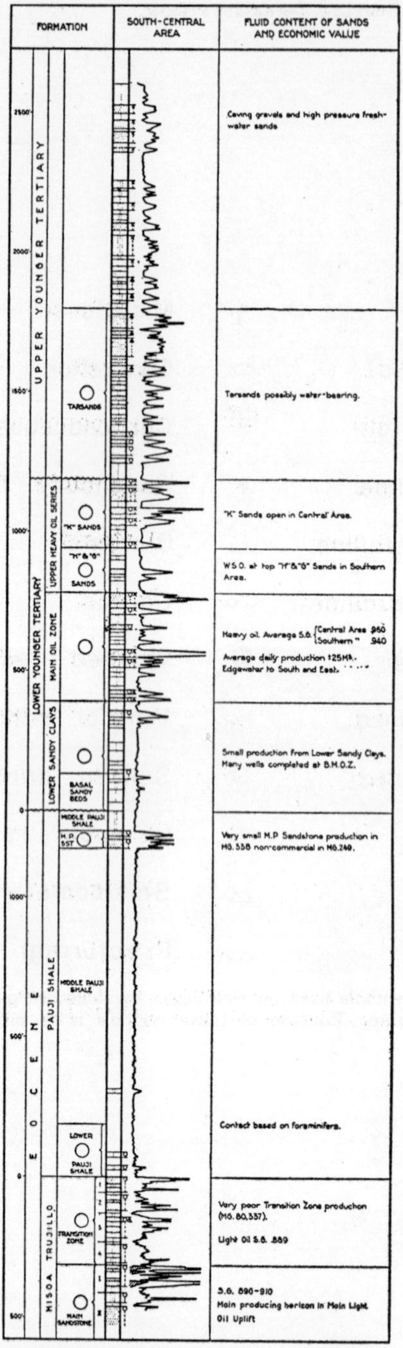

FIGURE 436. Type well section of Eocene and Miocene in south-central part of Mene Grande field, Venezuela. Such compilations should be prepared for all fields. (From Caribbean Petroleum Co. Reproduced permission Am. Assoc. Petroleum Geologists.)

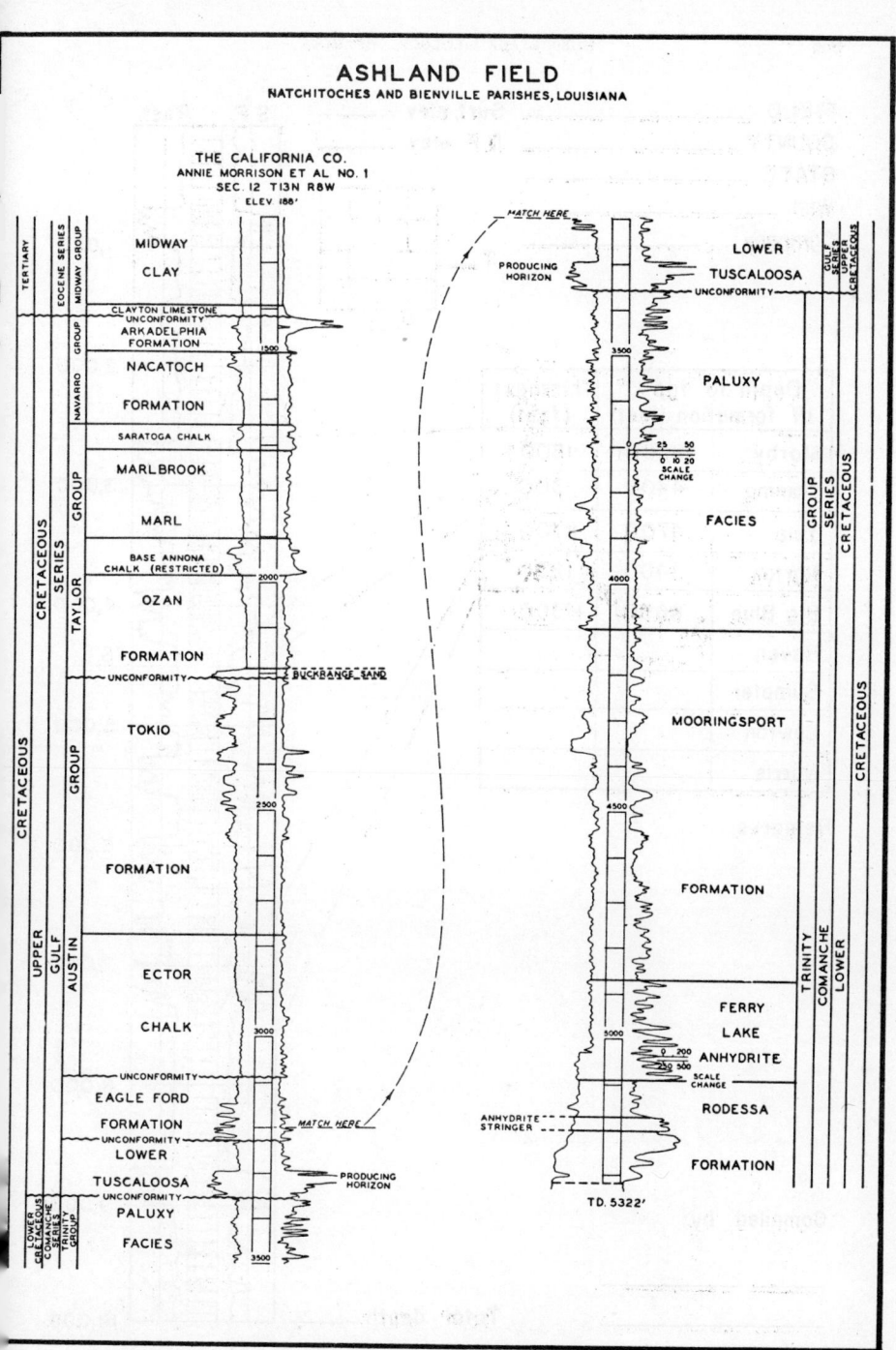

FIGURE 437. Correlation between an electric-log profile and formational units. (From Shreveport Geol. Soc.)

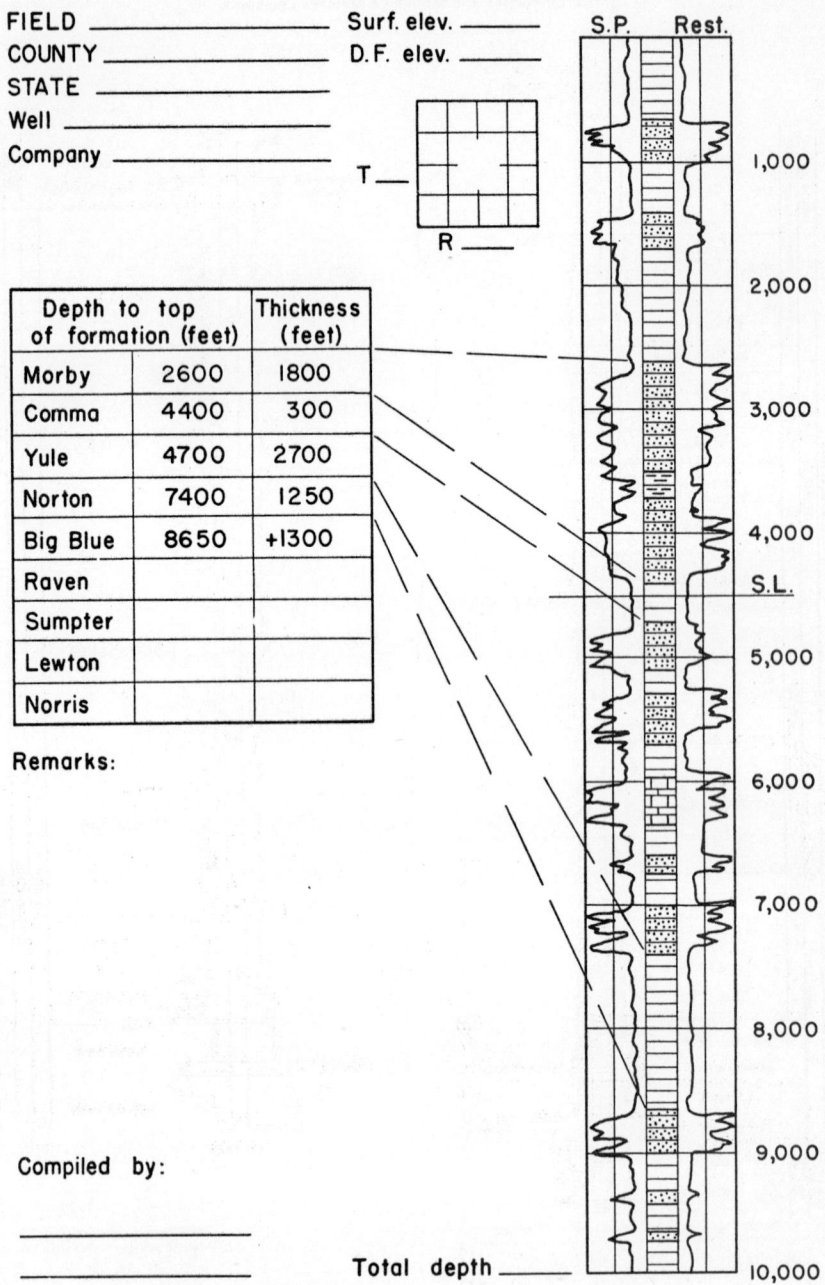

FIELD _____ Surf. elev. _____

COUNTY _____ D.F. elev. _____

STATE _____

Well _____

Company _____

T ___

R ___

S.P.   Rest.

1,000

2,000

Depth to top of formation (feet)		Thickness (feet)
Morby	2600	1800
Comma	4400	300
Yule	4700	2700
Norton	7400	1250
Big Blue	8650	+1300
Raven		
Sumpter		
Lewton		
Norris		

3,000

4,000

S.L.

5,000

Remarks:

6,000

7,000

8,000

Compiled by:

9,000

_____

_____   Total depth _____

10,000

FIGURE 438. Compilation chart; adaptable when assimilating data on an oil field. (Suggested by A. J. Gude, III.)

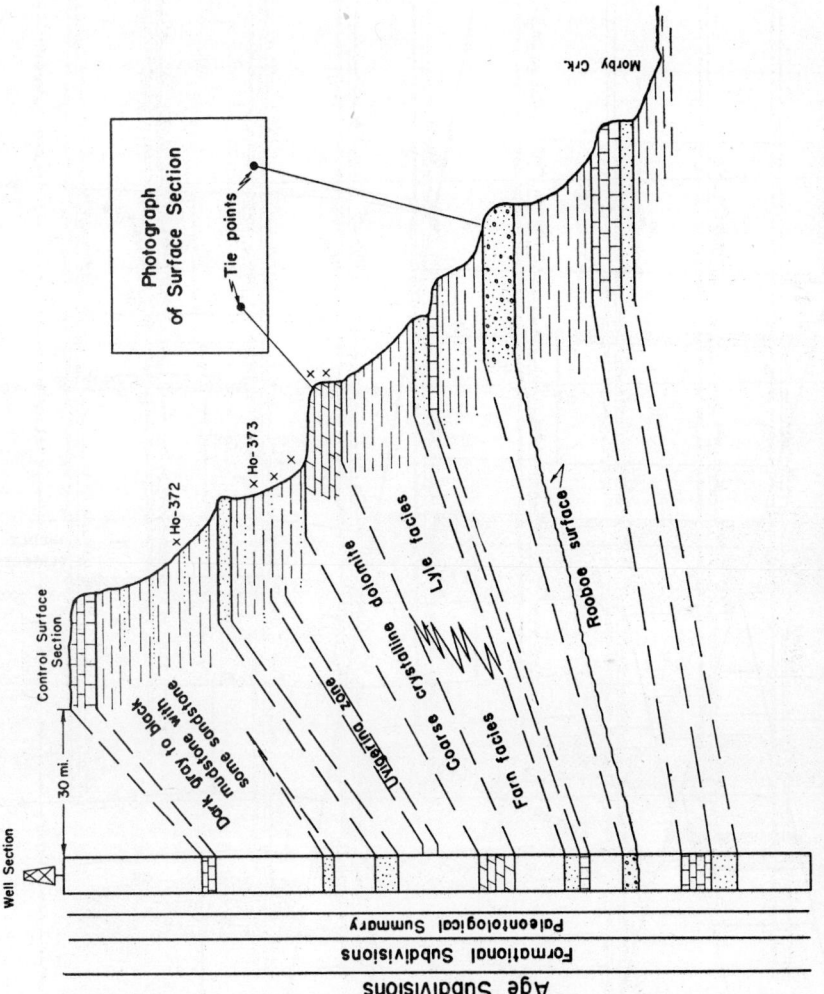

FIGURE 439. Composite drawing showing relationship of well and surface sections.

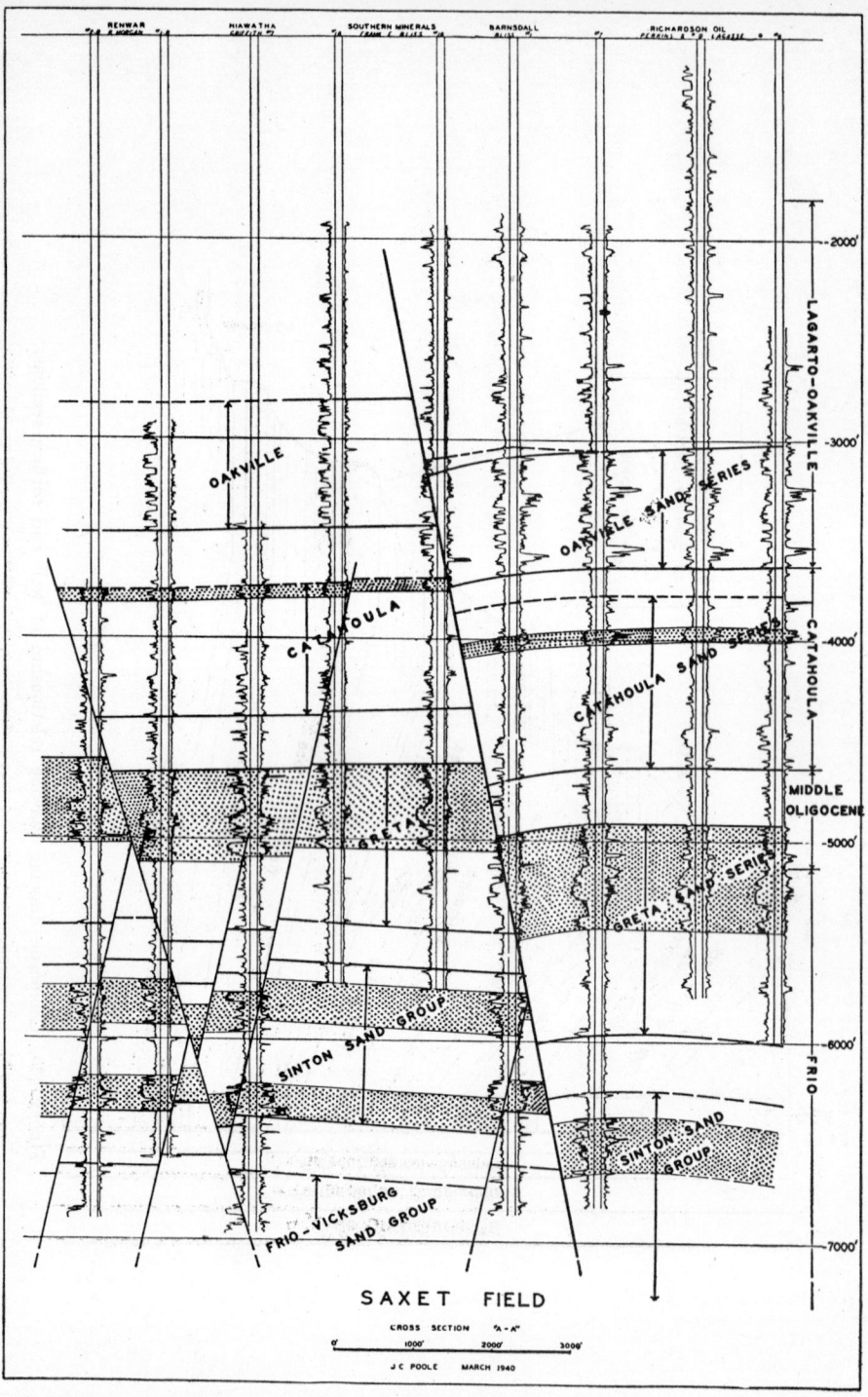

FIGURE 440. Chart showing complex faulting and divisions of sand series and groups with relation to geologic time divisions. Such conditions are deciphered by use of electric logs and micropaleontology. (From Poole. Reproduced permission Am. Assoc. Petroleum Geologists.)

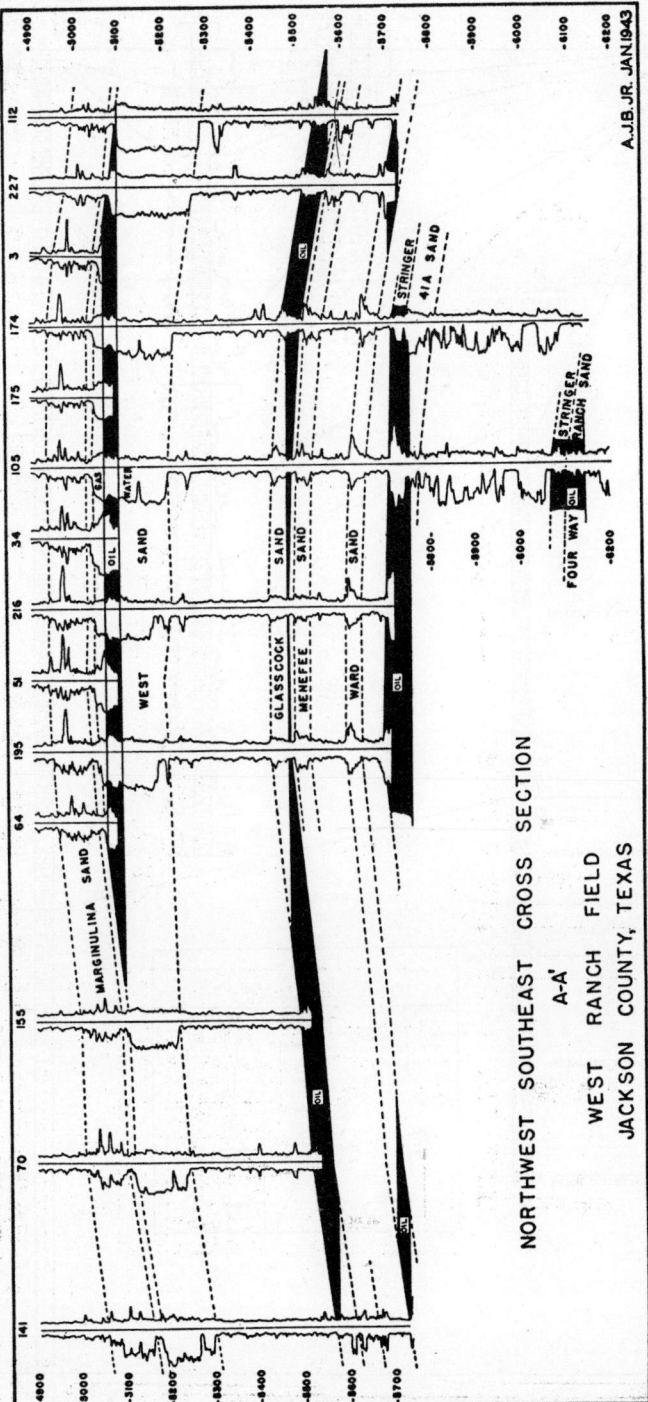

FIGURE 441. A standard method of illustrating distribution of oil and gas and their stratigraphic relationships. (From Bauerschmidt. Reproduced permission Am. Assoc. Petroleum Geologists.)

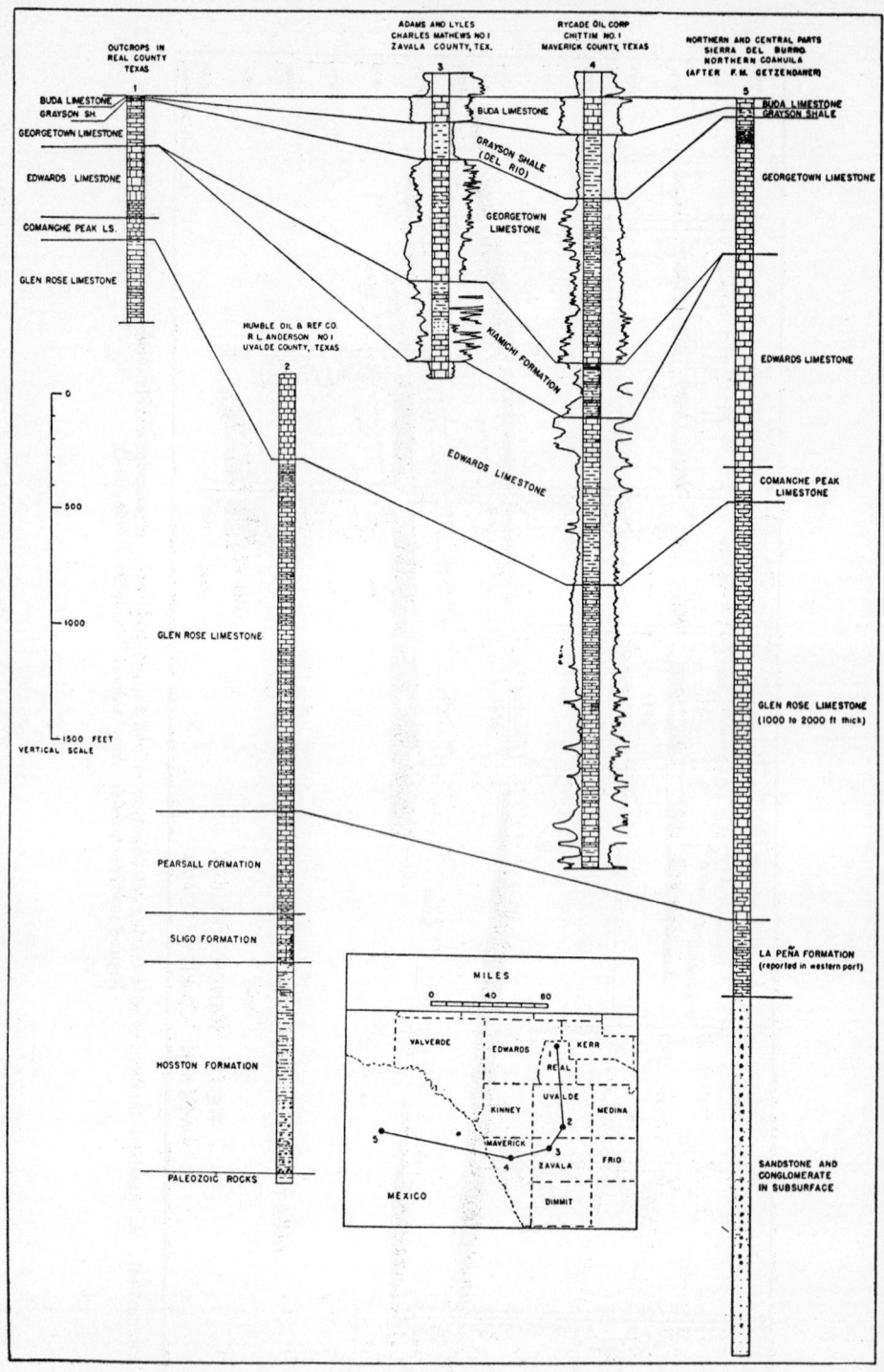

FIGURE 442. Correlation of well sections with control surface sections. An index map should always accompany such compilations. (From Imlay. Reproduced permission Am. Assoc. Petroleum Geologists.)

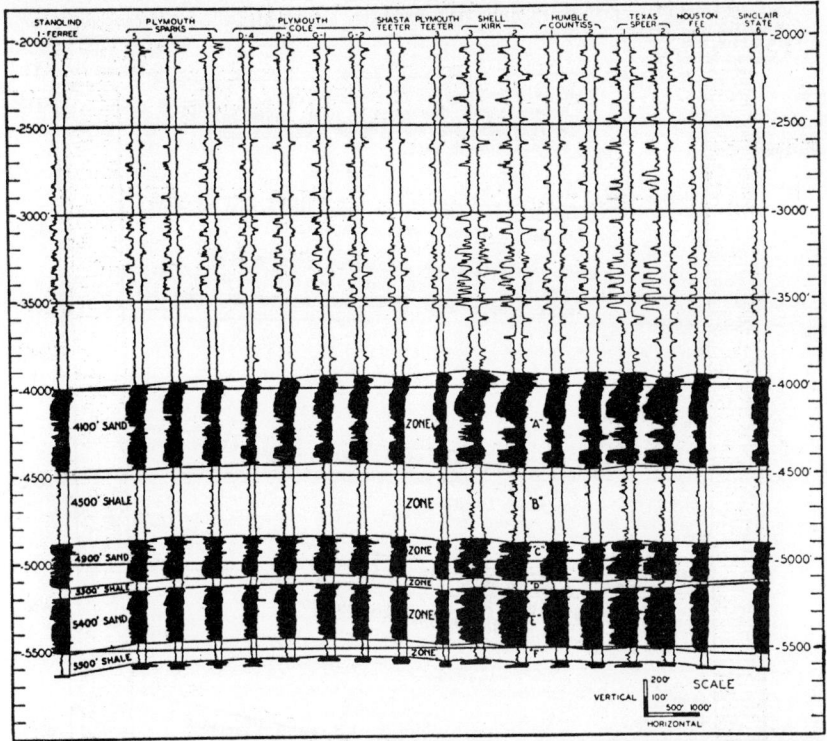

FIGURE 443. North-south section *(A–A′)* across East White Point field, San Patricio County, Texas, showing detail correlation and variation of Oligocene strata by use of electric logs. (From Martyn and Sample. Reproduced permission Am. Assoc. Petroleum Geologists.)

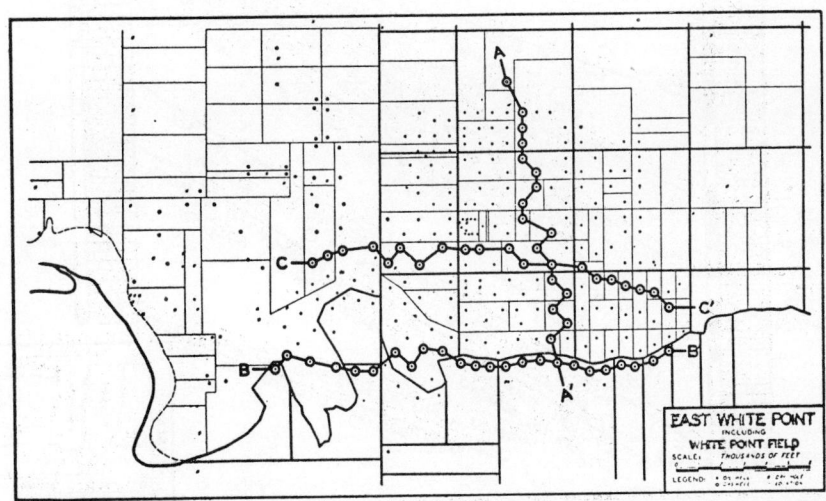

FIGURE 444. Index map showing position of section *(A–A′)* of figure 443. (From Martyn and Sample. Reproduced permission Am. Assoc. Petroleum Geologists.)

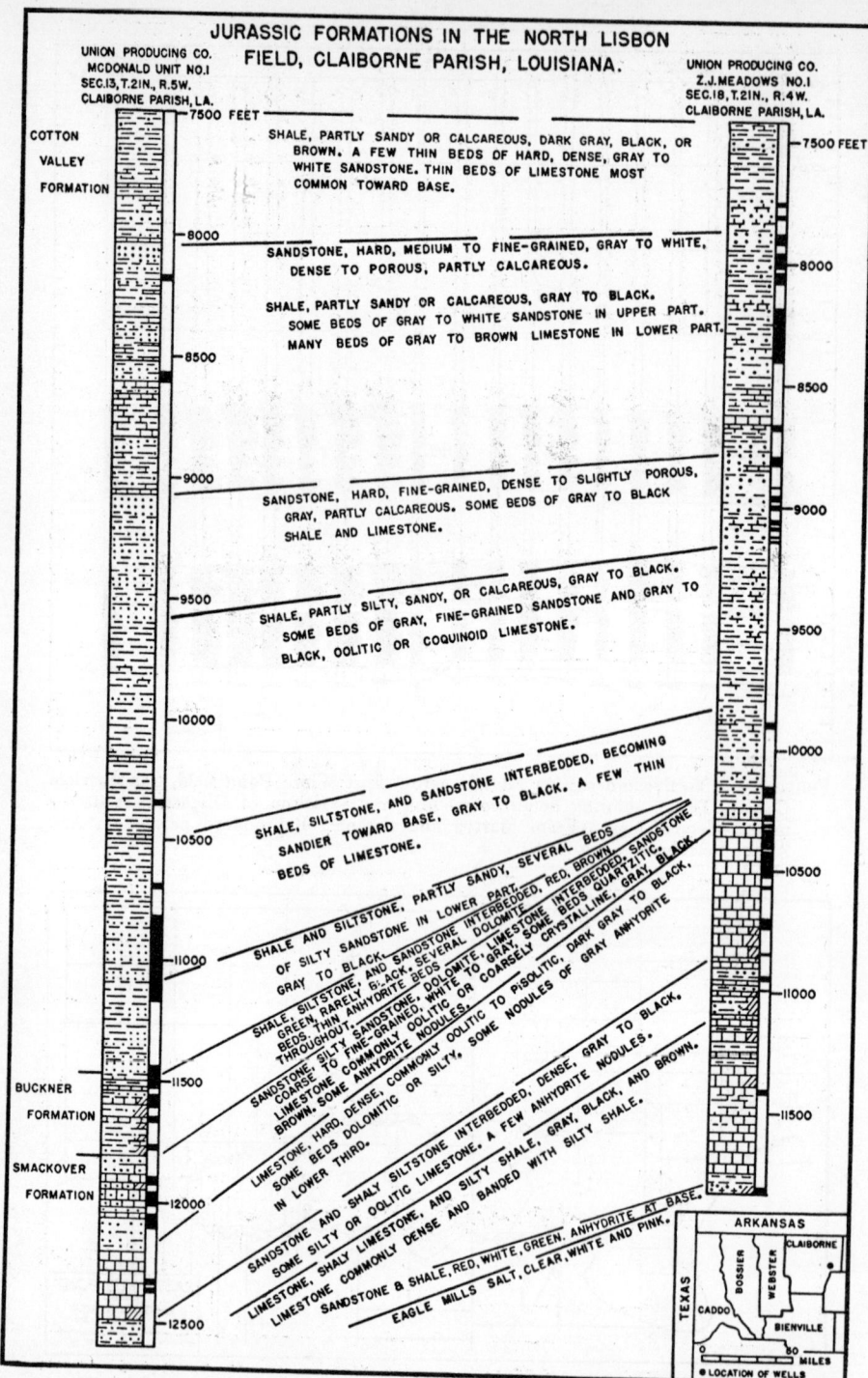

FIGURE 445. Columnar subsurface sections of Upper Jurassic formations in North Lisbon field, Claiborne Parish, Louisiana. (From Imlay. Reproduced permission Am. Assoc. Petroleum Geologists.)

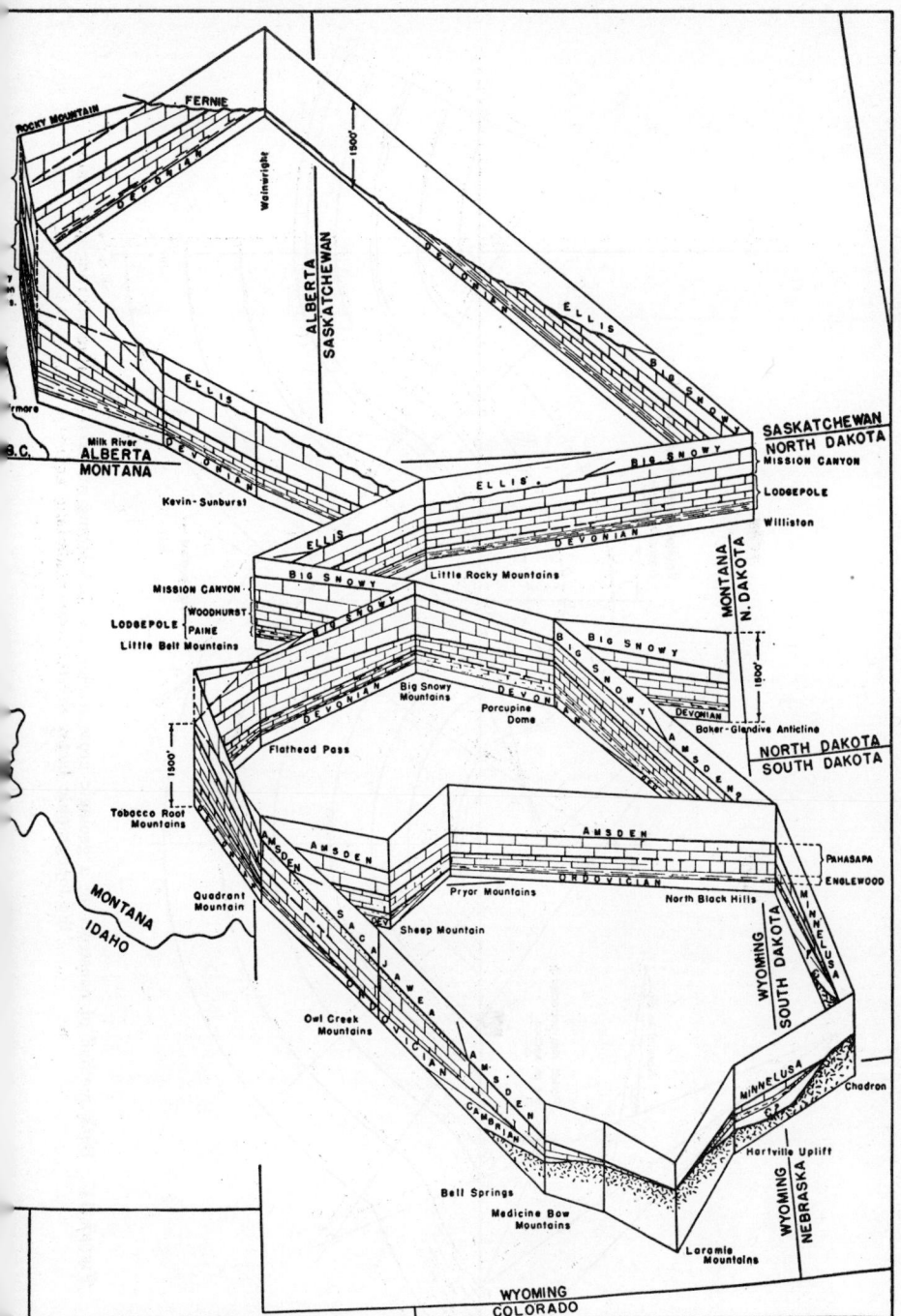

FIGURE 446. Panel diagram showing surface and well sections of Madison group and adjacent formations of northwest Great Plains area. (From Sloss and Hamblin. Reproduced permission Am. Assoc. Petroleum Geologists.)

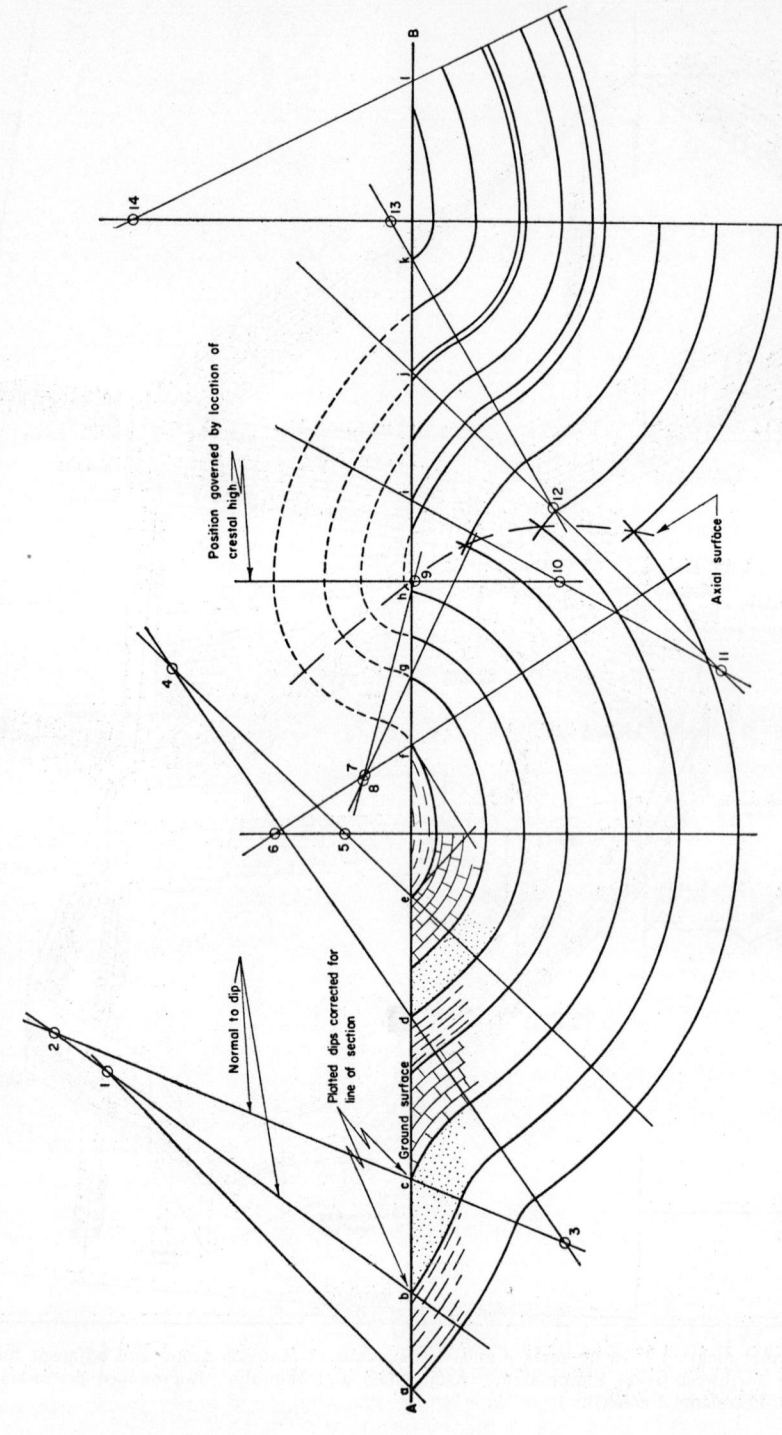

FIGURE 447. Busk method of constructing a geologic cross section and developing an axial surface. In rough terrain the topographic factor must be considered.

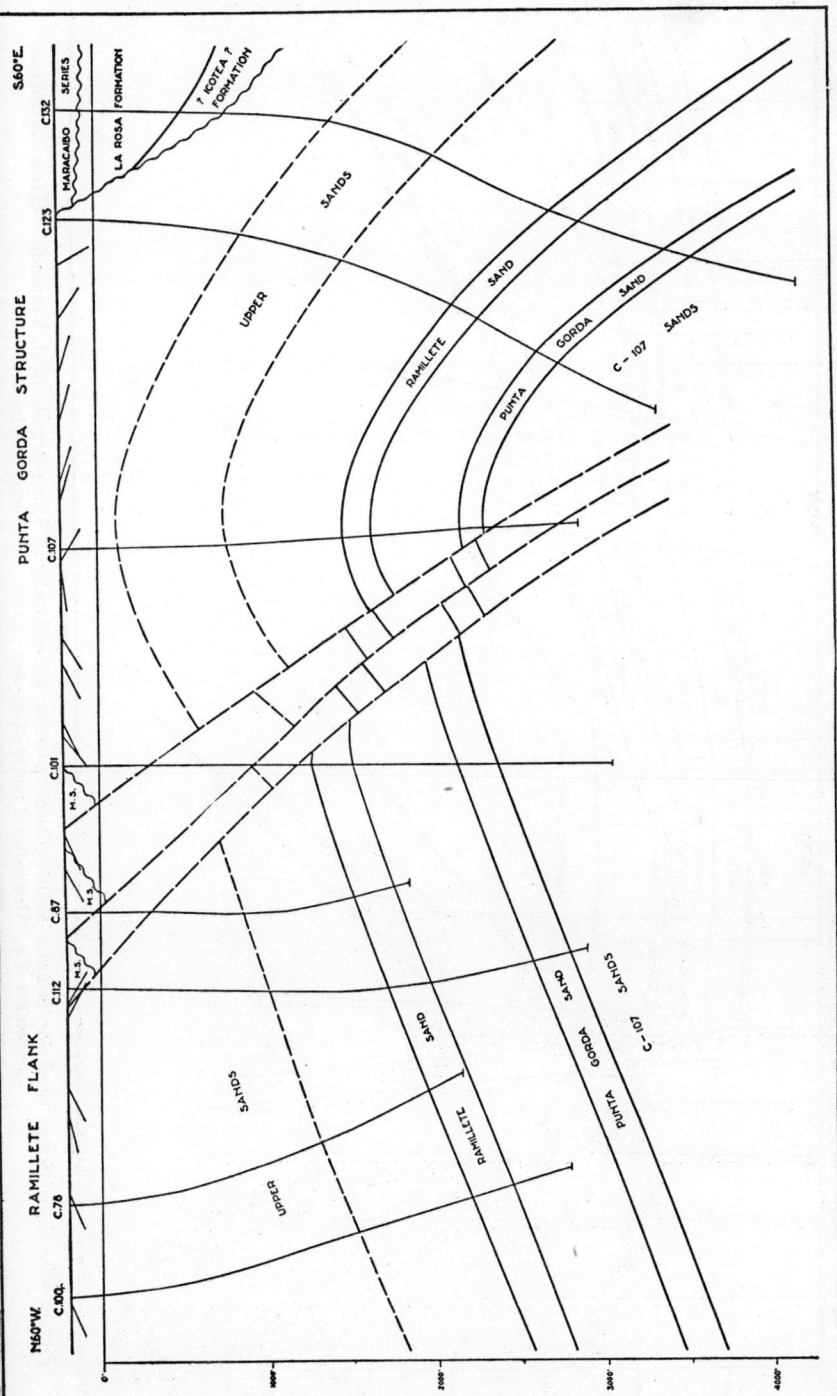

FIGURE 448. Cross section involving directional drilling on a complexly faulted structure. Section through Ramilette flank and Punta Gorda structure of Concepcion field, Venezuela, according to P. de Schumacher. (From Caribbean Petroleum Co. Reproduced permission Am. Assoc. Petroleum Geologists.)

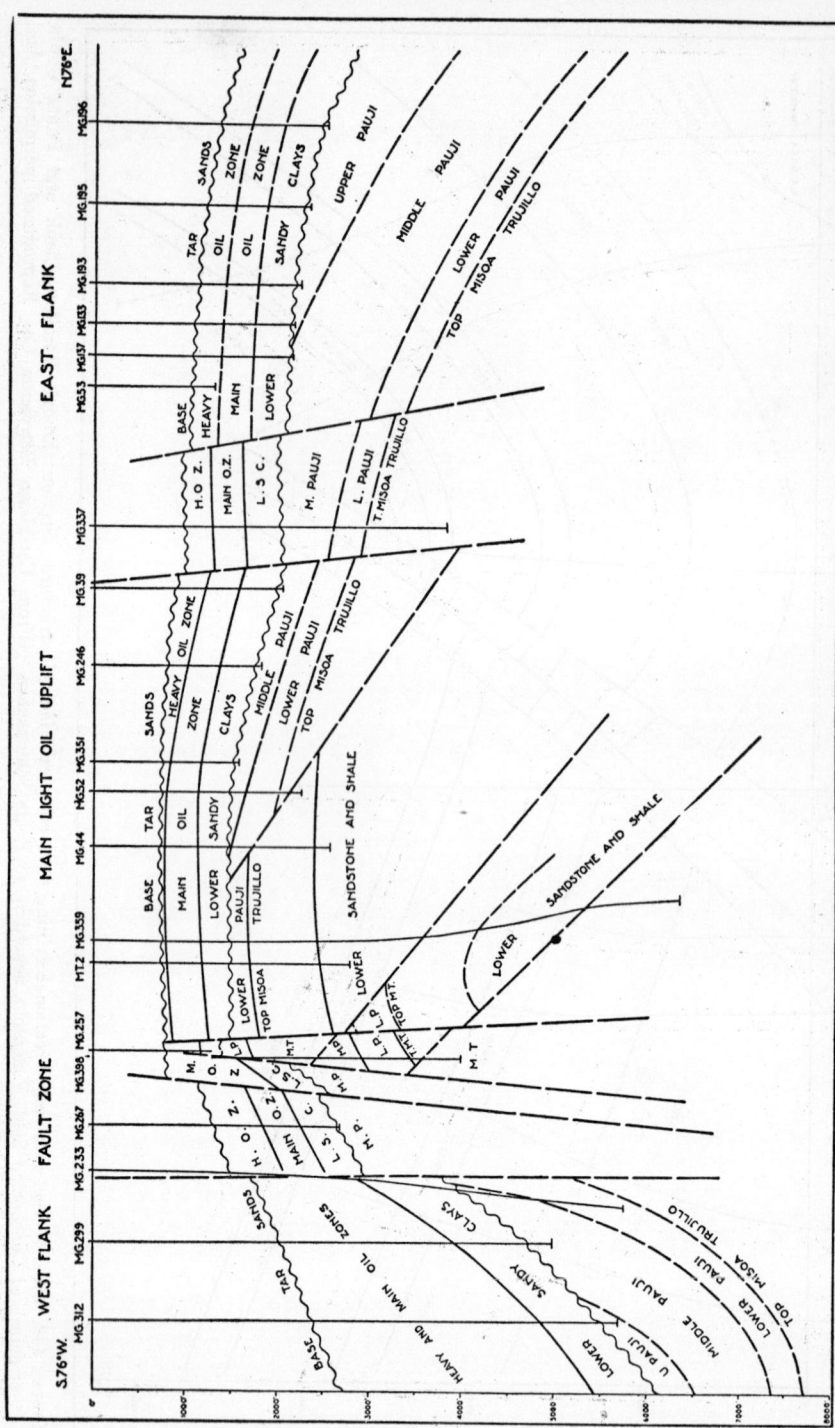

FIGURE 449. Cross section involving complex faulting below a major unconformity in Mene Grande field, Venezuela. (From Caribbean Petroleum Co. Reproduced permission Am. Assoc. Petroleum Geologists.)

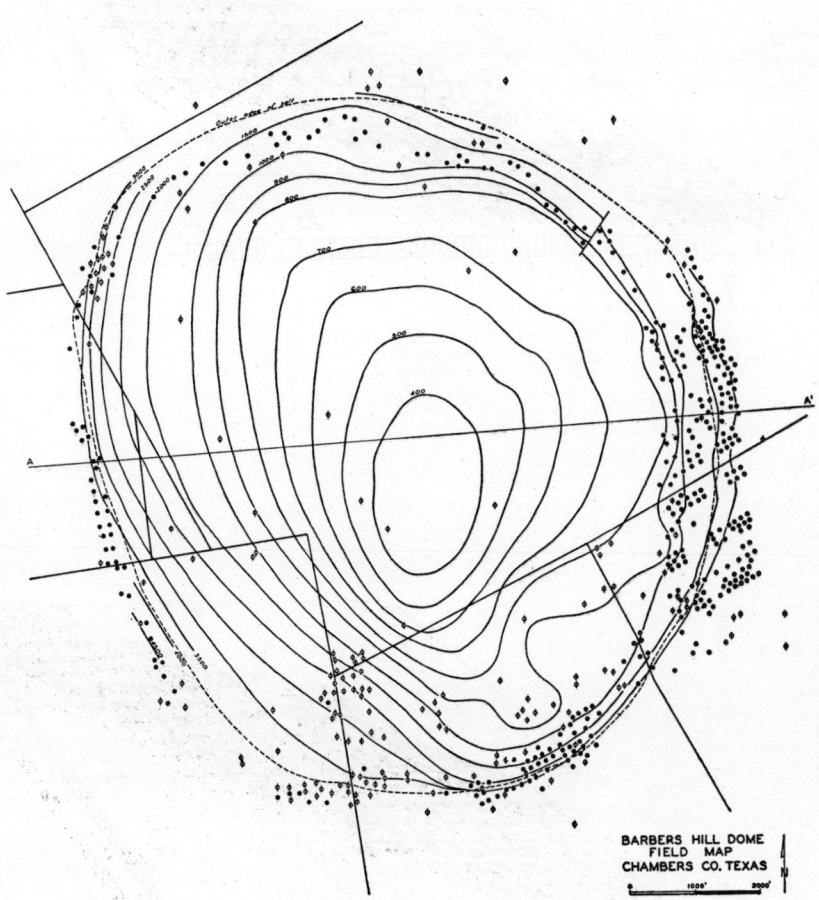

FIGURE 450. Plan structural view of Barbers Hill dome, Chambers County, Texas.. (Reproduced permission Houston Geol. Soc.)

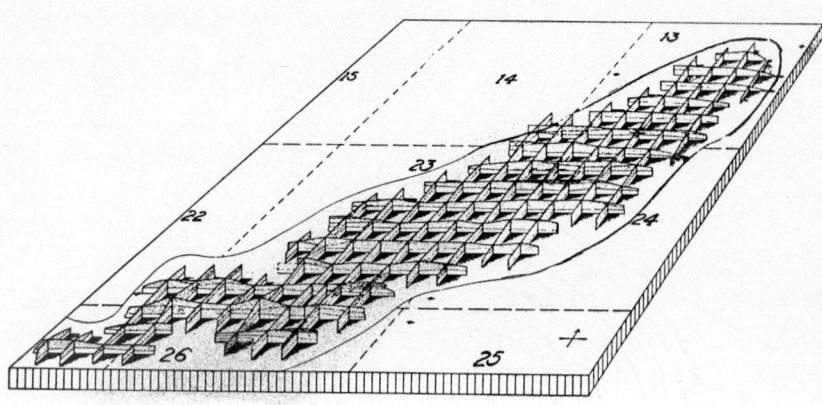

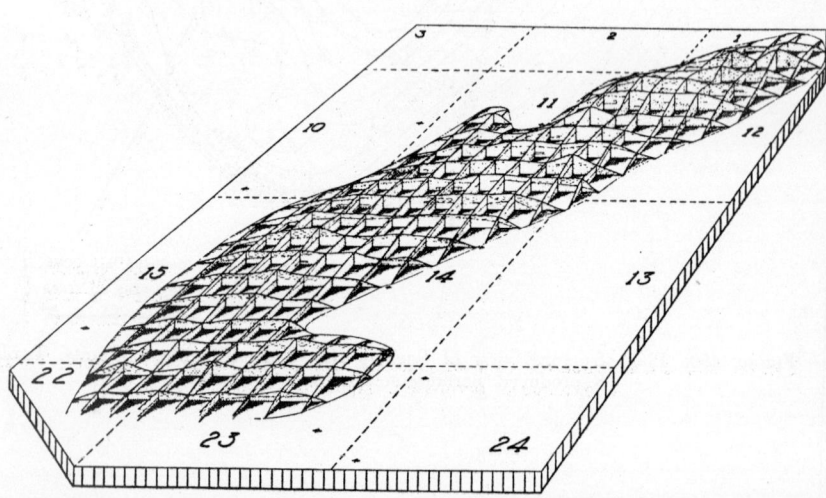

Figure 451. Use of block diagrams in representing variations in Bartlesville sand thicknesses in Burkett (upper figure) and Madison (lower figure) fields of Kansas. (From Bass. Reproduced permission Am. Assoc. Petroleum Geologists.)

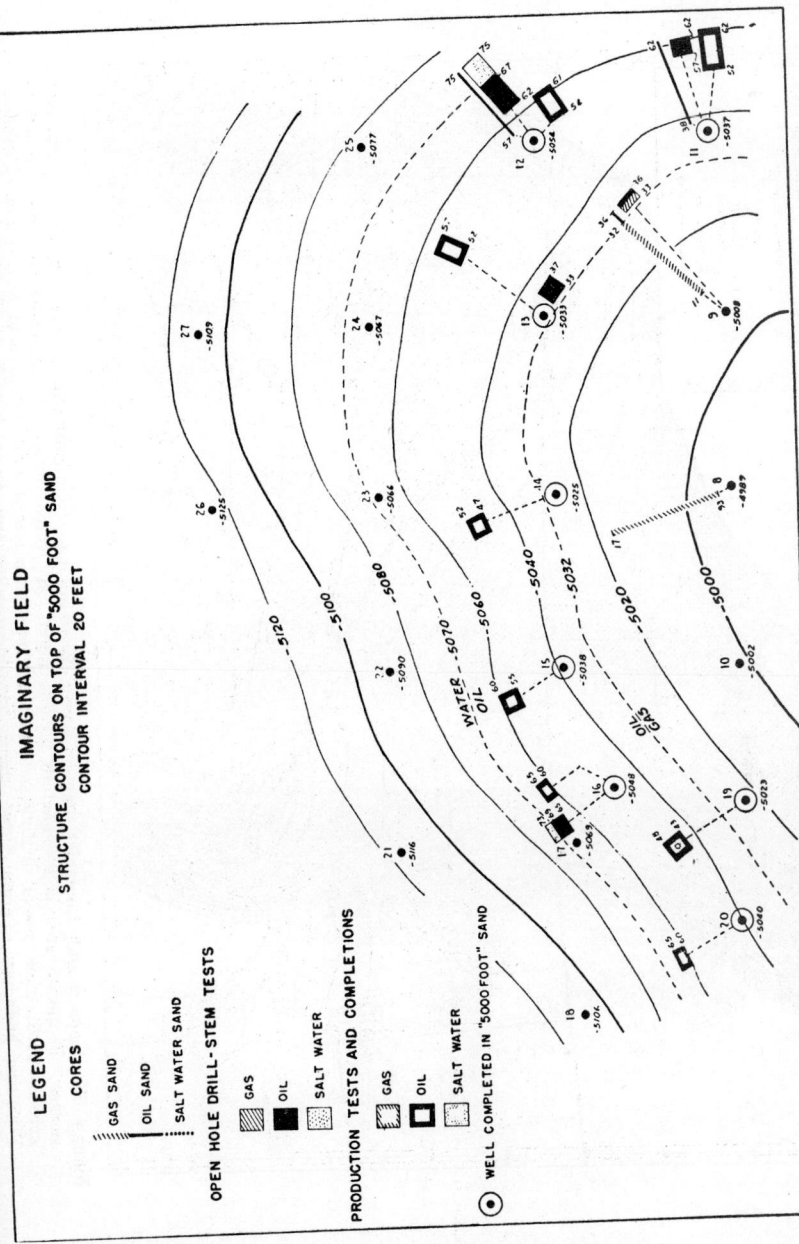

FIGURE 452. Graphical method of recording cores, drill-stem tests, production tests, and production histories of wells on structure-contour maps of oil and gas reservoirs. Posting of data aids in determining fluid contacts and in following expansion of gas cap or water encroachment. Color symbols may also be used to represent the data. Subsea depths of cored, tested, and producing intervals within reservoir being mapped are projected directly to map at positions indicated by structure contours. (From Alexander. Reproduced permission Am. Assoc. Petroleum Geologists.)

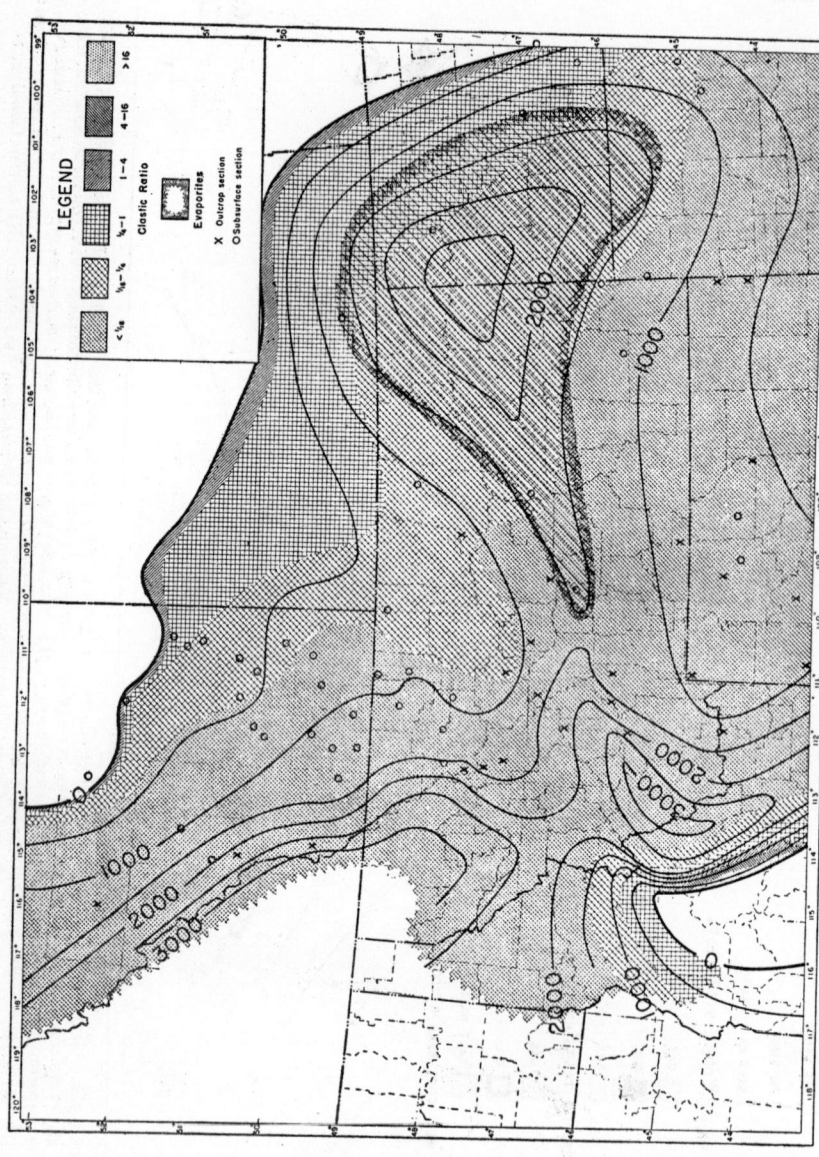

FIGURE 453. Isopach and lithofacies map of Lower Mississippian (Kinderhookian, Osagian, and Meramacian series). Evaporite area is intended to include points at which ratio of evaporites to other nonclastics exceeds 1:10; more recent drilling indicates southern extension of evaporate area to near center of Montana-Wyoming boundary. (From Sloss. Reproduced permission Am. Assoc. Petroleum Geologists.)

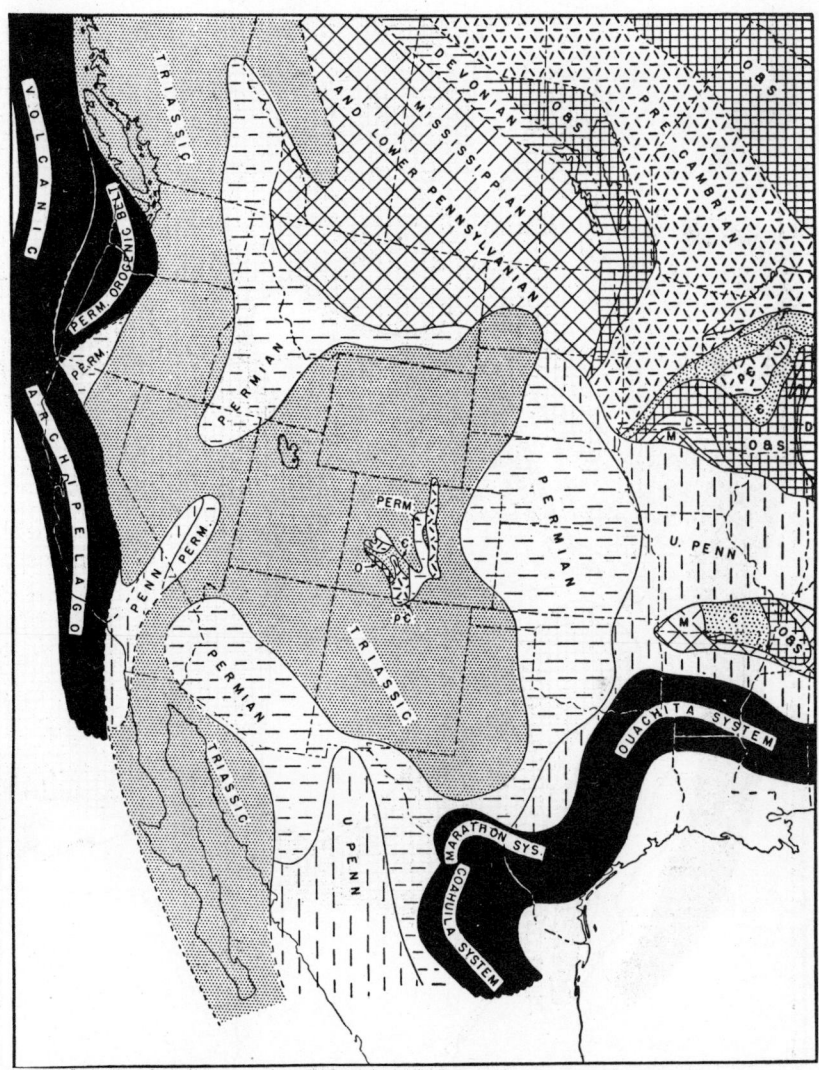

FIGURE 454. Paleogeologic map at close of Triassic. Maps of this type enable the
geologist to decipher geologic history and trends more accurately. (From
Eardley. Reproduced permission Am. Assoc. Petroleum Geologists.)

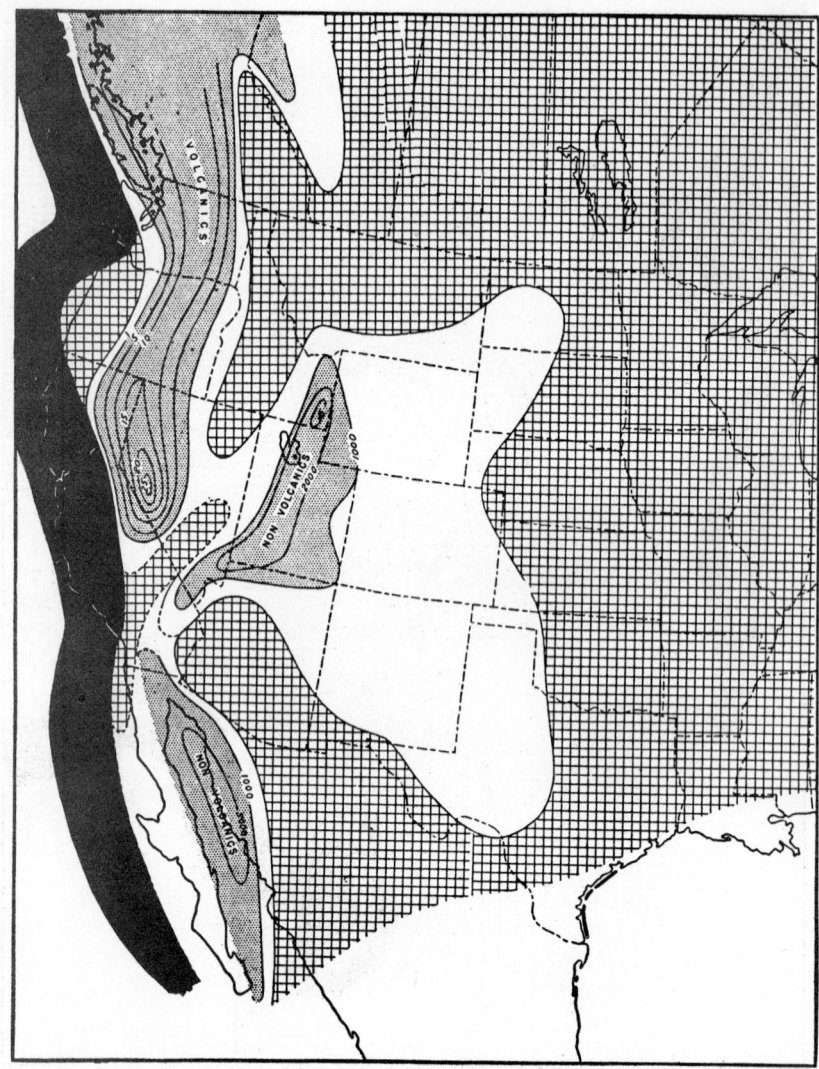

FIGURE 455. Paleotectonic map of Triassic. Black areas are orogenic belts; cross-ruled areas are epeirogenic uplifts that were subjected to erosion; white areas were covered by less than 1,000 feet of sediments; and stippled areas sank more than 1,000 feet and received more than 1,000 feet of sediments. (From Eardley. Reproduced permission Am. Assoc. Petroleum Geologists.)

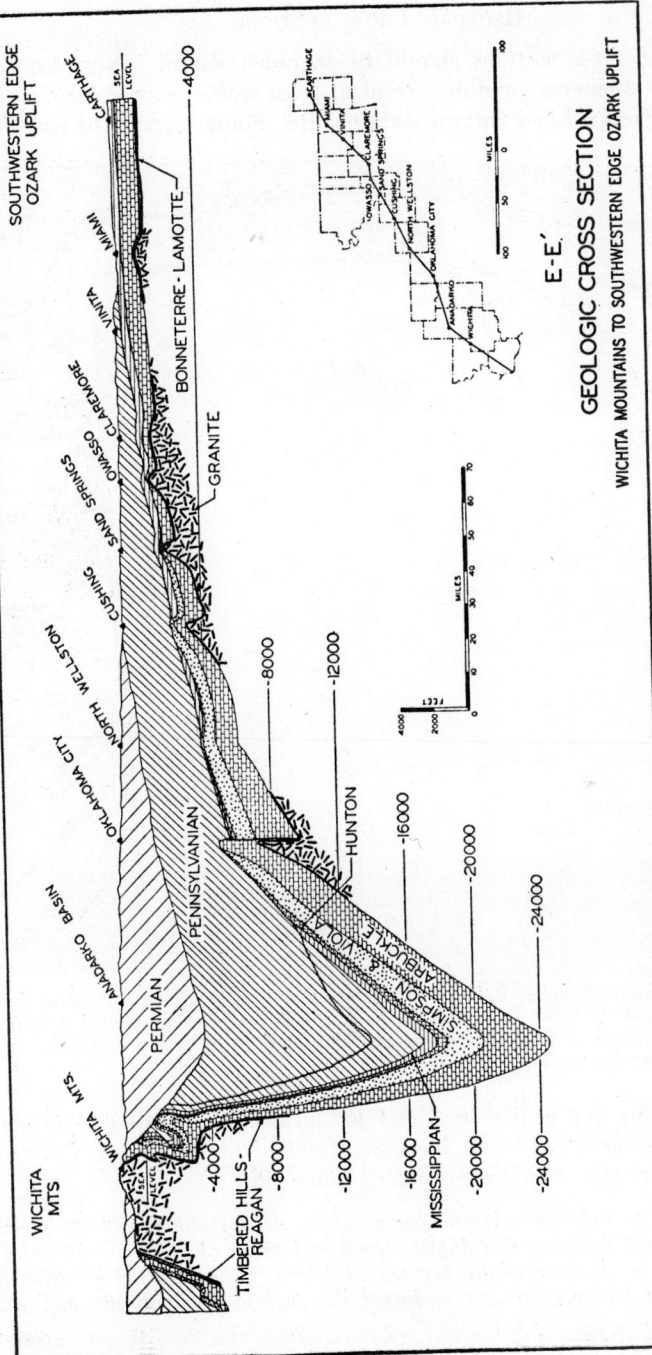

WICHITA MOUNTAINS TO SOUTHWESTERN EDGE OZARK UPLIFT

GEOLOGIC CROSS SECTION

E-E'

FIGURE 456. Effect of vertical exaggeration of scale. This type of drawing fails to exhibit true structural relationships; however, such drawings are sometimes necessary to emphasize geologic phenomena. (From Bartram, Imbt, and Shea. Reproduced permission Am. Assoc. Petroleum Geologists.)

Geologic Cross Sections

Geologic cross sections should be included within subsurface geologic reports whenever possible. An ideal cross section should incorporate both an exaggerated and natural-scale profile. Many features of structure

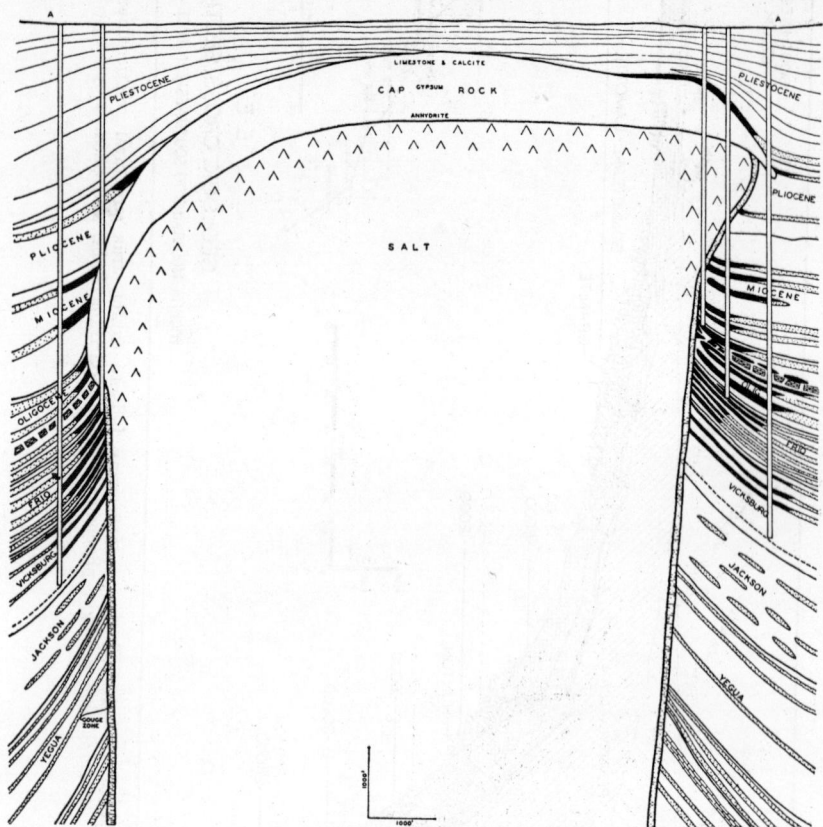

Figure 457. Cross section of Barbers Hill dome, a typical overhang salt plug; note truncated and upturned sediments on flanks of intrusion; also arched overlying section. In many areas of salt intrusion, faulting is extreme and complex. (Reproduced permission Houston Geol. Soc.)

and stratigraphy that cannot be illustrated on a natural-scale section may be shown in exaggerated form; however, it must be remembered that such sections distort true conditions. Suter [1] comments:

Obviously in extreme cases of exaggeration, the tectonical and the stratigraphical picture becomes meaningless, even in regions of moderate tectonics; moderate dips can become almost vertical, *et cetera*. Geologists not accustomed to exaggerated sections are apt to forget the fact of exaggeration and will

[1] Suter H. H., *Exaggeration of Vertical Scale of Geologic Sections*: Am. Assoc. Petroleum Geologists Bull., vol. 31, no. 2, pp. 318-339, Feb. 1947.

STRATIGRAPHIC SECTION OF NORTHWEST GERMANY

AGE			FORMATION			LITHOLOGY
QUATERNARY			ALLUVIUM DILUVIUM		O	SAND AND GRAVEL BOULDERS, GLACIAL TILL
TERTIARY	UPPER TERTIARY		PLIOCENE 0-1000'			SAND, CLAY
			MIOCENE 200-1000'			SAND, CLAY LIGNITE
	LOWER TERTIARY		OLIGOCENE 400-800'		☼	SAND, CLAY LIGNITE
			EOCENE 150-1100'		● ☼	SAND, LIGNITE
			PALEOCENE 100-600'			MARL, SHALE, CHALK
CRETACEOUS	UPPER CRETACEOUS		SENONIAN 500-660'		●	CHALK AND MARL
			EMSCHERIAN 600-900'			MARL
			TURONIAN 500-900'			LIMESTONE
			CENOMANIAN 100-200'			MARL
	LOWER CRETACEOUS	GAULT 200-400'	ALBIAN	UPPER		SHALE
				LOWER		
			APTIAN			
		NEOCOM 0-600'	BARREMIAN			SHALE
			HAUTERVIAN		O	
			VALENDIS		●	SHALE AND SANDSTONE
		WEALDEN 0-600'			●	SHALE SANDSTONE COAL
JURASSIC	MALM 0-1200'	PORTLAND	SERPULIT		●	LIMESTONE
			MUNDER MERGEL			MARL, GYPSUM
			EIMBECKHAUSER			
			GIGAS			
		KIMM	KIMMERIDGE			LIMESTONE
		OXFORD	KORALLENOOLITH		●	
			HEERSUMER			
			ORNATEN			SHALE
	DOGGER 0-600'	UPPER	MACROCEPHALEN		●	SANDSTONE, SHALE
			CORNBRASH			
		MIDDLE	PARKINSONI		●●	SHALE, SANDSTONE
			CORONATUS			
		LOWER	MURCHISONE			
			OPALINUS			
			JURENSE			
	LIAS 0-900'	UPPER	POSIDONIEN			BITUMINOUS SHALE
		MIDDLE	AMALTHEES			SHALE
			CAPRICORNU			
		LOWER	ARIETITES			SANDSTONE, SHALE
			ANGULATUS			
			PSILONOTUS		O	
			RHÄT			
TRIASSIC	KEUPER 500-1300'		GIPSKEUPER		●	MARL SANDSTONE GYPSUM
			KOHLENKEUPER			DOLOMITE SANDSTONE
	MUSCHEL KALK 600-900'	UPPER				LIMESTONE
		MIDDLE				DOLOMITE, GYPSUM
		LOWER				LIMESTONE
	BUNTSANDSTEIN 2300'-5500'	UPPER				MARL, GYPSUM ROCKSALT
		MIDDLE	BAUSANDSTEIN			SANDSTONE, SHALE
		LOWER	ROGENSTEIN			
PERMIAN	ZECHSTEIN 660'-4300'	UPPER			O ☼	ROCK SALT POTASH SALT GYPSUM ANHYDRITE SHALE
		MIDDLE	STINKSCHIEFER			BIT. LIMESTONE
			HAUPTDOLOMITE		● ☼	DOLOMITE
			ANHYDRIT			ANHYDRITE
			BLASENSCHIEFER		●	DOLOMITE
			ZECHSTEINKALK			LIMESTONE
		LOWER	KUPFERSCHIEFER			BITUM. SHALE
			CONGLOMERATE			CONGLOMERATE
	ROTLIEGENDES		ROTLIEGENDTONE		●	RED CLAY SANDSTONE ANHYDRITE AND SALT
			HASELGEBIRGE			
CARBONIFEROUS						SHALE SANDSTONE COAL

O OIL SHOW   ☼ GAS   ● OIL

FIGURE 458. Generalized stratigraphic section showing relationship between producing intervals and formations. (From Reeves. Reproduced permission Am. Assoc. Petroleum Geologists.)

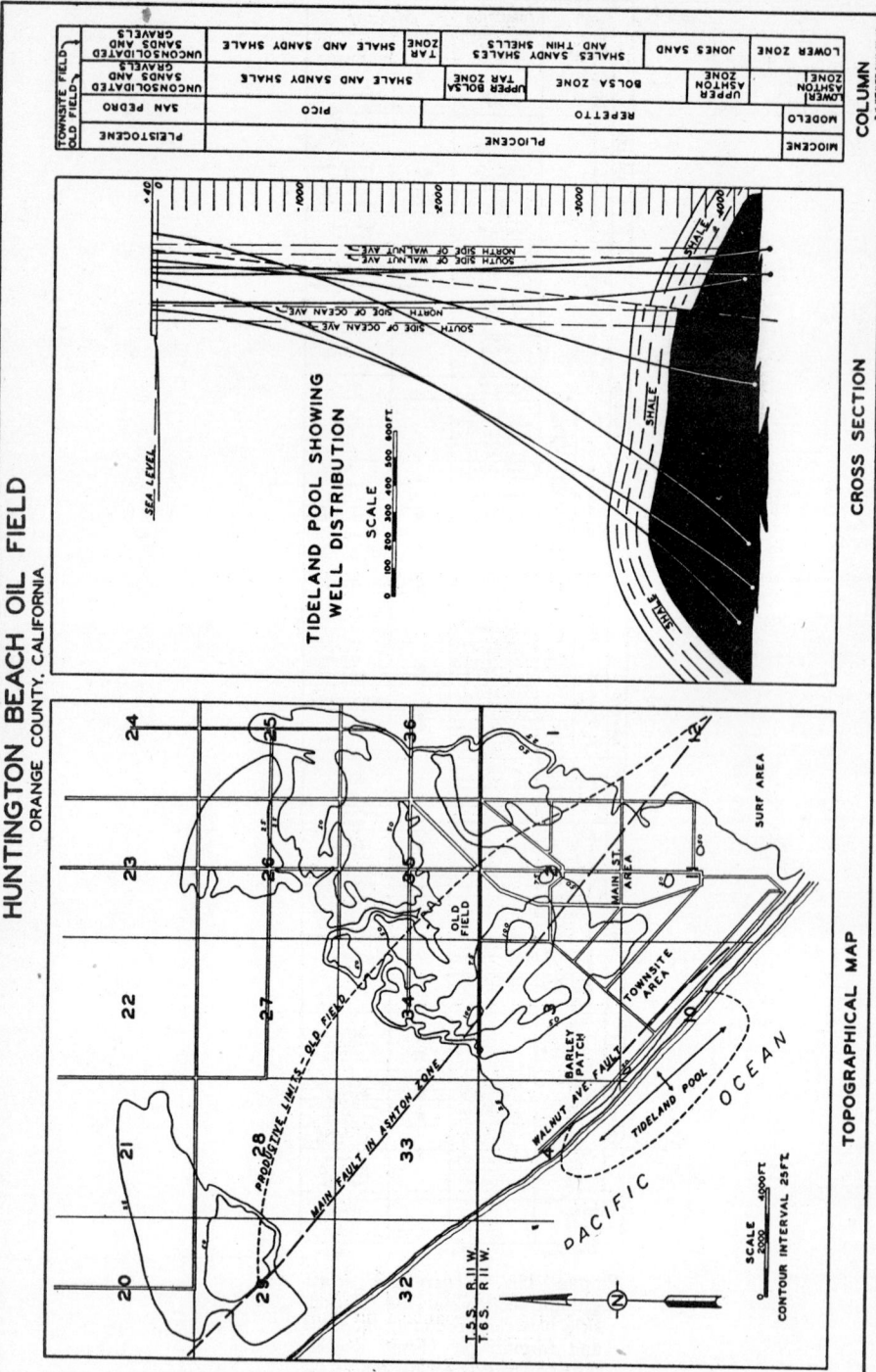

FIGURES 459 and 460. A well-organized compilation drawing illustrating directional drilling. (From Weaver and Wilhelm. Reproduced permission

HUNTINGTON BEACH OIL FIELD
ORANGE COUNTY, CALIFORNIA

TIDELAND POOL SHOWING
WELL DISTRIBUTION

SCALE
100 200 300 400 500 600 FT

SEA LEVEL

SOUTH SIDE OF WALNUT AVE.
NORTH SIDE OF WALNUT AVE.

NORTH SIDE OF OCEAN AVE.
SOUTH SIDE OF OCEAN AVE.

SHALE

SHALE

CROSS SECTION

TOPOGRAPHICAL MAP

PACIFIC   OCEAN

SURF AREA

MAIN ST AREA

TOWNSITE AREA

TIDELAND POOL

BARLEY PATCH

WALNUT AVE. FAULT

OLD FIELD

PRODUCTIVE LIMITS - OLD FIELD

MAIN FAULT IN ASHTON ZONE

T.5.S. R.11 W.
T.6.S. R.11 W.

SCALE
2000   4000 FT

CONTOUR INTERVAL 25 FT.

N

20   21   22   23   24
25
26   36
27
28   34   35
29   33   32

TOWNSITE FIELD	OLD FIELD		
UNCONSOLIDATED SANDS AND GRAVELS	UNCONSOLIDATED SANDS AND GRAVELS		
SAN PEDRO			
PICO			
REPETTO			
MODELO			

COLUMN			
UNCONSOLIDATED SANDS AND GRAVELS			
SHALE AND SANDY SHALE			
TAR ZONE			
SHALE AND THIN SHELLS			
SHALES SANDY SHALES AND THIN SHELLS			
JONES SAND			
LOWER ZONE			

SHALE AND SANDY SHALE	UPPER BOLSA TAR ZONE	BOLSA ZONE	UPPER ASHTON ZONE	LOWER ASHTON ZONE

PLEISTOCENE   PLIOCENE   MIOCENE

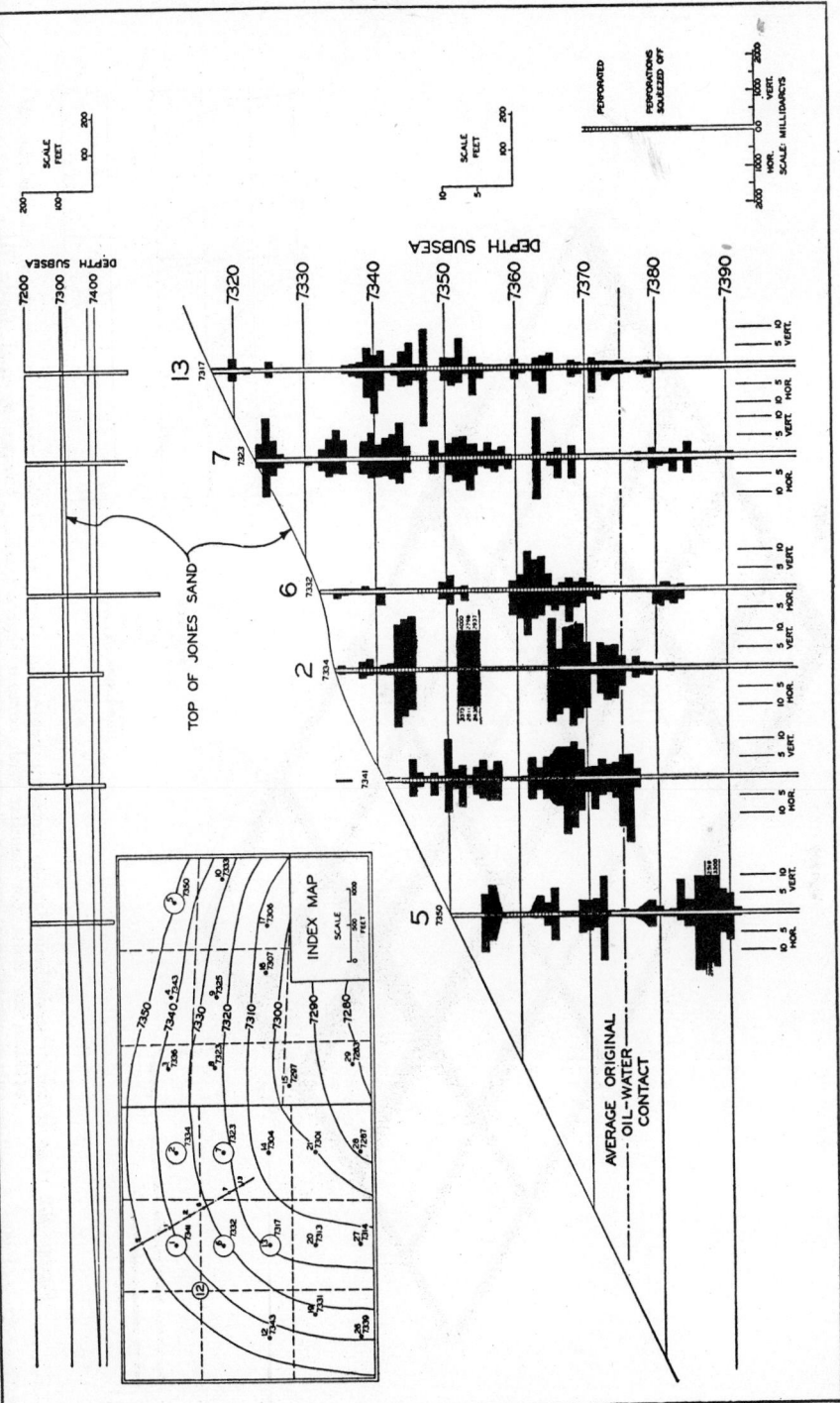

FIGURE 461. A method of showing relative permeabilities. (From Elliott. Reproduced permission Am. Assoc. Petroleum Geologists.)

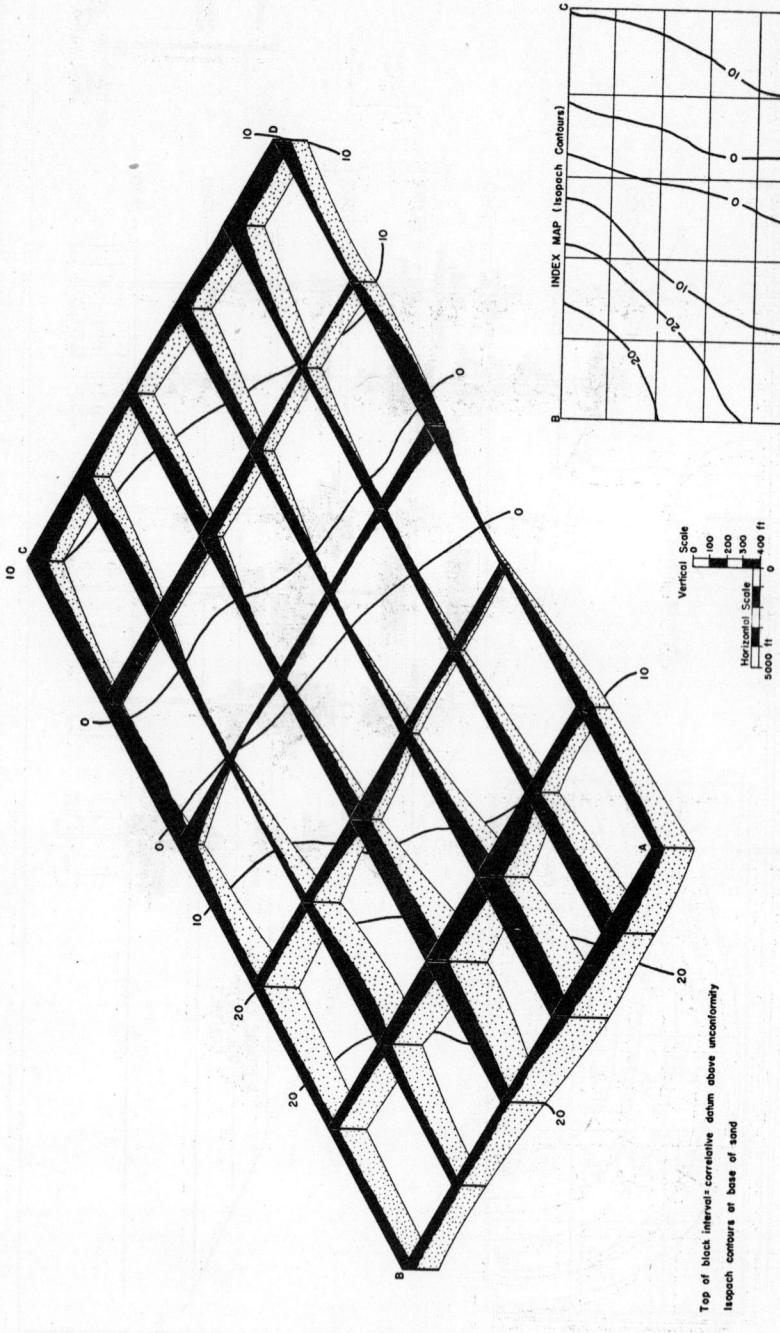

Top of block interval: correlative datum above unconformity

Isopach contours at base of sand

INDEX MAP (Isopach Contours)

Vertical Scale

Horizontal Scale

FIGURE 462. Conversion of isopachous data to isometric projection. Drawings of this nature permit rapid visualization of problem for those not familiar with contour representation. (Drawn by A. J. Gude, III.)

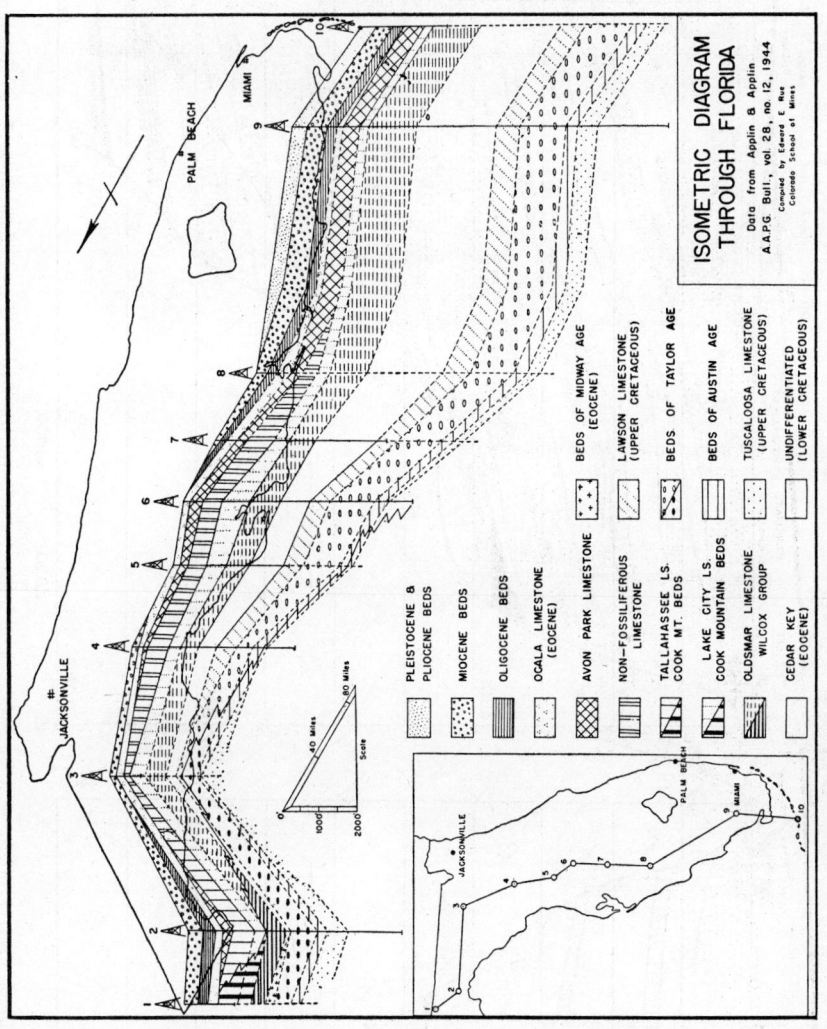

FIGURE 463. Application of isometric construction in stratigraphic compilation.

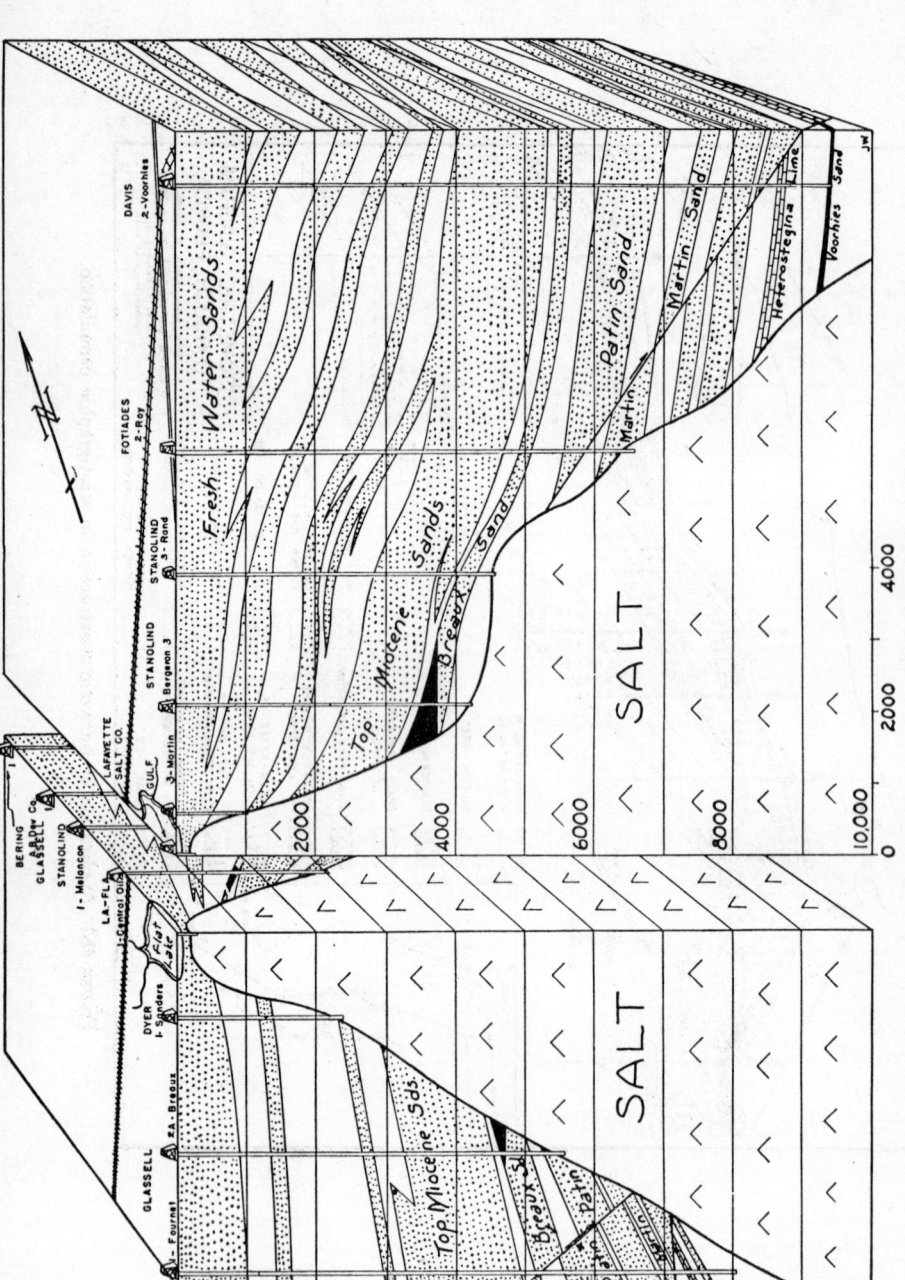

FIGURE 464. Block diagram of Anse la Buth salt dome looking northwest, St. Martin Parish, Louisiana. Such diagrams readily portray subsurface conditions in the third dimension. (From Bates and Wharton. Reproduced permission Am. Assoc. Petro-

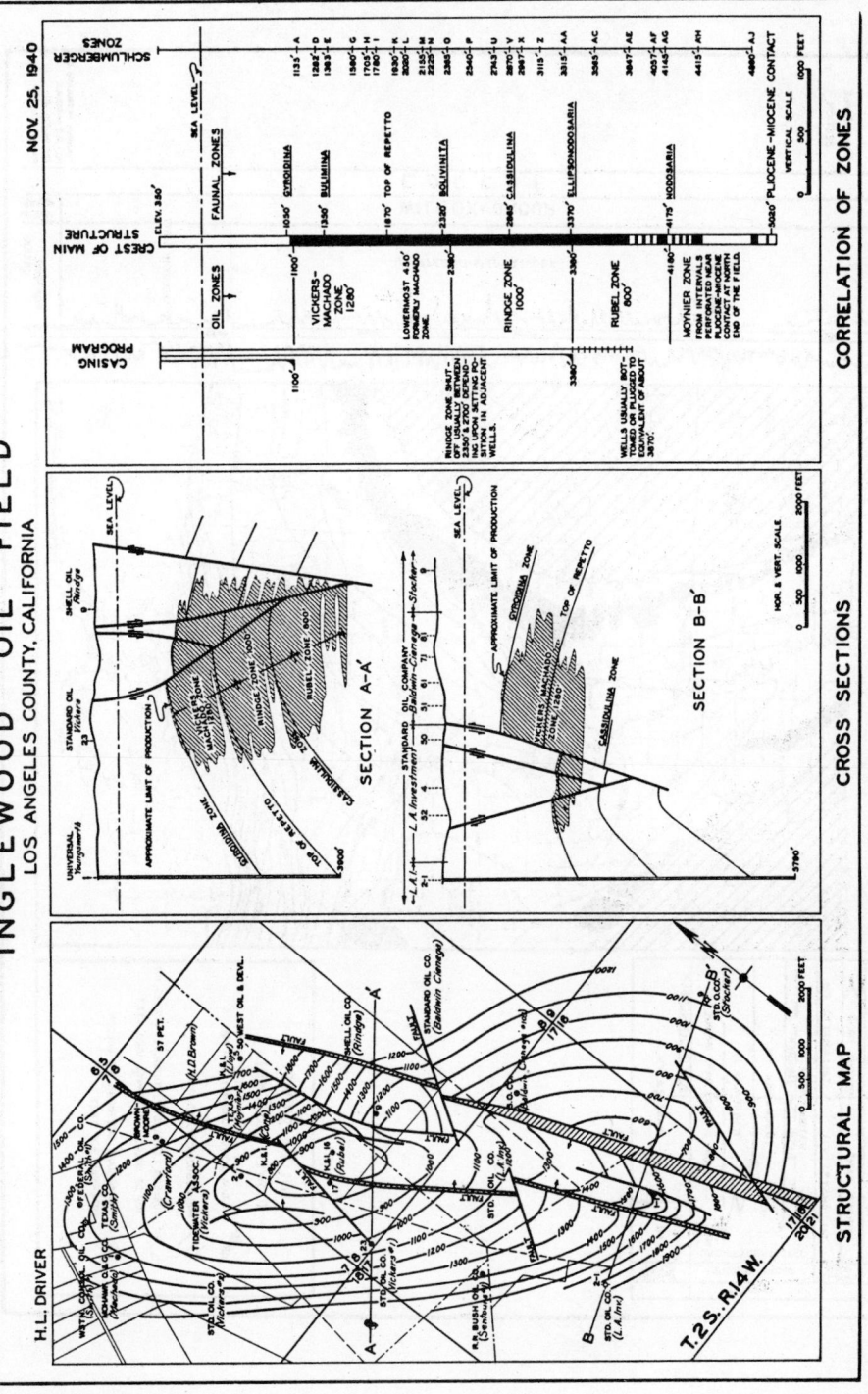

FIGURE 465. Well-organized drawing illustrating structural and stratigraphic conditions of an oil field. (From Driver. Reproduced permission California Div. Mines.)

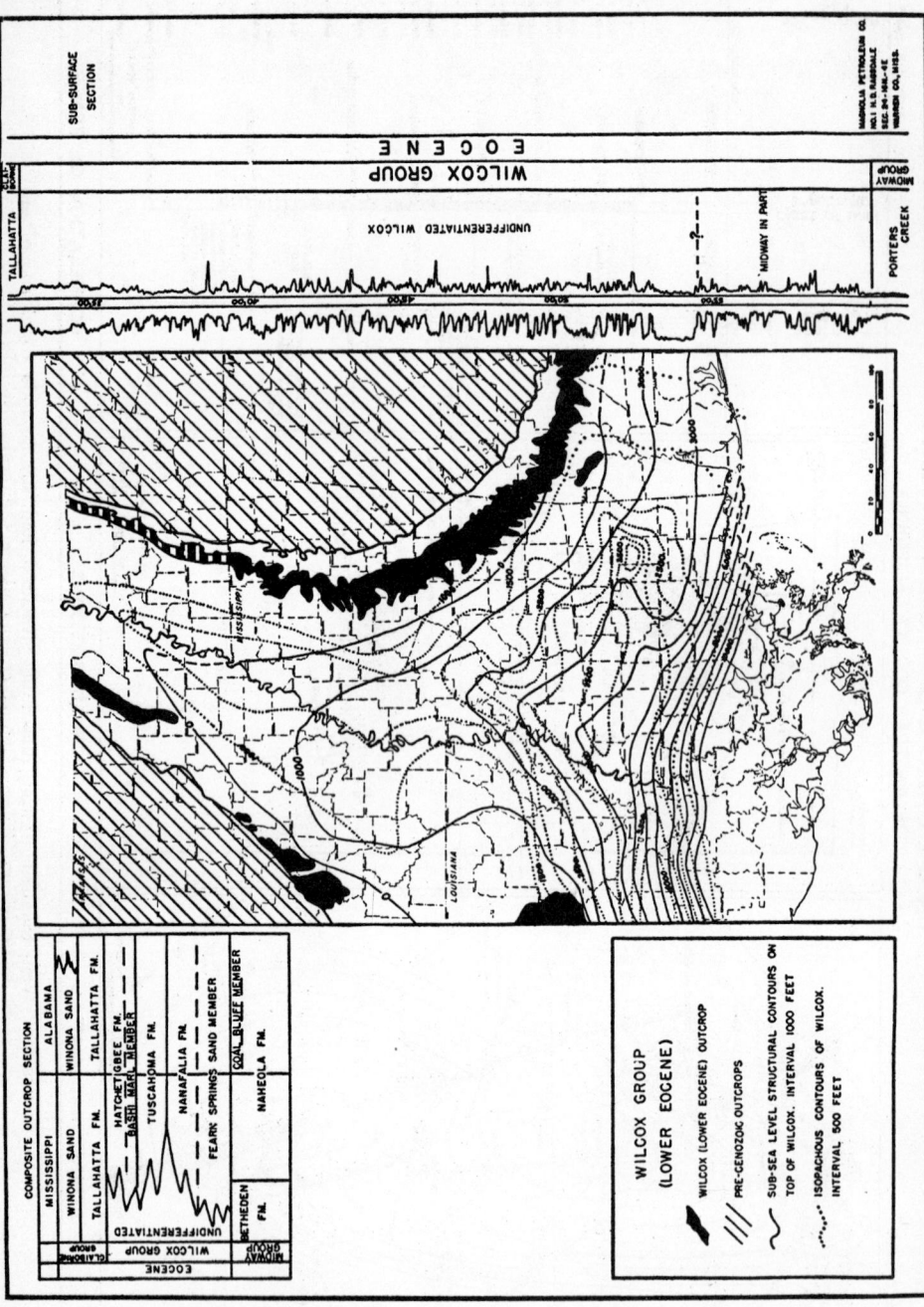

gain a mental picture of acute structural relief when, in fact, the tectonic relief may be very mild. . . . Exaggeration of vertical scale affects primarily the vertical dimensions of a geological form but it also affects, in a certain way, the horizontal dimension. In the vertical direction the picture is actually expanded; in the horizontal direction it is apparently contracted. . . . Vertical expansion causes distortion of dip and thickness of beds and this distortion varies with true dip and ratio of exaggeration. . . . The principal horizontal or lateral effect of exaggeration of vertical scale is to bring every element of a section closer together.

Prime factors to consider in preparing geologic cross sections are the lithologic plot (in part or in whole), the name and age of the formations, the natural as well as the exaggerated profile, proper captioning, a plan map showing the line of section, the vertical and horizontal scale, and the legend.

Control stratigraphic columnar sections should accompany the cross section as they contribute to a better understanding of the stratal sequence and relationships of the units involved. Unconformities and lithofacies boundaries should be shown. An excessively long cross section may be broken or interrupted.

The Busk method (fig. 447) of constructing geologic cross sections is frequently used by geologists. This method is applicable to many structural problems although its limitations must be recognized. Requisites for its use are adequate dip control, parallel folding, and constancy of formational thickness.

Figures 436 to 466 are included for the purpose of suggesting various methods of compiling and presenting subsurface data.

QUESTIONS

1. Why should geologic reports contain numerous graphic representations?
2. What data should be recorded on well-log headings?
3. When should the percentage log be used?
4. Review the abbreviations commonly used in giving lithologic descriptions on well strips.
5. Give the colored-well symbols for the following lithologies: dolostone, arkosic sandstone, ferruginous sandstone, arenaceous shale, tuff, and arenaceous limestone.
6. Study figure 432 which typifies data to be included on a final well log.
7 Discuss the percentage and interpretive log strip on plate 12.
8. Give the symbols for the following: drilling well, producing well, abandoned gas producer, strong-gas show, salt water, slight-oil show, faint-oil smell, coarse crystalline, detrital chert, pyrite, microfossil, medium-grained sandstone, and insoluble residue.
9. Why should geologic cross sections be drawn to natural scale?

CHAPTER 11

SUBSURFACE MAPS AND ILLUSTRATIONS

JULIAN W. LOW

The term "subsurface map" may be somewhat confusing in that nearly all types of both surface and subsurface maps display features that are actually concealed at the surface of the ground by soils, alluvium, and other types of overburden. The geologic formation or horizon contoured on a surface-structure map may lie beneath other formations over the greater part of the map area, as shown in the cross section in figure 467. In this figure, points 1, 2, and 3 are outcrops of the datum horizon where direct instrumental observations can be made. Point 4 is an out-

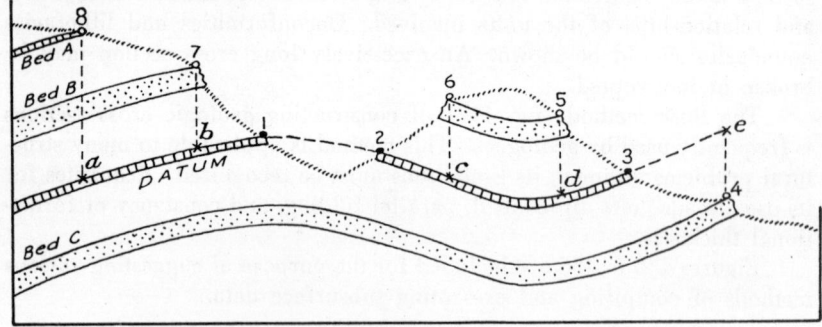

FIGURE 467. Geologic cross section compiled from surface data.

crop of bed *C*, stratigraphically below the datum, and 5 6, 7, and 8 are outcrops of beds *B* and *A*, stratigraphically above the datum. Control points *a*, *b*, *c*, *d*, and *e* are computed from instrumental observations obtained on the outcrops. The position of the datum surface from 1 to 2 and 3 to *e* is restored above the actual surface of the ground, and at all other places the datum is covered by other formations.

Figure 468 is a subsurface cross section. Neither the datum nor the two key beds *A* and *B* crop out at the surface. The control points, *a*, *c*, and *e* are determined from the logs of wells which penetrate the datum bed. Control points *b* and *d* are computed from the drilled points on key bed *A*. The similarities in the section shown in figures 467 and 468 are obvious. Indirect methods are used in the construction of both; yet one is a surface section and the other a subsurface. The principal difference in these sections and in surface and subsurface maps is that the surface map or section is constructed from surface data; that is, from outcrops. The subsurface section or map is constructed from data supplied by wells that have penetrated recognizable formations.

Preparation of Subsurface Data

The subsurface map can be only as good as the data used in its preparation. In surface mapping it is usually possible to observe the structural or stratigraphic behavior of formations over a considerable area around an instrument or rod station. The geologic interpretation of soils, topography, plant ecology, springs, and other natural conditions can aid materially in bridging over areas where the bedrock formations are concealed from direct observation. In contrast to the areal control available on the outcrops, there is only point control for subsurface work. For this reason it is necessary to prepare the data from wells with considerable care.

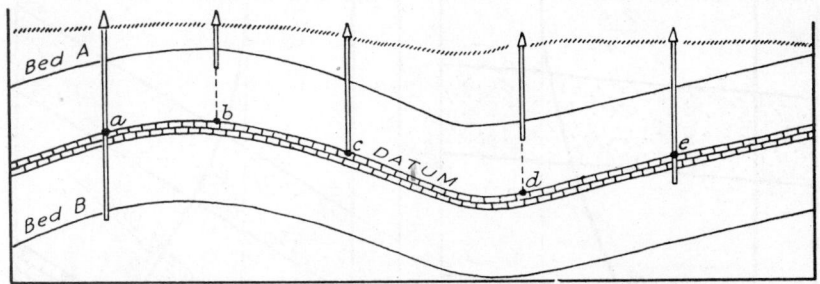

Figure 468. Geologic cross section compiled from subsurface (well) data.

Reduction of Datum Elevation

The elevation on the datum bed is the algebraic difference between the surface elevation of the well and the drilled depth to the datum. Thus, if the surface elevation is 5,000 feet and the depth to the datum is 4,000 feet, the datum elevation is 1,000 feet. If the depth is 6,000 feet, the datum elevation is minus 1,000 feet (1,000 feet below sea level).

If the drill hole is crooked, the apparent vertical depth to the datum will be either too great or too small. In figure 469, A shows a hole that has drifted down-dip and penetrated the datum at point a. Using the actual drilled interval from the surface to point a and the surface location of the well, the datum would appear to be at b. While the actual dip is to the west, the crooked hole produces an erroneous effect of east dip. In B of figure 469 the hole has drifted in a direction up the dip. In this case the actual drilled interval is less than the vertical depth of the datum bed. The effect is an apparently steepened dip between the two wells.

Unless a direction survey (chapter 6) has been made, it is impossible to adjust the log of a crooked hole to obtain a correct datum point.

Dips or dips and strikes determined from cores can aid the subsurface geologist greatly; but they can also lead him far astray in his interpretations. It is hazardous to use core dips indiscriminately in subsurface mapping. The core dips should always be adjusted by means of hole deviation or directional surveys when available. When straight-hole determinations have not been made, core dips should be used with caution. In figure 470, *A* shows a straight hole drilled on a sharply folded anticline. Both core dips and drilled stratigraphic intervals increase with depth. In figure 470, *B* shows a hole that drifts down the dip on a monocline, with erroneous increases in core dips and drilled intervals similar to those on the flank of the anticline. Deviation surveys used in conjunction with the core dips reveal the true subsurface conditions so that the formations can be correctly mapped.

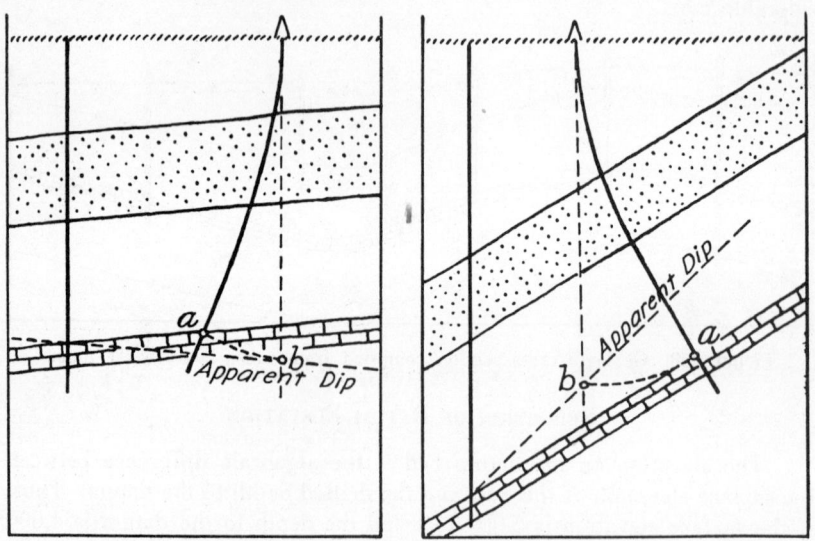

FIGURE 469. Effects of crooked holes on datum elevations and interpreted dips.

The two wells represented in *B* of figure 471 would suggest a similar convergence in the drilled intervals. There is one important difference, as shown by core dips: the dips above and below the inclined part of the hole are the same. Core dips from the straight hole on the right are constant throughout the apparently converging interval. Thus, where regional conditions are well-known through an adequate distribution of subsurface control, it is sometimes possible to infer correctly that a portion of one hole is crooked, even though a deviation survey has not been made.

Now examining figure 471 further, *A* shows a gradual thinning of two formations in the central portion of the stratigraphic succession. Core dips from the two wells would show a gradual increase with depth through the converging portion of the section. Dips below the thinning portion are steeper than those above, but remain constant as deep as the strata are parallel.

It is outside the scope of this chapter to describe in detail the causes and effects of crooked holes. The preceding examples are only a few of countless conditions that test the ingenuity of the subsurface geologist. These examples will serve to show that all data obtained from wells, such as formation tops, dips, thicknesses, and others, are subject to critical examination and balancing, one against the other, before they may be used with confidence in mapping.

STRUCTURAL CONTOUR MAP

There is no fundamental difference between a surface structural map and a subsurface one. Both attempt to show by means of contours the

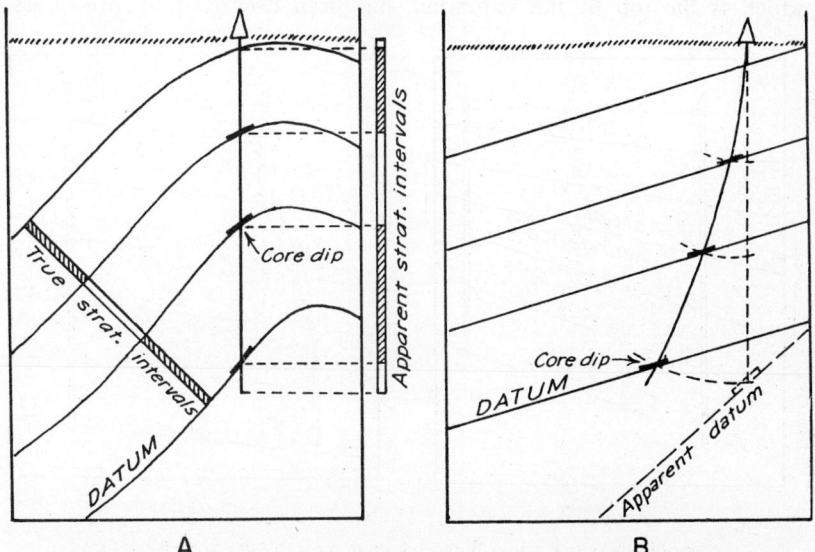

FIGURE 470. *A*—Core dips and stratigraphic intervals increasing with depth on sharply folded anticline. *B*—Migration of hole downdip resulting in erroneous core dips and stratigraphic intervals.

configuration of a selected continuous stratigraphic horizon, commonly called the "datum," or "datum horizon." As stated earlier, the principal difference is in the kinds of data used in their construction.

The subsurface structural map is almost or quite dependent on wells for the necessary control. Ordinarily, the elevations on which the contours are drawn are obtained by the simple process of subtracting the depth to the datum horizon from the surface elevation of the well, the latter being established at the point from which depth measurements are made. This point in most cases is the rotary table or the rotary bushing of the drilling rig.

The block diagram, *A*, in figure 472 illustrates an anticline that is typically eroded in the surface formations. In the same figure the block is separated, *B*, to show a buried major nonconformity, the presence of which is suggested nowhere in the surface geology.

The structural contour map in figure 473 is a surface map, as the datum bed crops out over a wide area, and all structural data needed for the map are readily available on the surface. The structural map in figure 474 is constructed solely on data supplied by the 24 wells that penetrated the Paleozoic formations below the Jurassic unconformity. The block diagram shows four of these wells, three of which are in the vertical planes of the block.

In the central part of the Cambrian structure map the actual datum, which is the top of the Cambrian, has been destroyed by pre-Jurassic

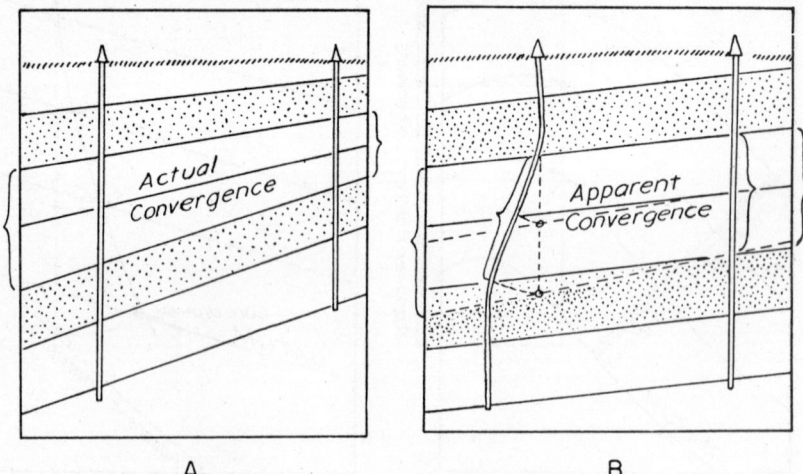

FIGURE 471. *A*—Actual convergence of section. *B*—Apparent convergence of section as a result of crooked hole.

erosion. This is shown by the fact that wells 8, 10, 16, 18 and 19 entered the Cambrian directly beneath the Jurassic without first penetrating the systems that overlie the Cambrian toward the edges of the map. Although the contours within this area of truncation are drawn to the elevations at which the Cambrian is encountered below the unconformity, they do not represent the true structural configuration of the strata. In order to do so, it would be necessary to reconstruct the original thickness of Cambrian throughout the area of truncation and raise the elevations accordingly. Even this process cannot be employed in the center of the area where Cambrian beds are absent, for there is no method by which the depth of erosion of the pre-Cambrian rocks can be determined.

Disconformities are frequently not recognizable in well samples or on resistivity and radioactivity logs, and, because of this, so-called regional structural maps may not truly represent the regional structure. The departure of the contouring from that of a true structural picture is controlled by the thickness of section eroded away, the relief of the unconformity, and the average rate of regional dips. This discrepancy is greatest where the dips are low and the relief on the unconformity is relatively

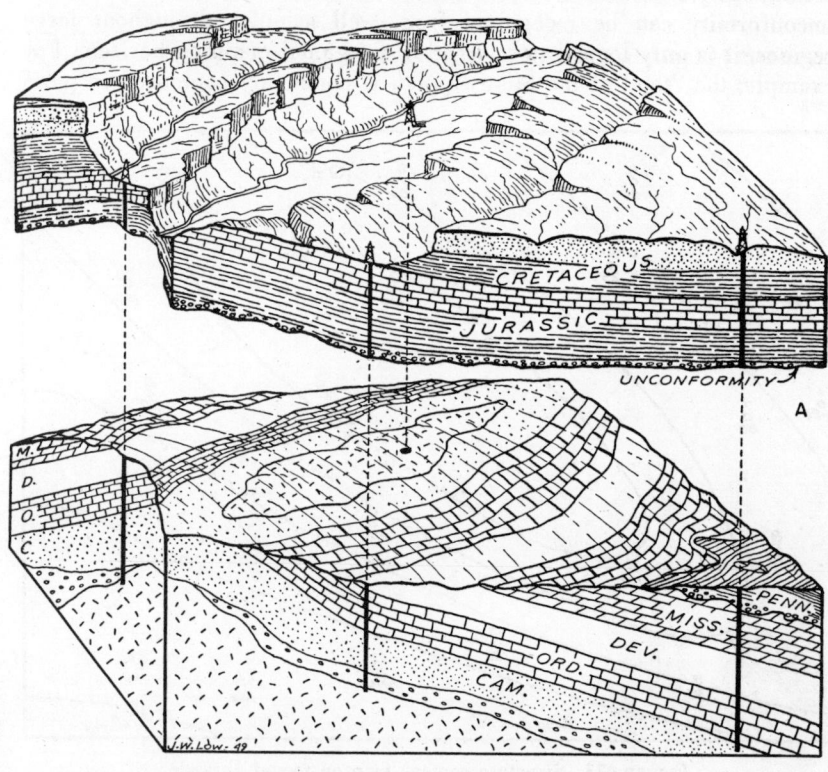

FIGURE 472. Typical eroded anticline involving a major unconformity in subsurface.

high. It is purely a matter of chance that certain wells strike the horizon of the unconformity at high points and others at low. Figure 475 shows a case where four wells fail to reveal the presence of a structure because the wells on the high part of the structure happened to strike locally low points on the unconformity, while those structurally lower encountered locally high points on the erosion surface.

Sometimes a subsurface disconformity can be recognized by the types or condition of the rocks in the drill cuttings; but very often the sample work is not done with sufficient care to differentiate these materials, or the diagnostic materials may be absent at the particular point where the well was drilled. A question likely to arise is "Why contour an unconformity when so many hazards exist?" There are two principal reasons: One is that a vast amount of oil has been found in the zones of unconformities and directly associated with them. The other is that in many regions the geologic section is lacking in stratigraphic horizons that can be traced continuously over wide areas; and, if the formation immediately under the unconformity can be recognized from well samples throughout large regions, it is only logical to select it as a regional mapping horizon. For example, the "top" of the Mississippian is commonly contoured in parts

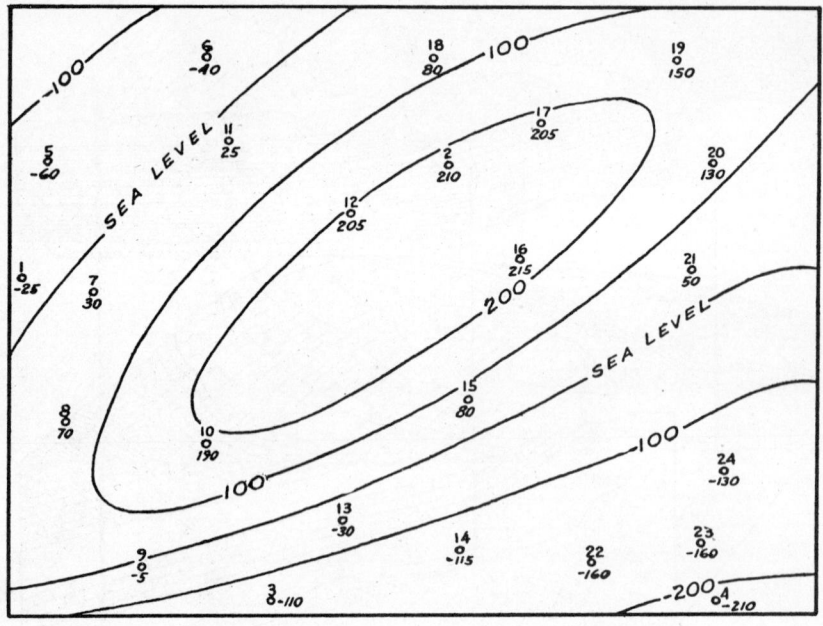

FIGURE 473. Structure contour map on top of Jurassic.

of the Midcontinent area, despite the fact that it is an erosion surface, because it can be easily recognized in well samples and drillers' logs.

In the construction of subsurface structural maps of oil fields that have not been entirely defined, it is often of the greatest importance to work out carefully even minor details such as the exact character of faulting in order to avoid drilling of unnecessary dry holes and to make certain that all potentially productive locations are tested.

The structure map in figure 476 shows a partly developed oil field with a number of oil and gas wells and dry holes. A normal fault dipping to the southwest cuts across the southwest end of the anticline. The structural datum is the top of the producing horizon.

While a fault is commonly represented on maps as a single line, a normal fault with a low-dipping plane invariably results in a zone where the datum surface is absent. This zone is called a "datum gap." The breadth of the datum gap is determined by the degree of dip in the strata, the dip of the fault plane, and the amount of throw. The datum gap can be worked out on a subsurface structure map if there are sufficient datum points to control the general contouring of the structure and at least three wells that have penetrated the fault plane in a triangular arrangement (not in a straight line).

It is first necessary to determine the dip and strike of the fault plane

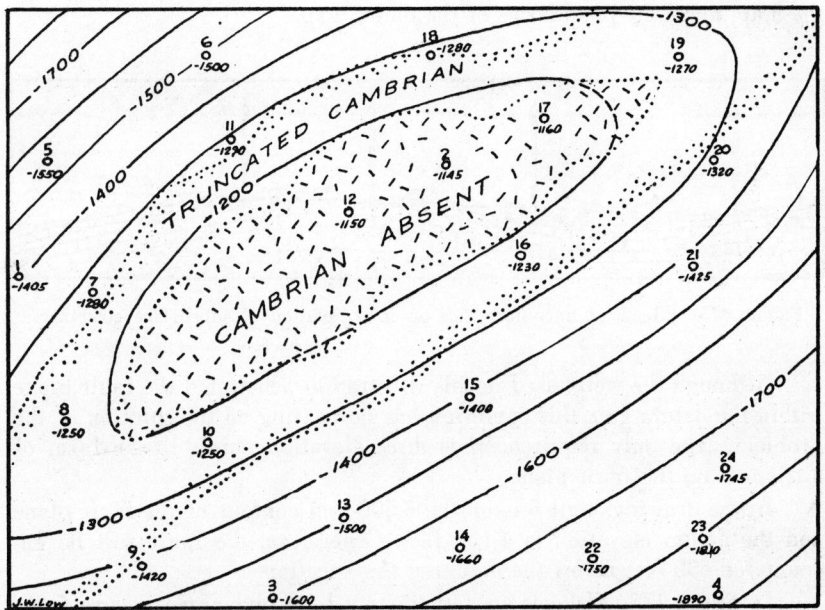

FIGURE 474. Structure contour map on top of Cambrian.

from the three or more wells. The method is shown in the fault-plane detail of the figure cited. The procedure is exactly the same as that used in obtaining a three-point dip and strike on a bed with the plane table. In the illustration wells numbered 1 to 5 have been plotted in their correct relative positions in a separate drawing in order better to illustrate the method. In actual practice, the determination of the dip and strike of the fault plane would be made directly on the map.

Referring again to the fault detail drawing: Wells 1, 2, and 3 penetrated the fault plane at elevations 4,100, 4,600, and 4,300 feet respectively. These wells are joined by straight lines. The difference in elevation (on the fault plane) between 2 and 3 is 300 feet. Divide the line into three equal parts. The difference in elevation between each of these points is

100 feet. The difference in elevation between wells 2 and 1 is 500 feet; therefore, five spaces are laid off along that line. Now, starting with the first mark on each side of the highest well (4,600 feet), strike lines are drawn as shown in the figure. These lines are contours on the fault plane, and the rate of dip of the fault plane is revealed by the spacing of the contours. This process assumes a true plane, which may not actually be correct.

When the fault-plane contours have been drawn on the map, the structural contours are carefully sketched to points where they intersect fault-plane contours of the same values. These intersections, as shown on the map, mark the boundaries of the datum gap.

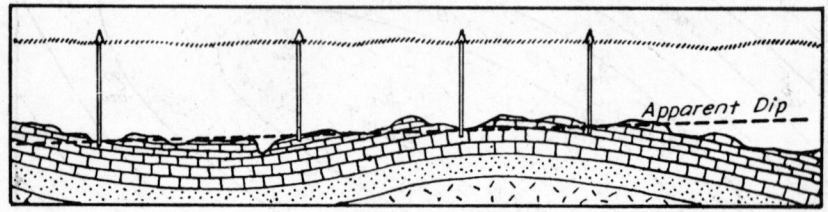

FIGURE 475. Effects of unconformities on interpretation of subsurface structure.

Although the wells used in this illustration penetrated the fault plane within the datum gap, this condition has no bearing on the solution of the problem. The only requirement is three elevation points in the form of a triangle on the fault plane.

In the drawing, well 5 is on the 3,500-foot contour of the fault plane and the datum elevation is 4,150 feet. Therefore, the fault will be encountered 650 feet below the datum at this location.

In figure 477, A shows an anticline cut by a high-angle reverse fault. The productive area on the upthrown block is ruled. The seven wells that drilled through the fault plane and encountered the datum beneath the fault are encircled. Three of these wells produce oil from the downthrown block. Saturated portions of the producing formation above and below the fault are shown in the cross section, C, of figure 477. The map, B, in figure 477 deals with the part of the datum in the downthrown block. The heavy dashed line in the central part of the area is the upper trace of the thrust sheet. The heavy dashed line in the northwest quadrant is the lower trace, and the area between these lines represents the horizontal displacement of the datum bed.

The fine dashed lines numbered 2,300 to 3,200 are contours on the fault plane, which is shown as a true plane, since all of the contours are straight lines. The solid-line contours are on the datum bed below the fault plane. The upper numbers at the wells are datum elevations below the fault, and the lower numbers (in parenthesis) are elevations of the fault plane. As in *A* of figure 477, the productive area is ruled.

The curving of the datum contours beneath the thrust sheet, northwest of the "hidden" syncline, suggests that some folding had taken place prior to the faulting; therefore, the possibility of accumulation of oil in the upper edge of the faulted flank beneath the fault might be anticipated, if the upturned edge of the reservoir had been adequately sealed by the fault.

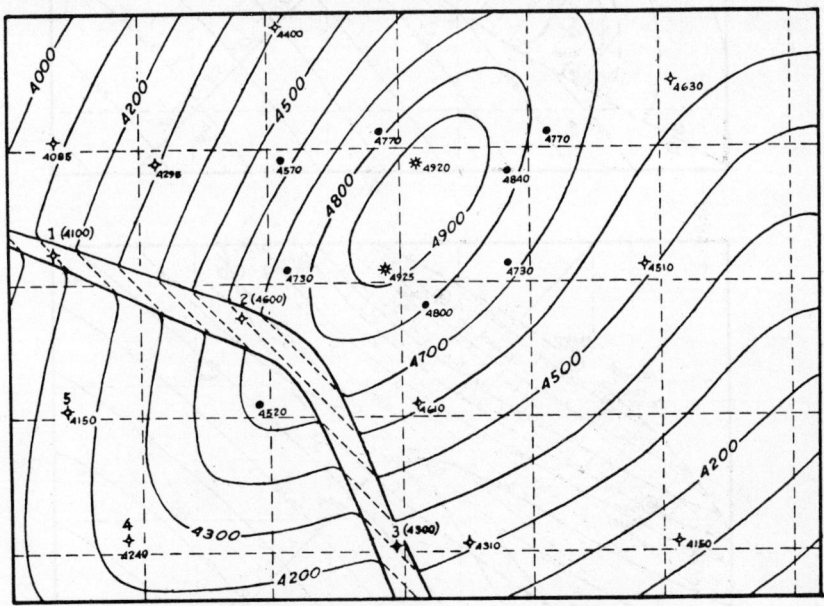

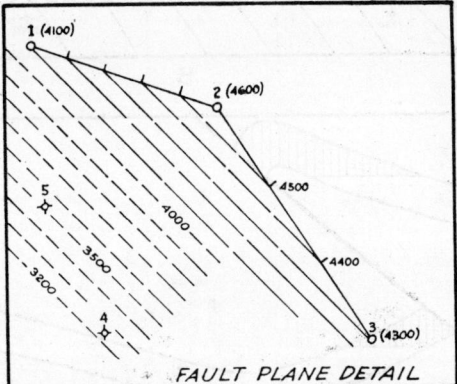

FIGURE 476. Structural contour map showing a datum (fault) gap and the contoured fault plane.

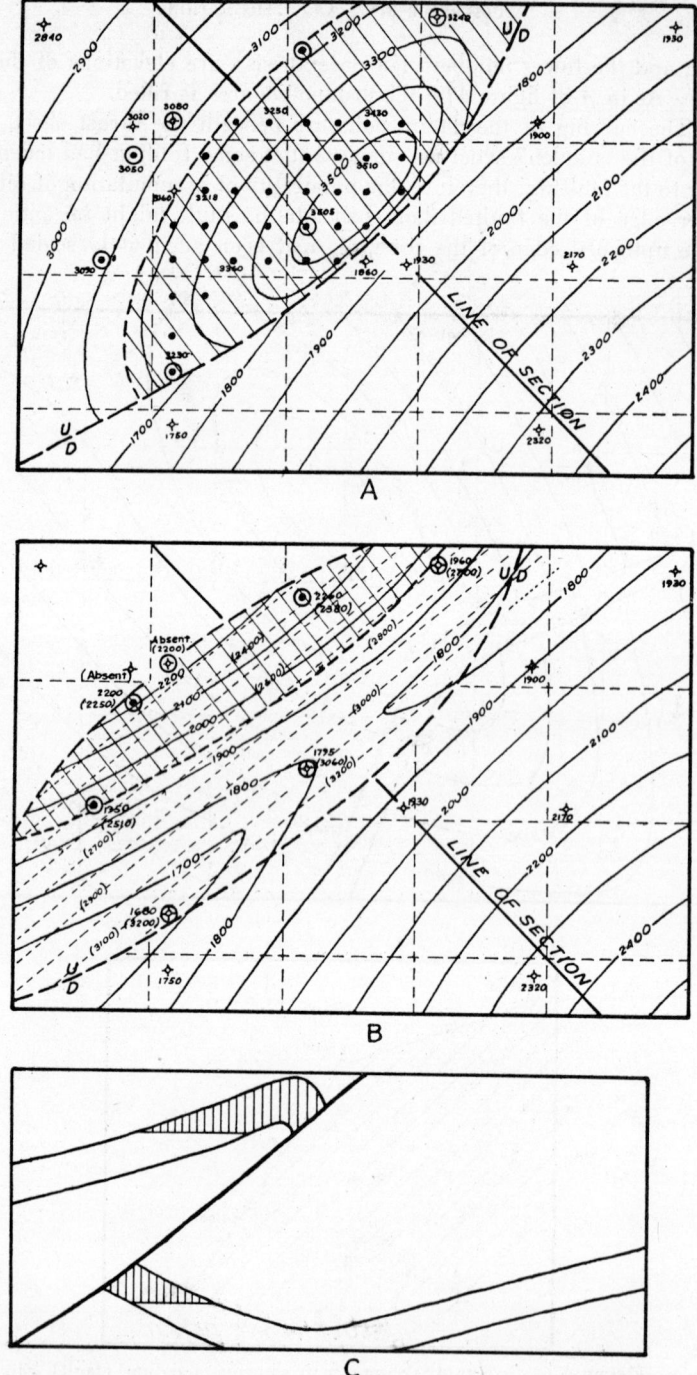

FIGURE 477. *A*—Structure contoured above a high-angled reverse fault. *B*—Contours on and below fault plane. *C*—Geologic cross section showing dip of fault.

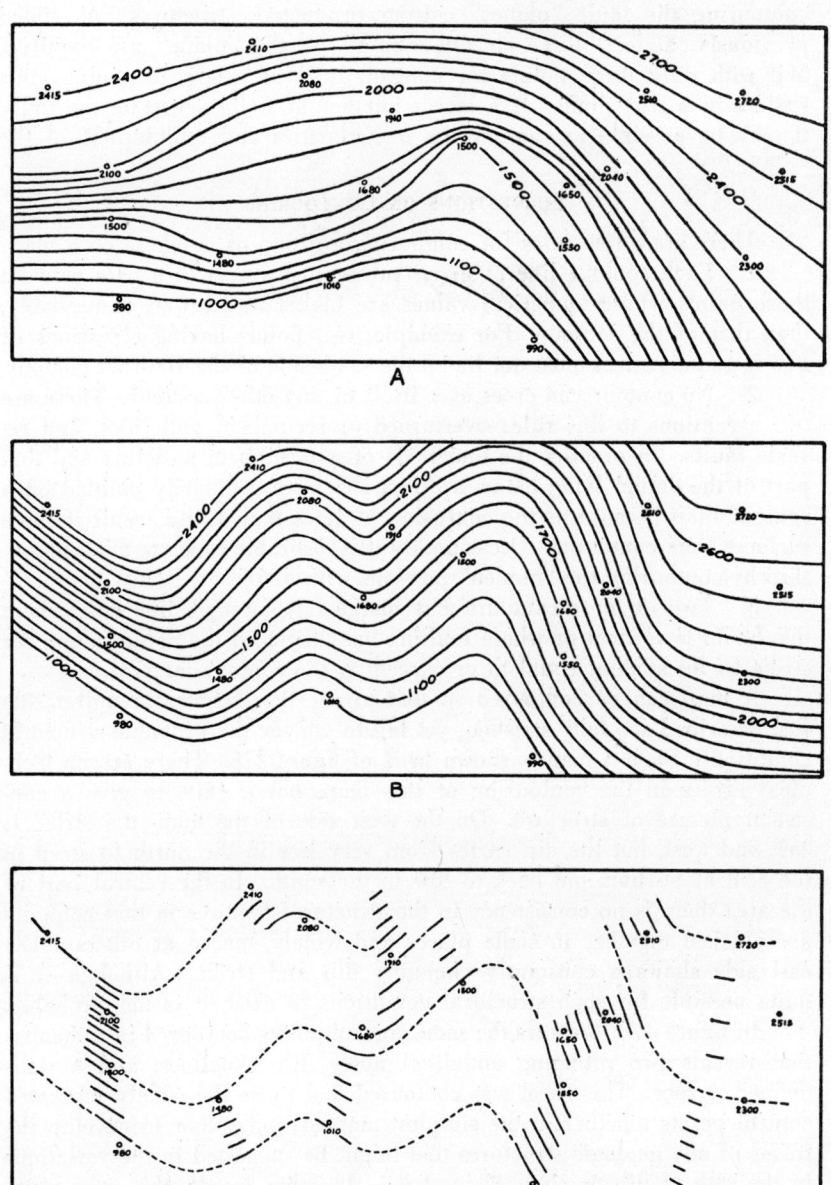

FIGURE 478. Contrast of careless and orderly methods of contouring the same data.

The two examples of faulted structures illustrate the importance of contouring the fault "planes" cutting productive structures. As stated previously, at least three elevation points on the "plane" are required, and with only three points for control, it is necessary to contour this surface as a true plane. If a larger number of wells penetrate the fault, it may be possible to contour the irregularities and undulations of the "plane."

SUGGESTIONS ON CONTOURING

There are a few rules for contouring a group of numbers on a map:

1. Each contour line of given value must everywhere pass between those points whose numerical values are higher and lower, respectively, than that of the contour. For example, two points having elevations of 110 feet and 95 feet must not lie on the same side of the 100-foot contour.

2. No contour can cross over itself or any other contour. There are two exceptions to this rule: overturned or recumbent anticlines, and reverse faults. In practice the underside of a recumbent anticline and that part of the datum lying below a thrust sheet are ordinarily omitted on a contour map because of the confusion of lines that would result if these surfaces were contoured. Occasionally it is desirable to show the relationship by contouring the "hidden" portions with dotted or dashed lines.

3. Two or more contours may merge into a single line only where the datum is vertical or where faulting has displaced the datum along the strike by an amount equal to or exceeding the contour interval.

A map can be contoured so that all of the technical requirements just described are fully satisfied, yet fail to convey the probable structural conditions. Such a map is shown in *A* of figure 478. There are no technical errors in the contouring of this map, but it fails to give a consistent picture of structure. On the west side of the map, the strike is east and west, but the dip varies from very low in the north to steep in the central portion and back to low in the south. In the central part of the area there is no consistency in the structural features in that contours are pinched together in some places and widely spaced at others. The east side shows a constantly changing dip and strike. Although it is quite possible for such structural conditions to exist, it is not probable.

In figure 478, *B* shows the same control points contoured in a manner that reveals two plunging anticlinal noses, two synclines, and a well-defined terrace. This sheet was contoured, not to tie the widely separated control points together in the simplest manner, but rather to develop the forms of any geologic structures that might be suggested in the variations in the rate of dip or changes in strike. In other words, this map bears the unmistakable marks of geologic interpretation of the data.

A knowledge of the general character and form of structures in the region aids greatly in correctly interpreting the subsurface structure where the well control is sparse. When the character of folding is known, an attempt should be made to contour the widely scattered points so that the features shown bear out the regional trends or tendencies. Often the subsurface geologist is called upon to construct structure maps where little is known about the regional trends. However, there are usually some clues in the datum elevations themselves. A common but often erroneous assumption is that most of the higher wells are on the highest parts of local structures, and most of the lower ones are on the lowest points of the structures. When starting to contour the subsurface map, it is better not to be too strictly constrained by the few scattered eleva-

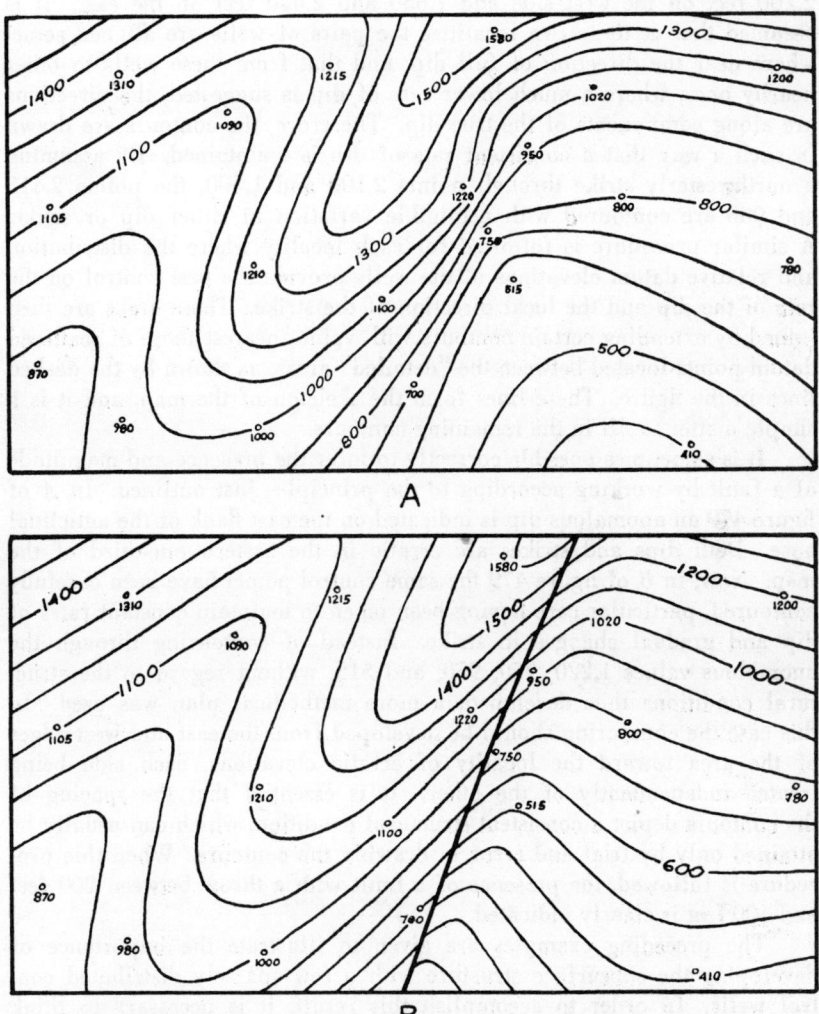

FIGURE 479. Examples of simple and interpretative contouring of the same data.

tions on the sheet. In some places the actual structural elevations probably exceed those of the highest wells and at others are less than those of the lowest wells. As long as the technical requirements of contouring are adhered to, the geologist has considerable license, and he should endeavor to present a consistent and feasible picture.

Figure 478, *C* illustrates the method by which the map, *B*, in that figure was constructed. A cursory inspection of the datum values of the wells shows that the regional strike is roughly east and west over most of the area. A high rate of dip is shown between elevation 1,500 and 2,100 feet on the west side and 1,650 and 2,040 feet on the east. It is assumed that at these two localities the pairs of wells are aligned somewhere near the direction of full dip, and that from these wells to other nearby ones, where a much lower rate of dip is suggested, the directions are along components of the true dip. Therefore, the contours are drawn in such a way that a consistent rate of dip is maintained. By assuming a northwesterly strike through points 2,100 and 1,500, the points 2,415 and 980 are contoured with negligible variation in either dip or strike. A similar procedure is followed for each locality where the distribution and relative datum elevations of the wells provide the best control on the rate of the dip and the local direction of the strike. These areas are then joined by extending certain contours with values nearest those of scattered datum points located between the "detailed" areas, as shown by the dashed lines in the figure. These lines form the skeleton of the map, and it is a simple matter to fill in the remaining contours.

It is sometimes possible correctly to infer the presence and magnitude of a fault by working according to the principles just outlined. In *A* of figure 479 an anomalous dip is indicated on the east flank of the anticlinal nose. Both dips and strikes are erratic in the eastern one-third of the map. Now, in *B* of figure 479 the same control points have been carefully contoured, particular care having been taken to maintain constant rates of dip and gradual changes in strike. Instead of contouring through the anomalous values 1,220, 750, 950, and 515, without regard to the structural conditions thus developed, a more methodical plan was used. In this case the contouring should be developed from the east and west edges of the area toward the locality of erratic elevations, each side being treated independently of the other. It is essential that the spacing of the contours depict a consistent structural condition, which can usually be attained only by trial and error in drawing the contours. When this procedure is followed, the presence of a fault with a throw between 200 feet and 400 feet is clearly indicated.

The preceding examples are given to illustrate the importance of developing the subsurface structure with a few sparsely distributed control wells. In order to accomplish this result, it is necessary to think geologically—to visualize the various structural forms as if they were solid models and to contour these forms in a manner that will withstand critical geologic analysis. As mentioned earlier, it is a very elementary task to contour a sheet technically correct. The geologist must go further: his map must be technically correct and geologically feasible.

PEG MODELS

The peg model is a device sometimes used to illustrate and study structural and stratigraphic conditions during the development of oil fields. It enables the geologist to view the behavior of the formations in three dimensions and is often a most useful tool in solving knotty structural problems. Peg models are not ordinarily used in broad regional

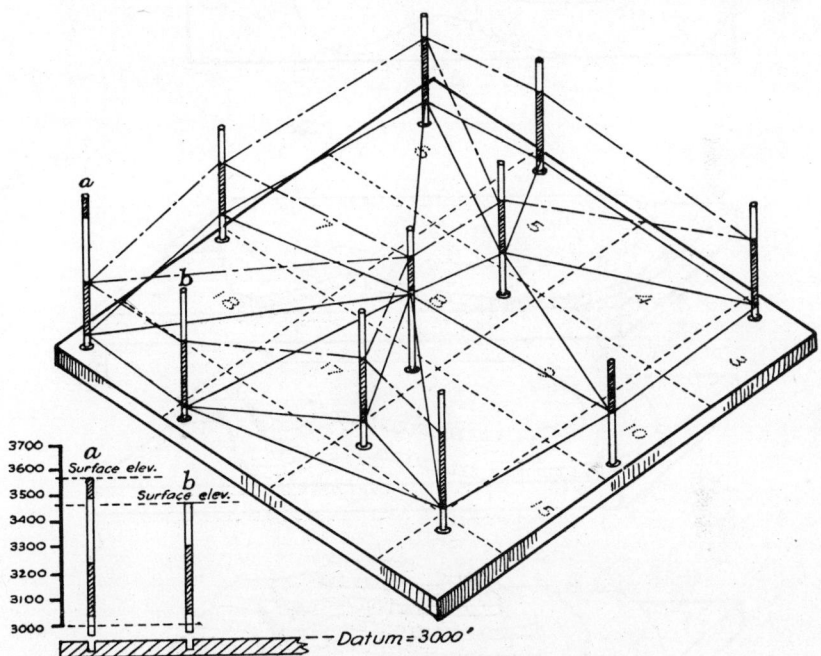

FIGURE 480. Peg model of a simple dome.

work, especially where wells are widely separated, but are better adapted to detailed subsurface studies of localized areas where well control is abundant. The principal objection to peg models is that they are bulky, occupy much floor space, and cannot be readily moved.

Figure 480 shows a peg model of a simple dome. The base into which the pegs are inserted should be made of wood not less than one inch thick. This base is painted, usually white or buff, and section and township lines and other desirable surface-map features are drawn on it to the scale selected for the model. At the location of each well, a one-fourth-inch hole is drilled almost but not entirely through the base. Ordinarily, a one-

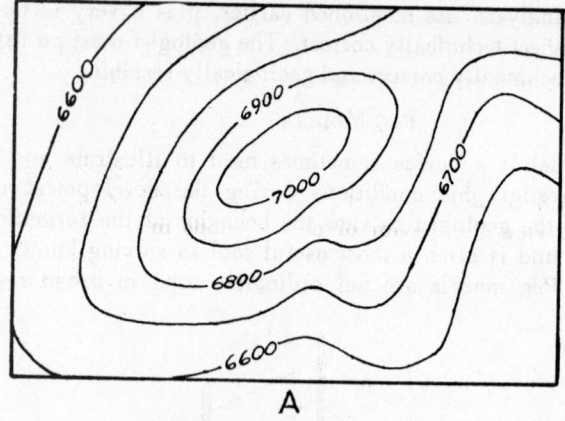

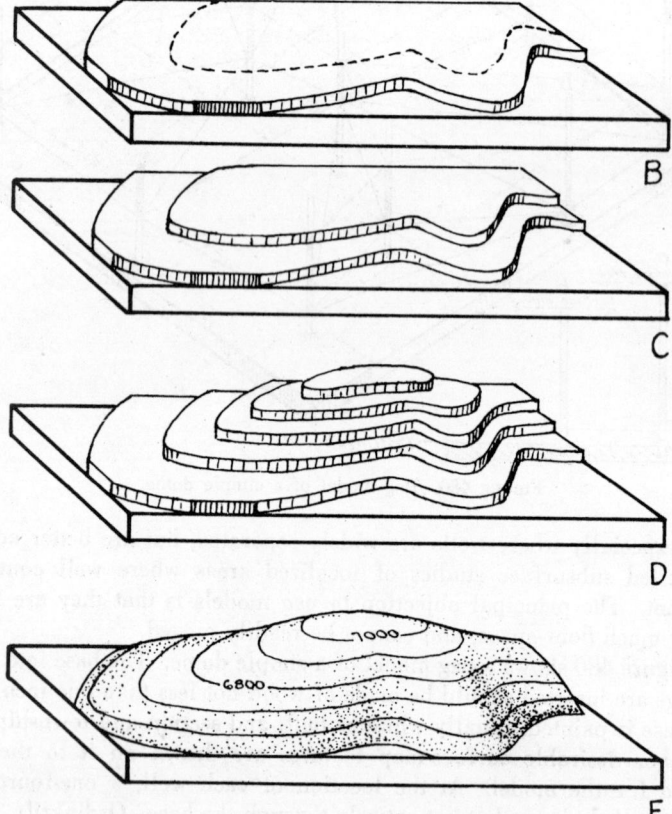

FIGURE 481. Solid relief model of a structure.

fourth-inch iron rod is used for the pegs, which, in turn, are to represent the wells.

The plane of the board is the elevation datum for the model. This assumed elevation should be somewhat below that of the lowest bottom-hole elevation in the field. Thus, if the lowest bottom-hole elevation is 2,300 feet above sea level, then the model might be constructed with the base elevation at 2,000 feet.

The rods are cut to lengths equal to the surface elevations of the wells on the vertical scale selected, plus an amount equal to the depths of the holes into which they will be mounted. The formations penetrated by the wells are represented on the rods by bands painted in different colors.

After the rods are cut, painted, and set in their respective locations, the formation tops from well to well are shown by strings colored the same as the formations which they represent, as indicated in figure 480. When all the rods are connected with the colored strings, the planes of the formations can be seen in their approximate relationships to one another. Of course, this illustrative method does not permit showing the curved surfaces of the structures, but it does suggest them in the sloping planes of which the strings are elements.

SOLID MODELS

The solid relief model of a structure can be made in much the same way as a topographic relief model, if there is a good structural contour map of the area.

The contour map is used as a pattern. Since the map will be cut during the construction of the model, several copies should be on hand. Sheets of pressed-fiber wallboard such as Celotex are cut with a coping saw along each contour line, as shown in *B* of figure 481. When the first sheet is mounted on the base, the next-higher contour is transferred to it, as shown by the dashed line in the figure cited. This may be done with one of the uncut maps and carbon paper. The transferred line serves to fix the position of the next fiber-board sheet, as in *C* of figure 481. When all the contours have been cut and the sheets are firmly nailed down, the skeleton of the model will look like *D* of figure 481. This skeleton should be partly waterproofed with several coats of shellac, after which it may be covered with plaster of paris, papier-mâché, or a similar material. Care must be taken to use only a sufficient thickness of the covering material barely to cover the edges of the fiber-board contours, or the accuracy of the model may be impaired.

If it is desired to show the structure contours on the finished model, pins may be stuck in the edges of the blocks before the covering material is applied. The pins will later serve as guides for sketching the contours and can easily be pulled out when they have served this purpose.

The solid model is used principally as an aid in teaching. Unlike the peg model, it cannot readily be altered to incorporate structural data that might become available at a later date, and for this reason it is not generally used for other than illustrative purposes.

SECTION MODELS

The section model is a series of parallel cross sections drawn or painted on any thin, rigid, boardlike material (fig. 482). These sections in turn are set in properly spaced slots in a solid base. A thick, hard cardboard is satisfactory for the sections, although transparent material such as cellulose acetate or cellulose nitrate is sometimes used because it is then possible to view one cross section through another. The section model is more practicable than the solid model in that any one of the sections can easily be removed and revised without affecting the others.

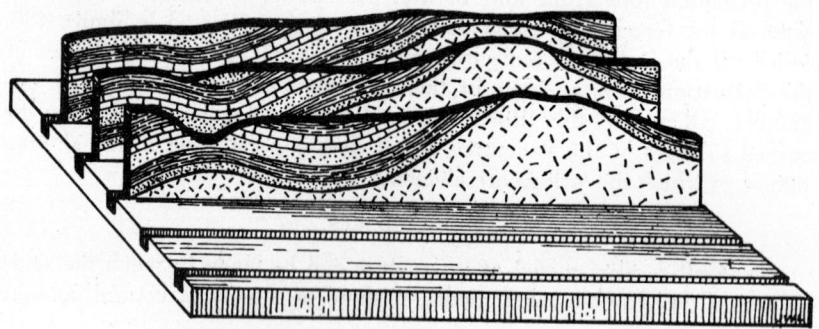

FIGURE 482. Models of parallel structural cross sections.

MISCELLANEOUS MODELS

Various types of working models have been made for the purpose of studying the behavior of fluids in permeable reservoirs. Others have been constructed to determine the deformation of rocks under different kinds of stresses. The models discussed earlier are similar to maps in that they are built as an aid to visualizing geologic conditions as they exist today. In contrast, the working models attempt to determine the series of events that bring about these conditions.

ISOPACH MAPS

An isopach or isopachous map is one which shows by means of contours the variations in stratigraphic thickness of a stratum, formation, or group of formations. As in the case of structural maps, isopach maps may be either surface or subsurface, depending upon the class of data used in their construction. The subsurface isopach map is based primarily upon formation thicknesses determined from well cuttings.

While contours must be drawn to agree with thicknesses plotted on

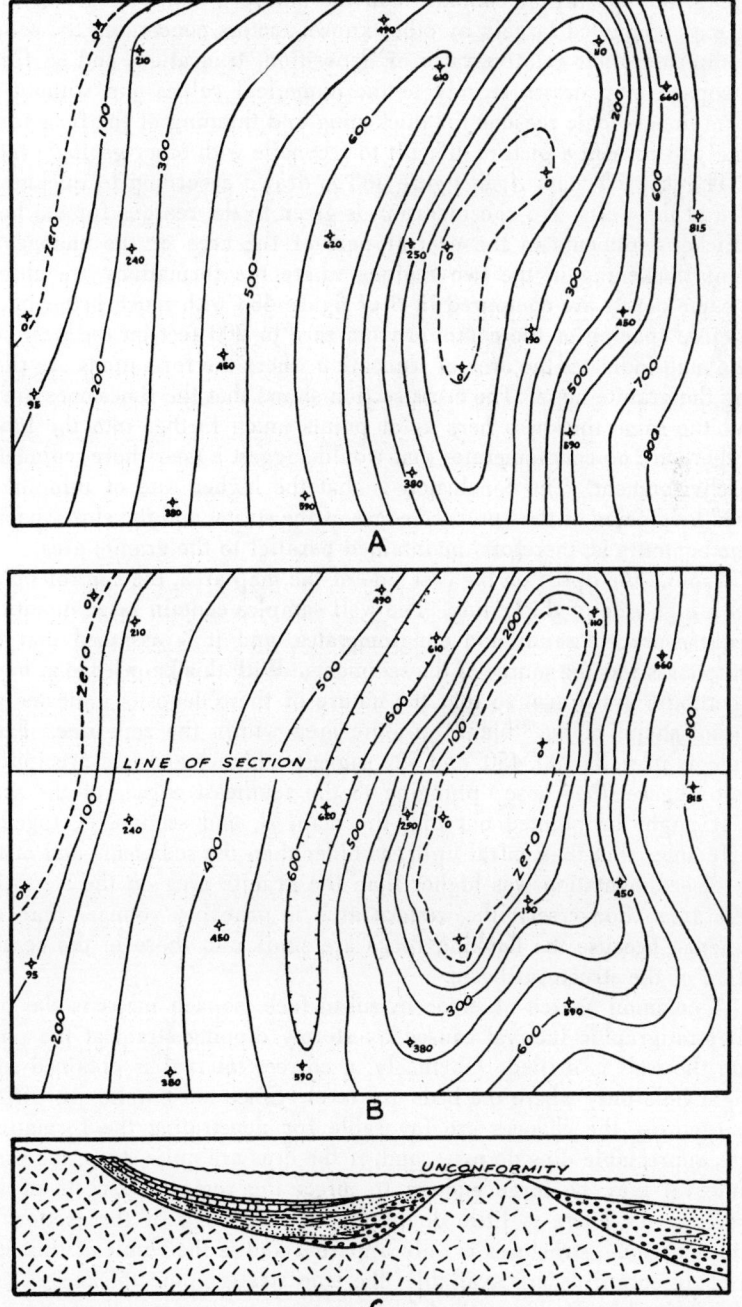

FIGURE 483. Two interpretations of contouring the same isopach data.

the map, the spacing of contours and the nature of thickening and thinning may be guided largely by other known factors concerning the source of sediments, their relative rates of deposition, truncation, and so forth. An isopach map drawn strictly to the numerical values and without regard to the geologic reasons for thickening and thinning of the formations is likely to present a picture difficult to reconcile with other geologic facts.

The isopach map, *A*, in figure 483 is drawn according to thicknesses shown at the wells. No consideration is given to the reasons for the body of thicker sediments in the central part of the area or the changes in rate of thickening in the two regions where the formations are absent. The same points are contoured in *B* of figure 483 with much better effect. The close spacing of the contours from zero to 200 feet on the west side of the map indicates the area of truncation where the formations are tilted along the granite mass. The cross section shows that the limestones are of about the same thickness here as at points much farther into the basin, and there are no conglomerates that would suggest a near-shore-sedimentation environment. The conclusion is that the higher rate of thinning is caused by erosion of the upturned edges of the strata, and the close spacing of the contours is, therefore, maintained parallel to the granite area.

Around the uplift on the east side of the map area, the control points show a high rate of thickening. The well samples contain large quantities of coarse arkosic sands and conglomerates, and it is assumed that the granite mass was the source of the sediments. With this knowledge at hand, the contours are drawn so that the nature of these deposits indicates the size and shape of the "blank" granite area within the zero line. Finer sediments in wells 190, 450, and 590 suggest a higher and less precipitous terrain; hence, the "nose" plunging to the southeast corner of the area.

It might be pointed out that the map, *B*, and section, *C*, together clearly show that the central uplift is older than the sediments and at the time of sedimentation was higher than the granite area on the west side of the area. Conversely, the western arch is probably younger than the sediments, because the flanking rocks are similar to those in the central portion of the structural basin.

A common source of error in subsurface isopach maps is the too-great stratigraphic interval caused by steeply dipping strata at the point where the well is drilled. Obviously, a correct interval is obtained in a straight hole only where the beds are level. Since most wells are drilled on structures, the chances are favorable for penetrating the formations where appreciable dips do exist, and, if the dips are quite steep, the error in interval may be large enough to affect the regional aspects of the isopach map. There is little doubt that the thinning of the section on the tops of some structures is only apparent and is the result of this condition. If core dips are available, the true stratigraphic thicknesses can be determined by Busk's method for obtaining stratigraphic thicknesses in inclined strata.

Although the subsurface maps representing drilled thicknesses are commonly called "isopach maps," a more precise term is "isochore." An isochore map is one that shows by contours the drilled thicknesses of the formations, without regard to the true stratigraphic thicknesses. The term is not ordinarily used but is mentioned here simply because it does come up occasionally in geologic literature.

Isopach maps are interesting to draw and frequently reveal intriguing and perplexing problems, but too often their many practical uses are not fully realized or employed in subsurface work. Isopach maps are

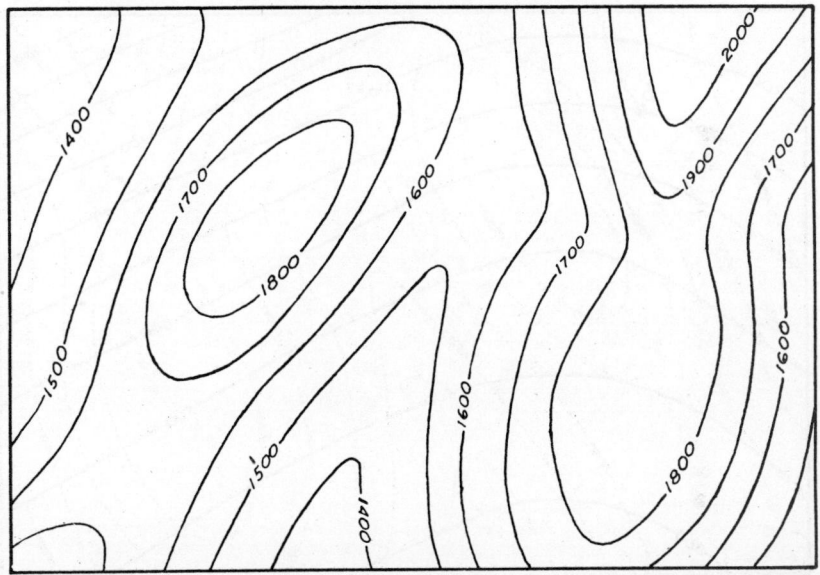

FIGURE 484. Subsurface structural map on top of Pennsylvanian.

generally used for the purpose of predetermining drilling depths to specific horizons in wildcat wells. They are also used as a means of locating buried structures in regions where formations habitually become thinner over the crests of the structures. A third common use is in estimating the elevation on a datum bed below the total depth of a well that has penetrated a higher known stratigraphic horizon. But there are many other practical uses, some of which are described below.

Figure 484 is a subsurface structural map on the top of the Pennsylvanian. In the northwest quadrant is an anticline with somewhat more than 100 feet of closure. The southward-plunging anticline on the east side is open on the north end. These structural contours are shown as dotted lines in figure 485. The thickness of the Pennsylvanian is shown by solid isopachous contours. Obviously, the 900 feet of convergence over the map area will have a profound effect on the form of the structure at the base of the Pennsylvanian or the top of the Mississippian. The procedure for reducing the Pennsylvanian structure to the Mississippian is as follows.

Point *a* in figure 485 is the intersection of the 1,500-foot structure contour and the 500-foot isopach contour. At this point the top of the Mississippian is 500 feet below the Pennsylvanian datum or at an elevation of 1,000 feet. At point *b* the Mississippian is 700 feet below the

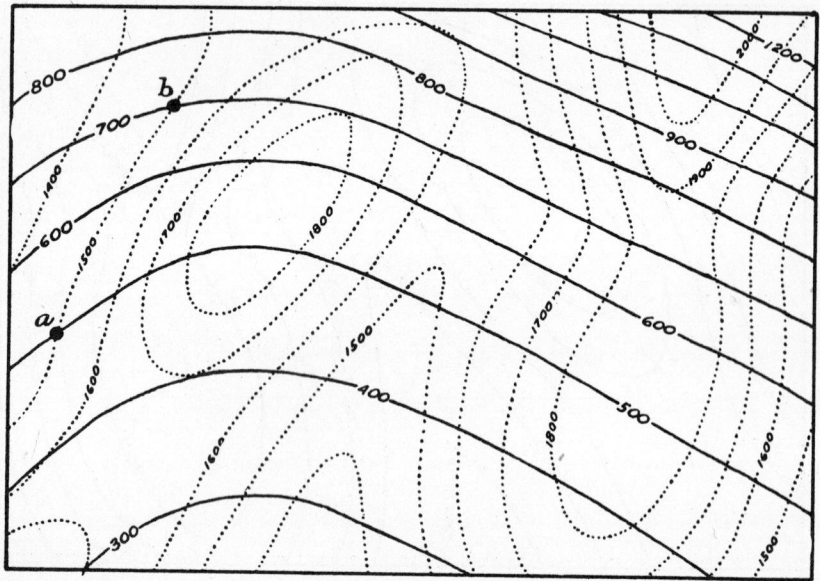

FIGURE 485. Isopach data of Pennsylvanian superimposed on structural map (fig. 484).

structure datum, and the elevation is, therefore, 800 feet. All intersections are reduced to Mississippian elevations in this manner, and these values are then contoured, as shown in the Mississippian structure map in figure 486. Now, if it is desired to determine the structure on the top of the Devonian, the Mississippian isopach is superimposed on the Mississippian structure, as in figure 487, and the process just described is repeated. The result of this step is shown in figure 488, the Devonian structure map. This figure shows the original Pennsylvanian structure map superimposed on the underlying Devonian structure map in order that the two may be compared directly. It should be pointed out that the closed Pennsylvanian structure is a northeasterly-plunging nose at the Devonian horizon. The south-plunging nose on the east side of the area is a north-plunging nose on the Devonian, and if the map were extended on the south, a large closure would be evident in the Devonian.

In figure 489 the combined thickness of the Pennsylvanian and Mississippian can be determined at any contour intersection simply by subtracting the lower value from the higher value, if both data are either above sea level or both are below sea level. Where one datum is above sea level and the other below, the contour values are added to obtain the thickness.

This method of reducing structural maps from higher to lower horizons should be applied wherever the rate of convergence (in feet per

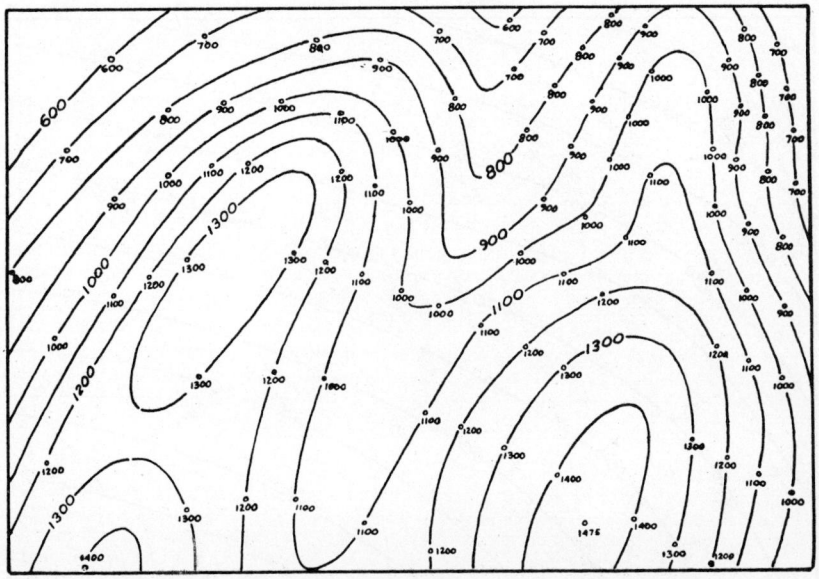

FIGURE 486. Subsurface structural map on top of Mississippian.

mile) between the structure datum and the prospective oil horizon approaches the rate of dip (in feet per mile) on the flanks of the structure.

In some regions persistent and sharply defined seismic-reflecting horizons are encountered several thousands of feet above the prospective oil-producing formations. Because of the fact that much of the wave energy is reflected here, it is sometimes impossible for the little remaining energy of the shot to reach the lower horizons and be reflected back to the seismometers in sufficient strength to produce usable records. Therefore, a detailed seismic structure map may be obtained on the upper horizon, but sparse data or none on the lower. Now, if a few deep wells provide the necessary convergence data, an isopach map can be constructed from the subsurface information, and, by means of this map, the seismic struc-

ture can be reduced to the prospective formation through the methods just
described.

Isopach maps of oil reservoir rocks, together with porosity determinations from cores, make it possible to calculate the volume of oil in a
structure. This method is most applicable to sandstone reservoirs where
there is little cementing or other interstitial material. Conditions of porosity and thickness of saturation are normally less predictable in limestone reservoirs, and for these reasons it is difficult to make accurate volumetric determinations.

In figure 490, *A* shows a structure contoured on the top of the pro-

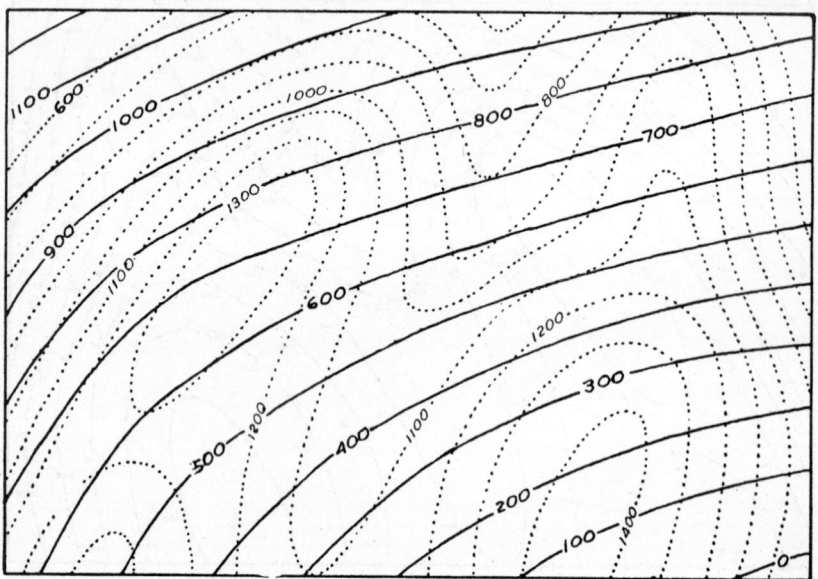

FIGURE 487. Isopach data of Mississippian superimposed on structural map (fig. 486).

ducing formation. A few dry holes have been drilled on the flanks of the
structure below the oil-saturated portion of the reservoir, and by means
of these dry holes the oil-water contact is established at a structural elevation of 660 feet. This oil-water contact is shown by the heavy dashed
line on the map and also in the cross section, *B*, in figure 490.

Since the thickness of the oil column is less than the thickness of the
reservoir rock, the computation of the volume of saturated sandstone is
quite simple, because the isopach map of the saturated rock is exactly
the same as the structure map with only the contour values being changed.
It is clear in the cross section that the "extra" structural contour (oil-
water line) of 660-foot elevation is the same as the zero isopach contour
for the saturated zone. Likewise, the 700-foot structural contour becomes

the 40-foot isopach line, the 800 becomes the 140, and so forth. The thickest part of the zone is on the top of the structure at an elevation of 1,050 feet, and the thickness here is 390 feet.

The computation of volume from an isopach map is as follows:

The area contained within each contour is determined with a planimeter, or, if a planimeter is not available, the area can be subdivided into rectangles and right triangles ,as shown in *C* of figure 490. These tracts are scaled, and areas are computed according to the scaled dimensions. The outline in the figure is the isopach zero line (660-foot structure contour). The same procedure is repeated for the next-higher contour, which in this

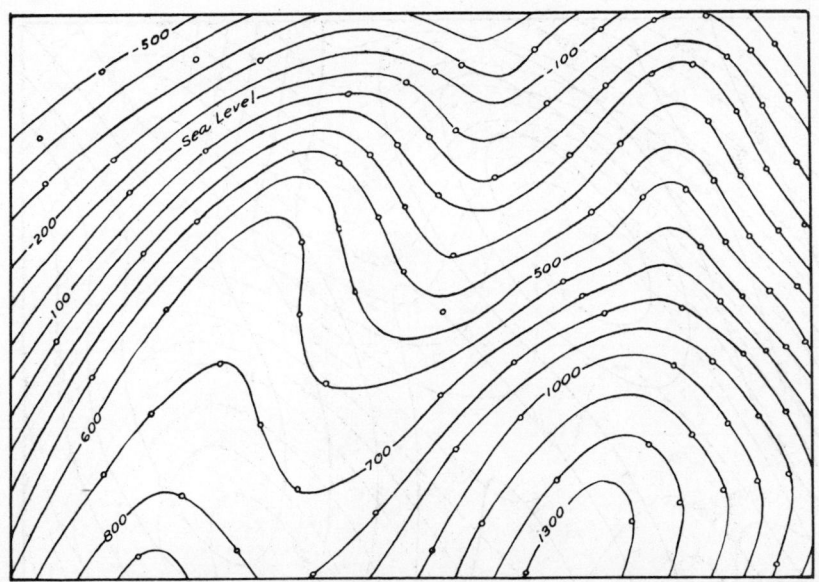

FIGURE 488. Subsurface structural map on top of Devonian.

case is the 40-foot isopachous contour. The volume of rock between these two planes is

Area within zero contour + area within 40-foot contour × (40 ÷ 2).

Since this gives the volume for only that portion between the zero and 40-foot contours, the process must be repeated for the segment between the 40-foot and 140-foot, and so on to the highest contour.

In figure 491, *A* shows the structure just discussed, but, as indicated in the cross section, the reservoir rock is uniformly 200 feet thick. The procedure for determining the volume of this reservoir is considerably different from that previously described. In this example it is assumed that the elevation of the oil-water contact plane is known and has been mapped according to the dashed line on the map. The uniform thickness of the reservoir is known from a few wells in the region that have penetrated it.

It can be seen in *B* of figure 491 that, if the folded reservoir bed were flattened, its actual breadth would be greater than that shown on the structure map. The first step, then, in computing the volume is to determine the actual area of the reservoir, rather than the area within the oil-water line as it appears on the map. This can be done graphically by first constructing several cross sections and one or two longitudinal sections, all on a natural scale: i.e., with no vertical exaggeration.

A series of horizontal lines are drawn across the profile. Using the

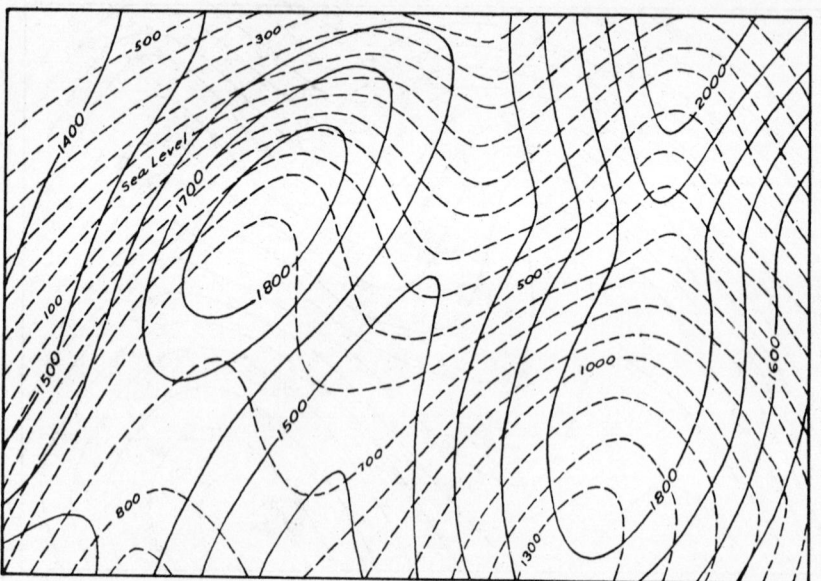

FIGURE 489. Devonian and Pennsylvanian structure maps combined.

intersections of these lines with the line of the profile as centers, short circle arcs are drawn upward from point to point, as shown in the figure. The sum of the distances, *a* to *d* and *a'* to *d'*, is a close approximation of the surface distance over the fold. In practice, the distance over the fold may be measured directly, or the profile may be drawn on profile paper. Wherever the dip is constant, the extension can readily be computed since the distance along the sloping surface is the hypotenuse of a right triangle whose other two sides are the difference in elevation and the horizontal distance between the points (scaled on the structure map).

When a number of zero points have thus been located on the map, a new zero (oil-water) line is sketched. This is the outline of the reservoir shown in *C* of figure 491. The bordering band of wide ruling is that portion of the reservoir cut by the water plane. The thickness of the oil-satu-

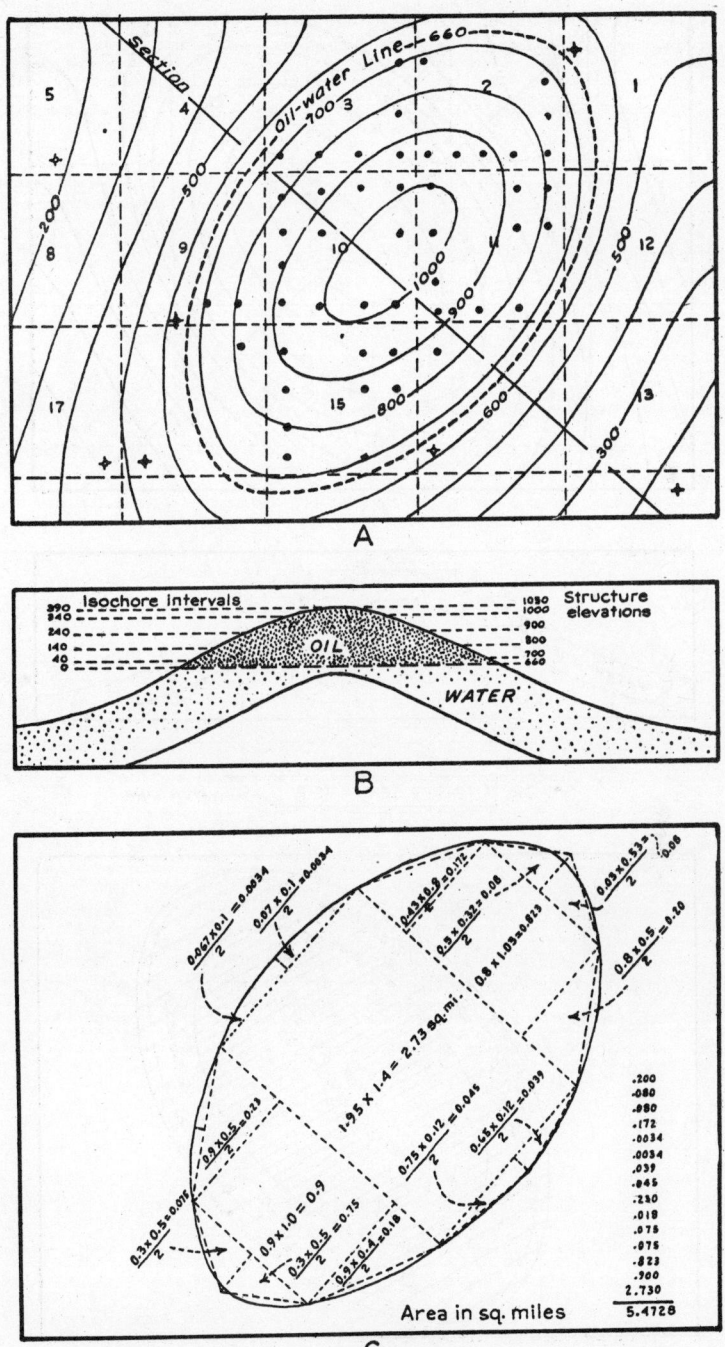

FIGURE 490. Method of computing oil-reservoir volume (case I).

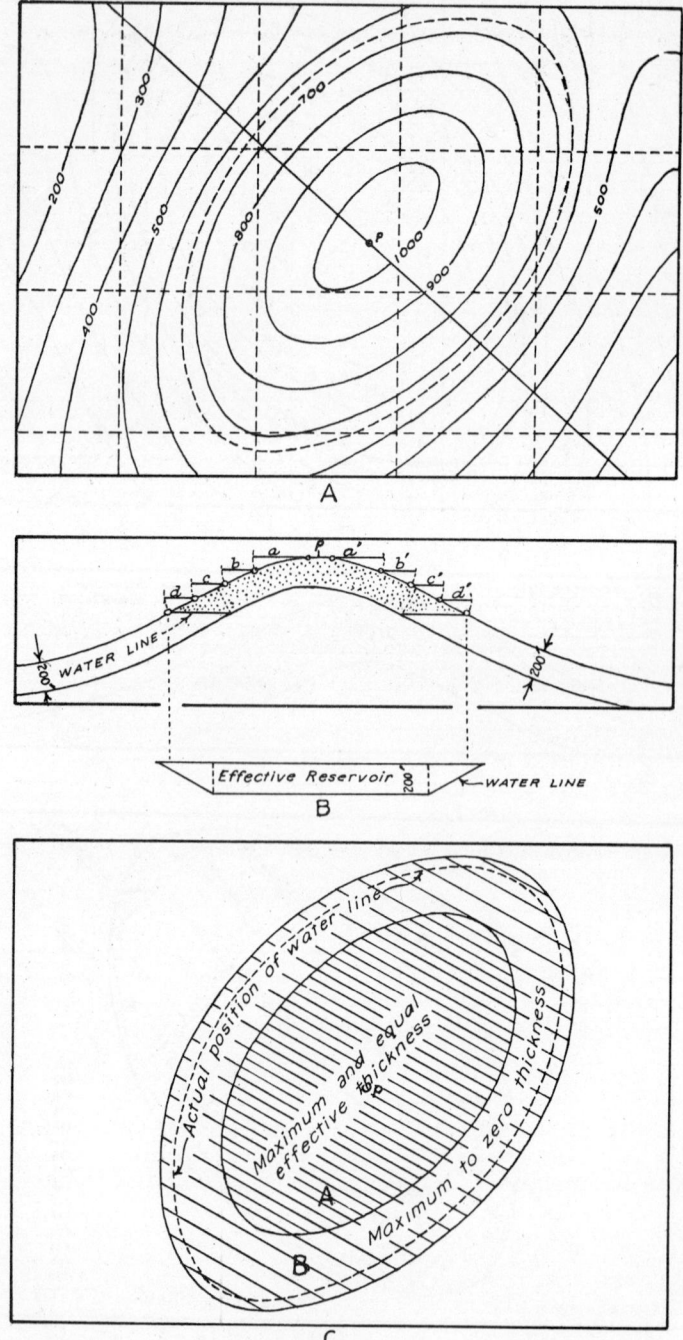

Figure 491. Method of computing oil-reservoir volume (case II).

rated reservoir in this band increases from zero at the outer edge to 200 feet at the inner. The volume is, therefore, the area (in square feet) × 100. The closely ruled area is that part of the reservoir where the thickness is everywhere 200 feet, and the volume within that area is the area (in square feet) × 200.

The two examples will serve to illustrate the use of isopach maps in special adaptations to determine volumes. There are cases where the plane of the oil-water interface is inclined, and others where the reservoir bed varies in thickness across the structure; but it is necessary only to show these variations by isopachous contours to compute the volume of the reservoir. These irregular conditions require some adjustment in procedure, but not in principle.

When the volume of the reservoir rock is obtained, the volume of the contained fluid is determined by multiplying by the percentage porosity, as ascertained from laboratory tests on representative cores.

From the foregoing it is evident that the volume of any stratum, such as a coal seam or a bed of gypsum, can easily be calculated from an isopach map of that stratum.

Other uses of isopach maps will be discussed in connection with paleogeologic and facies maps.

PALEOGEOLOGIC MAPS

A geologic or areal geologic map is one that shows the present distribution of consolidated rocks at the surface and immediately below the soil or unconsolidated mantle. A paleogeologic map shows the distribution of formations at a surface which existed at some specific time in the geologic past. Such a surface is shown in the lower block of figure 472. In figure 492, *A* is the areal geologic map of the upper block and *B* is the paleogeologic map of the pre-Jurassic surface in the lower block.

As might be presumed, paleogeologic maps are constructed from information supplied by wells. The four wells shown in the block diagram mentioned penetrate pre-Cambrian, Devonian, and Pennsylvanian beds beneath the pre-Jurassic unconformity. The remaining twenty wells of *B* in figure 492 encounter rocks of different ages beneath the unconformity, and it is upon this type of information that the map is constructed.

Several factors control the relative breadth of the bands or areas that appear on the paleogeologic map. Among these are the relative thicknesses of the formations, the rates of thinning, the relative rates of dip in the different formations and the actual degree of dip, the character of the eroded surface, and the amount and character of folding subsequent to truncation. It is well to keep these conditions in mind when drawing a paleogeologic map, because it may be necessary to interpolate several geologic boundaries between two control points, and any one of the conditions listed above might have a pronounced effect on the map position of the boundary lines.

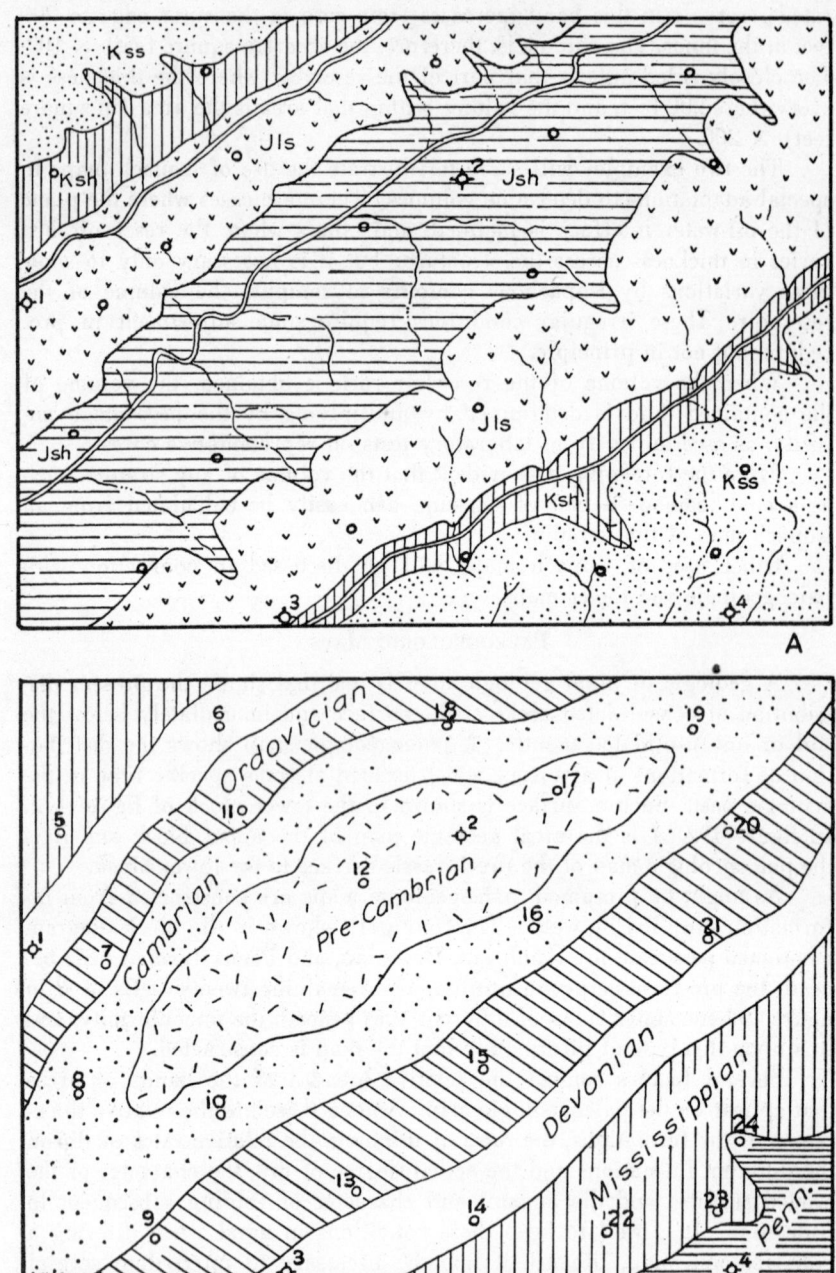

FIGURE 492. Paleogeologic maps. A—Surface areal map. B—pre-
Jurassic areal geology.

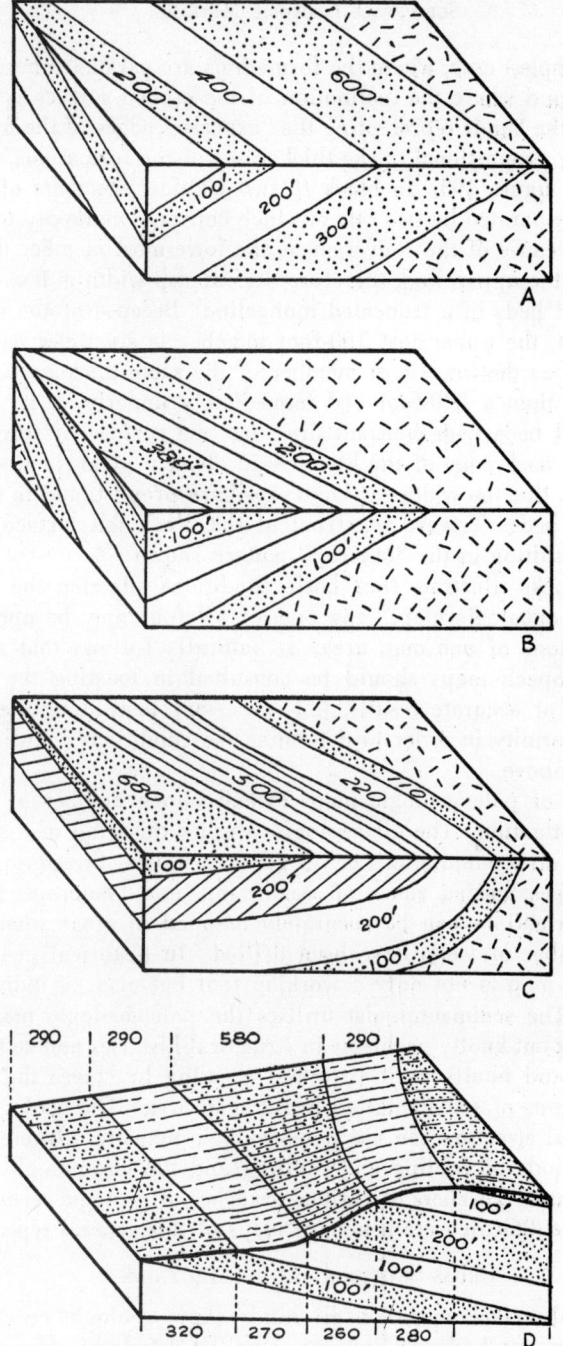

FIGURE 493. Block diagrams showing relationship of
outcrop bands to dip and topography.

In the simplest case, where the formations are parallel, where the dips are constant, and where the inclination of the eroded surface is constant, the widths of the bands representing the "exposed" edges of the formations will be exactly proportional to the thicknesses of the formations, as shown in block *A* of figure 493. In block *B,* two 100-foot members of constant thickness are separated by two others which converge markedly toward the "outcrop." Because of the convergence, the lowermost member dips more strongly than the upper, and, therefore, its outcrop width is less. Block *C* shows parallel beds in a truncated monocline. Because of the difference in rate of dip, the uppermost 100-foot member is six times as wide on the "outcrop" as the lowermost member of the same thickness and some-what broader than a 200-foot bed immediately underlying it. Block *D* shows parallel beds with constant dips, but the surface of truncation is variable. The back edge of the block is an element of an inclined plane, and along this line the widths of "outcrops" are proportional to the thick-nesses of the beds. Toward the front of the block the surface becomes terracelike, resulting in the "outcrop" pattern shown.

These blocks illustrate four basic conditions affecting the construc-tion of a paleogeologic map. Any one, or all four may be operative in different portions of one map area. It naturally follows that structural as well as isopach maps should be consulted in locating the geologic boundaries. For accurate results, it is necessary to contour the "plane" of the unconformity in order to determine the conditions shown in block *D* mentioned above.

The uses of paleogeologic maps extend into a number of fields of geologic investigation. The oil geologist applies this kind of mapping in the search for oil accumulations below unconformities. Stratigraphic traps of various types, buried zones of weathering, and "outcrop" trends of productive formations can be accurately mapped in areas where a con-siderable number of wells have been drilled. In historical geology, the paleogeologic map is not only a working tool but also an indispensable illustration. The sedimentologist utilizes the paleogeologic maps as an aid to working out knotty problems in structural histories and source areas of sediments and finally as illustrations showing by stages the progress and interruptions of sedimentation in the map area. The geologist work-ing on regional structure can use paleogeologic maps to advantage in de-termining periods of folding and faulting and the chronologic develop-ment of structure. Teachers of various branches of geologic science would find their tasks difficult were it not for maps of this general type.

Cross Sections and Projections

While subsurface maps of all kinds show geologic conditions in essentially horizontal planes, sections show the details of stratigraphy or structure in vertical planes. Neither the map nor the section, alone, tells the whole geologic story; and, for this reason, any exhaustive subsurface investigation of an area must use both sections and maps. Sections fall into two general groups, structural and stratigraphic, although there are many types that incorporate both stratigraphic and structural features.

In figure 494, *A* is a structure section plotted on a natural scale. This type of section does not attempt to show any structural features between the wells, such as might be indicated on structural maps of the area, but does show the difference in elevation of the formations at the wells.

In figure 494, *B* is a stratigraphic section along the same line, also plotted on a natural scale. In a stratigraphic section one continuous strati-

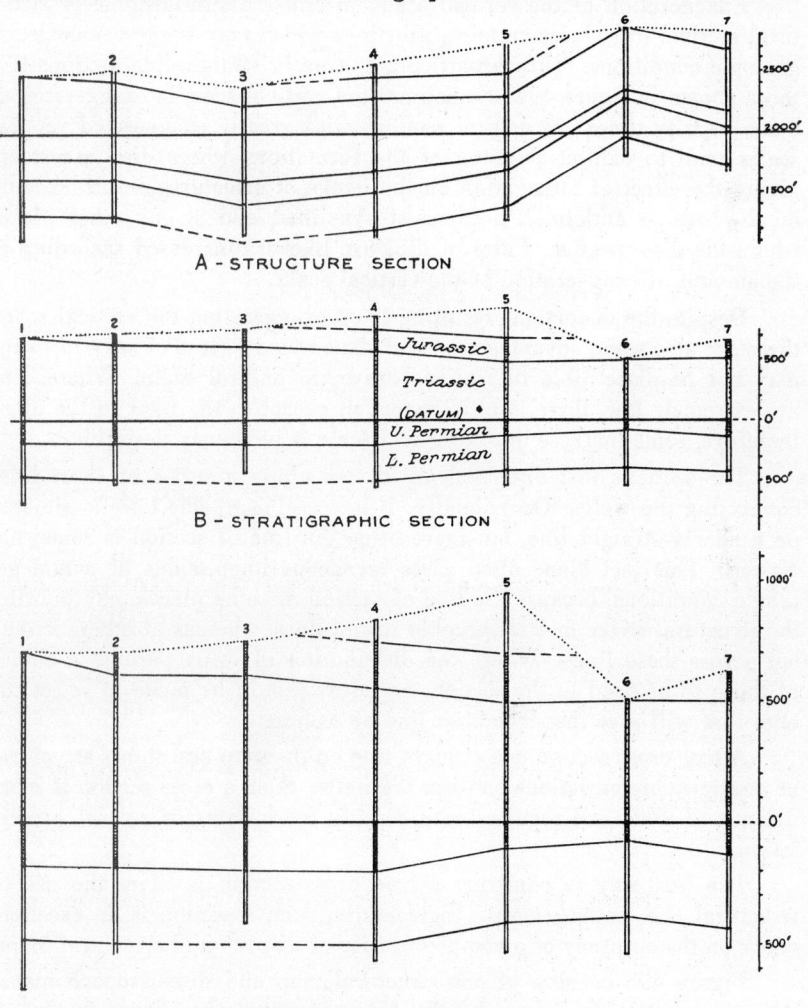

A - STRUCTURE SECTION

B - STRATIGRAPHIC SECTION

C - STRATIGRAPHIC SECTION
(DOUBLED VERTICAL SCALE)

FIGURE 494. Structural and stratigraphic sections.

graphic horizon is selected as a reference line or datum; this line is drawn straight across the sheet. All other formational boundaries are referred to this line according to the thicknesses of the formation. Thicknesses are usually indicated by numbers starting with zero at the datum horizon. In the drawing cited above, the datum is the top of the Permian, and thickness numbers start at this point. Figure 494, *C*, shows the same stratigraphic section, but with the vertical scale two times as large as the horizontal.

Exaggeration of the vertical scale, in either a stratigraphic or structural section, introduces certain distortions which may suggest nonexistent geologic conditions. This is particularly true in straight-line sections like those shown in figure 494. Sometimes the vertical scale is exaggerated as much as fifty times. Structural sections with greatly exaggerated vertical scales tend to exhibit thinning of the formations where dips are steep, giving the effect of attenuation on the flanks of structures and thickening on the tops of anticlines, bottoms of synclines, and at any other places where the dips are flat. Rates of dips are likewise increased according to the amount of exaggeration of the vertical scale.

Despite the distortions resulting from exaggerating the vertical scale, there are also some advantages. Local structures in areas of very low dips may not be discernible in sections drawn to natural scale. Where dips are extremely low, there is little distortion except in the rates of the dips; therefore, some increase in the vertical scale is obviously desirable.

The sections just discussed are drawn along a series of short lines connecting the wells. Occasionally, it is possible to select wells situated on a nearly straight line, but more often the line of section is somewhat zig-zag. This fact alone often gives erroneous impressions of actual geologic conditions, because the line of section at some places may parallel the structural strike or stratigraphic strand lines, whereas at others it may cut across these lines. Where the distribution of wells permits a choice of those to be used in the section, an effort should be made to select the ones that will give the straightest line of section.

A true cross section is a straight line on the map and shows structural or stratigraphic variations between the wells. Such a cross section is more difficult to draw and requires considerably more data for correct presentation.

The best way to construct a true cross section involves the use of structural and isopach maps. Incidentally, such a section is an excellent check on the accuracy of a complete series of isopach and structural maps.

Figure 495 consists of one structural map and three isopach maps; the cross section is drawn on the basis of data taken from the maps. The method employed is as follows:

The line of section is drawn on each of the maps. In the case illus-

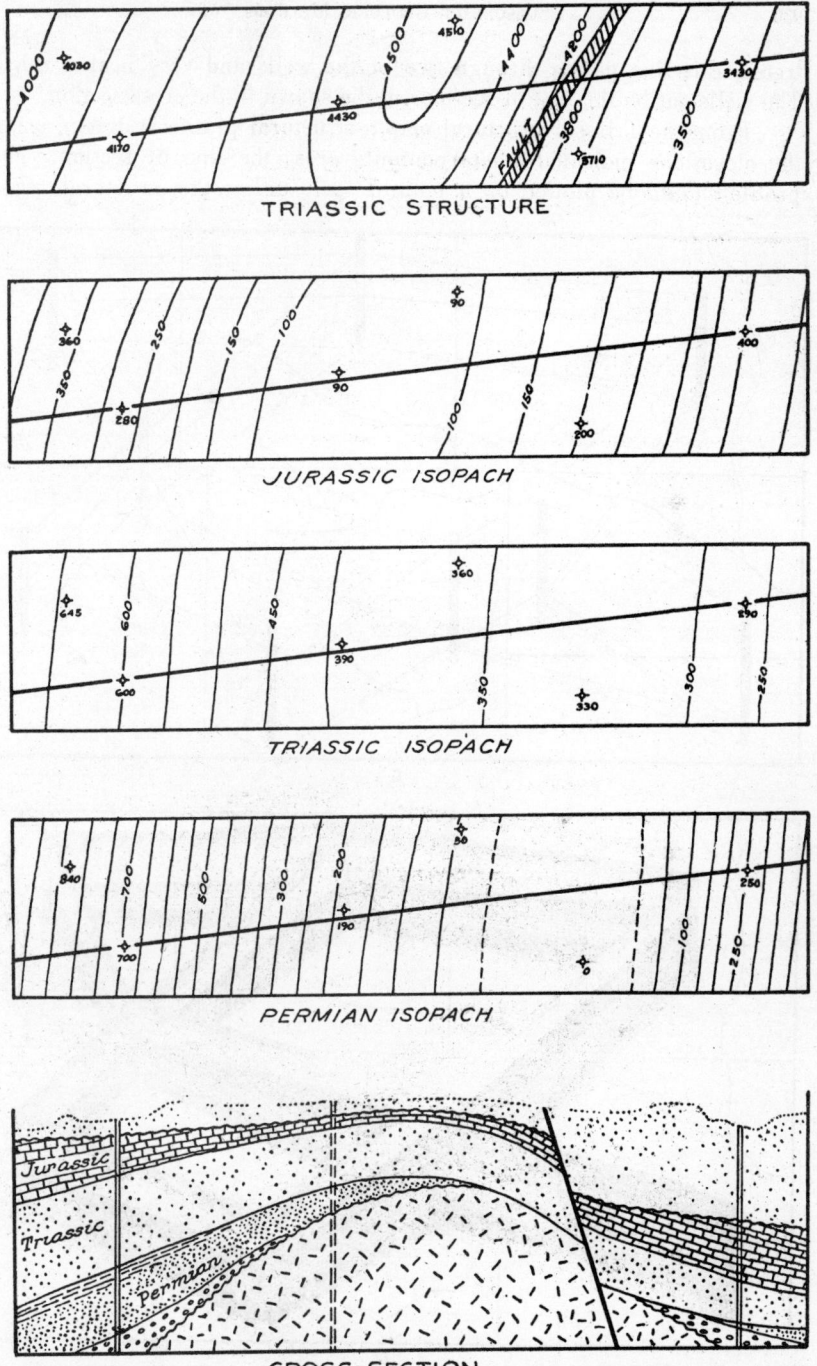

FIGURE 495. Structural and isopach maps and cross section based on data taken from maps.

trated, this line passes through two of the wells and very near a third. The wells cut by the line of section can be shown in the cross section.

From the Triassic structural map a structural profile is drawn, using the elevations indicated where contours cross the line of section. The profile should be plotted on a natural scale unless the structural relief

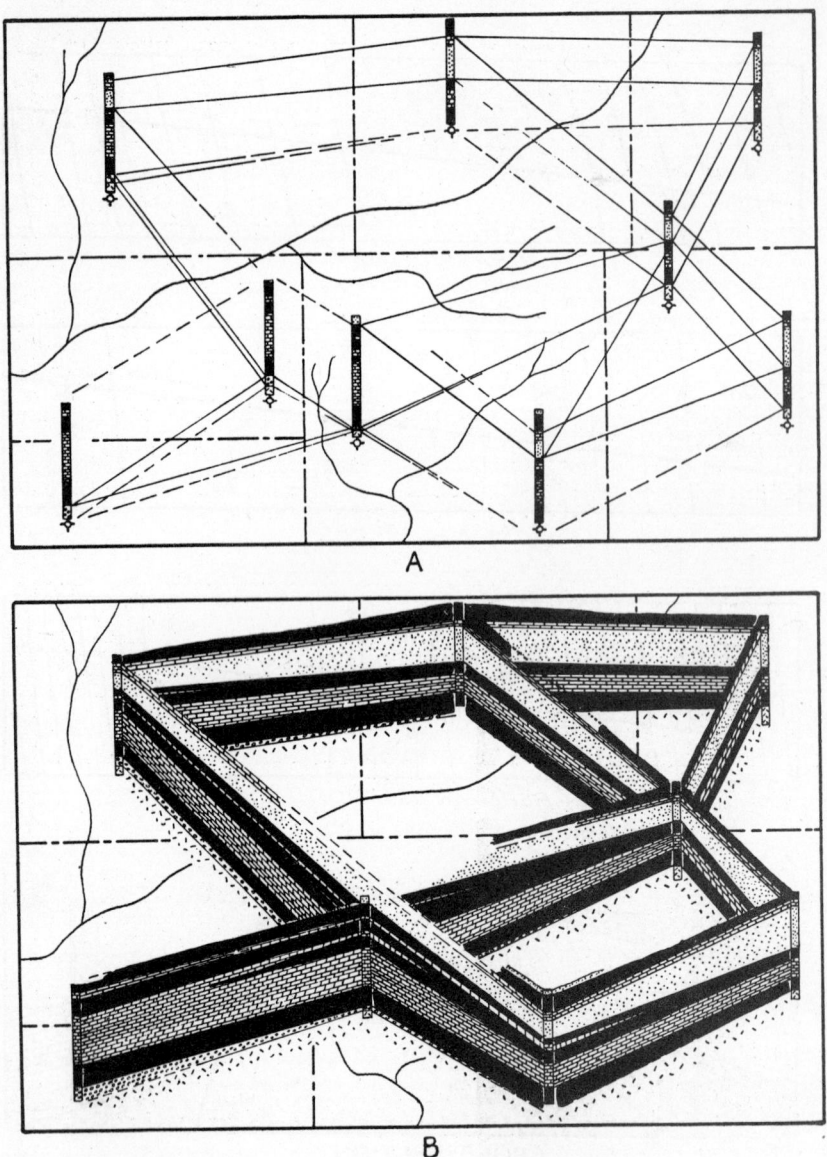

A

B

FIGURE 496. *A*—Log map. *B*—Panel map.

is extremely low. This profile is, in a sense, the structural datum of the cross section. It is also the reference line from which all succeeding geologic boundaries are drawn.

Now, from the Jurassic isopach map, thicknesses of the Jurassic along the line of section are obtained where isopachous contours cross the line. The thicknesses are plotted above the structural profile. The top of the Jurassic is then drawn through these points. The same procedure is followed with the Triassic and Permian isopach maps to draw the contacts stratigraphically below the structural reference line.

The surface profile can be taken from a topographic contour map.

Figure 496, *A,* is a log map. It consists of well logs plotted to any adaptable scale in their respective locations on the map. The example shows the bases of the logs at the map location, but they may be plotted with the tops of the logs or any selected continuous stratigraphic horizon on the logs at the respective map points. When the logs are plotted, they may be joined by formational correlation lines, as shown.

Figure 496, *B,* is a panel map of the same area as that shown in the log map. In the panel map it is possible to show changes in lithologic facies, pinch-outs, and other stratigraphic conditions occurring between the wells.

As only the front panels are shown in their entirety, they should be drawn first. In other words, the lowermost panels on the page are drawn, then the next higher, and so on to the top of the drawing. Panels joining wells along north and south lines are omitted, for they would appear only as single lines on the map.

The stratigraphic isometric projection is a special adaptation of the panel map. Figure 497, *A,* is a base map with a few principal streams and well locations. Figure 497, *B,* is the isometric projection made from this map.

In order to construct an isometric projection it is necessary to have the map contained in a rectangular grid, unless the land lines, as in the case illustrated, provide such a grid. This grid, which may be drawn to any scale, regardless of the scale of the map, serves only the temporary purpose of placing map features correctly on the perspective drawing. Instead of a grid, coordinate scales were used in the figure, and parallels were drawn from the section lines.

The isometric projection is referred to as a 20° or 30° projection, depending on the perspective effect desired and the construction necessary to produce this effect. An isometric projection is always less than 45°; for a 45° projection is simply a map rotated 45° on the sheet, and there is no foreshortening as in any perspective drawing.

Referring again to figure 497, *B,* the upper corner of the projection, is the northeast corner of the map. The east and west sides of the map are drawn at an angle of 30° to the horizontal (in a 30° projection). Thus, the northeast and southwest 90° angles of the map become 120°, and the southeast and northwest corners become 60° in the projection. All north-south and east-west lines are parallel, and scaled distances between map points along these parallel lines are the same as on the map. Scaled distances in any direction except parallel to these lines are either greater or less than those on the map. Obviously then, in order to transfer map information to the projection, it is necessary to do so by means of coordinate measurement. Point *a* on the map is 10.5 units (any scale) east of the northwest corner of the map and 2.8 units south. Point *á*

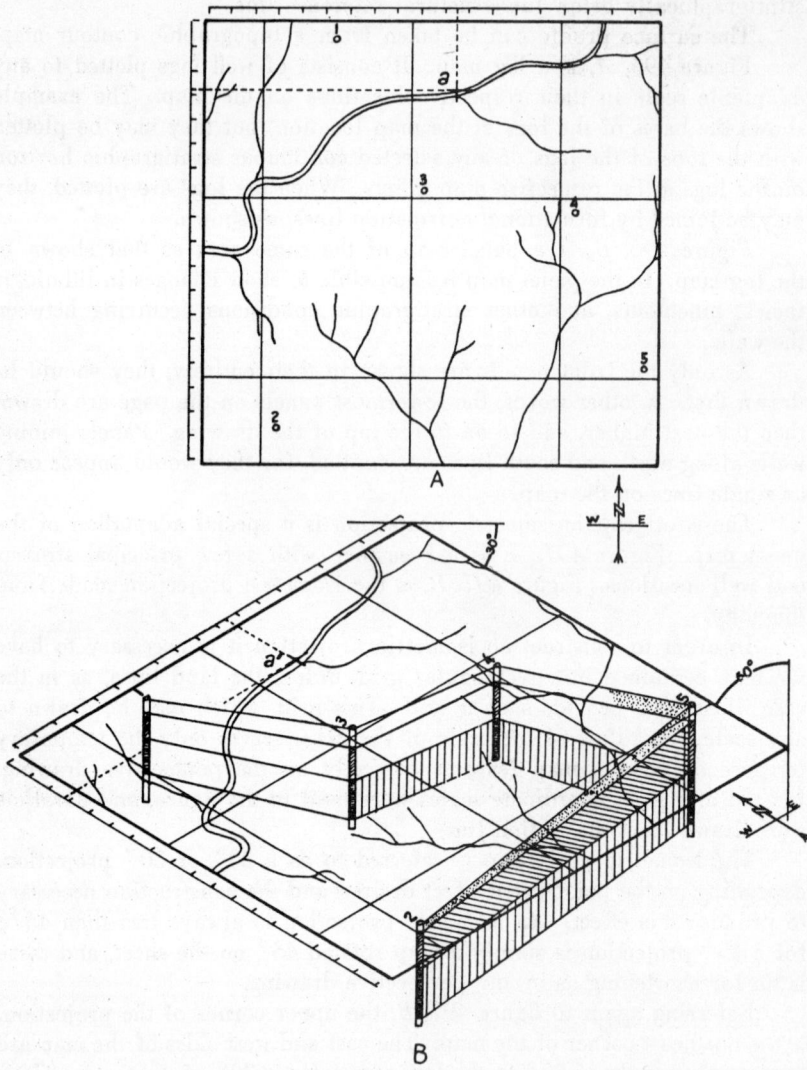

FIGURE 497. Stratigraphic isometric-projection drawing.

on the projection is the same number of units along corresponding lines. Any point on the map can be accurately located on the projection by scaling along coordinates; and this procedure must be followed in locating a sufficient number of control points, such as the confluences of streams, road intersections, and well locations, to insure accurate sketching between these points.

When the wells are located on the projection, logs are plotted at the locations on a vertical scale adapted to the scale of the projection so that the desired effect is attained. The isometric base is considered as a level plane. Therefore, if it is assumed to be at sea level and the logs are placed so that the sea-level point on the log is adjacent to the map location, then the panels will represent the structure. Of course, if the bottoms of the wells are above sea level, then the plane of the projection should be at some datum plane high enough to cut all of the wells.

The plane of the projection may also be considered a stratigraphic datum, i.e., the top of a formation; and all the logs are then placed with this horizon at the location point. Another common practice is to draw subsurface geology entirely below the plane of the projection and surface geology above. The principal objection to this method is that the drawing may have neither a structural nor a stratigraphic datum. The panel projection in figure 497, *B*, is drawn with the plane at the top of a formation; this, therefore, is a stratigraphic projection. The panels have not been completed in order to permit a clearer view of some of the map features.

BLOCK DIAGRAMS AND OTHER ILLUSTRATIONS

It is of the utmost importance that the geologic concepts developed as a result of studies in structure or stratigaphy be shown in some manner that is most comprehensible to those who have only occasional contact with the projects. Maps and cross sections sometimes fail in their purpose of conveying to others certain complex geologic conditions, mainly for the reason that each is two-dimensional, one in the horizontal plane, the other in the vertical. Block diagrams effectively combine the features of both maps and sections and are, therefore, an indispensable mode of illustration.

Geologic block diagrams are constructed according to certain principles of projection and perspective. Space here does not permit going into all the details of block diagrams: only the fundamental principles necessary for constructing the simplest illustrations can be given.

The two upper blocks in figure 498 are examples of the simplest projection. All opposing sides are parallel to each other, and, because of this feature, they can readily be drawn with a drafting machine or a triangle and straightedge. Distances along the front and back edges and all lines parallel to these edges are drawn to the scale of the map. Distances along the sides and parallel to the sides may or may not be to the scale of the map; but, in any case, the scale is constant along these lines. This type of block is sometimes called a "parallelogram block," but it is essentially an isometric projection. The block may be drawn with any desired degree of tilt. The high-angle block, as illustrated, should be used in illustrations requiring considerable details in the horizontal or map plane. Low-angle blocks are more effective where it is desired to emphasize the vertical sections in two directions. The high-angle block may be drawn so that the scale is the same along the two horizontal coordinates; but the low-angle kind should use a somewhat smaller scale along the front-to-back lines in order to produce a more realistic effect of perspective. The scale is reduced along this coordinate in the section lines drawn on the low-angle block in the figure cited.

Because of the simplicity of the isometric block, it is the one most

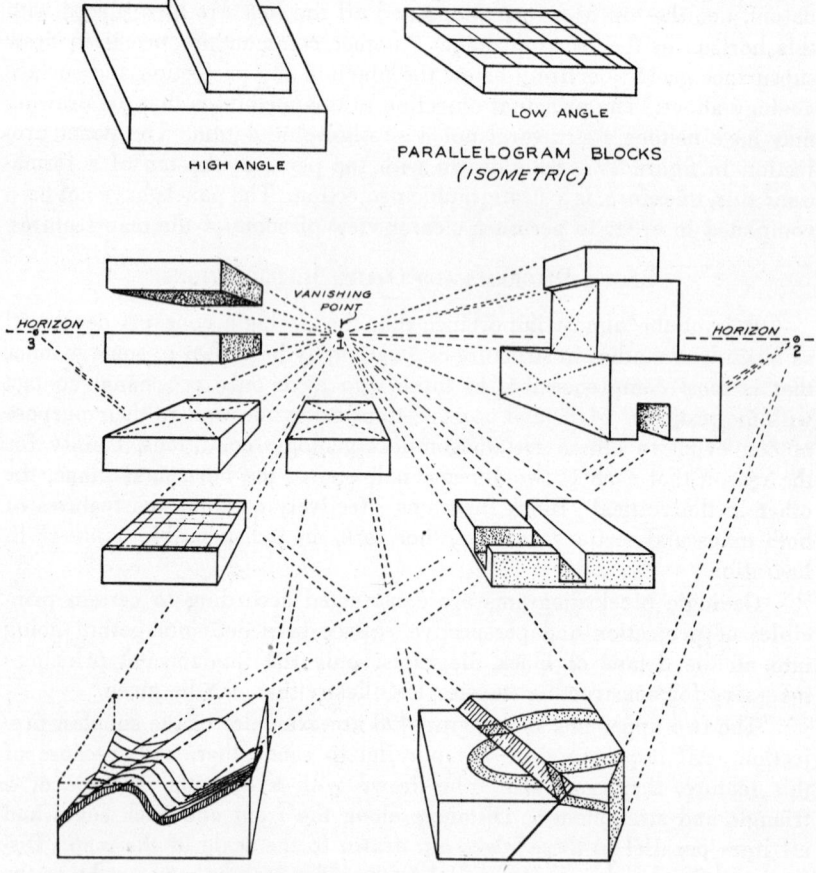

FIGURE 498. Geologic block diagrams.

commonly used in geologic illustrations. A number of examples appear in this chapter to demonstrate various features. Since this type of block is not a true perspective figure, the distant or upper end appears to be larger than the front or lower end, and the block, therefore, appears somewhat awkward and distorted. Despite this fact, it is the most generally useful of the block diagrams.

The lower set of blocks in figure 498 is drawn in one-point perspective. In this construction, lines forming the sides of the block and all others parallel to these converge into a single point called the "vanishing point." The vanishing point lies on the horizon, which, in turn, is level with the observer's viewpoint. Any pair of parallel lines not parallel to the construction lines of the block also converge to a point on the horizon but to the right or left of the vanishing point of the block. The blocks shown in the figure present one face without perspective distortion directly toward the observer. This is a departure from a true or natural perspective drawing, except in the one case where the vanishing point lies directly above or below the block. Such a view does not expose the sides of the block. All other positions of the block would, in natural perspective, require some convergence in the frontal face; but since this would unnecessarily complicate the drawing of geologic features, the front of the block is made a true rectangle, and the sides and top, quadrilaterals.

The perspective block, although somewhat more difficult and tedious to draw, is also more natural in appearance. Effects of towering heights and deep depressions and low or high vantage points are readily attained by mechanical drawing methods.

Figure 498 shows blocks in various positions relative to the observer. The uppermost blocks are above the horizon and, therefore, above the observer's position. The block in the upper left is placed so that the bottom is exposed to view. The stack of three blocks on the right is drawn so that the base of the stack is somewhat below the observer's eyes, and the top at a considerable height above. Note that the second block in the left-hand series lies on the horizon, and, consequently, the eyes are exactly in the plane of the upper surface. The two blocks below, which are successively lower, expose more of the upper surface.

The optical centers of the faces are shown in several instances at the intersections of the diagonals. It is quite apparent that the center of the block is always to the rear of the scaled midpoint. This is illustrated in the sectionized block on the left where the spacing between section lines is progressively less from front to back. This fact must be kept in mind when geologic features are transferred from maps to blocks in one-point perspective.

The geologic diagrams at the bottom of the figure illustrate the use of secondary vanishing points. The block on the left shows an anticline and two synclines, the axes of which are parallel and trend diagonally across the block. Since the block is drawn in perspective, these parallel lines must also be shown with the same degree of convergence; therefore, it is necessary to select a new vanishing point on the horizon (2 in the figure). When the folds are constructed according to the convergence of lines into this point, the perspective in the geologic features will be the same as that in the block.

The lower right diagram utilizes three vanishing points, as indicated by the construction lines. Point 3 is the focus for the lines bounding the fault plane, and 2 controls the lines cutting off the corner of the block.

Figure 499 shows a structural map and a one-point perspective block of the same area. Both have been shaded with a pencil to emphasize the structural relief. A contour map shaded in this manner is called a "shadow-graphic map." There are two methods for attaining the shadow effect. The simplest is by hand shading, as mentioned. In both the map and block it is assumed that the source of light is the upper-left-hand corner. All structural surfaces facing this corner receive the greatest amount of light, and those sloping toward the lower-right-hand corner bear the heaviest shading. High points have high lighting; low areas, dark shadows.

A more cumbersome method consists in first soaking a contoured map until it can be moulded into ridges and depressions conforming to features shown by the contours. This may be done by working the softened map over a mass of wet papier-mâché. When the modeling is completed, the shadow-graphic map is obtained by photographing from directly above with one source of light, preferably from the upper-left-hand corner. Obviously, this is a much more tedious method than shading the map with a pencil.

Facies Maps

In stratigraphy it is axiomatic that the lithologies of any formation change in some manner from one part of a basin to another. The degree of variation and rate of change may be small or large, depending upon the physical conditions of the basin and adjacent terrain, the chemistry of the waters, the climate, and many other factors that determine the type of sediment laid down. Therefore, a single formation may be a coarse conglomerate at one locality, a sandstone or shale at another, a limestone at a third, and all three lithologies at some places. Facies changes, although perhaps not so drastic as the example given, are the usual and normal condition; and it is often of the utmost importance to the stratigrapher to determine the characters of the variations and where they occur and then to have some means of showing the change on maps.

A large number of methods have been devised by geologists to illustrate facies changes on maps. Some of these have been mentioned earlier,

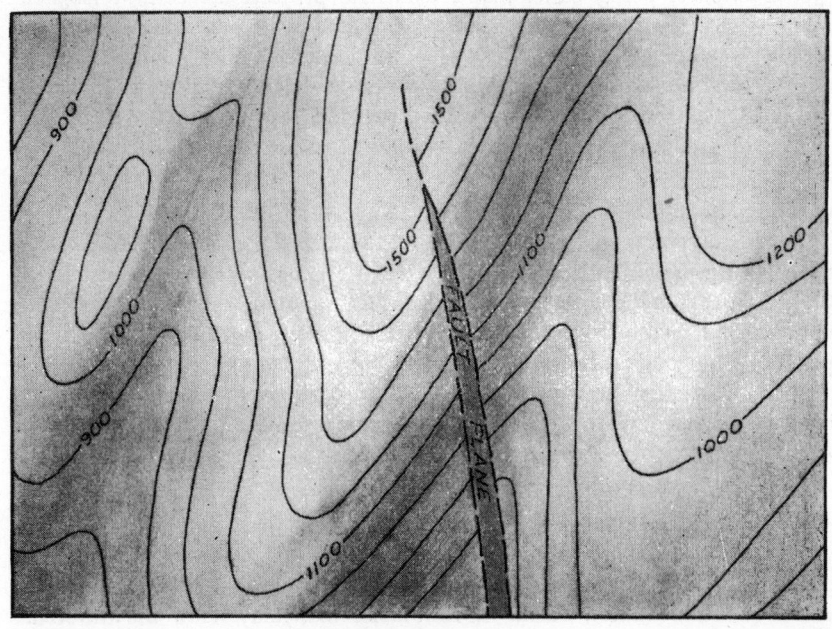

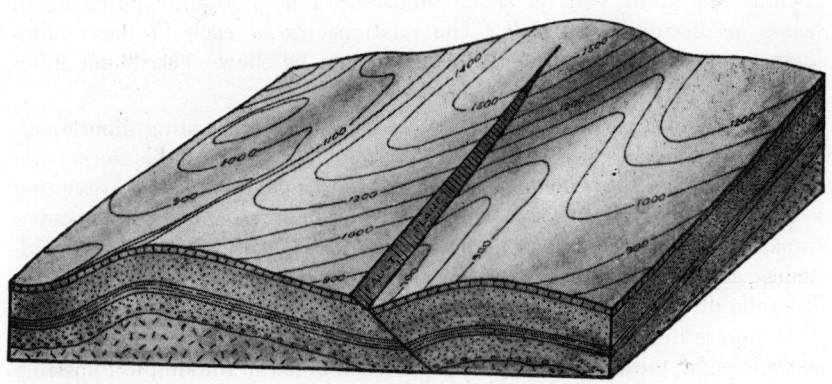

FIGURE 499. *A*—Shadow-graphic structure map. *B*—One-point
perspective block diagram.

the commonest being panel maps, isometric projections, certain isopach maps, and cross sections; but these methods fail in one way or another to give a complete and continuous picture where facies change rapidly and the section as a whole is highly diversified. Other methods invented for special conditions or for a specified problem lack general applicability.

Most attempts to show facies changes on maps have been concerned with qualitative data, rather than quantitative. However, comparatively recently, greater effort has been directed toward lithologic analysis and lithologic mapping on a quantitative basis, and the results of this work have been gratifying. Since the scope of this book is somewhat restricted to the more practical aspects of subsurface mapping, it is necessary to omit certain methods which occasionally or locally do have practical applications but which are not generally useful in lithologic mapping. The methods described below illustrate an approach to the problems which may, in turn, stimulate further endeavor along similar lines.

Lithofacies Maps

"Lithofacies map" is a comparatively new term in the geologist's vocabulary. It denotes a map that shows by one means or another the changes in lithologic facies of a formation, group, or system of rocks within a sedimentation basin. A lithofacies map may show the different facies in either a qualitative or quantitative way. Each is important in its own way. The definition of the term is broad enough to include a rather wide variety of maps dealing with facies changes.

It is difficult, or impossible, to show by conventional maps what and where facies changes take place in a highly diversified section, such as rapidly alternating beds of shale, sandstone, limestone, and anhydrite in lenses or discontinuous beds. The relationships of each of these lithologic groups to the others in the section can be shown clearly on lithofacies maps.

Figure 500 shows a well log consisting of alternating limestones, shales, sandstones, and evaporites. The total thickness of this succession is 478 feet. To the right of this log are four columns, each representing a lithologic class of rocks. The first column contains only the sandstones transferred from the well log; the second, only the limestones; the third, shales; and the fourth, evaporites—all plotted in their correct thicknesses. The total thicknesses are 163, 160, 110, and 45 feet, respectively; of course, their sum is the thickness in the original log. Now, in the column on the extreme right, these lithologic units are recombined in the simplest possible manner: i.e., all of the units of one class, regardless of the thickness of the individual members, are plotted as if they occurred as one thick bed. Thus, the thirteen members of the original log are reduced to four in the analytic log.

In figure 501, *A* is a stratigraphic cross section showing normal facies

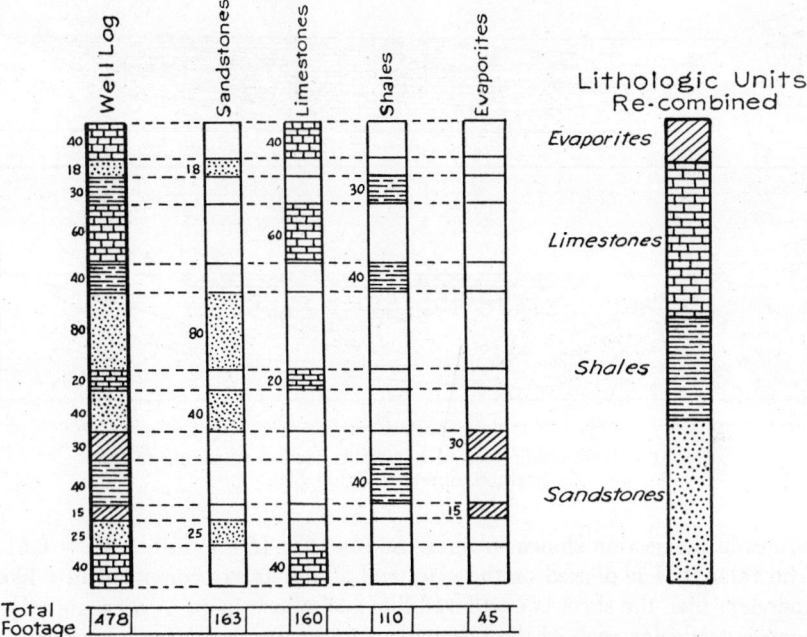

FIGURE 500. Lithologic breakdown of a well log into four main rock types.

changes from the edges toward the center of the basin. By the method described above, the complex nature of the stratigraphy is simplified (for a specific purpose) to the three-unit section shown in *B* of figure 501.

To digress momentarily, it is not necessary to replot a log in the simpler form. The thicknesses of all the individual beds of limestone, for example, are tabulated and then totaled for use on maps. A convenient form for this purpose is shown in figure 502. When the lithologic breakdown of the well logs or surface sections has been made, the results are recorded in the appropriate column as aggregate thicknesses, ratios, or percentages, depending upon the mapping units to be used. Such a table provides a permanent record of all computations of lithologic proportions.

The lithologic values obtained by the process just described may be used on maps in a great variety of ways, a few of which are discussed briefly below.

Ratio Maps

The ratio contour map shows the ratio of the aggregate thickness of one lithologic class to that of the remaining classes that go to make up the complete section. For example, a sandstone ratio map shows the ratio of sandstones to all other rock types. Thus the ratio value of sand-

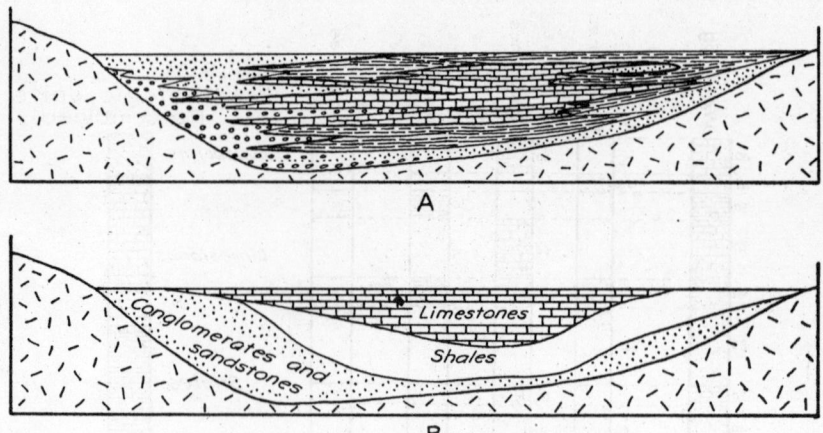

FIGURE 501. Simplification of a complex section according to the
method shown in figure 500.

stones in the section shown in figure 500 is 163: 160 + 110 + 45 = 0.51.
The value 0.51 is plotted on the map, and all others are computed in a like
manner; then the sheet is contoured like an isopach map. A separate ratio
map is made for each of the lithologic classes considered in the analysis.

Percentage Maps

The percentage contour map is very similar to the ratio map, except
that the number used for contouring is the percentage value of sandstones,
for example, in the total thickness of the formation. Thus, in the case of

the log in figure 500, this number is $\dfrac{163}{478} = 0.34$. Although the numerical

values of the contours will be different from those on the ratios map, the
appearance of the map should be the same.

STATE: *COLORADO*   FORMATION: *PERMIAN*

NAME & LOCATION	TOT. THICK.	CARBONATES		EVAPOR.		FINE CLASTICS		COARSE CLAST.		REMARKS
		LIME.	DOLO.	GYP.	SALT	SHALE	SILTST.	SAND.	CGLM.	
Ames-#1 Smith 6-21N-43W	3045	130	520	45	15	1560	365	510	—	*Sample Log by John Jones*
Std.Oil-#4 Black 5-7S-16W	1560	380	210	—	—	460	310	200	—	*Drillers Log*

FIGURE 502. Lithofacies-data sheet.

Isolith Maps

The ratio and percentage maps show lines of constant relationship of the thickness of one lithologic group to the total thickness of the remaining groups or to the total thickness of the formation, respectively. The isolith map differs from these in that actual thicknesses of each class are contoured. In a sandstone isolith map the total aggregate thickness of all the sandstones in the formation is the numerical value contoured on the map. For the case cited above this number is 163. As in the series of ratio or percentage maps, one isolith map should be made for each of the lithologic classes in order to provide a means for a thorough sedimentation analysis.

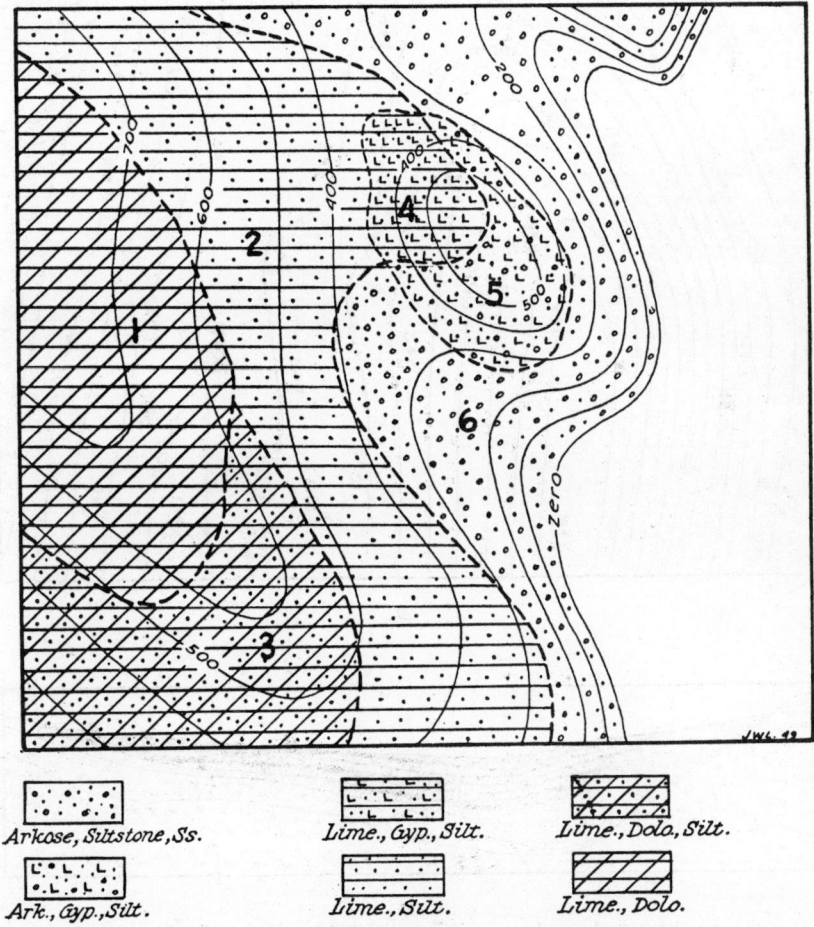

FIGURE 503. Combination isopach and lithofacies map.

Isofacies Maps

A single lithologic facies may include several rock types. For example, one facies of a formation may consist of alternating shales and sandstones, whereas another facies of the same formation may be made up of shales, limestones, and evaporites. Since the ratio of one type of sediment to another varies within a given facies, some difficulty may develop in attempting to map such a facies by the ratio or percentage method. The complex facies may comprise only a small portion of the formation, in which case the remaining beds would tend to mask the effects on the percentage maps. A type of map called an "isofacies map" is designed to show the change in lithologic characters within a facies. The method

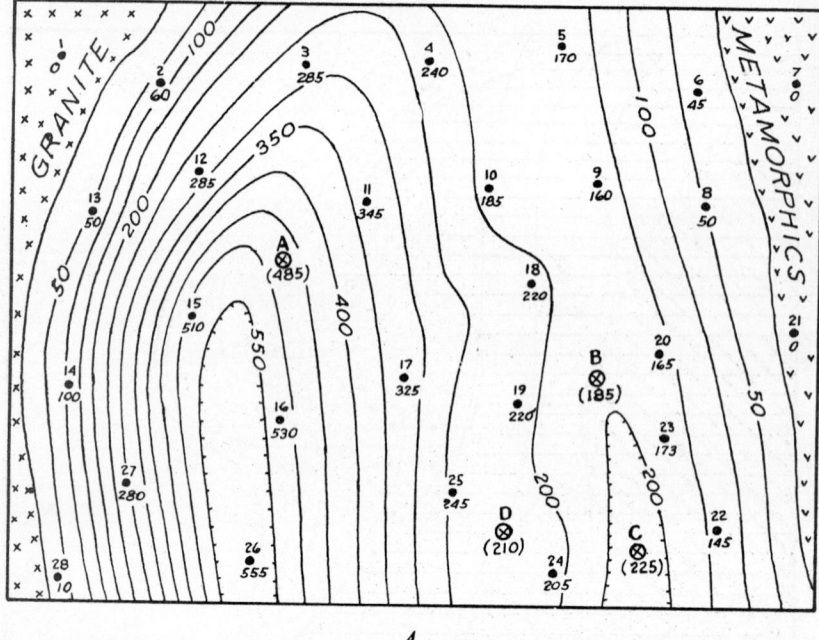

A

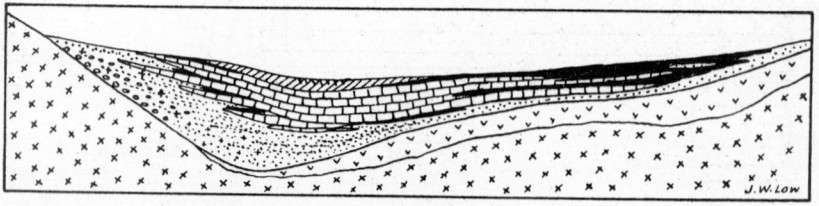

B

FIGURE 504. *A*—Isopach map of a sedimentation basin. *B*—Cross section showing complex relationships of formations.

is simple and is essentially a special adaptation of those described above. It consists in first delimiting the area and stratigraphic interval of the facies. Now, this portion of the formation can be treated as if it alone were a formation, and the required isopach or isolith maps constructed accordingly.

A form of generalized isofacies map is shown in figure 503. The contours show the total thickness of the formation, and the areas shown by different patterns within the heavy dashed lines are the various facies of the formation. These facies are as follows:

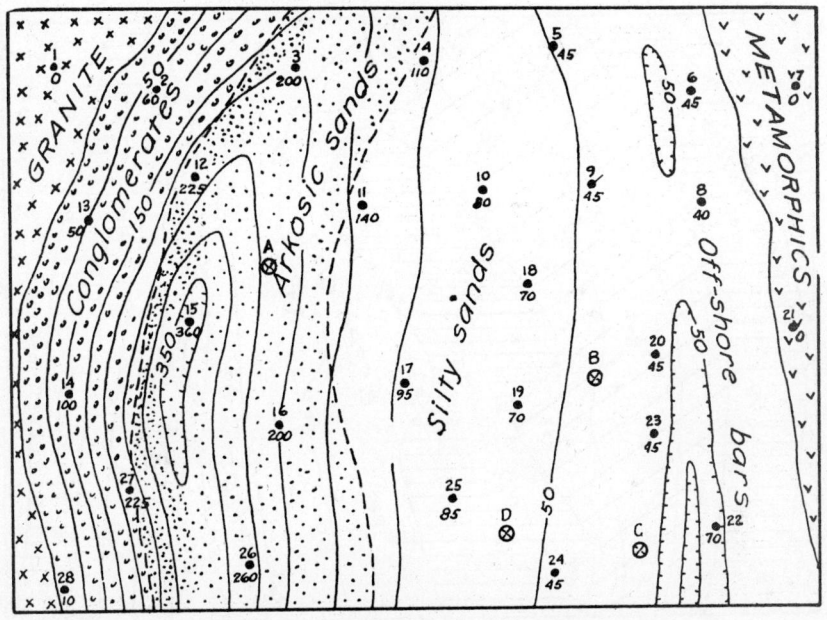

FIGURE 505. Sandstone isolith map of basin shown in figure 504.

Area 1—Limestones and dolomites
Area 2—Limestones, siltstones, and sandstones, interbedded.
Area 3—Limestones, dolomites, sandstones, and siltstones.
Area 4—Gypsum, limestones, and sandstones.
Area 5—Gypsum, arkosic sandstones, and siltstones.
Area 6—Arkoses, sandstones, and siltstones.

Figures 504 to 508, inclusive, are a series of isolith maps of a portion of a sedimentary basin. Figure 504 is an isopach map of the total thickness of the formation, and the cross section shows the stratigraphic relationships of conglomerates, sandstones, shales, limestones, and evaporites. This type of section is somewhat too complex to analyze lithologically on one lithofacies map by percentages or ratios. For this reason an isolith map is made for each of the principal lithologic classes: coarse clastics, fine clastics, precipitates, and evaporites. The control points used in contouring are wells whose logs have been broken down according to these lithologic classes and tabulated on the form shown in figure 502.

Figure 505 is the isolith map of coarse clastic rocks: i.e., sandstones, arkoses, and conglomerates. The thicknesses shown represent the aggregate thickness of all rocks falling in this classification, regardless of the thickness of the individual beds in which they occur. Thus, of two control points, each indicating a thickness of 100 feet, one might be made up of

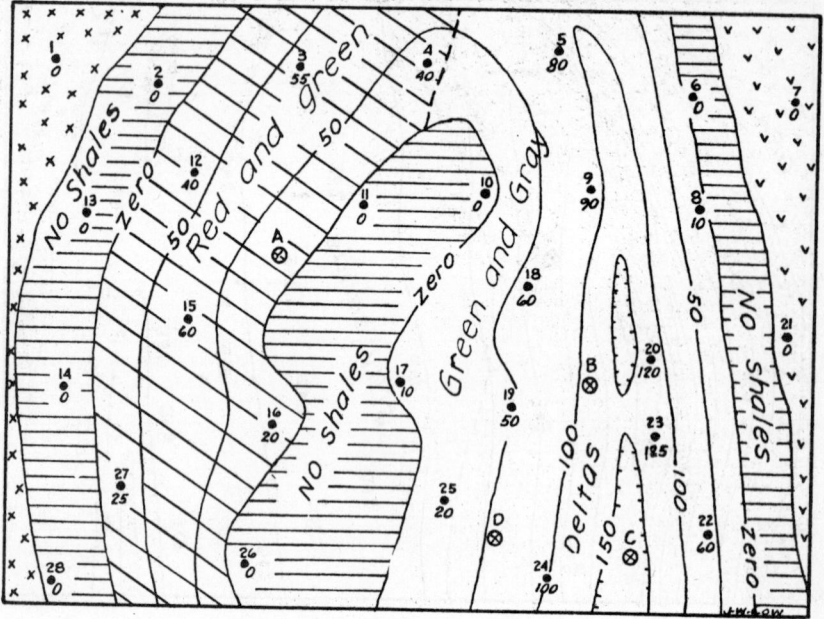

FIGURE 506. Shale isolith map.

two 50-foot members, the other of five 20-foot beds. They have the same values on the isolith map.

The sandstone isolith map shows not only the aggregate thicknesses of the coarse clastics, but also the areas where certain kinds of sandstones predominate. The shale isolith shows the aggregate thicknesses of the shales and a further differentiation on the basis of color. Likewise, the limestone and evaporite maps indicate areas of different types. There are many other ways of differentiating within each of the lithologic groups, and, depending on local conditions, it may be advantageous to set up other main classes. These examples are given to indicate how lithologic facies may be drawn on maps in both a qualitative and quantitative manner.

It has been necessary here to differentiate areas by contrasting pat-

terns in black. This can be done much more effectively by colors. The best
practice is to represent the variations within a given lithologic class by
colors restricted to a definite range of hue and tone; for example, the subdivisions of limestones on a limestone isolith map should be shown by
various shades of one or two basic colors chosen to represent the carbonate
group; on the sandstone isolith, by shades of the basic colors selected to
represent coarse clastic rocks. This procedure obviously must be varied
when the facies separation is made on the basis of rock color. The only
logical map presentation in this case is to show the rock color facies by

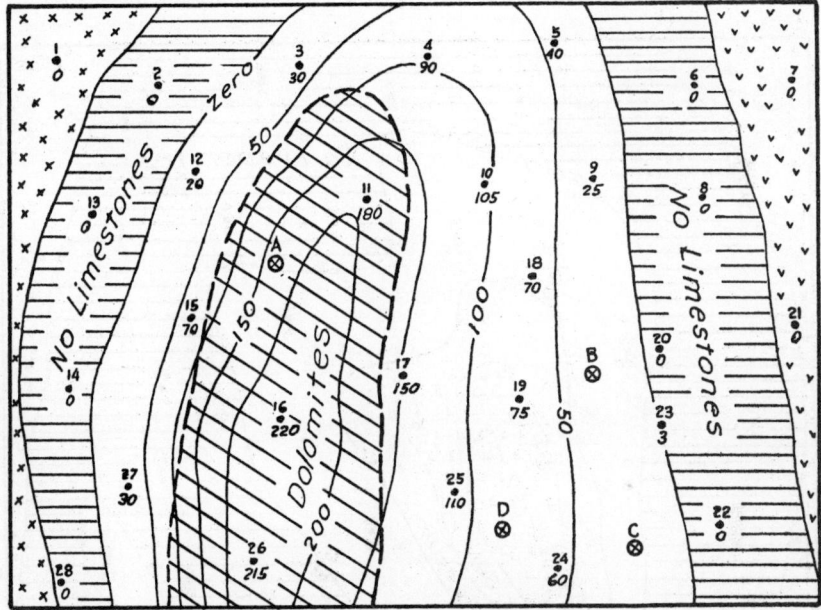

FIGURE 507. Limestone isolith map.

a similar color on the map; e.g., red rocks with red colors and gray with
gray colors.

The chart in figure 509 will serve as a guide for coloring lithofacies
maps. When colors are used to represent lithologies, it is possible to show
gradations from one facies into another by alternate bands of color; for
example, where a sandstone facies grades laterally into a limestone facies,
the area of gradation would be shown by alternate bands of blue and
yellow or brown. If the proportions of sandstone to limestone are two to
one, respectively, then the yellow bands should be twice as wide as the
blue. This plan can be extended to include several color bands.

Figure 510 is a lithofacies map drawn in black and white, with certain
patterns representing colors, which, in turn, represent the different rock
types. For very simple drawings like these the patterns are satisfactory;

but, where many lithologies are involved, the inevitable similarities of patterns in black and white make the map difficult to read.

Lithofacies maps of various kinds are almost indispensable in stratigraphic investigations in regions where the lithologic character of formations change radically from one locality to the other. Isolithic contour maps, used in conjunction with maps showing mineralogic compositions of the sediments, provide one of the best means of accurately locating with only a few points of control the source areas of clastic sediments, the areal extent of marine, continental, or fluviatile environments, and many other features vital to a thorough understanding of sedimentation processes.

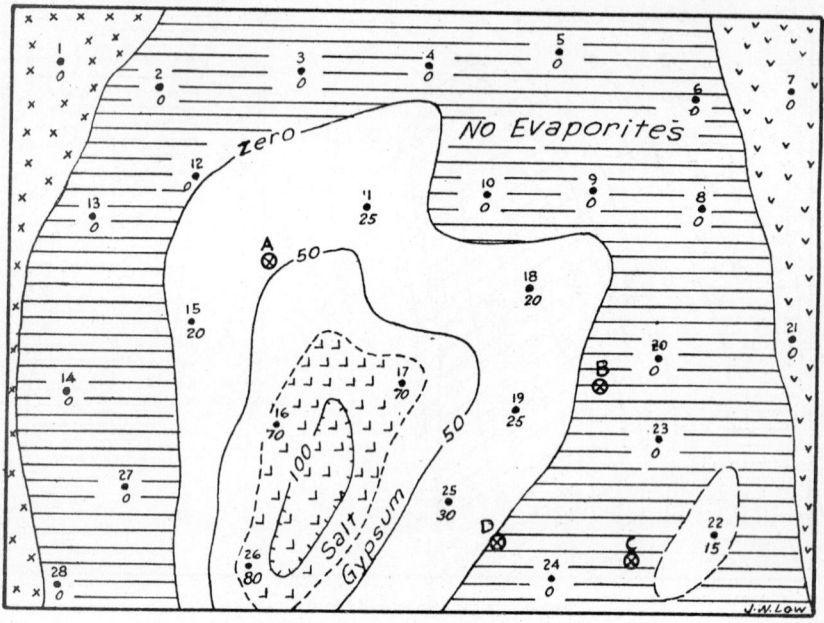

FIGURE 508. Evaporite isolith map.

In regions where contemporaneous strata are predominately carbonates at one locality, shales and siltstone at another, and sandstones and conglomerates at a third, an ordinary isopachous map can provide only a small part of the information needed by the stratigrapher. On the other hand, isolithic maps permit each of these rock groups to be appraised separately and without the distracting effect of having to deal simultaneously with the remaining rock groups of the complete section. Obviously, the shorter the time range and the thinner the stratigraphic interval studied, the more precise the results will be. But the different kinds of facies maps are also an excellent method for studying thick series of strata which, because of monotonous repetition of similar units, cannot be subdivided into thinner mappable units.

An isolithic map, as explained earlier, shows the aggregate thicknesses of beds in a specific lithologic class, but it does not necessarily differentiate the rocks within this class on a mineralogic, chemical, or physical basis. There are a number of ways in which this differentiation may be shown. One simple method is to show such a breakdown by percentage contours superimposed on the contoured isolithic map. Thus, on a sandstone isolith map the percentage of arkosic, silty, or carbonaceous sands or maximum grain sizes may be shown by color bands or a second set of contours. The thicknesses of rocks, alone, sometimes fail to reveal

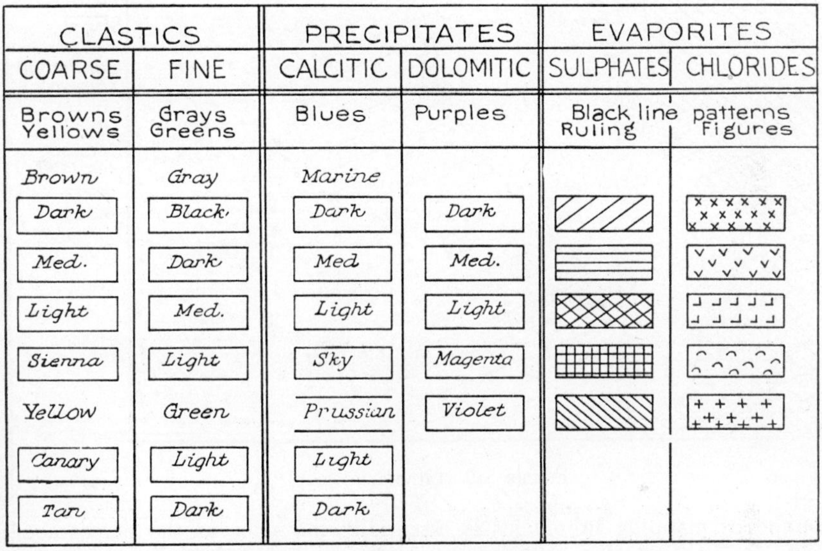

CLASTICS		PRECIPITATES		EVAPORITES	
COARSE	FINE	CALCITIC	DOLOMITIC	SULPHATES	CHLORIDES
Browns Yellows	Grays Greens	Blues	Purples	Black line Ruling	patterns Figures
Brown	Gray	Marine			
Dark	Black	Dark	Dark	/////	x x x x x x x x x x x x x x x
Med.	Dark	Med	Med.	≡≡≡	v v v v v v v v v v v v
Light	Med.	Light	Light	⊠⊠⊠	⌐ ⌐ ⌐ ⌐ ⌐ ⌐ ⌐ ⌐ ⌐ ⌐
Sienna	Light	Sky	Magenta	▦▦▦	⌒ ⌒ ⌒ ⌒ ⌒ ⌒ ⌒ ⌒ ⌒
Yellow	Green	Prussian	Violet	▨▨▨	+ + + + +₊+₊+₊+₊
Canary	Light	Light			
Tan	Dark	Dark			

FIGURE 509. Guide for coloring lithofacies maps.

the sources of the sediments or the conditions under which they were deposited; but this information, together with that suggested above, often clarifies an otherwise cloudy picture.

A word of caution should be given in the interpretation of isolithic maps. Apparent discrepancies in the locations of apparent highs and lows within large basins are bound to occur if the predominance of one lithologic class in one locality is replaced by that of another class at other localities. For example, on a coarse-clastics isolithic map, the thickest deposits are likely to lie close to the source materials near one edge of the basin and will thin out toward the central part of the basin. Conversely, the limestones and evaporites are likely to reach their maximum development farther toward the center. Therefore, coincidence of the thin and thick areas on the two maps should not be expected. This is one of the reasons that the isolithic maps are of such importance in the study of stratigraphy. From the foregoing it is clear that a set of isolithic maps with their opposing thin and thick sections will tend to cancel one another out in an over-all isopach map of the same stratigraphic interval. In other words, the isopach map may be generally featureless, whereas the series of isolithic maps reveals much of the stratigraphic information sought.

W. C. Krumbein, of Northwestern University, describes the ratio method of lithofacies mapping in the American Association of Petroleum Geologists Bulletin 10, Volume 32 (1948). As mentioned earlier, the ratio

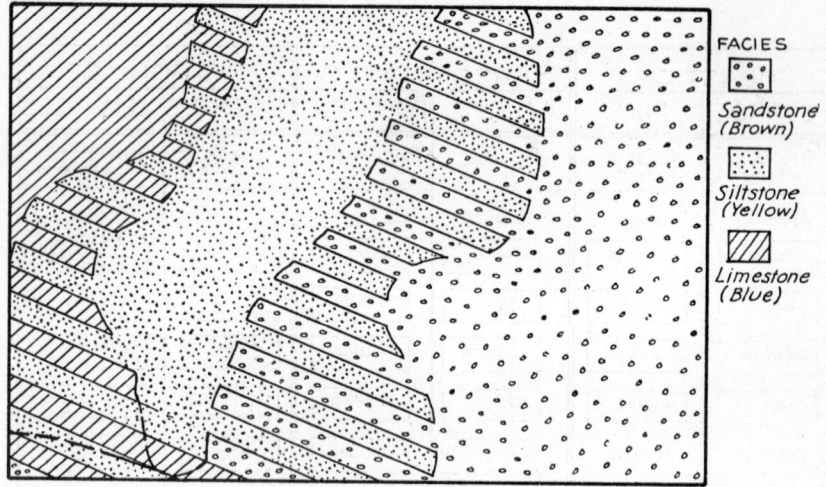

FIGURE 510. Lithofacies map.

method of mapping lithofacies is essentially the same as the percentage method. However, the relative spacing of contours may vary greatly.

Figure 511 is an isopach of a group of rocks consisting of sandstones, shales, limestones, and anhydrites. This isopach is used as a guide for drawing the lithofacies map of figure 512. Table 32 is a summary of the control points, showing the total thicknesses, thicknesses of clastic and non-clastic rocks, the clastic percentages and clastic ratios. The values in the last two columns are obtained from the following formulas:

$$\text{Clastic ratio} = \frac{\text{sandstones} + \text{shales}}{\text{limestones} + \text{anhydrites}}$$

$$\text{Clastic percentage} = \frac{\text{sandstones} + \text{shales}}{\text{sandstones} + \text{shales} + \text{limestones} + \text{anhydrites}}$$

In figure 512 the clastic percentages are shown by solid contours. The clastic ratios are shown by dashed lines, and ratio intervals are indicated by shading. The 50-percent contour is also the ratio 1.0 contour.

In general the two methods are very similar, the principal difference being that the percentage map is contoured on a regular contour interval

TABLE 32

Well No.	Total	Thickness clastics	Non-clastic	Clastic & non-clastic %	ratio
1	140	130	10	91	13.0
2	190	160	30	84	5.3
3	220	160	60	76	2.4
4	270	170	100	63	1.7
5	270	160	110	59	1.4
6	340	150	190	44	0.8
7	360	180	180	50	0.8
8	370	180	190	49	0.9
9	380	170	210	45	0.8
10	450	160	390	35	0.4
11	520	180	340	37	0.5
12	530	190	340	36	0.6
13	530	200	330	38	0.6
14	210	160	50	76	3.2

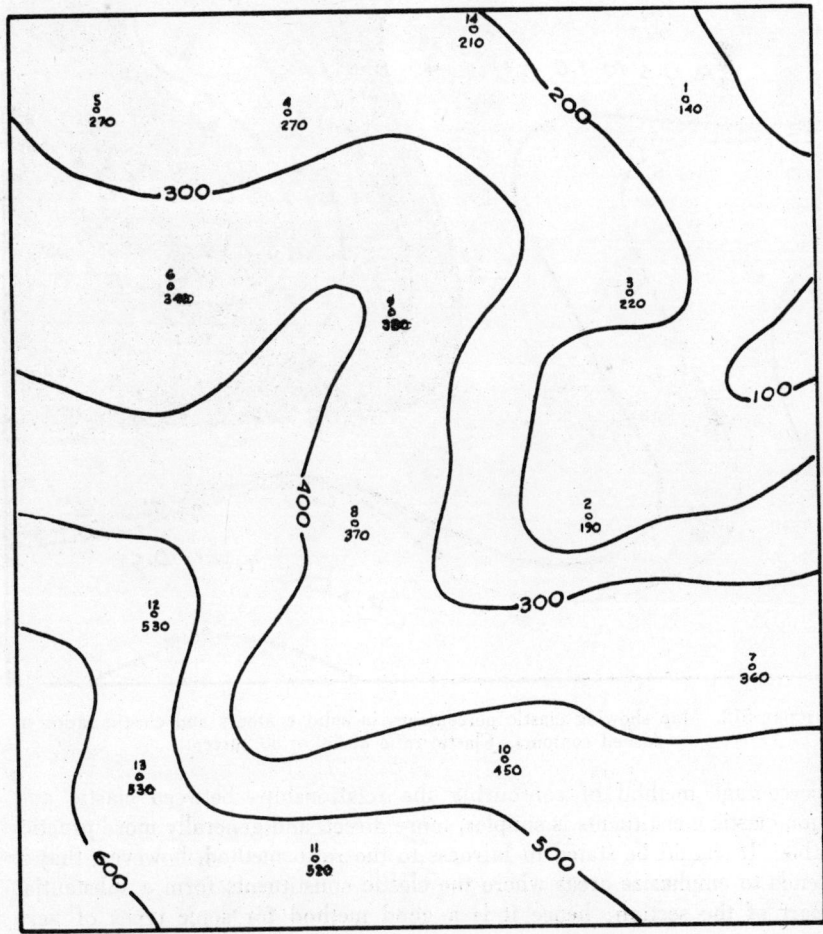

FIGURE 511. Isopach map used as a base for clastic-ratio map.

of 10 per cent, whereas the intervals of clastic ratios are 0.5, 1.0, 5.0, and
10.0. If the ratio-contour interval were constant arithmetically, the contour spacing would decrease at an exceedingly high rate toward the higher
ratios. Therefore, as Krumbein points out, the contour interval should be
determined by a logarithmic or geometric means. For this reason the

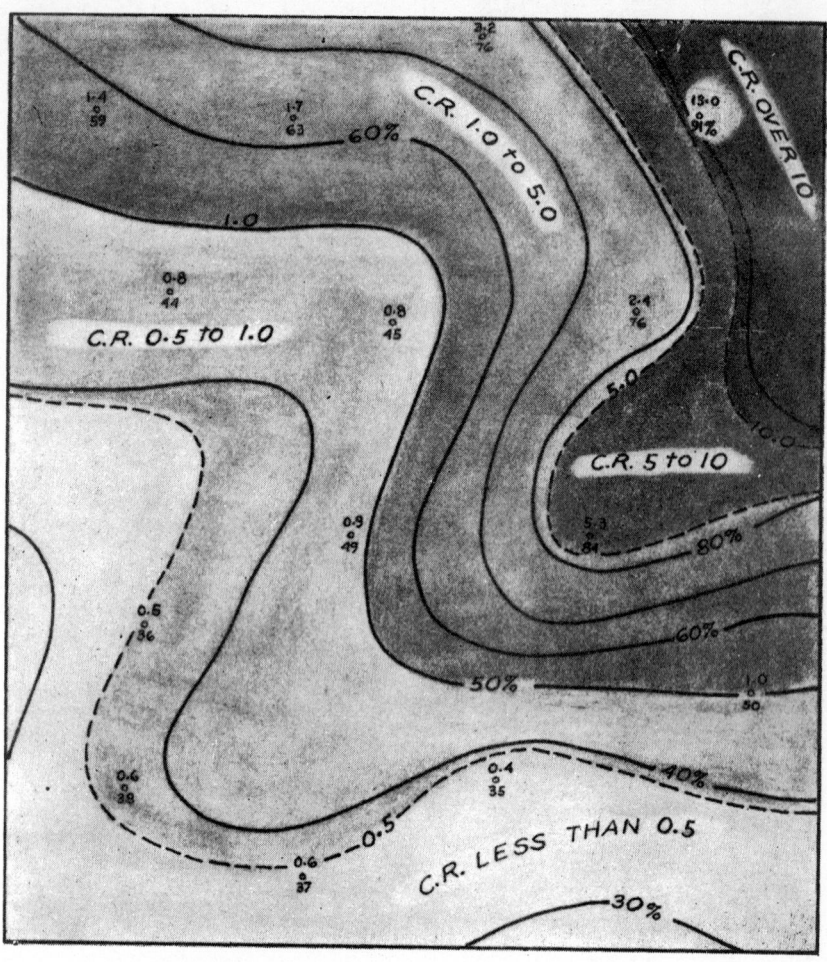

Figure 512. Map showing clastic percentages in solid contours and clastic ratios in
dashed contours. Clastic ratio of 1.0 = 50 percent.

percentage method of contouring the relationships between clastic and
non-clastic constituents is simpler, more direct, and generally more practicable. It should be stated in fairness to the ratio method, however, that it
tends to emphasize areas where the clastic constituents form a substantial
part of the section; hence it is a good method for some types of very

generalized broad regional work. Conversely, where control points are numerous, and the data reliable, the percentage maps are better.

Practical Uses of Lithofacies Maps

A number of the preceding pages have been devoted to lithofacies maps, how they may be constructed, and what they represent. In most instances the lithofacies map, regardless of the type, is only one phase of

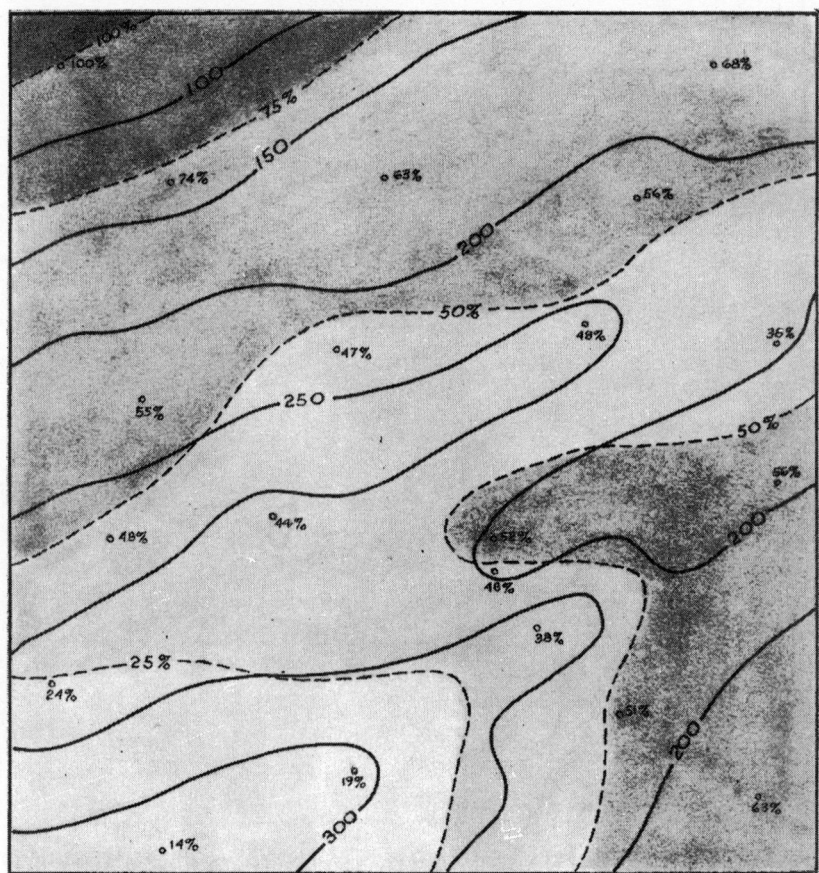

FIGURE 513. Sandstone isolith contours with shading according to percentage of sands over 25 feet thick.

regional geologic investigation and analysis. These maps do not depict a complete geologic story, but, rather, present complex stratigraphic data in a simplified form so that broad concepts can be developed. All the various lithofacies maps discussed reveal only the *averages* of geologic phenomena. None is precisely specific where the stratigraphic section is complex. Maps contoured on the bases of percentages, ratios, or aggregate thicknesses of lithic classes do not take into account the manner in which the individual units occur. As stated before, ten 5-foot units have the same contouring value as one 50-foot unit. There are means of further classifying a lithofacies map so that it yields the practical information desired.

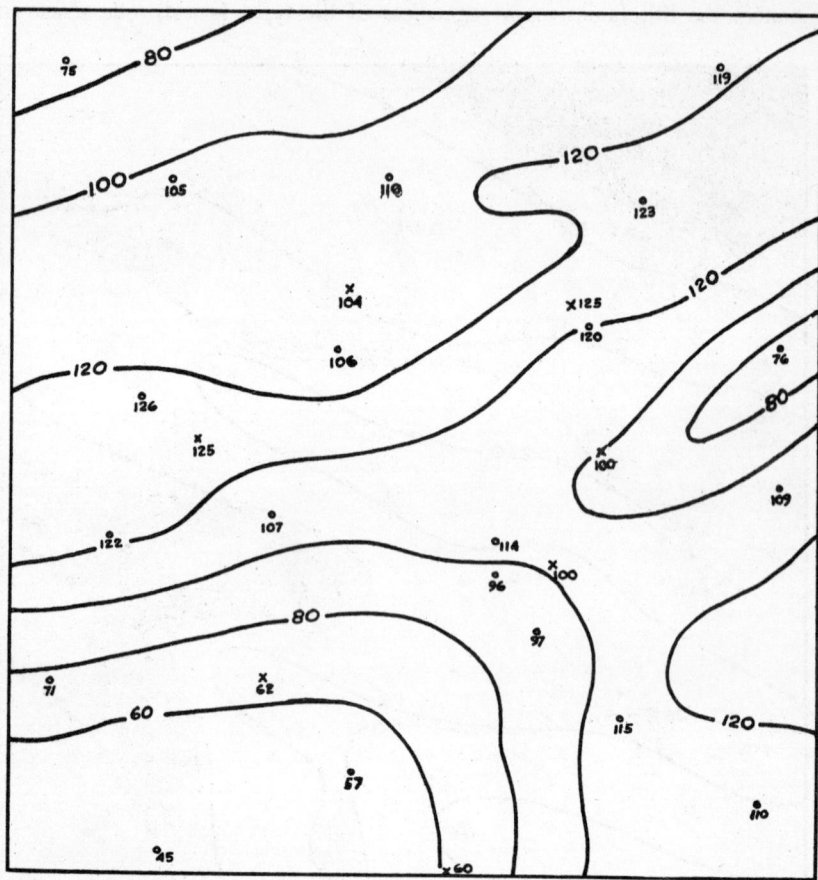

FIGURE 514. Map showing aggregate thicknesses of reservoir sands over 25 feet thick.

The following example will illustrate a direct practical application of a lithofacies map.

The problem is to determine the amount of effective reservoir sand and its relative worth over a broad area. It is assumed here that individual sands less than 25 feet in thickness, even though several such units occur in the section, are not worth exploitation. Since these thin units might coalesce to form more attractive reservoirs, locally, they must be given some consideration.

The first step is to construct a sandstone isolith map. In tabulating thicknesses from the logs, extremely thin beds are omitted. Sandstones containing a great deal of clay or silt may also be ignored, particularly if it is known that such characters are persistent over the region. Figure 513 shows a sandstone isolith prepared in this way. This isolith map includes sandstones as thin as five feet; but since the objective is to classify sands on the basis of a 25-foot thickness, the next step is as follows:

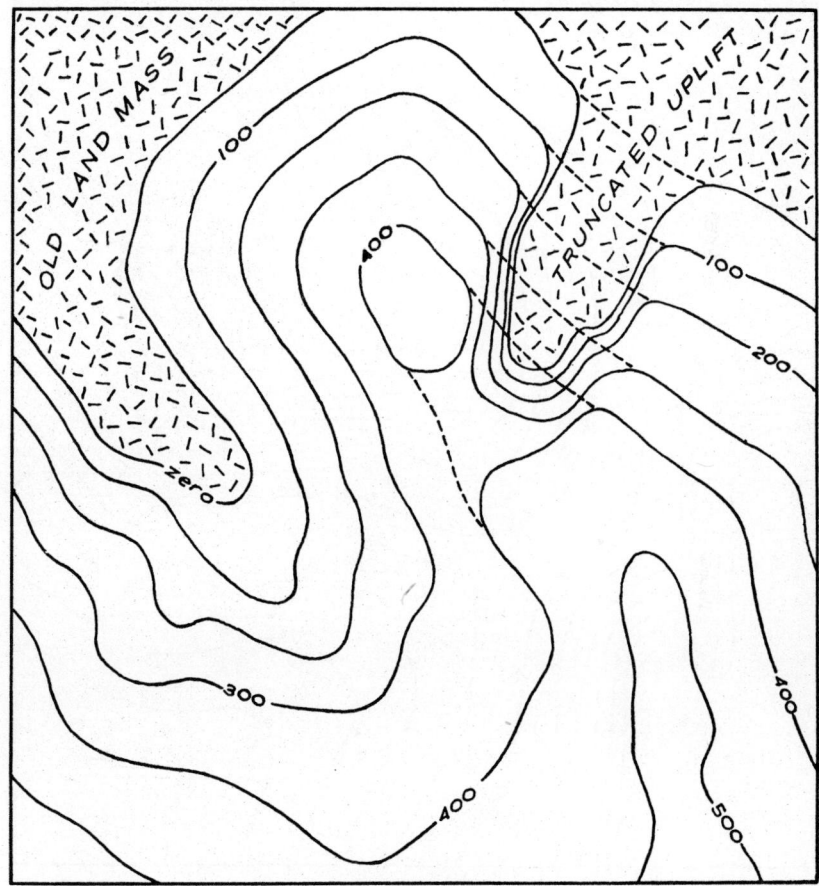

Figure 515. Isopach map showing reconstruction of thickness contours over a young, truncated uplift.

First, compute the percentage by total thickness of all sands that occur in beds 25 feet or greater in thickness. Thus, if the total thickness of sandstones in a well is 150 feet, occurring in beds of thicknesses 30, 25, 40, 35, 10, 5, and 5 feet, the effective reservoir sands are the first four, with a combined thickness of 130 feet. This is 87 percent of the total aggregate thickness of sandstones. These percentage values are posted on the isolith map, and percentage contours are sketched as shown by dashed lines in the figure. In the case illustrated the interval used is 25 percent. The map is somewhat easier to read if color or shading is used between the percentage contours.

A sheet of tracing paper is now placed on the map and fastened with

FIGURE 516. Isopach map used in construction of spalinspastic map of figure 518.

drafting tape. The total thickness of effective sand is computed on the basis of total aggregate thickness and percentage of effective sand, and these values are plotted on the overlay. The overlay is then contoured as in figure 514. This map shows by contours the thickness of effective reservoir sands that might be expected over the entire region.

Approximately the same result can be attained if only those sands adjudged to be of adequate thickness are considered in tabulating values from the logs. In this case the isolith map itself is the effective reservoir map. It should be pointed out, though, that this artificial classification might yield values which would be difficult to contour or which would give erroneous trends. By following the procedure set forth all significant sands are dealt with in the isolith map so that more reliable thickness trends are established. The overlay map is then contoured according to the general grain of the sandstone-isolith map. It is assumed that the thick

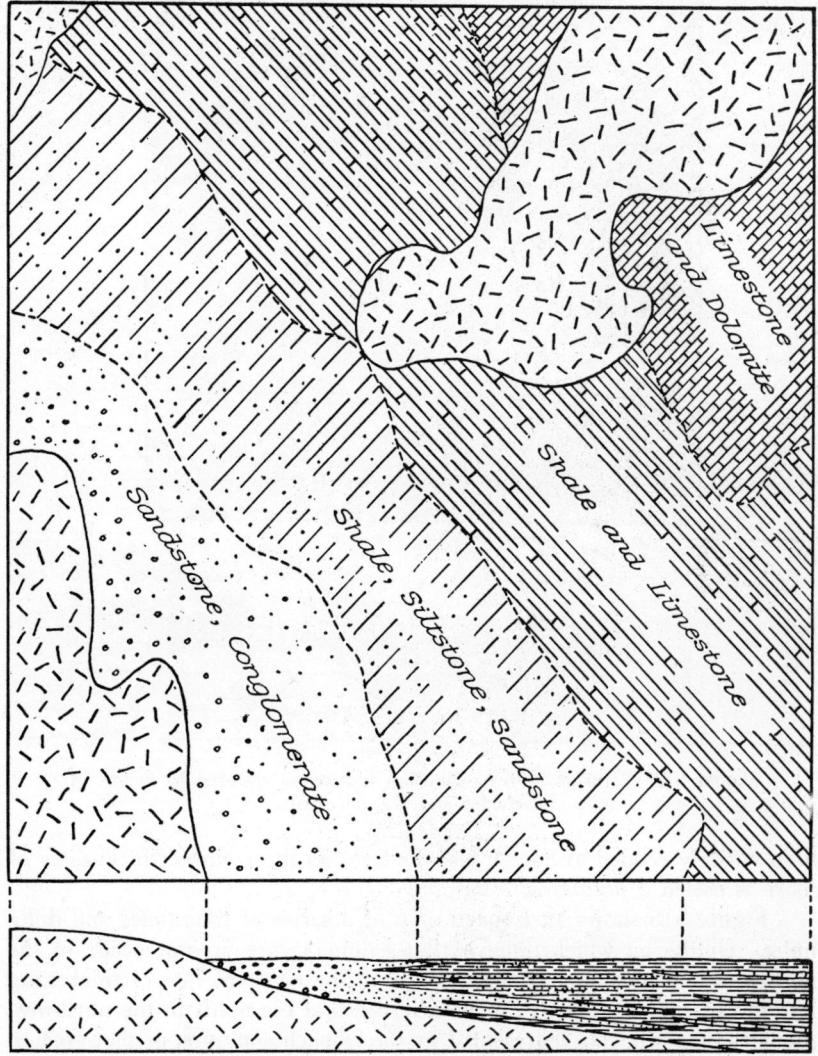

FIGURE 517. Generalized lithofacies map showing granitic uplifts of different ages.

and thin trends of the selected reservoir sands will correspond in a general way to the pattern of distribution of the entire sand content of the section.

In order to decipher the geologic history of a region in which major uplifts and severe truncation have occurred, it is often necessary to reconstruct by isopach contours the original thicknesses of strata that have

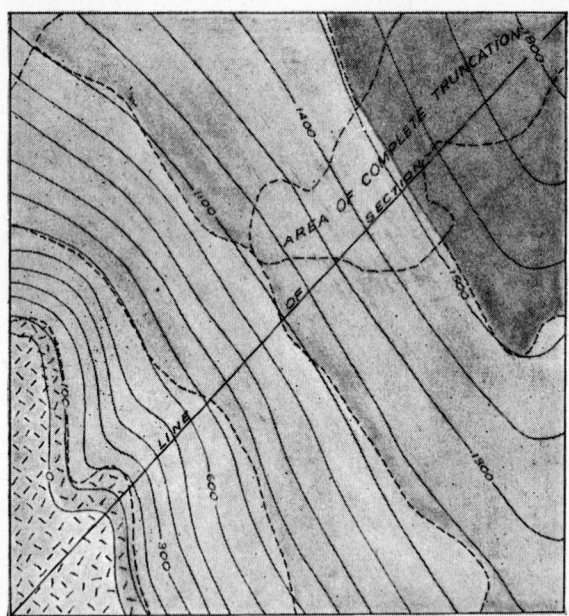

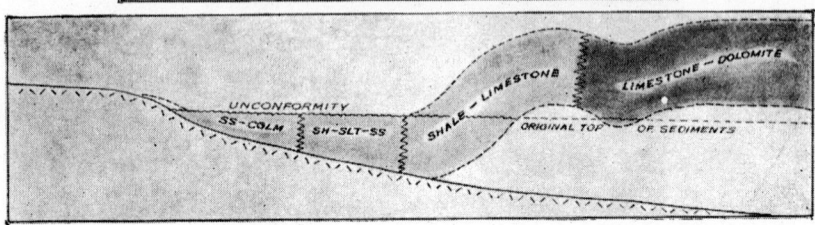

FIGURE 518. Palinspastic map showing thicknesses restored by means of lithofacies and isopach maps.

been entirely eroded away. A map that shows the restored thicknesses of rocks is called a *palinspastic* map.

Figure 515 shows an isopach map of a series of limestones and dolomites. Uplifts on which none of the sediments are present occur in the northwest and northeast quadrants. From the constant rate of thickening into the basinal regions, it can be inferred that the uplift in the northwest predates the sediments represented by the isopach contours, or that erosion has been uniform over the entire region. The axis of this uplift extends far to the southeast from the granite area.

The contours tell a different story about the uplift in the northeast quadrant. The extremely high rate of thinning along the flanks and the abrupt termination of the axis show that the sediments were truncated and probably extended across the nose at an earlier date; therefore, this uplift is younger than the sediments. The restored thickness contours are shown as dashed lines, which represent the thicknesses of the limestone series before truncation.

The relative ages of the two uplifts shown in figure 516 cannot be determined from the behavior of isopach contours representing the flanking sedimentary rocks. Therefore, a lithofacies map is compiled, as shown in figure 517. It can be seen that the facies trends conform to the general outline of the granite area in the southwest corner, but are unaffected by the one in the northeast. From the lithofacies it can be assumed that the uplift in the southwest predates the sedimentary series and contributed clastic materials. The northeastern uplift obviously is younger than the sediments.

Figure 518 shows a restoration of thicknesses based on the isopach and lithofacies maps cited above. In constructing this map, it is assumed that the changes in thickness will correspond in a general manner to the change in facies. The cross section at the bottom shows the restored formations after uplift but prior to complete truncation.

Coloring of Geologic Maps

Many kinds of maps require coloring, and since the maps that are made by commercial organizations are not reproduced in color by printing processes, they must be colored by hand methods. When several copies of a map are needed, the coloring of the prints may develop into a burdensome task. For this reason, it is desirable to be familiar with a number of different methods in order to select the one which will best satisfy the needs of the project.

Maps are usually colored for practical reasons, not merely for embellishment. The main purpose in using colors is to increase the legibility so that significant features can be seen at a glance. Because of this fact, color contrasts are more desirable than color harmony with low contrasts of tone or hue. On the other hand, "clashing" colors should be avoided if pleasing effects are desired.

The U. S. Geological Survey and various state surveys use only certain colors for rocks of specific geologic ages: one color range for Paleozoic rocks, another for Mesozoic, and so forth. Subdivisions within a main group are designated by patterns, such as ruling, stippling, and others. Since it is not practicable in most cases to use patterns on hand-colored maps, the color standards set up by the Geological Survey cannot be maintained. However, some consistency can be practiced, and if the colors are well chosen, the result in the finished maps will justify the discrimination exercised in the selections.

For paleogeologic maps the coloring is more effective if the darkest tones or most striking hues are used for the oldest rocks, the palest pastel tints being reserved for the youngest systems or formations, as the case might be. As an example, the following associations would be effective:

Cretaceous ....shades of yellow
Jurassic ....shades of brown, generally medium to light
Triassic ....pink, orange
Permian ....light red
Pennsylvanian ....light and dark grays
Mississippian ....shades of blue
Devonian ....light green
Silurian ....lavender
Ordovician ....reddish-purple to bluish-purple
Cambrian ....dark greens of different hues
Pre-Cambrian ....dark reds, some pattern

A similar arrangement can be determined when the colors are to represent a number of units within one system. However, an effort should be made to avoid representing a geologic system on one map by a certain color, and on another map by a different color.

In the different types of lithofacies maps, the main rock types should always be shown by certain colors. It is advisable to follow as closely as possible the color system used to represent lithologies on colored well logs, which, in turn, varies among oil companies. This practice greatly simplifies the interpretation of the lithofacies map by those already familiar with the color adaptation in lithologic logging. Some variations are necessary in mapping, particularly where colors of the rocks, as in shales, are important features of the facies; but in general, the associations given in figure 509 are adequate.

When colors are to be used on the map, some consideration must be given to the kind of prints that are to be made from the line tracing. Thin-paper translucent prints, such as "sepias," are not satisfactory. The best prints for coloring are blue-line or black-line Ozalids on medium- to heavy-weight rag-stock paper. Van Dyke positive prints on heavy paper can be hand colored with good results, but they are somewhat more difficult to work with than the Ozalids. Photostats are still more difficult because of the natural gloss and hard, impervious surface. Linens are not suitable for coloring.

Several methods of hand coloring are in use, each having peculiarities that are advantageous under certain circumstances.

Crayon Pencils

These wax-base colored pencils are applied as evenly as possible over the surface of the sheet, with particular attention being given to boundary margins. When all areas of one color have been covered, a fine, even tone of clear color is achieved as follows: It is necessary to have at hand a few paper charcoal stumps (blenders) of sizes 4 to 8, which can be purchased at any artists' supply store. Dip the stump in white gasoline, benzine, or dry cleaning fluid; and, after the excess on the surface has soaked in, rub the stump over the penciled area with a light, fast circular motion. The solvent dissolves the wax and carries the pigment into the absorbent paper. This process produces a water-resistant color on the map. When the solvent in the map paper has evaporated, a second application of color and solvent may be applied if darker tones are desired. A dark tone can be graded imperceptibly into a lighter one in this manner; or two colors may be so graded, one into the other. After the wax-pencil coloring has been treated with solvent, it has no tendency to rub off.

Indelible Pencils

The indelible colored pencil, sometimes called a "water color pencil," is soluble in water but is relatively unaffected by the aromatic solvents used with the wax-base pencils. The method of application is identical to that described above, except that the blender is used dry. The indelible pencil spreads easily and rapidly with either a dry stump or a wad of facial cleansing tissue or blotting paper.

For maps that are to serve only a temporary use, the indelible pencil is better than the crayon pencil. The indelible pencil is especially useful in preliminary work which may have to be revised, for the color can be removed with a soft rubber eraser even after blending with a stump. On the other hand, it rubs off on clothing or other maps, and changes to brilliant hues on contact with even small amounts of water. Perspiration from the hands will cause unsightly blotches on the map; therefore, when there is danger of this blotching, colored portions should be covered while work is in progress.

Water Color

Transparent water colors may be applied as a wash on Ozalid prints, but they are difficult to use and may cause appreciable shrinkage and distortion of the map scale. The air brush is an efficient and generally satisfactory method, although it may cause some shrinkage. When using the air brush one should mask all the map except the portion that is to be colored. Heavy wrapping paper is used for this purpose. For lithofacies maps, slotted stencils can be cut from stiff paper or cardboard. The stencil is laid over the exposed part of the map and the air brush, set on a wide spray, colors the map in bands. When the first color is dry, the stencil is offset one space and the second alternating color is applied. If three color bands are required, two stencils are needed. They are placed one on top of the other and are shifted until the desired exposure of the map surface is attained.

Printers Inks

Colored printers inks provide a means of coloring large areas with a uniform tone with no evidences of overlap. Since the inks have a grease base, there is no scale change in the map. The method is easy and fast, and the results are nearly as flawless as printing.

The viscous printers ink is thinned with about three parts of mineral spirits to one part of ink. Thorough mixing is essential.

Apply the color with a sable artist's round brush of size 6 to 12, depending on the size of the area to be colored. Cover all the surface, but there is no necessity for spreading the ink evenly. When an area of five or six inches square has been covered, lay a double-thickness of facial cleansing tissue flat on the moist color and pat it down so that the excess ink is absorbed. Make a pad or wad of the tissue and wipe and rub all "free" color from the map. When extending the color, overlap the edge previously colored. There will be no visible overlap when this is wiped with the cleansing tissue.

Better results can be expected if the border areas are colored first with a medium-sized brush not too heavily charged with ink. A larger brush may be used in the central parts. Apply the darkest colors first, and gradually work up to the lightest. If this procedure is followed, no overlaps will show, even where two different colors are involved.

There is no satisfactory means of removing the printers ink from the paper. Therefore, considerable care must be taken that the color is applied correctly. Maps colored by this method may be soaked in water for cloth mounting without danger of disturbing the colors. Colored areas will "take" colored pencils within a few minutes, but should be given several hours or days to dry before ink lines are attempted.

The printers inks can be mixed to obtain an infinite variety of shades. Pastel tints are obtained by mixing the colors with the white "transparent base." Colors must be mixed before they are applied to the map because, as suggested earlier, the paper becomes charged with the first color applied and is, thereafter, resistant to further applications. Within reasonable limits, the viscosity of the color has no effect on the hue or tone on the map: i.e., the same color is obtained with either a thin or thick mixture.

Pencil Shading of Isopach Maps

Shadowgraphic maps and the methods of drawing them have been described. Similar use of a soft pencil can add much to the over-all legibility of isopach maps, as shown in figure 519. Thick areas are darkest, thin ones lightest. About four different tones are most effective. The texture of the map paper determines to a considerable extent the hardnesses of pencils that should be used. On medium-weight Ozalid black-line paper, hardnesses of 2B to 2H are satisfactory. The toning is accomplished by rubbing the graphite with a charcoal stump blender, as described for indelible colored pencils. The change in tone should be made along contours at equal thickness intervals. In the figure cited, this interval is 500 feet, although the map is contoured on a 100-foot interval. The difference in this type of shading and that employed in shadowgraphic maps is that the latter suggests an illuminated model, whereas the former is shaded strictly on the basis of contour values.

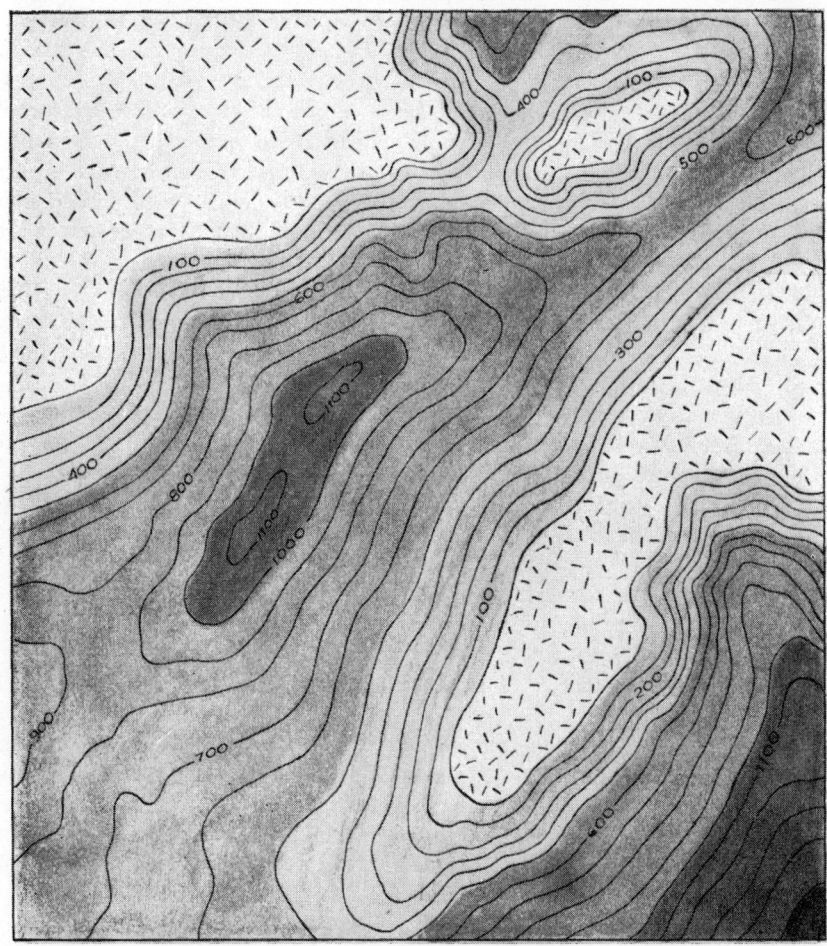

FIGURE 519. Isopach map shaded with pencil to emphasize areas of thick section.

REPRODUCTION OF MAPS

Most maps and other geologic drawings must be reproduced by one means or another. Better prints can be expected if the draftsman is aware of the peculiarities and limitations of the more common processes of reproduction. An exceptionally well drawn map may lose much of its effectiveness in reproduction if certain principles in drafting are neglected.

The so-called photo-copy processes depend upon the transmission of light through the medium upon which the map is drawn. The lines and figures drawn on the map prevent the light from passing through and exposing those portions of the sensitized printing paper. It is, therefore, desirable for the lines to be opaque and the medium transparent or highly translucent, and the quality of prints will depend largely on the degree to which these conditions are met. Obviously the best reproductions will be obtained from drawings made in black drawing ink on transparent material, such as cellulose acetate. Lines drawn with colored, waterproof inks may look well on the original map, but are liable to be indistinct or discontinuous on the print because these inks are not opaque to the intense lights of the printing machines. If colors must be used on originals to be reproduced by photo processes, the water-soluble types are more satisfactory. When extremely thin lines are necessary on acetate or similar, thick materials, it is better if they are inked on the back side where they will be in direct contact with the printing paper. Thin lines on the upper surface are likely to "burn out" in printing because of diffraction along the edges of the lines.

Since the necessary exposure time is longer with more opaque media, such as thick vellum, or tracing paper, the opportunity for light leakage along thin lines is greater.

Photocopies can be made from penciled originals; but it is essential to maintain considerable contrast between the opacity of the tracing material and the lines. For this reason thin tracing paper with a "toothy" texture combined with pencils which make very black lines give the best results. It is somewhat difficult to make reproducible maps with penciled lines on acetate and tracing linen.

Good photostats can be made from copy that is too weak to reproduce by light transmission processes. Better reproductions are obtained from originals having high line and background contrasts. Generally speaking, prints by Ozalid and Van Dyke methods are better than photostats for hand coloring by methods described earlier. Photostat prints are more receptive to colored pencils if they are rubbed with drafting pounce or some other very fine abrasive. Of the processes discussed above, only the photostat permits a reduction or enlargement of the original scale.

Maps and drawings for publication should be drawn to a slightly larger scale than that desired in the reproduction. This is done so that the inevitable irregularities occurring in lines, figures, or lettering will be reduced or eliminated in the reduction of scale. Over-all proportions, however, are not improved in the reproduction; in some cases, faulty proportions of map features are even more evident in the reduced reproduction.

Plain black and white drawings, that is, black lines on a white or blue background, are the most economical to reproduce. These drawings can be reproduced by several different processes, the principal ones being offset, engraving, lithograph, or metal-lithograph. The line drawings in this chapter are reproduced by the zinc plate, or offset process.

Shaded drawings, such as figures 499 and 519 are half-tone reproductions. All reproductions of photographs are made by one type or another of the half-tone process. The half-tone methods are appreciably more expensive than black and white line work. Drawings shaded by stippling, ruling, or hachuring, as in the central blocks of figure 498, can be reproduced by black and white methods.

Reproductions in color are expensive, and should be avoided when black and white or half-tone methods can be substituted. A separate plate must be made for each primary color used. In the printing, the paper must be run through the press for each of the color plates.

DRAFTING OF MAPS

The effective drafting of geologic maps of all kinds is too important to be neglected entirely, even though the subject cannot be fully discussed here. There are two main reasons for drawing subsurface maps, and which is the more important depends largely upon circumstances.

The first reason is obvious to those engaged in the technical phases of subsurface investigations: the various types of subsurface maps are in a sense geological tools. The technician must appraise the various data which he has processed by means of contours so that trends, gradients, anomalies, and other phenomena are developed. Subsurface geology is three-dimensional, and *must* be developed by a method which takes the three dimensions into account. Only contoured maps can do this in a continuous manner. Contour maps are quantitative, and any geologic condition that can be reduced to numbers can also be contoured. Examples are thicknesses, elevations, grain sizes, porosities, permeabilities, temperatures, percentages and ratios of any two or more rock constituents or extraneous materials contained in the rocks. In addition to mapping geologic conditions quantitatively it is often necessary to show the chemical, mineralogical, or physical characters of the rocks by a means which depicts only the qualitative aspects, as in certain lithofacies maps. It is only when all the properties are mapped that the subsurface geologist can grasp the complete geologic picture.

The second function of geologic maps is often as important as the first, although it affects the geologist only indirectly. This use is the presentation of mappable geologic phenomena to those having only slight acquaintance with the subject. Often the executives who control the operations of organizations know little about the details of subsurface methods and lack the technical training and experience necessary for an understanding of the geologist's working maps. Since the men who manage and control the operations of exploratory companies must be kept informed on the geologic developments, it devolves upon the geologist to construct maps that convey the essence of his work in a direct manner, without the details so necessary in the working maps. Maps that are to be used for the purposes described above should emphasize the ideas and conclusions of the geologist, not merely present or evaluate geologic data.

The methods of coloring maps have been described. But other features of map construction are equally important. It is difficult to show a number of different classes of geologic data on one sheet without some confusion of lines or areas. Therefore, the legibility of the map should be the principal guide as to how much can be shown to advantage. A number of factors influence the legibility of a map:

(1)   Standard symbols—the U. S. Geological Survey has published sheets of standard symbols for geologic maps. These symbols should be used wherever applicable because they are more likely to be understood by everyone using the map. If it is necessary to invent a symbol to indicate some subsurface feature, this sheet of standard symbols should be consulted to avoid using a figure which is standard for some other feature. The size, form, and weight of symbols should be kept uniform, except in special cases where a variation in the size or mass of the symbol denotes a corresponding variation in the size or importance of the feature shown.

(2)   Line weights—the careful grading and uniformity of line weights have much to do with the legibility and general appearance of the map; for example, line weights should grade downward from state boundaries, county boundaries, townships, sections, etc., despite the fact that certain of these are further distinguished by various sequences of short and long dashes.

(3)   Lettering—the lettering on most maps is done largely by LeRoy or Wrico lettering sets. These guides are capable of producing letters in a wide variety of sizes and line weights in both slanting and vertical styles. The presentation of the map is greatly enhanced when the choice of letter sizes and weights is made judiciously. Before any lettering is done, the features should be classified, and proper templets selected for each. The same templet will produce different-appearing letters when pens of varying sizes are used. Likewise, the spacing between letters is important in the appearance of a name. It is good practice to employ wide spacing for linear features and compressed spacing for locations of small areal extent. Use slant letters for surface hydrographic features, such as streams or lakes, and for descriptive or explanatory notations. Vertical letters in upper and lower case are used for geologic and geographic names. Large letters made with fine pens are less troublesome in obscuring control points or figures than are small letters in heavy lines. Lettering on vertical lines should read from the bottom upward. Lines that are even slightly inclined to the left at the top are lettered from the top downward. In other words, the letters should never be even slightly upside down.

(4)   Legends and Explanations—all symbols whose meanings might

be misconstrued should be fully explained in a legend. This applies also to colors and colored lines which are not identified by notations where they occur. The best location for the legend is near the title of the map. It is bad drafting practice to place portions of the legend at several locations in the margins.

Using a name or a short explanation to identify a feature on the body of the map is better than using a symbol which must be explained in the legend. It is tiresome for the map reader to have to refer frequently to a legend in order to understand the map. Explanatory notes at the approximate location of the feature to which they refer can hardly

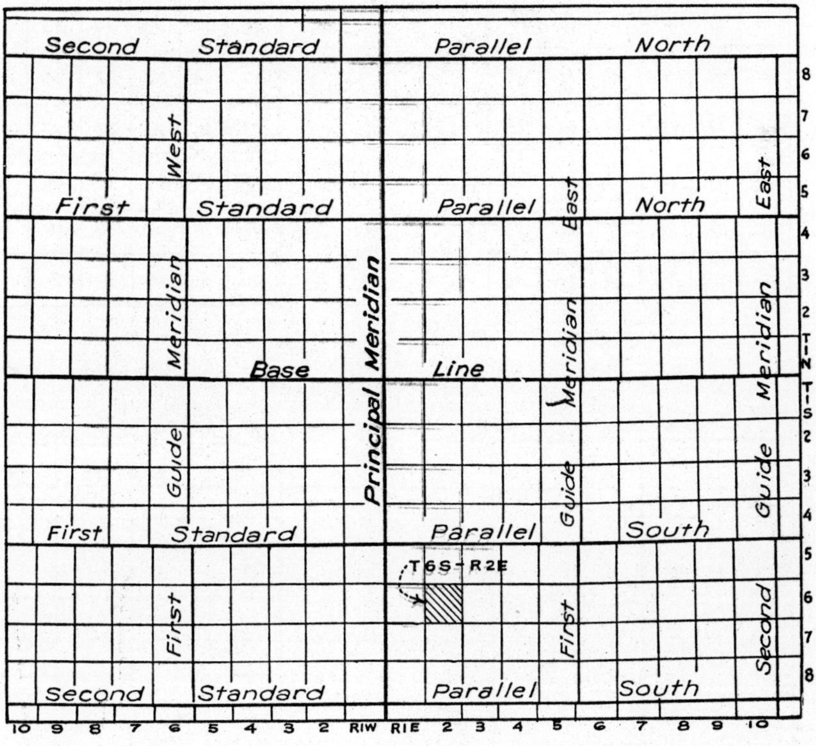

FIGURE 520. Plat illustrating principles of township and range designations.

be misconstrued and are less diverting than the see-saw reference to a legend.

(5) Geographic references—on a map which is well-planned and correctly drafted, it is not difficult to describe the location of any feature shown or to plot accurately new points of control. In other words, the map base should contain all the reference lines necessary for these operations. In subsurface work it is annoying to work on a map whose base consists of only geographic coordinates (meridians and parallels).

Practically all well locations are referred to township, range, and section lines; and if these lines are not shown on the map, it is difficult to determine the correct locations. On very small-scale maps, township lines may be undesirable; but, since county lines are tied in to the "land net," they may be shown. It should be borne in mind that geographic coordinates are not lines surveyed and marked on the ground, but, rather, are the framework of the map projection. The subsurface geologist is con-

TOWNSHIPS

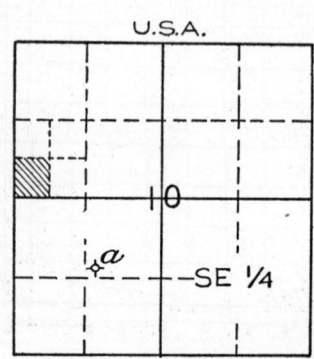
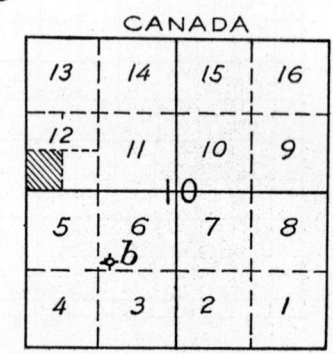

U.S.A.

6	5	4	3	2	1
7	8	9	10	11	12
18	17	16	15	14	13
19	20	21	22	23	24
30	29	28	27	26	25
31	.32	33	34	35	36

CANADA

31	32	33	34	35	36
30	29	28	27	26	25
19	20	21	22·	23	24
18	17	16	15	14	13
7	8	9	10	11	12
6	5	4	3	2	1

SECTIONS

U.S.A.   CANADA

FIGURE 521. Plats showing subdivisions of United States and Canadian townships and sections.

cerned primarily with legal subdivisions of state and federal surveys; and all points of control will be located on the map by means of these surveyed lines.

Figure 520 illustrates the basic principles of the township system of land surveys. There are many deviations from these principles, but ordinarily only small areas are involved. In Canada all townships are numbered north, beginning with township 1 at the international boundary. All ranges are east and west of one principal meridian located in the eastern side of Manitoba. Guide meridians are spaced about 30 townships apart and are numbered 1st, 2nd, 3rd, etc., meridians west.

Figure 521 shows the numbering of sections in United States and Canadian townships, and the subdivisions of the sections. Townships in Canada, like those in the United States, are composed of 36 sections, each approximately one mile square and containing 640 acres. The numbering of the sections is shown in the figure cited.

The Canadian system of designating the subdivisions of a section is more convenient than ours. The 1/16 section (40-acre) tracts are numbered from 1 to 16 according to the Canadian plan of numbering sections in the township. These subdivisions are called *Legal Subdivisions*, abbreviated Lsd. Thus, the shaded portion in the section diagram is described as $SW\frac{1}{4}$ of Lsd. 12. According to our system the same tract is described as the $SW\frac{1}{4}$ of the $SW\frac{1}{4}$ of the $NW\frac{1}{4}$, or SW SW NW. The location of a well situated at *a* is described as follows: SW SW NE $SW\frac{1}{4}$ of section 10. In Canada the description (at *b*) is: SW SW Lsd. 6 of section 10.

For a number of years it has been common practice in western United States to make the footage location of a well according to multiples of 330 feet from surveyed land lines. Thus, a location described as 1650 feet from the west line and 1650 feet from the south line of section 10 would be in the SW SW NE SW 1/4 of the section, or the location of *a* in figure 521.

Well "spotting" templets, made from acetate of about 0.020 inch thickness, greatly facilitate the posting of maps. The templet is made with inked lines corresponding to the land lines shown on the map, with holes punched at the normal positions of well locations. The templet is registered over the land lines on the maps and the well is "spotted" by inserting a sharp pencil in a hole at the correct location.

Because of the fact that the surveying of public lands over the years has not been carried out in a continuously systematic manner, there are many duplications in the numbering of townships and ranges. Therefore, in addition to the township and range, the state, and in some cases the county, must also be known before a location on the map can be made. For this reason, the state and county should be given in the location of a well.

QUESTIONS

1. What is meant by the term, "subsurface map"? Upon what data is this type of map prepared?
2. What is the effect of crooked holes on datum elevations and interpreted dips?
3. What is the effect of crooked holes on convergence interpretation?

4. What is the effect of unconformities on the interpretation of subsurface structure?
5. What is meant by a "fault gap" and how is it shown on a structure contour map?
6. What basic rules apply to contouring procedure?
7. Is it true that "squeezed' contours may suggest a fault?
8. What is the principle and purpose of peg models? of solid relief models?
9. Do section models have a definite application in regional geologic work? Why?
10. Define "isopach map." State its use.
11. Define "isochore map." State its use.
12. Study the maps showing combination of structure and isopachous data. Be prepared to explain.
13. For what purposes are paleogeologic maps used and how are they prepared?
14. What is the relation between width of outcrop bands to dip and topography?
15. How are structural and stratigraphic sections prepared?
16. What is the difference between a panel and isometric projection diagram? How are these diagrams used?
17. Define, "facies map," "lithofacies map." For what purposes are such maps used?
18. What is meant by ratio map, percentage map, isolith map, and isofacies map?
19. May isopach and lithofacies data be represented by a single map? What is the purpose of such combination?
20. Discuss a lithofacies-data sheet.
21. Study the color control for lithofacies maps.
22. Define "clastic ratio," "clastic percentage."
23. Discuss the practical use of lithofacies maps.
24. What colors are commonly used on geologic maps to represent the rocks of the various geologic periods?
25. Discuss the various types of pencils and inks used for coloring geologic maps.
26. Comment on fundamentals of map reproduction.
27. What is the difference between the United States and Canadian systems of subdividing sections?
28. Study the section "drafting of maps" and be prepared to discuss.

CHAPTER 12

SUBSURFACE METHODS AS APPLIED IN MINING GEOLOGY

TRUMAN H. KUHN

ORIGIN, CHARACTERISTICS, AND CONTROLS

Many geologic factors enter into the concentration of ore minerals within the earth's crust. In fact, so many conditions must be met to produce an economic mineral accumulation that an ore deposit has been termed an "accident of geology," and such accidents underlie considerably less than one percent of the earth's surface.[1] To prospect a region most efficiently it is important that origins, characteristics, and controls be understood.

Sedimentary Ore Deposits

Sedimentary ore deposits are formed by normal sedimentary processes and possess all of the characteristics of sedimentary rocks. They may be products of mechanical concentration such as placers, or they may be products of chemical concentration. The chemical products can either be soluble salts such as halite and the borates or insoluble products such as some iron, aluminum, and clay deposits.

Igneous Ore Deposits

The origin of deposits associated with igneous activity is not definitely understood and agreed upon, but most geologists consider that ores result from differentiation [2] of a magma. One product of differentiation is the usual igneous-rock series, and another is the ore-bearing solutions that accumulate in the top of stocks. During solidification the ore-bearing solutions are forced into the chilled hood of the batholith or into the surrounding country rock. The bulk of introduced deposits (epigenetic) is collectively termed "hydrothermal deposits," and these have been further divided into "hypothermal" (high temperature and pressure), "mesothermal" (intermediate temperature and pressure), and "epithermal" (low temperature and pressure) classes.[3]

According to Buddington,[4] a relationship exists between kinds of metallization and kinds of igneous rocks. Thus, platinum, chromium, and diamond are genetically associated with ultrabasic igneous rocks. Tin and tungsten are genetically related to granite. Copper, lead, and zinc are almost always related to intermediate rock groups. A spatial rela-

---

[1] Lovering, T. S., *Minerals and World Affairs*, p. 5, New York, Prentice Hall, Inc., 1943.
[2] Differentiation is the splitting up of an originally homogeneous magma into different or unlike fractions.
[3] Lindgren, Waldemar, *Mineral Deposits*, New York, McGraw-Hill Book Co., Inc., pp. 444-694, 1933.
[4] Buddington, A. F., *Ore Deposits of the Western States*, (Lindgren volume), pp. 350-385, Am. Inst. Min. Met. Eng., 1933.

tionship also exists between mineral deposits and the source pluton. Butler [5] and Emmons [6] have pointed out that the great bulk of epigenetic deposits are found either in the top of stocks or above the stocks in the country rock. Thus, if erosion has truncated the stocks and the main mass of the batholith is exposed, most deposits genetically related to that igneous mass have been removed. Many deposits occur in igneous rocks that are older than the source rocks, in which case the enclosing rock is the host and not the source rock.

Rising, reactive liquids or gases will follow and be controlled by pre-existing structures. They react more readily with certain rock types than with others and will leave behind evidences of their passage. The general geologic approach to the study of epigenetic deposits is based on the controls affecting the passage of liquids and gases and effects of these liquids and gases on wall rocks.

Structural control [7] [8] [9] refers to the controls on rising liquids by preexisting (premineral) structures. These controls may act as guides or as barriers. A consideration of structural control involves a study of the regional pattern as well as a detailed study of an individual mining district or mine. Billingsley and Locke [10] have suggested that the location of important mining districts is, in part, determined by orogenic belts which are continental in scope. They claim that orogenies have strengthened or made the crust sufficiently competent to support the deep-seated faults which are necessary to tap concentrations of metal-bearing material. Butler [11] has pointed out that a zone of structural disturbance and igneous activity surrounds much of the Colorado Plateau and that many major ore deposits of the Rocky Mountain region are found in this disturbed zone. In a more restricted area, Butler [12] has shown that much of the Utah production is from mining districts located near the intersection of the westerly-trending Uinta axis and the northerly-trending Wasatch belt. Specifically, in individual deposits, premineral structures are of extreme importance. Faults are good premineral structural controls in that some form zones of greater permeability, which act as channelways or conduits aiding and controlling the passage of ore-bearing solutions. In a similar manner, the more open zones at the crests of certain folds, at contacts of dissimilar rocks, and along bedding planes, act as guides for rising solutions. Slates, shales, clays, and faults with considerable gouge act as

[5] Butler, B. S., *Relation of Ore Deposits to Different Types of Intrusive Bodies in Utah:* Econ. Geol., vol. 10, pp. 101-122, 1915.
[6] Emmons, W. H., *Principles of Economic Geology*, pp. 185-190, New York, McGraw-Hill Book Co., 1940.
[7] Newhouse, W. H., editor, *Ore Deposits as Related to Structural Features*, Princeton University Press, 1942.
[8] McKinstry, H. E., *Mining Geology*, pp. 277-342, New York, Prentice-Hall, Inc., 1948.
[9] Butler, B. S., *Ore Deposits of the Western States*, (Lindgren volume), pp. 198-326, Am. Inst. Min. Met. Eng., 1933.
[10] Billingsley, Paul, and Locke, Augustus, *Structure of Ore Districts in the Continental Framework:* Am. Inst. Min. Met. Eng. Trans., vol. 114, pp. 9-64, 1941.
[11] Butler, B. S., *Ore Deposits of the Western States*, (Lindgren volume), pp. 215-222, Am. Inst. Min. Met. Eng., 1933.
[12] Butler, B. S., et al., *Ore Deposits of Utah:* U. S. Geol. Survey, Prof. Paper 111, 1920.

barriers to rising solutions damning the ores below premineral features. In referring to the barriers, Mackay [13] contends that impounding structures or rocks do not act as perfect barriers. If that were the case, stagnant solutions would be trapped and very little ore deposited. Mackay argues that the solvent passes through the relatively impermeable barrier after the metals have been deposited.

Sedimentary rocks act as host to numerous deposits of igneous origin, but the control exerted by the sediments on localization of the deposit is not fully understood.[14] In many replacement deposits, the control is structure modified by stratigraphy. Selective replaceability is not confined to sedimentary rocks. The Smuggler vein, Telluride, Colorado, passes through both andesite, which is productive, and an overlying rhyolite, which is essentially barren.[15]

Rising solutions, in addition to the depositing of ore and gangue minerals, penetrate into the wall rock and form wall-rock-alteration minerals.[16] [17] [18] [19] [20] Detailed work has shown that in certain districts a direct relationship exists between alteration and metalliferous deposition.

Numerous types of epigenetic deposits are mined and in general the exploration techniques differ for each.

The more important types of epigenetic deposits may be defined as follows:

A vein is a tabular or sheetlike body of minerals occupying or following a fracture or a group of fractures.

A disseminated deposit is an irregularly shaped body in which the ore minerals occur as small individual grains and in small seams or veinlets.

A pipe is a vertically elongated deposit of circular or elliptical cross section.

A replacement deposit is one in which the host rock has been partly or completely replaced by the ore and gangue minerals. The resulting deposit may be tabular, sheetlike, pipelike, or irregular and is more or less restricted to one horizon, bed, formation, or structure.

Metamorphic Ore Deposits

Some ore minerals are formed by metamorphic processes, and if the deposits are of sufficient size and grade they can be termed "metamorphic

---

[13] Mackay, R. A., *The Control of Impounding Structures on Ore Deposition:* Econ. Geol., vol. 41, no. 1, pp. 13-46, 1946.

[14] Rove, O. A., *Some Physical Characteristics of Certain Favorable and Unfavorable Ore Horizons:* Econ. Geol., vol. 42, pp. 57-77, 161-193, 1947.

[15] Emmons, W. H., *op. cit.,* p. 217.

[16] Butler, B. S., *Influence of the Replaced Rock on Replacement Minerals Associated with Ore Deposits,* Econ. Geol., vol. 27, no. 1, pp. 1-24, 1932.

[17] Lindgren, Waldemar, *op. cit.,* pp. 444-694, 1933.

[18] Lovering, T. S., *Rock Alteration as a Guide to Ore: East Tintic District, Utah:* Econ. Geol. Mono. 1.

[19] McKinstry, H. E., *op. cit.,* pp. 233-242, 1948.

[20] *Applied Geology:* Colorado School of Mines Quart., vol. 45 (75th Anniversary volume), no. 1B, Jan. 1950.

ore deposits." They will have all the characteristics of a metamorphic rock.

Supergene Deposits [21] [22]

All deposits formed by primary processes may be acted upon by ground water producing characteristic downward changes through the leached, oxidized, and supergene sulphide zones.

EXPLORATORY PROCEDURES

After accumulating all known data on the geology, grade, and size of a deposit, a mining geologist considers various approaches to prove the value of a given property. Generally, he thinks first in terms of geology; of the type and origin of the deposit and the possible ore controls and guides. After such factors are weighed, further work may be indicated to obtain more geologic data. Such additional work could involve geologic mapping of the surface and accessible underground workings, sampling, laboratory study, and geophysical investigations. It also may be advisable to carry out exploratory work to examine the deposit further in order to obtain more geologic data and more information on grade and size. The geologic data accumulated not only is important in evaluating the property but also aids the mining engineer in laying out mining methods.

Geologic Guides and Controls

Structural Control—Of the various guides and controls used as aids in locating ore perhaps the most important is structural. Regional control is well explained by Fowler [23] in the Tri-State area where he has shown that ore bodies are closely related to structural disturbances and that where the beds are relatively undisturbed little or no ore is present. Fowler also states:[24]

After initial discovery, drilling of 600 holes under informed geologic guidance, in lieu of most of 30,000 actually drilled, could have prospected an area of 37 square miles sufficiently to define the mining area and many important features in and around it.

Any fracture formed before mineralization is important as a control in that the fault may be a pervious zone along which solutions can travel. Fault surface deviations and irregularities can increase the permeability, and detailed study of a vein may show that ore bodies are localized at changes of dip and strike, at intersections of veins, or at intersections of veins and fractures. Even where no detailed control within a vein can be established, it may be possible to locate channelways and show that ore

---

[21] Bateman, A. M., *Economic Mineral Deposits*, pp. 243-289, New York, John Wiley and Sons, Inc., 1942.
[22] Emmons, W. H., *op. cit.*, pp. 105-140.
[23] Fowler, G. M., *Tri-State Geology:* Eng. and Min. Jour., vol. 144, no. 11, pp. 73-79, Nov. 1943.
[24] Fowler, G. M., *idem*, p. 79.

bodies are located near those channelways. In other veins no control or localization other than the premineral host fracture is apparent.

Folding represents another type of structural control. In competent rocks, crests of folds are more open than the flanks; and, when aided by an impermeable cap rock, deposits are restricted and localized near the crests. In the saddle-reef deposits of Bendigo and Ballarat, Victoria, Australia, ore occurs more abundantly near the crest line, but it also extends a short distance down the flanks. In tight, overturned folds, principally in metamorphic rocks, deposits occur along the flanks of some folds but the controls might be considered as faults developed because of intense folding.

Structural petrology [25] is a newly applied tool, being used to help

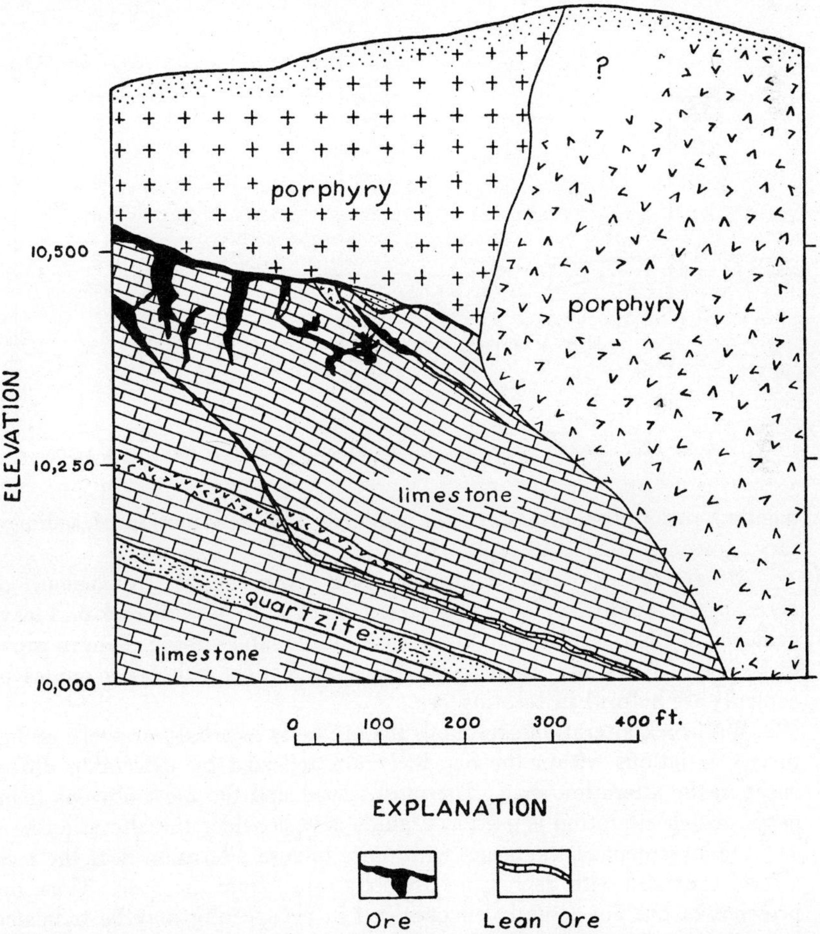

EXPLANATION

Ore   Lean Ore

FIGURE 522. Section showing ore in limestone below relatively unreplaceable and impermeable porphyry, Oro La Plata mine. (After Emmons.)

[25] Fairbairn, H. W., in Newhouse, *op. cit.*, pp. 265-267.

unravel structure in igneous and metamorphic rocks and thus aid in the search for new ore. By a study of planar and linear structures and spatial relations of the rock minerals, the fabric of a rock can be analyzed and the more or less hidden structures determined. These igneous and metamorphic structures can play about the same role in ore deposits as do the more obvious deformations in sedimentary rocks. Such work involves very detailed field studies followed by microscopic studies. (See "Petrofabric Analysis," this volume.)

Contacts of dissimilar rocks may be considered under structural control. In general, contacts are more permeable than the rocks on either side and in the absence of fractures, solutions more readily pass along the contacts than across them. Deposits occur at contacts of dissimilar sedi-

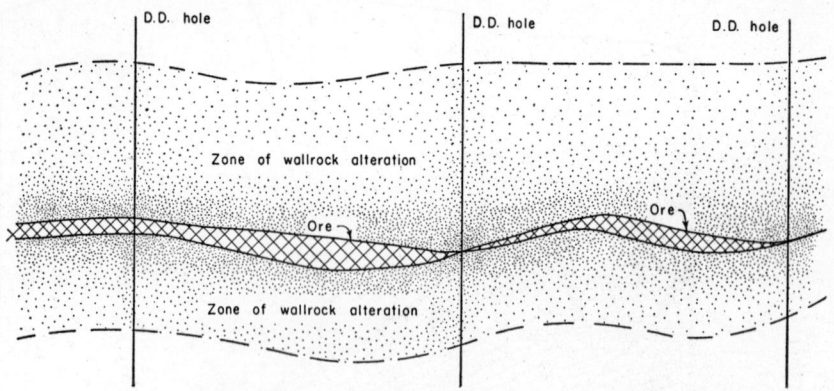

FIGURE 523. Sketch showing relation of wall-rock alteration to a vein.

mentary rocks, dissimilar igneous rocks, or any combination of sedimentary, igneous, and metamorphic rocks (fig. 522).

Mineralogic Guides—Solutions rising from below (hypogene) or descending from the surface (supergene) react with the wall rock and leave evidence of their passage. In many cases, these mineralogic changes prove to be useful tools and when applied in conjunction with other guides or controls are helpful in locating ore.

Wall-rock alteration surrounds the ore body as a halo or shell, and at places variations within the ore body are reflected by detectable differences in the alteration zone. The most useful and the most obvious manner in which alteration is used as a guide is to consider the alteration halo as an enlargement of the target with more intense alteration near the metallized zone and with decreasing intensity away from the zone. Thus, the presence of ore not actually encountered in prospecting may be indicated by alteration (fig. 523). It should not be inferred that alteration necessarily means ore. Solutions passing through a fracture and depositing less than an inch of pyrite may alter the wall rock for a number of feet on both sides of the fracture. Conversely, a vein containing several feet of copper ore may show less wall-rock alteration than a one-inch pyrite seam.

The foregoing type of alteration is caused by rising hypogene solutions. Descending supergene solutions act on the hypogene deposits and the enclosing country rock and also bring about typical changes.[26][27] The bases of supergene action are several: pyrite in the presence of air and water will oxidize to form sulphuric acid; dilute sulphuric acid will react with most minerals in the hypogene zone, take them into solution, and carry them down toward the water table; pyrite below the water table will precipitate copper and silver. If sufficient pyrite is present, all common metals with the possible exceptions of gold and lead may be entirely leached from the outcrop. As a result of the leaching action, characteristic but complex phenomena can develop and conclusions may be reached as to the character of the original deposit.[28][29] Below the leached zone, the metal-bearing supergene solutions may react with a carbonate to form azurite, malachite, or smithsonite. If the solution reaches the water table and comes in contact with pyrite, then chalcocite, covellite, or argentite may be precipitated.

Geochemical and Biogeochemical Guides—The geochemist is concerned with minute traces of metals that work their way into the soil, either as a result of hypogene mineralization or as a result of supergene action. These solutions may cause characteristic alteration or may deposit minute amounts of the metals in the overlying and surrounding rocks. Supergene action affects the primary deposits, and the metals are moved from their original position. Gold, tin, and lead may travel mechanically, and copper, zinc, and silver may be removed in solution. By systematic sampling of the soil and ground water, traces of metal may be found.[30][31][32] A study of the movement of the soil and ground water may indicate the source of the metals and lead to an ore body.

The biogeochemist, who is interested in the relation between vegetation and ores, reasons geologically in the same manner as the geologist. The distribution of specific types of vegetation has been studied with reference to ore deposits; and, although the work is not conclusive, it appears that in a few districts, there is a distinct relationship between vegetation and ore deposits. As a variation of the distribution of specific types

[26] Emmons, W. H., *The Enrichment of Ore Deposits:* U. S. Geol. Survey Bull., 625, 1917.

[27] Lindgren, Waldemar, *op. cit.*, pp. 813-877, 1933.

[28] Locke, Augustus, *Leached Outcrops as Guides to Copper Ores*, Baltimore, Williams and Wilkins, 1926.

[29] Blanchard, R., and Boswell, P. F., A Group of Papers on Various Types of Limonite Boxworks: Econ. Geol., vols. 22, 25, 29, 30, 1925-1935.

[30] Sergeev, E. A., *Geochemical Method of Prospecting for Ore Deposits:* from Materials of the U.S.R.R. Geol. Inst., Geophysics, Fasc. 9-10, pp. 3-55, 1941.

[31] Hawkes, H. E., *Annotated Bibliography of Papers on Geochemical Prospecting for Ores:* U. S. Geol. Survey, Circ. 28, Aug. 1948.

[32] Hawkes, H. E., *Geochemical Prospecting for Ores: A Progress Report:* Econ. Geol., vol. 44, no. 8, pp. 706-712, 1949.

of vegetation, ash from plants, twigs, and leaves has been analyzed for metal content. Some exacting work has been done along these lines, but, because of the lack of sufficient data, the results are still inconclusive. Like geochemical sampling, biogeochemical sampling shows great promise, and more work is being carried out at the present time.[33]

Stratigraphic Control—As a guide in layered rocks, stratigraphy is important in that certain beds are more easily replaced by mineralizing solutions and are better hosts than others. No clear and definite reason can be given for the preference of epigenetic ores for one particular type of rock or one particular formation, but certainly permeability, as reflected by relative brittleness, and composition, as reflected by chemical reactiveness, are important. On a purely statistical basis, it can be demonstrated that certain formations, as the Leadville in Colorado, the Homestake in South Dakota, and the Boone of the Tri-State region, are host rocks for the great bulk of the ore in those areas. In other areas as Bisbee, Arizona, much of the ore occurs in one formation but is not restricted to that formation. In either case, if such facts are known, initial prospecting certainly will be aimed toward the better host rocks, but until the district is well explored it cannot be assumed that the ore is restricted to those certain horizons. Many exceptions to the foregoing statements show that all favorable guides, of which stratigraphy is but one, must be examined by the mining geologist in his search for ore.

It is obvious that in sedimentary ore deposits, stratigraphy is of prime importance, and any prospecting must be based on sedimentation studies. The Tertiary gold-bearing channels of California, the potash deposits of New Mexico, and the phosphate deposits of Idaho all are examples of deposits of which the sedimentary features must be known before there can be efficient prospecting.

Geologic Mapping

Surface geologic mapping, as carried out by the mining geologist, differs very little from surface mapping performed by other geologists. Such necessary equipment as a Brunton compass, plane table, aneroid barometer, and airplane photographs are employed by all geologists. The principal difference between a mining problem and most other types of geologic mapping is in the choice of scale. It is not at all uncommon in mining geology to map the surface using a scale of one inch equals 100 feet. Detail is important, and it is only on such maps that detailed representation is possible. When mapping the general relations of the district, the geologist may use a scale as small as one inch equals 1,000 feet.

Underground geologic mapping (see chap. 13) has been well described,[34][35] and in principle is similar to surface mapping. The scale, of

[33] Warren, H. V., and Delavault, R. E., *Further Studies in Biogeochemistry:* Geol. Soc. America, vol. 60, no. 3, pp. 531-560, 1949.

[34] Forrester, J. D., *Principles of Field and Mining Geology:* pp. 331-348, New York, John Wiley and Sons, 1948.

[35] McKinstry, H. E., *op. cit.*, pp. 1-34, 1948.

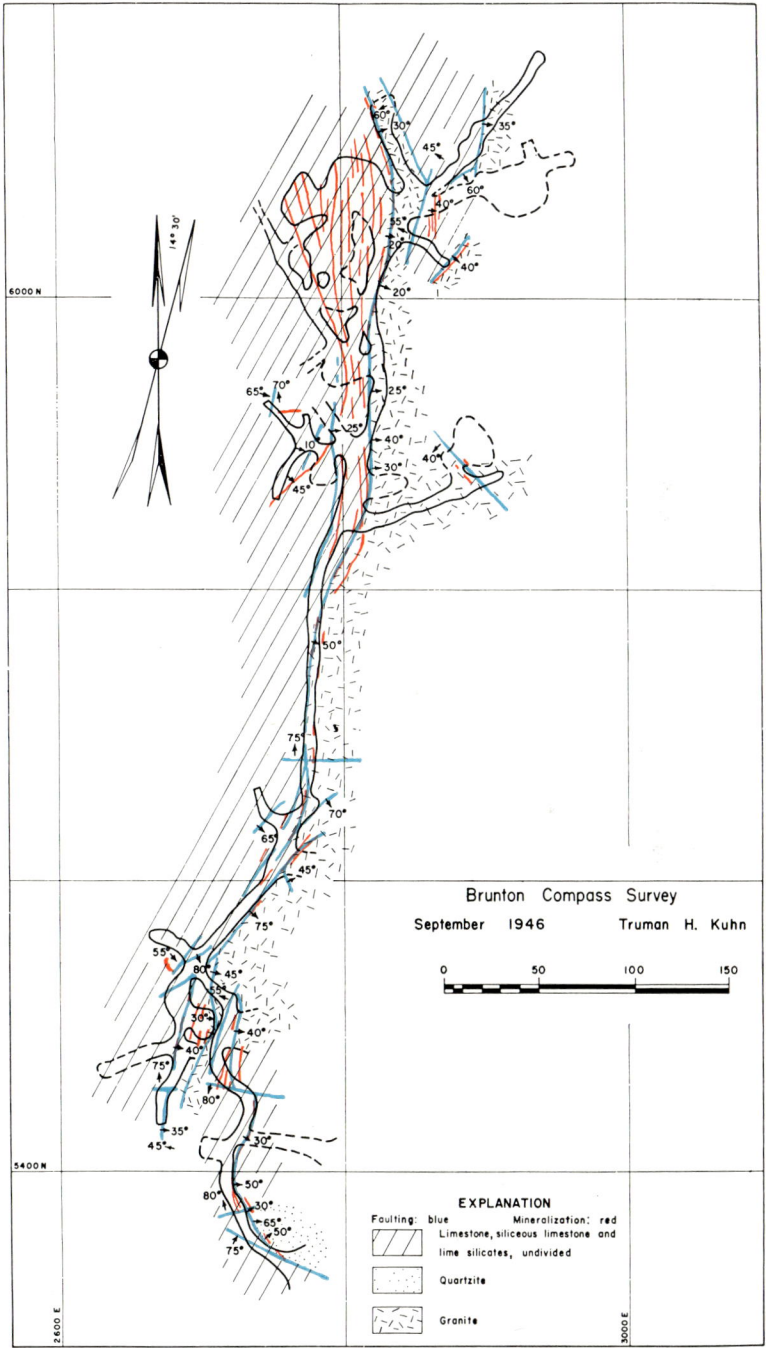

Brunton Compass Survey

September 1946   Truman H. Kuhn

EXPLANATION

Faulting: blue   Mineralization: red

Limestone, siliceous limestone and lime silicates, undivided

Quartzite

Granite

PLATE 13. Geologic map, Leader mine, Pima County, Arizona, showing conventional colors and symbols.

course, differs. Depending on the complexity of the deposit, the scales most commonly used are 20, 40, or 50 feet to the inch. An underground geologic map also differs from a surface map in that either a waist-high or breast-high horizontal mapping plane is assumed, and all data are projected to that plane. The mapping height varies with the practice of different geologists, but, if the plane is consistent and noted, either is acceptable. Under certain conditions, the mapping level may be the back or top of the mine working. Underground mapping has an advantage over surface work in that the vertical dimension normally is better revealed by underground workings, but the geologist, because of restricted mine openings, is hampered when attempting fully to develop the plan maps. As in surface mapping, conventional color and pattern symbols are used (pl. 13).

Geophysics

For a number of years geophysics was considered, without much success, as a possible aid in finding new ore bodies. One of the reasons for failure was the lack of understanding between the mining geologist and the geophysicist. The mining geologist knew little of geophysics and the geophysicist knew even less of geologic problems encountered in mining. In recent years, however, the picture has been changing. Mining companies are realizing that geophysics can be an important tool of the mining geologist. Research has been started by various companies and government agencies, so that at the present time geophysical staffs have been established who are devoting all of their time and effort to the application of geophysics to mining. Such organizations certainly are useful now and will be more important in the future in aiding the geologist to find new ore. The use of geophysical methods is described in chapter 14 of this volume. Numerous papers and summaries [36] [37] discuss the use, applications, and limitations of mining geophysics.

Laboratory Methods [38]

The laboratory investigation of ore deposits involves the study of two types of minerals; those that are nonopaque in thin section and those that are opaque. Nonopaque minerals are studied in transmitted light using a petrographic microscope, and opaque minerals are studied in reflected light using a reflecting or ore microscope. Many problems, such as mineral content and size of mineral grains, are readily solved in the laboratory, while other questions, particularly those relating to origin of the specimen, may be more difficult to answer.

Petrographic examination of fragments crushed to about 125–150

[36] Heiland, C. A., *Geophysical Exploration*, New York, Prentice-Hall, Inc., 1940.
[37] McKinstry, H. E., *op. cit.*, pp. 115-132, 1948.
[38] McKinstry, H. E., *op. cit.*, pp. 113-161, 1948.

microns (screen size 100–120) and immersed in oils of varying indices [39] [40] is a very satisfactory means of identifying nonopaque minerals. In thin sections, minerals also can be identified,[41] and in addition, grain-relation studies can be made. All wall-rock-alteration products should be investigated under the petrographic microscope to determine changes that have taken place within the rock and properly to classify the alteration as to kind and intensity. Structural-petrology problems entail orientation studies of the individual minerals, which necessitate the use of the universal stage under the petrographic microscope.

To prepare an ore for study under the reflecting microscope one surface of the opaque mineral must be given a high polish. Because of the ease of handling and storage, many laboratories mount the ore specimens in bakelite or sealing wax before polishing, but the ores also can be studied if the specimen is mounted in clay when polished. After a flat surface is obtained, either by grinding on a coarse lap or with a rock saw, the specimen is further treated on metal laps using powders of increasing fineness to remove most of the pits. Polishing may be done in a number of ways but cloth-covered laps or lead laps are generally employed.[42] Identification under the reflecting microscope is based on a number of factors, mainly color, hardness, polarized-light effects, standard etch tests, and microchemical tests.[43] The metallurgical microscope, after calibration by use of stage and ocular micrometers, can be used to determine grain sizes. Grain size is of specific importance to the metallurgist, who must know the grinding size necessary to free the various ore minerals. Mineralographic studies [44] [45] may help determine the origin of a deposit and materially aid the mining geologist in his search for ore; hypogene or supergene mineralization may be indicated, or age-relation studies may show a definite relationship between ore and ages of fracturing.

Other laboratory techniques such as fire and wet assaying, chemical analysis, spectrographic analysis, X-ray-diffraction analysis, differential-thermal analysis, and heavy-mineral determinations are used in conjunction with petrographic and ore microscopes in determining the kind and amount of minerals and metals present in the ore, gangue, and wall rock.

Exploration and Sampling Methods [46] [47]

No matter how complete the geologic data and how accurate the interpretation, mining properties cannot be proved on geology alone.

[39] Rogers, A. F., and Kerr, P. F., *Optical Mineralogy*, pp. 135-136, New York, McGraw-Hill Book Co., Inc., 1942.
[40] Larsen, E. S., and Berman, Harry, *The Microscopic Determination of the Non-Opaque Minerals:* U. S. Geol. Survey Bull. 848, 1934.
[41] Rogers, A. F., and Kerr, P. F., *op. cit.*, pp. 3-7, 1942.
[42] Short, M. N., *Microscopic Determination of the Ore Minerals:* U. S. Geol. Survey Bull. 914, pp. 4-44, (1948 reprint), 1940.
[43] Short, M. N., *op. cit.*, pp. 59-292.
[44] McKinstry, H. E., *op. cit.*, pp. 141-155, 1948.
[45] Bastin, E. S., *et al.*, *Criteria of Age Relations of Minerals with Special Reference to Polished Sections of Ores:* Econ. Geol., vol. 26, no. 6, pp. 561-610, 1931.
[46] McKinstry, H. E., *op. cit.*, pp. 35-114, 1948.
[47] Forrester, J. D., *op. cit.*, pp. 349-441, 1946.

Geology may aid in finding ore, may show that ore exists, and may indicate a certain size of deposit, but to satisfy the investor and, incidentally, the geologist, the deposit must be explored and samples taken. Depending upon the kind and size of deposit, the exploration work may consist of trenching, stripping, sampling from drill holes, or underground mining operations.

Trenching and Stripping—As is often the case, the surface outcrops may be covered at critical spots with vegetation or rock detritus. If the cover is not too great, structure, rock, and possible ore of any type may be uncovered by a trench dug across areas of interest. Trenches are treated as outcrops, and the location and geology are shown on the surface maps. If the cut exposes ore, sampling across the ore zone may be necessary; in which case a uniform sample can be taken either by deepening the trench or sampling the wall of the trench. It should be remembered that the exposed rock is near the surface and most surface ores and rocks have been attacked and changed by ground waters.

Overburden can also be removed with power machinery, most commonly with a bulldozer.[48] A good operator can economically remove small shrubs and trees and several feet of overburden in a very short time. Such exposures can be mapped as surface outcrops and sampled by taking cuts at random, on a grid, or by trenching at a known angle across the general trend of the exposed structure. In sampling an area uncovered with a bulldozer, caution must be exercised to be certain that incompletely removed soil does not dilute the sample.

Drilling—Auger and Hammer Holes: Drill holes put down from the surface can explore greater depths than can be reached by trenching or stripping. If relatively shallow holes in soft material such as clay or concentrator tailings are needed, a hand auger is satisfactory. Two men can easily drill 50 feet in this type of material, and 100-foot holes are not uncommon. In auger drilling, as it is possible to recover 100 percent of the cuttings, the material removed is an accurate sample of the hole. Depending upon the depth of hole and character of cuttings, samples may be taken at regular or irregular intervals, or the entire hole may be considered as one sample.

Short holes from the surface to prospect any type of deposit may be put down by pneumatic hammers to a depth of 30 to 50 feet. Sampling this type of hole is more difficult in that the cuttings must either be washed out if drilling is done wet or must be blown out if drilling is done dry. In either case, difficulty may be encountered in removing and collecting the cuttings for accurate sampling.

Churn-Drill Holes: Churn drilling[49] is a common, noncoring method of prospecting by means of vertical surface holes and is effective for exploring relatively widespread sedimentary deposits, disseminated epigen-

---

[48] *New Prospectors in the Hills:* Mining World, vol. 11, no. 3, p. 38, Mar. 1949.
[49] McKinstry, H. E., *op. cit.,* pp. 171-182, 1948.

etic deposits, and vein and replacement bodies of not too steep a dip or rake. For most geologic examinations, the cuttings obtained from the six- to twelve-inch churn-drill hole are sufficiently large to identify rock types, minerals, and ores. Logging to the nearest foot is possible. In churn drilling, care must be taken to avoid erroneous samples. Contamination of the sample by material from the side of the hole can be partly avoided by keeping the casing near the bottom of the hole. Because there is a ten- dency for gold and other heavy minerals to settle, care also must be main- tained in removing the cuttings or sludge from the bottom of the hole. The usual dart-type bailer will not clean the bottom 10 to 15 inches of hole; and, if settling is suspected, other bailers must be used to remove cuttings more completely. Samples are taken at least every five feet in ore horizons; and, where abrupt changes are expected, this interval may be reduced to two feet. Rarely is all of the sludge kept for a sample. Most commonly, the cuttings are emptied into a splitter, which uniformly reduces the amount to one-half, one-fourth, one-eighth, or one-sixteenth of the original volume. The retained portion may be further split, one half being sent to the assay office and the other half examined and saved by the geologist. As churn-drill holes are sufficiently large for instruments, common oil-industry practices such as electric logging and the measure- ment of radioactivity can be, but rarely are, employed.

Diamond-Drill Holes:[50] [51] The various methods outlined above, under certain conditions, will accurately sample the rock below the surface, but for many purposes a diamond drill is more desirable. By diamond drill- ing it is possible to obtain cores that can be more carefully examined and to drill shallow or deep holes either vertically or at any desired inclina- tion. Because of greater flexibility both as to angle of hole and maneuver- ability, the diamond drill can be employed for exploring narrow, steep veins, irregular deposits, and small pipe deposits, as well as for prospect- ing the same types of deposits as the churn drill.

In diamond drilling a circular, hollow bit set with diamonds is mounted at the end of a hollow rod or series of rods, which are clamped in the drill. The rods and bit are rotated and advanced by a screw or hydraulic feed with the rate of rotation and advance controlled by a series of changeable gears in the drill. Cuttings are removed from the hole by water pumped through the rods. The bottom rod contains a core barrel 5, 10, or 20 feet long, which receives and holds the rock core that passes through the bit. When the barrel is full or when the bit "blocks," the rods are pulled, the core barrel is removed, and the accumulated core is taken from the barrel for examination. Standard core sizes range in diameter from $\frac{7}{8}$ inch (EX) to $2\frac{1}{8}$ inches (NX). In solid homogeneous rock 100-percent core recovery is common, but with increased fracturing, alteration, and friability the core recovery decreases. However, a good,

[50] Forrester, J. D., *op. cit.*, pp. 398-411, 1946.
[51] McKinstry, H. E., *op. cit.*, pp. 82-105, 1948.

careful driller who is more concerned with core recovery than with footage drilled will recover a high percentage of core even in bad ground. Recovery in bad ground is important, because so often ore is found associated with fractured and altered rock, and the purpose of most drilling is to obtain representative samples for assay and study.

The geologist records all possible information from a diamond-drill core. He not only logs geologic data such as depths to contacts, rock descriptions, ore descriptions, and inclination of bedding, structural planes, schistosity, and veins, but he also notes the conditions of the core and the amount of core recovery. The last two observations may indicate zones of structural weaknesses. The desired information obtained from diamond drilling includes dip, strike, width and grade of the ore body, and structural and stratigraphic data. Obviously, one drill hole will not yield all the desired information. Under suitable circumstances with sufficient drill holes it is possible to compute the attitude of veins, faults, and bedding.[52] With the dip and strike known and the inclination of the hole measurable, the width or thickness of veins and beds can be determined. The assumption that drill holes have the same inclination at depth as at the collar may lead to incorrect calculations. Drill holes deviate for a number of reasons; and, if deviation is suspected, a drill-hole survey may be necessary.[53] [54] Many deposits are of irregular grade and thickness, and the variation and inaccuracy in sampling can be minimized by closer spacing of drill holes.

To sample the material from diamond-drill holes, both the sludge and the core are assayed, and, if warranted, a weighted average can be calculated which is computed on the basis of the relative amounts of sludge and core. Very seldom is all of the core assayed. Most commonly the core is split longitudinally with one half going to the assay office and the other half being retained for study and record.

In laying out an exploration program, limitations of the diamond drill should be understood; if they are, diamond drilling offers a relatively fast and inexpensive means of exploring unknown ground. For an effective program, a trained geologist should be on the job. The geologist certainly should see that the general drilling schedule is carried out, that the driller is recovering the maximum amount of core, that the driller's records are accurate, and that the samples are properly handled and tagged. In addition, the geologist should have authority to modify the program to fit the pattern developed or changed by the continued drilling and increased geologic knowledge.

Calyx-Drill Holes: A method somewhat similar to diamond drilling is shot drilling (calyx drilling), in which chilled-steel shot is used for cutting rather than a diamond bit. Calyx holes are vertical but can be drilled up to five feet in diameter. Since calyx holes are drilled vertically,

[52] McKinstry, H. E., *op. cit.*, pp. 100-103, 1948.
[53] Forrester, J. D., *op. cit.*, pp. 409-411, 1946.
[54] McKinstry, H. E., *idem*, pp. 97-98, 1948.

their exploration use is restricted to sampling relatively flat-lying deposits and providing access to underground-mine workings.

Underground Drilling: The foregoing discussions apply to drill holes from the surface, but diamond drilling and hammer drilling also are extensively used underground where holes may be drilled up or down at any angle. Coring underground with diamond drills is very little different from coring on the surface except that air or electric drills must be used in place of those powered with gasoline. For holes drilled upward all cuttings can be recovered readily, and under favorable conditions hammer drills and diamond drills with noncoring bits can be advantageously used. The purpose of drilling underground is the same as for surface drilling.

Mining Operations [55]—If geologic information indicates such a course, and particularly if drilling shows favorable ore or even strong possibilities of ore, exploratory mining operations may be instituted. Initial mining operations may include sinking a vertical or inclined shaft, which might follow or stay near the ore body. In country with relief, the original underground workings may be nearly horizontal, either starting at the outcrop and drifting with the ore or cross cutting to the ore body and then drifting until the limits are reached. Exploratory work should be closely watched by the geologist, who maps, records, and analyzes all geologic data and sees that the sampling is adequate. Normally one level is not sufficient, and, to develop further, additional levels will be drifted out following the ore and limiting all possible extensions of the ore on each level. To prospect a level thoroughly it might be necessary to drift through barren rock or a barren vein in an attempt to find the continuation of a known ore shoot or to find a new ore body. The different levels will be connected by shafts or winzes, but wherever possible, because of lower mining costs, connections and mining will be upward from a lower level to a higher. Development, exploration, and mining methods differ for dissimilar kinds of ore bodies.

Sampling—Sampling means the taking of a small representative portion of a mass in such a manner that it will accurately depict the entire body.[56 57 58] Sampling techniques are not restricted in use to the mining geologist. Every geologist has occasion to sample a geologic body, hoping that in laboratory studies he will obtain a clearer picture of the larger feature. Since all deposits differ in characteristics from place to place, one sample will not accurately represent the entire body. To obtain more representative results, a number of samples should be taken; but there is an economic and practical limit to the amount of sampling possible, and the actual number of samples will depend to a large extent on the

[55] Jackson, C. F., and Hedges, J. H., *Metal-Mining Practice:* U. S. Bur. Mines Bull. 419, pp. 35-58, 1939.
[56] Forrester, J. D., *op. cit.*, pp. 349-367, 1946.
[57] McKinstry, H. E., *op. cit.*, pp. 35-69, 1948.
[58] Jackson, C. F., and Hedges, J. H., *idem*, pp. 35-58, 1939.

judgment of the sampler after the deposit has been studied. Indiscriminate sampling without regard for geologic conditions and the extent and grade of ore should be avoided. Wherever feasible, individual samples should represent material of only one general grade or character. Not only are more accurate determinations of mineral content possible, but the assay maps will show distribution by grade and width, and the content-distribution pattern may indicate otherwise unrecognizable guides and controls.

Several techniques can be used to obtain good samples. As stated in preceding paragraphs, cuttings and cores from drill holes can accurately represent the rock through which the hole passes. Another common method of sampling is to cut across the body along a predetermined channel, removing proportional amounts of the differing material. In addition to sampling in definite zones, grab samples are taken either according to a fixed pattern or more or less at random. Grab sampling is less accurate than channeling but serves a useful purpose for preliminary work or for checking other samples. Grab samples underground are taken from the piles of broken rock, ore chutes, or ore cars. Before taking a sample care must be exercised to see that the walls or workings are clean and represent the rock to be tested. Leached zones and accumulations of soluble salts must be removed along with all dust and mud.

A sample too large to be taken to an assay office economically or conveniently may be reduced to one-half, one-fourth, or one-eighth. By the use of a mechanical splitter fractions can be obtained that are of uniform grade. In most prospecting work, however, mechanical splitters are not available, and the sample must be divided by coning and quartering. As the term "coning" implies, the rock is first coned by pouring or shoveling all the sample onto one spot on a smooth, clean surface. The cone is uniformly flattened, the resultant mass is divided into quarters, and opposite quarters are combined to make a sample one-half the volume of the original. If further reduction is necessary, the procedure may be repeated one or more times.

Representation and Correlation of Data

Considerable data may be accumulated on a particular property, but unless the data can be evaluated and correlated they are of little value. One important factor and aid in correlation is proper representation of data.[59][60] Standard methods of showing underground geology include plan maps of all levels, cross sections, and longitudinal sections. The level map is most important in that sections can be constructed from a set of level maps. In mines where the geology is complicated, stope maps generally are made. These are of most value if made at regular intervals between levels and to be usable must be related in plan and elevation to

[59] Forrester, J. D., *op. cit.*, pp. 445-497, 1946.
[60] McKinstry, H. E., *op. cit.*, pp. 162-198, 1948.

the level maps. In mines of simple geology, the stopes may be mapped geologically only at critical points. For use in conjunction with level and stope maps, the geology of all shafts, raises, and winzes is recorded, and then by using all sources of information more reliable sections can be made.

To supplement level and stope maps, cross sections are prepared showing geologic relations in a vertical plane as contrasted to the horizontal plane of the level map. For a simple vein one or two cross sections may show all necessary information, but in more complex districts, sections developed at 50-foot intervals may be needed. Most sections differ from level maps in that information is not plotted directly but must be projected to a plane. Should it be through a shaft, raise, winze, drill hole, or mapped stope, more direct information can be obtained, and the cross section will be more accurate. Sections must be constantly revised to fit all new data. A cross section, even though from projected data, is an extremely important aid for visualization of three-dimensional relations and thus aids in the search for ore.

To give a third picture, all geology and mine workings are projected to a longitudinal section, which generally is a vertical plane approximately parallel to the strike of the vein. As veins are rarely regular in dip and strike, some distortion is produced in projecting geology and mine workings to a vertical plane. On longitudinal sections the relation of ore bodies or ore shoots to the vein and wall rock are very well shown and structural and stratigraphic controls may become apparent that were unnoticed in other sections or in plan. Because the wall rock on opposite sides of an ore structure may differ, the reference wall of the longitudinal section must be specified (fig. 524).

A combination of plan maps and cross and longitudinal sections shows the geology in three major planes, but in more complicated mines three-dimensional representations are often helpful. To show general relations isometric or block diagrams can be used (fig. 525). For detailed representation models are more informative. Numerous kinds of models have been made,[61] but the one most used by geologists is the transparent type with the geology and mine workings shown on either horizontal or vertical sheets of glass or plastic. Backed by a strong light source, it is possible to look through the transparent sheets and see the exact relations as they are known in the mine.

Numerous other ways of representing ore structures have been attempted. One of the most useful is a method by which contours of a vein are made which refer to a datum plane that can be horizontal, vertical, or inclined.[62] For showing subtle changes in dip or strike and the relation between ore distribution and the structure of the vein, an inclined

---

[61] McKinstry, H. E., *op. cit.*, pp. 180-182, 1948.
[62] Conolly, H. J. C., *A Contour Method of Revealing Some Ore Structures:* Econ. Geology, vol. 31, pp. 259-271, 1936.

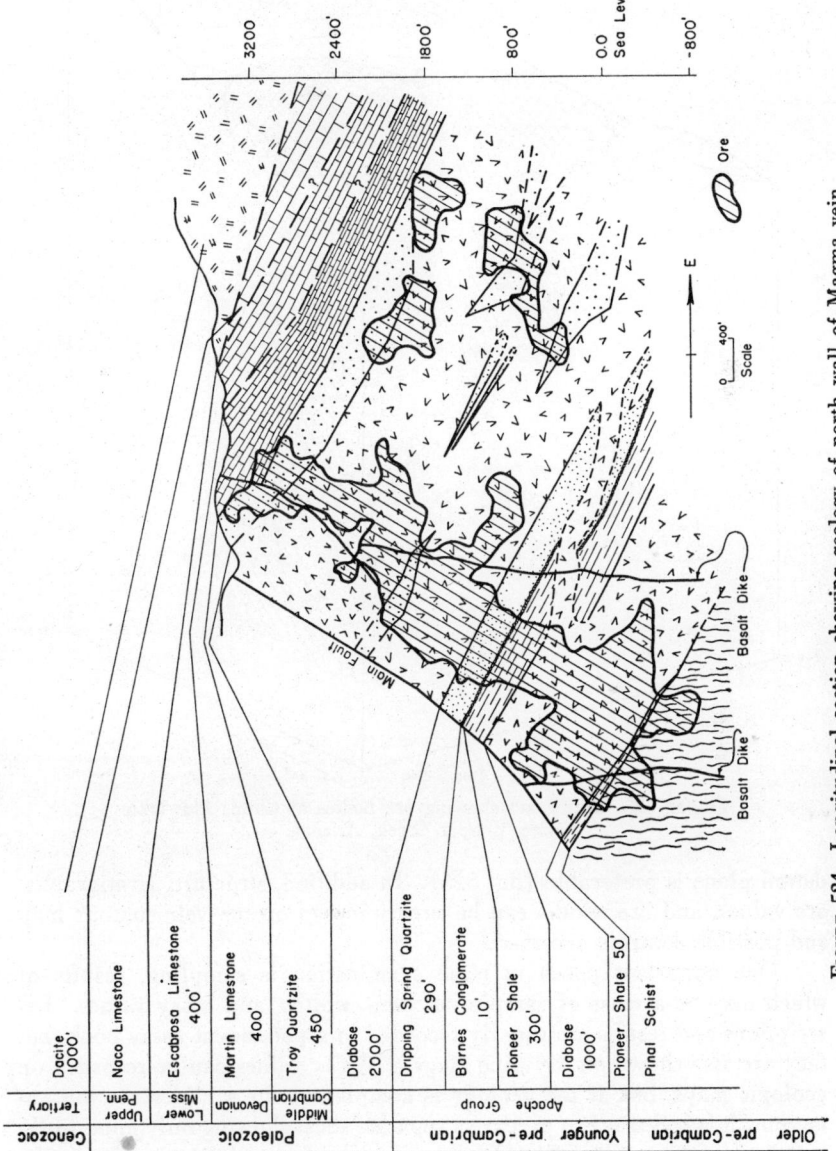

		Dacite 1000'
Cenozoic	Tertiary	
	Upper Penn.	Naco Limestone
	Lower Miss.	Escabrosa Limestone 400'
	Devonian	Martin Limestone 400'
	Middle Cambrian	Troy Quartzite 450'
Paleozoic		Diabase 2000'
Younger pre-Cambrian	Apache Group	Dripping Spring Quartzite 290'
		Barnes Conglomerate 10'
		Pioneer Shale 300'
		Diabase 1000'
		Pioneer Shale 50'
Older pre-Cambrian		Pinal Schist

3200'
2400'
1800'
800'
0.0 Sea Level
-800'

Ore

E

0   400'
Scale

Main Fault

Basalt Dike

Basalt Dike

FIGURE 524. Longitudinal section showing geology of north wall of Magma vein.

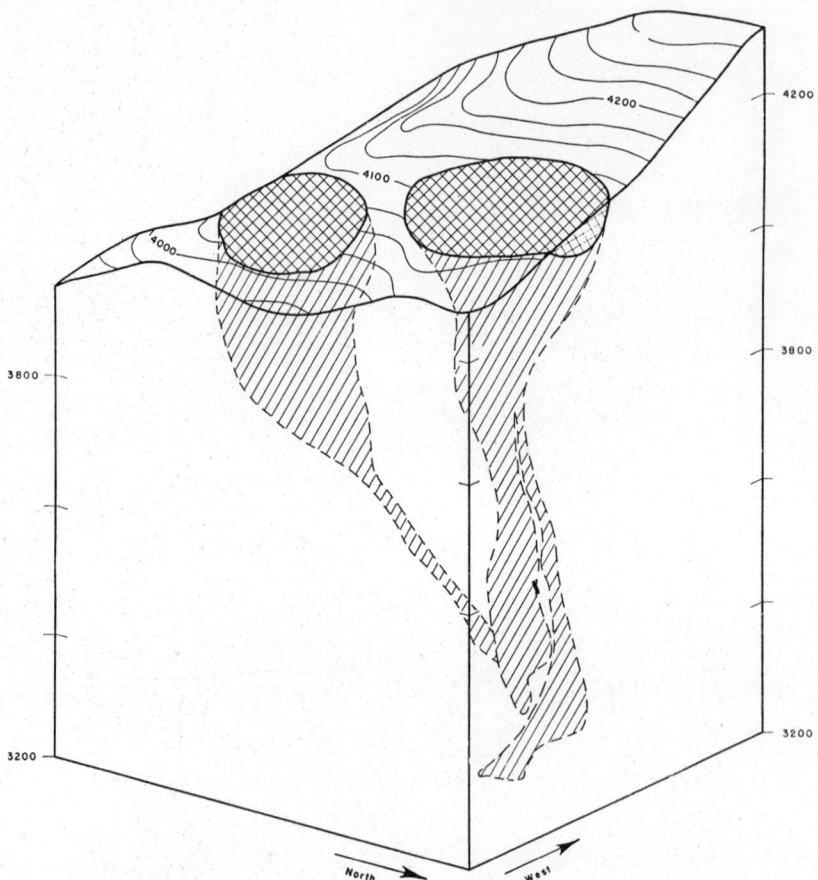

FIGURE 525. Sterogram showing ore bodies at Childs-Aldwinkle
mine, Copper Creek, Arizona.

datum plane is preferable (fig. 526). In addition, structure, stratigraphy,
ore values, and ore widths can be superimposed on the vein-contour map
and possible controls observed.

One important phase in mine examination is sampling, results of
which may be a mass of sample numbers, widths, and assay values. Descriptions and results are usually recorded in a permanent assay book, but
they are also shown on the mine maps.[63] Assay values can be recorded on
geologic maps, but, if the geology is complex or the samples are too numerous, it is necessary to prepare a separate sheet showing only mine workings and assay results. For clarity and ease of correlation, assay results
may be represented by color. The values also may be plotted near the
drift, raise, or stope with one coordinate approximately parallel to the
vein and the value coordinate at right angle to the vein trend. To account

[63] McKinstry, H. E., *op. cit.*, pp. 172-173, 1948.

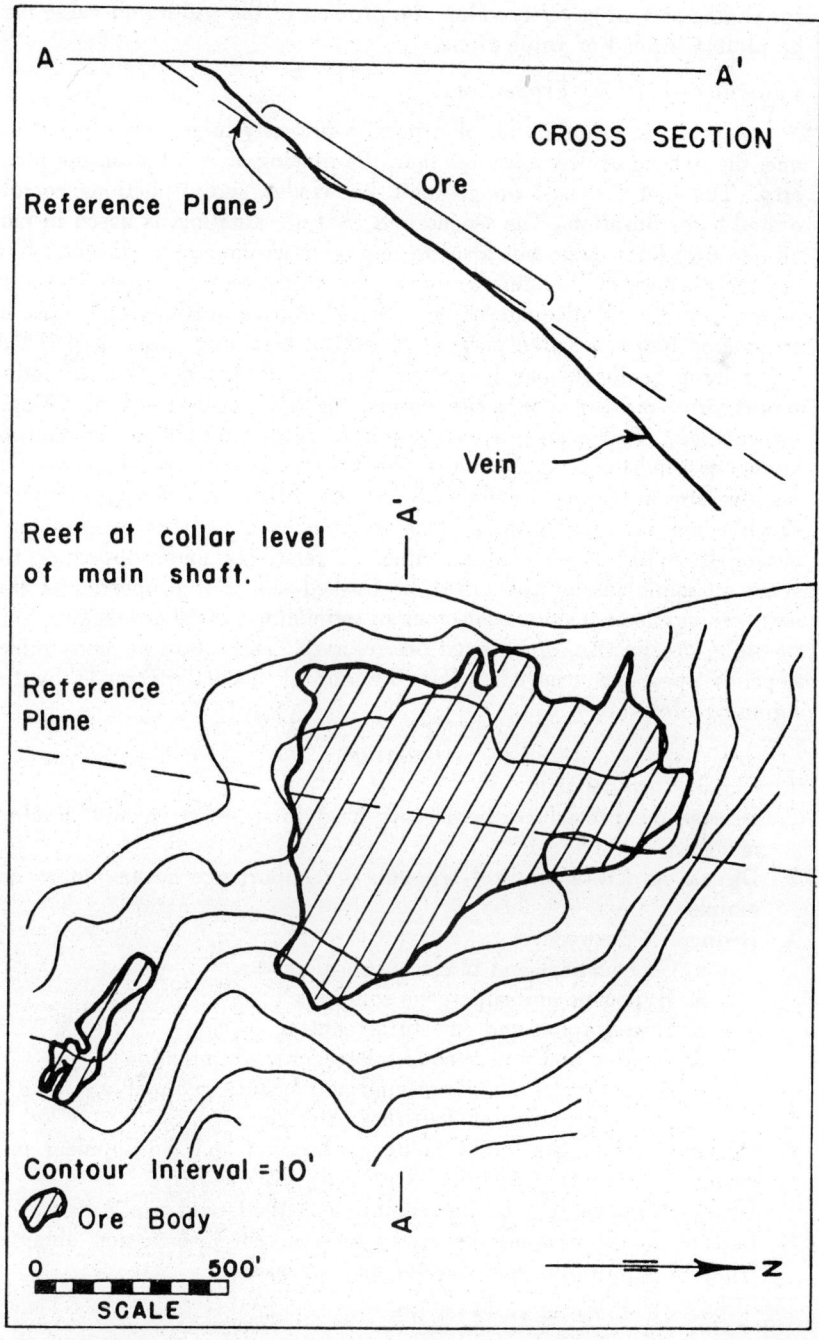

FIGURE 526. Sketch showing vein contours referred to an inclined plane. Ore bodies localized at change of dip. (After Conolly, *Economic Geology*.)

for widths as well as assay value, the product of the width and value may be plotted instead of value alone.

Valuation of Mining Properties

A geologic examination of a mining property might have two aims: one, the finding of ore, and the other, the placing of a value on the property. The first is based on geologic knowledge and deductions corroborated by exploration. The second aim, that of valuation, is based in part on geologic knowledge but also on numerous economic conditions. It is not the purpose of this chapter to discuss valuation or appraisal; recent books cover in an adequate manner the many ramifications that must be considered before a value may be placed on a mining property.[64][65][66][67] The mining geologist may be called upon to prepare the final valuation report, or he may work with the mining engineer, furnishing him geologic information. The basis for all valuation is the grade and amount of known ore and the possible future production. The measured, proved, or positive ores are easily accounted for in the valuation, but rarely is much ore blocked out in a property that is being examined for possible purchase. In most cases the mine either is relatively undeveloped or for some economic reason appears to be worked out. For properties of this sort, geology is of prime importance in estimating future production from possible, prospective, or inferred ore reserves. The future of many mines depends upon the accuracy of predictions or deductions made by the mining geologist.

Questions

1. Classify ore deposits.
2. Discuss the procedures commonly used in exploration and development work.
3. Discuss structural control stressing its importance as an aid to ore search.
4. Distinguish between:
   a. Geochemical and biogeochemical guides
   b. Hypogene and supergene solutions
   c. Stratigraphic and structural control
   d. Opaque and non-opaque microscopic examination
   e. Scales used for underground and surface geologic mapping.
5. Discuss the importance of wall-rock alteration.
6. Outline a geologists duties in an exploration and development program.
7. Define *sampling*, and outline sampling methods and precautions.
8. Discuss the illustrations necessary to represent information obtained from an exploration and development program.

[64] McKinstry, H. E., *op. cit.*, pp. 459-502, 1948.
[65] Forrester, J. D., *op. cit.*, pp. 491-573, 1946.
[66] Jackson, C. F., and Hedges, J. H., *op. cit.*, pp. 58-76, 1939.
[67] Parks, R. D., and Whitehead, W. L., *Examination and Valuation of Mineral Property*, 3d ed., Cambridge, Mass., Addison-Wesley Press, Inc., 1949.

CHAPTER 13

SUBSURFACE AND OFFICE REPRESENTATION IN MINING GEOLOGY*

SIGURD KERMIT HERNESS

The problems of geologic field and office map representation have been present since at least the time of Agricola, who lived in the sixteenth century. It was only towards the close of the nineteenth century that practical techniques of detailed mapping came into use in the industrial exploitation of mineral deposits. Winchell's mapping system, developed towards the close of the nineteenth century and still in use at Butte, Montana, was one of the first applications of detailed mapping to ore exploration. Since that time, the men interested in hard-rock geology have been too preoccupied with the theoretical and physiochemical abstract theories to continue research in bettering such techniques, and as a consequence the Winchell system is still in general use without modification.

The techniques and system evolved as a result of the writer's research has departed to a large extent from the Winchell system as used at Butte. The system has been gradually and carefully developed on the basis of logic and of trial-and-error methods. It is believed that this system permits far more detailed representation than any now in use and that the results are a saving in mapping time and expenditures. Because of the greater detail possible and the maximum utilization of graphic and colorimetric representation of qualitative and quantitative values it is believed that greater utilization of geologic data with consequent increased ore finding efficiency is obtained. It is hoped that continued research in representation technique will eventually introduce a better understanding of geologic processes such as the genesis and localization controls of ore deposits.

It is desirable that men who have been trained in this system will initiate its installation at properties where they accept employment, and that they will make efforts to improve upon the system. It is expensive to change mapping systems at large mines, once certain mapping techniques have been long established; therefore, when a mining property is in its infancy or youth, it is important that great caution be taken that every detail of a proposed mapping system will work towards the most efficient utilization of geologic data. Any geologist who accepts employment should assume that the property will eventually develop into a large-scale operation (even though the contrary is usually true) and must, therefore, exercise great care and judgment in the initiation of mapping technique

---

* This paper is a preliminary and partial release of material from a Doctor's thesis on mining-geology representation.

in order that future geologists at the property will not be hindered by inefficiency in ore finding.

In this paper it has been the intention to emphasize that geologic representation is of two types carried out in connection with two phases of work leading to ore finding: (1) field and underground representation techniques leading to efficient and detailed collecting of geologic data, and (2) office representation as related to the efficient and complete utilization of such data to predict the localization of ore. The importance of the first phase is fairly obvious. Perhaps there is more of a tendency to underrate the importance of the latter phase of representation. Many geologists seem to think that a map is a map, and that the map layout, scheme of colors or symbols, scale, and types of materials used in the preparation are of minor importance so long as the data have been collected and are available. As an illustrative example of the relative importance of office map representation the following is cited: During a two-year period at a certain mining property, the writer's ore-finding efficiency, utilizing ordinary methods of office map representation, was 12 tons of new ore added to reserve for each foot of sill development driven. During the next year, uitlizing improved representation on supplementary maps, the ore-finding efficiency was 54 tons of new ore per foot of development.

OBJECTIVES OF GEOLOGIC REPRESENTATION

The Purpose of Geologic Mapping

The main purpose of geologic mapping is to find mineral deposits, arrive at solutions to engineering problems, and increase the general knowledge of geologic processes.

Inasmuch as the controls of ore deposition are usually subtle, techniques of underground, field, and office representation of geologic data must be capable of expressing graphically subtle qualitative and quantitative variations of texture, composition, and structure in their relative spatial relationships.

All megascopically visible features should be mapped and described. These features include:

1. Physiography (topography and drainage).
2. Man-made features such as excavations, timber, cribbing, lagging, fill or gob, roads, buildings, installations, and survey points. All should be indicated and described on the note sheet.
3. Structural features including primary flow structure, bedding planes, fissures and faults, veins, and joints.
4. Compositional components such as the type of rock, color and facie variations, mineral components, degree or rate of gradation of one rock type to another, superimposed metamorphic or hydrothermal effects, vein fillings, and composition of soil or overburden.
5. Textural components such as grain size, variation in grain size,

orientation or attitude of grains or minerals, and texture of soil or overburden.

There must be no sorting of data to eliminate "unimportant" facts. The relative importance of unimportance of data cannot be determined until after the entire mine, district, or area has been mapped. Often the importance of certain apparently insignificant data is realized 20 or 50 years later, by which time the workings have probably become inaccessable for remapping. Because mine excavations seldom remain open long, the mining geologist is responsible as the "custodian" of all data, and he must make as completely unprejudiced and impartial a representation of fact as his technique will permit. At the time that he is mapping he should give no thought to the interpretation and correlation of the things he sees, but should completely detach himself from the interpretative problems on hand and map fact mechanically. In this connection Harrison Schmitt[1] says: "Most mining geologists with a number of years of experience in mapping believe that all details capable of being mapped should be recorded, including those which at first appear to be of remote significance. *They always become significant when integrated and plotted on the office maps.*" Wilson [2] likewise says: "It is rare that the significant features, that is, those which may help find ore, are all definitely known. It is only after details, no matter of how much present unimportance, have been mapped, studied, and correlated that the essential ones can be selected."

Geologic Technique and Mineral Exploration

Many college graduates become employed by large companies, and a year of intensive training after employment is usually necessary to finish the graduate's preparation for the effective application of geologic technique to ore finding. The school that takes responsibility for a more complete training in applied techniques will have little difficulty in placing their graduates, once the extent of such training becomes known to the mining industry.

Then, too, such training will result in increasing utilization of geologic technology by mining companies, and especially by the smaller mining companies. Up to the present time, these smaller companies have been discouraged when employing untrained geology graduates because of the poor results whenever they have done so. The past failure of geology in connection with the small mine operator has resulted from the operator's inability to complete the training in techniques of the men he hires. Men who have been trained by the larger companies are reluctant to accept positions with small operators.

The other alternative of the small mining company is to hire high-priced "experts" or consultants, who often are nothing less than respecta-

---

[1] Schmitt, Harrison, *On Mapping Underground Geology:* Eng. and Min. Journ., vol. 137, p. 557, 1936.
[2] Wilson, Philip D., *Report on the Collection, Recording and Economic Application of Geological Data:* Am. Min. Congress, 25th Ann. Convention, 1924.

ble racketeers who prey on the small operators and investors. Some of these are men of "respectability" and prominence who have never found a pound of ore but have achieved prominence through some means or another—such as publication of numerous papers on abstract theory. They usually give nothing for their high fees except a nonoperating report setting forth generalities that the operator already knows. The writer knows of one consultant who was paid a fee of $30,000 for a five-day examination at a mining property without even going underground. The report contained nothing of value but a prominent name on the title page.

Past experience, therefore, has tended to lower the prestige of geologists with the small mine operator, who today sees no reason to increase his operating overhead by the employment of nonproductive personnel. And yet the small operator is the very person who can least afford the luxury of being without competent geologic technology. A large company can more afford to make a mistake in planning ore-finding development than can the small company. To a small company one costly mistake means the end of the enterprise. Numerous operating failures result from lack of ore-finding technology, and thus the vicious circle is completed.

The difference between efficient, effective, economic, research application of geology and ineffective geologic application is the difference between skilled and unskilled field and office representation of geologic data. The degree of success in finding ore is usually a factor of the man's ability to map detail and make effective representation of the facts collected so that they can be integrated and readily interpreted. Good representation makes it possible to predict localization of ore from a study of empirical or statistical relationships of geologic data and without necessarily knowing the theoretical factors or processes that brought about this localization. In this respect ore finding is based on statistical odds and very often on empirical criteria.

Geologic Technique and Theory

The literature is saturated with undocumented and interpretative maps and charts of all types, theories, and hypothesis; but very little detailed and objective factual data is available for study. Therefore, conclusions and theories are arrived at on the basis not of factual data but on the weight of authority.

For a man to become a proficient surgeon he must study medical theory, but he must also take intensive training in the techniques of surgery. The medical profession is loaded with technique and that is why this branch of science has progressed so rapidly. Today mining geology is in a primitive stage of development in comparison to what this science will eventually become, once we become less pretentious about our knowledge of geologic processes, get our feet on the ground, and devote some time to the less "brainy" study of ore finding and fact-finding techniques. The results of geologists' inability and incompetency in gathering factual data is that little progress has been made during the last half century in the solution and understanding of the processes active in the bringing about of genetically interrelated orogenic effects such as intrusion, differentiation, metamorphism, hydrothermal effects, and ore deposits.

The best-equipped research laboratory available for such study is the field where abundant data is available to the geologist who possesses the training, skill, and initiative to gather and integrate this material. We only need to sharpen the tools that we work with, and great progress and revitalization of hard-rock geology is inevitable. At this moment the tool that needs to be sharpened is geologic-representation technique.

ESSENTIALS OF GEOLOGIC REPRESENTATION

Any effective field-underground and office-representation technique or system in terms of purpose and specifications should measure up to the following:

1. Techniques are such that the average mining geologist can be trained to acquire effective skill in their utilization. If these are overly tedious or slow, mapping and representation is retarded.

2. Geologists using a given system are calibrated in order that their representation will conform in qualitative and quantitative intensity so that their collective efforts are integrated into one set of maps. Two or more geologists working together in one organization are then capable of calibration to a degree that independent mapping of any given area will produce nearly identical representation of geologic values. In cooperative mapping there is no visible break or change in representation where one geologist terminates and the other continues. After a system has been decided upon, there is little place for individualism in geologic representation except in the degree of neatness and precision of note recording and plotting.

3. Such a system or technique is capable of representing all megascopically visible features.

4. Representation legend is based on *systematic* and logical chromatic or geometric sequence and arrangement rather than haphazard use of colors and symbols in order that:

   a. Representation colors and symbols are easily remembered even by non-geological personnel.

   b. The general geology is evident from casual inspection of the maps by the optional ignoring of slight variations in symbol or color utilized to show minor facie variation. By logical grouping of representation symbols or color in such a manner that similar lithologic values are represented by similar colors, it is possible to interpret readily the map in detail or in generalities.

   c. Qualitative geologic values are as accurately represented graphically as these values can be estimated or determined in the field.

   d. Quantitative geologic values are represented as accurately as they are determined in the field.

   e. It is possible to show gradational trends as well as abrupt changes; where obscure and uncertain field relationships exist, the representation must be equally obscure and uncertain.

   f. Subtle values are magnified so that they are readily apparent on the maps.

   g. Qualitative and quantitative values represented colorimetrically or graphically are cumulative in terms of compositional trends, process, time, and criteria of ore localization. Thus, all hydrothermal effects are represented by similar colors so that highly hydrothermalized areas are readily apparent and the *total* hydrothermal effect is readily evaluated.

   h. An office-map system is such that economic or assay metallization representation is quantitative, qualitative, graphic, cumulative. Metallization may be readily integrated with the geologic representation.

5. All field-subsurface and office representation is accomplished on a basis of factual data. McKinstry[3] defines a map as

> a record of geological facts in their correct space relations—facts, be it noted, not theories. There must always be sharp distinction between observation and inference. You can see a contact where it is exposed but you cannot see it under covered ground. . . . This failure to distinguish between fact and inference is a criticism that can justly be leveled at some of the otherwise impeccable maps published by government surveys . . . The geologist who fails to distinguish fact from inference on his maps is inconsiderate both of other workers and of his own reputation. As successors cannot tell which localities supplied the evidence, they must search the whole area for exposed contacts. They must either accept all of the work, fact and theory alike, or reject it all and start the mapping from scratch . . . Thus the map should be drawn in such a manner that either the man who made it or some one else will later be able to eliminate all the interpretation, preserve all the observations and build up an entirely new interpretation on the same set of facts . . . It is his (the geologist's) duty to offer interpretations at every opportunity, for no one is in a better position to draw inferences than the man who has made the maps and studied the ground. It is not speculation or imagination that is to be discouraged; it is merely the failure to recognize and to indicate the element of uncertainty, a failure which carries not only mechanical but also psychological dangers. The misconception that one's theoretical interpretation is the only one possible is likely to be exposed by the propensity of nature to contrive an interpretation that the geologist had not foreseen.

6. Any portion or portions of an area that has been examined by a geologist is accounted for even though there are no outcrop exposures available for mapping. A good working theory of mapping is to account for all area which has been examined in the field by posting some symbol or color representing the mapped portion on the note sheet. If the area is covered by soil, a symbol representing soil cover is posted

---

[3] McKinstry, Hugh E., *Mining Geology*, Prentice-Hall, p. 1, 1948.

on that portion of the note sheet; otherwise, there is no way of knowing which areas have been examined and which have not been examined.

7. An efficient mapping system permits individuals or departments to accomplish their duties cooperatively with a minimum of confusion and friction, and a maximum of efficiency. Professional draftsmen post geologic data directly from note sheets prepared by the geologist with little or no oral or written supplementary instruction. The ground or excavation lines and other survey data are posted by the engineering department or surveyor. Sampling done either by a sampling department or the shift bosses is posted as assay data from sample sheets. It is therefore imperative that every attempt be made to install a system that is adaptable to cooperative map preparation.

8. It is desirable that the number of maps be kept to a minimum with no duplication, unnecessary scales, or overlap in area. The writer recalls examining a mining property that had closed down for apparent lack of ore. He was given a piano box crammed with maps of the mine. No one had the slightest idea of the geologic and assay data available in these miscellaneous odds and ends of maps compiled by various operators during sixty years. When *all* the data had been replotted, it was represented on fifteen sheets each 20″ x 40″. This integrated information indicated a half-million-ton ore reserve of good grade already developed; and the operators were executing development away from this ore on barren vein.

9. Efficient working maps are prepared with the objective that revision and correction can be readily made.

10. Sectional representation is integrated with plan-map data. Because there is no planning for efficient utilization of sections, much cross- and longitudinal-section data are collected and lost in the files. At some mining properties there are hundreds and thousands of subsurface sectional note sheets that are haphazardly oriented and filed.

11. One condition of a good mapping system, often overlooked by mining companies, large and small alike, is that notesheets and maps are filed in a manner that geologists and draftsmen alike can readily and with minimum effort locate notes or maps of any particular area or place.

12. Coordinate values are carefully determined so that error and confusion will be reduced to a minimum. Coordinate values are large enough so that all points in the district are designated in terms of north and east, never as west or south values. In other words any point within the mineralized area is capable of location in positive or plus values and never in terms of minus values; the zero value lies well outside and to the south and west of the mineralized area regardless of property boundaries. It is these zero areas that are the danger areas for "boners." In most cases 50,000 north and 50,000 east is about right for the central portion of a district.

13. Scales are such that detail is shown but not so large that areas covered

are too small for effective interpretation and correlation. The use of many scales is avoided, and the use of "odd" scales, such as $1'' = 30'$, $1'' = 80'$, is discouraged. For operating geologic mapping and representation, scales of $1'' = 50'$ and $1'' = 200'$ will usually suffice. For district maps, smaller scale maps such as $1'' = 1000'$ and $1'' = 5000'$ are useful.

14. Office maps are not excessively large. The initial cost in tracing cloth is greater than if smaller sheets are used. Large maps are difficult to handle and are soon damaged or destroyed by use. It should be possible to reach any portion of the area without creasing the bottom portion of the map over the edge of the table. Map size is standardized for all scales so that blank map stock is procured or prepared at minimum cost, and systematic filing by coordinate value is facilitated.

15. The note sheet is of standardized size in order that sufficient area is covered on each sheet. Frequent transfer or changing of sheets underground or in the field is unnecessary. These sheets are not so large that they are difficult to handle. A letter-size sheet has been found to be the optimum size.

16. Note sheets and maps are prepared so that coordinate lines are parallel to the edges of the sheets. Nothing presents such an untidy appearance as a map with the coordinates at an angle to the margins of the sheet. Then, too, it is difficult to orient directions of structure because most people are accustomed to thinking in terms of the cardinal directions parallel to the edge of the sheet.

17. Map and note-sheet areas conform to a district wide grid system so that no overlap in the area exists. The office-map area is an even multiple of the note-sheet areas. Systematic layout of this type creates efficiency.

18. The material used in geologic representation is chosen with durability in mind. It is poor economy to post geologic data on tracing paper if the posting is to cost hundreds or thousands of dollars in labor. Operating maps should always be on tracing cloth of good quality. Underground and field pencils are waterproof (not "indelible"), and are capable of maintaining fine points. Colored inks used on maps are waterproof.

19. Maps and note sheets should present a pleasing appearance and should be executed with a sense of artistry. The finding of ore is an art as well as a science.

20. Good geologic notes and maps must have sales appeal; the geologist uses his maps not only to work out his concepts but to sell them to mine operators. It is important that geologic notes by their very appearance inspire confidence. Nothing is more impressive evidence of the ability of a geologist than a well-executed note sheet or map. The training and skill of a mining geologist is apparent from a very casual inspection of his notes and maps. Most experienced mining

geologists judge other mining geologists' ability by their skill in taking notes. Some men say this ability is not important; however, these men have one other thing in common—they have never found a pound of ore.

The System

Mapping Scales

The writer has found that the following scales integrate best into a mapping system and are adequate for all purposes.

1. Extreme detail .... $1'' = 1'$
2. Detail .... $1'' = 10'$
3. Operating detail .... $1'' = 50'$
4. Semi-detail correlation maps... $1'' = 200'$
5. Detail reconnaissance and district maps.......................... $1'' = 1,000'$
6. Sub-regional maps.. $1'' = 5,000'$
7. Regional maps .... $1'' = 10,000'$
8. Smaller scale regional and provincial maps are also useful for district evaluation of regional metallization and geologic trend or pattern. Provincial geologic and metallization representation often results in a better evaluation of local pattern with consequent increase in ore-finding efficiency.

It will be noted (with two exceptions) that the above scales increase progressively in a ratio of $1:5$. The exceptions are the ratios between the $1'$ and $10'$ scale units and the $50'$ and $200'$ scales; in the former case the ratio is $1:10$ and in the latter case $1:4$.

The first two detailed scales mentioned under 1 and 2 are used only when necessary to show extreme detail significant in the interpretation of time or structural relationships.

There are so many advantages in using the $50'$ scale detailed unit over the $40'$ scale unit that the latter scale is obsolete. Some argue that a $40'$ scale permits more detail than $50'$ scale mapping; however, it is contended that, inasmuch as the purpose of a larger scale is the representation of small features of less than $1'$, the difference in width of $1'$ using the $50'$ and $40'$ scale is too slight to introduce any practical increase of detailed representation.

The difference in area which is mapped on a note sheet of constant size is in the ratio of $12:20$; thus, almost twice the amount of area is mapped on a sheet of given size if mapping is done to $50'$ scale than is mapped on the same sheet using the $40'$ scale mapping unit. Since more mapping can be accomplished on each sheet, fewer transfers of traverse from one sheet to another are necessary. If the map "quadrangles" cover standardized unit areas such as $1000' \times 2000'$, the utilization of 50′ scale representation results in smaller-sized office maps. Such "quadrangles" require a 25″ x 50″ sheet if posted to 40′ scale and a 20″ x 40″ sheet if posted to 50′ scale. The smaller 50′ scale 20″ x 40″ map books are much more convenient to handle; consequently wearing longer before replacement becomes necessary, and the cost of tracing cloth is 40 percent less. At many mining properties stope-book floor plans are prepared with a plan map for every 8′ or 10′ vertical interval. Such stope books sometimes contain 200 or 300 pages, and cover the entire area of the mine. It can be appreciated, therefore, that the saving of expenditure for cloth is a large item.

If 200′ scale representation is utilized for detailed correlation maps, 100′ scale maps serve no useful purpose. In actual practice the detail that can be represented on 100′ scale maps is not much more than can be shown on 200′ scale maps. If plotting is on a sheet of constant size, 100′ scale maps cover an area only one-fourth as large as that covered by the latter scale.

Mapping and representation scales of intermediate values between 200′ and 1000′ scale can readily be dispensed with. It seems that in mapping and representation practice, a leveling off in the ratio of the detail representation possible to the rate of scale increase takes place above the 200′ scale value. It is therefore possible to show nearly the same detail in 1000′ scale mapping as in 400′ or 500′ scale mapping.

Mapping on smaller scales than $1000′ = 1″$ is often useful to determine the general geology of an area and to delineate the "hot" areas for more detailed mapping. Data useful for mineral exploration may be secured from published maps and plotted on small-scale regional maps. Local geology can be better evaluated if the regional geologic, metallogenic, or tectonic environment of an area is known and clearly indicated on maps.

Coordinate System

Coordinate values are first determined when initiating a geologic representation system. As stated previously, a coordinate system must be such that all points within a district may be located in terms of positive (north and east) values. The north and east zero values must, therefore, be well outside the mineralized area. In order to avoid negative values, numbers less than five digits are never assigned to the initial point. It is desirable to derive local coordinate values from a national, provincial, or state-wide system that has previously been established by federal or state agencies or bureaus. These coordinate systems are of such regional nature that values are too large to be used with convenience in actual local practice; it is, therefore, advisable to drop the first few digits and limit local values to the last five or six digits. Thus, if the regional value of a point is 5,654,432N and 6,476,451E, this point can be assigned a local value of 54,432N and 76,451E and used as the initial point for the local system.

Areal Map Grid System

In order to bring about efficiency and coordination of surveying, field mapping, sampling, office map preparation, ore-reserve calculation, indexing, and filing, it is necessary to subdivide the entire district into systematic rectangular subdivisions. All map layout, indexing, and filing of data are based on the grid system described below. This grid system is based on the assumptions that letter-size field note sheets containing an 8" x 10" map area and office sheets containing 20" x 40" map areas are used. These standardized sheets are used for both plan and sectional representation.

Note Sheet Grid—The size of rectangle included in an 8"x10" note sheet area at 1" = 50' scale is 400' × 500'. This area is the basic unit area or "quadrangle" into which the entire district is subdivided. These areas are laid out in a grid system with the long dimension of the rectangle oriented north.

The boundaries of these unit areas are determined by the coordinate system. The north and south boundaries of the unit areas are made to coincide with ordinate values whose last three digits are alternately 000 and 500; thus, the border local ordinate values are always evenly divisible by 500 and increases as follows: 47,500N–48,000N–48,500N–49,000N– etc. The east and west boundaries of the unit areas are made to coincide with ordinate values which are evenly divisible by 400. Thus, the sequence may be 52,000E–52,400E–52,800E–59,600E–60,000E–60,400E– etc.

A rectangle that is four unit 50' scale areas long and four unit areas wide, contains sixteen of the 50' scale unit areas, and is 1,600'x2,000'. This is the unit area used for 200' scale mapping on the letter-size standardized note sheet. The north and south regional ordinate boundaries of these 200' scale unit areas are always evenly divisible by 2,000. The east and west boundaries are evenly divisible by 1,600. The boundaries of every fourth 400'x500' rectangle coincides with the boundaries of this larger unit area. (fig. 529)

If the 200' scale rectangles are combined in multiples of five in north-south and east-west directions, a 1,000' scale unit results in a map area of 8,000'x10,000'. The north and south regional values of the boundaries are evenly divisible by 8,000; the east and west ordinate boundary values are divisible by 10,000.

In a similar manner 1,000' scale unit areas are combined to form 5,000' unit mapping areas, which are 40,000'x50,000'. The 50-scale 400'x500' unit rectangles are subdivided where necessary into 10-scale 8'x100' rectangles for representation of extreme detail. These areas may be further subdivided into 1' scale rectangles of 8'x10' size.

Sectional Grid System—The same type of letter-size note sheet is used for both plan and sectional representation. Whenever possible all sectional representation is made or projected on vertical planes parallel to ordinate lines. All sections are prepared as if facing in either a northerly or easterly direction. Sections are prepared in a manner that lateral boundaries coincide with the boundaries of the above-described plan unit-area rectangles of the same scale. This manner of preparation is so that sectional data may be filed with the plan data of the same unit area. Sections must not laterally overlap from one "quadrangle" to another because confusion in indexing and filing will result.

East-West Sectional Grid—This grid, which is identical to the plan

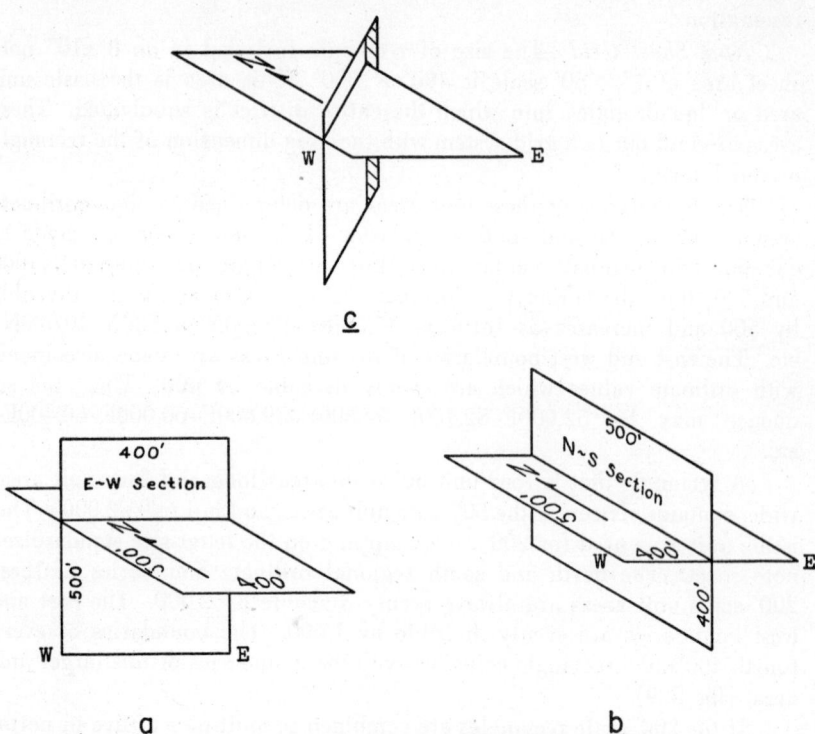

FIGURE 527. (a) Relationship of east-west sectional unit area to plan unit area; (b) relationship of north-south sectional unit area to plan area; (c) correlation of diagonal sectional boundaries with plan area boundaries.

grid, is erected in a vertical plane of east-west orientation in such a manner that the long dimension of the rectangle is vertical. The east and west borders of the sectional map areas are made to coincide with the east and west borders of the unit plan-areas. The elevations of the top and bottom borders of the 50' scale-unit sectional areas are made to coincide with elevations evenly divisible by 500 (fig. 527a).

North-South Sectional Grid—A grid in a vertical plane of north-south orientation is erected with the short dimension of the rectangle vertical.

The north and south borders of the sectional-unit areas coincide with the north and south borders of the unit plan-areas. The elevations of the top and bottom borders of the 50' scale-unit-sectional areas are made to coincide with elevations evenly divisible by 400 (fig. 527b).

Oblique Sectional Grid—Vertical sections, diagonal to the ordinate lines, are prepared in a manner similar to the north-south sections. The northwest or southwest boundaries of the sectional unit areas are made to coincide with the west (or, if necessary, with the north or south) unit plan-area boundaries which are traversed by the lines of section. That portion of any note-sheet area which extends across the boundary into an adjacent unit-plan area is not utilized; a new sheet is prepared for this portion of the sectional area. The elevations of the top and bottom of the rectangles are made to coincide with elevations evenly divisible by 400 (fig. 527c).

Office-Map Grid System—The office-map grid system is based on the same principles as the note-sheet grid system. Ten unit note-sheet areas of any standard scale combine to form a unit office map area. The office-map grid system is based on the combination of ten 8"x 10" note-sheet areas to form a map area 20"x 40". If two east-west tiers, of five 50' scale note-sheet areas in each tier are combined, the resulting 50' scale office-map rectangle is 1,000' wide in a north-south direction and 2,000' long in an east-west direction. (See fig. 530.) Standardized sheets of 20" x 40"map area are used for all scales of office representation—sectional and planar. Large-scale office-map areas are even multiples of small-scale map areas (fig. 531).

The north and south borders of the 50' scale unit office-map areas are made to coincide with north-regional-ordinate values evenly divisible by 1,000. The east- and west-ordinate values are evenly divisible by 2,000.

The ordinate values of the north and south boundaries of 200' scale map areas are divisible by 4,000, and the east- and west-boundary ordinate values are divisible by 8,000.

Similarly, the ordinate values of the north and south boundaries of 1,000' scale map areas are divisible by 20,000 and the east- and west-boundary ordinate values are divisible by 40,000.

North-ordinate values of the 5,000' scale-map boundaries are divisible by 100,000, and the east-ordinate values are divisible by 200,000.

Sectional Office-Map Grid System—In most respects the office-map sectional grids are the same as the previously described note-sheet sectional grids.

East-West Sectional Grid—East-west office sectional unit-areas are oriented so that the long or 40" dimension of the rectangle is in an east-west direction, and the east and west borders of the sectional map areas coincide with the borders of the plan map areas of the same scale. The elevations of the top and bottom borders of the sectional rectangle are always evenly divisible by 1,000 (fig. 532).

North-South Sectional Grid — North-South sectional unit-areas are oriented so that the short or 20″ dimension of the area is in a north-south direction, and the north and south borders of the sectional rectangles coincide with the north and south borders of the plan rectangles. The elevations of the top and bottom of the rectangle are divisible by 2,000 (fig. 532).

Diagonal Sectional Grid—These are laid out in a manner similar to the note-sheet diagonal grid. The short or 20″ dimension of the office-map rectangle is vertical and the elevations of the top and bottom are evenly divisible by 1,000.

Specifications of Data and Map Sheets

All essential field and office data and map sheets used may be printed from four plates. Survey sheets, field and underground note sheets, and ore-reserve listing and classification sheets are printed on ledger paper. Assay-record sheets are the same as the field-note sheets, but the printing is on tracing paper instead of ledger paper. Office geological map sheets are printed on cloth. Sheets for the representation of metallization, ore-reserve blocks, geochemical or biochemical data, vegetation, or geophysical data are printed from the same plate as that used for the office geological maps and are prepared on clear plastic material.

Survey Sheets—The survey sheet is dual purpose in that it is used both as a field recording sheet and office survey ledger sheet. The printing is on 32-pound ledger paper twice the size of a letter sheet, so that, when folded at the center, it may be placed in either a letter-size, spring-back field book (of identical type to that used for geology field-note sheets), or in an office post binder. When thus folded each sheet makes two oppositely facing pages. Data collected in the field and all preliminary calculations are entered on the left-hand page. Latitude, departure, and elevation data are entered on the right-hand page. To insure permanency, it is desirable to retrace all field data in India ink. The horizontal lines on the sheet are of such spacing that all the calculations may be typewritten for clarity in reading. A key-hole punching along the folded edge permits removal of any sheet without taking the binder apart or removing the top sheets. In the upper right-hand corner is an entry space for coordinate indexing; this indexing indicates survey data source by 400′x500′ unit areas (fig. 528).

Field- and Underground-Map Sheets—A letter-size field-note sheet with a 1″ grid covering an 8″ x 10″ map area has been found to promote the maximum efficiency in mapping and filing. The coordinate grid is printed in black on 28-pound ledger paper. On the north 8″ border of the map grid is a 1″ margin, which fits into the binder of a spring-back field book; on the east border is a $\frac{1}{2}$″ margin in which location, elevation, and filing data are entered in appropriate places provided for such information. The south and west margins of the sheet are kept to a minimum

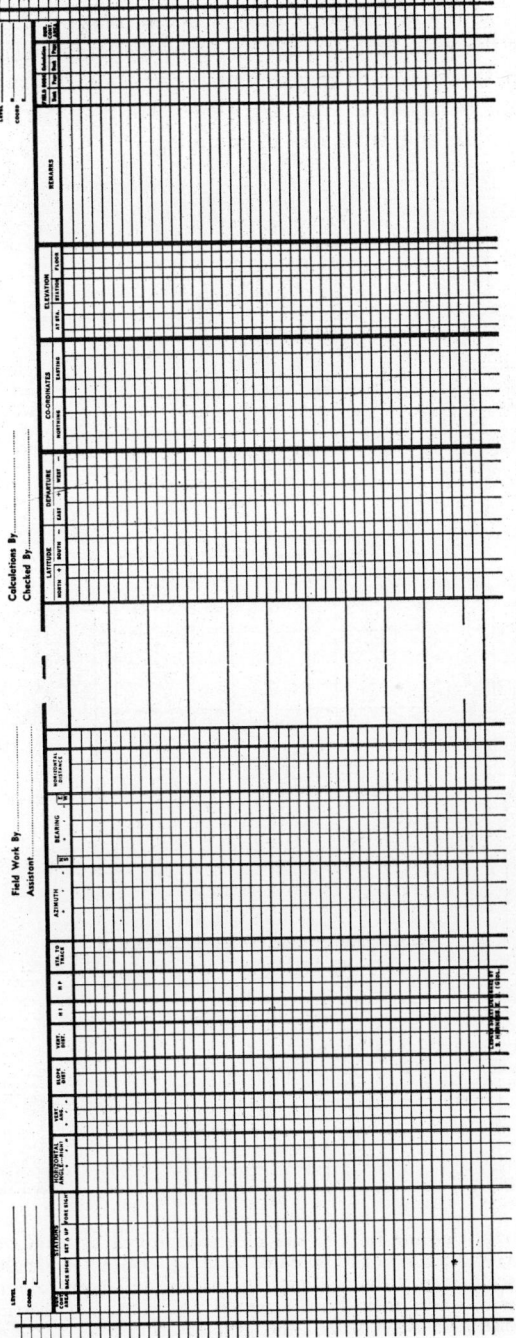

FIGURE 528. Field-recording and office-ledger survey sheet.

of less than $\frac{1}{16}''$ so that the narrow west or south margin of one sheet may be lapped over the wide east or north margin of the adjacent sheet. The narrow overlap facilitates the transfer of surveys and excavation when a map-area margin is reached. Identical sheets are used for plan and sectional representation and all scales of mapping.

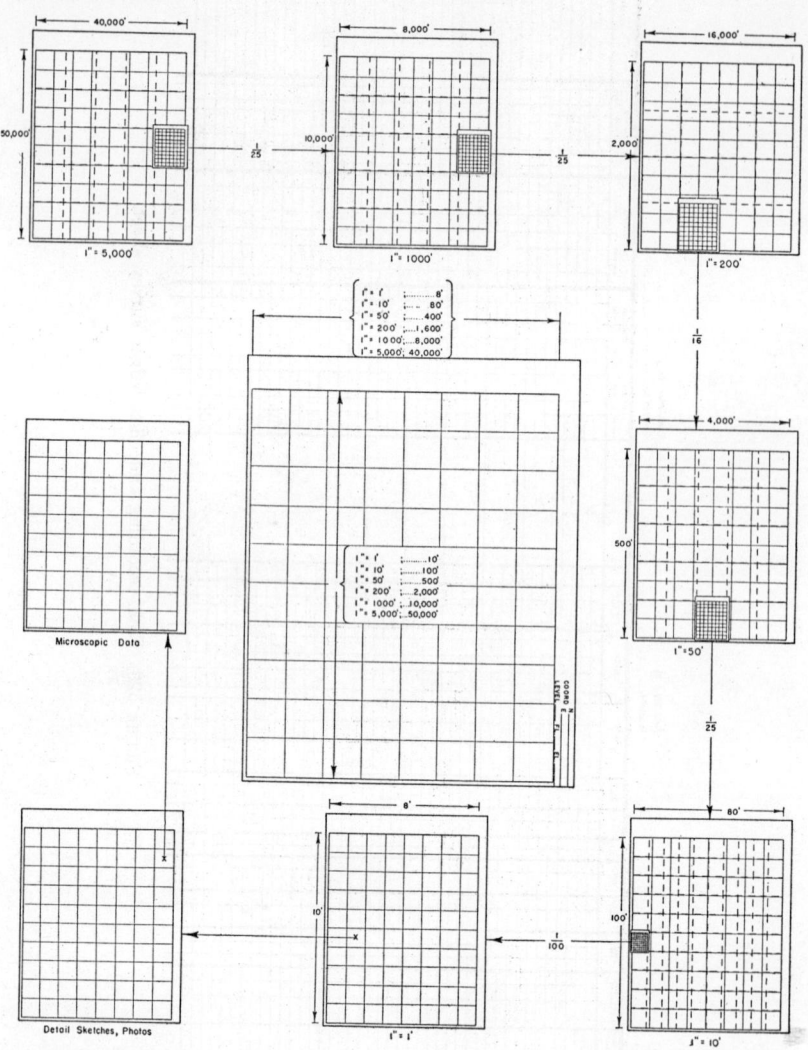

FIGURE 529. Diagram showing relative size of areas covered by standard 8½"x11" notebook sheets using various standard mapping scales. It should be noted that smaller-scale map areas are even multiples of larger-scale map areas; this is true only when critical scale values are used. (If metric system is used, letter-size field sheet with 1-cm. grid covering a 20 x 25-cm. map area is used. Metric scale ratios are 1/500, 1/2,000, 1/10,000.)

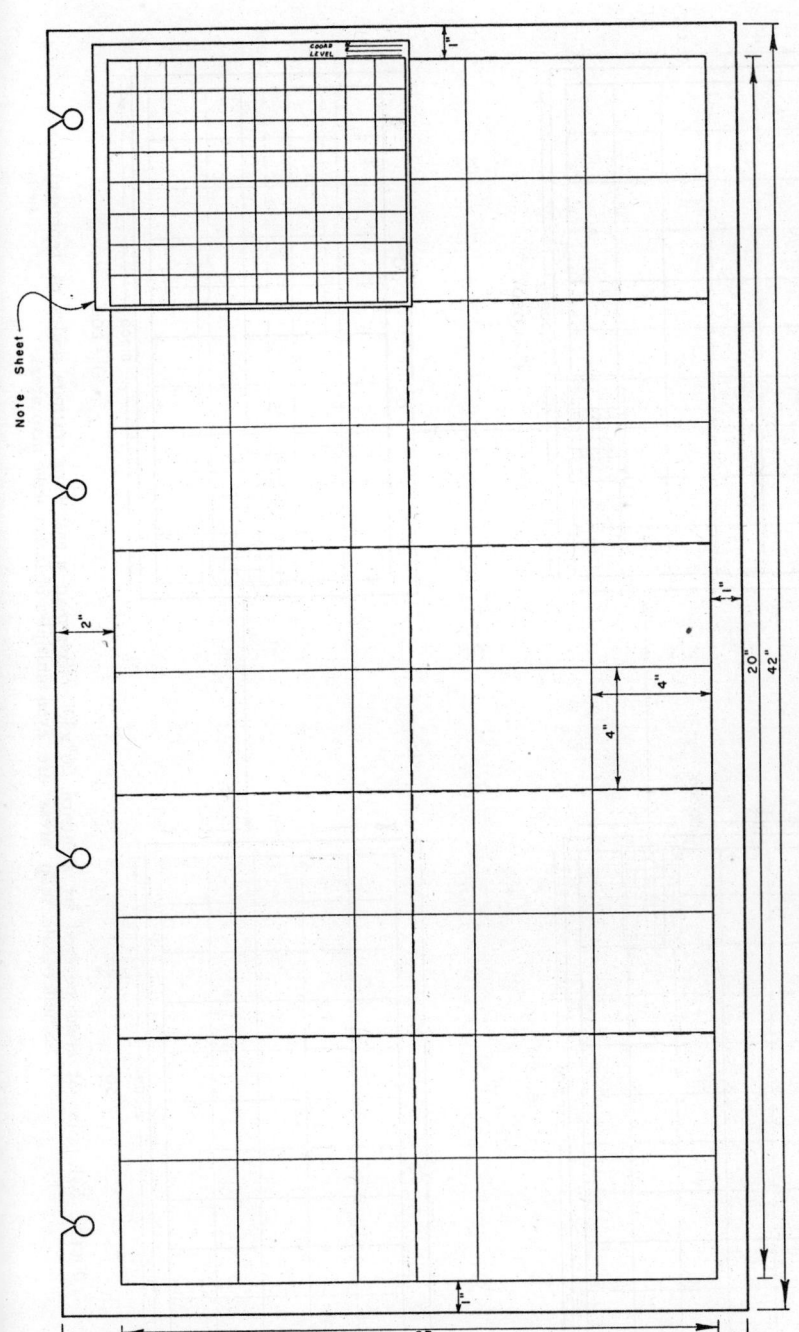

FIGURE 530. Relationship of 8″x10″ note sheet area to standardized 20″x40″ office map area. Office map specifications are indicated. (If metric system is used, map-area size is 100 x 50 cm.)

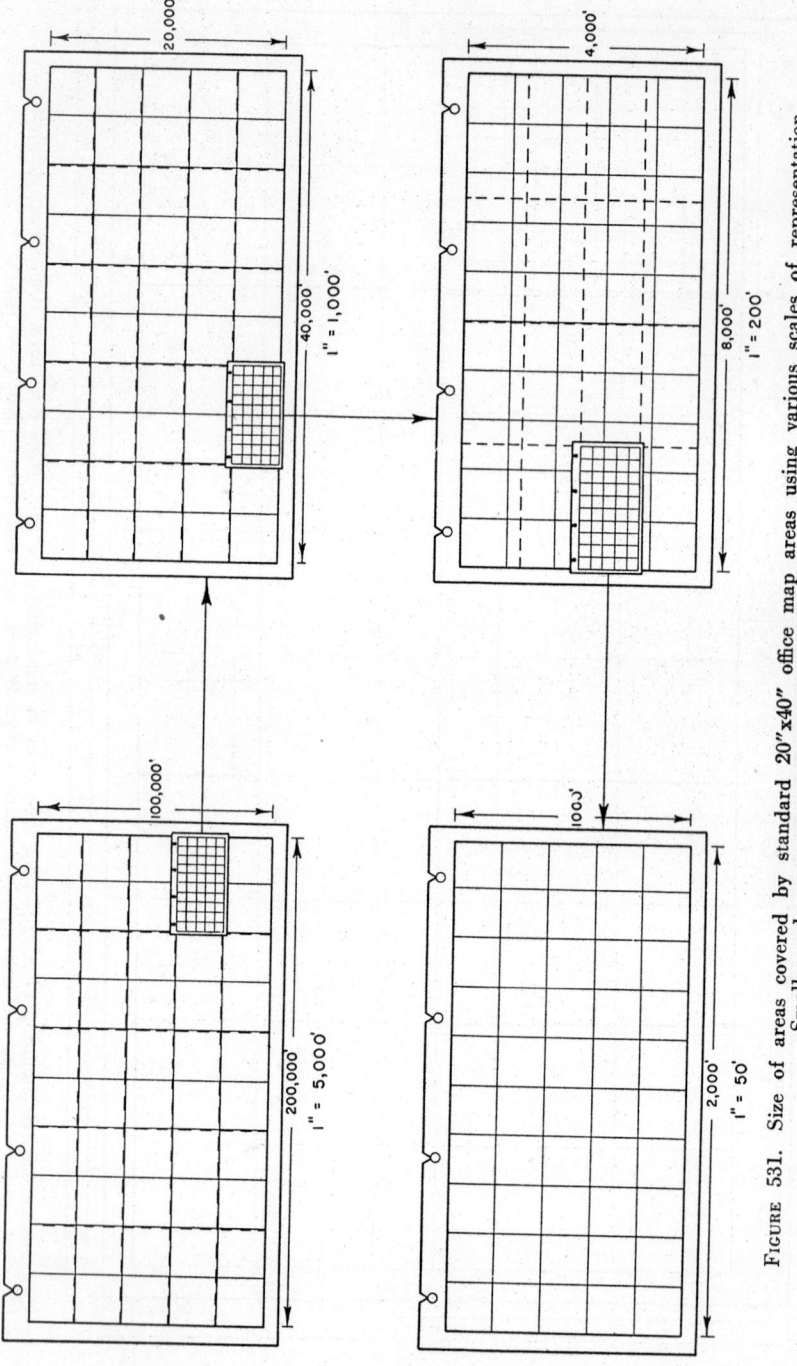

FIGURE 531. Size of areas covered by standard 20"x40" office map areas using various scales of representation. Smaller-scale map areas are even multiples of larger-scale map areas.

The area that can be mapped on one sheet follows.

$1'' = 1'$ is $8' \times 10'$
$1'' = 10'$ is $80' \times 100'$
$1'' = 50'$ is $400' \times 500'$
$1'' = 200'$ is $1600' \times 2000'$
$1'' = 1000'$ is $8000' \times 10,000'$
$1'' = 5000'$ is $40,000' \times 50,000'$
$1'' = 10,000'$ is $80,000' \times 100,000'$

All scales of note-sheet areas are even multiples or subdivisions of larger- or smaller-scale note-sheet areas. Detailed sketches, photographs, and microscopic data are posted on this type of sheet.

Assay-Record Sheets—The sheets are identical to note sheets except that they are printed on tracing paper. They may be stapled on the geologic note sheet as overlays.

Office-Map Geologic Sheets—Office maps have an over-all size of $23'' \times 42''$, with a $4''$ coordinate grid covering a $20'' \times 40''$ map area. The ordinate lines are printed or ruled in black on the glazed or reverse side of good-quality tracing cloth. There is a $2''$ margin along the north or $42''$ border, and this margin is punched with four $3/8''$ holes to fit a $24'' \times 43''$ post binder. Along the other borders of the map area there is a $1''$ margin. Location, elevation, and filing data are entered in the lower right-hand margin of the sheet. Ordinate values, engineering data, and culture are posted on the reverse or glazed side of the sheet. The lettering is first penciled on the unglazed side of the sheet and then "mirror" lettered on the glazed side with a LeRoy lettering device and reverse lettering guides. The unglazed side is reserved for the representation of geology. Identical sheets are used for plan and sectional representation and all map scales (figs. 530, 531, and 532).

The area covered by each sheet at various scales of representation is indicated below:

$1'' - 50'$ ....$1000' \times 2000'$
$1'' - 200'$ ....$4000' \times 8000'$
$1'' - 1000'$ .... $20,000' \times 40,000'$
$1'' - 5000'$ ...... $100,000' \times 200,000'$
$1'' - 10,000'$ ....$200,000' \times 400,000'$

The $23'' \times 42''$ map size is conducive to efficiency because the entire area may be reached without much shifting of person or sheet. The map books handle with ease and consequently maps wear longer.

Metallization-Ore Reserve-Map Sheets—The sheets are identical to the office geologic-map sheets except that the printing is on clear plastic material ($.005''$ cellulose acetate). These data are superimposed as an overlay on the geologic maps. These maps are filed in the same post binder alternately with the geologic maps so that every geologic map is overlaid by a plastic metallization and ore-reserve sheet. Since the superimposed overlays contain data supplementary to the geologic maps, there is no purpose in posting engineering, excavation, cultural, or geological data on these sheets. The metallization data is posted on the plastic colorimetrically with acetone-base inks. Ore-reserve blocks are posted with diluted acetone-base inks to facilitate periodic revision.

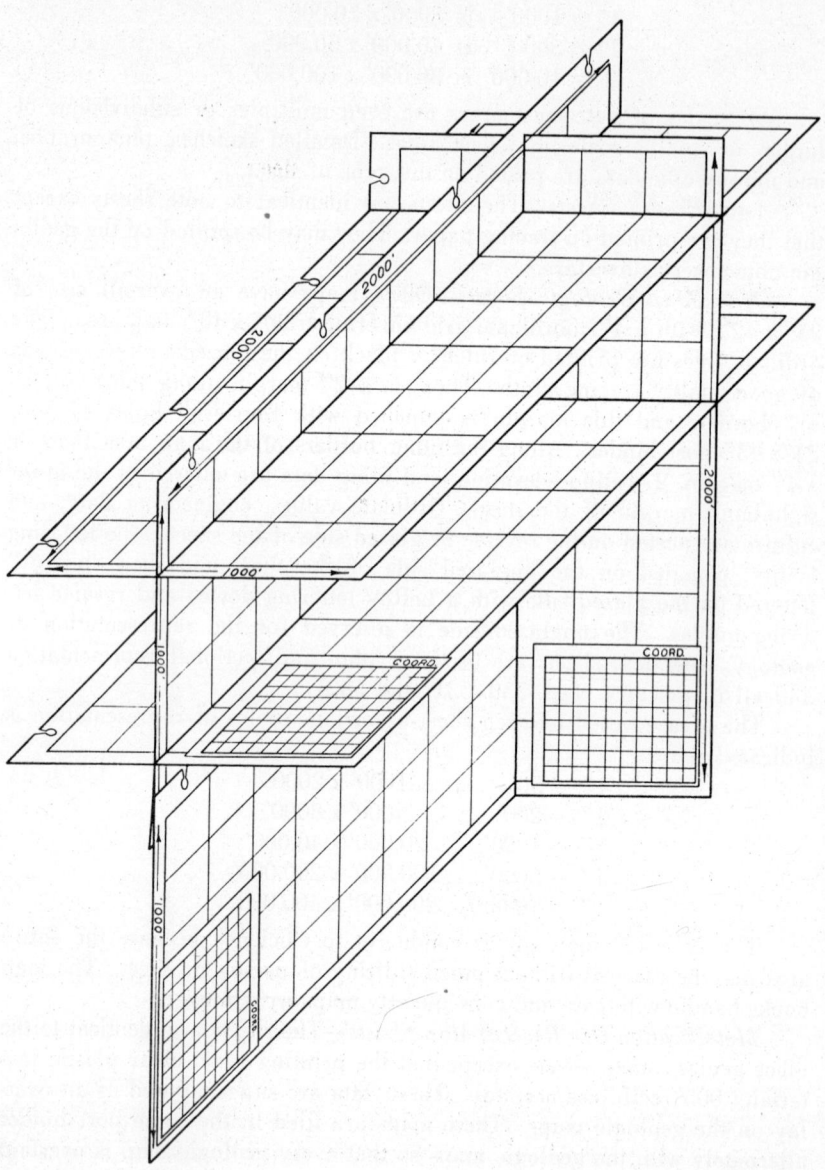

FIGURE 532. Three-dimensional representation of planar and sectional map correlation. Field note-sheet-area subdivisions are also indicated.

Geochemical, Biochemical, Geophysical, and Vegetation Data—The data are posted separately on plastic overlays identical to the metallization sheets.

Ore-Reserve Listing and Classification Sheets—Ore-reserve sheets are printed on 32-pound ledger paper. The sheet is 31″ long and 8½″ wide. When it is folded and inserted into a letter-size post binder, there are two opposite facing pages. The left-hand page is letter size and the right-hand page is infolded once to letter-size dimension. The extreme left tabulation headings provide entry places for general ore-reserve block data such as block number, location, width and grade, area and length, type of vein, and type of ore-reserve map representation. The tabulation to the right contains the gross-ore-reserve listing of tonnage, and grade times tons equals the products. The main or central portion of the sheet is for

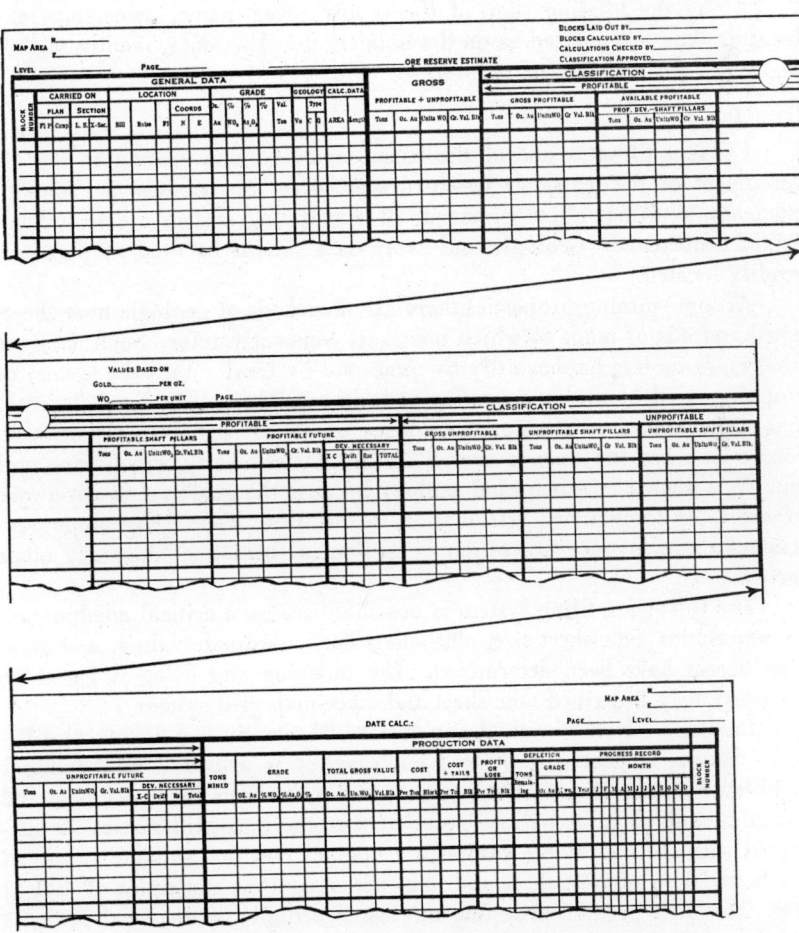

FIGURE 533. Ore-reserve listing and classification sheet.

classification of the reserve on the basis of economic profitability and availability. The infolded right-hand portion of the sheet contains tabulation headings for production or depletion data to be entered when an ore-reserve block is mined. In the upper right- and left-hand corners of the sheet are entry spaces for coordinate indexing to indicate the location of listed blocks within a 400' x 500' unit rectangle area (fig. 533).

Three-Dimensional Representation Sheets—Blank 23"x 42" office-map size sheets are used. The representation is oriented with the vertical axis parallel to the long dimension of the sheet. The plan area covered by each block diagram is 8"x 10" note-sheet size.

Reports—Reports are on standard $8\frac{1}{2}$" x 11" note-size paper. A place for coordinate indexing by rectangle-grid area is provided in the upper right-hand corner of the cover page as indicated in figure 535d.

Along the binding edge of the report cover, name, ownership, and location data are entered as on the note sheets. This data, readily visible, facilitates accessibility of reports.

Indexing and Filing

Efficient filing is one of the basic considerations leading to the development of the mapping system described in this discussion. Without efficiency in filing, all other merits of a mapping system are to a large extent neutralized. Geologic and assay data should be filed so that it is readily located.

At some mining properties there are thousands of geologic note sheets and hundreds of maps to which one must constantly refer. Such data are filed more or less haphazardly by mine and by level. With the system of mapping used at these properties, it is impossible to file data efficiently and systematically. Often an abundance of data is not properly utilized, for no one has the time to search for it. Consequently, much time and energy is wasted in remapping geology because the geologist is not aware of available long-buried information in the files. Poor filing alone contributes more to poor utilization of geological personnel than any other factor.

The following filing system is possible because a critical combination of map scales, note-sheet size, office-map size, coordinate values, and map-grid layout have been determined. The indexing and filing is based on the previously described note-sheet and office-map grid system.

Indexing—Note Sheets, Assay Record Sheets, Survey Sheets, Ore Reserve Sheets, Reports—A printed entry space is provided in the upper right-hand corner of the letter-size note and data sheets for coordinate and elevation indexing. It will be recalled that the map-grid system is integrated with the coordinate values in a manner that the ordinate values of the boundaries of note-sheet and map areas are even multiples of 400 or 500. This fact permits dropping the last two digits of the local ordinate value in indexing; thus, if the local ordinate value of the north boundary of a data sheet is 50,500 and the ordinate value of the east boundary is 40,800, the coordinate indexing is as indicated in figure 534a.

It will also be recalled that it was recommended that local values be derived from a provincial coordinate system. In local use the first few digits are dropped and local values are limited to the last five digits. For the purpose of regional indexing of geological data the regional values are recombined with the local value in the manner illustrated in figure 534b.

In the indexing of surface sheets the word *surface* is entered in the

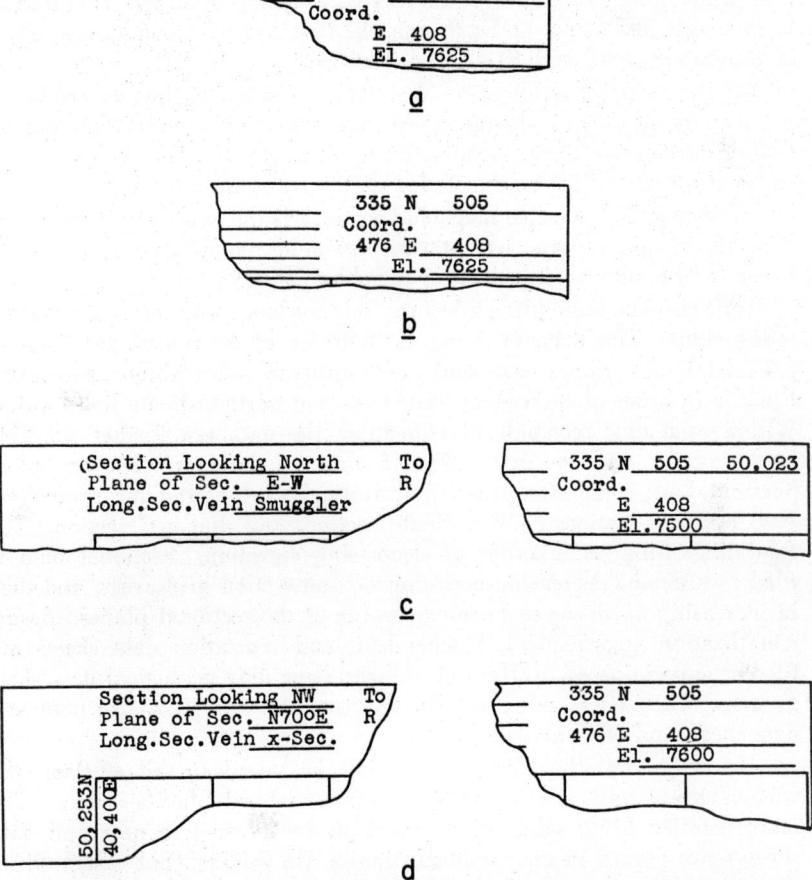

FIGURE 534. Manner of indexing field subsurface note sheets. *(a)* Local ordinate values of east and north boundaries and elevation of mapping plane entered on note sheet; *(b)* regional values recombined with local values; *(c)* longitudinal projection indexing in upper left and right corners of a field subsurface note sheet; *(d)* cross-section location, orientation, and filing index data entered on a note sheet.

space provided for elevation. Entry spaces are also provided for mine level and floor. Supplementary location data is entered in appropriate spaces provided for that purpose along the remainder of the margin of the sheet.

Sections are indexed in a similar manner to that of plans. Primary indexing is by planar area in which the section is prepared. In addition, however, the exact ordinate value of the line or plane of section is indicated. In the space provided the elevation of the top boundary of the section is entered as shown in figure 534c.

The indexing shown on figure 534c indicates an E-W longitudinal section of the Smuggler vein with the line of section at 50,023N. It is located in rectangle 505N and 408E; that is, the east and west boundaries of the section are 40,400E and 40,800E, respectively.

If the section is oblique to the ordinate lines, the line of section is not expressible as an ordinate value; and other means must, therefore, be devised to indicate exact position. Position is indicated by the coordinate value of a vertical line on the section as shown in figure 534d.

The indexing of office maps and sections is the same as for letter-size data sheets and reports, except that the index entry spaces are in the lower right-hand margin corner of the sheet.

Filing—The basic filing system is in accordance with the unit-rectangle index value. The primary filing is in order of increasing sequence of provincial or regional east- and north-ordinate index value. Secondary filing is in order of increasing *local* east- and north-ordinate index value. Within each unit rectangle classification, the data are further sub-filed in accordance with the orientation of plane of the representation—plan, horizontal or composite projection, vertical or longitudinal projection, N–S ordinate section, E–W ordinate section, and diagonal section. The final plan filing is in order of decreasing elevation. Sectional data is filed in order of decreasing elevation of unit-section grid areas, and then in increasing north- or east-ordinate value of the sectional planes. Assay, metallization, geochemical, biochemical, and vegetation data sheets are filed as superimposed overlays on the corresponding geological data sheet or map. The system is essentially the same for the letter-size note and data sheets and office maps.

In order that the filing system will be practical and efficient, the various filing units are separated by index-tabbed Manila sheets. The more detailed filing subdivisions (such as local-unit-area note and data sheets) are placed in index-tabbed Manila file folders. Subsidiary filing divisions of data contained in the folders are further separated by index-tabbed sheets. The system of tab indexing is described below.

Letter-size field and data sheets are filed together in steel drawer filing cases. It is desirable that the cases are fire-proofed and if possible located in a fire-proof vault. In the event of loss, usually less than 30

percent of subsurface data can be replaced by remapping; mine workings
seldom remain open long.

Regional filing is based on 100,000′ square area units. A provincial

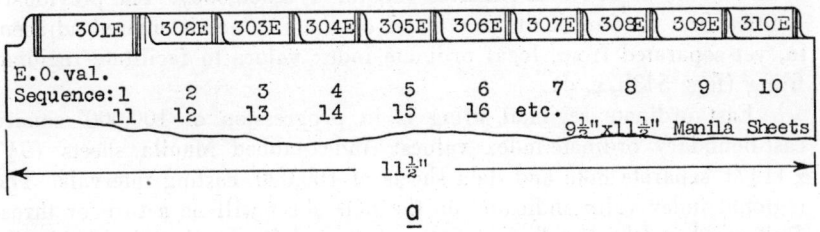

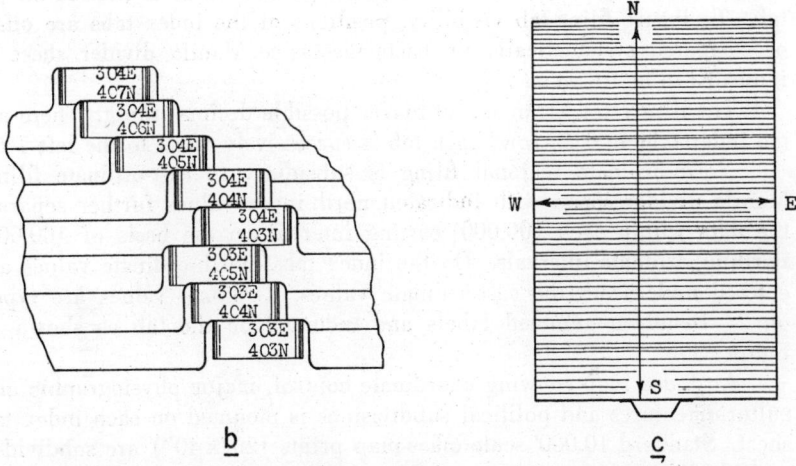

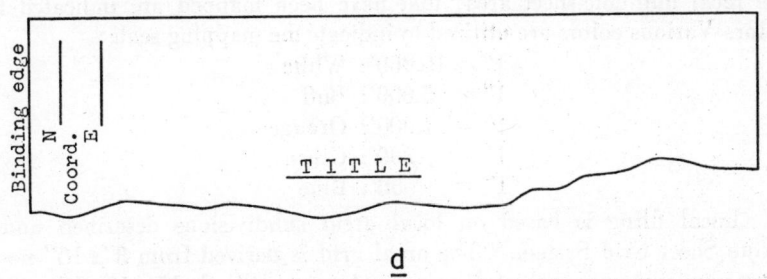

FIGURE 535. *(a)* Regional east ordinate filing tab arrangement (each successive tab
to right represents increase of 100,000 feet in east-ordinate value); *(b)* regional
filing arrangement, first in order of increasing east, ordinate values and second
in order of increasing northing; *(c)* plan view of filing drawer showing ordinate
arrangement of index tabs; *(d)* manner of report indexing for filing.

index wall map adjacent to the filing cases facilitates locating of file data. Ordinate lines and values of the 100,000′ grid system are posted in India ink on a U.S.G.S. geological wall map of the United States (Scale: 1/2,500,000) or of North America (Scale: 1/5,000,000). The previously omitted first portion of the provincial ordinate values is entered adjacent to, yet separated from, local ordinate index values to facilitate regional filing (figs. 543b, c, d).

East ordinate regional filing is in progression of 100,000′ square east-boundary ordinate-index values. Index-tabbed Manila sheets (9½″ x 11½″) separate note and data sheets at 100,000′ easting intervals. The regional index value indicated on the note sheet will be a two- or three-digit number (the last five digits are reserved for local use). A Manila index sheet is inserted to indicate a last-digit change of the regional index value: thus, 335, 336, 337, etc. The easting-index value is entered on the tab. To insure filing-tab visibility, positions of the index tabs are offset or staggered systematically on each successive Manila divider sheet as indicated in figure 535a.

The illustrated arrangement makes possible decimal filing. There are ten index tabs in each row; each tab is successively offset to the left 1.1″.

North-ordinate regional filing is subsidiary to east-ordinate filing. Manila divider sheets with indicated north-index values further separate the data within each 100,000′ easting interval on the basis of 100,000′ northing-ordinate intervals. On the index tabs, north-ordinate values are entered underneath the east-ordinate values. Ordinate values are typed on 2″ Dennison gummed labels and mounted on the tab as shown in figure 535b.

An index map showing coordinate control, major physiographic and cultural features and political subdivisions is mounted on each index tab sheet. Standard 10,000′ scale office-map prints (20″x 40″) are subdivided to 100,000′ square areas, cut, and mounted on the index divider sheets; or U.S.G.S. topographic sheets may be cut and "edited" to the proper 100,000′ square areas and mounted on the index sheets. On these maps the local unit-note-sheet areas that have been mapped are indicated by color. Various colors are utilized to indicate the mapping scale:

$$1'' = 10,000': \text{White}$$
$$1'' = 5,000': \text{Buff}$$
$$1'' = 1,000': \text{Orange}$$
$$1'' = 200': \text{Green}$$
$$1'' = 50': \text{Blue}$$

Local filing is based on local areal subdivisions described under "Note Sheet Grid System." The areal grid is derived from 8″x 10″ note-sheet map areas, when the following scales are utilized: $1'' = 50'$, $1'' = 200'$, $1'' = 1,000'$, $1'' = 5,000'$. Data mapped to any of the above scales is placed in the same file drawer. All planar and sectional data representing any one unit rectangle are placed in a 9″x 11½″ Manila file folder. The local ordinate index values of the contained data sheets are indicated on the filing tab in a manner similar to provincial file indexing. The north- and east-ordinate index numbers are typed on "Fangold" gummed labels which are mounted superimposed on the folder tab. These labels are obtainable in a number of colors, and the mapping scale of the material contained in the folder is indicated by the color of the index label. The colors are standardized to conform with scale colors used to indicate mapped areas on the 100,000' square area index maps which are mounted on the regional file-index sheets.

East-ordinate filing is in order of increasing 400' east-ordinate values The tabs are staggered for purpose of visibility (fig. 536a).

North-ordinate filing is in order of increasing 500' north-ordinate values as shown in figure 536a.

Plan data sheets are filed in order of decreasing elevation and precede filing of sectional data.

For level or elevation filing a letter-size Manila sheet is inserted so that it overlies each level sheet. To this Manila sheet is attached a $\frac{1}{4}''$x $\frac{1}{2}''$ "Mak-Ur-Own" plastic index tab with typed insert indicating level or sill. These tabs are obtainable in various colors. Levels of approximately the same elevation in different mines are indexed with the same tab color. Tabs for the various levels are staggered or offset to promote visibility and the offset should be systematic and in progressive conformity with elevation as indicated in figure 536b.

A full-sized template key sheet like the one illustrated in figure 536b, should be prepared to determine tab position on new level index sheets. Sectional data are filed in secondary sequence in the same folder with the corresponding plan-area sheets.

Orientation classification is accomplished by index-tabbed Manila divider sheets separating the various sectional data which are filed in the following sequence:

> Horizontal or Composite Projections.
> Longitudinal Vertical Projections.
> North-South Ordinate Sections.
> East-West Ordinate Sections.
> Diagonal Sections.

Sectional classification is typed on gummed labels and superimposed on the Manila divider-sheet tabs.

The sub-filing within each sectional classification is in order of decreasing vertical interval. Plastic-tabbed index sheets (described under "level indexing"), separate sectional data of different vertical intervals. The position and tab color are determined by the level-elevation template sheet previously illustrated. The sectional data are again sub-filed in order of increasing north- or east-ordinate value of the section plane.

The following information is recorded on overlay sheets: assay,

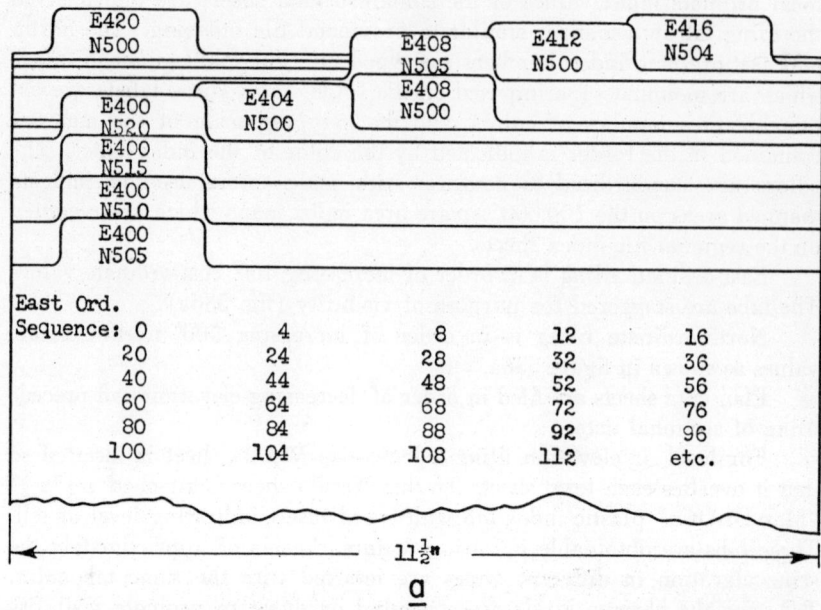

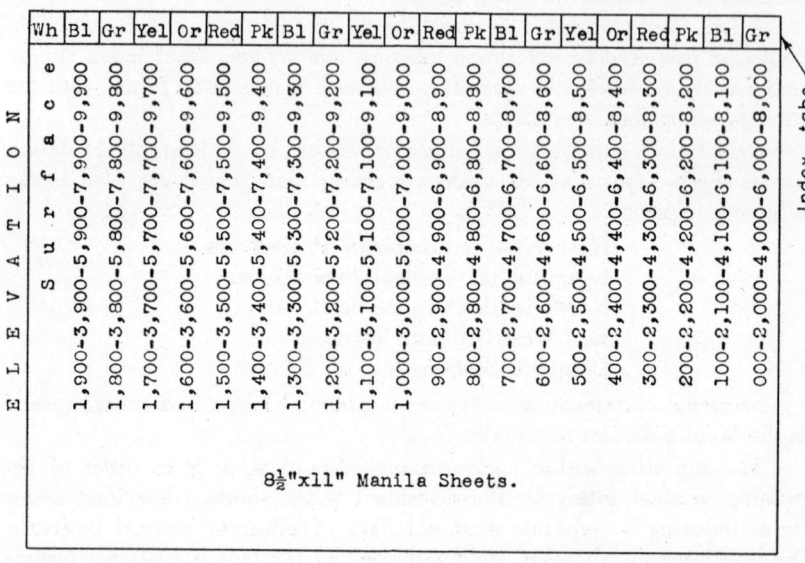

Figure 536. (a) Local ordinate filing arrangement of tabbed Manila note-sheet folders; (b) color and position of plastic elevation index tabs.

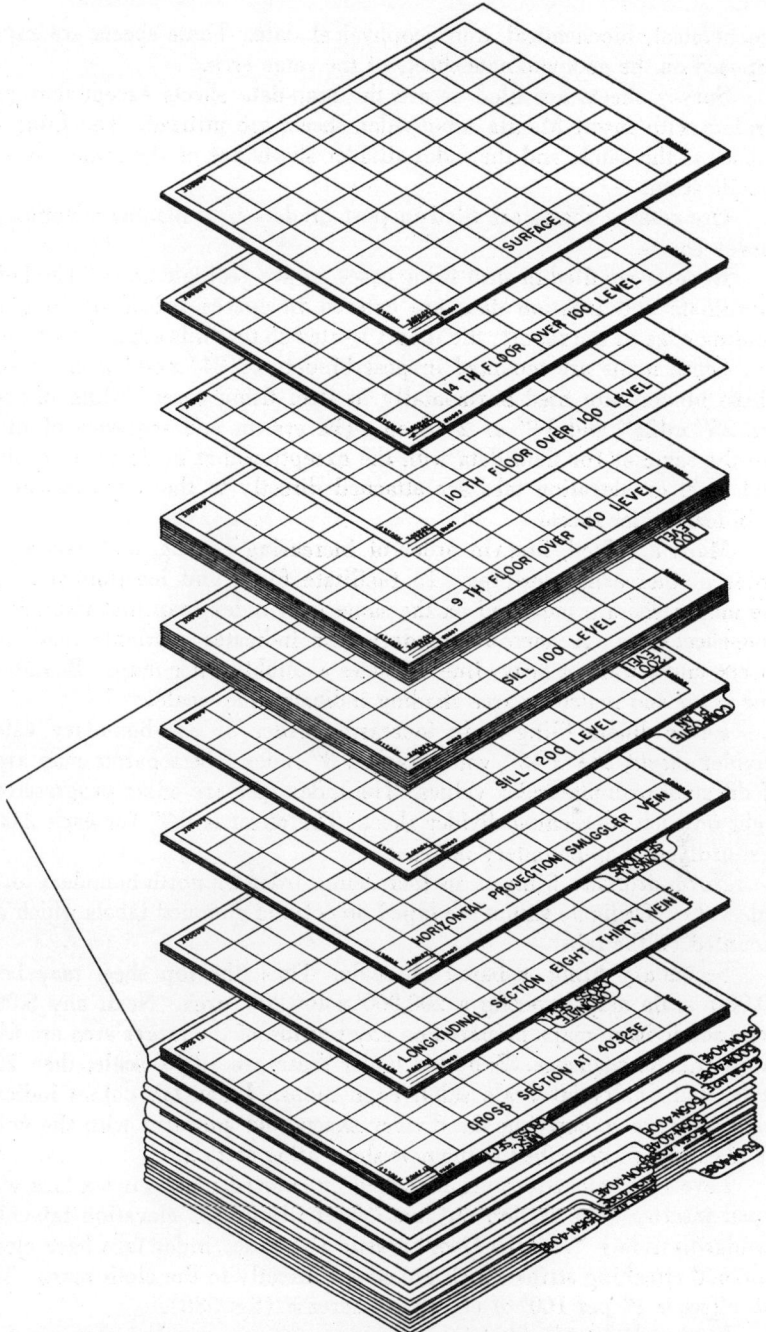

FIGURE 537. Exploded diagram illustrating sequence of note sheet filing. Primary filing may be in order of increasing east-ordinate values or in order of increasing north-ordinate values.

geochemical, biochemical, and geophysical data. These sheets are super-imposed on the geologic note sheets of the same area.

Survey sheets are filed as are the map-data sheets except that post binders with insert Manila tab-divider sheets are utilized. The filing sequence is the same, and the index divider sheets are of the same size and specification.

Ore reserve sheets are filed in post binders in a manner identical to survey sheets.

Reports are filed in steel filing cases, as are geologic notes. The index coordinate and location data are entered in spaces provided along the binding edge of the report; the report is filed so that this edge is at the top.

Office maps are mounted in post binders of 24″ x 44″ over-all size. These binders are filed horizontally in steel filing cases Filing drawers are 25″ x 42½″ and 1½″ or 2″ deep. The system and sequence of filing are the same as for field data with the exception that no folders are used and level or elevation tabs are attached directly to the maps instead of to index-divider sheets.

Maps are filed, first, in order of increasing easting, and, second, in order of increasing northing. To facilitate filing and location of maps, the map areas are indicated on the same wall index map that is used for note-sheet data. A narrow boundary line indicates available field note sheets, and a wide or heavy line indicates available office maps. Boundary-line color and matching-area shading indicates map scale.

East-ordinate filing is in increasing order of east-boundary value. Divider sheets, 23½″x 42″, with attached 3″ index tabs separate map areas of different ordinate index values. The index tabs are offset progressively right on each succeeding divider sheet. The offset is 0.8″ for each 2,000′ east-ordinate map-boundary increase.

North-ordinate filing is in increasing order of north-boundary ordinate value. Ordinate values are typed on colored gummed labels which are mounted on the tabs.

Secondary filing is based on scale. Thus, the top sheet may be a 10,000′ scale map covering a 200,000′ x 400,000′ area. Next, any 5,000′ scale subdivision maps prepared on any portion of the larger area are filed in coordinate sequence. Sub-filed under these are 1,000′ scale, then 200′ scale, and finally 50′ scale subdivision maps. Index tab colors indicate scale of maps underlying the divider sheets and conform with the color standardization described in the note-sheet filing section.

Levels and elevations are indicated by colored plastic index tabs with typed inserts. (See section on "note sheet filing" for elevation tab-color standardization.) "Mak-Ur-Own" plastic (3″ wide) index tabs have cloth-gummed attaching strips which are sewed directly to the cloth maps. The tab offset is 1″ per 100′ of elevation decrease (fig. 538).

Sectional data is filed in secondary sequence to the corresponding plan maps (those traversed by the section). Tabs indicate sectional classification. The sequence is identical to note-sheet filing. Assay, geochemical, biochemical, and vegetation data are represented on clear plastic overlay sheets, which are superimposed on the corresponding geologic plan or sections.

Legend

The legend shown in figures 539, 540, 541, and 542 is the result of eleven years of research and trial application. The objective has been simplicity and logical, systematic, and graphic representation of all megascopically mappable features. Representation is the same for plan and

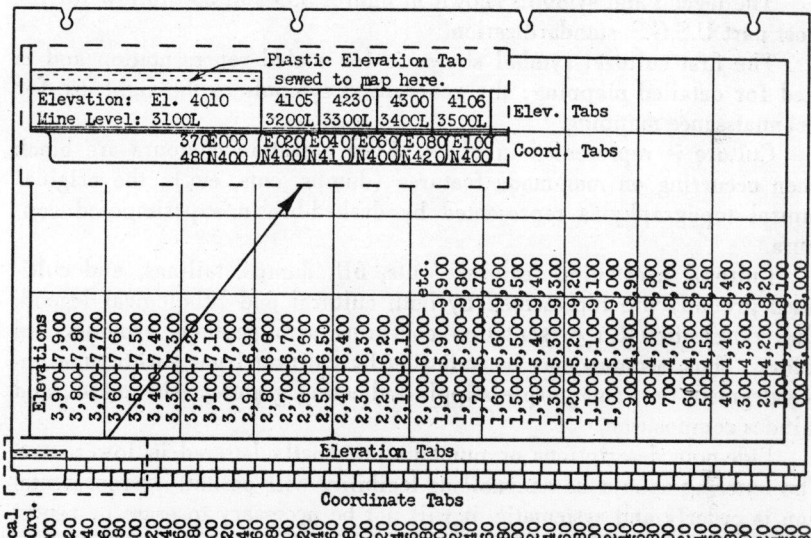

FIGURE 538. Position of plastic elevation office map tabs and coordinate areal-divider-Sheet filing tabs.

section. (A plan map is a horizontal section.) Unless otherwise specified the legend is the same for note sheets and office maps.

Culture and Physiography—These features are mapped concurrently with geology unless previous surveys exist. It is imperative that culture and physiography be completely represented so that spatially correlated geological data may be readily oriented and located. Such nongeologic data are also important in planning economic development and exploiting mineral deposits.

The location, size, and composition of mine dumps, old fill, and caved subsurface workings aid in determining the location, size, and composition of veins in inaccessible workings. Stope or raise holings, slumps, and caves help to deduce position, attitude, and economic tenor of veins. The amount of timber or lagging on subsurface maps is a criterion of fault location and magnitude. Surface installations (mills, smelters, tailings, dumps, roads, etc.) indicate localization and size of past operations, and the recovery processes utilized.

All fragmentary information must be integrated to determine mineralogy, magnitude of deposits, and past mine economics. This type of integration facilitates better interpretative geology and consequent future planning.

Physiography usually reflects geology, and, when locally calibrated to geology, is a valuable and useful aid to geologic correlation and interpretation.

The legend and symbols shown in figures 539 and 540 follow for the most part U.S.G.S. standardization.

The first cultural symbol shown is for scaled representation and is used for detailed mapping; the generalized symbols to the right are for reconnaissance mapping.

Culture is represented in black. Even elevation contours are black when occurring on man-made features (dumps, cuts, etc.); the original natural topography is represented by dashed-brown superimposed contours.

Cultural excavation features (cuts, fill, dumps, tailings, and cultivated ground) are represented by both cultural and lithological legend. The compositional representation conforms with geologic symbols and colors. Thus, a dump may be indicated by cultural legend with superimposed green, red, and purple stipple indicating granite, vein quartz, and sulfides composition.

Side-note descriptions or numbers are neatly lettered in lower case. The lettering should be as small as legibility will permit. If representation is orderly and systematic, it will not be necessary to erase or repost previous side-note observation to make room for new data. An area or point (described by side note) is indicated by a line parallel with ordinate lines. When it has become habit to post side-note lines in an east-west or north-south direction, the geologist will find that he has more room on the sheet for recording of detailed observation.

All culture, ordinate lines and values, and engineering data are posted on the glazed or reverse side of the tracing-cloth maps.

Geology—Mining-geologic representation facilitates prediction of ore localization. The following geologic legend graphically and decisively represents structure, texture, composition (primary and superimposed metamorphic or hydrothermal effects), and stratigraphy. A better understanding of the geologic-representation system will result if the representation objectives are considered.

    1. Ore Localization Factors:

        a. Genetic or Source Control: The location of a cupola determines source point and consequent primary pattern of hydrothermal effects. Zonation or annular pattern of magmatic and

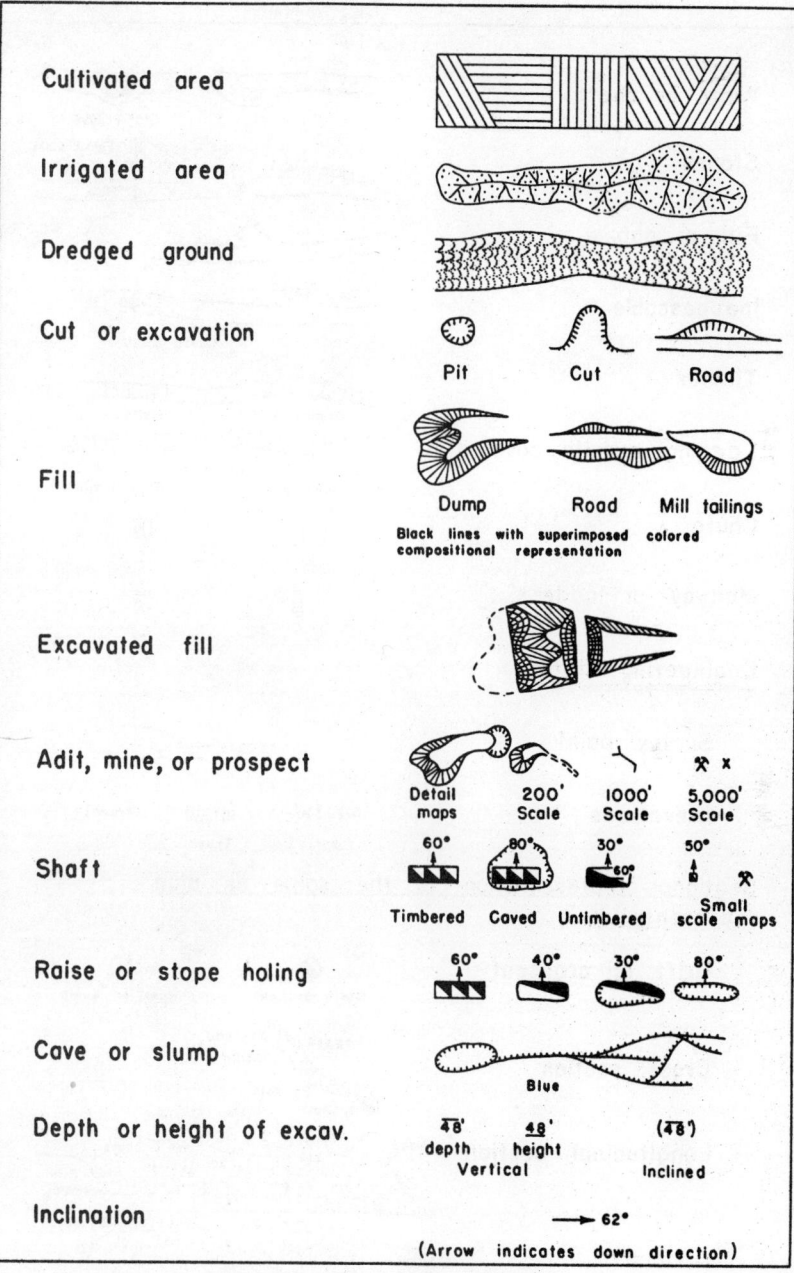

FIGURE 539. Surface cultural legend.

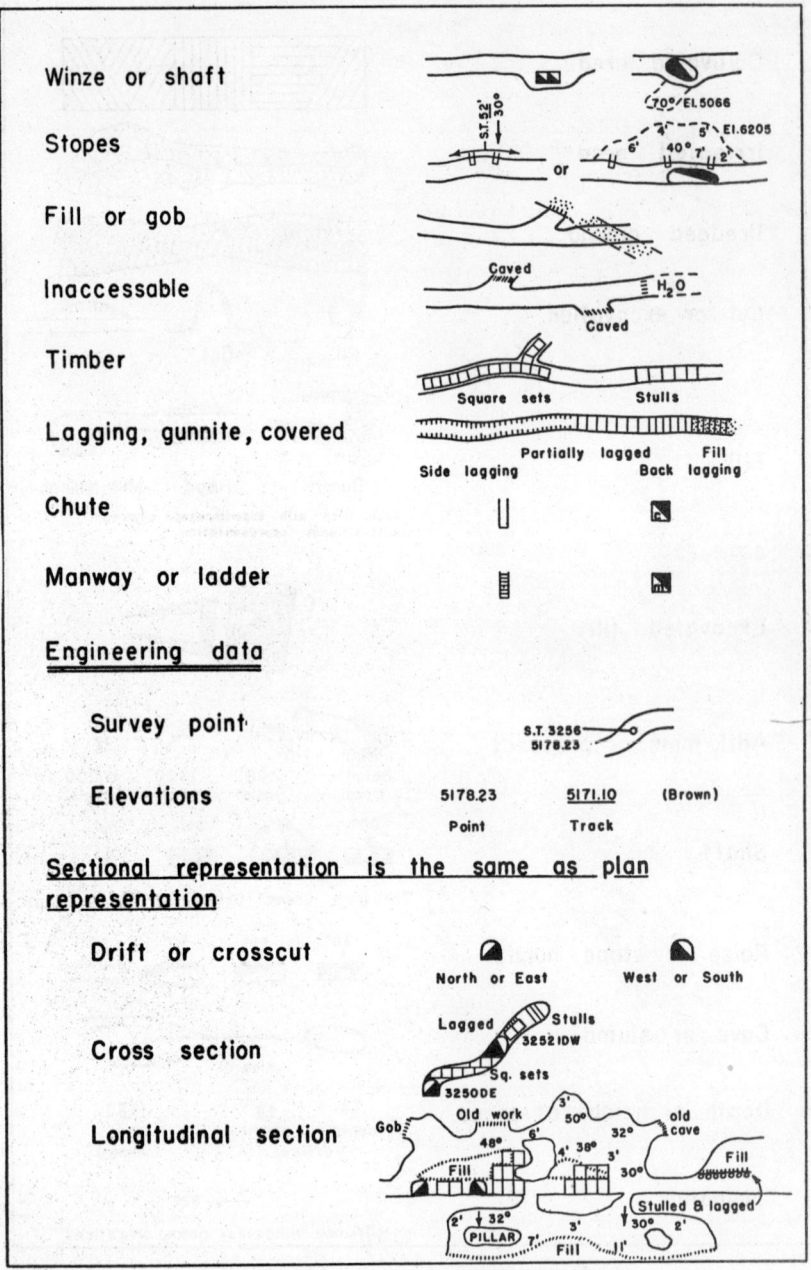

Winze or shaft

Stopes

Fill or gob

Inaccessable

Timber

Lagging, gunnite, covered

Chute

Manway or ladder

Engineering data

Survey point

Elevations

Sectional representation is the same as plan representation

Drift or crosscut

Cross section

Longitudinal section

FIGURE 540. Subsurface cultural legend.

deformational effects (intrusive, metamorphic, or hydrothermal facie), represented by effective colorimetric geologic mapping facilitates source point determination.

b. Structural Control: Structural location, shape, and attitude exercise a controlling effect on ore localization.

c. Lithologic Control: Ore deposits are localized where certain wall-rock types occur. Subtle and slight variations in facies may decisively affect ore localization. The lithologic localization may be determined by

Compositional Variations, indicated by color (magmatic effects) and pattern (sedimentary rocks)

Textural Variations, represented independent of composition.

d. Stratigraphic Controls: Ore is often found in equivalent stratigraphic units. Although lithology of the unit may regionally vary from quartzite to limestone, there is an unmistakable tendency for ore localization within the particular stratigraphic unit. No physical or chemical factor has been determined explaining this ore localization. Stratigraphic representation is independently superimposed on lithologic representation.

2. Simplicity of Representation: Similar effects are represented by similar colors or texture symbols. Texture symbols are suggestive of the textures represented. Texture symbols and colors are in gradational systematic sequence corresponding to the textural and compositional classification of magmatic rocks and effects. (fig. 541)

3. Flexibility: Lithology is indicated by combinations of color and texture. Symbols and colors may be proportionately blended to graphically and colorimetrically represent slight facies variations, gradations, and composite textures and compositions.

4. Permissible Detail: A primary objective of "hard rock" geologic mapping is a maximum representation of detailed data.

5. Time Sequence: The representation legend shown in figure 541 is based on a systematic colorimetric arrangement parallel to normal time sequence.

Structure Representation—Compositional structure-dip symbols are posted contiguously to the lithic representation whenever possible.

Planar structure is indicated by double-barbed dip arrow and posted contiguously to and attached directly to a bed, fault clay, vein, igneous-flow layer or other lithic bands. Planar mineral or clot orientation is indicated by flow symbol.

Linear structure is indicated by half-barbed arrows showing the bearing and inclination of linear structures (striations, flutings, fold-axis lines, minerals, clots, and xenoliths). In representation, crystalline-rock-

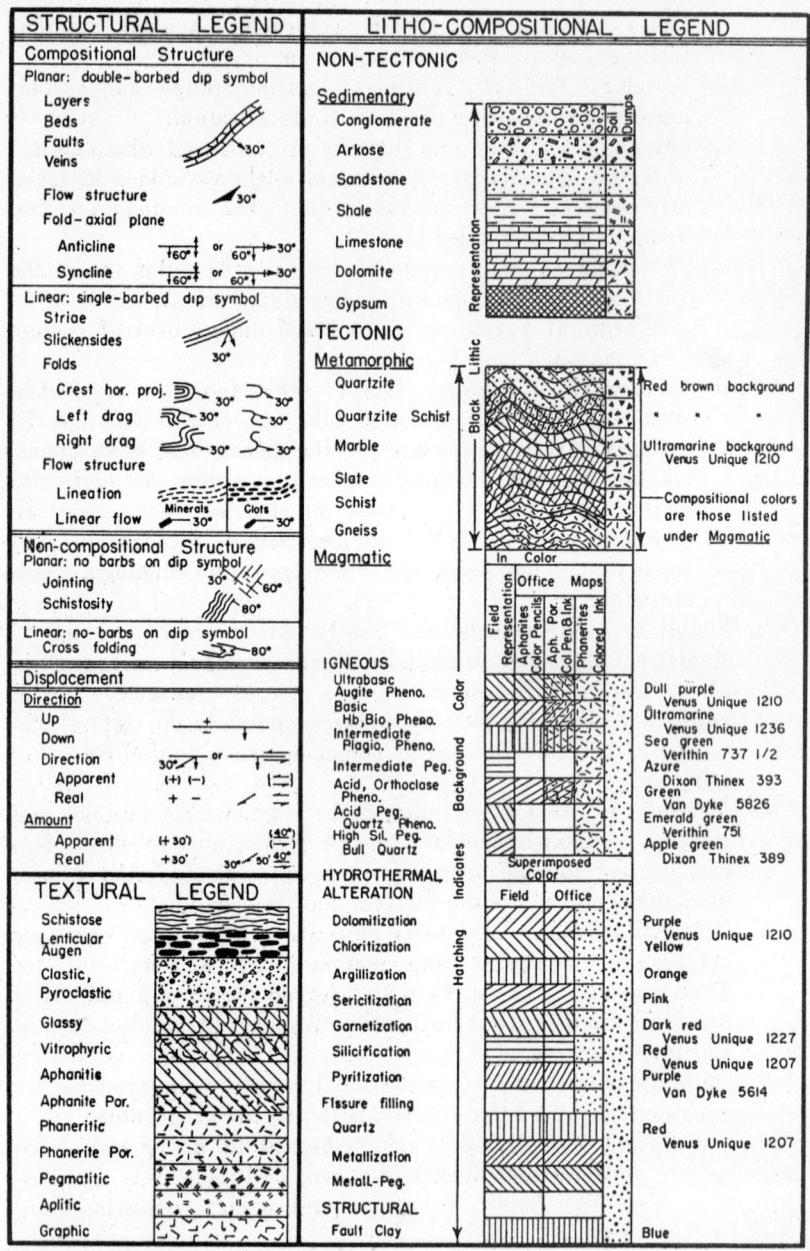

FIGURE 541. Three classifications of geologic legend. Color stratigraphic
representation optional.

mineral-texture symbols are oriented to show predominate lineation direction; half-barbed arrows are posted contiguously to show horizontal projections and inclinations of maximum elongation lines (fig. 541).

Non-Compositional structure (schistosity and jointing) is indicated by black lines superimposed on the lithic representation. Schistosity and jointing inclination is represented by unbarbed dip symbols. The representation is quantitative in that spacing of lines indicates spacing of joints. Two or more systems of joints may be represented in superposition.

Direction and amount of displacement are indicated whenever apparent. On plan maps, positive or up, and negative or down displacement are indicated respectively by plus and minus signs. On sections, movement toward and away from the observer are indicated respectively by plus and minus signs. Half-barbed arrows show relative directional displacement. Parenthetically enclosed symbols denote apparent displacement direction. Amount of displacement is indicated numerically in inches or feet. Parenthetically enclosed numbers denote apparent displacement.

Texture—Texture is represented by symbol (fig. 541). Textural quantitative, qualitative, and gradational values are shown graphically and relatively by size, shape, spacing, and attitude of symbols. Simple and composite textural combinations are accurately and graphically represented by combinations of symbols.

The symbol is usually a pictorial representation of indicated texture. Banded, layered, or bedded rocks (sedimentary, metamorphic, igneous) are indicated by parallel lines conformable with attitude and relative width of bands. The clastic symbol (to denote sediments, metamorphics, and pyroclastics) is modified to show relative size, shape (angularity or sphericity), and spacing of clastic material. Solid color represents either soil (if gray) or aphanitic texture.

Sedimentary rocks are represented on plan and section by black symbolic lithic-mapping units. Lithic symbols are proportionately blended and superimposed to show either abrupt or gradational facies variations.

Metamorphic rocks are represented by black symbols with superimposed color to indicate magmatic or hydrothermal effects. Closely spaced black lines indicate schistose or micaceous-rock cleavage; the relative line spacing indicates schistose layers. Crystalline or gneissic textures are shown in the same manner as igneous rocks. Facies banding is indicated by compositional color banding. Metacrysts are indicated by phenocryst or phanerite symbols.

Igneous rock textures are indicated by standardized compositional colored symbols. Two basic media, solid pencil or air brush color indicating aphanitic texture and dash symbols indicating phaneritic texture, are sufficient to represent most igneous rocks. The colored-dash symbols superimposed on the solid-color background indicate felsite porphyries. Relative size, attitude, and spacing of phenocrysts or crystals are indicated by symbols. A modified dash symbol represents pegmatitic and graphic textures.

Since colored pencil permits less detail than pen and ink office representation, it is not always possible to represent symbolically texture on the field-note sheet. Small igneous rock masses (narrow dikes, gneissic rocks, lit-par-lit, etc.) are represented on the field-note sheet in solid color with supplementary textural descriptive side notes. These rocks are then graphically represented in detail on the office maps by textural symbols.

Composition—Compositional representation of non-tectonic rocks is shown by black lithic-symbol patterns; tectonic compositional effects (magmatic, metamorphic, hydrothermal, and related deformational effects) are represented by color, as indicated in figure 541. Sedimentary-rock composition is indicated by lithic symbol. The black lithic-symbol patterns are superimposed and proportionately combined to show composite compositional facies.

Metamorphic rocks are represented by color (to show metamorphic, magmatic, or hydrothermal effects) superimposed on black sedimentary lithic symbols. Thus, marble is represented by a dull ultramarine superimposed on black, a limestone symbol; quartzite is represented by red-brown superimposed on black sandstone dots or stipple. Micaceous bands are represented as black lines because in superimposition on other colors, fine structural detail is possible and black is compositionally neutral; it is difficult to determine mica-schist genesis in the field because schists are derived from either sedimentary or igneous rocks. The "crystalline" metamorphics or gneisses are represented in the same manner as igneous rocks; in the field no attempt is made to distinguish primary-igneous from secondary-metamorphic gneiss. Interlayered micaceous-banded gneiss is represented by black lines (quantitatively and pictorially indicating amount, attitude, irregularity, and crinulation of schistose material) superimposed on the corresponding igneous compositional color. Gneisses of phaneritic texture are indicated by structurally oriented compositional color banding (see "Igneous rocks" which follows).

Magmatic effects (igneous rocks, hydrothermal alteration, veins) are represented in color conforming to a compositional chromatic progression. An orderly, systematic, compositional color scheme permits continuous recording (rather than time-consuming, incomplete "spot" side-note recording) of variable composition without previous lithic-mapping, unit-reconnaissance determination for use in any given area. Equivocal compositional values are indicated by superimposed, chromatic, compositional representation utilizing possible compositional color extremes.

Igneous rocks are represented in chromatic progression ranging from ultramarine (basic parent rocks) through various shades of green (intermediate rocks) to yellow green (acid rocks). Alteration is shown in chromatic progression ranging from yellow (early pneumatolytic) through orange (intermediate) to red (late hydrothermal). Metallization color is brilliant purple, a resultant of red and blue.

The following criteria determine representational colors and their use. Colors used in superposition are either opaque or transparent, depending on purpose and desired effect. Metallization color is somewhat opaque, so that sulphides (indicated as either bands or disseminations) may be superimposed on background of either red (quartz) or green (granite). Alteration colors are transparent, so that when superimposed on colors indicating igneous rocks, the primary rock compositional color is partially visible.

Colors are meticuously selected, so that when they are superimposed, the resultant effect clearly indicates superposition, rather than one resultant color indicating a primary unaltered rock. Thus, if yellow (alteration) is superimposed on green (granite), the resulting effect must not be a pegmatite yellow-green.

The colors are in a chromatic progression from dull colors for early effects toward increasingly brilliant colors for late magmatic effects. Color-brilliance progression is parallel with increasing merit of ore-localization criteria. Siliceous differentiates are represented in more brilliant colors. Quartz veins are shown in bright red, and metallization by brilliant purple. A zonation of increasing brilliance (from the periphery toward an ore-area center) results. Favorable areas are, therefore, readily apparent.

Colors denoting similar compositional effects permit qualitative integration and are consequently quantitatively cumulative. Orange, pink, red, and purple are colors similarly based on the presence of red component and are, therefore, cumulative. When each color represents a hydrothermal phase or type, the red component of all the colors will tend to integrate total hydrothermal effect.

Igneous rocks of basic composition are indicated by ultramarine. The increasing addition of yellow component for increased acidity produces greens of various shades, grading from blue-green to yellow-green. The various shades represent either over-all composition or phenocrysts. Ultramarine represents a basic rock; augite or olivene, phenocrysts; blue-green is used to indicate either intermediate composition or plagioclase phenocrysts. If a map area is posted with small, blue-green, dash texture having like-colored, larger, interspersed, dash symbols, a phanerite of intermediate composition and porphyritic texture is indicated. If the large interspersed dash symbols are ultramarine, the phenocrysts are basic; and if yellow-green, the phenocrysts are quartz. A green pencil-colored background superimposed with ultramarine clastic symbols, and large blue-green and small yellow-green dash symbols indicates aphanite porphyry of rhyolitic composition containing basic clastic material, large plagioclase and small quartz phenocrysts. Since it is not always possible to indicate such detail graphically or pictorially in the field, textural detail and composite composition (indicated on the field sheet by side-note description) are later translated into graphic, pictorial representation on the office maps.

Compositionally banded or layered igneous rocks (gneisses, extrusives, etc.) are represented, relatively, on the field sheet by compositional, colored bands. This layering is similarly indicated on office maps, except that phanerites are represented by appropriately colored bands of dash symbols indicating compositional banding.

Pegmatites, aplites (micropegmatites), and graphic textures are indicated by modified symbols of relative size.

Hydrothermal effects are represented by a chromatic sequence ranging from yellow to red, and then to brilliant purple. The legend is adaptable to local conditions by substituting other alteration minerals and effects in place of those indicated on the chart.

Alteration representation is quantitatively proportional to amount and degree of hydrothermal effect, and is shown by color overlay (blanket alteration), stipple (disseminated alteration), or lines (joint or fissure alteration). Gradational effects are indicated by corresponding gradational-color intensity. Two or more colors may be applied in superposition to indicate two or more alteration effects. On office maps, an artist's air brush may be used to apply coarse, alteration-color stipple in superposition indicating alteration effects singularly and collectively.

Vein quartz and other gangue minerals are shown in red, the intensity of red indicating degree of replacement. The intensity of red is increased by superimposed, black (6H) pencil lines. Silicification is indicated by lightly applied red.

Metallization (sulphides, oxides, etc.) is indicated pictorially by brilliant purple stipple (disseminations) or scaled lines (veins). Small veinlets (less than 3″ wide), stringers, and mineralized zones are shown by generalized representation (small purple lines of proper attitude); the amount of colored area is proportional to the amount of mineral.

Supplementary, descriptive side notes indicate width and type of mineralization. Most minerals are listed by chemical formula:

$Qtz^{+++}, FeS_2^{++}, \underrightarrow{PbS, ZnS}, CaCO_3^{-}, CuFeS_2^{=}$, etc. An arrow indicates observed paragenetic sequence; e.g., sphalerite is later than galena. Plus and minus signs show relative amount of each mineral.

Deformational effects such as faults are shown in blue. The intensity of blue indicates the fault-clay fineness, and the consequent relative amount of displacement. Gritty fault clays are indicated by pale blue; sticky, gummy clays by dark blue (derived by retracing fault lines with a sharp 6H pencil). Fault clays and breccias (3″ and up) are plotted to scale on 50′ scale maps. Shear zones are represented by faint blue lines of either regular or irregular orientation conforming to observed attitude. Breccia is indicated by blue background with superimposed stippled colors representing breccia fragments.

Unconsolidated material such as soil is mapped. The map must be a recording of observed facts to be of value in mineral exploration; it must be possible to reinterpret data at any time. All area traversed should be examined and mapped whether outcrops are present or not. Areas where the bedrock is covered should be clearly indicated and accounted for, so that clear distinction between observation and interpretation and examined or unexamined areas will result. Inferred contacts are never represented on detail maps even for short distances. A factual compositional representation of over-burden usually makes apparent the location of subsoil contacts (veins, dikes, lithology, and general features); contact lines are undesirable if no outcrops are present.

Soil, gravel, mine dumps, tailings, and mine or stope fill are represented texturally and compositionally. Soil representation should blend with outcrop representation along the boundary area, because in the field the demarcation between outcrop and overburden is usually indefinite and gradational.

Fine soil is represented in gray; if fragments or clastic materials are observed in the soil, relative amounts, composition, and lithology of such fragments should be qualitatively and quantitatively indicated by superimposed color dots (clastic symbol). Representation symbols are shown on the accompanying legend chart. The compositional colors are those used to represent outcrop lithology; soil containing kaolinized granite, vein quartz, and limonite is indicated by gray with independently superimposed green, orange, red, and purple dots. An air brush may be used on office maps to apply fine-textured gray background as well as the coarse-colored stipples.

Man-made deposits (dumps, mine fill, and tailings) are potential ore reserves and are represented with the same care of geology in place. With increased metal prices, lower handling and processing costs, and increased metallurgical recovery, man-made deposits, especially if old, are now profitably exploited. Much production of Butte copper is from old "gob" (mine fill) and tailings. The "gob" is mapped with the same meticulous care used in maping virgin ground. The amount and type of vein material and ore-fragment attitude are indicated on geologic note sheets. The same colorimetric legend is utilized for mapping man-made deposits as for unbroken rock. The representation is so stippled as to indicate fragment composition and degree of angularity. Detail compositional side notes are supplementary and essential.

Detail representation of man-made deposits facilitates geologic determination of inaccessible ore deposits and mine excavations. Careful mapping of mine dumps will give indications of the length of inaccessible mine tunnels, lithology, distance to veins, type of ore, etc.

Stratigraphic Representation—Stratigraphy is usually not mapped in the field by mining geologists. The stratigraphic "contact" units are determined and indicated on office maps by superimposing very light pastel colors over the lithologic representation. It is sometimes difficult to determine the lithologic factors localizing ore deposits; empirical stratigraphic controls are in some cases determined by stratigraphic representation on mining geology maps. The colors used should be dull (olive greens, gray-blue, and brown) and superimposed very lightly.

Assay, Metallization, and Ore Reserves—Assay-metallization data are represented to facilitate integration with geologic data. Compositional values are represented by a progressive chromatic series. (fig. 542) Black represents "barren" vein or waste. Colors used are chosen for each particular area or deposit. If copper is the chief economic metal present, copper percentages are indicated by color progression from yellow to red (inclusive). A color break is made for a 1% grade variation. Copper content is indicated, thus: minus 1%, black; 1%–2%, yellow; 2%–3%, orange; 3%–4%, brick-red; 4%–5%, red; and 5% plus, scarlet. Similarly, lead and zinc content may be indicated by blue-red (purple) and yellow-blue (green) color progressions, respectively. The "stronger" or more brilliant color is added with increasing grade of ore, and the resultant chromatic progression is quantitatively cumulative.

Assay filing-sheet records are posted on letter-size, tracing-paper, overlay sheets which are stapled superimposed on the geologic note sheets; cut-sample assay values are posted in colored inks. If only two or three economic metals are present, the assay is posted adjacent to the point where the sample was taken. If veins are closely spaced, or if ore is complex, a number (in color to indicate approximate ore-grade classification) is posted at the sample point; width and grades are numerically indicated on an attached listing sheet. Car samples are represented on the excavated area by appropriate grade-color pencil shading or hatching.

Assay office-map representation is posted directly on the geological maps if the veins are not too closely spaced. Assays are indicated in the same manner as side notes. A narrow side-note line (east-west or north-south) is posted (with a ruling pen and straightedge) from the point where the sample was taken to 2″ from the edge of the excavation; sample data for each location are posted on the outward projection of side-note line; all figures for any one metal will be in a column, usually curved to conform with the mine-excavation trend. The columnar position indicates the metal given (fig. 543).

Sample data are indicated in the following assay order: width, ounces gold, silver, percent copper, lead, zinc, etc. If grades for certain metals are missing, a dash line is substituted for the value.

Wd.	Au	Cu	Pb	Zn
3.1′	0.09	....	4.2	8.9
4.2′	0.15	3.1	8.3	7.2
2.6′	0.17	4.1	....	9.9

Metal-grade values are posted in colors conforming to grade (indicated in the legend chart). The predominant economic metal is posted in yellow-red chromatic progression. Minor accessory metals are indicated by black.

Metallization maps are prepared on clear, plastic, overlay sheets.

Quantitative representation is by line symbol (and circle) indicating width (fig. 542). On plan maps a simple semi-decimal system, based on number and width of lines, indicates width of veins. Representation, posted with ruling pen and French curve, is superimposed on the vein geologic representation. Circular symbols represent metallization width on longitudinal section, and accessory metals on plan; this representation, posted with a drop-circle pen, is adjacent to major metals linear representation. Qualitative representation is by color. Width symbols are posted in color to indicate grade. Ore reserve blocks are posted in semi-permanent colored ink on the metallization overlay maps; the color indicates grade

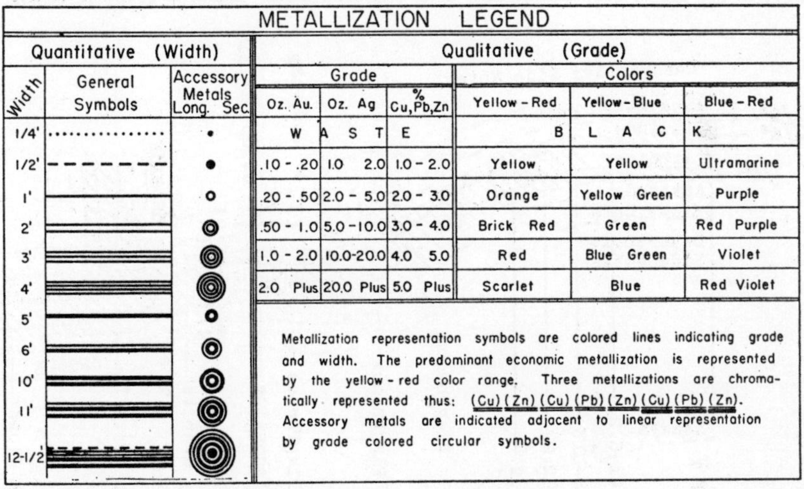

METALLIZATION LEGEND								
Quantitative (Width)			Qualitative (Grade)					
Width	General Symbols	Accessory Metals Long. Sec.	Grade			Colors		
			Oz. Au.	Oz. Ag	% Cu,Pb,Zn	Yellow – Red	Yellow – Blue	Blue – Red
1/4'	··············	•	W	A S T	E		B L A C	K
1/2'	− − − − − −	●	.10 − .20	1.0 2.0	1.0 − 2.0	Yellow	Yellow	Ultramarine
1'		o	.20 − .50	2.0 − 5.0	2.0 − 3.0	Orange	Yellow Green	Purple
2'		◎	.50 − 1.0	5.0 −10.0	3.0 − 4.0	Brick Red	Green	Red Purple
3'		◎	1.0 − 2.0	10.0-20.0	4.0 5.0	Red	Blue Green	Violet
4'		◎	2.0 Plus	20.0 Plus	5.0 Plus	Scarlet	Blue	Red Violet
5'		o						
6'		◉	Metallization representation symbols are colored lines indicating grade and width. The predominant economic metallization is represented by the yellow – red color range. Three metallizations are chromatically represented thus: (Cu)(Zn) (Cu)(Pb)(Zn)(Cu)(Pb)(Zn). Accessory metals are indicated adjacent to linear representation by grade colored circular symbols.					
10'		◎						
11'		◎						
12-1/2		◎						

FIGURE 542. Assay-metallization symbols as indicated, posted on plastic overlay sheets.

classification. When revision erasures are necessary, the plastic is resurfaced by "painting" with acetone.

Recording and Posting Specifications

Standardization of note-sheet recording and map posting is imperative, so that there is sufficient space for legible, factual, detailed recording and resultant effective utilization.

General Specifications—The subsurface mapping plane is shoulder-high; structures above and below this plane are projected to this elevation and represented at the mapping-plane intersection.

All culture and engineering data (contour pen excavation lines and LeRoy-lettered descriptive data) are posted on the reverse, or glazed side of tracing-cloth office maps. Descriptive geologic side notes and assays are lettered free-hand on the unglazed side.

Lithology—The lithic, graphic, geologic representation is usually superimposed on the entire subsurface-excavation area, and extends 1/8" outside the ground or excavation lines. If the excavation is lagged or covered so that visibility is entirely obscured, the lagging symbol is substituted for the lithic representation. Where partially lagged, the lagging symbol is superimposed on the lithic representation (fig. 543). This method facilitates later evaluation of the degree of visibility and consequent validity of data.

Structure Attitude—Attitudes are indicated adjacent to, but outside of, the lithic representation area. Attitudes are indicated by dip symbols appended directly to strike representations extended 1/4" from the excavation boundary (fig. 543).

Excavation Data—An area 1/4" wide, 1/4" from the excavation boundary, is set aside for excavation and cultural data. Lateral stoping extent,

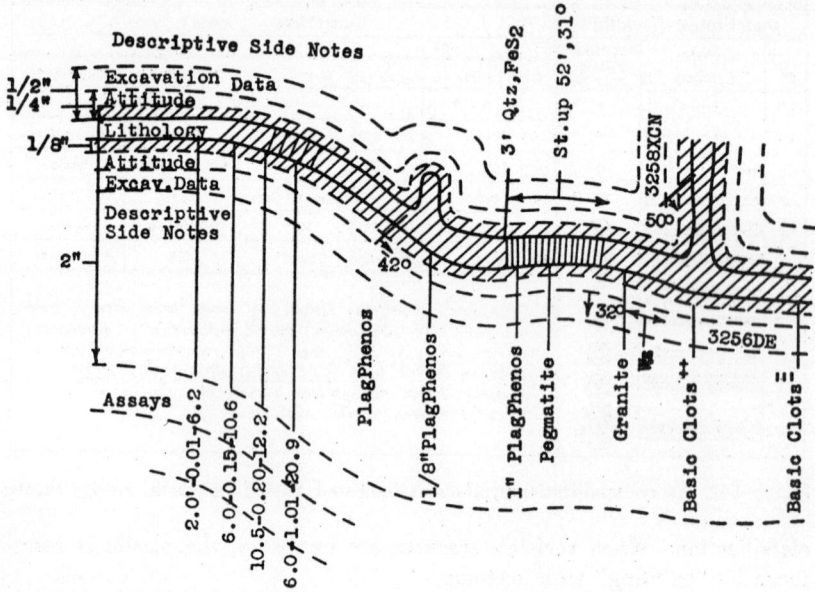

FIGURE 543. Note sheet and office-map posting specifications.

raise location, and level-to-level vein inclination (where determined by stopes and raises) are indicated on plan.

Side Notes—All descriptive side notes, whether they refer to geology, excavation, or engineering, are posted in 1½" wide area, ½" from the excavation boundary. Sides notes are necessary to indicate detail not graphically or colorimetrically mappable. Side notes are lettered neatly in lower-case as small as legibility will permit; they must be brief and descriptive. If the mapping is skillful, an auxiliary-data-recording book is unnecessary; all data is recorded on the field-note sheet. To conserve space on field sheets and office maps, all side-note descriptions and lines indicating described points are oriented parallel to either east-west or north-south oridinate lines rather than at right angles to the mine excavation trend. A line is sufficient to indicate the described point; it should be straight, not curved, with no unnecessary "jogs," or direction barbs.

On note sheets, textural and compositional gradations are indicated at gradational-interval-end points by side-note descriptions joined with a line.

Compositional or textural gradation ("fading") is also indicated as shown in the lower central portion of figure 543. Abbreviations and chemical formulae are utilized when possible. The abbreviations should be "natural" and non-ambiguous; e.g. chalcocite, Cu_2S; pyrite, $FeS_2^{\cdot}$; bornite, Born; covellite, CuS; quartz, Qtz; silicate or siliceous, Sil; augite, Aug; hornblende, Hb; feldspar, Feld; gouge, Gg; fault, Flt. Non-complex minerals are always designated by chemical formulae (pyrrhotite, Fe_{11}-S_{12}); silicates by first syllable (garnet, Gar); and other terms by omitting vowels (jointed, Jtd). Successive nouns are separated by commas; successive adjectives are not punctuated. Only descriptive material is included in side notes; e.g., "sooty" chalcocite is preferably over "secondary" chalcocite. Descriptions should indicate observations rather than deductions. Properly recorded observations facilitate deduction at any later date. Side notes are so recorded that a maximum of space is available for detail, and precise comprehension by persons other than geologists is possible.

Note Sheet Dating and Signing—The mapping date and geologist's initials are indicated on the reverse side of the sheet in the mapped-area space. The mapped-area position is determined by holding the sheet to the light.

Sectional Specifications—Cultural and engineering data posting is the same on plan and section. Whenever possible, sectional representation is on either north-south or east-west vertical plane orientation, as indicated in figure 544. Apparent dip corrections are made, therefore, before angular data is posted.

Cross-section data are readily integrated for ore finding if the section plane coincides with the even hundred ordinate lines. Nothing is gained by preparing sections at right angle to the vein strike; a section will probably be at an acute angle to cross-cutting faults also represented on the section.

The degree of cultural- or geologic-data probability and reliability are indicated thus:

1. Known Data:
   a. Known location.
      Data, located on the sectional plane, are posted as on maps.
      Projected data to the sectional plane are represented by long dashes. Spacing distance indicates projection distance. Distance and direction of projection are indicated thus: "25′ E."

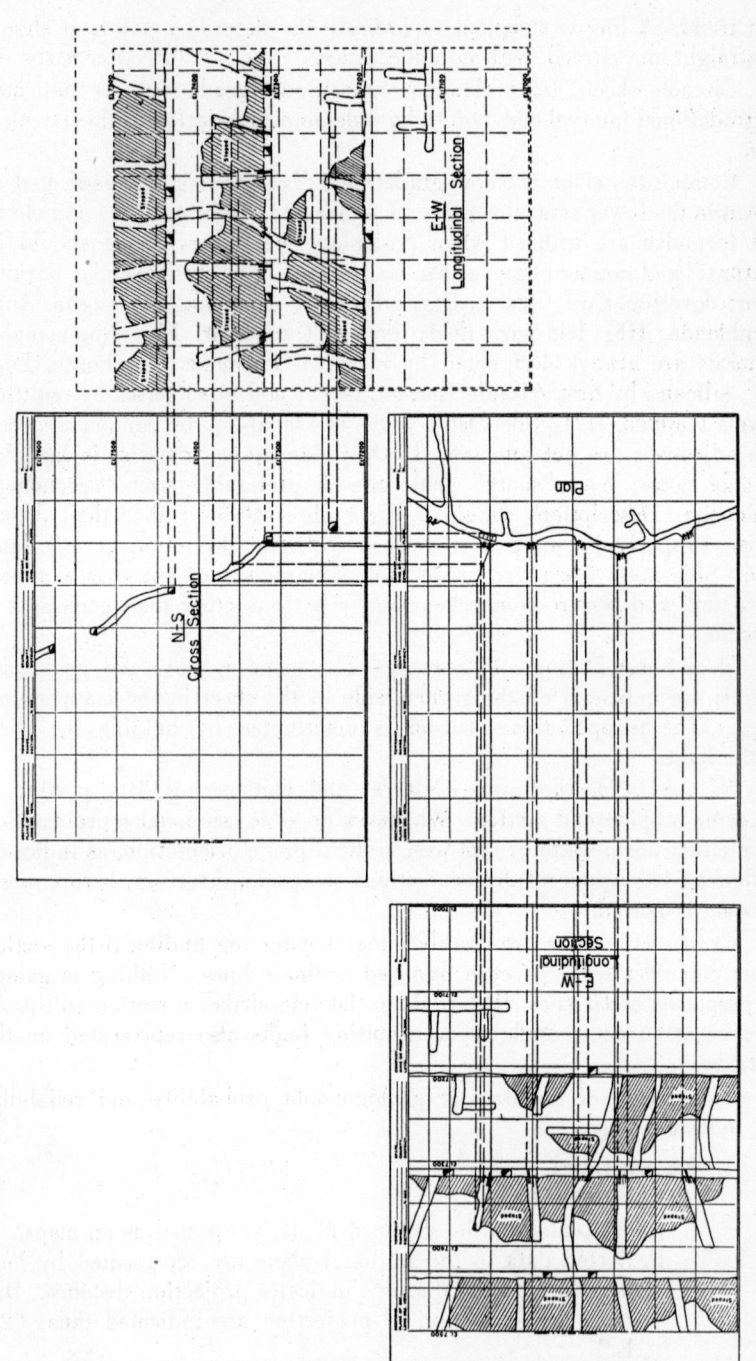

FIGURE 544. Correlation and manner of preparation of plan, north-south cross section, and east-west longitudinal projection. Sections are prepared underground by projection as indicated by dash lines on note sheets shown.

b. Approximate location and shape.

Data located on the sectional plane are represented by short dashes closely spaced.

Projected data to sectional plane are represented by short dashes; spacing distance indicates projection distance.

2. Doubtful or universal data:

Data located on the sectional plane are represented by closely spaced dots. Projected data are indicated by dots; spacing distance indicates projection distance.

The projection direction and inclination of geology is in a direction parallel to linear structure elements (flutings, mullion, striae, fold-axis lines, flow lines, etc.); the lines of least curvature on geologic structure usually coincide with net displacement or maximum elongation direction.

Longitudinal projections are on either horizontal or vertical planes. Vertical-section projection is convenient unless veins are at very low dip angles. Vertical plane projection implies vertical shortening distortion, and that ore volumes are products of length, horizontal width, and vertical distance rather than length, true width, and inclined distance. The projection plane is parallel to ordinate lines, even though the vein strike is diagonal to ordinate lines; veins are unpredictable warped "planes" varying laterally and vertically in attitude (frequently 30°). Since dip distortion is accepted, additional strike distortion will not seriously affect utilization of longitudinal projections systematically prepared on east-west or north-south planes. Ordinate-direction orientation of the projection plane simplifies longitudinal section preparation and facilitates better sectional integration, correlation, and filing. East-west or north-south, rather than true lengths of ore, are used in volume calculations; ton or volume factors indicated by numerical contours on the longitudinal section facilitates calculation.

Geology (composition, texture, and intersecting structures) of both vein walls may be represented on two separate longitudinal projections as on plan maps.

Vein configuration is represented by superimposed contours. The Connolly vein-contouring technique is utilized to show vein or orebody configuration. An auxiliary "datum" plane designated as zero "elevation" is erected nearly parallel to the vein. Vein-offset distances or "elevations" are measured; measure-point positions are projected to an east-west or north-south vertical longitudinal section; and the "elevation" values are contoured. Contours are superimposed on the texture-composition representation.

Mineralographic and Hand-Specimen Representation—(Indicated in figure 545.) Mineralogic legend and paragenetic sequence representation are combined. The same shading, hatching, or colors are utilized on both the legend-paragenetic sequence diagram and the micro-sketch or photo. A graphic scale is placed beneath the micro-sketch.

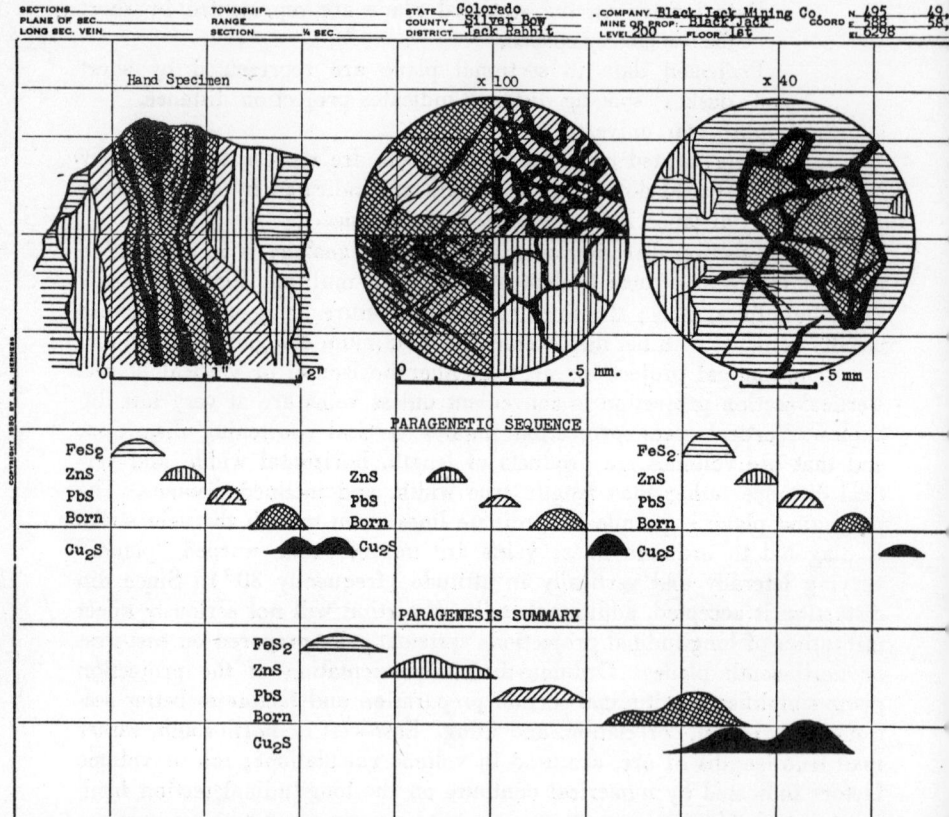

FIGURE 545 Coordination of mineralogic symbols on micro-sketches and paragenetic sequence re
resentation. Sketches are drawn on standard letter-size field subsurface note sheets.

QUESTIONS

1. What two geologic representation phases are important in mineral exploration?
2. What are five feature classifications which should be mapped?
3. What criticism can be made of published government survey maps?
4. Why is 50′ scale preferable to 40′ scale for detailed mapping?
5. What are "odd" mapping scales?
6. What mapping scales are recommended in this paper?
7. Why is the use of these scales advantageous?
8. Why is it desirable that coordinate values of the initial point in an area be five digit numbers?
9. What size office maps are recommended and why?
10. What are the advantages of using an 8″ x 10″ field-subsurface note sheet?

11. Describe note-sheet and office-map indexing.
12. Describe note-sheet and map filing and filing sequence.
13. Why are side-note lines oriented parallel to ordinate lines?
14. What are three classifications of geologic features which are represented on geologic maps?
15. What symbols are used to designate phaneritic, aphanitic, porphyritic, and pegmatitic textures for office-map representation?
16. What rock types and geologic features are represented in black?
17. For the representation of what types of rocks or effects is color used?
18. What is meant by "color progression"?
19. Transparent colors are used for what type of representation?
20. Why must color used in geologic representation be carefully chosen?
21. What precautions must be taken when selecting colors in superposition?
22. What are "cumulative" colors?
23. What color types are utilized to represent basic and acid rocks respectively?
24. List color representation progression from basic igneous rocks to metallization effects.
25. Why are these colors used?
26. How does office-map geologic legend differ from field-subsurface legend?
27. Why is soil cover indicated on note sheets and maps?
28. Why are dumps and other man-made deposits represented compositionally?
29. What are metallization maps?
30. How are metallization maps prepared?
31. How are quantitative and qualitative values, respectively, indicated on metallization maps?
32. Give two reasons why neat and orderly side notes are important.
33. How large should side note descriptions be?
34. What determines projection direction of geologic features?
35. In what manner and where are side notes recorded on note sheets and maps?
36. What type of abbreviations are used in side-note descriptions?
37. Describe gneissic rock representation on field-subsurface and office maps respectively.
38. What is the recommended reference projection-plane attitude used in longitudinal section preparation?
39. How is vein configuration represented on longitudinal projections?
40. How is paragenetic sequence and textural representation correlated?

CHAPTER 14

SUBSURFACE METHODS AS APPLIED IN GEOPHYSICS

HARRISON E. STOMMEL

Since the introduction of geophysical methods of prospecting in this country in the early 1920's, refinements in instrumentation, in field procedure, and in interpretation have made geophysics the foremost exploration technique available to the oil industry. In recognition of its successes the larger oil companies have created geophysical departments on a par with their geological departments. The location of new reserves has become of such major importance that most oil companies now rank the exploration division, usually headed by a vice president, with the producing, pipe-line, refining, and marketing divisions. Coffin [1] states:

> The exploration budgets of at least three companies whose geological and geophysical expense exceeds five million dollars yearly carry items for geophysical expense in excess of total geological expense in ratios ranging between 5 to 1 and 10 to 1.

Few petroliferous provinces exist where surface mapping alone gives evidence of subsurface structure. In this country, for example, the entire Gulf Coastal Plain, the Midcontinent, the Great Plains, Western Canada, and a part of California are blanketed by a thick series of younger sediments which are often unconformably related to the deeper potential-reservoir rocks. In such areas the combined efforts of subsurface geologist, the paleontologist, and the geophysicist are required to locate new accumulations. Finding another East Texas, Slaughter, or Yates field is considered very unlikely by the geologist or geophysicist, his efforts being directed to the location of smaller and deeper structures that were overlooked in the past or to prospecting in new, less favorable areas.

In the mining industry the future of geophysics looks exceedingly bright. Within the past several years a number of the country's largest metal producers have added geophysicists to their exploration staffs and are at present engaged in geophysical field work and research. Although geophysics was first applied to the location of mineral deposits in the 1600's, refinements in technique have not kept pace with the needs of the mining industry. This is not solely the responsibility of the mining geophysicist. He has been hampered by a lack of research funds, by the most complex geologic situations possible, and by the small size of the target.

It is rumored that new electrical techniques are being developed by the mining industry which show promise of much greater resolving power than has heretofore been achieved. Mining geologists today feel the need for additional tools to aid them in mineral exploration, because attention is

---

[1] Coffin, R. C., *Recent Trends in Geological-Geophysical Exploration:* Am. Assoc. Petroleum Geologists Bull., vol. 30, no. 12, p. 2014, Dec. 1946.

being focused on areas where outcrops are rare or unknown. The responsibility for providing these tools has fallen on the geophysicist and geochemist. Problems faced by the mining geophysicist were discussed at a recent meeting of the Society of Exploration Geophysicists.[2] Geophysics applied to mining in the United States is in the same position today that petroleum geophysics was a quarter of a century ago. The addition by the United States Bureau of Reclamation of geophysicists to its engineering and geologic staff indicates interest in applications of geophysics to civil-engineering problems.

It is beyond the scope of this chapter to discuss the detailed techniques involved in geophysical exploration, and only a general discussion of the topic is presented. For a complete treatment, the reader is referred to the several text books on geophysical methods of exploration.[3] [4] [5]

Statistics

During 1947 the drain on our oil resources was greater than ever before, yet, according to Eckhardt [6] the industry was able not only to maintain its reserves but to increase them. Lahee [7] shows that of the 3,471 new-field wildcat wells drilled in 1947, 394 became oil producers. It may be interesting to compare the effort of the various exploration techniques in these unproved areas: 10.5 percent of the holes drilled only on geologic information were producers; 16.7 percent drilled only on geophysical information were successful; 17.7 percent drilled on combined geophysical-geologic information were producers; and only 3.8 percent of the wells drilled on nontechnical information became productive. It is little wonder that the petroleum industry spent a minimum of $105,000,000 on geophysical exploration during 1947, not including the millions spent on research.

The number of geophysical crews in the field has shown a steady increase, averaging about 555 crews during 1947. This is double the number reported in 1942. Rapid expansion of seismic programs is chiefly accountable for the increase, as the number of magnetic and gravity parties in the field has decreased somewhat during the past several years. (See fig. 546.)

Geophysical Exploration

Geophysical exploration may be defined as "prospecting for mineral deposits and geologic structure by surface measurement of physical quantities." [8] In order to be considered a usable geophysical method the following five criteria must be satisfied:[9]

[2] Symposium on Mining Geophysics: Geophysics, vol. 13, no. 4, pp. 535-583, Oct. 1948.
[3] Heiland, C. A., Geophysical Exploration, New York, Prentice-Hall, 1940.
[4] Jakosky, J. J.. Exploration Geophysics. Los Angeles, Times-Mirror Press, 1950.
[5] Nettleton. L. L.. Geophysical Prospecting for Oil. New York. McGraw-Hill Book Co.. Inc.. 1940.
[6] Eckhardt, E. A., Geophysical Activity in the Oil Industry in the United States in 1947: Geophysics, vol. 13, no. 4, p. 529, Oct. 1948.
[7] Lahee, F. H., Statistics of Exploratory Drilling in 1947: Am. Assoc. Petroleum Geologists Bull., vol. 32, no. 6, pp. 851-868, June 1948.
[8] Heiland, C. A., op. cit., p. 3.
[9] Rust, W. M. Jr., Evaluation of New Geophysical Methods: Geophysics, vol. 10, no. 3, p. 331, July, 1945.

1. The method must involve physical measurements of some property of the earth.
2. The measurements must give reproducible results.
3. The process must be able to compete economically with existing methods.
4. The method must have a significant relation to the occurrence of oil (economic minerals).
5. The method must have a teachable interpretation.

The major methods of exploration may be roughly divided into those which measure variations in one of the earth's potential fields, magnetic

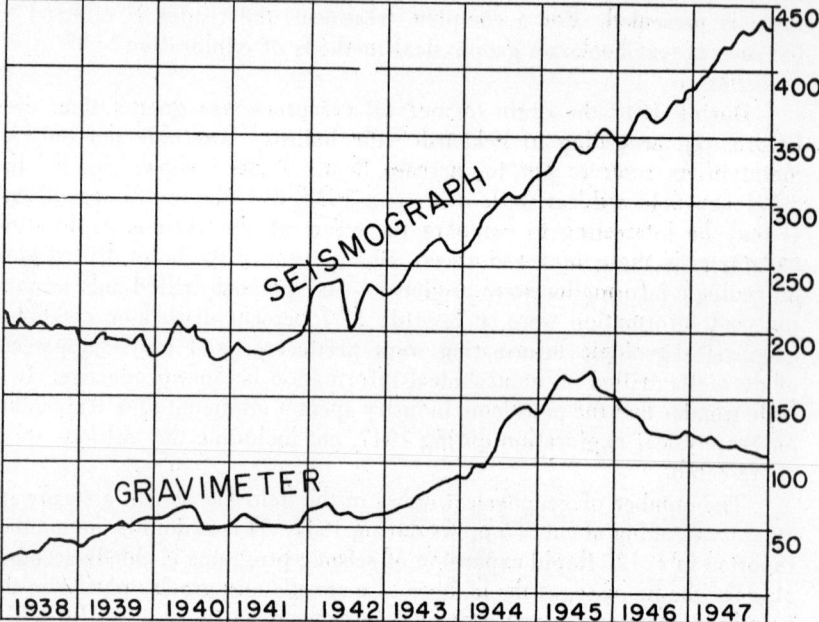

Figure 546. Number of seismograph and gravimeter parties in United States from 1938 to 1947. (After Eckhardt. Reproduced permission *Geophysics*.)

and gravitational; or into those in which a measurable reaction is produced at the surface by the application of an artificial force field, seismic and electrical. Inference may be made as to subsurface geologic conditions only when detectable differences in physical properties exist below the surface. The physical basis for the use of the magnetic methods is thus a variation in magnetism or susceptibility, for gravitational methods a variation in density, for seismic methods a variation in velocity, and for electrical methods a variation in electrical conductivity.[10] Generally speaking, the geophysicist is unable to give specific information concerning the

---

[10] Does not apply to the self-potential method.

presence or the absence of a given mineral or group of minerals. He is able, however, to indicate geologic conditions that are favorable to the accumulation of a given mineral or group of minerals. For example, it is impossible to predict the presence of petroleum by surface geophysical measurements; geophysics does, however, indicate favorable geologic structure for the accumulation of petroleum. Several exceptions to the foregoing statement should be noted. Gravitational methods will indicate accumulations of salt, peridotite, serpentine, etc., in certain areas where good geologic control is available. Under the same conditions the magnetometer will indicate the presence of serpentine plugs, concentrations of "black sand" in placers, and hematite deposits. The refraction seismograph on the Gulf Coast presents unmistakable evidence of salt uplifts. The self-potential method shows accumulations of the heavy sulphides.

The potential methods, magnetic and gravitational, although rapid and relatively inexpensive, are essentially reconnaissance tools. For a given magnetic or gravity map any number of subsurface conditions may exist that would produce the observed surface-intensity distribution; predications based on these methods may, therefore, be ambiguous. In the seismic and electrical methods control is maintained over the depth of penetration by operational technique, leading to a unique interpretation of subsurface geologic conditions.

Planning a Geophysical Program

Assuming that the prospective area is potentially petroliferous, geophysical programs generally follow a well-integrated plan. The "structural grain" may first be determined by one of the rapid reconnaissance methods. This may be accomplished by a number of air-borne-magnetometer profiles or by rather widely spaced magnetic or gravity observations (station density of two or three to the township). In the event that the "grain" is fairly evident and competition for leases is not keen, a more complete survey is generally begun at once with either the gravity meter or both magnetometer and gravity meter working the area. On a survey of this nature, gravity observations will be made with a density of about 85 stations to the township. Magnetometer observations are commonly spaced one mile apart or with a station density of about 48 stations to the township. In promising areas, more detailed work is usually desirable to substantiate an anomaly.

Several advantages accrue from the use of such reconnaisance instruments: (1) Prospects may be leased in a minimum of time, (2) Leases in the less-promising areas may be immediately dropped, and (3) Areas to be worked by more expensive seismic or core-drilling methods are reduced in size, with a better chance of locating structure for the exploration dollar spent.

With the location of either interesting magnetic or gravity anomalies the next step in the geophysical program is to shoot the magnetic or gravity anomalies with the seismograph. The general procedure today is to shoot continuous profiles spaced as closely as is deemed necessary.

Subsurface maps are carried on as many horizons as are convenient. In many areas resort must be made to dip shooting and the contouring of so-called phantom horizons. In other regions three-dimensional dip shooting is practiced, the final map being expressed in terms of strike and dip symbols. Various combinations of the foregoing seismic techniques are often practiced, depending upon the particular area involved. Then upon completion of the seismograph survey and evaluation of all data bearing on the area, recommendations are made by the geological and geophysical departments for the location of the test well.

If such a program as outlined above is followed, it is obvious that the credit for a discovery does not belong to any individual method, as would probably be indicated in an analysis of exploratory effort. Coffin [11] cites an example involving the Richardson and South Ellinwood fields in Barton County, Kansas, where credit for the discoveries was given the core drill and reflection seismograph. One year prior to the completion of the discovery wells, on the basis of a magnetically anomalous trend supported by limited gravity work, 50,000 acres were leased at a cost of ten cents an acre. After seismic and core-drill work on anomalous areas, the discovery wells were located.

Cost of Geophysical Surveys

When discussing costs it must be borne in mind that wide variations are bound to occur, physiographic conditions probably affecting the ultimate cost to the greatest degree. Other variables are climatic conditions, costs of materials and equipment, and salaries of personnel. The average cost of a magnetic survey, either air-borne or on the ground, with a density of 48 stations to the township will be approximately $1\frac{1}{2}$ cents an acre under average conditions. For a gravity survey with a station density of 85 stations to the township the costs will average about five cents an acre under normal conditions. Costs of seismograph surveys show the greatest variation, as the previously mentioned factors influence not only the cost but also the type of survey conducted. For instance, a reconnaissance survey with shot points located a mile apart might cost an average of 25 cents an acre, whereas a continuous survey in the same area might average two dollars an acre. It follows then that the amount and accuracy of subsurface information varies more or less with the cost of obtaining the information.

MAGNETIC PROSPECTING

Perhaps the oldest application of geophysics to prospecting for mineral deposits and buried geologic structure was the use of magnetic methods by the Swedes in 1640. During the three centuries that have elapsed since its inception, the magnetometer has proved to be a valuable exploratory tool. The discovery of several oil fields may be directly attributed to the magnetometer. In a score of others the magnetic results recom-

---

[11] Coffin, R. C., *op. cit.*, p. 2016.

mended the use of more accurate exploration methods, which were given credit for the location.

Soon after the discovery of the Hobbs field, New Mexico, on a magnetic high, any person who could raise the necessary thousand dollars to purchase a magnetometer became a magnetic "expert." Inadequate field technique and poor interpretation resulted in the drilling of hundreds of dry holes in structurally normal areas. Although several crews operated periodically during the following decade, it was not until World War II

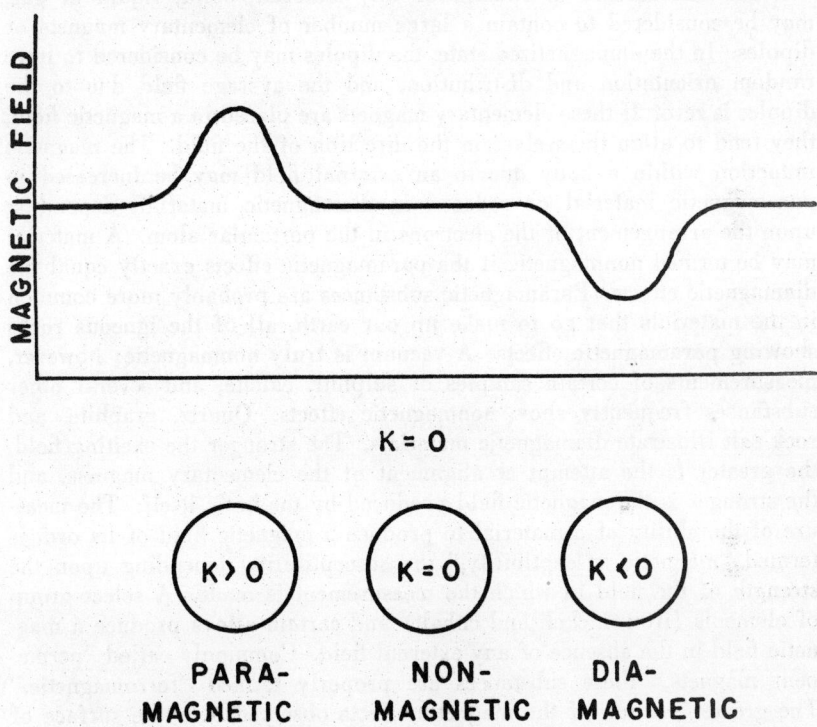

FIGURE 547. Effect on vertical component of earth's magnetic field of paramagnetic, nonmagnetic, and diamagnetic materials. Susceptibilities of paramagnetic and diamagnetic materials assumed to be equal.

that extensive magnetic programs were again included in the geophysical efforts of many companies. Peters [12] reports about 165 crew-months of ground magnetic activity during 1947 and about 17 crew-months of airborne-magnetometer work.

Basic Principles

Stated in simplest terms, magnetic prospecting consists in comparing the intensity of the magnetic field between observation points that are

[12] Peters, J. W., *The Role of the Magnetometer in Petroleum Exploration*: Mines Mag., vol. 39, no. 7, pp. 11-15, 1949.

located at or above the earth's surface. This natural-force field will consist of the field due to subsurface geologic features superimposed on the terrestial magnetic field. Variations in the distribution of magnetized material beneath the surface will alter the intensity of the magnetic field at the surface. Since sedimentary rocks are generally nonmagnetic and the basement rocks slightly magnetic, variations observed at the surface are ordinarily attributed to basement relief or to intrabasement variations in distribution of magnetic materials.

For convenience of illustration any material, solid, liquid or gas, may be considered to contain a large number of elementary magnets or dipoles. In the unmagnetized state, the dipoles may be considered to have random orientation and distribution, and the average field due to the dipoles is zero. If these elementary magnets are placed in a magnetic field, they tend to align themselves in the direction of the field. The magnetic induction within a body due to an external field may be increased in paramagnetic material or reduced in diamagnetic material, depending upon the arrangement of the electrons in the particular atom. A material may be termed nonmagnetic if the paramagnetic effects exactly equal the diamagnetic effects. Paramagnetic substances are probably more common in the materials that go to make up our earth, all of the igneous rocks showing paramagnetic effects. A vacuum is truly nonmagnetic; however, measurements of certain samples of sulphur, calcite, and several other substances frequently show nonmagnetic effects. Quartz, graphite, and rock salt illustrate diamagnetic materials. The stronger the exciting field, the greater is the attempt at alignment of the elementary magnets, and the stronger is the magnetic field produced by the body itself. The measure of the ability of a material to produce a magnetic field of its own is termed "magnetic susceptibility," the susceptibility depending upon the strength of the field in which the measurement is made. A select group of elements (iron, nickel, and cobalt) and certain alloys produce a magnetic field in the absence of any external field. Commonly called "permanent magnets," these substances are properly termed "ferromagnetic." The greater portion of the magnetic effects observed over the surface of the earth is due to the variation in the distribution of magnetite, which is a ferromagnetic substance.

In the earth's magnetic field, paramagnetic materials (susceptibilities greater than zero) tend to increase the normal field; nonmagnetic materials (susceptibilities equal to zero) have no effect on the normal field; and diamagnetic materials (susceptibilities less than zero) tend to decrease the normal field. (See fig. 547.)

Rock Properties

As a rule, basic igneous rocks show the largest susceptibilities, ranging as high as $16,000 \times 10^{-6}$ c.g.s. units. With an increase in silica content, the susceptibility decreases, values given in the literature ranging from 8 to $2,700 \times 10^{-6}$ c.g.s. units for granites. Sedimentary rocks seldom exceed

50×10^{-6} units, and many are diamagnetic, the susceptibility of gypsum and anhydrite, for example, varying from -1 to -10×10^{-6} c.g.s. units. In general it may be stated that two factors are chiefly responsible for the magnetization of rocks: magnetite content and geologic history (metamorphism, tectonic movements, lightning, etc.). The presence of magnetite in many of the igneous rocks suggests that the magnetization of the rocks is due to two sources. First, the contribution due to the "permanent magnetization" or remanent magnetization of the magnetite and, second, the magnetization induced in the rock by the earth's magnetic field. Designating polarization as I, susceptibility as k, the exciting field as H, and the remanent magnetization as I_p, one may write the relationship as follows:

$$I=kH+I_p$$

In most igneous rocks the remanent magnetization is the dominant factor contributing to the total polarization of the rock. The contribution due to remanent magnetization is impossible to determine in rocks lying thousands of feet below the surface, for not only is intensity involved but the direction of the force as well. This fact greatly limits the quantitative interpretation of magnetic results.

Inasmuch as irregularities in the distribution of magnetic materials give rise to the observed anomalies at the surface, it has become routine practice for many of the larger operators to make determinations of the susceptibility and remanent magnetization in the laboratory. Cuttings, cores, and outcrop samples are carefully studied and the results tabulated. Studies of this nature make more accurate quantitative interpretation possible and aid in the planning of field surveys.

The Earth's Magnetic Field

Magnetic measurements made over the surface of the earth would indicate that the central core of the earth is magnetic, and seismic results indicate that the core is metallic. However, the interior of the earth must be at a high temperature, temperatures much in excess of the Curie point, at which materials lose their ferromagnetic properties. The most reasonable theory proposes that the magnetic field is due to electric currents flowing through the central core. This theory is substantiated by recent measurements of telluric currents, an increase in the earth's magnetic field being accompanied by an increase in telluric-current activity.

The distribution of the magnetic field at the surface of the earth indicates that the earth approximates a polarized sphere with magnetic poles located near the axes of rotation. For sake of illustration, we may imagine the surface distribution of the earth's magnetism to be accounted for by a short bar magnet placed in the center of the earth. (See fig. 548.) By determining the total intensity and the direction of the magnetic vector, we may describe the magnetic field at any point on the surface of the earth. The strength of the earth's magnetic field is about 0.5 oersteds in the United States. The unit of measurement in magnetic prospecting is the gamma, which is equal to one-hundred-thousandth part of an oersted.

$$1 \text{ gamma} = \frac{1}{100,000} \text{ oersted or } 10^{-5} \text{ oersted}$$

In prospecting with the air-borne magnetometer, relative measurements are made of the total intensity; in prospecting with magnetic-field

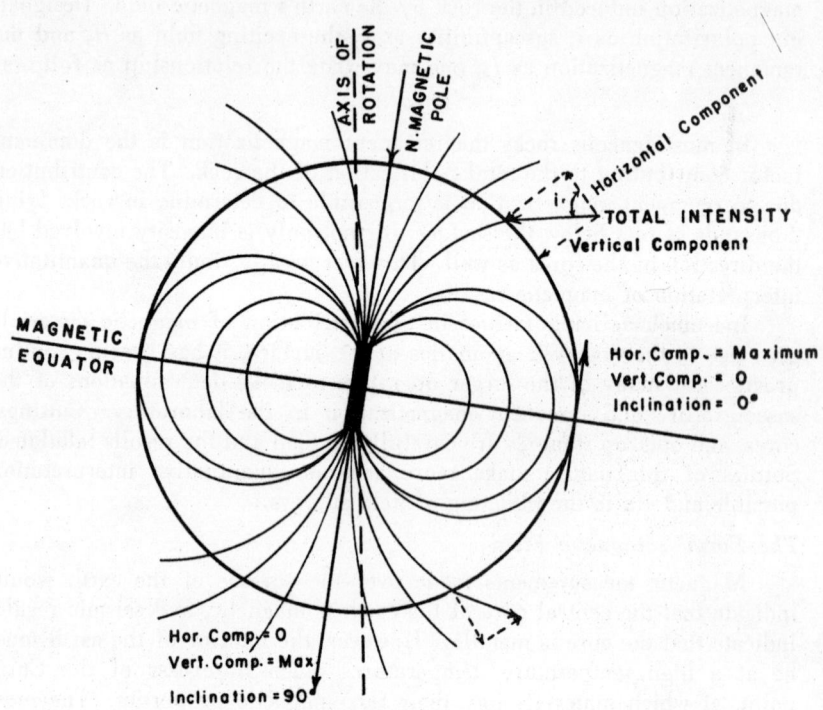

FIGURE 548. Simplified diagram of the earth's magnetic field.
(Adapted from Jakosky.)

balances, the vertical or horizontal components are compared. Certain techniques in mining exploration involve comparative measurements of the angle of inclination.

Normal Variation of Field—It will be noted that there is a normal increase in intensity of the vertical field as we approach the magnetic poles until at the poles maximum value is attained. Likewise, the horizontal component increases in intensity toward the magnetic equator and reaches its maximum value there. In large-scale magnetic surveys, corrections are applied to the field data to compensate for the normal variation of the magnetic-field strength with latitude and longitude. In the state of Oklahoma, for example, the normal increase of intensity for the vertical component toward the north is about 13 gammas per mile; toward the east it is about 4 gammas per mile.

Time Variations of the Earth's Field—Even at a fixed point on the earth's surface, the magnetic intensity does not remain constant, variations being observed daily, yearly, and over much longer periods of time. Certain short-period and daily changes are apparently related to sun-spot

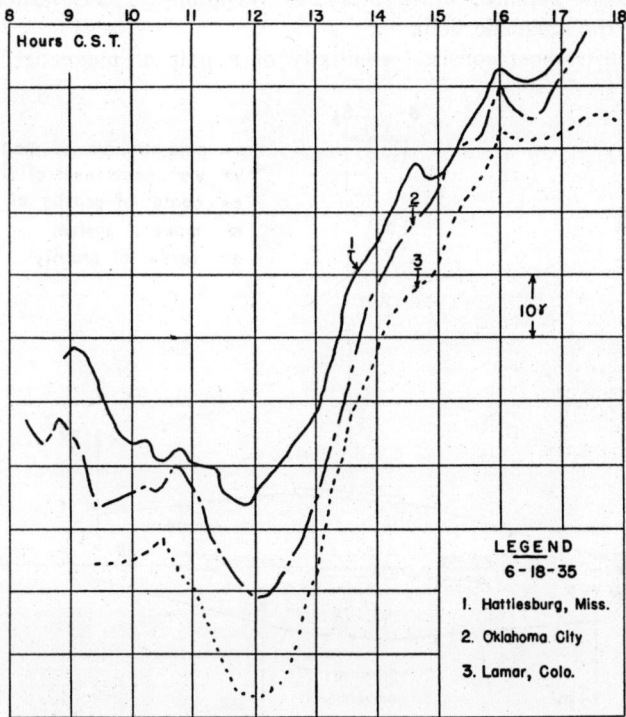

FIGURE 549. Diurnal-variation curves observed at different localities on the same day. Data obtained by observation of a fixed-base-station instrument. (After Vacquier.)

activity; variations of several hundred gammas within an hour are not uncommon during magnetic storms. Corrections must be made for the normal diurnal or daily variations of magnetic intensity with time before a survey may be utilized. The diurnal curves are obtained in the field either by reading a base instrument every ten or fifteen minutes or by returning the field instrument to a base or check station within every two-hour period. Observations made during periods of rapidly changing intensity are of no value. (See fig. 549.)

In summation, the magnetic-field intensity at any point at a given time consists of the field due to subsurface geologic bodies superimposed on the terrestrial magnetic field for that latitude and longitude.

Magnetic-Prospecting Instruments

None of the instruments used in magnetic prospecting today measures the magnetic-field intensity; without exception they compare intensities from point to point over the surface of the earth. The most widely used instrument, both in oil and mining prospecting, is the Schmidt-type magnetic-field balance, which measures variations of the vertical component of the magnetic field.

The instrument consists essentially of a pair of magnetized blades

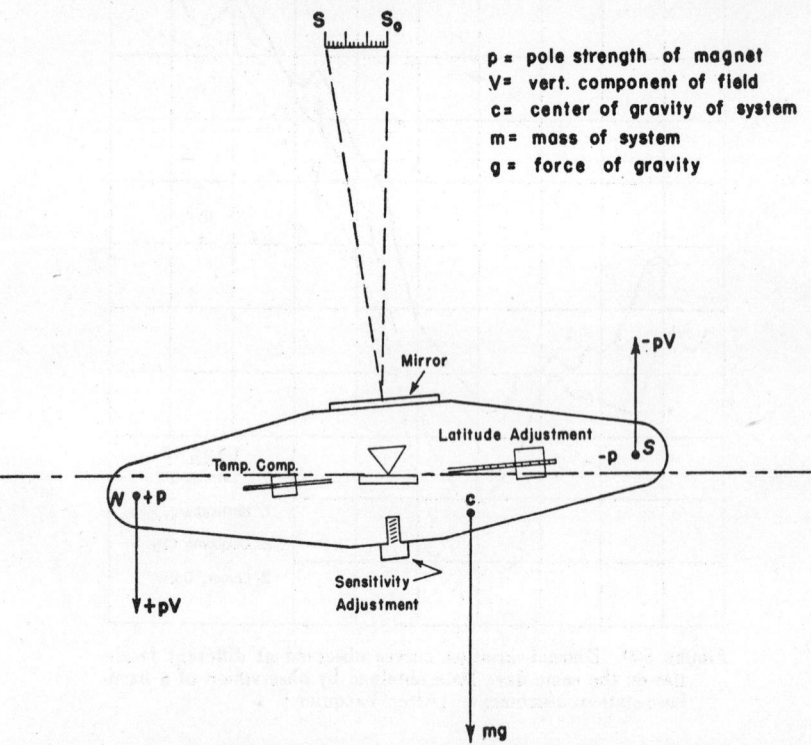

p = pole strength of magnet
V = vert. component of field
c = center of gravity of system
m = mass of system
g = force of gravity

FIGURE 550. Schematic diagram of compensated-vertical-magnetometer system showing position of latitude, temperature, and sensitivity-adjusting weights.

pivoted on a horizontal quartz-knife edge. A magnetic couple, depending on the strength of the vertical magnetic field, tends to rotate the blades about the horizontal axis. The magnetic couple is opposed by a gravity couple acting in the opposite direction. By observing the equilibrium position, where magnetic couple equals gravity couple, between two points on the earth's surface, we are able to determine the difference in vertical intensity between the points. Magnetometers may be read to approximately one or two gammas, depending upon the sensitivity of the instrument. Corrections for temperature fluctuations are generally applied to field observations, as variations in temperature change the instrument sensitivity by varying the moment of the magnetic system. (See fig. 550.)

Rapidly gaining in popularity for reconnaissance surveys is the first geophysical instrument to be used successfully in a moving aircraft, the saturable-core or flux-gate magnetometer. According to Muffly,[13]

. . . this type of magnetometer is basically a rod or strip of ferromagnetic material acting as an open magnetic core for one or more a.c. windings connected to exciting and indicating circuits which can measure the magnetization produced in the core by an ambient magnetic field.

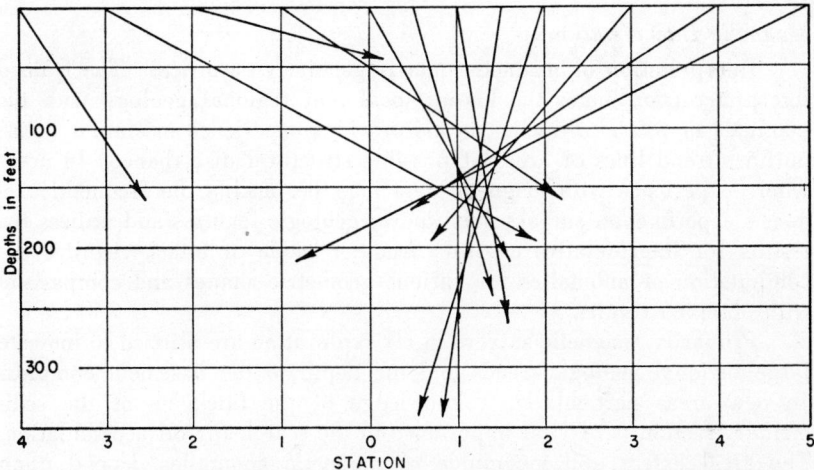

FIGURE 551. Magnetic vectors over an ore body at Falconbridge, Canada. (After Eve and Keys.)

The instrument has a sensitivity of about one gamma, is continuously recording, and measures the total magnetic-field intensity. The total-field recording of the saturable-core magnetometer requires changes in the interpretation technique ordinarily applied to field magnetic data of the vertical-component magnetometer. Using modern aircraft such as the DC-3, surveys may be successfully carried out over land, swamp, jungle, or sea in a minimum of time.

The chief problem confronting the operators of air-borne equipment is the accurate determination of position. War-born, high-frequency radio methods are being used for many of the surveys. The use of the air-borne magnetometer has not only speeded up field operations but has also increased the accuracy of the results. By making observations above the earth's surface, interference due to near-surface effects is largely eliminated.

13 Muffly, Gary, The Airborne Magnetometer: Geophysics, vol. 11, no. 3, pp. 321-334, July 1946.

Volcanic "float," telephone lines, pipe lines, fences, railroads, and other factors that disturb the conventional ground survey are rendered ineffective at higher elevations. Surveys are commonly flown at about 1,000 feet above the ground, this elevation largely eliminating near-surface effects and allowing an ample safety margin for the aircraft.

Of particular importance in mining surveys are measurements made with the Schmidt-type horizontal-component field balance. This instrument is similar in design to the vertical-component field balance with the exception that the magnetic system is placed in a vertical rather than a horizontal position. By using both vertical and horizontal data, vectors may be plotted that indicate quite accurately the location of the disturbing body. This technique is extremely valuable in underground surveys. (See fig. 551.)

Magnetic Interpretation

Interpretation of magnetic data is generally empirical. The trained interpreter coordinates the known local and regional geology with the magnetic map, and, drawing on his previous experience in similar areas, outlines trend lines or areas of possible structural disturbance. In areas where experience with magnetic data may be lacking, he frequently requests experimental surveys over known geologic features and utilizes the results for interpretative control. Another mode of attack requires the computation of anomalies for various geometric shapes and comparison with observed results.

Primarily, magnetic surveys in oil exploration are utilized to indicate large regional geologic trends and the depth to the basement complex. In new areas particularly, a knowledge of the thickness of the sedimentary section is of vital importance in the search for oil accumulation. The areal extent and magnitude of magnetic anomalies depend upon several factors, among which are depth, size and shape, and polarization or intensity of magnetization. The sharpness of magnetic anomalies depends almost entirely on the depth alone, comparatively sharp anomalies originating at shallow depth and broad anomalies at greater depths. By assuming certain simple geometric shapes, rules are available for estimating the depths to magnetized material from the extent of their surface magnetic effects.[14] Depth rules are simple to apply and often give important clues to the geologic structure. It should be understood that they do not give conclusive information, and they should be applied with reservations. In regions where the approximate depth to the basement complex is known, application of the depth rules may aid in distinguishing intrabasement variations from actual basement uplift.

Where the disturbing geologic body is roughly circular and the section small compared with the depth extent, the lines of vertical intensity will be nearly circles or short ellipses, and only a positive anomaly

14 Nettleton, L. L., *Geophysical Prospecting for Oil*, p. 224, New York, McGraw-Hill Book Co., Inc., 1940.

will be in evidence. If this condition exists, the assumption of a single magnetic pole is justified. If a rather broad negative anomaly encircles the sharper positive anomaly, assumption of a sphere will probably more closely fit the geologic conditions. If the isoanomalic lines are much longer in one dimension, a horizontal cylinder may be assumed. To determine approximate edge of the disturbing mass one customarily picks the inflection points on either side of the magnetic profile where curvature changes from concave downward to concave upward. In the case of the sphere this will give the approximate diameter and thus allow estimates to be made as to the depth below the surface to the top edge of the mass.

Depth Rules

Single pole...Depth to pole $= 1.305\ X_{\frac{1}{2}}$
Sphere ....Depth to center $= 2.0\ X_{\frac{1}{2}}$
Horizontal cylinder..................................Depth to center $= 2.05\ X_{\frac{1}{2}}$
Fault ....Depth to center $=\ X_c$

$X_{\frac{1}{2}}$ is the horizontal distance from the center of the anomaly to the point where the anomaly has fallen to one half its maximum value, and X_c is the distance from the center of the fault picture to the maximum or minimum of the curve.

The depth rules are subject to many errors, yet under the proper conditions they do give valuable information. More accurate interpretative techniques are available involving the comparison of magnetic effects from spheres, vertical cylinders, horizontal cylinders, and series of disks with the observed magnetic anomaly. For a discussion of these the reader is referred to Nettleton.[15] However, all geophysical methods involving one of the natural potential fields of the earth are subject to ambiguous interpretation. That is, a given magnetic anomaly may be due to a variety of geologic conditions. Even though we are able to calculate a certain distribution, depth, and polarization of magnetic material that will exactly satisfy the magnetic anomaly obtained in the field, it is still no guarantee that the geologic structure deduced is the correct one. By using other combinations of distribution, depth, and polarization, it is quite possible that equally good fits could be achieved. This ambiguity means that magnetic surveys should be supplemented by other types of subsurface control such as direct well control or seismograph data. The magnetic method does have the advantage over the gravity method in that most magnetic effects come either from within the basement or at the basement surface.

Regional Gradient—In many areas large magnetic anomalies covering hundreds of square miles are due primarily to polarization variations within the basement rocks. Generally they are in no way related to local structural conditions. These strong regional gradients act to mask mag-

---

[15] Nettleton, L. L., *Gravity and Magnetic Calculations:* Geophysics, vol. 7, no. 3, p. 293, July 1942.

netic anomalies due to structure. To the experienced geophysical interpreter this presents no problem, for he has learned to recognize the
small variations in regional gradient which indicate local anomalous
conditions. Persons not familiar with magnetic or gravity maps are
aided by the removal of the regional gradient. A number of methods for
eliminating extraneous effects are in common usage, the aim of all being
to leave only the anomaly due to local structure.[16] The term "residual" is
frequently applied to maps from which the regional has been removed.

Henderson and Zietz [17] have demonstrated that a definite relationship
may exist between residual maps and "second vertical derivative" maps.

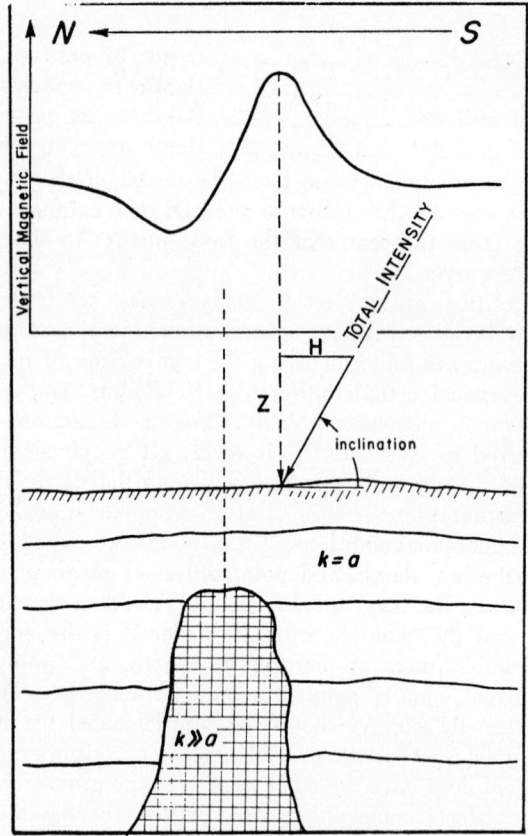

Figure 552. Shift of axis of magnetic anomaly from
center of structural disturbance.

[16] Nettleton, L. L., *Geophysical Prospecting for Oil:* pp. 126, 222, New York, McGraw-Hill Book Co.,
Inc., 1940.
[17] Henderson, Roland G. and Zietz, Isidore, *The Computation of Second Derivatives of Geomagnetic
Fields:* Geophysics, vol. 14, no. 4, p. 516, Oct. 1949.

The location of the maxima and minima as well as the zero contours may appear in exactly the same positions on both residual and derivative maps, the only variation being in the magnitude of the anomalies.

Shift of Axis—Magnetic anomalies associated with local structure generally originate in the uplifted basement rock under the structure. The center of the magnetic anomaly is often shifted laterally (to the south in the northern hemisphere) from the center of the structural disturbance, owing to the inclination of the earth's field. The amount of shift depends upon a number of factors including the depth to the basement rocks, the angle of inclination of the earth's field, shape, and size. This effect is illustrated in figure 552.

Horizontal and Total Field Interpretation—It is sometimes suggested in the literature that the combination of both vertical and horizontal magnetic data will afford the magnetic method the property of depth control. Depth control is also claimed for the magnetic method when the observations are taken at different elevations as with the air-borne magnetometer. Skeels writes,[18]

Since a corollary to Green's theorem holds for any potential function it also holds for magnetics; the magnetic potential and all of its derivatives are completely defined by the vertical intensity map. If two magnetic distributions can be found which satisfy the same set of vertical intensities and they always can be found, they will also satisfy the same set of horizontal intensity data. So combining horizontal with vertical intensity data does not give us a unique solution, as has been implied in some of the literature. Nor does observing magnetic effects at two or more levels resolve the ambiguity, for by means of the theorem it can be shown that if two magnetic distributions produce identical effects on one level they will also produce identical effects on all higher levels. Magnetic observations in a mine are severely restricted in areal extent; in such a case the addition of horizontal data to vertical data would greatly improve the interpretation. The more geological data the interpreter can bring to bear upon the problem, the closer will he be able to limit the range of possibilities, and the less ambiguous his interpretation will be.

Henderson and Zietz [19] have taken air-borne magnetic data at several levels for several regions in the country and demonstrated that it is possible to compute from the first-level data all essential information obtained at higher levels. They suggest that, if information is desired at several elevations, it would be more economical to compute the data for the higher elevations than to take it from a plane. Skeels and Watson [20] have calculated the horizontal components from vertical magnetic data over several magnetic features and have shown that the computed values agreed very well with the observed horizontal data.

[18] Skeels, D. C., *Ambiguity in Gravity Interpretation:* Geophysics, vol. 12, no. 1, p. 55, Jan. 1947
[19] Henderson, Roland, and Zietz, Isidore, *The Upward Continuation of Anomalies in Total Magnetic Intensity Fields:* Geophysics, vol. 14, no. 4, p. 534, Oct. 1949.
[20] Skeels, D. C., and Watson, R. J., *Derivation of Magnetic and Gravitational Quantities by Surface Integration:* Geophysics, vol. 14, no. 2, pp. 133-150, Apr. 1949.

Applications

Falconbridge Ore Body, Ontario—The vector diagram in figure 551 shows that the vectors obtained by combining horizontal and vertical magnetic data cross in a region about 150 to 175 feet below the surface of the ground. Although the intersection is only approximate, it does give an indication of the depth. The top of the ore is actually about 112 feet below the surface. If we allow for the depth of the pole beyond the edge

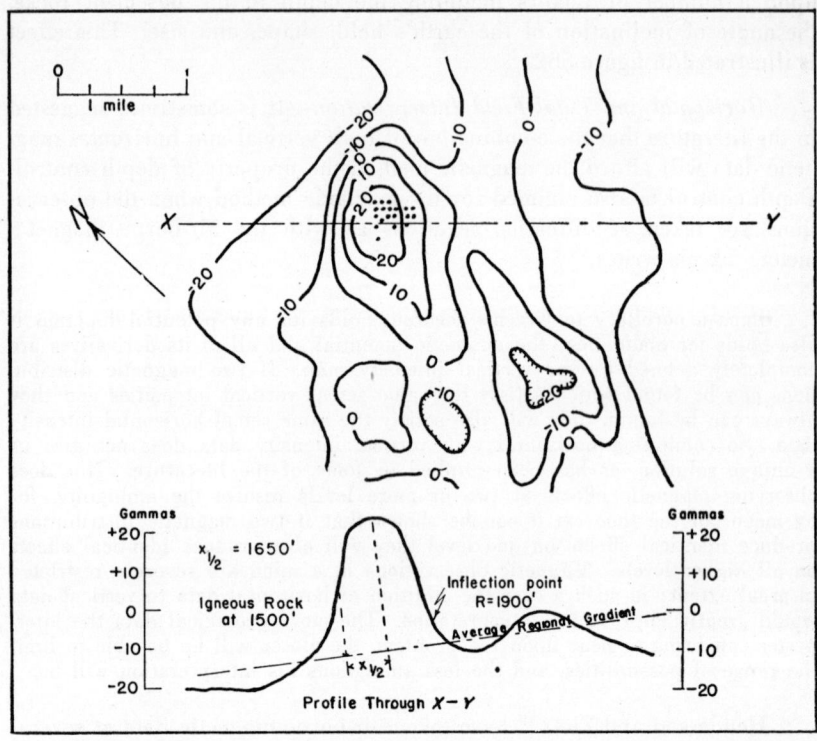

FIGURE 553. Magnetic effects over Yoast field, Bastrop County, Texas. Magnetic anomaly is due to serpentine intrusion at depth of about 1,500 feet. (Adapted from Collingwood. Reproduced permission Am.· Assoc. Petroleum Geologists.)

of the ore body, the magnetic results would agree more closely with the actual data. The ore consists in large part of pyrrhotite, a fact which accounts for the magnetic effects.[21]

Yoast Field, Bastrop County, Texas—One of the best applications of the magnetometer in prospecting for oil is afforded by the serpentine plugs of south Texas. The plugs offer excellent chances for the wildcatter,

---

[21] Eve and Keys, *Studies of Geophysical Methods:* Canadian Geological Survey Mem. 170, p. 39, 1930.

for, where the serpentine is porous, they are generally found to contain oil. As the plugs are relatively shallow and small in areal extent, magnetic surveys should be of a rather detailed nature in order that anomalies will not be passed over. Figure 553 shows the lines of equal magnetic vertical intensity and a profile across the Yoast field. Applying the

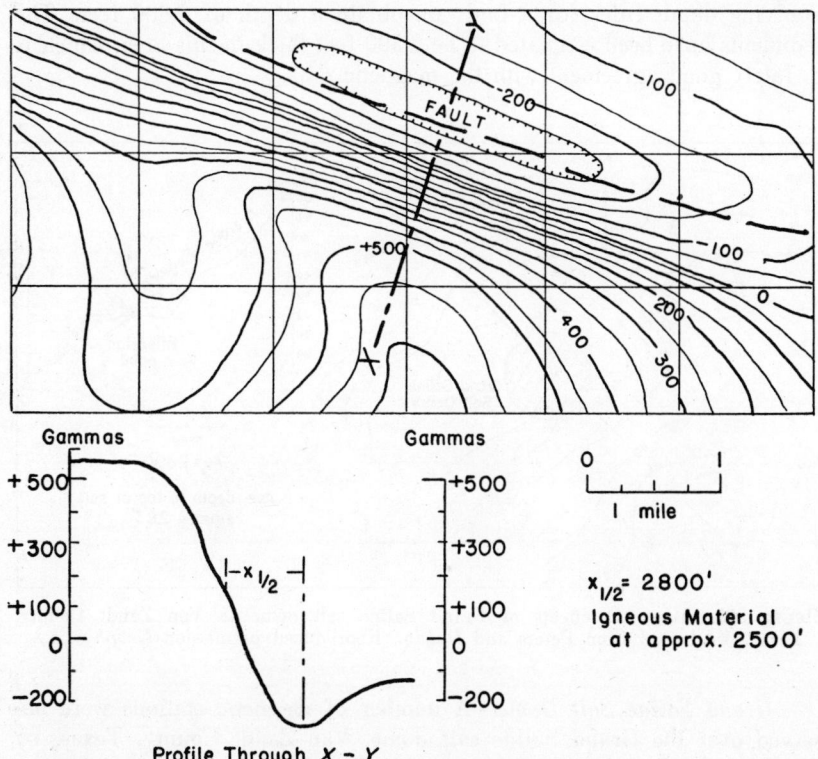

FIGURE 554. Portion of magnetic survey over Beckham County fault in Wheeler County, Texas. (Adapted from Stearn. Reproduced permission Am. Inst. Min. Met. Eng.)

depth rules to this anomaly one finds that $X_{\frac{1}{2}}$ equals 1,650 feet and the inflection point occurs at about 1,900 feet. Using the rules for a sphere, one finds the center to be roughly at 3,300 feet below the surface, which places the upper edge of the serpentine about 1,400 feet below the surface. The serpentine in the Yoast plug enters along a fault zone and probably represents a submarine extrusive at the end of Austin time. It lies approximately 1,500 feet below the surface of the ground and is bounded on top and sides by the Taylor formation, with the Austin chalk beneath.[22]

[22] Collingswood, D. M., *Magnetics and Geology of Yoast Field, Bastrop County, Texas:* Am. Assoc. Petroleum Geologists Bull., vol. 14, no. 9, pp. 1191-1198, Sept. 1930.

Magnetic Effects over Fault—Figure 554 is a portion of a magnetic survey in the southeast corner of Wheeler County, Texas. A normal fault with a throw of 300 to 500 feet parallels the strike of the buried Wichita Mountains. Named from the adjacent Oklahoma county, the Beckham County fault has been traced from well data for a number of miles, the magnetic results indicating an extension of this fault to the west. By applying depth rules for a fault we obtain a depth of 2,800 feet. The sediments have been estimated to be 2,500 feet thick in this area, which is in fairly good agreement with the magnetic data.[23]

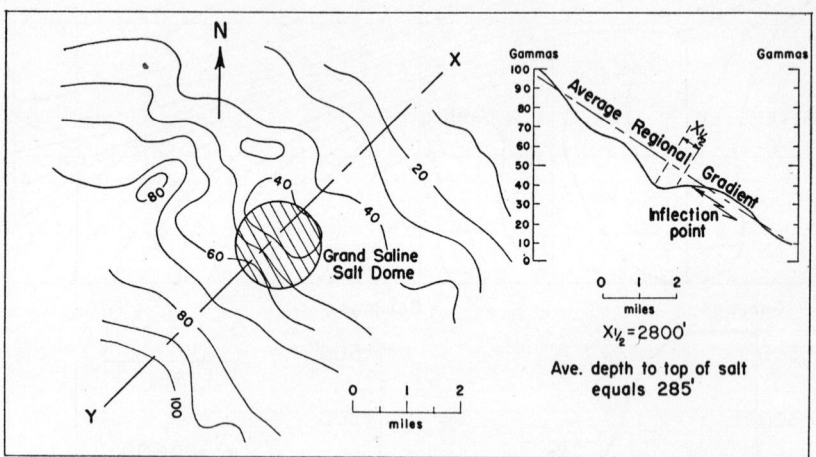

FIGURE 555. Magnetic effects of Grand Saline salt dome in Van Zandt County, Texas. (Adapted from Peters and Dugan. Reproduced permission *Geophysics.)*

Grand Saline Salt Dome—A number of magnetic stations were observed over the Grand Saline salt dome, Van Zandt County, Texas, by the Magnolia Petroleum Company. (See fig. 555.) After substracting the regional gradient, a weak negative magnetic anomaly of approximately 15 gammas existed over the salt dome. By applying the depth rules for a sphere we find that the center lies at a depth of approximately 5,600 feet. From the inflection point in the curve the radius would appear to be about 4,700 feet, which places the top edge of the salt mass about 900 feet below the surface.

Susceptibility measurements were made on the salt as well as the surrounding sediments. The average susceptibility of the salt was -0.56×10^{-6} c.g.s., and the average susceptibility of the surrounding sediments from the Wilcox formation at the surface to the Travis Peak at a depth of approximately 8,700 feet was 12×10^{-6} c.g.s. The susceptibility

[23] Stearn, N. H., *Geomagnetic Prospecting with the Hotchkiss Superdip:* Am. Inst. Min. Met. Eng., Geophysical Prospecting, p. 191, 1932.

contrast as measured is not sufficient to explain the magnitude of the anomaly, and either the sediments have retained some remanant magnetization or higher susceptibilities must exist at depth.

From gravity data the dome is a symmetrically shaped salt mass,

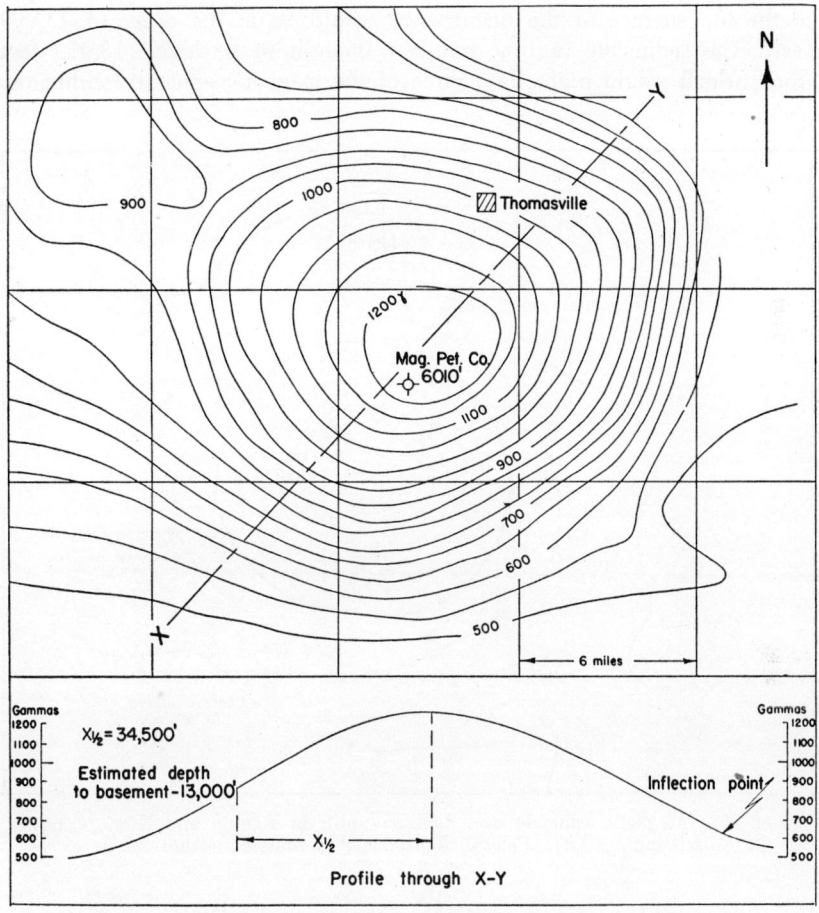

FIGURE 556. Thomasville magnetic anomaly in northern Clarke County, Alabama. (Adapted from Eby and Nicar.)

with its upper and lower bases at depths of about 215 feet and 15,000 feet and the radii from 4,000 to 2,500 feet.[24]

Thomasville Anomaly, Clarke County, Alabama—One of the better examples of magnetic anomalies that are not associated with structure is afforded by the large symmetrical magnetic anomaly lying in the northern

[24] Peters, J. P., and Dugan, A. F., *Gravity and Magnetic Investigation at the Grand Saline Salt Dome, Van Zandt Co., Texas:* Geophysics, vol. 10, no. 3, p. 376, July 1945.

part of Clarke County, Alabama. (See fig. 556.) The anomaly extends over an area of some 325 square miles with a magnitude of about 700 gammas. Rough estimates based on the depth rules for a sphere indicate that the center of the disturbing mass must lie at a depth of about thirteen miles. If 50,000 feet is allowed for the radius of this sphere, the depth to the top surface of the disturbance would be in the order of 15,000 feet. The sediments in this area are thought to be about 13,000 feet thick, which would place the source of the magnetic anomaly within the

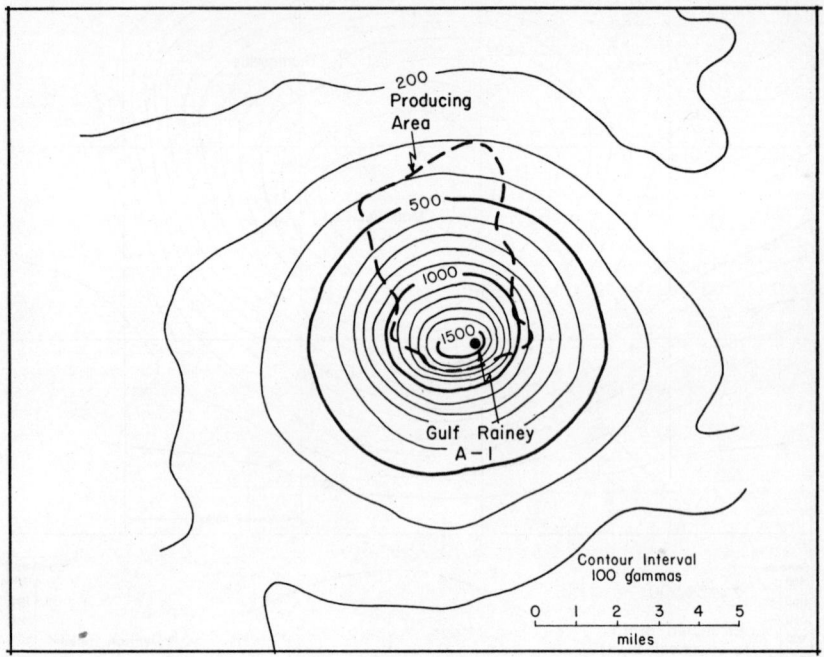

FIGURE 557. Magnetic anomaly over Jackson uplift in Rankin and Hinds Counties, Mississippi. (After Peters. Reproduced permission *Mines Mag.)*

basement complex. A well drilled by the Magnolia Petroleum Company in 1934 to a total depth of 6,010 feet gave evidence that a structural low existed over the anomaly, although regional well control was limited.[25]

Jackson Uplift, Mississippi—The Jackson gas field in Rankin and Hinds Counties, Mississippi, is expressed by a magnetic anomaly of some 1,200 gammas extending over an area of about 400 square miles. (See fig. 557.) The anomaly is roughly symmetrical, and depth rules for the sphere would place the center of the disturbance at about 8,500 feet.

[25] Eby and Nicar, *Magnetic Investigations in Southwest Alabama:* Alabama Geol. Survey Bull. **43,** 41 pp., 1936.

Allowing 9,500 feet for the radius, the igneous core should be encountered at 7,500 feet below the surface.

The Gulf Rainey A-1 entered igneous rock at 2,996 feet, and drilling ceased at 3,607 feet while still in igneous material described as a "perido-tite" in the upper part and a "lampophyre" in the lower part. Subsequent deeper wells have indicated that the igneous material encountered by the drill may be sills or slivers and that the actual igneous core has not

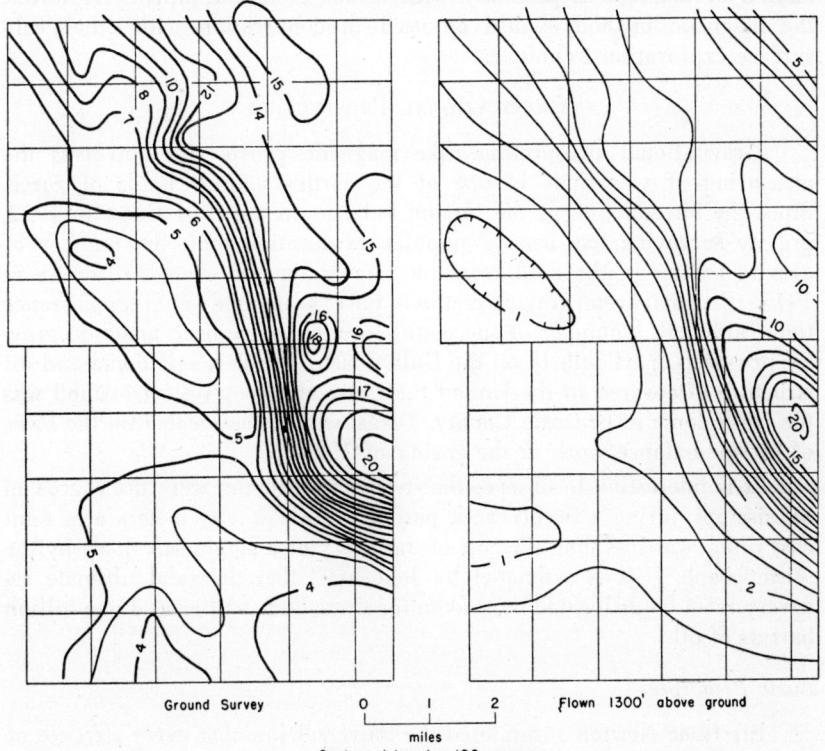

Ground Survey   0   1   2   Flown 1300' above ground

miles
Contour interval = 100 gammas

Figure 558. Portion of air-borne-magnetometer survey in Greer County, Oklahoma. Results of a ground survey in same area also shown. (After Balsley. Reproduced permission U. S. Geol. Survey.)

as yet been reached. Better agreement between magnetic data and sub-surface conditions would be achieved if the presence of a large igneous core at depth were confirmed. The igneous activity in this region apparently occurred during Tuscaloosa time and extended over southern Arkansas and central Mississippi.[26]

Mangum, Oklahoma—The results of an air-borne magnetic survey are

[26] Spraragen, L., *Magnetometer Survey of the Jackson Area:* Oil and Gas Jour., vol. 30, no. 26, pp. 14, 15, 83, Jan. 1932.

shown in figure 558. This area was flown by the United States Geological Survey at an altitude of 1,300 feet. The results of a conventional ground survey accompany the map of the air-borne survey to show the similarity between the methods. This area is located in Greer County, Oklahoma, near the Wichita Mountains.[27]

In summary, it can be seen that application of the simple depth rules will in many instances give a clue to the probability of the association of a magnetic anomaly with actual structural uplift. However, the magnetic method should be used in cooperation with other subsurface exploration techniques.

GRAVITATIONAL PROSPECTING

Gravitational prospecting, like magnetic prospecting, involves the measuring of variations in one of the earth's natural fields of force. Since the introduction of the torsion balance on the Gulf Coast in 1922, gravity surveying has been a popular exploration tool, the number of gravity parties in the field reaching a maximum of about 175 crews in 1945. Primarily the gravity methods have been used as reconnaissance tools, although a number of successful locations have been made on gravity prospects, particularly on the Gulf Coast. The first salt dome and oil structure discovered in the United States by any geophysical method was the Nash dome in Brazoria County, Texas, which was located on the basis of torsion-balance work in the spring of 1924.

It is interesting to observe that twenty salt domes were discovered in Mississippi during a twenty-week period by two gravity meters at a total cost which was less than the cost of discovery of a single salt dome by the seismograph.[28] It is estimated by Jakosky[29] that the total ultimate recovery from fields located by gravitational methods will exceed one billion barrels of oil.

Basic Principles

Sir Isaac Newton formulated the universal law that every particle of matter in the universe attracts every other particle. If two objects of mass m_1 and m_2 are separated by a distance r, and r is considerably larger than the dimensions of either object, there is a force of attraction between the two objects which is related directly to the masses and inversely as the square of the distance r. This expression may be written as follows:

$$F = \gamma \frac{m_1 m_2}{r^2}$$

---

[27] Balsley, J. R., *The Airborne Magnetometer:* U. S. Geol. Survey Geophys. Inv. Preliminary Rept. 3, p. 1, 1946.
[28] Eckhardt, E. A., *A Brief History of the Gravity Method of Prospecting for Oil:* Geophysics, vol. 5, no. 3, p. 232, July 1940.
[29] Jakosky, J. J., *Exploration Geophysics:* p. 6, Los Angeles, Times-Mirror Press, 1940.

The constant γ is independent of the composition of the masses involved and has a numerical value depending on the units used. If the mass is given in grams, the distance in centimeters, and the force in dynes, the constant γ is equal to 66.67×10^{-9} c.g.s. units.

The familiar force known as the "weight of an object" is due to the gravitational attraction of the earth on the object in question. Gravity, a force directed approximately toward the center of the earth, may be quantitatively expressed as a force per unit mass with units of dynes per gram, or the equivalent acceleration may be used with units of centimeters per second per second. The approximate value of gravity so expressed is 980 dynes per gram or 980 centimeters per second per second. Geophysicists commonly speak of gravitational acceleration in terms of gals, named after the famous renaissance scientist Galileo. The approximate value of gravity may thus also be expressed as 980 gals.

The force of gravity is everywhere present over the surface of the earth, its intensity at any one position being affected not only by matter located in the vicinity, but by all mass in the earth, on the earth, and in the solar system surrounding the earth. In gravity surveys, depending upon the nature of the survey, corrections may be made for irregularities in the distribution of matter beneath the surface of the earth (isostatic correction) and on the surface of the earth (cartographic or terrain correction), as well as for the relative positions of the sun and the moon (tidal correction). The corrections may be applied in addition to the corrections for the normal variations of gravity with latitude, elevation, and surface density, which are always required.

Gravity exploration depends on the existence of density contrasts between geologic bodies and the surrounding materials in a horizontal direction. Density contrasts, of course, exist between individual geologic horizons as we progress toward the center of the earth. It is only when a denser or lighter material is uplifted or intruded into the normal country rock that an anomaly is observed. Structures involving rocks of the same density will not produce a gravity anomaly. (See fig. 559.)

The force of gravity is defined as the rate of change of the gravitational potential in the vertical direction. Gravity, being a vector quantity, has both magnitude and direction. If it were not for surface and near-surface variations in the distribution of the masses that go to make up the outer shell of our earth, the gravity vector would be directed toward the center of the earth, assuming the earth to be a perfect sphere. Inhomogeneities due to either topography or variations in mass distribution beneath the surface cause the gravity vector to be deflected in the direction of the excess mass. (See fig. 560.) For example, near an igneous plug intruded into sedimentary rocks we should expect the gravity vector to be deflected toward the denser material. The magnitude of the vector increases as we approach the plug. The total force of gravity at a point near such an inhomogeneity would be the vectorial sum of the normal terrestrial gravitational force plus or minus the Newtonian attraction of the disturbing mass, depending on the densities involved.

The absolute direction of the gravity vector may not be measured in the field although comparative differences in direction may be deduced by field methods. Relative measurements of the "deflections of the vertical" could theoretically be used as an exploration tool; to be of local geological value, however, measurements would have to be made to an accuracy beyond the scope of portable equipment.

It should be recognized that measurements of the deflections of the

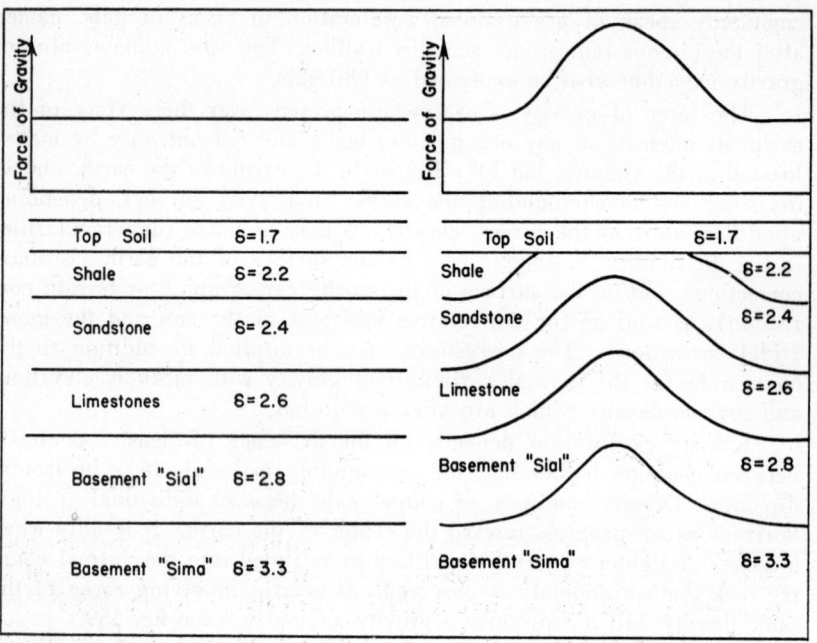

FIGURE 559. Gravity effects over a normal area in which no horizontal variation in density occurs compared with gravity effects over a disturbed area.

vertical are of importance in geodetic and large-scale geologic studies.

The force of gravity may be measured directly by the use of pendulums, although extreme care must be used in such observations to obtain even moderate accuracy. The absolute gravitational force has been measured with the greatest accuracy at Potsdam, Germany, and the value determined there has been carried to other points throughout the world by comparative-gravity measurements.

As the gravity vector is deflected by variations in distribution of mass, it is evident that the total force of gravity may be resolved into vertical and horizontal components. The horizontal component is an exceedingly small force, but it may be indicated by sensitive instruments such as the torsion balance. Gravity prospecting is accomplished with instruments that either compare the force of gravity between points on the earth's surface, the gravity meter or the pendulum; or measure the derivatives of gravity in a horizontal direction (gradient and curvature quantity), the torsion balance.

The unit of measurement in gravity prospecting.is the milligal which is roughly one-millionth part of the total gravity field. The measure

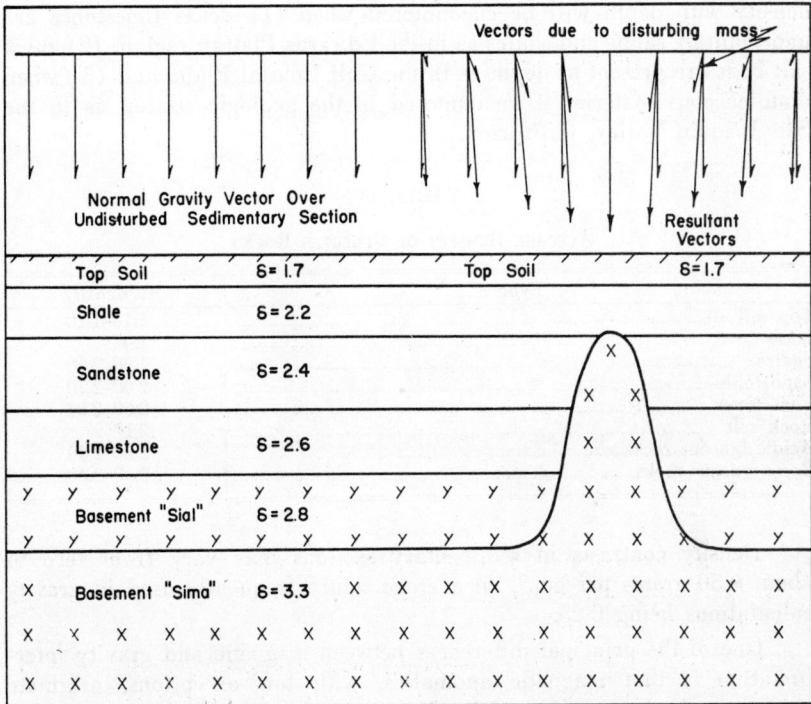

FIGURE 560. Gravity vectors over a normal area compared with a disturbed area showing change in magnitude and direction of vectors. (Size greatly exaggerated.)

of gravity gradient is the eotvos unit, which represents a rate of change of gravity equal to 10^{-9} gals per horizontal centimeter.

Rock Properties

Because geologic application of the gravity methods demands a horizontal variation in rock density, many oil companies are making systematic studies of this property from cores and outcrop samples. In general, it has been found that the density varies with geologic age, depth of burial, and lithology. For example, a decrease of silica content in an igneous rock or an increase in lime content of a sedimentary rock is generally accompanied by an increase in density. Older rocks generally have a higher density than younger rocks. This probably is caused by a higher lime content and greater compaction. Density shows a more or less consistent increase with depth of burial. Barton [30] found that the density increased at the rate of 0.07 grams per cubic centimeter per thousand feet of depth to about 8,000 feet on the Gulf Coast. This increase with depth is due in large part to greater compaction of the sediments with increasing load.

Several exceptions to the foregoing should be noted. A decrease in density with depth will be encountered when (1) dense limestones are underlain by sands and shales as in the Edwards Plateau region, (2) thick salt beds are present at depth as in the Gulf Coastal Plain, and (3) when diatomaceous material is encountered in the geologic section as in the San Joaquin Valley, California.

TABLE 33

AVERAGE DENSITY OF SELECTED ROCKS

Type	Av. density
Top soil	1.10–1.70
Loess	1.43
Shales	1.90–2.60
Sandstones	2.00–2.70
Limestones	2.60–2.84
Rock salt	2.17
Acidic igneous rocks	2.60–2.80
Basic igneous rocks	2.80–3.40

Density contrasts in sedimentary sections may vary from zero to about 0.50 grams per cm.^{3}, an average figure frequently used in gravity calculations being 0.25.

One of the principal differences between magnetic and gravity interpretation is that magnetic anomalies, with few exceptions, originate either at or in the basement rock, whereas gravity anomalies may represent horizontal variations in density at any point within the geologic column, including intrabasement variations. Gravity anomalies due to local structure are in many cases superimposed on larger anomalies caused by variations of density within the basement complex.

Earth's Gravitational Field

Any measurements of gravity or its derivatives involve the normal terrestrial gravitational field, with an additional effect caused by variations in density within the earth's crust. Since geologic inference may be drawn only from gravity variations associated with density changes in the under-

---

[30] Barton, D. C., *The Science of Petroleum*, New York, Oxford Univ. Press, vol. 1, p. 374, 1938.

lying rocks, it becomes necessary to remove from the observed gravity all influencing factors except those due to density irregularities.

Variation of Gravity with Latitude—If the earth were not rotating about its axis, it is quite possible that we would be living on a perfect sphere. However, owing to this rotation, centrifugal force has increased the equatorial radius some thirteen miles over the polar radius, thus forming an oblate spheroid. The acceleration of gravity at the equator is about five gals less than at the poles because (1) centrifugal force, which acts to oppose the gravitational force, is a maximum at the equator, and (2) the earth's surface at the equator is about thirteen miles farther from the center of mass than at the poles.

Equations have been developed [31] that indicate the normal gravitational force at sea level as a function of latitude for a homogeneous earth. Since the force of gravity varies from north to south, corrections are applied to all regional gravity surveys eliminating this normal variation from consideration. In the state of Oklahoma the normal increase of gravity to the north is roughly 1.1 milligals per mile.

Variation of Gravity with Elevation—An elevation correction must be applied to all relative-gravity data to compensate for the normal decrease in the force of gravity as the distance to the center of the earth increases. As an example, if a gravity-meter observation were made on the earth's surface and if another observation could be made in a balloon located one hundred feet above the earth's surface, it would be found that gravity had decreased by 9.4 milligals. This normal decrease of gravity with increase in distance to the earth's center is known as the "free-air effect."

If we could fill in the space between the balloon and the ground with a soil layer of infinite extent, we should find that our original gravity difference of 9.4 milligals has diminished to about 7.0 milligals, depending upon the density of the intervening soil layer. The attraction of the soil layer, which opposes the decrease of gravity with altitude, is known as the "Bouguer effect." The Bouguer attraction varies with the density of the soil and the thickness as follows:

$$\text{Attraction in mg.} = 0.012276 \delta h$$

where h is thickness in feet and δ is density.

The elevation correction as ordinarily used in gravity surveys combines both the free-air and Bouguer corrections with due regard for the density of the surface rocks within the topographic extremes. The combined correction is usually carried to a sea-level datum plane for convenience. This correction equals $(0.094 - 0.012276 \delta) h$ milligals per foot and has a value of 0.070 milligals per foot for a surface density of about 1.9 grams per cm.^{3} In rugged terrain where the assumption of an infinite

---

[31] Heiland, C. A., *op. cit.*, p. 97.

layer for the Bouguer effect would be in error, separate corrections are applied to gravity data.

Geologists examining gravity-meter maps should be aware of the importance of the elevation-correction factor and the need for terrain

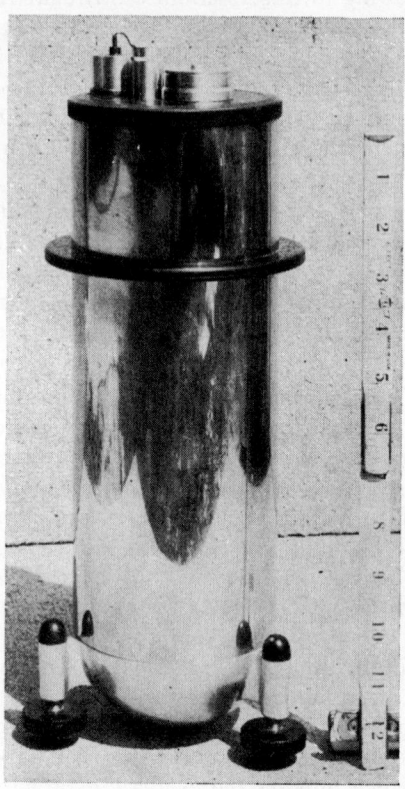

FIGURE 561. Modern portable gravity meter enclosed in thermos bottle for thermal insulation. Weight of meter is six pounds. (Reproduced permission Houston Technical Laboratory.

corrections in certain areas. In many areas, particularly those of rugged terrain, the use of the proper elevation factor and terrain corrections make data easier to interpret. Wrong factors cause the final map to appear ragged and may cause "false" anomalies to appear on topographic features.

Anomalies occurring on "hanging" or "stub" lines, which have not been rerun by the surveyor and the meter operator, should be eliminated from consideration unless detailed work in the area substantiates the "hanging" line. Many "anomalies" found to lie on these unchecked lines disappear with additional work.

Tidal Effects—Corrections for the attraction of the celestial bodies are generally not applied to gravity data, unless unusual accuracy is required or special field methods demand such corrections. The maximum amplitude of tidal effects is about 0.3 mg., and the change occurs with a period of about 24 hours. Normal operational procedure, in which the gravity meter returns to a check station every two hours, automatically compensates for these effects as instrumental "drift."

Gravity Instruments

The history of gravity prospecting in the United States records the use of the torsion balance, the pendulum, and the gravity meter. Although the gravity meter does not have the resolution powers of the torsion balance, its greater speed, portability, and consequent economy of operation have offset the advantages offered by the torsion balance. Gravity explorations conducted in rugged, mountainous terrain in the past several years with the gravity meter would have been almost impossible with the torsion balance because of the extreme sensitivity of the balance to the disturbing influence of terrain irregularities and near-surface density variations. Mott-Smith [32] has compared the influence of mass irregularities on the torsion balance and gravity meter and concluded that shallow density variations would make torsion-balance data practically worthless. The discussion of gravity methods is, therefore, more or less restricted to the gravity meter. As can be seen from figure 546, there are about 105 gravity-meter crews in operation in this country today.

A gravity meter is essentially an accurate weighing device in which the gravitational force on a mass is offset by the tension in a spring. Gravity meters may be either direct-reading devices, in which the final resting place of the mass is observed (Mott-Smith meter), or they may be null-type instruments, in which the mass is returned to a given position on the scale (LaCoste-Romberg meter). In null-type instruments the restoring force required to bring the mass to its zero position is read from an appropriate indicator and converted to gravity units.

The extreme sensitivity of the gravity meter may be appreciated from the following illustration. If a scale could be built that would accommodate 3,000 tons of coal (corresponding to the force of gravity on the earth) with a sensitivity such that the addition or subtraction of one ounce of coal would be measurable (corresponding to about 0.01 milligal), we should have a scale of comparable accuracy to the modern gravity meter. In order to attain such a high degree of accuracy, instruments must be carefully constructed of materials with the lowest temperature coefficients and further protected from thermal variations by enclosure in a thermostatically controlled "oven."

Modern instruments are light in weight (25-45 pounds) and readily portable. Manufacturers are now offering portable meters enclosed in

---

[32] Mott-Smith, L. M., *Gravitational Surveying with the Gravity Meter:* Geophysics, vol. 2, no. 1, pp. 30-32, Jan. 1937.

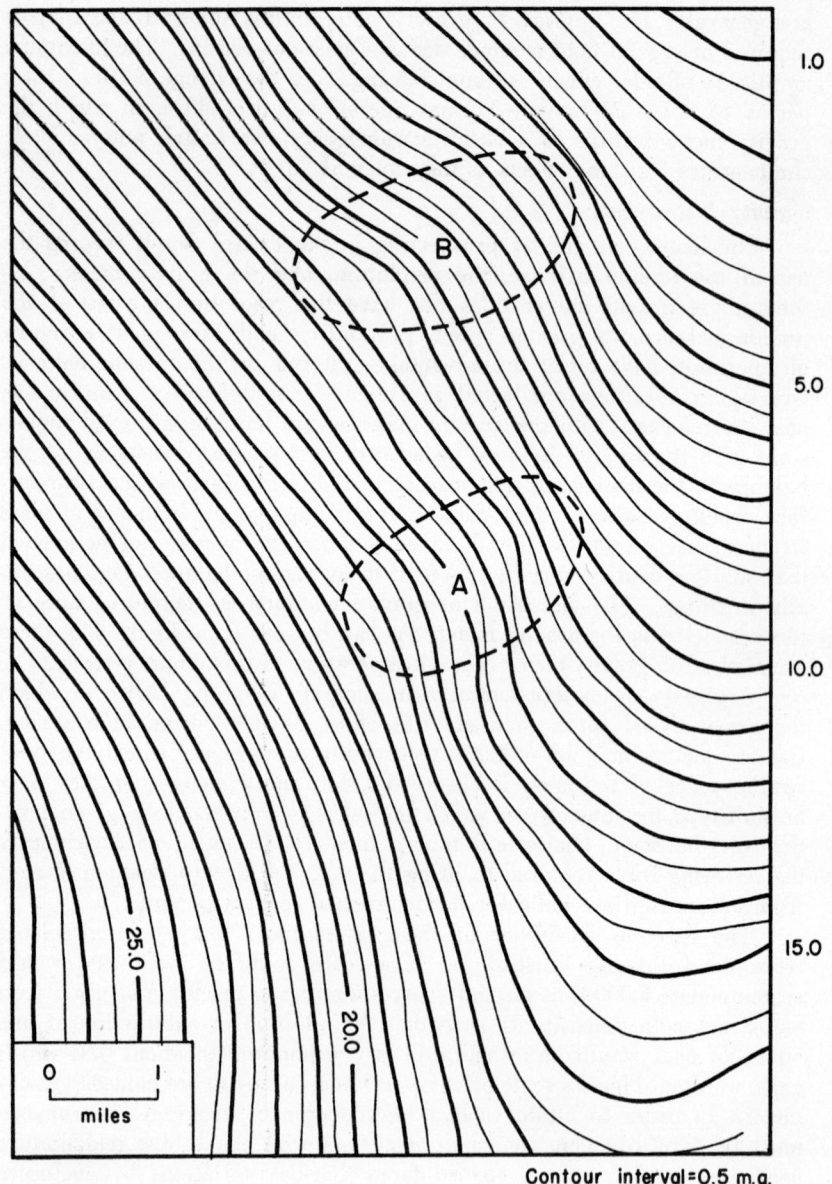

FIGURE 562. Gravity contours of an area in west Texas showing anomalies at *A* and *B* due largely to density contrasts between red beds and limestone at depth of approximately 4,000 feet. Both anomalies outline known structures. (After Coffin. Reproduced permission Am. Assoc. Petroleum Geologists.)

a thermos bottle for thermal control and weighing only a few pounds (fig. 561).

Gravity Interpretation

The results of gravity surveys are generally shown in the form of a contour map. The values plotted on the map are corrected for latitude, elevation, and Bouguer effects, and for terrain and tidal effects where necessary. This final map then represents the surface distribution of gravity due only to density variations in the rocks beneath the surface.

The interpretation of gravity results, like magnetic results, is usually qualitative. Knowledge of the geology is even more important in the interpretation of gravity data, as anomalies may be caused by density variations at any point within the geologic section, including the basement. By a careful analysis of the stratigraphic column, zones that are capable of producing gravity anomalies may be noted, reducing to a large extent the ambiguity of gravity methods. For example, in the Permian Basin of west Texas and New Mexico the gravity anomalies associated with local uplift are due in large part to the density contrasts between the relatively light sands, shales, and salt of the Dockum and Ochoa groups against the dense limes and dolomites of the Guadalupe and lower groups. Magnetic anomalies associated with structure in the same area are caused by actual uplift of the magnetic basement complex. It is possible in this area, therefore, to have gravity anomalies exist with or without corresponding magnetic anomalies. (See figs. 562, 563.)

Interpretation is always aided by a knowledge of the gravity effects over known structural features in the same region. It is also possible to calculate the magnitude and surface distribution of the gravity effects to be expected for simple geometric forms and to compare these results with the observed data. Another interpretation technique requires the calculation of the gravity effects over known structures or fields by the use of the gravity integrator. A knowledge of the density contrasts is particularly desirable in such quantitative computations.

Gravity surveys indicate large regional geologic trends as well as outline areas of local structural disturbance. The areal extent and magnitude of gravity anomalies depend upon a number of factors, among which are the depth, size and shape, and the density contrasts involved. The sharpness of gravity anomalies is related to the density contrast and the depth at which such contrasts occur. Comparatively sharp anomalies originate at shallow depths, whereas broad anomalies may arise from great depths. Observe that broad anomalies do not necessarily originate at great depths. It is quite possible to obtain a broad anomaly of large areal extent from density contrasts occurring quite near the surface. In fact, from any gravity profile it is possible to calculate almost an infinite variety of depths, shapes, and density contrasts that would satisfy the observed data. It is therefore apparent that unique solutions in gravity interpretation are possible only when a great amount of additional data

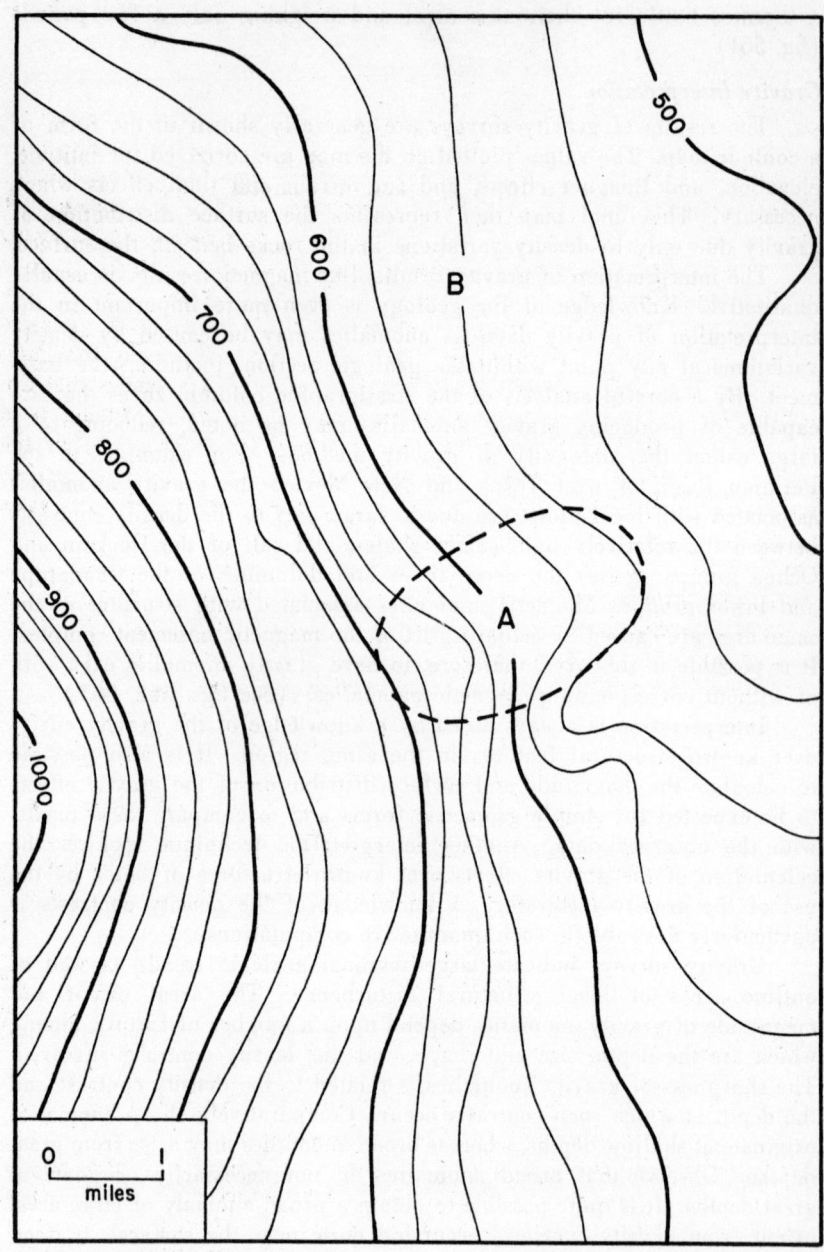

Contour interval = 25 gammas

Figure 563. Magnetic contours of same area shown in figure 562. Magnetic anomaly is present at *A* but absent at *B*. Pre-Cambrian formation producing *A* anomaly is 3,000 feet deeper at *B*. (After Coffin. Reproduced permission Am. Assoc. Petroleum Geologists.)

is available in the form of seismic maps, well logs, and density information from well cores.

When used judiciously, depth rules for certain simple geometric
shapes may prove of value to estimate the depth to the surface of the
density contrast causing the anomaly. The rules should be used with
the mental reservation that ambiguity is inherent in interpretation of gravity data in terms of structure. They are similar to those mentioned in the
section on magnetic technique, in that the distance from the center of the
anomaly to the point at which the anomaly has fallen to one-half of its
maximum value is multiplied by a factor to give the depth. Where the lines
of equal gravity or isogals (corrected for regional gradient) are roughly
circular, we may assume either a sphere or a vertical cylinder, depending
upon which shape is more probable geologically. For buried ridges and
anticlines the horizontal cylinder may be selected. As before, the inflection
points on either side of the gravity profile may be selected to give an idea
of the size of the disturbing body.

Depth Rules

Vertical cylinder	depth to top $\quad = 0.56\ X_{\frac{1}{2}}$
Sphere	depth to center $= 1.30\ X_{\frac{1}{2}}$
Horizontal cylinder	depth to center $= 1.00\ X_{\frac{1}{2}}$
Fault	depth to center $= X_c$

$X_{\frac{1}{2}}$ is the horizontal distance from the center of the anomaly to the
point where the anomaly has fallen to one half its maximum value, and
X_c is the distance from the fault trace to the point at which the gravity
has fallen to one half the total change from the fault trace to infinite
distance. The depth rules above may be determined from curves presented
by Nettleton.[33] Interpretative techniques based on spheres, vertical cylinders, horizontal cylinders, and series of disks are included in the above-
cited paper by Nettleton.

Although gravity data alone will not permit a unique solution, they
do limit the depth to the density interface causing the anomaly. It is
obvious that an anomaly of small areal extent must be caused by density
contrasts close to the surface. Thus we can place the maximum depth from
which a given anomaly could arise. It is also possible by mathematical
means to determine the shallowest depth from which a structure could
cause a given anomaly for a given density contrast. Between these two
limits it is impossible to predict the depth, size, or shape of the mass
causing the anomaly without the benefit of additional data. Quantitative
interpretation becomes even more involved when consideration is given
the several surfaces of density contrast that may be present in the geologic
column.

Reliable quantitative interpretation of gravity information requires

[33] Nettleton, L. L., *Gravity and Magnetic Calculations:* Geophysics, vol. 7, no. 3, pp. 296-300, July 1942.

(1) accurate field data, (2) the proper removal of the regional gradient and other disturbing effects, and (3) sufficient additional control to limit the depth from which the anomaly may arise.

Minimum and Maximum Control

Structural uplifts may be expressed by either gravity minima or maxima, depending upon the geographic location. In the Gulf Coast for example, uplifts are indicated by gravity minima (excluding maxima observed over cap rocks of piercement salt domes and igneous plugs); in Oklahoma or the Rocky Mountains, however, structural uplift is indicated by gravity maxima. The approximate limits of salt or minimum control

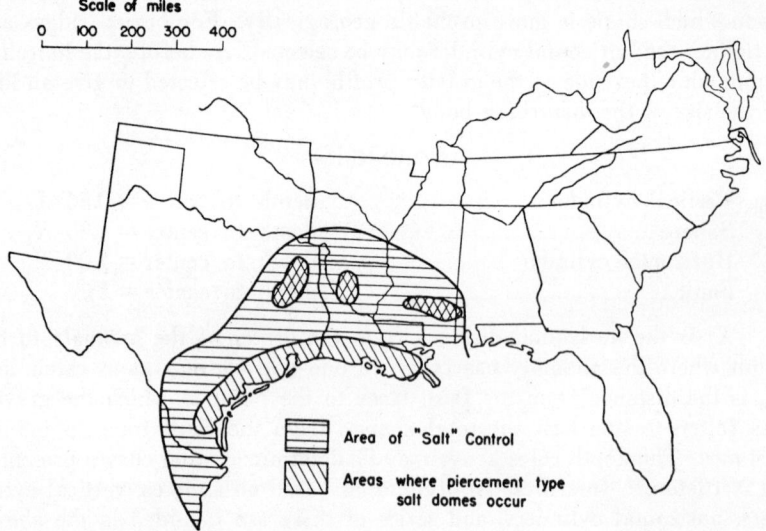

FIGURE 564. An approximate outline of an area of minimum gravity control in the Southern States.

are shown in figure 564. In the rest of the United States, with a few local exceptions such as portions of the San Joaquin Valley in California, maximum control exists. Gravity minima associated with uplifts are caused by the presence of a thick bed of light material at depth, such as the Louann salt of Permian age, which consists of more than a thousand feet of rock salt and extends from southwest Texas to western Alabama. Local thickening or actual intrusion of this salt bed into overlying sediments is associated with most of the oil fields in the area underlain by the salt and thus produces minimum gravitational effects.

In other areas the warping of a dense limestone at depth or the actual uplift of the basement itself will produce gravity maxima over structures.

Regional Gradient

It is an extremely rare occurrence to find a gravity anomaly not influenced by nearby local structure or by density contrasts within the basement itself. Frequently density variations within the basement act to mask gravity anomalies due to local structure. As was mentioned in the section on magnetic interpretation, the experienced interpreter is able to recognize the small variations in regional gradient that betray the presence of local structure. Before quantitative work may be done on the anomaly, all disturbing effects must be removed. The accuracy of the final interpretation depends upon the degree to which the interference is removed. A number of methods are in use to prepare the "residual" map,[34] [35] but none gives a guarantee of removing the disturbing or regional effects and nothing more.

The work of Henderson and Zietz [36] on the second derivatives of magnetic maps would also be applicable to gravity maps. A definite relationship may exist between gravity-residual maps and "second vertical derivative" maps. The location of the maxima and minima as well as the zero contours may appear in exactly the same positions on both maps, the only variation being in the magnitude of the anomalies.

Torsion-Balance Data

It is sometimes suggested that, as the torsion balance gives data of higher order (derivatives of gravity), it will render unique solutions to gravity problems. Skeels [37] has shown that all derivatives of potential are dependent upon one another and that additional data such as gradients and curvatures act only to supplement the gravity picture. It is theoretically possible, in fact, to calculate all of the torsion-balance quantities from a gravity-meter map. Skeels and Watson [38] have calculated the gravity-curvature quantities from gravity data and compared it with observed curvatures obtained with the torsion balance for several geologic features. They have also demonstrated that it is possible to calculate the "deflection of the vertical" from conventional gravity maps.

The use of the torsion balance is indicated in areas where sufficient coverage with the gravity meter is not obtainable. During the war, for example, only main thoroughfares across military reservations were kept open to civilian traffic, and torsion-balance stations made along these roads improved the gravity picture through the restricted areas.

Gravity operators today frequently find large areas in which they are refused permission to conduct surveys. In many instances, company files contain torsion-balance maps made in these same areas ten or fifteen

[34] Nettleton, L. L., *Geophysical Prospecting for Oil*, p. 22, New York, McGraw-Hill Book Co. Inc., 1940.
[35] Griffin, W. R., *Residual Gravity in Theory and Practice:* Geophysics, vol. 14, no. 1, pp. 39-56, Jan. 1949.
[36] Henderson, Roland G. and Zietz, Isidore, *The Computation of Second Derivatives of Geomagnetic Fields:* Geophysics, vol. 14, no. 4, p. 516, Oct. 1949.
[37] Skeels, D. C., *Ambiguity in Gravity Interpretation:* Geophysics, vol. 12, no. 1, p. 52, Jan. 1947.
[38] Skeels, D. C., and Watson, R. J., *Derivation of Magnetic and Gravitational Quantities by Surface Integration:* Geophysics, vol. 14, no. 2, pp. 140-147, Apr. 1949.

years previously. By combining such data with gravity-meter information around the area, a satisfactory picture may be obtained.

Applications

    Grand Saline Salt Dome, Texas—A residual gravity map of the Grand Saline dome in Van Zandt County, Texas, is shown in figure 565. This salt plug is a piercement-type dome rising to within 215 feet of the surface and is capped by a very thin layer of limestone and anhydrite. The gravity minimum observed is roughly symmetrical, with a magnitude of about 3.8 milligals centered on the salt mass. From the areal extent of the anomaly it is apparent that the structure must lie close to the

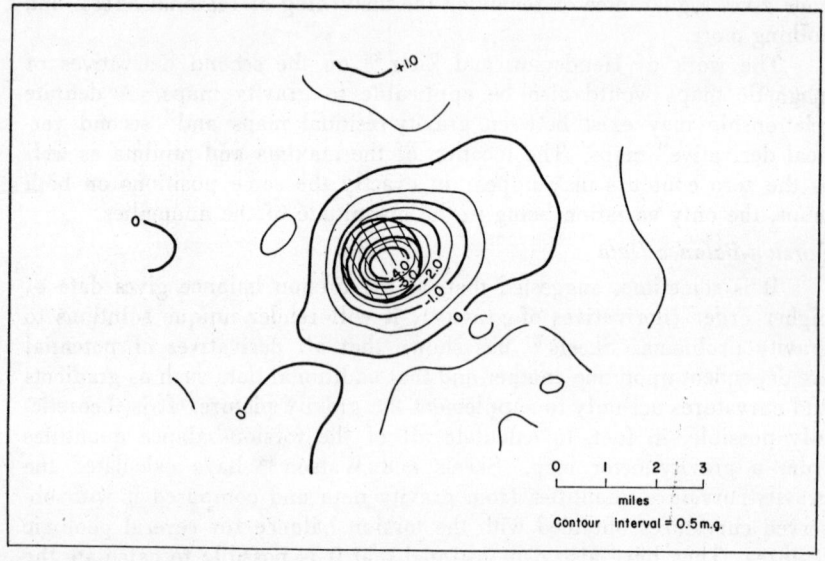

FIGURE 565. Residual gravity map of Grand Saline salt dome, Van Zandt County, Texas. Position of upper salt mass is shown by cross hatching. (After Peters and Dugan. Reproduced permission *Geophysics.*)

surface. Density determinations made on the salt and on the surrounding sediments give a maximum contrast between the salt and the sediments of 0.75 grams per cubic centimeter.[39]

    Hawkins Field, Texas—A gravity-meter survey over the Hawkins field in Wood County, Texas, is shown in figure 566. The structure has probably resulted from the arching of formations through the uplift of a large, deep-seated salt mass. A pronounced gravity minimum is noted over the field, which by observation of the areal extent of the anomaly indicates a deep density interface. The producing Woodbine formation has been uplifted approximately 1,200 feet above its normal depth for the area, and geologic evidence would indicate that the structure is even more pro-

---

[39] Peters, J. P., and Dugan, A. F., *op. cit.*, p. 382.

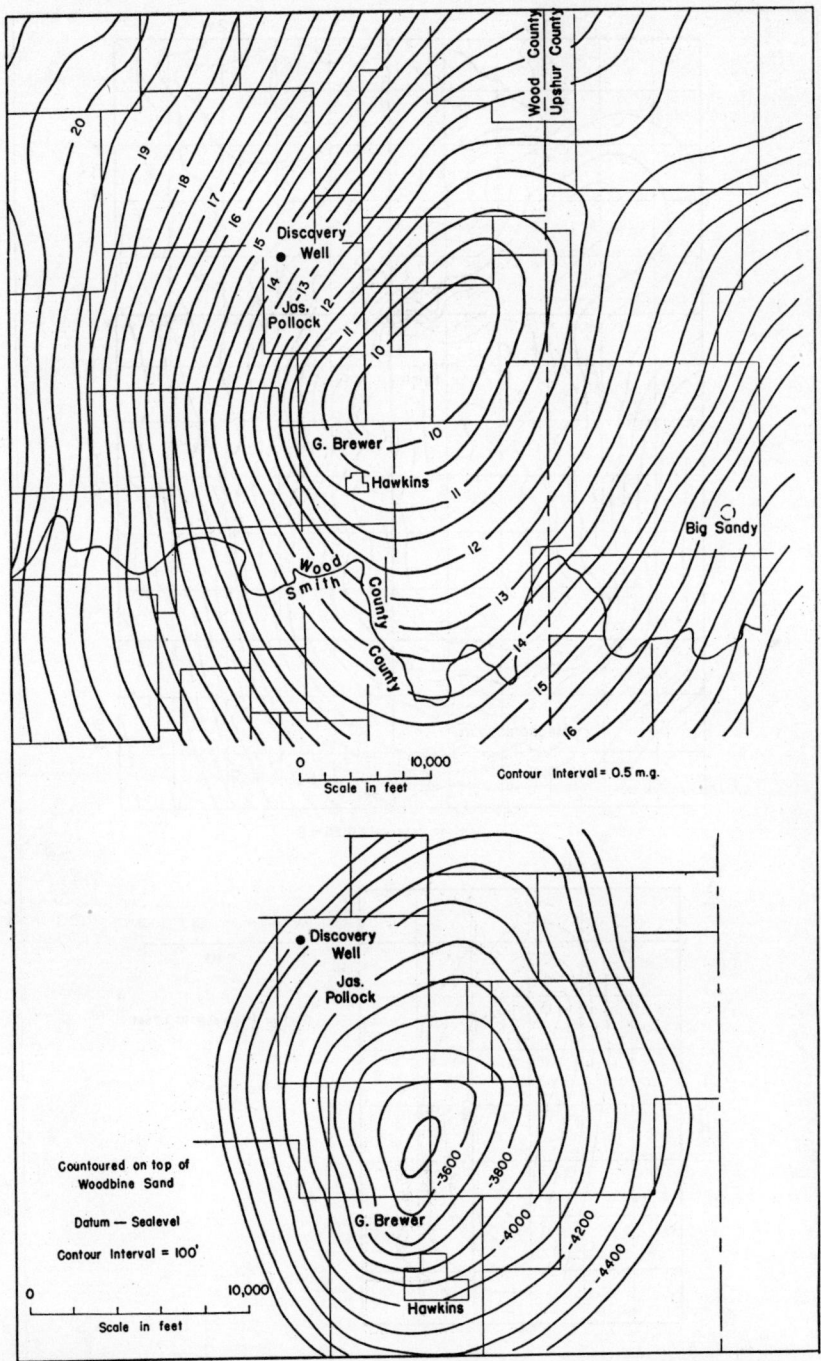

FIGURE 566. Observed gravity contours over Hawkins area, Wood and Smith Counties, Texas (above). Generalized contours on top of Woodbine sand are shown below.

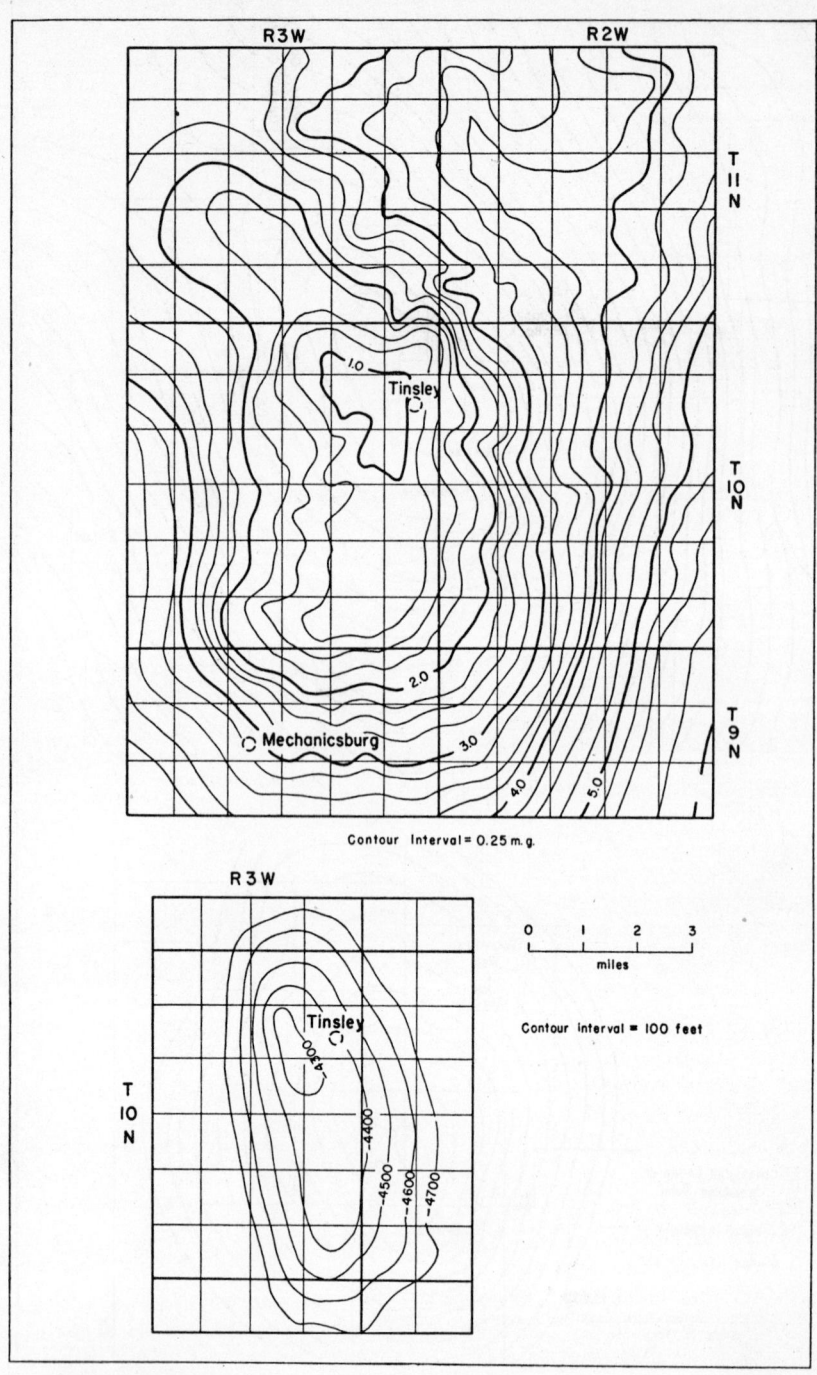

FIGURE 567. Observed gravity contours over Tinsley field, Yazoo County, Mississippi (above). Generalized contours on top of producing horizon are shown below.

nounced with depth. The gravity expression of the Hawkins field is characteristic of many of the fields in the East Texas Basin.

Tinsley Field, Mississippi—The structure of the Tinsley field in Yazoo County, Mississippi, is also related to the movement of the deep-seated salt mass. (See fig. 567.) The Tinsley anticline is an elongated structure with several major faults across the crest of the structure and is expressed by a marked gravity minimum. From the areal extent of the anomaly it is apparent that a great volume of salt is involved in the movement and that the salt mass does not approach the surface. It is interesting to observe that the Jackson dome, lying about forty miles to the

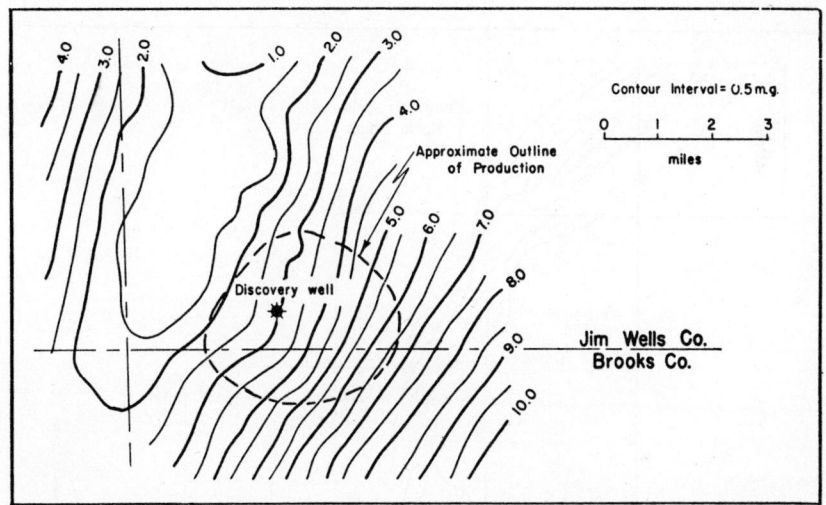

FIGURE 568. Observed gravity contours over La Gloria field, Jim Wells and Brooks Counties, Texas. Producing area is outlined. (After Wooley. Reproduced permission *Geophysics.*)

southeast, gives rise to large gravity maximum. Such gravity data give valuable information as to the cause of structural deformation, the maximum of course being due to the intrusion of a large mass of igneous rock.

La Gloria Field, Texas—The La Gloria field in Jim Wells and Brooks counties, Texas, was first indicated by torsion-balance work in the area.[40] The gravity-meter map (see fig. 568) indicates a northeast-southwest minimum axis extending along the west edge of the area. The structural uplift is indicated by a change in the gradient of gravity on the east side of the regional axis. This map shows the masking effect of strong regional gradients, the anomalous area being indicated by only a very slight change in gradient. Although a producing sand at $-5,650$ feet has a closure of about 400 feet, the gravity anomaly has a magnitude of only 0.4 milligals.

[40] Wooley, W. C., *Geophysical History of La Gloria Field, Jim Wells and Brooks Counties, Texas:* Geophysics, vol. 11, no. 3, pp. 292-302, July 1946.

Many of the fields along the Gulf Coast are indicated by such slight changes in gradient, and the outlining of structurally anomalous areas becomes extremely difficult. Qualitative inspection of the map would indicate that the large minimum axis must be caused by density variation at great depth or by regional thickening of the mother salt bed, and that the structural uplift is related to a salt uplift rising toward the surface.

Kettleman Hills-Lost Hills Trend, California—Figure 569 shows a residual gravity map of a portion of the San Joaquin Valley, California. Kettleman north dome is expressed by a gravity-maximum closure. Towards the southeast a second gravity-maximum closure is encountered which corresponds to Kettleman middle dome. The gradient again flattens

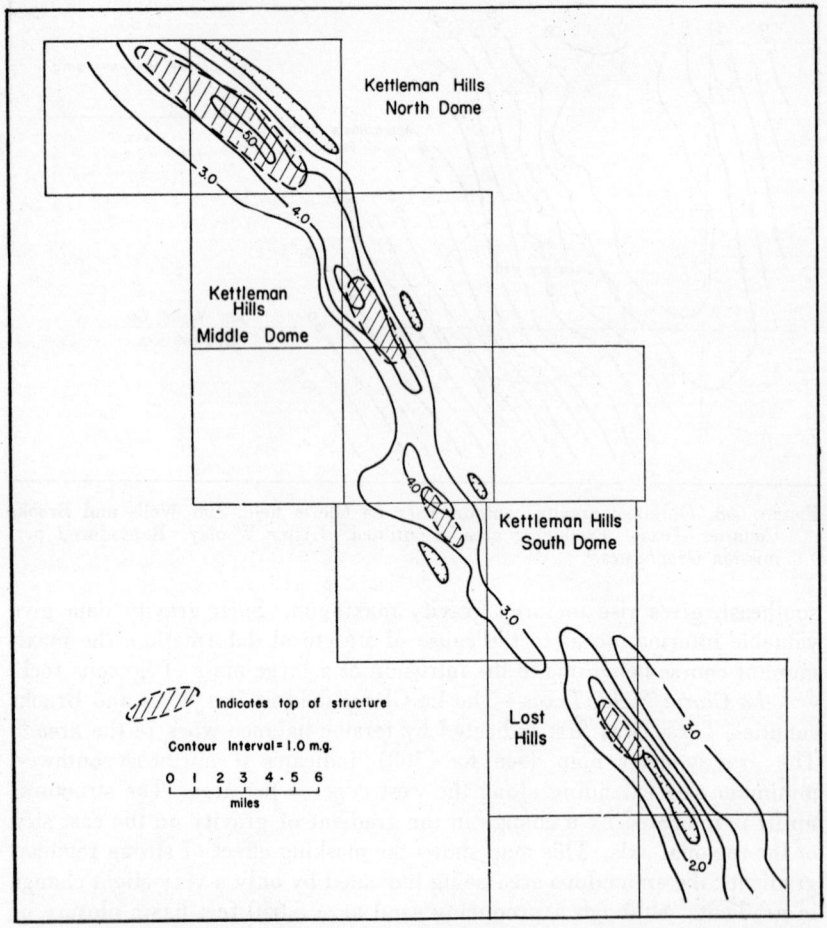

FIGURE 569. Residual-gravity contours over structures in San Joaquin Valley, California. Highest closing contour on each structure is indicated. (Adapted from Boyd. Reproduced permission *Geophysics.*)

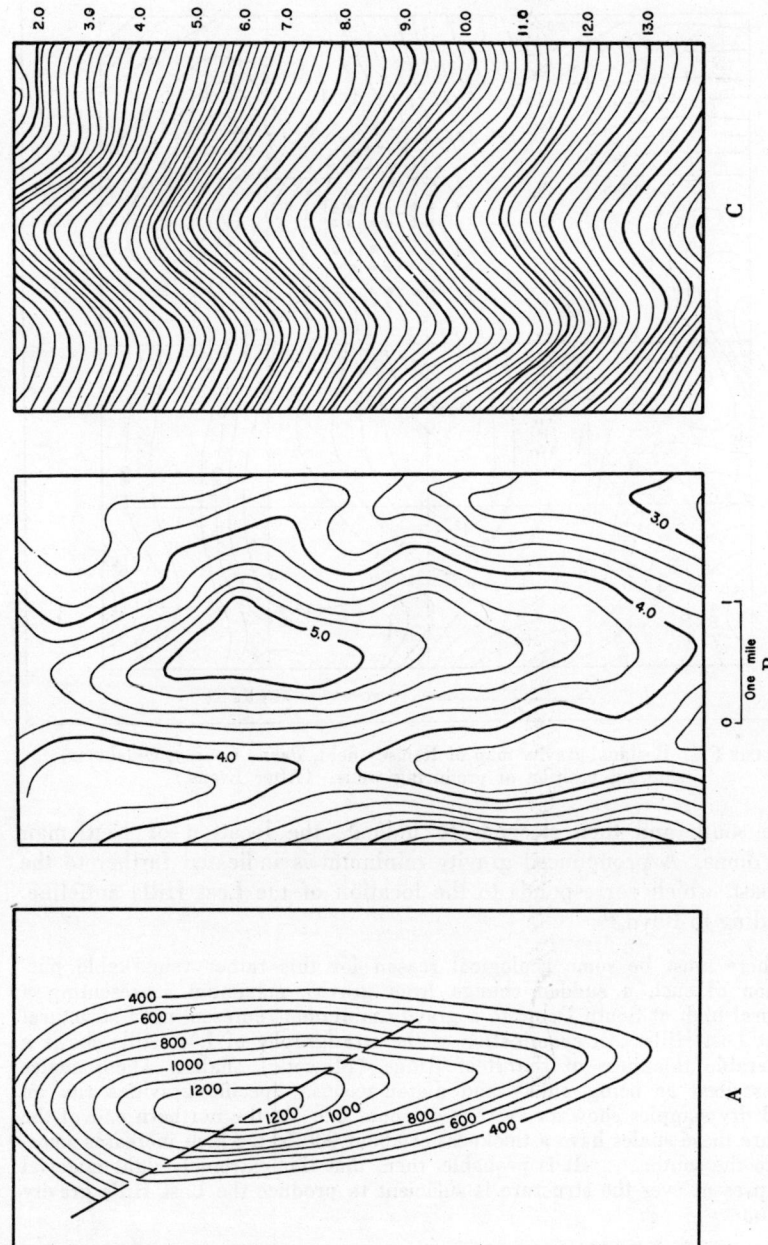

FIGURE 570. *A*—Subsurface structural map contoured on Muddy sand, Wellington field, Larimer County, Colorado. Contour interval 100 feet. *B*—Residual-gravity map over Wellington field. Contour interval 0.2 milligals. *C*—Observed-gravity map showing strong regional gradient. Contour interval 0.2 milligals. (After Wilson. Reproduced permission *Geophysics*.)

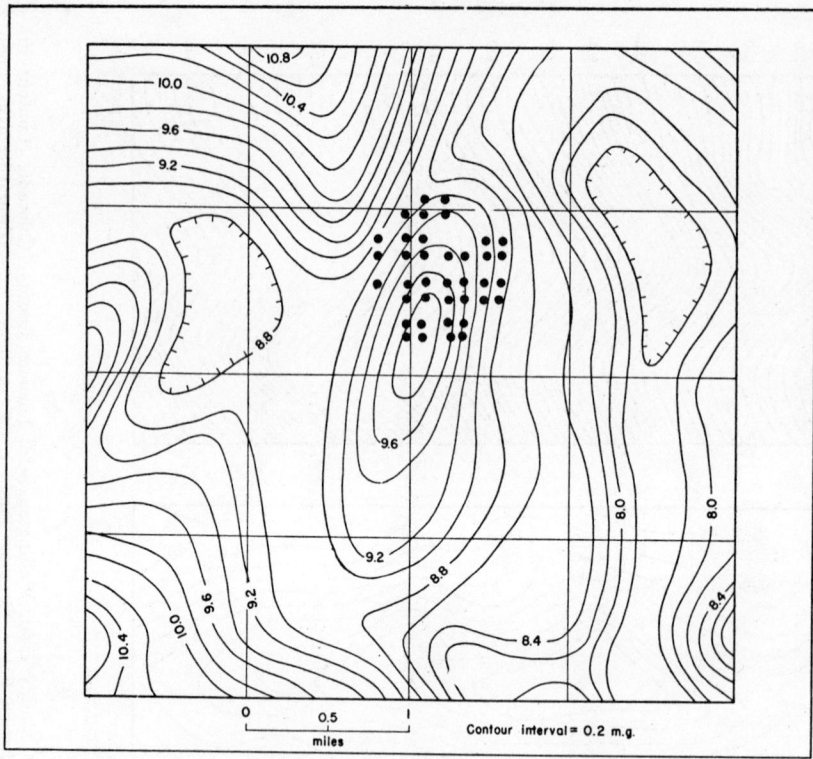

FIGURE 571. Residual-gravity map of Ramsey field, Payne County, Oklahoma,
showing position of producing wells. (After Evans.)

to the south and then steepens to indicate the location of Kettleman
south dome. A pronounced gravity minimum is indicated farther to the
southeast, which corresponds to the location of the Lost Hills anticline.
According to Boyd,[41]

There must be some geological reason for this rather remarkable phenomenon of such a sudden change from gravity maximum representing a
structural high at South Dome to a gravity minimum representing a structural
high at Lost Hills. An examination of the stratigraphy of Lost Hills shows a
considerable thickness of the Reef Ridge (Miocene) shales. These shales
are described as being punky and diatomaceous. Specific gravities run on
several dry samples show an average of about 0.9. At the northern end of the
structure these shales have a thickness of about 900 feet, which increases somewhat to the south. . . . It is probable, then, that this extremely light material
being present over the structure is sufficient to produce the Lost Hills gravity
minimum.

Wellington Field, Colorado—The strong gradient of gravity shown

⁴¹ Boyd, L. R., *Gravity-Meter Survey of the Kettleman-Hill-Lost Hills Trend, Cailfornia:* Geophysics,
vol. 11, no. 2, pp. 121-127, Apr. 1946.

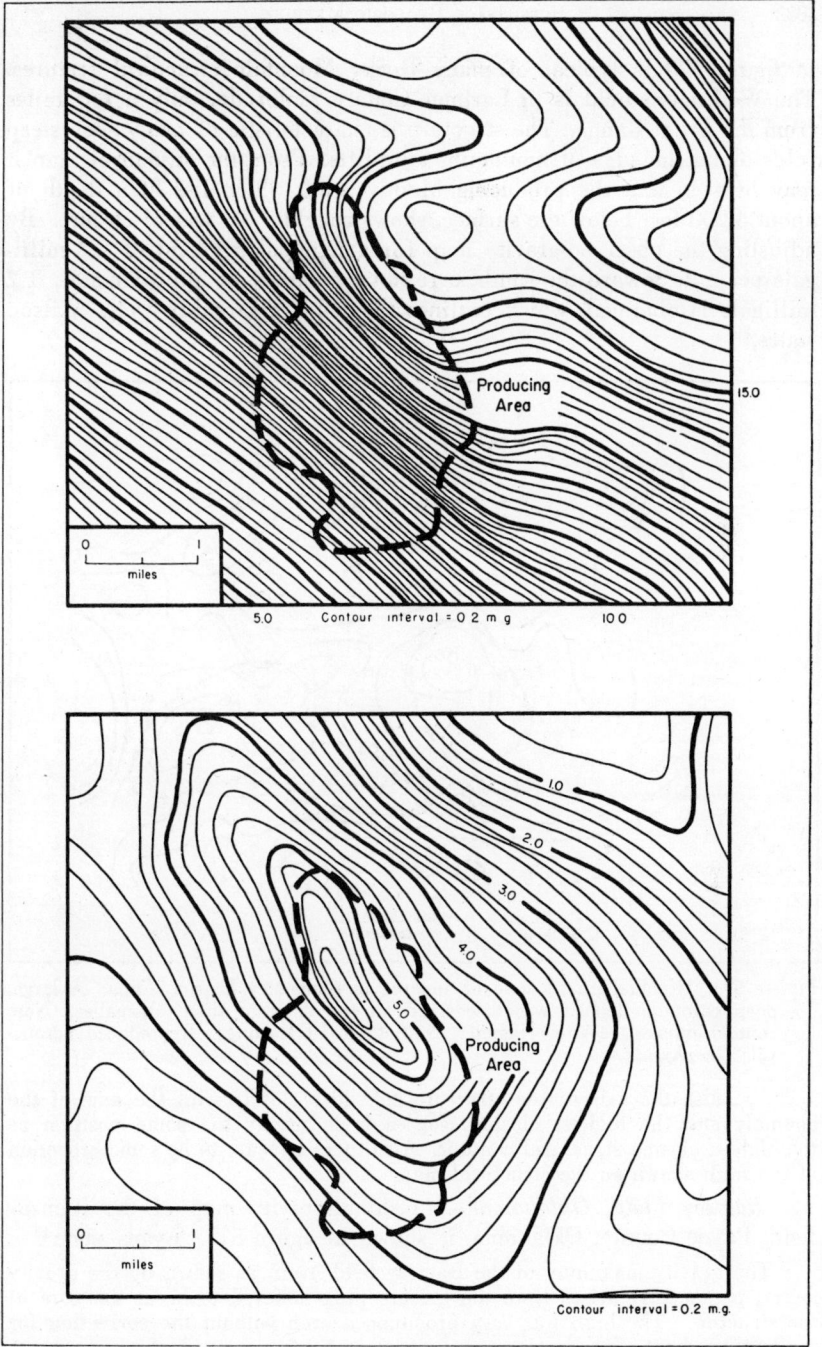

FIGURE 572. Observed-gravity contours over Altus field, Jackson County, Oklahoma (above); residual contours below. Contour interval 0.2 milligals. Producing area outlined. (After Coffin. Reproduced permission Am. Assoc. Petroleum Geologists.)

in figure 570 is typical of many Rocky Mountain structural features. The Wellington field is in Larimer County, Colorado, only a few miles from the Front Range. The structure is characteristic of many such steep folds along the edge of mountainous uplifts. From the structural map it may be seen that the producing Muddy sand, which lies at a depth of about 4,250 feet below the surface, shows roughly 700 feet of closure. By adjusting the observed-gravity map for a regional gradient of 2.2 milligals per mile toward the south, a residual anomaly of approximately 1.2 milligals is obtained, which outlines the field remarkably well. Wilson states,[42]

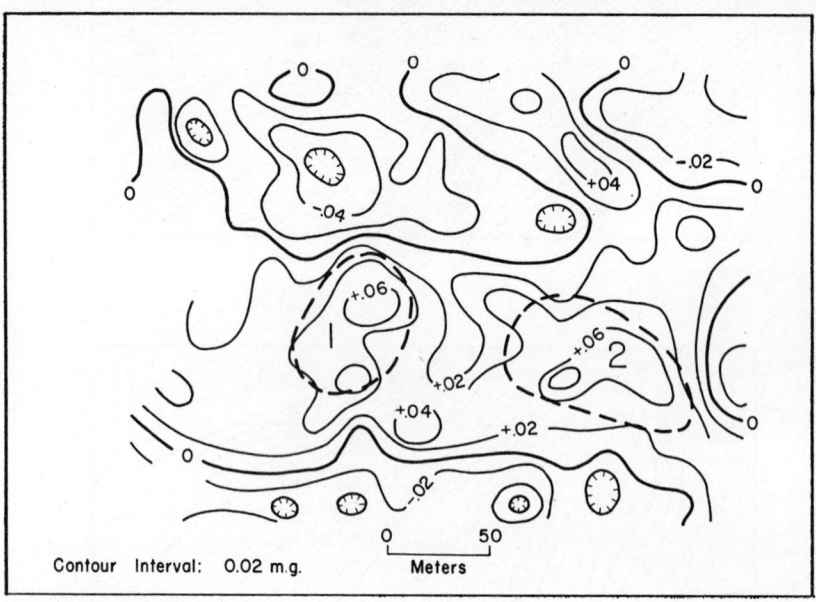

Contour Interval: 0.02 m.g.

FIGURE 573. Residual-gravity map of an area in Camaguey district, Cuba. A large, deep chromite deposit was discovered on the basis of these anomalies. Note contour interval. (After Hammer, Nettleton, and Hastings. Reproduced permission *Geophysics.*)

. . . also the axis of the structure coincides closely with the axis of the anomaly and the highest closing isogam is in nearly the same position as the highest closing structural contour. Also there appears to be some reflection of the fault shown on the structural map.

Ramsey Field, Oklahoma—A residual-gravity map of the Ramsey field, Payne County, Oklahoma, is shown in figure 571. Evans says:[43]

The gravity maximum of the Ramsey field itself, as shown by the gravity meter, presumably arises from the granite plug believed to be at the core of the structure. The high was very pronounced even without the correction for regional gradient.

[42] Wilson, J. H., *Gravity-meter Survey of the Wellington Field, Larimer County, Colorado:* Geophysics, vol. 6, no. 3, pp. 264-269, July 1941.
[43] Evans, J. F., *Correlating Gravity Maximum with Oil Structure in Ramsey Field:* Oil and Gas Jour., vol. 32, no. 26, p. 37, Nov. 1939.

Although it is possible that granite lies at the core of this structure, it is more probable that density contrast within the sedimentary section accounts for the gravity anomaly. Large density contrasts are known to exist between the Permian-Pennsylvanian series and the Mississippian and older sediments, which are predominantly limestones.

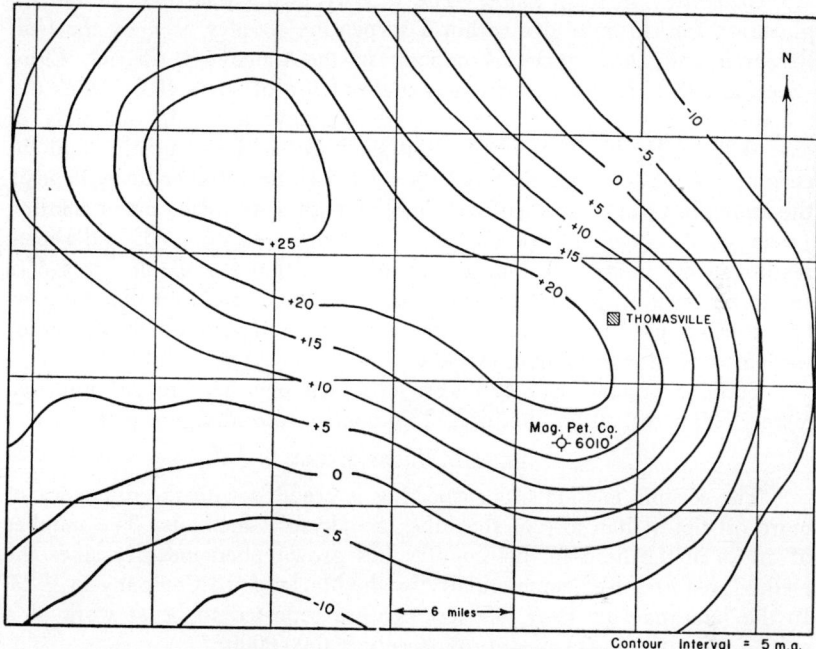

FIGURE 574. Thomasville gravity anomaly in northern Clarke County, Alabama. (Courtesy Magnolia Petroleum Company.)

Altus Field, Jackson County, Oklahoma—The Altus structure in Jackson County of southern Oklahoma was discovered by geophysical exploration and subsequent core-drill information. The field lies to the west of the Wichita Mountains on the trend of the buried Amarillo granite ridge. The structure, which lies on an igneous uplift, is clearly indicated on the residual-gravity map. (See fig. 572.) Although the observed-gravity map does not give the striking picture presented by the residual map, owing to the masking effect of the strong northerly regional gradient, the presence of the structure is betrayed by the pronounced change in gradient. The field had five active wells in 1936 with 340 proved acres and was producing from a limestone at about 1,200 feet.

Thomasville Anomaly, Clarke County, Alabama—Further evidence of an intra-basement variation in mineralogic composition at the Thomasville anomaly in Clarke County, Alabama, is offered by the gravity map shown in figure 574. From the gravity data it would appear that the causative intra-basement feature is elongated in a southeast-northwest direction. From the magnetic map (figure 556) it may be seen that the center of the magnetic "high" is shifted about three miles to the south of the gravity "high." It is apparent from the areal extent of this anomaly that it must be due to a density contrast within the basement complex.

Chromite Deposits, Cuba—The gravity meter was used in the exploration for chromite ore within a serpentine country rock by the Gulf Research and Development Company in the Camaguey district, Cuba. (See fig. 573.) It was found that a mass of ore of some 40,000 tons distributed from the surface down to 70 feet gave a gravity anomaly of only 0.35 milligals. The density contrast between the serpentine and the chromite was 1.5. Since this was considered to be a rather large deposit, the anomalies to be expected over smaller bodies would be much smaller. Twenty-meter spacing between gravity stations was adopted, and repeat readings were made with the gravity meter until the probable error of a single observation was 0.016 milligals. When it is realized that the contour interval used was 0.02 milligals, the precision with which the survey was carried out is even more evident.

Stations numbering 5,320 were observed over an area of approximately 520 acres, and a total of 113 anomalies were designated.[44]

SEISMIC PROSPECTING

The seismic method of prospecting is credited with the discovery of more oil fields than any of the other geophysical methods. The number of crews in the field in this country has grown phenomenally since the method was first used commercially by the Marland Oil Company in 1923. In the later part of 1947, 450 crews were reported to be at work at a cost to the oil industry of approximately $90,000,000.

From 1924 to 1929 the Gulf Coast was intensively worked and reworked by the fan-shooting method, which resulted in the discovery of a number of the shallower salt domes. In a period of several months a dozen domes were discovered for the Louisiana Land and Exploration Company. This feat is even more remarkable when we consider the swampy, marshy terrain in which the work was conducted and the instruments available at the time.

After considerable research the technique of reflection shooting was inaugurated with the discovery of three Oklahoma oil fields in rapid succession during 1929. By 1937 there were 250 reflection crews in the field compared with the four at work in 1929. The present high level of seismic activity will undoubtedly continue until the day that a proved direct method of oil finding is introduced.

Principles

The reflection seismograph is one of the few geophysical methods that does not involve determining potential distribution or derivatives of

---

[44] Hammer, S., Nettleton, L. L., and Hastings, W. K., *Gravimeter Prospecting for Chromite in Cuba:* Geophysics, vol. 10, no. 1, pp. 34-49, Jan. 1945.

potential of either natural or artifically created force fields. For example, in the electrical-resistivity methods, the depth of penetration is dependent upon the electrode spacing at the surface, whereas the depth of penetration of the reflection seismograph is in no way related to the interval between the shot point and the detectors. In the magnetic and gravity methods, the magnitude and distribution of the anomaly at the surface are dependent on the size, the contrast of physical properties, and the depth. Reflections may be obtained from great depths with about the same ease that they are obtained from shallow depths.

The seismic methods are similar to optics in regard to the physical phenomenon involved, in that they both deal with a type of energy propagated in the form of waves. Such physical quantities as velocity, fre-

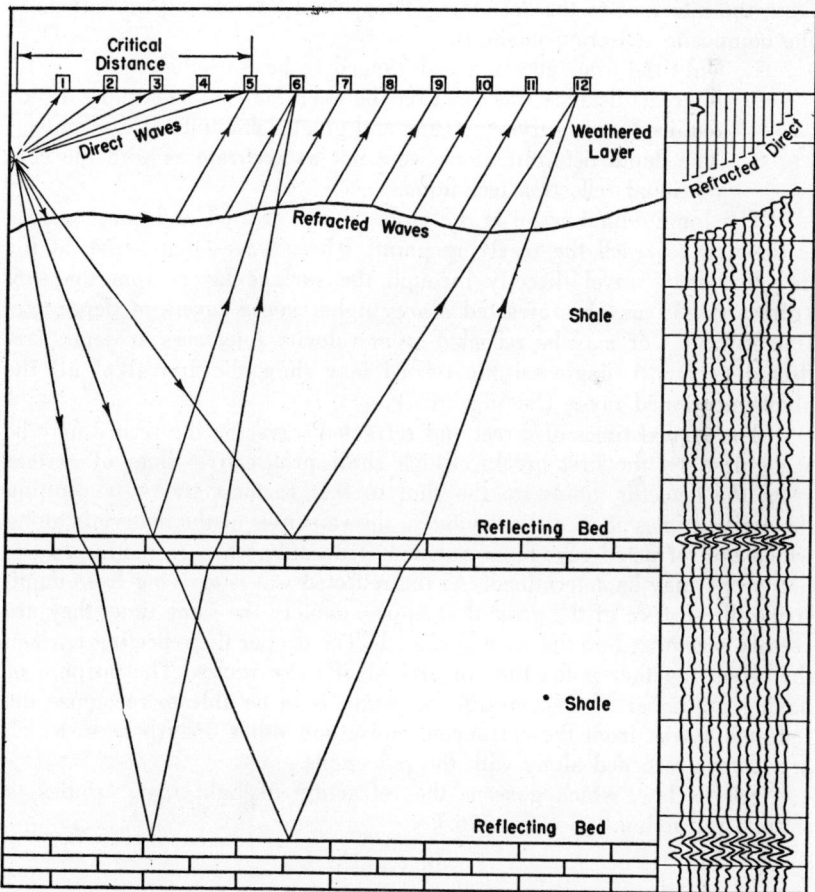

FIGURE 575. Schematic diagram of seismic-wave paths showing direct, refracted, and reflected waves and recorded arrivals. In order to record direct waves distance between shot point and detectors would be much less than depth to base of weathered layer.

quency, intensity, phase, and direction, with their associated or derived phenomena such as travel time, wave length, absorption, refraction, and reflection, are common to both. Owing to the rapid rate of propagation of light, however, only quasi-stationary phenomena are investigated in optics; in seismic work certain of the derived quantities such as travel time, refraction, and reflection are of prime importance.

When an artificial-force field is created in the earth by the explosion of dynamite, several types of waves are generated, among which are the longitudinal, transverse, Rayleigh, and Love types. The only wave utilized in present-day prospecting is the longitudinal or compressional wave.

Ricker and Lynn [45] have recently described the use of composite reflections in mapping geologic structure. A composite reflection consists of longitudinal waves to the reflecting interface and transverse waves from the interface to the detectors. They list the following limitations to the composite reflection method:

1. Only the first reflecting bed seemed to be workable.
2. A great distance was required between shot and detectors which required more surveying time and greater amounts of dynamite.
3. The depth determinations were not as accurate as with the conventional reflection technique.

The longitudinal wave is the fastest wave created and consequently is the first to reach the receiving point. These waves that arrive at the detectors may travel directly through the surface layer from the shot (direct rays), may be refracted along higher speed layers at depth (refracted rays), or may be reflected from velocity interfaces at depth (reflected rays). A single seismic record may show the arrival of all the above-mentioned rays. (See fig. 575.)

The arrival times of direct and refracted waves on the record may be recognized by the first breaks, which show progressive times of arrival from the detector closest to the shot to that farthest way. By plotting the arrival times against the distances, the velocities of the materials along which the refracted rays have traveled (true only when refracting bed is flat-lying) may be determined. As the reflected waves arriving from depth strike the surface of the ground at approximately the same time, they are readily recognized on the seismic record. The deeper the reflecting horizon the closer together is the time of arrival of these waves. The purpose of having a number of detectors in the set-up is to be able to recognize the reflected waves from the extraneous noise and other disturbances, which are always recorded along with the reflections.

Snell's law, which governs the refraction of light rays, applies to seismic refraction. (See fig. 576.)

$$\eta = \frac{\sin i}{\sin r} = \frac{V_1}{V_2}$$

---

[45] Ricker, Norman, and Lynn, R. C., *Composite Reflections:* Geophysics, vol. 15, no. 1, pp. 30-49, Jan. 1950.

where $\eta =$ index of refraction
$i =$ angle of incidence
$r =$ angle of refraction
$V_1 =$ velocity in upper medium
$V_2 =$ velocity in lower medium.

Of primary importance in refraction-seismograph methods is the critical angle of incidence where the angle of refraction becomes 90°. A ray striking the interface at the critical angle travels along the interface and returns to the detectors at the surface at the same angle. Rays striking the interface at angles less than the critical may be both refracted and reflected. Those striking at angles greater than the critical are reflected only. The angle of incidence of a reflected ray is equal to the angle of

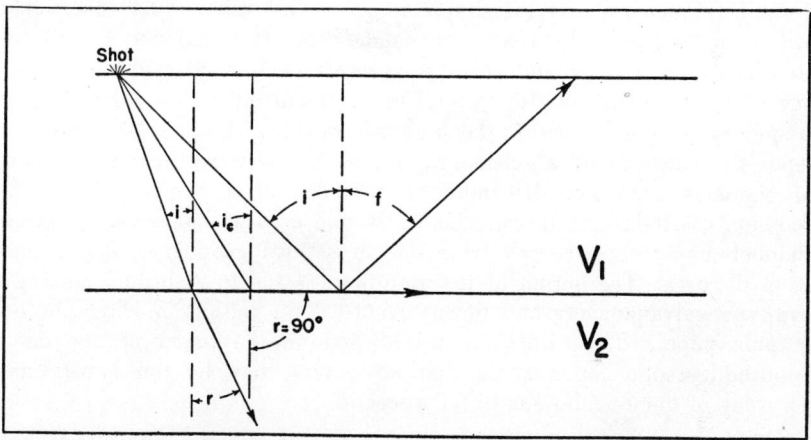

FIGURE 576. Three of the many wave paths that rays may travel after a seismic disturbance at or near the surface.

reflection, as in optics. In order that the seismic methods may be used as a prospecting tool, variations in the physical properties of the rocks must occur with depth.

In all modern seismic methods there are a shot point and a number of receiving points, the disposition of the receptors depending upon the method used and local conditions. Energy is introduced into the ground by the explosion of dynamite, which may be placed from a few feet to hundreds of feet below the surface in drilled shot holes, or from the explosive force of dynamite suspended above the ground. The latter method, called the "Poulter method," is at present in the development stage.[46] As shot-hole drilling represents one of the largest expenditures on a seismograph crew, shooting above the ground should prove attractive

[46] Kastrop, J. E., *The Poulter Method of Geophysical Seismic Exploration:* World Oil, vol. 128, no. 9, pp. 53-60, Jan. 1949.

to many operators, particularly in areas such as the Edwards Plateau of west Texas. The creation of an elastic-wave train within the earth by the explosion of dynamite has been described by Morris.[47]

The instant of the explosion is transmitted to the recording unit by either radio or wire and is placed on the record. The distance between the receiving points and the shot may be a number of miles in the refraction method or may be only several hundred feet in the reflection method. At each receiving point one or more geophones or detectors convert the mechanical energy of the seismic waves into electrical energy. These electrical impulses are carried to the recording unit by wire and fed into specially constructed amplifiers, which not only act to increase the intensity of the impulse but also act as discriminators. Filters are included in the amplifiers, which allow the operator to emphasize the desired frequencies and discriminate against all others. As reflections ordinarily fall into the frequency band from 30 cycles per second to 70 cycles and as, "ground roll" is generally below 20 cycles per second, the disturbance due to ground unrest from either the surface low-frequency seismic waves or the higher-frequency wind "noise" may be reduced. The size of a seismograph unit is indicated by the number of channels or traces that may be recorded at any given time. In common use today are instruments of 24 and even 48 channels. A given channel may receive energy from one or several geophones, depending upon the area. The output of the amplifiers is fed to recording moving-coil-type galvanometers and thence recorded on rapidly moving photographic paper. Time lines are established on the paper at one one-hundredth-second intervals so that any event may be timed with an accuracy of one one-thousandth of a second.

Many companies prefer "mixed" records, in which a certain portion of the energy from each channel is fed to the following channel. This is accomplished within the recording unit by the use of a mixer tube in the amplifiers or by a resistance network, in which a portion of the output is fed to the next channel. Results similar to those obtained from multiple geophones are obtained by this method. Mixing and multiple geophones have the advantage of lowering the disturbance level on the record due to ground roll and the like and of strengthening the reflections recorded. In faulted areas and in regions of appreciable dip, erroneous interpretations may result from the use of mixed records, owing to the effect of the mixing on the "step-out times" or "dip times." Most companies today will shoot mixed records, but correlations are not attempted on reflecting horizons that do not appear on the unmixed or pure record. Frequently reflections that do not actually exist may be "created" by the use of mixing.

---

[47] Morris, George, *Some Considerations of the Mechanism of the Generation of Seismic Waves by Explosives:* Geophysics, vol. 15, no. 1, p. 61-69, Jan. 1950.

Rock Properties

Both refraction and reflection of seismic waves are controlled by the physical properties of the rocks involved. The velocity of a longitudinal wave is determined by the following relations between the elastic constants and the density:

$$V_L = \sqrt{\frac{k + 4/3n}{\delta}}$$

where k is the bulk modulus
n is the shear modulus
δ is the density

It has been found that a number of factors influence the seismic velocity of rocks. Both the mineralogic composition and the crystalline structure affect the wave speed in igneous rocks. The velocity increases with a decrease in silica content and with an increase in the size of the mineral grains.

In sedimentary rocks the velocity varies with the depth, geographic location, lithology, local structural position, and geologic age. In general, velocity increases with depth owing to the increase of the value of the elastic moduli with pressure of the overlying strata. The geographic location influences the velocity because of the effects of metamorphism, severe folding, and regional variations in lithology. In the San Joaquin Valley of California it is noted that the velocities for corresponding depths or for corresponding reflection times are very much higher near the margins of the valley than in the central portion. The mineralogic composition, size, porosity, interstitial fluids, and degree of cementation of sediments all tend to influence the velocity, increases in velocity being observed when (1) the lime content of sands or shales increases, (2) the porosity decreases, (3) the interstitial fluid content of unconsolidated rocks increases, (4) the interstitial fluid content of consolidated rocks decreases, and (5) the degree of cementation increases. Velocities are generally higher over a structure owing to the induration of sediments lying above uplifts and the increase in the elastic constants. In general, velocity increases with geologic age. Two factors are probably involved: (1) the older sediments are more deeply buried, and (2) the younger sediments show lesser degrees of consolidation and cementation. It has been observed that most sedimentary and metamorphic rocks are anisotropic; that is, the velocity of the material is higher parallel to the bedding planes or schistosity than at right angles thereto. Although in certain shales the velocity may be as much as 50 percent higher parallel to the stratification than at right angles, an average figure would probably be about 10 percent.

Because the success of the seismic method depends on the accurate determination of the velocity, extreme care is exercised in obtaining

TABLE 34
Velocities of Selected Rocks

Type of rock	Velocity (ft. per sec.)
Alluvium	1,000– 2,000
Clay, sandy clay	6,000– 8,000
Shale	6,000–13,000
Sandstone	8,000–13,000
Limy sandstone	12,000–14,000
Limestone	7,000–21,000
Rock salt	14,000–25,000

values. Although several methods may be used for velocity determinations, the direct technique, in which a detector is lowered to known depths in a well, is favored. From such data the average velocity to any horizon, the interval velocity, and the time-depth relationship may be determined. By computing the mathematical formula governing the velocity from the time-depth relationship, seismologists are able accurately to predict the depth and disposition of subsurface strata.

Other parameters of importance in discussing rock properties are the transmission characteristics of the rocks. Significant in this connection are (1) acoustic impedance, (2) spreading and dispersion, and (3) absorption and dissipation of energy. The average rate of flow of seismic energy through the ground depends upon the amplitude and frequency of the wave and the acoustic impedance of the medium. Materials for which the acoustic impedance is high transmit more energy per unit area than those in which the value is low, if we assume the same values of amplitude and frequency for the wave. This factor also controls the reflection and refraction of seismic waves. If, for example, the acoustic impedances of two adjacent media are in the ratio of two to one (about the maximum difference encountered in stratified beds), the amplitude of the refracted or transmitted wave is two-thirds of the normal-incident wave, and the amplitude of the reflected wave is one-third of the normal-incident wave. Also of importance is the phase of the reflection: that is, whether at a given reflecting horizon a compression is reflected as a compression or as a rarefaction. For a compression arriving at the surface, the first movement of the ground is upward; for a rarefaction, the first movement of the ground is downward. If the acoustic impedance of the upper medium at a reflecting interface is larger than that of the lower medium, a phase shift of π occurs, and the compression is reflected as a rarefaction. If the acoustic impedance of the upper medium is smaller than that of the lower, no phase shift occurs, and a compression is reflected as a compression. Numerically, the specific acoustic impedance is the product of the longitudinal velocity times the density referred to unit dimensions.

The intensity of seismic waves decreases with the distance from the shot point, varying inversely with the surface areas of the advancing wave fronts. This factor is known as "geometric divergence" or "spreading." Scattering of energy occurs as the seismic wave front advances because of reflection and refraction of energy on prominent irregularities in the medium. It has been noted that scattering accentuates the low-frequency content of waves and attentuates the high frequencies as the wave propagates. The effects of dispersion of seismic energy in prospecting seismology are not very well known. It has been observed that in stratified ground there is a change of velocity with frequency, which is not observed in homogeneous or nearly homogeneous ground.

Energy absorption is one of the losses accompanying the decrease of seismic intensity with distance. The damping properties of the materials through which the waves are propagating largely determine the rate of attenuation of the energy. Because the attenuation factor increases with frequency increase, low frequencies are accentuated and high frequencies attenuated. Born [48] has shown that an increase in the moisture content of consolidated rocks also increases the attenuation at high frequencies, in that the damping properties of the rocks are increased. The propagation of waves may also be affected by the dissipation factor, which determines the ability of a material to sustain vibrations and depends upon the internal resistance to elongations and contractions. From this brief discussion it may be seen that a number of factors are acting to attenuate seismic energy, several of which are discriminatory to the higher frequencies. Several of the factors affecting the propagation of seismic waves are discussed by Clewell and Simon.[49] They show that although the earth acts as a low-pass filter for refracted energy, it acts as a bandpass filter for reflected energy. The low-frequency "cut-off point" is in large part determined by the thickness of the reflection horizon.

Methods of Prospecting

Prospecting with the seismograph may be divided into the refraction technique, in which travel-time data are obtained for the elastic waves that have been refracted at boundaries separating material of different elastic properties, and the reflection method, in which time data are obtained for those elastic waves reflected from subsurface boundaries. Several arrangements of shot points and detectors are employed in the refraction technique, depending upon the purpose of the survey and local conditions. The fan-shooting method resulted in the discovery of a large number of the shallower Gulf Coast salt domes in the 1920's and has been employed recently in the search for favorable structure off-shore. The method consists in comparing the travel times from a single shot point to a number of detectors placed at approximately equal distances from the shot point and arranged in the form of a fan. Rays traveling

---

[48] Born, W. T., *The Attenuation Constant of Earth Materials:* Geophysics, vol. 6, no. 2, p. 138, Apr. 1941.
[49] Clewell, D. H., and Simon, R. F., *Seismic Wave Propagation:* Geophysics, vol. 15, no. 1, pp. 50-60, Jan. 1950.

through abnormally high- or low-velocity regions may be identified by comparing the times to each detector in the fan with the time for a corresponding distance determined in a structurally normal area. As the depth of penetration of the rays increases with the distance from the shot point, spreads of as great as ten miles are sometimes used, requiring up to a ton of dynamite. By shooting a number of such fans and noting the rays that show abnormal velocities, the approximate location of salt domes may be determined.

The fan-shooting technique has been adapted to the tracing of buried river channels that have been scoured in more resistant rocks. By noting rays that show an abnormally low velocity, the Imperial Geophysical Experimental Survey in Australia has successfully applied the method in prospecting for favorable placer grounds.

In the profile method of refraction shooting, measurements are made of the travel times of waves that are refracted along high-speed subsurface boundaries. A number of arrivals of refracted energy may be observed, depending upon the distance between the shot point and the detectors and the number of high-speed refracting horizons within the subsurface that are penetrated by the rays. As the order of arrivals of refracted energy is dependent upon the shooting distance, depths, dips, and velocities of the refracting horizons, reliable identification of a refracting horizon can be made only on the basis of well control. At the optimum shooting distance for any refracting horizon, interpretations are inherently more accurate, although information from the other refracting horizons is frequently used. The refraction method does not evaluate small anomalies because of the reduced accuracy of the method; however, it may be of great value in locating large anomalies and investigating regional dips over large areas. This technique and modifications of it are being used at present in the exploration of the Edwards Plateau of west Texas and in the Florida Peninsula.

The reflection method, where adaptable, gives an actual subsurface map of geologic horizons. A number of advantages make it the preferred exploration tool in most regions. Among these advantages might be mentioned the greater resolving power of the method, the use of smaller explosive charges, the fact that the depth of penetration is not controlled by the dimensions of the effective beds, and that data may be obtained for a number of subsurface boundaries with a single shot. Two general types of reflection shooting are practiced: the correlation and the dip-shooting methods. Correlation shooting is employed where the reflecting beds are persistent and readily identifiable. For reconnaissance work, shot points may be located a mile or so apart with the detectors placed to afford the maximum character and amplitude to the recorded reflections. A spread in which the geophones and shot points are spaced regularly along the entire length of a traverse line, the interval between shot points being intercepted by a constant number of detectors, is used in areas of complex geology and where more accurate control is desired. This technique is known as "continuous profiling." Several variations of this technique are in common use, but, when maximum detail is desired or reflections are difficult to correlate between records, overlapping continuous spreads are preferred. Dip shooting must be practiced in areas where correlation is doubtful or impossible because of the lack of persistent reflecting horizons, a condition which exists in much of the Gulf Coast and in parts of California. It is also indicated in regions where dips in excess of a few degrees are expected. Frequently both dip data and correlations are used together in order to give a better indication of subsurface conditions. Dips are determined by utilizing the "step-out" times (the difference in time between two detector positions for a given reflection).

All seismic data must be corrected for elevation differences and variations in the thickness of the weathered zone or low-velocity layer. Several correction techniques are available, detailed descriptions of the corrections being found in the text books on geophysical exploration. Frequently extreme variations in the thickness of the weathered layer require a revision of operating procedure. Olson [50] reports unusual variations of the weathered layer due to the irregular distribution of lenses of unconsolidated mud and vegetation in the Tucupita area of Venezuela. By mounting geophones on individual pipes driven ten to fifteen feet into the ground, the effects were eliminated in certain portions of the area. Gaby and Solari [51] mention a similar situation in San Joaquin County, California, where the effects of the erratic distribution of the surface material were eliminated by burying the geophones to a depth of 28 feet.

Interpretation Methods

The object and the fundamental problem in seismic exploration are to obtain an accurate picture of subsurface conditions from data obtained at the surface. A computation method is selected for a given area to give the greatest degree of accuracy with the minimum expenditure of time and labor. The assumption must always be made that there is a conformable relationship between the geologic strata and discontinuities of the physical properties of the geologic section. Although velocity discontinuities are generally correlated, West [52] has shown that reflections may be obtained from density contrasts alone. (See figure 577).

It is always necessary to make an assumption concerning the attitude of the velocity zones that lie between the surface and the reflecting hori zon. Alcock [53] mentions several possibilities. (See figure 578.)

1. The velocity zones are all horizontal. This assumption is most •

[50] Olson, W. S., *Geophysical History of Tucupita Oil Field, Venezuela* in *Geophysical Case Histories*, vol. 1, p. 615, Tulsa, Soc. Exploration Geophysicists, 1948.

[51] Gaby, P. P., and Solari, A. J., *Geophysical History of the McDonald Island Gas Field, San Joaquin County, California* in *Geophysical Case Histories*, vol. 1, p. 605, Tulsa, Soc. Exploration Geophysicists, 1948.

[52] West, S. S., *The Effect of Density on Seismic Reflections:* Geophysics, vol. 6, no. 1, p. 45, Jan. 1941.

[53] Alcock, E. D., *Selection of Computational Methods for Seismic Paths:* Geophysics, vol. 8, no. 3, p. 297, July 1943.

frequently made and leads to rather simple mathematical relationships.

2. The velocity zones are all parallel to the reflecting horizons.

3. The velocity zones at the surface are horizontal and dip at a rate that varies as a function of the depth. This assumption undoubtedly fits the actual subsurface conditions most accurately, but it unfortunately leads to complex mathematical relationships that are difficult to apply.

4. An average velocity is assumed from the surface to the reflecting horizon. The ray paths thus become straight lines, and this greatly simplifies calculations.

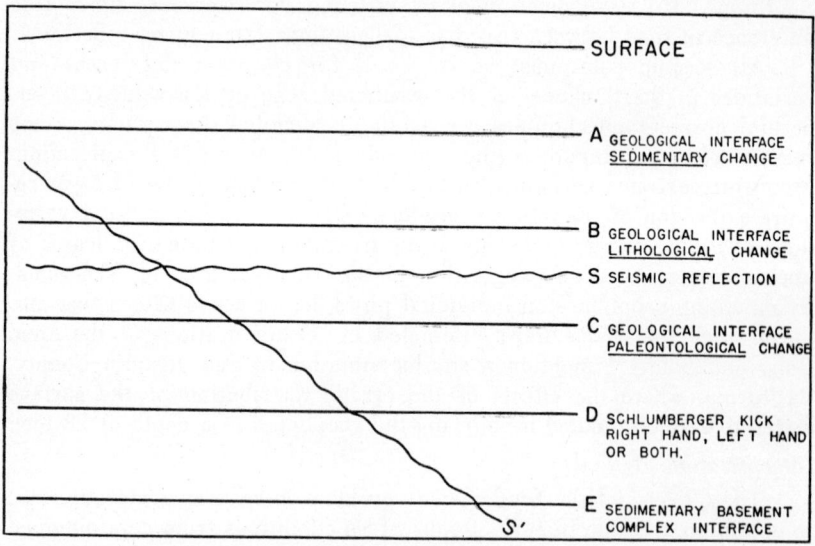

Figure 577. A seismic reflection *(S)* may correlate with any of the horizons *(A, B, C, D,* or *E),* or it may *not* precisely coincide with *any* of them. The important thing is that S is conformable with *A, B, C, D,* or *E,* if weathering corrections and velocity changes are properly accounted for. A seismic event such as *S'* seldom, if ever, will occur. (After Handley. Reproduced permission *Oil and Gas Jour.)*

The assumption to be made concerning the attitude of the iso-velocity surfaces depends upon the geology of the area concerned. For example, on the Gulf Coast where an extremely thick section of geologically young, incompetent sediments are arched upward or pierced by flowage of a deep salt layer, the assumption may be made that the iso-velocity surfaces are parallel with the reflecting horizons. This is reasonable since the uplift in many cases occurred contemporaneously with or after deposition of the sediments and is reflected through the sedimentary section to the surface. The reflecting horizons are approximately parallel not only to each other, but to the iso-velocity surfaces. This assumption leads to the use of extremely simple computing devices.[54]

---

[54] See U. S. Patent no. 2,535,220—G. M. McGuckin, *Apparatus for Solving Seismic Problems.*

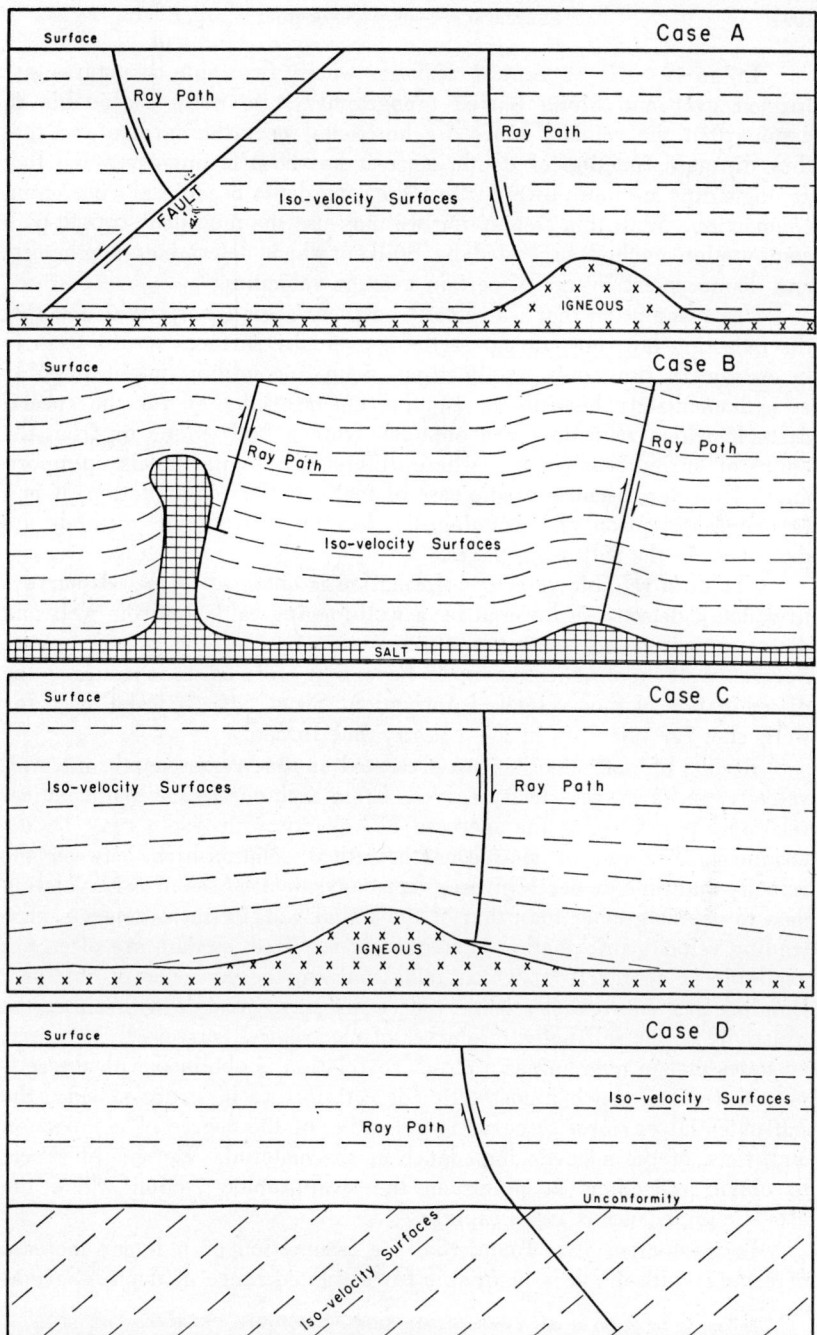

FIGURE 578. Several possible dispositions of iso-velocity surfaces with respect to reflecting horizons and associated ray paths for normal ray. Case A—Iso-velocity surfaces all horizontal. Case B—Iso-velocity surfaces parallel to reflecting horizons. Case C—Iso-velocity surfaces dipping as a function of depth. Case D—Iso-velocity surfaces at an angular unconformity.

In areas such as central Kansas, where favorable structures are formed over and along buried topography, it is more reasonable to assume that the velocity zones are horizontal near the surface and that they dip as a function of depth, since it has been frequently noted that geologic dips are more severe with increasing depth because of differential compaction. With this assumption we may use the modified straight-path computation method suggested by Stulken which determines depth, dip, and displacement by use of certain average velocities.[55]

The assumption that iso-velocity surfaces are all horizontal and that the reflecting horizons transgress the iso-velocity surfaces would seem to be extremely limited in application. Such a condition might possibly arise immediately beneath an angular unconformity or for the special cases in which reflections are obtained from a fault plane or from the flanks of buried topography (where differential compaction is unimportant). However, because of the ease of mathematical computation, it is a favorite assumption of seismologists. Various curved-path methods are described on the following pages.

The most reliable velocity information is obtained from well surveys in which a detector is lowered by a well-logging cable into the well and direct measurements are made of the travel time of the seismic waves. Several well-shooting associations have been formed to encourage the dissemination of such velocity information. Swan [56] has indexed the many wells shot for velocities in this country and abroad.

Results of well shooting are presented as time versus depth, interval-velocity, and average-velocity curves. From such curves the mathematical relationship governing the increase of velocity with depth may be determined. The use of simple mathematical relationships between the velocity and time or depth gives a very convenient means of extrapolating data to depths greater than that of the actual velocity measurements. Extending velocity information is frequently required, as data are often not available at depths from which reflections can be consistently obtained. Handley has shown that seismic reflections may correlate with either the resistivity or the self-potential curves of electric-well surveys.[57] However, he states that in many areas a closer correlation is obtained with the self-potential curve. Such a correlation is certainly to be expected since the self-potential or porosity curve is indicative of the degree of compaction and, thus, of the acoustic impedance of the material. We should expect to obtain reflections at points in the stratigraphic section where the acoustic impedance changes rapidly.

In most areas it is found that the assumption of a linear increase of velocity with depth is justifiable for a limited range of depth. Several

[55] Stulken, E. J., *Effects of Ray Curvature Upon Seismic Interpretation:* Geophysics, vol. 10, no. 4, pp. 472-486, Oct. 1945.

[56] Swan, B. G., *Index of Wells Shot for Velocity:* Geophysics, vol. 9, no. 4, p. 540, Oct. 1944; vol. 11, no. 4, p. 538, Oct. 1946; vol. 14, no. 1, p. 58, Jan. 1949.

[57] Handley, E. J., *Can Geophysical Reflections be Correlated with Geological Horizons?:* Oil and Gas Journal, vol. 47, no. 44, pp. 84-87, Mar. 1949.

methods have been described in the literature for fitting empirical data to mathematical expressions of type $V = V_o + kZ$ (linear increase of velocity with depth). Hafner [58] has described a method in which observed time versus depth curves are matched with a set of master curves, the constants in the equation being read from the master curve that best fits the observed data. Another method proposed by Legge and Rupnik [59] is based on a least-squares determination of the velocity function. It is frequently observed that the rate of increase of velocity is greater at shallow depths. Sparks [60] suggests that the velocity function describing a hyperbolic increase of velocity with depth will best fit the observed data. He presents a method for obtaining the constants for the hyperbolic relation

$$V = \frac{V_o (1 + aZ)^2}{(1 + 2bZ + abZ^2)}$$

A discussion of the several mathematical expressions that may best describe the observed velocity data in many areas is given by Mott-Smith.[61]

After selecting the velocity function that best fits the empirical data, it becomes necessary to describe the path of the reflected ray in terms of this velocity. When seismic rays pass through a medium in which the velocity is variable, the phenomenon of refraction requires that they must deviate from a straight line. Completely and rigorously to describe the path of the ray would require a detailed knowledge of the wave velocity in the medium between the reflecting beds and the surface. Obviously, since we have smoothed our observed data to fit a mathematical relationship, it becomes impossible exactly to describe the path. It should be noted that curved-ray paths are described by instantaneous-velocity functions rather than average-velocity functions. By knowledge of the ray path, it becomes possible to determine the depth, dip, and horizontal offset of the reflecting point. Rice has assumed iso-velocity surfaces to be horizontal and an asymmetric structure to show variations of depth and displacement for various computational methods.[62] For a parabolic increase of velocity with depth, the over-all results of using the curved-path method and the modified straight-path method proposed by Stulken [63] are quite similar. As the computations required to obtain these data are involved and tedious to perform, charts are frequently constructed from which the information may be readily obtained. Simplified computational techniques for the linear increase of velocity with depth $(V = V_o + kZ)$

[58] Hafner, W., *The Seismic Velocity Distribution in the Tertiary Basins of California:* Seismological Soc. America Bull., vol. 30, no. 4, pp. 309-326, Oct. 1940.

[59] Legge, J. A., Jr., and Rupnik, J. J., *Least Squares Determination of the Velocity Function* V=Vo+kZ *for Any Set of Time-Depth Data:* Geophysics, vol. 8, no. 4, pp. 356-361, Oct. 1943.

[60] Sparks, N. R., *A Note on Rationalized Velocity-Depth Equation:* Geophysics, vol. 7, no. 2, pp. 142-143, Apr. 1942.

[61] Mott-Smith, Morton, *On Seismic Paths and Velocity-Time Relations:* Geophysics, vol. 4, no. 1, pp. 8-23, Jan. 1939.

[62] Rice, R. B., *A Discussion of Steep Dip Computing Methods Part I:* Geophysics, vol. 14, no. 2, pp. 109-122, Apr. 1949, and *Part II,* Geophysics, vol. 15, no. 1, pp. 80-93, Jan. 1950.

[63] Stulken, E. J., *Effects of Ray Curvature Upon Seismic Interpretations:* Geophysics, vol. 10, no. 4, pp. 472-486, Oct. 1945.

are described by Slotnick,[64] Soske,[65] and Jakosky.[66] If the velocity increases exponentially with depth $(V = V_o \exp (aZ)^n)$, the interpreter may use the charts prepared by Slotnick.[67] In certain areas the velocity relationship is best described by the assumption of a linear increase of velocity with time, which becomes a parabolic function $(V = V_o \sqrt{1 + kZ})$ when expressed in terms of depth. Houston [68] has prepared charts that simplify the calculations associated with this velocity function.

Several mechanical devices have been described in the literature enabling the seismologist to plot data that are accurately disposed as to dip; depth, and horizontal offset on the cross section.[69][70] Several slide rules have been proposed for the linear increase of velocity function.[71][72]

To avoid the laborious computation methods involving the curved ray, approximations are often employed in which the actual curved-ray path is replaced by a straight-ray path. The average-velocity-approximation method assumes that the velocity between any reflecting bed and the surface is constant and is equal to the average velocity to the horizon. The paths of the ray are therefore straight, and the depth for the normal ray is simply:

$$d = \tfrac{1}{2}Vt$$

where V is average velocity
t is total corrected reflection time

This method is frequently used in areas of simple geology where correlations are possible; however, it is definitely not recommended when dips are to be determined. The modified straight-path approximation is sometimes applied when dips are of importance. The method is not applicable in regions where the dip exceeds a few degrees, nor where the depth and horizontal offset of the reflection point are critical factors.

We have discussed only the variation of velocity with depth; however, one should not overlook the possible effects of lateral variations of velocity. Stulken [73] shows that in certain areas, such as the San Joaquin Valley of California, the velocity may vary laterally as much as 100 feet per second per mile. (See fig. 579.) Within a distance of some eighteen miles the velocities at a constant reflection-arrival time show a difference of over 2,000 feet per second. A map computed with a single time-depth

[64] Slotnick, M. M., *On Seismic Computations with Applications, I:* Geophysics, vol. 1, no. 1, p. 13, Jan. 1936. *On Seismic Computations with Applications, II:* Geophysics, vol. 1, no. 3, p. 299, Oct. 1936.
[65] Soske, Joshua, *Computing Seismic Reflection Data by a Simple Consistent Method:* Mines Mag., vol. 32, no. 10, pp. 489-495, Oct. 1942.
[66] Jakosky, J. J., *Exploration Geophysics*, pp. 490-498, Los Angeles, Times-Mirror Press, 1940.
[67] Slotnick, M. M., *op. cit.*, I. p. 17; II, p. 302.
[68] Houston, C. E., *Seismic Paths, Assuming a Parabolic Increase of Velocity with Depth:* Geophysics, vol. 4, no. 4, pp. 242-246, Oct. 1939.
[69] Wolf, A., *A Mechanical Device for Computing Seismic Paths:* Geophysics, vol. 7, no. 1, pp. 61-68, Jan. 1942.
[70] Daly, J. W., *An Instrument for Plotting Reflection Data on the Assumption of Linear Increase of Velocity:* Geophysics, vol. 13, no. 2, pp. 153-157, Apr. 1948.
[71] Fillipone, W. R., *Depth—Displacement Slide Rule:* Geophysics, vol. 11, no. 1, pp. 92-95, Jan. 1946.
[72] Mansfield, R. H., *Universal Slide Rule for Linear Velocity vs. Depth Calculations:* Geophysics, vol. 12, no. 4, pp. 557-575, Oct. 1947.
[73] Stulken, E. J., *Seismic Velocities in the Southeastern San Joaquin Valley of California:* Geophysics, vol. 6, no. 4, pp. 327-355, Oct. 1941.

relation would contain a cumulative depth error of over 100 feet per mile. This would cause sufficient distortion to make a major structure appear as a shelf. Several methods of correcting the effects of lateral variations of velocity have been proposed,[74][75] the correction to be applied to the data being read from correction maps created for certain horizons on the basis of well-velocity information. Gaby [76] proposed a method of com-

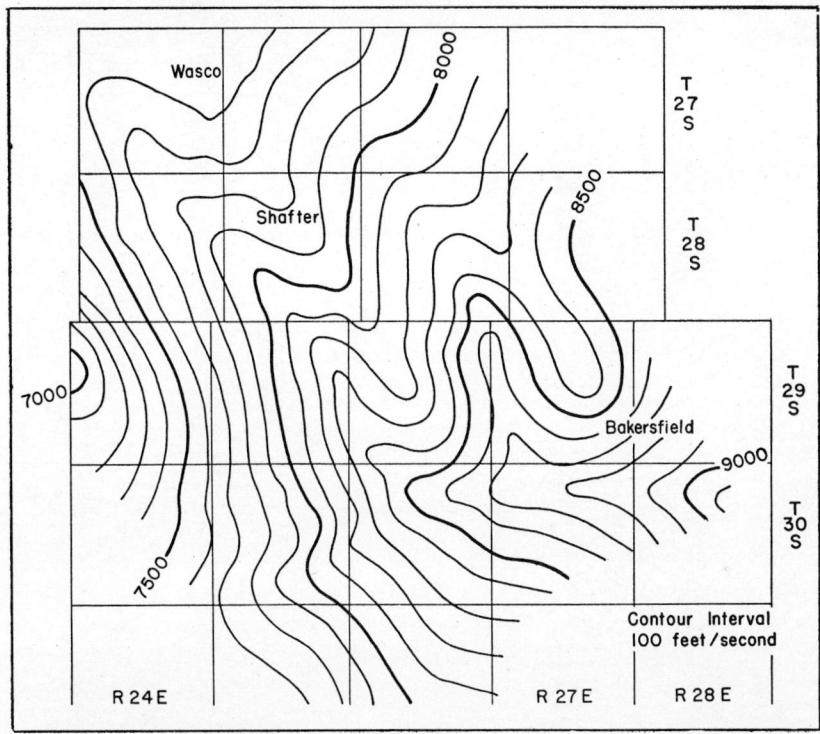

FIGURE 579. Average velocity at a constant depth of 7,500 feet below base of weathering in southeast portion of San Joaquin Valley, California. (Adapted from Stulken. Reproduced permission *Geophysics.)*

putations that enables one conveniently to reassign velocity scales for the express purpose of maintaining agreement between seismic interpretations and velocity revisions. For an excellent discussion of the interpretation of data obtained by well shooting, the reader is referred to the papers by Dix.[77]

[74] Stulken, E. J., *op. cit.*
[75] Navarte, P. E., *On Well Velocity Data and Their Application to Reflection Shooting:* Geophysics, vol. 11, no. 1, pp. 66-81, Jan. 1946.
[76] Gaby, P. P., *A New Type of Seismic Cross Section Wherein Accuracy of Representation is Rendered Insensitive to Velocity Error:* Geophysics, vol. 10, no. 2, pp. 171-185, Apr. 1945.
[77] Dix, C. H., *The Interpretation of Well Shot Data, I:* Geophysics, vol. 4, no. 1, pp. 24-32, Jan. 1939; *II:* Geophysics, vol. 10, no. 2, pp. 160-170, Apr. 1945; *III:* Geophysics, vol. 11, no. 4, pp. 457-461, Oct. 1946.

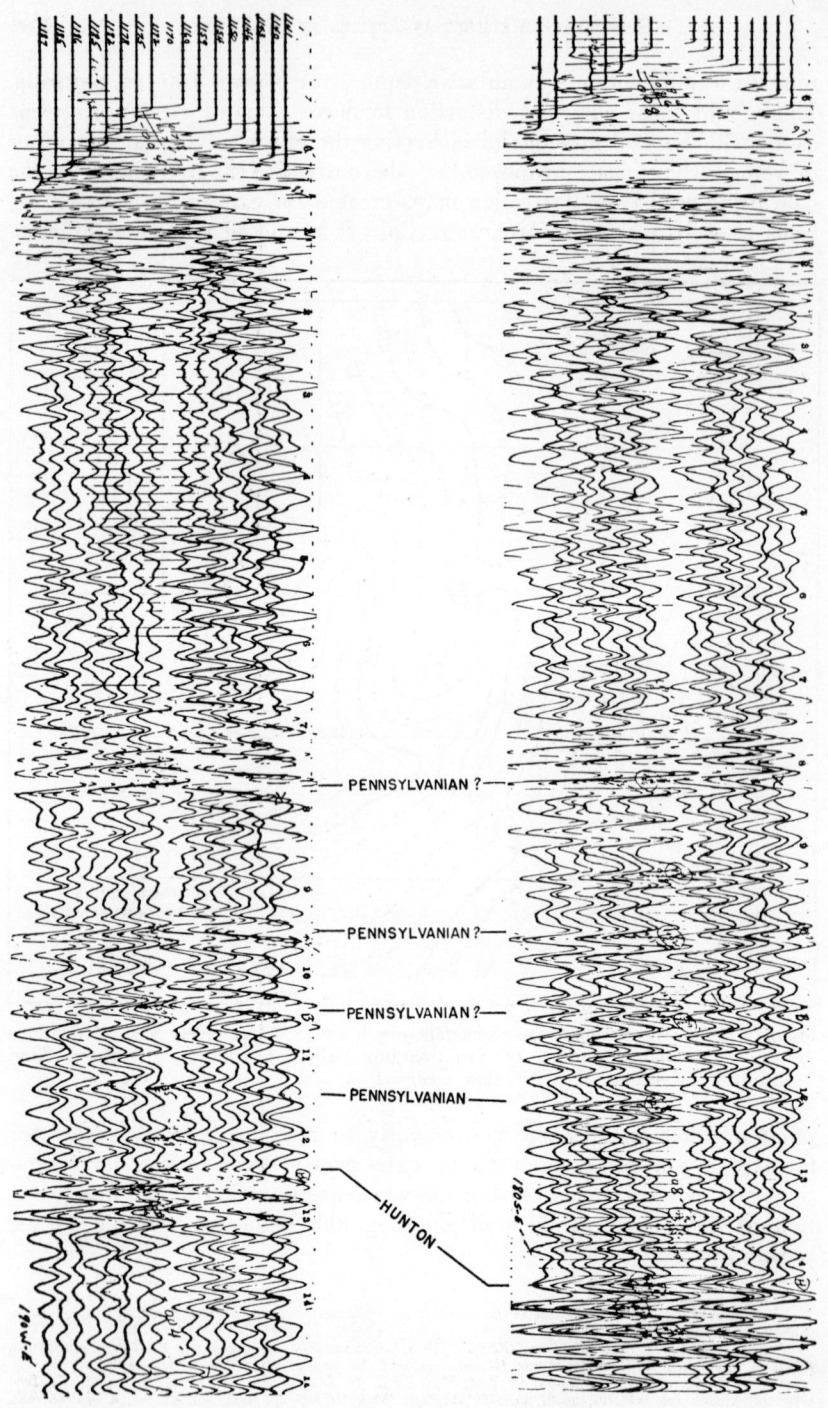

FIGURE 580. Seismic-reflection records in Anadarko Basin, Oklahoma. Shot points 1½ miles apart; fault shown below Pennsylvanian has a throw of approximately 1,000 feet. (Courtesy Magnolia Petroleum Co.)

Correlation shooting with distances of 2,000 to more than 5,000 feet between shot points may be practiced in those areas in which certain horizons yield a similar and distinctive character of reflection that may be readily identified by the interpreter. No effort is made to determine the dip, as the areas in which this technique is applicable are those of low relief. By picking corresponding phases of reflections that possess character and abnormal amplitude, it becomes possible to trace and map reflecting horizons over wide areas. Extreme care must be taken to insure that the proper "legs" or phases of a reflection are picked from record to record. (See fig. 580.) Gaby [78] has described a number of the criteria by which reflections may be correlated between records. Computations are generally simple, as they may be based on a straight-line ray path. Depths may be plotted directly on the contour map, which obviates the necessity for the cross section.

Continuous shooting is indicated in regions where correlations become somewhat uncertain or maximum detail is required. Since only short intervals exist between subsurface control points, reflections may be correlated with greater certainty. By using interlocking or overlapping continuous spreads, the correlations become almost mechanical. The continuous methods are commonly used in regions of steep or variable dip as well as in those areas where persistent reflecting horizons are absent. The more rigorous computation methods involving the curved ray are generally chosen in order to present dip and depth data as accurately as possible. The computed dips and depths of strata are shown on cross sections. True dips are only indicated when the line of traverse is perpendicular to the strike. In those regions where persistent reflecting horizons are absent, subsurface relief may be expressed by use of the phantom horizon. At an arbitrary point on the cross section, a traverse or phantom is drawn by averaging (paralleling) the dips for a reasonable distance on either side of the phantom horizon. By planning the survey so that frequent ties are obtained to previous work, errors that have accumulated in the loop may be adjusted out. Large portions of the Gulf Coast and parts of California have been surveyed by this technique.

Seismic Applications

The application of the seismograph to exploration problems may best be illustrated by reviewing the discovery history of several oil fields. The reader is referred to the volume entitled *Geophysical Case Histories* [79] for a complete discussion.

Cameron Meadows Dome, Cameron Parish, Louisiana—The first geophysical work was done in the area of the Cameron Meadows dome in Cameron Parish, Louisiana, in 1926, using mechanical seismographs. Although the Seismos Company report noted abnormal conditions, the data

[78] Gaby, P. P., *Grading System for Seismic Reflections and Correlations:* Geophysics, vol. 12, no. 4, pp. 590-617, Oct. 1947.
[79] *Geophysical Case Histories*, vol. 1, 671 pp., Tulsa, Soc. Exploration Geophysicists, 1948.

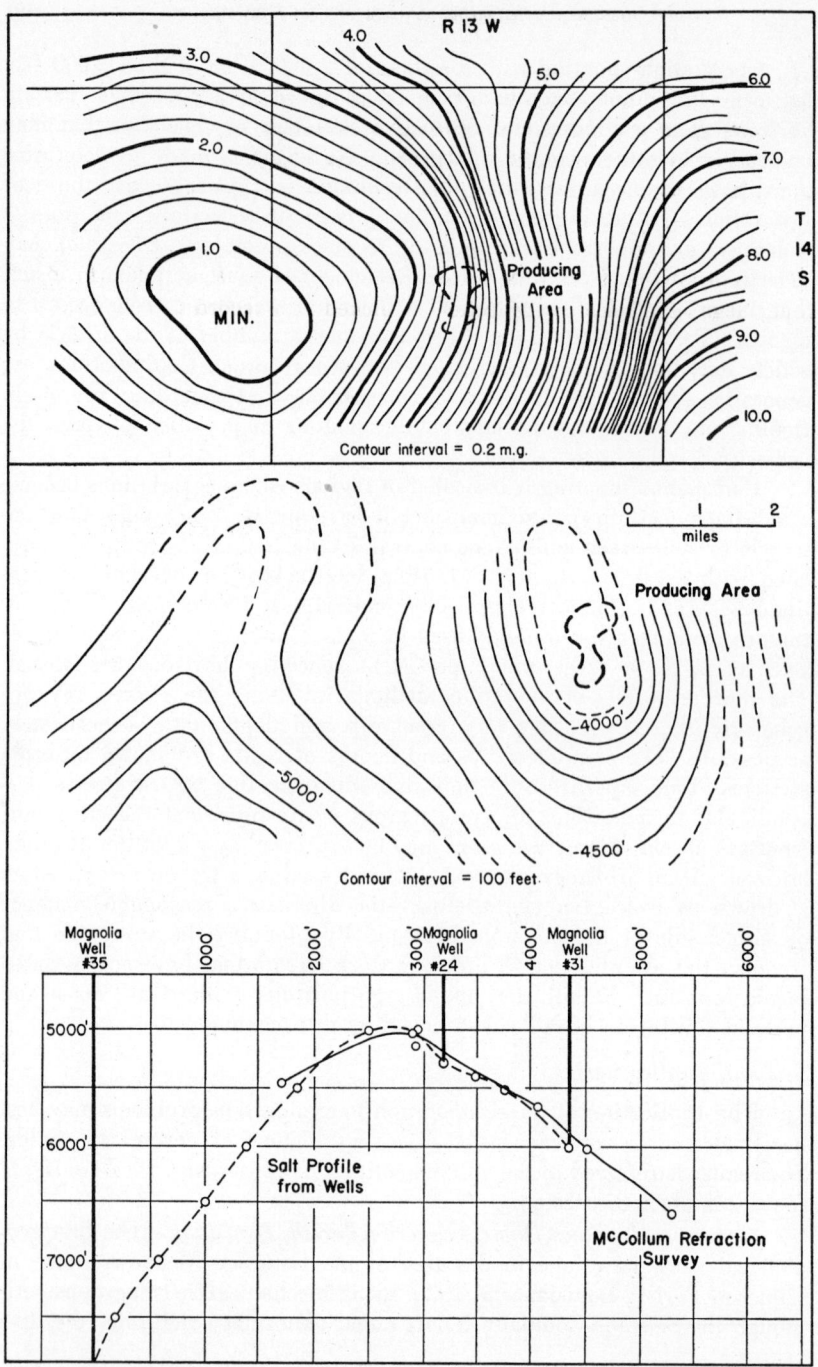

FIGURE 581. Gravity-meter survey, dip-reflection survey, and McCollum refraction profile at Cameron Meadows dome, Louisiana. (After McGuckin. Reproduced permission *Geophysics.)*

were incomplete. Later geophysical work included a torsion-balance survey and several refraction and reflection surveys, which indicated the presence of a piercement-type salt dome. The results of a dip survey made in 1933 after three producing wells had been drilled are shown in figure 581. This survey served to guide further drilling in the area. In order to investigate the possibility of additional structural deformation in the area favorable to the accumulation of oil, a detailed study was made in 1942 using the McCollum profiling technique. This patented method consists in lowering a detector in a well on the dome to a point at or near the salt contact and observing travel times of seismic impulses generated at shot points spaced along radical lines extending outward from the well. A comparison of the salt outline determined by drilling with the results of the refraction survey may be seen in the figure.[80]

. *Lake St. John Field, Concordia and Tensas Parishes, Louisiana*— The Lake St. John field is located on a structure produced by a deep-seated salt uplift. A reconnaissance gravity survey in this region first indicated the presence of a possible structure. As may be seen in figure 582, a closed gravity minimum of considerable extent and moderate intensity lies in close proximity to the producing area. A dip-reflection survey made to check the gravity anomaly showed the possibility of closure in the area. A more detailed seismic survey in 1940 indicated the Lake St. John structure to be a large anticline which trended northwest-southeast. Maps made on several horizons showed increasing closure with depth. Contours near the top of the Lower Cretaceous are shown in the figure. On the basis of this work the first well was drilled to the Wilcox and logged shows of oil. However, it was later found to be in a broad subsurface saddle between two highs. Additional detailed work led to the location of the discovery well. Seventy-eight wells are now producing in the area with 13,000 proved acres.[81]

Odem Area, San Patricio County, Texas—Although eight dry holes had been drilled in the Odem area, San Patricio County, Texas, a reflection-seismograph survey located several structures that subsequent drilling proved to be major oil and gas fields. As the early drilling had shown the area to be regionally high, a reflection-seismograph survey was begun in 1938. Owing to an unfavorable leasing situation in the Odem area proper, the crew was shifted to the Riverside area to the south after the discovery of abnormal dips. The Riverside structure was located after very little work, and subsequent drilling confirmed the seismic structure shown in figure 583. During exploration of the Riverside area, sufficient lease had expired in the Odem area to warrant seismic investigations. Continuous profiles shot across the structure revealed several hundred feet of closure. Owing to the lack of continuous reflecting horizons, it

[80] McGuckin, G. M., *History of the Geophysical Exploration of the Cameron Meadows Dome, Cameron Parish, Louisiana:* Geophysics. vol. 10, no. 1, pp. 1-16, Jan. 1945.

[81] Smith, N. J., and Gulmon, G. W., *Geophysical History, Lake St. John Field, Concordia and Tensas Parishes, Louisiana:* Geophysics, vol. 7, no. 3, pp. 369-383, July 1947.

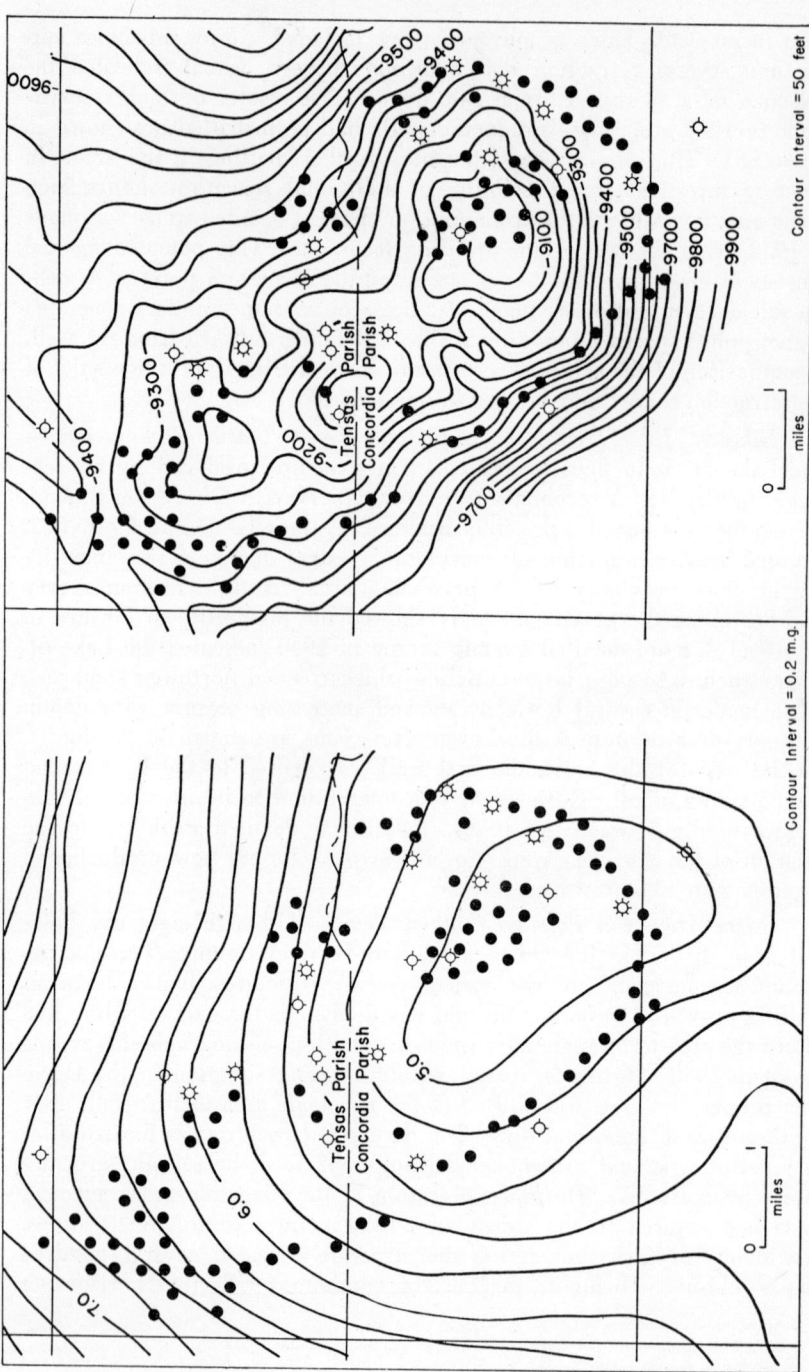

Contour Interval = 0.2 m.g. Contour Interval = 50 feet

FIGURE 582. Gravity map and combination dip-correlation seismograph map on the horizon near top of Lower Cretaceous at Lake St. John field, Louisiana. (After Smith and Gulmon. Reproduced permission *Geophysics.*)

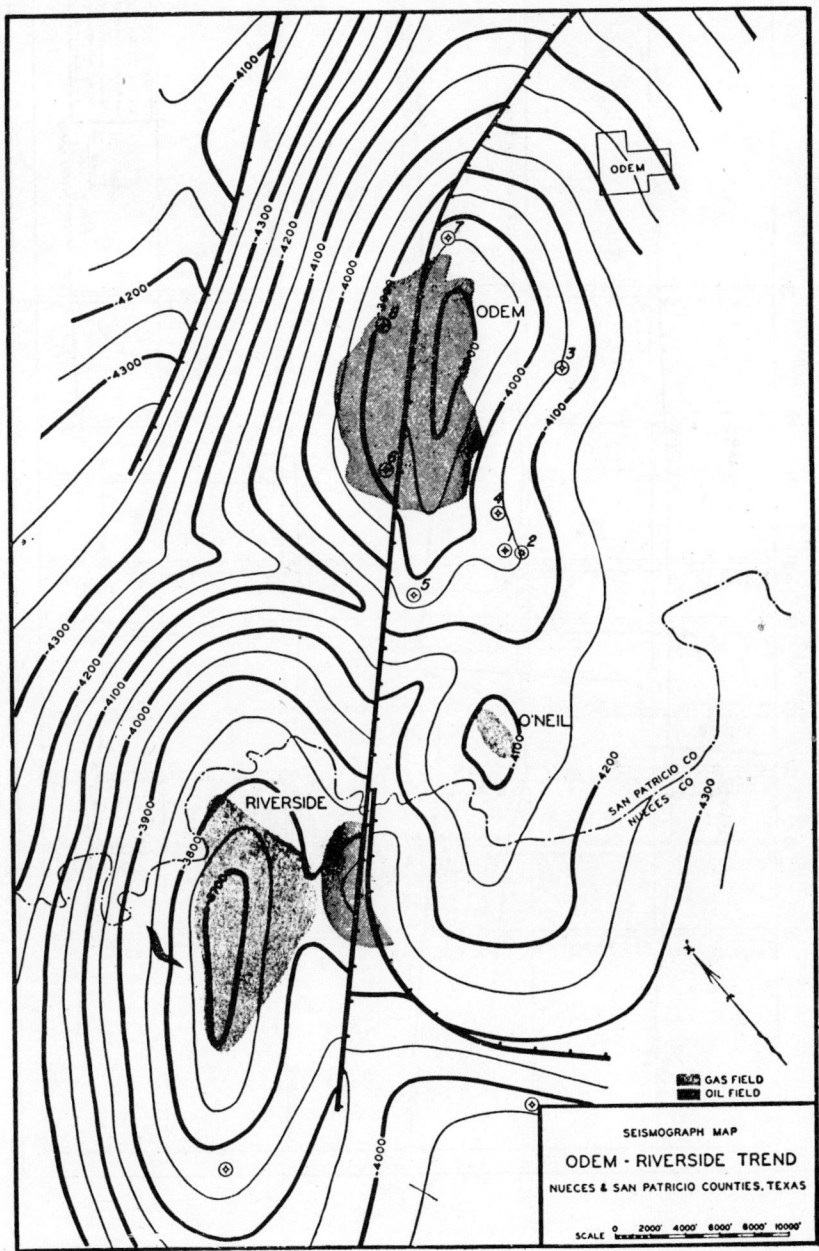

FIGURE 583. Seismograph map of Odem-Riverside area, Texas, showing relationship of seismic closure to proven productive area. (After McCarver and West. Reproduced permission *Geophysics.*)

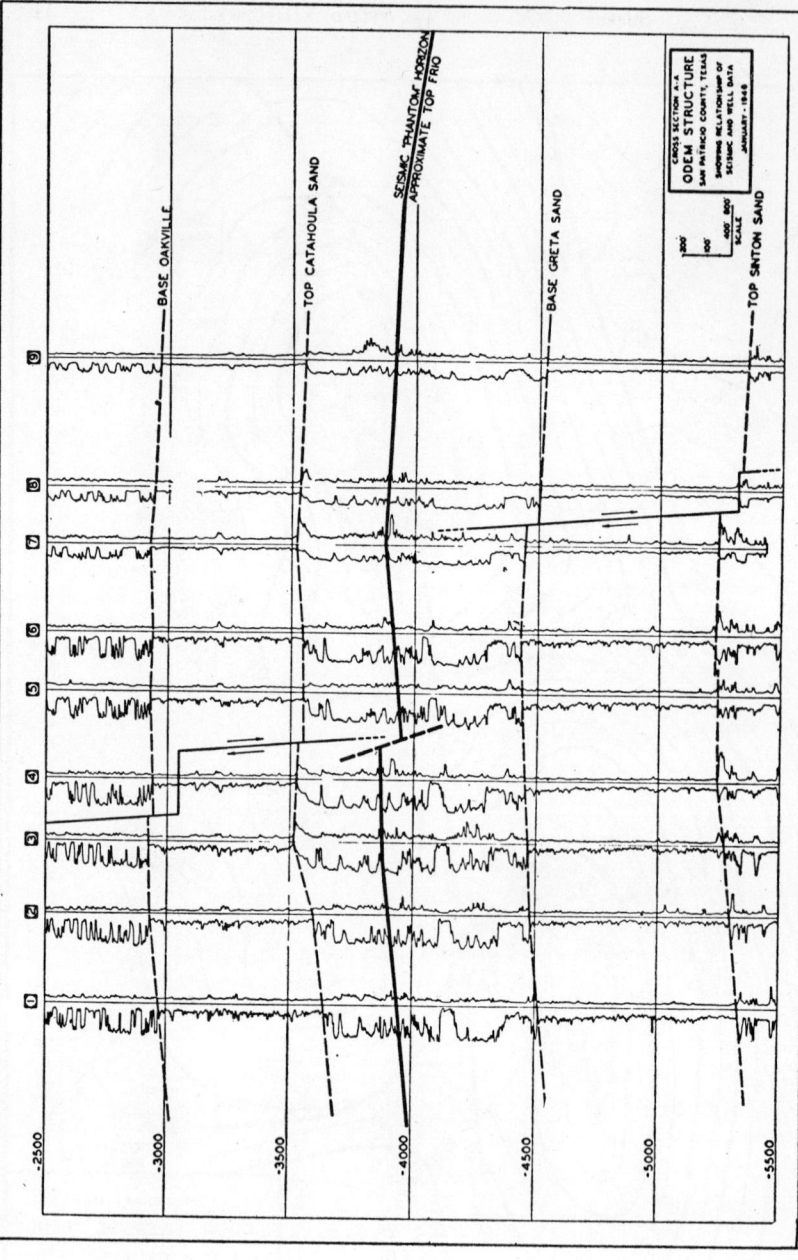

FIGURE 584. Subsurface cross section across Odem structure comparing seismic-reflection horizon with well data. (After

was necessary to contour phantom horizons. Figure 584 is a combined seismic and subsurface cross section, which compares the phantom horizon with well data. This seismic program discovered the Odem and East Riverside oil fields and the Riverside and O'Neil gas fields.[82]

La Gloria Area, Jim Wells and Brooks Counties, Texas—Favorable torsion-balance results in the La Gloria area of Texas initiated a correlation seismograph survey in 1936. Two years later the area was again shot; this time the continuous dip-profiling technique was used. The dips indicated by the reflections were plotted on cross sections, a phantom hor-

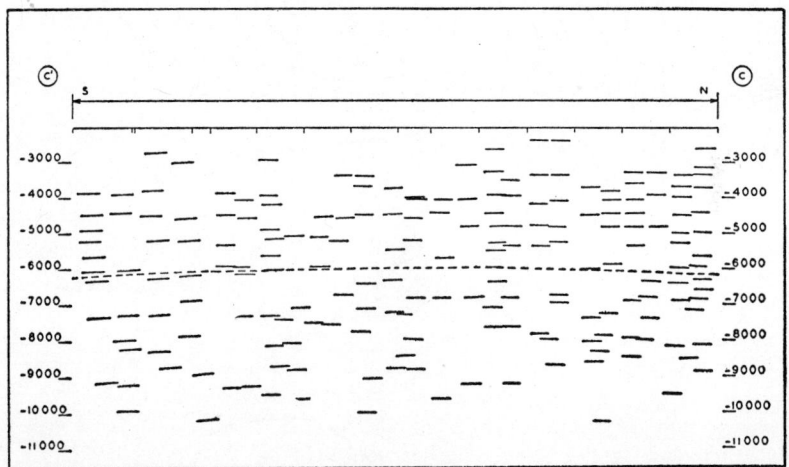

FIGURE 585. Reflection-seismic-dip cross section across La Gloria field, Texas. Phantom horizon shown by dotted line. (After Wooley. Reproduced permission *Geophysics.)*

izon was drawn on each section, and the traverses were closed and adjusted. Figure 585 is a typical dip section across the La Gloria field. A contour map prepared on the basis of dip data showed about 250 feet of closure. (See fig. 586.) After the completion of the dip survey, a continuous correlation survey was made over the closed structure for comparison with the dip map. The results of this survey agreed closely with the dip map, as may be seen in the figure. The discovery well was located from these maps and now produces from sands in the Frio formation of Oligocene age.[83]

Tucupita Oil Field, Venezuela—The Tucupita field is located in eastern Venezuela and was discovered on seismograph work done during the years 1939 to 1941. Continuous spreads were used with jump correlations being made across gaps caused by faulting and zones of no reflections. Good correlation was obtained, even though the section contained

---

[82] McCarver, H., and West, L. G., *The Geology and Geophysics of the Odem Oil Field, San Patricio County, Texas:* Geophysics, vol. 12, no. 1, pp. 13-29, Jan. 1947.
[83] Wooley, W. C., *op. cit.*

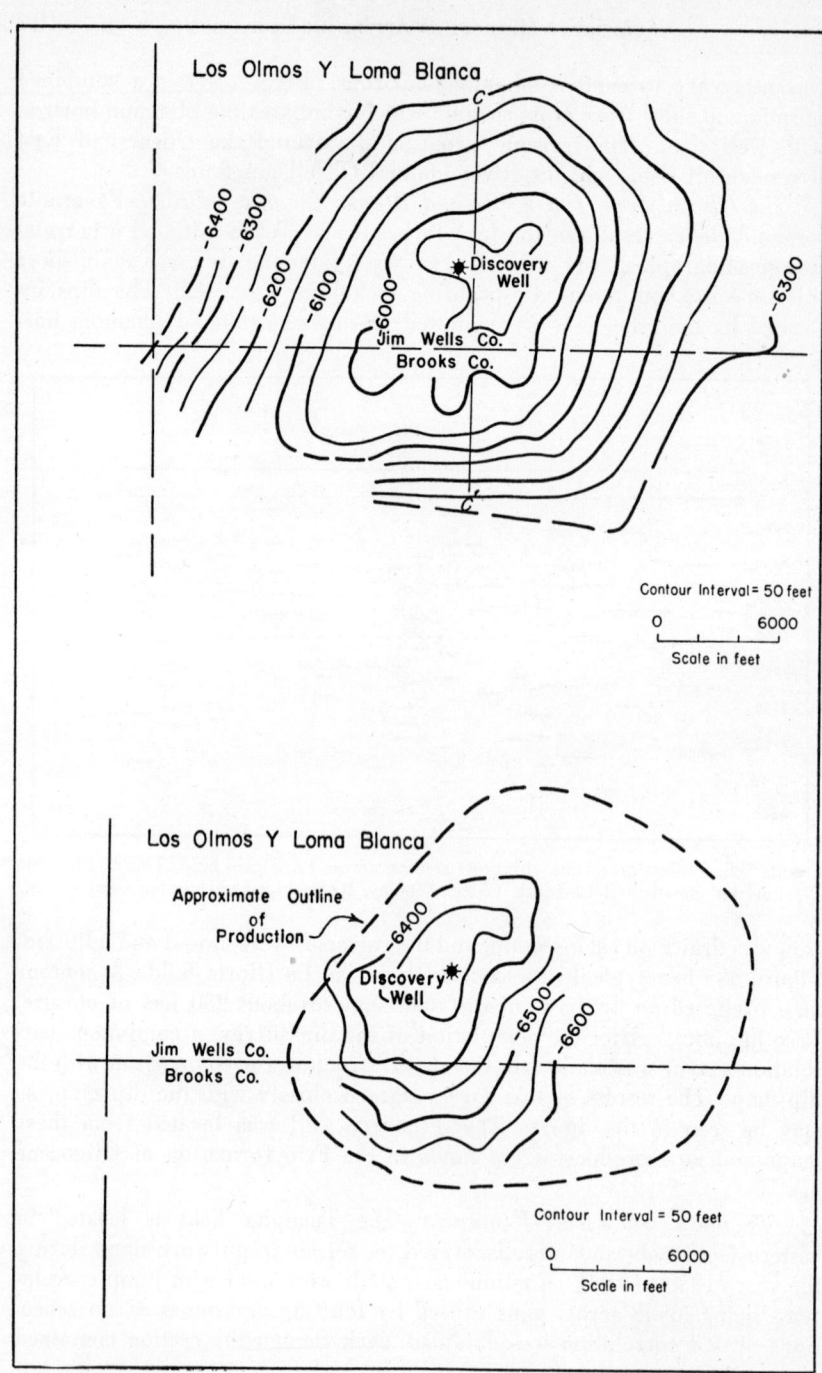

FIGURE 586. Dip-reflection map on −5,950-foot phantom horizon at top. Correlation reflection-seismic map on −6,350-foot horizon below. La Gloria field, Texas. (After Wooley. Reproduced permission *Geophysics*.)

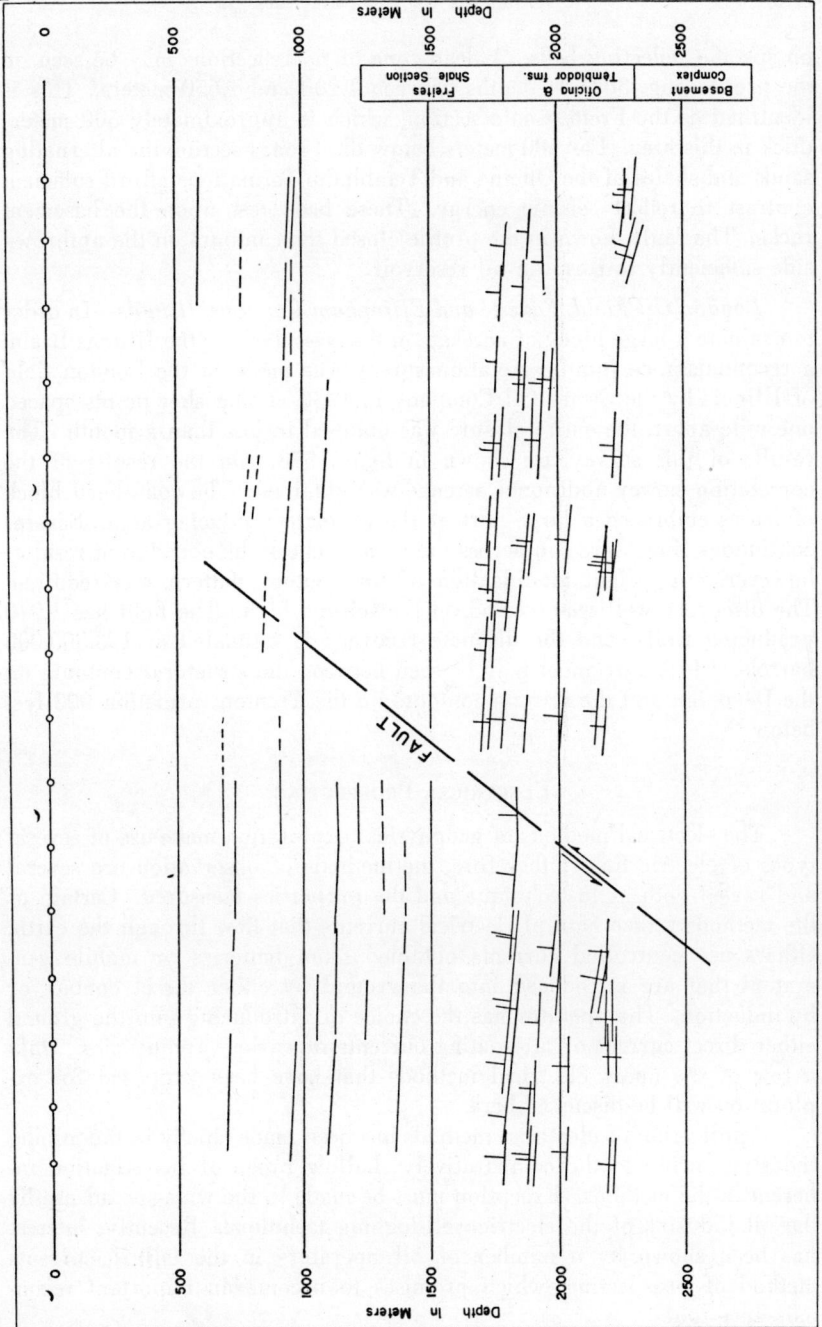

FIGURE 587. Cross section from reflection-seismograph survey at Tucupita field, Venezuela. (Adapted from Olson. Reproduced permission *Geophysics*.)

no specific reflecting beds. A dead zone of no reflections may be seen on the profile (fig. 587) at depths between 1,200 and 1,650 meters. This is identified as the Freites-shale section, which is approximately 500 meters thick in this area. For 500 meters below the Freites section the alternating sands and shales of the Oficina and Temblador formations afford sufficient contrast to reflect seismic energy. These beds rest upon the basement rocks. The fault shown in the profile closed the contours on the upthrown side sufficiently to form an oil reservoir.[84]

Loudon Oil Field, Fayette and Effingham Counties, Illinois—In order to evaluate a large block of acreage on the west flank of the Illinois Basin, a reconnaissance jump-correlation survey was made in the Loudon field of Illinois by the Carter Oil Company in 1936. Using shot points spaced one mile apart, the entire closure was mapped in less than a month. The results of this survey are shown in figure 588. On the results of the correlation survey additional acreage was obtained. The completed block of leases embraced a large part of the ultimate producing area. Several continuous lines were run across the area to check the correlation results; however, only slight modification of the contour pattern was required. The discovery well was located on the seismic high. The field has 1,940 producing wells, and the ultimate recovery is estimated at 193,500,000 barrels. Close agreement may be seen between the structural contours on the Devonian and the seismic contours on the Trenton formation 900 feet below.[85]

ELECTRICAL PROSPECTING

The electrical methods of geophysical exploration make use of several types of electric fields; therefore, the methods of observation are several and varied both as to technique and the properties measured. Certain of the methods utilize natural electrical currents that flow through the earth. Others use controlled currents obtained from batteries or mobile generators that are introduced into the ground by either direct contact or by induction. The operator has the choice of introducing into the ground either direct current or alternating currents of various frequencies. Only a few of the many electrical methods that have been proposed for exploration will be discussed here.

Application of electrical methods has been made chiefly in the mining industry, owing to the comparatively shallow range of investigation inherent in the methods. Exception must be made to the wide-spread use by the oil industry of the electric-well-logging technique. Recently, interest has been shown by a number of oil operators in the telluric-currents method of prospecting, which promises to become an important reconnaissance tool.

---

[84] Olson, W. S., *op. cit.*, pp. 611-618.
[85] Lyons, P. L., *Geophysical Case Histories*, vol. 1, pp. 461-470, Tulsa, Soc. Exploration Geophysicists, 1948.

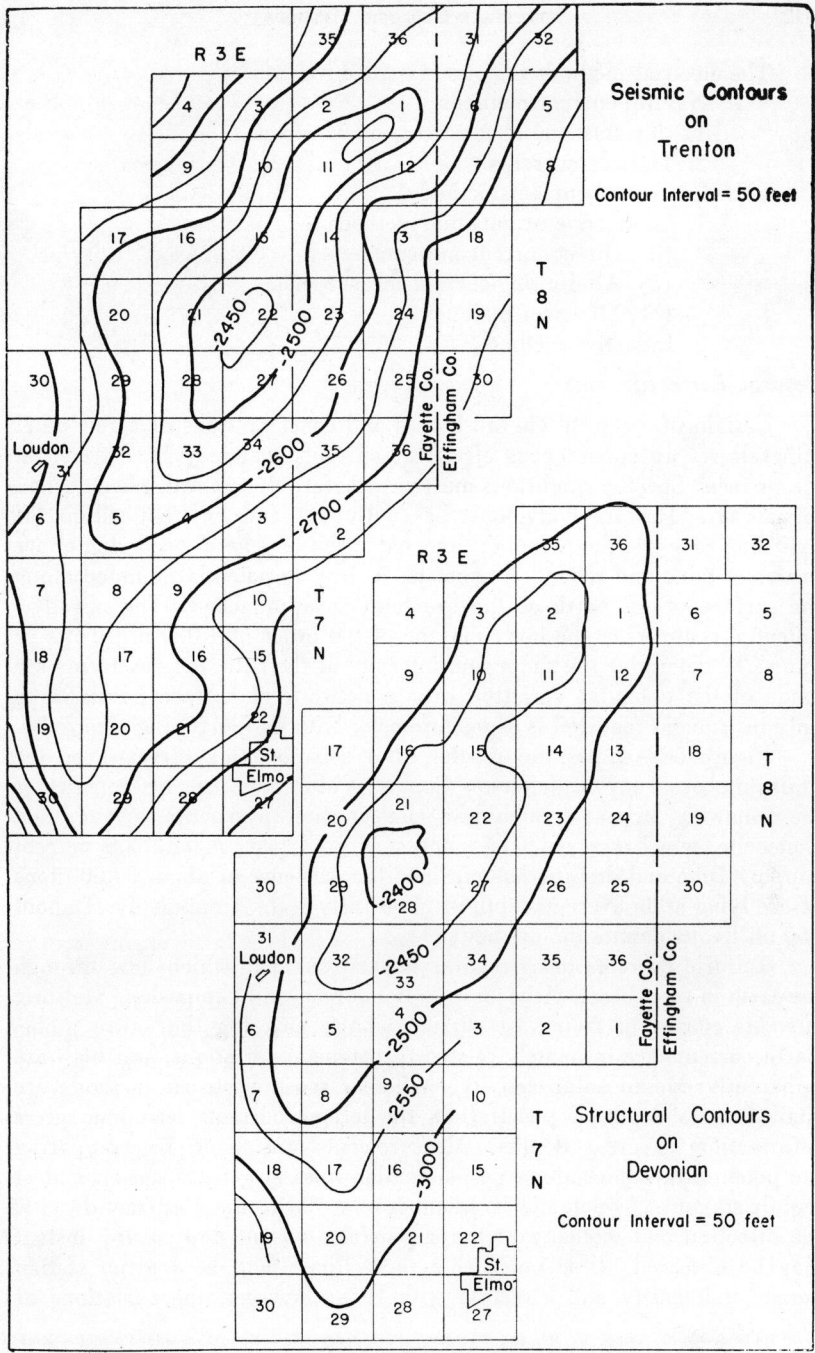

FIGURE 588. Trenton structure by seismograph work prior to drilling compared with subsurface contours on Devonian from well data, London field, Illinois. (After Lyons. Reproduced permission *Geophysics.*)

The electrial methods may be classified as:
I. Natural-current methods
    1. Currents due to electrochemical effects
    2. Telluric currents
II. Controlled-current methods
    1. Conductive or galvanic methods
       (*a*) Direct-current measurements
       (*b*) Alternating-current measurements
       (*c*) Transient methods
    2. Inductive methods.

Natural Earth Currents

Certain of the sulphide ore bodies and concentrations of a few other minerals set up spontaneous electrical currents that may be detected at the surface. Specific conditions must exist before this natural phenomenon may occur. The ore body must be continuous and lie both within the oxidizing zone and the reducing zone, oxidation in general occurring above the water table and reduction beneath. A drop in potential is noted along the surface of the earth as the ore body is approached. The negative-potential center over the body may be of the order of 1,000 millivolts or more. To determine the shape and intensity of the field, measurements are made of the potential variation over a network of surface points. The only instrument required is a potentiometer with nonpolarizing electrodes.

Figure 589 shows the results of a self-potential survey over the Malachite ore body in Jefferson County, Colorado. The ore consists of the following primary sulphides: chalcopyrite, pyrrhotite, pyrite, and chalcocite, which assayed 0.05 ounces of gold per ton and 3.5 percent copper. Diamond drilling has outlined the existence of about 34,000 tons of ore lying at an average depth of about fifty feet. An anomaly of about 160 millivolts defines the ore body.[86]

Telluric currents are irregular natural currents which flow through the earth in vast sheets. According to Boissonnas and Leonardon,[87] telluric currents consist of four vast current whorls covering the entire globe. Earth currents are intimately related to magnetic variations, and both are apparently due to solar activity. Whereas most electrical methods are limited as to depth of penetration, the telluric-currents technique gives information to great depths. Measurements are made by comparing the potential gradient along perpendicular lines at a fixed station and at mobile stations. As potential gradient is related directly to current density, the direction and intensity of the subsurface current flow at any instant may be observed. It is noted that the telluric field at a given station varies in intensity and direction with time; however, the variations of

---

[86] Heiland, C. A., Tripp, R. M., and Wantland, Dart, *Geophysical Surveys at the Malachite Mine, Jefferson County, Colorado:* Am. Inst. Min. Met. Eng. Tech. Pub. 1947, pp. 2-4, Feb. 1946.

[87] Boissonnas, E., and Leonardon, E. G., *Geophysical Exploration by Telluric Currents, with Special Reference to a Survey of the Haynesville Salt Dome, Wood County, Texas:* Geophysics, vol. 13, no. 3, p. 389, July 1948.

potential at points many miles apart are very similar in form. The pref-
erential direction of the currents is determined along with the ratio of
the intensity at the base station to the intensities at the mobile stations.
The magnitude and direction of telluric currents is governed partly by
variations in the electrical resistance of the earth's crust; they avoid
rocks of high resistivity and tend to flow over or around them, thus

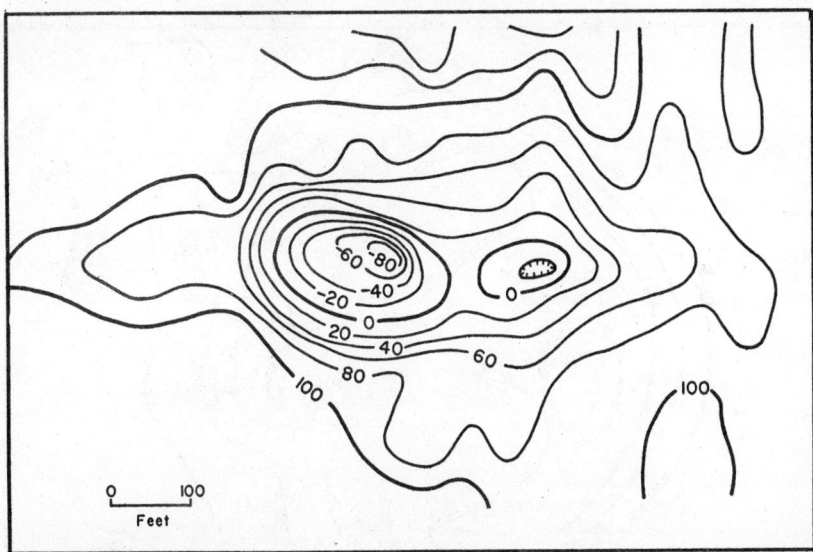

FIGURE 589. Self-potential anomaly at Malachite mine, Jefferson County, Colorado.
(After Heiland, Tripp, and Wantland. Reproduced permission Am. Inst. Min.
Met. Eng.)

varying the surface-potential gradient. Amplitude ratios are known to be
high over uplifts and low over synclines and basins.

An experimental survey was conducted over the Haynesville salt
dome in Texas. (See fig. 590.) Salt, a nonconductor of electricity, deflects
the current over and around the salt plug, causing a high current density
over the plug. The same effect would exist over basement uplifts. Ratios
of current density of as great as one to five may be noted on the Haynesville map. Remarkable correspondence between the outline of the salt
plug and the contours of equal magnitude of the telluric field is shown.

Controlled Currents

In the several controlled-current methods the ground is energized
either by means of electrodes inserted into the ground (galvanic methods),
or the current may flow in the ground by electromagnetic induction from
alternating currents flowing in lines, loops, or coils at or above the surface
of the ground (inductive methods). The power may be provided either
by batteries or by mobile generators.

When electrical energy is supplied to homogeneous and isotropic ground at two points, the current flow lines and the equipotential lines (at right angles to current lines) are symmetrically disposed about the electrodes. The current flow lines and consequently the equipotential lines are disturbed by the presence of either unusually low- or high-resistivity bodies. In the equipotential-line method the paths of the lines of equal

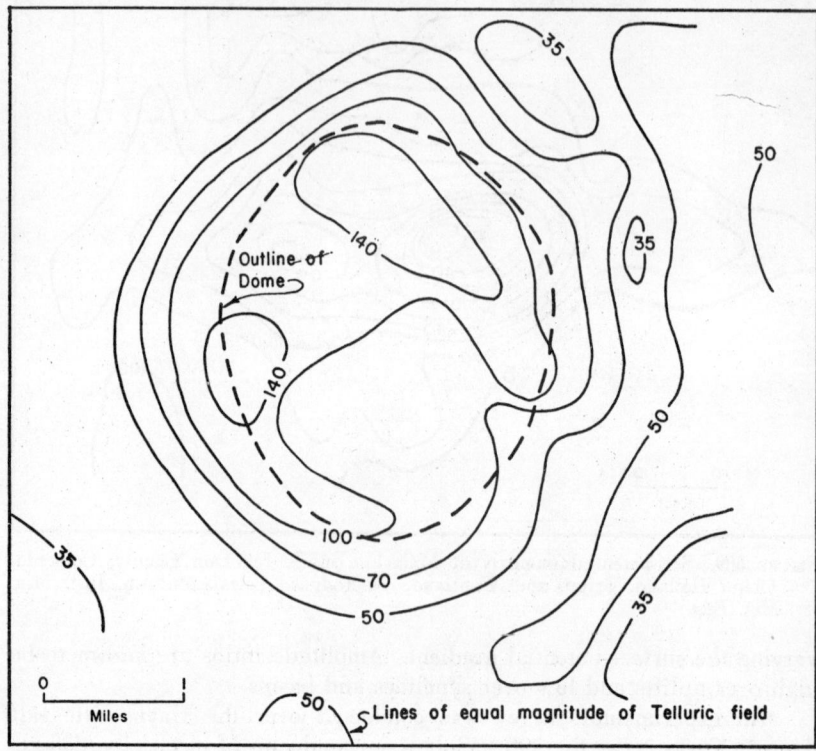

FIGURE 590. Results of telluric-current exploration at Haynesville, Texas, salt dome. (After Boissonnas and Leonardon. Reproduced permission *Geophysics.*)

potential are traced by probing the ground and locating points of the same potential. The indicating instrument may be either a galvanometer when direct current is used or an amplifier and head set when audio-frequency alternating currents are used. By noting areas in which the equipotential lines are abnormally disposed, predictions may be made as to the presence of conducting or nonconducting bodies. The Russian and Swedish geophysicists have had considerable success with this technique.

If current measurments are made in the power-electrode circuit and potential measurements made at two other points, it becomes possible to calculate the apparent resistivity of the ground for any electrode arrangement. Measurements may be made with either direct current and non-polarizable electrodes or with commutated direct current and metal electrodes. One common electrode system is the Wenner configuration in which the four electrodes are equally spaced at intervals a along a straight line. For this particular arrangement the ground resistivity (apparent) may be calculated from the following relationship: $\rho = \dfrac{V}{I} 2\pi a$. The resistivity obtained by such surface measurements applies to a volume of ground that depends on the electrode spacing. The greater the spacing between electrodes the greater the depth of penetration. As a rule of thumb, the depth of penetration is approximately equal to the spacing a in the Wenner-Gish-Rooney·arrangement. As the spacing increases, apparent resistivities are obtained from deeper and deeper strata. A plot of apparent resistivity versus electrode separation will thus show different values as the current reaches beds of different resistivities. Interpretation

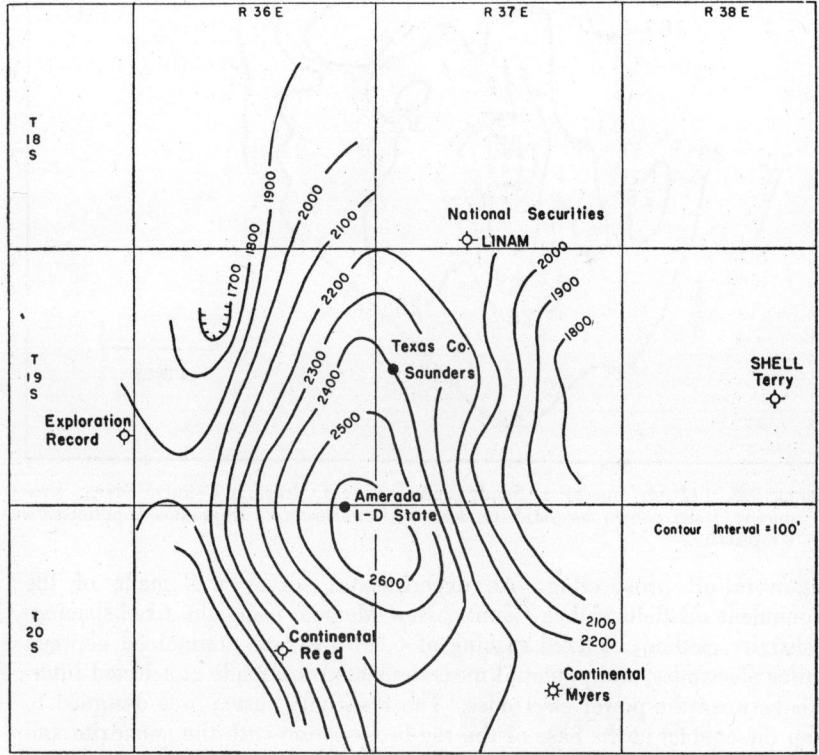

Figure 591. Resistivity map on base of red beds corrected for lateral changes in conductivity, Monument oil field, New Mexico. (After England. (Reproduced permission *Geophysics.)*

of such apparent-resistivity curves gives valuable depth information. This method is particularly well suited to location of the water table.

By maintaining a fixed electrode separation and taking observations along a traverse line, a resistivity map may be constructed. This method is applicable to a number of mining problems such as the location of ore bodies, faults, and vein extensions. Figure 591 illustrates an appli-

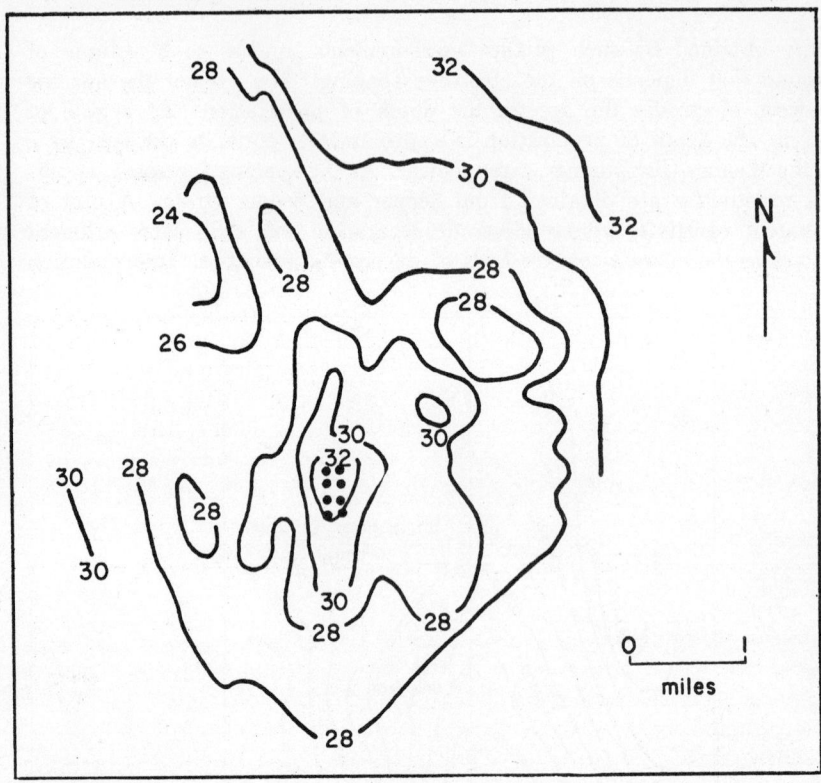

FIGURE 592. "Eltran" survey at Sandy Point oil field, Brazoria County, Texas. Producing wells shown as solid circles. (After Rosaire. Reproduced permission *Geophysics.*)

cation to oil prospecting. An experimental survey was made of the Monument oil field in Lea County, New Mexico, using the fixed-spacing-resistivity method. A fixed spacing of 6,000 feet was maintained between power electrodes, and potential measurements were made at selected intervals between the power electrodes. The resistivity survey was designed to map the contact at the base of the red-beds section with the anhydrite and salt lying immediately beneath. The red beds are relatively good conductors of electricity while the anhydrite and salt are almost insulators. By using existing well data and applying corrections for the lateral variation of resistivity in the red-beds section, a contour map on the base of the red beds was obtained.[88]

A modification of the resistivity method is the potential-drop-ratio technique, in which the potential differences are not measured absolutely as in the resistivity method but in the form of a ratio of successive differences. Sharper indications are obtained on vertical-formation boundaries, and in favorable sedimentary sections the determination of horizontal stratigraphic boundaries is more readily accomplished. A description of the instrument, field procedure, and interpretation of results is presented by Heiland.[89] This method is being used at present by Pemex to locate anomalous areas associated with large faults that extend to the surface. Rummerfield[90] reports that "Considerable success has been attained in making well locations by this method."

By using a direct current interrupted at ten-second intervals, Karcher and McDermott[91] proposed a method of deep electrical prospecting known as the "eltran technique." By measuring the form of the resulting potential when a sharp current pulse is introduced into the ground, deductions may be made as to the electrical properties of the subsurface. According to Nettleton[92] there is considerable controversy as to the real value of the electrical-transient methods. It appears that the method has no greater depth range than the conventional resistivity methods, and results obtained by the method should correspond with those observed by conventional resistivity techniques. An "eltran" survey over the Sandy Point field in Brazoria County, Texas, is shown in figure 592. About fifty feet of subsurface closure is recognized in the producing area.

CONCLUSION

Since geologic structure must be inferred from the results of geophysical surveys, it behooves the interpreter to familiarize himself with all geologic information concerning the area in which he is working. It is sometimes suggested in the literature that geophysical interpretations should be made on a mathematical and physical basis only without recourse to geologic data. However, it must be remembered that mathematical analyses require certain assumptions concerning the disposition of physical properties of the subsurface materials. It remains for the interpreter to make only assumptions that are geologically feasible in order to lead to a particular solution which is probable in that area. Future oil fields will be found through the combined efforts of the geologist, paleontologist, and geophysicist working together in a spirit of cooperation.

[88] England, C. M., *A Resistivity Survey of the Monument Oil Field:* Geophysics, vol. 8, no. 1, p. 20, Jan. 1943.
[89] Heiland, C. A., *Geophysical Exploration*, pp. 744-757, New York, Prentice Hall, 1940.
[90] Rummerfield, B. F., *Oil Exploration in Mexico:* Mines Mag., vol. 38, no. 12, p. 35, Dec. 1948.
[91] Karcher, J. C., and McDermott, Eugene, *Deep Electrical Prospecting:* Am. Assoc. Petroleum Geologists Bull., vol. 19, no. 1, pp. 64-77, Jan. 1935.
[92] Nettleton, L. L., *Geophysical Prospecting for Oil*, p. 374, New York, McGraw-Hill Book Co., Inc., 1940.

Questions

1. May the "divining rod" which has been used in prospecting for water, petroleum, and minerals be classified as a geophysical method? Discuss.
2. Which general class of geophysical methods lacks the ability to render unique geological solutions? Discuss reasons for the ambiguity. In which general category would geothermal measurements be classified?
3. List several economic minerals which may be located directly by geophysical methods and indicate the proper instrument for each mineral.
4. Outline a complete exploration program for a remote region in Brazil where little is known of the regional geology.
5. Show distortion of a magnetic field by the insertion of paramagnetic, nonmagnetic, and diamagnetic bodies.
6. By reference to the typical daily diurnal curves in figure 549, plan a magnetic survey using only one field instrument so that maximum error in correcting diurnal variation by assuming linear diurnal variation will always be less than five gammas. Although base checks should be made at least every two hours, large errors may occur at certain portions of the day unless the survey is carefully planned.
7. By inspection of figure 549, indicate source of error in using published daily magnetic variation curves issued by government observatories at Tucson, Arizona, and Cheltenham, Maryland.
8. For a Schmidt type compensated magnetic system operating in a vertical field of 0.5 oersteds, calculate the pole strength of the magnets (assume both blades equal) if the system is to be in equilibrium. Assume that the gravitational force is equal to 980 dynes and is acting on a mass of 5 grams at center of gravity which is located one centimeter from axis of rotation. Distance between magnetic poles is seven centimeters.
9. Calculate the force of attraction in dynes between two spheres separated a distance of 100 centimeters between centers if the spheres consist of lead (mass = 50 grams, density = 5.6) and aluminum (mass = 100 grams, density = 2.4).
10. List factors which influence the force of gravity on the earth.
11. Using depth rules for a sphere, calculate depth to the top of the disturbing mass in figure 565.
12. Using depth rules for the horizontal cylinder, calculate depth to top of disturbing mass in figure 571.
13. Calculate the critical angle i_c if velocity of V_1 is 2000 feet per second and V_2 is 6000 feet per second. See figure 576.
14. Discuss differences between correlation and dip methods of seismic

prospecting. Suggest regions where correlation shooting will be successful.

15. Discuss differences between refraction and reflection methods of prospecting. List advantages of the reflection method.

16. What is meant by the "weathered layer"?

17. Discuss relationship between the geologic history of an area and the assumptions to be made concerning dispositions of the iso-velocity surfaces.

18. Calculate throw of the Hunton Lime in figure 580 if average velocity function from datum is $7,000 + 0.7z$ feet per second, and if corrected reflection time to the Hunton for shot-point 180 S-E is 1.422 seconds and for shot-point 190 W-E is 1.233 seconds. Assume vertical reflections.

19. Sketch equi-potential lines between a source and sink of current for homogeneous ground. Insert a conductive body between power electrodes and again sketch distribution of equi-potential lines.

20. Indicate applications of resistivity measurements to problems of the mining and petroleum industries, and in civil engineering work.

CHAPTER 15

GEOLOGIC TECHNIQUES IN CIVIL ENGINEERING

GEORGE L. ROBB

Geology in its application to civil engineering is primarily concerned with the rocks, soils, and ground water that make up the surface and shallow subsurface of the earth's crust. The civil engineer builds his structures on rock and soil foundations and uses the soil and rock as construction materials. He adapts his structures to existing terrain and topography, surface and underground water, geologic structures, lithology, and stratigraphy at the construction site. All combine to create the problems that he must surmount if his structure is to be efficient, permanent, appropriate, and economical. On this common ground, that is, topography and geology, the engineer and the geologist must meet. Engineering structures are custom-made, and their efficiency and economy depend largely on how well they have been adapted to the peculiarities of the site and the valuable construction materials.

Engineering geology is thus born of the engineer's need that someone interpret these geologic conditions expertly—not abstractly but in terms of their engineering significance—not in abstruse geologic terms but in words that the engineer can understand and apply specifically to given problems. The engineering geologist must interpret the conditions of the crust with such clarity and discernment that the engineer can design and build the kind of structure that is most appropriate to and compatible with the natural conditions of the site.

A summary of the functions and responsibilities of the engineering geologist and his relationship to the engineer can be summed up in the following paragraphs by Rhoades: [1]

In essence, it is the function of the engineering geologist to interpret the character of structure sites and prospective natural construction materials, thus supplying information essential to the engineer and fulfilling his function of developing plans and specifications most effectively, reconciling the engineering objectives with the natural conditions. The functions are inseparable.

It is the engineer's responsibility to define what kind of information he needs concerning materials and surface and subsurface conditions. It is the engineering geologist's responsibility to obtain and interpret that information. The burden of geologic interpretation rests with the geologist. The burden of engineering interpretation and application rests with the engineer. The geologist must assimilate the data and present conclusions and recommendations to the engineer in a concise, practical form. Such conclusions are of value to the engineer only to the extent that they have recognized all the pertinent

---

[1] Rhoades, Roger, *Geology in Civil Engineering:* Address before Second Pan-American Congress of Mining Engineering and Geology, Rio de Janeiro, Brazil, Oct. 1946.

factors of the geological situation and have interpreted their meaning in the light of the engineering problems.

The first step in any problem of engineering geology is to elucidate all the geologic conditions that may pertain to the engineering problem. The engineering geologist, like any other geologist, studies faults, folds, joints, stratigraphy, petrography, geomorphology, and ground water. But the elucidation of the geologic conditions is only the first step for the engineering geologist. The next step is to interpret the significance of these conditions in terms that are intelligible and useful to the engineer. The objective of engineering geology is to translate the geologic conditions into answers to such questions as these: Will this canal leak, and how much? Will this rock dissolve or break down through the agencies of weathering, and how rapidly? Will this material stand on steep slopes, and how steep can they be? Can this material be used in construction, and how can it be used to best advantage? Will this foundation settle, how much, and how rapidly? How will these soils react to a pile foundation?

The writer wishes to thank the members of the geology section of the United States Bureau of Reclamation for their assistance and comments on various parts of this paper. Roger Rhoades, assistant head, Research and Geology Division, Bureau of Reclamation, has offered valuable criticism during the preparation of this chapter.

General Geologic Techniques

The area to be investigated for most construction sites is frequently rather restricted. The engineering geologic study of a site may best be considered as two separate phases. The first phase should be a study of the geologic conditions in the immediate construction area, more commonly called "site geology." The second phase should be a consideration of the geologic conditions on a regional scale, from which conclusions may be drawn to answer or explain the geologic conditions at the construction site.

One of the first responsibilities of an engineering geologist should be the preparation of a geologic surface map. This geologic map should include all available information on the attitude of the exposed formations, structural relationships, stratigraphy, and lithology, as well as topographic conditions. Information on the geologic map should be further amplified by means of geologic cross sections or stratigraphic columns so that the subsurface conditions in the area concerned may be clearly defined. Where bedrock exposures at the construction site are meager, it may be necessary to make interpretations based on observations of geologic conditions some distance from the site. The farther the distance from the actual site, the less valuable the subsurface geologic predictions become. An important purpose of the regional geologic study is to determine the uniformity or variability of a particular stratum or series of strata throughout the area. By knowing the foregoing, exploration programs can be designed to obtain the maximum amount of subsurface geologic information with a minimum amount of exploration.

Engineering geology is concerned not only with the technical geologic description of the material at a particular site but with the physical properties of the material as well. The designation of a geologic formation as sandstone, shale, or some other type of lithology is not sufficient for engineering purposes. It is important to know the load-carrying capacity in pounds per square foot, the coefficient of friction, cohesion values, the modulus of elasticity, and the permeability. To obtain this information on physical properties, the engineering geologist must make use of other chemical and physical sciences. In this way he can obtain actual laboratory data on the particular materials in question in order to obtain the answer to these pertinent problems.

All of the more common geologic techniques are employed in engineering geology. The lithology, stratigraphy, geomorphology, structure, and ground water are all integral parts of an engineering geologic report. A complete understanding of the basic geologic information must be the basis for a sound interpretation of the engineering implications of a particular site.

In listing the requirements of the engineering geologist, Berkey [2] has said:

He must have the principles of geology so well in hand and feel so sure of them in their application to the actual ground as it is that he is not in the least disturbed at finding everything of a geologic nature belonging to a particular project materially different from anything he has ever seen.

Lithology

The lithologic description of a geologic formation should contain all pertinent data regarding the grain size, the mineralogy of the individual grains, the type and amount of cementation material, as well as a summary term such as "shale" or "sandstone," to complete the description. The actual formation names as given in geologic literature are of minor importance to the engineer. He is more interested in an accurate word description of the material or geologic formation as found at the location where he must build his structure.

To the engineer "shale" implies a hard, durable rock having considerable strength. The term "shale" does not adequately describe a number of formations that are composed of poorly indurated clays and silts. Such additional descriptions as soft or hard, and fissile or nonfissile; the

    [2] Berkey, C. P., *Responsibilities of the Geologist in Engineering Projects:* Am. Inst. Min. Met. Eng. Tech. Pub. 215, p. 5, 1929.

type of structure, whether bedded or massive; and, if possible, the mineralogy of the grains are all helpful to the engineer. Another pitfall in lithologic descriptions is the use of the term "sandstone." For example, along the Front Range of the Rocky Mountains the Fox Hills sandstone is in many places a very well-cemented material and forms prominent hogbacks. In areas of North and South Dakota, however, where the Fox Hills formation has been recognized, it is a weakly cemented to uncemented sand. The term "sandstone" fails to bring out the poor quality of this geologic formation, which, to the engineer, is a very unfavorable foundation material. An accurate description of the geologic formations occurring at the site is one of the basic requirements of sound engineering geology.

Care must be taken in describing formations exposed in outcrops. A number of clay-shale formations tend to air-dry on exposure, thus exhibiting a more competent nature than when found in their unweathered, unexposed condition. Many sandstones have a tendency to case-harden on exposure to weathering agencies, consequently exhibiting a much harder outcrop surface than in their natural condition. For this reason some subsurface exploration, even though a small amount, should be carried out at any site to determine whether the surface exposures of certain geologic formations are a true indication of their natural characteristics. Caution must be used in predicting the character of a geologic formation from a knowledge of its character at some other location. Normal geologic quadrangle maps may give a very excellent picture of the geologic formations present in an area; however, their quality and physical properties cannot be told from the average geologic map. One of the long-range phases of engineering geology is to conduct a large-scale mapping program that will more specifically delineate the physical properties of particular geologic formations as shown on existing maps.

Stratigraphy

Stratigraphy basically is not so important in engineering geology as in other forms of geologic study. The engineering geologist must, however, study the stratigraphic sequence of sedimentary rocks in any geologic investigation in order that intelligent predictions may be made as to the possible occurrence of various lithologic types. As has been pointed out previously, the occurrence of geologic outcrops at a particular construction site may be so few or of such quality that definite information cannot be obtained. Thus, geologic interpretations must be made from areas outside the immediate construction area. An understanding of the normal stratigraphic sequence in the area will permit the engineering geologist to make more accurate interpretations as to the types of rock likely to occur at the construction site.

An important phase of the exploration program in its early development is the drilling of a sufficient number of holes to establish definitely the stratigraphy at a particular site. This is not necessarily done for any great depth below the site but for the most part on the near-surface formations.

One of the most difficult tasks that an engineering geologist faces is convincing the engineer of the need for drilling at least one or two deep test holes in a construction area. Naturally the engineer is only concerned with the first few feet into bedrock. Several cases have developed where objectionable material has later been found at relatively minor depths below the original exploration. A knowledge of the general stratigraphy of the area will permit the engineering geologist intelligently to specify the depth to which exploration must be carried to delineate fully the subsurface geologic problems in the area.

Geomorphology

A knowledge of physiography and geomorphology is helpful in that they are indicative of particular processes of erosion. Land forms, when identified, may give a clue to the underlying rock or the depth of surficial material. Examples might be pediment surfaces, river terraces, alluvial fans, landslides, or the more common forms of erosion such as may be found in areas where outcrops of sandstone or shale are found. One of the first indications that an engineer uses in selecting the location of a particular site is the topographic relief of the area in which he is interested. Oftentimes this topographic relief is the direct result of some land form that may be explained by geologic processes. Thus, on the initial reconnaissance investigation, geologists may be helpful in selecting a site that does not have some particularly objectionable subsurface characteristic.

Structural Geology

The geologic structure, i.e., faults, folds, joints, and dipping strata, has a decided influence on engineering features. A clear picture of the geologic structure will permit the engineer to design his structure with due allowance both for good or bad conditions.

Faulted areas are usually zones of weakness or potential leakage and are to be avoided if possible. Careful investigation is necessary fully to determine the character of each fault encountered. The age of the fault and the possibility of renewed movement must be determined. The dip and strike of the fault plane with relation to the intended structure have an important bearing on strength. A fault dipping in the same direction as the ground slope provides a potential slide plane, but one dipping in the opposite direction has little detrimental effect on the foundation strength. The leakage problem in faulted areas may be influenced by the following factors: the type of formation through which the fault plane passes, the age of the fault, and the amount of movement. For example, ancient faults in limestone are apt to be recemented and offer little trouble.

Faulted sandstone may be brecciated and unless recemented may have many open channels. Faults in shale or claystone have a tendency to fail by plastic flow and therefore usually are relatively impermeable.

Extreme caution must be used when dipping strata are involved in foundation areas. Many landslides have occurred where soft, incompetent beds dipping in the direction of the ground surface become saturated or overloaded as a result of construction. Excavations that remove the support from dipping strata are potentially dangerous. Heavy structures such as retaining walls or building foundations may slide downdip if not properly anchored or keyed into the foundation.

Joints may provide zones of structural weakness if the trend is in the same direction as the stress pattern of the structure. More often, however, joints provide paths of leakage. Jointed rocks are difficult to excavate, as the limits of the excavation are controlled by the joint system and not by the arbitrary limits set by the engineers.

Ground Water

Ground-water investigations are those in which ground-water hydrology is studied in conjunction with the climate and the properties of the rock and soil as they influence the hydrology of an area. A discussion of the measurement and calculation of ground-water reservoirs would be too lengthy for this chapter. It is sufficient to say that porosity, permeability, hydraulic gradient, discharge, and recharge must be taken into account. Most of these are similar to petroleum-reservoir calculations.

There are many other applications of ground-water geology to engineering geology. The construction of a dam and reservoir will influence the ground-water levels in the vicinity of the structure. The effect of the change is often deleterious to water-supply sources, sewage disposal, arable soils, foundations, and the like. Attention to the ground-water level should be given during the investigational stages, the construction, and the operation of a structure, so that the status and future effect of ground-water conditions may be fully appraised.

Ground-water conditions may influence the actual design and construction procedures. Problems relating to the load capacity of saturated foundation materials may be encountered. Ground-water conditions may affect construction-material sources. Artesian uplift pressure may affect load distribution. The buoyant effect of a raised water table on watertight structures may cause damage. Slope stability due to saturation must be considered. Mineralized ground water may have a deleterious effect on the materials with which a structure is constructed.

Excavations for tunnels, bridges, canals, or structure foundations that extend below the water table may involve ground-water disposal or dewatering problems. The possibility of damage to ground-water reservoirs by lowering water table or the cutting of recharge channels or beds must be investigated.

Exploration

Subsurface exploration at a construction site is usually necessary to furnish specific design information. The extent of this exploration is determined by the size of the structure, the type of foundation present, and the amount of overburden cover. The more complex the subsurface geologic conditions, the more extensive the exploration program must be to clarify fully all pertinent foundation conditions.

An exploration program must specify not only the location of the drill holes but also the type of data to be obtained from the holes. These data include core recovery, an important factor in determining the soundness of the rock, and water-test information, which is necessary to determine permeability and leakage. Locating, sampling, and determining the extent of shattered zones, faults, and cavities that may transect good rock are of primary concern during foundation exploration. These zones might effect unequal or excessive consolidation, sliding, or large water losses and might even cause complete failure of a structure. Frequently the zones of primary interest are those where core is the most difficult to obtain. Therefore, it is essential that every effort be made to recover cores from such zones.

Depending on the type of information desired, rotary core drilling, churn drilling, wash boring, test pitting, tunneling, and geophysics have been used to investigate subsurface geologic conditions. Geophysical applications will be discussed under specialized techniques. Churn-drilling and wash-boring methods have a rather limited application, because core is not recovered in a form amenable to testing. Wash boring, which has the additional disadvantage of being unable to penetrate through large boulders or heavy, gravelly soils, makes true bedrock determinations difficult.

Test pits may be used for foundation exploration when the overburden cover is slight, or when large block samples are required for test purposes. Test pits have the additional advantage of allowing visual examination of the materials in place. Recently large-diameter, 12- to 48-inch power augers have been used to eliminate test pits in some tests.

Exploratory tunnels have been used in numerous explorations, particularly where large concrete dams have been under consideration. These tunnels are primarily used to obtain actual rock specimens for laboratory testing and for performing field tests in strength and permeability. The advent of more adaptable rotary-type core drills has minimized the use of tunnels for foundation exploration.

By far the largest percentage of exploration work has been done by means of various types of rotary core-drilling equipment. A type of core-drilling equipment widely used is a bottom-discharge core barrel that cuts a $2\frac{1}{8}$-inch core. The designation of this size core barrel in the X series would be "NX." The X series is a standard size range, which will take cores ranging in diameter from $\frac{7}{8}$ inch for the EX size to $2\frac{1}{8}$ inches for the NX size. (See table 35.)

The need for obtaining cores on which laboratory tests may be made has given an impetus to the development of core barrels of a larger diameter than the NX size. A four-inch core is about the minimum size that can be effectively used, but a six- to twelve-inch core would be preferable. Calyx drills or shot drills ranging up to 84 inches in diameter

TABLE 35

NOMINAL DIMENSIONS OF THE X-SERIES CORE-DRILL EQUIPMENT

Size designation		Casing O.D. (inches)	Casing coupling		Casing bit O.D. (inches)	Core-barrel bit O.D. (inches)	Drill rod O.D. (inches)	*Approx. diameter of hole made by core-barrel bit (inches)	Approx. diameter of core (inches)
Casing, casing coupling, casing bits, C.B. bits	Rod, rod couplings		O.D. (inches)	I.D. (inches)					
EX	E	$1\frac{13}{16}$	$1\frac{13}{16}$	$1\frac{1}{2}$	$1\frac{31}{32}$	$1\frac{7}{16}$	$1\frac{5}{16}$	$1\frac{1}{2}$	$\frac{7}{8}$
AX	A	$2\frac{1}{4}$	$2\frac{1}{4}$	$1\frac{29}{32}$	$2\frac{3}{16}$	$1\frac{31}{32}$	$1\frac{5}{8}$	$1\frac{7}{8}$	$1\frac{1}{8}$
BX	B	$2\frac{7}{8}$	$2\frac{7}{8}$	$2\frac{3}{8}$	$2\frac{11}{16}$	$2\frac{5}{16}$	$1\frac{29}{32}$	$2\frac{3}{8}$	$1\frac{5}{8}$
NX	N	$3\frac{1}{2}$	$3\frac{1}{2}$	3	$3\frac{9}{16}$	$2\frac{11}{16}$	$2\frac{3}{8}$	3	$2\frac{1}{8}$

* For a closer figure assume hole one thirty-second inch larger than bit.

have been used. These drills usually are used in competent rocks and are less expensive than excavating a shaft by conventional methods.

It may be of interest to explain why large-diameter cores are required for the testing of undisturbed foundation samples. Two tests are especially important, both requiring large-diameter cores from which test specimens are cut. The first of these is the triaxial shear test, which is used to determine the angle of internal friction (coefficient of friction) and the cohesive properties. When fine-grained formations are being tested, three or more $1\frac{3}{8}$ x $2\frac{3}{4}$-inch cylindrical specimens are cut from a single short section of the sample. After allowance for trimming away the disturbed outer surface of the core and for the trimming loss between each specimen, there is little leeway even with a six-inch core. The second important test requires undisturbed specimens $4\frac{1}{4}$ inches in diameter to determine the rate and amount of consolidation that may be expected under specific loading conditions.

These and other tests also provide data on pressures developed by the pore fluid during consolidation from superimposed loads and on percolation rates of water through the material. Obviously, rather large cores are required for these tests, and it is essential that the samples be recovered with as little disturbance and change in water content as possible.

Another method of taking undisturbed samples of overburden or foundation materials is with the Denison sampler. The Denison sampler

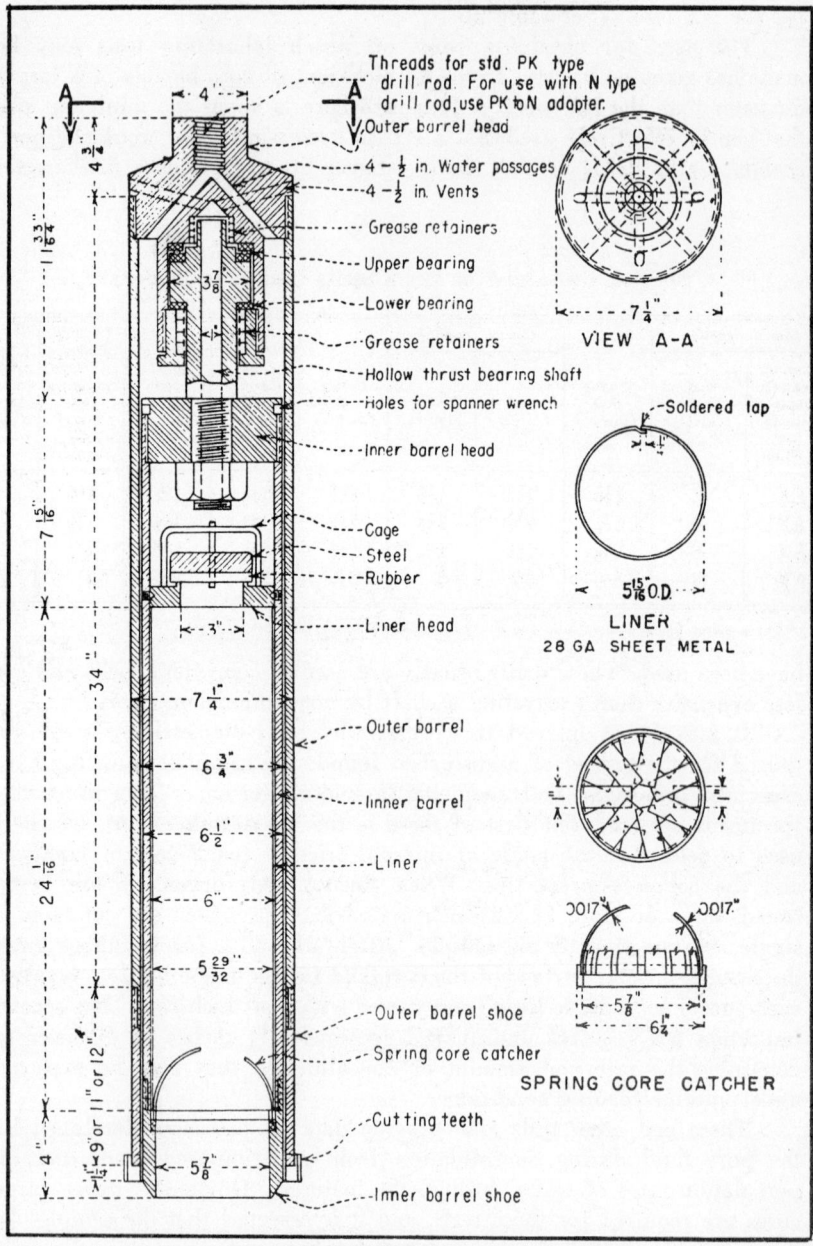

FIGURE 593. Denison double-tube core barrel, designated by H. L. Johnson, U. S. Engineer Office, Denison, Texas, October 1939.

was developed at the United States engineer office at Denison, Texas, by H. L. Johnson. This sampler is basically a double-tube core barrel, which cuts a core approximately six inches in diameter. (See fig. 593.) Five advantageous features of the Denison sampler are as follows:

1. The size of the sample is satisfactory for testing.

2. Provision is made for a sample container of sheet metal, which acts as a removable inner-barrel liner.

3. The inner nonrotating barrel trims the core as the sampler penetrates into the ground and thus anchors the inner barrel to prevent rotation and erosion of the core.

4. A spring core catcher retains noncohesive fine-grained materials as the sampler is withdrawn from the hole.

5. A check-valve assembly vents fluids or water from the inner barrel as the core is received. This valve also seals the inner barrel at the top when the core barrel is lifted.

The Denison sampler is by no means an all-purpose core drill. Coarse materials or those finer grained materials that contain gravel cannot be successfully sampled. Materials most adaptable to sampling by this method are fine-grained cohesive materials and extremely fine damp sand.

The Denison core barrel is forced into the material by a combination of pressure and the rotary cutting action of the core bit. The drilling mud is pumped through the drill rod between the inner and outer barrels and then upward along the exterior of the outer core barrel to remove the cuttings. This drilling mud forms a skin on the drill hole, and thus holes can usually be drilled without casings. Holes drilled below the ground-water surface should not be pumped or bailed dry because the external ground-water pressure on the mud casings will very likely cause caving; furthermore, the flow of water through the bottom of the hole is likely to disturb the structure of the material.

Numerous types of drive-sampling devices have been developed for obtaining undisturbed type samples in cohesive and plastic materials. This sampling procedure is not suitable for brittle, highly compacted, cemented materials or cohesionless materials. Drive sampling is not used as a means of boring or drilling an exploration hole but rather to obtain representative samples while the main drilling is made by other means. Drive sampling consists in forcing a sampling tube or barrel into the material without rotation. Usually a quick, steady drive for the full length of the sample is the best means of obtaining an undisturbed type sample. The most common method is by means of repeated blows of a driving head. Samplers with thick walls in general increase disturbance by causing more displacement and compaction of the samples. Thin walls and sharp cutting edges that taper on the outside are very important.

SPECIALIZED TECHNIQUES

The mapping of the surface geologic features is a very important and indispensable phase of engineering geology, but the fact should be remembered that it is only the first in a series of investigations. If the geologist is to obtain information that can be resolved into quantitative terms of rock or soil behavior, he must apply every possible investigational tool. The specialized tools of the petrographic laboratory, the materials laboratory, and geophysics may be required to provide the necessary data. Frequently the final answers to an engineering geologic problem represent the combined viewpoints of the geologist, the petrographer, the soils technician, the geophysicist, and others whose specialties can be applied to some phase of the problem.

Application of Petrography to Engineering Problems [3]

Petrography serves the engineer in several ways: first, a detailed study of rocks assists the engineering geologist in establishing the geologic structure and interrelation of formations at construction sites; second, petrography assists in determination of the engineering properties of the rock materials in place at the site and in materials to be used in the construction.

The engineering geologist requires the application of petrography to obtain the maximum geologic information from limited exposures or relatively few drill cores so that preliminary estimates will be as valid as possible, and so that future explorations might be the most appropriate. For example, the geologist may wish to know if the alteration observed at the surface was caused by weathering and hence is limited in depth or by hydrothermal processes and is likely to continue downward. The decision would greatly influence preliminary estimates of exploration and construction costs, for if hydrothermal, deep-seated alterations were indicated, the advisability of extensive preconstruction exploration would be established. At Anderson Ranch dam site the intense hydrothermal alteration rendered the granite incoherent to depths in excess of 300 feet below the original surface, and serious landsliding has occurred during excavation of the site.

Precise description and identification of rock formations aid the geologist in problems of correlation. Thus, the presence or absence of zones of shearing and faulting can be established by correlation of strata across the site. For example, at Canyon Ferry dam site on the Missouri River near Helena, Montana, the sedimentary formation constituting the foundation and abutments contains thin sills of altered andesite, distinguishable only after petrographic analysis, continuity of which proved the absence of significant faulting beneath the river alluvium.

Petrography is a valuable tool for determination of the properties

---

[3] Mielenz, R. C., *Petrography and Engineering Properties of Igneous Rocks:* U. S. Dept. Interior, Bur. Reclamation, Eng. Mon. 1, pp. 8-9, 1948.

of rocks, either when applied independently or as a means of selecting those quantitative tests necessary to measure specific properties. The latter function is the more important. Most tests to measure properties of rock materials are expensive and time consuming, and they usually require carefully selected samples so protected that at the time of testing they truly represent the character of the rocks and materials in place, especially with regard to moisture content and fractures. Consequently,

Figure 594. Rock stresses in walls of Prospect tunnel are determined by cementing one-inch strain-gage rosettes on rock and then relieving stress by drilling around gages with six-inch-diameter core drill. This tunnel forms a part of the Colorado-Big Thompson project.

it is wise to determine the necessity for certain tests before they are requested. For example, testing the quality of an aggregate by performing tests on concrete containing the aggregate usually can be avoided by the application of physical and chemical tests and petrographic examination of the aggregate. Tests of concrete need be performed only when the results of the physical and chemical tests of the aggregates are anomalous, or if the petrographic examination indicates adverse properties not evaluated by the aggregate tests. Determination of volume increase of rock materials with wetting is significant only if clay minerals of the montmorillonite type (bentonitic) are present. The presence and abundance of these clays can be determined quickly by petrographic and X-ray-diffraction analysis. If the number of samples available for test is so great as to preclude testing of all, petrographic examination to select specimens representative of the group frequently will simplify the test program without sacrificing the significance of the results.

Engineering petrography and geology in coordination serve to relate the properties of the individual specimens subjected to laboratory tests to

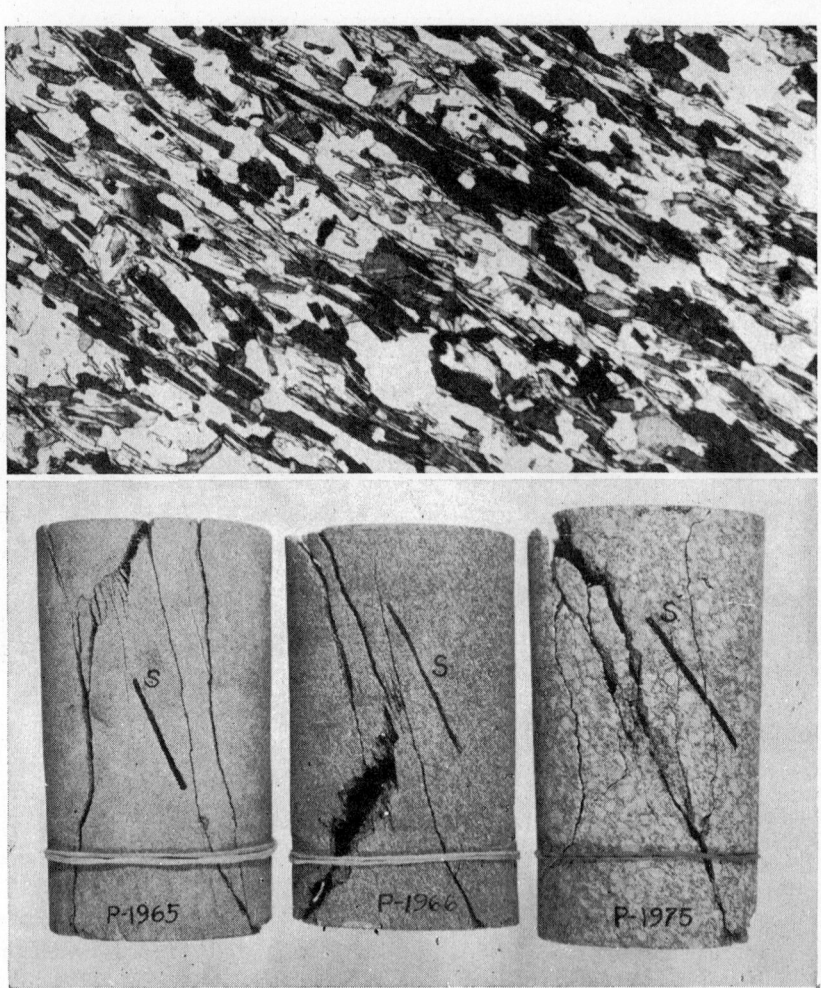

FIGURE 595. Petrographic examination of thin slice of metamorphic rock shown in upper photograph (X50) revealed that mineral grains were arranged in parallel planes and led to prediction that rock would fail along these planes. Crushing tests conducted on cores of this rock (lower photograph) confirmed this prediction. Line marked S on side of cores was predicted line of failure.

the properties of rock formations in place. For example, the petrographer and geologist may be called upon to decide to what extent the fractures, joints, and planes of shear in the rock in place should be cause for reducing below the measured strength of rock specimens the strength of the rock mass; or to what extent the swelling clays might remain stable by virtue of their impermeability; or to what degree discontinuities such as joints, bedding planes, and faults would augment the known permeability of the rock itself. Not infrequently, the results of tests of rock specimens, however carefully selected, are more deluding than instructive. Only experience, geologic and petrographic skill, and good judgment will permit translation of test results to the data required for engineering design. The petrographer can assist the engineer in design, construction, and maintenance problems. By working with the engineering geologist the petrographer facilitates selection, exploration, and subsurface geologic investigation of construction sites. Through application of petrography to materials testing, specific tests to be applied and samples to be tested can be selected with minimum hazards of inefficiency. Petrography is effective in predicting the engineering properties of rocks because those properties are determined by the texture, structure, and composition of the rocks, characteristics that can be discerned by petrographic methods.

Soils Mechanics

The never-ending demand for exact information concerning the physical properties of foundation materials has introduced the application of soils mechanics and earth-testing procedures into geologic investigations. In the design of his structures the engineer is confronted with the problem of stability of the material upon which he must place his structure. The principal properties to be determined in a foundation study are consolidation, shearing resistance, and permeability.

Settlement of a foundation may effect a structure in several ways. Most engineering works are constructed to precise measurements. Thus, even a small settlement may throw various features out of line or grade with each other. Differential settlement may actually cause failure of certain key parts of a structure. Settlement may result from direct vertical consolidation, lateral deformation or shear, or a combination of the two.

Consolidation tests must be made on poorly cemented, friable sandstones and poorly indurated shales or claystones. These materials while in their dry state as found exposed appear to be highly competent and capable of withstanding extreme loads. However, one must consider that in the construction of a dam or other water-diverting structure these apparently competent materials will probably become saturated and tend to lose their properties as rock and behave more like natural soil types. Conversely, some shales that appear to be incompetent on the surface or when subjected to water in an unconfined state may be competent when properly confined.

Shearing resistance, which is the property of a material to withstand loading without objectionable lateral deformation, is, for the engineer, a most important consideration. The original basic assumptions on stability may be stated as follows:

1. A force applied to a granular mass is transmitted throughout the mass by contact pressure between the individual particles.

2. The loading will cause the particles to rearrange, and as they rearrange they will tend to slide with respect to each other.

3. This sliding is resisted by frictional resistance, which is a function of the coefficient of friction between the particle surfaces and the pressure between the surfaces.

4. The sliding is also resisted by mechanical interlocking of the aggregation of particles and by cohesion between the particles.

These basic assumptions would serve for the evaluation of shear tests if all materials were free-draining. However, when materials of relatively low permeability are tested or when drainage restrictions are imposed, the test results vary with such conditions as those of testing, sample thickness, drainage restrictions, density, and moisture content. The principal cause of these variations is variations in induced pore pressures within the test specimens. For this reason pore-pressure measurements are taken and the laboratory-test data are reduced to zero-pore-pressure conditions to aid in the analysis.

All geologic formations, with the exception of a few, are composed of material having a definite particle size and shape. Void spaces occur between the particles or grains. These voids may be filled with air, water, or a combination of both. This void filling is called pore fluid.

When a load is applied, the initial effect is an increase in contact pressure between grains and a subsequent reduction in volume by readjustment. If the material is 100 percent saturated with water, a slight load will produce a fluid condition where all the load is carried by the water. Usually the void spaces are filled with a mixture of air and water so that the loading will cause a reduction in void space and will compress the air, building up a pressure within the material. This pressure is called "pore pressure." Pore pressure opposes the applied loading and thus reduces the contact pressure between the soil particles. That part of the load carried by the fluid space does not increase the shearing resistance of the mass because the fluid has no resistance to change in shape. Because all soils are more or less permeable, pore pressure once produced will decrease with time as the pore fluid is extruded. One of the important factors in pore-pressure determinations, therefore, becomes the rate at which this pore pressure is dissipated. If the load is applied faster than the pressure is dissipated, failure through flow may result.

The permeability of a foundation of an engineering structure has several major implications. The first is that of structural stability. If the flow of water through the foundation actually dissolves the foundation rock, causes removal of the finer rock particles, or causes uplift pressures to be built up, structural failure may result. Another major consideration is one of economies; that is, how much water can be lost at a certain structure before it becomes economically unfeasible.

In the drilling program for most foundations a requirement is made that all holes be water-tested at various suitable pressures at successive depths in the drill hole. Careful measurements are made of the water loss, the length of time of the test, and the pressure used so that calculations may ultimately be made as to the permeability of the material tested. It can be seen that joints, cracks, or other structural weaknseses in a rock formation may greatly influence the permeability as measured from pressure tests in bore holes. These weaknesses are usually sealed off or filled during construction operations by high-pressure grouting.

Geophysics

In the investigation of a dam site and in many other problems of engineering geology, certain types of information can often be obtained by geophysical methods at less cost and more quickly than by drilling or other means; for example, such data as the depth to firm bedrock or the extent of a buried gravel deposit. Geophysical exxploration is a blend of physics and geology, for it involves measurements that are interpretable in terms of local subsurface geology, which are made with suitable instruments at the ground surface. Although the methods and equipment used in the present-day geophysical explorations are extremely accurate, the interpretation of the data obtained by these methods may be complicated by the diversity of subsurface conditions that are usually encountered. However, geophysical methods are rapid and economical and, under proper conditions of control, can be used to great advantage in locating rock profiles quickly and inexpensively during preliminary investigations and in establishing the continuity of strata between borings during more detailed explorations.

Reconnaissance of a site should be made by experienced personnel to determine the geophysical procedure best adapted to the local conditions and the specific problems to be investigated. Since identical subsurface conditions are seldom encountered over wide areas, and since each site presents special problems, a stereotyped program cannot be applied. It is very important that the geologic and engineering problems involved at a given site be thoroughly understood by those undertaking the geophysical survey. Most geophysical explorations in connection with engineering geologic problems pertain to the near-surface material, which normally is corrected for or eliminated from standard geophysical explorations as conducted in the petroleum industry. For this reason, as the areas under consideration in engineering geology are rather small, each geophysical-exploration program must be tailor made to fit the conditions that are expected to be encountered.

The most widely used methods of geophysical exploration in the engineering field are the seismic-refraction and the electrical-resistivity techniques. These two methods depend on artificial or man-made fields of force. The two other forms of geophysical exploration have been used to a minor extent: namely, the magnetic and gravimetric methods. In these last two methods, the force field, in which measurements are made, is set up by nature: i.e., the earth's magnetic and gravitational fields.

The matter of force fields is brought out at this time because it is basic to the functioning of geophysical methods and their ability to give results interpretable in geologic terms. In field problems amenable to geophysical solution, there must exist contrasts in the physical properties of rocks or formations that give rise to detectable differences in the particular field of force involved. For example, in seismic-refraction measurements for bedrock depth, the bedrock must transmit seismic waves at a higher velocity than the overburden. In like manner, to be detectable with a magnetometer, a dike must be more magnetic than its surrounding rocks. Figures 596 through 599 depict general and ideal application for the various types of geophysical methods named in the preceding paragraphs.[4]

Sequence of Engineering Geologic Investigations

Most engineering geologic investigations are carried out in successive stages: namely, reconnaissance, site selection, preliminary design and estimate, final design, and specifications. The amount and type of investigation required in each stage cannot always be determined beforehand. Exploration should be limited to the essentials necessary to obtain the information required for any particular stage. It is important that data be assembled during or immediately after the work is finished on each stage, while it is fresh in the geologist's mind and before the material, such as cores, has been affected by exposure to air or moisture. The stages of exploration will be discussed in the following paragraphs.

Stage A: Reconnaissance

Stage A or the reconnaissance is what the name implies. It will, in most cases, determine whether it is possible to build economically the desired structure in the area chosen. Usually, the reconnaissance examination for investigation is conducted on a number of possible sites. Reconnaissance examination will involve the utilization of all data, detailed or general, already existing for the area, such as topographic and geologic maps and reports by state or federal agencies or private industries. A field examination will be made by experienced geologists and

---

[4] Wantland, Dart, U. S. Bur. Reclamation, Denver, Colorado, personal communication.

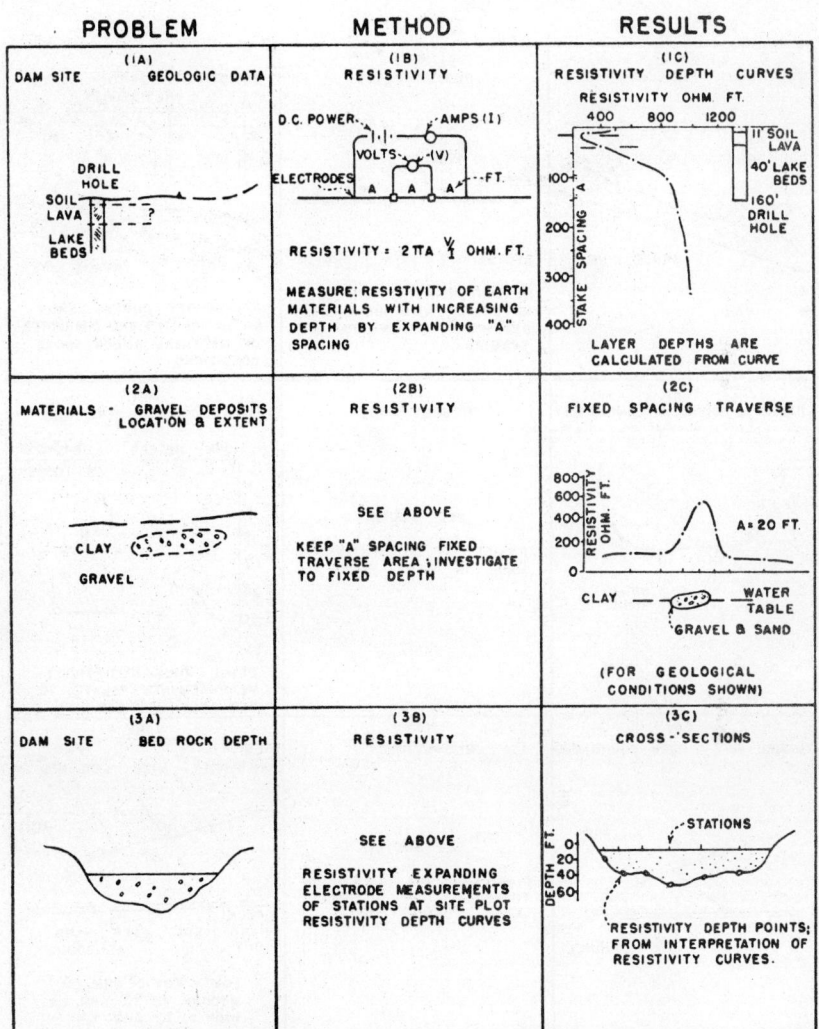

FIGURE 596. Resistivity methods applied to engineering geology.

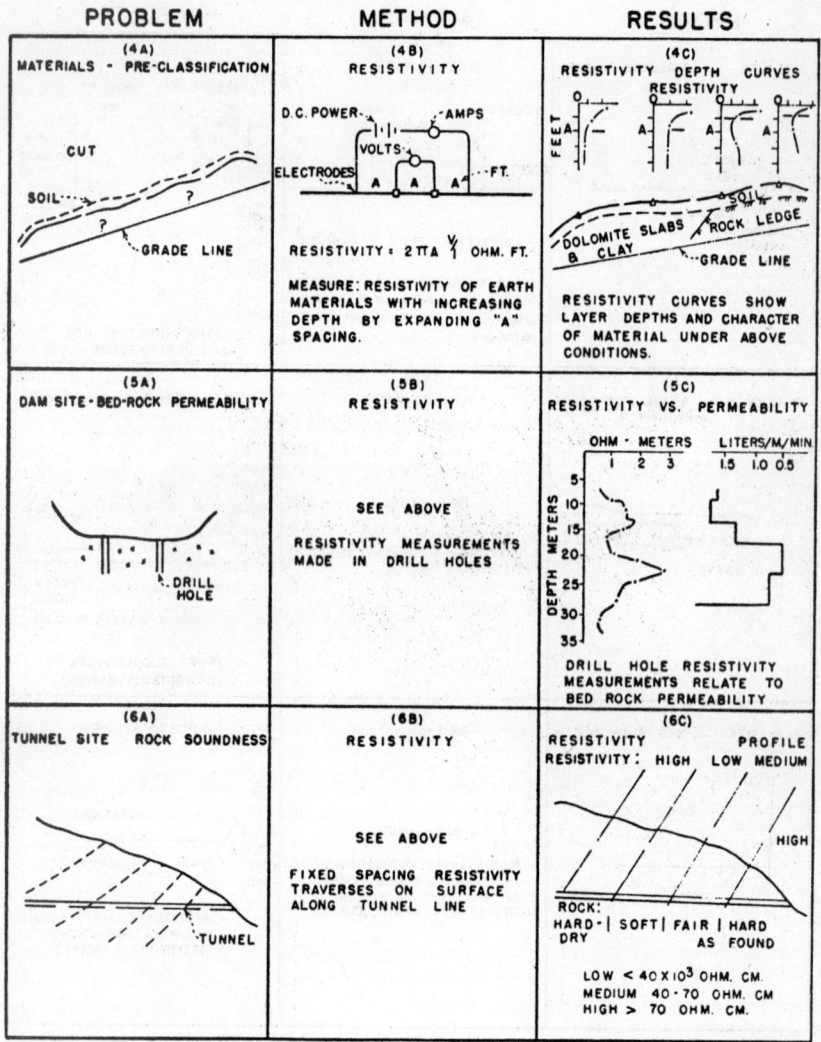

FIGURE 597. Resistivity methods applied to engineering geology.

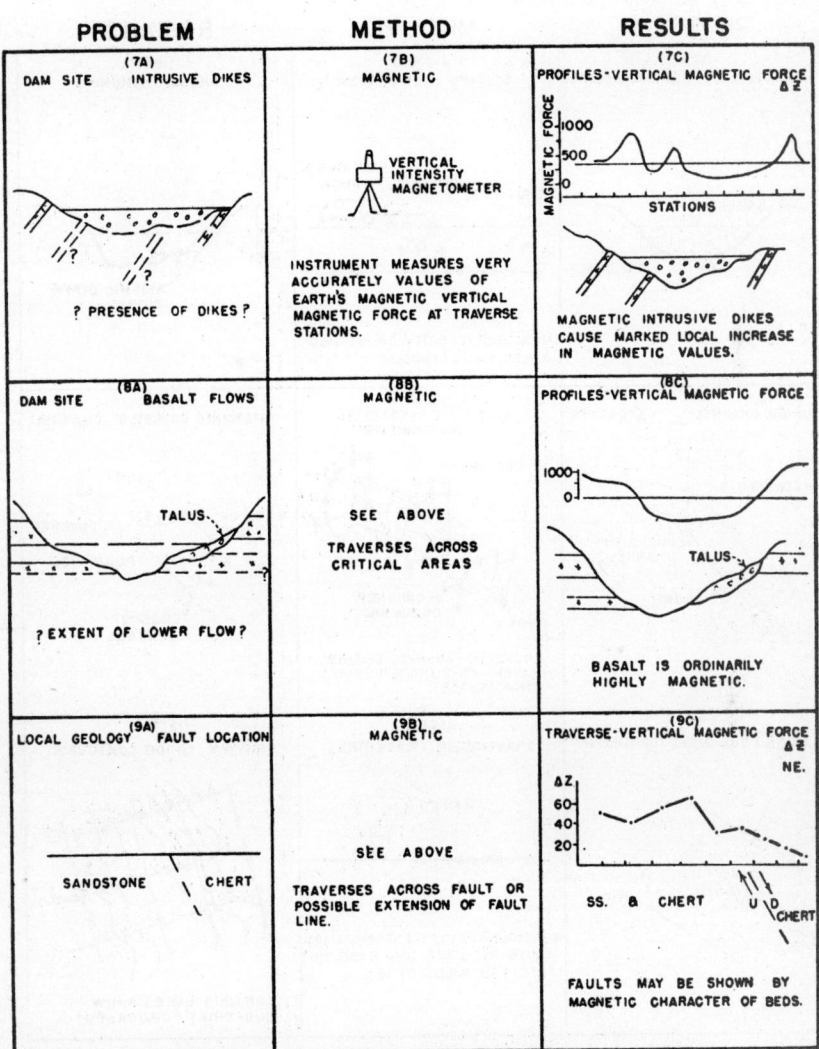

FIGURE 598. Magnetic methods applied to engineering geology.

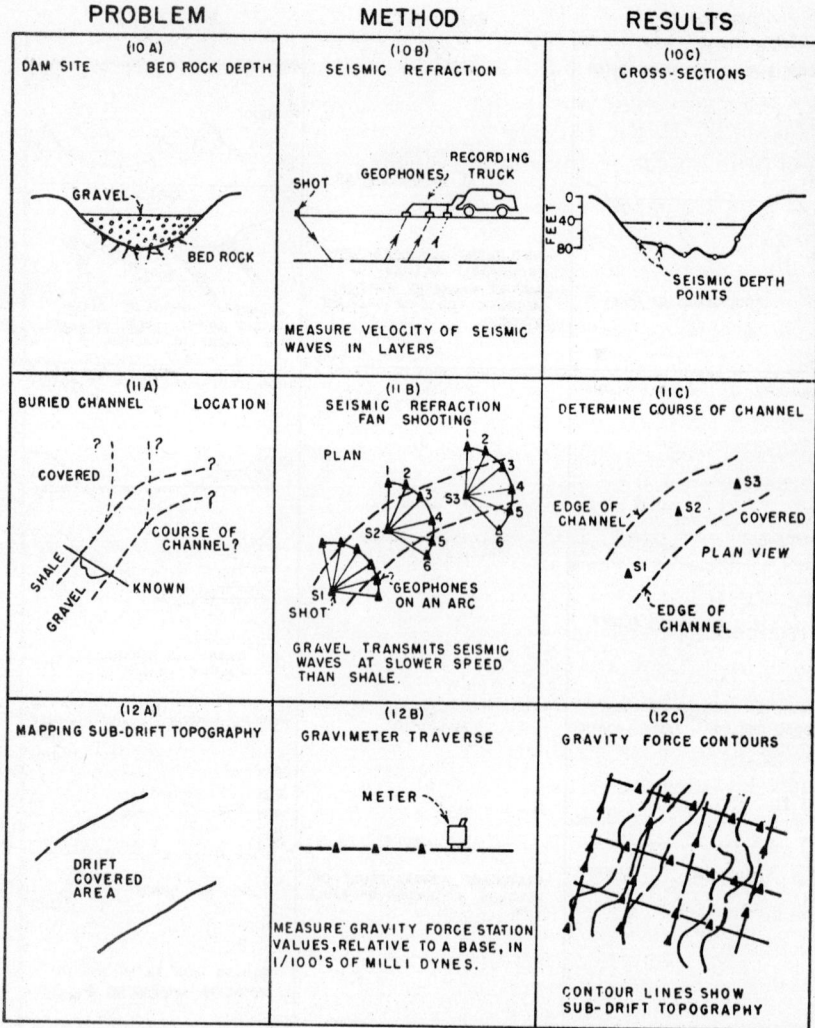

FIGURE 599. Seismic methods applied to engineering geology.

engineers, not only of the local areas of possible sites, but also of the surrounding regions, to detect, if possible, any features such as faults and contacts that are obscured at the site but are capable of being projected into it. No detailed mapping is included in stage *A* exploration. Sufficient time should be allowed that a thorough examination can be made and no essential data overlooked. A reconnaissance geologic report should be as complete as possible and should set forth all that is known of the geology and engineering aspects of prospective sites. In the case of a dam, the feasibility of the reservoir site should also be discussed. Reports should include maps and photographs from other sources when available.

Stage B: Selection

When there is any doubt as to the choice of site after the reconnaissance examination, sufficient additional work should be done to select the site to be used. Selection may depend upon (*a*) a comparison of two or more sites in rough cost estimates; (*b*) a comparison of the details of the geology as to bedrock areas, the type and quality of the rock for foundation material, the areas covered by overburden, the thickness of overburden, and its adaptability for use in construction in areas of required excavation; and (*c*) a comparison of the availability and quality of the construction materials.

When additional exploration is necessary to resolve the choice of site, it is desirable that the stage *A* data be expanded and supplemented by the following additional work: (1) Detailed geologic maps covering the alternative sites should be made. These will allow comparison of the geology of the sites and will serve as a guide to necessary drilling and other testing, if such are needed in the choice of sites. Detailed geologic maps of small areas such as dam sites constitute one of the least expensive steps in the exploration program and one of the most useful in all subsequent investigations. If stage *B* is not necessary, the map can be relegated to stage *C*, confined only to one site after the completion of stage *A*. (2) Surface profiles or rough topography for comparison of quantities should be submitted. (3) Limited test sampling and laboratory testing should be done as necessary for comparing the suitability, quantity, and costs of construction materials. (4) Limited core drillings, exploration tunnels, and the like are necessary for the comparison of foundation conditions, bedrock profiles, and the layout of appurtenant structures. (5) Geophysical surveys should be conducted, when pertinent, to determine the depth to bedrock and to obtain a rough classification of materials, especially if the site is covered by overburden that conceals all or most of the bedrock.

The stage *B* geologic report may be appended to and added as a supplement to the reconnaissance report or as a revision thereof. The data obtained from this exploration stage should be formulated into an appendix to the reconnaissance report and should include drill- and test-pit logs, geologic cross sections when feasible, and outcrop or geologic maps. The importance of such a map should be emphasized, for, although an outcrop map is better than nothing, it cannot replace a properly made geologic map.

Stage C: Preliminary Designs and Cost Estimates

The over-all purpose of stage C is to provide a basis for requesting authorization and appropriation. Most of the main points relative to stage C explorations are covered in paragraphs relative to stage B above, but the stage C investigations are carried on with a view to providing all data necessary for the preliminary design and estimate and are more complete than those limited to the purpose of site selection. The purpose of carrying out the steps necessary in stage C is to develop more specifically the surface and subsurface conditions at the proposed location of appurtenant structures, as well as at the main structure, and to establish more definitely the character, quantity, and cost of construction materials. The extent of stage C explorations will depend on the completeness of stage A and stage B explorations, the complexity of subsurface geologic conditions, and the importance of the structure.

The preparation of an adequate geologic map again cannot be too strongly emphasized. It should be one of the first steps in the stage C exploration, especially if such a map was not prepared previously. A geologic map should be made before or while topographic map is made. The geologic map should show (1) contacts of overburden in bedrock areas, contacts between types of overburden, such as slope wash, talus, alluvial and glacial drifts, and landslides; contacts between kinds of rocks in the outcrop area; and formation boundaries within outcrop areas; (2) dips and strikes in sedimentary or metamorphic rocks taken at frequent intervals; (3) geologic columns with detailed descriptions of the sedimentary section if possible; (4) structural contours in sedimentary rock where appropriate; (5) the physical condition of the rock exposed, as shattered, crushed, sound, hard, or soft, and the degree of weathering; (6) a surface trace of fault planes or zones, the amount and direction of displacement, and the effect on the attitude and condition of rock; (7) the prominent joint systems, directions, spacing, and effect on rocks; (8) the estimated thickness of the overburden where this has not previously been determined by drill holes and test pits, which are to be shown by figures and circles on maps; (9) the lines of geologic cross sections; (10) the location of all drill and auger holes, exploration tunnels or shafts, and test pits; and (11) a legend or explanation using formation or rock symbols or conventional signs.

Stage C explorations should be sufficiently complete to permit the development of geologic cross sections.

Stage D: Geologic Report

The stage *D* geologic report should contain a comprehensive coverage of all data on hand at the time of its preparation. In addition to data acquired in stages *A* and *B* it should contain more of the detailed information obtained in stage *C*. The text should include (1) a brief account of the regional geology; (2) a discussion of the kind of rock and the effects of faulting, shearing, weathering, and like influences at locations shown as such on the map, as well as other structural features, such as dip, strike, jointing, and unconformity; relations of possible leakage grouting programs; the amount of exploration necessary; and data obtained in tests on undisturbed foundaiton materials as they have been logged; (3) a description of cross sections; (4) a statement of geologic conditions at sites of proposed appurtenant structures; (5) a discussion of such matters as percolation tests and ground-water conditions; (6) a description of materials for various construction purposes, their origin, hauling distances, and quality, and the results of laboratory testing; (7) and reference to special conditions affecting the preliminary design and estimate, such as landslides, vertical cliffs, the occurrence of bentonitic materials, and similar matters.

The following illustrations should be included: (1) a combined topographic and geologic map; (2) a sheet or sheets of geologic cross sections showing geologic interpretations as derived from all surface and subsurface data available; (3) a map showing bedrock contours where appropriate; (4) logs of drill holes, the test pit, and exploration tunnels; (5) maps of prospective borrow areas showing locations of test pits and cross sections, from which may be estimated the quantity and types of material; (6) maps of profiles of sections showing the results of geophysical work; and (7) photographs, both ground and aerial.

Stage E: Final Design and Specifications

Upon the completion of stage *D*, exploration and investigations, the designing engineer and geologist usually will undertake discussions relative to the amount and type of exploration remaining to be done in order to prove the competency of the geologic conditions at the specific site of the more important structures appurtenant to the main feature. Additional exploration will probably be determined and controlled by the engineers rather than by the geologists. During this stage the geologists should prepare and maintain up-to-date information, which can be submitted to the designer at periodic intervals, upon the completion of the final design stage of exploration. An expansion of the geologic report is usually necessary, and this will contain all of the information up to the time of the preparation of the report.

Construction

During construction new information will be available as excavations progress. This information should be recorded with appropriate changes

made in the map, new cross sections constructed or old ones revised, and
any other new data recorded for use, in that further questions may arise
during construction, operation, or maintenance, or subsequent repairs or
alterations. The construction phase is a geologist's proof of predictions
made during the previous stages. The geologist should usually be on
hand during the construction phase in order to check the interpretations
made in the light of the evidence uncovered, in order that revisions may be
made in the design if the need should become apparent. Frequently
economy can result from interpretations that may prove the foundation
area better than originally anticipated.

EXAMPLES

The following are only a few examples that point up the need for,
and the application of, various laboratory and investigation techniques.
Up to the present time, a large number of the problems solved by engineering geology have been related to failures or impending failures of
existing structures. The present engineering geologist has gained considerable knowledge from these past failures and has a better idea of
what to search for in the investigational stages of new projects.

Davis Dam

Investigations disclosed a large fault zone in the area of the proposed spillway and powerhouse foundation for Davis dam. The fault
consisted of a zone of gouge and brecciated granite-gneiss rock, which was
an average of forty feet in thickness, with dips from 12 to 37 degrees.

The design problems were resolved into the strength of the foundation rock, the amount of the fault zone that would require remedial
treatment or removal, and the most appropriate location for the required
structures.

An exploration program was laid out and further amplified or modified as the work progressed. This exploration included NX core holes,
six-inch core holes, test pits or shafts, and 36-inch calyx holes. Field bearing tests, laboratory bearing tests, and petrographic investigations were
all employed. A geologic peg model was constructed and maintained up to
date in order to correlate the data derived from the exploration program.
(See fig. 600.)

The general outlines of the proposed structures were etched on
plastic. Exploration holes were represented by metal pegs, which were
cut to length so that the top of the peg represented the top of the hole.
Different rock types were represented by means of colors on the pegs,
with white representing the difference between the bottom of the hole
and the chosen datum. The portions of the hole showing fault breccia
were tied together with string. White string represented the top and
bottom of the major fault zone. Dark string represented a minor fault

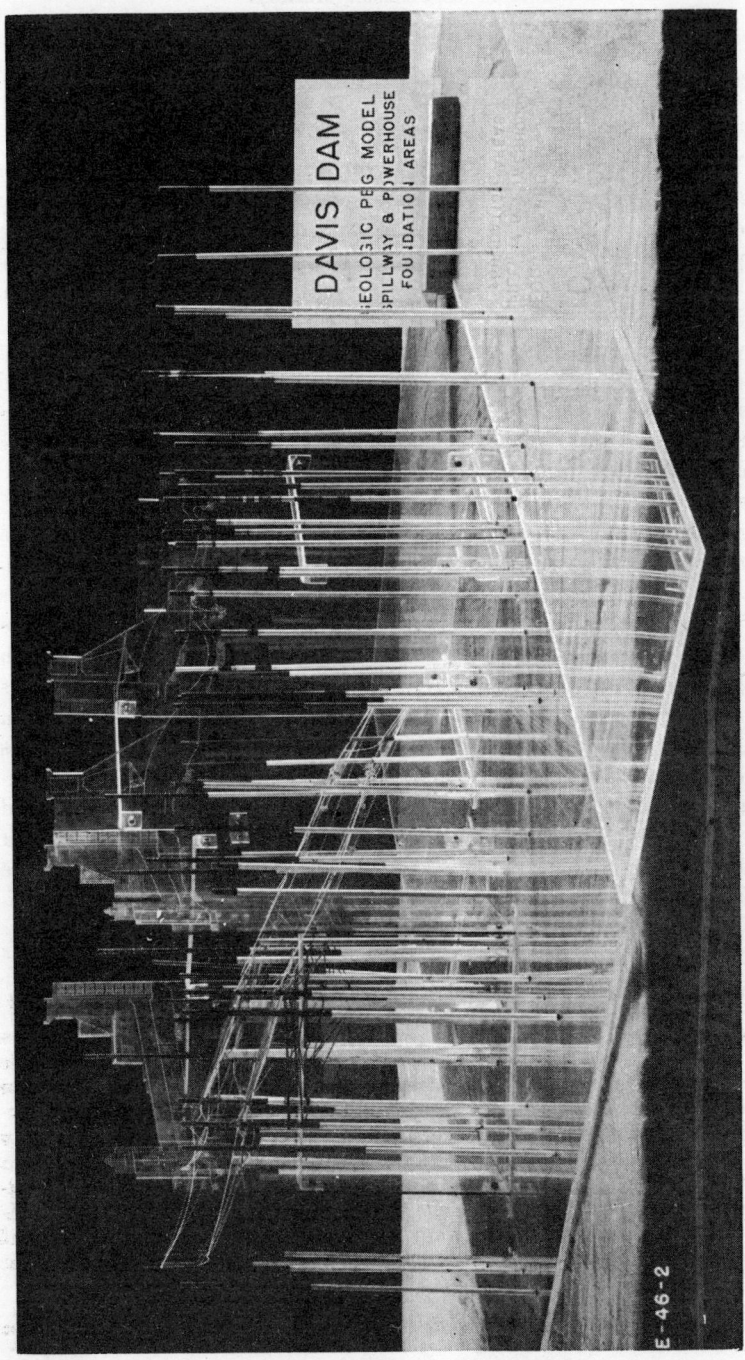

FIGURE 600. Geologic peg model of Davis dam on Colorado River. Pegs represent exploration holes. Foundation outlines of major structures are etched on plastic. White string eliminates top and bottom of major zone; dark string outlines a minor fault.

zone. By this means an accurate relationship between the fault zone and the structure foundation was shown.

Keyhole Dam Site

The Keyhole dam site is located on the Belle Fourche river near Moorcroft, Wyoming. The river at an earlier stage aggraded a deep channel through sediments of the Lakota formation. This channel has subsequently been refilled with gravel, sand, and clay. One of the proposed types of structure for the site was a slab-and-buttress-type dam. It became necessary to know if the river-fill materials were of sufficient density to provide adequate support for the buttress foundations. Normal means of exploration were not deemed advisable. The alluvium contained large-size gravel and cobbles, which precluded the use of normal drilling and sampling methods. The water table was near the surface, making the dewatering of a test pit some 50 feet in depth a costly operation.

It was decided to explore the area by means of a gravimetric survey. The sediments are essentially horizontal, and sufficient drilling had been done to outline the river channel in the area to be studied. A gravimetric survey, as applied to the regional exploratory problems, makes use of the following two corrections to bring the data to a usable base level: (1) a correction for elevation differences or a free-air correction and (2) a correction for the density of the near-surface alluvia by means of the Bouguer reduction. The first of these can be readily obtained from the surface elevations. The second, namely, the Bouguer correction, is the one of interest to this problem since this correction is made by assuming various densities for the material between an arbitrary datum and the ground surface. The correct average density will give a straight line through the corrected points, which may dip in a direction of the regional dip and may reflect the dip of the strata; or, in the case of horizontal strata, it may be horizontal.

The theory behind this correction has been discussed by Nettleton,[5] who makes an interesting comment:

The analysis of a density traverse consists primarily of plotting profiles of the elevation and of the gravity values with the usual reduction for latitude and free-air corrections and with different curves for the Bouguer correction made with different densities. Under favorable conditions a quite definite selection can be made of the density which comes closest to giving reduced gravity values on a straight line across the topographic feature. Frequently there are some stations which do not fit into a smooth curve for any density having departures of a few tenths of a milligal (milligal numerically equal to a millidyne). Our experience indicates that these irregularities are caused by real inhomogeneities in the material, as checked observations have confirmed the gravity differences.

The foregoing clearly sets forth the idea on which the field test

[5] Nettleton, L. L., *Determination of Density for Reduction of Gravimeter Observations*: Geophysics, vol. 4, no. 3, pp. 180 ff., July 1939.

here reported is based: namely, that small gravity differences may reflect a density variation in the material beneath a particular station. The density value used in the Bouguer reduction, which gives the corrected gravity stations on a profile line the smoothest character, is the average density of the materials above the datum plane.

Malheur Siphon

The Malheur siphon is a feature of the irrigation system in the state of Oregon. It is approximately 23,000 feet long, and the portion under consideration consists of welded steel pipe 80 inches in diameter, supported at intervals by concrete piers. Within three years after completion the piers were being displaced laterally and vertically, the displacement being of such magnitude that the siphon kinked in one place and threatened to fall from its support in other places. The field geologist reported that shales of the Payette formation formed the foundation of the structure, and he obtained undisturbed samples for study by the petrographic and earth-materials laboratories, where it was found that the shales contained bentonite and would swell conspicuously when wet. Further field studies disclosed that the shales continued downward to great depth, and the laboratory undertook to determine the swelling pressure so that a calculation could be made of the depth at which the swelling pressure would be equaled by the superincumbent load; any remedial measures to be adopted would necessarily have to be applied down to this depth. It was found that the swelling pressure equaled the weight of the empty siphon plus sixty feet of overburden, or the full siphon plus thirty feet of overburden. Inasmuch as the topography would not permit gravity drainage to such depths, it was decided to investigate the possibility of cutting off the source of the water that was saturating the shales. Observation holes were drilled over the surrounding area to permit a detailed analysis of the ground water. A contour map of the water table disclosed the direction of ground-water flow and indicated that ground-water recharge was occurring through leakage from an adjacent canal. The materials laboratories designed an impervious asphaltic lining for the canal, and after its installation no further displacement of the siphon has been reported. Opportunities for this kind of field-laboratory cooperation are becoming more frequent and important with the use of consolidated materials as foundations.

Palisades Dam Site Intrusive

At Palisades dam site in southeastern Idaho, preliminary drilling disclosed in the left abutment a tilted tabular mass of andesite 300 feet thick, underlain and overlain by beds of clay, silt, and conglomerate. Preliminary designs contemplated three large tunnels in the left abutment, two for outlet works and one for a spillway. The geologist was given tentative locations and elevations of the proposed tunnels and was requested to determine whether the tunnels could be kept exclusively in the andesite.

The solution of the problem lay in determining the regularity of the lower surface of the andesite body. It was reasoned that if the andesite had been emplaced as an extrusive flow during the deposition of the surrounding sedimentary series, the lowest surface would be roughly regular and little or no drilling would be necessary to establish its position with respect to the tunnel lines. On the other hand, if the andesite was intrusive in the sedimentary series, the lower surface could have any configuration whatever; hence the relationship of the lower surface of the andesite to the tunnel line could be determined only by extensive drilling. Field studies were inconclusive, but petrographic analysis of thin sections of the andesite and the adjacent sediments demonstrated that the andesite was intrusive. The additional drilling could not be eliminated from the exploration program.

Columbia Basin Landslides

Grand Coulee dam impounds Lake Roosevelt, which is 150 miles in length. The lake fills a gorge of the Columbia River, which during glacial times was dammed by ice and received lake sediments that were deposited in bars to thicknesses of hundreds of feet. These sediments, known as the "Nespelem formation," were subsequently trenched and now form terraces, which stand high above the lake's surface. Some large prehistoric landslides had occurred before the reservoir was filled, but new slides occurred in considerable numbers after the lake was impounded, especially when the lake was drawn down from a high to a lower level.

Farms, highways, and a railway would be endangered if slides occurred at certain locations on the reservoir rim. To determine the extent to which such areas should be condemned and reserved from use, it was necessary to classify the entire reservoir rim in terms of relative landslide potentialities. A geologic survey classified the different kinds of materials exposed on the edges of the reservoir, measured and mapped the old and new slides in both plan and profile, identified different types of Nespelem silt, and obtained samples for laboratory study.

Although the immediate engineering problem of how much land and which land to remove from use was quickly solved, the laboratory studies are continuing in the interest of analyzing further the mechanics of landslides in general. Petrographically, the Nespelem sediments are ultrafine rock flour of fluvioglacial origin; the particles are clay-size, but mineralogically they are mainly quartz and feldspar. The permeabilities, angles of internal friction, apparent cohesions, pore pressures, and related properties have been determined for the various facies of the formation.

One aspect of the Nespelem sediments has intrigued our analysts.

The sediments in some places sometimes possess natural-moisture contents that exceed 50 percent. Yet, after they are oven dried, no more than 25 percent of moisture can be added to them before they lose all solidity and assume a semiliquid consistency. How was the initial water retained? The answer, when obtained, will doubtless also explain an interesting phenomenon observed in the field: the natural material will stand indefinitely on vertical cuts if undisturbed and can be observed to have done so in the scarps of high terraces; yet, if dug with a power shovel or otherwise drastically disturbed, it will flow and assume an angle of repose of two or three degrees. This disturbance apparently releases the excess water which, through the action of time and pressure, it has bound to itself, and it becomes a semiliquid slurry. Research continues in the hope of unraveling facts fundamental to the explanation of the natural mechanisms of consolidation of sediments and to the factors involved in the stability, in the engineering sense, of sediments in general.

Questions

1. What are the responsibilities of the engineering geologist to the engineer?
2. Why are physical properties of rocks and materials important?
3. Why must subsurface exploration be conducted at most engineering sites?
4. What might be the significance of a zone of rock where core recovery was very poor?
5. What methods of subsurface exploration are employed in engineering geologic examinations?
6. Why are large diameter cores desirable, particularly in unconsolidated or poorly consolidated material?
7. In what way does petrography serve the engineer?
8. Why is soils mechanics considered an engineering geologic tool?
9. What is the normal sequence for conducting engineering geologic examinations?
10. What geologic features of a site should be included on an engineering geologic map?

CHAPTER 16

SOURCES OF WELL INFORMATION

Well Sample Library, Colorado School of Mines
Golden, Colorado

The well-sample library at the Colorado School of Mines was originally under the auspices of the Board of Industrial Development and Research of the State of Colorado. It is now operated as a function of the school.

The library now has on file samples from 1,531 wells. The geographic distribution of these wells is as follows:

Arizona	3	Nebraska	12
California	3	New Mexico	157
Canada	16	North Dakota	5
Colorado	462	Oklahoma	16
Idaho	6	South Dakota	8
Kansas	100	Texas	74
Louisiana	1	Utah	56
Mexico	1	Washington	1
Montana	233	Wyoming	377

Samples from the wells are vialed, systematically filed, and card indexed. More than 225,000 samples are involved in the collection.

Several well lists have been published and may be obtained free upon request. In addition to these lists an issue of the *Quarterly* of the Colorado School of Mines entitled "Selected Well Logs of Colorado" by Barb [1] is available from the Department of Publications of the school.

It is a violation of policy for the samples to leave the depository once they have been vialed and cataloged. Space in Berthoud Hall, the geology building, is available for those interested in sample examination. Microscope, lamp, and other equipment must be supplied by the examiner.

In addition to ditch-cutting samples, electric logs from Colorado, Oklahoma, Kansas, and Texas wells are available for study.

Paleontological Laboratory
Midland, Texas

The primary function of the Paleontological Laboratory is to furnish paleontologic data to oil operators who desire more detailed information for their stratigraphic work. Emphasis is placed on the identification

---

[1] Barb, C. F., *Selected Well Logs of Colorado:* Colorado School of Mines Quart., vol. 41, no. 1, Jan. 1946, $2.00.

and use of the Fusulinidae, important Permian and Pennsylvanian index fossils.

The laboratory offers two types of service:

1. Reports on important tests in the west Texas-southeastern New Mexico district are issued at regular intervals. These reports comprise paleontologic data and some formation or zone tops based on fossils and lithology.

Subscribers have access to (*a*) paleontologic data prepared during the period of their subscription, except "tight well" data, which are reported after samples are released; (*b*) fusuline thin sections, fossils, microfossil slides, a comprehensive library, card catalogs, faunal lists, etc. More than 30,000 indexed fusuline thin sections and microfossil slides are on file.

2. Special investigations are conducted in paleontology or stratigraphy, surface or subsurface. This work is on a consulting basis.

Regular reports are issued on the first and fifteenth of each month. Normally the paleontologic data involve twelve to fourteen tests each month, covering 30,000 to 50,000 feet of section.

An attempt has been made to describe lithologies using color terms from the Rock Color Chart prepared by the Rock Color Chart Committee and distributed by the National Research Council. An attempt is also being made to apply the tentative grade scale for carbonate rocks proposed by Deford.[2]

SOUTH DAKOTA SCHOOL OF MINES AND TECHNOLOGY
RAPID CITY, SOUTH DAKOTA

The Department of Geological Engineering at the South Dakota School of Mines and Technology has on file samples and electric logs for nearly all of the significant wells in the Dakota Basin, as well as numerous wells in adjacent parts of Nebraska, Wyoming, and Montana. These are available for study at any time, and space is available for the visiting geologist.

An up-to-date list of wells and electric logs may be obtained upon request.

KANSAS GEOLOGICAL SOCIETY
WICHITA, KANSAS

The Kansas Geological Society, a nonprofit organization, sponsors the Kansas Well Log and Sample Bureau.

The well-log service consists of all the current completions mailed each week to the subscribers at a monthly rate of $12.50.

If only old logs of a certain district are desired, prices are as follows: 1 to 25 logs, 25 cents each; 26 to 50 logs, 20 cents each; 51 to 100 logs, 15 cents each; and over 100 logs, 10 cents each.

---

[2] Deford, R. K., *Tentative Grade Scale for Carbonate Rocks:* Am. Assoc. Petroleum Geologists Bull., vol. 30, no. 11, pp. 1921-1928, Nov. 1946.

Logs can be supplied of wells drilled not only in Kansas, but also in Colorado, Nebraska, Iowa, and Missouri. Several thousand time-logs are available at 25 cents each. Electric logs may be purchased by subscribers at seventy cents each, wildcat logs at eighty cents each, and occasional logs at ninety cents each.

Subscriptions to the Kansas Well Sample Bureau are $55.00, $30.00, and $15.00 a month. Miscellaneous samples are four cents a sample. Cuts of well samples are available only for all current wells. The bureau does not maintain a sample rental library and thus does not retain any sample sets after the cut has been made.

Guidebooks of the following field trips and cross sections are available:

1. Guidebook, Tenth Field Conference (1936), $4.00.
2. Guidebook, Fourteenth Annual Field Conference (1940), $5.00.
3. Guidebook, Fifteenth Annual Field Conference (1941), $5.00.
4. Cross Section of the Central Part of the United States, $4\frac{1}{2}$ feet by 9 feet, colored $8.00, plain $5.00.
5. Cross Section from Western Missouri to Western Kansas, $2\frac{1}{2}$ feet by 8 feet, colored $6.00, plain $3.00.
6. Cross Section from Granite Ridge to Southeastern Nebraska to the Salem Oil Field in Illinois, 2 feet by 8 feet, colored $4.00, plain $2.00.

These cross sections can be furnished mounted on cloth at additional cost.

KANSAS SAMPLE LOG SERVICE
407 EAST FIRST STREET, WICHITA, KANSAS

The personnel of the Kansas Sample Log Service consists of two geologists, one draftsman, and one part-time utility man. The geologists log their interpretative sample determinations on a conventional log strip. These data are then transcribed to a tracing log strip (scale 1 inch = 100 feet), from which ozalid prints are made. The logs are printed each Friday and mailed to subscribers Saturday.

The letter of agreement of the Kansas Sample Log Service includes the following:

1. Detailed microscopic analyses of samples will be made on all important wildcat wells in Kansas.
2. Samples will be obtained from the Kansas Well Sample Bureau.
3. Logs will be plotted on standard log strips (1 inch = 100 feet), and conventional symbols will be used. The logs will also show the detailed description of the formations and the correlation of the key markers.
4. Samples will be examined currently, and logs will be mailed to clients each week.
5. This service will be offered at a monthly rate of $150.00 payable the first of each month.

6. Upon 30 days' notice the service may be discontinued by either party.

Summary of Operations From January 1, 1948 to December 31, 1948

Total number of logs..437
Average number of logs monthly..36.4
Total footage 599,416
Average footage monthly 49,951
Average footage of each log..1,372
Total number of sample tops reported..2,834
Average number of sample tops each log..6.5
Total cost of service at $150.00 monthly...$1,800.00
Average cost per log to subscribers...$4.12

Petroleum Information, Inc.
Continental Oil Building, Denver, Colorado
(Casper, Wyoming)

Petroleum Information, Inc., was established in 1928 as an oil- and gas-data service covering the states of Wyoming, Colorado, Montana, Utah, North and South Dakota, and Idaho, and western Nebraska and northern New Mexico.

A complete Rocky Mountain reporting service covering all drilling and leasing operations in the region is published, as well as special news bulletins through the week. The reports are accurate as to geologic data, and maps of special areas are a weekly feature of the service. The reports are in standard use by oil operators interested in the region. Well-completion-data cards are issued as a part of the confidential reporting service.

An annual "Résumé of Rocky Mountain Oil and Gas Operations" is published by the company, listing statistical and current information on activities. Well completions, production data, and geophysical and other information are contained in the volume.

The company has a large collection of drillers' and sample logs on wells drilled in the Rocky Mountain region. Copies of these logs are available at a nominal charge. Area and structure maps of the region are sold at the Denver office. These are predominantly of maps of known surface structure and producing fields and oil and gas state maps of the region.

Copies of electric logs of wells are reproduced by the company at the Casper office; some of the logs are available for general sale, and some are restricted to the use of operating companies in the Rocky Mountain region.

Land information is furnished through the company, and special land-block books are available for sale to subscribers of the reporting service. Land-block outlines are frequently carried in the weekly report service.

Geologic and engineering studies are handled by the company on a consulting basis, with the work being directed from the Casper office.

WELL-SAMPLE LIBRARY, UNIVERSITY OF TEXAS
AUSTIN, TEXAS

The well-sample library of the Bureau of Economic Geology at the University of Texas now offers increased facilities for study of 2,000,000 individual samples representing 25,000 oil, gas, and water wells from every section of the state. This collection, in its new location at the Off-Campus Research Center of the University of Texas, weighs approximately 200 tons, covers 11,050 square feet of floor space, occupies 16,000 cubic feet of shelf space, and in length of geologic sections represented measures approximately 10,000 miles.

Extensive provisions have been made and are being projected for accommodating visiting geologists who wish to conduct full-scale research programs at the well-sample library. The collection is housed in a three-story brick building 50 by 110 feet. Research and processing rooms are equipped with adequate facilities for washing, drying, and examining samples. Plans have been made to construct individual research rooms, where visitors may keep their equipment and enjoy a reasonable amount of privacy.

Every effort has been made to provide an easily accessible system for filing and locating material available.

For every well filed, a serial-number card is made listing the county, operator, fee owner, county and survey locations, sample range, and contributors. In addition, each well is cross-indexed according to county, operator, and fee owner. Finally, serial numbers and depth ranges of wells for which accurate locations are known are plotted on county maps.

A general index of all oil, gas, and water wells in the library collections to be published soon will be of interest to the oil industry. All wells in the collection will be listed in this index by county, and, within each county, alphabetically by company and fee owner. The sample ranges of every well will be noted. Included in the index will be a list of wells for which duplicate samples are on hand. These duplicate samples may be obtained for permanent possession by educational institutions, oil companies, or individuals upon request. The index will also include a list of certain surface samples on file at the library for examination and study.

In general, the well-sample library is designed to promote a greater knowledge of the geology of Texas by bringing together an extensive range of material concerning geologic problems. More specifically, it offers a permanent file of subsurface samples, which can be worked and reworked as new discoveries are made and new techniques developed.

The well-sample library also offers an initial field of research for many special studies. The department of geology of the University of

Texas draws upon this collection as a basis for subsurface studies covered by various courses.

Chemical and Geological Laboratories
Casper, Wyoming

The geologic department of the Chemical and Geological Laboratories offers the following services:

Stratigraphic well logging: Detailed descriptions of the lithology and paleontology and other pertinent information, together with their stratigraphic correlations based on microscopic examination of well cuttings. Cuttings of current wildcat wells from Wyoming, Montana, North and South Dakota, northern Utah, and Colorado are described on strip logs and standard typewritten sheets.

Routine field work: This service consists of supervision of drilling wells.

General consultation: Consulting in all phases of geologic work.

Reservoir evaluation: Comprehensive study both geologic and from accumulated engineering data based upon the review of primary and secondary core analysis, subsurface-fluid analysis, actual production figures, and extended estimated future development. The service is available for individual oil and gas lease and partial or complete field or total holdings.

Casper sample cut: Samples from current drilling wells in Wyoming, Montana, North and South Dakota, and northern Colorado and Utah are available.

Denver Sample Log Service
Denver, Colorado

The Denver Sample Log Service endeavors to serve the oil industry in the following operations:

Well Sample Processing: The operators of all tests that are drilled in Colorado, Utah, and western Nebraska are contacted for the release of their samples. After acquisition, the samples are washed, dried, and packaged, and then distributed to companies, institutions and individuals that desire a set for examination and library storage. In this service the organization functions as a clearing house for formation cuttings and relieves the companies and institutions of the time and expense involved in the acquisition, preparation and distribution processes. Sample cuts from currently drilled wells are available on a cost per sample basis.

Sample Analysis: Approximately 40,000 feet of samples from tests drilled in Colorado, Utah, northern New Mexico and western Nebraska are examined microscopically each month. The geologic data observed are recorded in strip log form by conventional graphic lithologic symbols and typewritten abbreviated descriptions. When available, electric logs are used in conjunction with the cuttings as an aid in the interpretation of the stratigraphic sequence.

Library: In addition to the sample processing and analysis services, a library is maintained which contains cuttings from many tests drilled in all of the Rocky Mountain states. These are rented on a cost per set basis and are utilized by those geologists that do not have space for library storage or whose other duties confine them to offices so that they cannot study samples at depositories lacking shipping facilities.

Arkansas Geological Survey
Little Rock, Arkansas

The well-sample library of the Arkansas Geological Society has only recently been organized. The principal interest is in wells of the Gulf Coastal Plain area. Most of the samples now available are from wells in eastern and northeastern Arkansas. Most of the cuts of samples have been contributed by the companies. In certain key wells in areas of particular interest the office has supplied geologic services in return for sample sets.

The laboratory has facilities for working and drying samples and a storage room. An assistant is available for processing the samples.

New Mexico Bureau of Mines and Mineral Resources
Socorro, New Mexico

The New Mexico Bureau of Mines and Mineral Resources lists wells by township, range, section, operator, and lease. Pool wells in southeastern New Mexico are listed only by field, county, and number of sample sets from each pool. Additional information on these pool wells will be given upon request.

All samples are available for examination at any time in Socorro. They may also be borrowed by reliable companies or individual geologists with the understanding that they will not be cut, washed, or damaged in any way. In fairness to all, the bureau requests that no set of samples be kept longer than thirty days at any one time.

INDEX

This index consists of two parts: one listing authors, the other subjects. Since Dr. LeRoy was called to South America on a summer assignment before the page proof was ready, the index was prepared and supervised by Stanley Reichert and W. Alan Stewart of the Department of Geology, Colorado School of Mines.

AUTHOR INDEX

Barstow, O. E.
   Deep-Well Camera 664
Bryant, C. M.
   Deep-Well Camera 664
Caran, J. G.
   Core Analysis 295
Carlsten, Kirk
   Oriented Cores 521
Cooke, S. R. B.
   Spectrochemical Sample Logging.. 487
Crawford, J. G.
   Water Analysis 272
De Ment, Jack
   Fluoroanalysis in Petroleum
   Exploration 320
Doll, H. G.
   Induction Logging and Its Appli-
   cation to Logging of Wells Drilled
   with Oil-Base Mud 393
   The Microlog 399
Fancher, G. H.
   The Porosity and Permeability of
   Clastic Sediments and Rocks........ 685
Ford, R. D.
   Electric Logging 364
Frey, M. G.
   Magnetic Core Orientation............ 596
Gabelman, J. W.
   Micro (Petrographic) Analysis.... 172
Gill, R. J.
   Composite-Cuttings-Analysis
   Logging 495
Hamilton, R. G.
   Application of Dipmeter
   Surveys 625
Hardin, P. N.
   The Electric Pilot in Selective
   Acidizing, Permeability Determin-
   ations, and Water Locating.......... 676
Hassebroek, W. E.
   Hydrafrac Treatment 723
Herness, S. K.
   Subsurface and Office Represen-
   tations in Mining Geology............ 989
Hills, J. M.
   Sampling and Examination of
   Well Cuttings 344
Ireland, H. A.
   Insoluble Residues 140
Johnson, F. Walker
   Shale Density Analysis.................. 329

Johnson, G. W.
   Duties and Reports of the Sub-
   surface Geologist 810
Johnson, J. H.
   Calcareous Algae 95
Kerr, P. F.
   Multiple-Differential Thermal
   Analysis 240
Kuhn, T. H.
   Subsurface Methods as Applied
   in Mining Geology 969
Kulp, J. L.
   Multiple-Differential Thermal
   Analysis 240
Landua, H. L.
   Coring Techniques and
   Applications 609
Langton, Arthur
   Well Logging by Drilling-Mud
   and Cuttings Analysis 449
LeRoy, L. W.
   Comments on Sedimentary Rocks 71
   Driller's Logging 475
   Graphic Representations 856
   Micropaleontologic Analysis 84
   Settling Analysis 193
   Shape Analysis 199
   Size Analysis 184
   Stain Analysis 193
   Stratigraphic, Structural, and
   Correlation Considerations 12
   Well Acidization 750
Low, J. W.
   Subsurface Maps and
   Illustrations 894
Mercier, V. J.
   Radioactivity Well Logging.......... 419
Moore, C. A.
   Electron-Microscope Analysis........ 202
Murdoch, J. B., Jr.
   Controlled Directional Drilling.... 504
   Oil-Well Surveying 548
Nichols, P. B.
   Drilling-Time Logging 478
Owsley, W. D.
   Oil-Well Cementing 746
Payne, L. L.
   Design and Application of
   Rock Bits 643
Rittenhouse, Gordon
   Detrital Mineralogy 116

Robb, G. L.
  Geologic Techniques in Civil
  Engineering 1120
Schieltz, N. C.
  X-Ray Analysis 211
Shepard, G. F.
  Drilling-Time Logging 455
Sloss, L. L.
  Spectrochemical Sample Logging 487
Sources of Well Information............1150
Stewart, W. A.
  Unconformities 32
Stommel, H. E.
  Subsurface Methods as Applied
  in Geophysics 1038
Stratton, E. F.
  Application of Dipmeter
  Surveys 625
  Electric Logging 364

Sutter, H. F.
  Drilling Fluid Chemistry.............. 713
Tapper, Wilfred
  Caliper and Temperature
  Logging 439
Todd, J. D.
  Valuation and Subsurface
  Geology 792
Torrey, P. D.
  Secondary Recovery of
  Petroleum 775
Tripp, R. M.
  Geochemical Methods 760
Wagner, W. R.
  Micro (Petrographic)
  Analysis 172
  Petrofabric Analysis 157
Wallace, W. A.
  Formation Testing 731

SUBJECT INDEX

Bedding, 162
  map symbols, 162
Cable-tool samples, 345
Calcareous algae, 95
Caliper logging, 439
  equipment, 441
  uses, 443
Civil engineering, geologic techniques
  in, 1120
  applications of petrography to, 1130
  construction, 1143
  cost estimates, 1142
  examples, 1144
  exploration, 1126
  final design and specifications, 1143
  general gelogic techniques, 1121
  geologic report, 1143
  geomorphology, 1124
  geophysics, 1135
  ground water, 1125
  lithology, 1122
  magnetic methods applied to, 1139
  preliminary designs, 1142
  reconnaissance, 1136
  resistivity methods applied to, 1137,
    1138
  seismic methods applied to, 1140
  selection, 1141
  sequence of investigations, 1136
  soils mechanics, 1133
  specialized techniques, 1130
  stratigraphy, 1123
  structural geology, 1124
Clastic sediments and rocks, 685
  capillary pressure, 708
  characteristics of, 685
  degree of consolidation, 687
  permeability, 694
  permeability measurement, 699

Clastic sediments and rocks—Cont.
  permeability, relative, 699
  porosity, 688
  porosity and permeability, 685
  porosity measurement, 692
  structural configuration, 687
Clay-stain results, 198
Cleavage, 162
  map symbols, 162
Composite-cuttings-analysis logging, 495
  application, 501
  method, 496
  plotting, 500
  theory, 495
Conodonts, 105
Core analysis, 295
  applications, 318
  connate water, 304
  core sampling, 296
  definition, 296
  residual core water, 304
  residual-fluid saturations, 302
  residual oil and condensate, 302
Core data, 317
  graphic presentation, 317
Core-data interpretation in production
  calculations, 314
  gas reserves, 317
  maximum recoverable oil, 315
  oil in place, 314
  recoverable gas, 317
  solution-gas-expansion recovery, 316
Core-data interpretation in reservoir
  characteristics, 308
  bleeding cores, 311
  chalk and serpentine, 313
  condensate sands, 310
  conglomerate, 313
  gas sands, 309

Core-data interpretation in reservoir
characteristics—Cont.
oil sands, 310
oil-saturated limestone, 311
transitional zones, 313
water sands, 313
zones of secondary migration of
water, 314
Core samples, 299
physical characteristics of, 299
types of, 299
Core sampling, 296
Coring techniques and applications, 609
conventional coring, 609
core-analysis data, 624
correlation with electric logging, 623
diamond coring, 612
geologic and development
problems, 619
limitation in geologic problems, 623
production work, 621
reverse-circulation coring, 615
side-wall coring, 617
types of coring, 609
wire-line coring, 615
Correlations, 60
difficulties, 69
indicators, 68
magnitude of, 68
methods of, 64
purposes of, 64
Cuttings samples examination, 352
oil stain, 356
permeability, 356
porosity, 356
use of in limestone-reservoir
studies, 358
Deep-well camera, 664
camera, 667
equipment, 666
future developments, 673
operating equipment, 670
principle of operation, 664
stereoscopic pairs, 676
Deflection tools, orientation of, 519
bottom-hole orientation, 519
drill-pipe alignment, 519
Detrital mineralogy, 116
choice of methods, 117
color, 121
composition, 125
grain roundness, 135
grain size, 121
heavy minerals, 131
megascopic and binocular
examination, 119
minor minerals, 138
orientation, 125
porosity and permeability, 129
rock types, 128

Detrital mineralogy—Cont.
shape and roundness, 122
structure, 128
subsurface samples, 118
surface texture, 124
texture, 127
thin sections, 139
Diatoms, 101
Dipmeter, 625
areas of application, 627
basic principles, 625
field procedure, 627
Dipmeter surveys, 628
applications, 625, 629
areas of application, 627
field procedure, 627
interpretation, 628
selection of levels, 629
Directional drilling, 504
applications, 533
directional well planning, 527
directional well types, 525
equipment, 504
orientation of deflecting tools, 519
Directional-drilling applications, 533
business and residential areas, 540
faults, 546
inland waters, 539
mountainous terrain, 539
multiple well, 544
offshore drilling, 544
offshore structures, 533
relief wells, 537
salt domes, 537
Directional-drilling equipment, 504
knuckle joint, 510
removable whipstock, 505
spudding bit, 516
Directional well planning, 527
Directional well types, 525
Disconformity, 33
Driller's logging, 475
interpretation, 477
terminology, 475
Drilling fluids, 713
chemical treatment of, 718
general comment on, 723
lost circulation, 722
properties of drilling muds, 714
temperature effects on, 721
types of, 722
Drilling-time logging, 455, 478
definitions, 456
drilling rate, 459
drilling time, 459
drilling-time logs, 462
geolograph, 482
Log-O-Graf, 486
Martin-Decker recorder, 486
types of recorders, 482
value in diagnosing lithology, 457

Drilling-time logs, 462
  application, 462
  geolograph, 468
  method of preparing, 468
  plotting, 471
Electric logging, 364
  application of, 379
  resistivity, 376
  spontaneous potential, 364
Electric logs, 345
Electrical prospecting, 1110
  controlled currents, 1113
  natural earth currents, 1112
Electric pilot, 676
  equipment, 677
  in permeability surveys, 681
  in selective acidizing, 679
  in water locating, 684
Electron microscope, 202
  description, 203
Electron-microscope analysis, 202
  bed identification, 205
  correlation of clays, 207
  correlation, geologic
    applications in, 202
  correlation, long-range, 208
  correlation, possible uses in, 205
  crude oil, 208
  paleontologic studies, 210
  specimen mounting, 205
Facies, 22
  definition, 24
  examples of ancient facies
    changes, 30
  examples of modern facies
    changes, 30
Faults, 51
  criteria for recognition of, 52
  definition, 51
  geologic symbols, 160
  gravity fault, 51
  subsurface recognition of, 52
  thrust fault, 51
Feldspars, 196
  identification of, 196
Fish scales, 113
Fluorescence exploration, 325
Fluoroanalysis in petroleum
  exploration, 320
  fluorescence exploration, 325
  fluoroescence microscopy, 325
  fluoroadsorption analysis, 324
  fluorographic exploration, 327
  fluorography, 324
  fluorologging, 326
  fluorometry, 323
  methods of, 323
  quenching analysis, 325
Fluorochemistry of petroleum, 321
Fluorographic exploration, 327
Fluorolog, 328

Fluorolog—Cont.
  section of, 328
Fluorologging, 326
Folds, 161
  geologic symbols, 161
Foliation and cleavage, 162
  map symbols, 162
Foraminifera, 86
Formation testing, 731
  applications, 739
  equipment, 733
  pressure-recording-device chart, 745
  pressure-recording devices, 744
  procedure, 732
  types of testing packers, 737
Geochemical methods, 760
  analytical techniques, 764
  correction factors, 768
  future of, 772
  historical development, 760
  interpretations, 769
  results, 769
  theory, 761
Geologic cross sections, 884
  block diagram, 890
  directional-drilling, 886
  isometric construction in stratigraphic
    compilation, 889
  isometric projection, 888
  outcrop area and subsurface distribu-
    tion of Wilcox group, 892
  relative permeabilities, 887
  stratigraphic section, 885
  structural and stratigraphic condi-
    tions of oil field, 891
Geolograph, 468
Geophysics, subsurface methods as
  applied in, 1038
  cost of geophysical surveys, 1042
  electrical prospecting, 1110
  geophysical exploration, 1039
  gravitational prospecting, 1060
  magnetic prospecting, 1042
  planning geophysical program, 1041
  seismic prospecting, 1084
  statistics, 1039
Grain roundness, 135
Graphic representations, 856
  block diagrams, use of, 878
  busk method, 874
  colored well-log symbols, 856
  columnar subsurface sections, 872
  compilation chart, 866
  complex faulting and divisions of
    sand series and groups with re-
    lation to geologic time divisions, 868
  complex faulting below major
    unconformity, 876
  cores, drill-stem tests, production
    tests, and production histories
    of wells recorded on structure-

Graphic representations—Cont.
  contour maps of oil and gas
    reservoirs, 879
  correlation between electric-log
    profile and formational units, 865
  correlation charts, 857
  correlation of well sections with
    control surface sections, 870
  directional drilling on complex
    faulted structure, 875
  electric profiles for plotting litho-
    logic logs, 861
  final well log, 860
  geologic cross sections, 884
  isopach and lithofacies map, 880
  lithologic graphic symbols, 859
  oil and gas distribution and
    stratigraphic relationships, 869
  paleogeologic map, 881
  paleotectonic map, 882
  plan structural view, 877
  panel diagram, 873
  relationship of well and surface
    sections, 867
  scale, vertical exaggeration of, 883
  standard well-log strips, 858
  subsurface oil, gas, and water
    symbols, 862
  symbols, misc., 863
  well logs, 856
  well symbols, 862
Grass seeds, 111
Gravitational prospecting, 1060
  applications, 1074
  basic principles, 1060
  control, minimum and maxi-
    mum, 1072
  earth's gravitational field, 1064
  instruments, 1067
  interpretation, 1069
  regional gradient, 1073
  rock properties, 1063
  torsion-balance data, 1073
Heavy minerals, 131
  errors in analysis of, 135
  mineral identification, 133
  preparation of sample, 131
  use of data on, 134
Heavy-mineral separation, 174
Hydrafrac treatment, 723
  application, 725
  procedure, 724
Induction logging, 393
  geometry of, 397
  resistivity measurements, 394
Insoluble residues, 140
  chart for, 144
  data, plotting of, 154
  description of, 143
  descriptions, 154
  preparation of, 140

Insoluble Residues—Cont.
  terminology for, 146
  use of, 150
Lineations, 163
  map symbols, 163
Magnetic core orientation, 596
  applications, 608
  cores from deflected wells, 606
  experimental results, 606
  magnetic field of core barrel
    and drill pipe, 605
  residual magnetic minerals,
    distribution in core, 604
  rock movements, 604
  secondary residual magnetism, 603
  secular variation of earth's
    magnetic poles, 601
  sediment, coarseness of, 600
  theory, 597
  variable geologic factors, 600
Magnetic prospecting, 1042
  applications, 1054
  basic principles, 1043
  earth's magnetic field, 1045
  instruments, 1048
  interpretation, 1050
  rock properties, 1044
Micrologging, 399
  equipment, 402
  field examples, 417
  interpretation of, 404
  interpretation, quantitative, 418
  principle of, 399
Micropaleontologic analysis, 84
  calcareous algae, 95
  conodonts, 105
  diatoms, 101
  fish scales, 113
  Foraminifera, 86
  grass seeds, 111
  ostracods, 94
  otoliths, 108
  Radiolaria, 105
  spores and pollen, 108
Micropaleontologic laboratories, 114
Micropaleontologic studies, 114
Micro (petrographic) analysis, 172
Mining geology, subsurface and
  office representation in, 989
  areal map grid system, 999
  coordinate system, 998
  data and map sheets,
    · specifications of, 1002
  field-recording and office-
    ledger survey sheet, 1003
  filing, 1010
  geologic representation,
    essentials of, 993
  geologic representation,
    objectives of, 990

Mining, geology, subsurface and·
office representation in—Cont.
  geologic technique and
    mineral exploration, 991
  geologic technique and theory, 992
  indexing, 1010
  legend, 1019
  litho-compositional legend, 1024
  mapping scales, 997
  metallization legend, 1031
  purpose of geologic mapping, 990
  recording and posting
    specifications, 1031
  structural legend, 1024
  subsurface cultural legend, 1022
  surface cultural legend, 1021
  system, 997
  textural legend, 1024
Mining geology, subsurface
  methods in, 969
  biogeochemical guides, 975
  drilling, 979
  exploration methods, 978
  exploratory procedures, 972
  geochemical guides, 975
  geologic guides and controls, 972
  geologic mapping, 976
  geophysics, 977
  igneous ore deposits, 969
  laboratory methods, 977
  metamorphic ore deposits, 971
  mineralogic guides, 974
  mining operations, 982
  representation and correlation
    of data, 983
  sampling, 982
  sampling methods, 978
  sedimentary ore deposits, 969
  stratigraphic control, 976
  structural control, 972
  supergene deposits, 972
  trenching and stripping, 979
  valuation of mining properties, 988
Multiple-differential thermal
  analysis, 240
  allophane, 255
  alunite, 259
  alunite-jarosite mixtures, 267
  alunite-kaolinite mixtures, 266
  apparatus, 241
  applications, qualitative, 250
  artificial mixtures, 261
  carbonates, 260
  dickite, 251
  endellite, 255
  halloysite, 255
  hydrous oxides, 258
  jarosite, 259
  kaolinite, 253
  kaolinite-dickite mixtures, 268
  kaolinite-goethite mixtures, 263

Multiple-differential thermal
  analysis—Cont.
  kaolinite-montmorillonite
    mixtures, 269
  kaolinite-quartz mixtures, 264
  kaolinite-sericite mixtures, 265
  minerals, 3-layer lattice, 257
  montmorillonite-sericite
    mixtures, 270
  procedure, 245
  theory, 247
Nonconformity, 34
Offlap, 55
Oil-well cementing, 746
  procedure, 746
  requirements for, 747
Oil-well surveying, 548
  applications, 584
  equipment, 549
  recording, 581
Onlap, 55
Oriented cores, 591
  equipment, 591
Ostracods, 94
Otoliths, 108
Overstep, 55
  complete, 55
  definition, 55
Permeability, 299
  clastic sediments and rocks, 685
  cuttings sample, 356
  definition, 299
  detrital mineralogy, 129
  spontaneous potential,
    effect on, 370
  thin sections, 181
Petrofabric analysis, 157
  applications, 168
  definition, 157
  field work and its presentation, 158
  laboratory techniques and
    presentation of data, 164
Petrographic laboratory, 180
  flow sheet of, 180
Pollen, 108
Porosity, 299
  clastic sediments and rocks, 685
  cuttings sample, 356
  definition, 299
  detrital mineralogy, 129
  spontaneous potential,
    effect on, 370
  thin sections, 181
Radioactivity logs, 345
Radioactivity well logging, 419
  applications, 427
  correlation with rock types, 438
  general considerations, 436
  instrumentation, 421
  interpretation, 422
Radiolaria, 105

Report writing
  (cf. Subsurface geologist)
Resistivity, 376
  electrode spacing, 377
  resistivity curves,
    characteristics of, 378
  terminology, 377
Rock bits, design and
  application, 643
  bits for formations on the harder
    side of medium-soft, 654
  bits for medium-hard to hard,
    nonabrasive formations, 655
  compressive strength of
    various rock types, 658
  drillability of formations, 659
  drilling hard to hard,
    abrasive formations, 656
  hard and hard, abrasive
    formation bits, 652
  medium-hard formation bits, 648
  medium-hard to hard, nonabrasive
    formation bits, 650
  modifications of OWS,
    type OWC, 651
  OSC-1 tricone, 645
  OSC tricone, 647
  OSQ-2 tricone, 648
  OWS tricone, 650
  soft to medium-soft
    formation bits, 653
  W7 and W7R tricones, 652
  weight from tandem drill
    collars, 656
Rotary samples, 346
  catching of, 350
  contamination from upper beds, 346
  cuttings, loss of, 348
  cuttings, powdering of, 347
  elutriation, 348
  methods of obtaining
    correct depths, 348
Sample logging, 344
  cable-tool samples, 345
  cuttings samples, 352
  plotting, 360
  rotary samples, 346
  use of, 362
Secondary-recovery methods, 778
  air and gas injection, 778
  mining, 781
  vacuum, 778
  water flooding, 779
Secondary recovery of
  petroleum, 775
  costs and results, 785
  history of operations (U. S.), 776
  limiting factors, 788
  methods, 778
  operations in Rocky Mtn.
    states, 786

Secondary Recovery of Petroleum—Cont.
  secondary oil reserves, 781
  susceptibility of oil fields to, 781
Sedimentary rocks, 71
  classification of, 72, 73
  color of, 82
  diagenesis of, 82
  structures in, 80
  texture of, 79
  types of, 71
Sedimentary rock types, 71
  breccias, 71
  carbonaceous rocks, 78
  conglomerates, 71
  dolostone, 76
  evaporites, 77
  limestone, 76
  miscellaneous, 78
  mudstone, 76
  sandstones, 74
  shale, 76
  siltstones, 75
Seismic prospecting, 1084
  applications, 1101
  interpretation methods, 1093
  principles, 1084
  prospecting methods, 1091
  rock properties, 1088
  velocities of selected rocks, 1090
Settling analysis, 193
Shale-density analysis, 329
  apparent specific gravity, 330
  application of results, 334
  bulk density, 330
  compaction, 331
  grain density, 330
  natural density, 330
Shale-density determination, 331
  method of, 331
  sources of error in, 333
Shape analysis, 199
Size analysis, 184
  computations, 190
  cumulative-frequency curve, 190
  histogram plot, 188
  plotting of data, 188
  preparation of sample, 184
  simple-frequency curve, 188
Spectrochemical sample logging, 487
  densitometry, 493
  instrumentation, 491
  interpretation, 493
  sample treatment, 489
  sampling, 489
  standards, 492
Spontaneous potential, 364
  effect of bed thickness, 373
  effect of drilling mud, 370
  effect of porosity and
    permeability, 370
  effect of special muds, 373

Spontaneous potential—Cont.
  effect of variation in .
    hydrostatic pressure, 376
  factors influencing resistivity
    of drilling fluids, 370
  fresh-water-bearing formations, 366
  salt-water-bearing formations, 366
Spores and pollen, 108
Stain analysis, 193
  benzidine test, 197
  calcite and aragonite, 195
  calcite and dolomite, 195
  clay-mineral stain tests, 197
  copper nitrate method, 195
  crystal-violet test, 198
  Fairbanks method, 195
  Lemberg method, 196
  malachite-green test, 198
  potassium ferricyanide method, 196
  safranine y test, 198
  silver chromate method, 195
Stratigraphic geology, 13
  subdivisions of, 13
Stratigraphic units, 14
  nature and classes of, 14
  rock units, 15
  time-rock units, 15
  time units, 15
Stratigraphy, 12
  definition, 12
Subsurface geologist, 1
  duties of, 5
  requirements of, 1
  training of, 7
Subsurface geologist, duties
    and reports of, 810
  duties on exploratory well, 810
  report writing, 823
Subsurface geologist, duties
    on exploratory well, 810
  caliper logging, 818
  core descriptions, 813
  core sampling, 814
  coring, 811, 818
  ditch sampling, 810
  ditch samples, 818
  drilling rate, 811
  electric logging, 818, 819
  formation testing, 815, 819
  gas samples, 817
  general, 820
  mechanical logging, 819
  mud and water tests, 819
  oil and gas samples, 820
  preparation of core samples for
    analysis and shipment, 813
  radioactive logging, 818
  recording data, 821
  salinity test, 817
  tests for oil, 817
  well log, 810

Subsurface geologist, examples
    of report outlines, 847
  on exploratory (wildcat) well, 847
  on oil field, 850
  on petroliferous province, 853
  on well in proved field, 848
Subsurface geologist, report
    writing, 823
  completion log report, 834
  Creole well reports, 825
  examples of outlines of reports
    on subsurface geology, 847
  publication, 825
  sand data report, 844
  weekly chronological
    well report, 825
  weekly well-sample report, 832
Subsurface geology, 1
  definition, 1
  future of, 9
  relationship to other sciences, 2
Subsurface laboratory methods, 84
Subsurface logging methods, 344
  caliper logging, 439
  composite-cuttings-analysis
    logging, 495
  driller's logging, 475
  drilling-mud and cuttings
    analysis, 449
  drilling-time logging, 455, 478
  electrical logs, 345
  electric logging, 364
  induction logging, 393
  micrologging, 399
  radioactivity logs, 345
  radioactivity well logging, 419
  sample logging, 344, 345
  spectrochemical sample
    logging, 487
  temperature logging, 444
Subsurface maps and illustrations, 894
  block diagram, one-point
    perspective, 937
  block diagrams, etc., 933
  block diagrams showing relation-
    ship of outcrop bands to
    dip and topography, 925
  block diagrams showing
    unconformity, 899
  coloring of geologic maps, 957
  coloring of lithofacies maps, 947
  contouring, suggestions on, 906
  cross sections and projections, 926
  drafting of maps, 963
  facies maps, 936
  isofacies maps, 942
  isolith maps, 941
  isopach maps, 912
  lithofacies maps, 938
  lithofacies maps, uses of, 951
  log map, 930

Subsurface maps and
illustrations—Cont.
miscellaneous models, 912
paleogeologic maps, 923
panel map, 930
peg models, 909
percentage maps, 940
preparation of subsurface data, 895
ratio maps, 939
reduction of datum elevation, 895
reproduction of maps, 961
section models, 912
shadow-graphic structure map, 937
solid models, 911
spalinspastic map, 956
stratigraphic and isometric-
projection drawing, 932
structural contour map, 897

Subsurface problems, 5

Subsurface techniques, 10
comparative use of, 10

Temperature logging, 444
instrumentation, 444
interpretation, 446
uses, 449

Thin sections, 174
of fragments, 178
of heavy concentrates, 178
preparation of, 174
study of, 179
correlation, 183
porosity and permeability, 181
storage of, 183
vug and opening studies, 183

Traps, oil and gas, 55
Clapp's classification of, 56
Wilson's classification of, 57

Unconformities, 32
chemical sediments, 44
continental, 36
criteria for recognition of, 40
definition, 33
evidences of erosion or
weathering, 42
genesis of, 36
marine, 37
nomenclature, 33
recognition of, 39
structural features, 41

Unconformities, importance of, 45
oil reservoirs, 46
ore deposits, 49
stratigraphic, 45

Valuation and subsurface
geology, 792
checking the value, 808
engineering valuations, 795
figuring the reserve, 796
gas properties, 805

Valuation and subsurface
geology—Cont.
hypothetical case, 798
market value, 794
proration, 793

Water analysis (Rocky Mtn.
oil fields), 272
classification, 272
Colorado group, 277
Cretaceous, Lower, 281
Cretaceous, Upper, 275
Devonian and older, 292
Frontier formation, 279
Green River formation, 274
Jurassic, 283
Mississippian, 292
Montana group, 275
Muddy (Newcastle) sand, 279
Pennsylvanian, 285
Permian, 285
Shannon sandstone, 278
surface waters, 273
Tertiary, 274
Triassic, 285
Wasatch formation, 274

Well acidization, 750
dolomite, 750
examples of, 753
limestone, 750
methods of locating
porous zones, 755
procedure, 756
production increase by, 754
sandstone, 751
shooting, 758
structural, stratigraphic, and
lithologic influence on, 753

Well information, sources of, 1150
Arkansas Geol. Survey, 1156
Chemical and Geol.
Laboratories, 1155
Colorado School of Mines
Well Sample Library, 1150
Denver Sample Log Service, 1155
Kansas Geol. Soc., 1151
Kansas Sample Log Service, 1152
New Mexico Bur. Mines and
Mineral Resources, 1156
Paleontological Laboratory, 1150
Petroleum Information, Inc., 1153
South Dakota School of Mines
and Technology, 1151
Texas Univ. Well-Sample
Library, 1154

Well logging by drilling-mud
and cuttings analysis, 499
application, 453
correlation of data, 452
methods of analysis, 450
theory, 450

X-radiation, 226
  measurement of lines in pattern
    and conversion to d values, 228
X-ray analysis, 211
  advantages, 215
  apparatus for recording
    diffraction patterns, 217
  applications, 239
  identication of minerals and
    components of mixtures, 231

X-ray analysis—Cont.
  identification, special
    problems of, 237
  limitations, 215
  position and type of film, 225
  powders, 221
  preparation and mounting
    of specimen, 221
  scope, 215
  single crystals, 221